Dictionary of
Agricultural and Environmental Science

Dictionary of
Agricultural and Environmental Science

Frederick R. Troeh

Roy L. Donahue

Iowa State Press
A Blackwell Publishing Company

Iowa State Press
2121 State Avenue, Ames, Iowa 50014

Orders:	1-800-862-6657
Office:	1-515-292-0140
Fax:	1-515-292-3348

Web site: www.iowastatepress.com

Printed in India

First edition, 2003

Library of Congress Cataloging-in-Publication Data

Troeh, Frederick R.
  Dictionary of agriculture and environmental science / Frederick R. Troeh and Roy L. Donahue.—1st ed.
      p. cm.
  ISBN 0-8138-0283-0 (alk. paper)
  1. Agriculture—Dictionaries. 2. Environmental sciences—Dictionaries. I. Donahue, Roy Luther, 1908-1999 II. Title.
  S411 .T76 2003
  630'.3—dc21

2002000824

The last digit is the print number: 9 8 7 6 5 4 3 2 1

Contents

Preface

This dictionary is intended to facilitate communication in agricultural and environmental sciences. Both of these broad sciences deal with the environment but in very different ways. People working in these areas need to be able to communicate with each other clearly and effectively. Too often, however, the tendency is to develop special terms and usages that are understood within a particular group but that fail to convey their unique meanings to persons outside the group.

Roy Donahue was an educator and writer in the field of soil science and had broad interests in all fields of agriculture and environmental science. He taught for many years at Michigan State University and accepted long-term international assignments in India and Africa. He undertook this project to help bridge the gap that has often existed between the terminologies used in these two broad subject areas, as well as the gaps that often exist between scientists and practitioners in both areas. He spent approximately ten years compiling terms and definitions from many sources. His emphasis was on special and unique terms and usages, including unusual words and short phrases along with special usages of common words. Words and usages that were not included in standard dictionaries were of special interest to him as he compiled his material.

Dr. Donahue asked me to complete the project. As a professor of agronomy at Iowa State University with a background in soil science and an emphasis on soil conservation, I, too, had worked abroad in Uruguay, Argentina, and Morocco. I was able to provide an additional viewpoint and the advantage of being an experienced computer user. I edited and revised Dr. Donahue's work, verifying the meanings by consulting various dictionaries, textbooks, other books, and online references, especially the compilation of hundreds of dictionaries compiled on the OneLook site. I made sure that all of the definitions were consistent with materials already published, but they were rewritten for a consistent style and to avoid plagiarism. In the process, I reviewed volumes of environmental and agricultural literature, expanded the material by adding thousands of entries, and designated the parts of speech and derivations for most of the words. It should be noted that the derivations are provided mainly to clarify the source and meaning of the words and do not constitute complete etymologies.

Many of the definitions are not unique to agricultural and environmental sciences but are needed to show that the word or phrase might be used in ways other than its unique scientific meaning. Dr. Donahue's illustrations were mostly photographs and artwork from government sources. I used many of them and supplemented them with photographs, my own artwork, and clip art. All photographs and drawings not otherwise credited on the Figure Credits page are mine.

Dr. Donahue initiated this project and his input into the work was considerable, but regrettably he did not live to see this dictionary published. The final form therefore became my responsibility. Users' comments and suggestions are welcome and will be taken into consideration for any future editions that may be undertaken. In spite of all the effort that went into this dictionary, there are certainly many more terms and usages that could be added. Such a project can never be regarded as final and complete, but the current form should be useful to many people.

Frederick R. Troeh
Ames, Iowa
August 2002

Figure Credits

The figures chosen to illustrate this dictionary are from a variety of sources. About half of them are the work of Dr. Troeh in the form of original photographs, artwork, or graphs (sometimes based on data or figures from some of the sources listed below). Ten photographs were taken by Dr. Donahue. The remainder come from the following sources:

Individuals:
    Julian P. Donahue—Banyan tree
    Frank Shuman (University of Illinois)—Fertilizer (effect), Fertilizer placement

Universities:
    Texas A&M University—Asexual reproduction, Silverfish insects
    University of Missouri—Ant (termite and ant), Cockroaches
    University of Puerto Rico—June bug

U.S. government agencies:
    U.S. Agency for International Development—Emission (smoke), Family planning
    U.S. Bureau of Reclamation—Clouds
    U.S. Council on Environmental Quality—Chlorofluorocarbon release
    USDA Agricultural Research Service—Field bindweed, Canada thistle, Cocklebur, Lambsquarters, Morning glory, Pigweed, Ragweed, Rhizobium nodules, Velvetleaf, Wasp, Wild oats
    USDA Natural Resources Conservation Service—Gully erosion, Manure (pollution potential), Mollisol, Spodic soil profile, Strip cropping, Wetland
    U.S. Department of Agriculture—Cabbage looper, Climatic types, Codling moth larva, Drained land, Dryland agriculture, Lady beetle, Pheromone, Scabies (hair on fence), Ticks, Trichogramma
    U.S. Department of Energy—Global warming
    U.S. Department of the Interior—Aquifer
    U.S. Environmental Protection Agency —Acidity scale, Aerating wastewater, Air pollution sources, Asbestos hazard, Water lily
    U.S. Fish and Wildlife Service—Texas longhorns
    U.S. Geological Survey—Agonic line, Volcanic dust

USDA Yearbooks of Agriculture:
    Grass (1948)—Foxtail millet, Lupine, Millets, Orchardgrass, Reed canarygrass
    Insects (1952)—Bark beetle, Boxelder bug, Carpenter ants, Centipede, Chinch bug, Colorado potato beetle, Corn borer larva, Cucumber beetle, Drone fly, Ground beetle, Heel fly, Mosquitoes, Pacific Coast tick, Praying mantid, Seed corn maggot, Stinkbug, Sweet potato weevil
    Trees (1949)—Chestnut leaf and nut, Honeylocust, Locust (trees), Spruce, Sugar maple
    U. S. Agriculture (1992)—Kenaf

International organizations:
    Food and Agriculture Organization of the United Nations—Erosion, Shifting cultivation, Terraces
    World Bank—Vetiver grass

Abbreviations Used in This Book

ac	acre		km	kilometer
ac-in.	acre-inch (water measure)		km/h	kilometer per hour
atm	atmosphere		kW	kilowatt
ATP	adenosine triphosphate		kWh	kilowatt-hour
Btu	British thermal unit		L	liter
bu	bushel		lb	pound
C	Celsius		m	meter
cal	calorie		m^2	square meter
cc	cubic centimeter		m^3	cubic meter
cfs	cubic foot per second		meq	milliequivalent
Ci	curie		mg	milligram
cm	centimeter		mi	mile
cm^2	square centimeter		mi^2	square mile
cm^3	cubic centimeter		min	minute
cv	cultivar		mL	milliliter
dB	decibel		mm	millimeter
e.g.	for example (*exempli gratia*)		mmol	millimole
F	Fahrenheit		MPa	megapascal
ft	foot		mph	mile per hour
ft^2	square foot		N	normal
ft^3	cubic foot		nm	nanometer
ft-lb	foot-pound		Pa	pascal
g	gram		pk	peck
gal	gallon		ppm	parts per million
ha	hectare		pt	pint
ha-cm	hectare-centimeter		qt	quart
hr	hour		sec	second
Hz	hertz		SI	Standard International units (Système International d'Unités)
in.	inch			
$in.^2$	square inch		V	volt
$in.^3$	cubic inch		W	watt
K	Kelvin		yd	yard
kcal	kilogram calorie		yd^2	square yard
kg	kilogram		yd^3	cubic yard

A

A *n.* **1.** an excellent grade; top quality. **2.** formerly the chemical symbol for argon (now Ar).

Å *n.* See Angstrom.

aardvark *n.* (Dutch *aardvarken*, aardvark) a large burrowing mammal (*Orycteropus afer*) about 2 ft (60 cm) high at the shoulders, with strong digging claws and long snout, tongue, ears, and tail. A native of Africa that feeds mostly on ants and termites. Also called anteater.

abattoir *n.* (French *abattoir*, slaughter place) slaughterhouse.

abdomen *n.* (Latin *abdōmen*, abdomen) **1.** the belly; the lower body cavity of a human or of many animals, containing the intestines, bladder, and reproductive organs (in females); separated from the thorax by the diaphragm. **2.** the part of the body that encloses this cavity. **3.** the posterior body segment of an arthropod, containing the reproductive organs.

abiosis *n.* (Greek *a*, without + *biosis*, way of life) absence of life.

abiotic *adj.* (Greek *a*, without + *biōtikos*, pertaining to life) **1.** without living organisms. **2.** not biotic; any component of the environment that is not a living organism; e.g., air, light, plant nutrients, water.

ablactate *v.* (Latin *ablactare*, to wean) to wean, as a calf from its mother.

ablate *v.* (Latin *ablatus*, carried away) **1.** to remove, as by surgery or erosion. **2.** to be worn away.

ablation *n.* (Latin *ablatio*, carrying away) **1.** removal of an organ or abnormal growth by surgery. **2.** the process of wearing away; gradual removal of a surface layer, as by erosion or sandblasting.

abloom *adj.* (Anglo-Saxon *a*, in + *blōstma*, bloom) blooming; producing blossoms.

Abney level *n.* a handheld instrument used to measure the slope of a surface. It has a split line of sight: one side has a line to be aimed parallel to the surface, and the other side has a mirror reflecting a leveling bubble. A scale indicates the slope in percent or degrees. The internal mirror enables the user to see the bubble at the top while sighting through the body of the instrument.

abnormality *n.* (Latin *anormalus*, not normal) an observable unusual condition or behavior in a person, animal, plant, or thing; something that deviates from

Abney level

the norm, especially in an undesirable way. Some abnormalities are symptoms that indicate a specific disease, mineral toxicity, mineral deficiency, or other problem.

abomasum *n.* (Latin *ab*, away + *omasum*, bullock's tripe) the 4th of the four stomachs of a cow or other ruminant; the true stomach, where digestion occurs.

aborigine *n.* (Latin *aboriginē*, first inhabitant) **1.** a descendant of the original or earliest known inhabitants of a nation or region; the native plants and animals of a region. **2.** an Australian Aborigine.

abortion *n.* (Latin *abortio*, miscarriage) **1.** termination of a pregnancy by removal of an embryo or fetus from the uterus (called a spontaneous abortion or miscarriage if it occurs by natural cause). **2.** abandonment of a project or mission. **3.** failure to develop, as in the abortion of an illness at an early stage.

abortive *adj.* (Latin *abortivus*, relating to abortion) incompletely developed, as a fetus born prematurely, a seed that does not germinate, or an action that is not completed.

abrasion *n.* (Latin *abrasio*, scraping away) **1.** a scrape, scratch, or cut that breaks the skin or other surface. **2.** wear of parts in a machine by friction. **3.** the wearing away of rocks by wind or water.

abscess *n.* (Latin *abscess*, a going away) a local collection of pus, commonly caused by bacterial infection.

abscission *n.* (Latin *abscissio*, cutting off) **1.** a natural separation, as the fall of a leaf, flower, or fruit. **2.** sudden termination by trimming.

abscond *v.* (Latin *abs*, from + *condere*, to hide) to go away hastily and hide, especially to avoid capture or injury. Example: honeybees that leave their hive when pesticides are being used nearby.

absinthe *n.* (Latin *absinthium*, wormwood) **1.** a bitter green liquid extracted from wormwood (*Artemisia absinthium*, family Asteraceae); the principal component is absinthin ($C_{30}H_{40}O_6$). **2.** a bitter drink with a licorice flavor containing 60% alcohol made from the wormwood extract. Its continued use can cause mental deterioration. It is banned in many countries.

absolute humidity *n.* the amount (weight or liquefied volume) of water vapor present in a unit volume of air.

absolute temperature *n.* a temperature measured from absolute zero, normally using either the Kelvin scale or the Rankine scale.

absolute zero *n.* the theoretical temperature of zero thermal energy for all substances; equivalent to 0 K (–273.16°C) or 0°R (–459.69°F). See Kelvin scale, Rankine scale.

absorption *n.* (Latin *absorptio*, sucked in) **1.** the passage of a substance into and dispersion through another, such as oxygen dissolved in water or nutrient ions entering plant roots. **2.** passage through a membrane into another part or substance, such as absorption of nutrients into the blood stream. **3.** assimilation; taking in, as the absorption of water by a seed or a sponge.   **4.** deep mental concentration. See adsorption, sorption.

abstract *adj.* (Latin *abstractus*, drawn off) **1.** generalized; theoretical; not a specific object or concrete reality. **2.** a characteristic that is difficult to define or to illustrate with a specific example; e.g., an abstract quality like righteousness. **3.** difficult to understand. *n.* **4.** a summary; a condensed expression of a larger concept or work, as an abstract of an article. *v.* **5.** to draw out the main points; to write a summary. **6.** to divert or steal attention. **7.** to generalize the qualities or characteristics of a body or group.

abutment *n.* **1.** a retaining wall such as a seawall or the end of a bridge, especially one made of masonry. **2.** the junction of adjoining parts, as the abutment of a part against another.

abyss *n.* (Greek *abyssos*, bottomless) **1.** an extremely deep place; generally very large and not measurable. **2.** the deep part of the ocean. **3.** a topic that cannot be understood or explained adequately, as the abyss of time.

abyssal plain *n.* the surface of a relatively flat part of the deep ocean floor.

acacia *n.* (Greek *akakía*, Egyptian thorn) **1.** a shrub or small tree (*Acacia* sp., family Fabaccac), mostly of warm, semiarid climates (e.g., Australia), that bears clusters of small yellow or white flowers; used as ornamentals and as a source of gum arabic. **2.** any of several similar species, including some locusts. **3.** gum arabic.

acaricide *n.* (Greek *akari*, mite + Latin *cide*, killer) a chemical for killing acarines. Also called miticide. See acarine, miticide, pesticide.

acarid *n.* or *adj.* (Greek *akari*, mite) a mite or pertaining to mites.

acarine *n.* (Greek *akari*, mite) any of a large number of arachnids (order Acarina); any mite or tick.

accelerated erosion *n.* erosion that is accelerated by human activity to a rate that is faster than geologic erosion and soil formation. See geologic erosion.

acceleration of gravity *n.* the acceleration of a body falling without friction under the influence of gravity. On Earth, the acceleration of gravity is 32.17 ft/sec^2 or 9.8 m/sec^2. Symbol: g.

accession *n.* (Latin *accessio*, an approach or addition) **1.** the action or process of coming into something, as accession to a title or position. **2.** an addition to one's holdings, as an accession of new property or new books for a library. **3.** acceptance or agreement, as accession to a demand or to a treaty. **4.** joining, as accession to a confederacy or other group. **5.** onset, as the accession of a disease.

access tube *n.* a tube installed vertically in soil to provide a place to insert a radiation source and detector, especially a neutron probe to determine the moisture content of the soil or a gamma-ray probe to measure soil density at various depths.

acclimation or **acclimatization** *n.* the physiological and behavioral adjustments of an organism to changes in its environment, especially to changes in temperature and/or altitude.

acclivity *n.* (Latin *acclīvitus*, steep slope) an upward slope; rising ground; e.g., a hillside viewed from the bottom, as opposed to declivity.

accretion *n.* (Latin *accretio*, increase) **1.** growth by gradual acquisition from an external source. **2.** growth of snowflakes or ice crystals by contact and freezing of water droplets.

accumulative pesticide *n.* a chemical that tends to concentrate in the food chain of animals or in the environment. See pesticide.

acetaminophen *n.* a widely used pain reliever and fever reducer ($C_8H_9NO_2$), named for its acetate, amine, and phenol components, that is less likely than aspirin to cause upset stomach. Its excessive use can cause liver damage.

acetylene *n.* a colorless hydrocarbon gas (C_2H_2) with a triple bond between its two carbons (HC≡CH); a highly combustible fuel used for oxyacetylene cutting and welding, for lighting, and in the synthesis of many organic compounds. Also called ethene.

acetylene reduction *n.* a process used to evaluate the activity of nitrogenase and thereby determine whether nitrogen is being fixed by a microbe.

achene *n.* (Latin *achenium*, not to gape) a small, dry, one-seeded fruit; e.g. the small spots on the surface of a strawberry.

achondroplasia *n.* (*a*, non + Greek *chondros*, cartilage + *plassein*, to form) a hereditary form of dwarfism resulting from a single gene that causes stunting by ossification of the cartilage in the long bones.

acid *n.* (Latin *acidus*, sour) **1.** a substance capable of releasing hydrogen ions (H^+). *adj.* **2.** having acid properties; a pH less than 7.0. **3.** an igneous rock high in silica (containing crystals of quartz).

acid deposition *n.* either dry or wet deposition of acidic particles or drops on exposed surfaces; e.g., acid rain. See acid precipitation.

acidification *n.* (Latin *acidus*, sour + *ificare*, to make + *atio*, action) the process of making something acid or adding to its acidity.

acidity *n.* (Latin *acidus*, sour + *itas*, condition) **1.** the condition of having a pH less than 7.0, often evidenced by a tart or sour taste or a corrosive effect on exposed surfaces. **2.** the degree of acid conditions; e.g., strong acidity in soils and water (below pH 4.0) is not a favorable environment for most higher (seed bearing) plants because it may cause changes in the microbial population, deficiencies of plant nutrients (especially calcium, magnesium, nitrogen, phosphorus, potassium, or molybdenum), or toxicities related to excessive concentrations of metallic ions (aluminum, iron, manganese, zinc, or other heavy metal ions). See alkalinity. Note: "Natural" rain has a pH of 5.6 caused by dissolved carbon dioxide forming carbonic acid (H_2CO_3).

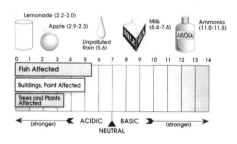

Acidity scale

acid mine drainage *n.* runoff or seepage water acidified by oxidation of impurities in coal or other mined materials (especially by oxidation of sulfur and nitrogen compounds to form sulfuric and nitric acids).

acidophilus milk *n.* a food product resulting from the culturing of milk with the bacterium *Lactobacillus acidophilus*. It is used to modify the bacterial population in a person's intestines and is commonly known as acidophilus-cultured buttermilk or sweet acidophilus.

acid precipitation or **acid rain** *n.* raindrops or other precipitation with pH less than 5.6. The principal chemicals responsible for acid precipitation are the oxides of nitrogen (NO_x) and sulfur (SO_2). See acidity.

acid wood *n.* wood used for making charcoal, acetic acid (CH_3COOH), and methanol (CH_3OH).

aclinic line *n.* (Greek *aklinés*, not bending) an imaginary line around the Earth near the equator where the Earth's magnetic fields are parallel with the Earth's surface and where a magnetic needle will not dip. See agonic line.

acoustics *n.* (Greek *akoustikos*, to listen) **1.** the physical science that deals with sound and sound waves. **2.** the characteristics of a room that determine how clearly a sound can be heard.

ACP *n.* See Agricultural Conservation Program.

acquired characteristic *n.* a change in a physiological or morphological characteristic caused by environmental factors; not hereditary. See Lamarackism, Lysenko.

acre *n.* (German *acker*, acre) a measure of land area equal to 43,560 ft^2 (4047 m^2 or 0.4047 ha) (abbr. ac). There are 640 ac in a square mile.

acre-inch *n.* a common measure of irrigation water; the volume required to cover an area of 1 ac to a depth of 1 in. (3630 ft^3 or 102.8 m^3).

acre-furrow slice *n.* the tilled part of 1 ac of land, often assumed to be about 6⅔ in. (17 cm) thick and to weigh 2 million pounds (1000 tons or 907.2 metric tons). See furrow slice.

acronym *n.* (Greek *akros*, at the end + *onym*, name) an abbreviation (usually the first letter from each of several words) that can be pronounced as a name. Example: AID = Agency for International Development.

actic *n.* the seashore between high tide and low tide.

actinic rays *n.* (Greek *aktinos*, ray or beam) short electromagnetic rays (beyond the violet end of the spectrum) that produce chemical changes. Example: X rays.

actinium *n.* (Greek *aktinos*, ray or beam + *ium*, element) element 89, atomic weight 227.0278; a radioactive (about 150 times as radioactive as radium), silvery-white heavy metal that glows with a blue light in the dark.

actinomycetes *n.* (Latin *actinomycetes*, ray fungus) a nontaxonomic term applied to a group of organisms with characteristics intermediate between the simple bacteria and fungi. Actinomycetes are common soil organisms that are very active in the composting of organic residues and give off an odor of rich soil under moist conditions. Most actinomycetes are unicellular microorganisms that produce a slender branched mycelium without internal cell walls and sporulate by segmentation of the mycelium or, more commonly, by segmentation of special hyphae. The group includes many but not all organisms belonging to the order Actinomycetales. Some species (e.g., *Streptomyces* sp.) are a source of antibiotics.

actinorhizal plant *n.* any of about 200 plant species nodulated by the nitrogen-fixing actinomycete, *Frankia*; e.g., plants of the genera *Causarina*, *Allocasuarina*, *Gymnostoma*, and *Alnus*.

activated carbon or **activated charcoal** *n.* a highly adsorbent form of carbon used to remove odors and toxic substances from liquid or gaseous emissions. In waste treatment, it is used to remove dissolved organic matter from wastewater.

activated sludge *n.* sludge that results when primary effluent is mixed with bacteria-laden sludge and then agitated and aerated to promote biological activity.

active dune *n.* a sand dune that has not been stabilized. Windblown sand moves over it and causes the dune to migrate slowly downwind.

active gully *n.* a gully that has not been stabilized. Its banks are not vegetated, and erosion occurs whenever water flows through it.

active ingredient *n.* the toxic component in any pesticide product that kills or otherwise controls target pests. Pesticides are regulated primarily on the basis of active ingredients.

active layer *n.* the layer of soil that thaws each year in cold climates where the soil is underlain by permafrost. Also known as suprapermafrost.

active site *n.* the part of an enzyme molecule that interacts with a substrate to convert it into a product.

acute exposure *n.* a single exposure to a toxic substance that causes severe biological harm or death. Acute exposures are usually characterized as lasting no longer than a day.

acute toxicity *n.* **1.** the ability of a substance to cause poisonous effects resulting in severe biological harm or death soon after a single exposure or dose. **2.** any severe poisonous effects resulting from a single short-term exposure to a toxic substance. See toxicity.

adapt *v.* (Latin *adaptare*, to fit or adjust) to adjust to a new environment, a new set of conditions, or a new purpose.

adaptation *n.* (Latin *adaptātio*, fitting) **1.** the process of adjusting to a new environment by an individual, group, or organization or by a species as an evolutionary process, especially one that affects the genome and phenotype of a species. **2.** alterations in form or structure of a species by natural selection in response to a new environment. **3.** a sensory response such as the enlargement or reduction in the opening of the pupil of an eye when light intensity changes. **4.** the changed product resulting from adjusting to a new situation; e.g., a movie that is an adaptation of a novel. **5.** a change in behavior in response to social pressure.

addax *n.* (Latin *addax*, addax) a large, pale antelope (*Addax nasomaculatus*) with long, twisted horns, native to northern Africa.

adder *n.* (Anglo-Saxon *naeddre*, snake) **1.** any of several venomous snakes (family Viperidea), especially the common European snake (*Vipera berus*), whose bite causes painful swelling but is rarely fatal to humans. **2.** any of several harmless American snakes, including the North American milk snake (*Lampropeltis triangulum*) that sometimes enters buildings in search of mice, and the North American puffing adder (*Heterodan contortix*), also called blowing adder, flat-headed adder, hognose snake, spreading adder, or sand viper. **3.** the venomous Australasian snake known as the death adder (*Acanthophis antarctius*). **4.** the venomous African puff adder (*Bitis arietans*). **5.** either the Indian banded adder (*Bunganis coeruleus*) or the venomous Indian asp or serpent (*Naje haje*). **6.** the venomous horned adder (*Cerastes cornutes*) of the Near East. **7.** certain European fish, including the sea adder (*Spinachia spinachia*) and the pipefish (*Nerophisi* sp., family Syngnathidae).

additive *n.* (Latin *additivus*, added) **1.** anything added to a diet, ration, food, fertilizer, pesticide, or other product to increase its efficiency or to give it more desirable characteristics; e.g., molasses, urea, and/or anhydrous ammonia added to forage when it is ensiled to increase its feeding value. *adj.* **2.** cumulative; a group of items with one or more properties that can be added together.

addled egg *n.* an egg with its yolk mixed with the white; sometimes a rotten egg; generally considered inedible.

adenine *n.* (Greek *adenin*, gland chemical) a purine base alkaloid ($C_5H_5N_5$) that is a fundamental component of nucleic acids and combines with ribose to form adenosine. It is extracted from tea or synthesized for use in certain medicines.

adenosine *n.* (Greek *adenosin*, gland sugar) a white, crystalline organic compound ($C_{10}H_{15}N_5O_4$) formed by combining adenine and ribose to form a water-soluble nucleoside that combines with phosphate groups to form ADP and ATP.

adenosine diphosphate (ADP) *n.* a nucleotide that serves in the energy transfer process when glucose is oxidized in plant or animal tissue. It is composed of adenine, ribose, and two phosphate groups.

adenosine triphosphate (ATP) *n.* the most important nucleotide in the energy transfer processes of both plants and animals. It is formed by an enzymatic reaction that adds a third phosphate group with a high-energy bond to adenosine diphosphate.

adenovirus *n.* (Greek *adenoeidés*, adenoid + Latin *virus*, poison) any of 49 or more 20-sided viruses (family Adenoviridae) originally identified in adenoid tissue that cause respiratory diseases (including common colds, bronchitis, and pneumonia) and conjunctivitis (pink eye) and may cause tumors in animals. Altered forms are used in gene therapy against tumors and cystic fibrosis.

adephagous *adj.* (Greek *adēn*, enough + *phagein*, to eat) having a gluttonous appetite, especially certain gluttonous beetles.

adhesion *n.* (Latin *adhaesio*, a clinging) **1.** the action or process of attaching or being attached to something. **2.** the attraction between unlike materials that are in contact; e.g., water adheres to any surface that it wets.

adhesive *n.* (Latin *adhaesio*, clinging + *ivus*, quality of) **1.** a substance; e.g., glue, that sticks to something. *adj.* **2.** tending to stick and hold. **3.** having a sticky coating, as adhesive tape or an adhesive bandage.

adiabatic *adj.* (Greek *adiábatos*, cannot be crossed) taking place without gain or loss of heat. Adiabatic compression results in warming, and adiabatic expansion results in cooling.

adiabatic rate *n.* the average rate of cooling of an air mass as it rises; in summer = 1°F/330 ft or 1°C/180 m; in winter = 1°F/400 ft or 1°C/220 m.

adipose *adj.* (Latin *adeps*, fat + *osus*, full of) **1.** fatty or pertaining to fat. *n.* **2.** accumulations of animal fat in the body.

Adirondack Mountains *n.* a forested mountain range of about 5000 mi^2 (13,000 km^2) in northeastern New York State; the site of many health resorts. Mount Marcy, the highest peak, is 5344 ft (1629 m) high. The principal tree species in the Adirondacks are spruce and maples.

adjustment *n.* (Latin *ajuster*, to make conform + *mentum*, act) **1.** a small change to make something fit properly. **2.** a change made in response to a particular condition or situation; e.g., changes in range management practices as warranted by specific conditions of the herbage or water facility, including changes in kinds or classes of animals, stocking rates, or seasons of use. **3.** the changes an organism makes in response to its environment. See adaptation.

adjuvant *n.* (Latin *adjuvant*, to help) any solid or liquid substance added to a product such as a pesticide or fertilizer to increase its effectiveness or efficiency. Examples include carriers, diluents, emulsifiers, solvents, spreaders, stickers, or sometimes another pesticide or fertilizer.

ad lib *adj.*, *adv.* (Latin, short for *ad libitum*, at pleasure) **1.** impromptu, as in an unrehearsed statement, speech, or musical piece. **2.** as desired; without restriction; commonly used to express the availability of animal feed on a free-choice basis.

administrative order *n.* a legal document directing an individual, business, or other entity to take a specific action or to refrain from an activity. For example, the Environmental Protection Agency may use it to describe violations of environmental laws and corrective actions to be taken. Such orders may be issued, for example, as a result of an administrative complaint whereby the respondent is ordered to pay a penalty for violations of a statute, and they can be enforced in court.

adobe *n.* (Arabic *at tob*, the brick) **1.** unburned, sun-dried bricks made of a mixture of mud and straw. **2.** clay and silty deposits used for making sun-dried bricks, usually in dry areas such as the desert basins of southwestern U.S.A. and Mexico. *adj.* **3.** made of adobe bricks; e.g., an adobe hacienda. Also spelled adaubi, adabe, dobe, dobie, dobby, doby, dogie.

adopt-a-highway *n.* a program whereby a group (usually an organization or a business) agrees to remove litter periodically along a designated section of a highway as a public service project.

ADP *n.* See adenosine diphosphate.

adsorption *n.* (Latin *ad*, at + *absorptio*, sucked in) **1.** bonding of molecules of gas, liquid, or dissolved solids to a surface by either chemical or physical attachment, normally weak enough for desorption or exchange to occur. **2.** an advanced method of treating wastes by using activated carbon to remove organic matter from wastewater. See sorption, absorption.

adularia *n.* See potassium feldspar.

adulterants *n.* (Latin *adulterant*, altered) chemical impurities or substances that by law do not belong in a food, pesticide, or other product.

advection *n.* (Latin *advectio*, carried or brought) the horizontal flow of a fluid (especially an air mass) and the accompanying transfer of properties such as temperature, pressure, and humidity.

advection fog *n.* fog produced by the cooling of an air mass as it flows across a cold surface.

advective frost *n.* frost produced by cold air moving downslope and accumulating in a "frost pocket."

adventitious *adj.* (Latin *adventicius*, coming from outside) growing from an unusual place; an adventitious root or stem looks out of place.

adventive *adj.* (Latin *adventus*, coming from elsewhere) **1.** not native to the ambient environment, usually rare and exotic. *n.* **2.** a non-native plant, animal, or human. See feral.

adze *n.* (Anglo-Saxon *adesa*, ax) a hand tool like an ax with its blade turned crosswise and curved; used to cut a flat surface.

AEC *n.* the U.S. Atomic Energy Commission, since renamed the Nuclear Regulatory Commission—an independent agency with statutory responsibility for atomic energy regulations.

aedes *n.* a mosquito (*Aedes aegypti*) that carries the virus of urban yellow fever and dengue fever and possibly filariasis and encephalitis.

AEI *n.* See agri-environmental indicator.

aeolian or **eolian** *adj.* (Greek *aiolos*, wind) **1.** eroded and deposited by wind, usually a silty or sandy soil material. See loess, dune. **2.** the erosive action of wind.

aeration *n.* a process that exposes materials to air, often by mechanical mixing or bubbling devices. It is commonly used to support life or to promote biological degradation of organic matter in wastewater. The process may be passive (as when waste is exposed to air) or active (as when a mechanical device makes wastewater aerobic).

Aerating wastewater

aeration pore space *n.* the pore space in a soil that drains readily after a rain. Soils generally need to have about 10% aeration pore space to provide adequate oxygen for the roots of most plants.

aerenchyma *n.* (Greek *aer*, air + *énchyma*, infusion) tissue composed of thin-walled cells with large internal spaces where air can circulate in certain aquatic plants.

aerial *adj.* (Greek *aerios*, of the air) **1.** above ground; surrounded by air; e.g., the stem, branches, and leaves of a plant. **2.** as seen or done from above, as aerial photography. **3.** occupants of the atmosphere, as birds are aerial creatures. **4.** operating above ground, as an aerial ski lift. **5.** ethereal; light and graceful, as aerial sounds or music. **6.** growing in the air, as aerial roots. *n.* **7.** an antenna.

aerobic *adj.* (Greek *aero*, air + *bios*, life) **1.** needing, requiring, or tolerating the presence of air or gaseous oxygen for life or other processes. **2.** the presence of air in an environment. See anaerobic.

aerobiology *n.* the study of microorganisms carried by wind.

aerology *n.* the study of the upper atmosphere, especially through observations made with balloons or airplanes. See meteorology.

aeropathy *n.* (Greek *aer*, air + *pathos*, disease) any sickness caused by change in atmospheric pressure, such as air-decompression sickness.

aerophobia *n.* (Greek *aer*, air + *phobos*, fear) the fear of drafty air or airborne matter.

aerophyte *n.* (Greek *aer*, air + *phyton* plant) an epiphyte.

aeroplankton *n.* (Greek *aer*, air + *planktos*, wandering) small plants, bacteria, pollen grains, spores, and animals that are suspended in the atmosphere. See plankton, phytoplankton, zooplankton.

aerosol (Greek *aer*, air or gas + *sol*, an abbreviation for solution) *n.* **1.** a colloidal system consisting of tiny solid particles and/or liquid droplets dispersed in air or any other gas. Examples are fog—colloidal-sized droplets of water dispersed in air; smoke—colloidal-sized carbon and associated materials suspended in air; and, in medicine, a bactericidal solution sprayed in colloidal-sized droplets to sterilize the air, usually for inhalation therapy. **2.** finely atomized spray or smoke with particles ranging between 0.1 and 50 microns in size. The particles may be produced by blasts of heated air, exhaust gases, rapid volatilization of a liquefied gas containing a nonvolatile chemical solution, or the release of a pressurized liquid through a small orifice. Insecticides, antibiotics, germicides, and deodorants are often used in aerosol form.

aerosphere *n.* (Greek *aer*, air + *sphaira*, round body) the atmosphere surrounding the Earth.

aesthetic or **esthetic** *adj.* (Greek *aisthētikos*, perceptible by the senses) beautiful to behold; pleasing in an emotional rather than an intellectual sense.

aesthetic insults *n.* damage done to the environment through mistreatment or neglect.

affinity *n.* (Latin *affinitas*, adjacent) **1.** a relationship established by marriage rather than shared descent. **2.** a close relationship, especially a similarity that implies a common origin, as the affinity of Latin languages. **3.** a special attraction, often a mutual attraction; e.g., that between a man and a woman who love each other. **4.** the chemical property that causes the atoms of certain elements to form stable compounds with the atoms of another group of elements.

affluent *n.* (Latin *affluent*, rich) a tributary stream.

afforestation *n.* (Latin *af*, addition to + *forestis*, unenclosed woods + *atio*, act of making) the establishment of trees where they never grew before. See reforestation.

aflatoxin *n.* (Latin *A. flavus*, a fungus + *toxicum*, poison) a potent carcinogen produced by fungi (*Aspergillus flavus* and *A. parasiticus*). These fungi survive over winter in soil or on crop residues and develop rapidly during high-moisture grain storage. They enter kernels of corn, peanut, rice, and other grains through injuries caused by insects or cracks resulting from careless mechanical harvest. Caution: never feed moldy grain to livestock or poultry or eat moldy foods, because they may contain aflatoxins that can affect people, domestic livestock, and poultry.

Africanized honeybee *n.* a hybrid honeybee produced in Brazil in the mid 1950s by the mingling of domestic honeybees with African honeybees imported with the objective of producing more honey in the Tropics. However, it is more aggressive than other species of honeybees and stings repeatedly. The stings of droves of bees can be deadly. The Africanized honeybees entered Texas from Mexico about 1990 and have been expanding northward.

African sleeping sickness *n.* See trypanosomiasis.

African swine fever *n.* a highly contagious viral disease in swine; similar to cholera. It causes high fever, internal hemorrhaging, skin discoloration, and a very high mortality rate.

aftosa *n.* (Spanish *aftosa*, aphthous fever) foot-and-mouth disease; a viral disease capable of infecting all cloven-hoofed animals and, rarely, humans.

Ag *n.* (Latin *argentum*, silver) the chemical symbol for silver.

agamic *adj.* (Greek *a*, not + *gámos*, marriage) **1.** reproducing without mating. **2.** capable of reproduction without fertilization.

agave *n.* (Greek *agaué*, noble, brilliant) a desert plant (*Agave* sp., family Amaryllidaceae) native to the Americas, with thick spiny leaves and a tall flower stalk, especially the century plant (Sp. *maguey*). Fibers from the leaves are used to make rope and paper, and extracts from its sap are used to make soap and are fermented to make alcoholic beverages such as pulque or distilled to make mescal or tequila.

age *n.* (French *aage*, time or lifetime) **1.** the length of time an organism or object has existed since its birth or creation. **2.** a period during the existence of an organism or thing marked by certain characteristics, as middle age of a person or the carboniferous age of geologic time marked by certain types of rocks and fossils. **3.** the contemporary time or world, as the present age. **4.** a level of mental, physical, or emotional development. **5.** the latter period of the life of an individual, as old age. *v.* **6.** to mature, ripen, or grow old. **7.** to cause something to mature, ripen, or grow old.

Agency for International Development (AID or **USAID)** *n.* an agency of the U.S. Department of State that provides agricultural and other assistance in developing countries.

agent *n.* (Latin *agent*, doer) **1.** an independent being capable of making decisions. **2.** a person or entity that acts on behalf of another; a representative. **3.** a causative force. **4.** a chemical that causes a reaction to occur. **5.** an organism that causes or transmits a disease.

Agent Orange *n.* a powerful herbicide containing 2,4-D and 2,4,5-T with a contaminant, dioxin, thought to cause cancer. It was used to defoliate crops and forests during the Vietnam War. See dioxin.

agglomerate *v.* (Latin *agglomerare*, to accumulate and form a ball) to grow larger by collecting a cluster of particles or a mass of droplets. Examples: miscellaneous rock fragments may agglomerate to form a rock, or tiny droplets of water may agglomerate to form raindrops, sleet, or snowflakes.

agglomeration *n.* (Latin *agglomerare*, to accumulate and form a ball + *ātiōn*, action) **1.** the cluster formed as fragments agglomerate. **2.** the process of forming clusters of fragments.

agglutination *n.* (Latin *agglutinatus*, glued together + *atio*, action) **1.** a joining, as by adhesion or being glued together. **2.** the clumping together of fat globules in milk or of bacteria in immune sera. **3.** the action of antibodies called agglutinins causing bacteria, other microbes, or cells suspended in a fluid to stick together. **4.** clotting of red blood cells. **5.** the formation of words by combining morphemes.

agglutinin *n.* (Latin *agglutinans*, sticking) an antibody produced as the response of an animal body to an infection or to the injection of microorganisms. Agglutinins cause the organisms responsible for their formation to agglutinate.

aggradation *n.* (Latin *ag*, add to + *gradi*, to walk + *atio*, action) **1.** the process of building up a land surface by gradual deposition of sediments on lowlands and on gentle slopes below steeper slopes. **2.** the growth of a permafrost area.

aggregate *v.* (Latin *aggregare*, to add to) **1.** to collect, unite, or bring together into a mass. *adj.* **2.** made up of many particles that can be separated by mechanical means. *n.* **3.** the fragmented mineral material (sand, gravel, slag, crushed stone, etc.) that is mixed with cement or bituminous material to form mortar or concrete. **4.** a fertilizer or grain sample composed of several subsamples of material being analyzed. **5.** a group of soil particles that cohere together. Stable aggregates may become peds or parts of peds.

aggregate stability *n.* the strength of soil aggregates; a function of the amount and type of clay, humus content, soil reaction, microbial activity, etc., determining resistance to disruption when aggregates are exposed to water, especially to the impact of raindrops.

aging cycle of lakes *n.* the life cycle of lakes, whereby they gradually fill in with sediment and plant debris (over decades or millennia, depending on the size and surroundings of the lake) until they become swamps and eventually organic soil. See peat, muck.

agonic line *n.* a line passing through all the points on the Earth's surface where the magnetic declination is zero; i.e., where the needle of a compass points exactly north and south.

agoraphobia *n.* (Greek *ageirein*, to assemble + *phobos*, fear) a fear of open places, crowds, or public places; especially that occurring in people or animals kept in confinement for a long period. Among horses, it is also called picket-line-bound; an affected horse is called a barn rat.

Agonic line

agrarian *adj.* (Latin *agrarius*, of the land) **1.** pertaining to land, land ownership, and land use. **2.** promoting agriculture and agricultural interests. **3.** rural; agricultural.

agrarian reform *n.* an effort to improve land ownership policies, usually by dividing large ownerships into smaller units owned by those who work the land.

agribusiness *n.* a coined word referring to the full assemblage of operations related to the business of agriculture. It includes the interrelationships among farmers and all of the businesses that provide services to farmers; e.g., farm machinery manufacturers and vendors and a major portion of the chemical industry as a supplier of agrichemicals.

agrichemicals *n.* a coined word for agricultural chemicals, including all chemical materials used in agriculture, such as fertilizers, liming materials, herbicides, insecticides, and fungicides.

agric horizon *n.* a diagnostic subsurface horizon in Soil Taxonomy. An illuvial horizon that is formed by accumulation of silt, clay, and humus in the zone just below the plow layer of soils that have been tilled for long periods.

agricolist *n.* a farmer.

agricolous *adj.* pertaining to farmers or farming.

agricoltor *n.* a farmer.

agricultural chemicals *n.* See agrichemicals.

Agricultural Conservation Program (ACP) *n.* a U.S. government program to assist farmers by sharing costs related to specific objectives, especially to reduce soil erosion and sedimentation; incorporated into the Environmental Quality Incentives Program (EQIP) in 1996. See Farm Service Agency, Agricultural Stabilization and Conservation Service.

Agricultural Extension Service *n.* See Cooperative Extension Service.

Agricultural Hall of Fame. *n.* See National Agricultural Center and Hall of Fame.

agricultural land *n.* any land that is used for agricultural purposes, such as raising crops or keeping animals to be used for food, work, or pleasure.

agricultural pollution *n.* the polluting effects of all of the liquid, solid, and gaseous wastes produced by producing crops, raising livestock, and other forms of agriculture. This includes soil sediments, fertilizer and pesticide residues and losses, animal manures, and dead animals. See nonpoint source, point source.

Agricultural Research Service (ARS) *n.* a federal agency in the U.S. Department of Agriculture responsible for conducting research on agricultural problems and food products.

Agricultural Stabilization and Conservation Service (ASCS) *n.* an agency established in the U.S. Department of Agriculture in 1961 as a successor to the Agricultural Adjustment Administration (AAA), which was established in 1933, and followed by the Production and Marketing Association (PMA) and then the Commodity Stabilization Service (CSS). Until it was replaced by the Farm Service Agency in 1994, the ASCS handled agricultural production and marketing programs and the Agricultural Conservation Program that helped finance approved soil and water conservation structures.

agriculture *n.* (Latin *agri*, field + *cultura*, culture) **1.** the occupation, art, and science of raising crops and livestock (farming and ranching). **2.** The entire food production industry, including the supply of materials used in farming, the practice of farming and ranching, the marketing of agricultural products, and the information services available to farmers and ranchers.

agri-environmental indicator (AEI) *n.* a measurable environmental characteristic that is suitable for evaluating the condition of an environment as a basis for policy decisions.

agrobiology *n.* (Latin *agri*, field + Greek *bios*, life + *logos*, word) the science that studies plant life and plant nutrition.

agroecology *n.* (Latin *agri*, field + *oeco*, house + Greek *logos*, word) the science of applying a combination of agronomic and ecological principles to put production agriculture on an environmentally sound, sustainable basis. See agroecosystem.

agroecosystem *n.* (Latin *agri*, field + *oeco*, house + *systema*, placed together) an artificial agricultural environment managed by humans; e.g., a cropped field, an orchard, or a pasture, understood as an ecosystem. See ecosystem.

agroforestry *n.* (Latin *agri*, field + *forestis*, unenclosed woods + *aria*, place for) a system of growing annual and other agricultural crops as intercrops with established shrubs and trees, ranging from home gardens with trees to fields shaded by multilayered tree crops, and ranging from fields with an occasional tree to areas that are dominantly trees with small areas of annual crops. See alley cropping, vetiver.

agrology *n.* (Greek *agros*, tilled land + *logos*, word) the scientific study of agricultural crop production; agronomy.

agronomy *n.* (Greek *agros*, tilled land + *nomia*, law) the integration of the sciences of crop production and soil management. See agrology.

agropastoral system *n.* (Latin *agri*, field + *pastoralis*, like a shepherd) an agricultural production system that emphasizes livestock production and relies heavily on the use of pastures and good pasture management.

agropyron mosaic *n.* a viral disease that causes stunting, spots and stripes on leaves, and leaf necrosis in wheatgrass (*Agropyron* sp.).

agrostology *n.* (Greek *agrōstis*, kind of grass + *logos*, word) the scientific study of grasses and grasslike plants; a branch of systematic botany.

agrypniaphobia *n.* (Greek *agrypnos*, sleepless + *phobos*, fear) a human fear that if something can keep you awake, it will keep you awake. Examples: worrying that everything is wrong with the environment or that coffee will keep you awake. See caffeine.

A horizon *n.* the upper layer of a soil that is darkened by organic coatings on the soil particles; commonly called topsoil, but topsoil often includes additional layers that are not technically A horizons.

AID *n.* See Agency for International Development.

AIDS *n.* acquired immune deficiency syndrome. A terminal disease caused by a retrovirus, characterized by a deficiency of certain leukocytes, and resulting in increasing susceptibility to other diseases. AIDS is spread by transfer of body fluids, especially by blood transfer via intravenous injections and by sexual contact.

ailurophobia *n.* (Greek *ailouros*, cat + *phobos*, fear) an abnormal fear of cats.

air *n.* (Greek *aēr*, the lower atmosphere) the mixture of gases that compose the Earth's atmosphere. Dry air is composed of about 78% nitrogen, 21% oxygen, 1% argon, and trace amounts of other gases and particulates. See atmosphere.

airborne particulates *n.* total suspended particulate matter found in the atmosphere as solid particles or liquid droplets. The chemical composition of particulates varies widely, depending on location and time of year. Airborne particulates include windblown dust, emissions from industrial processes, smoke from the burning of wood and coal, the exhaust of motor vehicles, and pollen from plants that may cause hay fever.

air classification *n.* a process used in a resource recovery plant to sort waste materials that have been shredded. A current of air lifts low-density combustible materials while denser noncombustible materials fall into a chute.

air contaminant *n.* any extraneous particulate matter, gas, or combination thereof present in the atmosphere. See air pollutant.

air curtain *n.* a method of containing oil spills on water. Air bubbling through a perforated pipe causes an upward water flow that slows the spread of oil. It can also be used to stop fish from entering polluted water.

air drainage *n.* **1.** the gravity-driven flow of cold air down a valley or a slope, or of warm air up a valley or slope. **2.** the relative movement of air about a plant, a home, or a livestock facility. Free air circulation reduces frost hazard and dissipates pollutants.

air glow *n.* a faint light caused by ionic radiation in the upper atmosphere (much dimmer than the aurora).

air mass *n.* a widespread body of air that gains certain meteorological or polluted characteristics; e.g., a heat inversion or smog.

air mass weather *n.* a weather system that may persist for several days, caused by the influence of a large air mass.

air pollutant *n.* any extraneous substance in air capable, if in high enough concentration, of harming humans, animals, vegetation, or materials. Pollutants may be in the form of solid particles, liquid droplets, gases, or a combination of these forms. Usually, they fall into two main groups: (1) those emitted directly from identifiable sources and (2) those produced in the air by interaction between two or more primary pollutants or by reaction of one or more pollutants with normal atmosheric constituents. The following are considered air pollutants: airborne solids, sulfur compounds, volatile organic chemicals, nitrogen compounds, halogen compounds, radioactive compounds, and odors. Hazardous indoor air pollutants include radon, lead paint, asbestos fibers, fumes from a wood stove or kerosene heater, and such common items as pesticides, cleaners, glues, and disinfectants.

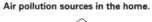
Air pollution sources in the home.

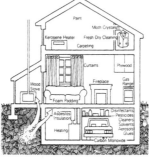
Air pollution sources

air pollution episode *n.* a period of abnormally high concentration of air pollutants that can cause illness and death, often associated with low winds and temperature inversion. See episode, pollution.

air pressure *n.* **1.** atmospheric pressure. **2.** the pressure of air against any surface, whether confined or unconfined. Commonly expressed in Pascals, atmospheres, bars, inches or centimeters of mercury, lb/in.^{2}, or kg/cm^2.

alachlor *n.* a herbicide used mainly to control weeds in corn and soybean fields.

alar *adj.* (Latin *alar*, winged) **1.** winged or wing-shaped. *n.* **2.** trade name for daminozide ($C_6H_{12}N_2O_3$), a plant growth regulator that makes apples redder, firmer, and less likely to drop off trees before growers are ready to pick them. It is also used to a lesser extent on peanuts, tart cherries, Concord grapes, and other fruits.

Alaskan pipeline *n.* a 789-mile (1270-km) pipeline that crosses Alaska from north to south. After oil was discovered in 1968 at Prudhoe Bay on the Arctic Ocean, it was a controversial news item in the U.S.A. and around the world in the 1970s because the proposed pipeline would cross the environmentally sensitive Alaskan tundra. Environmentalists successfully argued the following:
- A pipeline carrying *hot* oil, if buried, would melt the permafrost and the line would collapse and pollute the environment.
- The pipeline could be supported above ground but if so built it would obstruct animal migration. (This method of construction was chosen.) See permafrost, tundra.

albatross *n.* (Portuguese *alcatraz*, sea eagle) **1.** any of several large seabirds (family Diomedeidae) with long wings (some wingspans as long as 11 ft or 3.3 m) that enable them to soar for hours at a time. They have white bellies, spotted brown or white backs, and webbed feet and are sometimes called great gulls. **2.** an extremely heavy burden, especially of guilt or responsibility, that makes progress nearly impossible.

albedo *n.* (Latin *albedo*, whiteness) the amount of sunlight reflected from a surface as a percentage of the total amount of light falling on that surface. The average albedo for the Earth is 30%. Typical values are fresh snow, 75%–95%; grass, 10%–30%; and forest, 3%–10%. The high albedo of snow can cause snow blindness.

albic horizon *n.* an eluvial diagnostic subsurface horizon in Soil Taxonomy. A light-colored, leached layer at least 1 cm thick that usually underlies a dark-colored surface horizon and overlies a less permeable horizon such as an argillic horizon, a spodic horizon, or a fragipan.

albino *n.* (Latin *albus*, white) **1.** a human or animal with inherited white skin, white hair, and pink eyes resulting from the absence of melanin. **2.** a plant with white color resulting from a lack of chlorophyll.

albumen *n.* (Latin *albus*, white) **1.** the white of an egg. **2.** albumin.

albumin *n.* (Latin *albus*, white + *in*, protein) a class of water-soluble proteins that contain about 2% sulfur. Albumin is soluble in water, dilute salt solution, and 50% saturated ammonium sulfate and is coagulated by heat. It is found inmilk, blood, egg white, muscle, and vegetables. Egg white (albumen) is about 20% albumin, consisting of a glycoprotein also known as ovalbumin.

alburnum *n.* (Latin *alburnus*, whitish) sapwood; the soft, light-colored, new wood between the bark and the heartwood of a tree or shrub, consisting of cambium and phloem cells. Native Americans ate the alburnum of some species of trees during famines; e.g., that of Ponderosa pine (*Pinus ponderosa*) of the northwestern U.S.A.

alcohol *n.* (Arabic *al*, the + *kohl*, powdered antimony) **1.** the family name of a group of organic compounds composed of a hydrocarbon chain plus a hydroxyl group; includes methanol, ethanol, isopropyl alcohol, and others. **2.** ethanol (C_2H_5OH). See ethanol. **3.** a beverage that intoxicates because it contains ethanol; e.g., whiskey, brandy, vodka.

alcohol-alizarin test *n.* a test for milk freshness. Ethyl alcohol (C_2H_5OH) and alizarin ($C_{14}H_8O_4$) are added to the milk being tested. A reaction color of lilac red indicates normal milk, yellowish brown indicates sour milk, and violet indicates mastitis milk. See mastitis.

alcoholic fermentation *n.* the anaerobic transformation of glucose and other simple hexose sugars into ethanol. Carbon dioxide is a by-product.

alcoholic slime flux *n.* a white, frothy slime that seeps from the bark or sapwood near the base of a diseased tree. It is rich in starches, sugars, and proteins and is fatal to a tree if sustained over long periods.

alcohol test *n.* a test for milk quality. Equal parts of milk and ethyl alcohol (C_2H_5OH) are mixed to detect abnormal milk or milk with unusual salt balances. The test tube remains clear with normal milk, whereas the curd clings to the glass if the milk is abnormal.

aldehyde *n.* (short for alcohol dehydrogenated) an organic chemical with a double-bonded oxygen (a CHO group) at the end of a carbon chain or, in formaldehyde, simply CH_2O.

aldose *n.* an aldehyde sugar.

Aleochara beetle *n.* a tiny rove beetle (*Aleochara bimaculata*) that attacks and eats cabbage maggots and other fly maggots much larger than itself. These beetles destroy as many as 80% of the cabbage maggots in a field. They lay eggs in the soil where cabbage maggots are likely to occur so their larvae can find and consume the cabbage maggot larvae. See cabbage maggot.

aleukia *n.* (Greek *a*, absence of + *leukós*, white + *ia*, condition) **1.** absence of leukocytes in the blood of a person or animal. **2.** alimentary toxic aleukia in people and some animals is caused by ingesting a mycotoxin produced on damp grain crops by one or more of the following genera of fungi: *Actinomyces, Alternaria, Cladosporium, Fusarium, Penicillium, Piptocephalis, Rhizopus, Thammidium, Tricoderma, Trichothecium,* and *Verticillium*. See mycotoxin.

alfalfa *n.* (Arabic *al fisfisa*, fresh fodder) lucerne (*Medicago sativa*, family Fabaceae); a perennial leguminous plant with purple flowers, widely used for forage (as hay, silage, or pasture with careful management to avoid bloat) and often considered the oldest and best of all hay crops, combining high yield and high protein content. It is subject to diseases and insects, including leaf spot, black stem, bacterial wilt, verticillium wilt, potato leafhopper, clover leaf weevils, and alfalfa weevils.

Alfisol *n.* (from the chemical symbols Al and Fe + soil) a soil order in Soil Taxonomy composed of weathered soils that have moderate accumulation of clay in their B horizons (subsoils). They occupy 14% of the U.S. land area.

alga (*pl.* **algae**) *n.* (Latin *alga*, seaweed) any of several single-celled, rootless, photosynthetic plants (phylum Thallophyta) that grow in colonies in sunlit waters in proportion to the supply of the most limiting nutrient (often phosphorus). Algae are primary food producers that serve as food for fish and small aquatic animals. Also called seaweed, kelp, or pond scum.

algae, blue-green *n.* See blue-green algae.

algal bloom *n.* a spurt of rapid algal growth in a body of water; often defined as more than 500 algae per milliliter of water. Algal blooms generally affect water quality adversely by lowering the dissolved oxygen in the water and giving water a bad taste and odor. Decay of an algal mass depletes the oxygen content of water, produces putrid odors, and can cause hazardous changes in local water chemistry. Note: The word "bloom" in this usage is figurative; algae do not produce blossoms.

algalization of rice fields *n.* the use of blue-green algae on rice fields to supply symbiotic atmospheric nitrogen to the rice crop. See azollae, blue-green algae.

algefacient *n.* (Latin *algere*, to be cold + *faciens*, making) something that cools. Examples: a cool drink or a cold wave.

algicide *n.* (Latin *alga*, seaweed + *cida*, killer) any chemical used to kill algae, commonly copper sulfate ($CuSO_4$).

algin *n.* (Latin *alga*, seaweed + *in*, carbohydrate) a colloidal, gelatinous carbohydrate found in the cell walls of kelp (brown seaweed) and used in medicines, ice cream, jellies, paper sizing, and plastics. It makes ice cream and gelatin smooth and noncrystalline. Its acid form ($C_6H_8O_6)_n$ is called alginic acid, and its salts are alginates.

algology *n.* (Latin *alga*, seaweed + *logica*, word) the scientific study of algae.

alidade *n.* (Arabic *alidādah*, a rule) **1.** a scaling device used on a surveying instrument, including a vernier scale and a pointer. **2.** a surveying instrument designed to rest on a plane table to read distance (by means of a stadia rod) and indicate direction. It is used to locate points on a map that represent positions on a landscape.

aliphatic *adj.* (Greek *aleiphat*, oil or fat + *ic*, like) the type of structure exhibited by organic compounds that have carbon chains but do not have aromatic ring structures; e.g., the aliphatic acids (acetic, butyric, propionic, etc). See aromatic.

alkali *n.* (Arabic *al*, the + *qali*, saltwort ash) a chemical compound that can react with and neutralize acids, forming water and a salt. Alkalis are composed of hydroxyl groups (OH^-) and metallic elements, such as sodium, potassium, calcium, and magnesium, or the ammonium radical.

alkali bee *n.* a small wild bee (*Nomia melanderi*, family Nomiinae) that nests in the soil in salty valleys in western U.S.A. The females make small individual nests in fine-grained soils with good drainage, high moisture storage capacity, sparse vegetation, and low organic content. Most nesting sites are found on low hummocks and gentle slopes where a high evaporation rate has left salty conditions and sparse vegetation. In suitable sites, they form aggregations of nest burrows that number thousands or even millions of nests and occupy 1 ac or more of ground. Alkali bees are valuable to growers of legume seeds, especially for the pollination of alfalfa.

alkali chlorosis *n.* a yellowing or blanching of plant leaves caused by excessive salts and/or high pH in the soil.

alkali disease *n.* **1.** selenium poisoning of livestock causing emaciation, hair loss, deformed hooves, and blind staggers. It is caused by grazing on selenium accumulator plants (e.g., *Astragalus* sp.) growing on alkaline alluvial soils that are high in selenium. The selenium concentration in the forage can be reduced by sulfur fertilization. See selenium. **2.** tularemia.

alkali soil *n.* **1.** soil with a pH higher than 7.0, especially one higher than 8.5. See alkaline soil. **2. black alkali** sodic soil; soil that has a pH higher than 8.5, usually caused by more than 15% exchangeable sodium on the cation exchange complex. Black alkali soils generally have dissolved organic matter that gives them a dark color, dispersed clays, crusted surfaces, very low permeability, and little if any plant growth. See sodic soil. **3. white alkali** saline soil; called white alkali because white salts crystallize on the surface during dry periods. See saline soil.

alkaline soil *n.* any soil that has a pH above 7.0 is technically alkaline, but the term usually means a soil with a pH near 8 or above. A pH from 7.8 to 8.2 commonly indicates an excess of calcium carbonate and limited availability of phosphorus, potassium, iron, manganese, zinc, copper, and boron. Many common field and garden crops may not grow well because of nutrient deficiencies. See alkali soil, saline soil, sodic soil.

alkalinity *n.* (Arabic *al*, the + *qali*, saltwort ash + Latin *itās*, state) **1.** the condition of being alkaline. **2.** the degree of alkaline condition, as high alkalinity means that the pH is much above neutral (usually more that 8.5).

alkaloid *n.* (Arabic *al*, the + *qali*, saltwort ash + Greek *eidos*, form) any of a group of about 2000 complex ring compounds that contain nitrogen (normally in heterocyclic rings). Some alkaloids are toxic or carcinogenic, some are used as medicines, and some cause strong reactions when used by humans. Examples include nicotine in tobacco, opium from poppies, quinine from chinchona bark, and caffeine from coffee and tea.

alkene *n.* any of a series of aliphatic hydrocarbons (C_nH_{2n}) with a chain structure containing a double bond; e.g., ethylene ($H_2C=CH_2$); also called olefin or olefine.

Allegheny mound ant *n.* a mound-building ant (*Formica exsectoides*, family Formicidae) common in northeastern U.S.A. Each mound houses some tens or hundreds of thousands of ants. The ants clear the surrounding area of trees, shrubs, and other vegetation that might shade the mound within a radius of 20 to 30 ft (6 to 9 m).

allele or **allelomorph** *n.* (Greek *allēlo*, one of two or more + *morphē*, form) one of two or more forms of a gene. An individual normally has two alleles for each trait. They occupy corresponding positions on comparable chromosomes inherited from each parent and are passed on to progeny in accordance with Mendelian law.

allelopathy *n.* (Greek *allēlon*, of each other, + *pathos*, suffering) a growth-repressing influence of one plant on another by the secretion of a toxic substance. Fresh plant residue seems to be more toxic than weathered residue. Some crop residues and plants that have been demonstrated to be toxic to specific plants are

- corn residues on corn or soybean seedlings
- soybean residues on corn seedlings
- quackgrass on beans and soybeans
- broomsedge on black locust and loblolly pine trees
- ragweed and timothy on sugar maple trees
- tall fescuegrass on black locust and sweetgum trees
- bracken fern on wild black cherry trees
- black walnut and butternut on most pines and hardwood trees
- rye on lambsquarter and pigweed
- wheat straw on morning glory.

See amensalism.

allergen *n.* (Greek *allos*, other + *ergia*, activity + *genes*, born) any substance (usually a protein; e.g., ragweed pollen) that produces an allergic reaction in humans. Such allergic reactions are commonly known as hay fever or allergic rhinitis.

allergenic *adj.* (Greek *allos*, other + *ergia*, activity + *ikos*, of) capable of causing sensitive individuals to exhibit an allergic reaction.

allergy *n.* (Greek *allos*, other + *ergia*, activity) an immediate or delayed hypersensitive state involving antibodies that cause the organism to react adversely when exposed to specific allergens. The allergen may be pollen, dust, food, heat, cold, bacterial products, drugs, or plants such as poison ivy. Common foods that may produce allergic reactions in some persons include milk, eggs, and strawberries. The sensitivity is usually hereditary and can often be overcome by treatment with low doses of the antigen. Local honey can sometimes relieve the symptoms of allergies caused by plant pollen. See histamine,.

alley cropping or **alley farming** *n.* a system of raising crops between rows of trees and shrubs planted on the contour on steep slopes with erodible soils. Three commonly used shrubs are *Leucaena leucocephala*, *Gliricidia sepium*, and *Sesbania sesban*. *Sesbania sesban* gives the most rapid early growth, but *L. leucocephala* outperforms it beginning the second year. See agroforestry, vetiver.

alliaceous *adj.* (Latin *allium*, garlic + *āceus*, having the nature of) tasting and/or smelling like garlic or onion; cepaceous.

alligator *n.* (Spanish *el lagarto*, lizard) **1.** a large, dangerous, warm-climate amphibian (*Alligator mississippiensis* of southeastern U.S.A. or *A. sinensis* of eastern China, family Alligatoridae) with a broad snout and a long tail; related to the crocodile. **2.** the scaly leather made from the skin of an alligator, or shoes or other objects made from such leather.

alligatorweed *n.* (Spanish *el lagarto*, lizard + Anglo-Saxon *wēod*, weed) an aquatic weed that choked many of the waterways in Florida before it was controlled by the release of a South American moth (*Vogtia malloi*) in 1971.

allochthonous *adj.* (Greek *allos*, other + *chthon* land + Latin *osus*, like) not native to the area where it now exists. See autochthonous for contrast.

allodial land or **allodium** *n.* (Latin *allod*, free and clear + *ium*, material) land held free of any superior claim and without any rent or other payment in service; a freehold estate.

allogamy *n.* (Greek *allos*, other + *gamy*, marriage) cross-fertilization of flowering plants.

allogenic *adj.* (Greek *allos*, other + *genos*, kind + *ikos*, like) **1.** related but somewhat dissimilar. **2.** plant succession whereby a plant community is replaced by another in response to a change in the nonbiotic factors of the environment; e.g., improved drainage; fertilization.

allomorph *n.* (Greek *allos*, other + *morphē*, form) one of two or more forms of the same compound; e.g., quartz, cristobalite, and tridymite are allomorphs of SiO_2.

allophane *n.* (Greek *allos*, other + *phanés*, to seem, because heat causes allophane to lose color) an amorphous hydrous silicate clay mineral without a structure identifiable by X-ray diffraction. Allophane is abundant in certain young soils formed in volcanic ash.

allopolyploid *n.* or *adj.* (Greek *allos*, other + *polýs*, many + *ploos*, fold or multiple) a polyploid containing genetically distinct sets of chromosomes inherited from two or more different species.

allosome *n.* (Greek *allos*, other + *soma*, body) **1.** an unusual chromosome that has peculiarities of behavior or changes in size or shape. **2.** a sex chromosome.

allothigenic *adj.* (Greek *allothi*, elsewhere + *genes*, produced) produced elsewhere, as a rock fragment that has been moved from its place of origin; e.g., by a glacier, an ice raft, a landslide, or a mudflow.

allotment *n.* (French *allotement*, share) **1.** a share of something; the part granted to an individual. **2.** a permit to graze a specified number of cattle or sheep (or some of each) on a specific area; e.g., a range allotment. **3.** an allowance to plant a certain number of acres of a crop regulated under a government program; e.g., an acreage allotment.

allotrope *n.* (Greek *allos*, other + *trópos*, change) one of the two or more forms of an element differing in either physical or chemical properties or both. Ozone (O_3) and molecular oxygen (O_2) are allotropes of the element oxygen (O); graphite and diamond are allotropes of carbon; phosphorus has white and red allotropes, sulfur has several allotropes, etc.

allspice *n.* **1.** a tropical American tree (*Pimenta officinalis*, family Myrtaceae). **2.** the berries or the spice derived from the allspice tree used in pickles, relish, and certain medicines; named allspice because the flavor of the spice is like that of a mixture of cinnamon, nutmeg, and cloves. Also called pimento allspice.

alluvial *adj.* (Latin *alluvius*, washed against + *alis*, like) pertaining to water deposition and/or materials deposited by moving water.

alluvial cone *n.* a landform composed of soil and rock detritus deposited in a fan shape where a tributary stream enters a larger valley. It is similar to an alluvial fan, but alluvial cones have steeper slopes and coarser stone fragments. See alluvial fan.

alluvial fan *n.* a landform composed of soil and rock fragments moved by flowing water and deposited in the form of fans where small streams enter a larger valley or other open area. Alluvial fans are most common in arid regions with high relief. Several fans may coalesce to form an alluvial slope that commonly blends into a terrace, piedmont plain, or floodplain below. The soil on alluvial fans tends to be coarse textured, porous, and somewhat droughty. Alluvial fans are often used as sites for orchards because they have good water and air drainage. Also known as delta fan.

alluviation *n.* (Latin *alluvion*, an overflowing + *atio*, action) the process of depositing alluvium on bottomlands, alluvial fans, or other positions where decreased slope and/or spreading of water slows the water velocity and causes it to deposit sediment.

alluvion *n.* (Latin *alluvio*, an overflowing) **1.** land added by accretion; i.e., by deposition from a sea, lake, or river; by recession of a shoreline; or by the natural shifting of a river channel. It belongs to the owner of the adjacent land. **2.** alluvium. **3.** a flood.

alluvium *n.* (Latin *alluvius*, washed against) material deposited by streams that is usually composed of eroded soil containing a mixture of mineral and organic particles. Such deposits usually have lenses and strata of varying textures, representing changes in the flow rate and channel locations of the streams that deposited them. Alluvium most commonly refers to recent deposits on flood plains; more generally, it also includes older deposits on terraces and other areas that may be far removed from any current stream and no longer subject to flooding. Also called alluvion.

aloe *n.* (Greek *aloē*, aloe) any of 200 or more species of stemless, stoloniferous, clump-forming, succulent, cactuslike plants (*Aloe* sp., especially *A. vera* and *A. barbadensis*, family Liliaceae) native to southern Africa and some other areas. Aloe is commonly grown as a houseplant. Extracts from the leaves provide drugs that serve as laxatives, tonics, promoters of topical healing, etc. It has similar growth characteristics to agaves, but agaves have fibrous leaves that are used for making cord and rope.

alp *n.* (Latin *alpes*, high mountain) **1.** a high mountain peak, especially one of the Swiss Alps. **2.** a mountain meadow in Switzerland. **3.** an obstacle; something difficult to surmount or attain.

alpaca *n.* (Spanish *alpaca*, alpaca) **1.** a hoofed ruminant mammal (*Lama pacos*, family Camelidae) native to the Andes mountains of South America, domesticated in Peru and adjoining countries. Alpacas and llamas are believed to have been domesticated from guanacos and are related to camels. **2.** the long, soft wool of the alpaca.

alpenglow *n.* (German *alpenglühen*, with glow) a reddish glow that often covers mountain summits in the dim light of early morning or late evening.

alpenhorn *n.* (German *alpen*, Alps + *horn*, horn) a long wooden horn used by herdsmen and other mountain people, especially in Switzerland to call cattle.

alpestrine *adj.* (German *alpestris*, of the Alps) **1.** of high mountains, such as the Alps. **2.** subalpine; growing below the timber line on high mountains. See alpine.

alpha diversity *n.* the degree of diversity among the species present in a community or agroecosystem in a particular area. See beta diversity.

alpha-helix or **α-helix** *n.* a spiral structure cross-linked by hydrogen bonds; typical of part or all of many protein molecules. These linkages add to the strength of fibers such as hair or wool.

alpha-mesosaprobic zone *n.* an area in a heavily polluted stream where organic wastes are decomposing both aerobically and anaerobically.

alpine *adj.* (Latin *alpinus*, of the Alps) **1.** of the Alps or their inhabitants. **2.** in any mountainous area at a high elevation. **3.** above the timber line, either covered by grass and bushes (alpine tundra) or unvegetated.

Alps *n.* a high mountain range in Europe that extends about 750 miles (1200 km) from Italy into France, Switzerland, and Austria. The highest peak is Mont Blanc (15,771 ft or 4807 m).

alsike clover *n.* a common, domesticated, European clover (*Trifolium hybridum*, family Fabaceae) used as a hay or pasture legume. Sprawling when mature, resembling common red and white clovers, with pinkish white blossoms. The clover tolerates soil pH as low as 5.0. Also called alsike, Swedish clover, or bastard clover.

alter *v.* (Latin *alterare*, to change) **1.** to change the style, size, or other characteristic of something. **2.** to castrate a male or spay a female animal. Also called cut, emasculate, or geld. **3.** to prune a bush or other plant. **4.** to take the comb out of a beehive.

alternate grazing *n.* grazing two or more pastures alternately so that the forage grows back before it is grazed again. Also called rotation grazing.

alternate host *n.* a plant or animal that serves as host to a disease organism for only a part of its life cycle. Example: cedar/apple rust or wheat/barberry rust. See cedar-apple rust.

alternation of generations *n.* reproduction involving a haploid, sexual generation that produces a diploid, asexual generation, which is followed by another haploid generation. Thus, many common characteristics are found only in every second generation.

alternative agriculture *n.* nontraditional agriculture; alternative methods used in lieu of traditional farming and ranching, often with strong emphasis on environmental conservation and avoidance of agricultural chemicals. See organic farming, sustainable agriculture.

alternative energy *n.* energy produced by means such as solar energy cells, wind power, and biomass sources without use of conventional sources such as petroleum products, coal, and large hydroelectric plants.

altimeter *n.* (Latin *alti*, high + Greek *metron*, measure) an instrument for measuring altitude, especially that of an airplane; commonly an aneroid barometer that determines elevations indirectly from changes in atmospheric pressure or a radio altimeter that measures the time required for a radio wave to bounce off a surface and return.

Altiplano *n.* (Spanish *altiplanicie*, high plain) a high plateau in the Andes Mountains of Argentina, Bolivia, and Peru. The Altiplano supports some livestock grazing on its grassland vegetation and raises some short-season crops. Lake Titicaca is located on the Altiplano at an elevation of 12,500 ft (3810 m). See paramo.

altitude *n.* (Latin *altitudo*, height) **1.** the elevation above a reference level, especially above mean sea level. **2.** the angle of sight between the horizon and a heavenly body.

altitude cooking *n.* allowances such as increased cooking time used to compensate for lower boiling temperatures at high elevations. The boiling point of water decreases by 1°F for every 550 ft in elevation (1°C for every 300 m).

altitude sickness *n.* nausea, shortness of breath, and other disorders caused by reduced oxygen availability at high altitudes. Also called mountain sickness.

alum *n.* (Latin *alumen*, alum) **1.** potassium aluminum sulfate [$KAl(SO_4)_2 \cdot 12\ H_2O$]; used as an astringent, as an emetic, as a component of baking powder, as a soil amendment to acidify alkaline soil in greenhouse cultures, as a dye or a leather-tanning agent, and as a water purifier. **2.** any double sulfate of a monovalent cation and a trivalent cation. **3.** erroneously used for simple sulfates such as aluminum sulfate [$Al_2(SO_4)_3 \cdot 18\ H_2O$], chromium sulfate, or iron sulfate (also known as pearl alum or pickle alum).

alumina *n.* (Latin *alumen*, alum) aluminum oxide, Al_2O_3, occurring naturally as corundum.

aluminosilicate *n.* (Latin *alumen*, alum + *silex*, flint + *ate*, mineral) a mineral that contains aluminum, silicon, and oxygen; e.g., feldspars, amphiboles, pyroxenes, micas, clay minerals. Aluminosilicates include the most abundant minerals (except quartz) in rocks and soils.

aluminum or **aluminium (Al)** *n.* (Latin *alumen*, alum + *ium*, element) **1.** element 13, atomic weight 26.98; an abundant, widely distributed, low-density, metallic element, commonly occurring as Al^{3+} ions in various silicate minerals in soils and rocks. Some acid soils with pH 5 or below contain sufficient soluble and exchangeable aluminum to stunt or kill sensitive plants. *adj.* **2.** containing or made of aluminum; e.g., an aluminum pot.

aluminum dross *n.* a by-product of refining aluminum ore that contains aluminum oxide (Al_2O_3), aluminum nitride (Al_2N_2), and various other impurities. Dross is also relatively high in magnesium, copper, manganese, and zinc. Dross with high aluminum content may be toxic to plants if used as a soil amendment.

aluminum magnesium boride *n.* the second hardest substance known (after diamond) since its discovery was announced at Iowa State University in 1999; a likely replacement for cubic boron nitride for making tools and instruments.

alunite *n.* (Latin *alumen*, alum + *ite*, mineral) a hydrous aluminum and potassium sulfate mineral [$KAl_3(SO_4)_2(OH)_6$]. Large deposits occur in Utah and other western states. Heat treatment makes the potassium soluble in water and suitable for use as a potassium fertilizer. As marketed in the western U.S.A., it averages about 6.5% K_2O equivalent.

alutaceous *adj.* (Latin *alūtacius*, softened leather) **1.** leatherlike; covered with minute cracks like the human skin. **2.** colored brown, like certain leathers.

alveolus (pl. **alveoli**) *n.* (Latin *alveus*, concave + *olus*, little) **1.** a small cavity, pit, sac, or cell. **2.** a tiny, thin-walled air sac at the end of a bronchiole in the lung where gases are exchanged through the walls between air and blood. **3.** a cell in a honeycomb. **4.** a pit in the wall of a stomach. **5.** a tooth socket. **6.** the tiny, balloon-like structure of the mammary gland where milk is secreted. **7.** a terminal lobule (acinus) in a racemose gland. **8.** a pore or pitted perithecium in a fungal fruiting body. **9.** a pit of a carduaceous (thistle family) plant. **10.** a land surface with small, shallow pits similar to kettle holes.

alveus *n.* (Latin *alveus*, concave) **1.** the bed of a river. **2.** an anatomical duct that carries the flow of a body fluid.

AM *n.* arbuscular mycorrhizae. See endomycorrhiza.

amalgam *n.* (Arabic *al*, the + Greek *malagma*, softening agent) **1.** an alloy that includes mercury as one of its important components. **2.** an alloy composed mostly of silver with some mercury and other metals for use as a dental filling. **3.** any mixture or blend of distinct or unusual components.

amaranth *n.* (Greek *amaranton*, unfading flower) **1.** an imaginary flower that never fades or dies. **2.** any of a large number of plants (*Amaranthus* sp., family Amaranthaceae), including several weeds such as tumbleweed (*A. albus*), prostrate pigweed (*A. blitoides*), smooth pigweed (*A. hybridus*), red-root pigweed (*A. retroflexus*), and spiny amaranth (*A. spinosus*); some ornamental plants such as love lies bleeding (*A. caudatus*), prince's feather (*A. cruentus*), and Joseph's coat (*A. tricolor*); and some species that have been or in the future may be grown as crops. **3.** a purplish red dye ($C_{20}H_{11}N_2O_{10}Na$) used to color foods, pharmaceuticals, and fabrics.

amaryllis *n.* (Latin *Amaryllis*, a shepherdess) any of several bulbous plants (*Hippeastrum* sp., especially *H. puniceum*, family Amaryllidaceae); a houseplant with large red or pink lilylike blossoms. Also called belladonna lily.

Amazon ant *n.* a red ant (*Polyergus* sp., subfamily Formicinae) that captures and enslaves other ants (usually of *Formica* sp.).

amber *n.* (Latin *ambra*, ambergris) **1.** the fossilized resinous sap of coniferous trees used for jewelry and mouthpieces on the ends of pipe stems. Amber readily acquires a charge of static electricity when rubbed. Often amber contains preserved insects, leaves, and other organisms that make it more attractive for jewelry. Amber is usually yellowish to brownish and ranges from transparent to translucent, with flecks of orange, red, and white. *adj.* **2.** having the yellowish brown of amber. Geologic synonyms: succinite, bernstein, and electrum.

ambergris *n.* (Arabic *anbar*, ambergris + French *gris*, gray) a gray waxy secretion from sperm whale intestines found floating in the ocean or on the seashore and highly valued for making perfume.

ambience or **ambiance** *n.* (Latin *ambient*, gone around + *antia*, quality) the mood or atmosphere of a special place or situation.

ambient *adj.* (Latin *ambient*, gone around) **1.** the current condition of the surrounding environment, usually expressed in terms of temperature, pressure, and humidity. **2.** surrounded by the outside air.

amblotic *n.* (Greek *amblōsis*, abortion + *ikos*, having to do with) tending to cause abortion.

ambrosia *n.* (Greek *ambrosios*, food of the gods) **1.** mythical food of Greek and Roman gods that was supposed to give immortality to persons who ate it. **2.** a dessert with a delightful taste or smell, usually containing oranges, pineapple, and coconut. **3.** a genus of ragweed whose pollen causes hay fever.

ameba *n.* See amoeba.

amebiasis *n.* (Greek *amoibe*, change + *iasis*, condition) illness caused by infection with amoeba (especially *Entamoeba histolytica*). When the intestines are affected, it is called amoebic dysentery.

amelanism *n.* (Greek *a*, not + *melanos*, black + *ismos*, condition) lack of natural dark pigment in the skin of an animal or person. See melanism.

amelia *n.* (Greek *a*, not + *melos*, limb) failure to develop one or more arms or legs.

amelioration *n.* (French *a*, to + Latin *melioratus*, make better + *atio*, action) **1.** improvement; the process of making things better. **2.** the things that are done immediately following an accident to limit its consequences and to reduce their severity.

amensalism *n.* (Latin *a*, not + *mensa*, table + *isma*, action) a negative effect of an organism on another with no impact on the organism that produces the effect; a form of allelopathy. See allelopathy, symbiosis.

ament *n.* See catkin.

American bittersweet *n.* a twining woody vine (*Celastrus scandens*, family Celastraceae) with orange berries that contain red seeds, remain on the vine throughout the winter, and are poisonous to livestock. It is found from the North Carolina coast to New Mexico and northward through South Dakota into eastern Canada. Commonly used as an indoor winter decoration. Also called climbing bittersweet, false bittersweet, stafftree, or waxwork.

American chestnut *n.* a large tree (*Castanea dentata*, family Fagaceae) that was prominent in colonial times but was nearly exterminated by the chestnut blight (*Endothia parasitica*) first reported in New York City in 1904. American chestnut trees were found from Maine to Alabama and as far west as Michigan. Only a few American chestnut trees remain now, mostly in isolated orchards and ornamental plantings near the Pacific coast. The nuts they bear are edible and sweet. Chestnut wood is used in making crossties, fence posts, furniture, poles, pulp, tannin, etc. Also called chestnut. Note: The Chinese chestnut (*Castanea mollissima*) is resistant to chestnut blight.

American cockroach *n.* a house-dwelling winged insect (*Periplaneta americana*, family Blattidae), commonly 1 to 2 in. (2 to 5 cm) long with a hard, flat, brown body, slender legs, and long feelers that is especially common in kitchens, bathrooms, and basements. Active at night, cockroaches feed on garbage, food scraps, book bindings, and paper. They are also common in greenhouses, where they feed on seedlings and girdle plants.

American dog tick *n.* any of several bloodsucking ticks (*Dermacentor variabilis* and related species, family Ixodidae) that carries and transmits Rocky Mountain spotted fever, tularemia, and bovine anaplasmosis. It hatches in bushes and attaches itself by burrowing its head into the skin of any animal that comes close enough. Also called wood tick.

American elder *n.* a shrub (*Sambucus canadensis*, family Caprifoliaceae) grown for its flowers and fruit. The leaves, buds, and young shoots can kill cattle and sheep that eat them, but the ripe berries are eaten by birds and are used in making wine and jelly. Also known as elderberry.

American Society for Testing and Materials (ASTM) *n.* an organization based in Philadelphia that sets standards for materials, products, and services and specifies methods for sampling and testing such items, including many that are important environmentally.

amictic lake *n.* (Latin *a*, not + *mixtus*, mixing) a lake that does not turn over as a result of seasonal temperature inversions. Also called oligomictic lake if thermally stable, as in many tropical lakes, or meromictic if permanently stratified by chemical differences, such as a bottom layer of saline water. See dimictic lake, monomictic lake, polymictic lake.

amine *n.* (ammonia + Latin *inus*, made of) a chemical structure with one or more of the hydrogens of ammonia (NH_3) replaced by a bond to an organic structure, as RNH_2 or RNHR' where R and R' represent organic groups. See amino acid.

amino acid *n.* any of a group of 20 naturally occurring, nitrogen-containing building blocks of protein plus several hundred nonprotein amino acids. They are organic compounds whose molecules contain both an amine ($-NH_2$) and a carboxyl ($-COOH$) functional group attached to their α carbon. These groups react, releasing a water molecule and forming the peptide bond ($-CONH-$) that links amino acids in protein molecules. Of the 18 amino acids contained in food proteins, 8 are essential for health and must be contained in human diets, as the body is incapable of synthesizing them. In this group are tryptophan, leucine, lysine, isoleucine, valine, phenylalanine, threonine, and methionine. Two, histidine and arginine, are semiessential in that they are synthesized but in inadequate amounts, and the others are nonessential, as they can be synthesized by the body. See peptide.

amitosis *n.* (Greek *a*, without + *mitos*, thread + *osis*, condition) cell division by nuclear cleavage without the formation of chromosomes.

ammonia *n.* (Latin *ammoniacum*, from *sal ammoniac*, an ammonium salt) a chemical (NH_3) consisting of 82.25% nitrogen and 17.75% hydrogen on a weight basis. A colorless, pungent gas at ambient atmospheric pressure that can burn the eyes, skin, and lungs if handled improperly, it is commonly stored in liquefied form under pressure or dissolved in water (sometimes called aqua ammonia). It is used in refrigeration, as a laboratory reagent, and as a nitrogen fertilizer. It is the cheapest source of nitrogen fertilizer, but it must be injected into the soil about 5 in. (13 cm) deep when applied. The ammonia (NH_3) molecules are soon converted to ammonium (NH_4^+) ions that are adsorbed and held by clay and humus. However, over a period of days, weeks, or months (depending on temperature, moisture, and aeration), bacteria convert the ammonium ions to more mobile nitrate (NO_3^-) ions. The nitrate ions move with percolating water and may contaminate the groundwater.

ammonification *n.* (Latin *ammoniacum*, ammonia + *ficatio*, making) **1.** the release of ammonium ions during the decomposition of organic matter by soil microbes. **2.** the addition of ammonia, especially as a component of fertilizer.

ammonium nitrate *n*, a white, water-soluble, crystalline salt (NH_4NO_3) containing 34% nitrogen by weight; used as a nitrogen fertilizer. It is also a strong oxidant that becomes explosive in contact with combustible materials and, for that reason, must be handled carefully. Two major explosions involving ammonium nitrate were (1) an accidental explosion that killed 561 persons at a pier in Texas City on April 16, 1947, and (2) the intentional bombing of the federal building in Oklahoma City that killed 168 persons on April 19, 1995.

amoeba or **ameba** *n.* (Greek *amoibē*, change) microscopic, single-celled, amorphous protozoan organisms (order Amoebida). Some are pathogenic to humans and animals.

amoebic dysentery *n.* dysentery caused by protozoa (*Endamoeba histolytica*), usually including intestinal ulcers. See dysentery.

amorphous *adj.* (Greek *a*, without + *morphos*, shape) without a definite or characteristic shape; lacking crystalline structure.

ampere (A or **amp)** *n.* the standard unit for measuring the flow of an electrical current, named for Andre Marie Ampere (1775–1836), a French physicist; the amount of current flow produced by 1 volt through a resistance of 1 ohm (1 ampere = 1 coulomb per second).

amphibian *n.* (Greek *amphi*, both + *bios*, life + Latin *ianus*, animal) **1.** a frog, salamander, or toad; a cold-blooded animal that obtains oxygen underwater by means of gills in the early stages but develops lungs and breathes air in the later stages of life. **2.** a vehicle that can travel on either land or water. **3.** an airplane that can alight and take off from either land or water.

amphibious *adj.* (Greek *amphibios*, living a double life) capable of living in water as well as on land or of functioning on either land or water. See amphibian.

amphibole *n.* (Greek *amphibolos*, ambiguous) a silicate mineral with a double-chain structure that produces crystals with a fibrous texture and 124° angle between faces; e.g., hornblende or asbestos. Compare pyroxene.

amphicarpic or **amphicarpous** *adj.* (Greek *amphi*, two + *karpos*, fruit) bearing two different types of fruit, either of different shape or of different time of maturity.

amphicarpogenous *adj.* (Greek *amphi*, two + *karpos*, fruit + *genous*, producing) fruit that is fertilized above ground and then buried to grow and mature below ground; e.g., the peanut.

amphidiploid *n.* (Greek *amphi*, two + *diploos*, twofold) an allotetraploid; a hybrid plant or animal having cell nuclei that contain two nonmatching diploid sets of chromosomes, one set derived from each parent.

amphigean *adj.* (Greek *amphi*, two + *gē*, earth) native of both the Old and New Worlds; existing around the Earth; e.g., cattails (*Typha* sp., family Typhaceae) are amphigean.

amphimixis *n.* (Greek *amphi*, two + *mixis*, mingling) **1.** sexual reproduction; the uniting of sperm and egg from separate individuals. **2.** crossbreeding.

amphiphilic *adj.* (Greek *amphi*, two + *phile*, friends) able to interface with a water environment on one side and a nonpolar lipid environment on the other.

amphiphyte *n.* (Greek *amphi*, both + *phyton*, plant) a plant with amphibious characteristics that grows in the border zone of wetland around a body of water.

amphoteric *adj.* (Greek *amphoteros*, both) able to react either as an acid or as a base.

amplitude *n.* (Latin *amplitudo*, breadth) **1.** the breadth or width of something, especially of an oscillation such as a light wave or radio wave, usually measured from the central position to an extreme. **2.** the range of ecological conditions an organism can tolerate and continue to exist. See hardy.

amu *n.* See atomic mass unit.

amylase *n.* (Greek *amylon*, starch + *ase*, enzyme) any of several digestive enzymes that hydrolyze starch into sugar molecules.

amylolysis *n.* (Greek *amylon*, starch + *lysis*, dissolution) the process of breaking down starch into sugar molecules.

amylopectin *n.* (Greek *amylon*, starch + *pēktos*, congealed) a polysaccharide component of starch; composed of chains of glucose molecules joined together through α-1,4 linkages with an α-1,6 linkage providing a branch point after each chain of about 25 glucose units. See starch.

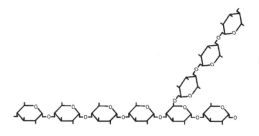

A fragment of amylopectin

amylose *n.* (Greek *amylon*, starch + *ose*, protein hydrolysis product) a polysaccharide component of starch composed of long chains of glucose molecules joined together through α-1,4 linkages. The chains roll into a helix shape that forms a tube stabilized by hydrogen bonding that can enclose other molecules; e.g., iodine can enter the tubes and form a blue complex that is often used as a test to detect the presence of either starch or iodine.

anabatic *adj.* (Greek *anabatikos*, climbing) ascending; upward moving, as a local breeze caused by the sun heating the land surface at low elevations, warming the air, and creating air currents that move up valleys and slopes. See katabatic.

anabiosis *n.* (Greek *anabiōsis*, coming back to life) **1.** revival of an organism that appeared to be dead. **2.** suspended animation during unfavorable living conditions.

anabolism *n.* (Greek *anabolē*, rising up + *ismus*, action) constructive metabolic processes in plants, animals, and humans that incorporate simple substances into more complex components of living tissue. See catabolism.

anadromous *adj.* (Greek *anadromos*, running upward) swimming upriver to freshwater spawning grounds to reproduce after having spent most of their adult life in the sea. Example: salmon. Opposite of catadromous.

anaerobic *adj.* (Greek *an*, not + *aēro*, air + *bios*, life) **1.** anoxic; lacking or severely deficient in free oxygen (air). **2.** a life or process that occurs in, or is not destroyed by, the absence of oxygen. See aerobic.

anaerobiosis *n.* (Greek *an*, not + *aero*, air + *biōsis*, way of life) the sustaining of life by anaerobic processes in the absence of air or gaseous oxygen.

anaphylactic shock *n.* a sudden violent attack of symptoms such as severe shock, collapse, or even death from exposure to or injection of a serum or protein (usually of an animal or person that has developed hypersensitivity from previous exposure).

anaplasmosis *n.* (Greek *ana*, without + *plasma*, something molded + *osis*, condition) a blood disease of cattle that may also affect other ruminants. It is caused by a parasite (*Anaplasma marginale*, phylum Protozoa) transmitted by ticks, destroys red blood cells, and makes the blood watery. Symptoms are anemia, labored breathing, high fever, and cessation of milk flow, followed by jaundice in the hairless areas. Death rates vary, and death may occur in one to several days after symptoms appear. Recovery of surviving animals is slow. Also called yellow teat disease or gall sickness. A much milder form of the disease prevalent in Africa is caused by *A. centrale* or *A. marginale* var. *centrale*.

anasa wilt *n.* a wilt disease caused by the squash bug feeding on cucurbits plants, without any parasitic microorganism involved.

anastomosis *n.* (Greek *anastomōsis*, opening) **1.** a branching, interlacing, and intercommunicating stream that produces a netlike or braided pattern. **2.** interconnected blood vessels. **3.** netted, interveined leaves marked by cross-veins forming a network (sometimes the vein branches meet only at the margin of the leaf). **4.** a joining of two parts that are normally separate; e.g., reconnecting an intestine when a segment has been surgically removed.

anatomy *n.* (Greek *anatomie*, a cutting up) **1.** the study of the physical structure and form of plants, animals, or humans. **2.** the structure of an organism and its component parts.

anchored dune *n.* a sand dune stabilized by growing vegetation.

anchor ice *n.* ice that forms at the bottom of an unfrozen stream or lake, often on stones. Also called ground ice or bottom ice.

ancylostomiasis *n.* See hookworm.

Andisol *n.* (*ando*, volcanic + French *sol*, soil) an order in Soil Taxonomy that includes young soils with a high content of allophane. Andisols develop from black volcanic ash and similar materials. They occupy about 2% of the land area in the U.S.A. See allophane, Soil Taxonomy.

androconia *n.* (Greek *andro*, male + *konios*, dusty) minute scales on the wings of some male butterflies.

androecium *n.* (Greek *andro*, male + *oikion*, little house) the assemblage of stamens and associated parts of a flower; the pollen-producing part.

androgenous *adj.* (Greek *andro*, male + *genous*, producer) **1.** producing male offspring. **2.** a plant that has stamens but lacks female parts.

androgynous *adj.* (Greek *androgynos*, hermaphrodite) **1.** having both male and female parts (stamens and pistils) on the same plant. **2.** hermaphrodytic; having both male and female characteristics in the same animal or person.

Andromeda *n.* (Greek *Andromedē*, a Greek goddess) **1.** in Greek mythology, the daughter of Cepheus, king of Ethiopia, and Cassiopeia. **2.** a constellation in the northern sky, farther from the North Pole than Cassiopeia. **3.** a genus of plants (family Ericaceae) that bears masses of white or rose-colored flowers.

anecic *adj.* feeding on leaf litter and mixing it with soil while feeding and burrowing into the soil; e.g., certain large earthworms that are typically pigmented on the anterior half of their bodies and make long-lasting vertical burrows in the soil.

anemia *n.* (Greek *an*, negative + *haima*, blood) a weakened condition in humans and animals caused by too few red blood cells, too little hemoglobin in the blood, or both. It can be caused by excessive bleeding, iron or vitamin B_{12} deficiency, or failure to produce enough red blood cells.

anemochore *n.* (Greek *anemos*, wind + *chorein*, to spread) an ecological term denoting a plant with seeds distributed by wind, such as milkweed.

anemometer *n.* (Greek *anemo*, wind + *metron*, measure) **1.** an instrument that measures wind speed; a wind gauge. **2.** the **Dines anemometer** measures differential air pressure for translation into wind velocity.

anemone *n.* (Greek *anemo*, wind + *one*, female) **1.** any of several plants (*Anemone* sp., family Ranunculaceae) with white or colored cup-shaped blossoms that have 5 or more petal-like sepals surrounding several golden stamens. **2.** sea anemone.

anemophilous *adj.* (Greek *anemo*, wind + *philos*, loving) pollinated by pollen carried by wind.

anemoscope *n.* (Greek *anemo*, wind + *scopion*, to look at) a device that indicates and/or records wind direction; e.g., a weather vane.

aneroid barometer *n.* (Greek *a*, not + *nēros*, liquid + *eidos*, form) a closed container with an elastic surface connected to a needle that indicates atmospheric pressure.

anestrus *n.* (*an*, not + Latin *oestrus*, gadfly, frenzy) a period of sexual inactivity between two estrus cycles of a female mammal.

angiosperm *n.* (Greek *angio*, container + *sperma*, seed) a flowering plant; any plant having seeds that develop in an enclosed ovary. See gymnosperm.

angle of repose *n.* the steepest slope of a soil or loose stony material that is stable against mass movement. Also called critical slope. See quick clay.

angler *n.* (Greek *ankylos*, bent + *er*, person) **1.** a person who uses a fishing pole, line, and hook to catch fish. **2.** a person who uses dubious schemes. **3.** certain saltwater fish (*Lophius americanus* and others of family Lophiidae) that prey on other fish, attracting them with a wormlike lure that dangles from their heads.

angleworm *n.* (Greek *ankylos*, bent + Latin *vermis*, worm) an earthworm that either is used or could be used by an angler as bait to catch fish. Also called fishworm.

Angstrom (Å) *n.* a unit of length equal to 1 ten-thousandth of a micron or 1 ten-billionth of a meter (one 254-millionth of an inch); named for A. J. Angstrom (1814–1874), a Swedish astronomer and physicist. It is used for small measurements such as wavelengths of light, radii of atoms or ions, and the spacing of layers in mineral structures. Also called angstrom or angstrom unit.

anhydrous *adj.* (Greek *anydros*, without water) **1.** without water, especially without any water of hydration; e.g., removing water from gypsum ($CaSO_4\cdot 2H_2O$) produces anhydrite ($CaSO_4$). *n.* **2.** anhydrous ammonia, especially when used as a fertilizer.

anhydrous ammonia *n.* pure ammonia under sufficient pressure to liquefy it; commonly injected into soil as a nitrogen fertilizer or used as a raw material to produce other nitrogen compounds.

animal *n.* (Latin *animal*, breathing) **1.** any member of the kingdom Animalia; a nonphotosynthetic organism that lives by ingesting, eating, and digesting food, typically free to move from one location to another of its own volition. Excludes plants and bacteria; generally does not include humans. **2.** a mammal, as opposed to a bird or fish.

animal agriculture *n.* agriculture based on the use of domestic animals, especially for power but also for food, fiber, and other products. Animal agriculture serves human needs by providing power for transportation, tillage, and other purposes; nutritious food products; and a variety of useful by-products including clothing, pharmaceuticals, and cosmetics. Three-fourths of the protein, one-third of the dietary energy, most of the calcium and phosphorus, and substantial quantities of essential vitamins and minerals in the American diet are from animal products.

Animal and Plant Health Inspection Service (APHIS) *n.* an agency in the U.S. Department of Agriculture that inspects agricultural products as they are exported and imported.

animal husbandry *n.* the science or practice of keeping, breeding, and tending domesticated animals, especially meat or dairy animals and horses. Also called animal science.

animal rights *n.* the ethical claim that animals have rights similar to those of humans to be well treated and to be protected from abuse and exploitation.

animal traction *n.* the use of animals (e.g., horses, mules, oxen, donkeys, or camels) to pull implements or vehicles.

A camel–donkey team at work

animal unit month (AUM) *n.* one cow (or the equivalent in other livestock, such as 1 horse, 1 mule, 5 sheep, 5 goats, or 5 hogs) grazing on pasture or rangeland for 1 month; used as a means of evaluating the carrying capacity of pasture and rangeland. Also called animal month, cow month.

anion *n.* (Greek *aniōn*, going up) an ion with a negative charge; one that is electrically attracted to an anode. Anions in soil tend to be more mobile than cations because clay and humus usually have negative charges that attract cations but repel anions. Examples: chloride (Cl^-) and nitrate (NO_3^-).

anise *n.* (Latin *anisum*, anise) an annual herb (*Pimpinella anisum*, family Umbelliferae) native to Egypt that is used as a spice and for medicinal purposes. It tastes like licorice and reduces gas in human stomach and intestines.

anisotropy *n.* (Greek *aniso*, uneven + *tropos*, turn) **1.** the characteristic of having unequal properties (e.g., strength, hardness, light transmission, or electrical conductivity) in different directions (e.g., along different axes of a crystal or along vs. across the grain of wood). **2.** the characteristic of a living organism that has unequal responses to external stimuli.

ankylostomiasis *n.* See hookworm.

annual *adj.* (Latin *annual*, yearly) **1.** pertaining to an entire year, as annual growth or annual precipitation. **2.** occurring once each year, as an annual feast. *n.* **3.** a yearly publication; a yearbook. **4.** a plant that completes its life cycle from seed to maturity and death in one year (often including a dormant period). Also called a therophyte. See biennial, perennial.

annual ring *n.* a growth ring common in trees and some other organisms consisting of a layer of large cells produced during a period of rapid growth (e.g., during a favorable spring season) and a layer of small, dense cells produced during a period of slow growth (e.g., during a cold or dry season). The age of a tree is often determined by counting the annual rings in its trunk.

annular eclipse *n.* an eclipse of the sun with a ring of sunlight showing around the moon.

annulation *n.* (Latin *annual*, yearly + *atio*, action) a circular formation, as the annual growth rings visible in a tree stump or the rings around the body of a worm.

anode *n.* (Greek *anodos*, way up) an electrode with a positive charge; one that attracts anions and repels cations; opposite of cathode.

anoxia *n.* (Latin *an*, not + French *oxy*, oxygen) absence of oxygen; suffoca-tion. See hypoxia.

ant *n.* (German *āmete*, ant) any of about 5000 species of social insects (family

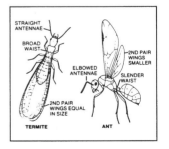

Termite and ant

Formicidae) that live in colonies including a winged queen, winged males, and wingless female workers. Ant bodies have three distinct sections and are usually black or red. See Amazon ant, anthill, carpenter ant, fire ant, termite.

antagonism *n.* (Greek *antagónism*, contention against) **1.** active opposition; hostility. **2.** the interaction of two chemicals having an opposing or neutralizing effect on each other. **3.** a chemical interaction with an opposing or neutralizing effect that prevents an expected biological response. **4.** exceptionally strong competition in the processes involved in cation absorption by plant roots between ions of different elements but the same charge, such as K^+ and Rb^+ or Ca^{2+} and Sr^{2+}.

Antarctica or **Antarctic Continent** *n.* a continent surrounding the South Pole that is almost entirely covered by glacial ice. It has a land area of about 5,500,000 mi^2 (14,000,000 km^2) and holds about 91% of all the ice on Earth. This cold mass has a strong influence on global air circulation, winds, and weather. See Antarctic Circle, Arctic region.

Antarctic Circle *n.* an imaginary circle parallel to the equator at about 66°33′ south latitude, marking the northern limit of the southern region that experiences 24-hr days of sunlight in the summer and 24-hr nights in the winter. See Arctic Circle, Arctic region.

ant cow *n.* any aphid that ants use to obtain honeydew. See honeydew.

anteater *n.* **1.** a tropical mammal (family Myrmeco-phagidae) with a long snout that uses its strong claws to open burrows and its long, sticky tongue to feed on ants and termites. **2.** any of several other species of mammals that feed on ants and termites. See aardvark.

antelope *n.* (Greek *antholops*, fabulous beast) **1.** the pronghorn (*Antilocapra americana*, family Antilocapridae), a deerlike ruminant native to the Great Plains of North America. About 3 ft (90 cm) tall at the shoulder with branched horns, it is noted for its grace, beauty, and speed. **2.** any of about 50 species of ruminants (family Bovidae) native to Africa and Asia that somewhat resemble the pronghorn, including the gazelle, the impala, and the gnu. **3.** leather made from antelope skin.

antenna *n.* (Latin *antenna*, a sail yard) **1.** an aerial; a wire or set of wires used to send or receive electromagnetic signals (radio waves). **2.** one of the pair of jointed sensory appendages on the head of an insect or other arthropod. **3.** (figurative) a means of intercepting rumors or other messages.

anthelmintic *adj.* (Greek *ant*, anti + *helmith*, certain worms) **1.** capable of killing parasitic gastrointestinal worms. *n.* **2.** a drug that kills gastrointestinal worms.

anthesis *n.* (Greek *anthēsis*, full bloom) the period in flowering plants from the opening of the flower to the setting of fruit; full flower.

anthill *n.* an earth mound built by ants around the entrance to their nest.

anthocyanin *n.* (Greek *anthos*, flower + *cyanine*, blue dye) any of a group of organic compounds built around a flavonoid ring with attached sugars and other groups. Anthocyanins provide most of the red, purple, and blue of flowers and the orange, red, and brown of autumn leaves and fruits.

anthoecology *n.* (Greek *anthos*, flower + *oikos*, house + *logos*, word) the study of flowers in relation to their environment.

Anthony pig *n.* the smallest pig (runt) in a litter; dedicated to Saint Anthony, the patron saint of swine.

anthophilous *adj.* (Greek *anthos*, flower + *philos*, loving) **1.** being fond of flowers. **2.** living and feeding among flowers; e.g., anthophilous bees.

anthracite *n.* (Greek *anthrakitēs*, resembling coal) high-grade, hard, lustrous black coal that is low in volatiles and burns with little flame or smoke, producing the most heat per unit of weight of any coal. It is used mostly for heating residences and commercial buildings. Lower-grade coals, in decreasing order, are bituminous, lignite, and peat.

anthracite culm *n.* waste from anthracite coal preparation, especially in Pennsylvania; a mixture of coarse rock fragments, small pieces of coal, and silt-sized coal particles.

anthracnose *n.* (Latin *anthrac*, carbuncle + French *noso*, disease) a fungal plant disease that produces lesions as it attacks many plants, including apple, black walnut, cotton, grapes, melons, and sycamore.

anthracosis *n.* (Greek *anthrak*, coal + *osis*, disease) a form of pneumoconiosis caused by inhaling coal dust. Also called black lung.

anthrax *n.* (Latin *anthrax*, carbuncle) an infectious disease caused by a bacterium (*Bacillus anthracis*) that attacks especially cattle, sheep, and goats but also can infect horses, pigs, dogs, foxes, rabbits, rats, mice, guinea pigs, and humans, but not birds. Anthrax occurs in all parts of the world but mostly in tropical and subtropical regions. Symptoms of attack are pustules, high fever, throat swelling, and enlarged spleen, with *Bacillus anthracis* in the blood. Anthrax produces spores that are hard to destroy and may remain viable for 10 years or longer in soil. Earthworms are suspected of having carried the anthrax organism from deeply buried animals to the soil surface. Anthrax may be used in germ warfare. See apoplectic anthrax, pulmonary anthrax.

anthropic *adj.* (Greek *anthrōpos*, human) pertaining to or caused by humans.

anthropic epipedon *n.* a surface soil horizon defined in Soil Taxonomy as having acquired a color dark enough for a mollic epipedon and a concentration of more than 250 ppm available phosphorus because of long-term human use.

anthropocentrism *n.* (Greek *anthrōpos*, human + French *centriste*, of the center) the assumption that humans are the focal point of the universe; viewing everything from the human aspect.

anthropogenesis *n.* (Greek *anthrōpos*, human + *genesis*, origin) the study of human origins and the development of the human race.

anthropogenic *adj.* **1.** dealing with anthropogenesis. **2.** generated by humans. Example: a compact soil layer created by tillage.

anthropoid *adj.* (Greek *anthropoeides*, having human shape) said of animals that resemble humans, such as apes, gorillas, chimpanzees, and orangutans.

anthropology *n.* (Greek *anthrōpos*, human + *logos*, word) the study of the origins, development, physical and biological characteristics, culture, customs, and beliefs of humans.

antibiosis *n.* (Greek *anti*, against + *biosis*, way of life) a negative interaction between two organisms; one of them is harmed by the other.

antibiotic *adj.* (Greek *anti*, against + *biōtikos*, pertaining to life) **1.** harmful to life, especially to bacteria. *n.* **2.** a chemical substance of microbial origin used in the treatment of an infectious disease to control the causal organism. The term was first applied by S. A. Waksman (1888–1973) of Rutgers University (Nobel Prize in 1952) to describe compounds such as streptomycin. Most antibiotics are produced by fungi, actinomyces, or bacteria.

antibody *n.* (Greek *anti*, against + Anglo-Saxon *bodig*, body) a protein manufactured by the body of an animal or person as a defense against a disease antigen. Antibodies are so specific that they can differentiate between antigens produced by different species or even by different individuals of the same species. Vaccination causes antibodies to be produced in response to disease antigens that have been killed, weakened, or otherwise altered. See antigen.

anticline *n.* (Greek *antiklin*, leaning against each other) a geologic structure produced by uplift along a central line so that rock layers slope downward on both sides of the anticline. The opposite is a syncline. See syncline.

anticyclone *n.* (Greek *anti*, opposite + *kyklōn*, revolving) a large high-pressure air mass, usually associated with fair weather. It has descending air and surface winds that blow spirally outward in a clockwise direction in the Northern Hemisphere and counterclockwise in the Southern Hemisphere. Weather systems normally follow each other as a series of cyclones and anticyclones. See cyclone.

antifungal *adj.* (Latin *anti*, against + *fungus*, fungus + *alis*, suitable) **1.** able to kill fungi or suppress their growth and reproduction. **2.** useful against fungal infections.

antigen *n.* (Latin *anti*, against + *gen*, production of) a large molecule foreign to the body that stimulates production of an antibody in an animal or human. Synthetic or denatured antigens are used as vaccines to cause the body's natural defense system to produce antibodies that will resist a specified disease. See antibody.

antimony (Sb) *n.* (Greek *anti*, against + *monos*, alone) element 51, atomic weight 121.75; a soft bluish white metal that burns readily when heated and that is used in alloys, semiconductors, batteries, paints, etc.

antioxidant *n.* (Greek *anti*, against + *oxys*, sour + Latin *antem*, does) **1.** a chemical that inhibits oxidation. **2.** an organic compound, commonly an enzyme or coenzyme such as vitamin C or E, that helps prevent oxidation damage by free radicals in human or animal tissue. **3.** a chemical added to prevent oxidation and thus extend the life of fats, vegetable oils, animal feed, or synthetic rubber.

antiseptic *adj.* (Greek *anti*, against + *sēptikos*, rotten) **1.** sterile; free from microbes. **2.** exceptionally clean. *n.* **3.** a substance used to kill pathogenic microbes; e.g., alcohol.

antler *n.* (French *antoillier*, antler) **1.** the branched bony growth produced and shed annually from the upper part of each side of the head of most animals of the deer family (especially males), as distinguished from permanent horns grown by ungulate animals. **2.** any of the branches of such growth, sometimes called prongs or points.

ant lion *n.* a nocturnal insect (family Myrmeleontidae) that has four delicate wings, resembling a damselfly. It lays its eggs in sandy materials and is most common in dry climates like southwestern U.S.A. Its larvae prey on ants and other insects and are known as doodlebugs. See doodlebug.

apatite *n.* (Greek *apatē*, deceit + *ite*, mineral) a calcium phosphate mineral [$Ca_5(PO_4)_3(F,Cl,OH)$] that occurs in rocks and is the principal structural component of bones and teeth. Apatite is mined as rock phosphate for use in fertilizer, mostly after treatment with acid to change it into a more soluble superphosphate compound. Fluoride is added to water and toothpaste to convert the apatite in tooth enamel to the fluorapatite form because it is harder and more resistant to decay than is chlorapatite or hydroxyapatite.

ape *n.* (Anglo-Saxon *apa*, ape) **1.** any of several large primates (order Primate, suborder Haplorhini), including the lesser apes (gibbons) and the great apes (chimpanzees, gorillas, and orangutans); the animals most closely related to humans. **2.** any primate, including the above plus monkeys and prosimians. **3.** a person who imitates or mimics others. **4.** (slang) a person who has bad manners, is clumsy, or otherwise uncouth. *v.* **5.** to mimic or imitate the actions of another.

aphelion *n.* (Greek *aphélion*, off sun) the position of a planet or other celestial orbiter when it is most distant from the sun. Earth reaches its aphelion in early July. See perihelion.

aphid *n.* (Latin *aphis*, a bug) a small plant louse, a soft-tissued insect (family Aphididae) that sucks the sap of various plants and helps to spread certain viral diseases. Aphids secrete a honeydew that is eaten by ants, bees, and wasps.

APHIS *n.* See Animal and Plant Health Inspection Service.

aphis lion *n.* the lacewing larva (*Chrysoperla* sp., family Chrysopidae), a helpful predator on many destructive insects.

aphotic *adj.* (Greek *a*, not + *phōs*, light + *ikos*, like) dark, without light, especially referring to the ocean depth where light is insufficient for photosynthesis. See euphotic zone.

aphthous fever *n.* foot-and-mouth disease.

Apiaceae *n.* family name adopted in 1972 for hollow-stemmed, dicotyledonous, flowering plants with blossoms that form an umbel; e.g., celery and parsley. Formerly called Umbelliferae.

apical dominance *n.* the growth habit of a tree or other plant with a single apex; the result of an auxin produced by the uppermost growing point that inhibits the growth of buds below it.

apiculture *n.* (Latin *api*, bee + *cultura*, care) beekeeping or the study of beekeeping.

apiology *n.* (Latin *api*, bee + *logos*, word) the scientific study of bees. Also called melittology.

apiotherapy *n.* (Latin *api*, bee + Greek *therapeia*, healing) the use of bee venom for treatment of illness. Also called melissotherapy.

apiphobia *n.* (Latin *api*, bee + *phobos*, great fear) an abnormal fear of bees and/or bee stings. Also called melissophobia.

aplasia *n.* (Greek *a*, not + *plasis*, formation) absence of an organ or tissue as a result of defective development.

aplite *n.* (German *aplit*, simple) fine-grained granite composed mostly of feldspar and quartz.

apogamy *n.* (Greek *apo*, away from + *gamia*, marriage) development of a plant without fertilization; apomixis.

apogee *n.* (Greek *apo*, away from + *geō*, Earth) the position in the orbit of the moon or other satellite when it is farthest from the Earth. See apsis, perigee.

apomixis *n.* (Greek *apo*, away from + *mixis*, mixing) asexual production of seeds in plants; apogamy. The formation of viable embryos without union of male and female gametes, as in dandelions or Kentucky bluegrass.

apoplectic anthrax *n.* a form of anthrax that may cause the death of cattle, sheep, or goats within a few hours. See anthrax.

aposematic *adj.* (Greek *apo*, away from + *sēmatos*, a sign + *ikos*, of) coloration and/or toxins to ward off enemies. The striped larva of the monarch butterfly and the stripes on a skunk are examples of aposematic coloration. Also called warning coloration.

appendage *n.* (Latin *appendere*, to hang + French *age*, place of) **1.** an addition or extension; an auxiliary or subsidiary part. **2.** a body part such as an arm, leg, or tail that extends out from the trunk, an antenna on the head of an animal, or a blade or leaf on a plant.

apple *n.* **1.** (German *apfel*, apple) a firm, somewhat rounded tree fruit that has a thin, usually red but in some varieties yellow or green peel, a white interior, and a few small seeds in its core. The most common varieties are produced by *Pyrus malus*, a plant of the rose family. Apples are eaten raw, cooked to make pies and other dishes, or squeezed to make apple juice, cider, or vinegar. Apple skin and flesh are a mild laxative, and apple seeds contain enough cyanide for a cupful to kill a person. Idioms: The apple of one's eye = someone or something that is highly cherished. Apple-pie order = perfectly arranged. The Big Apple = New York City. **2.** the deciduous tree that produces such fruit.

apple cider *n.* apple juice obtained by squeezing apples. It may be consumed fresh, pasteurized so it remains "sweet cider" for a longer time, or fermented to produce "hard cider."

apple rust *n.* a serious fungal (*Gymnosporangium juniperi virginianae*, family Pucciniaceae) disease that lives part of its life cycle on eastern red cedar and part on apple trees. It causes rust spots on the leaves and lesions on twigs and fruit followed by loss of leaves and dwarfing of fruit. See cedar-apple rust, hawthorn rust.

applesauce *n.* **1.** apples cut into small pieces, sweetened, often spiced with cinnamon, and stewed to make a soft dessert or appetizer. **2.** slang for nonsense, bunk.

apple scab *n.* a parasitic fungal (*Venturia inaequalis*, family Pleosporaceae) disease of apples that causes lesions on leaves, blossoms, and fruit of apples; dwarfs the fruit; and causes premature fruit dropping.

apple scald *n.* brown discoloration of the skins and flesh of apples during or after storage resulting from the release of alcoholic esters. It is especially prevalent on green-skinned apples and apples harvested before they are mature.

apposition *n.* (Latin *appositio*, placing with) **1.** the action of placing things together or near each other or of placing one thing on another. **2.** thickening of a cell wall by the deposition of an additional layer or layer of particles. **3.** the position or structure resulting from such action.

appropriative right to water *n.* a doctrine of rights to the use of water legally based on established prior use for the designated purpose; e.g., for domestic use, industrial use, or irrigation. It is prevalent in the western half of the U.S.A. Also called prior appropriation, as opposed to riparian right based on ownership of adjacent land. See riparian right.

apricot *n.* (French *abricot*, apricot) **1.** a yellowish orange stone fruit that resembles a plum or a peach. **2.** a tree (*Prunus armeniaca* and related species, family Rosaceae) that produces such fruit. Native to China, it grows in areas with cold winters and a continental climate. **3.** a yellowish orange color similar to that of the fruit.

apsis *n.* (Greek *hapsis*, a wheel) either the nearest or the farthest point in the orbit of a celestial body around a larger body. See aphelion, perihelion, apogee, perigee.

apteryx *n.* (Greek *a*, without + *pteryx*, wing) a nearly extinct genus of New Zealand birds with long, slender bills, hairlike feathers, rudimentary wings, and no tails; a kiwi.

aqua ammonia *n.* ammonia dissolved in water; often used as a fertilizer or as a household cleaner.

aquaculture or **aquiculture** *n.* (Latin *aqua*, water + *cultura*, care) **1.** the propagation and commercial rearing of fish and other animal species in a controlled water environment. See pisciculture. **2.** growing plants in water without soil; hydroponics.

aquamarine *n.* (Latin *aqua*, water + *marinus*, of the sea) **1.** a variety of beryl valued as a gem stone that is a transparent light blue green. **2.** a color between blue and green like that of the gem (similar to the color of seawater).

aquarium *n.* (Latin *aquarius*, of water) **1.** a transparent tank, bowl, or other container holding water where fish or other aquatic animals or plants are grown. **2.** a building with transparent tanks or containers for public viewing of living fish and other aquatic animals and plants.

aquatic *adj.* (Latin *aquaticus*, in water) **1.** living, growing, or taking place in or on water. *n.* **2.** a plant or animal that lives in or on water; e.g., water lilies or beaver. **3. aquatics** games or exercises that take place in or on water.

aqueduct *n.* (Latin *aqua*, water + *ductus*, drawing off) a bridgelike structure, usually made of masonry with several short spans, having a canal on top to carry water, especially one that supplies water to a city. The Romans were famous for building aqueducts. A similar structure for a road or railroad is called a viaduct.

aqueous *adj.* (Latin *aqua*, water + *osus*, characterized by) **1.** having a water base. **2.** formed by the action of water, as shale is an aqueous rock.

aquic *adj.* (Latin *aqua*, water + *icus*, characteristic of) a soil moisture regime in Soil Taxonomy; wet enough to cause reducing conditions in the soil and too wet to grow upland plants without artificial drainage. See wetlands, soil moisture regime.

aquiclude *n.* (Latin *aqua*, water + *clude*, to close) a buried sedimentary rock that does not transmit a significant flow of water even though it may be porous. See aquifer, aquifuge.

aquiculture *n.* See aquaculture.

aquifer *n.* (Latin *aqua*, water + *fer*, bearing) an underground stratum of porous rock, gravel, or sand containing or conducting usable amounts of water. See aquiclude, aquifuge.

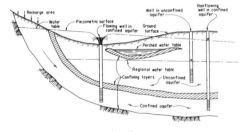

Aquifer

aquifuge *n.* (Latin *aqua*, water + *fuge*, driven away) a nonporous underground formation that neither holds nor transmits usable amounts of water. See aquiclude, aquifer.

Ar *n.* the chemical symbol for argon (the previous symbol was A). See argon.

arable *adj.* (Latin *arare*, to plow + *abilis*, able) suitable to be plowed to raise cultivated crops. Arable land is used to grow field crops, tree crops, and horticultural crops and for pasture. Some of it is fallowed periodically, and some is used for other purposes such as roadways and building sites. Also called tillable.

arachnid *n.* (Greek *arachnē*, spider) a wingless, carnivorous arthropod with 8 legs (class Arachnida, phylum Arthropoda), including predacious and parasitic spiders, mites, ticks, and scorpions.

aragonite *n.* (Spanish *Aragon*, a Spanish province + *ite*, mineral) the orthorhombic form of crystalline calcium carbonate ($CaCO_3$). Aragonite has the same chemical composition as calcite but is denser (2.94 g/cc) and harder (3.5 to 4.0). See calcite, calcium carbonate.

arbor *n.* (Latin *arbor*, tree) **1.** a small, shady structure where vines are grown; a shady, restful place. **2.** a tree; a woody plant larger than a shrub. **3.** a spindle, axle, or bar that supports a rotating cutting wheel.

Arbor Day *n.* a day to plant trees and to appreciate the environmental benefits of trees. National Arbor Day in the U.S.A. is the last Friday in April. As of the 1990s, all states in the U.S.A. as well as the District of Columbia, Guam, and the Virgin Islands celebrate Arbor Day, and the practice has spread to many other countries. Dates of celebration, however, differ in various states and countries to coincide with the most suitable planting season for trees.

arboriculture *n.* (Latin *arbor*, tree + *cultura*, to cultivate) the art and science of raising trees and shrubs, especially to provide shade and beauty.

arborvitae or **arbor vitae** *n.* (Latin *arbor vitae*, tree of life) an evergreen tree (*Thuja* sp., family Pinaceae) that has soft, aromatic, scalelike leaves on branchlets. Native to North America and Asia and popular as an ornamental tree or shrub. See eastern arborvitae.

arbuscular *adj.* pertaining to or shaped like a shrub or dwarf tree.

arbuscular mycorrhizae *n.* See endomycorrhiza.

archaea *n.* (Greek *archaios*, ancient) microbes such as methane producers, extreme halophiles, and certain thermophiles that are now classified as a separate kingdom, making them one of three major divisions of life forms: archaea, bacteria, and eukaryotes (based largely on each division having its own distinct form of RNA sequencing and on the composition and structure of cell membranes and cell walls). Also called archaebacteria.

Archimedean principle *n.* a body immersed in a fluid displaces a volume of fluid equal to the submerged volume of the body and is buoyed up by a force equal to the weight of the displaced fluid. Also called Archimedes' principle. See buoyancy.

Archimedean screw *n.* an ancient water-lift device consisting of either a large screw built into an inclined cylinder or of a pipe wrapped in a spiral around an inclined

Archimedean screw

axle. Rotating the assembly lifts water up the incline. The action is similar to that of an auger except that the screw is attached to the inside of the cylinder and the whole cylinder rotates. Also called Archimedes' screw.

archipelago *n.* (Italian *arcipelago*, the Aegean Sea) **1.** the Aegean Sea. **2.** a large body of water that contains many islands. **3.** a group of oceanic islands that are related to each other; commonly the peaks of a mountain chain rising from the ocean floor. Such chains are typically formed where one of Earth's plates moves across a hot spot in the mantle. See plate tectonics.

arctic *adj.* (Greek *arktikos*, extremely cold; of the bear) **1.** frigid; extremely cold. See polar climate. **2.** related to the far north, especially the area north of the timber line. **3. Arctic** the area north of the Arctic Circle.

Arctic Circle *n.* an imaginary circle parallel to the equator at about 66°33′ north latitude, marking the southern limit of the northern region that experiences 24-hr days of sunlight in the summer and 24-hr nights in the winter. See Arctic region, Antarctic Circle.

Arctic Ocean *n.* the ocean surrounding the North Pole, north of Asia and North America.

Arctic region or **Arctic zone** *n.* **1.** the area between the Arctic Circle (66°33′ north latitude) and the North Pole. **2.** the region north of the 50°F (10°C) July isotherm or north of the tree line. See Arctic Circle, Antarctica.

are *n.* the base unit of areal extent in the metric system, equal to 100 m^2 (1075.84 ft^2).

area *n.* (Latin *area*, open space) **1.** a surface, expressly that of a piece of land, often used with quantitative units; e.g., an area of 100 ft^2 or of 5 mm^2, or qualitative usage; e.g., a small area or a large area. **2.** a vicinity; e.g., the New York area or the London area. **3.** a section dedicated to a particular purpose, as the dining area of a home or the business area of a city. **4.** a subject or field of study, as the area of philosophy or the scientific area.

area source pollution *n.* a diffuse source of air and/or water pollution such as a watershed or airshed. Also called nonpoint source pollution. See point source pollution.

Arecaceae *n.* the family name adopted in 1972 for palm trees; formerly called family Palmae. See palm.

arenaceous *adj.* (Latin *arena*, sand + *āceus*, of) **1.** having a sandy nature. **2.** derived from sand. **3.** growing in sand.

Argas n. a genus of ticks that infest poultry.

Argentine ant *n.* an ant species (*Iridomyrmex humilis*, family Formicidae) that can kill bees in hives, trees, livestock, and humans. They occur in southeastern U.S.A. and in other countries.

argillaceous or **argillic** *adj.* (Latin *árgill*, clay + *āceus*, of) having a claylike nature; containing clay.

argillic horizon *n.* a diagnostic subsurface horizon in Soil Taxonomy. A mineral soil horizon that shows evidence of illuviation, characterized by coatings of oriented clay films on pore and ped surfaces and usually by a clay content at least 1.2 times as much as that in an overlying eluvial horizon.

argon (Ar) *n.* element 18, atomic weight 39.95; an inert gas constituting about 1% of the ambient atmosphere. Argon is used in lightbulbs and in vacuum tubes.

arid *adj.* (Latin *aridus*, to be dry) **1.** dry, parched, lacking sufficient moisture to be productive. See arid climate, Aridisol, humid, subhumid, semiarid. **2.** barren and unproductive from lack of moisture. **3.** dull and uninteresting.

arid climate *n.* a very dry climate with only enough precipitation for widely spaced desert plants. The upper limit of annual precipitation ranges from about 10 in. (250 mm) in cool regions to about 20 in. (500 mm) in the tropics. See arid, arid region.

aridic *adj.* (Latin *aridus*, to be dry + *icus*, of or like) too dry for plant growth more than half of the growing season. Water is available for plant growth for fewer than 90 consecutive days during the growing season when the soil temperature at 50 cm is above 8°C.

Aridisol *n.* (Latin *aridus*, dry + *solum*, soil) in Soil Taxonomy, a soil with arid climate, having no more than 3 months of growing season that is both warm enough and moist enough for plant growth. Aridisols occupy 8% of the land in the U.S.A. and 19% of the world. See arid climate, Soil Taxonomy.

arid region *n.* an area where the precipitation is less than the potential evapotranspiration. Arid regions may have either arid or semiarid climate.

armadillo *n.* (Spanish *armadillo*, armed a little) any of several burrowing mammals (*Dasypus* sp., family Dasypodidae) that carry a jointed armor protective covering; native to South America, Central America, and Texas.

armyworm *n.* the larva of a moth (especially *Leucania unipuncta*, family Noctuidae). Such larvae constitute a serious pest because they move in hordes and destroy whatever vegetation is in their path. See fall armyworm.

aromatic *adj.* (Greek *aromatikos*, fragrant) **1.** producing a strong aroma, usually a pleasing sweet or spicy fragrance. **2.** characteristic of certain organic compounds that contain benzene rings and commonly have an aroma (as contrasted with aliphatic compounds that have chains but no ring structures).

arrowhead plant *n.* any of several marsh plants (*Sagittaria* sp., family Alismataceae) with arrow-shaped leaves and small white blossoms.

arroyo *n.* (Spanish *arroyo*, stream) a small, steep-banked stream course that is usually dry except for short periods after a rain (used mostly in southwestern U.S.A.).

arsenic (As) *n.* (Latin *arsenicum*, arsenic) element 33, atomic weight 74.92; a semimetal that is listed as carcinogenic by the U.S. Environmental Protection Agency. Many of its arsenates are very poisonous. Arsenic trioxide (As_2O_3) is a toxic white powder that must not be inhaled, swallowed, or even touched by skin. Arsenic is used in metallurgy to harden copper and lead alloys and in the manufacture of certain kinds of glass. An artificial isotope, ^{76}As, is used as a radioactive tracer in toxicology.

artesian well *n.* a well that taps groundwater under sufficient pressure to cause the water to rise above the land surface (named after the French province of Artois, where there are many such wells).

arthropod *n.* (Greek *arthro*, joint + *pod*, foot) a member of the phylum Arthropoda, including insects, centipedes, millipedes, mites, ticks, spiders, and crustaceans, altogether about 80% of the species in the animal kingdom. Arthropods have segmented bodies, exoskeletons, and jointed legs.

artichoke *n.* (Arabic *al kharshuf*, the artichoke) **1.** a plant (*Cynara scolymus*, family Asteraceae) of the thistle family that produces flower heads with numerous large brachts. **2.** the unopened flower head of the artichoke plant. The flower head is cooked; the thick basal portion of each bracht and the fleshy core are eaten as a vegetable. See Jerusalem artichoke.

artificial insemination *n.* a procedure that uses a syringe to place semen collected by artificial means into the reproductive tract of the female. The advantage is that semen from a superior male can be used to breed several hundred females each year.

As *n.* the chemical symbol for arsenic.

asbestos *n.* (Greek *a*, not + *sbetos*, quenched) a fibrous magnesium silicate mineral, mainly the chrysotile form of serpentine [$Mg_6Si_4O_{10}(OH)_8$], but also including an amphibole known as riebeckite, that is fire resistant and an excellent insulator against both heat and electricity. It was widely used as a fireproofing and insulating material and as a constituent of floor coverings, roofing materials, building siding, pipes, and brake linings before it was recognized as carcinogenic. Dust containing asbestos fibers can cause asbestosis, a lung cancer condition in humans.

ascarid *n.* (Greek *askaris*, threadworm) any of a family of parasitic intestinal roundworms (family Ascaridae,

Asbestos hazard

especially *Ascaris* sp.) that infests several animal species and causes colic and diarrhea in humans.

ascorbic acid *n.* vitamin C ($C_6H_8O_6$). A white, crystalline, water-soluble material that is often added to foods as an antioxidant; also used for the prevention or treatment of scurvy. Also called cevitamic acid.

ASCS *n.* See Agricultural Stabilization and Conservation Service.

aseptic *adj.* (Greek *a*, not + *sēptikos*, rotted) **1.** unlikely to putrefy because living organisms are absent. **2.** absence of pathogens. See antiseptic.

asexual reproduction *n.* any form of reproduction that does not involve male and female germ cells, usually referring to vegetative reproduction such as cuttings from a stem to start a new plant. Other examples of asexual reproduction are shown in the figure below.

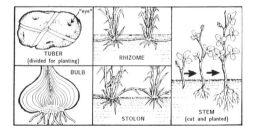

Asexual reproduction

ash *n.* (Anglo-Saxon *aesce*, base) **1.** the residue from burning, usually a gray powder composed mostly of oxides of metallic elements such as calcium, magnesium, and potassium. **2.** powdery angular fragments of glassy rock material from a volcanic eruption deposited in a thin layer over a wide area. See cinder, tephra. **3.** a hardwood tree or shrub (*Fraxinus* sp., family Oleaceae) having pointed pinnate leaves and clusters of small green flowers; used as timber and shade trees. The most important species are black ash

(*F. nigra*), blue ash (*F. quadrangulata*), European ash (*F. excelsior*), green ash (*F. pennsylvanica*), pumpkin ash (*F. tomentosa*), and white ash (*F. americana*). **4.** wood from an ash tree. Ash is typically straight grained, tough, and elastic; used in furniture and for tool handles. Young resilient ash is a favorite wood for baseball bats. *v.* **5.** to burn something until only ashes remain.

ashfall *n.* the ashes that fall over an area from a particular event, such as a volcanic eruption.

aspartame *n.* aspartyl phenylalanine ($C_{14}H_{18}N_2O_5$), produced by linking two amino acids that occur naturally in both plants and animals. Neither is sweet alone, but the compound appears to be 200 times sweeter than sucrose. On July 15, 1981, the U.S. Food and Drug Administration cleared aspartame for use as a nonnutritive sweetener. It has been cleared for use in cold cereals, drink mixes, instant coffee and tea, gelatin, pudding fillings, dairy products, and dessert toppings.

ASPCA *n.* American Society for the Prevention of Cruelty to Animals.

aspect *n.* (Latin *aspectus*, appearance) **1.** general nature of something; the visible appearance or expression of an object. **2.** a viewpoint or interpretation of a topic. **3.** the compass direction faced by a ground slope or a building; e.g., a southern aspect means facing south. **4.** the angular distance between two heavenly bodies as seen from the Earth. **5.** a verb form indicating the duration or completeness of an action.

Aspergillus *n.* (Latin *aspergillus*, a fungus) any of several fungi (*Aspergillus* sp.) occurring in many different molds, especially moldy feed and moldy silage, including the fungus that causes aspergillosis. See mycotoxins.

asphalt *n.* (Greek *asphaltos*, pitch) **1.** a thick black or brownish black residue composed mostly of bitumens left from petroleum after gasoline and other products have been distilled away; used to make cements, fluxes, sealants, shingles, and pavement. **2.** a naturally occurring deposit of similar nature. **3.** a mixture of asphalt with gravel used as pavement, especially for highways.

asphyxiation *n.* (Greek *asphyxia*, cessation of pulse + *atio*, condition) **1.** a systemic condition of animals or people caused by oxygen deprivation and resulting in anoxia and carbon dioxide accumulation in the body; a serious hazard in garages, silos, and manure pits with insufficient circulation of air. **2.** a systemic condition of plants, brought on by severe oxygen deficiency and characterized by seeds not germinating, blighting of plants, and breakdown of fruits and tubers.

aspirin *n.* **1.** acetyl salicylic acid; a white crystalline solid ($CH_3COOC_6H_4COOH$) used as a painkiller, anti-inflammatory agent, and blood thinner; so named because it was discovered in spirea blossoms. **2.** an aspirin tablet.

ass *n.* (Latin *asinus*, donkey) **1.** the donkey (*Equus asinus*); a sure-footed domesticated beast of burden; related to the horse, but smaller and with longer ears; also called burro. Cross-breeding asses with horses produces mules. **2.** any of several species of *Equus*; e.g., wild asses of Asia and North Africa. **3.** a derogatory term for a person behaving stubbornly or foolishly. **4.** (slang) the posterior of a person or animal.

assimilation *n.* (Latin *assimilatio*, making similar) **1.** the process of incorporating new individuals or components into an existing body. **2.** the process of absorption of nutrients or food by plants, animals, or humans. **3.** the ability of a body of water to purify itself of pollutants. **4.** the blending of the sound of a letter into that of another; e.g., the loss of the p sound in cupboard.

astatine (At) *n.* (Greek *astatos*, unstable) element 85, a radioactive synthetic halogen heavier than the four naturally occurring halogens. Its half-life is 8.3 hr for its most stable form (^{200}At).

Asteraceae *n.* the family name adopted in 1972 for herbaceous plants with composite blossoms that usually have a cluster of tiny flowers surrounded by a ray of petals and a whorl of bracts; e.g., asters, daisies, dandelions, marigolds, sunflowers, and thistles. The older name is family Compositae.

asteroid *n.* (Greek *asteroeid*, starlike) a solid body smaller than a planet that orbits the sun. Most asteroids orbit between Mars and Jupiter, but some (known as Apollo asteroids) have elliptical orbits that cross Earth's orbit. See Ceres.

asthenosphere *n.* (Greek *asthenēs*, weak + *sphaîra*, ball or spherical layer) the dense, somewhat plastic rock mass underlying the tectonic plates that make up the crust of the Earth and extending inward to where it surrounds the molten core.

ASTM *n.* See American Society for Testing and Materials.

astragalus *n.* (Greek *astragalos*, a vertebra) **1.** the upper ankle bone that supports the tibia, also known as tarsus. **2.** a genus of about 1000 species of leguminous plants that are also known as milk vetch. Some *Astragalus* species are called locoweed because they accumulate selenium from the soil and, when eaten, are poisonous to sheep, cattle, and horses. See locoweed.

astrobiology *n.* See exobiology.

astronomical unit (AU) *n.* the average distance between the Earth and the sun, equal to about 93 million miles (150 million km or 1.6×10^{-5} light years); used for expressing distances within the solar system.

astronomy *n.* (Latin *astronomia*, star law) the science that studies the universe and its components.

ataraxia *n.* (Greek *ataraktos*, unmoved) a calm, peaceful, tranquil state; being in harmony with nature and the environment.

atavism *n.* (Latin *atavus*, grandfather) an individual's apparent inheritance of characteristics from remote ancestors that were not apparent in intervening generations; throwback or reversion to an earlier type.

atmometer *n.* (Greek *atmos*, vapor + *metron*, measure) an instrument for measuring the rate of evaporation of water into the atmosphere. Also called an evaporometer.

atmosphere *n.* (Greek *atmos*, vapor + *sphaira*, sphere) **1.** the air; the gaseous envelope surrounding the earth to a distance of about 22,000 miles (35,000 km), composed largely of 78% nitrogen, 21% oxygen, and 1% argon, with 0.03% carbon dioxide and variable content of water vapor. **2.** a standard unit of pressure representing the mean atmospheric pressure at sea level at 45° latitude, equivalent to a 29.92-in. (760 mm) column of mercury, a 33.9 ft (10.33 m) column of water, 14.696 lb/in.^{2}, or 1.033 kg/cm^2. **3.** the mood, tone, or distinctive quality of a place or a work of art. See air, hydrosphere, lithosphere, biosphere.

atmospheric pressure *n.* the pressure of the air on a surface. See atmosphere, barometric pressure.

atoll *n.* (Malayalam *atolu*, atoll) a coral island or circle of islands that is not associated with a larger land mass but is built on an underwater structure, usually an extinct volcano. Atolls are roughly circular with an internal shallow lagoon of quiet water. Bikini in the South Pacific is a famous atoll.

atom *n.* (Greek *atomos*, undivided) **1.** the smallest particle of an element capable of expressing the chemical properties of the element, composed of a nucleus containing protons and neutrons surrounded by orbiting electrons. **2.** the tiniest possible amount of something.

atomic energy *n.* energy released in nuclear reactions. Of particular interest is the energy released when nuclei are split into smaller pieces (fission) or when nuclei are joined together at hundreds of millions of degrees of heat (fusion). Atomic energy is more correctly called nuclear energy.

atomic mass unit (u or amu) *n.* the usual unit for describing the mass of atomic particles; one-twelfth of the mass of the ^{12}C isotope of carbon, or 1.66054×10^{-24} g. On this basis, a proton weighs 1.0073 amu, a neutron weighs 1.0087 amu, and an electron weighs 5.486×10^{-4} amu.

atomic number *n.* the total number of protons in the nucleus of an atom of any element. It is also the total number of electrons orbiting around the nucleus of a nonionized atom.

atomic pile or **atomic reactor** *n.* an apparatus that uses a controlled nuclear fission reaction to release heat for the production of electricity.

atomic time *n.* time as measured by the vibrations of an atom. Currently those of cesium atoms are used as the international standard. A leap second is added when needed to keep atomic time in agreement with universal time.

atomic weight or **at. wt.** *n.* the average relative weight of an atom of an element as compared with an arbitrary standard (e.g., oxygen = 16.000), usually a standard that makes the atomic weight approximately equal to the number of protons plus neutrons in the nucleus. Isotopes of the same element have different atomic weights because the number of neutrons differs from isotope to isotope; a quoted atomic weight is the average for the element unless a specific isotope is named.

ATP *n.* See adenosine triphosphate.

atrazine *n.* a white, crystalline organic herbicide ($C_8H_{14}N_5Cl$), so named because it has an amino group attached to a triazine; the most common herbicide for use on corn and one of the most common pollutants in water.

atropine *n.* (Greek *atropos*, not turning + *inos*, like) a poisonous alkaloid ($C_{17}H_{23}O_3N$) derived from belladonna plants (*Atropa* sp.); used as an antidote for organophosphate poisoning, to dilate the pupil of the eye, and to relieve spasms.

attainment area *n.* an area considered to have air quality as good as, or better than, the national ambient air quality standards as defined in the Clean Air Act. An area may be an attainment area for one pollutant and a nonattainment area for others.

attenuate *v.* (Latin *attenuat*, to reduce) **1.** to dilute, thin down, enfeeble or reduce. **2.** to reduce the virulence of an organism; to destroy the power of disease-producing bacteria by chemical action or viruses by passing them through unnatural hosts, etc. This principle is used to develop certain bacterins and vaccines.

attenuation *n.* (Latin *attenuat*, to reduce + *atio*, action) **1.** the process by which a compound, such as a toxin, is reduced in concentration over time, through adsorption, degradation, dilution, and/or transformation. **2.** the weakening without distortion of a radio signal.

Atterberg limits *n.* the water contents of a soil when it has barely enough water to be plastic (the plastic limit) or to flow as a liquid (the liquid limit), and the plasticity index (the liquid limit minus the plastic limit).

attractant *n.* (Latin *attrahere*, draw toward + *antem*, thing that) something that attracts. Specifically, any substance that attracts or draws insects or other animals by a sense of smell. For example, a substance identified as gyptol, obtained from the terminal segments of female gypsy moths, is an *attractant* for male gypsy moths. See pheromone.

attrition *n.* (Latin *attritio*, wearing away) **1.** a gradual reduction in numbers or mass. **2.** the wearing away by friction or abrasion. **3.** the wear and tear that rock particles in transit undergo through mutual rubbing, grinding, knocking, scraping, and bumping causing a gradual reduction in size. **4.** in business, a reduction in number of workers resulting from retirements, resignations, and deaths.

atypical *adj.* (Greek *a*, not + *typikos*, type + Latin *alis*, like) differing from the form, state, or situation usually found under similar circumstances; not typical; often abnormal.

Au *n.* (Latin *aurum*, shining dawn or gold) the chemical symbol for gold.

Audubon, John James *n.* an early American naturalist (1785–1851) renowned for his paintings and descriptions of North American birds.

Audubon Society *n.* See National Audubon Society.

auger *n.* (Anglo-Saxon *nafogār*, nave drill) **1.** a tool for boring holes, especially a hand tool larger than a gimlet used for boring holes in wood. **2.** either a hand or mechanized tool for drilling holes in the ground (e.g., post holes).

Aujeszky's disease *n.* a lethal viral disease of cattle, less commonly affecting dogs, cats, pigs, humans, etc. It is characterized by intense itching, paralysis, and convulsions. It was named in 1902 for Aladar Aujeszky (1869–1933), a Hungarian scientist. Death occurs within 36 to 48 hr after symptoms become evident. Also called bovine pseudorabies, infectious bulbar paralysis, or mad itch.

auk *n.* (Norse *alka*, auk) any of several black-and-white birds (*Alca* sp., family Alcidae) up to about 20 in. (0.5 m) tall that spend most of their time in the water, using their wings and webbed feet to swim. They live on fish, go ashore only to breed, and are native to the northern seas. Note: The great auk (*Pinguinus impennis*) has been extinct since 1844. See penguin.

auklet *n.* (Norse *alka*, auk + *let*, little) any of several small auks of the north Pacific; e.g., the crested auklet (*Aethia cristatella*).

AUM or **aum** *n.* See animal unit month.

aurochs *n.* (German *auerochs*, ox) a bovine species (*Bos primigenius*, family Bovidae) that was the ancestor to western cattle (*B. taurus*). They were once widely distributed from northern Africa through Europe and the Middle East into northern Asia but became extinct in 1627.

austral *adj.* (Latin *australis*, southern) **1.** from the south; southern. **2.** Australian. **3.** the Argentine monetary unit.

Australopithecine *n.* or *adj.* (Latin *australis*, southern + *pithēcus*, ape) a prehistoric hominid (*Austraopithecus* sp., family Hominidae) that lived in southern Africa about 1 to 4 million years ago. It was first identified in fossil remains in 1924 by anthropologist Raymond A. Dart. See hominid, hominoid.

autarky or **autarchy** *n.* (Greek *autarchia, autarkeia*, self-sufficiency) **1.** a national policy of self-sufficiency, especially in an economic sense. **2.** people and domestic animals eating only what is locally grown. This is a fact of life for millions of poor people around the world. The hazards of autarky are the following:

- It may be impossible to have a balanced diet because not all foods will grow locally.
- Deficiencies may develop because soils may have inadequate amounts of certain essential nutrients for people and domestic animals; e.g., iodine deficiency leads to goiter.
- Toxicities may occur in foods and feeds because of an excess of certain minerals; e.g., excess selenium causes selenosis, excess fluorine causes fluorosis, and excess molybdenum causes molybdenosis.

autochthon *n.* (Greek *autochthōn*, sprung from the land) **1.** the earliest known plants, animals, or humans to particular place; aborigines. **2.** a person who still lives at his/her birthplace; a native. **3.** an indigenous plant or animal.

autochthonous *adj.* (Greek *autochthōn*, sprung from the land + Latin *osus*, characterized by) **1.** indigenous; something native to or formed in the place where it is found. **2.** medical: remaining in the part of the body where it originated, as an autochthonous cancer. **3.** an abnormal obsession or schizophrenic characteristic originating in an individual. See allochthonous for contrast.

autoecious *adj.* (Greek *auto*, self + *oikos*, house + Latin *osus*, having) having all stages of a parasitic life cycle on a single host plant; e.g., autoecious rust.

autoecology *n.* (Greek *auto*, self + *oikos*, house + *logos*, word) the study of the ecology of a single organism; e.g., the interactions of an individual plant with various components of its environment, as distinguished from synecology.

autogamous *adj.* (Greek *auto*, self + *gamos*, marriage) self-fertilizing; a flower fertilized by its own pollen, as opposed to allogamous.

autogenic succession *n.* succession due to biotic reaction wherein the replacement of an association by another results chiefly from transformations induced by

the plants themselves; e.g., a pine forest being replaced by deciduous forest.

autoimmunity *n.* (Greek *auto*, self + Latin *immunitas*, exemption) an immune reaction by a human or an animal producing antibodies that damage some of its own tissues and may produce clinical disease.

autolysis *n.* (Greek *autos*, self + *lysis*, dissolving) **1.** self-digestion, as a result of certain diseases or as spontaneous decomposition after death. **2.** the natural softening process of fruits or vegetables after picking or of meat after slaughtering. **3.** spontaneous disintegration of bacteria by their own enzymes.

autopass *n.* a device that substitutes for a gate, constructed to allow automobile traffic but to prevent cattle or sheep from getting out of a pasture. Also known as cattlegap or cattleguard.

autophyte *n.* (Greek *autos*, self + *phyton*, plant) an organism that manufactures its own food, as a photosynthetic plant. See photosynthesis.

autopolyploid *n.* (Greek *autos*, self + *polys*, many + *ploos*, fold) an organism that has more than the normal two sets of chromosomes (usually a tetraploid, i.e., having four sets of chromosomes), all derived from the same parent species.

autotroph *n.* (Greek *auto*, self + *trophé*, nourishment) an organism that produces carbohydrates from carbon dioxide and water without using food from other organisms (in contrast to heterotrophs). Autotrophs are of two general types: photoautotrophs (photo-lithotrophs), whose energy is derived from sunlight; and chemoautotrophs (chemolithotrophs), whose energy for growth and reproduction comes from oxidation of inorganic materials. Important among the photoautotrophs are all higher plants, algae, and a few genera of bacteria. Principal chemoautotrophs of interest in food and agriculture (with their substrates and products) are *Nitrosomonas* sp. (ammonium to nitrite), *Nitrobacter* sp. (nitrite to nitrate), and *Thiobacillus* sp. (sulfur to sulfate). Also called lithotroph.

autotrophic *adj.* (Greek *auto*, self + *trophé*, nourishment + *ikos*, having) self-nourishing; able to use light energy or chemical energy to synthesize carbohydrates from water and carbon dioxide.

autotrophic nutrition *n.* (also known as lithotropic nutrition) the process of an organism manufacturing its own food from inorganic sources, using carbon dioxide as the sole carbon source.

autumn *n.* (Latin *autumnus*, fall) **1.** the period from the autumnal equinox (about September 22 in the Northern Hemisphere or March 21 in the Southern) to the winter solstice (about December 22 or June 21). **2.** the fall season; the months following summer but before winter, commonly regarded as September, October, and

November in the Northern Hemisphere or March, April, and May in the Southern Hemisphere. **3.** the latter part of a lifetime, the period following maturity and including old age. **4.** a cooling period that usually follows maximum development and activity of a plant or animal; maturity, dormancy and/or death may occur during this period; does not necessarily coincide with the calendar autumn.

auxin *n.* (Greek *auxein*, to increase) a plant hormone; a substance that promotes the growth of plants.

available nutrient *n.* an ion of an element that is essential for plant growth and that is present in or can readily become part of a pool of ions that can be absorbed by any nearby plant root.

available water *n.* water in soil that plants can use; water held at tensions between field capacity (about one-third atmosphere) and the wilting point (about 15 atmospheres).

available water-holding capacity *n.* the amount of available water that a soil can store, either per unit of depth (e.g., 2 in./ft or 0.16 cm/cm) or in the solum or root zone, normally expressed in inches or centimeters. Available water-holding capacity varies with soil texture and is also affected by soil density, structure, organic matter content, etc. Some typical values are shown in the table below.

soil texture	in./in. or cm/cm
sand	0.10
sandy loam	0.15
loam or silt loam	0.17
clay loam or silty clay loam	0.19
clay	0.17

Note: clay holds as much total water as any soil, but much of it is held too tightly to be available.

avalanche *n.* (French *la valanche*, avalanche) **1.** a large mass of snow and/or ice, sometimes accompanied by other material, moving rapidly down a mountain slope. Avalanches are usually classified by the type of snow involved as climax, combination, damp snow, delayed action, direct action, dry snow, hangfire, and windslab avalanches. **2.** a large quantity of something that arrives suddenly, as an avalanche of mail. **3.** a rapidly increasing chain reaction, as when the impact of a few electrons causes the release of succeeding generations with ever-increasing numbers.

avalanching *n.* the process whereby a few saltating sand grains cause progressively larger numbers to saltate and cause more severe wind erosion in the downwind direction.

aviary *n.* (Latin *aviarium*, a place to keep birds) a large enclosure for keeping birds.

avicide *n.* (Latin *avis*, bird + *cida*, killer) a chemical for killing unwanted birds.

aviculture *n.* (Latin *avis*, bird + *cultura*, growing) the raising of birds. See aviary.

avidin *n.* (Latin *avidus*, desiring + *in*, protein) a protein contained in raw egg white that can bind with biotin and cause dermatitis. It can also inhibit the growth of certain bacteria.

avitaminosis *n.* (Latin *a*, not + *vita*, life + *amin*, amine + *osis*, condition) a disease of humans or animals that results from a vitamin deficiency. See vitamin.

avocado *n.* (Nahuatl *ahuacatl*, testicle) **1.** a tropical tree (*Persea americana*, family Lauraceae) that bears an edible fruit. **2.** the green-skinned, pear-shaped stone fruit with a yellow or light green, buttery flesh that is often used in salads. Also called alligator pear.

Avogadro's law *n.* equal volumes of all gases under the same temperature and pressure contain the same number of molecules. Also called Avogadro's principle and named after Amadeo Avogadro (1776–1856), the Italian physicist who stated it as a theory.

Avogadro's number *n.* the number of molecules contained in the gram-molecular weight of a substance, equal to 6.0225×10^{23}. Symbol: N. Also called Avogadro constant.

avoirdupois *n.* (French *avoir du pois*, to have weight) the weight system based on pounds (abbr. lb; 1 lb = 453.6 g) in common use in the U.S.A.; originally from Britain.

awn *n.* (Latin *agna*, bristlelike fiber) a long, wiry, bristlelike plant appendage, especially that growing from the glumes in spikelets of many varieties of barley, oats, rye, wheat, and other grasses.

axenic culture *n.* a pure culture; free of contaminating germs.

axis *n.* (Latin *axis*, an axle) **1.** the center of rotation of a body; e.g., a straight line between the Earth's North and South Poles, the center line of its daily rotation. The inclination of the Earth's spin axis from a line perpendicular to its orbital plane around the sun, an angle of 23.5°, causes seasonal changes during the year. **2.** a line that divides a figure into two equal or balanced parts. **3.** a line of symmetry through a three-dimensional body. **4.** a reference line, such as the *x*- or *y*-axis of a figure. **5.** the line of support from the roots through the stem and other parts of a plant.

axis deer *n.* an Asian deer (*Cervus axis*) with white markings on a reddish brown coat, found in India and nearby parts of Asia.

azalea *n.* (Greek *azaleos*, dry) any of a large number of short to tall, flowering shrubs (*Rhododendron* sp., family Ericaceae) that do well in dry sites. Azaleas are similar to rhododendrons but with smaller, thinner petals that fall sooner than those of rhododendrons. They produce blossoms of white, pink, orange, red, purple, and other colors.

azimuth *n.* (Arabic *as sumut*, the way) **1.** horizontal direction on the surface of the Earth, designated clockwise from north at 0° or 360°, east at 90°, due south at 180°, and west at 270°. **2.** a similar direction in astronomy measured clockwise with south as 0°.

azollae *n.* (Greek *azoto*, dry + *ollyo*, to kill). a rapidly growing, free-floating water fern (*Azolla* sp.) that grows in still waters of temperate and tropical regions (e.g., rice paddies) and forms a symbiotic relationship with nitrogen-fixing cyanobacteria (*Anabaena azollae*) commonly known as blue-green algae. *Azollae* grow with their roots hanging in the water or, in shallow water, into the underlying mud. They form a dense mat in rice paddies that suppresses weed growth, and the association can fix enough nitrogen to satisfy the needs of the rice, comparable to the amount of nitrogen fixed by a legume-*Rhizobium* association.

azonal *adj.* (Latin *a*, not + *zona*, belt) independent of zone. A designation in the 1938 U.S. system of soil classification for soils that lack distinct profile characteristics because they are very young, they are being eroded too fast to form profiles, or their parent material is highly resistant to change. (In contrast to zonal and intrazonal soils.)

Azorhizobium n. nitrogen-fixing bacteria (*Azorhizobium caulinodans*) that can fix nitrogen in the free-living state, in nodules, or on the stems of *Sesbania rostrata*.

Azotobacter n. (French *azote*, nitrogen + *bactérie*, bacteria) any of several free-living, motile, aerobic, nonsymbiotic soil bacteria (*Azotobacter* sp., family Azotobacteraceae) that were first isolated in 1901 by M. Beijerinck. They occur in most soils, and at least 5 species are capable of fixing atmospheric nitrogen (N_2): *A. agilis, A. beijerinckii, A. chroococcum, A. indicum,* and *A. vinelandii*.

azoturia *n.* (French *azote*, nitrogen + *uria*, in urine) a disease of horses marked by an excess of urea in the urine, usually occurring in a draft horse that overeats when it is rested for a few days Symptoms include dark urine, sudden perspiration, and paralysis of the hind legs. Also called azotemia, Monday morning disease, myochemoglobinuria, or paralytic hemoglobinemia.

Aztec *n.* **1.** a member of the pre-Columbian civilization of Mexico; an advanced civilization with its capital, Tenochtitlan, located on an island in a lake where Mexico City is now. The Aztecs were conquered by Hernando Cortez in 1519–1521 with his few hundred Spanish soldiers aided by armies from rebellious native tribes. **2.** a modern descendant of these people. See Inca, Maya.

Aztec tobacco *n.* a wild tobacco (*Nicotiana rustica*, family Solanaceae) of high nicotine content; the tobacco first grown by the Indians in eastern U.S.A., sometimes cultivated in flower gardens and used in the manufacture of insecticides. Also called emetic weed, Indian tobacco, wild tobacco, or eyebright.

azygous *adj.* (Greek *azygos*, not paired) single, not one of a pair, as certain bones, muscles, or veins.

B

B *n.* **1.** the chemical symbol for boron. **2.** a good grade, but less than excellent. **3.** second class.

babassu *n.* (Portuguese *babaçú*) a palm tree (*Orbignya speciosa*) that grows wild in the Brazilian rain forest. The nuts are edible and produce babassu oil used for cooking and for making margarine and soap.

babble *v.* (French *babiller*, to babble) **1.** to make incoherent sounds or indistinct words. **2.** to make continuous background noise, as the babble of a brook. **3.** to talk foolishly on a subject. *n.* **4.** the sound produced by someone or something that babbles.

babesiosis *n.* infection of red blood corpuscles by parasitic sporozoa (*Babesia* sp., family Babesiidae) transmitted to warm-blooded animals by ticks; e.g., the cause of fever and anemia in cattle (*B. bovis*), malignant jaundice in dogs (*B. canis*), and Texas fever in humans (*B. bigemina*). Named after Victor Babeş (1854–1926), a Rumanian bacteriologist. Also called piroplasmosis.

babirusa *n.* (Malay *babi*, pig + *rusa*, deer) a piglike wild animal (*Babirousa babyrussa*) of the West Indies with canine teeth that grow upward outside its mouth; sometimes hunted for food.

baboon *n.* (French *babouin*, baboon) **1.** a large monkey (*Papio* sp. and related genera) of Africa and Arabia, characterized by a doglike muzzle and a short tail. **2.** (slang) a careless, brutish person.

Bacillus *n.* (Latin *bacillum*, a wand) a large genus of aerobic, rod-shaped, spore-producing bacteria (*Bacillus* sp., family Bacillaceae). Several species are pathogenic to animals and/or humans:
- *Bacillus abortus*: contagious abortion (Bang's disease of cattle).
- *Bacillus alvei* (*B. larvae*): foulbrood of honeybees.
- *Bacillus anthracis*: anthrax in animals and humans. See anthrax.
- *Bacillus cereus*: food poisoning.
- *Bacillus dysenteria*: dysentery.

Several species are used to produce antibiotics:
- Bacitracin: *Bacillus subtilis*.
- Colistin: *Bacillus colistinus*.
- Gramicidin: *Bacillus brevis*.
- Polymyxin: *Bacillus polymyxa*.
- Subtilin: *Bacillus subtilis*.
- Tyrocidine: *Bacillus brevis*.
- Tyrothricin: *Bacillus brevis* and others.

Bacillus thuringiensis (**Bt**) *n.* a soil bacterium that produces crystalline proteins that act as an insecticide, attacking the armyworm, cabbage looper, cabbage worm, and gypsy moth; officially registered in the U.S.A. as a biopesticide since 1961. It is harmless to people, domestic animals, and plants.

backcross *n., v.,* or *adj.* (Anglo-Saxon *baec*, back + *cros*, cross) the crossing of a first generation hybrid with one of the parental types (breeds). The offspring are referred to as the backcross generation, or progeny.

backfire *n.* (Anglo-Saxon *baec*, back + *fyr*, fire) **1.** a prescribed burning against the wind in the path of an advancing fire to consume the fuel and thereby control the advancing fire. **2.** a premature explosion in a cylinder of an internal combustion engine, tending to make the engine turn backwards. **3.** an explosion that flashes into the intake manifold of an internal combustion engine. *v.* **4.** to have an unexpected result contrary to the desired outcome.

back furrow *n.* a ridge of soil made by a moldboard or disk plow throwing soil on top of unplowed soil or a previously plowed ridge rather than into an open furrow. This is the usual method of beginning to plow a field. After that, each succeeding furrow turns soil into the preceding furrow until the final pass. The open furrow left at the end is called a dead furrow.

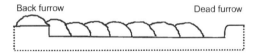

Back furrow   Dead furrow

Plow furrows

background radiation *n.* the radiation coming from sources other than the radioactive material to be measured; primarily produced by cosmic rays constantly bombarding the Earth from outer space.

backshore *n.* (Anglo-Saxon *baec*, back + *score*, to jut out) the usually dry part of an ocean beach above the level of normal high tide and extending inland to vegetation, dunes, or a cliff. See beach.

backwash *n.* (Anglo-Saxon *baec*, back + *waescan*, clean with water) **1.** water or air forced backward by the action of a propeller or other propulsion device. **2.** return flow of water on a beach after the advance of a wave. **3.** undesirable results from a past event.

backwater *n.* (Anglo-Saxon *baec*, back + *waeter*, water) **1.** a series of connected lagoons parallel to a coast, separated narrowly from the sea, and communicating with it by barred outlets. **2.** a currentless, stagnant body of water of the same trend as a river and fed from it at the lower end by back flows. **3.** water forced back upstream by a dam or other obstruction.

backwearing *n.* a process of landscape development postulated by Walther Penck in *Die Morphologische Analyse* (1927). Penck proposed a parallel retreat of slopes whereby a valley wall wears back toward a ridge rather than the summit wearing down, as proposed by W. M. Davis in his downwearing concept. Penck coined the term pediment for the gentle footslope that gradually lengthens as the valley wall retreats. See downwearing, mass wasting, pediment.

back woods or **backwoods** *n.* (Anglo-Saxon *baec*, back + *wudu*, forest) **1.** a sparsely populated wooded area. **2.** any remote area. *adj.* **3.** pertaining to a person or thing from a backwoods area; unsophisticated.

BACT (best available control technology) *n.* emission limitations based on the maximum degree of emission reduction achievable through application of available production processes, methods, systems, and techniques with appropriate consideration for energy, environmental, economic impact, and other costs. In no event does BACT permit emissions in excess of those allowed under any applicable Clean Air Act provisions.

bacteria (sing. **bacterium**) *n.* (Greek *baktērion*, little staff) **1.** microscopic primitive living plants, generally free of pigment, that reproduce by dividing in 1, 2, or 3 planes; bacteria occur as single cells, chains, filaments, well-oriented groups, or amorphous masses. Most bacteria do not require light, but a few species are photosynthetic and draw on light for energy. Most bacteria are heterotrophic and use organic matter for energy and for growth materials, but a few are autotrophic and derive bodily needs from inorganic materials. Bacteria are important in the decomposition processes that recycle nutrients from dead organisms. The self-purification potential of streams depends on aerobic bacterial action. Certain bacteria in soil, water, or air can also cause human, animal, and plant diseases. **2.** one of three major divisions of life forms—archaea, bacteria, and eukaryotes—based largely on each division having its own distinct form of RNA sequencing.

bacteria, facultative *n.* See facultative anaerobe.

bacterial gummosis *n.* exudation of gumlike liquids resulting from bacterial infection, especially from cut sugar cane stalks (caused by *Xanthomonas vasculorum*), bacterial canker of stone fruits (caused by *X. pruni*), and foot rot of citrus (caused by *Phytophthora* sp.).

bacterial wilt of alfalfa *n.* a wilt that is highly destructive on older stands of alfalfa; caused by bacteria (*Corynebacterium insidiosum*). Usually not serious until the third crop year, it causes yellowing and dwarfing of the stems and leaves, sometimes rotting of the crown, a yellow ring just under the skin, and a yellowing of the roots. The disease is most serious in wet areas with freezing conditions in winter.

bacterial wilt of corn *n.* a bacterial (*Bacterium stewartii*) disease that is most troublesome on sweet corn. The bacteria infect the whole plant and cause wilting by plugging the vascular bundles. The disease produces long, wavy streaks and a yellow exudate on the leaves. Plants that survive are stunted and produce abnormal ears. The disease attacks sweet corn, dent corn, popcorn, teosinte, and Job's-tears. It is more severe after a mild winter. Also called Stewart's disease.

bactericide *n.* (Greek *baktērion*, a little staff + Latin *cida*, killer) a chemical, such as an antibiotic, specific for killing bacteria. One of the first was streptomycin ($C_{21}H_{39}N_2O_{12}$).

bacteriocin *n.* an agent produced by certain bacteria that kills or inhibits the activity of other closely related species.

bacteriology *n.* (Greek *baktērion*, a little staff + *logos*, word) the biological science that studies bacteria.

bacteriolysis *n.* (Greek *baktērion*, a little staff + *lysis*, breakdown) **1.** destruction of bacteria either inside or outside the body of an animal. **2.** decomposition by bacterial action.

bacteriophage *n.* (Greek *baktērion*, a little staff + *phage*, eating) an ultramicroscopic, filterable virus composed of deoxyribonucleic acid and a protein coat. A particular bacteriophage can infect and can reproduce itself only in a specific strain or species of bacteria. It usually causes lysis, or explosive dissolution, of the infected bacterium, releasing the bacteriophage to infect other bacterial cells. Also called phage.

bacteriostasis *n.* (Greek *baktērion*, a little staff + *stasis*, standing) retardation of the life processes (growth or development) of bacteria without killing them.

bacteriostat *n.* (Greek *baktērion*, a little staff + *statēs*, stationary) a substance that inhibits bacterial growth.

badger *n.* **1.** a carnivorous, burrowing, nocturnal mammal (*Taxidea taxus*, family Mustelidae) with short legs, a heavy body, a round back with a black and gray coat, and a white streak running vertically on its face; native to North America; the symbol of the state of Wisconsin. **2.** a similar Eurasian species (*Meles meles*, family Mustelidae). **3.** the fur from badgers. *v.* **4.** (slang) to pester someone persistently, trying to get them to do something such as lower a price.

badlands *n.* a miscellaneous land type composed of an intricate maze of narrow ravines, sharp crests, and pinnacles resulting from severe erosion of soft geologic materials. Badlands are most common in arid or semiarid regions and are generally devoid of vegetation. A specific example on a large scale is the dissected region in the Black Hills in western South Dakota.

bagasse *n.* (Spanish *bagazo*, husks) residue after juice is extracted by crushing sugar cane stalks or sugar beets; used as fuel and in making paper, wallboard, insulating materials, and livestock feeds. Also denotes similar residues from sorghum and sisal.

bagworm *n.* the caterpillar phase of a moth (*Thyridopteryx ephemeraeformis*, family Psychidae) whose larvae surround themselves with baggy wrappings composed of silken threads, leaves, etc. They infest more than 100 plant species and eat the foliage in eastern and central U.S.A.

bahiagrass *n.* (Portuguese *Bahia*, a state in Brazil + German *gras*, grass) a palatable pasture and lawn grass (*Paspalum notatum*, family Poaceae) that originated in the American tropics. It forms a tough sod even on droughty soils and is sufficiently winter hardy for use in southern U.S.A. for forage or erosion control. The seed is low in germination percentage unless pretreated to soften the seed coat.

bait *n.* (Norse *beita*, split or crack) **1.** food or other enticement to lure birds, beasts, fishes, or insects into a trap to capture or destroy them. See pheromone. **2.** any kind of attractive enticement. *v.* **3.** to put bait someplace, as to bait a trap. **4.** to tease.

bait shyness *n.* the wariness of rodents, birds, or other pests that helps them avoid poisoned bait.

bajada *n.* (Spanish *bajada*, lower slope). **1.** the gently sloping surface of a coalescing group of alluvial fans that blend with a pediment to form a piedmont slope in a basin. Anglicized spelling is bahada. **2.** a series of coalescing alluvial fans along the base of a mountain range, underlain entirely by gravelly detritus that is ill sorted and poorly stratified. Syn.: compound alluvial fan, alluvial slope.

balanced fertilizer *n.* a soil additive containing suitable proportions of each mineral element needed for optimum plant growth. The appropriate composition varies according to circumstances and must be determined by a soil test, plant tissue test, field plot research, or trial and error.

balanced ration *n.* an appropriate daily food allowance for livestock or poultry, mixed to include suitable proportions of nutrients required for normal health, growth, production, reproduction, and well-being.

bald *adj.* (Greek *phaliós*, bald) **1.** uncovered, as a small, treeless area in a forested hilly or mountainous region, especially a crest, as a bald knob (southern U.S.A.). **2.** beardless, as bald wheat. **3.** hairless, as a bald head. **4.** not concealed, as a bald lie.

bald cypress *n.* a deciduous conifer (*Taxodium distichum*, family Pinaceae) native to swamps and other wet areas in southern U.S.A.; grown as an ornamental tree and used for railway ties and in outdoor construction for silos, boatbuilding, etc. Its wood is very durable in water or moist conditions. Also called taxodium, deciduous cypress, pond bald cypress. See wetland.

bald eagle *n.* a large North American bird (*Haliaeetus leucocephalus*) with a wing span of about 80 in. (2 m). The white feathers that cover their heads give adult birds a bald appearance. They feed mostly on fish but are also confirmed predators of livestock. The bald eagle is portrayed on the coins and in the coat of arms of the U.S.A. See eagle.

Bald eagle

bale *n.* (Latin *bala*, package) a large, compact bundle, usually wrapped with cords or wires to hold it together, as a hay bale, a bale of cotton, or a bale of trash. **2.** misery, woe, calamity. *v.* **3.** to pack something tightly and tie it to form a bale.

balk *n.* (Anglo-Saxon *balca*, bank or ridge) **1.** an unplowed area left between two plowed strips, by accident or on purpose to reduce erosion. **2.** refusal to do as commanded, as when a horse or mule stops and will not go on. *v.* **3.** to start doing something and then suddenly stop, as when a baseball pitcher starts to pitch, then improperly throws toward a base to try to get a runner tagged out. **4.** to refuse to act; e.g., to balk at cleaning up the environment.

balled and burlapped *adj.* covered with a ball of soil or other rooting medium and wrapped in burlap or other sacking as protection for the roots of a tree or other plant to be transplanted.

balling a queen *n.* the formation of a cluster of worker bees into a tight ball around an unacceptable queen bee, usually killing her.

balling hydrometer *n.* a triple-scale hydrometer designed to read the sugar content of a syrup or fruit juice based on the specific gravity of the solution. Also called Brix hydrometer. See Brix scale.

ball planting *n.* a method of transplanting that keeps enough compact soil on plant roots to provide nourishment and help their growth in a new location. See balled and burlapped.

balm *n.* (Latin *balsamum*, balsam) **1.** a soothing, fragrant ointment used for healing purposes or for anointings. **2.** anything that soothes, comforts, and leads to either physical or mental healing. **3.** an aromatic perennial herb (*Melissa officinalis*, family Labiatae) with small white flowers that produce a lemon fragrance; native to southern Europe and naturalized in Europe and the U.S.A. Also known as lemon balm. **4.** any pleasant odor; fragrance.

Balm of Gilead *n.* **1.** a small aromatic evergreen tree (*Commiphora* sp., especially *C. opobalsamum*) native to Africa and Asia Minor. **2.** a fragrant resin obtained from the tree and used to make perfume (named for its use in the region of Gilead in Biblical times). **3.** a hybrid American poplar tree (*Populus balsamifera* × *P. deltoides*, family Salicaceae) with large heart-shaped leaves and a pleasing aroma; grown as an ornamental tree in Canada and northeastern U.S.A. The wood is soft and is used for pulp, boxes, and excelsior. **4.** loosely used for balsam fir.

balmy *adj.* **1.** soothing; pleasing; acting as a balm. **2.** warm and pleasant, as balmy weather. **3.** foolish, strange-thinking.

balsa *n.* or *adj.* (Spanish *balsa*, raft) **1.** a tropical American tree (*Ochroma lagopus*) that produces wood of very low density. **2.** the wood of the balsa tree, often used to make rafts or model airplanes.

balsam *n.* (Latin *balsamum*, balsam) **1.** an exudate with a pleasing fragrance obtained from any of several species of trees (especially those of *Commiphora* sp.). See Balm of Gilead. **2.** a transparent resinous turpentine. See Canada balsam. **3.** a tree that produces balsam. See balsam fir. **4.** any of several plants that produce balsam (especially those of *Impatiens* sp.). **5.** any ointment with a balsamlike fragrance used in medicine or in ceremonies.

balsam fir *n.* **1.** an evergreen tree (*Abies balsamea*, family Pinaceae) with small needles and dark purplish cones about 2½ in. (7 cm) long; native to eastern Canada and northeastern U.S.A.; often used as a Christmas tree; the source of Canada balsam. **2.** wood from the tree, used in making cooperage and boxes.

bamboo *n.* (Malay *bambu*, bamboo) a woody, treelike perennial grass plant (family Poaceae, subfamily Bambusoidea) that grows in warm, often tropical climates. Bamboo plants are very tall grasses with jointed hollow stems that may die down to the ground or may remain living for many years; some attain heights of 100 ft (30 m) or more and diameters of 6 in. (15 cm) or more. Worldwide, more than 20 genera and 200 species are recognized. Among the more important genera are *Bambusa*, *Phyllostachys*, *Dendrocalamus*, and *Arindinaraia*. In India and many other tropical countries, they are often strong enough to be used in building construction. Smaller stems are used to make furniture, fishing poles, and musical instruments. In the tropics and subtropics, the succulent young shoots are used for food. Some varieties bear a drupelike fruit with a fleshy exterior. Plants of *Arindinaraia* sp. form the canebrakes of southern U.S.A. and provide year-round grazing for cattle.

bañado *n.* (Spanish *bañado*, basin) a flooded depression covered with dense vegetation that is shallow enough that cattle and horses can wade through it.

banana *n.* (Spanish *banana*, banana) **1.** a perennial, herbaceous, treelike, tropical plant (mostly classified as *Musa acuminata* or *M. paradisiaca*, family Musaceae; the taxonomy has not been clarified because there are hundreds of variations, and the wild species grow in remote areas). Banana plants grow 10 to 20 ft (3 to 6 m) high and produce bunches of fruit weighing up to 80 lb (36 kg) that are divided into a number of "hands," each containing several "fingers" that grow out and upward from their point of attachment. After the bananas are harvested, the entire plant is cut down. The next crop will be borne on a new plant that develops from a basal sucker arising from the crown of the parent plant. A new crop of fruit can be harvested approximately every 12 months. **2.** the elongated, somewhat curved fruit (botanically a berry) of a banana plant. The fruit is picked green for shipping and matures to a creamy white flesh covered by a yellow or red skin that is easily peeled away.

Harvesting bananas

band *n.* (French *bande*, band) **1.** a small group of animals, especially sheep or goats. **2.** a small orchestra or other specialized musical group. **3.** a group of people functioning together, often with ulterior motives, as a band of thieves. **4.** nomadic people wandering or migrating together. **5.** a contrasting stripe across an object or landscape or around an animal's body. **6.** a narrow strip of fertilizer or pesticide applied either on the surface or underground in or alongside a seed row. **7.** a binding strip wrapped around an object. **8.** a marker wrapped around the leg or ear of an animal or bird. **9.** a specified range of frequencies, as a radio band. *v.* **10.** to group together; unite to form a band. **11.** to mark an animal or bird by fastening a band with an identifying number or description to the ear, leg, or wing. **12.** to place a band or collar around a tree trunk or the stem of a plant to poison, repel, or trap insects. **13.** to place a line of fertilizer or pesticide in or near crop rows instead of spreading it uniformly over the cropped area.

bandicoot *n.* (Telugu, *pandi-kokku*, pig rat). **1.** a very large rat of India (as large as a rabbit) that is very destructive of rice fields and gardens; a carrier of rat-bite fever. **2.** an Australian ratlike marsupial (*Perameles* sp., family Peramelidae) that resembles the bandicoot of India.

Bang's disease *n.* a bacterial (*Brucella abortus*) disease that causes reproductive problems in cattle and humans. It is named after a Danish biologist, Bernard L. F. Bang (1848–1932). See brucellosis.

bank *n.* (German *bank*, bench) **1.** a long, narrow raised area. **2.** the rising ground bordering a lake, river, or sea. On a river, a bank is designated as right or left bank as it appears when you face downstream. **3.** a large area where an elevated sea floor is surrounded by deeper water but is safe for surface navigation; a submerged plateau or shelf, a shoal, or shallows. **4.** the crown or upper surface of a plow furrow slice. Also called furrow crown. **5.** a row of matched items, as a bank of switches. **6.** a financial institution that provides checking accounts, savings accounts, loans, etc. *v.* **7.** to pile earth around plantings to protect them against the cold, to preserve moisture, or to blanch their stems (celery). **8.** to pile a long mound of earth against a structure. **9.** to lower a wing of an airplane to maintain stability in a turn.

Bankhead-Jones Farm Tenant Act. a law Congress passed in 1937. It authorized the Farm Security Administration (later renamed the Farmers' Home Administration and subsequently made a part of the Farm Service Agency) to make loans to qualified farm tenants, sharecroppers, farm laborers, and others unable to obtain loans through commercial channels, thus enabling them to buy farms.

ban *v.* disallow on the market; e.g., DDT was banned in 1972 by the U.S. Environmental Protection Agency because it was declared toxic to humans and animals.

banner crop *n.* an excellent crop with unusually high yield or quality or both.

banquette *n.* (French, diminutive of *banc,* a small terrace). **1.** in the U.S.A., specifically, an earthen bank added to the base of a high levee away from the water to prevent seepage and sloughing of the levee. **2.** a sidewalk (in French-speaking areas of southern U.S.A.). **3.** an upholstered bench along a wall. **4.** a narrow ledge; a window seat.

bantam *n.* **1.** any dwarf or miniature breed of domestic fowl. *adj.* **2.** petite, lightweight, often feisty, as the bantam weight class in certain sports.

banteng *n.* (Malay *banténg*, banteng) a species of bovine animals (*Bos banteng*) of tropical south-eastern Asia.

banyan tree *n.* a tropical tree (*Ficus benghalensis*, family Moraceae) that germinates as an aerial plant. It sends its roots downward, and when they reach the surface, they grow into the soil. In the process, the roots may strangle the host tree. Also known as strangler fig tree. See bo tree.

Banyan tree

baobab tree *n.* **1.** a huge tropical tree (*Adansonia digitata*, family Bombacacae) with a swollen base up to 30 ft (9 m) in diameter. Some hollowed-out trunks are used as homes. It bears a large edible fruit called "monkey's bread" and produces fibers that are used in making rope, cloth, and paper. Native to Africa and long cultivated in India. **2.** a large tree species (*A. gregorii*) found in northern Australia that sometimes attains a diameter of 25 ft (8 m). Also called "cream-of-tartar tree" because of the pleasantly acid taste of the fruit.

bar *n.* (French *barra*, rod) **1.** a barrier, often with a uniform, long, slender form. **2.** a rectangular piece of something, such as a bar of soap or a gold bar. **3.** a mass of sand, gravel, or alluvium deposited on a part of the bed of a stream, sea, or lake or at the mouth of a stream, forming an obstruction to water navigation. **4.** a term used in a generic sense to include various types of submerged or emergent embankments of sand and gravel built on a sea floor by waves and currents, usually parallel to the shore. **5.** meteorology: a unit of pressure equal to 106 dyne cm^{-2}; equivalent to a mercurial barometer reading of 750.076 mm at 0°C (29.5306 in. at 32°F). It is equal to the mean atmospheric pressure at about 100 m above mean sea level, about 0.987 atm. **6.** the mouthpiece of a bridle bit. **7.** the landside of a plow. **8.** the straight line of a cattle brand (western U.S.A.). **9.** the part of a horse's hoof that is bent inward, extending toward the center of the hoof. **10.** ridges in the roof of a horse's mouth. **11.** the space in front of a horse's molars where the bit is placed. **12.** a pole used to close a fence gap. **13.** a place where liquor is sold by the drink. **14.** a counter used to dispense drinks or food. **15.** the legal profession, or any tribunal. **16.** a vertical line dividing a musical score into measures. *v.* **17.** to block the way.

barb (Latin *barba*, beard) *n.* **1.** a projection that resists withdrawal of an object that has penetrated a body, as the barb of a fishhook. **2.** a hairlike side branch of a feather. **3.** a hooked hair or bristle on a plant part; e.g., the hooks on the awns of grasses, especially barley. **4.** a pointed projection on a wire fence. **5.** mucous membrane projections under the tongues of horses and cows where submaxillary glands open. **6.** a breed of horses native to Barbary and introduced into Spain by the Moors; its strain is evident in all known present breeds. It was related to the Arab but smaller and coarser. **7.** a cutting remark.

Barbary fig *n.* an edible, spiny, multiple fruit produced by prickly pear cacti of northern Africa.

barbed wire or **barbwire** *n.* fencing made of two or more wires twisted to form one strand with barbs attached at short intervals to discourage animals, or sometimes people, from passing.

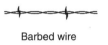

Barbed wire

barbel *n.* (Latin *barbula*, little beard) a threadlike growth on the lips or jaws of certain species of fish that is used as an organ of touch.

barberry *n.* (Latin *berberis*, barberry) a shrub (*Berberis* sp., family Berberidaceae) that is often grown as an ornamental plant or for birds to feed on its red or orange berries in the winter. The European, or common, barberry (*B. vulgaris*) and a few other species are the alternate hosts for the stem rusts (*Puccinia graminis*) of cereals, such as wheat, barley, and rye, as well as many other members of the grass family. However, many species of barberry, including the Japanese barberry (*B. thunbergii*), are not hosts for this rust.

bareback or **barebacked** *adv.* or *adj.* **1.** without a saddle; riding directly on the back of a horse, burro, or other animal. **2.** lacking feathers, as a featherless area on the back of a chicken or other fowl.

bare fallow *n.* in arid areas, land unsown for a season to store soil moisture but cultivated whenever necessary to prevent weed growth. It often results in a high yield for the following crop because plant nutrients (especially nitrogen) become available as organic matter decomposes. Conservationists discourage the practice because it exposes the land to excessive erosion. Also called summer fallow, naked fallow, black fallow, fallow.

barium (Ba) *n.* (Greek *barys*, heavy + *ium*, element) element 56, atomic weight 137.327; a soft, silvery white, divalent metal that behaves much like calcium; used in paint, X-ray diagnostic work, and glassmaking.

bark *n.* (Anglo-Saxon *barc*, bark) **1.** the exterior of a woody stem containing phloem tubes in the inner bark and usually some waterproof cork cells in the outer layer. **2.** a loud sharp sound made by a dog, or a similar coughing sound made by a person. **3.** a type of large sailing vessel having three or more masts. *v.* **4.** to remove the external layer from a tree or log, sometimes called debark. **5.** to cure or dye with bark extract; to tan. **6.** to enclose or surround with bark. **7.** to make a barking sound.

bark beetle *n.* any of several species of beetles (e.g., *Dendroctonus* sp., family Scolytidae) that bore under the bark of trees. Adults and larvae tunnel in the bark and sapwood of living, dying, or recently dead trees, making a network of passageways on the surface of the wood. Some species attack roots, twigs, cones, and solid wood. The beetles commonly are infested with fungi that cause tree diseases. See Dutch elm disease.

Adult bark beetle

bark grafting *n.* a method of grafting plants by slitting the bark and inserting a scion between the bark and the xylem of the stock. See grafting).

barkhan *n.* a type of sand dune having a crescent shape with the ends of the crescent pointing downwind.

bark mulch *n.* tree bark used as a mulch to protect the soil surface and prevent weed growth. It is porous, easily handled, and slowly decomposable. Bark can improve the tilth, structure, and aeration of clayey soils and can increase water absorption and penetration. As a mulch, it conserves moisture through weed control and reduced evaporation, improves granulation of surface soils, reduces topsoil erosion, and builds up organic matter and humus in soil with concurrent benefits to the soil microflora. Bark's darker color and lower susceptibility to termites makes it a better mulch than wood chips.

barley *n.* (Anglo-Saxon *baerlic*, barley) a small-grain cereal crop (*Hordeum* sp., family Poaceae, especially *H. vulgare*, the commonly cultivated barley) cultivated for human and animal consumption since prehistoric times. Barley is a cool-season crop that is more salt tolerant than any of the other small grains and more drought resistant than wheat. Barley ranks third in acreage of small grains in the U.S.A. The center of barley production in North America is the northern Great Plains of the U.S.A. and the adjoining provinces of Canada. The primary uses of barley are as human food (flour for dark, fairly heavy breads), feed for livestock, and malt (from sprouted grain) for making beer and whiskey. Malting barley comes from special cultivars that usually yield less but command a higher price than nonmalting cultivars.

barleycorn *n.* (Anglo-Saxon *baerlic*, barley + *corn*, grain or seed) **1.** a single grain of barley. **2.** barley as a grain. **3.** ⅓ in. (the approximate length of a grain of barley). **4.** John Barleycorn is used as a personification of whiskey or other strong liquor.

barley malt *n.* fermented barley used in the production of beverages, foods, and medicines.

barn *n.* (Anglo-Saxon *bern*, barley storage place) **1.** a farm building used to house domestic animals and/or to store hay; typically larger than other farm buildings. **2.** a large shed for storing farm implements or vehicles. **3.** a unit (10^{-24} cm^2) used to define the cross-sectional area of an atomic nucleus.

barnacle *n.* **1.** a saltwater shellfish (crustacean of subclass Cirripedia) that attaches itself to a rock, ship's hull, or other object and is difficult to remove. **2.** a person or thing that clings persistently to something.

barn owl *n.* an owl (*Tyto alba*, family Tytonidae) that feeds on rodents around barns and other such buildings. They are beneficial to the human environment.

barn swallow *n.* a migratory bird (*Hirundo rustica*, family Hirundinidae) native to North America. It is about 6 in. (15 cm) long and has a dark blue back, a rusty underside, and a deeply forked tail. It builds mud nests in barns and eats many flying insects.

barnyard or **barn lot** *n.* a plot of land adjacent to a barn; usually fenced to enclose livestock or fowls. Barnyard runoff can be a source of pollution.

barnyard fowl *n.* any of the common domesticated birds: chickens, ducks, geese, turkeys, and guineas.

barnyard grass *n.* a coarse annual grass (*Echinochloa crusgalli*, family Poaceae) that is often a weed in cultivated areas. When used for hay, special care must be given to the curing to prevent ergotism, as it is a host for the ergot fungus. Also called ankee, barn grass, cockspur grass, water grass, barnyard millet, Japanese barnyard millet. See ergot.

barometer *n.* (Greek *baros*, weight + *metron*, meter) **1.** an instrument that measures atmospheric pressure, commonly used in predicting weather. See hypsometer. **2.** an indicator of something changeable, as a barometer of public opinion.

barometric elevation *n.* an approximate elevation above mean sea level determined by the use of an instrument that measures atmospheric pressure.

barometric pressure *n.* the atmospheric pressure as measured with a barometer.

barrel *n.* (Latin *barillus*, large wooden container) **1.** a cylindrical container, originally of wood but also of metal; e.g., a 55-gal (0.208 m^3) drum. **2.** a large quantity of a liquid, a granular solid, or something abstract, as a barrel of fun. **3.** a liquid measure equal to 42 U.S. gal (5.6 ft^3, or 0.159 m^3) of petroleum products and most other liquids, but usually 31.5 U.S. gal (4.3 ft^3, or 0.119 m^3) of fermented liquors. **4.** a measure of fruits and vegetables equal to 105 dry qt (4.1 ft^3, or 0.116 m^3). **5.** the trunk of a domesticated animal. **6.** a cylindrical part, as the barrel of a fountain pen or a gun. *v.* **7.** to travel at high speed, as to barrel along.

barren *adj.* (French *baraigne*, unproductive **1.** unvegetated; unable to support vegetation; e.g., the Arctic barrens. **2.** having no offspring, seed, or fruit.

Barren Grounds or **Barren Lands** *n.* a large area of tundra in northern Canada, north of the forest line.

barrens *n.* a name used by settlers and explorers to describe land with sparse vegetation, open or scrubby tree growth. Frequently designated by the kind of growth, as pine barrens or oak barrens. Barrens may be caused by cold temperatures, low soil fertility, high winds, destructive water erosion, toxic wastes, low rainfall, and/or shallow soil. See bald.

barrier island *n.* an elongated island or group of islands paralleling a nearby mainland; commonly low-lying and composed of a few to many ridges and sand dunes.

barrier plants *n.* one or more lines of tall plants placed to protect a crop by reducing wind speed in a field. They are usually annual plants (e.g., corn, sorghum, sunflowers) that are planted before or at the same time as the crop to be protected. The protected crop is often a vegetable or fruit, such as broccoli, lettuce, tomatoes, or strawberries.

barrier reef *n.* a coral or rock reef that serves as a breakwater; separated from the shore by an elongated lagoon.

barrow *n.* **1.** a castrated male hog. **2.** a small push cart or a wheelbarrow. **3.** a hill or mound of earth, often marking a burial site. Also called a cairn or tumulus.

barrow pit *n.* in western U.S.A., a place where soil has been removed to serve as fill dirt elsewhere. Also called a borrow pit.

basal area *n.* **1.** the cross-sectional area of a tree, including its bark, at breast height (4.5 ft or 137 cm above ground level). **2.** the total basal area of trees per given land area (usually per acre). **3.** the total cross sectional area of plant stems per given land area, measured at 1 in. (2.5 cm) above soil level.

basal metabolism *n.* the minimum metabolic rate needed to sustain conscious life in an animal or human in the fasting and resting state. It is determined in humans at absolute rest in a thermoneutral environment 14 to 18 hr after eating. It is measured by means of a calorimeter and is expressed in calories per m^2 of body surface per hr.

basalt *n.* (Greek *basaltes*, dark, hard marble) a dark-colored, extrusive, fine-grained igneous rock containing plagioclase feldspar and ferromagnesium silicates (e.g., olivene, augite) but no quartz; the most common "lava rock."

base *n.* (Latin *bassus*, low) **1.** a support or foundation to hold something. **2.** the lower part of something, as the base of a plant. **3.** the primary component of something, as a solution with a water base. **4.** a headquarters or point of departure. **5.** the bottom side of a geometric figure, as the base of a triangle. **6.** mathematics: the number of symbols in a numeric system or a number that is raised to some power. **7.** chemistry: a substance that contains one or more ionizable OH$^-$ radicals or that can react with water to form OH$^-$ ions or a substance capable of accepting protons from a donor.

base exchange *n.* See cation exchange.

baseline *n.* (Latin *bassus*, low + *linea*, linen thread) **1.** a basic standard of reference; a guideline for future comparisons. **2.** the condition of an environment prior to the release of a hazardous substance. **3.** an east–west survey line used along with a principal meridian as a reference for legal descriptions in areas that are divided into townships and sections.

basement *n.* (Latin *bassus*, low + *mentum*, condition) **1.** a building story that is partly below the ground surface. **2.** the crystalline rocks that underlie the sedimentary rocks in a study area.

basic *adj.* (Latin *bassus*, low + *icus*, like) **1.** fundamental; serving as a base or basis, as a basic principle. **2.** acting as or having the chemical nature of a base; causing an alkaline pH. **3.** lacking quartz; e.g., a basic igneous rock such as basalt or gabbro. *n.* **4. BASIC** beginners all-purpose symbolic instruction code, a high-level computer programming language.

basic cation *n.* a cation that contributes to alkalinity or to the neutralization of acidity in a soil, especially Ca^{2+}, Mg^{2+}, K^+, and Na^+.

basidiomycete *n.* a fungus (phylum Basidiomycota) that bears its spores on a clublike structure after nuclear fusion and meiosis. Basidiomycetes include mushrooms, puffballs, wood and root rots, and rusts; some fungi in the group form beneficial root mycorrhizae.

basil *n.* (Greek, *basilikon*, royal) an aromatic annual herb (*Ocimum basilicum*, family Labiatae) native to the Middle East but now grown in most countries. Basil leaves are used as an herb for seasoning meats, soups, and salads and to bring luck.

basin *n.* (Latin *baccinum*, water vessel) **1.** a broad bowl-shaped container that can hold liquids; e.g., a wash basin. **2.** a low place that receives and holds runoff water. **3.** a largely enclosed sheltered area along a coast that can serve as a harbor. **4.** an extensive depressed land area with no surface outlet. (Use confined almost wholly to the arid West in U.S.A.) **5.** the drainage catchment area of a stream or lake. **6.** an amphitheater, cirque, or corrie. (Local use in the Rocky Mountains.) See watershed.

basin irrigation *n.* an irrigation system with level areas surrounded by earthen ridges to form basins. Irrigation water is turned into the basins and left until it infiltrates. Most such basins are relatively small, often the area allotted to one tree in an orchard. See irrigation.

bass *n.* (Anglo-Saxon *baers*, point or bristle) any of several species of scaly, spiny-finned game fish of both fresh and salt water. Bass range in size from Mississippi River rock bass about 6 in. (15 cm) long to sea bass 8 ft (2.5 m) long that weigh up to 700 lb (320 kg). They include rock bass and the smallmouth and largemouth black bass (family Centrarchidae), the white and yellow striped bass (family Moronidae), the Jewfish (family Epinephelidae), and the black sea bass (family Serranidae).

bastard *n.* or *adj.* (French *bast*, barn + *ard*, bold) **1.** not legitimate; born of unmarried parents. **2.** despicable; a mean person. **3.** a hybrid plant between two species that rarely cross.

bat *n.* (Swedish *backa*, bat) **1.** a flying mammal (order Chiroptera) that eats large numbers of insects every night; bats generally hang from the ceiling of a cave to sleep during the day and fly in swarms at night, finding their way by sound echoes like sonar. Bat bites are feared because bats can carry rabies. Sometimes called flying fox because a bat face resembles that of a fox. **2.** a club used to swing at a ball in baseball and several other games.

bathymetry *n.* (Greek *bathys*, deep + *metron*, measure) measuring depths and mapping the topography of the ocean floor or other large bodies of water.

battery *n.* (French *batterie*, something that beats) **1.** an electric energy source composed of two or more cells; or loosely, a cell (especially a dry cell). **2.** a group of similar objects that work together, as a gun battery on a ship. **3.** a series of confinement cages used for raising chickens, fattening chickens, or increasing the egg production of laying hens. Animal rights advocates consider this cruel treatment. **4.** an unlawful physical attack on a person.

baule unit *n.* the amount of a nutrient element (especially nitrogen, phosphorus, or potassium) needed to increase the growth of a plant or the yield of a crop by half of the remaining potential according to Mitscherlich's law of diminishing returns; named for Baule, a German mathematician who worked with Mitscherlich in the early part of the 20th century.

bauxite *n.* (French *Baux*, a town in France + *ite*, rock or mineral) **1.** weathered rock composed of one or more aluminum hydroxide minerals (boehmite–AlOOH, gibbsite–Al(OH)$_3$, or diaspore–HAlO$_2$) and impurities in the form of silica, clay, silt, and/or iron hydroxides; the principal ore of aluminum. Generally formed in tropical and subtropical latitudes under conditions of good surface drainage. **2.** used collectively for lateritic ores high in aluminum.

bay *n.* (French *baie*, reddish brown) **1.** a term frequently used to describe a reddish brown or chestnut horse with black mane, tail, and lower legs. **2.** the laurel tree (*Laurus nobilis*, family Lauraceae). The leaves are used as an herb for seasoning foods. **3.** a compartment in a barn or other large building for storage or for some special use, as a horse bay. **4.** a compartment in an airplane, as a cargo bay. **5.** a coastal inlet larger than a cove and smaller than a gulf. **6.** an indentation of a plain into a range of hills. **7.** a shallow swamp supporting dense tree and shrub vegetation and containing peat or muck soil. **8.** the deep and prolonged wail of a hound when it is on a hot scent. *v.* **9.** to howl like a hound following a hot scent.

bayou *n.* (French *bayou*, small stream) **1.** a backwater area with sluggish flow or stagnant water, often a former creek or river channel closed or partially closed at both ends. **2.** a marshy arm of a lake, often a sluggish inlet or outlet.

BC soil *n.* a soil profile with B and C horizons but with little or no A horizon, probably because of surface erosion or earth moving. See soil profile.

beach *n.* **1.** the sandy or gravelly zone of sedimentary material that borders a large body of water. The seaward limit of a beach—unless otherwise specified—is the mean low-water line. It extends landward from the low-water line to the place where there is a marked change in material or form or to the line of permanent vegetation (usually the effective limit of storm waves). A beach includes foreshore and backshore. *v.* **2.** to haul or run something onto a beach, as to beach a boat.

beach berm *n.* a nearly horizontal bench or narrow terrace formed by wave action in unconsolidated material on the backshore of a beach roughly parallel to the shoreline. Some beaches have no berm, others have more than one.

beach plum *n.* a dwarf seacoast shrub (*Prunus maritima*, family Rosaceae) native to the Atlantic Coast from Canada to Virginia; sometimes planted for erosion control on beaches. It has white flowers and an edible purple fruit that is used mostly for jams and jellies.

beak *n.* (Latin *beccus*, beak) **1.** the bill of a bird, consisting of the upper and lower mandibles (nib). **2.** the protruding mouth part structures of a sucking insect; proboscis. **3.** the protruding mandibles of some fishes or reptiles. **4.** any projection that has a form resembling that of a beak; e.g., the awn on the outer chaff of wheat.

beam *n.* (Anglo-Saxon *beam*, a piece of wood) **1.** a heavy piece of timber or metal used in the framework of a structure or machine. **2.** the central shaft of a plow that supports all its principal parts. **3.** the balance bar on a scale. **4.** in tanning, a sloping board used for dressing hides. **5.** a ray or a compact group of rays of electromagnetic radiation, as a beam of light or a radio signal. *v.* **6.** to emit a beam of light or other radiation. **7.** to smile broadly; to exhibit a happy countenance.

bean *n.* (Anglo-Saxon *bean*, bean) **1.** kidney-shaped (or in the broad bean, flattened) dicotyledonous seeds of certain leguminous annual or perennial herbs. The seeds are highly nutritious for people and animals and are valued for their high protein content, especially as sources of lysine and threonine (two essential amino acids). Beans are eaten as vegetables when pickled, cooked, or ground into meal. The dried seeds can be stored for a long time. **2.** the plants themselves (mostly *Phaseolus* sp.); bean plants may be climbing (pole beans) or trailing bushes (bush beans); the most common species (*P. vulgaris*) is known in England as kidney bean, in France as haricot, and in U.S.A. as navy beans or pea beans. The frijoles of Mexico and Spain are also forms of *P. vulgaris*. The bean of classical literature is a southern European vetch. See cowpea, faba, lima bean, soybean, tepary bean.

bean rust *n.* a fungal (*Uromyces phaseoli* var. *typica*, family Pucciniaceae) disease that causes reddish brown pustules on bean leaves. Each pustule produces thousands of spores that can blow to other leaves and produce new pustules in about 10 days, especially if the humidity is high. Bean rust spores overwinter on bean straw, not on the seed.

bean weevil *n.* a snout beetle (*Acanthoscelides obtectus*, family Bruchidae) that feeds on bean plants while they are small and then lays its eggs in the seed when the pods are formed. The larvae burrow into the dried beans and devour them. Bean weevils attack kidney beans, lima beans, and cowpea plants and beans, peas, and lentils in storage.

bear *n.* (Anglo-Saxon *bera*, large animal) **1.** a large, powerful, omnivorous mammal (family Ursidae) with coarse fur and very little tail, including black bears, brown bears, grizzly bears, and polar bears. **2.** any of several other smaller animals that resemble a bear. **3.** a strong person (usually a man) who is crude and clumsy. **4.** a person who believes that the stock market will fall in value (as opposed to a bull who believes it will rise). **5.** one who works hard at something in spite of difficulties, as a bear for punishment. *v.* **6.** to produce offspring, as to bear a child. **7.** to produce fruit. **8.** to withstand or endure, as to bear the pressure or to bear close examination. **9.** to carry, as to bear a load or to bear good news. **10.** to maintain, as to bear a good attitude or to bear malice. **11.** to express, as to bear testimony or to tell tales. **12.** to have, as to bear a resemblance or a mark or to bear title to property. **13.** to move or be located in a general direction, as to bear north.

bearberry *n.* an evergreen, trailing shrub (*Arctostaphylos uva-ursi*, family Ericaceae) with small evergreen leaves that are toxic and bitter tasting. Its branches root quickly and are planted to control erosion on steep slopes. Bearberry is also grown for its beautiful red berries in winter. Found in the U.S.A. and Eurasia. Also called kinnikinnick.

beard *n.* (Latin *barba*, beard) **1.** a growth of hair around the jaws and chin of a man. **2.** a similar growth of hair on an animal, as on the chin of a goat. **3.** the awns of a small grain, such as those on barley, rye, or wheat.

bearing *n.* **1.** demeanor or behavior; e.g., a person having a noble bearing. **2.** significance of a fact in regard to a topic under discussion. **3.** a machine part that supports a rotating shaft or wheel. **4.** (often plural: **bearings**) the direction of a line relative to the cardinal points of the compass, a ship, or other reference. **true bearing**: the horizontal angle between a ground line and either the south or north point of a geographic meridian. **magnetic bearing**: the horizontal angle between a ground line and the magnetic meridian. A magnetic bearing differs from a true bearing by the angle of magnetic declination of the locality. See declination, agonic line. *v.* **5.** carrying, producing, or bringing forth something, such as a crop or a baby. **6.** enduring, as bearing pain.

bearing tree *n.* See witness tree.

beast *n.* (Latin *bestia*, beast) a quadruped animal, especially a large mammal such as a cow or a horse.

beast of burden *n.* a domesticated animal used as a source of power, especially to carry or pull a load, as a horse, mule, ox, or donkey.

Beaufort wind scale *n.* a system for describing wind velocities invented by Admiral Sir Francis Beaufort (1774–1857) of the British Navy; originally based (1806) on the amount of canvas a full-rigged frigate of the early 19th century could carry at various wind speeds; a modified form has been widely used in international meteorology.

Beaufort Wind Scale for Land Use

Beaufort number and name	MPH	Knots	Wind effects observed on land	Terms used in U.S.A.
0	<1	<1	Calm; smoke rises vertically	
1 Light air	1–3	1–3	Direction of wind shown by smoke drift; but not by wind vanes	Light
2 Light breeze	4–7	4–6	Wind felt on face; leaves rustle; ordinary vane moved by wind	
3 Gentle breeze	8–12	7–10	Leaves and small twigs in constant motion; wind extends light flag	Gentle
4 Moderate breeze	13–18	11–16	Raises dust and loose paper; small branches move	Moderate
5 Fresh breeze	19–24	17–21	Small trees in leaf begin to sway; crested wavelets form on inland waters	Fresh
6 Strong breeze	25–31	22–27	Large branches move; whistling heard in phone wires; umbrellas used with difficulty	Strong
7 Moderate gale	32-38	28-33	Whole trees in motion; inconvenience felt walking against wind	
8 Fresh gale	39–46	34–40	Twigs break off trees; generally impedes progress	Gale
9 Strong gale	47–54	41–47	Slight structural damage occurs	
10 Whole gale	55–63	48–55	Seldom experienced inland; trees uprooted; considerable structural damage occurs	Whole gale
11 Storm	64–72	56–63	Very rarely experienced; accompanied by widespread damage	
12 Hurricane	>72	>64	Very rarely experienced; accompanied by widespread damage	Hurricane

beaver *n.* (Anglo-Saxon *beofor*, beaver) **1.** any of several amphibious rodents (*Castor* sp.) known for cutting down trees with their large, sharp teeth and using them to build dams; well equipped for swimming with webbed hind feet and broad tail. **2.** the fur of such animals, sometimes used for making coats or hats.

bebeeru *n.* See greenheart.

becquerel (Bq) *n.* the unit for the radioactivity of a source in the SI system; equivalent to the spontaneous release of 1 radioactive particle/sec. It is named for Antoine Henri Becquerel (1852–1908), the French physicist who discovered radioactivity in uranium.

bedbug *n.* a flat, oval, reddish, wingless insect (*Cimex lectularius* of temperate regions or *C. rotundatus* of the tropics, family Cimicidae) about ¼ in. (6 mm) long that feeds on human and animal blood, mostly at night.

bedding *n.* **1.** straw, leaves, sawdust, sand, peat moss, pine straw, or other material used to make a bed for an animal and to absorb urine and droppings. **2.** the process of land forming to produce a series of elevated, crowned parallel beds or lands separated by shallow ditches to improve soil drainage. **3.** growing flowers in large beds for a massed decorative effect. **4.** layering in sedimentary rock such as shale or sandstone.

bedding plane *n.* the contact surface between two distinct layers of a sedimentary rock.

bed load *n.* sand particles, stones, and other debris rolled along the bottom of a stream by the moving water, as distinguished from the silt and clay carried in suspension.

bedrock *n.* **1.** solid hard rock lying beneath the regolith (the soil layers and any unconsolidated geological deposits, such as glacial drift). **2.** a firm foundation. **3.** basic facts or principles.

bee *n.* (Anglo-Saxon *beo*, bee) **1.** any of a large number of flying insects (superfamily Apoidae), including the families Prospidae, Colletidae, Megachillidae, Xylocopidae, Ceratinidac, Nomadidae, Andrenidae, Anthophoridae, Bombidae, and Apidae. Bees are generally beneficial as pollinators, producers of honey and beeswax, parasites on other insects, etc. See honeybee, bumblebee. **2.** a neighborhood social gathering, especially in rural areas, to accomplish a task such as making a quilt, husking corn, or raising a barn.

beebread *n.* (Anglo-Saxon *beo*, bee + *bread*, crumb or morsel) flower pollen mixed with honey and stored in cells of the comb as food for bees.

bee dance *n.* movements of a worker bee as a means of communication to indicate the direction and distance to a source of nectar or pollen.

bee escape *n.* a device that lets bees pass in only one direction; usually inserted between honey supers and brood chambers to remove bees from the honey supers prior to harvesting the honey. See honeybee.

beef *n.* (French *boeuf*, beef) **1.** the flesh of cattle eaten as meat. **2.** a cow, steer, or bull raised to be eaten as meat. **3.** muscle; brawn. **4.** (slang) a complaint.

beefalo *n.* (French *boeuf*, beef + *alo* from buffalo) a cross between a domestic beef cow and an American buffalo (bison).

bee farming *n.* raising bees for the production of honey and beeswax and/or for the pollination of plants. See honeybee.

beefy *adj.* (French *boeuf*, beef + *y*, inclined to) **1.** a term used to designate plump, meaty features characteristic of a beef animal as contrasted to a dairy cow. **2.** a term used in poultry judging to describe (a) combs that are coarse and overgrown or (b) birds that are fat and coarse.

bee louse *n.* a relatively harmless insect (*Braula coeca*) that lives on honeybees, but its larvae can damage honeycombs.

bee martin *n.* a flycatcher (*Tyrannus tyrannus*), also called kingbird or eastern kingbird, that supposedly catches and eats bees in small numbers. It is about 7 in. (17 cm) long and has a brown back, white front, and a white band across the end of the tail.

bee pasture *n.* vegetation, within flying distance of a hive, providing nectar and/or pollen that are attractive to bees.

beer yeast *n.* yeast (*Saccharomyces cerevisiae*, family Saccharmycetaceae) used in baking and distilling beer. Also called brewer's yeast.

bee space *n.* a space 0.25 to 0.31 in. (6 to 8 mm) thick that is big enough to permit free passage for a honeybee but too small to encourage comb building. Bee space is allowed between parallel beeswax combs and between the outer comb and the hive walls when the hives are built to facilitate later removal of the combs to harvest the honey. See honeybee.

beestings *n.* (Anglo-Saxon *bysting*, colostrum) the first milk (colostrum) given by a cow after she has had a calf.

beeswax *n.* a waxy mixture of organic compounds secreted by special glands on the last four visible segments on the ventral side of the worker bee's abdomen; used by bees for building honeycomb. Its melting point is 143.6° to 147.2°F (62° to 64°C). It is used for making cosmetics, candles, comb foundation, salves, ointments, dental wax, modeling compounds, patterns, polishes, adhesives, crayons, ink ingredients, etc.

beet *n.* (Latin *beta*, beet) **1.** a biennial plant (*Beta vulgaris* and related species, family Chenopodiaceae) commonly grown in vegetable gardens or commercially for sugar production. **2.** the usually red fleshy root of the plant, eaten as a vegetable. **3.** the mostly white, large, fleshy sugar beet root.

beetle *n.* (Anglo-Saxon *bitel*, beetle) any of about 200,000 species of insects (order Coleoptera) with an upper pair of wings forming a hard case that folds over and protects the true wings. Certain beetles are very destructive; e.g., the Colorado potato beetle, Mexican bean beetle, and Japanese beetle.

bee tree *n.* **1.** a hollow tree containing honey and occupied by a colony of bees. **2.** a tree that attracts bees because its flowers are a source of nectar or pollen. See honeybee.

beggarweed *n.* any of several tick trefoils (*Desmodium* sp., especially *D. tortuosum*, family Fabaceae) of subtropical origin and adaptation; grown for forage and green manure to improve poor soils in southeastern U.S.A. Also called Cherokee tickclover.

bel *n.* a unit of relative power equal to the logarithm of the ratio of the power of two sounds or electric currents, named after Alexander Graham Bell (1847–1922), an American scientist. For practical use, 1 bel is commonly divided into 10 decibels. See decibel.

belladonna *n.* (Italian *bella donna*, fair lady) a perennial herb (*Atropa belladonna*, family Solanaceae) native to Eurasia; grown in the U.S.A. and sometimes found escaped, especially in eastern U.S.A. Its black berries are poisonous, and its sap yields the alkaloid atropine that is used to make a drug by the same name. Also called death herb, doftberry, deadly nightshade. See atropine.

bell glass or **bell jar** *n.* a bell-shaped glass covering used with its open side down: 1. in greenhouses for covering special plants and cuttings to reduce transpiration and/or to induce rooting, 2. in laboratories to produce a controlled environment, or 3. in displays to protect fragile objects.

bellwether *n.* See wether.

belt *n.* (Latin *balte*, belt) **1.** a band that encircles or crosses something. **2.** a flexible leather, plastic, or fabric band that buckles around the waist and is used to support clothing or tools. **3.** a geographical term denoting an area of similar soil and climatic environment, particularly suited to specific animals or crops, such as the Corn Belt. **4.** a band of hair or skin of a contrasting color, often white, around an animal's body, as in Dutch belted cattle and Hampshire swine. **5.** in machinery, a flexible band passed around two pulleys to transmit motion from one shaft to another. **6.** a ring or band around the midsection of a fruit caused by retarded growth, often from frost injury when the fruit was very young. This band is usually shrunken.

benchlands *n.* **1.** terraces or shelflike land features representing former floodplains of rivers or shorelines of lakes or seas. They are usually composed of a layer of alluvium covering an eroded surface. **2.** foothills near the base of a mountain.

bench mark or **benchmark** *n.* **1.** a marked point of known or assumed elevation used as a reference for determining the elevations of other points. Most bench mark elevations show the height above mean sea level. Official bench marks in the U.S.A. have iron pipes with brass caps engraved: "U.S. Geological Survey Bench Mark. Elevation _____ Feet." **2.** Any standard characteristic used to evaluate quality or value.

bench terrace *n.* a steplike earthen structure (most often a group of such structures) constructed across a slope to control runoff and erosion. Each terrace has a level top and a steep downhill face. Many ancient bench terraces have stone walls for their downhill faces.

beneficial element *n.* See functional element.

bentgrass *n.* any of several forage grasses (*Agrostis* sp., family Poaceae) that commonly have a creeping habit, including colonial bentgrass (*A. tenuis*), creeping bentgrass (*A. palustris*), and velvet bent (*A. canina*); often used in lawns and putting greens.

benthos *n.* (Greek *benthos*, depth of sea water) **1.** the bottom portion of a large body of water (also called benthic region or benthonic zone). **2.** bottom-dwelling organisms, including: a. sessile animals (e.g., sponges, barnacles, mussels, and oysters), some worms, and many attached algae; b. creeping forms (e.g., snails and flatworms); and c. burrowing forms (e.g., clams and most worms).

bentonite *n.* a fine-textured clay with a great ability to absorb water and swell; largely composed of the clay mineral smectite. Bentonite is commonly formed from the alteration in place of volcanic ash. It was named after Fort Benton, Montana, and U.S. Senator Thomas H. Benton (1782–1858). It is used commercially to seal reservoirs and in drilling fluids, catalysts, paint, plastic fillers, and waterproofing for paper or cloth.

benzene *n.* a volatile, flammable, carcinogenic aromatic hydrocarbon (C_6H_6) with a ring structure. It is a derivative of coal tar and is used to make chemicals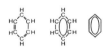

Benzene ring symbols

and dyes. It is a strong nonpolar solvent with properties that make it useful for degreasing bones and in fertilizer manufacture. Its toxic properties make it useful for destroying larvae, especially those of the screwworm, a blowfly.

benzene hexachloride *n.* the fully chlorinated form of benzene ($C_6H_6Cl_6$); used as an insecticide. There are 16 isomers with varying structural angles; the most potent isomer has been named lindane. See lindane.

benzoic acid *n.* an organic acid (C_6H_5COOH) found naturally in fruits such as cranberries, prunes, and plums and in trees such as cinnamon. It is used as a flavoring, an antiseptic, and an antimicrobial preservative.

bergamot *n.* (Italian *bergamotta*, bergamot) **1.** an aromatic herb with clusters of red flowers (*Monarda* sp., family Lamiaceae) native to northeastern U.S.A. An extract is used as a flavoring and a tea. The most common species (*M. didyma*) is called Oswego tea or horsemint. **2.** a small citrus tree (*Citrus aurantium bergamia*) with yellow pear-shaped fruit. **3.** the fragrant oil extracted from the rind of bergamot fruit; used in perfume.

beriberi *n.* (Singhalese *beriberi*, I cannot) a debilitating disease of peripheral nerves that causes anemia, paralysis, and heart failure; caused by a deficiency of thiamine (Vitamin B_1); prevalent in Asian countries where people eat large quantities of polished rice.

berm *n.* (French *berme*, bank of earth) **1.** a long, narrow earthen mound or ledge across the top, middle, or bottom of a slope. **2.** a long mound of earth (similar to a terrace ridge) erected across a slope to divert water and reduce erosion, sometimes accompanied by a ditch constructed along its upper side. **3.** the shoulder of a highway. **4.** the area of a beach that is smoothed by wave action. **5.** the strip between a ditch and its spoil pile, often used as a path or a road.

Bermuda grass *n.* a creeping, long-lived, perennial grass (*Cynodon dactylon*, family Poaceae) from the Old World (probably native of India) that is used for lawns and pastures in the tropics and subtropics, including southern U.S.A. It is a warm-season grass that will grow on most soils, but its spreading habit with runners as long as 3 or 4 ft (1 m or more) causes weed problems. Also called wiregrass, dog's tooth grass, and devil grass.

beryl *n.* (Latin *beryllus*, beryl) a cyclosilicate mineral ($Be_3Al_2Si_6O_{18}$) that is the principal source of beryllium. Emeralds and aquamarines are gemstone forms of beryl.

berylliosis *n.* (Latin *beryllus*, beryl + *osis*, disease) a disease that forms granular tumors in the lungs when beryllium salts are inhaled as dust or fumes or in the skin or other tissues when beryllium salts enter through lacerations or punctures.

beryllium (Be) *n.* (Latin *beryllus*, beryl + *ium*, element) element 4, atomic weight 9.012, forming Be^{2+} ions. A very lightweight metal that can easily become airborne and that can be hazardous to human health when inhaled, causing berylliosis.

best management practice (bmp) *n.* **1.** any of the best vegetative or mechanical practices known for controlling erosion and keeping land productive in a specific situation. **2.** a superior method of doing something, such as raising livestock, reducing pollution, maintaining good water quality, or producing something in a factory.

beta diversity *n.* the diversity between the plant and animal species inhabiting one area compared to those of another area in the same or another ecosystem. See alpha diversity.

beta (ß) particle *n.* an electron or positron emitted from the nucleus of an atom; more penetrating than an alpha particle, but less penetrating than a gamma ray.

beta-pleated sheet (β-pleated sheet) *n.* a secondary structure in many proteins composed of several peptide chains aligned side by side and linked by hydrogen bonds. The linkages add to the strength of fibers such as silk.

bezoar *n.* (Persian, *pād-zahr*, a counterpoison; antidote) **1.** precipitated mineral salts, mostly calcium carbonate or calcium phosphate formed as concentric shells around foreign objects (e.g., bullets, nails, stones, hair balls, wool balls, wood slivers) in the stomachs of deer, dogs, goats, mules, horses, lizards, and whales. **2.** a type of goat considered to be a likely ancestor of domestic goats.

bhang or **bang** *n.* See marijuana.

B horizon *n.* the zone of illuviation in a soil profile, generally the subsoil. B horizons form beneath an A, E, and/or O horizon and generally have lost soluble materials (such as carbonates) and accumulated resistant materials (such as sesquioxides), especially as coatings on soil peds. Blocky and/or prismatic structure is present in many B horizons.

biennial *adj.* (Latin *bi*, two + *annus*, year + *al*, pertaining to) **1.** happening once every two years, as a biennial celebration. **2.** living for two years, as a biennial plant that develops vegetative growth the first year and flowers, fruits, and dies the second year. Examples: beets, red clover. See annual, perennial.

bifurcate *v.* (Latin *bifurcare*, to fork) to divide into two branches.

big bang theory *n.* the hypothesis that the universe was created by a giant explosion and has been expanding ever since. Back-calculation based on the present distribution of galaxies and the rate of expansion indicates that the big bang would have occurred between 15 and 20 billion years ago.

Big Dipper *n.* a group of seven stars that outline a dipper in the Ursa Major constellation. The two outer stars of the bowl of the dipper point toward Polaris, the North Star. See Ursa Major.

big gun *n.* a large spray nozzle sometimes used at the end of a center-pivot irrigation system to irrigate the corners of the field.

Big gun irrigating the corner of a field

bighead *n.* **1.** a disease that causes a swollen head in certain animals, especially sheep. See tribulosis. **2.** (slang) egotism; a conceited opinion of oneself.

bighorn *n.* a large wild sheep (*Ovis canadensis*) native to the Rocky Mountains; the male carries a pair of assive, spiraling horns. Also called Rocky Mountain sheep.

bignonia *n.* **1.** a climbing shrub (*Bignonia* sp., family Bignoniaceae) native to tropical America and named for Abbé Bignon, a librarian for French King Louis XIV; grown for its showy trumpet-shaped blossoms. **2.** a vine, tree, or shrub of the bignonia family, including catalpa, princess tree, and trumpet creeper.

bilharziasis *n.* a parasitic disease that debilitates infected farmers, ranchers, and others in tropical areas throughout Africa, northern and eastern South America, southern Asia, and some western Pacific islands. It was named after a German physician, Theodor Maximilian Bilharz (1825–1862), but is also known as schistosomiasis, snail fever, and liver fluke fever. The pathogen is a worm (schistosome) 0.2 to 1.0 in. (5 to 25 mm) long that lives part of its life cycle in snails and part in people or other mammals. In humans, three species attack different parts of the body: *Schistosoma haematobium* attacks the bladder; *S. mansoni,* the intestines and liver; and *S. japonicum,* the spleen and liver. *S. japonicum* also causes infantilism in children. Symptoms of bilharziasis include itching, internal hemorrhaging, fever, chills, and general debility. According to the World Health Organization, bilharziasis is the world's most important parasitic disease. Schistosome eggs pass from humans or other mammals and hatch in wet areas, forming larvae that infest snails and grow into cercariae. The cercariae leave the snail and live in water, seeking any mammal. For example, they can burrow through the skin of a person swimming in such water and travel to the liver or other organ. In 4 to 6 weeks they mature, mate, and lay eggs, and the cycle is repeated. Control measures have focused on killing the host snails with synthetic organic molluscicides, but the expense is great.

bindweed *n.* (Anglo-Saxon *bindan*, binding + *weod*, weed) a twining, weedy plant with funnel-shaped white, pink, or blue blossoms, including the very troublesome field bindweed (*Convolvulus arvensis*, family Convolvulaceae), native to Europe but now widespread; sea bindweed (*C. soldanella*) of sandy

Field bindweed

seaside areas; and greater bindweed (*Calystegia sepum*), common in woods, hedges, and roadsides.

binomial *n.* (Latin *bi*, two + *nomen*, name) **1.** a two-word scientific name (genus and species) of a plant, animal, or human, in Latin; first proposed by Carolus Linnaeus (1707–1778). Examples: Corn = *Zea mays*; dog = *Canis familiaris*; and human = *Homo sapiens*. **2.** an algebraic expression with two additive terms, as $2x + 5y$ or $x^3 - 3x^2$. *adj.* **3.** having two names or two parts. **4. binomial distribution** = the statistical probability of one of two possible results occurring $1...n$ times out of n trials when the probability for each trial is constant. **5. binomial theorem** = Newton's theorem for raising a binomial expression to any power, as $(a + b)^2 = a^2 + 2ab + b^2$.

bioaccumulation *n.* (Greek *bio*, life + Latin *accumulatio*, increasing) the process whereby certain substances, such as pesticides or heavy metals, become more concentrated in tissues or internal organs as they move up the food chain; e.g., when small aquatic organisms are eaten by fish and the fish are eaten by large birds, animals, or humans. Also called biological magnification.

bioassay *n.* (Greek *bio*, life + French *essay*, test) the use of growth rates or other responses of living organisms to measure the concentration or potency of a substance, factor, or condition. The term is often used to mean cancer bioassay.

bioastronautics *n.* (Greek *bio*, life + *astron*, star + *nautikos*, ship), the science that studies and applies the physical and medical responses of organisms to space travel.

bioastronomy *n.* See exobiology.

bioavailability *n.* **1.** the availability of plant nutrients through biological processes. **2.** the availability of nutrients, vitamins, medication, etc., to an organism.

biocenology *n.* (Greek *bio*, life + *koinos*, common + *logos*, word) the biological science that deals with the reactions of organisms in a community to each other and to the environment.

biocenose or **biocenosis** *n.* (Greek *bio*, life + *koinōsis*, mingling) a self-sufficient organic community; a

group of naturally occurring organisms that live in the same area and interact with each other.

biochemical oxygen demand (BOD) *n.* See biological oxygen demand.

biochemistry *n.* (Greek *bio*, life + *chemia*, alchemy + *ry*, qualities) the science that deals with the chemistry of life, especially the structure, synthesis, and functions of biological molecules such as proteins and the genetic mechanisms that control the growth and reproduction of organisms.

biocide *n.* (Greek *bio*, life + Latin *cida*, killer) a poisonous substance that can kill living things, especially a pesticide.

bioclimate *n.* (Greek *bio*, life + *klima*, region) climatic factors that influence living organisms. For example, certain plants grow best in full sunlight, but others do better in the shade.

bioclimatic law, Hopkins *n.* the A. D. Hopkins (1857–1948) bioclimatic law states, "In the Spring in the Northern Hemisphere, all biological events occur 4 days later for each degree of latitude Northward, 5 degrees longitude Eastward, or for every 400 feet in elevation." This statement applies to the flowering of plants, the hatching of insects, and other biological events. See growing degree days.

biocontrol *n.* See biological control.

biodegradable *adj.* (Greek *bio*, life + Latin *degrad*, demote + *abilis*, able) susceptible to decomposition or disintegration by natural processes of living organisms, especially by microbial action.

biodiversity *n.* (Greek *bio*, life + Latin *diversitus*, diversity) the variety of living organisms occupying an area, including the number of species, variations within each species, and the habitat that sustains them.

bioengineering *n.* (Greek *bio*, life + Latin *ingeniare*, to design) manipulation of plant genetics to make plants resistant to specific pests; e.g., the production of corn strains that produce a Bt protein to kill the larvae of the European corn borer, thus making the plants resistant to the pest.

biofeedback *n.* a technique used to help a person control a body process or condition that is normally involuntary (e.g., blood pressure, pulse rate, depression, or anxiety) by relaxation techniques, electronic devices, or other manipulation.

biofertilizer *n.* a biological fertilizer; a plant grown to supply nutrients to crop plants; e.g., *Azolla* grown in rice paddies to supply nitrogen for the rice, or a green manure crop used to scavenge P and K from the soil and make it available to a succeeding crop.

biogas *n.* a combustible mixture of methane (CH_4), carbon dioxide (CO_2), and other gases produced by the controlled decomposition of manure, sewage, or other organic wastes.

biogenetics *n.* See genetic engineering.

biogeography *n.* (Greek *bio*, life + *geōgraphia*, geography) the study of the geographical distribution of plants, animals, and people and the causes thereof.

biohazard *n.* (Greek *bio*, life + Arabic *az-zāhr*, the death) **1.** the risk to human, animal, and plant health posed by an infective biological agent. **2.** an infective biological pathogen.

bioherbicide *n.* a living organism or natural product that controls weeds. Examples: a fungus (*Colletotrichum* sp.) that can kill certain weeds; a product from corn that can suppress weed growth.

biological control *n.* pest management methods based on the use of natural enemies of the pest, such as insects or plant diseases, that target the specific pest to be controlled; e.g.,
- chalcid flies to control harmful insects.
- ladybugs to control aphids.
- musk thistle weevils to control musk thistles.
- preying mantises to control insects.
- sweetclover residues to control cotton root rot.
- a fungus to control pathogenic fungi.
- geese to weed a rice paddy.
- sheep, goats, or donkeys to eat brush or thistles.

biological oxidation or **biological purification** *n.* the action of bacteria and other microorganisms feeding on and decomposing complex organic materials; occurs naturally in soil and water and is used in secondary treatment of sewage and other wastewater.

biological oxygen demand (BOD) *n.* the amount of oxygen needed by aerobic bacteria to break down the organic materials present in water. A high BOD indicates serious pollution. Also called biochemical oxygen demand, chemical oxygen demand. See BOD_5.

biology *n.* (French *biologie*, life study) **1.** the science of living organisms and life processes. **2.** the life processes characteristic of an organism. **3.** the assembly of plants and animals living in a specified region.

bioluminescence *n.* (Greek *bios*, life + Latin *lumen*, light + *escentia*, process) the production of light by living organisms; e.g., the phosphorescent light of a firefly (lightning bug). One mechanism of light production in living cells is by the oxidation of a heat-stable substance, luciferin, in the presence of a heat-sensitive enzyme, luciferase. Bioluminescence occurs in a variety of life forms, including certain algae, bacteria, fungi, protozoa, worms, mollusks, fish, and arthropods. Also called chemoluminescence.

biomagnification *n.* See bioaccumulation.

biomass *n.* (Greek *bio*, life + *maza*, barley cake) the total weight or volume of the living plants and animals in a given area, volume, or system at a specific time.

biome *n.* a major land area characterized by a group of ecosystems with climax plant and animal life forms determined by the environmental factors of climate and soil; e.g., tropical rainforest, tallgrass prairie, desert, Arctic tundra.

biometry *n.* (Greek *bio*, life + *metria*, measurer) **1.** the application of statistical analyses to data from the biological sciences. **2.** the estimation of human longevity by statistical methods.

biomonitoring *n.* **1.** the use of living organisms to test the suitability of treated sewage and other effluents for discharge into receiving waters. **2.** testing the quality of water downstream from a discharge. **3.** analysis of human blood, urine, tissues, etc., to measure the effects of exposure to hazardous chemicals. **4.** the use of canaries or other animals to test the suitability of the air for humans, especially in an underground setting such as a coal mine.

biophage *n.* (Greek *bio*, life + *phage*, eating) an organism that subsists on other living organisms; a parasite, pathogen, or predator.

biophilia *n.* (Greek *bio*, life + *philía*, liking) an innate liking for all living things.

biopsy *n.* (Greek *bio*, life + *opsis*, appearance) **1.** the removal of a sample of living tissue from a body for diagnosis. **2.** medical evaluation of excised tissue by microscopic study or other means.

bioregion *n.* (Greek *bio*, life + French *région*, land area) the area that is suitable for the growth of a particular ecological community.

bioremediation *n.* (Greek *bio*, life + Latin *remediatus*, remedial + *ion*, action) **1.** the use of bacteria and other organisms to purify polluted water. **2.** removal of pollutants from soil by biological means, such as the decomposition of toxic organic materials by microbes or the extraction of heavy metals by growing plants. See phytoremediation.

biosphere *n.* (Greek *bio*, life + *sphaira*, ball) **1.** the areas or volumes of land, water, and air that support life on Earth. **2.** the ecosystem encompassing all of Earth's living organisms. See ecosystem.

biostasy *n.* (Greek *bio*, life + *stasis*, standing state) a stable period of a land mass so covered by vegetation that physical erosion is prevented.

biostress *n.* (Greek, *bios*, to live + Latin, *strictus*, distress) a force, strain, mental or physical tension, or environmental problem that causes distress to living things.

biosynthesis *n.* (Greek *bio*, life + *synthesis*, putting together) the production of complex organic compounds by living organisms. See photosynthesis.

biota *n.* (Greek *bioté*, life) the assemblage of plants and animals living in a particular region.

biotechnology *n.* (Greek *bio*, life + *technē*, art + *logos*, word) **1.** manipulation of the characteristics of plants, animals, or microbes for practical purposes, such as the production of improved or specialized foods, medicines, or other products or processes. Sometimes used in a more restricted sense as synonymous with genetic engineering. **2.** genetic modification of animals to enhance certain characteristics or of humans to repair genetic defects. See genetic engineering.

biotic or biotical *adj.* (Greek, *biōtikos*, of life) pertaining to life or living organisms.

biotics *n.* the study of living things and their life processes.

biotite *n.* black mica [$K(Mg,Fe)_3(AlSi_3)O_{10}(OH)_2$]; a layer silicate that commonly occurs in igneous and metamorphic rocks; named after J. B. Biot (1774–1862), a French mineralogist.

biotope *n.* (Greek, *bio*, life + *topos*, a place) **1.** a small habitat with a uniform environment and the same fauna and flora throughout; a niche. See niche. **2.** the part of a host's body that is occupied by a parasite.

biotron *n.* (Greek, *bio*, life + *tron*, experimental chamber) a large room with environmental controls for light, temperature, humidity; used for growing plants, animals, and/or birds, often for research purposes.

bioturbation *n.* (Greek *bio*, life + Latin *turbāre*, to disturb + *atio*, action) stirring of soil by movement of living things such as ants or earthworms.

biotype *n.* (Greek *bio*, life + *typos*, impression) a group of plants, animals, or people with the same or closely related genetic composition.

biparous *adj.* (Latin *bi*, two + *parus*, bearing) **1.** bearing two offspring at a time. **2.** dividing into two branches or clusters, as in certain flowers.

bipolar *adj.* (Latin *bi*, two + *polaris*, pole star) **1.** having two poles, such as the North Pole and South Pole of the Earth or of a magnet. **2.** occurring in the regions near both poles. **3.** characterized by two opposite opinions or conditions.

Bipolaris *n.* a genus of fungi that produce mycotoxins. See mycotoxins.

birch *n.* (Anglo-Saxon *birce*, birch) **1.** any of several deciduous trees or shrubs (*Betula* sp., family Betulaceae) with serrated, pointed, oval leaves and smooth outer bark; sometimes used as ornamental trees or for erosion control. Native Americans used the bark to make birch bark canoes. The yellow birch (*B. lutea*) is the state tree of New Hampshire. **2.** the fine-grained wood from such trees; used to make furniture, boxes, flooring, railroad ties, etc.

bird *n.* (Anglo-Saxon *bridd*, bird) **1.** any member of the class Aves (phylum Vertebrata); a group of warm-blooded, egg-laying vertebrates with wings and bodies covered with feathers. **2.** domesticated fowl: chickens, ducks, geese, guineas, turkeys, etc. **3.** a clay pigeon. **4.** the feathered ball used to play badminton.

bird dog or **birddog** *n.* **1.** a dog trained to assist a hunter find birds and retrieve birds that have been shot; usually a dog of a breed that has been bred for this purpose. **2.** a person who specializes in locating special items or people. *v.* **3.** to follow closely, watch carefully, or seek after someone or something.

bird grass *n.* **1.** a lawn or turf grass (*Poa trivialis*, family Poaceae) used in temperate regions of North America. **2.** knotgrass (*Polygonum aviculare*, family Polygonaceae).

bird of prey *n.* a carnivorous bird that seizes small animals and carries them away; e.g., eagles, hawks, owls.

birdseed *n.* seeds used for feeding birds; often a type or mixture of seeds chosen to attract certain species of birds and placed in a bird feeder.

bird's-eye maple *n.* maple wood having a grain that when properly cut shows a pattern of dark eyelike markings (often used to make veneers).

birdsfoot trefoil *n.* a perennial legume (*Lotus corniculatus*, family Fabaceae) planted for forage, erosion control, or wildlife; tolerant of cold winters, dry summers, and moderately acid soil. The flower petals are yellow tinged with red; the pods resemble bird claws.

bird watching *n.* the observation, identification, and study of wild birds as a hobby.

birth *n.* (Swedish *byrth*, birth) **1.** the action of being born; the beginning of life as a separate individual. **2.** the initiation of a new concept, as the birth of an idea. **3.** ancestry or origin, as an American by birth.

bisequum *n.* (Latin *bi*, two + *sequentia*, sequence) a soil that has two sets of horizons, with one set above the other (e.g., ABAB), in the same deposit.

bisexual *adj.* or *n.* (Latin *bi*, two + *sexualis*, of sex) **1.** having both sexes present and functional in the same animal, human, or flower; e.g., earthworms.

Also called ambisexual. See hermaphrodite. **2.** being sexually attracted to both sexes; both homosexual and heterosexual.

bismuth (Bi) *n.* (Latin *bisemūtum*, white mass) element 83, atomic weight 208.98; a brittle metal with a low thermal conductivity that expands slightly when it solidifies; white with a slight pinkish tinge; used to make low-melting alloys, castings, thermocouples, etc.

bison *n.* (Latin *bison*, wild ox) **1.** the largest North American land animal now living; a shaggy, four-legged mammal (*Bison bison*, family Bovidae) with large forequarters and a massive head; commonly called buffalo;

North American bison

sometimes crossed with domestic cows to produce hybrids known as beefalo or cattalo. Bison once roamed in great herds across the American Great Plains. **2.** a related slightly larger and less shaggy animal (*B. bonasus*) native to Europe but now nearly extinct; also called wisent.

bitch *n.* (Anglo-Saxon *bicce*, female dog) **1.** a female canine (family Canidae), including dogs, wolves, foxes, etc. **2.** a derogatory term for an ill-tempered woman. *v.* **3.** to complain, especially at length in a grumpy manner.

bitterbrush *n.* a large, silvery shrub (*Purshia tridentata*) with yellow blossoms and bitter-tasting, three-toothed leaves that serve as browse for animals such as deer and antelope; grows along with sagebrush in arid regions of western North America.

bitter grass *n.* an aggressive allelopathic weed (*Paspalum conjugatum*) in annual cropping systems (e.g., cornfields) in warm climates (e.g., parts of Mexico).

bittern *n.* (Latin *būteō*, bittern) **1.** a bitter briny solution remaining after common table salt (NaCl) has been extracted from sea water; used as a source of bromides, iodides, etc., and sometimes as an herbicide. **2.** a freshwater marsh bird of the heron family (Ardeidae), including American bittern (*Botaurus lentiginosus*) and least bittern (*Ixobrychus exilis*).

bittersweet *n.* **1.** American bittersweet: a twining vine (*Celastrus scandens*, family Celastraceae) found from the North Carolina coast to New Mexico and northward through South Dakota into eastern Canada. Its decorative orange berries with red seeds are poisonous to livestock. Also called climbing bittersweet, false bittersweet, stafftree, or waxwork. **2.** a European vine (*Solanum dulcamara*, family

Solonaceae) with purple flowers and red berries that is a poisonous member of the nightshade family.

bitterweed *n.* (Anglo-Saxon *biter*, biting + *weod*, weed) a weed with a bitter taste, including
- a ragweed variety (*Ambrosia artemisiifolia*) that grows in America.
- certain plants of the family Asteraceae (*Picris* sp.).
- an annual herb (*Actinea ordorata*, family Asteraceae) found on overgrazed ranges in western U.S.A.; poisonous to sheep if they eat too much of it during a shortage of forage. Also called bitter rubberweed or Colorado rubberweed.

bitumen *n.* (Latin *bitumen*, bitumen) **1.** the solid or semisolid, dark-colored mixture of hydrocarbons and other residues left after various fuels and other liquids have been distilled from crude oil, tar, or similar materials; used as a sealant and/or adhesive, especially for road construction. **2.** asphalt as it occurs naturally.

bituminous coal *n.* the most abundant coal in active mines; a dense, black coal with bands of bright and dull surfaces; used mostly as a fuel for generating electricity, for heat and power in manufacturing plants, and for making coke.

biuret *n.* a white crystalline compound ($C_2H_5N_3O_2$) formed by thermal decomposition of urea. Biuret is toxic to some crops, especially citrus, and must generally be limited to 2% or less in fertilizer urea.

black alkali *n.* a sodic soil, so named because the pH is highly alkaline and soluble organic matter darkens the surface. See sodic soil.

black bear *n.* a North American bear (*Ursus americanus*) with a dense black coat and a dark brown face.

blackberry *n.* **1.** a viny shrub (*Rubus* sp.) with long, canelike, thorny stems. **2.** the black or purple berries (resembling raspberries) produced by the plant.

blackbird *n.* any of several bird species of mostly or entirely black color about 8 in. (20 cm) long (crows and ravens are also black but are larger than this), including
- Brewer's blackbird (*Euphagus cyanocephalus*): males are all black with some purplish or greenish sheen; females have dark brown colors.
- rusty blackbird (*E. carolinus*): common in swamps of southeastern U.S.A. in winter, Canada and Alaska in summer; similar colors to Brewer's blackbird but without iridescence.
- red-winged blackbird (*Agelaius phoeniceus*): males are mostly black and females mostly brown; both have red and yellow wing markings. They fly and roost in large flocks.
- tricolored blackbird (*A. tricolor*): males are similar to red-winged blackbird but with red and white

wing markings; females are brown and lack the wing markings.
- yellow-headed blackbird (*Xanthocephalus xanthocephalus*): males have black body and tail and a bright yellow hood; females are mostly brown with a yellow chest; common in some marshes of western U.S.A.

blackdamp *n.* chokedamp. See chokedamp, coal-mine gases.

black-eyed Susan *n.* any of several composite flowers with a daisylike blossom that has a dark center and, usually, yellow rays (especially *Rudbeckia hirta*). The state flower of Maryland; also called yellow daisy.

black fly or **blackfly** *n.* any of about a hundred or more species of gnats (Family Simuliidae) that are aquatic in the larval stage and that bite animals in the adult stage. The larvae produce a silklike thread to anchor themselves to objects in swift waters and use a pair of fan-shaped structures to ingest smaller organisms such as algae, protozoa, and diatoms. The adults are terrestrial; females feed on the blood of higher animals. The most common species that suck blood from animals and people are
- southern buffalo gnat (*Cnephia pecuarum*): this species has been known to kill horses by packing the nostrils and blocking breathing.
- turkey gnat (*Simulium meridionale*): a pest on poultry and other animals in southern U.S.A.
- blackfly (*Simulium pictipes*): a vicious blood-sucking gnat that prefers human blood; found in the northern and eastern parts of the U.S.A. and southeastern Canada.
- Adirondack blackfly (*Simulium hirtipes*): a pest, especially among loggers, hunters, campers, and fishers.

black frost *n.* a severe frost that kills and blackens vegetation, including staple crops, without a visible deposit of frost crystals.

blackhead disease *n.* a chronic intestinal and liver disease of poultry, especially of turkeys, caused by a protozoan parasite (*Histomonas meleagridis*). The disease causes listlessness, diarrhea with light green to brown droppings, and darkened skin on the head; it is commonly fatal to turkeys. Chickens are seldom seriously affected but may harbor the parasite. Most old poultry yards are contaminated with the organism. Also called enterohepatitis.

black ice *n.* a transparent, slippery layer of ice formed on pavement or other surfaces by freezing a thin layer of water.

blackleg *n.* **1.** a disease of cabbage and other crucifers caused by a fungus (*Phoma lingam*, family Sphaerioidaceae); a seed-borne disease that persists in the soil and causes sunken, dark-colored lesions on the stem near the ground or on the petioles and leaves. **2.** a potato disease caused by soft rot bacteria (including *Erwinia carotovora* and *E. atroseptica*). It causes a black, moist, soft rot on immature plants from the seed to the stem. It is most likely to be serious in cold, wet, or waterlogged soil. **3.** a stem rot disease (*Botrytis* sp. or *Rhizoctonia* sp.) of geranium or a bacterial leaf spot. **4.** an acute, infectious disease of young cattle and sometimes of sheep, goats, and swine; caused by a soil bacterium (*Clostridium chauvoei*) that infests the animals' legs and produces a high fever. Also called black quarter, quarter ill, felon, emphysematous anthrax or symptomatic anthrax, emphysematous gangrene.

black locust *n.* a leguminous tree (*Robinia pseudoacacia*, family Fabaceae) characterized by pinnate leaves and fragrant white (or pink in some varieties) blossoms. Its wood makes durable fence posts and erosion-stabilization timbers. It is used as an ornamental and for wildlife plantings. However, the leaves are poisonous to most livestock, and it is susceptible to the locust borer (*Megacyllene robiniae*).

black lung *n.* a disease caused by long-term inhalation of coal dust that blackens the interior of the lungs and regional lymph glands; a form of pneumoconiosis or silicosis. Also called anthracosis.

black nightshade *n.* a small poisonous annual herb (*Solanum nigrum*, family Solanaceae) with white flowers and black berries. A common weed pest, its leaves are toxic to bees and to livestock, causing salivation, vomiting, diarrhea, and bloating.

black rat *n.* a common rat (*Rattus rattus*) that is generally smaller than the Norway rat but has darker fur, larger ears, and a longer tail. Black rats are more common in the tropics, tend to be higher climbers than Norway rats, and are the most common carrier of plague. Both kinds are very destructive pests and vectors for many diseases. Also called Alexandrine rat, climbing rat, gray rat, or roof rat.

black rot *n.* any of several plant diseases exhibiting darkened tissue caused by fungi or bacteria. Some examples:

- black rot of apples, pears, quinces, and other hosts caused by a fungus (*Physalospora malorum*, family Pleosporaceae) that causes cankers on the twigs and limbs, leaf spots, and rotting of the fruit as it matures or while it is in storage. The fruit usually becomes shriveled and mummified and should be removed to prevent overwintering. Occurring widely east of the Rocky Mountains, the disease is sometimes called ring rot, blossom end rot, or brown rot. The symptoms on twigs and limbs are also known as twig blight, apple canker, black rot canker, or dieback. The leaf symptoms

are small dirty gray spots that are also known as leaf blight, brown spot, or frog-eye.

- black rot of bananas caused by a fungus (*Endoconidiophora paradoxa*) that is present wherever bananas are grown. On green fruit, it causes small black areas near the point of attachment to the stalk. It can cause fruit to drop from the bunches during ripening.
- black rot of cabbage, cauliflower, and other crucifers caused by a motile seed-borne bacterium (*Xanthomonas campestris*) that infects plants mostly during warm, humid weather and causes lesions along leaf margins. Cabbage leaves turn yellow with blackened margins and veins. Control practices include rotations, seed treatment, and shifting seed production to areas with low humidity during the seedling stage. Also called bacteriosis, bacterial rot, brown rot.
- black rot of grapes caused by a fungus (*Guignardia bidwellii*); a very destructive disease east of the Rocky Mountains. This fungus infects all green parts of grape plants, especially the newer growth. Tiny lesions grow and merge, forming tan or reddish brown spots that rot and turn black. Infected fruit may mummify and discharge ascospores the next spring.
- black rot of sweet potatoes caused by a fungus (*Ceratostomella fimbriata*) that can live in the soil for several years and can also survive on infected sweet potatoes in storage. It can be spread by wash water if the sweet potatoes are washed. Storage at 110°F (43°C) for a day kills the fungus.

black sage *n.* an aromatic perennial herb (*Salvia mellifera*, family Labiatae) that produces spikes of blue or white blossoms; occurs in Pacific coastal regions of California and Mexico. It is important as a bee plant and sometimes is grazed as forage.

black shank of tobacco *n.* a disease caused by a parasitic fungus (*Phytophtora parasitica*, var. *nicotianae*, family Pythiaceae). It causes damping-off of seedlings, blackened and dead roots, decay at the base of the stalk, and brown leaf blotches. The fungus is easily transported in plant material and soil, so cultivation causes infections to spread quickly.

black sigatoka *n.* a plague that kills banana plants. It appeared in Fiji in the 1920s and is slowly invading banana plantations around the world.

blacksnake or **black snake** *n.* **1.** a harmless dark-colored snake (*Coluber constrictor*) that is common in the U.S.A. It is long and slender, reaching lengths of 5 to 6 ft (1.5 to 1.8 m). **2.** any snake with a black color. **3.** a braided cowhide whip.

black spot *n.* a general name for many fungal and bacterial plant diseases that cause black spots on foliage or fruit, including

- apple scab caused by a parasitic fungus (*Venturia inaequalis*).
- bull's eye rot of apples caused by a fungus (*Neofabraea malicorticis*).
- avocado anthracnose caused by a fungus (*Colletotrichum gloeosporioides*).
- a disease of citrus caused by a fungus (*Phoma citricarpa*).
- peach scab caused by a fungus (*Cladosporium carpophilum*).
- internal black spot of bruised potatoes.
- a disease of roses caused by a fungus (*Diplocarpon rosae*).
- a disease of sugar beets caused by bacteria (*Pseudomonas aptata*).

black spot of beets *n.* See internal black spot.

blacktongue *n.* **1.** a disease of dogs caused by a deficiency of niacin (a B vitamin) that causes severe nervousness, ulcers in the mouth, and inflammation of the alimentary canal. Also known as pellagra in humans and dogs. **2.** a variation of anthrax that causes ulcerations on the tongues of cattle and horses.

black walnut *n.* **1.** a large deciduous tree (*Juglans nigra*, family Juglandaceae); native of eastern, central, and southern U.S.A. at higher elevations and higher latitudes. It grows best on fertile soils at a pH near neutrality. Roots of black walnut contain juglone, a substance toxic to many plants. **2.** a nut from the tree. The nuts are about 1 in. (2.5 cm) in diameter and slightly elongated, composed of a thick shell that parts into halves with an edible interior. The nuts are enclosed in a husk that contains a brown stain. **3.** the dense, hard, dark-colored wood from the tree, often used as a veneer; prized for making furniture, paneling, gunstocks, and novelties.

black water *n.* **1.** dark urine containing decomposing red blood cells resulting from any of several diseases. Sometimes called azoturia. **2.** effluent water that contains animal or human feces wastes. See gray water.

black widow spider *n.* a venomous spider (*Latrodectus mactans*). The female is about ½ in. (1.2 cm) long (larger than the male) and has a black body and a red to orange spot on the abdomen; its venom is a neurotoxin that can cause pain, weakness, and breathing difficulty in adults and fatal convulsions in children. It is named black widow because, after breeding, the female sometimes eats the male.

bladderwort *n.* (Anglo-Saxon *blaeddre*, bladder or blister + *wyrt*, root) any of several carnivorous plants (*Utricularia* sp.) of worldwide distribution that trap insects and crustaceans; mostly aquatic with threadlike leaves containing bladders that float, but some forms are terrestrial or epiphytic.

blanch *v.* (French *blanchir*, to whiten) **1.** to remove color; whiten; or bleach. **2.** to whiten a vegetable as it is growing by wrapping paper around the stalk and leaves or by mounding soil around the portion to be whitened, as celery is blanched. Also called etiolate. **3.** to briefly heat vegetables or fruit in boiling water, steam, or dry heat to inactivate enzymes before canning or freezing.

blast *n.* (Anglo-Saxon *blaest*, a puff of wind) **1.** a sudden gust of air, often making a loud noise. **2.** a sudden loud sound emitted by a horn, whistle, animal, radio, etc. **3.** an explosion. **4.** vigorous criticism. **5.** joyous revelry, especially at a party. **6.** a disease of sheep that causes flatulence. **7.** a disease that causes plant parts to dry or shrivel, such as blossom blast. **8.** a disease of rice (*Piricularia oryzae*).

blast cell *n.* an immature cell.

blaze *n.* or *v.* (Anglo-Saxon *blaese*, a torch or flame) **1.** burn vigorously with a bright flame. **2.** a passionate outburst, as a blaze of glory. **3.** shoot several times rapidly. **4.** mark a survey line or a trail by chipping off bark or painting spots on the bark of trees. **5.** open the way by cutting a trail through dense vegetation or pushing obstacles aside, either literally or figuratively. **6.** make known, proclaim, publicize. **7.** a white mark on the face of a horse or other animal.

bleak *adj.* (Norse *bleikr*, pale) **1.** bare, desolate, exposed to harsh conditions. **2.** depressing and dreary. **3.** an environmental disaster with a natural cause.

bleat *n.* or *v.* (Anglo-Saxon *blaetan*, to cry) **1.** a crying sound made by a sheep or goat. **2.** whining or complaining without any action, especially in a foolish manner.

bleeding heart *n.* **1.** a plant (*Dicentra spectabilis*) that has fernlike leaves and long clusters of red heart-shaped flowers. **2.** a person who feigns sympathy for someone or some cause.

blight *n.* (Norse *blikja*, turn pale) **1.** a plant disease or other cause that leads to rapid wilting and death of plants or plant parts. **2.** the continuing effects of blight. **3.** severe deterioration of an environment, especially urban blight.

blind inlet *n.* an inlet to a tile drain where water enters by seeping through a layer of coarse sand or gravel rather than directly into a pipe.

blind staggers *n.* erratic movement of cattle or horses poisoned by selenium as a result of eating locoweed. The animals lose hair and become lame, and their vision deteriorates. Their food intake declines, even to the point of starvation. See locoweed.

blind valley *n.* **1.** a feature in karst areas where a valley is closed at the lower end because its stream flows into an underground passage and disappears. **2.** a type of valley formed where a spring emerges from an underground channel. The resulting surface stream cuts a valley that is enclosed by steep walls at the head end. See karst.

blissom *adj.* in heat (a ewe).

blister beetle *n.* any of a large number of beetles (family Meloidae); also called meloid. Blister beetles feed primarily on leaves of plants and are very destructive on many garden and field crops, including tomatoes, potatoes, peas, beans, eggplant, melons, spinach, carrots, and chard. Many of these beetles secrete a body fluid containing cantharidin ($C_{10}H_{12}O_4$) that can raise blisters on a person's skin. Cantharidin is an ingredient in some topical medications for treating warts. The aphrodisiac known as Spanish fly is obtained from the dried bodies of a European species *(Lytta vesicatoria)*.

blister rust *n.* a fungal disease *(Cronartium ribicola,* family Melampsoraceae) of the five-needle white pines, with gooseberries and currants *(Ribes* sp.) as alternate hosts. The disease causes cankers and blisters on the bark of white pine trees in the spring.

blizzard *n.* **1.** a severe storm with high wind, heavy snow, and low temperature, often including whiteouts with near-zero visibility. **2.** an unusually large amount of something that overloads a system, as a blizzard of paperwork.

BLM *n.* See Bureau of Land Management.

bloat *n.* or *v.* (Scandinavian *blautr,* soft) **1.** a sometimes fatal condition of severe gas pressure in the stomach or intestines, usually in ruminant animals. It is caused by carbon dioxide and methane gas production after eating too much watery food or high-protein legumes too quickly. The condition in horses is also called colic. Also called swellhead. **2.** the swelling of the end of a can because of gas pressure. **3.** any severely distended condition, literally or figuratively, such as bloat in a bureaucracy.

blocky structure *n.* a soil structure composed of peds that are approximately equal in vertical and horizontal dimensions and that fit closely together in a complex pattern with surfaces that meet with either sharp corners (angular blocky structure) or rounded corners (subangular blocky structure); sometimes occurring as subunits within prismatic structure. Also see granular.

blood *n.* (Anglo-Saxon *blōd,* blood) **1.** a liquid tissue in humans and other vertebrates composed of plasma containing dissolved salts, sugars, fatty acids, and ionized organic substances and a heterogeneous suspension of red cells (erythrocytes) that transport oxygen, white cells (leukocytes) that fight disease, and platelets (cephalin and other proteins) that, along with fibrin, form clots when exposed to air. Blood is strongly buffered (at a pH of 7.3 to 7.5 in humans). **2.** a corresponding liquid tissue in nonvertebrate animals. **3.** temperament. **4.** ancestry, as having noble blood, sometimes called blue blood. **5. bad blood**: ill will between two persons or groups. **6. cold blood**: without feeling emotion, as murder in cold blood. **7. life blood**: the animating element of something, either literal blood or a figurative driving force. *adj.* **8. warm-blooded:** maintaining a nearly constant body temperature, as a warm-blooded animal **9. cold-blooded**: having a body temperature that varies with environmental conditions, as a cold-blooded animal.

blood poisoning *n.* a diseased condition with toxic matter or microbes present in the blood, accompanied by fever, chills, sweating, and possibly collapse and death; toxemia, septicemia, or pyemia.

blood pressure *n.* the pressure of blood circulating in a body, especially in the arteries. It pulses as the heart beats and is normally expressed as a fraction; e.g., 120/80, representing the systolic (highest) and diastolic (lowest) pressure expressed in equivalent millimeters of mercury; usually measured in the upper arm for consistency. Blood pressure varies with species, age, health, physical exercise, and emotional state. High blood pressure is considered a serious risk factor for a heart attack.

blood pudding *n.* a very dark sausage made mostly from animal blood, usually from a pig.

blood spot *n.* an egg defect that occurs mostly when a young hen starts to lay; detectable by candling. Such eggs can be used in livestock and poultry feed, but not for human consumption. See candle.

bloodsucking conenose *n.* an insect *(Triatoma sanguisuga,* family Pseudococcidae) with sucking mouth parts that inflict an intensely painful bite and inject a toxin that causes long-lasting swelling, faintness, and vomiting. Also called Mexican bedbug.

bloodworm *n.* (Anglo-Saxon *blōd,* blood + *wyrm,* worm) **1.** a red, segmented worm *(Polycirrus* sp. or *Enoplobranchus* sp.) used as fish bait. **2.** the larva of any of several midges *(Chironomus* sp.). **3.** any of several parasites *(Strongylus* sp. and *Elaeophora* sp.) that live in blood.

bloody scours *n.* a highly contagious, often fatal, intestinal infection that attacks small calves, pigs, and horses; characterized by diarrhea and bloody feces.

bloom *n.* (German *blume,* flower) **1.** a blossom of a flowering plant; a reproductive part of a plant. **2.** the state of flowering, as daisies in bloom. **3.** freshness, vigor, and health, as the bloom of youth. **4.** sudden appearance and fast growth, as an algal bloom. **5.** a white or gray powder that coats certain fruits (e.g., plums, grapes) and is easily rubbed off. **6.** a white, gray, or yellow powdery surface coating on stones, leather, metal, or new coins. *v.* **7.** to produce blossoms. **8.** to thrive and radiate well-being.

blossom *n.* (Anglo-Saxon *blōstma*, flower) **1.** an open flower of a plant in bloom, especially that of a fruit-bearing plant; e.g., a cherry blossom. **2.** the condition of flowering, as trees in blossom. **3.** the mixed sorrel and white color (similar to peach) of a bay horse. *v.* **4.** to produce open flowers. **5.** to develop vigorously; e.g., to blossom into a mature individual (often used figuratively).

blow fly or **blowfly** *n.* a fly (family Calliphoridae) that is capable of transmitting several diseases. It deposits its eggs on animal carcasses, on meat scraps, in wounds on living animals or birds, or on dung. The larvae hatch about a day later and produce maggots that eat the dead meat or exposed living tissue (e.g., the wool around the rump of sheep), or even live in the intestines of an animal that eats infested meat. They produce full-grown insects with black heads and thoraxes and steel-blue abdomens in 5 to 6 days. There are many species; some of the more common are also known as greenbottle and bluebottle flies.

blowout *n.* **1.** a rupture that allows air to escape suddenly, as from a tire. **2.** a depression made by wind erosion in loose soil, usually sand. **3.** a hole in the soil above a drainage tile line caused by soil washing into a broken tile.

blowout grass *n.* a grass (*Redfieldia flexuosa*, family Poaceae) that grows well on deep sandy soils in southwestern U.S.A. It is used to stabilize blowouts and control wind erosion in sandy soils, reseed droughty rangeland, and establish pastures.

blubber *n.* **1.** fat from whales and other marine mammals; used to produce whale oil. **2.** a thick layer of fat on a person, especially around the midriff. **3.** loud, unrestrained weeping and wailing. *v.* **4.** to weep and wail loudly. **5.** to be incoherent because of crying while speaking.

bluebell *n.* (French *bleu*, blue + Anglo-Saxon *bell*, to sound) any of several plants with small, blue, bell-shaped blossoms; e.g., the harebell (*Scilla nonscripta*) of the Old World and the lungwort (*Mertensia virginica*) of the U.S.A.

blueberry *n.* (French *bleu*, blue + Anglo-Saxon *berie*, berry) **1.** the nearly spherical, dark blue, edible berries produced by certain shrubs (*Vaccinium* sp., family Ericaceae), especially the highbush blueberry (*V. corymbosum*) of northeastern U.S.A., the rabbiteye blueberry (*V. virgatum*) of southeastern U.S.A., and the lowbush blueberry (*V. angustifolium*) often mistakenly called huckleberry. **2.** the shrubs that produce these berries.

bluebird *n.* (French *bleu*, blue + Anglo-Saxon *bridd*, bird) an attractive songbird (*Sialia* sp.) about 5 to 6 in. (13 to 15 cm) long. Males have a blue hood, back, wing, and tail; females and juveniles have browner colors, especially for the hood and back. The eastern bluebird (*S. sialis*) and western bluebird (*S. mexicana*) have red chests and a white belly; the mountain bluebird (*S. currucoides*) has a blue chest.

bluegrass *n.* (French *bleu*, blue + Anglo-Saxon *graes*, grass) **1.** any of about 200 species of *Poa* (e.g., *Poa pratensis*, Kentucky bluegrass) that grow mostly in temperate to cold regions and are used for lawn and forage grasses. They are more palatable but less productive than many forage grasses. **2. Bluegrass Region:** a region in the central part of Kentucky that is famous for raising horses on bluegrass pastures. **3. bluegrass music:** polyphonic folk music that originated in southern U.S.A.; typically played on stringed instruments, especially banjos and guitars.

blue-green algae *n.* primitive freshwater plants that are believed to have lived on Earth for about two billion years and are credited with changing the atmosphere from an early composition with a high concentration of carbon dioxide to its present composition low in carbon dioxide and high in oxygen. Traditionally, they were classified as a group called Myxophyceae (family Cyanophyceae) in the plant kingdom, but most taxonomists now call them Cyanobacteria and group them with bacteria in the kingdom Monera. They contain chlorophyl for photosynthesis but do not have true roots, stems, leaves, or seeds. They are prokaryotic (lacking a membrane around the cell nucleus), and they reproduce by simple fission of cells. The nitrogen-fixing capability of several of their genera combined with photosynthesis makes them the most independent of all life forms. Blue-green algae form a slimy filamentous mass known as pond scum in many bodies of water. Their nitrogen-fixing ability, as much as 40 lb atmospheric nitrogen/ac (45 kg/ha) each year, makes an important contribution to paddy rice production. Under dry conditions, blue-green algae combined with certain fungi form lichens that can grow on the surfaces of rocks and other objects. See cyanobacteria.

blue jay *n.* (French *bleu*, blue + *gai*, jay) a gregarious passerine bird (*Cyanocitta cristata*, family Corvidae) about 10 in. (25 cm) long that is common in eastern and central U.S.A. and southern Canada, especially in forested areas. It has a prominent blue crest, blue plumage on wings and back marked with black and white, and a white throat and breast with a black line between.

blue mold *n.* **1.** a fungal (*Penicillium* sp.) growth that forms a fuzzy bluish coating on stored food. Experiments with apples have shown that bruised fruit is highly susceptible and that infection can be transferred by wash water. **2.** a disease of spinach caused by a fungus (*Peronospora effusa*) and also called downy mildew. **3.** a tobacco disease common in plant beds in tobacco areas throughout the U.S.A.; caused by a fungus (*Peronospora tabicina*, family Peronosporaceae) whose resting spores overwinter in soil. Early indicators are seedlings with erect leaves. Infected plants that are a little older but still small have cupped leaves with round yellow spots and a moldy growth on their lower surfaces. It kills young plants; older plants may lose leaf tissue and then recover.

blue moon *n.* **1.** either the second full moon in the same calendar month (the usual interpretation during the last half of the 20th century) or the third full moon in a season that has four full moons (the older tradition); either of these events happens once in about three years. **2.** an unusually blue appearance of the moon resulting from atmospheric conditions on Earth. **3.** a very long time, as once in a blue moon.

blue racer *n.* a harmless blue-green snake; a subspecies (*Coluber constrictor flaviventris*) of the blacksnake; found in southeastern U.S.A.

blue-stain *n.* blue discoloration of light-colored wood such as pine; caused by fungal growth.

bluestem *n.* (French *bleu*, blue + Anglo-Saxon *stemn*, trunk) **1.** either of two important native prairie bunchgrasses with bluish leaf sheaths that were abundant in central U.S.A.; big bluestem (*Andropogon gerardii*, family Poaceae) grows up to 6 ft (1.8 m) tall, and little bluestem (*Schizachyrium scoparium*, formerly called *Andropogon scoparius*, family Poaceae) typically grows 1 to 3 ft (0.3 to 0.9 m) tall. **2.** any of several related or similar grass species; e.g., sand bluestem (*Andropogon hallii*, family Poaceae).

bluetongue *n.* a virus disease of sheep and sometimes cattle that is sometimes confused with hoof-and-mouth disease. Affected animals run high fevers, their mucous membranes redden in the nose and mouth, the animals salivate heavily, and their tongues swell. Many countries will not permit importation of animals from areas known to have bluetongue.

bluff *n.* (Dutch *blaf*, broad or flat) **1.** a high cliff, headland, or hill with a broad steep front. **2.** the high vertical banks of certain U.S. rivers. **3.** a bold act that misrepresents the truth. *v.* **4.** to proceed boldly in an effort to conceal weakness. **5.** to achieve by false assertion.

bmp *n.* See best management practice.

boa *n.* (Latin *boa*, water adder) **1.** any of several large nonvenomous tropical snakes (e.g., the anaconda, boa constrictor, python) that wrap around their prey and squeeze it to death. **2.** a long stole made of soft material such as feathers or fur.

boar *n.* (Anglo-Saxon *bār*, boar) a male swine capable of breeding a sow.

bobcat *n.* an American wildcat (*Lynx rufus*) with black spots on a brownish coat; named for its short (bobbed) tail.

bobolink *n.* a colorful North American songbird (*Dolichonyx oxyzivorus*) with a call interpreted to sound like its name.

bobtail *n.* short for bobbed tail; a tail that is either naturally short or has been docked, or an animal that has such a tail.

bobwhite *n.* a North American quail (*Colinus virginianus* and related species) with mottled brown, black, and white plumage and a call interpreted to sound like its name.

bodewash *n.* dried manure, especially of cattle or water buffalo; used for fuel in many developing countries where wood is scarce. Also called cowchips or buffalochips.

BOD$_5$ *n.* the amount of dissolved oxygen consumed in five days by the biological decomposition of organic materials in water; a measure of water quality. See biological oxygen demand.

bog *n.* (Gaelic *bog*, soft ground) **1.** a vegetated wetland, often an intermediate state between a lake and dry land, occurring in a depression or seep spot; generally characterized by either marsh (reeds, sedges, and other small plants) or swamp (water-loving shrubs and trees) vegetation, or peat deposits, and often either acidic or saline water. *v.* **2.** to slow down drastically; to impede progress either literally or figuratively.

bogged down *adj.* **1.** stuck in a muddy place; unable or barely able to move. **2.** so encumbered by problems, regulations, or other difficulties that very little progress is made.

bog spavin *n.* a diseased hock joint in horses caused by an accumulation of lymph fluid that enlarges the inner side of the hock joint and cripples the horse. Also called boggy in the hocks.

bogue *n.* a local term for a creek or slough, supposedly a Choctaw Indian word for stream or creek. It is used colloquially and in place names along the Gulf Coast; e.g., Bogalusa, Louisiana.

boiling point *n.* the temperature when a liquid boils. Water boils at 212°F (100°C) under a pressure of 1 atm (at sea level). Each 550-ft increase in elevation decreases atmospheric pressure enough to lower the boiling point of water 1°F; e.g., to 210°F at 1100 ft of elevation (or 1°C for 300-m increase in elevation).

bola or **bolas** *n.* (Spanish *bola* = ball) two or more (usually three) heavy balls (e.g., stones or metal) at the ends of an equal number of strong cords (commonly about 3 ft or 1 m long) that radiate from a center. Bolas are thrown by gauchos in South America to entangle the legs of running ostriches, cattle, or other animals.

bole *n.* (Norse *bolr*, tree trunk) **1.** the central stem or trunk of a tree, extending from the base to the top of some conifers but terminating in a crown in most deciduous species (syn. trunk, stem). **2.** (Latin *bolus*, clay) a type of reddish clay that is easily pulverized. **3.** a reddish brown color produced with such clay.

boll *n.* (Dutch *bol*, boll) the seed pod of cotton or flax.

boll weevil *n.* a beetle (*Anthonomus grandis*, family Curculionidae) whose larvae hatch in cotton bolls and cause heavy losses. The adult weevil may feed on okra, hollyhock, and hibiscus. Native to Mexico and Central America, the boll weevil has gradually expanded its range to the northern edge of the U.S. Cotton Belt.

bollworm *n.* (Dutch *bol*, boll + Anglo-Saxon *wyrm*, worm) insect larvae (*Heliothis armigera* or *Platyedra gossypiella*, family Phalaenidae) that feed on cotton bolls, corn, tobacco, tomatoes, beans, vetch, alfalfa, and other succulent plants and flowers. Also called corn earworm, tomato fruitworm.

bolly *n.* an unopened boll of frosted cotton.

bolly cotton *n.* the poor-quality lint obtained from bollies.

bolson *n.* (Spanish *bolsón*, big purse) an undrained basin in the arid regions of southwestern U.S.A. and Mexico; e.g., Death Valley. It includes a shallow pond of saline water in a nearly flat bottom flanked by alluvial fans and surrounded by hills and mountains.

bolt *n.* (Anglo-Saxon *bolt*, a catapult) **1.** a metal rod with a head at one end and threads at the other to be used as a fastener. **2.** a sliding rod or bar used to securely lock a closed door. **3.** a short section sawed from a log to be cut or split lengthwise into blocks, shingles, staves, etc. **4.** a log section used in making pulpwood or veneer. **5.** an arrow or other shaft to be shot from a bow. **6.** a flash of lightning. **7.** a length of cloth, usually about 40 yd (36 m) long, wrapped around a core. *v.* **8.** to fasten with bolts. **9.** to burst out and run away. **10.** to eat rapidly with little or no chewing. **11.** to flower or produce seed stalks suddenly, usually prematurely. **12.** to leave a political party or other group.

bolus *n.* (Latin *bolus*, clod of earth) **1.** a mass of chewed food ready to be swallowed; a cud. **2.** a large pill of veterinary medicine.

bone chewing *n.* the eating of bones of dead animals, usually by cattle, horses, or wild animals; usually indicating a deficiency of phosphorus or calcium. Also called pica, depraved appetite, licking disease. See pica, geophagia.

bone dry *n.* very dry; parched.

bone meal *n.* finely ground bones used in feed or fertilizer as a source of phosphorus and calcium. The principal kinds of bone meal products are:

- **green bone meal** made by drying and grinding fresh bones. It should not be used in animal feed because it might spread diseases such as anthrax and salmonellosis.
- **raw bone meal** made by boiling fresh bones in open kettles until they are freed of all adhering material; the bones are then dried and ground. It contains about 22% calcium (Ca) and 10% phosphorus (P).
- **steamed bone meal** made by cooking bones under steam pressure to remove excess meat and fat and make the bones easier to grind into a meal. It contains about 32% Ca and 15% P.
- **calcined bone meal** (bone ash) made by burning bones on a metal frame to sterilize them and rid them of all organic matter. The charcoal-like bone ash is friable and easily pulverized; it contains about 34% Ca and 16% P.

bone seeker *n.* a radioactive nuclide that is incorporated more in bones than in other tissues.

bonobo *n.* an African great ape (*Pan paniscus*, family Pongidae), also known as the pygmy chimpanzee, that lives in the Democratic Republic of the Congo; somewhat more slender and longer limbed than other chimpanzees. Its skin and hair color are black.

book farmer *n.* an advocate of scientific farming, as opposed to farming by rules of thumb and signs of the moon (often used in a derogatory sense). See moonarian.

booklouse or **book louse** *n.* any of several tiny wingless insects (*Liposcelis* sp., family Liposselidae, and others of the order Psocoptera or Corridentia) that damages books and papers. They thrive when the indoor humidity is high.

book lungs *n.* the gas exchange structures of most spiders and other arachnids, composed of a stack of plates resembling the pages of a book inside a chamber in the body of the insect.

book scorpion *n.* a small arachnid of the order Pseudoscorpionida that lives under stones and in old books. It resembles a scorpion but has no tail.

boom *n.* (Dutch *boom*, a tree or pole) **1.** a spar projecting from the foot of a mast to extend a sail on a sailing ship or to support a hoist for handling cargo. **2.** a spar projecting from a tractor or other machine to lift heavy loads. **3.** a long beam across the top of the mast of a hay stacker; used to hoist hay on or off a hay stack. **4.** a barrier made of a line of logs fastened end-to-end across a river to contain floating logs. **5.** a group of floating logs used to contain an oil spill. **6.** a long pipe or other structure used to apply fertilizer or a pesticide from a truck or tractor. **7.** a deep resonant sound. **8.** unusually strong building, sales, or other activity, as a housing boom. **9.** a sonic boom made by an airplane going faster than sound. **10.** an arm extending from an upright to support and position a microphone.

boot *n.* (French *bote*, boot) **1.** a high shoe or overshoe that encloses the foot and part or all of the leg of a person; high-heeled boots are typically worn by cowboys. **2.** feathering on the lower legs and feet of birds. **3.** a sheathlike structure that encloses the inflorescence of grass plants before it emerges as a head (in this condition, the plants are said to be in boot or in the boot stage). **4.** a tube that conducts seed to the shoe of a drill or planter. **5.** a protective shield around certain moving parts of a machine. **6.** a metal device that police sometimes lock onto a wheel to prevent a car from being driven away. **7.** (British) the compartment for carrying luggage in a car; the trunk. **8.** a kick. **9.** a new recruit in the U.S. Navy or Marine Corps; new recruits train in boot camp.

bora *n.* (Latin, *boreas*, north wind) a violent, cold, dry wind blowing down a mountain slope onto a lowland; named after the north winds that blow from the Alps in the former Yugoslavia to the Adriatic Sea; now used for such wind anyplace in the world.

borax *n.* (Latin *borax*, borax) sodium tetraborate $(Na_2B_4O_7 \cdot 10H_2O)$; a white, crystalline, water-soluble material used as a source of boron (B). The principal U.S. source is in Death Valley, California; also found in certain hot springs. Pure borax contains 11.34% boron on an elemental (B) basis, or 36.5% on a boric oxide (B_2O_3) basis. A less hydrated form $(Na_2B_4O_7 \cdot 4H_2O)$ is known as kernite. Borax is used as a boron fertilizer, cleaning agent, water softener, and soldering flux and in pottery glazes, certain alloys, flares, and nuclear reactor control rods. See boron.

border irrigation *n.* a form of surface irrigation using small ridges 10 to 100 ft (3 to 30 m) apart to guide water down a gentle slope. A line across the land between the ridges must be level so that water applied at the upper end will spread across the border and flow smoothly down its length.

boreal *adj.* (Latin *borealis*, northern) **1.** of high latitudes, including the needle-leaf forest and tundra areas of the Arctic region. **2.** pertaining to the north, especially to the north wind; frigid; bitterly cold.

borehole *n.* (Anglo-Saxon *borian*, to bore + *hol*, hollow) **1.** an exploratory well drilled in search of water, oil, gas, or minerals. **2.** a hole drilled to determine the nature of underground strata as an evaluation for construction plans or for geologic mapping.

Borlaug, Norman (1914–) *n.* an American crop breeder and director of Rockefeller Foundation wheat research in Mexico since 1944; credited with work on high-yielding varieties of wheat, maize, and rice that formed the basis of the Green Revolution. He received the Nobel Peace Prize in 1970 for this work. See Green Revolution.

boron (B) *n.* (Latin *borax*, borax + *on*, element) element 5, atomic weight 10.81. Boron commonly combines with hydrogen, oxygen, and other elements and occurs naturally in borax $(Na_2B_4O_7 \cdot 10H_2O)$, kernite $(Na_2B_4O_7 \cdot 4H_2O)$, tourmaline, colemanite, tincal, ulexite, and certain brines associated with hot springs. Boron is an essential micronutrient for plant growth, but excess boron is toxic to plants. Boron deficiencies cause distorted growth and rots that have often been interpreted as plant diseases. See borax.

borrow pit *n.* a pit formed where earth has been removed for use elsewhere as fill. These areas need special techniques (e.g., smoothing embankments, loosening compacted layers, often replacing some soil, fertilizing) and appropriate plantings to stabilize them and restore them to usefulness. Also called barrow pit.

boscage or **boskage** *n.* (French *boscage*, a grove) a natural growth of shrubs and trees; a thicket.

bosque *n.* (Spanish *bosque*, forest or woods) a wooded thicket in southwestern U.S.A., usually on a floodplain dominated by tamarisk, willow, and cottonwood.

boss *n.* (Dutch *baas*, master) **1.** a director and supervisor of a group of workers. **2.** a person with decision-making authority; one who dominates. **3.** an animal that leads a herd or that is first in the pecking order. **4.** a raised lump on the body or some part of an animal. **5.** an ornamental knob on a piece of furniture or a machine. **6.** a piece attached to each end of the bit of a horse's bridle where it extends outside the horse's mouth.

bot *n.* (Gaelic *botus*, belly worm) a larva of the botfly. See botfly.

botanical garden *n.* a garden where plants of scientific interest are grown for study and exhibition. The first botanical garden in the U.S.A. was established in St. Louis, Missouri, about 1860.

botanical name *n.* the Latin scientific name (genus and species) given to a plant. It is used for universality to avoid problems resulting from different common names for the same plant.

botanical pesticide *n.* a pesticide derived from plants; e.g., nicotine, strychnine, and rotenone.

botany *n.* (Greek, *botanē*, herb). **1.** the biological science of plants, including their growth and reproduction, parts, structures, and classification. Botany is divided into specialized disciplines such as plant anatomy, cytology, ecology, mycology, paleobotany, physiology, phytogeography, and taxonomy. It has practical application in agronomy, forestry, and horticulture. **2.** a generic term for Australian wool, such as a Botany wool suit.

botfly or **bot fly** *n.* a large, stout-bodied fly (families Oestridae and Gasterophilidae) that resembles a bee. The larvae are endoparasites of various mammals. The human botfly (*Dermatobia hominis*) deposits eggs on mosquitoes that carry them to humans. The sheep botfly (*Oestrus ovis*) is viviparous; it deposits larvae in the nostrils of sheep, and they feed in the frontal sinuses. The deer botfly (*Cephenomy* sp.) affects deer in a similar way. The ox warble flies (*Hypoderma bovis* and *H. Lineatum*) are serious pests of cattle; these species lay eggs on the legs of cattle, and the larvae penetrate the skin, migrate to the back, and develop in swellings (warbles) just under the skin. When fully grown, they burrow out through the skin and pupate in the soil. Ox warbles may seriously affect the health of cattle, and the holes made in the hide by the escaping larvae reduce the value of the skin for leather.

bo tree *n.* (Singhalese *bogaha*, bo tree) a large fig tree (*Ficus religiosa*, family Moraceae) that starts as an epiphyte. Such trees are sacred in India because Buddhism originated when Siddhartha Guatama, The Buddha, received inspiration under such a tree. Also called peepul or pipal. See banyan tree, fig.

botryose *adj.* (Greek *botrys*, a bunch of grapes + Latin *osus*, full of) growing in clusters, as flowers clustered like a bunch of grapes.

botrytis *n.* (Greek *botrys*, a bunch of grapes + Latin *itis*, disease) **1.** parasitic fungi (*Botrytis* sp., family Moniliaceae) that form clusters of spores that resemble bunches of grapes. Various species cause gray, moldy spots on flowers, leaves, and stems of plants, especially during wet weather. **2.** the subspecies name for cauliflower (*Brassica oleracea botrytis*).

bottle bill *n.* legislation that requires a returnable deposit on beverage containers such as plastic bottles and aluminum cans; a technique to encourage recycling and reduce littering.

bottled gas *n.* liquefied petroleum gas (often called LP gas) delivered in metal containers under pressure; usually propane (C_3H_8) in cold climates, butane (C_4H_{10}) in warmer climates, or a mixture of both. It is used where piped natural gas is not available for cooking, heating, lighting and fueling internal combustion engines, especially for pumping engines and farm tractors.

bottom *n.* (Anglo-Saxon *bodan*, ground) **1.** the lowest part of something, as the bottom of a bucket. **2.** the underside of an object. **3.** floodplain areas near a stream, often called bottoms or bottomlands. **4.** the soil or rock material beneath a body of water, as a stream bottom or lake bottom. **5.** the submerged part of a ship's hull. **6.** a chair seat. **7.** the part of a plow that cuts, lifts, and turns the soil, including the share, moldboard, and landside. **8.** stamina and endurance, as a horse with good bottom. **9.** the truth, as to get to the bottom of this matter.

bottom flow *n.* underflow; a density current that is colder or more loaded with sediment or dissolved salts than the overlying water, causing it to flow along the bottom of a body of water.

bottomland *n.* (Anglo-Saxon *bodan*, ground + *land*, land) nearly level land on valley floors; typically covered by alluvial deposits; commonly divided into first bottoms (floodplains) that are subject to periodic flooding and second bottoms (terraces) that represent former floodplains but are no longer flooded.

bottomland hardwoods *n.* forested wetlands that produce useful wood products and serve as an excellent environment for many species of wildlife.

botulism *n.* (Latin *botulus*, sausage + *ismus*, condition) food poisoning caused by an anaerobic bacterium (*Clostridium botulinum*) that produces toxic botulin, often in damaged or improperly processed canned goods. It may develop also when food and feed are stored too wet. Botulism can be fatal to humans and animals; it impairs vision, speech, and muscular function, ultimately leading to paralysis and suffocation. Treatment involves administration of an antitoxin to neutralize the toxin.

bougie *n.* (French *bougie*, a wax candle) **1.** a wax candle. **2.** a long slender medical instrument for dilating and/or medicating a natural body passage; e.g., a teat bougie; usually made of steel, rubber, or plastic. **3.** a slender medicated stick of gelatin to be inserted into the rectum or other body passage, where it melts from body heat.

bouillon *n.* (French *bouillir*, to boil) soup broth made from an extract of beef, chicken, or fish, often with vegetables; often dehydrated to make bouillon cubes.

boulder or **bowlder** *n.* (Swedish *buldersten*, noisy stone) a large, detached stone, often somewhat rounded by having been moved by gravity, running water, or a glacier; common in the soil of glaciated and mountainous regions of the U.S.A. In soil surveys, boulders are between 2 to 10 ft (60 to 300 cm) in diameter and up to 30 ft (10 m) long. In common usage, any stone too heavy to be lifted readily by a person (>10 to 12 in. or 25 to 30 cm in diameter) may be called a boulder.

boulder clay *n.* unstratified glacial deposits of compact gritty clay containing embedded gravel, stones, and boulders.

boundary *n.* (Latin *bodina*, a limit) **1.** the limit or border of a piece of property or a territory. **2.** a line, fence, ridge, river, or other indicator showing the outer limit of a property or territory; usually something that can be shown as a line on a map. **3.** a figurative limit, as the boundary of one's imagination.

boundary layer *n.* a layer adjacent to a surface that has different properties than the rest of the body it represents; e.g., the layer of air next to a leaf surface where there is little or no air movement and the relative humidity is 100%, or the water in a layer adjacent to a surface and attracted to that surface sufficiently to retard or prevent its flow.

boundary tree *n.* a distinguishable tree marking a point on a property line as defined in a metes and bounds survey; usually blazed or otherwise marked. See metes and bounds, witness tree.

bound water *n.* soil water strongly adsorbed on surfaces of clay and organic colloids and therefore not readily available to plants. Bound water is slightly denser than free water, and its freezing point is depressed. Also called hygroscopic water.

bourbon cotton *n.* a perennial shrub or small tree (*Gossypium purpurascens*, family Malvaceae) grown for commercial cotton production in the tropics. Also called Siam cotton or Puerto Rico cotton.

Bouyoucos block *n.* a gypsum block used for monitoring soil moisture; named after Dr. George J. Bouyoucos, Michigan State University. Two wires form terminals embedded a fixed distance apart in the block and extend far enough to be attached to a modified Wheatstone bridge. The gypsum blocks are buried at one or more depths in the root zone, and readings of their electrical resistance are taken periodically. The drier the soil becomes, the greater the blocks' resistance to electrical current. This resistance reading is calibrated for each kind of soil into estimates of soil moisture tension or plant available water. They are often used to indicate when it is time to irrigate.

bovine *n.* (Latin *bovis*, ox) any animal of the family Bovidae, including:
- domestic cattle (*Bos taurus*).
- gaur (*Bos gaurus*).
- gayal (*Bos frontalis*).
- yak *Bos grunniens*).
- zebu (Brahman) cattle (*Bos indicus*).
- water buffalo (*Bubalus bubalus*).
- bison (buffalo) (*Bison bison*).

All bovines have four parts to their stomachs: rumen (paunch), reticulum (honeycomb), omasum (manyplies), and abomasum (true stomach).

bovine farcy *n.* a chronic and fatal disease of cattle caused by a microbe (*Actinomyces farcinicus*). Its symptoms include nodules that cause small swellings on the lower legs. Also called cattle farcy.

bovine genital trichomoniasis *n.* a veneral disease of cattle caused by a protozoan (*Trichomonas fetus*) that causes breeding difficulty, abortion, or fetal death.

box *n.* (Latin *buxus*, boxwood) **1.** a container made of wood, metal, plastic, paper, etc., usually rectangular in shape, and often fitted with a cover. **2.** an enclosure with seats for several people, as a jury box. **3.** an area outlined on a printed page to enclose a picture, diagram, special text, etc. **4.** a space designated for a baseball player to stand, as the batter's box or the pitcher's box. **5.** a small evergreen tree or shrub (*Buxus sempervirens*, family Buxaceae) used for hedges and ornamental garden plantings since the days of the Romans. It has emetic and purgative properties that can be lethal to cattle, horses, sheep, and pigs. **6.** the wood of *Buxus sempervirens* or similar hard, durable wood from several other species of trees. Such wood is often used for making fine furniture and musical instruments. **7.** the lowest grade of softwood lumber. **8.** a predicament (figuratively), as he got himself into a box.

box canyon *n.* a canyon with only one entrance, steep rock walls, and a zigzag course, making its interior appear like a complete enclosure.

box elder or **boxelder** *n.* a North American maple tree (*Acer negundo*, family Aceraceae) also known as the ash-leaf maple or three-leaf maple. It is a fast-growing medium-sized tree with gray or brown bark and compound, sharply toothed leaves. It is used as a shade tree and for making furniture, boxes, novelties, railroad ties, pulpwood, etc.

boxelder bug *n.* a red and black bug (*Leptocoris trivittatus*) that lays its eggs on the fruit of boxelder trees. It is generally harmless but becomes a pest when large numbers seek shelter in houses in autumn.

Boxelder bug

box stall *n.* a stall where a horse or other large animal can be kept without being tied.

boysenberry *n.* a hybrid (*Rubus* sp., family Rosaceae) of blackberry, loganberry, and raspberry that bears large, sweet, dark red berries. Named after Rudolph Boysen, the American horticulturalist credited with developing it.

Bq *n.* becquerel.

Br *n.* the chemical symbol for bromine.

BR *n.* See Bureau of Reclamation.

brace *n.* (Latin *bracchia*, arms) **1.** a piece that strengthens or holds the parts of a framework in place, helping to rigidify a structure. **2.** an appliance to protect a weak or broken member, such as a leg brace. **3.** two of the same thing, as a brace of partridges or a brace of suspenders. **4.** a pair of joined musical staves. **5.** one of a pair of marks { } used to indicate grouping. **6.** a cranklike handle to hold and turn a bit for drilling holes. *v.* **7.** to support or prop up something. **8.** to reinforce or stiffen by adding a brace. **9.** to prepare for an impact, to brace oneself, either literally or figuratively.

brace roots *n.* adventitious roots that grow from aboveground nodes of corn and some other plants, helping to brace the plant in an upright position. Also called adventitious roots, aerial roots, prop roots.

bracken *n.* (Swedish *bräken*, fern) a large, coarse, poisonous fern (*Pteridium aquilinum*, family Polypodiaceae) that is an aggressive noxious weed of worldwide distribution, especially in temperate regions. It has triangular fronds and creeping rootstocks. Also known as bracken fern and brake fern.

bracken sickness *n.* a sometimes fatal livestock illness caused by eating bracken ferns. Bracken contains thiaminase, an enzyme that destroys body thiamine, reduces bone marrow, and produces severe intoxication in monogastric animals such as swine.

bracket fungus *n.* a corky or woody fungus (order Aphyllophorales) with a shelflike fruiting body that protrudes from a tree trunk or branch.

brackish water *n.* a mixture of freshwater and seawater with a salt content too high for most domestic and irrigation uses (typically with 1000 to 4000 ppm total salts but sometimes given a range as wide as 500 to 30,000 ppm). See brine, mixohaline.

braconid *n.* (Greek *brachýs*, short + *ides*, family) any of a large group of parasitic insects (family Braconidae) of value to agriculture and the environment. Various species parasitize aphids and larvae of the insect orders Lepidoptera, Coleoptera, and Diptera.

bract *n.* (Latin *bractea*, gold leaf) a specialized plant part resembling a leaf but with unique shape and/or color (usually small; often bright colored); usually located at the base of a flower.

Bradyrhizobium *n.* (Greek *bradys*, slow + *rhizo*, root + *bios*, life) a genus of nitrogen-fixing bacteria composed of species that form nodules on the roots of certain legume plants, such as soybeans (*Glycine max* hosting *B. japonicum*). *Bradyrhizobium* species are separated from species of the *Rhizobium* genus on the basis of a slower growth rate.

brae *n.* (Norse *bra*, brow) a steep bank or hillside (Scottish).

brahma or **brahman** *n.* **1.** a type of cattle (*Bos indicus*, family Bovidae) with a large hump on the back and a large dewlap under its neck, developed in southern U.S.A. from stock imported from India. It is uniquely adapted to the environments of India, China, East Africa, and southern U.S.A. because it is tolerant of heat, humidity, insects, and diseases. In India it is known as Zebu. **2.** a large Asian chicken that has feathered legs and yellow skin and lays brown eggs.

braided *adj.* (Anglo-Saxon *bregdan*, to move quickly) **1.** woven together or interlaced, as braided hair. **2.** interconnecting, as the multiple channels of a braided stream.

brake *n.* (German *brake*, a machine to break flax) **1.** a friction mechanism used to slow a moving vehicle, bring it to a stop, and hold it in place. **2.** an area overgrown with briars, brush, etc., such as a cane brake. **3.** one of several tropical ferns (*Pteris* sp.), especially bracken. **4.** a place to check the flow of irrigation water. *v.* **5.** to apply a brake to a vehicle to resist its movement.

bramble *n.* (Anglo-Saxon *braembel*, bramble) **1.** any of several prickly shrubs (*Rubus* sp., family Rosaceae), including blackberries, dewberries, raspberries. **2.** any prickly shrub.

bran *n.* (French *bren*, bran) the coating on kernels of cereal grains such as wheat, barley, corn, oats, rice, and rye. Bran contains protein and minerals; the white interior of the kernels is mostly carbohydrate. Products that contain bran are commonly called brown or whole, as brown rice and whole wheat flour.

branch *n.* (Latin *branca*, paw) **1.** a part that separates from the stem or another branch of a tree, shrub, or other plant. **2.** any part that divides from a main part, as the branches of a deer's antler. **3.** a stream that is a tributary of another stream. Also called a fork. **4.** a subdivision of an organization or system, as a branch office or a branch of learning. **5.** a portion of a larger family, as a branch from the family tree. **6.** a place in a computer program with two or more alternate routines. *v.* **7.** to diverge from a larger body and form a branch. **8.** to expand by adding branches.

brand *n.* (German *brand*, a burn) **1.** a means of identifying a product or a piece of property with a recognizable mark, logo, or other unique characteristic (often certified by a government); e.g., a trademark used to identify grades of fertilizer. **2.** an ear tag or a burn scar in the flesh of an animal to identify ownership.

brandy *n.* (Dutch *brandewijn*, burnt wine) **1.** a distilled alcoholic liquor made from wine. **2.** a similar distilled liquor made from other fermented juices; e.g., cherry wine.

Brassicaceae *n.* (Latin *brassica*, cabbage + *aceae*, family) the family name adopted in 1972 for plants of the mustard family; herbaceous plants that have pungent sap, alternate leaves, and clusters of four-petaled blossoms; e.g., broccoli, cauliflower, cress, mustard, radishes, turnips. The former family name was Cruciferae.

bread *n.* (Anglo-Saxon *bread*, crumb or morsel) **1.** a kind of food made by baking a dough or batter that contains flour or meal mixed with milk or water and usually salt, sugar or honey, and sometimes additional ingredients; called leavened bread if it is made with yeast, baking powder, or another agent that produces bubbles in the bread, or unleavened bread if it has no leavening agent. **2.** food in general. **3.** money, or other means of financial support.

breadfruit *n.* **1.** a tree (*Artocarpus altilis*, family Moraceae) with deeply lobed leaves, native to certain Pacific Islands. **2.** the fruit of the tree—a large, roughly spherical, starchy fruit that may be eaten fresh, cooked, or baked in bread. Some varieties have seeds, but others are seedless.

breeder seed *n.* seed or other propagating material produced under the direct control of the plant breeder or institution that produced the cultivar it represents. Breeder seed is used to grow the plants that produce foundation seed. See foundation seed, certified seed.

breeze *n.* (Spanish *brisa*, trade wind) **1.** a blowing wind, especially when it is light or moderate. **2.** in the Beaufort wind scale, breezes are classified as light breeze, 4 to 7 mph (6 to 11 km/h); gentle breeze, 8 to 12 mph (13 to 19 km/h); moderate breeze, 13 to 18 mph (21 to 29 km/h); fresh breeze, 19 to 24 mph (31 to 39 km/h); and strong breeze, 25 to 31 mph (40 to 50 km/h). **3.** a task that is easy to accomplish. **4.** ash, cinders, or dust from handling or burning coal, coke, or charcoal. Such material is sometimes used to reduce the weight of bricks, cinder blocks, or concrete. *v.* **5.** to move or work quickly and smoothly with minimum effort.

brew *v.* (Anglo-Saxon *breówan*, to brew) **1.** to prepare an alcoholic beverage (e.g., beer or ale) by steeping (soaking), boiling, and fermenting a mixture of malt and hops. **2.** to prepare a beverage (e.g., tea) by soaking in hot water or boiling. **3.** to prepare food by boiling, as to brew a pot of soup. **4.** to concoct a plan (especially a devious one), as to brew some mischief. *n.* **5.** a single batch of beer, ale, or other brewed beverage. **6.** a beverage prepared by brewing.

brewer's yeast *n.* unicellular fungi (especially *Saccharomyces cerevisiae*) used for baking bread and fermenting alcoholic beverages. They oxidize sugars, producing carbon dioxide and ethyl alcohol. Also called baker's yeast. See yeast.

brick *n.* (Dutch *bricke*, a block) **1.** a rectangular solid building block usually made from baked clay but sometimes of concrete. The standard size is 4¼ in. × 8½ in. × 2 in. (10.8 cm × 21.6 cm × 5.1 cm). In a dry environment, bricks may be sufficiently stable when made by sun-drying blocks of wet clay; otherwise, they must be baked in a kiln. **2.** a product made into the shape of a brick, such as a brick of cheese or a gold brick.

Making sun-dried bricks

brine *n.* (Anglo-Saxon *bryne*, a salt liquor) water containing more than 2.6% salt by weight (often used for any water with a high salt content). See brackish water.

bristle *n.* (Anglo-Saxon *bristl*, bristle) **1.** a long, stiff, usually glossy hair suitable for making a brush, especially hair from the back of swine. **2.** a pubescence grown by a plant that resembles such hair. **3.** any of the hairs or fibers in a brush. *v.* **4.** to cause bristles or other hairs to stiffen and stand more erect than usual, usually as an indication of fear or anger. **5.** to raise one's head and assume a defiant posture. **6.** to be thickly populated with something that threatens, as a hillside that bristles with weapons.

bristlecone pine *n.* a small pine (*Pinus aristata*, family Pinaceae) that grows near the tree line in western U.S.A., especially in California. This species includes the oldest known living things on Earth—trees as old as 4600 years.

bristletail *n.* (Anglo-Saxon *bristl*, bristle + *taegel*, tail) any member of a small order (Thysanura) of primitive, wingless insects (including silverfish and firebrats) with a fringed tail at the end of the abdomen and long antennae on the head; the carrot-shaped body about ½ in. (13 mm) long is often covered with scales. Bristletails live in soil, decaying litter, or nests of other animals.

British thermal unit (Btu or **BTU)** *n.* the amount of heat required to raise the temperature of 1 lb of water 1°F (from 59.5° to 60.5°F at a pressure of 1 atm); equal to 1054.35 joules, 252 calories, or 778.26 foot-pounds. See calorie.

Brix scale *n.* a hydrometer scale used to indicate the concentration of sugar in a solution such as syrup or fruit juice; named after its originator, Adolf F. Brix, a 19th-century German inventor. See balling hydrometer.

broad bean *n.* See faba.

broad-spectrum *adj.* applicable to a wide range of individuals or circumstances, as a broad-spectrum herbicide that controls many different weeds.

broiler *n.* (French *bruiller*, to broil) **1.** a pan, oven, or other container or means used to cook by intense heat. **2.** a young chicken that is or will be cooked by intense heat.

bromine (Br) *n.* (Greek *brōmos*, stench + *ina*, element) element 35, atomic weight 79.904; a member of the halogen family that form ions with a valence of −1. Bromine compounds (bromides) are used as light-sensitive materials in photographic films and papers, and bromine is an alternative to chlorine for a water disinfectant.

bronchial asthma *n.* allergic sensitivity of human air passages caused by an air pollutant such as plant pollen, smoke, or animal fur.

bronco *n.* (Spanish *potro bronco*, untamed colt) a range horse of the western U.S.A., especially one that is somewhat wild, broken or unbroken.

brood *n.* (Anglo-Saxon *brod*, brood) **1.** a group of young animals growing up together, as a brood of chicks. *v.* **2.** to incubate a nest of eggs or to care for a group of young. **3.** to worry excessively. *adj.* **4.** kept for breeding, as a brood hen.

brook *n.* (Anglo-Saxon *broc*, a stream) **1.** a small freshwater stream, commonly one that makes ripples and sounds (as a babbling brook) as it flows over and around stones. *v.* **2.** to tolerate or accept (usually in the negative), as to brook no evil or to brook no restraint.

brown bear *n.* **1.** a variant of the American black bear (*Ursus americanus*) with a brown coat. **2.** a European

bear (*Ursus arctos*) that has a brown coat. **3.** any other large bear with a brown coat, as the Kodiak bear (*Ursus middendorffi*) or a brown grizzly bear (*Ursus horribilis*).

brown heart *n.* discolored and deformed interior parts of beets or turnips caused by a boron deficiency; initially attributed to a disease.

brown planthopper *n.* an insect pest that attacks rice. In the 1980s, it developed resistance to virtually all known insecticides and drastically reduced rice production in Asia.

brown rat *n.* the common rat (*Rattus norvegicus*, family Muridae); a very serious environmental pest that spreads diseases, consumes and contaminates food supplies, and is very difficult to eradicate. Also called Norway rat, house rat, sewer rat, or wharf rat.

brown rice *n.* rice with the outer hull removed but not polished white. It retains the protein and minerals of the bran layer that has been removed from polished (white) rice.

browntail moth *n.* a white moth (*Euproctis chrysorrhoea*) with a brown tuft in the tail position. Its larvae damage trees by eating their leaves and carry a toxin on their hair that causes a skin rash on humans.

browse *v.* (French *broust*, twig or shoot) **1.** to eat tender leaves, twigs, and shoots of trees and shrubs. **2.** to nibble on forage. **3.** to shop leisurely, often with no particular purchase in mind. **4.** to read scattered parts of a book to get an impression of its content. *n.* **5.** the leaves, twigs, and shoots eaten by animals.

brucellosis *n.* (Latin *Brucella*, a genus of bacteria + *osis*, diseased condition) a very serious, highly contagious bacterial (*Brucella* sp.) disease also known as Bang's disease, Malta fever, Mediterranean fever, or undulant fever. It causes abortions, infertility, low milk production, and fever in the following species:

• cattle, bison, elk, humans (*Brucella abortus*).
• dogs, humans (*Brucella canis*).
• goats, humans (*Brucella melitensis*).
• swine, humans (*Brucella suis*).

brush *n.* (French *brosse*, brushwood) **1.** a thick growth of woody bushes. **2.** a pile of trimmed branches. **3.** a brief fight or other antagonistic encounter. **4.** a tool for spreading paint or other liquid or for cleaning, polishing, or smoothing; usually including flexible bristles and a handle. **5.** a bushy tail, such as that of a squirrel. **6.** an electrical part that conducts electricity by contact with a moving part such as an armature.

bryology *n.* (Greek *bryon*, moss + *logos*, word) the branch of botany concerned with bryophytes (mosses and liverworts).

bryophyte *n.* (Greek *bryon*, moss + *phyton*, plant) a nonvascular plant (division Bryophyta); a plant such as a moss or liverwort that represents an earlier form than the vascular plants that are now dominant. Bryophytes need a humid environment and are less adaptable than vascular plants.

Bt *n.* a bacterium (*Bacillus thuringiensis*) that produces natural insecticides that are harmless to mammals; a source of genes used in genetic engineering to produce insect-resistant crop plants (e.g., corn, cotton, and potatoes).

bubble *n.* (German *bubbele*, bubble) **1.** a roughly spherical body of gas enclosed in a liquid or in a solidified mass. **2.** a balloonlike body consisting of gas surrounded by a liquid film. **3.** an inflated part, as a bubble in a piece of tubing. **4.** a rounded covering, usually transparent. *v.* **5.** to emit bubbles. **6.** to effervesce. **7.** to exhibit enthusiasm, as to bubble with energy.

bubble concept *n.* See emissions trading.

bubonic plague *n.* a contagious disease caused by a bacterium (*Yersinia pestis*, formerly called *Pasteurella pestis*, family Enterobacteriaceae) and known as black death; transmitted from rats to humans by flea bites. Symptoms include high fever, swollen lymph nodes, weakness, and sometimes hemorrhages. It caused the Black Death plague that ravaged Europe and part of Asia in the 14th century, killing about a quarter of the population. See black rat.

Buchanan Amendment *n.* the amendment to the Agricultural Appropriations Bill for fiscal year 1930 (Public Law 70-769) that provided for the establishment of 10 soil erosion experiment stations and 10 plant materials centers in the U.S.A.

buck *n.* (German *bock*, male goat or deer) **1.** a male antelope, deer, goat, hare, rabbit, or sheep. **2.** a man, usually young and outgoing (sometimes used in a strongly derogatory manner, especially regarding Native Americans and other minorities). **3.** slang for a U.S. dollar. **4.** slang used figuratively for responsibility, as passing the buck. **5.** the lowest in rank; e.g., buck private, buck sergeant. **6.** a sawhorse. *v.* **7.** to leap and twist, as an animal trying to dislodge a rider or other burden. **8.** to charge with the head lowered, as a goat. **9.** to cut felled trees into logs. **10.** to gamble or take a risk, especially if successful, as to buck the odds.

buckaroo *n.* a cowboy (western U.S.A.).

buckeye *n.* **1.** any of several trees or shrubs (*Aesculus* sp., family Hippocastanaceae) with palmate leaves, showy yellowish green, bell-shaped flowers and scaly bark. They are popular shade trees for ornamental plantings. The Ohio buckeye (*A. glabra*) is native to the U.S.A. and is the state tree of Ohio. **2.** a bitter, poisonous nut from any of these trees. **3.** a butterfly (*Precis lavinia*) with purple or red spots on dark brown wings.

buckeye poisoning *n.* poisoning caused in livestock by eating certain plants (*Aesculus* sp., family Hippocastanaceae). It causes inflammation of the mucous membranes, vomiting, depression, stupor, lack of coordination, twitching, and paralysis.

buckhorn plantain *n.* a broadleaf weed (*Plantago lanceolata*, family Plantaginaceae) that is a serious pest in lawns and fields in humid parts of the U.S.A. Also known as ribgrass, ribwort, or English plantain. See plantain.

buck moth *n.* a moth (*Hemileuca maia*, family Saturniidae) with gray wings and a white band. In the autumn, the caterpillar of this moth is a very serious pest that eats the leaves of willows and oaks.

bucksaw *n.* (German *bock*, male goat or deer + Anglo-Saxon *seax*, knife) a saw with a coarse-toothed blade stretched between the handles of a wooden frame; commonly used for sawing wood on a sawhorse.

buckshot *n.* (German *bock*, male goat or deer + Anglo-Saxon *sceot*, thrown or shot) coarse lead shot used in shotguns for killing large birds or other game. Lead shot has been banned because some waterfowl can get lead poisoning by ingesting spent shot.

buckskin *n.* (German *bock*, male goat or deer + Norse *skinn*, peel) **1.** the skin of a buck deer. **2.** a soft leather made from deer skin or simulated from sheep skin. **3.** shoes or breeches made from buckskin. **4.** a grayish brown horse, similar in color to a buck deer, with a black mane and tail. **5.** a disease affecting grapefruit, oranges, and other citrus; caused by a mite (*Phyllocoptes oleivorous*). It causes the rind to be tough and leathery. **6.** a viral disease that causes yellowing of the foliage of cherry and peach trees and causes the fruit to drop or fail to ripen.

buckwheat *n.* (Anglo-Saxon *boc*, beech + *hwaete*, wheat) a plant (*Fagopyrum* sp., especially *F. esculentum*, family Polygonaceae) with triangular seeds that are used to make flour or livestock feed. Buckwheat has been called the poor farmer's crop because

- it grows on low-fertility, acid soils.
- it matures in 12 weeks.
- it is a good honeybee plant.
- the grain contains 6% lysine, an essential but scarce vitamin.

However, some livestock and albino humans are allergic to the protein, a disease known as fagopyrism.

bud *n.* (German *buddich*, swollen) **1.** a knob at the end of or on a short branch from a stem; buds contain miniature leaves or flowers that are exposed when the buds open. **2.** an undeveloped shoot, stem, or flower. **3.** a bulge that is the beginning of asexual reproduction in simple organisms such as yeasts and sponges. **4.** a nickname for a male person, especially one with the same name as his father. **5.** short for buddy; a friend. *v.* **6.** to graft by inserting a bud from one plant in a slot cut in the stem of another plant. It uses a minimum of grafting material, only a single bud instead of a scion. Also called budding. See scion.

budding *n.* a method of grafting using a single bud rather than a stem or twig scion. Budding is a common method to multiply varieties of fruit trees, roses, and other ornamental trees and shrubs that cannot be produced from seed. It may also be used for top working trees of stone fruits (e.g., cherries, plums, peaches) that cannot be easily grafted with cleft or whip grafts. See grafting.

buffalo *n.* (Latin *bufalus*, buffalo or gazelle) **1.** the water buffalo; a wild ox (*Bos bubalis*, family Bovidae) from India now found in other warm Asiatic and African nations. **2.** a large, black, fierce animal (*Syncerus* caffer) of southern Africa, usually called the Cape buffalo. **3.** a common name for the American bison (*Bison bison*), also known as American buffalo. The bison has been crossed with cattle to produce hybrid animals called beefalo or cattalo. *v.* **4.** to manipulate someone by intimidation and pretense.

buffalo berry *n.* **1.** a North American shrub (either *Shepherdia argentea* or *S. canadensis*) with silvery leaves, yellow flowers, and edible red or yellowish berries. **2.** the berries of these shrubs.

buffalo chips *n.* dried buffalo dung, sometimes used as fuel. Cow chips (dried cattle manure) are used as cooking fuel in many developing countries where wood is scarce.

buffalo fly or **buffalo gnat** *n.* any of several small insects (family Simuliidae) of the lower Mississippi Valley whose innumerable bites may kill a domestic animal. The most common species are southern buffalo gnat *(Cnephia pecuarum)*, turkey gnat *(Simulium meridionale)*, and common blackfly *(Simulium hirtipes)*.

buffalograss *n.* (Latin *bufalus*, buffalo or gazelle + German *gras*, grass) a short, grayish green forage or lawn grass (*Buchloe dactyloides*, family Poaceae) native to the Great Plains of the U.S.A.; also called early mesquite. As a lawn grass, it requires only half as much water as Kentucky bluegrass.

buffalo wallow *n.* a shallow depression that retains water during the rainy season in the Great Plains of the U.S.A. Pioneers thought the wallows were bathing places for buffalo.

buffer strip *n.* a strip of close-growing vegetation planted to reduce erosion or control pollution. See strip cropping.

buffer zone *n.* an intervening area that offers protection from a potentially negative external influence; e.g., a noncultivated border protecting a nature reserve from a nearby agricultural field.

bug *n.* **1.** any of a large group of insects (order Hemiptera, especially suborder Heteroptera); true bugs crawl while in the nymph (immature) stage and fly or walk as adults; they have mouthparts that are adapted to piercing and sucking. Some species have four wings, others are wingless. **2.** any insect. **3.** (slang) a virus or other microbe causing a short-term illness, as an intestinal bug. **4.** an addiction or a person addicted to a craze, fad, or hobby; e.g., a camera bug. **5.** (slang) a mechanical, electrical, or other defect that interferes with the proper operation of a machine. **6.** a flaw in a computer program. **7.** a hidden listening device. *v.* **8.** to install a hidden listening device. **9.** to pester or annoy someone.

buhrstone *n.* (Danish *burre*, bur + *steen*, stone) **1.** a hard, siliceous rock suitable for making millstones. **2.** a millstone. Also spelled burstone or burrstone.

bulb *n.* (Latin *bulbus*, a bulbous root) **1.** an underground reproductive organ of daffodils, hyacinths, lilies, onions, etc. It contains rudimentary stems and roots covered by several layers of fleshy leaves that produce a swollen part in an underground stem. Bulbs can be dried and stored for replanting later. **2.** a corm, tuber, or rhizome of a crocus, dahlia, or similar plant with a swollen shape and a reproductive function. **3.** any rounded swollen part of an object; e.g., a light bulb or the bulb of a thermometer. **4.** a rounded end or protuberance in an anatomical structure; e.g., the root of a hair or the medulla oblongata. **5.** in photography, a setting that holds the shutter open as long as its release is depressed.

bulk blend *n.* or **bulk-blend** *adj.* a mixture of two or more solid materials; e.g., a bulk blend prepared at a bulk-blend fertilizer plant by mixing two or more single-compound fertilizer materials.

bulk density *n.* mass per unit volume, including any pore space in the material; often expressed in grams per cubic centimeter. The bulk density of dry soil is usually between 1.0 and 1.6 g/cc. See particle density.

bull *n.* (German *bulle*, bull) **1.** a mature male bovine that is capable of breeding. **2.** a mature male of several other large animal species; e.g., elk or whales. **3.** a person who believes that stock prices will increase. **4.** a very large man. **5.** (slang) untrue statements; nonsense. *v.* **6.** to force one's way through a crowd or a thicket. *adj.* **7.** strong or rising, as a bull market.

bulldog *n.* (German *bulle*, bull + Anglo-Saxon *docga*, hound) **1.** a breed of domestic dog with short hair, short legs, a stocky body, and a large head. **2.** a very stubborn, persistent person. *v.* **3.** to throw a steer or other large animal to the ground by seizing the horns or head and twisting the neck.

bulldozer *n.* **1.** a crawler tractor with a sturdy blade mounted in front of it to push earth, stones, trees, and other materials. **2.** a person who ruthlessly insists on things being done in a certain way.

bullfrog *n.* (German *bulle*, bull + Anglo-Saxon *frogga*, frog) a large frog (especially the North American *Rana catesbeiana*) with a deep croaking voice.

bull nurse *n.* a man who cares for cattle during shipment (in southwestern U.S.A.).

bullock *n.* (Anglo-Saxon *bulluc*, little bull) a castrated male of genus *Bos*; an ox; used in many developing countries to pull carts and farm machinery.

bull pen *n.* **1.** a small enclosure for confining bulls. **2.** the sale ring where livestock are shown during an auction. **3.** temporary quarters for a group of workers or prisoners. **4.** an area near a baseball diamond where relief pitchers warm up.

bull snake *n.* a harmless snake (*Pituophis* sp.) with yellow and brown or black markings. Also called gopher snake because they eat rodents.

bull terrier *n.* a breed of strong, lean, active dogs that originated as a cross between bulldogs and terriers, usually white in color.

bull thistle *n.* a tall, weedy, biennial thistle (*Cirsium vulgare*, family Asteraceae) with pink to purple flowers and sharp spines.

bull whip *n.* a long, heavy whip, usually made of braided rawhide.

bulrush *n.* **1.** any of several wetland plants (*Scirpus* sp., family Cyperaceae or *Typha* sp., family Typhaceae) often growing in ditches or marshes with shallow open water; produces tall round or triangular stems topped with a brown spikelet composed of many tiny flowers. Also called cattail or tule. **2.** papyrus (*Cyperus papyrus*, family Cyperaceae).

bumblebee or **bumble bee** *n.* any of several large, hairy, yellow and black bees (*Bombus* sp., family Apidae). Bumblebees are larger and have longer tongues than honeybees; they can get nectar and pollen from such deep-throated flowers as red clover, alfalfa, and squash and thus pollinate them.

bumblefoot *n.* difficulty in walking caused by a swelling on the underside of the foot of a chicken or other bird; often associated with an abscess around a foreign body or an injury.

bunch *n.* **1.** a group of several similar items; a cluster, such as a bunch of grapes or bananas. **2.** a large quantity of something, as a whole bunch. **3.** a hump, as a camel's bunch. **4.** a lump where a wound healed improperly, especially following castration of pigs. *v.* **5.** to arrange items in groups; to bunch together. **6.** to gather hay into piles. **7.** to skid logs together so they can be loaded on a truck.

bund *n.* an earthen dike built to control flooding along a river, especially in India.

bung *n.* (Dutch *bonghe*, stopper) **1.** a large stopper for a hole in the side or end of a barrel or cask. *v.* **2.** to place a bung in its hole.

bunker *n.* (Scottish *bonker*, chest) **1.** a large storage bin or chest, often fixed in place, as a coal bunker. **2.** a sand trap or other hazard on a golf course. **3.** a partially buried bomb shelter or other fortification.

bunker silo *n.* a trench silo dug into an embankment or hillside and used to store silage. See trench silo.

bunkhouse *n.* sleeping quarters for workers on a large farm, ranch, or camp.

bunt *n.* **1.** a blow by the head of a goat, sheep, or other animal. **2.** a short bounce of a pitched ball off a bat. **3.** a smut disease that stunts infected plants and replaces wheat kernels with black, foul-smelling, fungal (*Tilletia caries or T. foetida*) spores. The spores may be either soil borne or seed borne to cause future infection. Also called stinking smut.

bunting *n.* **1.** any of several species of small (mostly 4 to 6 in., or 10 to 15 cm, long), brightly colored songbirds (*Passerina* sp. and *Plectrophenax* sp., family Fringillidae), including indigo bunting, lazuli bunting, varied bunting, painted bunting, snow bunting, and McKay's bunting. **2.** a flag, or the lightweight cotton or woolen cloth used to make flags. **3.** long, brightly colored strips of cloth or paper used for holiday decorations. **4.** a baby's sleeping garment that covers all except the face.

buoy *n.* (French *boue*, a float) **1.** a floating object anchored in a body of water to mark an area designated for a particular purpose or to warn of rocks or other hazard to navigation. *v.* **2.** to lift or support in water by buoyancy. **3.** to lift up the human spirit; to overcome despondency.

buoyancy *n.* (French *boue*, a float + Latin *antia*, quality) **1.** the upward force exerted by water, air, or other fluid on a floating or submerged body. Buoyancy is equal to the weight of the fluid displaced by the body. See Archimedean principle. **2.** a cheerful, light-hearted spirit.

buprestid *n.* any of several bright metallic-colored beetles (family Buprestidae) that can be deadly to cattle when accidentally eaten with forage. The larvae damage wood by tunneling through it.

bur *n.* (Danish *burre*, bur) **1.** a prickly case that encloses the seeds of certain plants, such as burdock or bur clover. **2.** a burr.

bur clover *n.* an annual forage legume (*Medicago arabica* or *M. hispida*, family Fabaceae) with prostrate growth habit, trifoliate leaves, small yellow flowers, and spiny, coiled seed pods.

burdock *n.* (Danish *burre*, bur + Dutch *docke*, dock plants) weedy plants (*Arctium* sp., especially *A. lappa*, family Asteraceae) with large, heart-shaped leaves, purple flowers, and burs that stick to clothing and animal hair.

Bureau of Customs *n.* an agency of the U.S. Treasury Department with more than 350 customs ports and stations to oversee the entry of goods into the U.S.A. and collect duties and taxes due on imported merchandise. The Bureau of Customs cooperates with the USDA–Animal and Plant Health Inspection Service to prevent entry of plant and animal diseases.

Bureau of Land Management (BLM) *n.* an agency in the U.S. Department of the Interior established in 1946 as a combination of the General Land Office and the U.S. Grazing Service; responsible for the administration of 264 million ac of public land (about ⅛ of the U.S. land area), mostly in the 12 western states, including grasslands, forest, desert, and tundra. Its mission is to sustain the health, diversity, and productivity of public lands for the use and enjoyment of present and future generations.

Bureau of Reclamation (BR) *n.* an agency in the U.S. Department of the Interior established in 1902 to develop water resources in the 17 western states; best known for its more than 600 dams and reservoirs that provide water for 31 million people and for irrigation of 10 million ac (4 million ha) of land, and for its 58 electric power plants. Its mission is to assist in meeting water needs while protecting both the environment and the public interest through prudent water policies, conservation programs, wetland preservation, and dam operations.

burgeon *v.* (French *borjon*, bud) **1.** to bring forth buds; to sprout. **2.** to begin growing rapidly, as a disease in an animal. **3.** to flourish.

buried soil *n.* a paleosol; a soil that developed on a former land surface and was buried by subsequent deposition. Such soils are often covered by alluvial deposits, glacial till, volcanic ash, or even lava flows.Buried soils are often studied to interpret the climate and vegetation of the former environment or the duration of interglacial periods.

burl *n.* (Latin *burla*, bunch) **1.** a knot or lump in a piece of cloth or strand of yarn. **2.** a knobby growth on the trunk of a tree or other plant. **3.** the distorted grain in lumber produced by growth around a burl. Large burls are considered valuable for the interesting patterns produced when they are made into lumber or veneer, especially for making furniture.

burlap *n.* (Latin *burra*, coarse cloth + Dutch *lap*, patch) coarse cloth made of jute, flax, or hemp; used to make bags and decorative handcraft items and to wrap the roots of plants. See balled and burlapped.

burley *n.* a strain of thin-leaved cigarette tobacco that is very light colored when cured; grown mostly in Kentucky.

burned lime or **burnt lime** *n.* calcium oxide (CaO) produced by heating ground limestone ($CaCO_3$) until carbon dioxide (CO_2) is driven off; used in mortar and for a fast-acting liming agent to raise the pH of acid soils. Adding water turns it into slaked lime [$Ca(OH)_2$].

burned-over land or **burnt-over land** *n.* an area of land where a forest or range fire has burned recently enough to affect the present vegetation. Partially burned dead trees can make it easier for a new fire to spread for many years after the original fire.

burning index *n.* a means for predicting the combustibility of a forest and the rate a fire could spread. It integrates the effects of ambient environmental factors, including the growth stage and moisture content of vegetation, relative humidity, temperature, and wind velocity.

burro *n.* (Spanish *burrico*, ass) a donkey, especially one used as a pack animal. See ass, donkey.

burrow *n.* **1.** a tunnel dug by a mole, chipmunk, ground squirrel, prairie dog, rabbit, or other animal that digs through the soil. **2.** a hole dug for refuge in time of battle; e.g., a soldier's foxhole. *v.* **3.** to dig a tunnel through soil, rock, refuse, etc. **4.** to search by digging through something. **5.** to move through something, as to burrow through the forest.

burrowing nematode *n.* a worm (*Radopholus similis*, family Tylenchidae) that attacks the roots of citrus, banana, sweet potato, sugarcane, etc.

bur sage *n.* a shrub (*Franseria* sp., family Asteraceae) that is the characteristic vegetation of desert areas in southwestern U.S.A. Its pollen is a significant cause of hay fever.

bush *n.* (German *busch*, bush) **1.** a shrub or cluster of shrubs; a woody plant smaller than a tree, having many stems and branches near ground level. **2.** remote, uninhabited land; e.g., the Australian outback or parts of northern Canada.

bushed *n.* **1.** covered with bushes. **2.** tired; exhausted.

bushel (bu) (U.S.) *n.* (French *boissel*, a unit of measure) **1.** a dry measure equal to 32 dry qt; 4 pk; 2,150.42 in.^{3}; or 35.24 L. (Note: a British imperial bu = 2219.36 in.^{3} or 1.0321 U.S. bu). **2.** the weight of a commodity considered equivalent to a volume of 1 bu: ear corn = 70 lb; wheat, beans, and peas = 60 lb; seed corn, rye, flax seed = 56 lb; barley = 48 lb; and oats = 32 lb. **3.** a basket or other container having a capacity of 1 bu. **4.** a large amount, as a bushel of money.

bush fallow *n.* a secondary succession of forest regrowth on land that has been cropped under a system of shifting cultivation.

bushlands *n.* unsettled lands of the northern provinces in Canada.

bush sickness *n.* anemia in ruminants in New Zealand as a result of cobalt deficiency. Affected animals are slow growing, emaciated, and listless; heifers may fail to calve. See cobalt deficiency.

bushwhack *v.* **1.** to cut a pathway through thick brush. **2.** to pull a boat upstream by pulling on tree branches and bushes along the bank. **3.** to attack from ambush.

bushwhacker *n.* **1.** a hook used to cut brush. **2.** a backwoodsman or frontiersman. **3.** a guerilla fighter, especially a Confederate during the U.S. Civil War. **4.** someone who attacks from ambush.

buster *n.* **1.** a person who breaks (trains) horses or other animals to be ridden or for draft work. **2.** a person who breaks things or organizations; e.g., a trust buster. **3.** a nickname for a boy or man, often belittling.

butane *n.* (Latin *butyrum*, butter + *ane*, an alkane) a gaseous aliphatic hydrocarbon (C_4H_{10}) that boils at 31°F (–0.5°C); used for making synthetic rubber and as a liquefied petroleum (bottled) gas where the climate is not too cold.

butt *n.* (French *but*, goal or objective) **1.** a target; the object of an attack, either physical or verbal, as the butt of a joke. **2.** the thick, blunt end of an object; e.g., the butt of a rifle. **3.** the thick part at the base of a tree or other plant. **4.** the remaining part when something is consumed or cut away, as a cigar butt or a stump, especially a walnut stump. **5.** the upper half of a ham or shoulder cut up for meat. **6.** the buttocks (slang). **7.** a liquid volume measure equal to 126 U.S. gal (477 L). **8.** a large cask used to store wine or beer.

butte *n.* (French *butte*, hillock) an isolated hill or mountain surrounded by steeply sloping sides, especially one with a flat top similar to a mesa but smaller. Common in western U.S.A. in arid and regions.

butter *n.* (Greek *boutyron*, butter) **1.** a soft, yellow, fatty solid made by churning milk or cream, commonly used in cooking and as a spread for making sandwiches; must contain at least 80% fat to be marketed legally. **2.** other types of sandwich spreads; e.g., apple butter and peanut butter. *v.* **3.** to spread butter on bread or other food. **4.** to spread mortar on bricks or blocks for building a wall. **5. butter up**: to flatter, often with hope of obtaining a favor.

butter-and-eggs *n.* **1.** a weedy plant (*Linaria vulgaris*) of the figwort family with yellow and orange flowers, also called wild snapdragon. **2.** other plants with similar yellow and orange flowers.

buttercup *n.* **1.** a small plant (*Ranunculus* sp., family Ranunculaceae) with cup-shaped yellow flowers and deeply cut leaves. Also called crowfoot. **2.** the family of flowering plants that includes buttercups, anemones, clematis, columbines, wild crocuses, larkspurs, and peonies.

butterfat *n.* the fat contained in milk; the main constituent of butter; composed largely of glycerides of butyric, oleic, palmitic, and stearic acids. Also known as milkfat.

butterfly *n.* (Anglo-Saxon *buttorfleoge*, butterfly) **1.** an adult insect with four large wings that allow it to flutter through the air, often displaying bright colors and attractive patterns. Butterflies feed mostly on nectar from flowers. They are classified in five major families in the order Lepidoptera: Hesperiidae, Lemoniidae, Lycaenidae, Papilionidae, and Nymphalidae. See moth. **2.** a person who dresses brightly and leads a carefree life, especially a young woman. **3.** a swimming breaststroke used in races; the swimmer lunges forward by bringing both arms out of the water and forward, then down and backward in coordination with a dolphin (up-and-down) kick.

Butterfly

butterfly weed *n.* a weed (*Asclepias tuberosa*) of the milkweed family with clusters of bright orange flowers; used medically as a cathartic and to induce perspiration.

buttermilk *n.* (Greek *boutyron*, butter + Anglo-Saxon *meolc*, milk) **1.** the acidic fluid that remains when fat is removed from milk or cream by making butter. **2.** a similar milk product made by fermenting low-fat or skim milk with certain bacteria (*Streptococcus lactis* and *Leuconostoc citrovorum*). Buttermilk contains at least 8.5% nonfat milk solids and about 0.8% acidity.

butternut *n.* (Greek *boutyron*, butter + Anglo-Saxon *hnutu*, nut) **1.** a tall deciduous tree (*Juglans cinerea*, family Juglandaceae) of the same genus as black walnut but with lighter-colored bark and wood; also called white walnut. It is found from the East Coast to the Dakotas and from Georgia and Arkansas northward to New Brunswick. The sap kills most nearby vegetation. The nuts are edible, oily, and spicy. See allelopathy, black walnut. **2.** the wood from such trees; used for furniture, cabinet work, interior trim, etc. **3.** a backwoodsman of the early settlement period of eastern U.S.A. who wore clothing dyed yellowish brown with coloring from butternut bark, especially a Confederate soldier.

butternut squash *n.* a winter squash (*Cucurbita* sp., family Cucurbitaceae) with an elongated pear shape, yellow skin, and orange interior.

butter oil *n.* liquefied butterfat with milk solids removed; used in making process butter. It can be shipped without refrigeration. In India it is known as gee or ghee. See ghee.

butterwort *n.* (Greek *boutyron*, butter + *wyrt*, twig or root) a small stemless plant (*Pinguicula* sp., especially *P. vulgaris*) with blue flowers and broad, fleshy, sticky leaves that trap and absorb small insects; grows best in wetlands.

button *n.* (Old French *boton*, bud or button) **1.** a bud or an immature fruit. **2.** an onion set. **3.** a nipple, especially on a hog. **4.** an immature or stunted horn growth, as on a calf. **5.** a flattened, often circular object sewn on garments as a fastener that passes through a buttonhole.

buttonweed *n.* See velvetleaf.

buttress *n.* (French *bouterets*, buttress) **1.** a structure used for bracing, as a reinforcement at the corner of a wall or short protrusions on the outside of a retaining wall; originally intended to add strength, but sometimes used for appearance. **2.** an increased diameter of the butt of a tree (near the ground); often seen on trees such as cypress in the humid tropics and subtropics. *v.* **3.** to strengthen a structure with a buttress. **4.** to present additional evidence to strengthen an argument.

C

C *n.* **1.** the chemical symbol for carbon. **2.** an average grade. **3.** coulomb.

Ca *n.* the chemical symbol for calcium.

cabbage *n.* (French *caboche*, head) **1.** a cultivated plant (*Brassica oleracea*, variety *capitata*, family Brassicaceae) with fleshy leaves wrapped around each other in an edible head about 6 in. (15 cm) in diameter. Cabbage is a popular green vegetable in home gardens, and it is adapted to a wide variety of cool-season environmental conditions. **2.** the cooked or raw leaves of the cabbage plant, often called coleslaw when eaten as a raw, shredded salad.

cabbage butterfly *n.* a butterfly (*Pieris rapae*, family Pieridae) whose bright green larvae feed on, and do serious damage to, plants of the cabbage family, principally cabbage, broccoli, Brussels sprouts, and cauliflower, but will starve to death on most other plants.

cabbage looper *n.* a very destructive larvae (*Trichoplusia ni*, family Phalaenidae) that eats leaves of carnations, nasturtiums, and most garden vegetables, especially cabbage and other crucifers.

Cabbage looper

cabbage maggot *n.* the destructive larvae of a two-winged fly (*Hylemya brassicae*, family Anthomyiidae). The larvae tunnel through and feed on the roots of cabbage, cauliflower, broccoli, Brussels sprouts, beets, cress, celery, radishes, and turnips, causing the plants to wilt. See Aleochara beetle.

cabbage worm *n.* the larva (caterpillar) of a yellowish-white butterfly (*Pieris rapae*, family Pieridae) with several black spots on the wings. The velvety green caterpillars are the most damaging insect pest on the cabbage family, as they feed on the leaves of cabbage, collards, cauliflower, broccoli, and related crops. The eggs are laid usually on the underside of the leaves. In warm weather, they hatch within a week, and the caterpillars take about 15 days to mature. There are several broods each year. They may be controlled by soil bacteria (Bt-endotoxin) or by a chemical pesticide.

cabrito *n.* (Spanish *cabra*, goat + *ito*, little) a small goat or the meat thereof. See chevon.

cacao *n.* (Nahuatl *cacahuatl*, cacao seeds) **1.** a small (up to 30 ft or 9 m), long-lived, tropical evergreen tree (*Theobroma cacao*, family Sterculiaceae) native to American lowlands from Mexico to Peru; now grown in many tropical areas, especially in Ghana, Nigeria, and Cameroon. **2.** fruit of the tree, borne in pods that each contain 20 to 40 seeds, and used as the source of cocoa and chocolate. Also known as cocoa bean. See cocoa.

cactus (pl. **cacti**) *n.* (Greek *kaktos*, prickly) any of a large number of flowering plants (family Cactaceae) with succulent green stems and sharp spines or scales instead of leaves. Cactus is highly drought tolerant and found mostly in warm arid climates. It is native to North and South America but is now found in other arid regions as well.

cadaster *n.* (Italian *catastro*, land registry) a register, survey, or map of property as a basis for taxation.

cadastral map *n.* a map showing the boundaries of subdivisions of land, usually with the bearings and distances thereof and the areas of individual tracts, for purposes of description and recording ownership.

cadelle *n.* (Latin *catellus*, puppy) a small, shiny black beetle (*Tenebroides mauritanicus*, family Ostomidae). Both the adults and the larvae eat grain, especially that stored in bins.

cadmium (Cd) *n.* (Greek *kadmeia*, earth) element 48, atomic weight 112.40; a bivalent metal whose salts are poisonous to people and animals. A heavy metal used in photography, ceramics, and insecticides; for plating other metals; and in alloys with copper, lead, silver, and aluminum. Cadmium in sewage sludge is a limiting factor when sludge is applied to soils used for the production of crops because cadmium accumulates in the environment and is hazardous to people, domestic animals, and shellfish. It can be toxic and has been shown to be carcinogenic in laboratory animals.

caducity *n.* (French *caducité*, frailty) **1.** senility. **2.** the quality or state of being perishable.

Caenozoic *n.* See Cenozoic.

caffeine *n.* (German *kaffein*, coffee) a white, odorless, bitter alkaloid ($C_2H_{10}N_4O_2$) present in coffee, tea, cocoa, kola nuts, and certain drugs. Caffeine stimulates the central nervous system and also has a diuretic effect (increases urination). Caffeine reaches peak blood concentration about an hour after ingestion and takes up to three hours for half of the stimulating effect to dissipate. See xanthine.

cairn *n.* (Gaelic *carn*, a mound of stones) a pile of stones used to mark trails, survey corners, property lines, or graves.

cajete *n.* (Spanish *cajete*, a bowl) a pit used to catch runoff water from a terrace and allow it to soak into the soil in certain parts of Mexico. Any eroded soil in the runoff water settles in the cajete and is later collected and spread on the adjoining field.

cal *n.* See calorie.

Calabar bean *n.* the poisonous seed of a climbing woody vine (*Physostigma venenosum*, family Faba-ceae) native to western Africa; the source of the drug physostigmine. Also called ordeal bean because it was used as a test for witchcraft.

calabash *n.* (Spanish *calabaza*, calabash) **1.** a tropical tree (*Crescentia cujete*, family Bignoniaceae) with showy tubular flowers, flat-winged seeds, and large gourds that is native to tropical America. **2.** a gourd from the tree, often used as a bowl, as a pipe, or to drink maté. See maté.

calamity *n.* (Latin *calamitat*, calamity) serious trouble; disaster, especially a sudden mishap of natural cause.

calcareous *adj.* (Latin *calcarius*, of lime) containing sufficient calcium carbonate and/or magnesium carbonate to exhibit visible effervescence (bubbles of carbon dioxide) when treated with acid such as 0.1 N HCl. Calcareous rocks form by biological deposition and/or inorganic precipitation in saline water and commonly contain snail shells and other fossils. Calcareous soil horizons are common in arid regions and sometimes occur at the soil surface around the borders of areas where water ponds or formerly ponded in subhumid regions.

calcic *adj.* (Latin *calx*, lime + *icus*, containing) containing calcium carbonate. A calcic horizon in Soil Taxonomy contains more than 15% $CaCO_3$ and at least 5% more $CaCO_3$ than the C horizon.

calcicole *n.* (Latin *calcx*, lime + *colis*, cabbage) a plant that grows well in soil high in calcium. Synonyms: calciphyte and calciphile. See calcifuge.

calciferol *n.* vitamin D, the sunshine vitamin, especially D_2 ($C_{28}H_{43}OH$). A crystalline alcohol found in milk, eggs, fish oil, etc. Also called ergocalciferol.

calciferous *adj.* (Latin *calcx*, lime + *ferous*, bearing) forming or containing calcite ($CaCO_3$), as calciferous sandstone or a calciferous gland.

calciferous gland *n.* a gland that secretes calcium carbonate ($CaCO_3$). Such glands are found near the esophagus of certain lower animals such as earthworms.

calcification *n.* (Latin *calcx*, lime + *ficatio*, making) **1.** the accumulation of calcium salts, especially $CaCO_3$, in joints and tissues, often resulting in stiff joints or stones such as gallstones. **2.** gradual petri-faction of the original parts of a fossil animal or plant by replacement of tissue with calcium carbonate. See petrifaction. **3.** a soil-forming process emphasized in soil classification systems that preceded Soil Taxonomy. Characteristic of arid or semiarid climates, it produces a layer enriched in calcium salts (especially $CaCO_3$) either in or immediately beneath the solum of most Aridisols and some Mollisols. See podzolization.

calcifuge *n.* (Latin *calcx*, lime + *fugere*, to flee) a plant that thrives in acid soil low in calcium but high in iron. See calcicole.

calcination *n.* (Latin *calcinare*, to heat + *ation*, action) the heating of a substance to its temperature of dissociation. Examples: the heating of gypsum ($CaSO_4 \cdot 2H_2O$) until it loses its water of hydration and becomes anhydrite ($CaSO_4$), or the heating of limestone ($CaCO_3$) to drive off carbon dioxide (CO_2) and produce quicklime (CaO). In modern kilns, it takes about 5.5 million British thermal units (Btu) to produce 1 ton of quicklime by heating limestone. See British thermal unit.

calcify *v.* (Latin *calcx*, lime + *ificare*, to make) **1.** to deposit or secrete calcium salts, as when gristle (cartilage) hardens and becomes bone. **2.** to solidify a soil horizon or a rock layer by the crystallization of calcium salts, especially $CaCO_3$.

calciphile *n.* (Latin *calcx*, lime + *philos*, loving) a calcicole. See calcicole, calcifuge.

calcite *n.* (Latin *calcx*, lime + *ite*, mineral) the most abundant form of calcium carbonate ($CaCO_3$), it has rhombohedral crystals with a density of 2.71 g/cc and a hardness of 3. Its pure form is a transparent or white mineral, but small amounts of impurities can give it any of several colors. Transparent crystals of calcite exhibit very strong double refraction. See aragonite, calcium carbonate.

calcium (Ca) *n.* (Latin *calcx*, lime + *ium*, element) element 20, atomic weight 40.078; the sixth most abundant element in the crust of the Earth. Calcium, which occurs naturally as a divalent cation in rocks, soils, and organic compounds, is the dominant cation on the exchange complex of most soils, and is an essential element for both plants and animals. The most abundant mineral element in the human body, it is concentrated in bones and teeth. See apatite, calcium carbonate.

calcium carbonate *n.* an abundant mineral ($CaCO_3$) that crystallizes as calcite or aragonite and forms oolite concretions. The principal component of limestone, marble, chalk, snail shells, oyster shells, etc. Ground calcium carbonate is used extensively to neutralize acidity in soil and water.

calcium chloride *n.* a very soluble salt (CaCl$_2$) that will deliquesce (absorb water from the air and go into solution). It is used as a drying agent, as a component of road salt to melt ice or reduce dustiness, and as a preservative added to tomatoes, apples, and other fruits and vegetables prior to canning or freezing.

calcium cyanide *n.* a toxic chemical [Ca(CN)$_2$] used as a fumigant to control insects and rodents.

calcium deficiency *n.* a cause of weak bones and lack of blood clotting in humans and animals, of soft-shelled bird eggs, and of stunted growth in plants. In humans and animals, it can result from a deficiency of vitamin D that prevents calcium utilization or from a shortage of calcium in the diet.

calcium hydroxide *n.* a strong base [Ca(OH)$_2$] that is formed when calcium oxide reacts with water; used in cement, mortar, and plaster. Sometimes called slaked lime or hydrated lime.

calcium nitrate *n.* a salt [Ca(NO$_3$)$_2$] that is sometimes used as a nitrogen fertilizer. The calcium component prevents it from causing soil acidity.

calcium oxide *n.* the chemical (CaO) that remains when calcium carbonate (CaCO$_3$) is heated enough to drive off carbon dioxide; used in cement, mortar, plaster, and several manufacturing processes. Called lime, burned or burnt lime, or quicklime. See calcium hydroxide.

calcium sorbate *n.* one of the salts of sorbic acid (C$_5$H$_7$COOH) that has broad antimicrobial activity against yeasts and molds and is used as a fungicide and as a preservative for a wide variety of foods—cheeses, pickles, beverages, and baked products. It can be synthesized or obtained from the berries of mountain ash (*Sorbus americana*, family Rosaceae).

calcium sulfate *n.* a neutral salt (CaSO$_4$) known as anhydrite in its dry form or as gypsum (CaSO$_4$·2H$_2$O) when hydrated. It is mined mostly as gypsum. Much of it is partially dried to form plaster that recrystallizes and hardens when allowed to absorb water again. It is used in plaster, plaster of Paris, gypsum wallboard, and several food products. It aids in the rapid maturing of flour, stimulates carbon dioxide production by yeast, and acts as a firming agent in certain canned vegetables. In agriculture, gypsum is used as a soil amendment to reclaim sodic soils by providing calcium (Ca^{2+}) to replace sodium (Na$^+$).

calculus (pl. **calculi**) *n.* (Latin *calculus*, small stone) **1.** concretionary material (mostly CaCO$_3$) that forms small "stones" inside the body (e.g., in the kidneys or bile ducts) of humans and animals when calcium concentrations are too high in stagnant body fluids. See bezoar, concretion. **2.** a similar hard deposit that forms on the inside of teeth. **3.** a system of higher mathematics based on the rate of change in the value of a mathematical expression.

caldera *n.* (Spanish *caldera*, cauldron) a large, steep-sided, cauldronlike depression with a diameter at least three times its depth. It is formed in the summit of a volcanic mountain by an explosion that scattered volcanic ash and/or by internal collapse of the mountain. See crater, volcano.

calf (pl. **calves**) *n.* (German *kalb*, calf) **1.** an offspring of a cow, of other bovines, or of certain other large mammals, including elephants and whales. **2.** the muscular section on the back side of a person's lower leg.

calf crop *n.* the productivity of a cow herd; usually expressed in percentage, as the number of calves weaned per hundred cows bred.

calf diphtheria *n.* an acute infectious disease of young calves caused by an anaerobic organism (*Spherophorus necrophorus*, formerly named *Actinomyces necrophorus*) that is characterized by ulcers in the mouth and throat, drooling, loss of appetite, and labored breathing. The mortality rate is high. Also called calf diphtheroid.

calf scours *n.* acute, infectious, often fatal diarrhea common in calves less than 10 days old; to be distinguished from other types of diarrhea. Symptoms include yellowish white droppings with a strong disagreeable odor, listlessness, gauntness, poor appetite, and much sleep. Also called white scours, 3-day calf scours, infectious diarrhea, or calf septicemia.

calibration *n.* (French *calibre*, size + *ation*, action) the process of adjusting a machine to give an accurate output under prevailing conditions; e.g., setting a laboratory instrument to read the concentration of a known standard or setting a field spreader to meter the desired rate of application of fertilizer, lime, or pesticide.

caliche *n.* (Spanish *caliche*, flake of lime) **1.** evaporite deposits of sodium nitrate (NaNO$_3$) in the Atacama Valley in Chile. Once a major source of nitrogen fertilizer. **2.** a cemented layer formed at the base of certain soils in semiarid and arid regions by crystallization of calcium carbonate and other salts. Sometimes called hardpan or duripan.

California holly *n.* See Christmasberry.

caliper *n.* (French *calibre*, size) **1.** a device with two legs and a width-adjusting mechanism for use to measure either thickness or outside diameter (an outside caliper) or an internal dimension (an inside caliper). **2.** the mechanism in a disk brake that applies pressure to the brake pads on both sides of the disk to provide braking action. **3.** the diameter of a tree trunk as measured 6 in. (15 cm) above the ground for trees up to 4 in. (10 cm) in diameter or 12 in. (30 cm) above the ground for larger trees.

callus *n.* (Latin *callum*, hard skin) **1.** the protective covering that forms over a cut or broken surface on a plant. **2.** a hardened area of thickened skin formed as protection from rubbing. **3.** a hardened exudate formed around the ends of a broken bone as part of the healing process. *v.* **4.** to form a callus.

calorie *n.* (Latin *calor*, heat) a unit of heat. The small calorie (cal), commonly used in physics, was originally defined as the amount of heat required to raise the temperature of 1 gram of water 1 degree Celsius, from 14.5° to 15.5°C at a pressure of 1 atmosphere; since redefined as exactly 4.1840 joules. The large calorie (kilocalorie or Cal), commonly used to evaluate the energy content of food, is equal to 1000 cal or 4184 joules. See British thermal unit, therm.

Calvin cycle *n.* the part of the photosynthetic process that does not require light (named after Melvin Calvin, 1911–1997, a U.S. chemist who won the 1961 Nobel Prize in chemistry for having explained this cycle). Carbon dioxide reacts with ribulose diphosphate (a 5-carbon sugar) to form a 6-carbon sugar that splits into two 3-carbon molecules that are transformed and phosphorylated by ATP and NADPH (from the light cycle) into triose sugars. These activated triose sugars then react with each other, and the product is transformed into glucose.

calving *n.* **1.** parturition; the birth of one or more calves. **2.** the process of icebergs breaking from an ice field and floating away.

calyx or **calix** *n.* (Greek *kalyx*, covering) **1.** the outer parts of a blossom. The sepals as a group, usually green (outside of the corolla). **2.** a cavity or cuplike part in the body of an animal. **3.** a cup or chalice.

camas or **camass** *n.* (Chinook *qamas*, camas) any of several plants (*Camassia* sp., especially *C. quamash*, family Liliaceae) having pale blue flowers. Their bulbous roots were used by Indians of western U.S.A. to make a type of flour. See death camas.

cambic horizon *n.* a diagnostic subsurface horizon in Soil Taxonomy. Cambic horizons have blocky and/or prismatic soil structures and/or brighter colors than the overlying horizons but lack the strong evidence of illuviation required for argillic or spodic horizons.

cambium *n.* (Latin *cambium*, an exchange) a layer of cells where growth occurs in dicotyledonous plants. In grafting, the cambium of the scion must contact the cambium of the stock so they can grow together. **Vascular cambium** produces layer upon layer of xylem cells (e.g., wood) on its inside surface and layers of phloem cells (e.g., bark) on its exterior surface. **Cork cambium** forms cork cells that constitute the outer portion of the bark of a tree.

camel *n.* (Greek *kamēlos*, camel) a large, humped, ruminant mammal (*Camelus dromedarius*, with one hump, or *C. bactrianus*, with two humps, family Camelidae) that can survive for long periods under adverse desert conditions. Fat stored in their humps and water in their bodies enable camels to survive for several days without eating or drinking, and their keen senses of smell and sight enable them to locate water from a long distance. They prefer to eat grass but can eat plants of saline areas and can survive almost indefinitely by browsing on shrubs. These qualities plus padded feet that enable them to walk on hot sand make camels excellent for transportation and draft power in desert areas; they are therefore sometimes called ships of the desert. However, they object to moving in mud or on wet surfaces. Approximately 71% of the world's 17 million camels are found in the parts of sub-Saharan Africa with less than 16 in. (400 mm) of annual rainfall. Most of the Bactrian (2-humped) camels are in the higher, cooler deserts of China and Mongolia. Camels provide milk and edible but low-quality meat. They have long intervals between parturitions and are late maturing.

camellone *n.* a raised bed, commonly 50 to 100 ft (15 to 30 m) wide, 500 to 1000 ft (150 to 300 m) long, and 6 to 10 ft (2 to 3 m) high, that is built of soil excavated from bordering zanjas and used in parts of Mexico to grow corn, beans, squash, alfalfa, etc. See zanja.

campestral or **campestrian** *adj.* (Latin *campester*, flat) having to do with open fields or plains; e.g., of the Great Plains of North America.

Canada goose *n.* the most common wild goose (*Branta canadensis*, family Anatidae), a large water-fowl marked with white cheeks on a black head and neck and a mostly brown body. Its honking is well known as it migrates in large flocks with a V formation between northern Canada and southern U.S.A.

Canada potato *n.* See Jerusalem artichoke.

Canada rice *n.* See wild rice.

Canada thistle *n.* a perennial thistle (*Cirsium arvense*, family Asteraceae) that grows 16 to 48 in. (40 to 120 cm) tall, has spines on multi-lobed leaves, and produces purple to white flowers. Native to Eurasia, it spreads by both seed and rhizomes and has become a troublesome weed throughout the U.S.A. and elsewhere.

Canada thistle

Canada yew *n.* See ground hemlock.

canal *n.* (Latin *canalis*, channel) **1.** an excavated waterway used for irrigation and/or navigation. **2.** a tubular body structure in a plant or animal, as an alimentary canal. *v.* **3.** to excavate a canal.

cancer *n.* (Latin *cancer* or *cancri*, crab, ulcer) **1.** a tumor; a malignant growth of abnormal cells in humans and animals. Cancer tends to spread and is often fatal, but some cancers are treatable by surgery, radiation, and/or chemotherapy. **2.** a disturbing development indicative of evil in a society; corruption. **3. plant cancer** See crown gall.

candela (cd) *n.* (Latin *candela*, candle) the international standard of light intensity adopted in 1979; a light of 540×10^{12} Hz shining through a conical opening of 1 steradian with an intensity of 1/683 W. Also called candle.

candle *n.* (Latin *candela*, candle) **1.** a mass of wax or fatty combustible substance, usually cylindrical, with an embedded wick protruding from the top. A flame produced by lighting the wick emits light and heat as it gradually melts, volatilizes, and burns the wax or fat. **2.** a former standard of light intensity equal to 1/60 of that produced by a black body of 1 cm^2 at the temperature of freezing platinum (2046°K). See candela. *v.* **3.** to examine an egg by holding it in front of a light (originally that of a candle) to determine freshness and whether the egg has been fertilized.

canid or **canine** *n.* (Latin *canis*, dog) an animal of the dog family (Canidae), including dogs, coyotes, foxes, jackals, and wolves.

canine madness *n.* See rabies.

canker *n.* (Latin *cancer*, gangrene) **1.** an ulcer, especially a raw sore in the mouth. **2.** a lesion in the bark and underlying xylem tissue in a woody plant. **3.** an inflammation of the horny tissue of a horse's hoof that can lead to a crippling lesion, rotting, and a fetid discharge from the frog and hoof. **4.** social decay; corruption. *v.* **5.** to damage by rot or rust.

cankerworm *n.* the fall cankerworm (*Alsophila pometaria*, family Geometridae) and the spring cankerworm (*Paleacrita vernata*) are both larvae of moths. The larvae do serious damage to the foliage of apple, elm, oak, hickory, and maple trees. Also called inchworm or looper.

cannabis *n.* (Greek *kannabis*, hemp) hemp (*Cannabis sativa*), an annual herb whose fibers are used for making rope, paper, and cloth. The oil from its seeds is used to make paint, varnish, and soap. It is illegal to cultivate cannabis in the U.S.A. because its leaves and other parts are used to make marijuana and hashish. See marijuana.

canner *n.* **1.** a person who cans food products for preservation. **2.** a kettle used for canning food. **3.** a marketing classification for poor quality beef. **4.** in western U.S.A., a wild horse slaughtered for meat. **5.** berries or other fruits to be canned because they are too soft for shipment.

canners' alkali *n.* a mixture of sodium carbonate and sodium hydroxide used to loosen the skin of fruit such as peaches, making them easier to peel.

cannibalism *n.* (Spanish *canibal*, cannibal + Latin *ismus*, practice) **1.** the act of a human or animal eating the flesh of another of the same species; e.g., some chickens peck on other chickens and some pigs chew on the tails of other pigs (especially in overcrowded conditions). Usually, the victims are smaller, weaker individuals. In severe cases, the victim may be killed. **2.** the removal of parts from a machine (usually a disabled machine) for use on another machine.

canopy *n.* (Latin *conopeum*, mosquito net) **1.** a covering mounted above a person or object to provide protection or dignity. **2.** the upper layer of vegetation in a wooded area. The crowns of trees, shrubs, or other large plants that shade other vegetation.

cantaloupe or **cantaloup** *n.* (Italian *Cantalupo*, a papal summer estate) a small round or oval melon (*Cucumis melo cantalupensis*, family Cucurbitaceae) with a tan exterior and sweet-tasting orange flesh. Also called muskmelon.

cantharidin *n.* (Latin *cantharides*, beetles + *in*, chemical) the main active ingredient in Spanish fly; the lactone of cantharidic acid ($C_{10}H_{12}O_4$); a bitter-tasting chemical that blisters the skin.

cantharis (pl. **cantharides**) *n.* (Greek *kantharis*, beetle) a bright green blister beetle (*Lytta vesicatoria*) of southern Europe. When dried, it is known as Spanish fly and has been used in powdered form as an aphrodisiac or as a blistering plaster. The main active ingredient is cantharidin.

canyon *n.* (Spanish *cañon*, a long cane or hollow) a deep valley with steeply sloping walls, typically having rapids in the stream and cliffs in the walls, especially at the rim of the valley.

caper euphorbia *n.* a poisonous annual herb (*Euphorbia lathyrus*, family Euphorbiaceae) that is native to Europe and is sometimes grown in gardens. Its milky sap is a skin irritant that causes blisters and inflammation. Animals that eat the plant (usually as a component of dried hay) may suffer scours and swellings on their heads. See euphorbia.

capillarid worms *n.* nematode parasites (*Capillaria annulata* or *C. contorta*) that infest the crop and esophagus or the small intestines (*C. columbae*) of chickens, ducks, turkeys, pigeons, and some wild birds. Symptoms include weakness, emaciation, ruffled feathers, diarrhea, soiled vent, and a tendency to sit on the ground. Also called crop worms, hair worms, or threadworms.

capillarity *n.* (Latin *capillaris*, hairlike + *itas*, condition) the tendency of a liquid to be pulled up into a small tube or passageway if it wets the surface it contacts or for its level to be depressed if it does not wet the surface of the passageway. See meniscus.

capillary *n.* (Latin *capillaris*, hairlike) **1.** long and slender, resembling a hair. **2.** a microscopic blood vessel that carries blood between an artery and a vein. Capillaries have porous walls that allow chemicals (e.g., oxygen and carbon dioxide) to be exchanged between the blood and the surrounding tissue. **3.** a small pore in soil. *adj.* **4.** showing properties of capillaries, as capillary flow or capillary rise.

capillary action *n.* liquid movement that depends on the small size of capillaries, especially capillary flow and capillary rise.

capillary capacity *n.* the amount of water held in a soil by capillary attraction after gravitational water has drained away. It is usually expressed as a percentage of the oven-dry weight of the soil or as a percentage of the soil volume. Also known as field capacity.

capillary flow *n.* flow through very small passageways. Such flow is generally slow but powerful and almost equal in all directions if the media is uniform because the wetting forces that drive it are much stronger than gravitational forces. Also called capillary action.

capillary fringe *n.* the zone of soil saturated with water by capillary rise above the true water table.

capillary potential *n.* See soil water potential.

capillary rise *n.* the vertical distance that water rises in a soil by capillarity above the level it would have as a free water surface.

capillary water *n.* the water held in micropores and on particle surfaces strongly enough to resist gravitational movement (about ⅓ atm or greater) but not too strongly to be lost by evaporation (less than about 30 atm of tension). In soil, the capillary water held less tightly than the wilting point (about 15 atm of soil moisture tension) is considered to be available for plant growth.

capon *n.* (Latin *capo*, castrated cock) a rooster or other male bird that has been neutered either surgically or chemically so it will fatten faster for market. Caponized birds lose male characteristics

such as the bright red color and large size of their comb, crowing, and the mating instinct.

caponette *n.* (Latin *capo*, castrated cock + *itus*, little) a male chicken whose reproductive organs have been deactivated by the injection of an estrogenic hormone (stilbestrol). The testes of these animals shrink, the secretion of testosterone is inhibited, and there is a regression of the secondary sex characteristics (comb, wattles, earlobes, mating instinct, and crowing).

caponize *v.* (Latin *capon*, castrated cock + *izare*, to make) to neuter a rooster or other male bird, either surgically or chemically.

capuchin *n.* (French *capucin*, a monk who wears a cowl) **1.** a monkey (*Cebus capucinus*) about 1 ft (30 cm) tall that is native to Central and South America. It has cowl-like hair on its head, and its long tail can curl around and grip branches. Also called ringtail monkey.

capybara or **capibara** *n.* (Portuguese *capibara*, one who eats grass) the largest living rodent (*Hydrochaeris hydrochaeris*), about 2 ft (0.6 m) tall and 3 to 4 ft (0.9 to 1.2 m) long, with no tail. Native to the Amazon floodplain area, it feeds on herbs and floating grasses and produces succulent meat.

caragana *n.* (Latin *caragana*, Siberian pea shrub) a genus of deciduous trees and shrubs (*Caragana* sp., family Fabaceae) with compound leaves and pealike yellow flowers, native to central Asia, where it is used as browse for livestock. Siberian peashrub (*C. aborescens*) is planted in shelterbelts, for wildlife, and as an ornamental in the northern Great Plains of the U.S.A.

carapace *n.* (Spanish *carapacho*, carapace) the external skeleton that covers the upper side of certain animals such as turtles, crabs, and armadillos (the part of the shell that covers the underside of a turtle is called a plastron).

caraway *n.* (Arabic *al karawiyā*, caraway) a biennial herb (*Carum carvi*) with finely divided leaves and white or pinkish flowers. Its aromatic seedlike fruit is used in cooking (especially on bread or in cakes or salads) and in medicines for controlling stomach and intestinal gases.

carbamate insecticide *n.* a salt or ester of carbamic acid (H_2NCOOH) used as an insecticide. Carbamates are also toxic to humans, causing weakness, dizziness, and sweating. More severe exposure causes headache, salivation, nausea, vomiting, abdominal pain, diarrhea, and slurred speech.

carbohydrate *n.* (Latin *carbonis*, coal + Greek *hydōr*, water + Latin *atus*, combine with) a sugar, starch, cellulose, or gum with the general formula $(HCOH)_n$, originally interpreted as $C_n + nH_2O$ (n is a whole number between 3 and 7 for the basic units). Carbohydrates are produced by photosynthesis and comprise about 75% of the composition of plants. They represent much of the diet for animals and people. The recognized classes of carbohydrates are monosaccharides (4-, 5-, and 6-carbon sugars), disaccharides (pairs of 6-carbon sugars), trisaccharides, and polysaccharides (cellulose, starch, and pentosans).

carbon (C) *n.* (Latin *carbonis*, coal) element 6, atomic weight 12.011; a nonmetallic element whose atoms have the unique ability to form four covalent bonds, making it central to organic chemistry. Carbon is an essential element in all living things and is also an important element in many inorganic compounds such as calcium carbonate ($CaCO_3$) and carbon dioxide. It is a constituent of soil humus, coal, charcoal, limestone, and most shales. Diamonds and graphite are forms of carbon.

carbonaceous *adj.* (Latin *carbonis*, coal + *aceus*, having the nature of) composed largely of carbon (e.g., coal) or having to do with carbon. Carbonaceous materials include organic tissues from dead plants and animals and carbon-containing decomposition products therefrom.

carbonate *n.* (Latin *carbonatum*, carbonated) a chemical containing carbonate (CO_3^{2-}) ions, especially $CaCO_3$. Carbonates in soil can be detected by the effervescence (bubbles of carbon dioxide) produced when acid is applied (10% HCl is commonly used for this purpose).

carbonation *n.* (Latin *carbonis*, coal + *atio*, action) **1.** the process of carbonating a beverage by dissolving pressurized carbon dioxide (CO_2) in the water. **2.** a chemical reaction involving carbon dioxide or carbonate ions; commonly a weathering process that breaks down mineral structures, making the elements in soil minerals more available to plants and/or contributing to the process of soil formation.

carbon bisulfide or **carbon disulfide** *n.* a clear, highly flammable liquid (CS_2) used as a fumigant to kill insects in stored grains, as an extractant to degrease products such as cottonseed meal, and as an input in the production of cellophane and rayon. It is also toxic to mice, other rodents, and humans.

carbon cycle *n.* the cyclic transformation of carbon absorbed as carbon dioxide (CO_2) from the atmosphere for photosynthesis by plants and built into numerous organic compounds, possibly consumed by an animal and made part of its body, and ultimately returned to the atmosphere by respiration, decomposition, or combustion. There is also a mineral carbon cycle that converts carbon into carbonate

rocks that may eventually weather and release carbonate ions.

carbon dating *n.* the use of radioactive carbon (^{14}C) with an approximate half-life of 5730 years for determining the approximate age of buried materials such as wood, bones, or even soil—anything that contains organic carbon. The proportion of ^{14}C in atmospheric carbon dioxide is nearly constant because ^{14}C is produced continuously by cosmic radiation. Fresh plant and animal tissues contain this same proportion of ^{14}C, but radioactive decay gradually reduces the proportion of ^{14}C in dead tissue in accordance with the half-life principle. Thus, tissue that is 5730 years old is half as radioactive as fresh tissue, 11,460 years makes it one-fourth as radioactive, etc. The instruments used for measuring radioactivity can detect such differences in materials as old as 40,000 to 50,000 years. Isolation of the material being tested from more recent material is very important, as a small admixture can add enough radioactivity to indicate a much more recent date.

carbon dioxide (CO_2) *n.* a colorless, odorless gas that all animals and humans exhale and that is released by respiring plants or plant parts (e.g., roots). Normal air contains 0.0325% carbon dioxide by volume or 0.05% by weight. Atmospheric carbon dioxide is the source of carbon for photosynthetic plants to make sugars and other organic compounds. (See photosynthesis.) Carbon dioxide is frozen to make dry ice and is used in soft drinks and fire extinguishers. It is also a greenhouse gas suspected of being a major contributor to global warming. See global warming.

carbon fixation *n.* that part of the photosynthetic process wherein carbon from atmospheric carbon dioxide is made into simple organic compounds.

carbonic acid *n.* the partially ionized acid (H_2CO_3) formed when carbon dioxide dissolves in water. Rainwater normally has a pH of about 5.6 because it contains carbonic acid formed from atmospheric carbon dioxide. See acid precipitation.

carbon monoxide (CO) *n.* a colorless, odorless, highly flammable gas produced by incomplete oxidation of fuels that contain carbon. Unpolluted air contains only about 0.05 ppm carbon monoxide, but pollution (75% to 80% from internal combustion engines) builds it up to 5 to 10 ppm or higher in many urban areas. This problem has been declining gradually since automobile emission control standards were established in the 1970s. Carbon monoxide binds to the hemoglobin in blood 210 times as strongly as does oxygen and thereby reduces the ability of the blood to transport oxygen. Atmospheric carbon monoxide levels above 10 ppm cause headaches and sleepiness, and much higher levels caused by unvented combustion in closed areas can cause death. Soil bacteria are the principal agents that remove carbon monoxide from the atmosphere. They obtain energy by oxidizing carbon monoxide to carbon dioxide.

carbon to nitrogen (C:N) ratio *n.* the weight of carbon divided by that of nitrogen present in an organic material or a material such as soil that contains organic matter. Fresh organic materials with wide C:N ratios (e.g., sawdust at about 400:1 or wheat straw at about 80:1) decompose more slowly than those with narrower ratios (e.g., alfalfa hay at about 13:1) because nitrogen is often a limiting factor for microbial activity. However, conversion of organic carbon to carbon dioxide gradually reduces the ratio while concentrating the most resistant materials, so a narrow ratio in older materials generally indicates slow decomposition (e.g., soil humus with a C:N ratio near 10:1).

carbon partitioning *n.* the process of allocating photosynthate to the different parts of the plant.

carbon sink *n.* a site of long-term storage of large amounts of carbon. The biosphere, soil organic matter, peat and coal deposits, and carbonate rocks are all carbon sinks.

carbonyl group *n.* an oxygen double bonded to a carbon, either at the end of a chain as an aldehyde or inside a chain as a ketone.

carboxydotrophic bacteria *n.* any of several bacteria (e.g., *Pseudomonas carboxydovorans*, *Bacillus schlegelii*, and *Alcaligenes carboxydus*) able to oxidize carbon monoxide as a source of energy for growth. Such bacteria living in the soil are believed to be the most important natural means for reducing the carbon monoxide content of the atmosphere.

carboxyhemoglobin (COHb) *n.* a complex of carbon dioxide and hemoglobin. Human hemoglobin bonds to carbon dioxide about 210 times as strongly as it bonds to oxygen, so even a small amount of carbon dioxide can drastically reduce the ability of the blood to transport oxygen. See oxyhemoglobin.

carcass *n.* (French *carcasse*, a skeleton or body). **1.** the dead body of an animal, especially one slaughtered for meat. **2.** (slang) the body of a person, living or dead. **3.** the shell or skeleton of an incomplete machine or building, either in the process of construction or after having been disassembled.

carcinogen *n.* (Greek *karkinos*, cancer + *genēs* producer) any substance that can cause or contribute to the development of cancer, especially a substance resulting from human activity. See mutagen.

carcinoma *n.* (Greek *karkinōma*, cancer) a malignant epithelial tumor; a cancerous growth on either the skin or an internal body surface.

cardamom *n.* (Greek *kardamon* cress + *amōmon*, spice plant) **1.** a tall herb (*Elettaria cardamomum*, family Zingiberaceae) with large leaves and white flowers marked with blue and yellow. It is native to the East Indies but grown in southern Florida and anywhere in the tropics. **2.** its aromatic seed, used in medicine and as a condiment.

cardinal *n.* (Latin *cardinalis*, principal) **1.** a crested songbird (*Cardinalis cardinalis*, family Emberizidae). The male is bright red, whereas the female and young are brown marked with red and yellow. **2.** a high ecclesiastical official in the Roman Catholic Church. **3.** a bright red. *adj.* **4.** of prime importance.

cardinal point *n.* any of the four principal compass directions, north, east, south, and west, or astrologically, the direction of sunrise, sunset, zenith, and nadir. See agonic line, azimuth.

carding *n.* the process of separating wool fibers from each other and removing foreign matter in preparation for spinning wool into yarn.

caribe *n.* (Spanish *caribe*, cannibal) piranha.

caribou *n.* (Algonquian *khalibu*, pawer) any of several large reindeer (*Rangifer* sp.) native to the northern part of North America.

Carlsbad Caverns *n.* one of the world's largest systems of interconnected caves. Located in New Mexico, these dry, dead caves are presumed to have been formed millions of years ago by acidic water dissolving limestone. A chamber near the entrance measures 300 × 1500 ft and 300 ft high (90 × 450 × 90 m) and is claimed to be the largest room in any cave in the world.

carniolan bee *n.* a variety of honeybee that originated in Carniola, Austria. The bees are popular in the U.S.A. for their honey production and gentle behavior.

carnivore *n.* (Latin *carno*, flesh + *vorare*, to devour) **1.** a mammal that customarily eats flesh; e.g., bears, cats, dogs, and seals. **2.** any animal that eats flesh. **3.** a plant that traps and consumes insects. See herbivore, omnivore, carnivorous fungi, carnivorous plants.

carnivorous fungi *n.* any of about 150 species of fungi known to entrap nematodes by means of sticky knobs, fixed rings, constricting rings, or netlike appendages that hold a nematode until fungal hyphae invade its body and digest it. Other fungi prey on amoebae, bacteria, copepods, etc.

carnivorous plants *n.* insectivorous plants. As many as 450 species of plants have been reported to entrap and digest insects and some other members of the animal kingdom. Among the well-known higher plants are Venus's flytrap (*Dionaea muscipula*), pitcher plant (*Darlingtonia californica, Nepenthes* sp., and *Sarracenia* sp.), and sundew (*Drosera* sp. and *Drosophyllum* sp.).

carnotite *n.* a radioactive ore of uranium and vanadium that is an earthy yellow mineral [$K_2(UO_2)_2(VO_4)_2 \cdot 3H_2O$] named after A. Carnot (?–1920), a French mine inspector.

carob *n.* (Arabic *kharrub*, bean pod) **1.** an evergreen tree (*Ceratonia siliqua*, family Fabaceae) native to the eastern Mediterranean area. It bears long, flat, curving pods that contain a sweet pulp. **2.** the pods of the tree, used as fodder or to make candy or a nutritious drink. Sometimes called algaroba, carob bean, or St. John's bread.

carotene or **carotin** *n.* (Latin *carot*, carrot + *ene*, source) any of three yellow to red fat-soluble hydrocarbon ($C_{40}H_{56}$) pigments found in such foods as carrots, sweet potatoes, milk fat, and egg yolk and in green leaves. Carotene is called the precursor of vitamin A or provitamin A because animal livers can convert it into vitamin A. Carotene in green leaves is masked by the green of chlorophyll but is revealed in yellow, orange, red, and brown fall colors after cold weather destroys the green chlorophyll. See vitamin, xanthophyll.

carp *n.* (Latin *carpa*, carp) **1.** any of several bottom-dwelling freshwater fish (*Cyprinus* sp., family Cyprinidae) with large scales. Edible but bony, carp thrive in ponds and other quiet waters, commonly stirring up mud and making the water murky. **2.** any of several related types of fish; e.g., goldfish. *v.* **3.** to complain, censure, or find fault, especially habitually or in a nagging manner.

carpel *n.* (Greek *karp*, fruit + Latin *ellum*, little) an individual set of female fruit parts, either a separate pistil or one set in a compound group.

carpenter ant *n.* any of several black or brown ants (*Camponotus* sp., family Formicidae) that are pests in buildings, utility poles, posts, tree cavities, and other wooden objects. Carpenter ants seek soft wood (especially wood that has weathered and begun to decay) to make cavities where they rear their young. The ants remove wood and eject it in fibrous shreds while constructing these chambers, but they do not eat it. They feed on honeydew obtained from aphids and on animal remains and plant juices. The chambers of carpenter ants are clean and are cut across the grain of the wood. Carpenter ants are distributed over most of the U.S.A.

Carpenter ants

carpenter bee *n.* any of several black bees (large, *Xylocopa* sp.; or small, *Ceratina* sp.; family Andrenidae) that make nests by tunneling in dry wood. They resemble bumblebees and pollinate many species of plants.

carpenter moth *n.* a moth (either *Prionoxystus robiniae* or *P. macmurtrei*, family Cossidae) whose larvae are known as carpenter worms because they burrow beneath the bark of trees.

carpenter worm *n.* a larva of a carpenter moth. The larvae burrow beneath the bark of black oak, red oak, elm, locust, poplar, willow, maple, ash, and chestnut throughout central North America. They make small holes that are marked by sawdust and dark sap that discolors tree trunks.

carpet beetle *n.* any of several species of beetles (family Dermestidae) that are serious household pests because their larvae feed on clothing, bedding, carpets, etc., made of wool and other natural materials; especially the black carpet beetle (*Attagenus megatoma*) and the common carpet beetle (*Athrenus scrophulariae*).

carpoptosis *n.* abnormal drop of fruit from trees.

carrageen or **carragheen** *n.* a red marine algae (*Chondrus crispus*) that grows in tidal pools along rocky seacoasts and is the source of carrageenan; named after Carrageen in southeastern Ireland. When dried, it is called Irish moss.

carrier *n.* **1.** any person, animal, or plant that has a disease organism in its system that it may transmit to other plants, persons, or animals without itself suffering from the disease. See vector. **2.** any substance added to a chemical compound to dilute an active ingredient to facilitate its storage, shipment, or uniform distribution in the field. **3.** a material such as water used to carry a pesticide to the target. **4.** a heterozygous individual that shows the characteristics of its dominant allele but has an unexpressed recessive gene. **5.** a radio wave whose amplitude (am) or frequency (fm) is modulated to carry a signal.

carrier pigeon *n.* a homing pigeon; a pigeon trained to carry messages.

carrion *n.* (Latin *caronia*, carcass) decaying flesh of dead animals; the common food of buzzards, hawks, vultures, hyenas, opossums, and several other bird and animal species.

carrion beetle *n.* any of several beetles (family Silphidae) that feed on carrion and deposit their eggs in it.

carrion crow *n.* a European crow (*Corvus corone*, family Corvidae) that eats carrion.

carrion flower *n.* **1.** any of several climbing plants (*Smilax* sp., family Liliaceae) with small white flowers that smell like carrion. **2.** any of several south African succulent plants (*Stapelia* sp.) with variegated blossoms and a fetid odor.

carrying capacity *n.* the maximum number of a given type of animals that a specific ecosystem can support without causing environmental damage.

Carson, Rachel Louise *n.* a distinguished American marine biologist (1907–1964) and writer who authored *Silent Spring*, an early environmental book published in 1962 that publicized the problem of DDT accumulating in the environment and in the food chain.

cart *n.* (Irish *cairt*, a little car) **1.** a two-wheeled vehicle used to convey goods and/or people. Large carts usually are pulled by one or more animals (donkeys, mules, horses, oxen, etc.); small carts may be pulled or pushed by a person. **2.** the load carried on such a vehicle. *v.* **3.** to transport something in such a vehicle.

cartography *n.* (Latin *charta*, paper + *graphia*, writing) the art of producing maps and charts that graphically represent both natural and anthropogenic features of a surface such as that of the Earth; e.g., topography, geology, soils, roads and railroads, and buildings.

Carver, George Washington *n.* famous chemist, inventor, and agriculturalist (1864–1943), especially well known for his work with peanuts. The first black graduate of Iowa State University.

caryopsis *n.* (Greek *karyon*, nut + *opsis*, appearance) a small fruit with the pericarp adhering to the seed coat (typical of most grasses and grains, including corn and wheat).

cascade *n.* (Italian *cascare*, to fall) **1.** a series of small waterfalls produced by water flowing down a steep, rocky channel. **2.** a showering of sparks or some kind of material in layers resembling a cascade waterfall. **3.** a series of events that follow one another in rapid succession.

cascara *n.* (Spanish *cáscara*, husk or shell) a small tree (*Rhamnus purshiana*, family Rhamnaceae) native to the west coast of the U.S.A.; the source of cascara sagrada.

cascara sagrada *n.* **1.** a laxative extracted from the bark of the cascara tree. **2.** the bark itself.

cascarilla *n.* (Spanish *cáscarilla*, little husk or shell) **1.** a shrub (*Croton eluteria*, family Euphorbiaceae) from the West Indies. **2.** the aromatic bark from the plant; used as a tonic or to flavor tobacco.

cashew *n.* (Portuguese *acajú*, cashew) **1.** a medium to small evergreen tree (*Anacardium occidentale*, family Anacardiaceae) native to Brazil but extensively cultivated in India and Africa. Its flowers are pale pink with a pleasant odor. **2.** the kidney-shaped nuts produced by the tree at the end of an edible, pear-shaped, fleshy peduncle. The fresh nut is covered by three shells; the middle shell contains a black oil that can blister the hands. Wet lime is used on the hands to prevent blistering. The dry cashew nuts of commerce contain about 46% fat, 30% carbohydrates, 17% protein, 5% water, and 2% other components.

cassava *n.* (Spanish *cazabe*, cassava bread) also known as **manioc 1.** a woody shrub (*Manihot esculenta*, family Euphorbiaceae) that originated in the lowlands of South America and is now widely grown in humid tropical lowlands of Africa and India. **2.** the tuberous starchy roots of the shrub that are used to make tapioca. Some varieties contain large amounts of poisonous hydrocyanic acid (HCN), but this is destroyed by boiling. Cassava has a reputation of being a survival food in the humid tropics because of its ability to grow on soils of very low fertility, to remain in the soil without spoiling for months after it becomes edible, to continue being productive after some of the fleshy roots are harvested with a hand hoe, and to reproduce readily from stem cuttings. See tapioca.

cassowary *n.* any of several large flightless birds (*Casuarius* sp.) with a bony rudderlike growth on the top of their heads; native to Australia, New Guinea, and other nearby islands.

CAST *n.* See Council for Agricultural Science and Technology.

caste *n.* (Latin *castus*, pure) **1.** a hereditary social level; e.g., the Brahman, Kshatriya, Valsya, and Shudra castes of Hindu society. **2.** a rank or level in any system of rigid social divisions. **3.** any of the three types of bees (workers, drones, and queen) that comprise the adult population of a honeybee colony.

casting *n.* **1.** an object produced by filling a mold with molten material. **2.** the process of choosing actors for a theatrical production. **3.** the action of throwing a fishing line or other object. **4.** a fecal pellet (usually plural); earthworm castings are an excellent fertilizer. See scatology, frass.

castor bean *n.* **1.** a tall, annual herb (*Ricinus communis*, family Euphorbiaceae) native to tropical Asia and Africa that is grown for its striking foliage and the oil produced from its seeds. **2.** the large, beautiful, poisonous seeds of the plant, whose ingestion causes nausea, gastric pain, diarrhea, thirst, and dullness of vision. Large amounts can be lethal.

castor oil *n.* vegetable oil produced from the seeds of castor bean (*Ricinus communis*, family Euphorbiaceae); used as a cathartic, as a lubricant, and as a dressing for leather.

castrate *v.* (Latin *castrare*, to geld) **1.** to emasculate a male animal or human by removing the testicles; to geld. See spay. **2.** to eliminate the significance or effect of a document or object by removing a vital part. **3.** to extract the stamens from a flower so it will not produce pollen.

catabatic warming *n.* the warming that occurs when a large mass of air descends to a lower elevation after having been pushed across a mountain range. The resulting compression warms the air and lowers its relative humidity, thus causing a rain shadow.

catabolism *n.* (Greek *katabolē*, throwing down + *ismus*, action) metabolic processes that release energy by breaking down complex organic compounds into simpler ones (in opposition to anabolism). Also spelled katabolism.

cataclysm *n.* (Greek *kataklysmos*, flood) **1.** a sudden, violent upheaval that changes the appearance of a large area; e.g., a large flood or extensive volcanic activity with lava flows. **2.** a violent social or political upheaval.

catadromous *adj.* (Greek *kata*, down + *dramein*, to run) freshwater fish that swim down to salt water to spawn. Opposite of anadromous.

catalpa *n.* (Greek *katálpa*, head wing) any of several trees (*Catalpa* sp., family Bignoniaceae) native to central U.S.A. and characterized by long seed pods resembling those produced by beans and clusters of white blossoms. Also called Indian bean.

catalytic converter *n.* an air pollution abatement device used in motor vehicle exhaust systems since about 1975. The exhaust passes through a container filled with pellets coated with platinum and palladium to catalyze the oxidation of unburned hydrocarbons and other pollutants into carbon dioxide and water and to reduce nitrogen to its gaseous form.

catamount *n.* (Latin *cattus*, cat + *montis*, hill or mountain) any wild member of the cat family (Felidae), especially a cougar, lynx, or puma.

cataract *n.* (Greek *katarraktēs*, waterfall) **1.** a clouding of the lens of the eye that obstructs vision; can occur in the eyes of people and most animals. **2.** a large waterfall.

catastrophe *n.* (Greek *katastrophe*, an overturning) an exceptionally damaging event of rare occurrence (often with widespread effect); e.g., a cataclysm.

catastrophism *n.* (Latin *catastropha*, catastrophe + *isma*, action) the theory that the Earth's landscapes are molded mostly by sudden events such as earthquakes rather than by slow processes such as erosion (proposed by William Whewell, 1794–1866, a British philosopher and scientist).

catch basin *n.* a depression constructed to remove sediment and debris from water before it enters a tile drainage system or a sewer.

catch crop *n.* **1.** a fast-growing crop grown between two regular crops. **2.** a substitute crop planted after the regular crop has failed or cannot be planted because it is too late or conditions are unusual. See buckwheat.

catchment *n.* a watershed; the area draining into a stream, lake, or sewer. See drainage basin, watershed.

cat clay *n.* strongly acid soil resulting from aeration of materials containing sulfide minerals. The sulfides are oxidized to sulfuric acid when swampy areas are drained or submerged horizons are excavated.

category *n.* (Greek *katēgoria*, accusation or affirmation) **1.** a class or group in a system of classification. **2.** a level of generalization in a system of classification. Each category is composed of a set of classes that includes all of the individuals or items being classified.

catena *n.* (Latin *catena*, a chain) **1.** a series of related ideas or events, especially with a theological topic. **2.** A sequence of soils formed in similar parent material but in different topographic positions with differences in soil drainage. Also called a toposequence.

caterpillar *n.* (French *catepelose*, a hairy cat) the wormlike larva of a butterfly or moth. The larva has a head with strong cutting jaws followed by a body with about a dozen segments and several pairs of legs. Some are covered with hair, others not. They feed voraciously on the succulent parts of plants for several days, weeks, or months (depending on the species) until they metamorphose into the pupal stage and finally into adult butterflies or moths.

Caterpillar *n.* a trademark for a brand of crawler tractor; often used to describe other brands as well. See crawler tractor.

catface *n.* **1.** a blemish on the trunk of a tree representing a fire scar or other healing or healed wound. **2.** abnormal fruit growth of tomato caused by unfavorable weather or other growth disturbances. **3.** scars from insect stings or other pests causing irregular growth on an apple or peach.

catfish *n.* (Anglo-Saxon *cat*, cat + *fisc*, fish) any of more than 1000 species in about 30 families of scaleless fish (especially those of family Ictaluridae). All have long barbels (whiskers) around the mouth and sharp spines on the head.

cat flea *n.* a small biting insect (*Ctenocephalides felis*, family Pulicidae) that infests cats, dogs, and people.

catheter *n.* (Greek *kathetēr*, a tube) an elongated tubelike instrument to be inserted into body passages for the insertion or removal of fluids, e.g., to remove urine from the bladder, to insert semen into the cervix in artificial insemination, or to treat diseases of the middle ear. Catheters may be made of rubber, plastic, metal, or glass.

cathode *n.* (Greek *kathodos*, a way down) an electrode with a negative charge; one that attracts cations and repels anions. Opposite of anode.

cation *n.* (Greek *kation*, going down) an ion with a positive electrical charge; one that is electrically attracted to a cathode. Cations in soil tend to have low mobility because clay and humus usually have negative charges that attract and hold them. Examples: hydrogen (H^+), calcium (Ca^{2+}), and ammonium (NH_4^+). See anion.

cation exchange *n.* (formerly called **base exchange**) the replacement of cations adsorbed on soil colloids (silicate clays and humus) by cations from the soil solution carrying the same total electrical charge. Cation exchange is considered one of the most important phenomena of soils because it provides a means of storing plant nutrients such as Ca^{2*}, Mg^{2+}, K^+, NH_4^+, and most of the micronutrients and also provides a mechanism that buffers soil pH. The first recorded practical example of cation exchange occurred in 1852 when Thomas Way of England used a fine-textured soil to absorb ammonium ions (NH_4^+) from decomposing animal manure, thus reducing losses of ammonia (NH_3) to the atmosphere. See isomorphic substitution.

cation exchange capacity (CEC) *n.* the quantitative evaluation of cation exchange in soil or other media (e.g., artificial growth media, plant roots, or other materials); expressed in the S.I. system in centimoles of charge per kilogram (cmol kg^{-1}) and formerly expressed in milliequivalents per 100 grams (meq/100 g). The numerical value is the same with either the new or the old units.

catkin *n.* (Dutch *katteken*, little cat) a branch or stem with scaly unisexual flowers with no petals (e.g., those produced by willows or birches such as pussy willow), so named because the flowers resemble the tip of a cat's tail or the toe of a cat. Also called ament.

catnip *n.* (Latin *cattus*, cat + *nepeta*, an herb) a perennial mint (*Nepeta cataria*) with egg-shaped leaves that contain aromatic oils that is used in making sauces and teas and is a special treat for cats.

catsteps *n.* narrow, irregular terraces on steep hillsides formed by animal trails that follow the approximate contour of the land and/or by slippage of saturated soil.

cattail *n.* **1.** a tall reed (*Typha latifolia*, family Typhaceae) that is very common in marshy areas. Cattails grow about 5 to 6 ft (1.5 to 1.8 m) tall, have long, stiff, swordlike leaves, and produce innumerable tiny petalless flowers in prominent cylindrical brown spikes about 1 in. (2.5 cm) in diameter and 6 to 10 in. (15 to 25 cm) long. **2.** any of the *Typha* sp. reeds that grow in marshes. Also called bulrush.

cattalo *n.* the offspring of a domestic cow crossed with a bison (buffalo). Also called beefalo.

cattle *n.* (Latin *capitale*, property) **1.** bovine animals, especially mature cows, bulls, and steers. **2.** a collective term for all of the 4-legged animals kept for meat and milk production on a farm or ranch. **3.** a derogatory term for a crowd of people, especially when misled by an agitator.

cattle egret *n.* a white heron (*Bubulcus ibis*, family Aredeidae) with a wingspan of about 3 ft (90 cm). Native to Europe, Asia, and Africa, and first reported in Florida in 1952, these egrets are believed to have flown to South America and then to Florida. They have been officially reported since then in all of the 48 conterminous states of the U.S.A. Cattle egrets have yellow or orange bills and yellow to pink legs; their necks are shorter and thicker than other herons. They invariably move and feed in and around cattle, eating insects that prey on the cattle. They roost in flocks as large as 25,000.

cattle gap *n.* See cattle guard.

cattle grub *n.* the larvae of heel flies (*Hypoderma lineatum* and *H. bovis*, family Hypodermatidae, the common and northern U.S. species, respectively); a parasite that has been a serious pest to cattle from time immemorial. Cattle grubs develop from fly eggs deposited in the spring, usually on hair of the lower legs, especially just above the hoof. The larvae hatch after a few days and burrow their way through the skin into body tissues and move through the animal body for several months. Ultimately, they arrive at the back of the animal and produce cysts. After several weeks, they drop to the ground where they pupate and produce adult heel flies. See heel fly, warble.

cattle guard *n.* an opening in a fence with a slatted area that substitutes for a gate (or sometimes, stripes painted on a road to simulate slats), permitting vehicles to pass but not cattle or sheep. Also called an autopass or cattlegap.

cattle pass *n.* a passageway beneath a road to permit cattle to cross from one side to the other without obstructing traffic.

cattle plague *n.* See rinderpest.

caudal *adj.* (Latin *cauda*, tail) **1.** resembling a tail. **2.** located near the rear or tail end of an organism, as a caudal fin of a fish.

caustic *adj.* (Latin *causticus*, burning) **1.** capable of destroying human flesh by burning, searing, or dissolving it; e.g., lye (caustic potash or caustic soda). **2.** sharply critical and usually sarcastic, as a caustic response.

caustic potash *n.* potassium hydroxide (KOH), a very strong base used to make soap.

caustic soda *n.* sodium hydroxide (NaOH), a very strong base.

cauterize *v.* (Greek *kautēr*, branding iron + *izein*, action) to burn living tissue with heat, electricity, or chemicals to seal a wound or stop growth. See dehorn.

cave *n.* (Latin *cava*, hollow) **1.** an underground hollow place such as an opening in a hillside that offers at least partial shelter. **2.** a cavern. *v.* **3.** to make a hollow. **4.** (usually cave in) to collapse, as heavy pressure might cause a box to cave in or a person's resolve to cave in.

cavern *n.* (Latin *caverna*, a large cave) a large cave, especially one with a small opening but with an extensive system of underground rooms and connecting passages formed by flowing water dissolving calcium carbonate from limestone. Many caverns have beautiful formations of stalactites, stalagmites, columns, and flowstone.

caviar *n.* (Persian *khāviyār*, egg) salted fish eggs (roe), usually from sturgeon or salmon.

CCC *n.* **1.** See Civilian Conservation Corps. **2.** See Commodity Credit Corporation.

cd *n.* candela.

ceanothus *n.* (Greek *keanōthos*, a kind of thistle) any of several low-growing, hardy shrubs (*Ceanothus* sp., family Rhamnaceae) with blue or white flowers; native to North America and used as ornamental plants. The leaves have been used to make tea. Also called New Jersey tea or redroot.

cecropia moth *n.* a large, colorful silkworm moth (*Hyalophora cecropia*, family Saturniidae) native to eastern U.S.A., with a wingspread of 7 in. (18 cm). Each wing has a crescent-shaped spot edged in red. Its larvae feed on the foliage of deciduous trees.

cedar *n.* (Latin *cedrus*, cedar) **1.** a large evergreen tree (*Cedrus* sp., especially *C. libania*, the famous cedar of Lebanon) that has as many as 30 needles in a cluster and produces a highly prized aromatic, low-density, straight-grained wood. **2.** a tree that produces low-density aromatic wood similar to cedar; e.g., eastern red cedar (*Juniperus virginiana*, family Pinaceae), Barbados cedar (*Cedrela odorata*), or Australian red cedar (*Cedrela australis*). **3.** the wood of any of the cedar trees. See eastern red cedar.

cedar-apple rust *n.* a fungal (*Gymnosporangium juniperi-virginianae*) disease that alternately affects several species of the genus *Juniperus* (i.e., eastern and western red cedars, and horizontal and savin junipers) and a number of apple and crabapple varieties. Damage to cedars is usually not severe but may appear quite striking when the orange, spore-bearing tendrils emerge from the cedar galls in the spring. Apple trees are more seriously affected, with the leaves yellowing and falling prematurely. Fruit may also be infected. It is usually recommended that any cedar host within 2 miles of a commercial apple orchard be removed entirely. Carefully timed sprays are an alternative control method.

cedar-hawthorn rust *n.* a fungal (*Gymnosporangium globosum*, family Pucciniaceae) disease that alternately infests junipers and hawthorns, and rarely apples. See hawthorn rust.

cedar waxwing *n.* an attractive, crested, brownish gray American bird (*Bombycilla cedrorum*, family Bombycillidae) about 6 in. (15 cm) long with a black mask, a yellow fringe on its tail, and a yellowish belly; named for small red waxlike spots on the tips of secondary wing feathers of adult birds.

celestial body *n.* an object visible in the heavens; e.g., a star or planet.

cell *n.* (Latin *cella*, room) **1.** a small compartment or room, as a prison cell. **2.** one of the hexagonal spaces in a honeycomb. **3.** the simplest collection of matter that can live, composed of a nucleus surrounded by cell fluid and usually bounded by a cell wall. Microbes are composed of one or more cells; plants, animals, and humans are made up of thousands of cells that function in harmony. **4.** an electric device including a chamber, electrodes, and an electrolyte. **5.** a group organized for a social objective; e.g., to distribute information or proselytize.

cellar *n.* (Latin *cellarium*, pantry) **1.** a storage room for provisions such as wine or home-canned goods, usually underground beneath a building. **2.** an underground room for protection against tornadoes, hurricanes, etc.; e.g., a bomb shelter. **3.** in sports, the position of the team with the poorest record.

cellulase *n.* (Latin *cellula*, small cell + *ase*, enzyme) an enzyme secreted by certain bacteria and fungi capable of hydrolyzing glucose molecules from cellulose. See cellulose.

cellulose *n*. (Latin *cellula*, small cell + *osus*, full of) an inert, complex, transparent carbohydrate $(C_6H_{10}O_5)_n$ that makes up the bulk of the cell walls of plants and is composed of long chains of glucose molecules joined through β-1,4 linkages. About 2000 to 6000 glucose molecules polymerize into a glucan chain, and commonly 30 to 40 glucan chains linked by hydrogen bonds form a cellulose fiber. Cellulose is used for making building materials, paper, fabrics, and explosives. Ruminant livestock digest cellulose through the action of microorganisms in the rumen.

cell wall *n*. a somewhat rigid external structure that surrounds the plasma membrane of most cells. Bacterial cell walls are composed largely of peptidoglycan; plant cell walls are composed mostly of cellulose and other complex carbohydrates.

Celsius scale or **centigrade scale** *n*. a temperature scale used in the metric system whereby, at sea level, 0°C = freezing and 100°C = boiling. Named after Anders Celsius, 1701–1744, and adopted in the U.S.A. in 1948. For conversion to Fahrenheit, °F = 1.8 × °C + 32. See Fahrenheit scale, Kelvin.

cement *n*. (Latin *cementum*, stone fragments from a quarry) **1.** an adhesive used for fastening things together or mending broken objects. **2.** the binding material in a rock or groundmass. **3.** portland cement; the binding agent used to make concrete (produced from lime and clay fired in a kiln and finely ground). **4.** any unifying or binding agent. *v.* **5.** to bond or unite.

Cenozoic or **Caenozoic** *n*. or *adj*. (Greek *koinos*, new + *zoic*, of animals) the geologic era characterized by the development and ascendancy of mammals, from about 65 million years ago to the present.

Center for International Forestry Research (CIFOR) *n*. an international agricultural research center founded in 1992 with headquarters in Indonesia. Its research focuses on forest conservation and sustainable development.

center of origin *n*. an idea proposed by N. I. Vavilov (?–1943), a Russian plant geneticist, that most of the world's important crops were domesticated in a few centers of origin where the largest diversity still occurs (since modified because some species have developed wide diversity in areas where they have been introduced).

center pivot irrigation *n*. sprinkler irrigation using a line of sprinklers supported by towers with wheels. The line rotates around a central point where the water is supplied by a well or an underground pipeline. A typical line is a quarter-mile (400 m) long and irrigates a 125-ac (50 ha) circle in a half-mile (0.8 km) square (160 ac or 64 ha). Some systems irrigate the corner areas by means of a big gun on the end of the line or a trailing segment that swings out into the corners.

centigrade scale *n*. See Celsius scale.

centimeter-gram-second system *n*. the system of weights and measures commonly known as the metric system. See metric system, Système Internationale d'Unites.

centipede *n*. (Latin *centipeda*, a hundred feet) the common name for arthropods (class Chilopoda). Centipedes are elongated, flattened, worm-like animals with 15 or more body segments, each with a pair of legs. They are fast-moving predators that prey on many species of insects and spiders. They can use their poisonous front pair of claws to bite an animal or person.

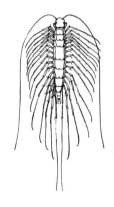

Centipede

century plant *n*. a desert agave (*Agave americana*) of tropical and subtropical America that requires 10 years or longer to mature. It produces a rosette of long, fleshy, pointed leaves at its base and a flower stalk that grows 20 to 30 ft (6 to 9 m) tall. Also known as maguey.

cepaceous *adj*. having the odor or flavor of onion or garlic; alliaceous; e.g., the breath of a person or the flavor of milk from a cow that has eaten onions or garlic.

cephalopod *n*. (Latin *cephalo*, head + Greek *podos*, with a foot) a mollusk (class Cephalopoda) with tentacles extending outward from its head, including the octopus, squid, and cuttlefish.

ceramic *n*. (Greek *keramikos*, made of potter's clay) **1.** an item made of fired clay, e.g., pottery or tile. *adj.* **2.** pertaining to such items and/or their production.

ceramics *n*. **1.** pottery and other ceramic products. **2.** the art and technology used to produce such items.

cereal *n*. (named after the Roman goddess Ceres) **1.** a grass cultivated for its edible seeds. Major cereals are corn, wheat, rice, barley, rye, sorghum, and oats. Minor cereals are millet, foxtail millet, proso millet, and teff. **2.** a food product made largely from grain, often eaten for breakfast.

Ceres *n*. (Latin *Ceres*, a Roman goddess) **1.** the Roman goddess of food grains. **2.** the largest known asteroid and the first one discovered (in 1801 by the Italian astronomer Giuseppe Piazzi), Ceres is about

578 miles (930 km) in diameter and orbits the sun in the asteroid belt between Mars and Jupiter.

cerium (Ce) *n.* (Latin *Ceres*, a Roman goddess + *ium*, element) element 58, atomic weight 140.12; a gray, metallic element that is the most abundant of the rare-earth group. Cerium is used to make arc lamps, colored glass, and ceramics and in photography and textiles.

cerrero *adj.* (Spanish *cerrero*, free and untamed) free and untamed, as certain wild horses in southwestern U.S.A.

certified milk *n.* nonpasteurized milk produced according to the regulations of an authorized medical milk commission.

certified pesticide applicator *n.* a person certified to be qualified to apply a restricted-use pesticide.

certified seed *n.* seed approved by a certifying agency (e.g., a Crop Improvement Association) as meeting established standards of germination, freedom from diseases and weed seeds, and varietal purity. Certified seed may be the progeny of foundation, registered, or certified seed that meets purity and quality standards.

cesium (Cs) *n.* (Latin *caesius*, bluish gray) **1.** element 55, atomic weight 132.91; one of the alkali metals that form ions with +1 charge. Cesium is a silvery-white metal that is liquid at room temperature. It is used in electronics and in atomic clocks. **2. cesium 137** an isotope of cesium having a mass number of 137. An important fission product that is a constituent of radioactive fallout, it has a half-life of 30 years and is used in cancer research and radiation therapy.

cesspit *n.* (Italian *cess*, privy + Latin *puteus*, pit) a compost pit where refuse, garbage, and night soil are dumped. See cesspool.

cesspool *n.* (Italian *cess*, privy + Dutch *poel*, deep place) **1.** a large, buried, porous tank or a pit lined with stones to receive household sewage or other liquid waste. The liquid leaches into the surrounding soil. Some of the solid material decomposes, and the rest accumulates as sludge that must be removed periodically. See lagoon, dry well, cesspit. **2.** a filthy place that smells like a cesspool. **3.** sometimes used figuratively to describe dishonest politics or other unsavory social activity.

cetacean *n.* or *adj.* (Greek *kētos*, whale + *acea*, animal) an aquatic mammal (order Cetacea), including whales and dolphins.

cetology *n.* (Greek *kētos*, whale + *logos*, word) the zoological study of whales and dolphins.

cevitamic acid *n.* ascorbic acid ($C_6H_8O_6$); also known as vitamin C.

CFC *n.* See chlorofluorocarbon.

cfs *n.* abbreviation for cubic feet per second, often used for measuring water flow, especially for irrigation; 1 cfs = 7.48 gal/sec, 28.32 liters/sec, 0.99 ac-in./hr, or 1.02 ha-cm/hr. In western U.S.A., 1 cfs is commonly used to irrigate 80 ac (32.4 ha) of cropland; it provides approximately 2 in. (5 cm) of water per week on 80 ac.

CGIAR *n.* See Consultative Group on International Agricultural Research.

chaff *n.* (German *kaff*, chaff) **1.** fine refuse from threshing grain or grass seed; mostly small husks from the kernels and broken pieces of straw. **2.** dry bracts from the flowers of certain plants. **3.** strips of metal foil or other material used to confuse radar signals. **4.** material of little or no value; something to be separated and discarded.

Chagas' disease *n.* a form of trypanosomiasis that exhibits an acute course in children but is chronic in adults and is often fatal. Named after Carlos Chagas (1879–1934), a Brazilian physician, it is widely distributed in Latin America. Chagas' disease is caused by a pathogen (*Trypanosoma cruzi*) that is transmitted by several insects (*Panstrongylus megistus*, *P. infestans*, and others, family Reduviidae). Reservoir hosts include armadillos, bats, cats, foxes, and guinea pigs. Chagas' disease is also known as American, South American, Brazilian, or Cruz trypanosomiasis; thyroiditis parasitaria; or careotrypanosis. See trypanosomiasis, nagana, tsetse fly.

chain *n.* (Latin *catena*, fetter) **1.** a series of rings, each linked to the rings on each side to form a flexible unit used for pulling, supporting, or tying objects or animals or for decoration; usually made of metal but also of paper, plastic, or other material. **2.** a series of things that are connected, as a chain of events. **3.** a series of similar natural features, as a chain of islands, lakes, or mountains. **4.** a measure of length equal to 66 ft (20.1 m) used in land surveys. See Gunter's chain. **5.** a measuring tape, usually 100 ft (30.5 m) long. *v.* **6.** to tie or bind with a chain. **7.** to enslave.

chain reaction *n.* **1.** a self-sustaining nuclear fission process. If a critical mass of fissionable material is present, the particles and energy produced by splitting a nucleus is enough to cause at least one other nucleus to split, so the reaction continues one nucleus after another. Atomic power plants to produce electricity use a controlled chain reaction whereby each nucleus that splits causes an average of exactly one other nucleus to split. Atomic bombs explode by reactions in which a nucleus splitting causes more than one other nucleus to split. **2.** a series of events wherein each event causes the next.

chalcedony *n.* (Greek *chalkēdōn*, after a town in Asia Minor) a translucent, microcrystalline form of quartz, often milky in color or grayish, including agate, jasper, chert, flint, opal, onyx, and petrified wood. Some forms are used as gemstones or carved into cameos and other ornaments. Also spelled calcedony.

chalcid fly *n.* any of several small insects (family Chalcididae) whose larvae are parasites on the eggs, larvae, or pupae of many harmful insects and thereby serve to control insect pests.

chalk *n.* (Latin *calx*, lime) **1.** a soft, saltwater, calcium carbonate ($CaCO_3$) deposit composed mostly of foraminifera shells; e.g., the deposit that forms the White Cliffs of Dover, England. **2.** a marker made from chalk, naturally white but often artificially colored. *v.* **3.** to rub with chalk. **4.** to apply chalk to a field to raise the soil pH.

chameleon *n.* (Greek *chamaileōn*, on the ground) **1.** a lizard capable of changing its skin color to match its background. The true chameleon (family Chamaeleontidae) is a native of Africa and India. **2.** the American chameleon (*Anolis carolinensis*), a member of the iguana family **3.** a changeable or fickle person who changes positions when challenged.

chamois *n.* (Latin *camox*, chamois) **1.** an agile European antelope (*Rupicapra rupicapra*) native to high mountain areas. **2.** a soft, pliable leather made from the skin of a chamois or of similar nature from some other source; often used to absorb water to dry a freshly washed car or other object.

champaign *n.* (French *champagne*, open field) a grassy plain; a flat expanse of open countryside.

chance seedling *n.* a plant produced from seed scattered by wind, animals, or birds. Some varieties of fruit (e.g., the Delicious apple) originated as chance seedlings. See sport.

changa *n.* a mole cricket (*Scapteriscus vicinus*, family Gryllidae) that eats the roots of garden crops, peanuts, and tobacco and is native to South America and the West Indies. Also called Puerto Rican mole cricket or mole cricket.

channel *n.* (Latin *canalis*, water pipe) **1.** the deep portion of a river or stream where the main volume or current of water flows. **2.** the part of a body of water deep enough to be used for navigation. **3.** a large strait, as the English Channel. **4.** the swale above a terrace ridge where water collects and is either held or conducted across the slope. **5.** the way something moves (often figuratively), as a trade channel. **6.** a way of access. **7.** a frequency band used for electronic communication. **8.** a long groove or flute. *v.* **9.** to cut a channel in something. **10.** to cause something to move by a certain route.

channelization *n.* the deepening (and often straightening) of a stream channel, thus enabling water to flow faster, reducing flooding, and draining nearby land. It may increase erosion, disturb fish breeding, and damage the stream's natural beauty.

chaparral *n.* (Spanish *chaparro*, evergreen oak) a dense (often thorny) thicket of shrubs or grove of small trees that is sometimes described as a fire climax community because it comes in after a fire and will be replaced if there are no more occasional fires.

chaps or **chaparajos** *n.* (Spanish *chaparajos*, chaps) leather leg coverings worn by cowboys over ordinary trousers from the waist to the feet (but with no seat) to protect their legs from cacti, thorns, and brush.

char *v.* **1.** to blacken by burning; to convert the surface or entirety of an object to charcoal. **2.** to become charred. *n.* **3.** the carbonaceous residue produced by incomplete combustion (pyrolysis) of organic material. Coal char is called coke; wood or bone char is called charcoal. Char will burn with little or no flame and is often used for cooking. Purified char is used to produce activated carbon for use as a filtering medium.

charco *n.* (Spanish *charco*, shallow pool) **1.** shallow pools of water along the channels of rivers where the floodwaters spread over adobe flats. Some charcos hold water from a few days to several weeks. **2.** a water-storage structure made of two parts, a desilting basin and a holding basin.

charcoal *n.* **1.** black carbonaceous material prepared by burning wood or bones slowly in an environment deficient of oxygen. Charcoal burns slowly with little or no flame. Most of it is used as a fuel, especially for cooking on outdoor grills. **2. activated charcoal** purified charcoal used to absorb odors and toxins from air or water. **3. animal charcoal** charred animal bones used as an absorbent and decolorizer.

charcoal rot *n.* **1.** a plant disease caused by a fungus (*Macrophomina phaseoli*, family Sphaeropsidaceae) that causes rotting of the roots and lower part of the stalks of over 30 different species of plants (e.g., soybeans). **2.** decay caused by a fungus (*Diplodia natalensis*, family Sphaeropsidaceae) that affects the mature fruit of limes (*Citrus aurantifolia*). The fruit develops brown spots, dries, and becomes filled with a hard mass of black mycelia. **3.** a warm-weather seedling blight that affects sweet potatoes, sorghum, peanuts, and a number of other plants and is caused by an imperfect fungus (*Sclerotium bataticola*) that forms brown lesions on the lower stems and upper parts of the root system.

chard *n.* (French *carde*, thistle) a garden beet (*Beta vulgaris*, var. cicla) grown for its large leaves and leafstalks that are used as a green vegetable. Also called Swiss chard.

charlock *n.* (Anglo-Saxon *cerlic*, charlock) a wild mustard (*Brassica kaber*, family Cruciferae), native to Eurasia, that bears yellow flowers and is a pernicious weed in grain fields. Animals that consume large amounts of its seed may suffer apathy, colic, diarrhea, enteritis, nephritis, and abortion. Also called carlick, corn mustard, wild mustard, or California rape.

chasm *n.* (Greek *chasma*, large gap) **1.** a deep cleft or canyon; a gorge. **2.** a deep recess in the floor of a cave. **3.** a large difference of opinion and sentiment.

chasmophyte *n.* (Greek *chasma*, large gap + *phyton*, plant) a plant growing with its roots in a crevice in a rock or on the surface of a rock.

chattel *n.* (Latin *capitale*, property) **1.** a piece of movable property; e.g., an appliance, an automobile, or furniture. **2.** a slave.

chatter *n.* **1.** rapid talk, often with little meaning. **2.** rapid speechlike sounds uttered by an animal or bird. **3.** a rapid clicking sound, as one's teeth may chatter in the cold. **4.** a noisy vibration of a sliding or rotating object. *v.* **5.** to make chattering sounds or speech.

chaussee *n.* (French *chaussée*, roadbed) **1.** a levee along a river or canal. **2.** a levee with a road on top of it in Louisiana.

cheatgrass *n.* a weedy annual grass (*Bromus tectorum*, family Poaceae) native to Europe that often becomes dominant on overgrazed or frequently burned rangeland in the Great Plains and inter-mountain areas of western U.S.A. Also called downy brome or downy chess.

check dam *n.* a small dam placed in a gully or other watercourse to slow the water flow and minimize erosion until the area is stabilized by perennial vegetation. A check dam may be constructed of brush, logs, stones, or other suitable material.

check plot *n.* a plot in field research that receives conventional treatment or no treatment to serve as a standard of comparison for the experimental plots.

chelate *adj.* (Greek *chēlē*, claw) **1.** having two opposable claws, as certain insects. *n.* **2.** an organic chemical that can enclose a divalent or polyvalent cation inside its structure. Natural chelates include chlorophyll, hemoglobin, and certain components of animal manures and soil organic matter. Some of the best-known synthetic chelating agents are ethylenediaminetetraacetic acid (EDTA), hydroxyethylenediaminetriacetic acid (HEDTA), and diethylenetriaminepentaacetic acid (DTPA). Chelates in the soil help keep micronutrient cations (iron, copper, man-ganese, and zinc) available to plants by slowing or preventing their conversion to insoluble and unavail-able oxides and hydroxides. See EDTA, sequester.

chemical *n.* (Greek *chemia*, alchemy + Latin *icalis*, concerned with) a substance with a specific composition, especially one that enters into a chemical reaction or is a product of the reaction of another chemical or chemicals. Every type of matter (solid, liquid, or gas) is composed of one or more chemicals.

chemical caponization *n.* causing a rooster or other male fowl to develop feminine characteristics by injecting or implanting a hormone (diethylstilbestrol = stilbestrol) under the skin of the neck. This makes the rooster a caponette by causing it to gain weight and lose its fighting instinct, and its comb, wattles, shanks, and skin become paler.

chemical control *n.* pest management methods that use chemical pesticides.

chemical dehorning *n.* arresting the growth of horns on young calves by applying chemicals to the horn buttons.

chemical energy *n.* The energy of chemical bonding between atoms in molecules. Some of it can be released as heat by chemical reactions such as combustion.

chemical fallow *n.* the use of herbicides to prevent vegetative growth on fallow land. See fallow.

chemical fertilizer *n.* an inorganic chemical that supplies one or more plant nutrients in a form that is or will become available for plant growth; e.g., diammonium phosphate.

chemical oxygen demand (COD) *n.* See biological oxygen demand.

chemigation *n.* the addition of fertilizers and/or pesticides to irrigation water to fertilize crops and/or control pests. It is often used with drip irrigation and sometimes with sprinkler irrigation. See fertigation.

chemiluminescence or **chemoluminescence** *n.* creation of visible light by chemical reaction without a source of heat. See firefly.

chemistry *n.* (Greek *chemia*, alchemy) science of the properties of matter; its composition, behavior, and changes.

chemoautotrophic *adj.* (Greek *chemo*, chemical + *auto*, self + *trophikos*, nourishing) characteristic of bacteria that produce organic matter by chemosynthesis.

chemocline *n.* (Greek *chemo*, chemical + *klinein*, to lean) the boundary in a lake at the base of the water that mixes periodically (the mixolimnion), separating it from the underlying water that does not mix and is therefore generally anaerobic (the monimolimnion). See meromictic.

chemolithotroph *n.* (Greek *chemo*, chemical + *lithos*, stone + *trophe*, nourishment) an autotrophic organism that uses carbon from carbon dioxide to synthesize organic compounds and obtains its energy by the oxidation of inorganic compounds; e.g., bacteria (*Thiobacillus ferrooxidans*) that oxidize ferrous iron (Fe^{2+}) to ferric iron (Fe^{3+}).

chemoprophylaxis *n.* (Greek *chemo*, chemical + *prophylassein*, to prevent) the use of chemicals (drugs, food products, etc.) for the prevention of certain diseases; e.g., the use of vitamin C to prevent scurvy.

chemosphere *n.* (Greek *chemo*, chemical + *sphaîra*, ball) that part of the atmosphere with maximum photochemical activity; a zone from about 12 to 50 or more miles (20 to 80 km) above the Earth's surface, where the ozone content is relatively high; mostly in the stratosphere. See ozone.

chemosterilant *n.* (Greek *chemo*, chemical + Latin *sterilis*, unfruitful) a chemical used to sterilize males that is used on domestic animals or for the control of certain insect pests.

chemosynthesis *n.* (Greek *chemo*, chemical + *synthesis*, making) production of organic compounds with carbon dioxide as the only carbon source by chemoautotrophic organisms that use energy obtained from the oxidation of certain inorganic substances; e.g., reduced forms of iron, nitrogen, or sulfur.

chemotaxis *n.* (Greek *chemo*, chemical + *taxis*, arrangement) movement of an organism in response to a chemical stimulus, either toward an attractant or away from a repellent. See klinotaxis.

chemotropism *n.* (Greek *chemo*, chemical + *tropos*, a turning) a change in the direction of plant root growth in response to chemicals such as fertilizers.

Chemotropism (fertilizer-increased root growth)

chemurgy *n.* (Greek *chemo*, chemical + *urgia*, work) the application of chemistry to the development of industrial (nonfood) uses for agricultural products; e.g., ethanol produced from corn for use as a fuel.

chenier *n.* (French *chêne*, oak + *ier*,concerned with) a long, narrow beach ridge or hummock about 5 or 6 ft (1.5 to 2 m) above the surrounding level in the coastal marshes of Louisiana. These ridges had oak trees as their dominant vegetation before the early settlers used them for home sites and farmlands.

Chernobyl *n.* a city in the Ukraine (in the former U.S.S.R.) where the world's worst nuclear accident occurred on April 26, 1986. One of the reactors suffered a core meltdown and a fire that produced a massive radioactive cloud. Many were killed, thousands were affected by radiation, a dense radioactive cloud crossed northern Europe, and the radioactive plume was detected for thousands of miles.

Cherokee *n.* **1.** a native American of an Iroquoian tribe called by the same name. **2.** the language of the Cherokees. *adj.* **3.** the tribe that once inhabited a region in what is now eastern Tennessee, the western part of the Carolinas, and parts of other nearby states, with surviving groups now located in Oklahoma and North Carolina.

Cherokee tickclover *n.* See beggarweed.

Cherokee rose *n.* a climbing rose (*Rosa laevigata*) with large, fragrant, white blossoms; native of China and now grown extensively in southern U.S.A.; the state flower of Georgia.

cherry *n.* (French *cerise*, cherry) **1.** any of several moderate-sized deciduous trees (*Prunus* sp., family Rosaceae) with pink and white blossoms that resemble roses, including several wild American species and domesticated species that came from Asia by way of Europe; grown in orchards and as ornamental fruit trees. **2.** the small, round (except for a dimple around the stem), yellow, red, or purple fruit from such trees, usually less than 1 in. (1 or 2 cm) in diameter, including sweet (*P. avium*), sour (*P. cerasus*), wild black (*P. serotina*), and other varieties, with a single hard pit in their centers. **3.** the hard reddish wood obtained from such trees that is used to make furniture. **4.** a bright red typical of the red fruit. *adj.* **5.** made from cherries or cherry wood, as a cherry pie or cherry furniture. **6.** having a bright red cherry color.

chess brome *n.* **1.** a weed (*Bromus secalinus*, family Poaceae) often found in grain fields that is especially troublesome in wheat because its seed mixes with the threshed wheat and lowers the grade. **2.** darnel ryegrass (*Lolium temulentum*).

chestnut *n.* (Latin *castanea*, chestnut) **1.** a tall, stately, deciduous tree (*Castanea* sp., family Fagaceae), including the American chestnut (*C. dentata*), the

European chestnut (*C. sativa*), the Japanese chestnut (*C. crenata*), and the Chinese chestnut (*C. mollissima*). The American chestnut was once one of the dominant species in eastern U.S.A. until the chestnut blight killed almost all of the trees. Asian varieties showing resistance to the blight are used as a partial replacement. **2.** the rounded, smooth-shelled nuts produced by chestnut trees. **3.** the wood from chestnut trees. **4.** a reddish brown. **5.** a reddish brown horse.

chestnut blight *n.* a disease of chestnut trees caused by a fungus (*Endotheca parasitica*, family Diaporthaceae) that attacks the bark and cambium of the trunk and branches of the American and European species. The cankers girdle the stem, eventually killing the tree.

Chestnut leaf and nut

chevon *n.* (French *chèvre*, goat) meat of a young goat. Also called cabrito.

chianophile *n.* a plant that tolerates a long winter with snow cover or one that requires snow cover during winter.

chickadee *n.* any of several species of small, friendly songbirds (*Parus* sp., family Paridae) with black caps and bibs and white cheeks.

chicken *n.* (Dutch *kicken*, chicken) **1.** any of several breeds of domestic fowl (*Gallus domesticus*, family Phasianidae) grown for their eggs, meat, and feathers, especially one less than a year old. See hen, rooster. **2.** the meat from a chicken. **3.** (slang) a cowardly person.

chicken hawk *n.* any hawk of a species that is reputed to prey on chickens, especially Cooper's hawk (*Accipiter cooperii*).

chicken mite *n.* a parasitic insect (*Dermanyssus gallinae*, family Dermanyssidae) that sucks the blood of chickens and other domestic and wild birds. It usually engorges itself with blood at night and hides in cracks around the roosts, nests, and walls of the poultry house in the daytime. It causes restlessness, anemia, weight loss, and reduced egg production, and has caused the death of setting hens. Also called bird mite.

chickpea *n.* **1.** a bushy annual pea (*Cicer arietinum*, family Fabaceae) that grows about 2 ft (60 cm) tall with pinnate leaves and small white or pink flowers. Also called garbanzo. **2.** the edible, round, yellowish brown peas that grow one or two in each pod of the plant and are widely used in soups, salads, sauces,

etc., in India, the Middle East, Southern Europe, Africa, and Latin America. Also called garbanzo bean. See falafel, hummus.

chickweed *n.* a weed (*Stellaria media*, family Caryophylaceae) native to Eurasia and found in fields, lawns, and gardens throughout the U.S.A. It has small leaves and white star-shaped flowers. Birds eat the leaves and seed. Also called birdweed, chickenwort, satin flower, starweed, starwort, tongue grass, or winterweed.

chicory *n.* (Greek *kikhora*, succory) a perennial herb (*Cichorium intybus*, family Asteraceae) with toothed leaves and bright blue flowers. Its leaves are used as a salad green (commonly blanched), and its roots are roasted and ground for use as a coffee additive. Imported from Europe, it is a weed along roadsides in the U.S.A. Also called succory, blue dandelion, blue sailor, or coffee weed.

chigger *n.* a tiny, red, bloodsucking, 6-legged larva of a mite (*Eutrombicula alfreddugesi*, family Trombiculidae); a parasite on people, domestic animals, and some birds and wild animals. Chiger bites cause inflamed spots and intense itching. Also called chigoe, red bug, or harvest mite.

chigoe *n.* a small, reddish brown sand flea (*Tunga penetrans*) of tropical America and Africa; also found in southern U.S.A. Impregnated female fleas cause painful sores by embedding themselves under the skin, most commonly on the feet. Also called chigger or jigger.

chilblain *n.* (Greek *cheilos*, lip + Anglo-Saxon *blegen*, a sore) itching and painful inflammation of fingers, toes, ears, or other exposed parts caused by exposure to cold and dampness.

Chile dodder *n.* a parasitic plant (*Cuscuta suaveolens*, family Convolvulaceae) on clover and alfalfa. Native to South America, it is common in imported red clover and alfalfa seed. See dodder.

chill hours *n.* the number of hours of temperatures below 45°F (7°C) required to induce dormancy for a particular species of fruit or nut tree. See vernalization.

chilling effect *n.* **1.** a lowering of the temperature of the Earth caused by volcanic eruptions increasing the amount of dust in the air. **2.** a condition or action that markedly reduces the enthusiasm of a person or group for a particular activity; e.g., bad news had a chilling effect on the project.

chimera *n.* (Greek *chimaira*, she goat) **1.** a mythological animal with a lion's head, a goat's body, and a serpent's tail. **2.** any grotesque imaginary creature; e.g., those commonly used during the Gothic period to adorn the walls of buildings such as the Cathedral of Notre Dame. **3.** in genetics, an individual plant or animal with two or more genetically different types of cells as a result of grafting, mutation, or radiation. Examples: an orange tree with tame branches grafted onto a sturdy wild rootstock, an apple tree that produces two different varieties, or peaches with fuzz on half of their skin.

chimpanzee *n.* (Bantu *kumpenzi*, chimpanzee) any of a genus of African great apes (*Pan* sp., especially *P. troglodytes*, family Pongidae) that are considered to be the closest living relatives to humans (98% gene similarity). A tropical forest to savanna dweller that is smaller than a gorilla or human and is covered with black hair. See bonobo, primate.

chinampa *n.* a raised bed 8 to 33 ft (2.5 to 10 m) wide and up to 330 ft (100 m) long built of lake sediments and organic materials. It is built like a small island or a peninsula extending from a lake shore, often on a floating platform, especially around the shores of Lake Texcoco in Mexico. The Aztecs built about 9000 ha of chinampas to help feed the people of Mexico City and other areas. Now less than 1000 ha remains, used mostly for growing vegetables and flowers.

chinch bug *n.* a tiny insect pest (*Blissus leucopterus*, family Lygaeidae) about 4 mm long that causes serious crop losses in the central part of the U.S.A., especially to small grains, corn, sorghum, and other grass species. The young bugs crawl from one

Chinch bug

plant to the next, so one method of control has been to use barriers such as a plow furrow treated with creosote accompanied by postholes about 2 ft (60 cm) deep. The chinch bugs fall into the postholes as they try to find a way around the line of creosote. They may also be controlled by direct spraying or dusting or by growing legumes or other immune crops.

Chinese mantis *n.* the largest mantis in North America (*Tenodera aridifolia sinensis*, family Mantidae), up to 4 in. (10 cm) long. It devours large numbers of other insects; the female may even eat her male partner after mating. It was brought to the U.S.A. from Asia about 1896. See praying mantis.

Chinese sumac *n.* a large, very fast-growing deciduous tree (*Ailanthus altissima*, family Simaroubaceae) with large compound leaves. It withstands smog and insects well in inner cities but is subject to wind damage. Male and female flowers are on separate plants, and the male flowers have a foul smell. Also called ailanthus or tree of heaven.

Chinese tallowtree *n.* a small tree (*Sapium sebiferum*, family Euphorbiaceae) with poisonous milky juice whose wood is used for engraving work. Its seeds have a waxy covering that is used in soap and candle making. Native of China and Japan, it grows in southern U.S.A.

chingma abutilon *n.* an annual herb (*Abutilon theophrasti*, family Malvaceae) known as a troublesome weed. See velvetleaf.

Chinook *n.* **1.** a native American of a tribe that formerly inhabited the area along the lower part and mouth of the Columbia River in the present state of Washington. **2.** the language of the Chinook tribe or, loosely, a pidgin language derived from it.

chinook *n.* (Salish Indian *tsi-núk*, Chinook Indian) **1.** a warm, dry, winter wind from the west or north blowing down the eastern slope of the Rocky Mountains or in the intermountain area. Snow cover melts rapidly under a chinook. Similar winds are called Santa Ana in California or foehn in the Alps. See foehn. **2.** a warm, moist southwestern wind blowing from the Pacific Ocean onto the northwestern coast of the U.S.A. or southwestern Canada.

chinook salmon *n.* the largest species of salmon (*Oncorhynchus tshawytscha*), averaging about 22 lb (10 kg) but ranging up to 100 lb (45 kg); native to northwestern U.S.A. and Canada. Also called king salmon, quinnat salmon, or tyee. See salmon.

chinquapin *n.* (Algonquian *chinquapin*, chinquapin) **1.** any of several bushy trees (*Castanea* sp., family Fagaceae) with oblong serrated leaves; e.g., Allegheny chinquapin (*C. pumila*) native to southeastern U.S.A. **2.** any of several evergreen trees (*Castanopsis* sp., family Fagaceae); e.g., golden chinquapin (*C. chrysophylla*) of far-western U.S.A. or other species from Asia. **3.** the edible nuts from any of these trees.

chip budding *n.* a form of asexual plant propagation using a small scion with a single bud. The scion is inserted into a cut in the stock and tied in place so the two will grow together. See budding.

chipmunk *n.* a small ground squirrel with black stripes on a brown body, either the eastern chipmunk (*Tamias* sp., especially *T. striatus*) or the western chipmunk (*Eutamias* sp.). Chipmunks are known for their chatter and for sudden rapid movement.

chiral *adj.* distinct from the mirror image; e.g., a molecular structure that is either right-handed or left-handed. The versions are mirror images that cannot be turned to identical configurations.

chirp *n.* **1.** a short, shrill sound made by certain birds, insects, etc. **2.** a short, squeaky sound made by a machine, often indicating a dry bearing or other problem. *v.* **3.** to make a chirping sound.

chisel *n.* (French *cisel*, chisel) **1.** a tool with a cutting edge at one end and a handle so it can either be pushed by hand or driven by a hammer to cut wood, metal, or other material. **2.** a farm implement with narrow blades that loosen the soil. This requires less power and results in less soil erosion than when a plow that lifts and turns the soil is used. *v.* **3.** to cut with a chisel. **4.** to obtain something by swindling or coercion without earning it.

chit *n.* **1.** a child, especially a pert young girl. **2.** a young animal. **3.** a plant shoot. **4.** a voucher promising to pay a small sum. **5.** a note (British). **6.** second- or third-grade rice.

chitterlings or **chitlins** *n.* the small intestines of young pigs or other animals used as casings for human food.

chive or **chives** *n.* (French *chive*, chive) **1.** a small bulbous perennial herb (*Allium schoenoprasum*, family Amaryllidaceae) related to garlic and onion. **2.** the long, slender leaves of the plant used as a food seasoning.

chlordane *n.* a persistent chlorinated hydrocarbon ($C_{10}H_6Cl_8$) insecticide that was once used for termite control but is now banned by the EPA for environmental reasons.

chlorinated hydrocarbons *n.* a class of persistent, broad-spectrum insecticides that linger in the environment and accumulate in the food chain. Among them are DDT, aldrin, dieldrin, heptachlor, chlordane, lindane, endrin, mirex, hexachloride, and toxaphene. Another example is TCE, used as an industrial solvent.

chlorinated lime *n.* See calcium hypochlorite.

chlorination *n.* **1.** the application of chlorine as a gas or liquid to drinking water, sewage, or industrial waste to disinfect or to oxidize undesirable compounds. **2.** the use of chlorine to remove impurities from gold or silver.

chlorine (Cl) *n.* (Greek *chlōros*, pale green + *ine*, a halogen) element 17, atomic weight 35.453; one of the halogens, the family of elements that react with a valence of –1. Chlorine is an essential micronutrient for plants as well as an essential element for animals and humans, but it is so abundant that it is rarely deficient and is more likely to be in excess supply, sometimes enough to be toxic. Chlorine is the most abundant anion in seawater and is the anion in table salt (NaCl), most potassium fertilizer (KCl), and many other compounds. It is commonly used as an oxidant and disinfectant.

chlorite *n.* (Latin *chloritis*, pale green) a partially expanded (1.43 nm layer spacing) silicate clay mineral composed of $Mg_3Si_4O_{10}(OH)_2$ layers (with some substitution of Al for Si) and interlayers of $AlMg_2(OH)_6$ that balance the resulting charge; found in certain soils and sedimentary rocks.

chlorofluorocarbon (CFC) *n.* any of a family of relatively inert, nontoxic, synthetic organic compounds containing both chlorine and fluorine; e.g., CCl_2F_2 (freon). Some of them are easily liquefied gases used in refrigeration, air conditioning, and aerosol propellants. Others are used in insulation, packaging, and as solvents. Because CFCs are not destroyed in the lower atmosphere, the gaseous forms drift upward to where they contribute to global warming through the greenhouse effect, and their chlorine components catalyze the decomposition of ozone in the stratosphere (it is estimated that one free chlorine radical can destroy up to 100,000 ozone molecules before it is converted to the relatively inactive chlorine dioxide form, ClO_2). Since 1975, the use of volatile chlorofluorocarbons is being phased out on a worldwide basis.

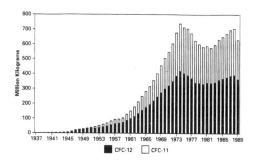

Global release of chlorofluorocarbons, 1937–1989

chlorophyll or **chlorophyl** *n.* (Greek *chlōro*, light green + *phyllon*, leaf) the green coloring that catalyzes photosynthesis in the chloroplasts of plant cells. Most green plants contain chlorophyll a ($C_{55}H_{72}O_5N_4Mg$) and chlorophyll b ($C_{55}H_{70}O_6N_4Mg$). Chlorophyll c occurs in brown algae and in diatoms (same formula as chlorophyll b). Chlorophyll d ($C_{54}H_{70}O_6N_4Mg$) is found in red algae. Chlorophyll enables plants to use light energy to synthesize carbohydrates from carbon dioxide and water. See photosynthesis.

chloropicrin *n.* (Greek *chlōro*, chlorine + *pikros*, bitter) a gas (CCl_3NO_2) used as a fungicide and insecticide to disinfect grains and cereal foods, to sterilize soil, and as a war gas (a severe skin and lung irritant). Also called nitrochloroform.

chloroplast *n.* (Greek *chlōro*, light green + *plastos*, formed) an organelle that contains chlorophyll in a plant cell; the site where photosynthesis occurs.

chlorosis *n.* (Greek *chlōro*, light green + *osis*, condition) an abnormal pale green, brown spotted, or yellowish color, especially of leaves. Chlorosis is caused by a deficiency of any of several nutrients, water stress, air pollution, disease, insects, or herbicides. The chlorotic color may be over the entire leaf or only between the leaf veins.

chlortetracycline *n.* an antibiotic ($C_{22}H_{23}ClN_2O_8$) derived from a soil microbe (*Streptomyces aureofaciens*). It is used for treatment of certain diseases and in livestock feeds as a growth stimulant.

chokecherry *n.* **1.** a small deciduous shrub (*Prunus virgiana*, family Rosaceae) sometimes planted as a component of a windbreak. Its leaves and fruit are toxic to people and animals. Poisoned animals may stagger, convulse, have difficulty breathing, bloat, and die. See wild black cherry. **2.** the bitter dark red fruit produced in clusters by the shrub.

chokedamp *n.* a poisonous but nonflammable and nonexplosive gas that forms in coal mines and is so low in oxygen that it causes choking. It is high in impure carbon dioxide (CO_2) that contains about 15% carbon monoxide (CO). Also called blackdamp.

cholecalciferol *n.* vitamin D_3 ($C_{27}H_{44}O$), a fat-soluble vitamin produced when animal or human skin is exposed to sunshine; stored in the liver. Vitamin D helps in the utilization of calcium and phosphorus to form strong bones. A deficiency in vitamin D causes rickets.

cholera *n.* (Greek *cholera*, a disease of the bile) a highly infectious disease caused by bacteria (*Vibrio cholerae*), often transmitted by contaminated drinking water. Cholera causes severe diarrhea, dehydration, and a high mortality rate and is endemic and sometimes epidemic among people in densely populated parts of Asia. See hog cholera, fowl cholera.

cholesterol *n.* (Greek *kholē*, bile + *stereos*, solid) the most abundant form of sterol in animal tissue. A white, fat-soluble alcohol ($C_{27}H_{45}OH$) that is a primary component of animal cell membranes, it is needed for the synthesis of bile acids, steroid hormones, and vitamin D. It is present in animal fats and oils, bile, blood, brain tissue, nervous tissue, the liver, kidneys, and adrenal glands. Cholesterol is transported from the liver through the arteries to other parts of the body by low-density lipoproteins (LDLs). Excess cholesterol can build up as fatty deposits in blood vessels and has been directly related to heart attacks. High-density lipoproteins (HDLs) have been shown to offset the effect of LDLs by helping to return excess cholesterol to the liver.

cholla *n.* (Spanish *cholla*, head) a very common, bushy, spiny cactus (*Opuntia* sp.) that grows in arid areas of southwestern U.S.A., Mexico, and elsewhere. Also called prickly pear.

chorion *n.* (Greek *chorion*, afterbirth) **1.** the outermost membrane around an unborn fetus in mammals, near the wall of the mother's uterus. **2.** a similar membrane inside the eggshell of a bird or reptile. **3.** a membrane around an insect egg.

C horizon *n.* unconsolidated material beneath the solum (true soil) and above the bedrock. Sometimes called parent material. See soil horizon.

chorography *n.* (Latin *chorographia*, ground graph) the process of systematically mapping and describing the geography of a region or district.

chow *n.* (Chines *kaú*, dog) **1.** (slang) food, especially a hearty meal. **2.** a medium-sized dog with a leonine head, a stocky body, and a thick coat with a ruff around the neck. A Chinese breed once raised expressly for eating and also used as a guard dog, for herding a flock, or for hunting. Also called chow chow.

Christmasberry *n.* an evergreen shrub (*Heteromeles arbutifolia*, family Rosaceae) native to California and used as Christmas decoration for its bright red berries and dense clusters of white flowers. Prussic acid in its leaves is poisonous to livestock, causing affected animals to stagger, have difficulty in breathing, hang out their tongues, bloat, and die. Also called toyon or California holly.

Christmas rose *n.* a perennial plant (*Helleborus niger*, family Ranunculaceae) with a palmate leaf, red-spotted peduncles, and white to purplish roselike flowers that blossom in winter. The plant is toxic to livestock, and its sap can cause severe dermatitis in human skin. Also called black hellebore or felon grass.

Christmas tree *n.* a tree decorated with ornaments and lights during the Christmas season; usually a wild evergreen tree or a tree grown in a plantation. Of the latter, the order of popularity of the tree species in the U.S.A. is Scotch pine, Douglas fir, balsam fir, black spruce, and eastern red cedar.

chroma *n.* (Greek *chrōma*, color) the relative intensity or saturation of a color; directly related to the purity of the dominant wavelength of light. It is rated in the Munsell color system on a scale from 0 (pure gray with no color) to 20 (pure color with no gray). See hue, value, Munsell color system.

chromatin *n.* (Greek *chrōma*, color + *in*, protein) dispersed genetic material in the nucleus of living cells that contains the genes and condenses into chromosomes during cell division. See chromosome.

chromatography *n.* (Greek *chrōmatos*, color + *graphe*, a representation) **1.** the science of colors or a textbook about such science. **2.** a means of chemical analysis based on differential rates of movement through a stationary medium; e.g., column chromatography, gas chromatography, or paper chromatography.

chromite *n.* ferrous chromate ($FeCr_2O_4$), a lustrous black mineral that is the commercial source of chromium.

chromium (Cr) *n.* (Greek *chrōma*, color + *ium*, element) a silvery metal, element 24, atomic weight 51.996; used in steel alloys and coatings to add hardness and resistance to corrosion (most stainless steel contains 10% to 20% chromium). It is considered essential in human and animal nutrition because it assists in glucose metabolism, but excessive amounts of chromium can cause lung cancer.

chromomere *n.* (Greek *chrōma*, color + *meros*, part) a microscopic beadlike structure on a chromosome; composed of aggregated chromatin.

chromosome *n.* (Greek *chrōma*, color + *sōma*, body) an elongated threadlike body that condenses in the nucleus of a cell prior to and during cell division. Chromosomes can be stained so they are visible under a microscope. Each species has a specific number of chromosomes containing the genes that carry the genetic code for the organism. Chromosomes occur in pairs except in sperm and eggs for sexual reproduction.

Chromosome numbers in humans and in domestic animals.

Animal/human	No. of chromosomes
Cat	38
Cattle	60
Chicken	78
Dog	78
Donkey	62
Goat	60
Horse	64
Human	46
Pig	38
Rabbit	44
Sheep	54

chromosphere *n.* (Greek *chrōma*, color + *sphaira*, round body) a red gaseous envelope that surrounds the photosphere of the sun and other stars. It erupts large quantities of hydrogen and other gases.

chronic poisoning *n.* cumulative poisoning as a result of many small doses of a toxic substance over a period of time.

chronobiology *n.* (Greek *chronos*, time + *bios*, life + *logos*, word) the study of the effect of time on living organisms, especially through rhythmic processes.

chronology *n.* (Greek *chronos*, time + *logos*, word) the science of measuring time in fixed periods and arranging events in their order of occurrence.

chronosequence *n.* (Greek *chronos*, time + Latin *sequentia*, series) a group of related soil series whose differences are functionally related to time as a factor of soil formation.

chrysalis (pl. **chrysalides** or **chrysalises**) *n.* (Greek *khrysallis*, pupa) **1.** pupa; the resting stage between the larva and adult stages of butterflies, moths, and some other insects. **2.** the cocoon that encloses and protects a pupa.

chrysanthemum *n.* (Greek *khrysanthemon*, golden flower) **1.** any annual or perennial bush or shrub of a large genus (*Chrysanthemum* sp., family Asteraceae) that originated in Eurasia but now grows in most parts of the world, including many ornamental species, some species grown as sources of medicinal products or insecticides (e.g., *C. coccineum*, florist's pyrethrum, and *C. cinerariaefolium*, Dalmatian pyrethrum), and others that are pernicious weeds. The most common weedy species are the oxeye daisy or white weed and the European mum (*C. leucanthemum*). **2.** a flower from a chrysanthemum plant. The blossoms are either single or double, range in size from small pompons to several inches across, and occur in many colors, especially white and various shades of yellow and red. Also called mums.

churchmouse threeawn grass *n.* a tufted perennial grass (*Aristida dichotoma*, family Poaceae), native to the U.S.A., that grows in dry, sandy soils and produces sharp awns that can injure grazing animals. Also called poverty grass.

chymosin *n.* rennin.

chytrid *n.* a class (Chytridiomycetes) of microscopic fungi that inhabit water or soil. Several plant diseases are caused by chytrids; e.g., clubroot of cabbage, powdery scab of potato, potato wart, cranberry gall, and beet root tumor.

CIAT *n.* Centro Internacional de Agricultura Tropical (International Center for Tropical Agriculture), an international agricultural research center founded in 1967 with headquarters in Cali, Colombia. It focuses on crop improvement and ecoregional approaches to developing agriculture in the lowland tropics of Latin America. Research covers rice, beans, cassava, forages, and pasture.

cicada *n.* (Latin *cicada*, cicada) any of hundreds of species of large insects (*Magicicada* sp. and *Tibicen* sp., family Cicadidae, order Homoptera) that are usually green with red and black markings; e.g., the harvest fly that matures in two years or the periodical cicadas that live mostly in the soil as nymphs that suck sap from tree roots but emerge in large broods of adults once every 13 years (most southern U.S. species) or 17 years (most northern U.S. species). The adults are active for a few weeks, long enough to make themselves heard by a shrill piercing sound and to lay eggs for the next brood.

cider *n.* (Latin *sicera*, strong drink) apple juice used either as a fresh (sweet cider) or fermented (hard cider) beverage or for making applejack brandy, wine, or vinegar.

cider yeast *n.* a yeast (*Saccharomyces malei*, family Saccharomycetaceae) used to make sparkling hard cider by slower than usual fermentation.

cienaga *n.* (Spanish *ciénaga*, marsh, bog, or miry place) **1.** a grassy wetland. The water table is at or near the soil surface in most of the area, depressions have standing water in them, and springs and small streams may flow for short distances in or from the area. Cienagas typically range in size from several hundred square feet to several hundred or more acres (from tens of square meters to hundreds of hectares). **2.** a seepy, marshy area on a hillside in southwestern U.S.A. See wetland.

CIFOR *n.* See Center for International Forestry Research.

cigar casebearer *n.* a small insect (*Coleophora occidentis*, family Coleophoridae) whose larvae weave tough, silken cases around themselves. They drag these cases with them while feeding on leaves, buds, and fruit of apple, cherry, pear, plum, and quince.

cilia *n.* (Latin *cilia*, eyelashes) **1.** short hairlike growths on the exterior of certain protozoa (known as ciliates) that move in unison to impart movement to the organism. Cilia are shorter and much more numerous on the cell than are flagella on a flagellated organism. See flagellum. **2.** short hairlike growth on the surface of certain plant cells, e.g., either around the fringe or across the entire underside of some leaves. **3.** eyelashes.

CIMMYT *n.* Centro Internacional de Mejoramiento de Maize y Trigo (International Maize and Wheat Improvement Center), an international agricultural research center founded in 1966 with headquarters in Mexico City. It focuses on crop improvement for maize, wheat, barley, and triticale.

cinchona *n.* a large genus of trees and shrubs (*Chinchona* sp., family Rubiaceae) native to South America that has opposite leaves and small panicled flowers. Quinine (used to alleviate muscle cramps and formerly to cure malaria) is obtained from the bark of several of its species. Named for Francisca Enriques de Ribera (?–1641), Countess of Chinchón.

cinder *n.* (French *cendre*, ashes) **1.** ash and/or partially burned residue from burned coal, wood, or other solid fuel. **2.** an ember that is burning without a flame. **3.** ashlike solidified magma ejected by a volcano in pieces 0.5 to 2.5 cm (0.2 to 1 in.) across. See tephra.

cinder cone *n.* a peak composed of cinders and other volcanic materials.

cinnamon *n.* (Greek *kinnamōmon*, cinnamon) **1.** a popular spice made from the sun-dried bark of several species of small trees or shrubs (family Lauraceae) in Sri Lanka (formerly Ceylon) and other tropical countries. The principal species are Ceylon cinnamon (*Cinnamomum zeylanicum*), Cassia or Chinese cinnamon (*C. cassia*), Saigon cinnamon (*C. loureirii*), and camphor tree (*C. camphora*). **2.** the tree that produces the spice. **3.** a yellowish or light reddish brown.

cion *n.* See scion.

CIP *n.* Centro Internacional de la Papa (International Potato Center), an international agricultural research center founded in 1971 with headquarters in Lima, Peru. It focuses on potato and sweet potato improvement, with special attention paid to ecoregional aspects of mountain area agriculture.

circadian rhythm *n.* (Latin *circa*, about + *dies*, day) the approximately 24-hr awake/sleep periods of humans and animals and the growth/droop periods of plants that synchronize with the light/dark periods in the day/night (diurnal) pattern.

circling disease *n.* an infectious disease of the central nervous system of sheep, swine, cattle, goats, humans, and several other species that causes affected animals to stagger and move in circles. Other symptoms include dullness, fever, eye inflammation, and paralysis that ends in death. It is caused by a bacterium (*Listeria* or *Listerella monocytogenes*). Also called listeriosis or listerellosis.

circumboreal *adj.* (Latin *circum*, around + *borealis*, northern) throughout the northern (boreal) region; e.g., a plant that grows in all the land areas of the far north.

cirque *n.* (Latin *circus*, a ring) **1.** a steep-walled, rounded basin carved into a mountainside by glacial erosion, often containing a small lake. **2.** a circular arrangement or ring.

cirrus *n.* (Latin *cirrus*, curl or tuft) **1.** a banded wispy cloud composed of small ice crystals that occurs at high altitudes (10,000 to 26,000 ft or 3000 to 8000 m in polar regions and 20,000 to 60,000 ft or 6000 to 18,000 m in tropical regions). **2.** a tendril of a plant or a slender appendage of certain invertebrates.

cistern *n.* (Latin *cisterna*, tank) **1.** a large tank for storing water or other liquid, especially an underground tank holding rainwater for use by a farm or ranch home. Water running off a roof may be caught by a gutter and directed into the cistern. **2.** a small body cavity or sac holding lymph or other fluid.

citric acid. a white tribasic acid crystalline substance present in lemons, limes, currants, gooseberries, raspberries, etc., that is usually obtained from the juice of lemons and limes and is used in artificial lemonade, food preservatives, medicines, and dyes. Also given as a diuretic.

citrinin *n.* a bacteriostatic yellow pigment ($C_{13}H_{14}O_5$) produced by a fungus (*Penicillium citrinum*).

citron *n.* (Latin *citron*, citrus tree) **1.** a large, thorny shrub or small tree (*Citrus medica*, family Rutaceae) grown in very warm areas for its highly aromatic fruit. Native to eastern Asia. Also called cedrat. **2.** the fruit of the tree, similar to a lemon but larger. Its thick rind is candied and used as a preserve.

citrus *n.* **1.** (Latin *citron*, a citrus tree) any of several trees or bushes (*Citrus* sp., family Rutaceae) grown in tropical and subtropical regions. **2.** the fruit of a citrus tree, also called citrous fruit, an excellent source of potassium and vitamin A and well known as a source of vitamin C that prevents scurvy. Citrus accounts for about 20% of all world fruits. Oranges comprise about 70% of all citrus, followed in production by grapefruit, lemons, limes, and tangerines. There are at least 16 principal species of *Citrus*, listed alphabetically in the table below.

Citrus fruits and their scientific names.

Common name	Scientific name
Celebes papeda	*Citrus celebica*
Citron	*C. medica*
Grapefruit	*C. paradisi*
Ichang papeda	*C. ichangensis*
Indian wild orange	*C. indica*
Khasi papeda	*C. latipes*
Lemon	*C. limon*
Lime	*C. aurantiifolia*
Mandarin (tangerine)	*C. reticulata*
Mauritius papeda	*C. hystrix*
Melanesian papeda	*C. jacrocarpa*
Orange (common, sweet)	*C. sinensis*
Orange (sour)	*C. aurantium*
Papeda	*C. micrantha*
Pummelo	*C. grandis*
Tachibana orange	*C. tachibana*

citrus belt *n.* the citrus-growing regions of the world, generally tropical or subtropical land at low elevation within 35° north or south of the equator.

citrus canker *n.* a bacterial (*Xanthomonas citri*) disease of citrus that was eradicated from the U.S.A.

until a new strain appeared in Florida in 1985, causing 12 million orange and grapefruit trees to be killed to avoid infecting the entire crop. It is so destructive that international trade in plant material from affected areas is restricted.

citrus mealybug *n.* a serious insect pest (*Pseudococcus citri*, family Pseudococcidae) of citrus and many ornamental and flowering plants. Biological control by an internal parasite (*Leptomastidea abnormis*) has been reasonably successful.

citrus oil *n.* oil from various species of citrus fruits (family Rutaceae) used to produce commercial oils for flavoring. Among the most important are
- sweet orange oil, produced from the fresh peel of the ripe common orange (*Citrus sinensis*)
- bitter orange oil, from the fresh peel of the ripe lime (*Citrus aurantiifolia*)
- lemon oil, from the fresh peel of the ripe lemon (*Citrus limon*)

citrus thrips *n.* a tiny insect pest (*Scirtothrips citri*, family Thripidae) that infests the buds, new growth, and fruit of citrus and other crops and weeds, especially in arid climates such as California and Arizona.

citrus white fly *n.* an insect pest (*Dialeurodes citri*, family Aleyrodidae) of citrus and other plants, especially the camelia. It sucks sap from the underside of leaves and causes stunting of the trees and fruit, especially in states bordering the Gulf of Mexico.

civet *n.* (French *civette*, civet) **1.** any of about 17 species of weasel-like carnivores (especially *Viverra* sp. of Asia and *Civettictis* sp. of Africa, family Viverridae) with a spotted coat, coarse hair, and rounded ears; often called civet cat. **2.** a dark yellow secretion from the musk gland of civets that is used as a fixative to make perfume last longer. **3.** the fur of civets. **4.** a skunk in western U.S.A.

Civilian Conservation Corps (CCC) *n.* a federal agency established in the U.S.A. in 1933 to provide work for unemployed persons (mostly young men) and supply workers for various conservation projects until it was terminated in 1942.

civilization *n.* (Latin *civilis*, citizen + *izatio*, action) **1.** an organized society with culture, arts, and science; not barbaric. **2.** the process of forming an organized society. **3.** the ensemble of civilized nations and societies.

clabber *v.* (Irish *clabar*, curdled) **1.** to curdle. *n.* **2.** milk that has curdled. See curdled milk.

claim *n.* (Latin *clamare*, to cry out) **1.** an assertion that something is true. **2.** an assertion of a right or supposed right or privilege. **3.** a right or supposed right to property, especially to obtain title to unsettled land, or the right to exploit mineral or oil resources discovered on land within the public domain. **4.** property that someone has claimed. *v.* **5.** to express a claim.

clam *n.* (Anglo-Saxon *clamm*, bond) **1.** any of a large group of hardshell (mostly bivalve) mollusks, especially those that are edible. **2.** a person who refuses to talk. **3.** (slang) a dollar or some other type of currency. *v.* **4.** to gather clams or dig for clams.

clambering monkshood *n.* a poisonous, perennial herb (*Aconitum uncinatum*, family Ranunculaceae) that grows in shady areas along streams in eastern U.S.A. Livestock poisoned by it suffer difficult breathing, weakness, bloating, and belching. Also called wild monkshood.

clammy locust *n.* an ornamental deciduous tree (*Robinia vitosa*, family Fabaceae) with attractive foliage and pink and yellow blossoms. Sticky secretions from glands in the leaves and bark are toxic and can be fatal to livestock.

clan *n.* (Irish *clann*, offspring or tribe) **1.** an early form of social group based on descent from a common ancestor and generally consisting of several families living in close proximity, following the same leader, and sharing the same family name. **2.** a part of a tribe with close family ties. **3.** a clique or group of people that work or socialize together, sharing a common interest.

clarification *n.* (Latin *clarificare*, to become clear + *atio*, action) **1.** removal of ambiguity; making a meaning clear. **2.** the removal of suspended solids by settling and/or filtration, sometimes aided by coagulating chemicals or centrifugal action; e.g., the purification of milk, fruit juice, or wastewater.

class *n* (Latin *classis*, class) **1.** a group of similar objects, especially those grouped in a system of classification. **2.** a group of people with something in common, e.g., all in the same grade at a school. **3.** a social stratum, as the middle class. **4.** distinguished or high quality, as a classy car or classy dress. **5.** the level of accommodation, as traveling first class. *v.* **6.** to classify; to place in a class.

classification *n.* (Latin *classis*, class + *ficatio*, making) **1.** the process of assigning items or organisms into classes having a defined range of characteristics within an organized system. The classes may be defined taxonomically, mathematically, or otherwise, depending on the purpose to be served. **2.** the class name assigned to an item or organism. The description of the class may be used as a partial description of the individual item or organism.

classify *v.* (Latin *classis*, class + *ficare*, to make) **1.** to place in classes according to a classification system. **2.** to designate a document or other source of information as secret or unavailable to persons without authorization.

claustrophobia *n.* (Latin *claustro*, closed + Greek *phobia*, fear) an abnormal fear in a person or animal of confinement in a small or narrow space, often causing debilitation.

clay *n.* (Anglo-Saxon *claeg*, glue) **1.** a mineral particle with a diameter of less than 0.002 mm (in soil science), less than 0.004 mm (in geology), or less than 0.074 mm (in engineering). **2.** a soil textural class containing more than 40% clay, less than 40% silt, and less than 55% sand (in soil science), more than 50% clay (in geology), or plastic material composed mostly of clay (in engineering). **3.** mineral material of small particle size suitable for making bricks, ceramics, or pottery.

clay mineral *n.* a mineral that forms in soil and typically occurs in clay size, especially those of the silicate family, including chlorite, halloysite, illite, kaolinite, montmorillonite, smectite, and vermiculite. Allophane (amorphous material) and sesquioxides may also be included.

claypan *n.* (Anglo-Saxon *claeg*, glue + *panne*, pan) a compact layer in the upper subsoil having a high clay content that impedes water percolation and air movement and restricts the growth of plant roots. It is formed by illuviation of clay eluviated from the topsoil and/or clay synthesis in place. Claypans are hard and brittle when dry and sticky and plastic when wet.

Clean Air Act *n.* a law passed in 1963 by the U.S. Congress asserting the need for the federal government to deal with interstate air pollution problems; 1970 amendments authorized the Environmental Protection Agency to establish air quality standards to be implemented by the individual states. See National Ambient Air Quality Standards.

clearcutting *n.* a timber-harvesting practice involving the removal of all the trees in an area at one time that is often criticized as a forest management technique because of the erosion, pollution, and wildlife habitat problems it can create, but supported by some because of economics and/or favoring the reproduction of certain tree species such as Douglas fir, black cherry, and certain species of pine.

cleft *n.* or *adj.* (Anglo-Saxon *clyft*, cleavage) a crack or divide formed by splitting; a fissure, as a cleft in a rock.

cleft graft *n.* a grafting method used on established trees. A branch is sawed squarely across, and the stub is split lengthwise to form a cleft. A scion is inserted into the cleft so that the cambium layers are in contact (often two scions, one on each side of the cleft). The cut surfaces should be covered with a thin coat of wax to reduce desiccation. See grafting.

click beetle *n.* any of various elongated, flexible beetles (family Elateridae) that curl up when threatened, often landing on their backs. After laying still for a time, a click beetle flexes its body suddenly and jumps in the air with a clicking sound. If necessary, it repeats the jumping process until it lands on its feet and can run away. Click beetle larvae are called wireworms.

cliff *n.* (Anglo-Saxon *clif*, cliff) a precipice or bluff; a nearly vertical rock face (including those in soft geologic deposits such as loess).

climacteric *n.* or *adj.* (Greek *klimaktērikos*, critical point) a crucial period, often including a maturity or climax of some activity (e.g., the peak respiration rate of mature fruit, especially apples) and/or a decline in some (e.g., sexual) activity.

climate *n.* (Greek *klima*, region) **1.** a long-term average of weather conditions. The typical annual cycle of weather events of a place or region, including especially temperature, precipitation, humidity, air pressure, wind, and cloudiness. The climate of an area has a strong influence on its soils, native vegetation, and potential use. **2.** prevailing attitudes and standards of a group of people, as a climate of discontent.

climate year *n.* a year with a beginning date selected for presentation of climatic data on temperature, precipitation, stream flow, etc.; e.g., the climate year of the U.S. Geological Survey, called a water year, from October 1 of a year to September 30 of the following year. Also called climatic year.

climatic type *n.* a unit of classification of climate based on one or more climatic elements such as temperature, rainfall, humidity, wind, and many others, including combinations and patterns; e.g., a Mediterranean climate has a cool, wet winter and a warm to hot, dry summer. Many different climatic types have been defined to serve a particular purpose or investigation.

climatic optimum *n.* a geologic period of unusually warm temperatures worldwide; e.g., the period between 7000 and 5000 years ago when temperatures were 2° to 3°C warmer than they are now.

climatic zone *n.* **1.** a zone with a distinct climate, often designated according to the needs of a particular type of plant or animal. **2.** any of the following five portions of the Earth's surface:

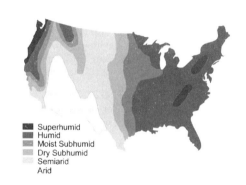

Superhumid
Humid
Moist Subhumid
Dry Subhumid
Semiarid
Arid

Climatic types based on humidity

- north frigid zone: between the North Pole and the Arctic Circle
- north temperate zone: between the Arctic Circle and the Tropic of Cancer
- torrid zone: between the Tropic of Cancer and the Tropic of Capricorn.
- south temperate zone: between the Tropic of Capricorn and the Antarctic Circle
- south frigid zone: between the Antarctic Circle and the South Pole

climatogenic *adj.* (Greek *klima*, region + *genesis*, birth) caused by climatic agents; e.g., periodic floods and droughts are both climatogenic.

climatology *n.* (Greek *klima*, region + *logos*, word) the study of climate and related phenomena, especially as it applies to large areas and extended periods. Weather conditions at a specific time and place are in the domain of meteorology.

climax vegetation *n.* a stable plant community in equilibrium with the environment; the culmination of a succession of plant communities that preceded it.

climbing bog *n.* a bog that gradually expands onto land that was previously dry, extending marshy conditions to higher land. The principal vegetation is usually sphagnum moss, and the climate is typically cool and humid with a short summer.

climograph *n.* (Greek *klima*, region + *graphos*, drawn) a chart with one climatic factor plotted against another; e.g., mean monthly temperature versus mean monthly precipitation.

climosequence *n.* (Greek *klima*, region + Latin *sequentia*, series) a set of related soils whose differences are related to climate.

cline *n.* (Greek *klinein*, to lean) a gradual change in the characteristics of plants, animals, or people along a geographic gradient; usually assumed to be genetic adaptation to environmental differences, e.g., temperature or moisture gradients or both.

clinical waste *n.* human and animal tissues, medicines, wound dressings, disposable medical tools, etc., that need special handling to avoid spreading disease, e.g., incineration or at least sterilization before being discarded.

clinometer *n.* (Greek *klinein*, to lean + *metron* measure) an instrument for measuring the slope or dip of a surface in percent or degrees.

clinosequence *n.* (Greek *klinein*, to lean + Latin *sequentia*, series) a group of related soils or plants whose differences are related to the slope gradient.

clitocybe root rot *n.* a fungal (*Clitocybe tabescens*, family Agaricaceae) disease of fruit trees and other woody plants across southern U.S.A. Its effects resemble those of Armillaria root rot and mushroom root rot.

cloaca *n.* (Latin *cloaca*, sewer) **1.** a sewer or cesspool, especially an ancient one. **2.** the common chamber in a bird, reptile, or amphibian where the urinary, digestive, and reproductive canals discharge waste material. It opens to the exterior though the anus.

cloche *n.* (French *cloche*, bell or bell jar) **1.** a bell jar. **2.** a bell-shaped jar or other transparent cover placed over an individual plant to protect it against cold temperatures. **3.** a bell-shaped woman's hat.

clod *n.* (Anglo-Saxon *clott*, rounded lump) **1.** an artificially produced mass of soil that is compact and coherent, variable in size from a fraction of an inch to several inches (from a few millimeters to a fraction of a meter), and produced by compression resulting from traffic, plowing, digging, or other manipulation, especially in soils with significant clay content when they are either wet enough to be plastic or dry enough to be brittle. See ped. **2.** a derogatory term for a person who appears clumsy and stupid.

clog *n.* **1.** an object that impedes or restrains movement; an encumbrance or blockage. **2.** a heavy wooden block strapped to a horse's hind foot to prevent it from kicking. **3.** a shoe or sandal with a thick sole, usually made of wood. **4.** an old-fashioned dance with the dancers wearing clogs. *v.* **5.** to plug or choke up, as to clog a drain. **6.** to overload a machine, especially a harvester, by feeding material so rapidly that it stops working.

clone *n.* (Greek *klōn*, young shoot) **1.** a plant produced from a vegetative part so it is identical in genetic character to the source plant; e.g., a strawberry plant produced from a stolon, a willow tree from a cutting, or Bermuda grass from rhizomes. **2.** an animal produced by artificially placing a substitute nucleus in an egg so its genetic character is controlled by the animal source of the nucleus. **3.** a group of organisms with identical genetics produced by cloning. **4.** a person, animal, or thing that mimics the appearance and/or behavior of another. *v.* **5.** to produce organisms, objects, or machines by cloning.

close breeding *n.* inbreeding among closely related plants, animals, or people, especially when done repeatedly to produce a line with consistent genetics. Also called inbreeding.

closed formula *n.* a statement of ingredients that gives guaranteed percentages of protein, fat, and fiber for the feed inside a container but does not specify the amount of each ingredient such as the percentage of each grain. See open formula.

closed pedigree *n.* hybrids produced by private seed companies using unrevealed combinations of inbred lines. Also called unpublished pedigree.

clostridium *n.* (Greek *klōstēr*, spindle + Latin *idium*, tiny) any of more than 200 species of spore-forming, obligate anaerobic, rod-shaped soil bacteria (*Clostridium* sp., family Bacillaceae) that are widely distributed in soils and intestinal tracts. Some of them fix atmospheric nitrogen (N_2) in soil; others cause tetanus and botulism. The best-known nitrogen fixer is *C. pasteurianum*, discovered in 1891 by S. Winogradsky.

clothes moth *n.* any of several species of moths (family Tineidae, order Lepidoptera) that are serious household pests because their larvae feed on clothing, bedding, carpets, etc., made of wool, silk, felt, hair, or feathers; especially the webbing clothes moth (*Tineola bisselliella*) and the case-making clothes moth (*Tinea pellionella*).

cloth house *n.* a structure similar to a greenhouse but with a cloth covering that provides reduced light and temperature compared with a glass-covered structure, while still providing protection from wind and insects. It is advantageous for certain flower crops (e.g., asters and chrysanthemums) and for maturing a certain quality of tobacco.

cloud *n.* (Anglo-Saxon *clud*, mass of rock) **1.** a visible water accumulation in the atmosphere consisting of water droplets or ice particles, often contaminated with smoke, dust, or volcanic gases. Clouds are usually classified into four groups and ten types: (1) high clouds (cirrus, cirrostratus, and cirrocumulus), (2) middle clouds (altostratus and altocumulus), (3) low clouds (stratus, stratocumulus, and nimbostratus), and (4) clouds of great vertical development (cumulus and cumulonimbus). See cirrus, cumulus, nimbus, stratus. **2.** an atmospheric mass of smoke or dust. **3.** a discolored area on an otherwise uniform surface, especially on

Stratocumulus clouds

a transparent surface. **4.** a large swarm of insects or other things moving together, as a cloud of locusts. *v.* **5.** to make cloudy or gloomy. **6.** to obscure sight of something, either literally or figuratively, as to cloud an issue.

cloudburst *n.* (Anglo-Saxon *clud*, mass of rock + *berstan*, to break) a sudden intense rainstorm, usually brief and commonly accompanied by wind, lightning, and thunder.

cloud seeding *n.* the injection of small crystals of silver iodide or dry ice into a cloud to serve as centers for water condensation in an effort to cause precipitation to end a drought or to reduce wind velocity in a hurricane. Results have been erratic.

clove *n.* (Latin *clavus*, nail) **1.** a natural segment of a bulb, as a clove of garlic. **2.** (often cloves) a fragrant spice from the dried flower bud of a myrtle tree (*Syzygium aromaticum* or *Eugeniaa caryophyllata*, family Myrtaceae) native to the East Indies but now grown extensively throughout the tropics and used whole or ground as a seasoning and in perfumes and medicines. **3.** in England, a weight unit for wool, cheese, etc.; about 7 or 8 lb (3½ kg).

clover *n.* an herb (*Trifolium* sp., family Fabaceae) with compound, usually trifoliolate, leaves. Symbiotic *Rhizobium* bacteria growing on clover roots fix atmospheric nitrogen. Several species make excellent forage crops and serve also as good bee plants. **2.** any of several leguminous plants of related genera; e.g., bur clovers (*Medicago* sp.), bush clovers (*Lespedeza* sp.), and sweet clovers (*Medicago* sp.). **3. in clover** an idiom for having enough wealth to live luxuriously.

clover leafhopper *n.* an insect (*Aceratagallia sanguinolenta*, family Cicadellidae) that infests both legumes (e.g., clovers and alfalfa) and grasses (e.g., small grains and forage grasses) and is especially destructive to new growth of clover either during the seedling stage or after cutting.

clover mite *n.* an arachnid (*Bryobia praetiosa*, family Tetranychidae) that lives on clover, alfalfa, grass, or weeds. It may become a nuisance to humans by invading houses in winter and infesting stored food, but it does not transmit disease or cause much damage.

clover root borer *n.* a small beetle (*Hylastinus obscurus*, family Scolytidae) that infests the roots of various legumes (clover, sweet clover, alfalfa, peas, and vetch) in northern U.S.A. and eastern Canada.

clover rusts *n.* any of several species of fungi (*Uromyces* sp., family Pucciniaceae) that attack clovers, forming pale-brown pustules on the underside of leaves and on petioles and stems. The species are autoecious (i.e., all of their growth stages occur on the same host plant) and are specific to one or a few types of clover: *U. elegans* attacks Carolina clover in southern U.S.A., *U. trifolii* var. *fallens* attacks red clover, *U. trifolii* var. *hybridii* attacks alsike clover, and *U. trifolii* var. *trifolii-repentis* and *U. nerviphilus* attack white clover.

clover seed chalcid *n.* a serious insect pest (*Bruchophagus gibbus*, family Eurytomidae) of alfalfa and clover seed. Its larvae develop inside the seed and consume the entire interior, leaving only an empty shell. Also called alfalfa seed chalcid.

clubroot of cabbage *n.* a disease of plants of the cabbage family caused by a sporulating fungus (*Plasmodiophora brassicae*, family Plasmodiophoraceae) that deforms, enlarges, and rots the underground parts of the plant and may cause serious damage before symptoms appear above ground (e.g., slow growth and wilting). Also called anbury, club, clubfoot, clump foot, dactylorhiza, or finger or toe disease.

cluster *n.* (German *kluster*, bunch) **1.** a large number of similar things grouped, arranged, or growing in close proximity; e.g., a bunch of grapes, a mass of bees clinging together in a hive, or a group of people. *v.* **2.** to gather objects or individuals together into a cluster.

clutch *n.* (Anglo-Saxon *clyccan*, to clench) **1.** a grip or hold on an object. **2.** a mechanical device that provides a linkage that can be interrupted. **3.** an emergency that tests one's ability or resolve. **4.** the set of eggs a bird lays for one incubation. Some birds lay a single egg (penguin), others lay two (pigeon), and others may lay 12 to 20 (partridges and most domestic fowls). *v.* **5.** to grip firmly or desperately.

cnidarian *n.* (Latin *cnidaria*, stinging cell) any of several mostly marine-dwelling invertebrate animals (phylum Cnidaria, formerly Coelentera) with stinging cells and a large central body cavity, including jellyfishes, sea anemones, hydras, etc. Also called coelenterate.

coagulation *n.* (Latin *coagulatus*, bound together + *ation*, action) clumping together of solids in a liquid medium, making them settle faster. Chemicals that supply soluble calcium (Ca^{2+}), aluminum (Al^{3+}), or iron (Fe^{3+}) cause coagulation of clay particles and other negatively charged solids in water. The coagulating effect of many cations on proteins accounts for the toxicity of such cations because it is impossible for coagulated proteins to act as enzymes. Coagulation of blood is known as clotting.

coal *n.* (German *kohle*, coal) a black or dark brown solid, containing more than 50% carbonaceous material by weight (more than 70% by volume), that is formed by the slow metamorphosis, compaction, and hardening of buried plant materials through several geologic eras, especially material from the Mississippian and Pennsylvanian periods of the Paleozoic Era, about 350 to 280 million years ago. This era is also called the Carboniferous Period for the abundant plant life that produced great peat deposits. With time and intense pressure, peat was converted to lignite (brown), bituminous (soft), or anthracite (hard) coal, with its heat-producing quality increasing in that order.

coal gas *n.* volatile fuel gas produced by the destructive distillation of certain bituminous coals. It is a mixture that typically includes, on a volume basis, about 50% hydrogen, 30% methane, 4% other hydrocarbons, 8% carbon monoxide, and small amounts of water vapor, carbon dioxide, nitrogen, and oxygen.

coal-mine gases *n.* hazardous gases that sometimes accumulate in poorly ventilated parts of coal mines include
- blackdamp or chokedamp: a mixture of mostly nitrogen (N_2) plus 10% to 15% carbon dioxide (CO_2) that is so low in oxygen that it causes choking.
- whitedamp: a mixture that includes toxic levels of carbon monoxide (CO).
- firedamp: a mixture that includes an explosive concentration of methane (CH_4).

coast *n.* (Latin *costa*, side) **1.** the strip of land adjacent to the sea from the shoreline inland to the first major change in type of landscape. *v.* **2.** to continue movement without inputting more energy, as a sled sliding down a hill or an auto or bicycle either rolling down a hill or slowing to a stop.

coastal zone *n.* lands and waters adjacent to the coast that exert an influence on the uses of the sea and its ecology, or, inversely, whose uses and ecology are affected by the sea.

cobalt (Co) *n.* (German *kobalt*, goblin) element 27, atomic weight 58.93; a transition metal used in some alloy steels and whose compounds are used as coloring agents in paints and ceramics. Cobalt is a trace element that is plentiful in meat, eggs, and dairy products. It is a constituent of vitamin B_{12} and is essential for the fixation of nitrogen by *Rhizobium* bacteria. Cobalt 60 is radioactive with a half-life of 5.25 years that makes it useful in agricultural research and for treatment of cancer.

cobalt deficiency *n.* a cause of pernicious anemia because a cobalt atom is required at the core of each vitamin B_{12} molecule, and vitamin B_{12} is essential for the production of red blood cells. Anemia causes a thin, emaciated appearance in the affected animal or person. Also called enzootic marasmus.

cobblestone *n.* **1.** a rounded stone between 3 and 10 in. (75 and 250 mm) in diameter. **2.** a rounded stone used for paving, as in a cobblestone sidewalk.

cobra *n.* (Latin *colubra*, snake) any of several hooded, poisonous snakes (family Elapidae) in Asia and Africa.

coca *n.* (Quechan *kuka*, coca) a shrub or small tree (*Erythroxylum coca*, family Erythroxylaceae) native to the Andes Mountains in South America. Its leaves are the source of cocaine.

cocaine *n.* (Quechan *kuka*, coca + Latin *ine*, made of) an alkaloid ($C_{17}H_{21}NO_4$) obtained from the leaves of the coca shrub. A local anesthetic, narcotic, and illicit drug that produces intense euphoria and damages the nervous system.

coccidia *n. pl.* (Latin *coccidae*, grains) an old name for Sporozoa, a group of microscopic, parasitic protozoa that invade organs of the digestive tract of animals and, rarely, of people. They pass through a complicated life cycle similar to that of malarial parasites within the host and produce minute, egglike cysts, called oocysts, which are usually discharged from the body of the host with the droppings. The disease is known as coccidiosis.

coccidioidomycosis *n.* (Latin *coccidae*, grains + Greek *mykēs*, fungi + *osis*, condition) a respiratory infection produced by a fungus (*Coccidioides immitis*) that resembles the early stages of tuberculosis. It is often accompanied by a skin rash. Also called desert fever.

coccidiosis *n.* (Latin *coccidae*, grains + Greek *osis*, condition) a disease of the digestive tract of birds, animals, and rarely of humans that causes large economic losses, especially in chickens. It is caused by coccidia (*Isospora hominis* in humans and *Eimeria* sp. and *Isospora* sp. in animals and birds).

coccidiostat *n.* (Latin *coccidae*, grains + *statio*, standing) an agent added to feed or drinking water to control coccidiosis.

cochineal cactus *n.* a tall, treelike cactus (*Nopalea cochenillifera*, family Cactaceae) native to Mexico and Central America that provides most of the food for cochineal insects.

cochineal insect *n.* any of several scale insects (*Dactylopius coccus* and others of family Dactylopiidae) native to Mexico that feed on cochineal cacti and whose red body fluid is a source of cochineal.

cock *n.* (French *coq*, cock) **1.** an adult male chicken or other fowl. Also called rooster or stag. See cockerel.

2. a conical pile of hay, manure, etc. **3.** a device used to regulate the flow of fluids through a pipe (also called a **stopcock**). It consists of a rotatable plug with a passageway that can be aligned with the pipe or turned to either restrict or block the passage of fluid. *v.* **4.** to pull back the hammer of a firearm so it can be discharged. **5.** to set the shutter of a camera ready to make an exposure. **6.** to turn at an angle, often in a jaunty manner, as to cock one's hat.

cockerel (French *coq*, cock, diminutive) *n.* a male chicken less than one year old. See cock.

cocklebur *n.* a troublesome, poisonous, annual weed (*Xanthium* sp., family Aster-aceae) with seeds enclosed in burs with spiny hooked barbs that get caught in fur and annoy animals. Eating the plant can cause nausea, weakness, and staggering. Also called clotbur.

Common cocklebur

cockroach *n.* (Spanish *cucaracha*, cockroach) any of several related insects (family Blattidae) that are among the most common and troublesome pests found in homes. Roaches are flat, quick moving, and nocturnal; avoid light by hiding in cracks and crevices; feed on food scraps, stored foods, and paper; reproduce rapidly; and adapt to a variety of conditions. They can carry organisms that cause dysentery, food poisoning, and diarrhea from garbage cans, sewers, and hospitals to human food. Their fecal droppings can contaminate both food and cooking utensils with bacteria from spoiled food eaten by the roaches. Large populations of roaches produce a very objectionable odor.

American cockroach
adult female

Brown-banded
cockroach
adult female

German cockroach
nymph

Oriental
cockroach
nymph

The four principal cockroach species

cocoa *n.* (Spanish *cacao*, cacao) **1.** brown powder made from the roasted and ground beans of the cacao tree (much of the fat is extracted in the process).

2. the beverage made from this powder mixed with hot milk or water and sugar. See cacao.

coconut or **cocoanut** *n.* **1.** a large, egg-shaped, hairy, brown seed produced by the coconut palm. It has a hard shell lined with white edible meat. It has a milky liquid in the center that is used as a beverage. **2.** the white meat from the seed, often shredded and used in baking and decorating cakes and pastries. See copra.

coconut palm *n.* a tall palm (*Cocos nucifera*, family Arecaceae) that is widely grown in the humid tropics at elevations below about 1000 ft (300 m), especially in lowlands near the seashore, and is used as a source of coconuts and of coconut oil.

cocoon *n.* (French *cocon*, eggshell) **1.** the silky outer covering spun by the larvae of certain insects before passing into the pupae stage. Silk fabric is made from the fibers unwound from cocoons of certain moths (especially *Bombyx mori*, family Bomby-cidae). **2.** other protective coverings such as those made by some spiders for their eggs. **3.** an airtight cover for equipment or machines placed in storage, usually formed by spraying polyvinyl chloride over a supporting web.

cocoyam *n.* a perennial food crop (*Xanthosoma sagittifolia* and *Colocasia esculent*) that grows in open areas of tropical rain forests. The edible part is a large root similar to a yam or a cassava.

cod *n.* **1.** an important food fish (*Gadus morhua*) of the North Sea and northern Atlantic. It is speckled gray or green with a white line on its side. It is commonly 2 to 4 ft (0.6 to 1.2 m) long and is the source of cod-liver oil. Also called codfish. **2.** related species, especially *G. macrocephalus* of the northern Pacific. **3.** the scrotum of a male animal, especially that part left after castration. **4.** a pod, a casing enclosing the seeds of a plant.

codling moth *n.* a small moth (*Carpocapsa pomonella*, fa-mily Tortricidae) whose larvae are the white or pinkish worms that are often found in apples and are also a pest of pears, quinces, English walnuts, and occa-sionally other fruit. Also called appleworm.

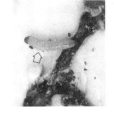

Appleworm
(codling moth larva)

cod-liver oil *n.* oil from the livers of codfish (*Gadus* sp., especially *G. morhuae*) that is used as a food supplement to supply vitamins A and D and to increase the number of red blood corpuscles. See cod.

coelacanth *n.* (Greek *koilos*, hollow + *akantha*, spine) a heavy, bottom-dwelling fish (*Latimeria chalumnae*, order Coelacanthiformes) of deep ocean waters. Sometimes called a living fossil because it was long thought to be extinct, its hollow spine indicates a closer relationship to land animals than that of other fish.

coelenterate *n*. See cnidarian.

coenzyme *n*. (Latin *co*, with + Greek *enzymos*, leavened) an ion or small organic molecule that functions with an enzyme and enables it to modify another molecule, usually by transferring a reactive group from a part of a molecule to another part or to another molecule. The coenzyme is also changed in the process but is changed back by a compensating process so the transfer can be repeated. Many coenzymes are either micronutrient cations or organic molecules synthesized from vitamins. Also called cofactor.

cofactor *n*. (Latin *co*, with + *factor*, doer) a small molecule that enables an enzyme to catalyze an oxidation-reduction reaction or a group-transfer reaction; either a metal ion (e.g., one of the micronutrients) or a coenzyme.

coffee *n*. (Arabic *qahwah*, coffee) **1.** an evergreen shrub or small tree (*Coffea* sp., family Rubiaceae, especially *C. arabica* and *C. liberica*) with opposite leaves, white flowers, and a berry-like fruit containing beanlike seeds that are dried in the sun and used green or roasted to make coffee beverages. It is native to Asia and Africa but is now grown throughout the tropical world, especially in Brazil, Colombia, Mexico, Guatemala, Indonesia, and Costa Rica. **2.** a beverage made from the beans of the coffee shrub and commonly used as a stimulant. It is useful in treating chronic asthma, headaches, and opium poisoning. The active ingredients are caffeine, coffee oils, sugars, proteins, and several volatile flavor oils. **3.** a dark brown color.

cohabitation *n*. (Latin *cohabitare*, to abide together) **1.** living together as husband and wife, especially those not formally married. **2.** two or more species of animals or plants living together.

coherence or **cohesion** *n*. (Latin *cohaerere*, to stick together + *entia*, state) **1.** the attraction or condition of sticking together, especially the molecular force that unites the molecules of a substance. **2.** the strength of the force holding particles together, as the cohesion of soil particles. **3.** the logical connection of words used in a communication or discourse.

coir *n*. (Portuguese *cairo*, a rope or cord) a tropical fiber obtained from the outer husk of coconuts (*Cocos nucifera*, family Palmae) that is used to make mats or as an inexpensive stuffing material.

coke *n*. **1.** a solid carbonaceous residue, containing fixed carbon and residual ash fused together, that is obtained by heating low-sulfur bituminous coal to about 2000°F (1100°C) to drive off the volatile components. It is used as a heating fuel and as a reducing agent for smelting iron ore in blast furnaces. **2.** a residue of the final decomposition process in cracking petroleum; high in carbon and low in hydrogen. Also called petroleum coke or catalyst coke.

cola or **kola** *n*. (Latin *cola*, cola) **1.** a western African tree (*Cola nitida* and *C. acuminata*, family Sterculiaceae) whose nuts are rich in caffeine and theobromine. When chewed, the nuts stimulate the body and relieve hunger. **2.** a carbonated soft drink made with an extract from the kola nut.

colchicine *n*. (Greek *kolchikon*, meadow saffron) a poisonous alkaloid ($C_{22}H_{25}O_6N$) extracted from seeds of crocus and other liliaceous plants (e.g., *Colchicum autumnale*) that is used to treat gout and to double the number of chromosomes in plants, often resulting in larger blossoms. See tetraploid.

cold *adj*. (German *kalt*, cold) **1.** deficient in heat; having a low temperature; chilled. **2.** uncomfortable for lack of warmth. **3.** unfriendly; lacking in affection. **4.** deliberate; unemotional; without passion or enthusiasm. **5.** lifeless; dead. **6.** not playing well in a game; far from the goal. **7.** a blue or green (in contrast to red as a warm color). *n*. **8.** the absence of warmth. **9.** a viral respiratory disorder that causes a watery nasal discharge, congestion, sore throat, coughing, and sneezing. Also called common cold, head cold, or coryza.

cold-blooded *adj*. **1.** poikilothermal; having variable blood temperature dependent on the temperature of the ambient environment; e.g., fish and snakes. See poikilotherm, ectotherm. **2.** ruthless; unsympathetic, as a cold-blooded killer. **3.** sensitive to cold temperatures.

cold cellar *n*. an underground storage room that keeps vegetables and fruits cool during warm weather.

cold frame *n*. an enclosure used to raise early plants or to harden plants in the fall. It is warmed by the greenhouse effect of the sun shining through its transparent glass or plastic cover. Some cold frames also use decomposing horse manure as a source of heat.

cold front *n*. the lead portion of a relatively cool air mass advancing to replace a warmer air mass. In addition to lower temperature, the front is marked by a trough of low pressure, usually accompanied by gusty winds. The colder air wedges under the warmer air and commonly causes short but violent thunderstorms and heavy rain. The skies clear after the front passes, and the weather is cooler and drier than before.

cold manure *n.* manure that ferments slowly and does not heat unduly while in storage, such as cattle, sheep, goat, or rabbit manure. See hot manure.

cold pit *n.* a covered pit extending down into frost-free soil and used for winter storage of bulbs and hardy plants. It is most common in Europe and in some developing countries. See cellar.

cold pocket *n.* a low area where cold air accumulates as it drains down from adjoining higher areas. Also called a frost pocket.

cold sterilization *n.* **1.** food processing that uses a cathode ray or electron-beam gun rather than heat to kill microbes or insects. **2.** sterilization of instruments by chemicals such as alcohol.

cold storage *n.* **1.** refrigeration; use of a stabilized low temperature (usually a few degrees above freezing) to store food or other perishable items. **2.** storage of plants, bulbs, or seed at a low temperature for purposes of vernalization.

cold test (seed germination) *n.* seeds are held at a constant temperature of 40°F (5°C) for three days or longer on moist soil or a wet cloth or blotter. The sample is then placed in an incubator to determine the percentage of germination.

coleoptile *n.* (Greek *koleon*, sheath + *ptilon*, soft feathers) the sheath that covers the embryonic shoot in a monocot seed and during the early development of a plant.

coliform bacteria *n.* a group of bacteria (*Escherichia* sp., *Klebsiella* sp., *Enterobacter* sp., and *Citrobacter* sp.) that normally inhabit the intestines of people and animals. Some strains are pathogenic, but most are not. Coliform bacteria are aerobic and facultative anaerobic, gram-negative, non-spore-forming bacilli that ferment lactose. *Escherichia coli* in drinking water is taken as an indication of human fecal contamination, and other coliform bacteria indicate animal fecal contamination. Their presence suggests that organisms capable of causing disease in humans and animals may also be present.

colitis *n.* (Greek *kolon*, colon + *itis*, inflammation) inflammation of the mucous membrane of the colon, typically accompanied by either constipation or diarrhea and by the passage of mucous and membrane shreds.

collagen *n.* (Greek *kolla*, glue) a fibrous protein of skin, tendons, bones, and connective tissue. Gelatin is derived from collagen by boiling animal tissues.

colligative property *n.* a characteristic of a solution that depends on the concentration of particles in the solution (each dissolved molecule or ion counts as a particle), including osmotic pressure, freezing-point depression, increase in boiling point, etc.

colloid *n.* (Greek *kolla*, glue + *eidos*, form) **1.** a finely divided material dispersed through another material, forming an aerosol, emulsion, gel, sol, etc. (any combination of solid, liquid, and gas except a gas in a gas). The dispersed material is in particles, droplets, or pockets that are larger than molecules but too small to segregate by gravity. **2.** tiny clay and humus particles (smaller than about 0.1 micron in diameter) that will not settle out if dispersed in water.

colloidal *adj.* (Greek *kolla*, glue + *eidos*, form + Latin *alis*, like) **1.** pertaining to a colloid or to its nature and properties. **2.** having a very small particle size; small enough to remain in suspension by Brownian movement without settling out, as colloidal clay.

colloidal peat *n.* a coprolitic, gelatinous organic layer formed as a bottom deposit in lakes, often underlying other types of peat. It has very low permeability and is unsuitable as a plant root medium. Also called sedimentary peat or coprogenous earth.

collop *n.* (Swedish *kalops*, slices of stewed beef) **1.** a small slice or piece of stewed meat. **2.** a small portion of anything. **3.** an area of grazing land capable of supporting a full grown horse or cow for a year.

colluviation *n.* the deposition of material on a toeslope where the flow velocity is reduced.

colluvium *n.* (Latin *co*, with + *alluvius*, washed against) mixed deposits of soil and rock material that accumulate near the base of steep slopes by mass movement from soil creep, frost action, slides, etc. Colluvium is unsorted; water serves as a lubricant but not as the transporting agent. See alluvium, residuum.

colmatage *n.* (Latin *culmen*, ridge + *aticum*, related to) sediment collected from low areas and spread on eroded areas, a laborious practice that is common in populous developing countries such as India and China.

A pile of colmatage

colonial *adj.* (Latin *colonia*, colony) **1.** pertaining to a colony or group of colonies; e.g., the 13 colonies that founded the U.S.A. **2.** living in a colony; e.g., the ants in an anthill.

colonization *n.* (Latin *colonia*, colony + *izatio*, action) **1.** the process of a plant or animal species entering and becoming established in an area previously unoccupied by that species. **2.** the entrance and establishment of humans of a specific national origin in an area outside their nation while remaining political subjects of their state of origin.

colony *n.* (Latin *colonia*, colony) **1.** the group of worker bees, drones, and a queen bee that live together in a hive. **2.** a cluster of microbes grown on a culture medium, usually developed from a single bacterium, spore, or other propagule. **3.** a group of people bound by ethnic ties, religious ideals, or other common interest that migrates to a new land for permanent settlement, especially those who claim the new land on behalf of their nation of origin. **4.** one of the groups of immigrants in the early European settlement of the Americas; e.g., any of the 13 colonies that became the origin of the U.S.A.

color or **colour** *n.* (Latin *color*, hue) **1.** the quality related to the dominant wavelength of light reflected or transmitted by an object and detected by the rods in the retina of an eye. The three subtractive primary colors are cyan, magenta, and yellow, and the three additive primary colors are red, green, and blue. Other colors are mixtures of these in various proportions. A balanced mixture of all three additive primary colors gives white, gray, or black, depending on whether nearly all, part, or almost none of the light is reflected. The distinct secondary colors of the spectrum, from long to short wavelengths, are red, orange, yellow, green, blue, indigo, and violet. These seven colors are sometimes known as the colors of the rainbow. See Munsell color system. **2.** the hue of one's skin, often related to race and commonly called white, yellow, brown, or black. Those other than white are sometimes collectively called colored. **3.** a pinkish flesh tone taken to mean either good health or embarrassment. **4.** vivid details in descriptions of objects or activities. **5.** a dye or pigment used to change the hue of a material or object. **6.** a trace of gold or other valuable material sought by a prospector. *adj.* **7.** something that has or shows color, as a color photograph. *v.* **8.** to add or apply color to an object or a depiction. **9.** to blush or otherwise show color. **10.** to express a biased version of something, as an explanation colored by one's personal viewpoint.

Colorado potato beetle *n.* a striped beetle (*Leptinotarsa decemlineata*, family Chrysomelidae) that feeds primarily on potatoes and to a lesser extent on several related plants. This native of Mexico has spread throughout the part of the U.S.A. east of the Rocky Mountains and to many other parts of the world. It is a major limiting factor in potato production wherever it occurs in a potato-growing area.

Colorado potato beetle

colorimeter *n.* (Latin *color*, color + Greek *metron* measure) a spectrophotometer; an instrument for measuring the intensity of light of a chosen wavelength. It is used extensively in laboratories, e.g., to determine phosphorus concentration.

colostrum *n.* (Latin *colostrum*, beestings) the milk produced for the first few hours or days following birth. It is high in protein, mineral content, and immune factors that can be passed to newborn offspring.

colt *n.* (Anglo-Saxon *colt*, a little donkey) **1.** a young male horse, donkey, zebra, or other member of the horse family less than 4 years of age (a young female horse is called a filly). **2.** a foal.

colter or **coulter** *n.* (Latin *culter*, knife) a sharp rolling disk or blade attached in front of a plowshare, drill shoe, or tool shank to cut crop residues and clods so the implement can work through residues without plugging.

columbium (Cb) *n.* (Latin *Columbia*, America + *ium*, element) an alternate name for niobium, especially prior to 1950. See niobium.

columnar *adj.* (Latin *columnaris*, columnar) **1.** formed of or in columns, as columnar data or text. **2.** having columns, as a building with a columnar façade. **3.** resembling columns, as columnar soil structure (prismatic soil structure with rounded tops on the prisms, often associated with sodic soil). Typical rounded and bleached tops on columnar soil structure are sometimes called biscuit tops.

Columnar soil structure with bleached tops

comb *n.* (Anglo-Saxon *camb*, comb) **1.** a strip of plastic, hard rubber, metal, etc., with long thin teeth used to clean and arrange the hair of a person or animal or to hold hair in place, often as an ornament. **2.** a toothed part of a machine that serves to clean, sort, or straighten material that passes by or through it. **3.** the fleshy, toothed, usually red crest on the heads of roosters and some other fowls. **4.** two series of back-to-back hexagonal cells in a beehive. Honey in such cells is called comb honey.

combine *n.* (Latin *combinare*, with pairs) **1.** a machine that both harvests and threshes as it moves through a field of grain or other seed crop. Also called a reaper. **2.** a collection of individuals or groups united for a common purpose; e.g., a cartel or syndicate. *v.* **3.** to harvest a crop with a combine. **4.** to bring together; to unite two or more parts into a whole. **5.** to unite with other persons or groups in a concerted effort.

combined sewer *n.* a water-disposal system that carries both sewage and storm water runoff (common in small, older systems). Normally, its entire flow goes to a waste-treatment plant, but a heavy storm may overload the system and cause an overflow, which might cause untreated mixtures of storm water and sewage to flow into receiving waters.

comfrey *n.* (French *confirie*, comfrey) **1.** a perennial herb (*Symphytum* sp., family Boraginaceae) with rough, hairy leaves and drooping clusters of purple, blue, yellow, rose, or white flowers; native to Eurasia. **2.** the common comfrey (*S. officinale*) is sometimes grown as an ornamental and may be eaten by livestock. Its root is used in medicines for coughs and intestinal disorders. It is known among organic gardeners as a "survival plant" that will thrive where others may fail.

commensalism *n.* (Latin *commensal*, with the table + *ismus*, act or condition) two or more species of animals, plants, or fungi living together in a nonparasitic host–guest relationship. The association benefits the guest(s) but has no noticeable effect on the host organism. See symbiosis.

commercial timberland *n.* forest land that produces, or is capable of producing, crops of industrial wood in excess of 20 ft³/ac (1.4 m³/ha) per year and is not withdrawn from timber utilization by statute or administrative regulation. Note: Inaccessible and inoperable areas are included.

comminution *n.* (Latin *comminutis*, reduced in size) **1.** pulverization; grinding a substance to a fine powder. **2.** mechanically shredding or breaking waste into small particles to make it more manageable. Comminution is used as the primary stage in wastewater treatment and in solid-waste management to make it easier to separate combustible materials, ferrous metal, other metals, etc., from each other.

Commodity Credit Corporation (CCC) *n.* an agency established in the U.S. government in 1933 to make loans to farmers as a means of price supports for selected crops and made a part of the U.S. Department of Agriculture in 1939. The price supports were terminated in 1998 by the Freedom to Farm Bill, but the agency continues to fund other programs as a part of the Farm Service Agency, including Food for Peace and promotion of U.S. agricultural exports for international markets.

common property *n.* property that belongs to everyone; e.g., national forests, air, oceans, and roads.

communal *adj.* (Latin *communalis*, of a community) **1.** shared by all the members of a group or community, as communal life. **2.** belonging to a group or community as a whole rather than to any individuals, as communal land or a communal project. **3.** involving two or more communities, as communal conflict.

communalism *n.* (French *communalisme*, commune system) **1.** government that emphasizes community authority. The state or nation is considered a federation of independent communities, often with internal ethnic or religious ties. **2.** stronger devotion to one's own ethnic or cultural group than to all of society. **3.** biotic interactions, either beneficial or harmful, among species of plants, animals, and humans that inhabit the same area.

community *n.* (Latin *communitas*, community or fellowship) **1.** a group of people residing in a particular area and commonly associating among themselves. **2.** an area where the people interact as a community; a neighborhood. **3.** a group of people united by a common interest even though they may not form a residential community; e.g., the business community or the religious community of a city. **4.** an association of groups or even nations that share a certain heritage or purpose, such as the European Economic Community. **5.** all of the living organisms that inhabit an ecosystem and interact with one another.

community ecology *n.* the study of interactions among organisms that share a particular location and interact with each other.

compaction *n.* (Latin *compactus*, bound together) **1.** the process of increasing density; of compressing material into a smaller volume. **2.** the condition of being compressed.

companion crop *n.* a crop planted along with another crop, especially a fast-growing annual crop (usually a small grain) planted with a slow-growing perennial forage crop. Companion crop is preferred to the term nurse crop because the relationship is somewhat competitive for water and plant nutrients rather than nursing (the seeding rate should be adjusted to avoid excessive competition).

compass *n.* (Latin *com*, with + *passus*, a pace) **1.** an instrument that indicates direction, usually by a magnetized needle free to turn in a horizontal plane, so it points toward the magnetic North Pole or South Pole. See agonic line, magnetic poles. **2.** a dip compass with a needle hung to move in a vertical plane for tracing magnetic iron ore. **3.** an instrument for drawing arcs or circles and transferring measurements. **4.** the line marking the perimeter of an area or property. **5.** a real or figurative limit, as the compass of a year or the compass of propriety. *v.* **6.** to surround, encircle, or comprehend.

compass plant *n.* a bristly leaved plant (*Silphium laciniatum*, family Asteraceae) with yellow blossoms that grows on prairies in central U.S.A. Its lower leaves are said to orient their vertical edges north and south. Also called rosinweed or polar plant.

compatibility *n.* (Latin *compatibilis*, suffer with) **1.** the ability of living organisms to live and function together harmoniously. **2.** the ability of a male and a female to mate and reproduce. **3.** the ability to function together philosophically, as groups with compatible ideologies. **4.** mechanical suitability; e.g., software that will run on a particular computer. **5.** chemical suitability for mixing; e.g., fertilizers that can be mixed with each other, pesticides that can be applied together, or a pesticide that can be added to a fertilizer (an adjuvant may improve chemical compatibility).

compensation point *n.* **1.** a condition of balance whereby living plants have no net gain or loss in energy; the light intensity and/or temperature required for the rate of photosynthesis to equal that of respiration. **2.** the water depth where photosynthesis and respiration are equal; the lower boundary of the euphotic zone. **3.** the minimum carbon dioxide concentration in the atmosphere in an illuminated, sealed container with plants growing inside. At 77°F (25°C), C_3 plants, such as soybeans, will lower the carbon dioxide concentration to about 50 ppm, whereas C_4 plants, such as corn, will reduce it to below 10 ppm.

compensatory payment *n.* a method of governmental support of commodity prices; a government agency pays farmers the difference between a guaranteed support price and the market price they receive for the commodity.

competition *n.* (Latin *competitio*, come together) **1.** rivalry. Two or more individuals or groups vying for the same objective. **2.** a contest, especially in an organized sport. **3.** the other participant or participants in a contest. **4.** the struggle for food, water, space, etc. between members of the same or different species when the local supply is inadequate for all. Example: Trees in a forest compete for light, water, and nutrients. See ecological interaction.

competitive crop *n.* **1.** a vigorous crop used to smother weeds or other undesired plants growing on a piece of land. Also called smother crop. **2.** crops with similar management schedules that compete for time in a farmer's work schedule.

competitive exclusion principle *n.* when two or more species of organisms compete for the same limiting resource, one will eventually win and the loser must adapt, move away, or die. Example: The red squirrel and the gray squirrel in the Ozarks. The gray squirrel won and the red squirrel moved away.

complementary enterprises *n.* two or more enterprises that contribute to each other; e.g., corn and hogs; or pasture, hay, and cattle.

complementary genes *n.* genes that interact to produce a different effect than either produces alone. Example: The C and P genes cause sweet peas to produce purple blossoms when both are present, but the blossoms are white when either C or P is present without the other.

complement fixation *n.* a biological reaction that binds an identifying substance to a surface; e.g., an antibody on an antigen or a marker on part of the immune system so other parts of the body defense system can tell the difference. This interaction enables tests for antibodies to be used as antigen indicators for the diagnosis of several diseases.

complete fertilizer *n.* a mixed or manufactured chemical fertilizer that contains significant, guaranteed concentrations of nitrogen, phosphorus, and potassium. Its label has three numbers showing the guaranteed percentages of total nitrogen (%N), available phosphorus (%P_2O_5 equivalent), and water-soluble potassium (%K_2O equivalent). It may also contain other fertilizer nutrients.

complete flower *n.* a flower containing at least four whorls of flower parts: at least one pistil surrounded by stamens, petals, and sepals. See perfect flower, incomplete flower.

complex slope *n.* a variable slope; land surface that changes gradient along the contour and/or up and down the slope. Such slopes have variable spacings between contour lines on a topographic map. Contour farming, strip cropping, and terrace construction are difficult or impractical on complex slopes.

Compositae *n.* the older name for family Asteraceae.

composite *adj.* (Latin *compositus*, put together) **1.** composed of two or more parts. **2.** belonging to the Compositae Asteaceae) family of plants. **3.** composed of more than one mathematical unit, as a composite number or a composite function. *n.* **4.** something that contains parts, as a composite structure. **5.** a composite plant (family Asteraceae). **6.** an assemblage or grouping of several elements, as a composite photo. *v.* **7.** to put components together; to make a composite.

compost *n.* (Latin *compositus*, composite) a mixture of organic residues, often interlayered with soil and manure. Lime and fertilizer may be included along with grass clippings, table scraps, etc. Compost is usually stored in a pile or bin. It is kept moist and occasionally mixed until the organic materials have disintegrated and it no longer heats excessively. It is then applied to a garden or other soil to improve its structure and to supply plant nutrients. Most organic materials can be used to make compost, but a few that should not be used are black walnut leaves and husks (see juglone) and seeds, rhizomes, or runners of pernicious weeds such as Bermuda grass and quack grass; pine needles and eucalyptus leaves (too slow to decompose); and diseased plants.

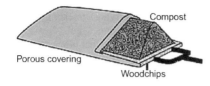

A compost pile with ventilation tubes

composted sewage sludge *n.* dried and composted solid materials removed in primary treatment of sewage. The process is faster if the sludge is mixed with wood chips or other chunky organic material that improves aeration. The composted sludge, like other compost, can be used for mulching and soil improvement.

concentrate *n.* (French *concentrer*, to center) **1.** a substance whose solid content has been increased, usually by removal of much of its natural water content; e.g., many drink mixes. **2.** animal feed high in energy, sometimes used with reference to a specified component such as protein concentrate. **3.** a chemical product that needs to be diluted before use; e.g., pesticides are often concentrated to reduce the cost of packaging and shipping. *v.* **4.** to increase in strength by removing diluents. **5.** to gather toward a designated place. **6.** to center one's efforts, attention, or thoughts on a particular project, idea, or topic.

concentration *n.* (French *concentrer*, to center + *atio*, state or action) **1.** the amount of active material in a unit volume or mass of diluent, solution, or mixture. **2.** close attention given to what one is doing or to one's surroundings.

conception *n.* (Latin *conceptio*, to conceive or understand) **1.** becoming pregnant; the process of combining male and female gametes. **2.** the early development of an idea; a mental image. **3.** a general notion or impression of something; e.g., a brief description of a project or an artist's sketch of a machine.

conception rate *n.* in animal breeding, the percentage of females that conceive from the first service.

conch *n.* (Latin *concha*, shell) **1.** the spiral shell of a gastropod mollusk of the family Strombidae. It has a screwlike point on one end and an open side with a wide outer lip. **2.** the mollusk that inhabits such a shell (some are edible). **3.** a concha; an anatomical structure whose form resembles a shell; especially an outer ear.

conchoidal *adj.* (Latin *concha*, shell + Greek *eidos*, form + Latin *alis*, having the form of) like the surface shape of certain conch shells; gently curved like the surface of broken glass.

conchuela *n.* an imported species of stink bug (*Chlorochroa ligata*, family Pentatomidae) common in southwestern U.S.A. that eats and injures leaves of peas, beans, cotton, tomato, pecan, citrus, and many other plants.

concolorous *adj.* (Latin *con*, with + *color*, hue + *ous*, full of) having a single uniform color over the entire body of an animal.

concrete *n.* (Latin *concretus*, solid) **1.** a stonelike building material composed of a mixture of gravel, sand, cement (especially portland cement), and water. The cement hydrates, and the concrete hardens as it dries. See cement. Note: When lime (CaO) is used as the cementing agent, the product is known as lime concrete. *adj.* **2.** made of concrete; e.g., concrete blocks. **3.** of or relating to real objects; not abstractions. **4.** something specific (not vague), as a concrete proposal.

concretion *n.* (Latin *concretio*, grown together) **1.** a localized concentration in soil of a chemical compound, such as calcium carbonate ($CaCO_3$), iron oxide (e.g., Fe_2O_3), or manganese dioxide (MnO_2), in the form of an aggregate or nodule that is often roughly spherical, different in color, and harder than the surrounding soil. **2.** a solidified inorganic mass formed in a body cavity; e.g., a kidney stone (to be distinguished from an ingested object that may cause "hardware" disease).

condemnation *n.* (Latin *condemnatio*, condemnation) **1.** censure; labeling as a wrongdoer, often accompanied by punishment. **2.** a judicial process to force an owner to sell real property, as when a governmental body exercises the power of eminent domain.

condensation *n.* (Latin *condensatio*, condensation) **1.** the process that changes a gas to a liquid. **2.** the liquid formed from a gas; e.g., dew. **3.** compaction; compressing the volume and increasing the density of a bulky material. **4.** shortening, as condensation of a novel.

condor *n.* (Quechan *cuntur*, condor) a large soaring American bird with mostly black plumage, white markings, and a bare red neck and head; either the California condor (*Gymnogyps californianus*, family Cathartidae) or the Andean condor (*Vultur gryphus*, family Cathartidae).

conducting tissue *n.* plant tissue that serves primarily for the movement of water and food; especially xylem for upward transport of water and nutrients from the soil and phloem for downward transport of carbohydrates.

conduit *n.* (French *conduit*, guide) **1.** any channel (e.g., a pipe or a ditch) intended for the conveyance of water or other fluid. **2.** a means of conveying anything, as an information channel. **3.** a metal or plastic pipe or channel used to protect electric wiring.

cone *n.* (Latin *conus*, cone) **1.** a geometric figure with a circular cross section whose diameter is proportional to the distance from the point of the cone. **2.** anything with a conical shape, as the cone of a volcano. **3.** one of the cells in the retina of the eye that detects both light and color (so named because of the conical shape of such cells). **4.** the seed-enclosing structure of certain evergreen trees and shrubs (e.g., pine trees).

conenose *n.* any of several hemipteran insects (*Triatoma* sp.) with a sucking beak that has a conelike base. They suck blood, sometimes transmit parasites, and are found in tropical America and southern U.S.A. See bloodsucking conenose.

cone of depression *n.* the roughly conical shape of the upper surface of a water table when water is pumped from a well.

cone penetrometer *n.* an instrument designed to evaluate soil strength by measuring the pressure required to force a rod with a cone-shaped tip into the soil.

confined water *n.* groundwater trapped in a permeable layer between two impermeable layers. It may be removed through a well (perhaps under artesian pressure) but is probably either nonrenewable or so slowly renewable that it is easily depleted. See connate water, fossil water.

confinement system *n.* a method of growing animals or birds in a controlled indoor environment, usually with large numbers in a small space.

conflagration *n.* (Latin *con*, with + *flagrare*, to burn + *tion*, action) a large fire that causes great destruction.

confluence *n.* (Latin *con*, with + *fluentia*, flowing together) **1.** the point of junction or the combined flow of two or more streams. **2.** the merging or assembling of a group of persons, things, events, or concepts.

conformation *n.* (Latin *conformatio*, forming) **1.** arrangement of parts; structure and form of an entity; configuration. **2.** external form; e.g., shape, sleekness, fullness. **3.** the arrangement of atoms in a molecular structure, especially any of the alternate conformations of an organic molecule resulting from rotation at bonds between carbon atoms.

confused flour beetle *n.* an insect (*Tribolium confusum*, family Tenebrionidae) commonly found in stored products such as flour, bran, grains, beans, peas, dried plant roots, dried fruits, nuts, baking powder, ginger, chocolate, and cayenne pepper. Also called bran bug.

congelation *n.* (Latin *congelatio*, congealing) **1.** the process of solidifying, thickening, or coagulating, especially of forming a gel. **2.** a mass formed by solidifying, often one that blocks a passage such as a blood vessel. **3.** freezing or frostbite.

congelifraction *n.* (Latin *congelatio*, congealing + *fractio*, breaking) disintegration of rocks into small fragments and individual mineral grains caused by water freezing in tiny cracks. Synomyms: frost riving, frost shattering, frost weathering, gelivation, and gelifraction.

congeliturbation *n.* (Latin *congelatio*, congealing + *turbidus*, stirred up + *ation*, action) soil mixing by frost action, including heaving and solifluction.

congeneric *adj.* (Latin *con*, with + *genericus*, kind) **1.** belonging to the same genus. **2.** similar in kind or class.

congenial *adj.* (Latin *con*, with + *genialis*, nature or kind) **1.** of a kindred nature; friendly; agreeable with one another; compatible. **2.** similar in tastes and temperament. **3.** suited to one's liking, as congenial work. **4.** plants able to cross-fertilize readily or to unite easily, as with stocks and scions suitable for grafting.

conifer *n.* (Latin *conus*, a cone + *ferre*, to bear) any of several trees or shrubs of order Coniferales, including cedar and cypress (family Cupressaceae), pine, fir, spruce, hemlock, and larch (family Pinaceae), yew (family Taxaceae), and redwood, sequoia, and bald cypress (family Taxodiaceae). Conifers produce their seeds in multiscaled cones ranging in length from 1 to 20 in. (3 to 50 cm). Their leaves are either needles or scalelike. Most conifers are evergreens that produce new leaves before all of the old leaves are shed. The two common exceptions, bald cypress and tamarack (a species of larch), are called deciduous conifers.

conilignosa *n.* forest vegetation dominated by trees and shrubs with needlelike foliage, such as pines and/or firs.

conjunctivitis *n.* (Latin *conjunctivis*, connecting membrane + *itis*, inflammation) inflammation of the conjunctiva (the mucous membrane that lines the eyelids and covers the front part of the eyeball), which affects humans and many species of animals. Conjunctivitis, keratitis, and some other eye diseases are sometimes called pinkeye.

connate *adj.* (Latin *connatus*, born at the same time) **1.** congenital; innate; existing together since birth. **2.** having a common origin; cognate. **3.** firmly united in a structure such as a compound leaf.

connate water *n.* fossil water trapped in pores when sediments are deposited. It may be highly mineralized and even briny, and may be tapped by a well, but will not be replaced. Synonyms: confined water, fossil water, and native water.

conservation *n.* (Latin *conservatio*, keeping) wise use that avoids waste and maintains resources for the future. It differs from preservation that excludes use of the resources. However, conservation has come to mean many different things to different users. A nature lover may see it as preserving pristine wild areas, whereas a hunter may be more concerned with maximizing the number of game animals, and a logger may see careful harvesting of mature timber as conservation. Soil conservation may emphasize erosion control in one setting and maintaining productivity in another, whereas water conservation probably means not only minimizing waste but also maximizing the amount available for one or more designated uses.

conservation district *n.* a special-purpose subdivision of state government locally organized to promote soil and water conservation within its boundaries (usually coinciding with county boundaries). Conservation districts have been formed in all 50 states, Puerto Rico, and the Virgin Islands. They include about 99% of all U.S. farms and ranches (most were originally called soil conservation districts).

conservation pool *n.* the water that cannot be drained from a reservoir because it is below the level of the outlet.

conservation practice *n.* any structure built or change made in manner of doing things for the purpose of conserving a resource, especially land treatments used to conserve soil, water, wildlife, or plant life.

Conservation Reserve Program (CRP) *n.* a U.S. government program established in 1985 to protect the most fragile part of U.S. cropland with vegetative cover; expanded in 1990 to include wildlife, water, and environmental protection emphases. It enabled landowners to enter into a 10-year agreement with the USDA to receive payments in exchange for maintaining perennial vegetation on about 36.4 million acres (15 million ha) of highly erodible soil. See sodbuster law, swampbuster law.

conservation tillage *n.* **1.** any tillage method that reduces soil erosion to an acceptable rate. Systems used for this purpose include no-till (crops are planted directly through the residues of the previous crop), reduced tillage (one or more of the conventional tillage operations are omitted, especially plowing and harrowing), strip tillage (only a narrow strip is tilled where the seeds are planted), and chisel plowing to replace moldboard plowing. **2.** any tillage system that leaves plant residues covering at least 30% of the soil surface when the next crop is planted.

consignment *n.* (French *consigner*, to deliver + *ment*, result or product) **1.** an arrangement whereby a consignor's property (e.g., produce or livestock) is entrusted to a consignee for sale or other designated action. **2.** property that is consigned.

consistency or **consistence** *n.* (Latin *consistere*, to stand together) **1.** reliability, steadfastness, continuity in behavior. **2.** harmony or uniformity among the parts of a thing. **3.** coherence and firmness; e.g., the plasticity and stickiness of a soil material. **4.** viscosity, as a watery consistency.

consolidated *adj.* **1.** (Latin *con*, with + *solidare*, to make solid) gathered or compressed into a compact mass. **2.** changed into a durable, hardened mass, often by drying and/or cementation, as consolidated rock.

consortium *n.* (Latin *consortis*, neighbor or partner) **1.** a community of different plant species, generally representing different phyla, that live together in close association. **2.** a group of businesses, educational institutions, or governments forming a partnership for a specific project or objective.

conspecific *n.* **1.** one of a group of individuals that belong to the same species. *adj.* **2.** pertaining to the same species.

constitution *n.* (Latin *constitutio*, establishment) **1.** the composition or physical characteristics of a body; factors that influence ability to live, grow, perform, and reproduce. **2.** the health and strength of a body. **3.** the act or process of establishing something. **4.** the fundamental principles or statement for establishing and governing an organization or nation, usually recorded in a document.

constraint *n.* (Latin *constringere*, drawing together) **1.** a limiting factor; a restriction on action or movement. **2.** self-control; repression of impulsive action.

Consultative Group on International Agricultural Research (CGIAR) *n.* an organization established in 1971 to support agricultural research on an international basis. See international agricultural research centers.

consumer *n.* (Latin *consumere*, to use up) **1.** any living thing whose food supply depends on the energy stored by other living things. People and animals are consumers, depending on green plants or other animals as their food sources. **2.** a person or organization that purchases goods and services for its own use; an end user.

consumption *n.* (Latin *consumptio*, process of consuming) **1.** use that consumes; e.g., eating, burning, wearing out, or any other manner of use that leaves the resource unusable. **2.** the rate a resource is consumed, as the fuel consumption of an automobile. **3.** an old name for tuberculosis.

consumptive use of water *n.* evapotranspiration; the total quantity of water taken up by vegetation (for both transpiration and growth) plus that evaporated from the soil. See evapotranspiration, duty of water.

contact dermatitis *n.* an acute inflammation of the skin caused by an allergic reaction to a toxic substance. Also called dermatitis venenata. See poison ivy.

contact pesticide *n.* an algacide, herbicide, or insecticide that kills pests when it touches them (in contrast to a systemic pesticide that must be eaten or absorbed from the soil).

contagious abortion *n.* See brucellosis.

contagious equine abortion *n.* a bacterial (*Salmonella abortivoequina*) disease among mares that can cause abortion at any stage of pregnancy, especially in the later stages.

container-grown *adj.* grown with a rooting medium in a container for ease of handling and transport, either for container planting or to be removed from the container for transplant. The container should be small enough for the roots to hold the rooting medium together at the time of transplant.

container planting *n.* a system of growing and planting seedlings in containers such as baskets, blankets, tubes, pots, or porous blocks. The entire container is planted along with the seedling in the field or garden.

contaminant *n.* (Latin *contaminatus*, defiler) any substance that pollutes or degrades the quality of air, water, soil, food, chemical, or other material; an extraneous chemical, biological, or radiological substance.

contamination *n.* (Latin *contaminare*, to defile + *ation*, being) **1.** a material whose presence degrades a larger mass of another material. **2.** a process that contaminates. **3.** the condition of being contaminated.

contemporary *adj.* (Latin *con*, with + *temporarius*, the time) **1.** of the same time; existing, living, or occurring simultaneously. **2.** modern; of the present age. *n.* **3.** a person who lived at the same time as another or is of the same age as another.

continent *n.* (Latin *continens*, containing) **1.** one of the seven large landmasses of the world: Africa, Antarctica, Asia, Australia, Europe, North America, and South America. **2.** the mainland (especially of Europe) as distinguished from associated islands or peninsulas.

continental climate *n.* climate affected primarily by a great landmass. It exhibits a wider range of temperatures than those of areas tempered by a nearby ocean.

continental drift *n.* the gradual shifting of the continents around the surface of the Earth. Until an estimated 200 million years ago, South America was joined to Africa and North America to Europe, but they have been drifting apart, widening the Atlantic Ocean and narrowing the Pacific Ocean since that time. Alfred Wegener, a German scientist, is credited with publishing in 1915 the first coherent theory (based on earlier suggestions) that the continents were once joined into a single landmass that he named Pangaea. See Pangaea, plate tectonics.

continental plate *n.* a large crustal segment that holds a large landmass and floats on the underlying layers of the Earth's crust. Earthquakes and volcanic activity are most common around the margins of these plates. See plate tectonics.

continental shelf *n.* the gently sloping floor of the ocean adjacent to most continental shorelines outward to an average depth of about 400 ft (120 m), where the slope becomes much steeper (often considered the true margin of the continent).

contingency plan *n.* a plan for an organized, coordinated course of action to be followed in case of a fire, explosion, or other accident, including building evacuation, handling of any toxic chemicals or hazardous wastes that might be released, and overall safety requirements; usually recorded in a formal document.

continuum *n.* (Latin *continuus*, held together) a connected series or set of elements with no discernible divisions, breaks, or gaps.

contour *n.* (Latin *con*, with + *tornare*, turn) **1.** the outline of an object, landmass, or other solid structure, especially one with a complex, curved shape. *v.* **2.** to make an outline showing the shape of a curved object. *adj.* **3.** working around a contour, especially following a level contour line, as in contour tillage.

contour curvature *n.* the shape of a curved contour line. The contour curvature around a nose or other projection is convex, whereas that inside a bowl is concave. See slope.

contour farming or **contouring** *n.* tillage operations and crop rows that follow contour lines so that small ridges on the soil surface will hold water, thus reducing runoff and erosion.

contour interval *n.* the vertical distance represented by the space between two adjacent contour lines.

contour irrigation *n.* the practice of using irrigation furrows that almost follow contour lines, so the water flows on such a gentle slope that it does not cause erosion.

contour line *n.* a line joining points on the land surface having the same elevation. A coordinated set of such lines drawn on a map depicting a constant vertical spacing outlines the topography of an area.

contour map *n.* a map showing contour lines that depict the topography of an area; a topographic map.

contract farming *n.* an arrangement whereby a farmer will sell and a purchaser will buy either all or a specified amount of a future crop at a specified price.

contract feeding *n.* a contractual arrangement for finishing cattle or other livestock for market. Typically, the stock raiser retains ownership of the livestock; the feeder furnishes the feed, equipment, and labor and is compensated on the basis of the increase in weight of the livestock.

contrail *n.* a white line of condensed moisture that trails behind an airplane when conditions are right. Also called a vapor trail.

control *n.* (French *contrôle*, verification) **1.** regulating the action of something or someone, especially restricting to less than the maximum rate or amplitude. **2.** the killing of a pest or limiting of its activity; loss prevention. **3.** a test unit given a standard treatment for comparative purposes in an experiment; e.g., a check plot. *v.* **4.** to check, verify, or regulate an activity. **5.** to reduce or limit a process; e.g., to control soil erosion.

controlled burning *n.* setting fire to vegetation or plant residues to fight a fire, reduce fire hazard, or accomplish specific silvicultural, wildlife, grazing, or cropping objectives. Also called prescribed burning.

controlled-release fertilizer *n.* fertilizer that releases one or more plant nutrients to the soil solution slowly over a period of several weeks or months. The fertilizer may have low solubility (e.g., dicalcium phosphate), be coated with a protective material (e.g., sulfur-coated urea), or be an organic material that takes time to decompose (e.g., manure or sewage sludge).

controlled traffic *n.* restriction of travel in a field so that all wheels follow the same paths. The intent is to limit compaction from tillage and other operations to the paths.

convection *n.* (Latin *convectio*, bringing together) transfer of heat or other energy by mass movement of a fluid medium such as air or water.

convective rainfall *n.* rainfall produced when solar heat warms an air mass and causes it to rise rapidly (often drawing moist air from adjacent areas), cool to a temperature below the dew point, and condense moisture that forms raindrops. The rainfall produced is usually localized, but it can be quite intense and can have thunder and lightning associated with it.

conventional tillage *n.* the established tillage practices of a region; e.g., plowing, disking, and harrowing followed by planting (and cultivating when row crops are grown).

convergence *n.* (Latin *convergere*, incline together) **1.** the trend or process of coming together; e.g., the narrowing of the space between nonparallel lines until they meet or the gradual diminishing of differences of opinion. **2.** a point of junction. **3.** the joining of air masses coming from different directions. **4.** the development of similar characteristics in unrelated plants or animals living in the same or similar environments.

convergent lady beetle *n.* a small, bright red and black beetle (*Hippodamia convergens*, family Coccinellidae) that parasitizes certain plant insect pests and their eggs and can be helpful in the control of such insects. See lady beetle.

converse adaptation *n.* adaptation of an introduced plant or other organism to its new habitat to such an extent that it crowds out competing species. Examples: Bermuda grass in warm climates, starlings, and house sparrows.

conversion *n.* (Latin *conversio*, complete change) **1.** a major change in appearance, belief, character, function, or use. **2.** a change in allegiance from one organization to another, especially the adoption of a religion or a change in religion or politics. **3.** an unlawful appropriation of property. **4.** a physical change, as water freezing into ice. **5.** a chemical change, such as the conversion of sugar into alcohol by fermentation. **6.** an internal atomic change, as the conversion of radioactive uranium into lead.

cool *v.* (German *kuhl*, cool) **1.** to reduce the temperature, either naturally or artificially, as for preservation or curing of milk or meat products. **2.** to reduce emotional intensity. *adj.* **3.** not warm, but only moderately cold. An air temperature between 50° and 60°F (10° to 15°C) feels cool to most people. **4.** calm and composed; not agitated. **5.** indifferent; not worried. **6.** (slang) an expression of agreement; approval of an idea or action.

coolant *n.* (German *kuhl*, cool + *ant*, something that does) a liquid or gas used to carry away the heat generated by power production in nuclear reactors, internal combustion engines, and/or various industrial and mechanical processes.

cooler *n.* (German *kuhl*, cool + *er*, something that does) **1.** either a room or a chestlike device for cooling and storing perishable products, such as milk, eggs, or meat. **2.** something that refreshes, as a cold drink. **3.** (slang) jail.

cooler shrinkage *n.* weight loss during storage in a cooler. An animal carcass typically loses about 2.5% of the original carcass weight.

cooling tower *n.* a structure that helps remove heat from water used as a coolant by dispensing the heat into the atmosphere.

cool-season plant *n.* a plant that thrives best in cool weather; e.g., certain grasses, small grains, garden peas, radishes, cabbage, onions, and Irish potatoes. See warm-season plant, vegetable.

co-op *n.* a cooperative society, residence, store, or other enterprise. See cooperative.

coop *n.* (Latin *cupa*, tub) a small airy cage or enclosure used to confine chickens, other poultry, or small animals; usually made of wire netting.

cooperage *n.* (Latin *cupa*, tub, cask + *aticum*, related to) barrels, casks, kegs, and tubs made by a cooper, usually of wooden staves and metal hoops. Tight cooperage is made from white oak or other nonporous woods to hold liquids such as whiskey. Slack cooperage is made from ash, elm, or other porous but tough wood to hold nonliquids such as rice or cranberries.

cooperative *adj.* (Latin *cooperatus*, working together) **1.** willing to cooperate. *n.* **2.** an enterprise organized, owned, and managed for the benefit of those using its services. Many such organizations are owned and used by farmers to facilitate their off-farm business activities—buying farm supplies, marketing farm products, furnishing electric and telephone service, etc. Essential features include democratic control and distribution of profits to the owner-users in proportion to their use of the services of the cooperative.

cooperative agreement *n.* a written document evidencing the intent of two or more parties to cooperate in carrying out an undertaking and indicating what each party agrees to do; e.g., an agreement between a farmer and a soil conservation district.

Cooperative Extension Service *n.* an agency now known as the Cooperative State Research, Education, and Extension Service; a means of extending the educational activities of the land-grant colleges and universities and the U.S. Department of Agriculture to farmers, ranchers, and other landowners and users that was created under federal legislation by the Smith-Lever Act of 1914 to help people identify their own problems and opportunities and then to provide practical research-oriented information to help them solve their problems. The Cooperative Extension Service has long been responsible for programs in four major areas: agricultural and natural resources, home economics, community development, and 4-H and youth development. Environmental education has been assigned more recently. See county agent; Knapp, Seaman Asahel.

coot *n.* (Dutch *koet*, coot) **1.** an aquatic bird (*Fulica americana* of North America or *F. atra* of Europe and Asia) about a foot (30 cm) long with webbed feet, a conical beak, and sooty black plumage with gray and white markings that is related to cranes, rails, etc. **2.** any of several other birds that swim and dive, especially the scoters. **3.** (slang) a cranky person, especially a foolish old man.

copepod *n.* (Greek *kōpē*, oar + *pous*, foot) a tiny crustacean (*Copepoda* sp.). Some are parasitic, and others occur in plankton. Some are intermediate hosts of certain tapeworms (*Diphyllobothrium* and *Dracunculus*).

copper (Cu) *n.* (Greek *kyprios*, Cyprus—an island that produced the best copper) **1.** element 29, atomic weight 63.546; a malleable, reddish metal that is an excellent conductor of electricity and a component of bronze and brass. It is an essential micronutrient for plants, animals, and humans. Organ meats, shellfish, nuts, and legumes are good sources of copper. **2.** a reddish color similar to that of the element. **3.** a coin made of copper. *adj.* **4.** made of copper; e.g., a copper tool, utensil, or wire.

copperhead *n.* **1.** a poisonous pit viper (*Agkistrodon contortrix*) with a copper-colored head; found in southern U.S.A. **2.** a contemptuous term for a Northerner who sided with the South in the U.S. Civil War.

coppice forest *n.* (French *copeiz*, fresh-cut wood) a forest of usually small trees cut periodically for fuel wood or other uses and regenerated by stump or root sprouts; e.g., willows, hybrid poplar, many oaks. See ratooning.

Coppiced trees along a highway

copra *n.* (Malay *koppara*, copra) the dried white meat from coconuts. Coconut oil is extracted from copra. See coconut.

copra oil *n.* coconut oil, extracted from copra and used in making candles, shampoos, soaps, shortening, synthetic rubber, etc.

coprogenous *adj.* (Greek *kopros*, dung + *genēs*, born) originating from or influenced by animal excrement; e.g., earthworm casts in soil.

coprogenous earth *n.* colloidal peat formed from feces of water animals. Also called sedimentary peat. See colloidal peat.

coprolite *n.* (Greek *kopros*, dung + *lithos*, stone) fossilized excrement of fishes, reptiles, or mammals.

coprophagous *adj.* (Greek *kopros*, dung + *phagos*, eating) living on dung; e.g., dung beetles are coprophagous.

coprophagy *n.* (Greek *kopros*, dung + *phagein*, to eat) **1.** the eating of its own feces by an animal. It is most common in wild animals, domestic rabbits, and horses. **2.** the eating of feces from one species of animal by another, e.g., dung beetles. Also, some state laws permit the sale of dried broiler litter and dried caged laying-hen litter as merdivorous livestock feed, reputed to be almost equal in feeding value to good-quality hay.

coprophilous *adj.* (Greek *kopros*, dung + *philos*, loving) living on animal excreta; e.g., many fungi, including certain mushrooms produced commercially.

copulation *n.* (Latin *copulatus*, fasten, link) **1.** sexual intercourse; the act that deposits male gametes (sperm) in position to reach the female gametes (eggs). **2.** the act of coupling or joining mating parts. **3.** the state of being coupled.

coquina *n.* (Spanish *coquina*, shellfish) **1.** a small clam (*Donax variabilis*) of the eastern and southern U.S. seacoasts whose shells often open into a butterfly shape. **2.** other similar clams. **3.** a soft limestone composed of shell fragments and coral; used as a building material, especially in Florida.

coquito *n.* (Spanish *coquito*, little palm tree) a type of palm tree (*Jubaea spectabilis*, family Arecaceae), native to Chile, with sugary sap that can be made into wine. The sap and the nuts are used as food. Also called wine palm.

coral *n.* (Latin *coralium*, coral) **1.** the calcareous skeletons of certain invertebrate organisms cemented together into a porous rock that commonly forms reefs, shoals, and atolls in tropical seas. **2.** the organisms (phylum Coelenterata, class Anthozoa and some Hydrozoa) that form coral rock. **3.** a yellowish red typical of coral reefs.

coral reef *n.* an offshore ridge of coral separated from land by a shallow lagoon. The upper surface is near sea level (or was at the time of its formation). The Great Barrier Reef off the northeastern coast of Australia is the world's largest coral reef, more than 1250 miles (2000 km) long, and contains more than 350 species of coral.

coral snake *n.* any of several small, poisonous snakes (*Micruroides* sp.) in tropical and subtropical areas. The snakes are commonly marked with coral, yellow, and black bands around their bodies.

corbiculum *n.* (Latin *corbiculum*, a little basket) the sac where pollen is carried on the hind tibia of bees. Also called pollen basket.

cordgrass *n.* (Latin *chorda*, cord + Anglo-Saxon *graes*, grass) any of several tall grasses (*Spartina* sp., family Poaceae) that grow up to 10 ft (3 m) tall in marshy areas, especially near coasts. Also called marsh grass.

cordillera *n.* (Spanish *cordillera*, mountain chain) a large mountain chain, especially the principal mountain chain of a continent, such as the Andes Mountains of South America or the Rocky Mountains of North America.

core *n.* (Latin *cor*, heart) **1.** the inner, central part of anything, especially the center of a fruit such as an apple. **2.** the heart of a nuclear reactor where energy is released. **3.** (figuratively) the central concept of a theory or plan.

coreopsis *n.* (Greek *koris*, bug + *opsis*, appearance) an ornamental plant (*Coreopsis* sp., family Asteraceae) with daisylike yellow, red, or crimson blossoms on wiry stems 1 to 3 ft (30 to 90 cm) tall and seeds that are shaped somewhat like a tick.

coriander *n.* (Greek *koriandron*) an annual herb (*Coriandrum sativum*, family Apiaceae) with strongly scented leaves and aromatic seeds that are used in cooking, especially for confections and bread.

Coriolis effect *n.* the tendency, caused by the rotation of the Earth, for fluids in the Northern Hemisphere to circulate with a clockwise rotation and those in the Southern Hemisphere to circulate counterclockwise. It is named after Gaspard Coriolis, the French engineer and mathematician who explained the effect in 1835. The Coriolis effect is easily overcome on a small scale by other forces but has a major effect on large-scale circulation in the atmosphere and oceans. Sometimes called Coriolis force.

cork *n.* (Arabic *korque*, cork) **1.** the thick, lightweight, resilient bark of mature cork oak trees that is used commercially for insulation, gaskets, floats, stoppers, etc. The bark can be stripped from the tree about once every 9 years, beginning when the tree is about

50 years old. Also called phellem. **2.** plant tissue similar to cork oak bark in structure, having many tiny cells of entrapped air. **3.** a bottle stopper, especially one made of cork, but sometimes used for those made of other materials. *adj.* **4.** made of cork. *v.* **5.** to put a stopper in a bottle or other container. **6.** to plug a hole or small passage.

cork oak *n.* a fairly small evergreen tree (*Quercus suber*, family Fagaceae) from Spain, Portugal, and northern Africa, whose thick, lightweight bark is the commercial source of cork. It is sometimes grown as an ornamental in southern U.S.A.

corktree, Amur *n.* a large, aromatic tree (*Phellodendron amurense*) with a rounded crown, low-spreading branches, and light gray, deeply fissured, corky bark. Native to northern China, it is hardy and can be used as an ornamental throughout most of the U.S.A.

corm *n.* (Latin *cormus*, tree trunk) a short, fleshy, vertical, underground stem of certain plants (e.g., crocus or gladiolus) that resembles a bulb but has more stem tissue and fewer scalelike leaves. A bud forms at the top and roots at the bottom.

cormorant *n.* (French *cormoran*, sea raven) **1.** any of several large seabirds (family Phalacrocoracidae) with long necks, webbed feet, dark plumage on their backs, and either light or dark fronts, depending on species. Their wingspans are mostly 3 to 5 ft (0.9 to 1.5 m). They dive beneath the surface and swim under water to catch fish and hold the fish in a pouch similar to that of a pelican. Also called sea raven or coal goose. **2.** a greedy or gluttonous person.

corn *n.* (Anglo-Saxon *corn*, a grain or seed) **1.** maize. A tall grass-family plant (*Zea mays*, family Poaceae) that produces grain on an ear (or sometimes more than one ear) near the middle of the plant. Typically a warm-season crop, it is planted in the spring and harvested in late fall. **2.** the grain shelled from the ears of corn plants; the world's third most important food and feed grain crop, after wheat and rice. **3.** the plant or grain of cereal crops such as wheat in England or oats in Scotland. **4.** a hard growth with a tender core at a pressure point on the foot of a person or animal.

Corn Belt *n.* a band of land with both soils and climate that are very well suited for corn production in northern U.S.A., including Iowa, Illinois, Indiana, and portions of nearby states. (Corn is also grown in many other areas but often as a minor crop rather than the dominant crop, as it is in the Corn Belt.)

corn borer *n.* the larva of a pyralid moth (*Pyrausta nubilalis*) native to Europe that bores passageways through the stalks, ears, leaves, and brace roots of corn and attacks many other large-stemmed flowering plants to a lesser extent (e.g., soybean, millet, oats, potatoes, peppers, and sorghum). Corn borers weaken the plants, cause lodging, and decrease disease resistance and yield. Also called European corn borer.

Larva of European corn borer

corncob *n.* the elongated woody core that holds the grain of an ear of corn.

corncrib *n.* a storage building for husked corn (still on the cob). It typically has a center driveway and bins on both sides with slatted walls that permit air movement to dry the corn. See granary.

corn leaf blight *n.* a disease caused by a fungus (*Helminthosporium maydis* for southern or *H. turcicum* for northern corn leaf blight) that produces tan elliptical lesions on the leaves of nonresistant corn varieties and can drastically reduce yield.

corn smut *n.* a fungal (*Ustilago zeae*, family Ustilaginaceae) disease that produces pea-sized or larger galls (up to 6 in. or 15 cm) on embryonic tissue of the leaves, stalks, and especially the ears of the plant, causing about 2% of the corn crop losses in the U.S.A. The galls appear greenish white on the outside but expose a powdery black spore mass when they rupture.

corolla *n.* (Latin *corolla*, little crown) the inner petals of a flower (inside the calyx).

corrasion *n.* (Latin *corrasus*, scraped off) abrasive erosion of soil and rock material by glaciers, sediment in running water, wind, or waves, and mass movement.

correction line *n.* a line where adjustments are made in the rectangular land survey in the U.S.A. The sections next to a correction line (especially those on the north or east side of a correction line) may be smaller, or occasionally larger, than 640 ac, and the sections on the north side may not line up with those on the south side of a correction line. See township.

correlation *n.* (Latin *com*, with + *relatio*, relation) **1.** a mutual relationship. **2.** a statistical measure of how closely two traits vary together, ranging from 0 to ±1. A correlation coefficient of +1.00 means that as one trait increases the other also increases in exact proportion. A correlation coefficient of –1.00 means that as one trait increases the other decreases in exact proportion. A correlation of 0.00 means that the two traits appear to be unrelated.

corrosion *n.* (Latin *corrosio*, gnawing) **1.** the process of wearing or eating away a material, especially by chemical reaction; c.g., by rusting. **2.** the product of corrosion, such as rust.

corrosive *adj.* (Latin *corrosio*, gnawing + *ivus*, tending to) **1.** able to react with and degrade the surface of an object or the skin of humans or animals; e.g., strongly acid (pH 2 or less) or strongly alkaline (pH 12 or more). **2.** (figurative) sarcastic; abrasive comments and attitude.

corrugation *n.* (Latin *corrugatio*, corrugation) **1.** a small furrow or ridge (often alternating furrows and ridges) in a surface; e.g., in a sheet of corrugated iron, plastic, or cardboard. **2.** a small channel used to guide water in an irrigated field.

corrugation irrigation *n.* surface irrigation that supplies water through small, closely spaced channels; commonly used for close-growing small grain or forage crops.

corundum *n.* (Hindu *kurand*, corundum) aluminum oxide (Al_2O_3). Second to diamond in hardness of all naturally occurring minerals, it is used as emery for grinding and polishing. It can be a gemstone in the form of a sapphire or a ruby. See aluminum magnesium boride for the hardest synthetic substance.

coruscation *n.* (Latin *coruscatio*) **1.** emission of a sudden gleam of light; sparkling or scintillating. **2.** a sudden, brilliant display of wit; e.g., a witty statement or retort.

cosmopolitan *adj.* (Greek *kosmos*, world + *politēs*, a citizen) **1.** worldwide; not limited to the viewpoint or characteristics of a region. **2.** comfortable anywhere; sophisticated. **3.** widely distributed, as a cosmopolitan plant or animal. *n.* **4.** a cosmopolite, one who has an international viewpoint and is free of local or national bias.

cosmos *n.* (Greek *kosmikos*, the universe) **1.** the universe exclusive of the Earth. **2.** order and harmony as represented by the universe.

cosset *n.* **1.** a lamb raised without the ewe; a pet lamb. **2.** *v.* to pamper a pet, as a lamb.

cost of production *n.* the average cost of inputs per unit of product, including raw materials and labor, an allowance for buildings and land, and sometimes a management cost. Farm cost of production may be expressed on a per-acre (or hectare), per-bushel (or quintal), per-marketable-animal, or weight (100 lb, 100 kg, or ton) basis. It may be calculated for an individual farm or for all the farms in a specified area, state, or nation.

cost sharing *n.* payment with public funds contributed through one or more government agencies to supplement the input of an individual or group for a project, especially payments made to promote soil, water, woodland, or wildlife conservation practices or for pollution control or recreational development.

cotton *n.* (French *coton*, cotton) **1.** an annual plant (*Gossypium* sp., family Malvaceae) widely cultivated in southeastern U.S.A. and other warm, humid climates for its soft white fibers that are used in making cloth; especially *G. hirsutum* (upland cotton) and *G. barbadense* (American Egyptian cotton). It is also the source of cottonseed oil, cottonseed meal, and gossypol. See gossypol. **2.** the soft white fibers produced by the seeds of cotton plants and contained in bolls. **3.** the balls, thread, fabric, and other items made from cotton fibers. **4.** any soft, fibrous substance produced by plants.

Cotton Belt *n.* originally the Southern states east of the Mississippi River, especially those along the Eastern and Gulf Coasts, where cotton was long the dominant crop, or even the only crop. With irrigation, the belt now includes southwestern U.S.A., but most farmers grow other crops as well as cotton.

cottonseed meal *n.* the fiber, hull, and other residues left after the lint (cotton) has been removed and most of the oil squeezed out of cotton seeds; used to feed livestock and as a fertilizer.

cottonseed oil *n.* oil extracted from cotton seeds that is used as a food product, as a cooking oil, and to make soaps.

cottonwood *n.* **1.** any of several large, fast-growing deciduous trees (*Populus* sp., family Salicaceae) used for shade trees and in shelterbelts. **2.** the soft wood produced by these trees and used to make boxes, crates, pulpwood, etc.

cottony maple scale *n.* a very destructive insect (*Pulvinaria innumerabilis*, family Coccidae) that attacks red maple and silver maple as well as several kinds of fruit trees and other woody plants. It turns leaves yellow, produces cottony masses on the undersides of branches and twigs, and kills infected branches.

cotyledon *n.* (Greek *kotylēdōn*, cavity) **1.** an embryonic leaf contained in a seed. Angiosperms (plants with seeds enclosed in an ovary) are classified as monocotyledons if their seeds have only one cotyledon (e.g., grasses such as corn) or dicotyledons if their seeds have two cotyledons (e.g., legumes such as peas and beans). Some gymnosperms (plants such as conifers whose seeds are not enclosed in an ovary) have seeds with several cotyledons. See monocotyledon, dicotyledon. **2.** a tufted area in the placenta of a ruminant animal.

couch grass *n.* See quack grass.

cougar *n.* (French *couguar*, puma) a large cat (*Felis concolor*, family Felidae) with a tawny coat on a long, slender body and a long tail; native to North and South America. Also called puma, mountain lion, panther, or catamount.

coulee *n.* (French *couler*, to flow) a term used mostly in western U.S.A. and Canada. **1.** a steep-sided lava flow, either molten or solidified. **2.** a steep-sided water channel. An arroyo that is dry in summer in southwestern U.S.A. or a steep-walled canyon in northwestern U.S.A. (the largest and best known, called Grand Coulee, is the site of Grand Coulee Dam).

coulter *n.* See colter.

Coulter pine *n.* a hardy pine tree (*Pinus coulteri*, family Pinaceae) native to southern California, a roundly pyramidal tree that grows 50 to 80 ft (15 to 24 m) tall and produces the largest pine cones (9 to 14 in. or 23 to 35 cm long) of any species.

A Coulter pine cone about 10 in. (25 cm) long that has opened to release its seeds

Council for Agricultural Science and Technology (CAST) *n.* an organization of professional societies, companies, and individuals that was established in 1972 with headquarters in Ames, Iowa. CAST organizes task forces to investigate agricultural issues and provides information to legislators and other decision makers, news media, members, etc.

Council on Environmental Quality *n.* a 3-person panel established by the National Environmental Policy Act in 1969 to monitor the environment and administer the environmental impact statement program. It was replaced in February 1993 by the White House Office on Environmental Policy.

country *n.* (French *contree*, region or country) **1.** a nation or its land and territory. **2.** a region or countryside, as a tract of arid country. **3.** the nation where one is a citizen or was born. **4.** the inhabitants of a nation. **5.** a rural area, as distinguished from a metropolitan area. *adj.* **6.** pertaining to a rural area, as country customs or a country road. **7.** containing mineral veins, as country rock.

countryside *n.* (French *contrée*, region or country + Anglo-Saxon *sīd*, broad) **1.** a rural area. **2.** a landscape, esp. in a rural area. **3.** the inhabitants of a rural area.

county *n.* (French *conté*, the jurisdiction of a count) **1.** a governmental subdivision of a state in the U.S.A. (the corresponding unit in Louisiana is called a parish). Many counties, especially in the central part of the U.S.A., are 24 miles square and contain 16 townships. Other counties, especially in mountainous areas, are irregular in shape and may be smaller or larger than the usual size. There are 3197 counties/parishes in the 50 states. **2.** a governmental subdivision of a territory in Great Britain, Canada, or New Zealand. **3.** the district of a count or earl.

county agent *n.* a professional in the Cooperative Extension Service employed to help diffuse practical information on agriculture and allied subjects among members of the farming community and other residents in a county. Also called, according to function, ag agent, agricultural agent, extension agent, farm and home advisor, home agent, home demonstration agent, 4-H agent, or youth agent.

covalent bond *n.* the chemical bond formed by two atoms sharing a pair of electrons. Also called coordinate covalent bond. Double bonds and triple bonds are formed when two atoms share 2 or 3 pairs of electrons, respectively.

cove *n.* (Norwegian *kove*, a closet) **1.** a small bay or open harbor; an indentation on the coastline of an ocean, sea, or lake. **2.** a sheltered nook, cave, or other recess in a hillside. The term is popularly applied to small areas of plains or valleys that extend into mountains or plateaus or to narrow strips of prairie land surrounded by forest. **3.** a steep, cirquelike area at the head of a valley. **4.** a channel or concave molding on a wall or ceiling, used for decoration or to accommodate wires, pipes, or fixtures. **5.** a ceiling arch.

cover *n.* (French *couvrir*, to cover) **1.** a protective layer over or around an item, e.g., a blanket or the front and back of a book. **2.** a removable upper surface, as the lid for a pot. **3.** plant growth, living or dead, that protects soil from erosion and provides food and protection for animals and birds. **4.** anything that conceals from sight, as the cover of darkness. **5.** an assumed identity or activity for someone who wants to hide who they are or what they are doing; a pretense. *v.* **6.** to place a cover over something. **7.** to hide something from view. **8.** to deal with an emergency; e.g., to work in place of another. **9.** to impregnate a female, as a bull covers a cow.

cover crop *n.* **1.** an annual grass or legume planted to protect a garden or field from wind and water erosion during a season or year when the main crop is not being grown. It is likely to be plowed or disked into the soil as green manure rather than being allowed to mature. **2.** a perennial grass or legume or mixture grown between and beneath the trees of an orchard or vines of a vineyard to reduce runoff, erosion, and soil compaction.

covert *adj.* (French *couvert*, covered) **1.** hidden, secret, concealed; protected; e.g., a covert operation. *n.* **2.** wildlife cover; e.g., a thicket. **3.** a small feather covering the base of a large feather in fowls.

covert escape *n.* vegetation planted along the edge of an open area to serve as a refuge for animals and birds.

covey *n.* (French *covée*, a brood or flock) **1.** a small flock of birds, especially a flock of quail or partridge. **2.** a small group of people; a party. **3.** (British) a pantry or closet.

cow *n.* (Anglo-Saxon *cū*, cow) **1.** a mature, female bovine (*Bos* sp.), especially one that has produced a calf (a younger female is called a heifer). **2.** a female of certain other large species; e.g., a female elephant or whale.

cowbane *n.* (Anglo-Saxon *cū*, cow + *bana*, to strike or wound) **1.** an herbaceous plant (North American *Cicuta maculata* or English *C. virosa*, family Umbelliferae) with small white flowers and very poisonous tuberous roots. Also called water hemlock. **2.** any of several other plants that are toxic to cattle; e.g., locoweed.

cowbird *n.* (Anglo-Saxon *cū*, cow + *bridd*, bird) a bird (*Moluthrus ater* or *M. aeneus*, family Icteridae) with black or gray plumage that resembles the Brewer's blackbird but has a shorter bill and is best known for laying eggs in other birds' nests for hatching and raising, often crowding out the offspring of the host species.

cowboy *n.* **1.** a man or boy who works with herds of cattle, usually on horseback, especially a cattle herder on a large ranch of western U.S.A. or Canada. Also known as a cowpoke or cowpuncher. **2.** a New Zealand boy who milks cows. **3.** a reckless driver or pilot.

cow chips *n.* dried cattle manure, commonly used as fuel in many developing countries where wood is scarce. See buffalo chips.

cowgirl *n.* a girl or woman who rides horses and works with cattle, especially on a ranch in western U.S.A. or Canada.

cowhand *n.* a cowboy or cowgirl.

cow heifer *n.* a young cow that has had only one calf.

cow month *n.* a mature cow grazing on pasture or range for 1 month. Also called an animal unit month (AUM).

cowpea *n.* **1.** an annual summer legume (*Vigna unguiculata*) with trifoliate leaves, elongated pods containing edible seeds, and indeterminant growth. Native to Africa or Asia, it is grown in southern U.S.A. and elsewhere with similar climate. **2.** the edible seed from the cowpea plant. Also called black-eyed peas or southern peas.

cowpea weevil *n.* a destructive weevil (*Callosobruchus maculatus*, family Bruchidae) that attacks cowpeas, other peas, and beans in storage. Eggs are laid in the cowpeas in the field; later, the larvae develop and eat the inside of the seed.

cow poke *n.* a metal or wooden device fastened around an animal's neck to discourage fence breaking by cattle. It has two hooked arms extending upward and downward with a barb on each arm pointing toward the animal to discourage it from pushing against a fence.

cow pony *n.* a horse trained to work with cattle, usually in western U.S.A.

cowpox *n.* a mild but contagious rash on the teats and udders of cows that is caused by a virus that was formerly used to vaccinate humans against the smallpox virus.

coyote *n.* (Spanish *coyote*, coyote) a carnivorous animal (*Canis latrans*, family Canidae) of the same family as wolves and dogs. The principal predator of sheep, goats, chickens, and rabbits, especially in western U.S.A. and western Canada.

coyote tobacco *n.* a poisonous plant (*Nicotiana attenuata*, family Solanaceae) widespread in dry, sandy soils in western U.S.A. Animals that eat it suffer from bloating, diarrhea, weakness, spasms, stupor, vomiting, and violent heart palpitations. Also called wild tobacco.

coyotillo *n.* (Spanish *coyotillo*, a little coyote) a thorny shrub (*Karwinskia humboldtiana*) of the buckthorn family found in Mexico and southwestern U.S.A. Its berries are toxic.

crab *n.* (Dutch *krab*, crab) **1.** a short-tailed crustacean (suborder Brachyura, order Decapoda) with a wide, somewhat flat body protected by a hard shell. Some live underwater and have gills, others live on land and can breath air, and some have a characteristic sideways movement called a crab walk. **2.** any of several other similar arthropods, such as the horseshoe crab. **3.** the Crab Nebula constellation. **4.** a sideways movement or adjustment in heading to compensate for drift caused by a strong wind or water current. *v.* **5.** to hunt or fish for crabs. **6.** to move sideways. **7.** to maneuver at an angle or in a zigzag to compensate for lateral drift.

crab apple *n.* **1.** a small tree (*Malus* sp., family Rosaceae) that bears small, tart apples. **2.** the tart apples produced by such a tree that are used to make jellies and preserves.

crabgrass *n.* (Dutch *krab*, crab + Anglo-Saxon *grues*, grass) a warm-season, coarse-bladed, annual grass (*Digitaria* sp., family Poaceae) with a ragged appearance. It spreads rapidly by stolons up to 4 ft (1.2 m) long that root at their nodes, making it a serious summer weed in lawns and fields.

crab louse *n.* an annoying body louse (*Phthirus pubis*, family Pediculidae) that inhabits hairy parts on humans, such as the pubic regions and armpits.

cracked stem of celery *n.* a growth disorder in celery affected by boron deficiency (aggravated by excess potassium and nitrogen) and characterized by brown stripes on the stalks, brown mottles on the leaves, broken and curled epidermis, and brown, dying roots.

cradle knoll *n.* a small mound of earth on a forest floor caused by the overthrow of a large tree; soil that was entrapped in the root system forms a mound. Many such events over several centuries can cover the land surface with knolls and pits that are sometimes called tree-tip mounds and tree-tip pits.

cradle region *n.* the area where a plant or animal species was first found and domesticated. Such areas provide biodiversity through an array of ancestral varieties and related species.

cranberry *n.* (German *kranbeere*, crane berry) **1.** a viny evergreen shrub (*Vaccinium* sp., especially *V. macrocarpon* in America or *V. oxycoccos* in Europe, family Ericaceae) that grows in acid bogs or wet acid soils, especially in northeastern U.S.A. Once called fen berry. **2.** the sour red berries produced by the vine. Cranberries are high in vitamin C and are often used to make a red sauce or jelly to brighten meals at Thanksgiving and Christmas.

cranberry bog *n.* a wet area that produces cranberries. Usually a low-lying area of peat, muck, or sandy soil that floods periodically and is strongly acid; common in Europe and northeastern U.S.A.

crater *n.* (Latin *crater*, mouth of a volcano) **1.** a roughly circular, steep-sided basin in the top of a volcano; relatively deep (deeper than one-third of its diameter) in relation to a caldera and sometimes holding a lake. **2.** a depression in the land surface caused by the impact of a meteorite. **3.** a depression left by an explosion, as a bomb crater.

Crater Lake *n.* a natural lake, known for its deep, clear, blue water, in the large caldera of an extinct volcano in southwestern Oregon.

crayfish *n.* (French *crevice*, crayfish) **1.** a small freshwater crustacean (*Astacus* sp. or *Cambarus* sp.) related to lobsters. Crayfish burrow in wet soil, leaving small mounds next to their holes. Also called crawfish or crawdad. **2.** spiny lobster, a saltwater crustacean.

crayfish (crawfish) land *n.* **1.** flat, wet, clayey soil with crayfish holes and tiny mounds. **2.** poor agricultural land in need of drainage.

crazy chick disease *n.* a disorder caused by vitamin E deficiency that causes lesions in the brains of baby chicks up to eight weeks of age. Symptoms include nervous tremors, difficulty walking, prostration, and body distortions. Also called encephalomalacia, nutritional encephalomalacia, or epidemic tremors.

crazyweed *n.* See locoweed.

cream *n.* (French *crème*, cream) **1.** liquid or semiliquid matter containing at least 18% milk fat that rises to the top of nonhomogenized milk. Also called sweet cream. **2.** a medicated ointment with a creamy consistency. **3.** the best part of something, as the cream of the crop. **4.** a yellowish white. *v.* **5.** to form a creamy mass, e.g., by stirring a mixture of butter and sugar. **6.** (slang) to win decisively.

creamery *n.* (French *crème*, cream + *erie*, a place for) a commercial establishment that buys milk and cream from producers and sells processed dairy products, including milk, cream, ice cream, and cheese.

creek *n.* (Dutch *kreek*, creek) **1.** a small stream of flowing water entrenched in its own valley; more stable and better vegetated than a gully, but smaller than a river. The exact size is inconsistent; a stream of water called a creek in one place may be large enough to be called a river in another. **2.** an indentation in the shoreline of a sea; e.g., an estuary. **3. up the creek** an idiom describing a hopeless situation.

creel *n.* (French *graïl*, griddle) a basket for holding fish, especially one made of wicker.

creep *n.* (Anglo-Saxon *creopan*, to creep or crawl) **1.** very slow movement; e.g., solifluction or a predator stalking its prey. **2.** an area with a small entrance that permits young animals (calves, piglets, etc.) to enter but keeps out mature animals. **3.** a derogatory description of a bothersome or disgusting person, e.g., a little creep. *v.* **4.** to move slowly, often difficultly, furtively, and/or close to the ground as in a crawl. **5.** to grow across a wall or along the ground, as a creeping plant. **6.** to enter or become apparent gradually, as the speaker's bias may creep into a lecture. **7.** to change shape gradually under sustained heat and pressure. **8.** to spread by means of stolons (surface stems) that root at the nodes; e.g., strawberry, Bermuda grass, crabgrass.

creep erosion *n.* See solifluction.

creep feeding *n.* a system for providing special feed for young domestic animals. The feed is placed in a creep that has a small entrance where the young can enter but mature animals cannot.

creep grazing *n.* an Australian practice similar to creep feeding. An area of pasture is closed to adult animals but provision is made for access by young stock.

cremation *n.* (Latin *crematus*, burned) incineration of a dead body. Human cremations are usually done in a crematory; the ashes are placed in an urn for burial or sometimes scattered in a special place.

creosote *n.* (German *kreosot*, flesh saver) a toxic, strong-smelling, dark brown to black, oily liquid that is irritating to skin. It is obtained by the anaerobic distillation of wood tar or coal tar and used as an antiseptic, fungicide, germicide, insecticide, and wood preservative.

creosote bush *n.* a shrub (*Larrea tridentata*, family Zygophyllaceae) with yellow flowers and resinous foliage that is highly tolerant of desiccation; native to arid regions in northern Mexico and southwestern U.S.A.

crepuscular *adj.* (Latin *crepusculum*, twilight) **1.** pertaining to dim light; twilight conditions. **2.** active at twilight and/or dawn; e.g., some insects, such as mosquitoes.

crest *n.* (Latin *crista*, crest) **1.** the pointed tuft of feathers or the comb on the heads of certain birds. **2.** the top of a mountain, hill, or ridge. **3.** the upper surface of a dam, dike, spillway, or weir. **4.** the highest stream flow of a river or creek in a runoff period. **5.** the foamy top of a wave, especially one that is breaking. **6.** a plume or emblem on a knight's helmet, armor, or shield. **7.** a ridge on a bone or on an animal's body, especially on the neck. *v.* **8.** to reach a high point. **9.** to form a crest.

cretaceous *adj.* (Latin *creta*, chalk + *eus*, composed of) chalky or containing much chalk.

Cretaceous *n.* a geologic period that began in the latter part of the Mesozoic era, about 135 million years ago, and lasted about 50 million years. The period when the dinosaurs became extinct, named for the chalk beds of southern England that formed during this period.

crevasse *n.* (French *crevasse*, crack or fissure) **1.** a deep fissure in a glacier formed as the ice moves. **2.** a deep cleft in a rock mass. **3.** a break in a levee along a river.

crevice *n.* (corruption of crevasse) **1.** a crack with a narrow opening; e.g., in the bark of a tree, in a board, or in the floor of a cave. **2.** a shallow fissure in bedrock that has caught and concentrated deposits of gold.

cricket *n.* (French *criquet*, a creaker) **1.** a jumping insect belonging to any of several genera (family Gryllidae). Some crickets are nocturnal and are known for the chirping sound the males make by rubbing their forewings together. Among the most known are the black field crickets (*Gryllus* sp.), the southern mole cricket (*Scapteriscus vicinus*), and the northern mole cricket (*Gryllotalpa hexadactyla*). The mole crickets are serious pests of field crops and gardens because they burrow in the soil and feed on young plant roots. **2.** a metal device that makes a clicking sound similar to the chirp of a cricket. **3.** a team game, popular in England, that is played with bats, a ball, and wickets. *adj.* **4.** fair, especially in the negative, as that's not cricket.

crimson clover *n.* a winter annual legume (*Trifolium incarnatum*, family Fabaceae) that is popular in southeastern U.S.A. as a cover crop and forage. Also called Italian, scarlet, or incarnate clover.

criollo *n.* or *adj.* (Spanish *criollo*, creole) **1.** Creole; a person born in Latin America with European (especially Spanish) ancestry, or one of the original French settlers in Louisiana (especially New Orleans). **2.** loosely, any native inhabitant of the Gulf Coast, Central American, or South American area settled by the Spanish. **3.** domestic cattle with a mixed heritage including the original criollo types (*Bos taurus* cattle) with 500 years of adaptation to tropical conditions, breeds derived from imports from India during the past century (*Bos indicus*), and recent imports from European and North American breeds.

critical level *n.* the highest concentration of a pollutant that will not cause the most sensitive ecosystems to suffer long-term harmful effects.

critical light period or **critical photoperiod** *n.* the minimum duration of daily light (or, more accurately, the maximum duration of darkness) that will permit certain long-day plants to grow and reproduce.

critical load *n.* the heaviest amount of deposition of an element that will not cause the most sensitive ecosystems to suffer long-term harmful effects.

critical mass *n.* the minimum amount of fissionable material required to achieve a self-sustaining chain reaction. For example, the critical mass of uranium 235 is 33 lb (15 kg).

critical period *n.* a time interval when the supply of a growth factor (e.g., water) has an unusually strong effect on plant growth and maturity.

critical slope *n.* See angle of repose.

critical velocity *n.* the velocity that corresponds to a basic change in some process. For example, the minimum velocity of water to cause erosion instead of sedimentation or the minimum velocity for a space vehicle to overcome the gravitational attraction of Earth or another celestial body.

critter *n.* (corruption of creature) (Slang) **1.** a cow. **2.** any animal or bird; a creature.

crocodile *n.* (Latin *crocodilus*) a large, amphibious, flesh-eating, tropical reptile (family Crocodylidae) that is a powerful swimmer and spends most of its time in the water but can also travel on land. It is related to the alligator but has a longer, narrower snout.

crop *n.* (Anglo-Saxon *crop*, top of a plant) **1.** any product growing on or harvested from a soil that has been worked, not including that from wild growth. **2.** the young animals and/or fowls produced in a season or year on a farm or ranch. **3.** a new group of anything, as a crop of recruits. **4.** craw; a pouch where food is stored in the esophagus of grain-eating birds until softened for digestion or regurgitated to feed young birds; the first of a bird's three stomachs. **5.** the honey sac in the alimentary tract of the honeybee. *v.* **6.** to cultivate land and grow a crop on it. **7.** to cut short (e.g., as a grazing animal bites off grass), to shear the hair or wool of an animal, or to cut off a tail or clip the ears of an animal (especially a dog).

crop bound *n.* distended; a domestic fowl's crop that retains too much food. Also called impacted crop.

crop diversification *n.* growing two or more (preferably several) kinds of crops in an area in an effort to avoid catastrophic crop failure from disease, insect, or weather problems.

crop improvement *n.* crossing plants with different genetics, selecting superior plants, and performing other manipulations to develop crop varieties that are more productive and/or higher in quality.

cropland *n.* land that has been cropped within the past two years, including harvested cropland, crop failures, tilled summer fallow, idle cropland used for pasture, land in orchards or vineyards, and land in soil-improving crops.

crop outlook *n.* a published forecast from the U.S. Department of Agriculture of the anticipated state or national harvest for a particular crop.

cropper *n.* See sharecropper.

crop residue *n.* any plant part that is not harvested or not used as a food product or raw product for processing. Crop residues that are left on the soil or returned to the soil help control soil erosion and return valuable plant nutrients and organic matter to the soil.

crop residue utilization *n.* use made of the plant parts left in a field after harvest, especially leaving them on the land for erosion control and soil improvement.

crop rotation *n.* alternation of two or more crops in succeeding years in a particular field. The primary purposes are to control erosion, weeds, insects, and diseases; to obtain fertility benefits such as a grain crop using the nitrogen fixed by a preceding legume; and to avoid growth suppression of a crop by residues of the same type of plant. Example: a 3-year rotation of corn, oats, and clover. The clover residues provide erosion control and a nitrogen supply for the next corn crop. See multiple cropping.

crop science *n.* the research, development of new varieties, teaching, and extension activities related to growing plants for food, fiber, and forage.

crop scout *n.* a person who checks fields regularly during the growing season to identify problems that need attention, such as nutrient deficiencies, weeds, damaging insects, or plant diseases.

crop yield *n.* the amount of marketable product harvested, usually expressed in weight or volume per unit area (e.g., bushels per acre, quintals per hectare, or tons per hectare).

crop yield index *n.* the average yield of a specified crop on a farm or other designated area expressed as a percentage of the average yield of the same crop in a reference area (usually the entire county, state, region, or nation).

crossbred *adj.* produced by crossbreeding.

crossbreed *n.* an individual animal (or plant in less common usage) produced by crossbreeding; a hybrid.

crossbreeding *n.* **1.** the fertilization of flowers of one type of plant with pollen from plants of a different variety or species. **2.** the insemination of a female of one type of animal with semen from a male of a different breed or race.

cross-compliance *n.* a government farm program requirement that a farmer receiving government assistance (e.g., price supports or loans for a particular crop) must also comply with the provisions of other programs (e.g., any other program crops grown on the farm).

cross-pollination *n.* sexual reproduction in plants resulting from the fertilization of a flower on a plant by pollen from another plant of the same species but different genotype.

crossroad *n.* **1.** a secondary road that crosses a main road. **2.** (or **crossroads**) a place where two or more roads cross each other. **3.** a center of activity; a place where people meet. **4.** a point of decision when the resulting action will preclude or make it difficult to change to the alternative.

crotalaria *n.* (Greek *krotalon*, a rattle) any of several hundred species of annual and perennial legumes (*Crotalaria* sp., family Fabaceae) of tropical origin that grow well on sandy soils with low fertility. Some species are used as forage or green-manure crops in southern U.S.A. Others (e.g., *C. juncea*) are considered possible sources of pulp. Some species produce toxins; e.g., arrow crotalaria (*C. sagittalis*) contains an alkaloid that can kill a horse.

croton *n.* (Greek *krotōn*, castor oil plant) **1.** any of several tropical herbs, shrubs, and trees (*Croton* sp., family Euphorbiaceae), including a tree of the Bahamas (*C. eluteria*) that yields cascarilla bark, an Asian plant (*C. tiglium*) that yields croton oil, and an annual weed (*C. texensis*) that grows from Alabama to Wyoming and can be toxic to livestock. **2.** any of several related trees and shrubs; e.g., garden croton (*Codiaeum variegatum*) grown as a houseplant or ornamental for its glossy, multicolored foliage.

croton oil *n.* a brownish-yellow oil with an acrid odor obtained from croton (*Croton tiglium*) seeds for external use as a pustulant or as a drastic cathartic.

crotovina *n.* a mass of contrasting material in a subsoil where a former animal burrow has been filled by material from the overlying soil.

crow *n.* (Anglo-Saxon *crawe*, a crow) **1.** a large, black nonmigratory bird (*Corvus* sp., family Corvidae) whose call is a distinctive caw. It is larger than a blackbird but smaller than a raven. **2.** the shrill cry of a rooster. *v.* **3.** to utter the cry of a rooster. **4.** to boast of an accomplishment or a triumph. **5.** (idiomatic) to **eat crow** means to admit that one was wrong; to **crow over** means to brag elatedly about besting another. **6. as the crow flies** means the distance in a straight line from one point to another.

crowfoot *n.* any of several small plants (especially *Ranunculus* sp., family Ranunculaceae) with mostly yellow flowers, including both ornamental and weedy species, named for their lobed leaves shaped somewhat like a crow's foot. It is toxic to livestock that eat it, causing muscle spasms. Also called tall buttercup.

crown *n.* (Latin *corona*, a crown or wreath) **1.** a decorative headgear worn by a monarch, often adorned with jewels. **2.** a sovereign ruler. **3.** a wreath or other head ornament used to mark victory or distinction. **4.** the uppermost part of something, especially a rounded top, as the crown of one's head, a hat, a hill, or a kernel of corn. **5.** an enlarged portion of a plant stem just above and just below ground level. **6.** the branches and leaves of a tree. *v.* **7.** to confer distinction, as to crown a winner or a king. **8.** to cut off the top part of a plant or tree, sometimes called topping. **9.** to form a rounded upper surface. **10.** to cover a tooth with a metal cap, usually made of gold.

crown cover *n.* a tree canopy; the upper level of foliage in a forest.

crown drip *n.* water from rain or melted snow that has been intercepted by a tree and that falls to the ground in large drops. It may cause erosion if the soil is not protected with understory vegetation or a litter layer.

crown fire *n.* a fire so hot that it jumps from one tree crown to the next and burns living trees and brush.

crown gall *n.* a malignant growth caused by bacteria (*Agrobacterium tumefaciens*) that affects 40 or more plant families and is economically important to sugar beets and nursery stock such as nut trees, pome fruits, and stone fruits. The bacteria enter plant stems through wounds, commonly near ground level, and cause nearby cells to divide and/or swell, forming a disorganized growth highly susceptible to secondary infections. Also called plant cancer.

crownvetch or **crown vetch** *n.* a sprawling perennial legume (*Coronilla varia*, family Fabaceae) commonly used as ground cover on roadbanks and other disturbed soils. A native of Europe, its clusters of pink and white flowers are so attractive that it is sometimes planted in flower gardens. However, it reproduces so rapidly by rhizomes and seeds that it soon smothers all other flowers. Also called axseed.

crowpoison *n.* a perennial herbaceous plant (*Amianthium muscitoxicum*, family Liliaceae; sometimes called *Zigadenus muscitoxicum*) that grows in low sandy soils, bogs, etc. Native to eastern North America, it grows from a bulb that contains a toxin that has been used in the manufacture of fly poison. It produces a raceme of white to greenish flowers. Also called stagger grass or fly poison.

CRP *n.* See Conservation Reserve Program.

crucifer *n.* (Latin *crux*, cross + *ferre*, to bear) any of a large family of dicotyledonous plants (family

Brassicaceae) with cross-shaped blossoms, pointed pods, and a strong odor (including broccoli, cabbage, cauliflower, radishes, and turnips). **2.** a person who carries a cross, e.g., in a religious ceremony.

Cruciferae *n.* the older name for plants of family Brassicaceae.

crude fiber *n.* the indigestible portion of a food product. Estimates are based on the part that remains insoluble following treatment with acidic and alkaline solvents. It is composed mostly of cellulose, lignin, and other components of cell walls.

crude oil *n.* a mixture of liquid hydrocarbons formed from plant remains (and possibly including some animal remains), subjected to heat, pressure, and metamorphosis, and accumulated in underground reservoirs. It is used as a source of gasoline and other fuels and products. Also called petroleum.

crude protein *n.* the amount of protein in livestock feed, either measured directly or estimated by measuring the nitrogen content and multiplying by 6.25 (N × 6.25 = crude protein), that is used as a measure of feed quality.

cruising timber *n.* the process of measuring trees and calculating the amount of lumber that could be harvested from a forested area.

crumb *n.* (Anglo-Saxon *cruma*, crumb) **1.** a small piece broken from bread, cake, cookie, or other food product. **2.** a porous aggregate in a granular soil.

crumbly *adj.* (Anglo-Saxon *cruma*, crumb + *ly*, like) **1.** easily broken into crumbs. **2.** friable; a desirable porous structure that makes a soil easy to work into a good seedbed. **3.** grainy; a texture defect of butter with fat particles that lack cohesion. Also called brittle. **4.** friable; a texture defect of dry ice cream that tends to fall apart when dipped. **5.** failing to mature; a problem of raspberries that have been damaged by certain pests or bad weather.

crumb structure *n.* a porous, granular soil structure described in several older soil description and classification systems. The peds are somewhat rounded, ranging from 1 to 5 mm in diameter, and have internal pore space. It is now considered a part of granular soil structure in Soil Taxonomy but is still used informally.

crust *n.* (Latin *crusta*, hard surface) **1.** a surface layer that is denser, harder, and more brittle than the material inside or beneath it; e.g., the surface of a loaf of bread or of a soil that has been pounded by raindrops. **2.** a cemented surface layer, as a soil surface made hard by calcium carbonate, gypsum, or other binding material. **3.** the outer rock shell of the Earth, ranging in thickness above the underlying mantle from about 3 miles (5 km) under much of the ocean to about 40 miles (65 km) under the Andes mountains. See Mohorovičić discontinuity.

crustacean *n.* (Latin *crusta*, hard surface + *aceus*, belonging to) an arthropod (class Crustacea) that lives mostly in an aquatic environment and typically has a hard shell covering its body. Crustaceans include barnacles, crabs, lobsters, and shrimps.

crustal movement *n.* the movement of segments of the Earth's crust that builds mountains, moves landmasses, and changes the size and shape of continents and oceans. See isostatic adjustment, plate tectonics.

cryic *adj.* (Greek *kryos*, cold) cold; used in Soil Taxonomy to name cold subgroups (soils that have mean annual temperatures between 0° and 8°C (32° to 46.4°F) and summer soil temperatures below 15°C (59°F).

cryobiology *n.* (Greek *kryos*, cold + *bios*, life + *logos*, word) the science that studies how humans and other warm-blooded organisms respond to cold temperatures.

cryology *n.* (Greek *kryos*, cold + *logos*, word) **1.** the science of cold objects; e.g., all forms of water formed below 0°C (32°F)—snow, ice, hail, and sleet. **2.** the study of cooling by refrigeration. **3.** glaciology (European usage).

cryophilic *adj.* (Greek *kryos*, cold + *philos*, loving) responding favorably to very low temperatures.

cryophyte *n.* (Greek *kryos*, cold + *phyton*, plant) a cryophilic plant; one that can grow on ice or snow. Examples: certain algae, mosses, and some species of willows.

cryopreservation *n.* (Greek *kryos*, cold + Latin *praeservatio*, preservation) **1.** preservation of organic materials (body tissues, etc.) by extreme cold; e.g., when an animal or person is frozen into a glacier. **2.** storing of cells (e.g., sperm, embryos, or oocytes), tissues, or other materials at extremely low temperatures to keep them metabolically inert and genetically stable; e.g., by cooling with liquid nitrogen (–196°C), dry ice (–79°C), or a freezer that can attain a similar temperature.

cryoturbation *n.* (Greek *kryos*, cold + Latin *turbinatus*, rotation) soil mixing by frost action, including frost heaving. Also called geliturbation.

cryptobiotic *adj.* (Greek *kryptos*, hidden + *biōtikós*, pertaining to life) **1.** leading a hidden or concealed life, as some species that live in caves. **2.** in a state of dormancy with no apparent signs of life.

cryptogam *n.* (Greek *kryptos*, hidden + *gamos*, mating) a nonflowering plant that reproduces by spores instead of by seeds; e.g., algae, ferns, fungi, and mosses.

crystal *n.* (Greek *krystallos*, ice or crystal) **1.** a single grain of homogeneous inorganic solid with an orderly atomic arrangement that is commonly indicated by external plane surfaces that form definite angles with each other, thus giving the crystal a regular geometric form. **2.** a transparent mineral; e.g., a clear block of ice. **3.** glass tableware, especially that of high quality. **4.** a transparent cover for a watch or other small object.

crystalline *adj.* (Greek *krystallos*, ice or crystal + *ine*, like) **1.** having an internal structure composed of crystals. **2.** like crystal; crystal clear.

crystallization *n.* (Greek *krystallos*, ice or crystal + *izein*, causing) the process of forming crystals, usually as a result of cooling or of a loss of water.

crystallized honey *n.* honey containing crystals of dextrose-hydrate. It can be liquefied by gentle heat, e.g., by immersing the container in hot water.

crystal system *n.* any one of the six systems used for classifying crystal shapes according to the inclinations and relative lengths of their crystallographic axes. The six systems are isometric, tetragonal, hexagonal, orthorhombic, monoclinic, and triclinic.

CSREES *n.* Cooperative State Research, Education, and Extension Service. See Cooperative Extension Service.

cub *n.* (Irish *cuib*, whelp) **1.** a young bear, fox, lion, tiger, wolf, whale, seal, etc. **2.** an inexperienced person; a novice or beginner, especially one who is young. *adj.* **3.** one who is learning a trade, as a cub reporter, or beginning in scouting as a Cub Scout.

cube *n.* (French *cube*, a cube) **1.** a 6-sided, 3-dimensional shape bounded by squares on every side; each side meets four other sides at 90° corners. **2.** any object that has the approximate shape of a cube; e.g., a sugar cube or a pellet of certain types of feed. **3.** a number or expression raised to the third power, e.g., X^3. **4.** a bulk package of butter, ranging in size from 63 to 80 lb (28.6 to 36.3 kg). **5.** (**cubé**) a shrub (*Lonchocarpus* sp., especially *L. nicou*, family Fabaceae) native to tropical South America. Its roots are a source of an insecticide, rotenone, and fish poisons known as cub, timbo, and barbasco. *v.* **6.** to cut into cubes. **7.** to raise a quantity to its third power.

cubic feet per second *n.* a common measure of water flow for irrigation water. Usually abbreviated cfs. See cfs.

cuckoo *n.* (French *coucou*, cuckoo) **1.** any of several species of slender, long-tailed birds (*Coccyzus* sp. in America and *Cuculus* sp. in Europe, family Cuculidae) with a call that sounds like its name.

European cuckoos deposit their eggs in the nests of other birds, but the American cuckoos raise their own young. *adj.* **2.** (informal) silly or crazy.

cuckoo bee *n.* a parasitic bee (family Nomadidae, subfamily Nomadinae) that does not gather pollen for honey and does not build its own nest but lays its eggs in nests of other bees. The cuckoo bee larvae are likely to kill many of the larvae of the host bees.

cucumber *n.* (Latin *cucumis*, cucumber) **1.** an annual vine plant (*Cucumis sativus*, family Cucurbitaceae) that is often grown in gardens for its fruit. **2.** the elongated green fruit with white flesh and small soft seeds that grows on such vines and is used in salads or to make pickles.

cucumber beetle *n.* either of two troublesome garden pests in humid regions, either the striped cucumber beetle (*Acalymma vittata*) or the spotted cucumber beetle (*Diabrotica undecimpunctata howardi*) that can kill young seedlings of cucumbers, squash, or melons; girdle the stems and eat leaves from older plants; and transmit bacterial wilt and mosaic diseases.

Striped cucumber beetle

cucurbit *n.* (Latin *cucurbita*, gourd) **1.** a gourd or a plant that produces gourds (order Cucurbitaceae), including melons, gourds, cucumbers, pumpkins, and squashes; especially any of about a dozen annual species with lobed, cordate leaves and large, fleshy fruit, including pumpkins and longneck squashes (*Cucurbita pepo*) and winter squash (*Cucurbita maxima*). **2.** a distillation vessel with a shape resembling that of a gourd.

cud *n.* (Anglo-Saxon *cudu*, chewed food) a wad of food that a ruminant has regurgitated from its rumen and is rechewing.

cuesta *n.* (Spanish *cuesta*, sloping land) a ridge composed of tilted rock layers with a smooth gentle slope on one side and a steep rough escarpment on the other side.

Cross section of a cuesta

cuirass *n.* (French *cuirasse*, leather breastplate) **1.** the part of a knight's armor that covers the torso. **2.** any similar armor, as that on a ship. **3.** a protective shell on an animal, e.g., a tortoise. **4.** a layer of stones or other cover that protects a landscape from erosion.

cul-de-sac *n.* (French *cul-de-sac*, bottom of a cavity) **1.** a dead end. A passage that ends abruptly, as a dead-end street or a blocked passage in a cavern. **2.** a sac or cavity, such as the cecum of the appendix.

cull *n.* (Latin *colligere*, collect) **1.** a fruit, vegetable, grain, animal, or object with sufficient defect in size, color, shape, or condition to render it of low value or not marketable. **2.** the lowest marketing grade of meat carcasses, dressed poultry, or lumber. **3.** a reject; any animal eliminated from a herd or fowl removed from a flock because of appearance, disease, low milk or egg production, etc. *v.* **4.** to sort, especially to remove plants, animals, or items of inferior quality. **5.** to cut a few trees from a stand without regard to silviculture.

cultigen *n.* (Latin *cultus*, cultivation + Greek *genēs*, born) a plant variety of unknown origin; not known to occur in the wild; presumably developed long ago by obscure plant breeders.

cultivar (cv) *n.* (shortening of cultivated variety) a cultivated variety of a crop, especially an improved variety developed by a plant breeder and given a name in a modern language. In taxonomic nomenclature, the cultivar name has a nonitalic, capitalized initial letter, is enclosed in single quotation marks, and is placed after the species name; e.g., *Betula pendula* 'Fastigiata.'

cultivation *n.* (Latin *cultus*, cultivation) **1.** tillage. Mechanical manipulation of the soil for growing crops, including plowing, disking, harrowing, and planting. **2.** loosening of the soil to kill weeds, especially between the rows of a growing crop. **3.** the process of growing crops. **4.** good culture; refined habits.

cultural control *n.* pest control by management practices other than chemical pesticides. Examples: use of disease-resistant varieties, introducing natural predators of pests, adjusting planting dates to avoid pests, and rotation of contrasting crops. See integrated pest management.

cultural eutrophication *n.* plant nutrients added to an aquatic system by human activity and causing excessive growth of algal slime and aquatic weeds. The plant growth may hinder or prevent certain uses of the water, and objectionable odors may result from decaying plant material.

cultural landscape *n.* an area that is managed by humans rather than left to natural ecosystems and that includes urban land, roadways, construction sites, cropland, and managed pastures and woodlands

cultural practice *n.* a technique or a combination of techniques used in tillage and in the planting, tending, and harvesting of plants as crops or for ornamental purposes.

culture *n.* (Latin *cultura*, cultivation) **1.** the art of raising crops or livestock, including tillage practices, propagation, and nurturing for improved growth. **2.** either the growing of microorganisms or the microorganisms grown in or on a special medium. **3.** the unique way people live and interact in a given society (community, tribe, or nation). Approximate modern synonym: lifestyle. **4.** human knowledge and practices that have been or were passed from generation to generation, especially intellectual and artistic works that are considered enriching and refining; e.g., the culture of a particular nation or ethnic group, either modern or ancient.

cultured buttermilk *n.* a milk product made by skimming (or partially skimming), pasteurizing, and souring milk with a culture of lactic bacteria. It contains at least 8.5% of nonfat milk solids and is thicker than buttermilk that is a by-product of butter making.

Culver's physic *n.* a perennial herb (*Veronica virginica* = *Leptandra virginica*, family Scrophulariaceae) native from Manitoba to Florida and Texas that is grown for its white or pale blue flowers. Its rhizomes contain leptandrin, a bitter glycoside that acts as a violent emetic and cathartic. Also called black root, Bowman's root, or Culver's root.

culvert *n.* (French *coulvir*, channel or gutter) a water conduit or passage beneath a road, railroad track, or other crossing. It is typically lined with a large metal pipe or other structure that can be covered with fill material to serve as a bridge across a small stream.

cumin *n.* (Latin *cuminum*, cumin) **1.** a small annual herb (*Cuminum cyminum*, family Umbelliferae) native to the Mediterranean region that grows about 6 in. (15 cm) tall and produces white or reddish flowers and an aromatic fruit. **2.** the dried fruit of the plant (often called seeds) that is used to flavor cheese, bread, kraut, pickles, and soups and as an ingredient of curry powder. Oil from the seeds is used to flavor liquors and make perfume.

cumulic *adj.* (Latin *cumulus*, a heap) **1.** thickened; a surface soil horizon that has been thickened by gradual deposition of sediment over many years, as on a floodplain. **2.** a subgroup in Soil Taxonomy for soils with a cumulic surface horizon.

cumulonimbus cloud *n.* a very tall cumulus cloud with a sharp outline, a dense base, and often with a flat "anvil" top. The parent cloud of a thunderstorm, it is often accompanied by strong surface winds, strong upward and downward air currents, and heavy precipitation. The frequent lightning and thunder with these clouds led to the popular names of thunderhead or thundercloud. See thunderstorm.

cumulus cloud *n.* (Latin *cumulus*, a heap) a puffy cloud, typically having rounded protrusions on top, a relatively flat base and a sharp outline. These clouds are bright white where sunlit, but otherwise white to gray, and include clouds commonly known as fair-weather clouds or picture clouds as well as larger, darker clouds that indicate rain.

curare *n.* (Portuguese *curare*, curare) a highly toxic substance that paralyzes motor control nerves. It is obtained by drying an extract from certain South American trees (*Strychnos* sp., especially *S. toxifera*) and used on arrows and blow darts by certain South American natives and in medicine as a relaxant to relieve spasm in tetanus, spastic paralysis, etc.

curculio *n.* (Latin *curculio*, a corn worm) **1.** a weevil (*Conotrachelus nenuphar*, family Curculionidae) that damages apples, peaches, pears, plums, and most other stone fruits. The adult beetle feeds on the leaves and flowers of the tree and then lays eggs in the fruit, causing the fruit to be grubby and misshapen and leaving wounds where the brown rot fungus can readily attack it. Its natural enemies are a small wasp (*Trichogramma*), several fungal diseases, and cold weather. Also called plum curculio. See Trichogramma. **2.** any of several similar weevils.

curd *n.* (Anglo-Saxon *crudan*, to press) **1.** a coagulated mass composed mostly of casein and fat that forms when milk is clotted by rennet, by natural souring, or by the addition of a "starter." It can be eaten as cottage cheese or used to make processed cheese. **2.** the edible white head of cauliflower.

curdled milk *n.* milk that has thickened and coagulated by souring or by the action of enzymes called rennet. Sometimes called clabber or clabbered milk.

cure *v.* or *n.* (Latin *cura*, care) **1.** to heal a wound or successfully treat an ailment and restore health. **2.** to preserve plant or animal products by drying, smoking, pickling, etc.; e.g., hay, tobacco, smoked ham, or pickles.

curie (Ci) *n.* a unit of radioactivity representing 3.700×10^{10} disintegrations per second, named to honor Polish-born French physicist Marie Sklodowska Curie (1867–1934).

curly wood *n.* lumber with a wavy, but uncrossing, grain characteristic of certain trees such as curly maple. Curly wood is highly prized for making attractive furniture and other wood products. See bird's-eye maple.

currant *n.* (French *raisins de Corinth*, raisins of Corinth) **1.** a shrub (*Ribes* sp., family Saxifragaceae) related to gooseberries, blueberries, and grapes that is the alternate host to the white pine blister rust on both eastern and western white pines. **2.** an important commercial berry produced by the shrub. The berries are seedless and black, red, or whitish.

current *n.* (Latin *currens*, running) **1.** a flowing stream, especially the main or fastest moving part of a stream of water or air. **2.** the velocity of flow. **3.** electric power as measured in amperes. **4.** a social or ideological trend of thinking, as a political current. *adj.* **5.** belonging to the present, as current affairs. **6.** common practice, as current usage. **7.** popular, as a current fad. **8.** newest, as the current issue of a periodical.

custom work *n.* work done under contract. The contractor usually furnishes the labor, equipment, and materials in return for a specified payment. Much grain harvesting, spraying and picking of fruit, and sheep shearing is done as custom work.

cutan *n.* (Latin *cutis*, skin) a clay coating on the surface of a soil ped as a result of illuviation.

cutaneous *adj.* (Latin *cutis*, skin + *eous*, like) having to do with the skin.

cuticle *n.* (Latin *cuticula*, little skin) **1.** epidermis; the outer layer of the skin of humans and other vertebrates. **2.** the hardened skin at the edge of a fingernail or toenail. **3.** the outer layer of the exoskeleton of an insect. **4.** a thin waxy layer of cutin and cellulose forming the outer wall of epidermal cells of aboveground plant tissue. **5.** one of the four major parts of an eggshell.

cutin *n.* (Latin *cutis*, skin) the waxy component of the cuticle on the exposed walls of epidermal cells of plant tissue. It is most noticeable where it can be polished to a high gloss, as on the surface of apples, cherries, and some other fruits.

cutover *adj.* cleared of forest, usually a removal of all trees from the area.

cutting *n.* **1.** a branch, shoot, bud, or other part of a plant that can be severed and used to generate a new plant or be grafted onto another plant. **2.** harvesting hay, grain, etc., by mowing. **3.** a crop of hay harvested by mowing (often designated as first cutting, second cutting, etc.). **4.** sawing or chopping down a tree. **5.** separating a horse or cow from the main herd. **6.** an article clipped from a newspaper or magazine. **7.** a recording, especially of a song on a record. **8.** any act of cutting, real or figurative. *adj.* **9.** designed to cut, as a cutting tool. **10.** penetrating, as a cutting wind. **11.** injuring someone's feelings, as a cutting remark.

cutting chute *n.* a narrow passageway for removing selected animals from a herd.

cutting cycle *n.* the time interval allotted from one major tree harvest to the next in a forest.

cutting horse *n.* a horse used by someone to remove selected animals from a herd.

cuttlefish *n.* (Anglo-Saxon *cudele*, cuttlefish + *fisc*, fish) any of several cephalopods (especially *Sepia* sp., family Sepiidae) with two long tentacles and eight arms covered with suckers. Its body is 8 to 10 in. (20 to 25 cm) long and has a hard internal shell. A cuttlefish can eject a jet of water to move itself or a dark, inky fluid to hide from danger.

cutworm *n.* (Anglo-Saxon *cyttan*, to cut + *wyrm*, serpent) a dingy yellow to black soft-bodied caterpillar (the larval stage of a moth) that feeds on tender plants. Some cutworms feed at or below ground level and may fell a small plant (e.g., the black cutworm, *Agrotis ypsilon*, family Phalaenidae). Others climb up plants (usually at night) and eat leaves or bore into fruits, e.g., the variegated cutworm (*Peridroma margaaritosa*, family Noctuidae), rated one of the most widely distributed and destructive of garden pests, or the granular cutworm (*Feltia subterranea*, family Noctuidae) that feeds on foliage near the soil as well as on tomato fruits in southern U.S.A. Cutworms attack most field crops, garden vegetables, flowers, and fruit trees.

cyanide *n.* (Greek *kyanos*, dark blue + *ide*, chemical) a radical composed of a nitrogen atom triple bonded to a carbon atom that carries one open bond (–CN) and unites with cations to form cyanide salts, e.g., potassium cyanide (KCN) or hydrogen cyanide (HCN). HCN is an extremely poisonous gas that forms hydrocyanic acid when dissolved in water and is sometimes used as a soil sterilant, insecticide, or rodenticide. It interferes with respiration because –CN combines with the iron in cytochrome oxidase. Sodium and potassium cyanides are sometimes used as rodenticides and also in the manufacture of certain plastics and in extracting gold from ore.

cyanobacteria *n.* (Greek *kyanos*, dark blue + *baktērion*, little staff) photosynthetic nitrogen-fixing organisms that can live either independently or as part of a symbiotic association, especially with a number of primitive plants such as liverworts, pteridophytes (*Azollae*), gymnosperms (cycads), and angiosperms (Gunnera). Also known as blue-green algae or pond scum. See blue-green algae.

cyanogen *n.* (Greek *kyanos*, dark blue + *genēs*, born) a component of various fruits, legumes, sorghum, cassava, etc., composed of cyanide linked through another carbon to a sugar molecule (sugar–CR_1R_2–CN). It can release toxic HCN when broken down, especially when plants wilt from frost or mechanical damage, and is most hazardous to ruminants because the rumen bacteria produce enzymes that can degrade cyanogens.

cycle *n.* (Latin *cyclus*, circle) **1.** one complete part of a repetitive series, as an electric current alternating from positive to negative and back to positive. **2.** a sequence of events that repeats during a stated time period; e.g., a daily cycle, a monthly cycle, or an annual cycle. **3.** a period of many years; an era or age. **4.** a bicycle, tricycle, unicycle, or motorcycle. **5.** a recurring biological pattern, as a growth cycle. *v.* **6.** to pass through one or more complete cycles. **7.** to ride or travel by bicycle or other cycle.

cyclone *n.* (Greek *kyklōn*, whirling) **1.** a large low-pressure air mass with upward-flowing air and winds that spiral counterclockwise and inward in the Northern Hemisphere or clockwise and inward in the Southern Hemisphere. The rising air often causes rain or snow. See hurricane, anticyclone. **2.** a hurricane in the vicinity of India or Australia. **3.** (nontechnical usage) a tornado or other violent windstorm rotating around a center of low pressure.

cyclone seeder *n.* a device, commonly driven by a hand crank, that drops seeds onto a spinner that scatters them in front of the operator.

cyclonic rainfall *n.* rainfall that results from air circulation in and around a low-pressure area formed where warm, moisture-laden air rises. Such a storm usually forms over an open ocean before moving across a landmass as a large system that can be seen in satellite photos and is commonly shown on weather maps. The leading edge forms a weather front.

cyclosilicate *n.* (Greek *kyklōn*, whirling + Latin *silex*, flint + *ate*, salt) any of a class of silicate minerals with a ring of tetrahedra as their basic structural element; e.g., beryl.

cyclostome *n.* (Greek *kyklōn*, whirling + *stoma*, mouth) a jawless fish with a circular, sucking mouth; e.g., hagfish and lampreys. Cyclostomes are the most abundant fish in the ocean, but most of them are small, transparent, luminescent species that live at depths between 1000 and 3000 ft (300 to 900 m).

cygnet *n.* (French *cygnet*, small swan) a young swan.

Cygnus *n.* (Latin *cygnus*, swan) **1.** the genus name of swans; large, white, long-necked aquatic birds. **2.** a northern constellation in the Milky Way supposed to represent a swan; also known as the Northern Cross.

cyma *n.* (Greek *kyma*, a wave) **1.** a cornice molding that has a sinusoidal cross section, partly concave and partly convex. **2.** a cyme. **3.** a wavy land surface with swells and sags or troughs; e.g., the surface of a Vertisol. See Vertisol.

cyme *n.* (Greek *kyma*, a wave) a flat or convex cluster of flowers wherein each flower has its own short stem, as in phlox and sweet William.

cynipid *n.* (Latin *cyniphes*, a gallfly) a small solitary wasp (family Cynipidae) that lays its eggs on a specific part of an oak tree, rose bush, or other plant (depending on the species of cynipid). The larvae cause the plant tissue to swell and form galls. Also called gall wasp.

cyst *n.* (Greek *kystis*, pouch) a sac, vesicle, or bladderlike structure in either plant or animal tissue, especially one filled with liquid or semisolid material. A cyst often forms around a foreign body or tiny organism, separating it from the host tissue.

cysteine *n.* an essential sulfur-containing amino acid ($C_3H_7O_2NS$) that is a component of most proteins. The linkage of two cysteines produces cystine.

cystine *n.* (Greek *kystis*, bladder + *ine*, amino acid) a cross-linked amino acid with a disulfide bridge formed by covalent bonding between 2 cysteine side chains of a protein or of neighboring proteins.

cystolith *n.* (Greek *kystis*, cyst + *lithos*, stone) **1.** a knobby crystalline deposit of calcium carbonate in a plant leaf cell. **2.** a concretion in the urinary bladder.

cytogenetics *n.* (Greek *kytos*, hollow + *genētikos*, generating) the study of the hereditary mechanism in cells, especially the roles of genes and chromosomes.

cytokinin *n.* (Greek *kytos*, hollow + *kinētikos*, putting in motion) any of several plant hormones (mostly purines) produced in the roots and translocated upward to stimulate cell division and differentiation and inhibit aging.

cytology *n.* (Greek *kytos*, hollow + *logos*, word) the scientific study of the life history, structures, and functions of plant, animal, and human cells, especially for the diagnosis of abnormalities and diseases.

cytoplasm *n.* (Greek *kytos*, hollow + *plasma*, something formed) the protoplasm between the nucleus and the cell wall of a microbial, plant, animal, or human cell.

cytosol *n.* (Greek *kytos*, hollow + *sol*, solution) the aqueous solution component of cytoplasm, containing many dissolved inorganic ions and organic compounds (e.g., salts, sugars, amino acids, and proteins). Insoluble particles and organelles are suspended in the cytosol.

cytosterility *n.* male sterility that is controlled by the cytoplasm rather than by chromosomal genes; a characteristic that facilitates the production of hybrid seed in certain lines of corn and some other plants. Cytosterility in the female parent line eliminates the need for detasseling.

D

D-activated sterol *n.* a livestock feed supplement that supplies vitamin D; produced by exposing a sterol fraction of plant or animal origin to ultraviolet light that converts the sterol into vitamin D_2 or D_3, respectively.

daffodil *n.* (Latin *affodilus*, daffodil) See Narcissus.

daily cover *n.* the practice of covering the material deposited in a landfill with soil at the end of each day to prevent fire and to control pests.

dairy *n.* **1.** a building where cows are milked and the milk is collected for processing. **2.** a plant that processes and distributes milk, cream, butter, cheese, and other milk products. *adj.* **3.** having to do with dairying; e.g., a dairy farm or dairy product.

dairy breed *n.* any of several breeds of cattle that are bred more for milk production than for meat; e.g., Ayrshire, Brown Swiss, Guernsey, Holstein, Jersey.

dairy cattle *n.* cows raised and kept for milk production.

dairy goat *n.* a goat kept for milk production. Goats tolerate hot, cold, dry, and mountainous conditions better than cows. Their milk is naturally homogenized with tiny fat globules and is easier to digest than cow's milk, especially for infants and invalids.

dale *n.* (Old English *dael*, dale or valley) a small valley, especially one with a broad area of gentle slopes.

dalles *n.* (French *dalle*, flagstone or gutter) rapids of a river flowing through a canyon; e.g., the Columbia River near The Dalles, Oregon.

Dallis grass *n.* a perennial pasture grass (*Paspalum dilatatum*, family Poaceae) named after A. T. Dallis, who grew it in Georgia in the late 19th century; native to South America and now widely grown in wet pasture land in southern U.S.A.

Dallis grass poisoning *n.* poisoning of cattle caused by ergot (*Claviceps purpurea*) growing on the grass seeds of Dallis grass.

dally *v.* (Old French *dalier*, chat) **1.** to loiter; to waste time. **2.** to frolic or play dangerously. **3.** to flirt or make love as a sport. **4.** to use a horse as an anchor when roping animals by looping the end of the lariat around the saddle horn.

dam *n.* (French *dame*, lady) **1.** a female parent, especially a domestic quadruped. **2.** an artificial structure, usually built of either compacted earth or concrete, that retains water for conservation, domestic use, flood control, irrigation, recreation, water power, etc. *v.* **3.** to obstruct the flow of water. **4.** to block or obstruct anything, physically or figuratively.

Damalinia sp. *n.* a biting louse (order Mallophaga) that is an external parasite on domestic and wild animals.

damp *adj.* (German *dampf*, vapor) **1.** moist; slightly or moderately wet; containing or coated with some water. See dank. **2.** having high humidity, as damp air. **3.** without enthusiasm, as a damp reception. *v.* **4.** to diminish movement; e.g., to reduce vibration or to deaden sound. *n.* **5.** any gas that accumulates in a mine; e.g., methane (CH_4) or hydrogen sulfide (H_2S). See stinkdamp.

damping-off *n.* a disease that rots young seedling stems at the soil level, often felling the plant; caused by any of several fungi (*Rhizoctonia solani*, *Pythium* sp., *Phytophtora* sp., or *Pellicularia* sp.). Also called foot rot.

damselfly or **damsel fly** *n.* a nonstinging insect (order Odonata, suborder Zygoptera) that captures other insects on the wing; related to the dragonfly but smaller and slower. The damselfly nymph has leaflike gill plates and an enormous lower jaw; it lives most of its life searching for food among submerged plants in still water. See dragonfly.

dandelion *n.* (French *dent de lion*, tooth of a lion) a serious weed (*Taraxacum officinalis*, family Asteraceae) in lawns and gardens; native to Europe. It has deeply notched leaves that are edible in the spring. This member of the sunflower family has a naked stem topped by a single yellow flower that soon turns into billowy white plumes of seeds that blow in the wind.

dander *n.* **1.** dandruff; small scaly particles of dead skin on the scalp. **2.** loose scaly particles from the skin of various animals or feathers of birds. **3.** anger; e.g., he got his dander up.

dank *adj.* (Swedish *danka*, moist) unpleasantly damp; humid and usually cold and clammy, as a dank cellar.

Darcy's law *n.* an equation for the velocity of saturated flow of water through an ideal porous medium, typically written as $V = kh/l$, where V is the apparent velocity of flow (the real velocity is faster because the water must go between and around the solid particles of the medium), k is the hydraulic conductivity (permeability constant) for the porous medium, and h is the difference in hydraulic head across a straight-line travel length, l. It is named for Henry Darcy of Paris, who formulated it in 1856 based on his work with water flow through sand beds. It can also be used for other liquids by determining appropriate *k* values.

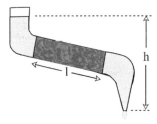

A setup for determining the *k* value for Darcy's law

dariloid *n.* sodium alginate, a gel that has a high capacity for absorbing water; obtained from the giant Pacific Coast kelp (*Macrocystis pyrifera*)*;* used to stabilize ice cream mixes so they will not form ice crystals.

darkling beetle *n.* a sluggish dark-colored beetle (family Tenebrionidae) that feeds only at night on plants and decaying plant matter. Its larvae are called mealworms. See mealworm.

dark matter *n.* matter that cannot be observed by telescopes but is postulated to exist to account for gravitational effects; probably consisting of gases and small nonluminous particles widely distributed through the universe.

dark of the moon *n.* that period of the lunar month when the moon is between the sun and the Earth so its lighted side is mostly hidden from view on Earth. Some folklore suggests that root crops such as potatoes should be planted during the dark of the moon and aboveground crops such as tomatoes should be planted in the light of the moon. See moonarian.

dark reaction *n.* **1.** any reaction that can take place in the absence of light, especially those that are part of the process of photosynthesis. **2.** the processes that fix carbon and synthesize sugar in the Calvin cycle of photosynthesis.

darnel *n.* (French *darnelle*, darnel) a weedy ryegrass (*Lolium temulentum*, family Poaceae) that often grows in grainfields and likely is the biblical tare.

Darwin, Charles Robert *n.* an English naturalist (1809–1882); the author of the theory of evolution by natural selection (especially in *The Origin of Species*, 1859).

Darwinism *n.* belief in the theory of evolution by natural selection as developed by Charles Darwin, holding that individuals endowed with properties that make them better adapted for survival and reproduction in a particular environment will be represented by larger numbers in later generations. This idea has been known as Darwinism, the survival of the fittest, evolution by natural selection. It is opposed by creationists. See creationism.

date *n.* (Latin *data*, given) **1.** a specified time, especially a day of the year. **2.** an appointment to meet someone. **3.** a companion for attending an event, especially one of the opposite sex. **4.** the sweet edible fruit of the date palm, usually in dried form with an elongated shape and a wrinkled surface; each date has an elongated seed that often is removed before marketing. *v.* **5.** to designate the time or age of something, especially of an artifact or an important historical event.

date palm *n.* a deep-rooted palm tree (*Phoenix dactylifera*, family Palmaceae) that produces edible dates; grown in arid and semiarid areas of northern Africa and eastward to Pakistan and India and in southwestern U.S.A. and Mexico.

Date palms

datum (pl. **data**) *n.* (Latin *datum*, a gift) **1.** the elevation of a point or level area (often mean sea level) taken as a reference for measuring other elevations, e.g., when making a topographic map. **2.** a quantitative evaluation; e.g., the number, quantity, or weight of an item; usually used in the plural, data. **3.** an assumed value that serves as a base for calculations.

datum plane *n.* **1.** the horizontal plane, real or imaginary, that serves as a zero level or other arbitrary reference level for ground elevations, soundings, or water surface elevations. **2.** an arbitrary reference surface used to minimize or eliminate local topographic effects on seismic time and velocity determinations in seismic mapping.

dawn *n.* (Anglo-Saxon *dagian*, to become day) **1.** daybreak; early light. **2.** the beginning of anything, as the dawn of civilization. *v.* **3.** to begin to be light. **4.** to begin to be understood (usually dawn on, as it dawned on him).

day-neutral plant *n.* a plant that is unaffected by shortening or lengthening of days. See photoperiodism.

days-to-pick *n.* the least number of days after a pesticide application before it is safe to harvest and eat the crop.

days-to-slaughter *n.* the least number of days after feeding an antibiotic before it is safe to slaughter the animal and eat the meat.

dbh *n.* diameter at breast height; used to measure tree size, usually at 4.5 ft (137 cm) above ground level. See basal area.

DDT *n.* dichlorodiphenyltrichloroethane ($C_{14}H_9Cl_5$), the first chlorinated hydrocarbon contact insecticide; a very effective broad-spectrum insecticide that became available in 1945 and was widely used thereafter. The EPA banned registration and inter-state sale of DDT for virtually all but emergency uses in the U.S.A. on June 14, 1972, because of its persistence in the environment and accumulation in the food chain. It accumulates in fatty tissues of people and animals and is very damaging to bird life.

deadfall *n.* **1.** a trap with a large weight poised to fall and crush the prey, usually a large game animal. **2.** a tangled mass of fallen trees.

dead furrow *n.* an open furrow left by a plow. See back furrow.

deadhead *n.* **1.** a log that is submerged or partially submerged in a stream or lake. **2.** poor-yielding blighted or dwarfed seed stalks. **3.** a dull person. **4.** a person traveling on a free ticket by commercial carrier. *v.* **5.** to not work on a return trip.

dead lake *n.* a lake overgrown with aquatic vegetation so there is little or no open water. This condition occurs naturally when a lake ages, but it is accelerated by eutrophication. In popular usage, the terms "dead," "dying," and "senescent" are often used interchangeably.

deadly nightshade *n.* See belladonna.

deadman *n.* **1.** a partially or completely buried log or stone that serves as an anchor for a guy wire. See guy. **2.** a safety switch that stops a machine from running when the operator releases it.

Dead Sea *n.* an extremely saline lake between Israel and Jordan that is 370 mi^2 (960 km^2) in area. It is fed by the Jordan River but loses water only by evaporation. Its shore is the lowest land elevation on Earth, approximately 1300 ft (400 m) below sea level.

dead zone *n.* an area of ocean where the water is toxic to fish and other aquatic organisms, usually from oxygen depletion.

dealate *adj.* (Latin *de*, away from + *alatus*, wing) wingless after mating; e.g., certain ants and termites whose wings are shed after a nuptial flight.

dealkalization *n.* the neutralization and removal of alkali from the soil by applying soil amendments (most commonly gypsum) and leaching. See sodic soil reclamation.

deamination *n.* the part of the decomposition process that removes amine groups from organic materials and converts them into ammonia; usually involves deaminase enzyme.

death camas or **deathcamas** *n.* **1.** a perennial weed (*Zigadenus paniculatus*, family Liliaceae) with nar-row leaves, small light green flower clusters, and a bulbous root containing a toxic alkaloid that is especially hazardous to sheep or to humans when it is mistaken for the edible camas root; found mostly in western U.S.A. **2.** a poisonous, bulbous herb (*Z. leimanthoides*) of southeastern U.S.A. that grows in wet areas; sometimes called pinebarren deathcamas.

death cup *n.*, a poisonous mushroom (*Amanita phalloides*) with a cuplike base; lethal when eaten by people. Also called death cap.

debacle *n.* (French *débâcle*, a breakup) **1.** a breakup of river ice. **2.** a flood of debris-laden water. **3.** a disastrous military defeat or other fiasco.

debeaker *n.* a device that cuts off the end of the upper beak of young (often day-old) chickens or turkeys to prevent pecking, cannibalism, and later egg eating.

debouchment or **debouchure** *n.* (French *déboucher*, to open into something + *ment*, result of) **1.** the place where a river or channel opens into the ocean, a sea, a lake, or other broad area. **2.** the place where a spring emerges. **3.** the place where a passageway connects with a larger passage or chamber.

debris *n.* (French *débris*, waste) **1.** the disorganized remains of something that has fallen to pieces or been destroyed. **2.** dead leaves, stems, and other plant residues left on the soil when a field is harvested. **3.** loose grains of mineral matter left when a rock disintegrates. **4.** soil and rock materials that accumulate at the debouchment of a stream or on the surface of a glacier as the ice melts. **5.** discarded litter along a highway or other public place.

debris cone *n.* a fan-shaped deposit of mixed stones and soil deposited where the velocity of a fast-flowing stream is greatly reduced; e.g., where a mountain stream flows into a major valley.

debris flow *n.* **1.** a landform produced where an earthen mass became so soggy that it flowed down a slope and onto a flat area below. (Debris flow generally describes a mass with a high content of gravel and stones; see mudflow for a similar mass of finer material.) **2.** the process that produces such a landform.

decalcification *n.* (Latin *de*, away from + *calcx*, lime + *ficatio*, making) **1.** leaching of calcium carbonate from a soil or rock material; a slow natural process of soil formation in humid regions. It causes the soil pH to drop from about 8.2 (where it is buffered by abundant calcium carbonate) to neutral or acid levels. **2.** the loss of calcium and density from bones of living animals and people; ultimately produces osteoporosis, a bone condition that is most prevalent in postmenopausal women and greatly increases their risk of fractures.

decapitation *n.* (Latin *decapitatus*, head cut off + *atio*, action) **1.** beheading; a common method of capital punishment in the Middle Ages. **2.** severe tree pruning that removes most or all of the branches. Trees that survive commonly produce a witches' broom type of growth consisting of many long, slender branches. Also called coppicing.

decare or **dekare** *n.* an SI (metric) unit of land area equivalent to 1000 m², 10 ares, 0.1 ha, or 0.2471 ac.

decastere *n.* an SI (metric) unit of volume equivalent to 10 steres, 10 m³, 13.08 yd³, or 353.16 ft³.

decay *n.* (French *decäir*, decline or decay) **1.** a gradual wasting away and disintegration into fragments; a loss of strength, integrity, or well-being. **2.** rot or the process of rotting, especially that resulting from the action of microorganisms. **3.** radioactive decomposition. **4.** deterioration of social and moral standards. *v.* **5.** to deteriorate, disintegrate, or become rotten, or to cause such conditions.

decibel (dB) *n.* **1.** one-tenth of a bel; a unit for evaluating the relative loudness of a sound; named after Alexander Graham Bell, 1847–1922. The rating in bels (tens of decibels) is equal to the logarithm of the loudness (power) of a sound divided by that of a standard barely audible reference sound (equivalent to 10^{-12} watts passing through 1 m²); an increase of 3 dB represents a doubling in loudness. Human hearing can detect differences of approximately 1 dB in sounds ranging from 0 to about 120 dB; sounds above 130 dB register as pain. The sound of a whisper may be 25 or 30 dB, a typical conversation about 60 to 70 dB, street noise often about 70 dB, and a clap of thunder about 100 dB. **2.** a similar scale for other forms of electronic energy; e.g., radio signal strength.

deciduous *adj.* (Latin *deciduus*, something that falls down) **1.** shedding leaves at the end of the growing season, as most nonevergreen trees and shrubs. **2.** characteristic of antlers or horns that fall off or of teeth that fall out. **3.** shedding of hair or molting of feathers.

declination *n.* (Latin *declinatio*, bending aside) the degrees of angle between the direction a magnetic needle points and true north, variable with geographic location. See agonic line.

declivity *n.* (Latin *declivis*, sloping downward) a downward slope; falling ground; e.g., a hillside

viewed from the top, as opposed to acclivity, an upward slope.

decomposer *n.* a microbe that obtains food energy and nutrients by decomposing dead organic materials of either plant or animal origin.

decomposition *n.* (Latin *de*, separation + *compositio*, putting together) **1.** the chemical breakdown of more complex substances into simpler ones by splitting molecules, hydrolysis, oxidation, and other chemical processes; e.g., the splitting of calcium carbonate molecules by heat:

$$CaCO_3 + heat \rightarrow CaO + CO_2$$

See disintegration. **2.** the breaking apart of any structure or assembly into its component parts.

decontaminate *v.* (Latin *de*, separation + *contaminatus*, contaminating) to remove or break down unwanted material (e.g., a toxic or radioactive substance) so it cannot do any harm or damage.

decreaser plant species *n.* a plant species that decreases as a percentage of the cover when rangeland is overgrazed. Often termed decreasers.

deer *n.* (Anglo-Saxon *deor*, a wild animal) any of at least 50 species of ruminant mammals (family Cervidae) with long legs, long necks, and small heads with long ears and prominent eyes, especially the smaller species (the larger caribou, elk, and moose are sometimes included). The males of most species carry multipointed antlers that they shed and regrow each year; female reindeer also carry antlers, but females of other species do not.

deerfly *n.* a troublesome fly (*Chrysops* sp., family Tabanidae) that attacks deer, cattle, horses, dogs, etc., especially around the neck and head; similar to a horsefly, but smaller. Females suck blood and may be a vector of tularemia.

deer tick *n.* a tick (*Ixodes scapularis*) that is a vector for Lyme disease. The tick has a 2-year life cycle including larva, nymph, and adult. The larva may be infected with Lyme disease by feeding on a dead infected mouse; the nymph and adult stages may transmit the disease to living animals or humans by their bites. Adult female deer ticks are orange-red with a black spot near the head; adult males are all black.

defecation *n.* (Latin *defaecatus*, cleansing from dregs) the evacuation of feces from the bowels; a bowel movement.

deferred grazing *n.* delayed grazing by livestock on a pasture or an area of rangeland until the plants have become well established, produced sufficient growth to support the livestock, or set seed. A deferred grazing system is usually more productive than continuous grazing, especially when combined with rotation grazing. The combination is called either rotational deferred grazing or deferred rotation grazing. See rotation grazing.

deficiency *n.* (Latin *deficiens*, lacking) **1.** a shortage or lack of anything important, as an ability, a material, or a product. **2.** inadequate nutrition, especially in terms of an essential element or vitamin (often designated as a nutrient deficiency or a vitamin deficiency).

deficiency symptom *n.* a visible indication that a nutrient is not available in adequate amounts to meet the needs of a plant or animal; e.g., chlorotic leaves on a plant or stunted growth in either a plant or animal.

deflation *n.* (Latin *de*, from + *flatus*, blowing) **1.** the process of allowing air or other inflatant to escape, as from a balloon, or the condition of having lost inflatant. **2.** a decline in prices and other economic indicators; opposite of inflation. **3.** soil erosion by wind.

deflorate *adj.* having shed its pollen or its blossoms.

defluorinated phosphate *n.* **1.** a product made by heating phosphate rock (apatite) to about 1500° to 1600°C to drive off fluorine in the form of HF. This converts phosphorus from the mineral apatite to tricalcium phosphate and improves the availability of the phosphorus for use as a fertilizer, in the production of other phosphorus fertilizers, or as an animal feed supplement (although it is still ranked below bone meal in value for most livestock).

defoliant *n.* (Latin *defoliare*, to remove leaves) a chemical (usually applied as a spray or powder) that causes the leaves to drop from a growing plant. For example, calcium cyanamide defoliant is sometimes dusted on cotton to hasten its ripening and make it ready for harvest. Agent Orange (a mixture of 2,4-D and 2,4,5-T) was used as a crop and forest defoliant during the Vietnam War.

defoliation *n.* (Latin *de*, from + *foliatus*, leafing) the removal or loss of leaves and other foliage.

deforestation *n.* removal of all of the trees and brush from a previously forested area, usually followed by converting the area to cropland or other new use. See forestation.

deformity *n.* (Latin *deformitas*, unshapeliness) any deviation from normal physical appearance, especially if deemed odd or ugly.

deglutination *n.* (Latin *deglutinatus*, ungluing) **1.** the ungluing of pieces that have been glued together. **2.** the separation of gluten from wheat or other cereal.

degossypolize *v.* to remove gossypol (a toxic phenolic pigment) from cottonseed meal; to reduce its content to not more than 0.04% free gossypol. See gossypol, cotton.

degradation *n.* (Latin *degradatio*, reduction in rank) **1.** a reduction in the rank, honor, or dignity of an individual. **2.** the process of degrading, especially when it occurs gradually. **3.** decomposition, especially of an organic compound. **4.** erosion of a landscape or other surface.

degree days *n.* See growing degree days.

degree of use *n.* utilization or consumption of forage on rangeland in respect to total plant growth. It may be expressed in qualitative terms such as ungrazed or lightly, moderately, closely, or severely grazed or as percentage of weight consumed for either an individual plant or the vegetation as a whole.

dehiscence *n.* (Latin *dehiscere*, to gape open + *entia*, action) bursting or splitting open along a seam; e.g., the opening of a seed pod or the reopening of a wound.

dehorn *v.* **1.** to remove the horns from cattle, sheep, or goats. **2.** to cauterize horn buds on young animals so that horns will not develop.

dehumidify *v.* (Latin *de*, from + *humidus*, moist) to reduce the amount of moisture; e.g., air can be dehumidified by means of a desiccant or by contact with a cold surface followed by reheating.

dehydrogenase *n.* any of several enzymes that catalyze the removal of hydrogen from organic compounds.

dehydroretinol *n.* See vitamin A.

deionized water *n.* water that has been purified by passage through ion exchange resins to remove both anions and cations. Many laboratories use deionized water as an alternative to distilled water.

delactation *n.* (Latin *de*, from + *lactatio*, giving milk) weaning of suckling young and cessation of milk production by the mother.

Delaney clause *n.* a clause proposed by James Delaney, U.S. Representative from New York, and passed as an amendment to the Food, Drug, and Cosmetic Act by the U.S. Congress in 1958. It established zero tolerance in food products sold in the U.S.A. for any chemical additive known to cause cancer in animals or humans. The meaning of a zero test effectively became more restrictive with time as analytical methods improved from those able to detect parts per million in 1958 to some that are now able to detect parts per quintillion. Consequently, the U.S. Congress repealed the Delaney clause in 1996 and replaced it with specific concentrations as tolerable standards for various chemicals.

delineation *n.* (Latin *delineatus*, marking) **1.** a line that marks a boundary or change on a map or diagram. **2.** a sketch or figure that illustrates an object or concept. **3.** the process of outlining or defining something.

deliquesce *v.* (Latin *deliquescere*, to become liquid) **1.** to absorb moisture from the air and go into solution. Calcium chloride ($CaCl_2$), for example, will deliquesce. **2.** to become liquid by melting (a solid) or when a certain maturity is reached (some fungi). **3.** to divide into many small branches.

dell *n.* (Dutch *del*, a dale or valley) a small valley, usually one that is wooded. See dale.

delousing *n.* the removal or extermination of lice, commonly by the use of insecticides.

delta *n.* (Greek *delta*, 4th letter of alphabet) **1.** the 4th letter in the Greek alphabet (Δ, δ). **2.** something triangular in shape, especially a nearly flat alluvial deposit in a lake or sea at the mouth of a river. Current deltas have their surfaces near the water level, but some old "hanging deltas" occur on hillsides that were once the shores of glacial lakes. **3.** an increment of change in the value of a mathematical variable.

demarcation *n.* (Spanish *demarcación*, dividing line) **1.** a line or other marker forming a distinct boundary, as the line of demarcation between two countries. **2.** the process of marking a boundary.

demersal *adj.* (Latin *demersus*, submerged) living near the sea floor; applies to fish and other aquatic animals that live on or adjacent to the sea bottom and feed on benthonic organisms. See benthos.

demineralize *v.* **1.** to remove dissolved minerals from water. **2.** to decrease the mineral content of bones and teeth. See osteoporosis.

demographics *n.* (Greek *dēmos*, people + *graphe*, a writing) population statistics and characteristics, especially data regarding age, education, income, etc.

demolition waste *n.* debris from removal or remodeling of buildings of any kind.

demurrage *n.* (French *demorage*, delay) **1.** a delay in departure time; a detention. **2.** a penalty for a delay; e.g., a charge levied when a consignor or consignee holds railroad cars or other transport vessels beyond a designated time.

denature *v.* (Latin *denaturare*, to change the nature of) **1.** to change the natural characteristics of something; e.g., to coagulate egg white. **2.** to make a product unsuitable for human consumption but still suitable for other purposes, especially the addition of either methyl alcohol or pyridine to ethyl alcohol to make it poisonous to drink.

dendritic *adj.* (Greek *dendrītēs*, belonging to a tree) branching like a tree; e.g., a dendritic drainage pattern.

Dendritic drainage

dendrochronology *n* (Greek *dendron*, tree + *chronos*, time + *logos*, word) the use of the growth rings of old trees and wood as a time scale for weather and associated events in prehistoric environments. Sequences of relatively favorable and unfavorable growth years are identified by the spacing of the growth rings both in living trees and in dead but not rotten wood. Such sequences have been established for periods of thousands of years in some locations.

dendrology *n.* (Greek *dendron*, tree + *logos*, word) the botanical study and classification of trees and shrubs.

dengue *n.* (Swahili *dinga*, cramplike attack) a tropical disease transmitted by a mosquito (*Aedes aegypti*) that causes a fever, a skin rash, and painful joints that typically last for about a week in the acute form and require a lengthy convalescence. The disease is estimated to affect 30 to 60 million people in the tropics and subtropics. Also called dengue fever, dandy fever, breakbone fever.

denitrification *n.* (Latin *de*, away from + French *nitrifier*, to form nitrates + *icus*, containing + *atio*, action) **1.** the microbial process that reduces soil nitrates to gaseous forms of nitrogen, including nitrogen gas (N_2), nitrous oxide (N_2O), nitric oxide (NO), and nitrogen dioxide (NO_2). Denitrification occurs mostly in soils that have been aerated enough to produce nitrates and then become anaerobic from being waterlogged. **2.** the process of removing nitrogen or nitrogen compounds from a substance.

denitrifier *n.* (Latin *de*, away from + French *nitrifier*, to form nitrates) a microbe that participates in the process of denitrification; generally an autotrophic organism (e.g., *Thiobacillus* sp.) or a facultative anaerobe (especially *Pseudomonas* sp., *Bacillus* sp., and *Paracoccus* sp.).

dense *adj.* (Latin *densus*, thick) **1.** compact; either naturally heavy or compressed into a mass with high density. **2.** thick; crowded; as a dense stand of trees. **3.** obstructing light, as a dense fog or a dense photographic negative. **4.** difficult to comprehend or having difficulty comprehending.

density *n.* (Latin *densitas*, thickness) **1.** the mass of a substance per unit volume, most often expressed in grams per cubic centimeter. **2.** population; the number of persons, animals, birds, or plants of a specified type per unit area. **3.** the closeness of packing, as the density of fibers in wool or the density of threads in a fabric.

density, tree *n.* density of tree cover expressed in number of trees, basal area, or volume of wood per acre or hectare.

density current *n.* the flow of a denser fluid beneath a lighter fluid of otherwise similar nature; e.g., the flow of strongly saline water from the Mediterranean Sea along the floor of the Straits of Gibraltar into the Atlantic Ocean (causing a return flow of surface water in the opposite direction), or the flow of turbid water beneath clear water. See turbidity current.

density separation *n.* **1.** the process of separating a mixture of materials into two or more parts by flotation in a liquid; e.g., separation of heavy and light minerals from a crushed rock. **2.** the separation of a mixture of materials by flotation in a moving stream of liquid (usually water) or air; e.g., use of a wind tunnel to separate a stream of solid waste into a heavy fraction (metal, mineral matter, etc.) and a light fraction (paper and plastic that can be burned, preferably in an electric power plant).

dental cup *n.* a small depression that develops in the incisor teeth of horses as the white enamel wears down, revealing a dark-colored interior known as a dental star. This permits a skilled person to estimate the age of a horse.

dental pad *n.* the very firm gum in the front of the upper jaw of cattle, sheep, goats, and other cud-chewing ruminants. The dental pad substitutes for upper teeth.

dentate *adj.* (Latin *dentatus*, toothed) **1.** toothed. **2.** serrated; e.g., the margin of a serrated leaf, especially those with sharp points between indentations.

den tree *n.* a hollow tree that serves as a home for a mammal such as a squirrel or raccoon.

denudation *n.* (Latin *denudatus*, stripping off) **1.** baring, especially of land; e.g., the removal of all plant cover from an area or the removal of all the leaves from a tree. **2.** severe erosion that removes most of the soil and even substrata from an area.

deodorant *n.* (Latin *de*, from + *odorans*, smelling) any substance used to absorb, destroy, or mask offensive odors; e.g., activated charcoal, aluminum chlorohydrate.

deoxyribonucleic acid (DNA) *n.* a linear polynucleotide with double-helical strands containing multiple units composed of a purine or pyrimidine base, deoxyribose sugar, and phosphate; the material in the nucleus of any cell (plus some in the mitochondrion or chloroplasts of plant cells) that carries the genetic code for an organism. See double helix, ribonucleic acid.

Department of Energy (DOE) *n.* an executive department in the U.S. government created in 1977 to coordinate the multitude of energy-oriented government programs and agencies.

depauperate *v.* (Latin *de*, from + *pauperare*, to make poor) to impoverish; to deprive a plant, animal, or person of adequate nourishment, causing a starved or underdeveloped appearance.

depergelation (Latin *de*, away from + *per*, very + *gelat*, frozen + *atio*, action) *n.* the thawing of permafrost (e.g., beneath a building or where plant cover is removed). See permafrost, tundra.

depletion *n.* (Latin *depletus*, emptying) serious diminishment of the reserves of a resource; the consumption of a natural resource at a rate in excess of the rate of replenishment. For example, the current rate of consumption of petroleum products is depleting the resource, and many irrigation projects that pump water from aquifers are depleting the aquifers. See Ogalala, fossil water.

depopulate *v.* (Latin *depopulari*, to lay waste) to eliminate or drastically reduce a population, as by extermination or expulsion.

deposit *n.* (Latin *depositus*, laying aside) **1.** money or goods placed for safekeeping. **2.** a partial payment on a purchase. **3.** a refundable payment made as security until something is returned; e.g., a bottle deposit. **4.** material transported and left in a new position by a natural agent such as water, wind, ice, or gravity, or by human activity; e.g., an alluvial deposit, an ore deposit, or a pile of trash. **5.** a coating placed on something by treatment with a chemical in solution, often by means of an electric current. **6.** the dust or spray that remains on the plants when a pesticide is applied. *v.* **7.** to place something on deposit for safekeeping, surety, etc. **8.** to carefully place something; e.g., to deposit a baby in a crib. **9.** to form a natural deposit; e.g., by sedimentation in a body of water.

deposition *n.* (Latin *depositio*, putting down) **1.** removal from an office or position. **2.** testimony, especially a written statement by a witness who is not present. **3.** the process of making a deposit; e.g., the deposition of alluvium by water, of loess by wind, or a covering by forming a solid coating from a vapor (the reverse of sublimation). **4.** a deposit, layer, or coating thus formed.

depositional *adj.* (Latin *depositio*, putting down + *al*, of or by) formed by the accumulation of sediment, as by flooding. Stones, gravel, sand, silt, and other debris are dropped when a transporting agent is slowed (a river or stream, ocean waves, wind, glaciers, or gravitational mass movement). Alluvial fans, offshore bars, dunes, glacial moraines, mudflows, etc., are depositional features.

depraved appetite *n.* an abnormal craving (perhaps caused by a diet deficiency) in people, animals, or fowls for items not normally eaten, e.g., soil. See geophagia, pica.

depression *n.* (Latin *depressio*, pressed down) **1.** a low place in a land surface. **2.** a dent or hollow in the surface of a solid object or in the flesh of an animal. **3.** a sad feeling of hopelessness, weakness, and dejection. **4.** a severe decline in business activity, drop in employment and income, and falling prices in an area or a nation; e.g., the Great Depression that was precipitated by the New York Stock Market crash in October, 1929, causing economic sluggishness throughout the industrial world in the 1930s. **5.** any notable lowering, as of atmospheric pressure, dew point, sound level, or activity.

Dermacentor *n.* a genus of reddish brown wood ticks that transmits Rocky Mountain spotted fever, tick paralysis, and tularemia. Rodents (especially squirrels) are its first and second hosts; humans and large animals (cattle, horses, deer, etc.) are its third hosts.

dermal toxicity *n.* the injury caused when a chemical is absorbed through the skin. See contact dermatitis.

dermatitis *n.* (Greek *dermatos*, skin + *itis*, inflammation) inflammation of the skin from any cause (e.g., allergy, chemical exposure, disease, sunburn).

dermatology *n.* (Greek *dermatos*, skin + *logos*, word) the study of skin and skin diseases.

dermatophytosis *n.* (Greek *dermatos*, skin + *phyton*, a growth + *osis*, condition) any skin infection caused by a fungus; e.g., ringworm, athlete's foot. Also known as dermomycosis and epidermomycosis.

dermis *n.* (Greek *derma*, skin) skin, especially the layer beneath the epidermis.

derris *n.* (Greek *derris*, a covering) a tropical tree (*Derris* sp., family Fabaceae) cultivated from Malaysia to Indonesia. The insecticide rotenone comes from the roots of this tree. See rotenone.

DES *n.* See diethylstilbestrol.

desalinization or **desalination** *n.* (Latin *de*, away from + *salinus*, salty + *izatio*, making) **1.** removal of salt from seawater or other saline water (e.g., by flash distillation, reverse osmosis, or electrodialysis). **2.** leaching of salt from saline soil.

de Saussure, Theodore *n.* a 19th-century Swiss physicist who elucidated plant growth processes, including respiration, absorption of carbon dioxide

from the air for photosynthesis, and mineral and water absorption from soil.

desert *n.* (Latin *desertus*, deserted) **1.** an arid region with so little vegetation that it is incapable of supporting any considerable population. Desert conditions occur in four types of settings: (1) areas dominated by subtropical high pressure; e.g., the Sahara Desert, Arabian Desert, most of Australia, and northern Mexico and southwestern U.S.A. (2) rainshadow areas in the interiors of continents, such as the Chilean desert and the Gobi Desert. (3) coastal deserts influenced by cold ocean currents, as on the west coasts of California and Peru. (4) cold deserts in polar regions with perpetual snow cover. *adj.* **2.** arid and nearly barren, as desert land. **3.** uninhabited; e.g., a desert island.

desert crust *n.* a hard surface layer on an arid soil; cemented by calcium carbonate, gypsum, or other binding material.

desertification *n.* (Latin *desertus*, deserted + *fication*, process) conversion of a habitable area to desert, usually at the border of an existing desert; loss of vegetation, perhaps accompanied by drier climate. Human causes of desertification include accelerated erosion, overgrazing, salt accumulation from irrigation, increased runoff and other water loss, and loss of vegetation killed by pollution of soil, water, and air.

desert pavement *n.* a thin layer of gravel or small stones left on the land surface in desert regions after the removal of the fine material by wind erosion; also called reg.

desert varnish *n.* a dark sheen on exposed rock surfaces in arid regions; caused by the polishing action of sand-laden winds.

desiccant *n.* (Latin *desiccatus*, drying up completely) a material that promotes dehydration, removing moisture from air, plant or animal tissue, or other substance; e.g., calcium chloride ($CaCl_2$), silica gel.

desiccation *n.* (Latin *desiccare*, to dry up completely + *tion*, action) drying process; subjection to conditions that remove water.

designer bugs *n.* a popular term for microbes developed through biotechnology to degrade specific toxic chemicals in waste materials or in ground water; e.g., microbes that can degrade oil or a pesticide.

desilting area *n.* an area used to protect a reservoir, pond, lagoon, field, or other area from sedimentation; usually a nearly level area with dense vegetation that slows water flow enough to remove silt and other debris from the water. See silt pond.

despoil *v.* (Latin *despoliare*, to plunder) **1.** to plunder; to rob the property and other valuables of a community or other area. **2.** to seriously damage an environment.

destructive distillation *n.* removal of the volatile materials from a substance such as coal or wood by heating it in the absence of air to produce coke or charcoal, respectively. The volatiles (oils, tars, etc.) are captured by condensation or absorption and may be separated into fractions for a variety of uses.

desulfurization *n.* the removal of sulfur from fuels and flue gases; e.g., removal of sulfur from coal, gasoline, and other fuels to meet clean air standards and reduce acidity of rainfall. See acid precipitation.

detassel *v.* to remove the tassels from corn so it will not self-pollinate; a practice used to produce hybrid corn seed. Usually about four rows of an inbred variety are detasseled so they will be pollenized by two adjacent rows of a different inbred variety. See hybrid corn, cytosterility.

detention dam *n.* a dam built for short-term storage of streamflow or surface runoff; the stored water is released slowly as a flood-control technique. See retention basin.

detergent *n.* (Latin *detergens*, wiping off) **1.** any chemical compound that helps clean by lowering the surface tension of water. **2.** a synthetic cleansing agent that substitutes for soap, makes oil a better lubricant, or serves other related uses; commonly, a foaming soaplike compound made from alkyl sulfates and benzene sulfonates. Being synthetic, some detergents are not readily biodegradable; also, some detergents (especially the older ones) contain enough phosphorus to be major contributors to eutrophication of streams that receive sewage effluent. See eutrophication.

deterioration *n.* (Latin *deteriorare*, to make worse + *tion*, action) **1.** the loss of quality and desirability. **2.** degradation of condition and ability. **3.** the state or condition of having deteriorated.

determinate growth *n.* limited or restricted growth, as a plant that has a flower at the end of each primary and secondary axis. Such flowers all blossom at about the same time, and their fruit or seed ripen in a single crop. See indeterminate growth.

detoxify *v.* (Latin *de*, away from + *toxicum*, poison + *ficare*, to make) **1.** to remove the toxic (poisonous) quality of a substance; e.g., cottonseed is detoxified by heating it enough to destroy the gossypol toxin. See gossypol. **2.** to hold an intoxicated person in detention long enough for the body to metabolize the alcohol.

detritivore *n.* (Latin *detritus*, a rubbing away + *vorare*, to devour) any organism that obtains its nourishment from dead organic and/or fecal matter. See decomposer.

detritus *n.* (Latin *detritus*, a rubbing away) fine particulate debris; e.g., a deposit of weathered rock material that has been transported from the weathering site to another site.

deuterium (D) *n.* (Greek *deutero*, second) heavy hydrogen. Each heavy hydrogen atom has a neutron along with the proton in its nucleus, giving it an atomic weight of 2.

deuterium oxide (D$_2$0) *n.* heavy water; heavy hydrogen combined with oxygen.

deuteromycete *n.* (Greek *deutero*, second + *mykētes*, mushrooms) a fungus (e.g., *Penicillium* sp. and *Aspergillus* sp.) that does not reproduce sexually, or at least is not known to do so; a large group, including most of the wilts, some damping-off fungi, ringworm fungi, etc. Also known as Fungi Imperfecti.

development *n.* (French *développement*, development) **1.** the process of growth or other progressive change; e.g., child development or soil development. **2.** a specific stage in such a process of growth or stage, especially the best or final stage. **3.** events or discoveries that impact or reflect the current status of a person, group, or field, as recent developments in chemistry. **4.** a community of homes built within a relatively short period of time, as a housing development.

deviation *n.* (Latin *deviatio*, turning aside) any variation from normal in characteristics or behavior in plants, animals, or people. Some deviations result from mutations and can be inherited. See mutation, jumping genes.

devilfish *n.* **1.** the manta ray (named for the hornlike appearance of its pectoral fins). **2.** the octopus and other large cephalopods. **3.** other aquatic species that strike someone as being devilish; e.g., the gray whale or the angler fish.

devil's darning-needle *n.* a common name for a dragonfly. See dragonfly.

devolution *n.* (Latin *devolutio*, rolling down) **1.** passage through a series of stages. **2.** passing of property or rights from one person to a successor. **3.** transfer of power from one governmental entity to another, especially from a central government to a local government. **4.** biological evolution toward simplicity in either form or function.

De Vries, Hugo *n.* a Dutch botanist (1848–1935) who researched evening primroses (*Oentherala marckiana*) and their mutations. Author of *Intracellular Pangenesis* and *The Mutation Theory*.

dew *n.* (Anglo-Saxon *deaw*, dew) condensed water vapor that collects (usually at night) on surfaces of plants, stones, and other objects cooled by natural radiation to a temperature at or below the dew point. See dew point.

dewclaw *n.* a vestigial claw or false hoof on, or just above, the foot of a quadruped. The dewclaw does not touch the ground as the animal stands or walks.

dewfall *n.* the condensation of dew from atmospheric moisture. In some humid climates, dewfall received is equal to 5 in. (13 cm) of annual precipitation. See dew pond.

dewlap *n.* **1.** the pendulous skin on the underside of the necks of some animals, particularly certain members of the ox family, such as Zebu cattle. **2.** the wattle of chickens, turkeys, etc. **3.** the loose skin that forms an inflatable pocket on the necks of certain lizards.

dew point *n.* the saturation temperature (corresponding to 100% relative humidity) for the water vapor present in air; water droplets will condense on surfaces whose temperature is below the dew point or form fog if the air temperature drops below the dew point. As water condenses, the release of heat of condensation has a warming effect that opposes the cooling of the air at the dew point.

dew pond *n.* a depression in a rock surface that is replenished with water by the condensation of dew, mist, or clouds (after they have once been filled with water). Dew ponds have been identified in chalk formations at high elevations near Sussex Downs, England, and on Isle Royale in Lake Superior, U.S.A. They are also called mist ponds, cloud ponds, rock pools.

dew retting *n.* wetting of jute, hemp, or flax by prolonged exposure to dew or rain to loosen the fibers so they can be separated from the mass.

dew worm *n.* an earthworm; a night crawler.

dewy *adj.* covered with, dampened with, resembling, or characterized by dew.

dexiotropic *adj.* (Greek *dexios*, to the right + *tropos*, turning) turning or spiraling from left to right (clockwise), as the whorls in most gastropod shells.

dextrin or **dextrine** *n.* (Latin *dexter*, right) a soluble, gummy polymer of D-glucose $[(C_6H_{10}O_5)_n]$ formed by hydrolysis of starches; used in adhesives, as a thickener in foods, and as a binding agent in medicines. Named for its property of rotating polarized light to the right.

dextrorotatory or **dextrorotary** *adj.* (Latin *dexter*, right + *rota*, a wheel) rotating polarized light to the right (clockwise).

dextrose *n.* (Latin *dexter*, right + French *ose*, carbohydrate) a hexose sugar $(C_6H_{12}O_6)$; D-glucose, the form that rotates polarized light to the right; usually obtained commercially by the hydrolysis of starch with acids or enzymes. Also called corn sugar, grape sugar, starch sugar.

DHIA (Dairy Herd Improvement Association) *n.* an organization of dairy farmers headquartered in Columbus, Ohio, that maintains dairy production records, certifies the quality of dairy products, advocates dairy interests, and advises its members of new technology.

diacetyl *n.* the flammable yellow aromatic chemical $(C_4H_6O_2)$ that is a carrier for the aroma of food products such as butter, coffee, vinegar.

diachronic comparison *n.* a means of studying changes in visible features of the Earth by comparing aerial photographs taken at different times.

diagnostic antigen *n.* a biological agent prepared for use in the diagnosis of a specific disease; e.g., tuberculin injected under the skin of cattle or humans to determine whether they have tuberculosis. Inflammation of the skin around the injection site indicates the presence of tuberculosis antibodies.

diagnostic horizon *n.* a set of soil characteristics used in Soil Taxonomy to define certain soil classes. Diagnostic horizons that form at the soil surface are called epipedons (e.g., a mollic epipedon); those that occur deeper in the soil are called diagnostic subsurface horizons (including albic, agric, argillic, cambic, kandic, natric, oxic, sombric, and spodic horizons). See epipedon.

diaheliotropism *n.* negative heliotropism; the turning of the leaves or other parts of certain plants away from light, especially sunlight.

diallel crossing *n.* a method of evaluating two male animals (e.g., bulls) by comparing the two sets of progeny produced by breeding them to the same females (at different times).

dialysis *n.* (Greek *dialysis*, separation) **1.** passage of dissolved material through a membrane. See osmosis. **2.** removal of solutes by permitting ions to diffuse through a membrane. **2.** medical treatment using dialysis to remove waste products from the blood of persons whose kidneys have failed.

diamagnetic *adj.* (Greek *dia*, apart + *magnēs*, magnet) repelled by an iron magnet, as certain minerals or forms of bismuth or copper. See paramagnetic.

diamond *n.* (French *diamant*, diamond) **1.** a form of carbon crystallized under high pressure in the isometric system; the hardest substance known; a very valuable precious gem; also used as a cutting, drilling, or polishing material. See aluminum magnesium boride for the second hardest substance or corundum for the second hardest natural substance. **2.** a precious gem composed of carbon, cut to reveal its natural facets and show brilliant colors from refracted light. **3.** an uncut gemstone composed of carbon. **4.** a rhombus, a figure with four equal sides, usually placed with its diagonals oriented vertically and horizontally. **5.** a playing card of the suit marked by red diamond figures. **6.** a playing field for baseball (either the infield or the entire field). **7.** something or someone of great but unrealized value, as in the idiom, a diamond in the rough. *adj.* **8.** adorned with one or more diamonds; e.g., a diamond ring or a diamond tiara. **9.** using diamonds, as a diamond saw.

diamondback *adj.* **1.** having diamond-shaped markings on the back. *n.* **2.** a large venomous rattlesnake (either *Crotalus adamanteus*, the eastern diamondback, or *C. atrox*, the western diamondback).

diamondback moth *n.* a small brown and white moth (*Plutella xylostella*, family Yponomeutidae) with wings that have a diamond shape when folded. The larvae feed on many greenhouse plants and conifers and are very destructive.

diamondback terrapin *n.* an edible turtle (*Malaclemys terrapin*) native to coastal salt marshes from Massachusetts to Mexico.

diapause *n.* (Greek *diapausis*, a pause) a period of inactivity and arrested development of an organism, especially in certain immature insects, that serves as a survival mechanism during adverse environmental conditions; often triggered by the short days preceding winter.

diaphragm *n.* (Greek *dia*, through + *phragma*, a fence) **1.** the muscular wall between the thoracic cavity and the abdomen of mammals. It aids in respiration by enlarging the thorax as it contracts. **2.** a thin disk that separates two chambers and either vibrates to produce sound in a telephone or microphone or flexes to serve as a pressure regulator or part of a pump. **3.** a birth control device that covers the entrance to the uterus. **4.** a plate or assembly with a hole, usually of variable size, that controls the amount of light entering a camera or other optical instrument. **5.** a porous plate or membrane separating two liquids, often adjacent to the wall of a cell.

diarrhea *n.* (Greek *diarrhoia*, flowing through) abnormally frequent and watery fecal discharges by animals or humans caused by many diseases or unusual foods. Diarrhea can involve enough water loss to cause serious dehydration and death. Diarrhea in horses, cattle, pigs, etc., is called scours. See enteritis.

diastase *n.* (Greek *diastasis*, separation) an enzyme that catalyses the hydrolysis of starch into maltose and then into dextrose.

diastrophism *n.* (Greek *diastrophē*, distortion) movements that bend, fault, fold, and warp the crust of the Earth and thus produce ocean basins, continents, mountains, plateaus, and other landforms.

diatom *n.* (Greek *dia*, through + *tome*, to cut) a microscopic to barely visible, single-celled, photosynthetic form of algae that is abundant in water around the world. The cell walls of diatoms are siliceous shells marked with a pattern characteristic of the species (at least 10,000 species have been identified). These shells accumulate as thick deposits known as diatomaceous earth or diatomite on the floors of seas and lakes, especially where the water is cold.

diatomaceous earth (diatomite) *n.* a deposit of siliceous shells of diatoms commonly found on the floors of seas and lakes; used in the manufacture of mild abrasives, pottery, insecticides, and dynamite and as a filter to remove giardia cysts from water. See giardia.

diazinon *n.* an organic phosphorus insecticide ($C_{12}H_{21}N_2O_3PS$) with a ring structure containing four carbon and two nitrogen atoms; especially useful against flies.

diazotroph *n.* (Greek *di*, two + *azo*, nitrogen + *trophe*, food) a microbe that can fix atmospheric nitrogen.

dibble *n.* **1.** a small, pointed hand tool used to make holes in the soil for planting bulbs, seeds, or small plants. *v.* **2.** to use a dibble for planting.

dichlorophenoxyacetic acid (2,4-D) *n.* the first widely used selective herbicide ($C_8H_6C_{12}O_3$); a growth regulator applied in an amine or ester form as a foliar spray that has little effect on grasses but kills most broadleaf herbaceous plants by altering physiological processes and causing rapid, contorted growth. The amine forms are less volatile and safer than the ester forms, but the esters are effective at a lower rate of application.

dichotomous *adj.* (Greek *dichotomos*, cutting in two) dividing into two equal branches, especially repeatedly.

dicotyledon or **dicot** *n.* (Greek *di*, two + *kotylēdōn*, a cavity) a flowering plant with two seed leaves (cotyledons), including most deciduous trees and shrubs, legumes, and forbs. The cotyledons store carbohydrates that provide energy for initial growth and are pushed up out of the soil to serve as the first leaves of the plant. Dicotyledons have netted or reticulate venation in contrast to the parallel venation of a monocotyledon plant. See monocotyledon, cotyledon.

dicoumarin *n.* a chemical ($C_{19}H_{12}O_6$) extracted from spoiled sweetclover (*Melilotus* sp., family Fabaceae) or synthesized; used as a blood thinner to prevent blood clotting in animals and humans. Also called dicumarol.

***Dicrocoelium* sp.** *n.* a liver fluke of ruminants; a trematode.

***Dictyocaulus* sp.** *n.* a lungworm nematode parasite of cattle, horses, sheep, and goats.

diecious *adj.* See dioecious.

dieffenbachia *n.* a tropical plant (*Dieffenbachia* sp., family Araceae) native to South America and the West Indies; named after a German botanist, Ernst Dieffenbach (1811–1855). It commonly is grown indoors in temperate regions for the decorative effect of its large green leaves (often splashed with white) that are attractive but also poisonous. See dumbcane.

dieldrin *n.* a highly toxic, persistent, chlorinated insecticide ($C_{12}H_8OCl_6$) that is licensed for nonagricultural use only.

diestrus or **dioestrus** *n.* that portion of a female mammal's cycle between periods of estrus.

diet *n.* (Greek *diaita*, way of living) **1.** food habitually eaten by a person or animal. See ration. **2.** a food regime prescribed for losing weight or to overcome a medical problem. **3.** (figurative) habitual activity, as a steady diet of watching television.

dietary factors *n.* characteristics of a diet; e.g., the total amounts of vitamins, minerals, carbohydrate, fat, protein, and fiber, the ratio of saturated to unsaturated fat, and the method of preparation.

dietary fiber *n.* food components that are resistant to the action of normal digestive enzymes (e.g., cellulose, hemicellulose, pectin, lignin, and certain undigested proteins); food components that absorb water, increase fecal bulk, reduce constipation, and promote normal bowel function.

dietetics *n.* the study of the kinds and amounts of food required for good health.

diethylstilbestrol or **stilbestrol (DES)** *n.* a synthetic hormone ($C_{18}H_{20}O_2$) and source of several estrogens; used to increase growth rate in meat animals, to cause female mammals to come into heat, in chemical caponizing, and to pacify tom turkeys. Residues in meat are suspected of being carcinogenic.

differential grasshopper *n.* a large (1.5 in. or 4 cm long), very destructive grasshopper (*Melanoplus differentialis*, family Acrididae), especially in the Mississippi Valley. It eats many kinds of crops and transmits several viral diseases from one plant to another.

differential weathering *n.* **1.** differences in the rate of weathering of the various minerals in a rock. Differential weathering causes the composition of mineral matter to change as rock materials weather to form parent material and soil. **2.** topographic changes that occur over time because some rocks weather much faster than others.

differentiation *n.* (Latin *differentia*, difference + *tion*, forming) **1.** the development or growth of a cell, embryo, or immature organism into all the parts of a mature organism. **2.** evolutionary development of new kinds of organisms. See Darwinism. **3.** a mathematical process used in calculus to define the rate of change of a variable represented by an equation.

diffraction *n.* (Latin *diffractus*, broken in pieces) **1.** the bending of light rays as they pass through a medium; e.g., the splitting of a light beam into the colors of the spectrum by passing it through a prism or the formation of a rainbow by sunlight passing through water droplets in the atmosphere. See rainbow. **2.** similar bending of beams of other forms of radiation; e.g., X-ray diffraction used to determine layer spacings in crystals.

diffused runoff *n.* a thin sheet of runoff occurring fairly uniformly over a large area rather than being concentrated in a channel. See erosion, sheet erosion, nonpoint source.

diffusion *n.* (Latin *diffusio*, diffusion) **1.** movement of dissolved ions or molecules in either the liquid or gaseous phase by random motion. **2.** distribution in random directions, as light diffusion by passing through a nonuniform medium or reflection from an irregular surface. **3.** spreading by random contacts, as the diffusion of information or of a disease.

diffusion coefficient *n.* a factor that evaluates the rate of movement of an ion or molecule through a uniform medium such as air or water. The corresponding value through a nonuniform mass such as soil is called the effective diffusion coefficient.

digestion *n.* (Latin *digestio*, carrying apart) **1.** the metabolic breakdown of food into materials that can be absorbed as nutrients by animals and people. Digestion begins in the mouth and continues in the stomach and intestines. **2.** the biochemical decomposition of organic matter; e.g., digestion of sewage sludge in tanks, where the sludge decomposes into gases, liquids, and small solid particles. **3.** the breakdown of binding agents; e.g., treatment of wood with acids and alkalis to make paper pulp.

digestive tract *n.* the various parts of the body of a human, animal, or fowl through which food passes, including the mouth, esophagus, stomach or stomachs, crop and gizzard of fowls, small and large intestines, and anus.

digger wasp *n.* any of several wasps (superfamily Specoides) that dig nest holes in soil. Digger wasps capture and paralyze beetle larvae, then place them in the nest holes along with each egg as food for the wasp larvae.

digitalin *n.* (Latin *digitalinum*, digitalis) a poisonous glucoside ($C_{36}H_{56}O_{14}$) made from the seeds of purple foxglove (*Digitalis purpurea*); used to treat cardiac problems.

digitalis *n.* (Latin *digitalis*, foxglove) **1.** the genus of plants known as foxglove (*Digitalis* sp., family Scrophulariaceae). **2.** a drug made from 2-year-old leaves of the purple foxglove (*D. purpurea*) gathered at the flowering stage and dried before extracting the drug. It is used to strengthen the heartbeat while slowing its rate and is used as a diuretic.

dike or **dyke** *n.* (Anglo-Saxon *dic*, dike or channel) **1.** a low embankment, usually built of earth or stones, that acts as a barrier to prevent water flow onto a protected area; used to control flooding, siltation (e.g., around a mined area), pollution (e.g., around a pesticide handling area), etc. **2.** a drainage ditch and the bank formed by its excavation. **3.** a thin, intruded vein of igneous rock that cuts across layers of older rock. *v.* **4.** to build a dike to control water movement.

dilapidation *n.* (Latin *dilapidatus*, squandered) the state of being dilapidated; neglected; damaged but not repaired; weathered; often discolored and misshapen.

dilation *n.* (Latin *dilatare*, to spread out) **1.** enlargement; e.g., dilation of the pupils for an eye examination, or the dilation of water when it freezes. **2.** dilatation; abnormal enlargement of an organ or a passage in the body, or the enlargement of a body passage by medical treatment. **3.** speaking or writing in great detail.

dill *n.* (Anglo-Saxon *dile*, dill plant) **1.** an annual or biennial herb (*Anethum graveolens*, family Apiaceae) native to Europe. Its aromatic seeds and leaves are used for seasoning food, especially pickles. **2.** the seeds or leaves of the plant prepared for use as a spice. *adj.* **3.** flavored with dill, as a dill pickle.

diluent *n.* (Latin *diluens*, diluting) any inactive material used to dilute or carry an active ingredient, e.g., for an insecticide or fungicide; often powdered mineral matter or water. Also called carrier.

dilute *v.* (Latin *diluere*, to wash away) **1.** to make less viscous by adding water, alcohol, etc. **2.** to weaken the brilliancy, concentration, flavor, strength, etc., of something by adding a diluent; e.g., most pesticides are sold in concentrated forms that must be diluted before use. *adj.* **3.** present in low concentration; having been diluted.

dilution ratio *n.* the proportion between an active ingredient (chemical, food, pollutant, etc.) and its diluent; e.g., the volume or weight of a pesticide to be mixed with a specified volume or weight of diluent, or the volume of sewage effluent discharged into a stream relative to the volume of water already in the stream.

dimethylnitrosamine *n.* a carcinogenic liquid ($C_2H_6N_2O$) present in tobacco smoke and in certain foods.

dimictic lake *n.* (Greek *di*, twice + Latin *mixtus*, mixing) a body of water with cold and warm strata that invert or mix twice each year, once in the fall when the surface water temperature drops to match that of the bottom layer and once in the spring when the ice melts and the water temperature rises to about 4°C. See amictic lake, monomictic lake, polymictic lake.

dimorphism *n.* **1.** (Greek *dimorphos*, having two forms) the occurrence of two forms of the same species of plants (differing in leaves, flowers, or stamens, etc.) or animals (differing in color, size, or other property). **2.** the occurrence of two forms of a mineral with different crystal structures.

dingle *n.* (Anglo-Saxon *ding*, dungeon) a small secluded wooded valley.

dingo *n.* a wild dog (*Canis familiaris dingo*, family Canidae) in Australia, typically having a yellowish red color, erect triangular ears, a white-tipped bushy tail, and white feet; a serious predator of cattle, sheep, and fowls.

dinocap *n.* a phenolic fungicide used on fruit trees to control powdery mildew until it was restricted by the EPA in 1986 because it was found to cause birth defects in rabbits.

dinosaur *n.* (Greek *deinos*, terrible + *sauros*, lizard) **1.** any of a large variety of reptiles that dominated the Earth during the Mesozoic period but are now extinct. Dinosaurs ranged from a few feet to nearly 100 ft (30 m) long. Some walked on all four feet, but many walked on their two rear feet and balanced themselves with their long tail. **2.** something that is oversized and out-of-date, as that old machine is a dinosaur.

dinosaur extinction *n.* an event that occurred at the end of the Mesozoic era, about 65 million years ago. Not only the dinosaurs but also many other species disappeared, possibly because of a large meteorite striking the Earth. Mammals became the dominant species on Earth after that. See Permian extinction.

dinoseb *n.* a contact herbicide applied either before the crop emerged or during dormancy to control weeds in strawberries, soybean, alfalfa, cotton, etc., until it was banned by the EPA in 1986 because it posed risks of causing birth defects and sterility.

dioecious or **diecious** *adj.* (Greek *di*, two + *oikos*, house) having staminate (male) flowers and pistillate (female) flowers borne on different individual plants; e.g., asparagus, date palm, spinach. See monoecious.

dioestrus *n.* See diestrus.

dioxin *n.* any of several isomers of chlorinated hydrocarbon ($C_{12}H_4Cl_4O_2$) that are toxic by-products in the manufacture of certain herbicides and therefore trace contaminants in such herbicides, especially the isomer known as Agent Orange used as a defoliant during the Vietnam War. Dioxins are also formed when certain plastics and bleached papers are burned; they form some of the most persistent and hazardous air pollutants. Dioxin is a teratogen and possible carcinogen, especially in the 2,3,7,8,tetrachloro-dibenzo-*p*-dioxin form (known as TCDD).

diphtheria *n.* (Greek *diphthera*, membrane) an acute, contagious disease of humans and animals caused by a bacillus (*Corynebacterium diphtheriae*) that causes yellow skin spots or sores, difficulty in breathing, and release of a toxin into the blood. It was one of the leading causes of death of children until a vaccine was developed in the 1930s.

diphtheroid *adj.* **1.** of or similar to diphtheria. *n.* **2.** a disease that resembles diphtheria but is caused by a different microbe. **3.** a microbe that looks and acts like the bacillus that causes diphtheria (*Corynebacterium diphtheriae*) except that it does not produce the toxin.

diplodia boll rot *n.* a disease of cotton caused by a fungus (*Diplodia* sp., family Sphaerioidaceae); diseased bolls are covered with pycnidia (cases containing asexual spores) that make them appear black and smutty.

diplodia disease of citrus *n.* a disease of citrus and some other fruits caused by a fungus (*Diplodia natalensis*, family Sphaerioidaceae), characterized by dead bark on branches of all sizes, oozing gum, and stem-end rot of the fruit. *D. natalensis* also causes stem-end rot of watermelon.

diplodia disease of corn *n.* a virulent, infective disease of maize (corn) caused by a fungus (*Diplodia zeae* and *D. macrospora*, family Sphaerioidaceae); common in warm, humid corn-growing areas; characterized by a seedling blight and by a dry rot of the ears and stalks. Also called mold, moldy corn, dry rot, ear rot, mildew rot, diplodia dry rot, diplodia ear rot.

diplodia tip blight *n.* a disease of pine, fir, and spruce trees caused by a fungus (*Diplodia pinea*, family Sphaerioidaceae) that stunts new growth and causes browning of the needles, especially on the lower branches.

diploid *adj.* (Greek *diploos*, double) having two full sets of chromosomes (the norm for most cells), usually one set from each parent, in contrast to haploid (one set) or tetraploid (four sets). See haploid, tetraploid, polyploid.

dip needle *n.* a magnetic needle mounted vertically (rather than horizontally as in a compass) to indicate the inclination of the Earth's magnetic field. It will be vertical at the magnetic North Pole and magnetic South Pole, horizontal at the magnetic equator, and at intermediate positions everywhere else.

dipterous *adj.* (Greek *dipteros*, two-winged) having two wings, as a two-winged insect or a two-winged seed.

Dipylidium caninum *n.* a parasitic tapeworm that grows to about 1 ft (30 cm) long in the intestines of dogs and cats, with fleas or sometimes lice as the intermediate host. Humans can be infected if they swallow infected fleas or lice.

Dirofilaria immitis *n.* a filarial heartworm that grows to about 1 ft (30 cm) long as a slender, whitish worm, mostly in the right ventricle or pulmonary artery of dogs, especially in warm climates; also attacks cats and sometimes humans. It is transmitted by mosquitoes and possibly fleas.

dirt *n.* (Norse *drit*, excrement) **1.** soil, especially when out of place. **2.** any unconsolidated geologic deposit. **3.** anything common, filthy, or nasty. **4.** filthy and obscene writing or speaking. **5.** rumors of bad behavior; gossip.

disaccharide *n.* (Greek *di*, two + Latin *saccharum*, sugar + *ide*, chemical) a double sugar formed by linkage of 2 simple sugar molecules and the release of a molecule of water. The most common disaccharide is sucrose ($C_{12}H_{22}O_{11}$), formed by linking glucose with fructose (2 $C_6H_{12}O_6$ molecules).

disaster *n.* (French *desastre*, disaster) an event that causes great damage, including destruction of material things, much hardship, and/or loss of life. Some disasters have natural causes such as earthquakes and floods, others depend on human involvement such as a stock market crash or a petroleum spill causing an ecologic disaster.

discharge *n.* (Latin *discarricare*, to unload) **1.** material passed through an opening or sent away by ejection or emission. **2.** the rate of flow of water, silt, or other mobile substances passing along a conduit or in a stream. **3.** an exudate from a wound or an abnormal emission from a body opening; e.g., a nasal discharge. **4.** the unloading of cargo or burden. **5.** the fulfillment of a duty or the payment of a debt. **6.** removal from office or release from military service. **7.** firing of a weapon. **8.** release of electricity from a battery or capacitor, especially if it seriously depletes the charge. *v.* **9.** to perform, cause, or order a discharge.

disclimax *n.* (Latin *dis*, lack of + Greek *klimax*, ladder) relatively stable vegetative cover established as a result of disturbance of the natural ecology, especially by human influence; e.g., plants growing on surface mining residues that have been undisturbed for several years.

discontinuity *n.* (Latin *dis*, lack of + *continuare*, to continue + *itas*, condition) **1.** a break or gap in an otherwise continuous distribution; e.g., a gap in the geographic distribution of a taxon of plants or animals. **2.** a value where a mathematical function is not continuous; e.g., the zero value of a divisor that ranges from positive to negative values. **3.** a contact between two geologic formations where the underlying formation shows the effects of a period of erosion. **4.** changes in the interior of the Earth evidenced by seismic data that indicate differences in physical properties associated with depth, e.g., the Mohorovičić discontinuity (Moho).

discord *n.* (Latin *discors*, unlike) **1.** lack of harmony among the persons or things in an environment; clashing of colors, patterns, sounds, or ideas. **2.** disagreement; argument; strife, or even war.

discrete *adj.* (Latin *discretus*, separated) **1.** separate and distinct from others; detached. **2.** discontinuous; having separate parts.

disease *n.* (French *desaise*, disease) any health problem in humans, plants, animals, or insects; causes include pathogens (viruses, bacteria, fungi, and protozoa), parasites, mental or physical stress, congenital deficiencies, poor nutrition, unfavorable environment, or any combination of these.

disease, environmental *n.* any disease caused by environmental conditions affecting a plant, animal, or human, including unfavorable temperature, air and/or water pollution, nutrient deficiencies or toxicities, too much or too little light or water, too little oxygen.

disease control *n.* any procedure used to reduce or eliminate the spread of an infectious disease, including isolation of infected individuals, inoculations, control of insect vectors.

disease resistance *n.* the ability of individual plants, animals, or people to resist certain diseases; possible sources include individual genetic differences, previous exposure, inoculation, and robust health.

disharmony *n.* (Latin *dis*, not + *harmonia*, agreement) discord; an environmental situation with a disturbing excess or deficiency of a particular factor.

disinfectant *n.* (Latin *dis*, not + *infectus*, put into) a chemical or physical treatment that kills pathogenic organisms; applied mostly on inanimate objects. See sterilization.

disintegration *n.* (Latin *dis*, not + *integratio*, making whole) the mainly physical process of falling into many small pieces; e.g., weathering of granite into grains of sand and gravel. See decomposition.

disorder *n.* (French *désordre*, lack of order) **1.** randomness; disarray; lack of order. **2.** a disturbance in a public place, as an unorganized protest. **3.** an unhealthy and unnatural physical, mental, or behavioral condition of a person or animal. See disease. *v.* **4.** to disarrange items that were in order. **5.** to seriously disturb the physical or mental condition of a person or animal.

dispersal or **dispersion** *n.* (Latin *dispersus*, scattered + *alia*, act or process) **1.** spreading from one place into other areas. **2.** the division of light into a spectrum by passing it through transparent glass or plastic. **3.** the degree of spreading or separation, as the distribution of values around a statistical mean. **4.** breaking down of aggregates into individual particles and suspending them in water, air, or other fluid.

disperse *v.* (Latin *dispersus*, scattered) **1.** to separate, scatter, and disseminate. **2.** to vanish or cause to vanish, as a vapor plume disperses in the wind. **3.** to break aggregates into their individual component particles. **4.** to distribute fine particles and form a homogeneous suspension in a dispersion medium, such as clay in water. See colloid.

dispersed soil *n.* soil in which the clay particles are not aggregated, usually because the soil contains excess sodium. Dispersed soils tend to shrink, crack, and harden when they dry and to swell into a mass with low permeability when wet. This condition is suitable for paddy rice but unsuitable for other crop plants. See sodic soil, paddy.

disposal *n.* (Latin *disponere*, to place apart) **1.** the process of getting rid of unwanted materials, often regulated by laws. **2.** the process of moving hazardous materials (toxins, carcinogens, explosives, radioactive wastes, etc.) to an approved repository. **3.** arrangement in a specified order. **4.** settling of affairs; distribution of goods, e.g., the estate of a deceased person. **5.** availability, as the items at one's disposal.

dissection *n.* (Latin *dissectus*, cut up + *atio*, action) **1.** the process of cutting apart an animal body or plant structure to separate and view its component parts. **2.** erosion of a land surface into a system of valleys with hills between them. **3.** (figurative) detailed analysis (often as a refutation) of an argument or a personality point by point.

dissemination *n.* (Latin *dis*, apart + *seminare*, to seed + *atio*, action) **1.** dispersal; widespread distribution; e.g., the natural scattering of pollen, spores, or seeds. **2.** the spread of a contagious disease from one individual to another. **3.** broadcasting of crop seed in a field. **4.** the spread of information or energy, as the dissemination of radio waves.

disseminule *n.* propagule; any detachable part of a plant that can give rise to a new independent plant; e.g., a viable seed, a fungal spore, a cutting that will form roots.

dissipation *n.* (Latin *dissipatus*, scattered + *atio*, action) **1.** dispersal to the point of becoming undetectable. **2.** loss of energy with little or no work being accomplished; e.g., heat dissipation. **3.** misuse and waste that consumes a resource. **4.** intemperate living; alcoholism.

dissolved oxygen (DO) *n.* molecular oxygen (O_2) freely available in water, expressed either quantitatively (e.g., as mg/L or % saturation) or qualitatively (e.g., low, medium, or high). DO is vital to fish and other aquatic life and is needed for the decomposition of organic wastes. Traditionally, the level of DO has been accepted as the single most important indicator of a water body's ability to support desirable aquatic life. The optimum oxygen for most species of fish in surface water bodies is 5 mg/L; the minimum is about 3 mg/L. See BOD_5.

distal *adj.* distant from the center; remote; opposite of proximal.

distemper *n.* (Latin *distemperare*, to derange) **1.** an infectious catarrhal disease caused by a virus (*Tarpeia canis*) that causes a discharge from the eyes and nose accompanied by dullness, fever, and loss of appetite in dogs (most common and frequently fatal in puppies from 2 months to 1 year old), ferrets, minks, skunks, and raccoons. Also called canine distemper. **2.** panleukopenia in cats. **3.** strangles of horses.

distillate fuel oil *n.* any of a group of liquid hydrocarbons obtained by distillation of petroleum; commonly classified as No. 1, No. 2, or No. 4 fuel oil or No. 1, No. 2, or No. 4 diesel fuel; used mostly in space heaters, diesel engines, and generators of electric power.

distillation *n.* (Latin *distillare*, to drop or trickle + *atio*, action) **1.** movement of a liquid (accompanied by heat) by vapor transport; the liquid evaporates in one place and condenses in another. **2.** purification of a liquid (e.g., water) by evaporation and condensation. Almost all of the impurities remain in the evaporation chamber. **3.** the separation of the components in a liquid mixture according to their boiling points by boiling them in a chamber at successively higher temperatures and condensing the vapors in separate receptacles. The process may be repeated several times to improve the separation. However, distillation can never completely separate some liquid mixtures because they form a "constant boiling mixture" that vaporizes in the same proportion as the liquid, e.g., a mixture of 95% ethyl alcohol and 5% water.

distilled water *n.* water that has been purified by distillation. See deionized water.

distillers' grains *n.* the residue left from the production of distilled liquors and other alcoholic products (e.g., ethanol) from barley, corn, rye, and other cereals; commonly used to feed livestock. See ethanol.

disturbance *n.* (Latin *disturbare*, to disturb + *antia*, action or condition) **1.** a commotion, disorder, or breach of peace; a disturbing action or condition. **2.** something that creates such an action or condition. **3.** an event or condition that is sufficiently unusual to alter a community or ecosystem through its effects on one or more of the component species. **4.** a cyclonic storm, especially a small area of inclement weather. **5.** a minor earthquake; a small area of crustal movement.

ditch *n.* (Anglo-Saxon *dic*, ditch) **1.** a long trench dug in the ground to control water movement for drainage, irrigation, or interception of runoff or seepage. *v.* **2.** to dig a ditch or ditches, especially when used to provide drainage. **3.** to abandon something (perhaps by leaving it in a ditch).

ditch rider *n.* a person employed to keep watch on irrigation ditches, measure and adjust water flow, and assure the delivery of the contracted amount of water to each farm.

diuresis *n.* (Greek *diourein*, to urinate) increased amount of urine.

diuretic *n.* (Greek *diourētikos*, promoting urine production) a product that increases urine production; used to treat excess water retention.

diurnal *adj.* (Latin *diurnalis*, daily) **1.** daily; occurring every day; e.g., opening of blossoms on a plant during the day and closing at night, or the reappearance of a celestial body on a daily cycle. **2.** active during daylight, as many insects, birds, animals, or persons. See nocturnal.

divergence *n.* (Latin *divergentia*, turning apart) **1.** a point or time of separation. **2.** a separation that increases with distance or time.

diversification *n.* (Latin *diversus*, diverse + *facere*, to make + *atio*, action) **1.** doing several activities or making several products; e.g., producing several kinds of complementary crops and/or livestock to enhance income on a farm or ranch. **2.** the process of diversifying; changing from an emphasis on one or two activities or products to a greater variety.

diversion *n.* (Latin *diversio*, turning in a new direction) **1.** turning in a new direction; e.g., redirecting resources from one purpose to another. **2.** a ditch that diverts water from its natural channel into an irrigation system or a pond or away from the head of a gully or wet area. **3.** an entertaining pastime. **4.** a strategy to attract attention to one activity so another activity will not be noticed.

diversity *n.* (Latin *diversitas*, difference) **1.** non-uniformity; differences occurring among a group of individuals. **2.** a measure of population variability; e.g., the number of plant or animal species found in a designated area. See alpha diversity, beta diversity.

diverted land *n.* land that is removed from production of major crops and devoted to approved conservation uses, especially when done under contract with the U.S. Department of Agriculture.

divide *n.* (Latin *dividere*, to separate) the ridge that separates two watersheds (catchments).

diviner *n.* (Latin *divinare*, to foresee) **1.** a person who claims to receive knowledge or insight from divine or mysterious sources. **2.** a person who claims to determine the location of water, oil, gas, or ore deposits with a divining rod; a dowser. See divining rod, water witch.

divining rod *n.* a forked stick (or wire) used by a diviner to locate underground water, oil, gas, or ore deposits. The two branches are held tightly in the diviner's hands, and the point is supposed to dip when held over ore, oil, or water deposits, depending upon the specialty of the diviner. Also called dowsing rod, wiggle stick. See water witch.

division *n.* (Latin *divisio*, dividing) of plants: propagation using segments of parent plants, e.g., by cutting plant crowns, rhizomes, stolons, stem tubers, or tuberous roots into sections. Each section must have at least one head, eye, bud, or stem; e.g., potato tubers cut into pieces and planted.

division box *n.* an irrigation control structure used to divide and regulate the flow of irrigation water into two or more ditches.

DNA *n.* See deoxyribonucleic acid, ribonucleic acid.

dobe *n.* or *adj.* (Arabic *al tob*, the brick) **1.** short for adobe. **2.** a soil that is high in clay and hard to work because it is hard when dry and sticky when wet.

dobe house *n.* (short for adobe house) a dwelling made of dobe (often clay mixed with straw), either as sun-dried mud bricks or tamped into a form; an ancient form of construction used in the Middle East, northern Africa, and by early settlers in arid regions of southwestern U.S.A. See adobe.

dobsonfly *n.* a large predatory insect (*Corydalus cornutus*, family Corydalidae) that eats other insects. Dobsonfly larvae are called hellgrammites.

docile *adj.* (Latin *docilis*, easily taught) tame, teachable, easily trained and managed.

dock *v.* (Icelandic *dokkur*, stumpy tail) **1.** to cut off most of the tail of an animal, especially a sheep. **2.** to bring a ship into a berth at a port. **3.** to connect a space vehicle to a space station. **4.** to reduce the price paid for a product because of contamination or poor quality. **5.** to reduce the wages of a worker; e.g., for misbehavior or low productivity. *n.* **6.** the flesh and bone part of an animal's tail, as distinguished from the hair. **7.** the tail stub remaining after much of the tail has been cut off. **8.** the area where a ship berths at a port. **9.** a platform for loading or unloading trucks or other vehicles, often adjoining a store or warehouse. **10.** the place occupied by a prisoner on trial in a courtroom. **11.** a serious perennial weed pest (*Rumex* sp., family Polygonaceae, especially *R. crispus*, known as curly dock for its curled leaf margins) introduced to America from Eurasia. It produces clusters of small greenish flowers intermingled with leaves and has a long taproot.

dockage *n.* (Icelandic *dokk*, hollow + French *age*, pertaining to) **1.** foreign material that can be removed from grain by sieving or other cleaning devices, including soil, weed seeds, stems, chaff, straw, grain other than the grain in question, and underdeveloped, shriveled, and small pieces of kernels. **2.** the amount subtracted from the weight of product sold to compensate for foreign matter. **3.** the amount subtracted from wages as a penalty for low productivity or misbehavior. **4.** the charge for the use of a dock by a ship.

dodder *n.* (Anglo-Saxon *dodder*, dodder plant) a leafless parasitic plant (*Cuscuta* sp., family Convolvulaceae) that grows entangled on host plants such as alfalfa and various clovers. Dodder stems have suckers that penetrate the stems of host plants to obtain food and water. Seeds of dodder are about the same size as alfalfa and clover seeds, so they cannot be separated by sieving. Other names for dodder are strangle weed, love vine, devils' hair, goldthread, and hairweed.

doe *n.* (Latin *dama*, doe) an adult female deer, antelope, goat, rabbit, or other animal with a male that is called a buck.

DOE *n.* See Department of Energy.

dog *n.* (Anglo-Saxon *dogga*, hound) a domesticated, carnivorous animal (*Canis familiaris*, family Canidae) often kept as a household pet or a watchdog and used to herd livestock (especially cattle and sheep), to hunt game, to guide the blind, and in law enforcement to detect drugs and track persons by using the sense of smell.

dog days *n.* **1.** about six weeks of hot, sultry weather in July and August (in the north temperate zone) when the dog star (Sirius) rises and sets at about the same time as the sun. **2.** a period when a person feels lethargic.

dog fennel *n.* a common weed (*Anthemis cotula*) with composite yellow and white flowers and an unpleasant odor; native to Eurasia but now present in many American pastures, hay fields, and grain fields. Also known as mayweed.

Dokuchaev, V. V. *n.* the Russian considered the father of modern soil science because, late in the 19th century, he proposed that soils are natural bodies with properties that depend on the climate, living organisms, subsoil, and age of the soil.

doldrums *n.* **1.** an equatorial area in the oceans with warm, rising air masses and gentle surface winds that often left sailing ships becalmed. **2.** a listless state of mind; depressed spirits.

dollarspot *n.* a disease of turfgrass, particularly of the bent grasses, caused by a fungus (*Sclerotinia homeocarpa*, family Sclerotiniaceae); characterized by small spots of turf that turn dark, then brown, and later straw-colored.

dolomite *n.* a double carbonate of calcium and magnesium ($CaCO_3 \cdot MgCO_3$) named after D. de Dolomieu (1750–1801), a French mineralogist. It is somewhat harder, less soluble, and less reactive than calcitic limestone that does not contain magnesium.

dolomitic limestone *n.* limestone that has a significant component of magnesium carbonate ($MgCO_3$) but less than that of dolomite. Finely ground dolomitic limestone is used like calcitic limestone as a soil amendment to raise soil pH; it is especially useful on soils that are deficient in magnesium, e.g., to cure grass tetany. Also called magnesian limestone.

dolphin *n.* (Latin *delphinus*, dolphin) **1.** an oceanic mammal (family Delphinidae) related to whales but smaller, generally quite intelligent and trainable; the best-known species have a beaklike snout. See orca. **2.** a large, slender fish (*Coryphaena hippurus* or *C. equisetis*) found in warm or temperate seas. **3.** the constellation Delphinus located in the northern sky west of Pegasus.

domain *n.* (Latin *dominium*, ownership) **1.** realm; a region under one management or government. **2.** a region characterized by a certain type of landscape, vegetation, or wildlife. **3.** a field of expertise of an individual. **4.** a region of uniform magnetic polarity in an object that has several such regions. **5.** a set of mathematical values for a given function or expression.

dome *n.* (Latin *domus*, house) **1.** a structure with a rounded shape resembling a section of a sphere, as a domed roof. **2.** a geological structure with rocks sloping away in all directions from a raised center. **3.** a mountain peak with a rounded top. **4.** (slang) a person's head, or the rounded upper part thereof.

domestic *adj.* (Latin *domesticus*, of the house) **1.** pertaining to the home and home life. **2.** tame, as a domestic animal. **3.** internal affairs; within a nation; e.g., domestic trade.

domestic animal *n.* an animal tamed and kept by humans as a pet, for work, or to produce food, clothing, or pharmaceuticals, especially one that has been developed into a breed with characteristics notably different than those of its wild relatives.

domesticate *v.* (Latin *domesticare*, to tame) to tame a wild animal, fowl, or plant for human use, often over a period of many generations of careful breeding, selection, and handling.

dominance *n.* (Latin *dominans*, ruling) **1.** a tendency to claim priority, take control, and give orders. **2.** a tendency of a gene to control a characteristic rather than its allele. See recessiveness. **3.** the ability of certain plant or animal species to crowd out others.

dominant species *n.* the species that has the most impact on both the biotic and abiotic features of its community.

donkey *n.* **1.** a domesticated animal (*Equus asinus*) that resembles a horse but is smaller and has longer ears; widely used as a beast of burden. Also called ass, burro. **2.** a derogatory term for a stupid or obstinate person. **3.** a donkey engine; a small auxiliary engine (originally a steam engine).

Donkeys transporting grain to market

Donnan equilibrium *n.* an equation that helps describe the distribution of ions in a system when some ions are held in one area but others can migrate freely; e.g., two compartments separated by a membrane that permits dissolved ions to pass but holds all colloidal particles on one side. The equation says that the proportion of each ion is the same throughout the system if the concentration is expressed as the first power for single-charged ions, as the square root for double-charged ions, as the cube root for triple-charged ions, etc.

donor *n.* (Latin *donator*, a giver) **1.** a person who makes a financial contribution. **2.** a person who provides blood, bone marrow cells, or an organ for transfusion or transplant, either while living or at the time of death. **3.** a person, animal, or plant that provides either male or female genetic material; e.g., an egg, sperm, or pollen donor.

doodlebug *n.* **1.** the larvae of the ant lion. Doodlebugs prey on ants and other small insects by building cone-shaped traps 1 to 2 in. (3 to 5 cm) across and hiding at the bottom of the cone. See ant lion. **2.** an old, stripped-down automobile converted into a homemade tractor.

Doppler effect *n.* a change in the apparent frequency of waves of sound, light, radar, or other radiation when the source and the observer are moving closer together (increasing the frequency) or farther apart (decreasing the frequency); used to measure velocities of nearby objects, weather systems, stars, etc. Named after Christian Johann Doppler (1803–1853), an Austrian mathematician.

dormancy *n.* (Latin *dormiens*, sleeping) **1.** a resting state induced by the presence of growth inhibitors and/or absence of growth promoters in a plant and/or by external environmental factors such as excessive drought, cold, heat, or toxins. It may affect all or only part of a plant (e.g., the flower buds). **2.** a physical (e.g., hard seed coat) or physiological (e.g., presence of a growth inhibitor) condition of a viable seed that prevents its germination even though temperature and moisture are favorable.

dormant spray or **dust** *n.* a pesticide applied to perennial plants (e.g., trees in an orchard) during their period of dormancy (e.g., in early spring before they leaf out).

dorsal *adj.* (Latin *dorsalis*, pertaining to the back) **1.** pertaining to or located on or near the back side of an organism or organ; opposite of ventral; e.g., a dorsal fin on a fish or a dorsal view of an organism. **2.** pertaining to the underside of a leaf.

dorsum *n.* (Latin *dorsum*, the back) the entire upper surface of an animal, a person, or a body part; e.g., the back of the hand.

dosage *n.* (Greek *dosis*, gift + French *age*, pertaining to) **1.** the amount in a dose of medicine. **2.** the rate of pesticide or fertilizer prescribed or applied.

dote *v.* (Dutch *doten*, to behave foolishly) **1.** to show senility and feebleness of old age. **2.** to express foolish or excessive affection. **3.** to decay (wood) or wither (a plant).

dothichiza canker *n.* a disease of poplars caused by a fungus (*Dothichiza populea*, family Sphaerioidaceae) and characterized by dark, sunken, elongated lesions on the trunk, branches, and twigs.

dothiorella rot *n.* a rot of harvested citrus fruit caused by a fungus (*Botryosphaeria ribis*, family Dothioraceae), usually characterized by a rot producing small spots or an area of leathery, drab brown, pliable decay beginning at the stem end of fruit in storage. It usually cannot be detected early enough to cull the fruit in the packing house.

double-cropping *n.* growing two successive crops on the same field in one year; e.g., soybean followed by wheat.

double-cross process *n.* the production of hybrid seed from four inbred lines; the process involves producing two single-cross lines by crossing pairs of the inbred lines and then producing commercial hybrid seed by crossing the two single-cross lines.

double helix *n.* the basic structure of DNA; two strands of protein linked like the sides of a ladder and twisted into a helix. The sequence of nitrogenous bases (adenine, thymine, guanine, and cytosine) that form the links carries the genetic code. See deoxyribonucleic acid.

Double-helix structure of DNA

dourine *n.* (French *dourine*, dourine) a sexually transmitted parasitic disease of horses and mules caused by protozoa (*Trypanosoma equiperdum*). Symptoms include swelling, blisters, skin lesions in the genital area, and paralysis of the hind limbs. Dourine has been eradicated from many areas by the use of artificial insemination. Also called genital glanders, equine syphilis, mal de coit.

dove *n.* (Norse *dūfa*, dove) **1.** any of many swift-flying, small-headed birds (family Columbidae) with pointed wings and either a fanned or a pointed tail. See pigeon. Note: a white dove is commonly used as a symbol of peace, holiness, or innocence. **2.** a person who advocates peaceful relations with other nations (opposite of a hawk). **3.** a gentle person. **4.** a soft gray color.

dovecote *n.* (Norse *dūfa*, dove + Anglo-Saxon *cote*, chamber) a small building with pigeon holes to serve as nesting sites for pigeons or doves; usually raised above the ground.

downstream *adj.* or *adv.* (Anglo-Saxon *dun*, hill + *stream*, to flow) **1.** in the direction the stream is flowing, as downstream effects of runoff. **2.** in a later stage of a process or procedure, as downstream on a production line. **3.** in the direction of transcription, as in the synthesis of RNA from DNA or of protein from RNA.

downwarp *n.* (Anglo-Saxon *dun*, hill + *wearp*, to throw) a large sunken area with gentle slopes at the edges, as where the weight of incoming sediment has caused an area to gradually sink and bend at the margins rather than produce faults.

downwearing *n.* a theory of landscape development proposed by William M. Davis in the 1890s; the concept that landscapes are gradually worn down by erosion toward a rather featureless plain that Davis called a peneplain. See backwearing, mass wasting, pediment, peneplain.

downy andromeda *n.* an evergreen flowering shrub (*Andromeda glaucophylla*, family Ericaceae) found in acid bogs and uplands of northeastern U.S.A. and southeastern Canada. It can be poisonous to sheep when they eat it (likely only when other browse is unavailable).

downy brome or **downy chess** *n.* See cheatgrass.

downy mildew *n.* any of a group of plant diseases of angiosperms caused by fungi (*Peronospora* sp., *Plasmopara* sp., *Rhysotheca* sp., *Bremia* sp, *Bremiella* sp., *Pseudoperonospora* sp., *Basidiophora* sp., and *Sclerospora* sp. of family Peronosporaceae plus *Phytophthora* sp. of family Pythiaceae). Downy mildew causes yellowing and/or a white powdery growth on a wide variety of plants (e.g., alfalfa, grapes, onions, tobacco), especially on the underside of young leaves.

dowser *n.* See diviner.

dowsing rod *n.* See divining rod.

Doyle rule *n.* a method of estimating the number of board feet that can be sawed from a log:

board feet = length in ft × [(diam in in. − 4)/4]^2

dracontic month *n.* See nodical month.

draff *n.* (Norse *draff*, dregs) refuse feeds obtained as a by-product of the distillation of grains; malt after brewing; commonly fed to swine.

draft *n.* (Dutch *dragt*, draw or pull) **1.** a preliminary form of a document, still subject to revision. **2.** a sketch or drawing. **3.** a current of air; e.g., an upward flow through a chimney or a cold air current in a room. **4.** a device to regulate the air flow through a chimney. **5.** the horizontal force required to pull an implement. Also called directional pull. **6.** a good-sized drink, especially of an alcoholic beverage. **7.** a withdrawal of money from a bank or supplies from a store. **8.** conscription for military service. **9.** the depth of water required to float a vessel. **10.** the fish caught in a net. *adj.* **11.** used to pull, as a draft animal. **12.** drawn from a keg or cask, as draft beer. *v.* **13.** to compose a preliminary form of a document. **14.** to make a sketch. **15.** to conscript for military service or some kind of work.

draft animal *n.* an animal used to pull heavy loads; e.g., a horse, mule, ox, donkey, or camel.

drag *v.* (Anglo-Saxon *dragan*, to draw or pull) **1.** to pull an object or animal, usually slowly against strong resistance. **2.** to move slowly with great effort. **3.** to search for something; e.g., by pulling a grappler across the bottom of a river. **4.** to pull a heavy, rigid implement (often made of planks) across the soil surface to crush clods and smooth the surface before planting. **5.** to pull someone into a discussion. **6.** to inject a new topic, especially an unwanted one. **7.** to prolong, as to drag on interminably. *n.* **8.** the friction force resisting the movement of a solid body through a fluid or of a fluid past a solid body. **9.** a device used to drag across a surface for searching or smoothing. **10.** a tedious chore or topic. **11.** a pulling or sucking action; e.g., a drink of liquid or an inhalation of air.

draggle *v.* **1.** to make something wet and dirty by dragging it through mud or wet vegetation. **2.** to straggle; to get behind.

dragnet *n.* (Anglo-Saxon *draegnet*, dragnet) **1.** a net that is pulled through the water to catch fish or along a surface to catch small game or other objects. **2.** an organized search to find someone, especially a criminal.

dragon *n.* (Greek *drakōn*, a serpent) **1.** an imaginary large, scary creature, often represented as a winged reptile that breathes smoke and fire. **2.** any extremely large serpent. **3.** a violent person. **4.** a watchful chaperone guarding a woman. **5.** any of several plants (e.g., green dragon, *Arisaemia dracontium*) whose blossoms have a long, slender spadix and a shorter green spathe. **6.** the constellation Draco.

dragonfly *n.* any of a large group of beneficial insects (many families and perhaps thousand of species, Order Odonata, suborder Anisoptera) that catch and eat mosquitoes and many other harmful insects as they fly. Larger and faster than damselflies, dragonflies rest with their wings extended rather than folded like the damselfly. See damselfly.

drain *v.* (Anglo-Saxon *dreahnian*, to drain or filter) **1.** to flow away slowly. **2.** to remove a fluid from an area or chamber, as to drain water from a field or oil from a crankcase. **3.** to exhaust a supply of energy, as to tire a person or discharge a battery. *n.* **4.** a system of pipes, ditches, or other conduits for carrying away water or other fluid. **5.** something that exhausts energy or resources, as a heavy load on an electric circuit.

drainage *n.* (Anglo-Saxon *dreahnian*, to dry out + French *age*, pertaining to) **1.** the withdrawal of excess fluid from an area, especially the removal of water either from the land surface or from within the soil or the flow of air across a land surface. **2.** the fluid that drains from an area, especially water. **3.** the fluid that drains from a wound or infection; e.g., nasal drainage caused by a cold. **4.** of agricultural land: a system of ditches, tile lines, or other means to remove water from land so it can be cropped. About 100 million ac (40.5 million ha) of land in the U.S.A. have been drained for agricultural purposes. See wetland.

drainage basin *n.* the area that supplies water to a particular stream or river. Also called watershed, catchment.

drainage class *n.* any of seven soil drainage classes that indicate the presence or absence of excess water relative to the use of the land for agriculture: excessively drained, somewhat excessively drained, well drained, moderately well drained, somewhat poorly drained, poorly drained, and very poorly drained.

UNITED STATES NET TOTAL
100 MILLION ACRES
(40.5 MILLION HECTARES)

1 DOT=10,000 ACRES

Drained land in the U.S.A.

drain field *n.* **1.** a set of buried drain tiles used to carry excess water from the soil in a field. See wetland. **2.** a set of buried tile lines used to deliver septage water from a septic tank to the soil.

drake *n.* (German *drako*, drake) **1.** a male duck. **2.** a mayfly. **3.** a small cannon of the 17th or 18th century.

dram *n.* (Latin *drachma*, dram) **1.** a unit of weight equal to 60 grains (3.89 g) in the apothecary system or 27.34 grains (1.77 g) in the avoirdupois system. **2.** fluid dram: a unit of volume equal to ⅛ fluid ounce or ½ tablespoon. **3.** a small drink of an alcoholic beverage. **4.** a little bit of anything.

dredge *n.* (Anglo-Saxon *dragan*, to draw or pull) **1.** a machine with scooping or suction equipment designed to remove earth from the bottom of a body of water, either for mining ore or to open a channel for navigation. Note: dredging is subject to regulation under the Clean Water Act because it disturbs and can do serious damage to the ecosystem, especially to aquatic life exposed to muddy water, siltation, and possibly heavy metals or other toxins. **2.** a sturdy net on a frame designed to harvest shellfish from the bottom of a body of water. **3.** a seeding mixture composed of two or more grains, especially oats and barley. *v.* **4.** to remove earth or harvest shellfish from the bottom of a body of water. **5.** to sprinkle a powdered substance (e.g., flour, corn meal, or bread crumbs) on a food product (e.g., meat to be cooked).

dregs *n.* (Norse *dregg*, dregs) **1.** sediment (e.g., coffee grounds), or lees. **2.** the remainder after the valuable part of something has been removed, sometimes figurative, as the dregs of society. **3.** small scraps or remnants.

dressing percentage *n.* the carcass weight of an animal as a percentage of its live weight. Typical dressing percentages of cattle and sheep are about 50%, and of hogs, about 70%.

dried food *n.* any of many dehydrated foods (eggs, milk, meat, various fruits and vegetables). The removal of much of the water (often about 95%) prevents microbial attack and allows many foods to be stored for long periods without refrigeration.

dried kelp *n.* brown seaweed (*Macrocystis* sp. and related genera) dried and ground into a meal or burned for use as an iodine source for livestock and humans. Its ash contains 0.15% to 0.20% iodine.

drift *n.* (Anglo-Saxon *drifan*, to drive) **1.** a wind deposit of snow, sand, etc., or a glacial deposit. **2.** deviation from a path or trajectory, as that caused by a wind or water current. **3.** aimless movement, as that of grazing livestock, restless people, or an unpowered floating object. **4.** general meaning, as the drift of a conversation. **5.** a gradual change in direction or philosophy, as a drift toward a political extreme. **6.** a metal punch with a long taper used for aligning or enlarging holes or to drive a piece loose. **7.** a passageway in a mine. *v.* **8.** to move aimlessly, driven by wind, water, or circumstances. **9.** to be deposited in drifts, especially by wind. **10.** to deviate from a planned course.

drill *n.* (Dutch *drillen*, to make a hole) **1.** a tool that makes holes by rotating; either the drill bit or the machine that holds and turns it. **2.** practice for learning a set routine or a set of facts. **3.** a machine that places seed at prescribed intervals and a set depth below the soil surface. **4.** a marine snail (*Urosalpinx cinera*) that bores holes in oyster shells and eats the oysters.

drilling mud *n.* a mixture of clay (usually bentonite or ground barite) with water or diesel oil and perhaps some chemical additives; used in drilling deep wells by pumping it down the center of the drill, out the tip, and back up along the outside of the drill shaft. It cools and lubricates the bit, seals the hole to prevent entrance or escape of water, oil, or gas, and carries cuttings to the surface.

drilosphere *n.* the soil volume and the microflora included in it that are influenced by earthworm activity.

drip irrigation *n.* delivery of water to individual plants by means of a branching system of plastic tubing; the water is usually released through an emitter that meters its flow. Such a system requires considerable equipment and time to install but is very efficient in operation. Also called trickle irrigation or microirrigation.

dripstone *n.* a limestone formation produced by evaporation of water containing dissolved carbonates, most commonly in caves, where it occurs in various formations. Also called travertine.

driving rain *n.* heavy rain accompanied by a strong wind.

drizzle *n.* (Anglo-Saxon *dreosan*, to fall) precipitation in the form of numerous small droplets less than 0.02 in. (0.5 mm) in diameter; heavier than a mist but lighter than a shower.

dromedary *n.* (Latin *dromedarius*, dromedary) a single-humped camel (*Camelus dromedarius*), as contrasted to the two-humped Bactrian camel; well adapted to desert conditions and commonly used as a beast of burden, especially in Arabia and northern Africa. See camel.

drone *n.* (Anglo-Saxon *dran*, drone) **1.** a male bee; hatched from an unfertilized egg; larger than the worker bees but smaller than the queen. Its chief function is to fertilize the queen on her mating flight. Drones have no sting and do not gather nectar for honey. See honeybee. **2.** a pilotless aircraft. **3.** a person who loafs and lives off other people or whose work is unmotivated and uninspired. **4.** a monotonous sound, as the drone of an engine.

drone fly *n.* a fly (*Eristalis tenax*, family Syrphidae) that resembles a honeybee drone in size, color, and action. Commonly known as syrphid flies, sweat flies, hover flies, and flower flies. The larvae of this and closely related species can live in highly polluted water. The adults feed on flower pollen and nectar and thus are important in cross-pollination of many plant species.

Drone fly

droop *v.* (Norse *drūpa*, to drop) **1.** to hang downward; to sag from dehydration or lack of nourishment, as a plant, animal, or person droops in the hot sun. **2.** to tire and weaken from exhaustion or dejection. **3.** to hang down naturally, as the droopy ears of a beagle.

drop *n.* (Anglo-Saxon *dropa*, a drop) **1.** a small body of liquid with a rounded surface. **2.** a very small quantity, especially of a liquid. **3.** a lozenge or other small rounded piece of medicine or candy; e.g., a cough drop. **4.** a fall; an abrupt downward movement. **5.** a place where the surface elevation suddenly decreases. **6.** a place where something can be left, as a book drop. **7.** a release of persons or goods by parachute. **8.** a structure in a ditch or other conduit used to let water fall to a lower level and dissipate the kinetic energy thus released. **9.** fruit that falls to the ground. **10.** a disease of vegetables caused by fungi (*Sclerotinia sclerotiorum*, *S. minor*, or *S. intermedia*, family Sclerotiniaceae). **11.** a decline in price or value. *v.* **12.** to fall or allow to fall. **13.** to lower, as a rope, a ladder, or the muzzle of a gun. **14.** to give birth, as to drop a calf. **15.** to kill an animal so it falls immediately. **16.** to decline in value or status. **17.** to make an incidental remark, as to drop a hint. **18.** to send a message, as to drop a note.

drop band or **drop herd** *n.* a group of ewes or nanny goats separated from the main herd during the season of bearing young.

droppings *n.* animal or bird excrement. See manure, guano.

dropsy *n.* (Greek *hydrōps*, dropsy) **1.** an older term for edema; swelling caused by abnormal accumulation of watery fluid in body tissue or cavities. **2.** an abnormal watery enlargement of succulent plant parts.

drosometer *n.* (Greek *drosos*, dew + *metron* a measure) an instrument for measuring the amount of dew that condenses on an exposed surface. Annual dew totals of 0.5 to 2 in. (12 to 50 mm) are common in middle latitudes.

drosophila *n.* (Greek *drosos*, dew + *philos*, love) fruit fly (especially *Drosophila melanogaster*); a genus of flies common around ripening or decaying fruit; used extensively in experimental genetics because it has large chromosomes, multiplies easily, and requires only a few days from one generation to the next.

drought or **drouth** *n.* (both forms derived from Anglo-Saxon *drugath*, dryness) a time when below-normal precipitation significantly reduces the growth and yield of the usual crops of the area.

drought-resistant *adj.* able to survive and reproduce in spite of dry conditions. Plants have three kinds of mechanisms for drought resistance—endure, escape, or evade drought. Cacti endure drought by protecting massive internal water storage with a well-sealed exterior, whereas mesquite does so by shedding its leaves and becoming dormant. Desert ephemerals escape drought by growing and maturing very rapidly during brief periods while water is available. Cereal grains evade drought by growing large root systems and limiting transpiration by having small leaf areas or closing stomata.

droughty soil *n.* soil that stores little water and therefore causes the usual crops of the area to suffer from lack of water. The soil may be shallow, sandy or gravelly, or have such low permeability that little water is absorbed.

drove *n.* (Anglo-Saxon *draf*, drove) **1.** a large group of animals of one species moving together, as a drove of cattle, oxen, sheep, or swine. **2.** a crowd of people in motion. **3.** a broad chisel used for dressing stone.

drowned topography *n.* a coastline where previously exposed land has become submerged, either by the land sinking or the water level rising. See estuary, fjord.

drum *n.* (Danish *drummel*, drum) **1.** a percussion instrument consisting of a hollow chamber (usually a cylinder) with a membrane stretched across one or both ends. **2.** the sound produced by a drum. **3.** an animal's organ that produces a deep booming sound. **4.** a large cylindrical container, as an oil drum. **5.** a cylinder that revolves as part of a machine, as the drum of a seed cleaner. *v.* **6.** to play a drum. **7.** to tap or beat rhythmically, especially with the fingers. **8.** to tap on the sides of a hive while transferring bees from one hive to another. **9.** to drill someone by repetition of a concept or idea, as to drum it into his head.

drumlin *n.* (Irish *druim*, narrow ridge + *ling*, small) an elongated hill shaped by an overriding glacier and covered by compact, unstratified glacial drift; usually its longer axis is aligned in the direction the glacial ice moved. Drumlins are common in parts of the glaciated region of the U.S.A., especially in parts of New York, Michigan, and Wisconsin.

drunken forest *n.* a forest with trees leaning or fallen in unusual directions as a result of wind, landslide, and/or heaving (freezing and thawing) of soil.

drupe *n.* (Latin *drupa*, an overripe olive) a fruit (e.g., peach, apricot, plum, cherry) that has a hard, woody covering surrounding its seed and separating the seed from the moist flesh of the fruit. Also called a stone fruit.

dry *adj.* (Anglo-Saxon *dryge*, dry) **1.** having no apparent water; not wet or moist, especially on the exterior. **2.** having lost much internal moisture, as dry soil or dry wood. **3.** measurements based on or pertaining to a dry condition, as oven-dry weight, or dry measure. **4.** arid; receiving little precipitation, as a dry region or a dry year. **5.** yielding no water, oil, or milk, as a dry hole (well) or a dry cow. **6.** thirsty. **7.** serving no alcoholic beverages. **8.** uninteresting, as a dry lecture. **9.** without humor or sympathy, as a dry tone. **10** having little sweetness, as a dry wine. *v.* **11.** to remove water, as to dry the dishes. **12.** to dehydrate food or other products for preservation or to reduce shipping cost. **13.** to cause a cow to stop giving milk.

dry basis *n.* a means of evaluating weights of materials with variable water contents by adjusting them to a consistent dry basis, usually the oven-dry weight, either by drying and weighing the material or by calculating the appropriate correction. For example, soil water is expressed as a percentage of the oven-dry weight of the soil.

dry deposition *n.* deposition of dust, salts, or other material in the absence of rainfall.

dry farming *n.* See dryland agriculture.

dry haze *n.* a slightly obscuring atmosphere caused by dust particles rather than water droplets.

dry ice *n.* **1.** solid carbon dioxide (CO_2); so called because it sublimes (changes from a solid to a gas without becoming liquid); used as a dry refrigerant. It sublimes at $-109°F$ ($-78°C$). **2.** ice so cold that it is dry to the touch.

drying by sublimation *n.* the drying of food or other product under a vacuum while it is frozen, which prevents spoilage by bacterial action. Also called freeze-drying.

dryland agriculture *n.* crop production (often a small grain) without irrigation in an area where water is the main limiting factor, usually a semiarid region with 10 to 20 in. (250 to 500 mm) of rainfall per year. Summer fallow may be used to accumulate soil water from one year for use by the next year's crop. Also called dry farming or rainfed agriculture. See fallow.

dry lightning *n.* lightning accompanied by little or no rain. The fire hazard is much higher with dry lightning than it is if potential fuels are wet by rain.

dry matter *n.* or *adj.* the plant material remaining after free water has been driven off by drying in an oven (typically at 70°C for 48 hr).

Major areas of dryland agriculture in the U.S.A.

dry off *v.* **1.** to reduce the water content of a plant, bulb, tuber, or corm to make it become dormant for storage. **2.** to end the lactation period of a cow, usually in preparation for the birth of a calf.

dry pergelisol or **dry permafrost** *n.* soil that has a mean annual temperature less than 32°F (0°C) but is too dry to form ice crystals.

dry rot *n.* decay of nearly dry wood (moisture may have been absorbed from the air). In an advanced stage of dry rot, the wood can be easily crushed to a dry powder.

dry wash *n.* a stream channel that rarely has any water in it because the climate is arid.

dry water *n.* a gel-like product containing about 98% water plus a vegetable product that turns it into a gel. It can be applied to soil next to a plant, where it will gradually release water as microbes consume the vegetable matter.

dry weight *n.* the weight of a material after it has been dried in an oven; e.g., plant material that has been dried for 48 hr at 70°C or soil that has been dried for 24 hr at 105°C. Also called oven-dry weight.

dry well *n.* **1.** a well that provides little or no water. **2.** a pit, usually filled with stones, used to dispose of water (e.g., runoff from a roof or drainage from a tile line) by allowing it to seep into the surrounding soil. **3.** an outdated substitute for a septic tank and drain field. See cesspool.

dual-use range *n.* an area of rangeland used for grazing by two or more kinds of livestock (e.g., cattle and sheep) at the same time, or at least during the same year.

duck *n.* (Dutch *duicken*, to duck or dive) **1.** any of many species of waterfowl (six of the eight tribes of the family Anatidae of order Anseriformes) with a wide variety of colors and patterns. Ducks have broad flat bills, short legs, and webbed feet; they are smaller and have shorter necks than geese and swans. Sometimes they are identified by sex by calling the male a drake and the female a duck. Wild ducks often migrate in large flocks and are widely hunted for sport and food. **2.** the meat of ducks used as food. **3.** a sturdy cotton fabric, often used to make tents, bags, etc., as well as clothes. **4.** an amphibious vehicle, especially one for military use.

duckling *n.* a young duck.

duckweed *n.* any of several small aquatic plants (family Lemnaceae, especially *Lemna* sp.) that form a green scum floating on freshwater; so called because it is eaten by ducks. It is widespread, especially in temperate regions, and includes the smallest known flowering plant.

duct *n.* (Latin *ductus*, water passage) a passageway for liquids, air, pipes, wires, etc.

dude ranch *n.* a ranch operated as a vacation resort to provide a ranching experience for nonagricultural people.

duff *n.* **1.** the organic covering on a forest floor composed of leaves, twigs, and other organic residues in all stages of decomposition from fresh to rotten. **2.** a type of steamed pudding. **3.** the buttocks.

duff hygrometer *n.* an instrument for determining the degree of fire hazard in a forest by measuring the moisture content of duff.

dugout *n.* **1.** a hollowed-out log used as a boat. **2.** a shelter recessed partly or entirely into the side of a hill, often crudely finished. **3.** a pond made by excavation, usually on level bottomland. **4.** a place where baseball players sit (usually below ground level).

dug well *n.* a well excavated with picks and shovels to reach below the water table; usually walled with bricks or stones; often fitted with a windlass to lift a bucket of water.

duiker *n.* (Afrikaner *duiker*, diver) a small, timid African antelope (*Cephalophus* sp., *Sylvicapra* sp., and related genera) about the size of a rabbit; prized as a meat dish across much of Africa south of the Sahara Desert.

dulosis *n.* (Greek *doúlōsis*, enslavement) the practice of certain ants (*Formica* sp. and *Polyergus* sp.) of enslaving other ants.

dulse *n.* (Gaelic *duileasg*, dulse) an edible red seaweed (*Rhodymenia palmata*); a marine alga with large wedge-shaped fronds.

dumbcane *n.* an ornamental plant (*Dieffenbachia seguine*, family Araceae) native to the West Indies; grown in greenhouses and as a houseplant for its blotched-leaf foliage. Chewing a leaf causes the tongue to swell so one cannot speak for a few hours. See dieffenbachia.

dummy *n.* (German *dum*, stupid + *y*, acting) **1.** a model of something, often used for display (as a clothes dummy) or to visualize something to be built. **2.** a person who acts in place of another without revealing the representation. **3.** a frame used with an artificial vagina to collect semen from a bull or other male animal for use in artificial insemination. See artificial insemination. **4.** a stupid person or animal, especially one that seldom speaks or one that has suffered brain damage. **5.** the partner of the winning bidder in certain card games, or the exposed cards dealt to that person.

dumpy level *n.* a surveying instrument with a telescope mounted on a leveling base; used for establishing a level line of sight, e.g., to determine elevations or to make topographic maps. It can measure horizontal angles but not vertical angles. See theodolite, transit.

dune *n.* (Dutch *duna*, dune) a mound or ridge of loose sand (or occasionally of sand-size aggregates) deposited by wind. Dunes are common where loose, bare sand occurs along shorelines of coasts or lakes, in river valleys, and in desert areas. Active dunes gradually drift downwind until they are stabilized by vegetative cover or windbreaks. Dunes may form from aggregated soil when unprotected tilled or otherwise bare soil dries out and is exposed to wind.

Sand dunes

dune grass *n.* grass (e.g., European beachgrass, *Ammophila arenaria*, and American beachgrass, *A. breviligulata*, family Poaceae) that will grow in droughty conditions, such as on sand dunes.

dung *n.* (German *dung*, dung) manure; excrement; the feces of animals, birds, or humans. See guano, manure.

dung beetle *n.* any of several scarab beetles that live in, breed in, or feed on dung. See tumblebug.

duramen *n.* (Latin *duramen*, hardness) See heartwood.

duripan *n.* (short for durable hardpan) a hard subsurface soil horizon cemented by illuvial silica into a hard mass that resists slaking in water; typically forms at about the normal depth of water penetration in an arid or semiarid region; usually calcareous.

durum wheat *n.* **1.** a group of hard wheat varieties that produce flour used mostly to make pasta. **2.** the grain from such varieties; used to make macaroni, spaghetti, etc.

dusk *n.* (Anglo-Saxon *dox*, dark colored) **1.** moderately dark colored; dusky. **2.** the darkening period in the evening. **3.** gloom, sadness. See dawn, twilight.

dusky *adj.* **1.** dimly lit; somewhat obscure. **2.** somewhat dark or dull in color, as dusky brown. **3.** gloomy; forlorn.

dust *n.* (German *dust*, dust) **1.** small particles of dry mineral matter (mostly soil) carried by air currents and deposited almost anywhere; typically of silt size, but including some clay and fine sand. **2.** a pesticide applied as a powder. **3.** powdered residue such as that produced by sandpaper. *v.* **4.** to sprinkle lightly with powdered material. **5.** to wipe the dust away from a surface, as in house cleaning.

Dust Bowl *n.* an area in the southern U.S. Great Plains region, including parts of southeastern Colorado and nearby parts of Kansas, Oklahoma, and northwestern Texas. This wheat-producing region sometimes is subject to severe droughts and wind erosion. Such an episode in 1935 carried an immense cloud of dust from the Great Plains all the way to Washington, D.C., and helped persuade Congress to establish the Soil Conservation Service. A similar episode occurred in the 1950s.

dust devil *n.* a dust-laden whirlwind; like a miniature tornado but often only a few feet in diameter. See snow devil.

dust mulch *n.* a powdered surface soil, usually produced by shallow cultivation; once supposed to prevent soil moisture from rising by capillary action and evaporating but no longer supported as a soil moisture–conserving tillage practice because the tilled soil loses too much water.

dustproof *adj.* protected so that dust will not interfere with successful operation; e.g., a watch or other delicate mechanism sealed so that dust cannot enter.

dust storm *n.* a strong wind that picks up and carries large clouds of dust, common in deserts and arid or semiarid plains. See sandstorm.

Dutch elm disease *n.* a lethal fungal disease (*Ceratostomella ulmi*) that killed nearly all the native elm trees in the U.S.A. during an approximately 50-year period. It was first observed in Holland in 1921 and immigrated to the U.S.A. in 1930. The leaves on elm trees wilt, turn yellow, and fall, and a brown discoloration occurs in affected wood. The disease is transmitted by a bark beetle. Also called elm blight. See elm bark beetle.

duty of water *n.* the total amount (depth) of irrigation water required to produce a specific crop, including transpiration, evaporation, deep percolation, surface runoff, and seepage from ditches.

dwarf *n.* (Anglo-Saxon *dweorg*, dwarf) **1.** an unusually small person, usually with somewhat abnormal body proportions, especially with short limbs (a small but normally proportioned person is a midget). **2.** an animal or plant that is unusually small for its species, especially one that has been intentionally bred and/or grown to be small (e.g., by grafting and pruning a plant). **3.** an imaginary small person, usually a highly skilled and/or magical old man. **4.** a star that has run out of nuclear energy and shrunk in size; either a white dwarf that still has enough heat to be luminous or a black dwarf that has cooled. **5.** a serious viral disease that stunts the canes and yellows the leaves of trailing blackberries, especially in the states along the U.S. Pacific Coast. **6.** a viral disease that affects plum trees, often called prune dwarf. *adj.* **7.** unusually small; diminutive. *v.* **8.** to cause something to be small. **9.** to cause something to appear small; to overshadow.

dwarfism *n.* the condition of being a dwarf, either hereditary or caused by disorders of the endocrine glands or of the bones. Dwarfism combined with impaired mental development is called cretinism.

dwell *v.* (Anglo-Saxon *dwellen*, to delay) **1.** to reside; to make one's home in a place. **2.** to remain in a given place or condition for some interval of time. **3.** to linger over an idea in deep thought or to discuss or write at length about it. *n.* **4.** the part of a cycle when a machine part is motionless, as the dwell in the ignition system of an engine.

dwelling *n.* a building or other shelter that serves as a residence.

dynamic equilibrium *n.* **1.** a mechanical balance of forces that produces no change in velocity or form of movement of an object or system. Also called kinetic equilibrium. **2.** a chemical equilibrium wherein two or more reactions balance each other and produce no net change. **3.** the condition in a river or stream wherein the sand and silt load is constant. **4.** the condition in an aquifer when the recharge water is equal to the discharge water. **5.** relative stability in the conditions in an ecosystem or other environment; life processes and other reactions are balanced so the prevailing conditions are constant. See steady state.

dynamite *n.* (Greek *dynamis*, power) **1.** a violent explosive containing a combustible carbonaceous substance (e.g., wood pulp) and a strong oxidant (usually nitroglycerine or ammonium nitrate); used to break apart or move resistant objects; invented and named by Alfred B. Nobel (1833–1896). **2.** a person, thing, or piece of information that produces or may produce spectacular activity or reaction. *v.* **3.** to use dynamite to destroy or move something. *adj.* **4.** powerful, exciting, marvelous.

dynamometer *n.* (Greek *dynamis*, power + *metron* to measure) a device that measures the power (force exerted over time or distance) of a machine or animal or the force required to do work (e.g., the resistance of soil to penetration by tillage implements).

dysentery *n.* (Greek *dys*, bad + *enteron*, intestine) **1.** an inflammation of the intestines, especially of the colon, that causes cramping and frequent, watery (often bloody) stools. The cause may be a chemical irritant, bacteria, protozoa, or parasitic worms. Amoebic dysentery is caused by *Entamoeba histolytica*; bacillary dysenteria is caused by bacteria (*Shigella* sp.). See shigellosis. **2.** the discharge of fecal matter by adult bees within a hive. Common contributing conditions are nosema disease, excess moisture in the hive, starvation, and low-quality food.

dysfunction *n.* (Greek *dys*, bad + Latin *functio*, performed) **1.** failure of an organ or other body structure to perform as it should. **2.** the part that fails to perform or exhibits impaired function. **3.** failure to act or behave in a socially acceptable way.

dysprosium (Dy) *n.* (Greek *dysprositos*, hard to get at) element 66, atomic weight 162.50; a very soft rare earth element with a bright silver luster that occurs along with other rare earth elements in several minerals.

dystocia *n.* (Greek *dys*, bad + *tokos*, birth) abnormal or difficult labor; difficulty in delivering the fetus and placenta.

dystrophic *adj.* (Greek *dys*, bad + *trophia*, nutrition) low nutrient content in water (especially in a lake or pond) that therefore supports few or no fish; opposite of eutrophic. See oligotrophic.

dystrophy *n.* (Greek *dys*, bad + *trophe*, nourishment) defective nutrition.

E

E *n.* east.

eagle *n.* (French *aigle*, eagle) a large flesh-eating bird of prey belonging to the falcon family (Accipitridae). See bald eagle.

ear *n.* (Anglo-Saxon *ēare*, ear) **1.** the organ of hearing for animals and humans. The ears of domestic and wild animals indicate the mood of the animal or conditions of the ambient environment. **2.** a well-developed ability to evaluate sound, as an ear for music. **3.** a projection that resembles an ear, as the ears on a sack of grain. **4.** the spike of a cereal plant, especially an ear of corn. *v.* **5.** to develop ears, as a field of corn. **6.** to hear or listen to someone or something.

early soil *n.* a soil that dries out and warms up earlier than others in the spring. It usually has enough slope and/or elevation for good surface drainage and a sandy texture for good internal drainage. A southern slope also favors early warm-up.

ear mange *n.* a disease caused by the ear-mange mite (*Otodectes cynotis*, family Demodicidae), which infests the skin inside the ears of cats, dogs, rabbits, and foxes, causing great irritation, impaired hearing, and/or deafness.

earmark *v.* **1.** to identify livestock by notching the ear in a specified pattern. The more modern technique uses ear tags impregnated with insecticides. **2.** to designate for a specific purpose. *n.* **3.** a mark of identification, usually in the ear of an animal. See brand.

earth *n.* (Anglo-Saxon *eorthe*, earth) **1.** or **Earth**; the world where we live; the third planet from the sun. **2.** land (lithosphere), as contrasted with atmosphere, biosphere, and hydrosphere. **3.** soil or any other loose, weathered mineral matter that once was rock.

Earth Day *n.* an annual event that originated April 22, 1970, with demonstrations by college students concerned about worldwide environmental issues.

earthen *adj.* (Anglo-Saxon *eorthen*, earthen) made from soil or other similar material, especially baked clay such as earthenware pottery.

earthly *adj.* (Anglo-Saxon *eorthlic*, earthly) **1.** worldly; having to do with things of the Earth rather than heavenly ideals. **2.** imaginable or possible, usually negated, as of no earthly use to anyone.

Earth orbit *n.* in orbit around the Earth, as the moon and satellites launched by humans.

earthquake *n.* (Anglo-Saxon *eorthdyne*, earthquake) **1.** an intense shaking that radiates from a specific location when slippage along a fault line releases tension built up by gradual shifting of plates in the Earth's crust. It is sometimes accompanied by volcanic activity. See plate tectonics, Richter scale. **2.** figuratively, a disruptive social upheaval.

earth science *n.* any science that deals with the Earth or one of its major physical components; e.g., geography, geology, meteorology, or pedology.

Earth's orbit *n.* the path the Earth follows around the sun that takes 365.2422 days per orbit; a slightly elliptical orbit (with a 100,000-year cycle between nearly circular and about 9% elliptical variation) that currently has the Earth about 3.3% closer to the sun in January than in July.

earthwork *n.* **1.** an engineering task involving reshaping a landscape by moving fill material and topsoil from one location to another. **2.** a fortification that consists largely of a pile of earth.

earthworm *n.* any of at least 1800 species of annelid worms that live in soil (especially *Lumbricus* sp.). Earthworms ingest soil material, digest part of the organic matter, and leave behind stabilized soil aggregates and channels that make the soil more permeable. Earthworms are bisexual, (hermaphroditic), but an individual cannot fertilize itself. Certain birds and fish commonly eat earthworms.

earthworm cast *n.* ingested soil material mixed with exudates and excrement, forming stabilized soil aggregates left behind by a passing earthworm.

earthy *adj.* (Anglo-Saxon *eorthe*, earth + *y*, characteristic of) **1.** composed of or having characteristics of soil, as an earthy smell. **2.** practical; a realistic individual or possibility. **3.** unsophisticated; crude, as an earthy comment.

east *n.* (Anglo-Saxon *east*, east) **1.** one of the four cardinal points of the compass (to the right when one faces north). **2.** in an easterly direction from one's location. **3.** a region located east of one's location. **4. the East** either the Orient (Asia and nearby islands) or the eastern part of the U.S.A. (east of the Mississippi River, especially the states east of the Allegheny Mountains).

east coast fever *n.* a disease caused by protozoa (*Piroplasma parva*) transmitted by the bite of the brown dog tick (*Rhipicephalus sanguineus*, family Ioxidae). The disease causes high fever and swelling of lymph nodes and is prevalent in cattle in eastern Africa. Also called Rhodesian fever.

eastern arborvitae *n.* an evergreen tree (*Thuja occidentalis*, family Pinaceae) with tiny scalelike leaves that is native to eastern North America and commonly used as an ornamental tree. Also called white cedar. See arborvitae.

eastern red cedar *n.* a large, hardy evergreen tree (*Juniperus virginiana*, family Pinaceae) that produces male and female flowers on separate plants. It is native to North America and grown as an ornamental and as a source of aromatic, straight-grained, low-density wood used to make cedar chest wood linings, shingles, pencils, wooden buckets, etc. Also called red cedar or Tennessee red cedar. The alternate host of cedar-apple rust. See cedar, cedar-apple rust.

eastern snow mold *n.* a disease of fine turf grasses (e.g., golf greens) and several field and range grasses (e.g., cheatgrass) that is caused by a fungus (*Fusarium nivale*, family Tuberculariaceae) and attacks during the winter, turning grass into matted pink or straw-colored spots with irregular circular outlines revealed when the snow melts. Also known as pink snow mold. See snow mold.

ebony *n.* (Latin *ebenius*, ebony) any of about 200 species of trees (*Diospyros* sp., family Ebenaceae) that have very hard, dense, dark-colored wood (mostly black but some red or green). Mostly native to Africa and Asia, ebony is used to make furniture, decorative woodwork, and golf-club heads. It produces yellow or orange fleshy fruit about an inch (2 to 3 cm) in diameter called persimmons. Also known as persimmon.

ecad *n.* a nonheritable plant form resulting from environmental effects. For example, leaves produced in the shade may be larger than those in full sunlight.

ecesis *n.* (Greek *oikēsis*, act of dwelling) the establishment of a plant or plants in a new environment; e.g., establishing vegetation on surface mine spoil.

Echinococcus *n.* a genus of tapeworm in dogs, wolves, and sometimes cats, whose larvae can form cysts (hydatids) in the liver, lungs, kidneys, etc., of most mammals, including humans.

eclipse *n.* (Greek *ekleipsis*, a failing or omission) **1.** a darkening of the Earth by the moon either partially or totally blocking the light of the sun (a solar eclipse) or of the moon by the shadow of the Earth (a lunar eclipse). **2.** any action of one object blocking the path of illumination for another object. *v.* **3.** to darken an object by blocking light. **4.** to surpass someone else's achievement greatly.

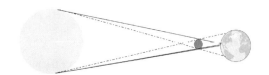

A solar eclipse (not to scale).
The small area would have a total eclipse;
the large area would have a partial eclipse.

ecliptic *n.* (Greek *ekleiptikos*, of an eclipse) **1.** the apparent annual path of the sun as seen from the Earth or of the Earth as if seen from the sun. **2.** a great circle drawn on a planet or moon representing the apparent path of the planet or star around which it orbits.

ecocide *n.* (Greek *oikos*, house + Latin *cida*, killer) ecological destruction of a large area by reckless mining or other exploitation of resources, by dumping of harmful wastes, or by deliberate destruction of vegetation.

ecoclimate *n.* (Greek *oikos*, house + *klima*, region) all of the meteorological factors influencing the living things in an environment. See climate.

ecocline *n.* (Greek *oikos*, house + *klinein*, to lean) **1.** a sequence of ecosystems along an environmental gradient such as along a slope or along a temperature gradient. **2.** a plant or animal species with characteristics that vary with location as an adaptation to an environmental gradient. **3.** the rate of change in the characteristics of a species that shows such an adaptation relative to the environmental gradient change.

ecodynamics *n.* (Greek *oikos*, house + *dynamikos*, power or force) the flow of energy and materials in the ecological environment.

ecofreak *n.* an enthusiast for environmental concerns (often used disparagingly).

ecogenetics *n.* (Greek *oikos*, house + *genesis*, origin + *tic*, of) the interaction of genetics with the environment; e.g., a genetic characteristic that influences the environmental adaptation of an individual.

ecogeography *n.* (Greek *oikos*, house + *geōgraphia*, geography) the study of the natural environment and how human beings and other living things are integrated into ecosystems as a function of geographical space.

E. coli *n.* See *Escherichia coli.*

ecological association *n.* same as plant type or forest type; e.g., oak-hickory forest type.

ecological climatology *n.* the study of plants, animals, and people in relation to their climatic adaptation and its effect on geographical distribution.

ecological density *n.* either the number or the biomass weight of organisms living in a unit area or unit volume of a specific habitat in an ecosystem.

ecological diversity *n.* the degree of heterogeneity in the composition of an ecosystem, including the variety of species present and their various activities and functions, genetic variability, spatial and trophic structures, and changes that occur over time.

ecological dominance *n.* the ability of one or more species in an ecosystem to exert major control on the nature of the ecosystem and the activity of associated species through their great abundance, size, or vigor.

ecological efficiency *n.* **1.** the amount of energy or nutrient actually used by an organism as a fraction of that contained in the food it consumes, usually expressed as a percentage. **2.** the characteristic ability of a living organism or a life process to make effective use of the energy received from a particular source. For example, the photosynthetic reaction in plants uses only a small percentage of the energy present in full sunlight.

ecological equivalent *n.* a species that occupies an ecological niche that is comparable to that occupied by another species in a different location with a similar environment.

ecological impact *n.* the effect of an action or activity (either human or natural) on living organisms and their environment.

ecological indicator *n.* a species or defined group of species whose presence or abundance indicates an important quality, either favorable or unfavorable, of an environment; e.g., the presence of coliform bacteria indicates fecal contamination.

ecological interaction *n.* the effect of two or more organisms on each other can be described as commensalism (one benefits, the other is unaffected), competition (one's gain is the other's loss), mutualism (beneficial to both species), neutralism (neither helpful nor harmful), parasitism (one living at the expense of another), or predation (one eats another). Symbiosis is a related term that is sometimes used for any of the above but most often in the sense of mutualism. See symbiosis.

ecological niche *n.* **1.** a combination of environmental factors that makes a suitable habitat for a particular life form. **2.** the place and function of a specific organism in an ecosystem, especially in relation to the resources it uses.

ecology *n.* (Greek *oikos*, house + *logos*, word) **1.** the relationships of organisms that live in an area to each other and to their ambient environment. **2.** the biological science that deals with the relationships between organisms and their environment.

economic entomology *n.* the study of when and how to control insect populations for the purpose of maximizing profit. Insect control is applied only if the population exceeds an economic threshold, and control practices are chosen to give the most beneficial results at minimum cost.

economic maturity *n.* a means of choosing the optimum time to harvest a tree or stand of trees, as determined by the age when the growth rate slows enough to cause the average annual profit over the life of the stand to begin to decrease. Also called financial maturity.

Economic Research Service (ERS) *n.* an agency in the U.S. Department of Agriculture that provides economic data and analyses for public and private decision makers regarding agriculture, food, natural resources, and environmental issues in rural America.

economic threshold *n.* the minimum degree of need that justifies addition of an input (more or better raw material, labor, management, or an additional item, e.g., a pesticide); the point where the value of the expected increase in production first becomes equal to or exceeds the cost of the input.

ecophene *n.* (Greek *oikos*, house + *phan*, appearance) the variations in physical appearance and behavior caused by the influences of different environments on a single genotype of a species of plants or animals.

ecoregion *n.* (Latin *oeco*, house + Latin *regionis*, ruling) **1.** an area of relatively uniform climate, physiography, hydrology, vegetation, and wildlife potential; a region suited to a particular ecosystem. **2.** a level of environmental classification ranging from ecodomains (most general) through ecodivisions, ecoprovinces, and ecoregions to ecosections (most specific).

ecosphere *n.* (Latin *oeco*, house + *sphaera*, round body) the biosphere; any place where plants, animals, people, and/or other organisms can live. This includes most of Earth's land and water areas and the lower part of the atmosphere.

ecosystem *n.* (Latin *oeco*, house + *systema*, placed together) any community of living organisms (any combination of species of plants, animals, people, and microbes) together with the environmental factors that interact with them. Examples: (1) a drop of water with its bacteria, protozoa, and other life-forms, (2) the tropical rainforest, (3) the Sahara Desert, (4) New York City, and (5) the Earth as a whole.

ecotone *n.* (Greek *oikos*, house + *tonos*, tension) the transition zone between areas covered by two distinctly different ecosystems, communities, or habitats; e.g., the transition between prairie and forest.

ecotourism *n.* promotion of good ecology by including ecological sites (nature reserves, traditional villages and farmlands, etc., that provide *in situ* sites of biodiversity) on tours along with conventional tourist sites.

ecotype *n.* (Greek *oikos*, house + *typos*, impression) a locally developed subdivision of a species that has unique physiological and/or morphological characteristics and genetics that differ from other populations as a result of the selective action of a particular environment (climate, soil, etc.) and that show adaptation to that environment.

ecraseur *n.* (French *écraser*, to crush or bruise) a surgical instrument used to sever a diseased part from an animal by looping a fine chain or cord around the part and gradually tightening the loop. It is also used for castrating large animals.

ectoderm *n.* (Greek *ektos*, outside + *derma*, skin) **1.** the outer of three layers of cells in a developing embryo; the precursor for the skin, hair, teeth, and nervous system. **2.** the outer layer of cells of a cnidarian or related organism.

ectodynamomorphic soil *n.* soil with properties produced or influenced mainly by factors other than parent material, i.e., by climate, living organisms, topography, and/or time. See endodynamomorphic soil.

ectogenic or **ectogenous** *adj.* (Greek *ektos*, outside + *genesis*, origin) growing parasitically on (outside) the body of a host.

ectomycorrhiza (pl. **ectomycorrhizae**) *n.* (Greek *ektos*, outside + *mykēs*, fungus + *rhiza*, root) a symbiotic association of a plant root and a fungus that forms a Hartig net in the space between cortical cells of the root (usually of a tree or shrub, especially conifers, beeches, and oaks), usually forms a white mantle over the root surface, and extends hyphae outward into the soil. The fungus absorbs water and nutrients from the soil for the plant and receives carbohydrates from the plant. See mycorrhiza, Hartig net.

ectoparasite *n.* (Greek *ektos*, outside + *parasitos*, one who eats at another's table) a parasite that lives on (outside) the body of its host; e.g., a leech or a tick.

ectophyte *n.* (Greek *ektos*, outside + *phyton*, plant) a plant that lives on the outside of another plant; e.g., moss on trees or dodder. Also called epiphyte. See dodder.

ectotherm *n.* (Greek *ektos*, outside + *thermē*, heat) an animal whose body temperature is controlled by the environment, commonly described as a cold-blooded or poikilothermic animal. Most nonmammals are ectotherms. See endotherm.

eczema *n.* (Greek *ekzema*, external boil or fermentation) an acute or chronic red inflammation of the skin of animals or humans that is accompanied by severe itching, papules, vesicles, and pustules, serous discharge, formation of crusts, and/or loss of hair. It may result from allergy, hypersensitivity to chemicals, photosensitization, or faulty feeding. Also known as facial eczema, mange, parakeratosis, or dermatitis.

edaphic *adj.* (Greek *edaphos*, soil) **1.** pertaining to the soil, especially to its influence on organisms. **2.** influenced by physical, chemical, or biological characteristics of the soil (rather than by climatic factors).

edaphology *n.* (Greek *edaphos*, soil + *logos*, word) the science that deals with the use of soil for plant growth, especially for crop production. See pedology.

edaphon *n.* any plant, animal, or microbe that lives in the soil.

eddish or **eadish** *n.* crop aftermath or regrowth after harvest; crop residue. See rowen.

eddy *n.* (Anglo-Saxon *ed*, turning + *ea*, water) **1.** a current of air or water running contrary to the main current, especially one moving circularly, as a whirlpool. See dust devil, snow devil. **2.** an opinion or event contrary to that usually expressed or occurring in public.

edema *n.* (Greek *oidēma*, a swelling) **1.** an abnormal accumulation of fluids causing swelling of body tissues or filling a cavity of a human or animal body. Also called dropsy. **2.** an abnormal accumulation of water causing swelling of plant cells.

edge effect *n.* the results of the special environment at the edge of an area; e.g., the difference in light and exposure may increase or decrease the yield of border rows. Many wildlife species are attracted to edges, as between fields and woodlands, because they offer a wider variety of food, shelter, and environment.

edge firing *n.* **1.** a controlled burn. Fires are set around the perimeter of an area and allowed to spread inward. **2.** yellowing followed by death of tissue around the margin of a leaf. This can be a symptom of potassium deficiency.

EDTA or **ethylenediaminetetraacetic acid** *n.* one of the most widely used sequestrants, commonly used as a micronutrient (copper, iron, or zinc) fertilizer in a metal acetate form [e.g., $Zn-C_2H_4N_2(CH_2COO)_4^{-2}$]. The EDTA molecule wraps around a divalent or trivalent cation and maintains the solubility of a cation that might otherwise become unavailable for plant growth. EDTA can also prevent vitamin breakdown, particularly of the fatty vitamins A, D, E, and K. See chelate, sequester.

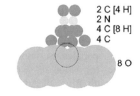

2 C [4 H]
2 N
4 C [8 H]
4 C

8 O

Structure of EDTA molecule.
H ions are too small to show;
chelated ion is in the black circle.

eel *n.* (Anglo-Saxon *ael*, eel) **1.** an elongated fish (order Apodes) that has a snakelike body with no ventral fins. **2.** any similar elongated fish; e.g., the lamprey.

eelgrass *n.* (Anglo-Saxon *ael*, eel + *graes*, grass) a flowering plant (*Zoster marina*, family Zosteraceae) with long grasslike leaves that grows submerged in shallow ocean water.

eelworm *n.* (Anglo-Saxon *ael*, eel + *wyrm*, worm) any small nematode (class Nematoda, phylum Aschelminthes), especially one that parasitizes plant roots.

eelworm disease *n.* plant disease caused by nematodes parasitizing the roots. See root knot.

effective precipitation *n.* the amount of water that actually becomes available for plant growth. It excludes losses such as runoff, evaporation, and deep percolation.

effervescence *n.* (Latin *effervesce*, boil + *entia*, action) **1.** the release of gas bubbles as a result of chemical action (not from heat); e.g., the carbon dioxide released when acid (commonly 1 normal HCl) is applied to limestone, calcareous soil, or other material that contains carbonates, especially calcium carbonate. **2.** the release of gas bubbles from a liquid when pressure is removed; e.g., when a bottle of carbonated beverage is opened. **3.** excitement; visible enthusiasm.

effete *adj.* (Latin *effetus*, exhausted) **1.** barren; spent; impotent; no longer able to produce progeny (animals or humans) or to bear fruit (plants). **2.** exhausted; worn out; degenerate.

effigy *n.* (Latin *effigia*, shape or form) **1.** a sculpted shape that represents some kind of figure, often crudely, as an effigy mound shaped like a bear or other animal. See Mound Builders. **2.** a crude figure used to ridicule a detested person. Such a figure is sometimes burned or hanged.

efflorescence *n.* (Latin *efflorescere*, to begin to bloom) **1.** the blossom of a plant. **2.** the flowering process. **3.** the period when blossoms are produced, especially the period of full bloom. **4.** a powdery deposit of soluble salts left by evaporation of water from a surface (e.g., on a wall, floor, porous container, or soil surface). **5.** the process of forming a mineral deposit by evaporation of water. **6.** a skin lesion, rash, or eruption; e.g., that produced by measles, smallpox, etc.

effluent *n.* (Latin *effluens*, flowing) **1.** a discharge or outflow of water or other liquid; e.g., a stream flowing from a lake or reservoir. **2.** the fluid discharge from domestic, industrial, and/or municipal waste collection systems or treatment facilities. **3.** the lava flowing through fissures from a volcano. See influent.

effluent limitation *n.* a restriction established by the Environmental Protection Agency on the quantity, rate, or concentration of a broad type of pollutants at a specific discharge point.

effluent standard *n.* a restriction established by the Environmental Protection Agency on the quantity, rate, or concentration of a particular pollutant in any wastewater discharge.

effluvium *n.* (Latin *effluvium*, outflow) a discharge of a vapor or other invisible stream of particles, often noxious or odorous.

eft *n.* (Anglo-Saxon *efeta*, newt) a small lizard (salamander) that is living on land. It is called a newt when it lives in the water.

egagropilus *n.* (Greek *aigagros*, wild goat + Latin *pilus*, hair) a hair ball or a hairy tumor in an animal's stomach. See bezoar, hair ball, stomach ball.

egestion *v.* (Latin *egestio*, discharging) **1.** the process of excreting or discharging waste material. **2.** excrement; waste material discharged from the digestive tract or, in an extended sense, any discharge from the lungs, skin, kidneys, or bowels.

egg *n.* (Anglo-Saxon *aeg*, egg) **1.** the reproductive body produced by a female bird, reptile, fish, or insect; typically enclosed in an oval calcareous shell or strong membrane. **2.** an ovum; a reproductive body produced by a female animal. Following fertilization by a sperm, it develops into an embryo. **3.** a female plant cell that, following fertilization, develops into a seed. *v.* **4.** to urge (often to do something questionable), as to egg on.

eggplant *n.* **1.** a bushy plant (*Solanum melongena*, family Solanaceae) grown in vegetable gardens as an annual with large, hairy, grayish green leaves and lavender star-shaped flowers with yellow centers. **2.** the large elongated or egg-shaped fruit from the plant with its shiny purple (or white, yellow, or brown in some varieties) covering over its edible off-white fleshy interior. It is eaten as a vegetable. **3.** a dark purple characteristic of the fruit.

eggshell *n.* **1.** the hard covering around an egg (of birds, certain insects, or reptiles), composed of a thin layer of crystalline calcite tied together by keratin fibers. **2.** a pale creamy white.

egg white *n.* the albumen of the egg. The part between the shell and the yolk that is nearly transparent but turns white when cooked.

egg yolk *n.* the yellow or orange central part of an egg.

Egyptian bee *n.* a nervous, ill-tempered honeybee (*Apis fasciata*, family Apidae) that is native to Egypt. It is a good honey producer that is prolific and useful in crossbreeding but is not suited to comb-honey production.

eider *n.* (Spanish *eider*, eider duck) any of several diving sea ducks (*Somateria* sp. and others, family Anatidae) of far northern areas of North America and Asia. Most are rather large, short-necked ducks that live in large flocks.

einkorn *n.* a primitive form of wheat (*Triticum monococcum*, family Poaceae) that includes the earliest varieties cultivated. It is diploid (14 chromosomes), whereas modern wheat varieties are tetraploid (28 chromosomes) or hexaploid (42 chromosomes).

einsteinium (Es) *n.* element 99, atomic weight 254; a synthetic trivalent cation element discovered in debris from a hydrogen bomb explosion in 1952 and named after Albert Einstein.

ejaculate *n.* (Latin *ejaculatus*, shot out) **1.** semen forced out of the body; especially that semen captured in an artificial vagina for use in artificial insemination. *v.* **2.** to eject or discharge, especially semen from a penis. **3.** to exclaim loudly and suddenly.

ejecta *n.* (Latin *ejecta*, ejected) **1.** material ejected; e.g., that cast from a volcano or a meteorite crater. **2.** excrement.

eland *n.* (Afrikaner *eland*, elk) either of two species of large African antelopes (*Taurotragus* sp.) with long, twisted horns. Also called Cape elk.

elapid *n.* any of a group of poisonous snakes (family Elapidae), including cobras and coral snakes.

elastrator *n.* a device used to castrate young calves or lambs or to dock their tails by stretching a specially made rubber ring over the scrotum or tail to stop the circulation of blood. This causes the scrotum or tail to atrophy and drop off.

elbow room *n.* enough space to move around unhindered, especially while working.

elder *n.* (Anglo-Saxon *eldra*, older person) **1.** an older person, either one of advanced age or older than another person (especially a participant in a discussion). **2.** an established, respected person in a tribe or community, especially a chief or other leader. **3.** an ecclesiastical office in certain churches; a presbyter. **4.** a woody shrub (*Sambucus* sp., family Caprifoliaceae). Its flowers contain a volatile oil used to dress wounds, as a laxative, as a diuretic, and as a diaphoretic. See elderberry.

elderberry *n.* a small red, black, or yellow berry produced by an elder bush. Birds like them, and they are sometimes used to make wine or jelly. See elder.

electrical storm *n.* a storm that produces numerous flashes of lightning, with or without rain.

electric eel *n.* an elongated cyprinoid fish (*Electroph electricus*, *Anguila* sp., or *Conger* sp., family Electrophoridae) that can give severe electric shocks.

electric fence *n.* a wire supported by insulators on widely spaced supports and connected to an electric source (usually one that emits brief high-voltage, low-amperage pulses). A single wire serves as a fence or enclosing device because any animal that touches the wire receives a sharp but harmless electric shock.

electric hygrometer *n.* an instrument that determines the relative humidity of the ambient atmosphere by measuring the electrical conductance of a hygroscopic material (e.g., a thin layer of lithium chloride).

electric prod(der) *n.* a handheld, battery-powered device that delivers a harmless electric jolt and is used to move or control livestock.

electrodialysis *n.* (Greek *ēlektron*, amber + *dialysis*, separation) a process for separating soluble materials and/or colloidal materials suspended in water from each other and/or from the water medium by means of an electric potential and a semipermeable membrane.

electrolyte *n.* (Greek *ēlektron*, amber + *lytos*, soluble) a substance (e.g., NaCl) that will conduct an electric current by the movement of ions either when dissolved in a solvent (especially water) or when fused by heat.

electromagnetic spectrum *n.* the entire range of radiant electromagnetic energy, including (from shortest to longest wavelengths) gamma rays, X rays, ultraviolet light, visible light, infrared light, and radio waves.

electromagnetic unit (emu) *n.* a unit used to express the magnetic effects of an electric current.

electromotive force (emf) *n.* **1.** the force that causes electrons (electricity) to flow or change direction. **2.** the maximum difference in electric potential between two plates of an electric cell, expressed in volts.

electron *n.* (Greek *ēlektron*, amber) a negatively charged elemental particle that orbits around the nucleus of an atom and carries a charge of 1.602×10^{-19} coulomb.

electronic *adj.* (Greek *ēlektron*, amber + *ikos*, of) **1.** working by controlled electron flow; e.g., an electronic device, machine, musical instrument, or circuit, especially those using transistors, integrated circuits, microchips, etc. **2.** having to do with electrons. **3.** controlled by a computer.

electronics *n.* **1.** the science that deals with controlled flow of electrons through semiconductors, gaseous media, vacuum tubes, etc., especially for the development of computers and other electronic devices. **2.** the electronic components of a machine or device.

electrophoresis *n.* (Greek *ēlektron*, amber + *phorēsis*, being borne) a process that characterizes the size and/or charge of large molecules or colloidal particles by their rate of migration through a medium under the influence of an electric field.

electrostatic precipitator *n.* a device that removes particulate matter (dust, smoke, etc.) from exhaust air by attracting particles to electrically charged screens or plates.

element *n.* (Latin *elementum*, first principle) **1.** a substance that cannot be decomposed into other substances by chemical means, especially one of the 92 naturally occurring chemical entities classified according to atomic number. Several more elements have been synthesized by bombarding some of the heavier nuclei with high-speed particles, but they are too unstable to persist in nature. **2.** a component part of an entity, a figure, or a mathematical set. **3.** a natural habitat. **4.** a group of people who share a common viewpoint or status (frequently implying a negative status; e.g., the criminal element).

elements *n.* weather; atmospheric forces.

elephant *n.* (Greek *elephantos*, elephant) **1.** a very large 5-toed mammal (either *Loxodonta africana* or *Elephas maximus*, family Elephantidae) noted for its long tusks (elongated teeth, the source of ivory), trunk (an elongated nose), and thick hide. Elephants are docile and easily trained to carry burdens or people and to push or pull heavy loads. They are also favorite circus performers. **2. white elephant** either an albino elephant or a costly possession that one does not use and would like to sell.

elephantiasis *n.* (Greek *elephantos*, elephant + *iasis*, condition) an advanced stage of filariasis in animals (especially horses) and people that is characterized by chills and fever and swollen lower legs and genitals resulting from lymphatic vessels being obstructed by nematodes (*Wuchereria bancrofti = Brugiamalayi*) carried by mosquitoes (*Culex* sp. and others). It is most common in tropical countries. See filariasis.

elephant seal *n.* a very large seal (*Macrorhinus proboscideus*) with a curved snout somewhat like an elephant's trunk. Native to the West Coast of North America and to the Antarctic Ocean, they grow as long as 20 feet (6 m) and are hunted for oil. Also called sea elephant.

elevation *n.* (Latin *elevatio*, lifting up) **1.** vertical distance above a specified reference, usually mean sea level or a nearby base level. **2.** a hill or other landscape feature that rises above its surroundings. **3.** the act or process of raising something to a higher level, either physically or figuratively (e.g., elevation of character or morality). **4.** the angle between a horizontal line and the line of view to an object (e.g., a star) or of sighting for a gun. **5.** a drawing that represents a view of an object or structure projected onto a vertical surface.

elicitor *n.* (Latin *elicere*, to draw out + *or*, one that) a substance or stimulus that elicits an immunologic response from a plant; often a product of a pathogenic organism or of cell wall degradation.

elixir *n.* (Arabic *al iksīr*, alchemical) **1.** a liquid used as a carrier for medication, usually a sweetened alcoholic solution. **2.** an alchemist's preparation intended to turn base metal into gold. **3.** a patent medicine promoted to cure almost any ailment and/or prolong life indefinitely (popular during the depression of the 1930s, sometimes called "elixir of life" or "snake oil").

elk *n.* (Anglo-Saxon *eolc*, elk) **1.** a large North American deer (*Cervus canadensis*). Also called wapiti. **2.** a moose (*Alces alces*) in Europe. **3.** the meat of an elk. **4.** the leather made from the hide of an elk.

Elk (BPOE) *n.* a member of the Benevolent and Protective Order of Elks, a fraternal organization.

elm *n.* (Anglo-Saxon *elm*, elm) a tall, hardy shade tree (*Ulmus* sp.) with serrated oval leaves, grown largely in humid and subhumid areas of the north temperate zone. The American elm (*U. americana*) was an important native species in the U.S.A. but was nearly wiped out by Dutch elm disease. Chinese elms and Siberian elms are resistant, and European elms have varying degrees of resistance to Dutch elm disease.

elm bark beetle *n.* a shiny dark reddish brown beetle (*Scolytus multistriatus*, family Scolytidae) that burrows through the bark of elm trees and is the principal vector for Dutch elm disease. See bark beetle, Dutch elm disease.

elm blight *n.* See Dutch elm disease.

elm lacebug *n.* an insect (*Corythucha ulmi*, family Tingidae) that infests the American elm and causes the leaves to yellow in early summer.

elm sawfly *n.* an insect (*Cimbex americana*, family Cimbicidae) whose larvae feed on the foliage of elm, willow, alder, basswood, birch, poplar, and maple trees, defoliating twigs and branches. The larvae are white with a black stripe on the back. The insects are common in northern U.S.A. eastward from Colorado. Also called giant American sawfly.

El Niño *n.* (Spanish *el niño*, little boy) a warm Pacific Ocean current that flows eastward toward the coast of Peru, alternating after a few years with a cold current known as La Niña that flows westward in the same area. The winds blowing across these currents influence the climate of large areas of North and South America. See La Niña.

eluvial *adj.* (Latin *eluvi*, washed out + *al*, pertaining to) removed from a soil horizon; e.g., soluble salts leached from the soil and colloidal clay formed in the A or E horizon but moved to the B horizon. The A and E horizons are eluviated and the B horizon is illuviated. See illuvial.

eluviation *n.* (Latin *eluvi*, washed out + *atio*, action) the process of removal of material from the A and/or E horizons of soil by weathering and soil formation, especially the downward transport of colloidal clay particles and of dissolved weathering products in percolating water. See illuviation, leaching, percolation.

emaciation *n.* (Latin *emaciatus*, wasted away + *tion*, action or condition) excessive leanness of animals or humans resulting from starvation or disease accompanied by a wasting away of both fat and muscle.

emasculate *v.* (Latin *emasculare*, to castrate) **1.** to castrate a male animal (remove the testicles). **2.** to remove the pollen-producing parts (the stamens and their appendages) from a flower.

embankment *n.* **1.** a ridge of earth or other material used as a dike or barrier. **2.** the process of building an embankment.

embarrass *v.* (French *embarrasser*, to obstruct) **1.** to cause someone to feel ill at ease and self-conscious, often by suggesting that they have done something wrong. **2.** to prevent someone from doing something; to cause difficulty or put them in debt; to complicate matters, often needlessly. *n.* **3.** a logjam that obstructs water flow in a river channel and may cause flooding in the surrounding area. Such obstructions have created lakes and wetlands, killed bottomland forests, and changed the character of soils and vegetation in bottomlands.

emblements *n.* (French *emblaement*, sown with grain) in law, a growing crop or the profit from growing a crop.

embodiment *n.* (Anglo-Saxon *em*, to put in + *bodig*, body + Latin *mentum*, condition) **1.** a visible expression of an idea or concept in a person, animal, or object; an incarnation. **2.** the process of embodying an idea or concept.

embolism *n.* (Greek *embolismos*, stoppage) **1.** a sudden blockage in a blood vessel by something that had been circulating in the bloodstream. **2.** an insertion; e.g., the extra day in a leap year.

embolus *n.* (Greek *embolos*, stopper) a clot, air bubble, or other obstruction that circulated with the blood until it blocked a blood vessel.

embouchure *n.* (French *emboucher*, to put to mouth) **1.** the mouth of a river or other opening from a restricted passage into a wider area; e.g., a narrow valley opening into a large one. **2.** the mouthpiece of certain musical instruments, or the adjustment of one's mouth to such a mouthpiece.

embryo *n.* (Greek *embryon*, ingrowing) **1.** a fertilized ovum in a womb or egg that eventually becomes a new individual person or animal. **2.** the part of a seed that grows into a new plant. **3.** a tiny ice crystal, dust particle, or other object that becomes the nucleus for an ice pellet or hailstone. **4.** the beginning stage of anything, as the embryo of an idea.

embryology *n.* (Greek *embryon*, ingrowing + *logos*, word) **1.** the science that deals with the formation, development, structure, and activities of embryos. **2.** the origin and development of embryos.

embryonic *adj.* (Greek *embryon*, ingrowing + *ic*, pertaining to) **1.** having to do with an embryo. **2.** in an early stage; undeveloped.

emergence *n.* (Latin *emergens*, rising out of + *entia*, action) a coming forth of something new, as a plant pushing its first leaves above the ground or an insect escaping from its cocoon or pupal case.

emergency spillway *n.* a discharge area for a pond or reservoir that functions only when the water level becomes higher than a design level and water flow exceeds the capacity of the usual spillway.

emergency tillage *n.* cultivation to roughen the soil for the reduction of wind erosion. Any implement that quickly produces a rough, cloddy surface may be used, including chisels, cultivators, and listers. A wide spacing is an advantage, so untilled strips are often left between the tilled strips.

emergent property *n.* a characteristic of a system that depends on interactions among its parts and cannot be observed or studied by considering the parts separately; e.g., a population involves not only the individual organisms that compose the population but also all the interactions among the individuals. At a community level, species may benefit other species as well as compete, symbiosis may occur, and one species may utilize resources that another cannot use, resulting in the whole community being more complex than the sum of its parts.

emergent species *n.* a species that grows taller than its neighbors and therefore receives full sunlight while its shorter neighbors grow in a reduced-light environment.

emergent vegetation *n.* aquatic plants that grow in shallow water with their roots in the underlying soil but with most of their vegetative growth above water. The plants may be erect, such as cattails (*Typha latifolia*) and bulrushes (*Scirpus* sp.), or may float on the water like water lilies (*Nymphaea* sp.).

Emerson, Ralph Waldo *n.* an influential 19th-century American poet (1803–1882) who advocated protecting natural resources.

emetic holly *n.* See yaupon.

emf *n.* See electromotive force.

emigration *n.* (Latin *emigratio*, removal) movement out of an area to establish residence elsewhere. See immigration.

eminent domain *n.* the right or power of government or a public agency to require an owner to relinquish private property for public use, subject to constitutional and statutory limitations (usually including reasonable compensation). The use of this power in the U.S.A. is limited by the "Due Process of the Law" clause in the 14th Amendment to the Constitution.

emission *n.* (Latin *emissio*, something emitted) **1.** a release or sending out of material, energy, or expression, as waste emission, light emis-

Smoke emission from coal-fired electric plants

sion, or propaganda emission. **2.** any material or product that is sent out. **3.** pollution discharged into a body of water or the atmosphere, especially from a point source such as a sewage outlet or a smokestack.

emission factor *n.* the amount of a pollutant produced per unit of raw material processed by a factory or per mile driven by a vehicle. Such factors are used to estimate total air pollution from a facility or in a geographical area.

emission inventory *n.* a listing of the sources and estimated amounts of various air pollutants discharged into the atmosphere of a community or other specified area.

emission standard *n.* the maximum amount of an air pollutant that a particular source is legally allowed to discharge, which depends on the nature of both the pollutant and the source, the location and history of the source, and the local air quality.

emissions trading *n.* an EPA policy that limits the total output of a particular pollutant but allows one or more sources to exceed their emission standards if the excesses are offset by reductions from other sources. Sources that reduce emissions substantially may sell their credits to others within a bubble concept. See bubble.

emit *v.* (Latin *emittere*, to send forth) **1.** to send forth a stream of gas, liquid, or solid particles. **2.** to release energy in the form of light, heat, or other radiation. **3.** to voice an opinion or utter a scream or other sound. **4.** to issue certain documents; e.g., paper money.

emitter *n.* (Latin *emittere*, to send forth + *er*, one that) **1.** a facility, machine, person, or animal that emits something, especially a pollutant. **2.** an outlet that controls the rate of water flow in a drip irrigation system.

emmer *n.* (German *amari*, wild wheat) an ancestral type of wheat (*Triticum dicoccon*, family Poaceae) that was domesticated early. The source of durum wheat.

Emmonsia *n.* a genus of saprophytic soil fungi (*Emmonsia parva* and *E. crescens*) whose spores, when inhaled, cause a lung disease in rodents and humans.

emphysema *n.* (Greek *emphysēma*, an inflation) **1.** a chronic lung disease that makes it difficult to expel air because the alveoli lose their elasticity. Cigarette smoking has been identified as one of the major causes. **2.** an abnormal accumulation of air in any body tissue or organ.

emu *n.* (Portuguese *ema*, emu) **1.** a large, flightless bird (*Dromaius novaehollandiae*) of Australia that resembles a shaggy ostrich and emits a deep booming sound. Emus may weigh up to 100 lb (45 kg), stand over 5 ft (1.5 m) tall, and run nearly 30 mph (50 km/h). The ostrich is the only living bird species that produces larger individuals than the emu. A nuisance to Australian farmers but raised for meat production in New Zealand, the United Kingdom, and the U.S.A. **2.** or **EMU** See electromagnetic unit.

emulsifiable *adj.* (Latin *emulsus*, milked out + *ificare*, to make + *abilis*, able) able to be emulsified; a liquid (usually an oily chemical pesticide) that can be mixed with water to form an emulsion.

emulsifier *n.* (Latin *emulsus*, milked out + *ificare*, to make + *er*, one that) a chemical (detergent, gum lecithin, agar, etc.) that helps produce a uniform suspension of one liquid in another; e.g., of an oily substance in water. This is important in the formulation of certain food products (e.g., cheese, ice cream, and margarine), cosmetics, and pesticides to assure uniform application.

emulsion *n.* (Latin *emulsio*, milked out) a colloidal suspension of one liquid as minute globules in another liquid; e.g., oil dispersed in water. Emulsions may also contain emulsifiers or other adjuvants; e.g., carbon tetrachloride added to a pesticide emulsion to increase droplet size and thereby decrease the hazard of drift.

encased knot *n.* a knot where the growth rings of the branch did not merge with those of the adjacent wood. An encased knot in a piece of lumber may fall out, leaving a hole.

encephalitis *n.* (Greek *encephalos*, brain + *itis*, inflammation) inflammation of the brain tissue in either humans or animals, causing various central nervous system disorders, either as a primary disease or as an effect of another disease. It is characterized by excitement, irregular movements, depression, and paralysis, and it can cause death. See meningitis.

encephalomalacia *n.* (Greek *encephalos*, brain + *malak*, soft + *ia*, condition) **1.** softening of the brain tissue. **2.** crazy chick disease, caused by vitamin E deficiency. The chicks suffer tremors and often fall grotesquely and wheel in circles. The brains of chicks that die from it show extensive lesions, especially in the cerebellum.

encroach *v.* (French *encrochier*, to hook in or seize) **1.** to infringe upon another's territory or rights. **2.** to advance gradually beyond one's designated limits.

endangered species *n.* an animal or plant species whose prospects of survival and reproduction are in immediate jeopardy. It is in danger of extinction from one or many causes: loss of habitat, overexploitation, predation, competition, disease, or even unknown reasons. Rare and endangered species are listed in the *Red Data Book* published by the International Union for Conservation of Nature and Natural Resources and in *Rare and Endangered Fish and Wildlife of the United States* compiled by the Bureau of Sport Fisheries and Wildlife, U.S. Department of the Interior. See threatened species, rare species.

Endangered Species Act *n.* a U.S. law passed in 1973 that makes the U.S. Fish and Wildlife Service, Department of the Interior, responsible for the listing of endangered and threatened species and prohibits federal agencies from providing permits or funds for projects that jeopardize the continued existence of such species.

endemic *adj.* (Greek *endēmios*, native) **1.** characteristic of or commonly occurring in a particular area or group; said of a plant, animal, disease, etc. **2.** found only in a particular area. See pandemic, epidemic, exotic.

end moraine *n.* **1.** an accumulation of debris pushed forward at the lower end of an active glacier. **2.** a hilly ridge of glacial debris marking a line where a glacier stopped moving forward and then receded. See terminal moraine, lateral moraine, ground moraine.

endocrine gland *n.* any ductless gland of a human or animal body that controls bodily processes by secreting hormones into the blood or lymph; e.g., pituitary, pineal, thyroid, parathyroid, thymus, adrenal, and pancreas.

endodynamomorphic soil *n.* a soil that is characterized primarily by properties inherited from its parent material. See ectodynamomorphic soil.

endogamy *n.* (Greek *endon*, within + *gamia*, marrying) **1.** the practice of restricting marriage partners to the same racial, ethnic, religious, community, or other group or to economic, educational, or other social peers. **2.** sexual reproduction between organisms with very similar germ plasm; inbreeding.

endogeic *adj.* (Greek *endon*, within + *geic*, born) living in the soil and feeding on roots and soil organic matter; e.g., certain unpigmented earthworms that process large amounts of soil, converting it into casts with a stabilized soil structure.

endogen *n.* (Greek *endon*, within + *genēs*, born) a plant that produces new growth throughout its stem rather than as external layers. A monocot plant. See exogen.

endogenous *adj.* (Greek *endon*, within + *genous*, produced) coming from within; produced internally in the cell or body.

endolithic *adj.* (Greek *endon*, within + *lithikos*, of stone) living within or penetrating deeply into weathered rock, coral, or other stony material; e.g. certain lichens.

endomycorrhiza (pl. **endomycorrhizae**) *n.* (Greek *endon*, within + *mykēs*, fungus + *rhiza*, root) a symbiotic association of a plant root and a fungus that forms structures such as vesicles and arbuscules inside plant root cells; hyphae extend outward through the cell walls and root surface into the soil. The fungi obtain carbohydrates from the plant and absorb water and nutrients from the soil for the plant. Also known as arbuscular mycorrhizae or vesicular arbuscular mycorrhizae. See ectomycorrhiza, mycorrhiza.

endoparasite *n.* (Greek *endon*, within + *parasitos*, a parasite) an organism living within, at the expense of, and to the detriment of another organism; e.g., a tapeworm in the intestinal tract of a person or animal, or a microbe that invades a body and causes disease.

endophyte *n.* (Greek *endon*, within + *phyton*, plant) **1.** a plant (e.g., a fungus) that lives inside another plant, usually as a parasite. **2.** a plant that lives inside the body of an animal; e.g., the fungus (*Acremonium coenophialum*) that causes fescue foot in cattle. See endoparasite.

endoplasmic reticulum *n.* the site of much organic synthesis in eukaryotic cells, either attached to the nuclear envelope or in Golgi bodies. Rough endoplasmic reticulum contains ribosomes that synthesize digestive enzymes and other proteins; smooth endoplasmic reticulum synthesizes enzymes that produce complex lipids.

endorheic *adj.* (Greek *endon*, within + *rheos*, flowing + *ic*, pertaining to) pertaining to an interior drainage basin; e.g., a body of water fed by land in the same property or state.

endoskeleton *n.* (Greek *endon*, within + *skeletos*, dried up) an internal skeleton; e.g., those of humans and other vertebrate animals.

endosmosis *n.* (Greek *endon*, within + *osmos*, impulse) passage of water into a cell by osmosis. See exosmosis, osmosis.

endosperm *n.* (Greek *endon*, within + *sperma*, seed) nutritive (nonembryonic, often starchy) triploid material that is produced early in the development of a seed from the combining of a pollen nucleus with the two polar nuclei of the embryo sac. In most legumes, the endosperm is absorbed by the embryo as it forms cotyledons. In grasses such as corn and wheat, the endosperm fills much of the seed alongside the embryo and nourishes the early stages of seedling development.

endotherm *n.* (*Greek endon*, within + *thermē*, heat) an animal that regulates its body temperature internally by adjusting the amount of heat generated and/or lost, commonly called a warm-blooded or homeothermic animal. Mammals and birds are endotherms. See ectotherm.

endothermic *adj.* (Greek *endon*, within + *thermē*, heat + *ikos*, of) adsorbing energy; a process that requires an input of energy, e.g., evaporation.

endotrophic *adj.* (Greek *endon*, within + *trophikos*, pertaining to food) growing within the cells of plant roots; e.g., an endomycorrhizal fungus. See endomycorrhiza.

endrin *n.* a highly toxic insecticide of the chlorinated hydrocarbon group ($C_{12}H_8OCl_6$). It is used for controlling grasshoppers and soil insects but has been banned in the U.S.A. because it has been shown to cause birth defects.

energy *n.* (Greek *energeia*, action) the capacity to do work, to produce or alter motion, to overcome resistance, and to bring about physical change. The seven forms of energy, capable of being changed from one to the other, are atomic, biotic, caloric, chemical, electric, kinetic, and potential.

energy cost *n.* the amount of energy required to accomplish a purpose; e.g., the amount of energy it takes to heat a house for a year, the amount of energy invested in tilling a field, the total amount of energy required to raise a crop, or the energy a lion must expend to catch a gazelle.

energy dissipator *n.* a device used to absorb and dissipate excess energy; e.g., a radiator in an automobile or a drop structure in a ditch.

energy efficiency *n.* the amount of work that can be accomplished from a given amount of energy by a particular process or machine. Data for transportation show that a bicycle makes the most efficient use of energy, followed by walking, train, bus, automobile, and airplane.

energy source *n.* any fuel or other means of providing energy. The principal energy sources in the U.S.A. are petroleum, natural gas, coal, nuclear power, hydroelectric power, wood, geothermal, solar, and wind.

Energy efficiency differs among these means of travel: walking, bicycle, rickshaw, oxcart, and automobile.

engender *v.* (French *engendrer*, to generate) **1.** to bring into being; to create. **2.** to procreate.

English sparrow *n.* See house sparrow.

enology *n.* See oenology.

enormous *adj.* (Latin *enormis*, huge) **1.** extremely large; beyond the ordinary size, amount, or degree. **2.** outrageous; extremely serious, e.g., an enormous error in judgment.

enrich *v.* (French *enrichier*, to enrich) **1.** to add wealth; to make richer. **2.** to add fertility by applying manure or fertilizer to soil. **3.** to add nutritional value by adding one or more minerals or vitamins to a food product.

enriched cereals *n.* processed cereals or cereal products (especially breakfast cereals) fortified with added minerals, proteins, and/or vitamins, including some that may have been depleted in milling and other processing.

ensete *n.* a relative (*Musa ensete*) of bananas with very large leaves, up to 18 ft (5.4 m) long by 3.5 ft (1 m) wide, and edible flower stalks that is native to Ethiopia.

ensile *v.* (French *ensiler*, to ensile) to store green fodder as silage in a silo. Silage is much higher in moisture than hay and is preserved by fermentation and acidification rather than drying.

ensnare *v.* (French *en*, to cause to be in + German *snare*, a trap) **1.** to catch in a trap (snare). **2.** to catch in a plot, often by deception; e.g., ensnared by lies.

enteritis *n.* (Greek *enteron*, intestine + *itis*, inflammation) inflammation of the lining of the intestines of animals or people, especially of the small intestine. It is a symptom of several infectious diseases characterized by frequent, watery (sometimes bloody), foul-smelling stools. See diarrhea.

enterotoxemia *n.* (Greek *enteron*, intestine + *toxikon*, poison + *haima* blood) **1.** a disease caused by a toxin produced in the intestine and found in the blood. **2.** a disease that causes high mortality among sheep in feedlots that is associated with overeating and bacteria (*Clostridium perfringens* and *Staphylococcus pyogenes* var. *aureus*). It is characterized by sluggishness, staggering, and sudden death. Also called pulpy kidney disease of sheep.

enterotoxin *n.* (Greek *enteron*, intestine + *toxikon*, poison) **1.** a toxic substance produced in food contaminated with bacteria (certain strains of *Staphylococcus aureus*) and stored for a few hours without refrigeration. The toxin resists breakdown from cooking. A small amount causes gastrointestinal distress for several hours but rarely if ever causes the death of an otherwise healthy person. **2.** a toxic substance produced in the intestine.

enthetic *adj.* (Greek *enthetikos*, put in) introduced from an outside source; exogenous; e.g., a contagious disease.

Entisol *n.* (French *récent*, recent + *sol*, soil) a soil with little or no profile development and no diagnostic horizons. Entisols are the least developed of the 12 soil orders in Soil Taxonomy. They comprise about 12% of the soil area in the U.S.A. and can occur wherever active erosion or deposition prevents formation of diagnostic horizons or where the soil horizons have been recently disturbed and mixed.

entomogenous *adj.* (Greek *entomo*, insect + *genēs*, born) growing on or in an insect (certain fungi). See parasite.

entomology *n.* (Greek *entomo*, insect + *logos*, word) the scientific study of insects (a branch of zoology).

entomopathogenic *adj.* (Greek *entomo*, insect + *pathos*, disease + *genēs*, born) attacking insects; bacteria, fungi, protozoa, or viruses that are pathogenic on insects.

entomophagous *adj.* (Greek *entomo*, insect + *phagos*, eating) insect eating, especially certain marsupials and parasitic insects.

entrainment *n.* (French *entraînement*, carrying along) drawing or carrying along, as the entrainment of silt particles in a stream of water.

environment *n.* (French *environnement*, environment) surroundings; the combination of all external conditions influencing the life, development, and survival of an organism.

environmental assessment *n.* a preliminary study to determine whether a project is required to have an environmental impact statement.

environmental biology *n.* ecology.

Environmental Defense Fund (EDF) *n.* an organization founded in the U.S.A. to promote environmental concerns through legal, economic, and scientific means.

environmental disease *n.* an illness caused by the surroundings of the affected organism; e.g., toxins in the air, lack of oxygen, or excess heat, cold, or moisture.

environmental impact *n.* any effects of a project (or anticipated effects of a proposed project) on the environment of the area and on the plant and animal species that inhabit that environment, especially any negative effects on rare or endangered species.

environmental impact statement *n.* a document required by the U.S. National Environmental Policy Act (PL 94-83) for major projects or legislative proposals that may have significant impact on the environment.

Environmental Protection Agency (EPA) *n.* an agency established in the U.S. government in 1970 by executive order of President Richard Nixon to formulate and enforce regulations limiting the release of pollutants that might adversely affect public health and/or the environment.

environmental quality *n.* an evaluation of the condition of the environment in a specified area, including such factors as degree and type of any pollutants present, natural beauty, ecosystems present, and any other factors that affect living conditions.

Environmental Quality Incentives Program (EQIP) *n.* a program established in 1996 in the U.S. Department of Agriculture to consolidate such previous programs as the Agricultural Conservation Program (ACP), the Great Plains Conservation Program (GPCP), Water Quality Incentive Projects (WQIP), and others providing cost sharing and other incentives for conserving soil and water. It is administered by the Farm Service Agency with technical oversight by the Natural Resources Conservation Service.

environmental resistance *n.* the inherent ability of an organism to tolerate various stresses, deficiencies, and threats in its environment.

environmental response team (ERT) *n.* a group of individuals specially trained and equipped to respond to releases of hazardous materials, e.g., from accidents that cause ruptures of tanks or pipelines.

environmental standard *n.* a requirement established by the U.S. Environmental Protection Agency or by a state agency specifying the maximum permissible concentration of a pollutant. State standards can be more strict but not less strict than the national standards.

environs *n.* (French *environs*, surroundings) surroundings, vicinity; especially the suburbs surrounding a city.

enzyme *n.* (Greek *en*, in + *zyme*, leaven) an organic catalyst produced by plants, animals, or microbes, usually identified by a name ending in -ase. Most enzymes catalyze a specific biochemical reaction; e.g., urease catalyzes the hydrolysis of urea. The six main groups of enzymes are hydrolases, isomerases, ligases, lyases, oxidoreductases, and transferases.

eolian *adj.* See aeolian.

EOSAT *n.* Earth Observation Satellite Company, a private company that works with NASA.

EPA *n.* See Environmental Protection Agency.

ephedra *n.* desert gymnosperm plants (*Ephedra equisetina*, *E. sinica*, and *E. vulgaris*, family Gnetaceae) that are a source of ephedrine.

ephedrine *n.* (Greek *ephedra*, horsetail + *inos*, like) **1.** an alkaloid ($C_{10}H_{15}NO$) extracted from ephedra plants or synthesized. It is used to decongest nasal passages in asthma, hay fever, and colds and to increase blood pressure. **2.** the horsetail plant.

ephemeral *adj.* (Greek *ephēmeros*, lasting only a day) transitory; short-lived; e.g., a brief fever, a flower that lasts only one day, or an insect such as a mayfly.

ephemeral gully *n.* a small channel that follows low points in the topography and is easily crossed and filled by tillage but also easily formed again the next time there is runoff. Similar channels that follow crop rows are called rills. See gully, rill.

ephemeral plant *n.* a plant that grows, sets seed, and dies within only a few days, typically in a desert environment where water is only briefly available; e.g., chickweed.

ephemeral stream *n.* an intermittent stream; one that carries flowing water for brief periods following rainstorms. Surface water is present fewer than 30 days during a year. See intermittent stream.

ephemeris *n.* (Latin *ephēmeris*, daybook or diary) **1.** a periodic table with a schedule of tides, phases of the moon, and/or positions of celestial bodies. **2.** an almanac containing astronomical tables. Some of these data are used in nonscientific forecasts as in astrology. See *Farmers' Almanac*, moonarian.

epicarcinogen *n.* (Greek *epi*, over + *karkinos*, cancer + *genēs*, born) an agent that increases the damaging effect of a carcinogen. See carcinogen.

epicarp *n.* (Greek *epi*, outer + *karpos*, fruit) the skin, rind, or outer layer of a fruit; the covering around the fleshy interior. Also known as exocarp.

epicenter *n.* (Greek *epikentros*, on the center) **1.** the location on the Earth's surface directly above the locus of an earthquake. **2.** the focal point or center of an activity.

epidemic *n.* or *adj.* (Latin *epidemia*, among the people) **1.** an outbreak of a large number of cases of a contagious disease in a short time, especially in humans in a single community or other relatively small area. See pandemic, epiphytotic, epizootic. **2.** any condition that arises suddenly and rapidly affects a large number of people, as an epidemic of terror.

epidemiology *n.* (Latin *epidemia*, among the people + *logos*, word) the study of epidemics as they affect populations, including the causes and distribution of diseases or other health-related events in human populations, factors (e.g., age, sex, occupation, or economic status) that influence epidemics, and the application of this study to control health problems.

epidermis *n.* (Greek *epi*, outer + *derma*, skin) **1.** the outer layer of cells in the skin of a person or animal. **2.** the outer layer of cells covering a plant or a seed.

epigamic *adj.* (Greek *epi*, outer + *gamos*, marriage) serving to attract a mate during the breeding season; e.g., the bright colors of some birds and animals.

epigeic *adj.* (Greek *epi*, surface + *geic*, born) living in leaf litter, fragmenting it, feeding on it, and converting it into stabilized organic matter in the process; e.g., certain small red or green earthworms that cannot burrow through the soil are epigeic.

epilimnion *n.* (Greek *epi*, upper + *limnion*, lake) the layer of warmer water overlying a colder and denser hypolimnion in a deep lake. The transition between the two is called a thermocline. Also called mixolimnion.

epinasty *n.* (Greek *epi*, upper + *nastos*, stamped down) a downward curl of a leaf or other part caused by its upper cells growing more rapidly than the lower cells. See hyponasty.

epipedon *n.* (Greek *epi*, surface + *pedon*, soil) a diagnostic surface soil horizon used to classify soils in Soil Taxonomy, including mollic, melanic, umbric, ochric, histic, anthropic, and plaggen epipedons. See diagnostic horizon.

epiphyll *n.* (Greek *epi*, surface + *phyllon*, leaf) a plant whose habitat is on the leaf of another plant but that does not draw nutrition from or harm the host plant (as an epiphyte lives on the trunk or stem of another plant).

epiphyte *n.* (Greek *epi*, surface + *phyton*, plant) **1.** a nonparasitic plant that grows on another plant or object but gets its nourishment from the air (from rain, dust, and volatiles); e.g., Spanish moss growing on an electric wire or a lichen on either tree bark or a rock. Also called an air plant. **2.** a parasitic fungus growing on the skin of a person or animal.

epiphytotic *adj.* (Greek *epi*, surface + *phyton*, plant + *otikos*, action or condition) rapidly spreading among plants; a plant disease comparable to an epidemic in humans or epizootic in animals.

episode *n.* (Greek *episodion*, an addition) **1.** a distinct event, especially an unusual one; one of several parts in a series, as a chapter in a story, a battle in a war, or an acute attack of a chronic disease. **2.** a disastrous event; e.g., an oil spill or other form of massive water or air pollution.

epizootic *adj.* (Greek *epi*, surface + *zo*, animal + *ōtikos*, action or condition) widespread occurrence of a disease among animals or insects; the equivalent of an epidemic among humans. See epidemic, epiphytotic.

epizootic aphtha *n.* See foot-and-mouth disease.

epoch *n.* (Greek *epochē*, fixed time) **1.** an extended period with certain consistent characteristics, as an epoch of peace or a geologic epoch. **2.** a significant date; the beginning of a new state of affairs. **3.** the position of a heavenly body at a specific time, or the time when it is at the specified position.

EQIP *n.* See Environmental Quality Incentives Program.

equable *adj.* (Latin *aequabilis*, can be made equal) nearly constant; having little variation, as an equable climate (typical of locations near the equator).

equator *n.* (Latin *aequator*, equalizer) **1.** an imaginary line around the Earth about 25,000 miles long that lies halfway between the North Pole and the South Pole and is designated zero latitude. The equator divides the Earth into Northern and Southern Hemispheres. **2.** a similar line around any sphere midway between its poles.

equatorial calm *n.* a zone near the equator that has mostly rising air and relatively little wind. See doldrums.

equestrian *adj.* (Latin *equestris*, horseman + *an*, pertaining to) **1.** related to horseback riding. *n.* **2.** a person (especially a man) who rides a horse. See equestrienne.

equestrienne *n.* (Latin *equestris*, horseman + *enne*, female) a woman who rides a horse. See equestrian.

equilibrium *n.* (Latin *aequilibrium*, balance) **1.** a constant condition wherein opposing forces are in balance. **2.** equality between opposing forces or powers. **3.** physical balance; e.g., the ability to maintain one's body in its position. **4.** mental and emotional stability. **5.** chemical balance. Any reaction in one direction is offset by an equal amount of the reverse reaction.

equine *adj.* (Latin *equinus*, of horses) having to do with horses, or characteristic of horses.

equinox *n.* (Latin *aequinoctium*, equal day and night) a day of the year when the Earth's poles are both the same distance from the sun and the extended plane of the equator passes directly through the sun. The equinox day and night are approximately 12 hours each worldwide. The vernal equinox (the official beginning of spring in the Northern Hemisphere) occurs on or about March 21, and the autumnal equinox (the official beginning of fall in the Northern

Hemisphere) occurs on or about September 23. Note: The seasons listed here are reversed in the Southern Hemisphere. See solstice.

equisetum *n.* (Latin *equisaetum*, horse bristle) a genus of rushes (*Equisetum* sp., family Equisitaceae) that contain a toxic substance that often poisons horses or sheep when they eat it in hay, especially horsetail (*E. arvense*) and giant horsetail (*E. telmateia*). In severe cases, the animals become very nervous, lose muscular control, stop eating, and die. Also called horsetail or scouring rushes (because they are sometimes used for scouring).

equivalent diameter *n.* the diameter of a spherical particle that would settle at the same rate as a particle of any shape being evaluated in a particle-size analysis (usually more than the smallest dimension and less than the largest dimension of a nonspherical particle).

equivalent weight *n.* the weight in grams of a substance that will combine with or replace 1 mole of hydrogen ions (H^+) or involve 1 mole of electrons in an oxidation-reduction reaction; the atomic or formula weight of the substance divided by its valence. Also called gram-equivalent weight; numerically equal to the newer preferred unit, moles of charge per liter.

era *n.* (Latin *aera*, fixed date) **1.** a significant period marked by certain characteristics; e.g., the jitterbug era. **2.** the time following a certain event, as the Christian era began with the birth of Christ. **3.** a major division of geologic time.

eradicant *n.* (Latin *eradicare*, to root out + *ant*, one that) **1.** any treatment that eradicates the causative organism of a disease after the disease is established. **2.** a chemical used to eradicate weeds. See herbicide.

eradicate *v.* (Latin *eradicare*, to root out) to remove or destroy completely; e.g., to pull or dig out all the roots of weeds, to eliminate all cases of a disease, to erase letters, words, or marks from a page, or to correct all the typographical errors in a manuscript.

eradicator *n.* (Latin *eradicatre*, to root out + *or*, one that) **1.** a chemical or process used to eradicate weeds, insects, or other pests. **2.** a person who eradicates pests. **3.** a material used to erase or cover mistakes or marks on a page, as ink eradicator.

erbium (Er) *n.* element 68, atomic weight 167.26; a soft, malleable rare earth metal that occurs along with other rare earth metals in several minerals and is used to color glazes for glass and porcelain.

eremacausis *n.* (Greek *ērema*, little by little + *kausis*, combustion) gradual decomposition of organic matter by oxidation upon exposure to suitable air, temperature, and moisture conditions. See subsidence.

eremium *n.* (Greek *erēmia*, a desert) a plant community in a desert.

eremology *n.* (Greek *erēmia*, a desert + *logos*, word) the field of science that deals with deserts.

eremophyte *n.* See xerophyte.

erg *n.* (Greek *ergon*, work) **1.** a metric unit for measuring work or energy; a force of 1 dyne acting through a distance of 1 cm. Also known as dyne-centimeter. **2.** a large area of land covered with sand; e.g., sandy areas in the Sahara Desert.

ergocalciferol *n.* vitamin D_2; a fat-soluble sterol that prevents or cures rickets and is needed for calcium metabolism to form strong bones and teeth. Also called calciferol.

ergonomics *n.* (Greek *ergon*, work + *nomos*, custom) a study of the interaction between workers and workplaces, especially as regards the relation between anatomy, physiology, and movements as they relate to machines and ways of increasing safety and reducing fatigue.

ergosterol *n.* a plant and animal sterol ($C_{28}H_{43}OH$) that is converted to vitamin D_2 by ultraviolet light striking either the source or the skin of a person or animal. Also called provitamin D_2 or ergosterin. See vitamin D, ergocalciferol.

ergot *n.* (French *ergot*, a rooster's spur) **1.** a disease of cereals (especially rye) and other grasses that is caused by a fungus (*Claviceps* sp., especially *C. purpurea*, family Clavicepitaceae) that replaces grain with sclerotia (black or dark purple, club-shaped structures composed of densely packed mycelia). Ergot reduces grain yield and causes ergotism when humans or animals consume contaminated grain or grain products. Also called clavus. **2.** the sclerotia from ergot used to produce medicines to reduce bleeding and induce muscular contractions, especially during childbirth. **3.** a small horny growth in the tuft of hair behind the fetlock joint of a horse. See ergotism.

ergotism *n.* (French *ergot*, a rooster's spur + *ism*, condition) a disease of people and animals caused by consumption of grain or grain products contaminated with mycotoxin produced by ergot. Ergotism causes excessive salivation, vomiting, muscular cramps, diarrhea, and/or constipation. Also called holy fire, St. Anthony's fire, bread madness, gangrenous ergotism, or dry gangrene. See ergot, mycotoxin.

ericaceous *adj.* (Latin *ericaceae*, heath) belonging to the heath family (Ericaceae) of plants.

ericoid *adj.* **1.** ericaceous. **2.** resembling heath.

ermine *n.* (French *ermine*, ermine) **1.** any of several northern weasels (*Mustela* sp., family Mustelidae) whose fur becomes white in winter. **2.** the soft white fur of the animals, especially when used to make fur coats or other garments.

erodibility *n.* (Latin *erodere*, to erode + *ibilitas*, ability) susceptibility to erosion. A silty soil has high erodibility because its particles detach easily and are small enough to be transported readily. See erosivity.

erodible *adj.* (Latin *erodere*, to erode + *ibilis*, able) susceptible to erosion by water, wind, or other abrasive or chemical agents; e.g., erodible soil. See erosive.

erosion *n.* (Latin *erosio*, wearing away) the wearing away of a surface by water, wind, or other abrasive agents and/or by chemical processes; e.g., soil erosion or dental erosion. Soil ero-sion is usually divided into geologic (natural) erosion and accelerated (human-induced) erosion. See gully erosion, rill erosion, sheet erosion.

Severe erosion

erosive *adj.* (Latin *erodere*, to erode + *ivus*, tendency) tending to cause erosion; e.g., the action of flowing water, wind, glacial ice, and acid leaching. See erodible.

erosivity *n.* (Latin *erodere*, to erode + *ivus*, tendency + *itas*, condition) erosiveness; the ability to cause erosion. The erosivity of water and wind depends greatly on their kinetic energy and is also affected by their intensity and duration. See erodibility.

erratic *adj.* (Latin *erraticus*, erring) **1.** unpredictable in occurrence or behavior. **2.** unusual; deviant; eccentric. **3.** out of place, as an erratic boulder transported by a glacier (identifiable because it is a different kind of stone than the local bedrock).

ERS *n.* See Economic Research Service.

eructation *n.* (Latin *eructatio*, a belching) **1.** the act of belching; the upward release of gas from the stomach. **2.** the gas released by belching. See flatulence.

eruption *n.* (Latin *eruptio*, sending forth) **1.** a spewing forth, as the release of ash and lava from a volcano or of water and gases from a geyser. **2.** the appearance of a new tooth erupting from a baby's gum. **3.** a rash, pimples, or pustules on the skin, as those caused by measles or chicken pox.

erysipelas *n.* (Greek *erythros*, red + *pella*, skin) **1.** an acute infectious disease that causes redness and swelling of the affected areas of skin or mucous membranes, sometimes including lesions. **2.** a contagious disease affecting mostly young swine that is caused by bacterial infection (*Erysipelothrix rhusiopathiae*). Symptoms include fever and red spots on the skin of the neck and body. It can be prevented by vaccination. Also called swine erysipelas.

erythroblastosis *n.* (Greek *erythros*, red + *blastos*, bud or sprout + *osis*, condition) **1.** the presence of erythroblasts in the bloodstream of humans, most commonly occurring in a fetus or very young infant as a consequence of anti-Rh agglutinin in an Rh-negative mother. **2.** a viral disease of chickens that is characterized by immature erythrocytes in the blood, causing yellowish combs and wattles, weakness, weight loss, diarrhea, prostration, and death.

erythrocyte *n.* (Greek *erythros*, red + *kytos*, hollow or cell) red blood corpuscle. A tiny disk-shaped body that contains hemoglobin and transports oxygen through the bloodstream.

escape *v.* (Norse *escaper*, to escape) **1.** to get away; to slip out of restraints or confinement. **2.** to evade capture. **3.** to become wild after having been tame (either an animal or a plant species). **4.** to flow through a crack or other opening, as air escapes from a punctured tire. **5.** to emit inadvertently, as a laugh escaped in spite of the solemn setting. *n.* **6.** the act of getting away or the fact of having gotten away. **7.** avoidance, as escape from responsibility. **8.** the loss of material that flows away through a puncture or crack. **9.** an animal or bird that has gotten away from its usual area of confinement. See feral. **10.** a plant cultivar that was developed for cultivation but is found growing wild.

escarpment *n.* (French *escarpement*, escarpment) a nearly vertical natural soil or rock face; e.g., a cliff produced by either erosion or a fault zone. The surface above the escarpment is usually gently sloping, whereas that below is often moderately sloping.

eschalot *n.* (French *eschallotte*, small scallion) a small winter onion (*Allium cepa*, family Liliaceae) propagated by a small clove (set). Also called shallot.

Escherichia coli *n.* a species of coliform bacteria (family Enterobacteriaceae) that occurs in large numbers in the gastrointestinal tract and feces of all warm-blooded animals and people and is used as an indicator of fecal contamination in water or food. Its presence in other parts of the body is not normal and can cause disease, especially if it is one of the more virulent strains. Often written *E. coli*. See coliform bacteria.

esculent *adj.* (Latin *esculentus*, food) **1.** edible, especially as a modifier of plants or plant parts.

n. **2.** a plant or plant part suitable for use as human food, especially a vegetable.

esculin *n.* a bitter, toxic substance in the bark of horse chestnut (*Aesculus hippocastanum*, family Hippocastanaceae).

escutcheon *n.* (French *escuchon*, shield) **1.** a shield adorned with a coat of arms. **2.** a protective metal shield, as that often placed around a keyhole. **3.** the part of the body covered by pubic hair, as the part of a cow just above and back of the udder where the hair turns upward instead of the usual downward direction. Also called milk mirror.

esker *n.* (Irish *eiscir*, ridge) a unique land feature of glaciated regions formed where a major stream flowed in or beneath glacial ice. When the glacier receded, it left a low, narrow, sinuous ridge with gravel and sand strata sloping in various angles and directions from deposition in glacial water. Eskers have good soil and air drainage and are often used as sources of sand and gravel. Also called osar or hogback. See drumlin, kame.

Eskimo *n.* See Inuit.

essence *n.* (Latin *essentia*, quality of being) **1.** an entity; something that exists. **2.** the fundamental nature of a being; stripped of unnecessary adornment. **3.** a concentrated extract carrying the fragrance, flavor, or other important characteristic of a plant, food, or drug, often dissolved in alcohol. **4.** a perfume. **5.** a spiritual entity.

essential element *n.* an element that a plant or animal must have to complete its life cycle. To be essential, an element must be involved in life processes, not just part of the environment, and it must not be replaceable by any other element. Essential elements include macronutrients that are required in relatively large quantities and micronutrients that are needed in much smaller quantities. See functional element, macronutrient, micronutrient.

essential host *n.* a plant or animal that is vital to the life cycle of a parasitic organism; e.g., many rusts that damage economically important plants can be controlled by removing their essential alternate host (e.g., barberry for wheat rust and cedar for apple rust) that they must have during part of their life cycle.

essential oil *n.* a volatile plant oil that carries an odor characteristic of the plant and that may serve as an insect attractant or as a defense mechanism. Essential oils are sources of many flavorings, perfumes, attractants, repellents, etc.; e.g., almond, camphor, lemon, peppermint, and rose.

estancia *n.* (Spanish *estancia*, mansion) a large landholding, especially a cattle ranch (southwestern U.S.A.).

estate *n.* (French *estat*, estate) **1.** a condition, class, rank, or stage of existence for a person or property. **2.** a relatively large acreage, usually indicating considerable wealth, including a residence and other structures, gardens, lawns, etc., often maintained by a number of servants. **3.** one's entire property, especially that left at the time of death.

ester *n.* (Latin *aether*, ether) **1.** an organic compound produced by reacting an alcohol with an acid (either an organic or an inorganic acid) and the release of a molecule of water. Many esters of organic acids have pleasant odors that are used in perfumes and flavorings. Most are good solvents, and some are used to make adhesives, lacquers, and enamels. **2.** an organic compound formed by reacting glycerol (a triple alcohol) with acids forming products such as fats and oils (esters of glycerol and fatty acids) and nitroglycerin (the ester of glycerol and nitric acid).

esthetic *adj.* See aesthetic.

estival or **aestival** *adj.* (Latin *aestivalis*, pertaining to summer) summery; pertaining to summer.

estivation or **aestivation** *n.* (Latin *aestivalis*, pertaining to summer + *tion*, action) **1.** passing the summer season in a particular activity. **2.** summer dormancy of certain animals that become inactive in response to high temperature and/or drought. See hibernation. **3.** the structural arrangement of a flower in the bud before it opens.

estrapade *n.* (French *strappare*, to break) the bucking of a horse.

estray *n.* (French *estraie*, straying) **1.** a person, animal, or thing that is out of place; one that has gone astray. **2.** a wandering, unclaimed domesticated animal.

estrogen *n.* (Latin *oestrus*, gadfly or frenzy + Greek *genes*, born) any of a group of female sex hormones capable of producing heat (estrus) in a female animal and of inducing feminine physical characteristics. Synthetic estrogens are used either for producing estrus in the female or for birth control, are fed to animals, and are used for chemical caponizing of male chickens.

estrus or **oestrus** *n.* (Latin *oestrus*, gadfly or frenzy) **1.** the recurrent period of sexual receptivity of a female mammal, also known as heat, corresponding to rut in a male. **2.** an overwhelming urge or frenzy.

estuary *n.* (Latin *aestuarium*, an inlet where the tide ebbs and flows) a coastal inlet with brackish water influenced by tides that mix seawater with fresh water. It is typically formed where the lower reaches of a river valley are drowned by the coastal area having sunk relative to sea level. See drowned topography.

ethanol *n.* (Latin *aether*, ether + *ane*, hydrocarbon + *ol*, alcohol) **1.** ethyl alcohol (C_2H_5OH), a transparent, colorless, volatile, combustible liquid that mixes readily with water. It is the intoxicating ingredient of beer, wine, whiskey, etc. and is commonly produced by fermentation of sugars such as glucose ($C_6H_{12}O_6$) with yeast (*Saccharomyces cerevisiae*). It is used medicinally to kill microbes and as a solvent for many medications. **2.** a mixture of 90% petroleum products and 10% ethyl alcohol used as a motor fuel. The alcohol increases the octane rating and makes the fuel burn cleaner. Also called gasohol.

ether *n.* (Latin *aether*, ether) **1.** ethyl ether, a colorless organic liquid [$(C_2H_5)_2O$] that is volatile and highly flammable, has anesthetic properties, is used as a solvent for organic substances, and was formerly used as an anesthetic. It has a sweet taste and an aromatic odor. **2.** any compound composed of two organic radicals joined through an oxygen atom. **3.** the pure (especially upper) atmosphere. **4.** outer space; the heavens. **5.** the medium supposed to occupy outer space according to ancient theories.

ethnology *n.* (Greek *ethnikos*, heathen + *logos*, word) anthropology that deals with the background, culture, and other characteristics of ethnic groups or language groups.

ethology *n.* (Greek *ēthologia*, depiction of character) **1.** the study of animal behavior in response to environmental influence, including factors that control intelligence and learning. **2.** applied ethics.

ethyl alcohol *n.* See ethanol.

ethylene *n.* (Latin *aether*, ether + *yl*, a chemical radical + *ene*, unsaturated hydrocarbon) a colorless, flammable, gaseous unsaturated hydrocarbon (C_2H_4) with a double bond between its two carbons that is used to hasten maturity of harvested fruit (especially pears and tomatoes), as a coloring agent for citrus fruits (especially oranges), as a fuel for blowtorches, and as an anesthetic. It is also polymerized to make polyethylene plastics.

ethylene dibromide *n.* a highly toxic liquid ($C_2H_4Br_2$) used as a fumigant for soil and grains until it was banned for most uses in agriculture because it causes cancer in laboratory animals. It was also long used as an antiknock compound for gasoline along with tetraethyl lead. Also called 1,2-dibromoethane.

ethylene glycol *n.* a sweet-tasting double alcohol [$C_2H_4(OH_2)$] widely used as a radiator antifreeze. Its sweet taste has contributed to its causing the death of dogs that consumed spilled antifreeze. Also called dihydroxyethane or 1,2-ethanediol.

etiolation *n.* (French *étiol*, making pale + *ation*, action) **1.** blanching (whitening) of a plant or plant part (e.g., cauliflower) by keeping it away from light.

2. paling of a person or animal as a result of illness or of insufficient light in their environment.

etiology or **aetiology** *n.* (Greek *aitia*, cause + *logia*, description) **1.** the study of cause and origin, especially of a disease or of diseases in general. **2.** the identification of the cause of an occurrence.

eucalyptus *n.* (Greek *eu*, well + *kalyptos*, covered) any of about 500 species of fast-growing evergreen trees (*Eucalyptus* sp., family Myrtaceae) with long, slender pendent leaves. Native to Australia and nearby islands, it is now widely grown in tropical to warm temperate climates and used for ornamentals, windbreaks, lumber, fuelwood, and as a source of eucalyptus oil and gums. Also called gum tree.

eucalyptus oil *n.* an aromatic essential oil derived from the leaves of some species of eucalyptus and used in antiseptics, expectorants, and perfumes.

eukaryote *n.* (Greek *eu*, well + *karyon*, kernel) **1.** any organism that has eukaryotic cells, including all higher animals and plants, protozoa, and fungi and all algae except blue-green algae. See eukaryotic cell, prokaryote. **2.** one of three major divisions of life forms: archaea, bacteria, and eukaryotes (based largely on each division having its own distinct form of RNA sequencing).

eukaryotic cell *n.* a cell that contains a distinct nucleus surrounded by a membrane and that also commonly contains organelles with specific biochemical functions. It is characteristic of multicellular plants and animals and of protozoa, fungi, and many algae. See cell, prokaryotic cell.

euosmia *n.* (Greek *eu*, good + *osmē*, smell) **1.** a pleasant odor. **2.** a normal sense of smell.

euphorbia *n.* any of many species of plants, shrubs, and trees of the spurge family (*Euphorbia* sp., family Euphorbiaceae), named after Euphorbos, a Greek physician, including poinsettia (*E. pulcherrima*), crown of thorns (*E. splendens*), Mexican fireplant (*E. heterophylla*), and some that are poisonous, emetic, and cathartic; e.g., an annual weed pest, sun euphorbia (*E. helioscopia*), whose milky sap can cause dermatitis in susceptible individuals. Milky euphorbia sap is a latex that contains hydrocarbons similar to those in gasoline and has been explored as a potential fuel source.

euphotic zone *n.* the upper layer of a body of water that enough light penetrates for plants to carry on photosynthesis [to a maximum depth of about 600 ft (180 m) or less]. See aphotic.

europium (Eu) *n.* element 63, atomic weight 151.965; a soft, ductile, rare earth metal that occurs along with other rare earth metals in several minerals and forms light pink salts.

eurybathic *adj.* (Greek *eurys*, broad + *bathos*, depth) able to tolerate a wide range of depth (pressure) in a water environment.

eurygamous *adj.* (Greek *eurys*, broad + *gamos*, mating) mating in flight, as many insect species do; e.g., the honeybee.

euryphagous *adj.* (Greek *eurys*, broad + *phagos*, feeding) able to consume any of a wide variety of foods. See stenophagous.

eurytherm *n.* (Greek *eurys*, broad + *thermē*, heat) a species that can live in environments with a wide range of temperature. See stenotherm.

eustasy or **eustacy** *n.* (Greek *eu*, good + *statikos*, standing) a global change of sea level, as that caused by either the formation or melting of continental ice sheets changing the volume of water in the oceans.

euthanasia *n.* (Greek *eu*, good + *thanatos*, death) **1.** a peaceful death, as easy and pain free as possible. **2.** a mercy killing to end pain and suffering.

euthenics *n.* (Greek *euthēn*, being well off + *ikos*, subject) the attempt to improve a race or breed (especially the human race) through regulation of the environment.

eutrophic *adj.* (Greek *eu*, good + *trophē*, nutrition) accumulating high concentrations of plant nutrients in a body of water (e.g., a lake or pond) that therefore produces a prolific growth of algae. Anaerobic conditions, murkiness, fish kills, and offensive odors result when the algae die and decay, especially in shallow, stagnant water. Much of the problem is attributed to nitrogen and phosphorus from waste materials (sewage etc.) and fertilizer nutrients in runoff and seepage waters. See dystrophic, oligotrophic.

eutrophication *n.* (Greek *eu*, good + *trephein*, to nourish + *atio*, action) **1.** nutrient enrichment; a process that contributes to nutrition. **2.** a condition in many stagnant bodies of water that leads to excessive vegetative growth (especially of algae known as pond scum). Decay of the vegetation depletes the water of oxygen, producing anaerobic conditions.

eutropic *adj.* (Greek *eu*, good + *tropos*, turning) turning of a plant toward the sun or other light source; e.g., a sunflower. Also called heliotropic.

evaginated *adj.* (Latin *evaginatus*, unsheathed) turned inside out; e.g., the head of a tapeworm larvae turned inside out so it can attach its suckers to the intestinal wall.

evaporation *n.* (Latin *evaporatio*, dispersion as a vapor) **1.** the process that changes a liquid to a gas. **2.** the loss or removal of water vapor from a body,

resulting in a loss of weight and increased concentration of solutes.

evaporation pan *n.* a container used by the U.S. Weather Service to measure evaporation; a circular tank 4 ft (122 cm) in diameter and 10 in. (25.4 cm) deep set in an open field on 2-in. (5-cm) supports. Water depth is measured and recorded at specified intervals to determine the rate of water evaporation, and water is added as needed to maintain a depth of 8 ± 0.5 in. (20.3 ± 1.25 cm).

evaporative cooling *n.* cooling air by evaporating water (often used as a form of air conditioning, especially in dry climates).

Evaporators providing chilled water to cool buildings

evaporator *n.* (Latin *evaporare*, to disperse as a vapor + *or*, one that does) a container where evaporation occurs. Evaporators are used to remove water from food or other materials and to absorb heat, as in a refrigeration system.

evapotranspiration *n.* (a combination of evaporation and transpiration) **1.** the movement of water from soil and bodies of water into the atmosphere by evaporation from exposed surfaces, either directly from the soil or water or indirectly through plant transpiration. **2.** the amount of water that moves by evapotranspiration, normally expressed in the same units as rainfall, either inches or centimeters of water evaporated in a specified period of time (e.g., per day, week, month, or growing season). It differs from consumptive use or water requirement only by the relatively small amount of water retained in plant tissue. See consumptive use of water, duty of water.

evening primrose *n.* any of several plants (*Oenothera* sp., family Onagraceae) whose yellow, pink, or white blossoms typically open in the evening.

everglade *n.* (Anglo-Saxon *aefre*, always + *glad*, smooth place) an area of wetland with many interlacing waterways between areas covered with water-loving tall grasses, reeds, etc.

Everglades *n.* an area of more than 4000 mi^2 (10,000 km^2) of partially wooded marshland with many interconnecting waterways flowing south out of Lake Okeechobee in southern Florida.

evergreen *n.* (Anglo-Saxon *aefre*, always + *grene*, green) a plant that remains green throughout the year either by retaining at least some green leaves at all times (e.g., fir, juniper, pine, and spruce trees and shrubs) or by having green stems and branches that contain chlorophyll (e.g., cacti, spurges).

eviscerate *v.* (Latin *eviscerare*, to disembowel) **1.** to remove the internal organs (entrails, lungs, heart, etc.) from a fowl or animal. **2.** to deprive of vital force, parts, or significance, as to eviscerate a novel.

evolution *n.* (Latin *evolutio*, unrolling) **1.** growth and development of an idea or practice. **2.** gradual change in genetics through mutations, genetic drift, and accommodation to the environment through natural selection. **3.** Darwinism. The theory that all living species evolved from earlier species through survival of the fittest. See Darwin, Charles Robert, regressive evolution, survival of the fittest. **4.** peaceful change. **5.** release of a vapor, as the evolution of bubbles of carbon dioxide.

evolve *v.* (Latin *evolvere*, to unroll) **1.** to develop gradually by a process of evolution. **2.** to emit or escape gradually, as heating the material caused it to evolve a gas.

ewe *n.* (Anglo-Saxon *eowu*, female sheep) a female sheep, especially one that is mature.

excess ice *n.* ice in a soil in excess of the amount that would be formed by freezing the water normally retained in the soil voids. Excess ice is present as ice lenses produced by water moving to them by capillary action as the soil freezes. Such ice lenses cause frost heaving.

exchangeable ion *n.* an ion adsorbed on a surface but subject to replacement by another ion or ions of equivalent charge.

excise *v.* (Latin *excidire*, to cut off) **1.** to remove by cutting, as to cut off a branch, cut out a tumor, or delete a passage from a text. **2.** to levy an excise tax. *n.* or *adj.* **3.** a tax levied on the production, sale, or consumption of a specified commodity (e.g., alcoholic beverages or tobacco).

excrement *n.* (Latin *excrementum*, refuse material) feces; manure; waste matter from the bowels.

excrescence *n.* (Latin *excrescere*, to grow out) **1.** an appendage or material that grows; e.g., a fingernail or hair. **2.** an unnatural growth caused by a disease, insect, toxin, or pressure; e.g., a bunion or a wartlike growth on the stem of a plant.

excreta *n.* (Latin *excreta*, separated) waste matter discharged from a human or animal body, especially sweat or urine, and sometimes including discharges from the nose or mouth and feces.

excretion *n.* (Latin *excretio*, separation) **1.** the process of discharging something from the body of a plant or animal. **2.** material discharged; excreta or certain liquid plant substances.

excurrent *adj.* (Latin *excurrens*, running forth) **1.** extending outward, especially giving passage outward, as some ducts. **2.** extending beyond, as the midrib of some leaves that extends past the rest of the leaf. **3.** extending from the base to the highest point without dividing, as the main stem of a spruce or hemlock tree.

exfoliate *v.* (Latin *exfoliare*, to strip of leaves) to lose or remove foliage, bark, scales, or other outer layer.

exfoliation *n.* (Latin *exfoliatus*, stripped of leaves) **1.** the natural peeling of an outer covering, such as the bark of certain trees (e.g., sycamores). **2.** the removal of a surface layer from a rock surface by mostly mechanical weathering (temperature fluctuations, ice wedging, roots growing in cracks, etc.).

exhaust *v.* (Latin *exhaurire*, to empty) **1.** to vent; to allow to escape, as to exhaust the combustion products from an engine or furnace. **2.** to use all of the available supply, as to expend all of one's resources or to deplete the fertility of a soil. **3.** to create a vacuum by exhausting all the air from a chamber. **4.** to consider thoroughly, as to exhaust all possible causes. **5.** to wear out and lose strength, as to exhaust one's energy and stamina or to run down a battery. *n.* **6.** a vent system to carry away combustion products. **7.** the combustion products that are vented.

exhaustible *adj.* (Latin *exhaurire*, to empty + *ibilis*, able) subject to depletion; limited in amount, so the supply can be used up; e.g., a resource such as petroleum or coal in contrast to a renewable resource such as wood.

exhume *v.* (Latin *ex*, out + *humare*, to inter) **1.** to dig up; to disinter buried remains. **2.** to uncover; to bring to light; to reveal or disclose. See paleosol.

exobiology *n.* (Greek *exō*, outside + *bios*, life + *logos*, word) the scientific study of the possibility of extraterrestrial life. Also called astrobiology or bioastronomy.

exocarp *n.* See epicarp.

exocytosis *n.* (Greek *exō*, outside + *kytos*, hollow or cell + *osis*, condition) the discharge from a cell of material produced by the cell. Also called reverse pinocytosis. See secretory vesicle, Golgi body.

exogen *n.* (Greek *exō*, outside + *genēs*, producing) a plant that grows by adding outer layers; a dicotyledon. See endogen.

exogenous *adj.* (Greek *exō*, outside + *genous*, produced) external; developing from the outside or on the exterior of an organism or object. See endogenous.

exoskeleton *n.* (Greek *exō*, outside + *skeletos*, dried up) an external skeleton, as the hard outer shell of crustaceans and many insects.

exosmosis *n.* (Greek *exō*, outside + *ōsmos*, thrusting) passage of water out of a cell; e.g., from plant root cells into the soil because the soil is very dry or contains excess salt, sometimes as a result of placing too much fertilizer too close to a seed. See endosmosis, osmosis, plasmolysis, salt index.

exothermic *adj.* (Greek *exō*, outside + *thermē*, heat + *ikos*, of) releasing energy; a process that produces heat; e.g., combustion.

exotic *adj.* (Greek *exōtikos*, foreign) **1.** not native; brought into the area from a remote place; e.g., a plant, animal, stone or other object, word, expression, or practice that originated somewhere else. **2.** strange, fascinating, and enticing because of unfamiliarity.

exploitation *n.* (French *exploitation*, exploitation) the selfish use of a resource that should be left for other use or for future use; exploitation is often detrimental to the value of the resource and to the environment.

exponential growth *n.* increasing according to a power series, e.g., 2^0, 2^1, 2^2, 2^3, 2^4 ... (= 1, 2, 4, 8, 16 ...), as opposed to arithmetic growth (e.g., 1, 2, 3, 4, 5 ...).

exposure *n.* (French *exposer*, to put out + *ure*, action or result) **1.** being uncovered; open to view. **2.** lack of protection from destructive environmental elements. **3.** the aspect of a slope or building; the direction it faces. **4.** the time when a camera shutter is open to take a picture.

exsanguinate *v.* (Latin *exsanguis*, bloodless + *atus*, making) to drain or draw blood from a human or animal.

exsanguination *n.* (Latin *exsanguis*, bloodless + *atio*, action) the withdrawal of blood; e.g., the action of a bloodsucking tick or female mosquito.

ex situ *adj.* (Latin *ex situ*, out of place) off-site, as opposed to in situ; e.g., ex situ conservation of animal species may be done by freezing semen and embryos, or conservation of plant species may benefit from ex situ seed banks.

extant *adj.* (Latin *exstans*, standing out) **1.** still in existence; not destroyed, lost, or extinct. **2.** protruding above a surface.

Extension Service *n.* See Cooperative Extension Service.

extermination *n.* (Latin *exterminatio*, destruction) the process or fact of complete elimination of a species.

external combustion engine *n.* an engine that separates the combustion process from the cylinder where mechanical power is generated; e.g., a steam engine.

externality *n.* (Latin *externus*, outside + *alis*, pertaining to + *itat*, condition) **1.** something external. **2.** an effect that occurs elsewhere, affecting someone else, as the downstream impact of erosion (a negative externality) or the aesthetic effect a visitor sees when a building is repaired and repainted (a positive externality).

extinct *adj.* (Latin *extinctus*, put out) **1.** no longer having living representatives on Earth, as an extinct species. **2.** ended; no longer in effect, as an extinct law.

extinction *n.* (Latin *extinctio*, putting out) **1.** the process or fact of elimination of a species. **2.** the ending of a condition, as the extinction of a fire or of a feud.

extirpate *v.* (Latin *extirpare*, to root out) to eliminate, especially by digging out the roots.

extravasate *v.* (Latin *extra*, beyond + *vas*, vessel + *atus*, making) **1.** to release or force out; e.g., to cause blood or plant sap to flow out of its normal vessels into the surrounding tissue. **2.** to spew out, as an erupting volcano.

extrinsic *adj.* (Latin *extrinsecus*, outward) **1.** external; coming from outside the body. **2.** nonessential; not part of the inherent nature of the body. See intrinsic.

extrusive *adj.* (Latin *extrudere*, to force out + *ivus*, tendency) **1.** formed by or pertaining to extrusion. **2.** formed of molten lava that flowed from a vent and hardened too quickly to form large crystals; e.g., basalt, andesite, or rhyolite.

exudate *n.* (Latin *exsud*, sweat + *atus*, making) a liquid or semiliquid substance that oozes from an injury on a person, animal, or plant; e.g., blood, pus, or sap.

F

F *n.* **1.** the symbol for filial generation; e.g., F_1, first filial generation after a cross; F_2, the second generation, etc. **2.** the chemical symbol for fluorine. **3.** a failing grade. **4.** farad.

faba or **fava** *n.* or *adj.* (Latin *faba*, bean) a large, edible broad bean (*Vicia faba*, family Fabaceae). Also known as broadbean, horse bean, Windsor bean.

Fabaceae *n.* the family name adopted in 1972 for leguminous plants, trees, shrubs, and vines; plants having compound leaves, keeled flowers, and fruit encased in pods that split along both sides; e.g., acacia, alfalfa, beans, clover, lentils, locust trees, peanuts. Formerly called family Leguminosae.

fabaceous *adj.* (Latin *fabaceus*, of beans) pertaining to beans or to the legume family of plants in general.

fabric *n.* (Latin *fabrica*, factory or fabric) **1.** any woven, knit, felted, etc., cloth or clothlike material. **2.** the positions and linkages of component parts, e.g., of the mineral grains in a rock or the particles in a soil. **3.** social structure, as the fabric of a society.

fabric pest *n.* any insect species (Tineidae and Dermestidae families) that lays eggs on clothing, linens, carpets, etc., made of wool, feathers, fur, and similar materials and whose larvae eat these substances. See clothes moth, carpet beetle.

facet *n.* (French *facette*, little face) **1.** a cut and/or polished plane surface, e.g., on a gem stone. **2.** an aspect or feature, as a facet of one's character. **3.** a segment of a compound eye, as in that of an arthropod.

facies *n.* (Latin *facies*, form) **1.** general appearance, especially that of the face. **2.** facial appearance that indicates a diseased condition or death. **3.** the characteristic appearance of a specific rock formation. **4.** a distinctive phase of a community or culture, especially one that is prehistoric.

factor *n.* (Latin *factor*, doer) **1.** something to be considered; an influence that helps determine a characteristic or outcome. **2.** one of two or more numbers that can be multiplied together to equal a number being factored; e.g., 2 and 5 are factors of 10. **3.** an agent who buys and sells commodities for others for a fee or commission. **4.** a gene. *v.* **5.** to divide a number into its component factors; for example, 12 can be factored into 2, 2, and 3. **5.** to conduct business for someone else.

factory *n.* (Latin *factoria*, factory) a place where goods are manufactured, including the buildings and machines needed. See cannery, creamery.

factory farm *n.* a large commercial farm, usually owned by a corporation (in contrast to a family farm) and operated as an investment for monetary profit.

facultative *adj.* (Latin *facultātivus*, facultative) **1.** giving or having the right to do or not do something. **2.** not obligate; capable of living in more than one way or under more than one condition: e.g., as either a saprophyte or a parasite, or in aerobic or anaerobic conditions.

facultative anaerobe *n.* a type of microbe that normally grows in the presence of oxygen but can adapt to anaerobic conditions; e.g., certain bacteria (*Klebsiella* sp., *Enterobacter* sp., and *Achromobacter* sp.) that fix atmospheric nitrogen under anaerobic conditions. Also called facultative bacteria.

facultative lagoon or **facultative pond** *n.* a lagoon or pond with an aerobic upper layer and an anaerobic lower layer; a deep lagoon or pond in a windy location is likely to be facultative.

fagopyrism *n.* (Latin *fagopyrum*, buckwheat + *isma*, condition) poisoning caused by consuming buckwheat.

fagot *n.* (French *fagot*, fagot) **1.** a small bundle of twigs used as kindling to start a fire or as a torch. **2.** any bundle or bunch; e.g., a bundle of iron rods.

Fahrenheit *n.* a temperature scale named after Gabriel Daniel Fahrenheit (a German physicist, 1686–1736). The freezing point of water is 32°F and the boiling point is 212°F (at 1 atm); 0°F can be produced by a mixture of equal weights of salt (NaCl) and snow. For conversion, °F = 32 + 1.8 × °C. See Celsius scale, Kelvin scale, Rankine scale.

fair *n.* (Latin *feria*, holiday or market) **1.** an event where farm products, equipment, etc., are exhibited on a competitive basis, usually with premiums offered for excellence, commonly in the summer or fall on a county or state basis. **2.** an exhibition and/or exposition of products; a form of advertising or a means of raising funds, especially for a charity.

fair condition *n.* **1.** denoting medium-quality plants or plant products; e.g., rangeland that is producing only 25% to 50% of its potential. **2.** medium condition of grazing animals.

fair market value *n.* an evaluation by an appraiser of the appropriate price of a property and terms of sale by a willing seller to an able, willing, and informed buyer. Also called market value, normal sale value.

fair weather *n.* pleasant weather with clear air, sunshine, and no rain; may have some white cumulus clouds that are also known as picture clouds.

fair weather clouds *n.* white cumulus clouds. Also known as picture clouds.

fairy ring *n.* a ring or circle (some as large as 50 ft or 15 m in diameter) of luxuriant dark green growth representing the border of a mat of fungal mycelia in the soil of an area of turfgrass or meadow; once imagined to be a dancing place for fairies. Mushrooms (especially *Marasmius oreades*) may grow on the periphery at certain seasons of the year. Also called fairy circle, fairy green.

falcon *n.* (Latin *falcon*, a hawk) **1.** the female of any of several birds of prey (especially *Falco* sp., family Falconidae) with hooked beaks and long, pointed wings. (The males are often called tercels; they are smaller and not as bold as the females.) Falcons are swift and agile in flight and often dive to catch prey. **2.** any hawk trained by a falconer to hunt and kill small game, especially the gerfalcon (*F. gyrfalco*) or the peregrine falcon (*F. peregrinus*).

fall *n.* (Anglo-Saxon *feallan*, to fall) **1.** autumn (when leaves fall from trees). **2.** a sudden descent under the influence of gravity. **3.** a collapse. **4.** something that has dropped to a lower level. **5.** the distance something has dropped. **6.** a downward slope. **7.** a decrease in value. **8.** a loss of status, power, prestige, or esteem. **9.** an animal birth or the number of animals born in a litter. **10.** a pin of a wrestling opponent.

fall armyworm *n.* the larva of a moth (*Laphygma frugiperda*, family Noctuidae) that spreads in large numbers (armies). The larvae eat the leaves of corn, alfalfa, cotton, peanuts, and many other cultivated and wild plants and bore into stalks and sometimes into ears of corn. The fall armyworm produces as many as six generations per year in warm climates (each of which may fly for several miles before laying eggs) but does not survive cold winters.

fallout *n.* **1.** radioactive particles from a nuclear explosion that are carried by the wind (sometimes for thousands of miles) until they fall from the atmosphere. They are most concentrated in a plume that extends downwind from the site of the explosion. **2.** the process of such particles falling from the atmosphere. **3.** an unintended or unplanned result of an event or action.

fall overturn *n.* an exchange between surface water and deep water that occurs in deep lakes as a result of cooler temperatures in the fall causing the surface water to cool and become denser. The denser water sinks, displacing warmer, less dense water from below.

fallow *n.* (Anglo-Saxon *fealga*, fallow land) **1.** land that is maintained in a tilled but unseeded condition for one or more seasons, usually to accumulate enough soil water in a semiarid region to raise a bigger crop the following year or to kill weeds. Also called summer fallow. *v.* **2.** to leave tilled land unseeded. *adj.* **3.** bare ground; tilled but unseeded. **4.** unused, as an idea lying fallow or a fallow (untrained) mind. **5.** a pale yellowish brown color.

fallow deer *n.* a small European deer (*Dama dama*) with a yellowish coat that has white spots in the summer.

false heat *n.* a semblance of estrus in a female animal when she is either pregnant or out of season, sometimes caused by diseased ovaries.

false hellebore *n.* any of several plants (*Veratrum* sp., family Liliaceae) that resemble hellebore, especially a North American species (*V. viride*) with clusters of yellow-green flowers; provides substances used in certain medicines and herbicides.

false indigo *n.* any of several North American legumes (*Baptisia* sp.) that produce long clusters of purple, cream, or white flowers.

false molt *n.* the loss of feathers by a bird at an unseasonable time caused by environmental conditions. See molt.

familism *n.* (Latin *familia*, family + *isma*, characteristic) a social pattern of people and some animals wherein family values and decisions are more important than individual interest or preference.

family *n.* (Latin *familia*, family) **1.** a husband and wife (or either as a single parent) plus any children they may have, especially those who live in the same household or maintain strong ties to the parent household. **2.** the offspring of one couple or one person. **3.** a group of closely related individuals, including parents, children, grandchildren, uncles, aunts, cousins, etc. **4.** a close-knit group of associates and assistants that work together. **5.** a group of people who share common interests or goals, as a church family. **6.** a group of related languages, as the Latin family. **7.** one or more genera of organisms (animals, plants, insects, or microbes) that are grouped for the next higher level of classification. **8.** all of the individuals of a certain species or strain in a plant community (often all descended from a single plant). **9.** a group of soil series in Soil Taxonomy at the categorical level between soil series and great soil group. **10.** a group of chemical elements that have certain characteristics in common; e.g., the halogen family.

family farm *n.* a farm of a suitable size to support a family and to be operated by that family alone or with minimal hired help.

family labor *n.* labor available within the family of an entrepreneur (especially that of a farmer) in addition to that of the entrepreneur. Its value is determined by what it would cost to hire outside labor to maintain an equivalent amount of business if family labor were not available.

family planning *n.* an effort to slow the rate of growth of the human population, largely by supplying information on birth control. The current population growth rate of nearly 2% per year is not sustainable for the long term (it was less than 0.1% prior to 1000 A.D.), and the problem is aggravated by a large proportion of the increase coming to families and countries with the least resources available to provide for the additional people. See zero population growth.

This family planning sign in New Delhi, India, says "Two children are enough."

famine *n.* (Latin *famina*, hunger) a catastrophic food shortage affecting a large area and causing large numbers of people to starve. Famines may be caused by crop failure (resulting from drought, insect pests, or plant diseases), natural disasters (earthquakes, flooding), or political problems (warfare, persecution).

fancy *adj.* (Latin *phantasia*, fantasy) **1.** of a grade indicating very high quality for many products, including vegetables, fruits, flowers, poultry, and livestock. **2.** adorned with designs or ornaments. **3.** whimsical; an image created in the mind. **4.** overly expensive, as a fancy price.

fanega *n.* **1.** a Spanish measure for grain equal to 55.5 L (1.57 U.S. bu) in some places, but variable with location. **2.** the land area that can be seeded with 1 fanega of wheat; 0.64 ha (1.6 ac) in Castilla but variable with location.

fang *n.* (German *fang*, capture or booty) **1.** a long, pointed tooth used by meat-eating animals to hold or tear prey; a canine tooth. **2.** a long hollow tooth that enables poisonous snakes to inject venom into their prey.

FAO *n.* See Food and Agriculture Organization.

farad (F) *n.* the unit of electrical capacitance in the SI system; the capacitance of a capacitor holding 1 coulomb of electricity with a potential difference of 1 volt between its plates. It is named after Michael Faraday (1791–1867), an English physicist.

farcy *n.* (French *farcin*, glanders) a chronic and often fatal form of glanders affecting the lymphatic glands, especially on the head and limbs; mostly a disease of horses but also communicable to other animals and humans. Also called farcin.

farding bag *n.* the rumen (first stomach) of a cow or other ruminant.

farina *n.* (Latin *farina*, meal or flour) **1.** a food product made by grinding a cereal grain, nut, bean, potato, cassava, etc., for consumption as a cooked cereal. **2.** potato starch.

farm *n.* (Latin *firma*, make firm) **1.** a rural property consisting of buildings and land used to grow crops (and often livestock) for market. Defined by the USDA as a unit that has $1000 or more gross sales of farm products annually (prior to 1978 the definition required only $250 of gross sales annually or an area of at least 1 ac with at least $50 of gross sales). **2.** a similar unit used to grow plants, animals, or fish, as a pig farm or a fish farm. **3.** a method of raising revenue by leasing land in districts, or the land so leased. **4.** historically, the rent paid for leasing property in England (often including or in lieu of taxes). *v.* **5.** to raise crops for market. **6.** to lease land to someone else for a fixed rental or for a percentage of the production. **7.** to hire a person for work or to care for someone or something.

farm accidents *n.* Accidents on farms are usually related to use of machinery, farm structures, and farm chemicals. The frequency of accidents is higher in farming than in most other occupations, partly because farmers work with a wide variety of machines and materials, sometimes with old equipment, and often work alone and without much of the training that would be required in industry.

Farm Bureau *n.* the American Farm Bureau Federation; the largest general farm organization in the U.S.A. and Puerto Rico. It is organized at the county, state, and national levels to analyze farm problems and formulate action plans, including lobbying legislators to achieve educational improvement, economic opportunity, and social advancement, while preserving individual opportunity and freedom. It is nonpartisan, nonsectarian, and nonsecret in character. Farm Bureau policies are determined and implemented in the counties by the members and on the state and national levels by delegate bodies. See Farmers Union, Patrons of Husbandry.

farm chemicals *n.* fertilizers and pesticides (especially herbicides and insecticides) applied to crops or livestock.

farm enterprise *n.* **1.** an individual farm, especially a family farm. **2.** a specific farm business activity, as growing a particular crop or a class of livestock.

Farmers' Almanac n. (now known as the *Old Farmers' Almanac)* an annual publication first published in 1792 (the oldest continuously published periodical in North America), founded by Robert B. Thomas, and headquartered in Dublin, New Hampshire. The new edition every September includes a calendar that gives weather predictions, moon calendars, planting charts, and tidal charts a year in advance. It is also liberally sprinkled with sayings and humor.

Farmers Home Administration (FmHA) *n.* an agency in the U.S. Department of Agriculture from 1946 (when it replaced the Farm Security Administration) to 1994 (when it became part of the Farm Service Agency). Its mission was to aid farmers financially, especially to provide loans and supervisory assistance to help tenant farmers become landowners and to provide emergency loans for recovery from drought, flood, or other disasters.

Farmers Union *n.* the Farmers Educational and Cooperative Union of America, also known as the National Farmers Union; founded in 1902 in Point, Texas, with the primary goal of sustaining and strengthening the economic interests and quality of life in family farm and ranch agriculture. Resolutions by an individual or group of members are presented for possible adoption at the local, state, and national levels. See Farm Bureau, Patrons of Husbandry.

farmland *n.* **1.** land in farms. **2.** cropland.

farm management *n.* decision making for the operation of a farm, either by the farmer or by a professional person or organization employed to make decisions regarding crops, livestock, equipment purchases, marketing, etc.

farm manager *n.* a person who makes decisions regarding the operation of a farm, either as a farmer or as a professional service.

farm planning *n.* **1.** the work of a farm manager analyzing the recent past, present, and projected future condition of a farm, making plans accordingly, and recording those plans for future action. **2.** the consulting work provided to farm managers by a farm planner from the National Resources Conservation Service or other government agency or private consultant.

farm pond *n.* a small artificial body of water (often a fraction of an acre; seldom larger than a few acres, or

1 or 2 ha) formed by damming a small stream (typically one that drains a watershed of 25 to 100 ac (10 to 40 ha). Farm ponds typically are used to provide water for livestock or for a small irrigation project and/or to raise fish; they often are designed by the Natural Resources Conservation Service.

Farm Security Administration *n.* an agency established in the U.S. Department of Agriculture by the Bankhead-Jones Act of 1937, replacing the Resettlement Administration (established in 1933) for the rural rehabilitation program, providing real estate loans up to 100% of the appraised value; designed to help tenant farmers become landowners by up to $12,000. Replaced in 1946 by the Farmers Home Administration.

Farm Service Agency (FSA) *n.* an agency established in 1994 in the U.S. Department of Agriculture by merging the Farmers Home Administration with the Agricultural Stabilization and Conservation Service. FSA supports farmers through production and marketing programs by financing part (often half) of the cost for building approved soil and water conservation structures and by making or guaranteeing loans to help new farmers get started or for established farmers to survive emergencies.

farrier *n.* (Latin *ferrator*, smith) **1.** a person (e.g., a blacksmith) who nails metal shoes on the hooves of horses, mules, or donkeys. **2.** a veterinarian, especially one who treats horses.

farriery *n.* **1.** (Latin *ferrator*, smith + *y*, action or place) the workplace of a farrier. **2.** the trade or practice of shoeing horses.

farrow *n.* (Anglo-Saxon *fearh*, pig) **1.** a litter of pigs or a young pig. *v.* **2.** to give birth to a litter of pigs.

farrowing house *n.* a structure designed to hold sows before and during the birthing process and, after birth, to hold sows and their litters.

fascine *n.* (Latin *fascina*, bundle of sticks) **1.** a fagot. **2.** a bundle of sticks or poles bound together and used to support a structure or a temporary roadway across a marshy area or ditch. **3.** a bundle of brushwood used as a facing to prevent erosion on a river bank, seawall, or other steep embankment. Fresh branches from a species that will sprout and grow (e.g., willow) to provide lasting protection in wet areas.

fat *n.* (Anglo-Saxon *faett*, loaded) **1.** a triple ester composed of three fatty acids (especially oleic, palmitic, and stearic acids) linked to glycerol; a triacylglycerol that is an oily solid at room temperature. Longer-chain fatty acids and those with higher saturation tend to raise the melting point. Also known as a triglyceride. **2.** any substance that looks, feels, and behaves like the solid triacylglycerols. **3.** a white or yellowish adipose tissue that serves for energy storage in plants (especially seeds) and animals. **4.** a food or cooking product derived from either plant or animal fat; e.g., shortening, lard. **5.** the most luxurious part, as the fat of the land. *adj.* **6.** containing fat, as a fat piece of meat. **7.** obese; overweight; enlarged from having stored excess fat. **8.** filled with an abundance of something, as a fat wallet or a fat envelope. **9.** big, wealthy, prosperous. **10.** stupid, unteachable, as a fat head. **11.** unlikely, as a fat chance.

fathom *n.* (Anglo-Saxon *faethm*, arm span) **1.** a length of 6 ft (1.83 m); used as a nautical measure of the depth of a body of water. *v.* **2.** to measure the depth of a body of water. **3.** to thoroughly understand something.

fatling *n.* (Anglo-Saxon *faett*, loaded + *ling*, little) a young animal that has been fattened for slaughter; e.g., a calf or a lamb.

fatty acid *n.* any of a number of organic acids with mostly straight chains of 2 to 24 or more carbon atoms. Fatty acids from plant sources are mostly unsaturated (have one or more double bonds in the chain), whereas those from animal sources are mostly saturated. The carboxyl (–COOH) group is at the end of the chain and behaves as a weak acid that can react with many other chemicals; e.g., fatty acids react with glycerol to form fats.

fault *n.* (Latin *falsus*, mistaken) **1.** an imperfection or defect. **2.** a blemish in the surface of an object. **3.** a flaw in one's character. **4.** blame; responsibility for a misdeed or for something that went wrong. **5.** a surface where rock layers are discontinuous because the body of rock has broken and one section has moved relative to another section.

fauna *n.* (Latin *fauna*, a Roman goddess) animal life of all kinds, including wild mammals, birds, fish, reptiles, insects, and even microscopic forms. See flora.

fava *n.* See faba.

favonian *adj.* (Latin *favonius*, west wind) gentle and mild, like the west wind.

favus *n.* (Latin *favus*, honeycomb) **1.** an infectious skin disease of humans and certain animals caused by a fungus (*Trichophyton schoenleini, T. violaceum,* or *Microsporum gypseum*) that causes itching of the scalp and the formation of yellow crusts with a honeycomb pattern and a moldy odor. Also called tinea favosa, honeycomb ringworm, or crusted ringworm. **2.** a similar disease of poultry caused by a fungus (*Achorion gallinae*) or of young cats or sometimes dogs (caused by *Achorion schoenleini*) that is transmissible to humans. **3.** a hexagonal

marble tile or slab used to form a honeycomb pattern in pavement.

fawn *n.* (French *faon*, offspring) **1.** a deer less than a year old. **2.** the pale yellowish brown coat color of a young deer. *v.* **3.** to show extreme courtesy or servile behavior. **4.** to behave affectionately or admiringly, as to fawn over a baby.

FDA *n.* See Food and Drug Administration.

Fe *n.* (Latin *ferrum*, iron) the chemical symbol for iron.

feasibility study *n.* an evaluation of the advisability of carrying out a proposed project, including an analysis of legality, practicality, costs, benefits, and probable environmental effects.

feather eating *n.* the action of poultry or other caged birds pulling out one another's feathers; a result of irritation by lice or quill mites, lack of exercise, faulty nutrition, or environmental stress; can lead to cannibalism. Also called feather pulling or feather picking.

feather pulling or **feather plucking** *n.* removal of the feathers from a bird in preparation for cooking it.

fecal *adj.* (French *fécal*, of feces) pertaining to excrement from the bowels of humans, animals, or birds.

fecal contamination *n.* pollution of food, water, or the environment with feces. See coliform bacteria.

feces *n.* (Latin *faeces*, dregs) excrement; waste material from the alimentary tract excreted through the anus of humans, animals, and birds.

fecund *adj.* (Latin *fecund*, fertile) **1.** fertile, capable of producing offspring, fruit, or continued growth; prolific. **2.** very productive or creative, as a fecund soil or the fecund years of a nation's history.

fecundate *v.* (Latin *fecundat*, make fruitful) to pollinate; to fertilize.

fecundity *n.* (Latin *fecunditas*, fertility) the ability of an individual to produce fertile eggs or sperm regularly.

Federal Extension Service *n.* See Cooperative Extension Service.

Federal Grain Inspection Service *n.* a function of the USDA Grain Inspection, Packers and Stockyards Administration (GIPSA), the agency that provides quality standards and a system for applying them to certify the purity of all grain exported from or imported into the U.S.A.

feed *n.* (Anglo-Saxon *fedan*, food) **1.** food, especially for livestock, including grain, forage, fodder. **2.** the quantity of feed supplied to one animal at one time.

3. a meal, especially one with abundant food. **4.** a mechanism or process to supply food or to supply fuel for combustion or raw material to a machine. **5.** the fuel supplied to a furnace or the raw material supplied to a machine. **6.** a radio or television connection to distribute a broadcast through a network. *v.* **7.** to furnish food or supplementary nutrients. **8.** to deliver, carry, or transport nutriments. **9.** to supply a fuel or raw material as needed for processing. **10.** to gratify; to satisfy a need. **11.** to assist, as to feed the ball to another player or to feed lines to an actor. **12.** to augment, as a tributary feeds a river.

feed additive *n.* a material added to feed to promote good health and/or rapid growth in livestock; usually an antibiotic (pioneered in 1949 by Dr. McGinnis of Washington State University and Dr. Jukes of Lederle Laboratory, who obtained more rapid growth of poultry by adding aureomycin to poultry feed), a tranquilizer, a growth hormone (e.g., diethylstilbestrol), or a vitamin or mineral supplement.

feedback *n.* **1.** a process or connection that returns some of the output of a device to the input of the same device, either desirably or undesirably (e.g., from a loudspeaker to a microphone); sometimes used to regulate the operation of a machine. **2.** a response to a process or activity, as feedback from a speech. **3.** the self-regulating response of certain biological systems.

feedlot *n.* a confined area for the controlled feeding of animals for fattening and finishing for market. In most states where they occur, large feedlots (those holding thousands of animals) are regulated, partly for animal disease and pollution control.

fee simple *n.* a legal title to property that is not restricted to any particular class of heirs; unrestricted ownership and inheritance.

fee tail *n.* a property title that is restricted to inheritance by a designated class of heirs (usually the owner's children).

feldspar *n.* (German *feld*, field + *spath*, mineral) any of several silicate minerals with silicon and oxygen in tetrahedra linked in a three-dimensional framework with large cations (Ca^{2+}, Na^+, K^+, and/or Ba^{2+}) in the large spaces between tetrahedra. See orthoclase, plagioclase.

felon grass *n.* See Christmas rose.

female *adj.* or *n.* (Latin *femella*, female) **1.** a person or animal that produces eggs that can be fertilized by the sperm of a male of the same species. **2.** a plant or the part of a plant (pistil) that is fertilized by pollen and produces seed. **3.** feminine; often associated with delicacy and gentleness. **4.** having an opening, as a nut with an opening for a bolt.

femto- a combining form for 10^{-15}; e.g., a femtosecond. A femtometer is called a fermi.

fen *n.* (Anglo-Saxon *fen*, fen or bog) a type of wetland that accumulates peat deposits. Fens are less acidic than bogs; much of their water comes from underground and is rich in calcium and magnesium. See wetland.

fence *n.* (Latin *defensus*, defense) **1.** a barrier surrounding a field or other enclosure, usually consisting of a series of posts connected by wires, rails, or poles to prevent the passage of large animals. **2.** a guard or protective piece; a part of a machine that controls or restricts movement, e.g., to keep hands away from a saw blade or to guide the movement of lumber being sawed. **3.** a person or place of business that deals in stolen goods. *v.* **4.** to build a fence around an area. **5.** to protect, defend, or prevent unwanted persons, animals, or things from entering. **5.** to sell stolen goods to a fence. **6.** to participate in the sport or a fight of fencing. **7.** to avoid giving a direct answer.

fencerow *n.* the untilled area next to a fence, where grasses and weeds are usually allowed to grow, often providing cover for wild animals and birds.

fennel *n.* (Anglo-Saxon *fenol*, fennel) **1.** a yellow-flowered herb (*Foeniculum vulgare*, parsley family) with feathery leaves. **2.** aromatic seeds of the plant; used for seasoning, especially in French or Italian cooking, and in medicine.

feral *adj.* (Latin *ferus*, wild) **1.** a tame species (e.g., horses or dogs) gone wild. **2.** naturally wild; untamed. **3.** brutal; ferocious. **4.** gloomy; deadly; mortal; e.g., having to do with a funeral.

fer-de-lance *n.* (French *fer-de-lance*, spear head) a large pit viper (*Bathrops atrox*) native to tropical areas in South America.

fermentation *n.* (Latin *fermentatio*, fermentation) **1.** an anaerobic chemical reaction catalyzed by enzymes produced by living microbes (bacteria or fungi, especially yeasts or molds). Fermentation produces many specific products (e.g., ethyl alcohol, acetic acid, bread, cheese, sour milk, certain antibiotics) according to the microbe used, the material being fermented, and critical conditions of heat, pressure, and light that are specific to the reaction. **2.** creation of a state of excitement or rebellion among people or animals.

fermenter *n* (Latin *ferment*, cause to rise + *er*, one that does) **1.** a bacterium, yeast, or other organism that causes fermentation. **2.** a closed container used to produce biochemical material or where cells containing recombinant DNA are multiplied. Laboratory units may be as small as a liter or less, whereas industrial units may hold tens of thousands of liters. **3.** or **fermentor** a container used to contain and promote a fermentation process (e.g., with temperature control), usually in a batch of 10 L or more for the commercial production of antibiotics or other products.

fermi (F) *n.* a very short unit of distance equal to 10^{-15} m; used to evaluate nuclear dimensions. See femto-.

Fermi, Enrico *n.* a nuclear physicist (1901–1954) who was born in Italy but worked in the U.S.A. In 1938 he won the Nobel Prize in physics for his work making radioactive chemical elements, and he achieved the first controlled nuclear reaction in 1942.

fern *n.* (Anglo-Saxon *fearn*, fern) any of a large group of nonflowering vascular plants (class Filicinae) that reproduce asexually by spores produced on their fronds (leaflike parts). Ferns range in size from minute to giant (tree-size) and are most abundant in warm, humid settings; they provided much of the vegetative material that was converted into large coal deposits and other plant fossils since the Devonian Period.

ferralitic *adj.* containing iron, as a ferralitic mineral or rock.

ferredoxin *n.* any of many small iron-sulfur proteins that function in photosynthesis, respiration, nitrogen fixation, etc., by transferring electrons from one enzyme system to another in plants and bacteria.

ferret *n.* (Latin *furetus*, little thief) **1.** a domesticated, weasel-like mammal (*Mustela*, sp., family Viverridae) about 14 in. (35 cm) long, with pale yellow fur and red eyes. It is of African origin but has been used in Europe and elsewhere to drive rats and rabbits from their dens and kill them. **2.** a narrow ribbon used to decorate cloth items. **3.** a piece of iron used by a glass blower to test molten glass to see if it is workable. *v.* **4.** to hunt with ferrets. **5.** to drive out of hiding; to investigate thoroughly; to uncover secrets.

ferromagnetic *adj.* (Latin *ferrum*, iron + *magnetis*, magnetic) having strong magnetic properties, most commonly including magnetite (Fe_3O_4), pyrrhotite ($Fe_{1\ or\ more}S$), and maghemite (Fe_2O_3). See paramagnetic, diamagnetic.

ferruginous *adj.* (Latin *ferruginus*, the color of iron rust) **1.** iron-bearing or iron-containing, as ferruginous ore. **2.** orange-brown in color; resembling the color of rust.

fertigation *n.* spreading fertilizer by dissolving it in irrigation water. Fertigation is often used with trickle irrigation, and sometimes with sprinkler irrigation, but rarely with surface irrigation because of the degree of difficulty of obtaining uniform distribution.

fertile *adj.* (Latin *fertilis*, fruitful) **1.** able to reproduce (as a person, animal, plant, seed, embryo, etc.).

2. highly productive; prolific; fruitful (as fertile land or a fertile imagination). **3.** able to supply all essential elements and water for abundant plant growth (as fertile soil).

Fertile Crescent *n.* a crescent-shaped area around the north end of the Arabian Peninsula that joins the Nile Valley on the west to the Tigris and Euphrates valleys on the east and includes the sites of ancient civilizations of both Egypt and Mesopotamia and most of the Bible lands.

fertilizer *n.* (Latin *fertilis*, fruitful + *izare*, making) a source of plant nutrients that can be applied to the soil or to plant leaves to supply one or more essential elements for plant growth. Fertilizers may be either organic (e.g., manure) or mineral (most commercial fertilizers); commercial fertilizers are required to have a label showing the fertilizer grade. In the illustration, the area on the left was unfertilized and produced stunted plants with no grain; the area on the right received 100 lb/ac (112 kg/ha) of nitrogen and produced a good corn crop. See soil amendment.

Unfertilized (left) and fertilized (right) crops

fertilizer grade *n.* the guaranteed amount of nitrogen, phosphorus, and potassium in a commercial fertilizer, shown on the fertilizer label as % total nitrogen, % available phosphorus (expressed as % P_2O_5), and water-soluble potassium (expressed as % K_2O).

fertilizer placement *n.* fertilizer may be broadcast on the soil surface and either left there or tilled into the soil, or it may be banded, either on the surface or injected into the soil. Before planting,

A village-made planter that places seed in one row and fertilizer in another

injected fertilizer such as anhydrous ammonia is often applied in a direction different from the way the crop will be planted so that the rows will not coincide; another form of injected fertilizer known as starter fertilizer is placed 2 in. (5 cm) to the side and 2 in. below the seed by the planter. In the illustration, a village-made planter uses two funnels to place seed and fertilizer in separate rows.

fescue *n.* (Latin *festuca*, stalk) **1.** any of a genus of cool-season grasses (*Festuca* sp., family Poaceae) native to Eurasia and America, grown on about 35 million ac (14 million ha) in the U.S.A. and used for lawn, turf, pasture, and hay, including tall fescue (*F. arundinacea*), hard fescue (*F.longifolia*), sheep or alpine fescue (*F. ovina*), meadow fescue (*F. pratensis*), red or Chewings fescue (*F. rubra*), etc. **2.** a pointer stick or straw used by a teacher.

fetch *v.* (Anglo-Saxon *fetian*, to fetch) **1.** to go after an object and bring it to a person, especially when done by a helper or a dog, as to fetch a tool or to fetch (retrieve) a bird that has been shot. **2.** to attract or bring, as to fetch a good price. **3.** to maneuver, as a boat. *n.* **4.** the unobstructed distance where wind of relatively constant direction and speed can generate waves on a body of water. **5.** the act of bringing or retrieving. **6.** the distance something is brought.

fetid *adj.* (Latin *fetidus*, stinking) having a foul odor; often a result of putrefaction.

fetter *n.* (Anglo-Saxon *feter*, shackle) **1.** a shackle (usually leg bands joined by a short chain) to restrain the feet of an animal or person. **2.** any limitation that prevents or drastically slows an activity. *v.* **3.** to shackle or restrain.

fetus or **foetus** *n.* (Latin *fetus*, offspring) a human or animal (mammal) developing in the womb, especially after it passes the embryonic stage and has an identifiable form and parts (after eight weeks for a human baby).

FFA *n.* See Future Farmers of America.

fiber or **fibre** *n.* (Latin *fibra*, filament) **1.** a slender filament of either natural (cotton, flax, etc.) or synthetic (nylon, rayon, etc.) origin. **2.** a mat or other object or mass composed of filaments. **3.** a strong or reinforcing factor, as moral fiber. **4.** elongated cells in animals or plants, as nerve fibers or vascular bundle fibers. **5.** plant material that adds bulk to a diet because it resists digestion. See dietary fiber.

fiber crop *n.* a crop grown for its fiber; e.g., cotton, flax, hemp, etc., used to make fabric or paper.

fiberglass *n.* any material composed of long, very thin glass filaments, as fiberglass insulation, fiberglass fabric, or fiberglass structural or covering materials composed of glass fibers embedded in resin.

fibrous *adj.* (Latin *fibra*, filament + *osus*, having) **1.** composed of or containing fibers. **2.** resembling fibers; long and slender; e.g., fibrous roots.

Ficus n. See fig.

Ficus religiosa *n.* See bo tree.

field bindweed *n.* a perennial weed (*Convolvulus arvensis*, family Convolvulaceae) native to Europe that is now prevalent through most of the U.S.A. and many other countries, spreading by both seeds and rhizomes. The vines twine around the stems of other plants and are difficult to eradicate. It has leaves up to 2 in. (5 cm) long with spreading basal lobes and white or pink bell-shaped flowers.

field capacity *n.* the water of a soil after downward movement by free drainage has nearly ceased (about 2 or 3 days after a rain); commonly assumed to have a soil water potential of ⅓ atm.

field crop *n.* a crop grown in fields, especially the major food crops of the world. The illustration gives the average annual world production of 10 major field crops.

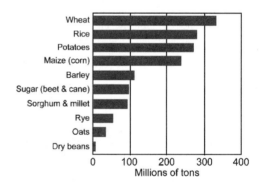

Average annual production of 10 world crops

field work *n.* **1.** work done by a farmer to raise a crop, etc. **2.** work done in the field by a scientist; e.g., archaeological excavation or travel to establish ground truth for interpreting aerial images.

fiery hunter *n.* a large, black ground beetle (*Calosoma calidum*, family Carabidae) that protects many crops and lawns by feeding voraciously on cutworms, caterpillars, and succulent larvae.

fig *n.* (Latin *ficus*, fig) **1.** a tree or shrub (*Ficus* sp., family Moraceae, especially *F. carica*) that grows in hot, dry climates and bears a multiple fruit known as a syconium. **2.** a soft, juicy, pear-shaped multiple fruit produced by a fig tree that may be eaten fresh or dried for storage and later consumption. **3.** a similar fruit produced by certain other species. **4.** an insignificant amount, as not worth a fig. **5.** Barbary fig; a cactus fruit. See Barbary fig.

fig, strangler *n.* See banyan tree.

fig wasp *n.* a tiny wasp (*Blastophaga psenes*, family Agaonidae) that enters the fruit of fig trees via a small opening and pollinates miniature flowers inside the fruit.

filamentous *adj.* (Latin *filum*, thread + *osus*, having) long and slender, like a thread or fiber or a fungal mycelium.

filaria *n.* (Latin *filum*, thread + *aria*, pertaining to) a common name for threadworms (class Nematoda) that cause filariasis; the species *Onchocerca volvulus* causes river blindness (onchocerciasis).

filariasis *n.* (Latin *filum*, thread + *aria*, pertaining to + *asis*, state of) a tropical disease caused by filarial worms (usually carried by mosquitoes) that results in swollen lymphatic glands and tissues, sometimes accompanied by swollen feet and legs; hence, the name elephantiasis for severe cases.

filbert *n.* See hazelnut.

filly *n.* (Icelandic *fylja*, female foal) **1.** an immature female horse, donkey, or zebra (a young male is called a colt). **2.** (slang) a young woman.

filter *n.* (Latin *filtrum*, felt) **1.** a substance or device used to remove solid particles larger than a stated size from a liquid or gas. **2.** a sheet of colored glass, plastic, or gelatin used to alter the color or amount of light that passes through it for artistic effects or photographic purposes. **3.** electronic circuitry that will pass certain radio frequencies but not others. *v.* **4.** to act as a filter. **5.** to pass through slowly, as when slowed by a filter.

filterable virus *n.* See ultravirus.

filter strip *n.* a strip of grass or other dense, permanent vegetation that filters runoff water before it enters a farm pond, lake, stream, diversion terrace, or other site that needs protection from siltation.

filth *n.* (Anglo-Saxon *fylth*, filth) **1.** refuse or other foul material that produces an offensive condition. **2.** vulgar language; obscenity. **3.** writing, speech, theatrics, or other activity that is obscene and offensive.

fiord *n.* See fjord.

fire *n.* (Greek *pyr*, fire) **1.** a process whereby a combustible material unites with oxygen, producing flames, light, and heat, either controlled (as in a stove or furnace) or uncontrolled (as a wildfire or a burning building). **2.** a mass of burning material. **3.** a colorful sparkle of light, as the fire in a diamond. **4.** a severe trial or ordeal, as trial by fire. **5.** expressing passion and enthusiasm, as on fire for a good cause. **6.** bullets from guns. *v.* **7.** to cause something to burn. **8.** to bake, as to fire pottery in a kiln. **9.** to shoot, as to fire a gun. **10.** to discharge from employment, as the boss will fire him.

fire ant *n.* any of several small (1- to 5-mm long) red or yellowish South American ants (*Solenopsis* sp., especially *S. geminata*, *S. xyloni*, or *S. ricteri*, family Formicidae) that are now common in southern U.S.A. and moving northward. Fire ants feed mostly on small insects, seeds, and plants; their bite is not only very painful to people and animals but also has caused some deaths. They build mounds in fields and pastures, some as tall as 3 ft (90 cm).

fire blight *n.* a serious bacterial (*Erwina amylovorus* = *Bacillus amylovorus*, family Bacteriaceae) disease of apple, pear, quince, and many other rosaceous hosts characterized by the rapid invasion of the cambium, causing blackening and death of blossoms, succulent leaves, twigs, branches, or entire limbs of susceptible varieties. Native to North America. Also called apple blight, blossom blight, twig blight, blight canker, body blight.

firebrat *n.* a bristletail insect (order Thysanura) that likes hot places such as near ovens or furnaces. Firebrats breed rapidly at temperatures between 90°F and 102°F (32°C to 39°C). See bristletail.

firebreak *n.* a naturally barren or, more often, an artificially cleared area that serves as a barrier to prevent or retard the spread of fire. Backfires can be started from firebreaks.

firebreak, living *n.* nonflammable green vegetation used as a firebreak; e.g., grass (well fertilized and closely grazed to keep it nonflammable), trees such as *Eucalyptus gmelina* in Australia (with their lower branches pruned by either natural or artificial means), and the semitropical iceplant from Africa that is widely grown in southern California.

firedamp *n.* a coal-mine gas containing an explosive concentration of methane (CH_4). See coal-mine gases.

fire danger *n.* in forestry, the environmental conditions that influence the likelihood of a fire being easily ignited and the likely difficulty of controlling a fire after it has started, including constant factors such as topography and presence of combustible material and variable factors such as humidity and wind.

fire-dependent *adj.* requiring fire for reproduction or long-term survival. Some cones open and release seeds following a fire; some seeds must pass through a fire before they will germinate. Some plants become senescent if they are not exposed to periodic fire.

fire farming *n.* the use of fire to clear patches of land for cropping; practiced mostly in tropical regions. Usually the land is abandoned after a few years and allowed to revert to forest or grass vegetation for a lengthy period to restore its soil fertility and productivity. Also called by many regional names, e.g., milpa or canuca in Latin America, caingin in the Philippines, langland in Indo-China, slash-and-burn or shifting cultivation in most English-speaking countries. See shifting cultivation.

firefighter (fireman in older usage) *n.* **1.** a member of a fire department whose primary work is to extinguish fires. **2.** a forester whose principal function is suppression of fires. Also called smoke chaser.

firefly *n.* flying males of any of about 1000 species of beetles (family Lampyridae or Elateridae) whose lower abdomen glows intermittently at night. The luminescent larvae and wingless females are known as glowworms. The light serves as a mating signal that helps males and females of the same species identify each other. Also called lightning bug. See chemiluminescence.

fire-resistant *adj.* able to resist the effects of fire. Some materials are relatively unaffected by fire; some trees have fire-resistant foliage and/or thick bark that protects the living interior.

fire-tolerant *adj.* able to survive in spite of being burned; e.g., a plant that can regrow from its crown following a fire.

firewood *n.* wood suitable for burning. Many people in developed countries use it primarily in fireplaces for an attractive setting; in developing countries, many people walk for hours to gather wood and carry it home for use in cooking and heating.

Using a donkey to carry firewood

firn *n.* (German *firn*, of last year) consolidated, granular snow that has not yet become solid ice, especially on the surface of a glacier.

firn line *n.* the line on the surface of a glacier between granular snow and solid ice. See snowline.

first bottom *n.* the floodplain area adjacent to a stream; land that is subject to periodic flooding. See bottomland, floodplain.

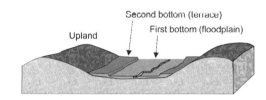
Alluvial deposits in a valley

first draw *n.* the water that flows immediately after a tap is opened. This water is likely to have the highest level of contamination with lead and other heavy metals dissolved from plumbing materials.

firth *n.* (Icelandic *firth*, fjord) an estuary; a long, narrow inlet of the sea.

fish *n.* (Anglo-Saxon *fisc*, fish) **1.** any of a large variety of cold-blooded, finned, vertebrate animals that live in water and obtain oxygen from the water through gills. **2.** the flesh of fish, used as food. **3.** a person who swims well or one who is unfortunate and/or easily baited.

fishery *n.* **1.** a fish hatchery; a place where fish are bred and raised until they can be released in a stream or lake. **2.** a place where fish may be caught. **3.** a fishing industry that catches, processes, and sells fish.

fish hawk *n.* See osprey.

fish ladder *n.* an inclined channel carrying water past a dam at a slow enough velocity or in a series of small steps so that fish (e.g., salmon) can swim upstream to reach their spawning grounds.

fishworm *n.* an angleworm.

fission *n.* (Latin *fissio*, cleaving) **1.** the act of dividing into two or more parts; e.g., bacteria reproduce by fission. **2.** the splitting of the nucleus of a heavy element such as uranium or plutonium into two or more parts, accompanied by the conversion of a small amount of mass into a large amount of energy and radiation. See atomic bomb, fusion.

fissionable material *n.* any nuclides capable of sustaining a neutron-induced fission chain reaction; e.g., uranium-233, uranium-235, plutonium-238, plutonium-239, plutonium-241, neptunium-237, americium-244, and curium-244.

fission fungi *n.* fungi (e.g., certain yeasts) that reproduce only by fission. See yeast.

fissure *n.* (Latin *fissura*, fissure) **1.** a crack or cleavage plane with very little space between the two surfaces. **2.** a natural groove or junction plane in an organ such as a brain. *v.* **3.** to form a fissure by cracking or breaking.

fit *v.* **1.** to be the right size, shape, and/or color, etc. **2.** to alter something so it will be the right size or shape. **3.** to match or correspond, as to fit actions to words or circumstance. **4.** to put together properly; e.g., to fit a piece in a puzzle. **5.** to supply or equip (also called outfit). **6.** to prepare a seedbed for planting. **7.** to prepare livestock for exhibition or sale. **8.** to notch a tree so it will fall properly, and to mark log lengths on a felled tree. **9.** to sharpen and set a saw. *n.* **10.** a match of two things that go together appropriately, as a glove that fits a hand. **11.** the way something fits, as a tight fit. **12.** an epileptic seizure, a sudden convulsion, or other uncontrollable spasm, as a fit of coughing. **13.** a sudden outburst of emotion. **14.** a brief burst of action. *adj.* **15.** appropriate. **16.** prepared. **17.** in good physical condition.

fixation *n.* (Latin *fixatio*, fastening in place) **1.** a locking in place. **2.** an obsession. **3.** a process or condition of repair or adaptation for a specific use. **4.** a photographic process that stabilizes the image on film and prints. **5.** a chemical process that converts a mobile material into an immobile form; e.g., fixation of atmospheric nitrogen (N_2) by bacteria in the soil (*Azotobacter, Clostridium, Rhizobium*, etc.) to form organic compounds. **6.** stabilization of a hazardous waste, usually by absorption or conversion into a solid form.

fixed carbon *n.* carbon converted from atmospheric carbon dioxide into plant biomass by photosynthesis (sometimes used as a measure of the productivity of an ecosystem by determining the rate of carbon fixation per unit area).

fixed groundwater *n.* water held in rocks or soils with interstices so small that it is either immobile or flows too slowly to serve as a source of water for pumping.

fixed nutrient *n.* a plant nutrient that has been added to the soil in an available form but that has been incorporated into either an organic or inorganic structure that holds it too tightly to be readily released back into solution.

fjord or **fiord** *n.* (Norwegian *fjörthr*, fjord) a narrow inlet of the sea bordered by steep slopes; formed by a mountainous coastline sinking relative to sea level, producing a drowned topography, most typically in Norway. See drowned topography, estuary.

flagellum (pl. **flagella**) *n.* (Latin *flagellum*, a whip) **1.** a whip or scourge. **2.** a long, stringlike appendage that either rotates or whips back and forth as a means of locomotion for certain bacteria, protozoa, and sperm cells. The two upper bacteria in the illustration have polar flagellation; the upper bacterium with three flagella at the end is lophotrichous (tufted). The lower bacterium has peritrichous (distributed around) flagellation. See cilia. **3.** the antenna of an insect (excluding the base). **4.** a stolon; a runner shoot.

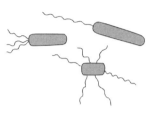

Bacteria with three types of flagellation

flagging *n.* **1.** flagstones, especially when used to make a walkway. *adj.* **2.** drooping, tired, weak. **3.** dwindling.

flaggy *adj.* containing flagstones (used to describe certain soils).

flagstone *n.* (Dutch v*lag*, flag + *steen*, becoming thick) **1.** a flat, relatively thin stone, especially one used to make a walkway. **2.** for soil survey: a piece of sandstone, limestone, slate, or shale 6 to 15 in. (15 to 46 cm) long with a flat, thin shape.

flail *n.* (Latin *flagellum*, whip) **1.** two sticks fastened end to end for threshing grain; one stick serves as a handle, and the other swings freely to thresh the grain. *v.* **2.** to thresh grain by hand with a flail.

flank *n.* (Latin *flancus*, side) **1.** the fleshy area on the side between the ribs and the hip of a person or animal. **2.** a section of beef from the flank area. **3.** the right or left side of a person, animal, group, or thing, especially that of a military formation. **4.** the pressure-bearing surface on the side of each tooth of a gear wheel in a machine. *v.* **5.** to occupy a position at one side, as to either defend or attack the flank of a military group. **6.** to mark the side of something, as stones that flank a pathway.

flatulence *n.* (Latin *flatulentus*, generating gas) **1.** the generation of gas in the digestive tract and its accumulation in the intestines (especially common following consumption of legumes such as beans and peas). **2.** an inflated ego; vanity; pompousness.

flatwoods *n.* any of several types of flat, wooded land in southeastern U.S.A., often designated by the kind of native vegetation as post-oak flatwoods, saw-palmetto flatwoods, etc. Flatwoods usually are either continuously or intermittently wet and are not well suited for agriculture.

flavone *n.* (Latin *flavus*, yellow) any of a group of organic compounds built around a flavonoid ring attached to glycogen. Mostly white to yellow in color, they serve as co-pigments in many flowers.

flavonoid *n.* (Latin *flavus*, yellow + *oides*, like) any of a group of organic compounds structured around a flavonoid ring, including anthocyanins and flavones.

flavor *n.* (Latin *flator*, stench or breath) **1.** a combination of taste, odor, and the way a substance feels in the mouth. **2.** flavoring; a substance added to change the taste of food. **3.** the essence or character of a setting, as the flavor of an event. *v.* **4.** to impart taste or flavor to food by adding spices, sugar, salt, etc.

flaw *n.* (Swedish *flaga*, chip) **1.** a defect or blemish, especially in the surface of an object. **2.** a weakness, as a flaw in one's reasoning. **3.** a sudden windstorm, usually brief, often accompanied by precipitation; a squall. *v.* **4.** to produce a defect or blemish; to damage.

F layer *n.* **1.** a partially decomposed (F for fermented) organic soil horizon, now designated as an O horizon. **2.** the uppermost level of the ionosphere that regularly reflects high-frequency radio waves; commonly subdivided into an F_1 layer composed mostly of oxygen ions at an altitude of about 100 to 150 miles (160 to 240 km) that fades away at night, and a denser, more stable F_2 layer composed mostly of nitrogen ions at an altitude of about 150 to 250 miles (240 to 400 km).

flea *n.* (Anglo-Saxon *fleah*, flea) any of about 1100 species of small, active, wingless, jumping, parasitic, blood-sucking insects (order Siphonaptera) that occur worldwide. Fleas and their wormlike larvae live on the blood of humans and many species of animals and birds, including dogs, cats, rats, and chickens. Various species transmit bubonic plague, typhus, certain tapeworms, etc.

flea beetle *n.* any of a large number of small jumping beetles (family Chrysomelidae) that damage crops by sucking sap from potato vines and tubers, tobacco plants (especially seedlings), corn, and several other plants.

fleawort *n.* See psyllium.

fledge *v.* (Anglo-Saxon *flycge*, fledged) **1.** to grow the feathers needed for flying. **2.** to raise a young bird until it can fly. **3.** to attach feathers, as to fledge an arrow.

fledgling *n.* (Anglo-Saxon *flycge*, fledged + *ling*, small) **1.** a young bird that has recently acquired the feathers necessary for flight. **2.** an inexperienced person attempting a new task.

flesh *n.* (Anglo-Saxon *flaesc*, flesh) **1.** the muscle and other soft tissue between the skin and the bones of a human or animal body. **2.** meat from animals used as food (often excluding that from fish or fowls). **3.** the edible portion of fruits and vegetables. **4.** the physical body of a person, excluding the spirit. **5.** family; kinfolk. **6.** humanity or human nature in general. **7.** all living things. **8.** the skin color of a person, especially the orange-pink color of a Caucasian. *v.* **9.** to feed meat, especially to prepare an animal or bird for hunting by feeding it meat. **10.** to fill out, especially to fatten. **11.** to add detail, as to flesh out a plan or a sketch. **12.** to remove the flesh from a skin or other parts of a carcass.

flesh fly *n.* any of many species of flies (family Sarcophagidae) that lay their eggs or place their larvae in living or dead flesh, dung, or other animal matter. Some are parasites on animals (e.g., *Wohlfahrtia vigil* on small children, rabbits, foxes, dogs, and minks) or other insects (e.g., *Sarcophaga kellyi* deposits its maggots on grasshoppers in flight), and some are scavengers.

flicker *n.* (Anglo-Saxon *flicorian*, flying) **1.** any of several woodpeckers (*Colaptes auratus*, family Picidae) native to North America, especially the common flicker (*C. auratus*), which has a brown back and wings with a yellow or pink underside, a spotted white front, a black bib, and a red crescent behind its neck. **2.** something that comes and goes very quickly, as a flicker of light. *v.* **3.** to move in a fluttering manner, as the flame of a candle flickers. **4.** to lighten and darken quickly, as a video image that flickers.

flint *n.* (Anglo-Saxon *flint*, flint) **1.** an impure form of quartz (SiO_2) that remains as a weathering residue from certain limestones and chalks; usually gray in color. Flint can be broken to produce a sharp edge. **2.** a fragment of flint shaped into a cutting tool or a spearhead or arrowhead. **3.** a fragment of flint that produces sparks when struck with a hammer; used to start a fire by flint and steel. **4.** a small piece of material used to produce sparks in a cigarette lighter, fire starter, or toy. **5.** (loosely) any material that is hard and brittle like flint.

flint corn *n.* a type of Indian corn (*Zea mays*, var. *indurata*) with hard kernels that are not dented at the end.

floating garden *n.* a platform floating on a canal or other body of water and supporting a garden (a common practice in Kashmir and China; also practiced in Mexico City, where it is known as chinampa cultivation).

floc or **floccule** *n.* (Latin *flocculus*, small mass) a small, fluffy clump of fine particles; e.g., an aggregate of smoke particles or of solids formed in sewage waters.

flocculation *n.* (Latin *flocculus*, small mass + *atio*, action) the process of forming floccules in a fluid (especially water or air) by aggregation of suspended particles as a result of either biological or chemical action (e.g., by adding a calcium compound to a soil suspension to make the clay settle). The floccules can be separated from the suspension by filtering or by allowing them to settle to the bottom.

floe *n.* (Norwegian *flo*, layer) ice floating on seawater, especially a large flat mass constituting an ice island. Drifting floes are a hazard to ships that pass by them. See iceberg.

flood *n.* (Anglo-Saxon *flod*, flood) **1.** water covering land that is usually dry; a result of unusually heavy rain and/or rapid snowmelt causing streamflow that exceeds the capacity of the channel. **2.** a rising tide that flows toward the shore; also called flood tide. **3.** a large stream of any material that can flow, as a flood of lava or a flood of light. **4.** a large outpouring, as a flood of tears or a flood of letters. *v.* **5.** to overflow across a floodplain or other flat area; to inundate. **6.** to fill an area, as to flood the road with cars or to flood the yard with light. **7.** to become flooded.

flood control *n.* an effort made to prevent or reduce flooding, e.g., by planting trees to slow runoff, building terraces to hold water in fields, or building levees to hold a stream in its channel.

flood crest *n.* the maximum depth of water flow past a site during a particular runoff event.

floodgate *n.* **1.** a control structure that restricts the passage of water in a stream or canal. **2.** a figurative restraint, especially one that fails, as the floodgates opened and crime was rampant.

flood irrigation *n.* any system of irrigation that covers the entire soil surface with water, including basin irrigation, border irrigation, level bench irrigation, and wild flooding. See irrigation, furrow irrigation, sprinkler irrigation, drip irrigation.

flood magnitude *n.* the size of a flood, usually described as the probable frequency of occurrence for a flood to reach a certain depth and extent at a specified site, e.g., a 100-year flood.

floodplain *n.* a nearly level area adjacent to a stream that is flooded when the stream overflows its banks. Also called first bottom or bottomland. Former floodplains at somewhat higher elevations are called terraces or second bottoms.

flora *n.* (Latin *Flora*, goddess of flowers) **1.** all plant life. **2.** all of the plants that grow in a specified area. **3.** all of the bacteria and fungi growing in a body, body part (e.g., intestine), soil, or other site. See fauna.

floral *adj.* (Latin *Flora*, goddess of flowers + *alis*, pertaining to) having to do with flowers, made of flowers, or like a flower.

florescence *n.* (Latin *florescens*, blossoming) **1.** the condition, process, or period of blooming; production of inflorescence. **2.** a period of growth, progress, and success, as the florescence of art.

floret *n.* (French *florete*, small flower) a small flower, especially one that is part of a larger compound flower.

floribunda *n.* (Latin *floribundus*, flowering freely) **1.** flowering abundantly. **2.** a variety of bush rose (*Rosa* sp.) that bears abundant clusters of large blossoms. **3.** a tall variety of ash (*Fraxinus floribunda*) from the Himalayas.

floriculture *n.* (Latin *Flora*, goddess of flowers + *cultura*, culture) a branch of horticulture dealing with the cultivation of flowering plants as ornamentals and for their blossoms.

floriferous *adj.* (Latin *Flora*, goddess of flowers + *ferous*, bearing) bearing flowers, especially flowering abundantly.

flotation *n.* (Anglo-Saxon *flota*, floater + *atio*, action) **1.** the process or condition of floating an object (e.g., a ship), usually on water. **2.** separation of materials of different densities (e.g., various ores or minerals) by mixing with a liquid of intermediate density so that some will float while others sink. **3.** separation of NaCl particles from KCl particles in sylvinite ore in a saturated solution of both. The mixture is conditioned with aliphatic amine acetate salts that stick to the KCl particles and cause them to adhere to air bubbles that are drawn through the solution to float the KCl so it can be skimmed off the top. **4.** the process of starting a business, issuing bonds, or obtaining a loan.

flour *n.* (French *fleur de farine*, flour of meal) **1.** wheat or other grain finely ground for use in baking and cooking, e.g., for making bread, cake, sauce, etc. **2.** anything that is pulverized to silt size. *v.* **3.** to make grain into flour by grinding. **4.** to coat with flour.

flour beetle *n.* a beetle (especially *Tribolium confusum* and *T. castaneum*, family Tenebrionidae) that infests flour, stored grains, and other food products.

flourish *v.* (Latin *florere*, to bloom) **1.** to thrive; to have vigor; to do well. **2.** to prosper. **3.** to grow luxuriantly. **4.** to behave dramatically, making sweeping gestures. **5.** to embellish writing by adding unnecessary lines or curves to the letters. **6.** to sound a trumpet. **7.** to brandish a weapon, tool, or baton.

flour mite *n.* any of several species of mites (*Acarus = Tyroglyphus* sp., family Acaridae) that infest flour, cereal products, sugar, dried fruits, cheese, etc. Also called cheese mite.

flour moth *n.* any of several small moths, e.g., the Mediterranean flour moth (*Anagasta kuehniella*) or the Angoumois grain moth (*Sitotroga cerealella*, family Gelechiidae) that are endemic in Europe and the U.S.A. They lay eggs in flour, on starchy grains, in breakfast cereals, etc.

flower *n.* (French *fleur*, a flower) **1.** a blossom; a reproductive part of a plant (male, female, or both). The four principal parts of a complete flower are sepals, petals, stamen (male reproductive organ), and pistil (female reproductive organ). **a. pistillate flower:** a blossom with only female reproductive parts. **b. staminate flower:** a blossom with only male reproductive parts. **2.** a plant that produces flowers. **3.** the best part or the best time of anything, as the flower of youth. *v.* **4.** to blossom. **5.** to cover or decorate with flowers. **6.** to mature.

flowering dogwood *n.* a small tree (*Cornus florida*, family Cornaceae) native to eastern U.S.A. (the state tree of Missouri, the state flower of North Carolina, and both the state tree and flower of Virginia). It is grown as an ornamental because it has white flowers in early spring and dark red foliage in late fall. In autumn, many birds eat its scarlet, drupe fruit. Also called American dogwood, boxwood, American cornelian tree.

Flowering dogwood

flowstone *n.* (Anglo-Saxon *flowan*, to flood + *stān*, stone) a calcium carbonate ($CaCO_3$) deposit formed in caves by the evaporation of seepage water flowing over rock surfaces. Related deposits hanging from the ceiling are stalactites, and deposits growing upward from the floor are stalagmites.

flue gas *n.* combustion gases from a furnace vented through a flue. See fly ash.

flue gas desulfurization *n.* a process used to scrub sulfur compounds from flue gas, most commonly by lime scrubbing. (See lime scrubbing.)

fluidized bed combustion *n.* a process designed to burn coal efficiently by first powdering the coal and then injecting it like a fluid into a stream of air so it burns while suspended in the air. Powdered limestone may be added to the mixture to reduce the emission of sulfur compounds.

fluke *n.* (Anglo-Saxon *floc*, flat) **1.** a flat worm (class Trematoda) that is an internal parasite of humans, vertebrate animals, and snails. The mature worms are usually seed-shaped, are found in the blood (*Schistosoma* sp.), or are attached by two suckers to the walls of body cavities in the liver, alimentary canal, or lungs. Discharged eggs produce intermediate stages that live in snails until they are deposited on grasses, where they grow into adult flukes that infect domestic animals and people through the food chain. The most common species is the liver fluke (*Fasciola hepatica*), which can cause a disease fatal to sheep and cattle. **2.** a flatfish (*Paralichthys* sp.); the flounder. **3.** either side of the tail of a whale or dolphin. **4.** the hook of an anchor or the head of an arrow, spear, or harpoon. **5.** an accidental stroke of good luck, such as winning a lottery or making an unlikely shot in billiards.

flume *n.* (Latin *flumen*, stream) **1.** an artificial channel to carry runoff water and/or logs down a slope. **2.** a trough (often elevated, as an aqueduct) to carry irrigation water or water for producing power, mining gold, etc. **3.** a natural channel carrying water through a gorge or other narrow space.

fluorapatite *n.* the fluoride form of apatite [$Ca_5(PO_4)_3F$]. Fluorapatite is harder and less reactive than the chlorapatite and hydroxyapatite forms. The addition of fluorides to drinking water and toothpaste is intended to convert hydroxyapatite in tooth enamel to fluorapatite.

fluorescence *n.* light produced by a phosphor (a substance such as fluorite, CaF_2) when it is acted upon by X rays or ultraviolet light. See phosphorescence.

fluoride *n.* (Latin *fluor*, flowing + *ide*, chemical) a compound that contains fluorine in a gaseous (e.g., HF), solid (e.g., CaF_2), or dissolved form. Small amounts (0.8 to 1.6 parts per million) that either occur naturally or are added to drinking water help reduce tooth decay; excessive amounts can lead to fluorosis (mottled or brown-stained teeth).

fluorine (F) *n.* (Latin *fluor*, flowing + *ine*, element) element 9, atomic weight 18.998; the most reactive element known; a halogen that forms gases (F_2, HF, CF_4, CCl_2F_2, etc.), reacts with almost any metal to form a solid salt, and occurs naturally in the minerals fluorspar (CaF_2), fluorapatite [$Ca_5(PO_4)_3F$], and cryolite (Na_3AlF_6). See chlorofluorocarbon, fluorocarbon.

fluorocarbon *n.* (Latin *fluor*, flowing + *carbon*, charcoal) any of a number of organic compounds analogous to hydrocarbons with one or more hydrogen atoms replaced by fluorine; e.g., Teflon (polymerized CF_2) or freon (CCl_2F_2). See chlorofluorocarbon.

fluvial *adj.* (Latin *fluvius*, river) **1.** pertaining to a river. **2.** produced by, growing in, or occurring in a flowing stream.

flux *n.* (Latin *fluxus*, flowing) **1.** the flow process of any fluid or fluidlike material. **2.** the rate of flow;

e.g., that of a fluid under known conditions, of electrons in a nuclear reactor, etc. **3.** the rate of change in a process, as the rate of conversion of atmospheric carbon dioxide into plant material by photosynthesis. **4.** a state of change. (The world has always been in a state of flux.) **5.** any persistent change, such as a tide. **6.** a substance that facilitates soldering or welding by floating away impurities.

fly *n.* (Anglo-Saxon *flyge*, fly) **1.** an insect with two transparent wings (family Muscidae), especially the housefly. **2.** any insect that flies, including bees, gnats, mosquitoes, moths, etc. Many flies are blood-sucking pests, disease vectors, or plant parasites, but others are valuable scavengers, pollinators, etc. **3.** a fabric flap, as that covering the opening of a tent or the zipper in a pair of pants. **4.** a fishhook disguised as an insect. *v.* **5.** to move through the air, as a bird, an airplane, or a kite. **6.** to move through the air (or space) after being thrown or propelled, as a ball or a spacecraft. **7.** to hang or flutter in the air, as a flag. **8.** to hurry or flee, as to fly away. **9.** to pilot an aircraft or to travel by aircraft. **10.** to move suddenly, as a door flies open. **11.** to change mood suddenly, as to fly into a rage.

fly agaric *n.* a poisonous mushroom (*Amanita muscaria*) with an orange or red cap with white spots; once used as a fly poison. Also called fly amanita or flybane.

fly ash *n.* fine particles that pollute the atmosphere as they are exhausted from a furnace along with combustion gases, especially from burning coal. Also called flue dust.

fly catcher *n.* **1.** any of several small birds (family Tyrannidae) that catch insects in flight. Some species (e.g., the scissor-tailed flycatcher, *Tyrannus forficatus*) have tails longer than their bodies. **2.** several other species of songbirds (family Muscicapidae, or Ptilogonatidae) that also catch insects in flight.

fly-free date *n.* the earliest date when susceptible wheat varieties can be planted without serious risk of infestation by the Hessian fly. See Hessian fly.

flying squirrel *n.* any of several tree squirrels (e.g., *Glaucomys volans* of eastern U.S.A.) with a wide layer of skin on each side of the body connecting the fore and hind legs that permits them to make long, gliding leaps from tree to tree.

fly speck *n.* **1.** a small dark spot composed of dried excrement from a fly. **2.** any tiny dark spot. **3.** a minute flaw. **4.** a trivial detail in an argument. **5.** a disease of apples caused by the imperfect stage of a fungus (*Leptothyrium pomi*, family Leptostromataceae), whose minute fruiting bodies resemble fly specks on the fruit.

flytrap *n.* **1.** a device for trapping flies. **2.** a plant that captures insects. See Venus' flytrap.

foal *n.* (Anglo-Saxon *fola*, young animal) **1.** the unweaned young of the horse, donkey, or zebra; a colt or filly. *v.* **2.** to give birth to a foal.

foam *n.* (Anglo-Saxon *fam*, foam) **1.** a frothy mass of tiny bubbles formed by agitating a liquid, passing a gas through it, or producing gas bubbles by a chemical reaction. **2.** frothy material produced from saliva or sweat, as when an animal is worked to exhaustion. **3.** a thick, frothy substance such as shaving cream, certain fire retardants, material used to make lightweight objects in a mold, etc. **4.** a solidified frothy mass, such as foam rubber. *v.* **5.** to produce or form foam. **6.** to cover with foam or to fill a mold with foam.

foaming agent *n.* a material that promotes foaming of a liquid, especially one that produces desired properties such as thoroughly wetting and covering plant leaves or other foamed surfaces.

foam suppressant *n.* a material that inhibits foaming.

fodder *n.* (Anglo-Saxon *fodor*, coarse food) dry plant material (e.g., hay, straw) used to feed livestock.

foehn *n.* (German *fōhn*, foehn) a wind blowing down from a mountain, especially from the Alps, and becoming warmer and drier in its descent. See chinook.

fog *n.* (Norwegian *fogg*, tall marsh grass) **1.** a cloud of very fine water droplets with its base at ground level. **2.** a cloud of smoke or dust thick enough to impair visibility. **3.** an aerosol spray; e.g., a spray used to apply a pesticide or a foliar fertilizer. **4.** mental confusion; addled thinking. **5.** an overall haze on a photographic negative or print. *v.* **6.** to envelop in a cloud or mist. **7.** to spray with an aerosol. **8.** to obscure an issue; to bewilder or confuse. **9.** to produce a hazy effect on a photographic negative or print; e.g., by exposure to stray light or X rays before development.

fog bank *n.* a fog mass as seen from a distance.

fog belt *n.* an area that is frequently covered by fog, commonly in a low-lying area downwind from a body of water.

fogger *n.* a device that applies pesticides (especially insecticides) in a mist under pressure, resembling a fog.

folacin *n.* See folic acid.

fold *n.* (Anglo-Saxon *fald*, pen or enclosure) **1.** a pen or other enclosure for holding and protecting sheep or other livestock, especially at night. **2.** a flock of sheep either in a pen or being herded. **3.** the members of a church or other group with shared beliefs. **4.** either the crease made by folding or a folded layer of paper, fabric, sheet metal, body tissue, rock strata, etc. *v.* **5.** to bend sharply, forming a crease. **6.** to bend repeatedly into a compact layered form. **7.** to bring close or intertwine, as to fold one's arms or as a bird folds its wings. **8.** to terminate or withdraw, as a business folds or as a card player folds a hand.

foliaceous *adj.* (Latin *foliaceus*, leaflike) **1.** leaflike. **2.** having leaves, as green plants. **3.** having many layers, as layered rocks; e.g., shale.

foliage *n.* (Latin *folium*, a leaf + *aticum*, related to) **1.** green, leafy plant material. See photosynthesis. **2.** ornamentation that represents leaves and associated plant parts; e.g., the bas relief on some buildings.

foliar fertilization *n.* spraying the leaves of plants with a suitable liquid fertilizer, usually to supply a micronutrient, especially one such as iron that may quickly become unavailable if applied to the soil. Note: the amount of material applied must be limited to avoid injuring the plants. See salt index.

foliar nematode *n.* a tiny (almost microscopic) worm (*Aphelenchoides olesistus*, family Anguillulinidae, and related species) that infests the leaves of many greenhouse plants and causes yellow mottles and necrotic (dead) areas that involve part or all of the leaf. Also known as nematodes or eelworms.

foliate *adj.* (Latin *foliatus*, leafy) **1.** like a leaf; in the form of leaves. **2.** having leaves; covered with leaves. Also called foliose.

folic acid *n.* a B vitamin ($C_{19}H_{19}N_7O_6$) found in the leaves of leguminous and other plants, wheat, yeast, and liver meal. Folic acid is essential for hemoglobin formation and for growth of humans and animals and is used medically to treat anemia. Also called pteroylglutamic acid, folacin, or vitamin M.

foliose *adj.* very leafy. See foliate.

folk *n.* (Anglo-Saxon, *folc*, the common people) **1.** people in general or persons of a specified type, e.g., country folk(s) or town folk(s). **2.** one's parents, or all the members of one's family. *adj.* **3.** relating to or originating with the general populace, as folk lore. **4.** traditional forms, often of unknown origin, as folk art.

follicle *n.* (Latin *folliculus*, a small bag or pod) **1.** a small cavity, depression, or sac, especially one containing something that grows, as a hair follicle. **2.** a dry seed pod that splits open along a seam. **3.** a cocoon.

follicle mite *n.* a mite (*Demodex folliculorum* and related species, family Demodicidae) that infests hair

follicles of humans and various other mammals and often causes intense itching and mange.

Folsom *adj.* pertaining to people who lived in the Great Plains of North America at the time of the last ice age (about 11,000 years ago); named after Folsom, New Mexico, where some of their stone projectile points were discovered in the mid-1920s. The prehistoric Folsom culture is characterized by Folsom points made of flint that have an elongated leaf shape with a flute (groove) on each side.

fomite *n.* an inanimate substance other than food that may harbor pathogenic organisms and transmit a disease from one individual to another.

food *n.* (Anglo-Saxon *foda*, food) **1.** any substance that a plant or animal can use as a source of nutrition to meet its needs for growth, body maintenance and repair, and energy, whether in solid, liquid, or even gaseous form. Also known as aliment or nutriment. **2.** any solid or mostly solid source of nutrition (when a distinction is made between food and drink). **3.** a specified type of food, as soft food or cat food. **4.** anything that promotes activity or growth, as a suggestion that provides food for thought.

food additive *n.* a substance added to food to maintain freshness, maintain or increase its nutritive value, make it easier to prepare, or make it more appealing.

Food and Agriculture Organization (FAO) *n.* a unit of the United Nations established in 1945 with headquarters in Rome, Italy; devoted mainly to resolving problems related to food production, distribution, and consumption and to raising living standards among rural people. FAO had 158 member countries in the year 2000 and provided technical assistance and training in all aspects of food and agricultural development, including worldwide collection, analysis, and dissemination of a wide range of data on food, agriculture, and rural development.

Food and Drug Administration (FDA) *n.* an agency in the U.S. Department of Health and Human Services that enforces the provisions of the Federal Drug and Cosmetic Act regulating the quality and safety of foods and drugs.

food-borne disease *n.* any disease spread by contaminated food. The most common food-borne diseases are amoebic dysentery, trichinosis, aflatoxicosis, mushroom poisoning, oyster poisoning, infectious hepatitis, bacillary dysentery, gastroenteritis, *Clostridium perfringens* (gas gangrene), botulism, and salmonellosis.

food chain *n.* a sequence of organisms that serve as food for one after the other from the lowest to the highest member of the chain (e.g., plants, worms, fish, birds, cats). At the base are the autotrophic plants (herbage) and along the way are various heterotrophic organisms (microbes, plants and animals), with various carnivores (including humans) at the top. See food web.

food extender *n.* a less expensive food product that can be added to a more expensive food product to increase the number or size of servings; e.g., a soybean product used to extend the supply of hamburger. See hamburger helper.

food grain *n.* relatively large grass seeds used to produce various cereals, flours, and other food products, especially the seed of wheat, rice, barley, oats, rye, corn, sorghum, and millet.

food intoxication *n.* poisoning in humans or animals (e.g., lambs and sometimes calves) following consumption of food that contains a toxin (often produced by bacteria such as *Clostridium* sp. or *Staphylococcus aureus*) that causes gastrointestinal distress and other problems.

food irradiation *n.* treatment of food or grain with radiation such as low-energy gamma rays (usually from cobalt 60 or cesium 137) to kill microbes, extend storage life, and suppress physiological processes such as ripening and sprouting. Sometimes called cold sterilization.

food poisoning *n.* an acute illness caused by either microbial toxins or infectious bacteria (especially *Salmonella* sp., *Campylobacter* sp., *Listeria* sp., *Clostridium* sp., and *Staphylococcus* sp.) ingested in food, especially gastro-intestinal problems such as diarrhea, but also including nervous system and muscular problems. Also called bromatotoxism or sitotoxism. See botulism.

food web *n.* the interrelated food chains of an ecosystem from the herbage of autotrophic plants through various heterotrophic organisms to the carnivores. See food chain.

fool's gold *n.* iron pyrite (FeS_2); a shiny mineral with a golden color that is often mistaken for gold (but is enough harder to be easily distinguished by a hardness test).

fool's parsley *n.* a foul-smelling, poisonous weed (*Aethusa cynapium*)

foot *n.* (Anglo-Saxon *fot*, foot) **1.** an appendage at the end of a leg (below the ankle) of a person or animal that serves for support when standing or moving. **2.** the end of a bed or grave nearest to the feet of an occupant. **3.** the bottom of a page or the end of a line or list. **4.** the part of a sewing machine that presses down on the upper surface of fabric being sewn. **5.** the basal part of something, as the foot of a chair or a mountain. **6.** the part of a boot, stocking, or other garment that covers a foot. **7.** a linear measurement

equal to 12 in. (30.48 cm), originating from the length of a man's foot.

foot-and-mouth disease *n.* an acute, contagious, viral disease of cattle and other animals with cloven hooves that causes high fever and blisters in the mouth and above the hooves. It can be communicated to other animals and humans. Also called aftosa, aphthous fever, epizootic aphtha, hoof-and-mouth disease, murrain.

foot-candle *n.* a standard measure of illumination; the light intensity of 1 candela (a standard international candle) falling on a surface 1 ft away and facing toward the candle; equal to 1 lumen per ft^2, 10.7639 lux, or 10.7639 lumens per m^2.

foothill *n.* a lower subsidiary hill (usually foothills) below a mountain, a mountain range, or a range of higher hills.

footing *n.* **1.** the part of a building foundation that rests directly on earth or rock. **2.** the principles that provide the basis for founding and running an organization. **3.** a firm support, especially as a place to stand. **4.** the condition of the ground surface as it relates to providing a nonslip contact for the feet of a person or animal. **5.** relationship, as a friendly footing.

foot maggot *n.* the larvae of certain species of blowflies (family Calliphoridae) or of the secondary screwworm (*Callitroga macellaria*) that infest wounds on the feet of sheep and other animals and may also attack humans. See myiasis.

foot mange *n.* a skin disease caused by a mite (*Chorioptes bovis*) that produces lesions in the legs and tails of animals, especially sheep and cattle. Also called aphis foot, chorioptic scab, or symbiotic scab.

foot pad or **footpad** *n.* **1.** the padded surface on the foot of an animal such as a dog or cat. **2.** a robber who travels on foot.

footpath *n.* a pathway intended for pedestrian use only.

foot rot *n.* **1.** a foot inflammation that affects many animals: a bacterial infection with pus formation between the toes, especially when sheep or cattle must stand in wet places. See pododermatitis. **2.** a disease that destroys the cambium and associated tissues at the base of the stem or in the roots of many different plants, ultimately causing loss of turgor and death of many plants. Nonwoody plants are likely to topple over (a condition known as damping-off). Foot rots are caused by several types of fungi (*Phytophthora* sp., *Citrophthora* sp., *Pythium* sp., family Pythiaceae, etc.). Also called root rot.

foots *n.* the sediment that settles from a liquid such as oil or molasses.

footslope *n.* a gently sloping area at the base of a hillside, valley wall, or other steeper slope but above the toeslope and bottomland (if present).

forage *n.* (French *fourage*, forage) **1.** either fresh or dried plant material consumed by livestock or wild animals as pasture, hay, or silage; especially the leaves and stems of grasses and legumes, usually at a growth stage prior to seed production. See roughage. *v.* **2.** to search for something, especially for food. **3.** to feed an animal with forage, as to forage a cow.

forage crop *n.* any crop used as forage, especially immature grasses and legumes.

forage legume *n.* any legume used as forage, especially alfalfa and various clovers.

foraminifer *n.* (Latin *foraminis*, hole) a marine protozoan (order Foraminifera) with a shell perforated by small holes with slender filaments projecting through the holes.

forb *n.* (Greek *phorbē*, food) a broad-leaved, herbaceous plant (excluding grasses, sedges, and plants with woody stems).

foredune *n.* a sand dune produced by onshore winds that pick up sand from the beach along the coastline of an ocean or large lake; often partially stabilized by grass and shrubs that keep the dunes in a narrow band along the coast.

forefoot *n.* **1.** one of the two front feet of a four-legged animal. **2.** the junction of the stem of a boat with the keel or the wooden piece that forms this junction. *v.* **3.** to capture a running animal by roping the front legs, usually from horseback.

Foreign Agricultural Service *n.* an agency of the U.S. Department of Agriculture that promotes the marketing of U.S. agricultural commodities in other countries, works to overcome trade barriers, and educates American producers in how to reach such markets.

foreign matter *n.* impurities that are not a natural part of a product and that detract from its quality; e.g., soil particles in grain.

foremilk *n.* **1.** the first ounce or so (25–50 mL) of milk withdrawn from the udder at the beginning of each milking, in contrast to middle milk and strippings. This milk should be rejected because it is of poor quality and may be high in bacteria. **2.** colostrum.

foreshore *n.* the part of the shore that is alternately covered by water and exposed to air, from the low waterline to the upper limit of wave wash. The foreshore is typically smoothed by the runup and return of waves.

forest *n.* (Latin *forestis*, unenclosed woods) **1.** an extensive area with a plant community dominated by trees and associated underbrush; wooded land. *v.* **2.** to establish a forest by planting trees; forestation.

forestation *n.* (Latin *forestis*, unenclosed woods + *atio*, act of making) establishment of a forest on an area, either afforestation on land that was not previously forested or reforestation of previously forested land.

forest fallow *n.* a regrowth of forest for the purpose of renewing the productivity of land in the slash-and-burn system of farming used in some tropical areas.

forest fire *n.* a fire that burns in forested land as a result of either natural causes (mostly lightning) or human causes. Forest fires have many effects in addition to the obvious destruction of trees, wildlife, and property that may be in their path. They are more destructive of some tree and wildlife species than others (e.g., they often kill more of the hardwood tree species than pines in southern forests). They also improve the reproduction of tree species that require strong sunlight (e.g., Douglas fir and longleaf pine). See burned-over land.

forest fire retardant *n.* a chemical applied to slow the spread of a forest fire; e.g., monoammonium phosphate ($NH_4H_2PO_4$) has proven to be effective in slowing the spread of forest fires and is also a fertilizer that supplies nitrogen and phosphorus to enhance tree growth.

forest growth rate *n.* a measure of the productivity of a forest, commonly expressed as annual growth in cubic feet per acre or cubic meters per hectare. The average forest growth rate in the U.S.A. is about 38 ft^3/ac (2.7 m^3/ha).

forest harvest *n.* removal of trees from an area for use in wood products; either clear-cutting (removing all of the trees) or selective cutting.

forestry *n.* (French *foresterie*, forestry) **1.** the management of timber and other resources on lands designated as forest or timberland. **2.** the science that studies tree growth and forest management.

forestry incentive program *n.* a tax reduction system that encourages planting and caring for trees. It is offered through the U.S. Forest Service and state forestry agencies under the Managed Forest Tax Incentive Program (which was the Managed Forest Tax Rebate Program before 1988).

Forest Service *n.* an agency in the U.S. Department of Agriculture created in 1905 to manage the National Forests and National Grasslands in the public interest as woodlands or grazing lands, as wildlife areas, for recreational use, and for watershed protection; to cooperate with state and private owners for protection and management of forest lands; to conduct research pertaining to forest resources; and to make the results of such research available to the public.

forge *n.* (French *forge*, furnace to heat metal) **1.** a fireplace used by a blacksmith to heat iron or other metal to be shaped, cut, welded, hardened, etc., to make metal items; usually equipped with a bellows or blower to increase the air supply for a hotter fire. **2.** a place where iron ore or pig iron is made into wrought iron.

A blacksmith forge and other blacksmith tools

fork *n.* (Latin *furca*, fork) **1.** a division into branches, as a fork in a tree or in a road, or the location of such a division. **2.** either of the branches where something divides. **3.** a tool with two or more tines or prongs used to hold and lift loose material such as hay or food. *v.* **4.** to divide into two branches. **5.** to hold, penetrate, or lift something with a fork.

formaldehyde *n.* a colorless, pungent, irritating gas (CH_2O) used as a disinfectant (insecticide and fungicide) for seeds, tubers, bulbs, soil, tools, and surfaces and in water solution as a preservative for biological specimens.

formation *n.* (Latin *formatio*, formation) **1.** the process of forming, making, or shaping something. **2.** an education. **3.** a special grouping or arrangement of similar or related items, as airplanes flying in formation. **4.** a geologic layer composed of a specified type of rock.

formulation *n.* (Latin *formula*, small pattern + *tion*, action) **1.** a specific combination of ingredients and form of a material (e.g., a fertilizer or pesticide) to be used for a specified purpose. **2.** the process of preparing a material by following a formula.

forsythia *n.* a shrub (*Forsythia* sp., family Oleaceae) with showy yellow early spring flowers; native to China and southeastern Europe. Named for William Forsyth (1737–1804), an English horticulturalist.

fossil *n.* (Latin *fossilis*, dug up) **1.** hardened (often petrified) remains or traces of plants or animals of former times. The oldest fossil on Earth is estimated to be 2.7 billion years old. **2.** a person who is old, out-of-date, and unlikely to change. *adj.* **3.** deposited or formed in a past geologic time (e.g., coal, petroleum, and connate water).

fossil aquifer *n.* an aquifer that is no longer being recharged because its water source has been cut off; e.g., much of the Ogalala aquifer in central U.S.A..

fossil fuel *n.* coal, petroleum, natural gas, or other combustible material formed from plant and/or animal remains deposited in a past geologic time.

fossilization *n.* (Latin *fossil*, dug up + *izatio*, process) **1.** the process of converting the body of a dead organism into a fossil, especially by replacing organic materials with minerals such as calcite ($CaCO_3$) and opal (SiO_2). See mineralization, petrification. **2.** becoming antiquated and out-of-date; e.g., new methods result in the fossilization of old ways of doing things.

fossil water *n.* water entrapped in a rock layer that has no surface connection for natural replacement of the water. Also called confined water, connate water, native water. See fossil aquifer.

foul *n.* (Anglo-Saxon *ful*, stinking) **1.** a violation of the rules, especially in a sporting event but also in other affairs of life. *adj.* **2.** offensive; repulsive, as a foul odor, foul air, or a foul environment. **3.** muddy or dirty, as a foul road. **4.** bad; stormy, as foul weather. **5.** clogged with foreign matter, as a foul passageway. **6.** profane; disrespectful, as foul language. *v.* **7.** to defile or soil. **8.** to entangle or clog, as to foul a propeller or a conduit. **9.** to violate the rules, as to foul an opponent.

foulbrood *n.* a putrefying disease of honeybee larvae caused by bacteria (*Bacillus alvei*).

foundation herd *n.* the animals used for breeding to produce or renew a herd (e.g., cows, bulls).

foundation seed *n.* the stage after breeder's seed and before registered and certified seed in the production of a controlled (usually new) variety of a crop. See certified seed.

Four-H (4-H) *n.* a national organization to acquaint youth with modern agricultural practices and home economics; sponsored by the U.S. Department of Agriculture. The 4-H organization uses a four-leaf clover as an emblem to symbolize its goal of improving the head, heart, hands, and health of its participants.

fourwing poisonvetch *n.* a poisonous plant (*Astragalus tetrapterus*, family Fabaceae) found in Nevada and southern Utah. Animals (mostly cattle and sheep) that consume the green or dried foliage become nauseated, weak, and shaggy, drag their feet in an irregular gait, lose appetite and flesh, and may die.

fourwing saltbush *n.* a salt-tolerant evergreen shrub (*Atriplex canescens*, family Chenopodiaceae) whose leaves and twigs are grazed by cattle, sheep, and goats. It grows on highly saline soils in arid regions, and its foliage is higher in crude protein, phosphorus, and carotene than that of associated grasses.

fowl *n.* (Anglo-Saxon *fugol*, fowl) **1.** a domestic bird, either live or cooked as food; e.g., a chicken, duck, goose, or turkey. **2.** originally, any bird. **3.** a restricted type of birds, as water fowl or wild fowl.

fowl cholera *n.* an acute bacterial (*Pasteurella multocida*) disease that causes diarrhea, hemorrhaging, listlessness, loss of appetite, increased thirst, rapid respiration, and accumulation of mucus and is often fatal to poultry. Also called chicken cholera, chicken septicemia, or pasteurellosis.

fowl gapeworm *n.* a roundworm (*Syngamus trachea*) that lives in and clogs the trachea (windpipe) of young birds, causing them to cough, sneeze, and gasp for air, often with the beak gaping open. The growing worms and associated mucus can cause death by choking in young chickens, turkeys, other fowls.

fowling *n.* hunting of wild fowl.

fowl leukemia *n.* a fatal infectious disease of the blood-forming organs in chickens and other fowls accompanied by atrophy of the bone marrow; caused by a virus (*Trifur gallinarum*).

fowl plague *n.* a highly acute, contagious disease caused by a virus (*Tortor galli*) that causes fowls to have difficulty breathing. Also called chicken pest or fowl pest.

fowl pox *n.* **1.** a contagious, cancerous, viral disease of birds (especially chickens) characterized by blisters under the skin (especially on the comb and wattles). Also called epithelioma contagiosum or sorehead. The most serious form is a chronic infection that also affects the mucous membranes of the eyes, mouth, and air passages; this form can be fatal and is also called diptheria.

fowl spirochetosis *n.* a disease of chickens, turkeys, ducks, and geese that causes a rapidly fatal blood infection by a pathogen (*Borrelia anserina*). It is spread by the fowl tick. Also called fowl tick or relapsing fever.

fowl tick *n.* **1.** an extremely hardy tick (*Argas persicus*, family Argasidae) that lives on blood and infests chickens and other fowls (especially on skin without many feathers). A vector for certain diseases, it can survive in uninhabited nesting sites or other bird habitats for three years or longer without eating

but is adversely affected by cold winters and wetness. Also called bluebug. **2.** fowl spirochetosis.

fox *n.* (Anglo-Saxon *fox*, fox) **1.** a doglike animal (*Vulpes vulpes* and related species, family Canidae) with a bushy tail and commonly a reddish brown color; known for its craftiness and for its appetite for rabbits and chickens. **2.** the fur pelt of a fox or a coat made from such pelts. **3.** a sly, deceptive person. *v.* **4.** to trick or deceive (especially to outfox an adversary).

fox fire or **foxfire** *n.* **1.** luminescence produced by various fungi growing on rotting wood. **2.** any of the fungi that produce luminescence when growing on rotting wood.

foxglove *n.* (Anglo-Saxon *foxes glofa*, foxglove) **1.** a tall perennial flowering plant (*Digitalis* sp., family Scrophulariaceae) with drooping bell-shaped flowers about 2 in. (5 cm) long on a long spike. Native to Europe, it is especially abundant in rocky areas in England. Foxglove is grown as a garden flower; although poisonous, it usually is not fatal to animals that eat it. **2.** the purple-flowered species (*D. purpurea*) whose leaves are the source of the cardiac drug, digitalis.

foxtail barley *n.* a biennial or perennial weed pest (*Hordeum jubatum*, family Poaceae) in pastures, meadows, roadsides, and waste areas of western U.S.A. The awns are irritants when they get in the eyes or noses or under the tongues of grazing animals.

foxtail millet *n.* an annual grass (*Setaria italica*) grown in the U.S.A., especially in the Great Plains, for supplemental hay, pasture, and green fodder. Its seed is used for wild bird food.

Foxtail millet

fractal *n.* (Latin *fractus*, broken + *alis*, like) a geometric figure or structure that has a similar appearance at any scale and whose physical properties (e.g., length or porosity) depend on the minimum size measured. For example, the length of a shoreline between two points depends on whether the measurement is straight across or follows major indentations, minor indentations, indentations between sand grains, or even finer detail.

fracture *n.* (Latin *fractura*, fracture) **1.** a break, especially a break in a bone or other support structure. **2.** the process of breaking, either literal or figurative, as a fracture in their relationship. **3.** the way something breaks, as glass has conchoidal fracture. *v.* **4.** to break or crack or cause something to break.

fractus *adj.* (Latin *fractus*, broken) scattered; small, ragged-edged cumulus clouds scattered across the sky are called cumulus fractus.

fragile land *n.* land that is easily degraded by humans using common farming, pastoral, or forestry practices and that often is difficult to restore if it has been degraded; e.g., highly erodible land or land with saline-sodic soil.

fragipan *n.* a compact layer at the base of the solum of many soils formed in loamy materials in humid climates, especially in soils with imperfect drainage in areas once covered by glaciers. Fragipans are high in silt, have very low permeability, resist root penetration, and are hard and brittle when dry and still brittle but softer and more difficult to detect when wet. See duripan.

fragmentary *adj.* (Latin *fragmentum*, fragment + *arius*, pertaining to) incomplete; present only as fragments, as fragmentary evidence or fragmentary remnants.

fragrance *n.* (Latin *fragrantia*, fragrance) **1.** the quality of having an odor. **2.** a sweet, pleasing scent. **3.** a perfume or other substance that has a sweet, pleasing scent.

frambesia *n.* See yaws.

francium (Fr) *n.* element 87, atomic weight 223; the heaviest of the alkali metal series (+1 valence). Francium, which is unstable (half-life 22 min), is produced by the decomposition of actinium or made artificially by bombarding thorium with protons.

frankia *n.* a filamentous bacterium or actinomycete (*Frankia* sp.) that forms nodules and fixes atmospheric nitrogen (N_2) symbiotically on the roots of a number of plants (mostly shrubs or trees of temperate climates; e.g., alder, Russian olive, *Purshia*, and *Casuarina*).

frass *n.* (German *frass*, insect damage) **1.** the excrement and wood fragments left by wood-boring insects. **2.** debris and excrement from insect larvae, especially in a water environment.

free drainage *n.* water movement through an unconfined permeable stratum. There may be an impermeable layer at some depth, but the water has an avenue of escape so the water table is sufficiently deep to drain the soil or other material that has free drainage. See aquifer.

free-living *adj.* living apart from other living organisms; neither parasitic nor symbiotic; e.g., *Azotobacter*.

freemartin *n.* a defective female calf born as a twin to a male calf. The defect (usually sterility) results from exposure to the male sex hormone and occurs in about 90% of such twins.

free radical *n.* **1.** a highly reactive atom or molecule with one or more unpaired electrons that are capable of very rapid reactions and of destabilizing other molecules to produce more free radicals in the body of a person or animal. Free radicals are neutralized by antioxidants. **2.** a similar atom or molecule in the atmosphere or other medium; e.g., O formed when O_3 (ozone) is split into O_2 and O. Certain free radicals help produce smog; some help preserve the ozone layer by destroying molecules of chlorofluorocarbons (CFCs). Most accelerate oxidation and combustion; others accelerate polymerization reactions.

free soil *n.* **1.** land that is easy to till (relative to associated land with a higher clay content). **2.** land where slavery is (was) not practiced, especially that part of the U.S.A. that did not have slavery before the Civil War. *adj.* **3.** opposed to slavery, as the Free-Soil Party (1848–1854).

freestone *adj.* easily removable pits; a characteristic of certain varieties of peaches, nectarines, and plums in contrast to other varieties with flesh that clings to the pits.

free water *n.* soil water that is free to move in response to gravity. This water can move downward into drier soil below; it will recharge the water table if it goes below the root zone. Also called drainage water, gravitational water, percolating water.

freeze *v.* (Anglo-Saxon *freosan*, to freeze) **1.** to solidify by loss of heat; especially to form ice from liquid water. **2.** to become very cold or to die from coldness. **3.** to rigidify or lock into a set position; to immobilize. **4.** to break by the expansion of ice, e.g., to freeze a water pipe. **5.** to become very formal or disdainful, as an expression that freezes. **6.** to immobilize by fright. *n.* **7.** a period of time with freezing temperatures. **8.** an action to stop an activity or to maintain prices, production, or other activity at a constant level.

freeze-drying *n.* See drying by sublimation.

freezing *n.* **1.** the act or process that turns a liquid into a solid by withdrawal of heat. *adj.* **2.** at or below the temperature required to solidify the liquid form of a material; e.g., the freezing point of water is 0°C (32°F). **3.** very cold (uncomfortable for a person).

freezing point depression *n.* the decrease in the solid–liquid equilibrium temperature that results from the presence of solutes. For water, the freezing point is depressed 1.86°C (3.35°F) for each mole of solute molecules or ions present in a liter. See colligative property.

freezing rain *n.* precipitation that falls as rain but freezes on contact with the ground or other surface.

frequency *n.* (Latin *frequentia*, multitude or crowd) **1.** common occurrence; happening often. **2.** the number of repetitions of an event in a specified time interval, as 60 Hz (cycles per second) or once per week. **3.** the fraction represented in a population, as 5 per 100.

fresh breeze *n.* **1.** a refreshing breeze. **2.** on the Beaufort scale, a wind whose speed is 19 to 24 mph (31 to 39 km/h). See breeze.

freshen *v.* (Anglo-Saxon *fersc*, fresh + *nian*, make) **1.** to make fresh; to clean or renew. **2.** to make water potable by removing salt. **3.** to start giving milk, as when a dairy animal gives birth and begins producing milk.

fresh gale *n.* on the Beaufort scale, a wind with a speed of 39 to 46 mph (63 to 74 km/h).

freshwater *n.* **1.** potable water; water that can be used for drinking, cooking, bathing, etc. Also called sweet water. **2.** water that has not been contaminated by pollution or excess salts (generally contains less than 0.2% dissolved salts). **3.** water bodies (lakes, ponds, and streams) that have an outlet and are not salty. See brackish water, salinity. *adj.* **4.** living in freshwater, as freshwater fish. **5.** working in or on freshwater, as a freshwater sailor.

friability *n.* (Latin *friabilis*, crumbling) **1.** ease of pulverization. **2.** consistence of moist soil (soil about midway between dry and wet). The terms used to describe the friability of soil are loose, very friable, friable, firm, very firm, and extremely firm. Friable soils crush easily, whereas distinct pressure is required to crush a firm soil.

friable soil *n.* soil that crumbles easily into small aggregates, is easy to work, and forms a good seedbed.

frigid *adj.* (Latin *frigidus*, coldness) **1.** very cold, as a frigid climate. **2.** a descriptive term used in Soil Taxonomy to name cold subgroups (soils that have mean annual temperatures between 0° and 8°C [32° to 46.4°F] and summer soil temperatures above 15°C [59°F]). See cryic, mesic. **3.** unfeeling; lacking warmth and enthusiasm, as a frigid response.

frigid zone *n.* a cold area around either of the Earth's poles. The North Frigid Zone lies between the North Pole and the Arctic Circle at 66° 33´ north latitude; the South Frigid Zone lies between the South Pole and the Antarctic Circle at 66° 33´ south latitude.

fringed tapeworm *n.* a tapeworm parasite (*Thysanosoma actinioides*) that resides in the liver and bile duct of sheep and related wild animals. Livers so infected are considered unfit for human consumption.

fringing reef *n.* a coral reef without a deep lagoon between it and the shore.

frit or **fritt** *n.* (Italian *fritta*, fried) **1.** a partly fused mixture of sand and flux used as a ceramic glaze or enamel. **2.** a partly fused sand mixture used for making glass or porcelain. **3.** powdered glass that contains a micronutrient to be used as a fertilizer.

frog *n.* (Anglo-Saxon *frogga*, a frog) **1.** any of several small, tailless, jumping animals (order Anura) with powerful hind legs, small front legs, and webbed feet. Frogs develop from tadpoles and become amphibians that range from those that live mostly in the water to those that live mostly on land or in trees. **2.** such an animal (family Ranidae) with aquatic habits and smooth, moist skin. Also called true frog or ranid. (Note: a toad is a similar but more terrestrial animal with a drier, warty skin.) **3.** a triangular mass of horny tissue in the middle of the sole of a foot of a horse or related animal. **4.** a device to keep railroad cars on the right tracks at a switch or an intersection. **5.** a small metal holder with many needlelike points to be placed in the bottom of a vase to hold flower stems in place. **6.** a noticeable hoarseness in one's voice, as a frog in the throat. **7.** (disparaging slang) a French person, especially one who likes to eat frogs' legs. **8.** a type of ornamental fastening for the front of a coat. **9.** a sheath that connects a scabbard to a belt.

frogbit or **frog's-bit** *n.* **1.** a floating aquatic plant (*Hydrocarus morsus-ranae*, family Hydrocharitaceae) with roundish, thick, spongy leaves; native to Eurasia. **2.** a similar plant (*Limnobium spongia*, family Hydrocharitaceae) with reddish brown spots on ovate or oblong leaves; native to tropical America.

frond *n.* (Latin *frondis*, bough) **1.** a leaf (usually large and compound) of a fern or palm. **2.** a leaflike part of a lichen or other small plant or animal that is not divided into stem and leaf.

front, climatic *n.* a transition zone between two different air masses, generally marking a change in weather. Basic frontal types are the cold front (when cooler air advances, replacing warmer air), the warm front (when warmer air advances, replacing cooler air), and the stationary front (when neither air mass is advancing appreciably).

frontal precipitation *n.* precipitation resulting from the convergence of two air masses. Generally, the warmer air mass rises over a wedge of the cooler air mass, creating a slow ascent and cooling; it often produces gentle precipitation with a relatively long duration over a large area.

frontogenesis *n.* (Latin *frontis*, forehead + *genesis*, generation) the strengthening of a climatic front with an increase in the temperature difference across the front.

frontolysis *n.* (Latin *frontis*, forehead + *lysis*, loosening) the weakening of a climatic front; a decrease in the temperature difference across the front.

frost *n.* (Anglo-Saxon *frost*, freezing) **1.** a mild freeze; an overnight temperature a little below 0°C (32°F). **2.** frozen condensation in the form of tiny ice crystals forming a white covering on exposed surfaces. **3.** a frozen condition in the soil, as frost down to 2 ft. *v.* **4.** to turn cold (below freezing). **5.** to cover with frost. **6.** to kill sensitive plants such as tomatoes by a killing frost. **7.** to cover a cake with frosting (icing). **8.** to etch or scratch a surface (especially of glass) so it diffuses light. **9.** to make someone angry.

frost action *n.* the effects of frost, especially when repeated many times; e.g., weathering by frost action.

frostbite *n.* or *v.* injury to body tissues by exposure to excessive cold. See chilblain.

frost control *n.* measures taken to reduce frost damage in a garden or orchard, including covering plants, turning on sprinkler irrigation, using fans to mix the air, or using smudge pots or other heaters.

frost-free days *n.* the number of days at a specific location between the average date of the last killing frost in the spring and the average date of the first killing frost in the fall. Commonly known as the growing season.

frost heaving *n.* the production of ice lenses in the soil, elevating the surface; perennial plant roots, building foundations, roads, and sidewalks may be broken; fenceposts may be lifted, etc. The ice lenses form at the frost line by freezing water drawn up by capillary action from wet soil below.

frost line *n.* **1.** the bottom of the frozen zone in soil. **2.** the maximum depth of frozen soil that occurs in a specific location. Water pipes should be buried below the frost line to protect them from freezing, and building footings should be below the frost line to protect them from frost heaving.

frost pocket *n.* a depression or other low area with poor air drainage that permits cold air from adjoining higher elevations to accumulate and produce frost.

frost table *n.* the contact surface between frozen soil below and thawed soil above as seasonally frozen ground thaws in the spring.

frost zone *n.* the soil layer that is subject to annual freezing and thawing. In cold climates with permafrost, it is also known as suprapermafrost or the active layer.

Average annual maximum depth of frost in the U.S.A. in inches.
Note: Mountainous areas are too variable to show detail on a map of this scale.

frozen zone *n.* the frozen part of the soil from either the soil surface or the frost table down to the frost line.

fructification *n.* (Latin *fructificare*, to bear fruit + *atio*, action) **1.** the process of bearing fruit. **2.** spore production by fungi or algae. **3.** the organs that bear fruit or spores.

fructose *n.* (Latin *fructus*, fruit + *ose*, sugar) a ketose sugar with the same formula as glucose and dextrose ($C_6H_{12}O_6$) but with its carbonyl group at the second carbon. It is very sweet and occurs naturally in honey, sweet fruits, corn, tubers of Jerusalem artichoke, etc. When linked to glucose, it forms sucrose. Because it does not require insulin for metabolism, fructose can be digested by persons with diabetes.

frugivore *n.* (Latin *frugis*, fruit + *vore*, eating) an animal that eats mainly fruit; e.g., a monkey.

fruit *n.* (Latin *fructus*, fruit or enjoyment) **1.** the contents of a mature plant ovary, including one or more seeds and accessory tissues. **2.** an edible pulpy plant part developed from a flower, either including or surrounding one or more seeds. **3.** the spores and accessory parts of algae, ferns, fungi, lichens, or mosses. **4.** all useful products of plant growth. **5.** the products of one's effort, as the fruits of honest labor. *v.* **6.** to bear fruit.

fruit fly *n.* **1.** a small fly (*Drosophila melanogaster* and related species, family Drosophilidae) that feeds on fermenting fruit and is often used in genetic studies because it reproduces in only nine days and has large chromosomes that are easy to study. **2.** any of many small black or green flies (family Tephritidae) that deposit their eggs in fruit that later serves as food for their larvae; e.g., the Mediterranean fruit fly.

fruitful *adj.* (Latin *fructus*, fruit + Anglo-Saxon *ful*, full of) productive; bearing fruit, offspring, or results, especially when the produce is abundant or the results are highly profitable.

fruiting body *n.* the spore-producing organ of an alga, fern, fungus, lichen, or moss.

frumentaceous *adj.* (Latin *frumentaceus*, of corn or grain) having to do with or made from a cereal (wheat, barley, oats, etc.) or resembling a cereal.

fruticose *adj.* (Latin *fruticosus*, bushy) branching repeatedly, as certain lichens that form shrublike structures.

FSA *n.* See Farm Service Agency.

fuchsia *n.* **1.** any of several bushy plants (*Fuchsia* sp., family Onagraceae) named after L. Fuchs, a German botanist (1501–1566). Fuchsias produce profuse, hanging, cylindrical blossoms from spring to fall; they are used as ornamentals, especially in hanging baskets. **2.** a purplish red color characteristic of many of the flowers.

fuel *n.* (French *fouail*, fuel) **1.** any combustible substance burned to produce heat or power; e.g., wood, coal, gasoline, diesel oil, natural gas, or cow dung. **2.** something that nourishes; food. **3.** a stimulant for action, ideas, or debate, as fuel for gossip. *v.* **4.** to supply with fuel. **5.** to stimulate a response, as to fuel a rebellion.

Dried cow manure to be used as cooking fuel

fuel cell *n.* a device that produces electricity from chemical energy, e.g., by the oxidation of hydrogen.

fuel oil *n.* a petroleum product (excluding gasoline) used for fuel rather than lubrication; e.g., kerosene, diesel oil, or jet engine fuel. Fuel oil is generally more viscous than gasoline but less viscous than most lubricating oils.

fuel rod *n.* a tube containing enriched uranium, plutonium, or other radioactive material used to generate energy by nuclear fission in an atomic reactor.

fuelwood *n.* wood used or intended for use in cooking, heating, or power production. It may come from trees grown for such use, or it may come from dead trees, trimmings from live trees or brush, or scraps from sawmills. Fuelwood is the principal renewable fuel used by humans.

fugacious *adj.* (Latin *fugacis*, apt to flee) short-lived; fading or falling early; e.g., a flower that lasts only a short time.

fugitive emission *n.* a gas, liquid, or other material that escapes from a factory or other source and becomes a pollutant.

fulgurite *n.* (Latin *fulgur*, lightning + *ite*, rock) glassy rock fused by lightning, especially a tube fused in sand or soil that has been struck by lightning.

full moon *n.* the time in each lunar month when the entire sunlit side of the moon is seen from Earth. Full moon occurs when the moon is on the side of the Earth directly opposite from the sun. Note: a lunar eclipse occurs at the time of a full moon when the sun-Earth-moon vertical alignment is exact. See eclipse, new moon, moonarian.

fulvic acid *n.* the portion of soil humus that can be extracted with an alkaline solution and that remains dissolved after the solution is acidified to pH 2 or below; mostly brown in color. See humic substance.

fumaric acid *n.* an unsaturated organic acid [$C_2H_2(COOH)_2$] produced in the citric acid (Krebs) cycle of respiration in most living organisms. It is derived from plants or synthesized and used to make resins, baking powder, and beverages and to extend the shelf life of powdered milk, nuts, potato chips, and butter.

fumarole *n.* (French *fumerolle*, smoke chamber) a fuming vent in a volcanic area.

A fumarole (Castle Geyser)
in Yellowstone National Park

fume *n.* (Latin *fumus*, smoke, steam) **1.** any smokelike or steamy emission, especially one that is harmful or offensive. **2.** an impatient, angry mood. *v.* **3.** to emit smoke or steamy vapor, as from a fire or a substance that is evaporating. **4.** to show impatience and anger.

fumed oak *n.* oak wood that has been exposed to ammonia (NH_3) fumes to make it darker in color.

fumigation *n.* (Latin *fumigare*, to smoke + *tion*, action) the use of volatile chemicals to control weeds, nematodes, insects, fungi, and bacteria, as in grain bins or soil.

functional element *n.* an element that is useful in life processes and that increases growth when present in appropriate amounts. Functional elements include all the essential elements but are often thought of as the additional useful elements; e.g., for plants: cobalt that is needed for symbiotic nitrogen fixation, sodium that often balances much of the negative charge, silicon that strengthens stems, and vanadium that may also be helpful in nitrogen fixation. Also called beneficial element. See essential element.

fungal or **fungous** *adj.* (Latin *fungosus*, spongy) pertaining to, caused by, or having the nature of fungi.

fungicide *n.* (Latin *fungi*, fungus + *cida*, killer) a pesticide used to control or kill fungi.

Fungi Imperfecti *n.* See deuteromycete.

fungistat *n.* (Latin *fungi*, fungus + Greek *stat*, static) a chemical or agent that inhibits the growth and reproduction of fungi but does not kill them.

fungivore *n.* (Latin *fungi*, fungus + *vorus*, eater) an organism that consumes fungi; e.g., a microarthropod or certain nematodes.

fungus (pl. **fungi**) *n.* (Latin *fungus*, mushroom) a heterotrophic, multinuclear organism that reproduces by spores and usually lacks mobility and has a filamentous structure with visible cell walls (classified now as kingdom Fungi but formerly as division Thallophyta in the plant kingdom). Fungi include molds, mildews, yeasts, mushrooms, puffballs, etc.; some are microscopic but others are easily visible. Fungi grow in soil, on decaying organic matter, or sometimes on or in living plants or animals. Most fungi are decomposers, but some cause plant or animal diseases. See mycorrhiza.

funnel cloud *n.* a funnel-shaped extension of rotating air reaching downward from a cumulonimbus cloud, often associated with a wall cloud. A funnel cloud becomes a tornado if it touches the ground or a waterspout if it touches a large body of water. Also called a tornado cloud, pendant cloud, or tuba.

furrow *n.* (Anglo-Saxon *furh*, furrow) **1.** a small channel cut through the soil by a plow. The soil usually is rolled over to fill the previous furrow. See back furrow, dead furrow. **2.** a rut, track, or crease that has an appearance similar to that of a plow furrow. *v.* **3.** to cut a furrow. **4.** to form wrinkles with furrows between, as to furrow one's brow.

furrow irrigation *n.* a form of surface irrigation that uses the channels between the rows of a cultivated crop to guide the water across the field. Furrows are generally larger than the corrugations that are similarly used for a pasture, hay, or small-grain crop.

furrow slice *n.* the part of the soil that is tilled, especially by a plow; usually the top 6 to 10 in. (15 to 25 cm) of soil. Also called plow layer. See acre-furrow slice.

fusarium *n.* (Latin *fusarium*, spindle location) any of a group of fungi (*Fusarium* sp., family Tuberculariaceae) that cause root rot and wilt in barley, rye, oats, corn, wheat, etc.; stem rot in many ornamental plants; and dermatitis in horses and some other animals. See mycotoxin.

fusion *n.* (Latin *fusio*, pouring out) **1.** the process or product of merging and blending materials, things, or styles. **2.** the process of melting. **3.** the merging of atomic nuclei, forming heavier nuclei and releasing energy; e.g., the combining of deuterium or tritium atoms in a star or in a hydrogen bomb.

Future Farmers of America (FFA) *n.* a youth organization for high school students in vocational agriculture; founded in 1928.

G

gabbro *n.* (Latin *glaber*, smooth) a dark, coarse-grained igneous rock composed mostly of labradorite feldspar and augite (essentially a coarse-grained basalt).

gabion *n.* (Italian *gabbione*, a large cage) a bundle of loose stones held together by wicker, metal, or woven wire that is used for building fortifications (the original use), foundations, or dams; e.g., a woven wire cage placed in a gully and filled with stones to control erosion.

gadfly *n.* **1.** a horsefly or botfly that irritates animals by biting or stinging. **2.** an irritating person who persistently criticizes others.

gadolinium (Gd) *n.* element 64, atomic weight 157.25; a malleable, ductile, highly magnetic, silvery white metal that tarnishes readily in moist air. It is useful in nuclear control rods because it has the highest thermal neutron capture cross section of any known element.

gaggle *n.* **1.** a flock of geese on the ground (not flying). **2.** a noisy, disorderly group of people. *v.* **3.** to honk like a goose.

galactagogue *n.* (Greek *galaktos*, milk + *agōgos*, leading to) something that stimulates or increases the flow of milk.

galaxy *n.* (Latin *galaxia*, galaxy) **1.** a large cluster of stars held together by gravitational attraction and orbiting around a common center; e.g., our solar system is part of the Milky Way Galaxy. **2.** a large gathering of important people or things.

galbanum *n.* (Greek *chalbanē*, galbanum) a bitter-tasting resin with a bad odor obtained from certain tall herbaceous plants (*Ferula* sp., especially *F. balbaniflua*, family Apiaceae) cultivated principally in Iran, Turkey, and Lebanon and used in incense, perfumes, and medicines and for flavoring food products.

gale *n.* (Danish *gal*, furious) **1.** a stong wind, capable of causing property damage. **2.** on the Beaufort wind scale, wind with a velocity of 32 to 63 mph (51 to 101 km/h).

gall *n.* (Latin *galla*, sore or gallnut) **1.** an abnormal swelling on a plant caused by insects, microbes, or chemicals. **2.** nerve; audacity; brazenness. **3.** bile; a bitter yellowish green fluid produced in the liver and stored in the gallbladder. **4.** anything bitter or distasteful. **5.** a sore spot worn by much rubbing, especially on a horse. **6.** a vexation or other persistent irritation or the cause of such irritation. *v.* **7.** to make or become sore by rubbing. **8.** to irritate, annoy, harass, or injure.

gallery *n.* (Latin *galeria*, gallery) **1.** a balcony or other raised area for exhibits or seating in a public building, especially the uppermost of such areas. **2.** the public; common people, especially those who occupy a gallery because those are the cheapest seats. **3.** a long, covered walkway or corridor, especially one open on one or both sides. **4.** an underground passageway in a mine or fortification. **5.** a passageway for insects or other macrobes in the soil, wood, or other material subject to burrowing.

gallfly *n.* any insect (especially *Cynips* sp.) that causes galls on plants by depositing its eggs in plant parts (often in leaves or stems).

gallinaceous *adj.* (Latin *gallina*, hen + *aceus*, like) belonging or having to do with ground-feeding fowls (order Galliformes) such as chickens, turkeys, quail, and pheasants.

gallium (Ga) *n.* (Latin *Gallia*, France + *ium*, element) element 31, atomic weight 69.723; a metal with a valence of +3; similar to aluminum but with a low melting point (29.78°C) and a very long liquid range (up to 2403°C). It expands slightly when it solidifies, is used in transistors and other semiconductors, and makes a brilliant mirror on glass.

gallon *n.* (Latin *galona*, a gallon) a measure of liquid volume. The U.S. gallon is equal to 4 quarts, 8 pints, or 231 in.³ (3.785 L), whereas the British imperial gallon is equal to 277.42 in.³ (4.546 L).

gambrel *n.* (French *gamberel*, legging) **1.** the hock of a horse or other animal. **2.** a stick or rod whose ends are inserted into the two hind hocks of a slaughtered animal to support it for butchering. **3.** a gambrel roof; i.e., a roof with a relatively flat part near the ridge and a steeper part below (like many old barns).

A barn with a gambrel roof

game *n.* (German *gaman*, glee) **1.** an amusing pastime involving competition, effort to accomplish a task according to specified rules, or demonstration of certain skills or knowledge, often including an element of chance. **2.** the equipment and rules needed for such a pastime. **3.** an event when individuals or teams compete in a contest. **4.** any task or occupation that involves competition, as the game of politics. **5.** a trick or joke. **6.** wild animals, birds, or fish that are hunted, caught, or otherwise taken for sport or profit.

game preserve *n.* an area reserved for private hunting or fishing.

game refuge *n.* an area closed to hunting to permit desirable wild animals and birds to increase.

gamete *n.* (Greek *gametē*, wife) a mature egg or sperm capable of combining with its counterpart to begin the formation of a new individual organism.

gametophyte *n.* (Greek *gametē*, wife + *phytos*, plant) a plant in its haploid stage. See sporophyte.

gamma ray *n.* radiant energy similar to an X ray or other electromagnetic wave but higher in frequency (shorter in length); produced by nuclear processes along with alpha and beta particle radiation.

gangue *n.* (German *gang*, going) the rock material surrounding a mineral or precious gem deposit.

gapes *n.* (Icelandic *gapa*, open mouth wide) **1.** a poultry disease causing birds to gasp for breath because their trachea is obstructed by parasitic gapeworms. **2.** uncontrollable yawning; being open-mouthed.

gapeworm *n.* a nematode (*Syngamus trachea*) that infests the trachea of poultry and other birds and causes the gapes.

gar *n.* (Anglo-Saxon *gar*, spear) **1.** any of several large North American freshwater fish (*Lepisosteus* sp.) that prey on other fish. They have a long beak with large teeth, and their bodies are covered with hard ganoid scales. **2.** needlefish (family Belonidae).

garbage *n.* (French *garbeage*, refuse removal) **1.** kitchen refuse; scraps of animal and vegetable matter left over from meals and food preparation (often fed to hogs or chickens, especially in the past). **2.** solid waste; trash of any kind. **3.** a worthless or inferior item. **4.** foolish talk; mistaken ideas. **5.** meaningless output from a computer.

garden *n.* (French *jardin*, garden or yard) a small plot of ground used to grow vegetables or flowers and other ornamental plants, often a place of enjoyment where the plants receive tender loving care.

garbanzo bean *n.* See chickpea.

garlic *n.* (Anglo-Saxon *gar*, spear + *leac*, leek) **1.** an onionlike perennial herb (*Allium sativum*, family Amaryllidaceae) with a strong odor and flavor. **2.** the bulbous base of the plant divided into several cloves (sections) and used as a flavoring for meat and vegetable dishes. Research indicates that eating garlic reduces cholesterol level in humans and reduces cancer and heart disease. **3.** wild onion.

garner *n.* (French *grenier*, garner) **1.** to reap or gather, especially for deposit in a granary or other depository. **2.** to build or acquire gradually, as to garner a reputation. **3.** to collect, especially to hoard a large amount of something. *n.* **4.** a granary or other storage place, or its contents.

garnet *n.* (Latin *granatum*, granular) **1.** any of a group of hard isometric minerals with independent tetrahedra (SiO_4) that are commonly produced in high-grade metamorphic rocks, the most common being almandite ($Fe_3Al_2Si_3O_{12}$). Other forms have Mg, Mn, or Ca substituted for some or all of the first (divalent) cation and/or Fe or Cr substituted for some or all of the second (trivalent) cation. Most forms are a deep red, but some are shades of yellow, orange, brown, green, or black. Garnet crushes to a gritty sand that is used to make sandpaper. Some garnet is used as a semiprecious gemstone. **2.** a deep red or reddish brown.

gas bacillus *n.* bacteria (*Clostridium* sp.) that cause gas to form in infected wounds.

gas chromatography *n.* a method of chemical analysis that uses a gas chromatograph, a column packed with charcoal or fine beads of silica or coated with a suitable gel. A chemical mixture to be analyzed is volatilized and carried through the column by an inert gas. Different chemicals accumulate in separate bands because they move at different rates depending on their attraction to the packing material. Each band can then be analyzed separately. See chromatography.

gasification *n.* conversion of a solid or liquid into a gas, especially the conversion of coal to manufactured gas, a fuel that is cleaner and more easily transported than coal. See methane.

gasohol or **gasahol** *n.* (combination of gasoline and alcohol) a blend of about 90% gasoline and 10% ethyl alcohol used as a fuel for automobiles. The alcohol increases the octane rating and makes the fuel burn cleaner. Also called ethanol.

gasoline, oxygenated *n.* gasoline with an additive that contains oxygen to make it burn cleaner, either ethyl alcohol (see gasohol) or methyl tertiary butyl ether (mtbe). Several cities and states in the U.S.A. require the use of oxygenated gasoline to reduce air pollution.

gastrolith *n.* (Greek *gastēr*, stomach + *lithos*, stone) a calcium carbonate or other concretion in the stomach

gastropod *n.* (Greek *gastēr*, stomach + *pous*, foot) any of many mollusks (class Gastropoda), including slugs, snails, whelks, etc., a small invertebrate that creeps on a single flexible foot. Many gastropods carry a spiral shell composed mostly of calcium carbonate.

GATT *n.* See General Agreement on Tariffs and Trade.

gaur *n.* (Sanskrit *gaura*, gaur) a wild bovine (*Bibos gauros*) in Malaysia that has been hunted both for meat and for its head as a trophy until it is near extinction. It is the largest of all cattle, weighing up to 3300 lb (1500 kg).

Gause's law *n.* an observation published in 1934 stating that two species cannot occupy the same ecological niche at the same time because one will compete better than the other.

gazelle *n.* (Arabic *ghazāl*, gazelle) any of several small, graceful antelopes (especially *Gazella dorcas*) of Africa and Asia. See antelope.

gecko *n.* (Malay *ge'kok*, gccko) any of a large number of small, tropical or subtropical, smooth-skinned lizards (family Gekkonidae) with large heads, stout bodies, and suction pads on their feet that can cling to smooth walls and other surfaces. They eat mostly insects. See tokay.

gee *interj.* **1.** the command for a horse or other draft animal to turn right. See haw, whoa. **2.** an expression of surprise.

Geiger counter *n.* a scientific instrument for measuring radioactivity (named after German physicist Hans Geiger, 1882–1945).

gel *n.* (Latin *gelat*, frozen) **1.** a semisolid material composed of small cells with solid walls surrounding liquid interiors. *v.* **2.** to form a gel. **3.** to become clear or real, as the plan gelled.

gelding *n.* (Icelandic *gelding*, gelding) a castrated male animal, especially a horse but sometimes a male of another species or even a human.

Gelisol *n.* (Latin *gelatus*, frozen + *solum*, soil) a soil order in Soil Taxonomy for soils of cold climates, including soils that either have permafrost within 1 m of the soil surface or have permafrost within 2 m and other cold features (broken horizons, oriented rock fragments, etc.) within 1 m of the surface.

geliturbation *n.* (Latin *gelatus*, frozen + *turbidus*, disturbed + *atio*, action) repeated frost heaving that mixes soil vertically, causes irregular or broken soil horizons, and forms distinctive polygonal surface patterns with oriented stones in cold climates. Also called cryoturbation.

gemma *n.* (Latin *gemma*, a swelling) a budlike growth that detaches and becomes a new organism by asexual reproduction, as in liverworts, certain fungi, and some protozoa.

gemmule *n.* (Latin *gemmula*, a small swelling) a small gemma (bud), especially on certain sponges.

gene *n.* (Greek *gen*, to produce) the basic hereditary unit of all living organisms. It is a segment of DNA in a chromosome with a specific sequence of four purine–pyrimidine bases (A = adenine, G = guanine, C = cytosine, and T = thymine) that serves as a pattern to control the development of a particular kind of protein, tissue, or other heritable trait. See deoxyribonucleic acid, genetic code, genome, jumping gene, mutation, transfer RNA.

gene bank or **germ-plasm bank** *n.* **1.** a place where rare plants or exotic animals are grown or their germ plasm (e.g., seeds or semen) is stored to preserve germ plasm that might otherwise become extinct. **2.** a coordinated group of genes, especially one that has been identified as a unique genomic sequence of a species or strain. **3.** a database of DNA sequences; e.g., the Gene Bank in Heidelberg, Germany, or the U.S. Genbank DNA sequence data bank. Also called a gene library.

gene mapping *n.* determining the relative locations of a group of genes that occur on a particular chromosome. Also called gene sequencing.

gene pool *n.* the total variability of all of the genes represented in a particular population.

General Agreement on Tariffs and Trade (GATT) *n.* a series of agreements dating from the end of World War II to the present intended to reduce tariffs and increase world trade. The GATT was signed by 115 countries in 1993, after 15 years of talks. It includes treating foreign products the same as nationally produced products, treating all trading partners equally except that developing countries are given certain preferences, and sharply reducing worldwide trade barriers such as tariffs. These agreements are now enforced by the World Trade Organization.

generalist *n.* (Latin *generalis*, common or general + *ista*, one who practices) **1.** a person who has knowledge relative to a broad range of topics, in contrast to a specialist who has studied a narrow topic in great depth. **2.** a species that has a broad ecological niche because it is able to tolerate a wide range of environmental conditions.

General Land Office *n.* an agency established in the Treasury Department by the U.S. Congress in 1812 to survey U.S. lands. It was combined with the U.S. Grazing Service in 1946 to form the Bureau of Land Management in the Department of the Interior.

generation *n.* (French *generacion*, generation) **1.** all of the individuals (people or an animal or plant species) that live at about the same time. **2.** the average length of time between the birth of one generation and that of their offspring (about 25 years for humans, one year for annual plants, etc.). **3.** one level in the lineage of an individual, family, or other group (the parents are one generation, the individual and contemporaries another, their children another, etc.). **4.** a similar concept in the stages of technological development, as a new generation of machines. **5.** the process of producing children, products, figures, or ideas.

gene splicing *n.* a technique for introducing genes from a different organism into a chromosome to produce a desired trait in the target organism.

genet *n.* (French *genette*, genet) **1.** any of several small, spotted African carnivores (*Genetta* sp., family Viverridae). **2.** the fur of such animals.

genetic *adj.* (Greek *gen*, to produce + *ikos*, of) **1.** controlled by genes (rather than by environment). **2.** pertaining to genes and/or the science of genetics. **3.** pertaining to natural origin and development. **4.** produced by active natural processes, as genetic soil horizons formed under the influence of the local climate and vegetation.

genetic code *n.* the sequence of nucleotides that provides biochemical control for translating information from genes through transfer RNA into the correct sequence of amino acids for synthesis of a particular peptide or protein.

genetic diversity *n.* **1.** a characteristic of a community of organisms with variations in the alleles of many genes, giving the members of a species a healthy variability that makes the population as a whole more adaptable to changes in ecological conditions and more resistant to diseases. **2.** variability in hereditary material that causes differences between the species in a community and between the individuals of any one species.

genetic drift *n.* a random change in the frequency of occurrence of gene alleles and their associated traits in a population (especially in a small population) over time.

genetic engineering *n.* the development and use of techniques for transferring genetic material from one organism to another. Genes from another source are inserted into existing plant or animal cells to produce progeny with desired characteristics more rapidly than conventional plant breeding and to accomplish transfers that cannot be done by conventional plant breeding methods. Also called biogenetics. See biotechnology, recombinant DNA.

genetic erosion *n.* loss of genetic diversity caused by selection and widespread use of a few species of plants and animals with little variability in their genetics.

genetics *n.* (Greek *gen*, to produce + *ikos*, of) **1.** the biological science that deals with the principles of heredity and explains resemblances and differences among related plants, animals, or people. **2.** the genetic composition of an organism or group.

genetic vulnerability *n.* susceptibility to great damage by disease or pest outbreaks because of widespread use of a few species of plants and animals with little genetic diversity.

genome *n.* (Greek *genos*, origin + *ōma*, mass or group) **1.** a set of genes inherited from one parent; one complete haploid set of the chromosomes in an organism. **2.** the complete genetic code of an individual.

genomics *n.* the complete characterization of the genetic composition of an organism, including the DNA sequence and the location and function of all of its genes.

genotype *n.* (Greek *genos*, origin + *typos*, blow or impression) **1.** the hereditary makeup of an individual, as distinguished from the phenotype (physical appearance). **2.** a specific combination of the alleles of a gene or genes that control one or more traits.

gentian *n.* (Latin *gentiana*, gentian) any of a large family of plants (especially *Gentiana lutea* and related species, family Gentianaceae) with 5-petaled flowers (typically blue, but yellow, white, and red also occur). **2.** a product made from the roots and rhizomes of the yellow gentian for use as a gastrointestinal tonic and an antibacterial and antifungal agent.

genus (pl. **genera**) *n.* (Latin *genus*, birth or kind) a unit of classification containing one or more species of plants, animals, or microbes; a subdivision of a family. See species.

geochemistry *n.* (Greek *geō*, Earth + *chēmia*, alchemy + *istēs*, one who practices + *ry*, study) the study of the elemental composition of the Earth and other celestial bodies, including the distribution of the various elements and their isotopes.

geochronology *n.* (Greek *geo*, Earth + *chronos*, time + *logos*, word) the science of Earth time. Determination of the age of geologic events, both by measurement (e.g., counting growth rings of trees or by radioactive residues) and by relative methods (e.g., fossil studies).

geochronometry *n.* (Greek *geo*, Earth + *chronos*, time + *metron*, measure) the measurement of geologic time; e.g., based on the rate of decay of radioactive elements.

geocorona *n.* (Greek *geo*, Earth + *korōnē*, crown) a shell of ionized hydrogen surrounding the Earth at the outer limit of the atmosphere; similar to the corona around the sun. Note: Geocorona has also been used as a synonym for the Van Allen belt, but this usage is discouraged to avoid confusion.

geode *n.* (Greek *geoidēs*, earthlike) a stone containing a cavity lined with crystals.

geodesic *adj.* (French *géodésique*, geodesic) **1.** relative to the curvature of a surface and its geometry. **2.** the shortest line joining two points on a curved surface; e.g., a great circle route. **3.** the form of a structure composed of a polygonal grid, as a geodesic dome.

geodesy or **geodetics** *n.* (Greek *geōdaisia*, Earth division) **1.** the measurement of the dimensions of the Earth or any large portion thereof, taking the surface curvature into consideration. **2.** the determination of the precise locations of points on the Earth's surface. **3.** the study of the Earth's movements and fluctuations, including measurement of gravity, tides, polar drift, etc.

geodetic coordinate *n.* a specification of location, usually using latitude and longitude, and sometimes including elevation.

geodynamic *adj.* (Greek *geō*, Earth + *dynamikos*, pertaining to energy and action) pertaining to physical processes within the Earth as they affect features at or near the surface.

geodynamics *n.* (Greek *geō*, Earth + *dynamikos*, pertaining to energy and action) **1.** the science that deals with geodynamic topics. **2.** the geodynamic processes.

geogeny *n.* (Greek *geō*, Earth + *genesis*, origin) the study of the origin of the Earth.

geographic information system (GIS) *n.* a computer-based system developed by the USDA Natural Resources Conservation Service to collect, analyze, manage, and display geographic data, especially data relevant to policy decisions on environmental issues.

geography *n.* (Greek *geō*, Earth + *graphia*, drawing) **1.** the study of features that occur on the surface of the Earth, including locations and shapes of land areas, climate, plant and animal life, nations, structures, people, and activities. **2.** the observable features of a specific region or place. **3.** a book (especially a textbook) about geography. **4.** the features of any complex entity, as the geography of the mind.

geologic era *n.* a major period in Earth's geologic history characterized by certain types of rocks and fossils. Four eras are recognized (from most recent to most ancient): the Cenozoic, Mesozoic, Paleozoic, and Precambrian. The four eras are divided into 14 periods, and the two most recent periods are subdivided into seven epochs.

geologic erosion *n.* natural erosion; erosion that occurred or that occurs without human involvement (estimated to average about 1000 lb/ac or 1000 kg/ha annually). Geologic erosion and tectonic shifts shape the land surface, forming valleys, shaping hills, filling low areas, etc. See accelerated erosion, erosion.

geologic time *n.* the sequence of geologic eras, periods, and epochs identified by geologists from the various rock layers and fossils found in the Earth's crust, currently estimated to span about 4.6 billion years.

Geological Survey (USGS) *n.* an agency of the U.S. Department of the Interior that collects environmental information (especially in the form of topographic, geographic, climatic, gravitational, and other maps) to enhance the quality of life, manage natural resources, and protect life and property against natural disasters such as earthquakes.

geology *n.* (Greek *geō*, Earth + *logos*, word) **1.** the science of the origin, history, and present attributes of Earth, especially as related to its rock structures. It is sometimes extended to a similar study of other celestial bodies. **2.** the rock structures and other geologic features of a given area or formation, as the geology of the Rocky Mountains. **3.** a textbook or other written work dealing with geology.

geomere *n.* a landscape area with a repetitive pattern indicating uniformity of material, structure, and function.

geomorphology *n.* (Greek *geō*, Earth + *morphē*, form + *logos*, word) the science that describes Earth's landforms and studies the processes that shape them.

geophagia or **geophagy** *n.* (Greek *geō*, earth + *phagein*, to eat) the eating of clay, chalk, or other earthy matter by people or animals. It is caused by either famine or psychosis. See pica.

geophysics *n.* (Greek *geō*, Earth + *physika*, science of nature) the science that studies the physical structure of the Earth by measuring earthquake waves, gravity, magnetism, heat flow, etc.

geophyte *n.* (Greek *geō*, earth + *phyton*, plant) a plant that propagates in the soil by means of underground buds on bulbs or rhizomes. Also called cryptophyte.

geoponics *n.* (Greek *geōponikos*, pertaining to husbandry) the raising of plants in soil, as distinguished from hydroponics.

geostationary satellite *n.* a satellite that orbits the Earth once each day and is positioned at a stationary location 22,300 mi (35,900 km) above a fixed point on the equator. It is especially useful for telecommunications and for meteorological purposes.

geosynchronous *adj.* (Greek *geō*, Earth + *synchronos*, of time) synchronized with the Earth; e.g., a geostationary satellite has a geosynchronous orbit.

geosyncline *n.* (Greek *geō*, earth + *synklinein*, lean together) a very large, sunken, troughlike area holding thick deposits of sediments whose weight contributes to further sinking. Compression and uplift turn a geosyncline into a folded mountain range.

geothermal energy *n.* heat energy from the Earth's interior, especially from a hot spot that approaches the land surface and produces hot water and steam.

geotropism *n.* (Greek *geō*, Earth + *tropus*, turning) movement or growth in response to gravity. Plant roots exhibit positive geotropism when they grow downward, and stems exhibit negative geotropism when they grow upward.

germanium (Ge) *n.* (Latin *Germania*, Germany + *ium*, element) element 32, atomic weight 72.61; a metal that is chemically similar to silicon. It is a very important semiconductor and is useful in optical glass because it has a very high index of refraction.

germicide *n.* (Latin *germen*, sprout or embryo + *cida*, killer) a substance that kills germs.

germination *n.* (Latin *germinatio*, sprouting forth) the beginning of growth, as a seed germinates by imbibing water, swelling, breaking its seed coat, and sending forth an embryonic stem and roots. See emergence.

germination test *n.* a test to determine the viability of seed. Results are usually expressed as a percentage. Note: All U.S. states have laboratories that test seeds for germination percentage and freedom from noxious weeds, in support of state laws.

germ plasm *n.* the reproductive material of a cell, especially that of a seed or embryo; the chromosomes and surrounding material.

gestation *n.* (Latin *gestatio*, carrying in the womb) **1.** the action or process of carrying young in the womb from conception to delivery. **2.** the development of young while they are in the womb. **3.** the gradual development of an idea or opinion in the mind of an individual. **4.** any slow, progressive development of a concept or plan. Social progress often requires a period of gestation.

ghee *n.* (Hindi *ghi*, butter) the liquid portion of melted, boiled, and strained butter made from cow or water buffalo milk. It stores well without refrigeration.

giant cane *n.* a bamboolike grass (*Arundinaria gigantea*, family Poaceae) native to North America. Its stems grow 12 to 25 ft (3.7 to 7.6 m) tall and produce rough, sharp-edged leaves nearly a foot

(30 cm) long. It grows in canebrakes from Virginia southward and provides fair grazing for cattle.

giant horsetail *n.* a poisonous perennial herb (*Equisetum telmateia*, family Equisetaceae) native to wet meadows in the Pacific Northwest of the U.S.A. Also called ivory horsetail.

giant kelp *n.* a very large seaweed (*Macrocystis* sp., especially *M. pyrifera*) that grows profusely along the California coast and in other warm, temperate waters and serves as food for both sea and land animals. It sometimes grows 200 to 300 ft (60 to 90 m) or more long.

giant wild rye or **giant wildrye** *n.* a coarse perennial bunchgrass (*Elymus condensatus*, family Poaceae) that commonly grows 10 ft (3 m) tall and is readily grazed by cattle and horses while the plants are young. It is most abundant on moist or wet saline soils of western U.S.A.

giardia *n.* flagellate protozoa (*Giardia* sp., especially *G. lamblia* in humans) that live in the intestines of people and animals and cause giardiasis. Named after A. M. Giard, a French biologist, it spreads in sewage sludge and septage, but its cysts can be effectively (99.9%) removed from drinking water by filtration through diatomaceous earth.

giardiasis *n.* an intestinal disease caused by giardia and characterized by intermittent severe diarrhea, abdominal pain, and flatulence.

gibberellin *n.* (Latin *Gibberella*, small hump + *in*, in) any of at least three hormones that stimulate growth in plants and fungi (so named because the first one was isolated from *Gibberella fujikuroi*).

gibbon *n.* (French *gibbon*, gibbon) any of the lesser apes (*Hylobates* sp., family Hylobatidae); a relatively small (up to about 2 ft or 0.6 m tall except for the siamang, up to 3 ft or 0.9 m tall), swift-moving tree dweller with no tail, whose long arms enable it to swing as far as 10 feet (3 m) or leap as far as 30 feet (9 m). Native to southeast Asia, it is covered with a gray, black, or brown coat and noticeable buttock pads. See primate.

giblets *n.* (French *gibelet*, stew of a fowl) organ meat from a fowl, especially the liver, heart, and gizzard.

gid *n.* (Anglo-Saxon *gydig*, insane) a fatal disease that attacks the brain and spinal cord of sheep (and sometimes of goats or, rarely, of cattle or humans), causing the sheep to walk unsteadily. It is caused by the gid bladderworm (*Coenurus cerebralis*), the larval stage of a tapeworm (*Multiceps multiceps*) of dogs. Also known as staggers, blind staggers, turnsickness, silage disease, or cornstalk disease.

gigantism *n.* (Latin *gigantes*, giants + *isma*, condition) **1.** growth of a part or all of the body of a human or animal to an abnormally large size that is believed to be caused either by genetics or by an overactive pituitary gland. Also called giantism or pituitary. **2.** unusually large and luxuriant vegetative plant growth, usually accompanied by delayed flowering and fruiting. Also called gone-to-stalk or gone-to-weed.

gilgai *adj.* undulating; a form of microrelief characteristic of Vertisols, in which low parts occur where soil falls into large cracks when the soil is dry, and high parts rise between the cracks when the soil wets and swells. The elevation difference ranges from a few inches to 3 ft or more (from several centimeters to 1 m or more).

gill *n.* (Danish *gaelle*, gill) **1.** the organ used by fish and other aquatic animals to absorb oxygen from water. **2.** the underside of the cap of certain mushrooms. **3.** a liquid measure in the U.S.A. equal to 4 fluid ounces (half a cup or 118 mL). **4.** (British) a ravine or small stream.

gilt *n.* (Icelandic *gylta*, gilt) **1.** a young female hog (one that has not yet produced a litter). *adj.* **2.** coated with gold or other material that produces a golden color.

gin *n.* (French *engin*, engine) **1.** a machine for removing seeds from cotton. **2.** any of several types of machine used for lifting heavy weights, driving piles, pumping water, etc. **3.** a type of trap or snare to catch wild animals. **4.** a distilled alcoholic beverage made from rye or other grain and flavored with juniper berries or certain other fruit extracts. *v.* **5.** to remove seeds from cotton with a gin. **2.** to snare small wild animals.

ginger *n.* (French *gingivre*, ginger) **1.** a reedlike perennial tropical plant (*Zingiber officinale*, family Zingiberaceae) native to Southeast Asia. **2.** any of several related or similar plants. **3.** a pungent spice made from the rhizomes of the ginger plant for use in cooking and as a medicine. **4.** a yellowish or reddish brown. **5.** liveliness or animation, as that animal (or person) has a lot of ginger.

ginkgo or **gingko** *n.* (Japanese *ginkyo*, gin silver + *kyo*, apricot) a deciduous shade tree (*Ginkgo biloba*, family Ginkgoaceae) with leathery fan-shaped leaves that is native to China. Female trees produce an edible but foul-smelling fruit, so male trees are preferred for ornamental use. They are noted for having high tolerance to air pollution. Also called maidenhair tree and described as a living fossil.

Ginkgo leaf ~3 in. (8 cm) across

ginseng *n.* (Chinese *jēn shēn*, ginseng) **1.** an herb (Asiatic, *Panax ginseng*, or American, *P. quinquefolius*, family Araliaceae) with a thick branched aromatic root **2.** the root of the plant used medicinally, especially as an aphrodisiac.

ginseng family *n.* a family (Araliaceae) of herbs, shrubs, and trees with fragrant leaves and small white or greenish flowers, including ginseng, ivy, and sarsaparilla.

giraffe *n.* (Italian *giraffa*, giraffe) a tall ruminant (*Giraffa camelopardalis*) with very long neck and legs and with dark spots on a tan coat. It is the tallest living animal, reaching heights up to 18 ft (5.4 m).

girasol *n.* (Latin *gyrare*, to turn + *sol*, sun) **1.** Spanish for sunflower. **2.** Jerusalem artichoke. **3.** French for a form of milky opal that is a pale blue.

girdle *n.* (Anglo-Saxon *gyrdel*, girdle) **1.** a band, belt, or anything that surrounds or encircles. **2.** an undergarment worn to shape the hips and waist. **3.** a cut through the bark and cambium layer all around a tree trunk (girdling will kill the tree unless there are branches below the girdle that can nourish the roots). *v.* **4.** to bind with a belt or other band. **5.** to encircle. **6.** to make a circular cut around a tree trunk through the bark and cambium layer.

girth *n.* (Norse *gerth*, girdle) **1.** the distance around something, especially around a body or object that is somewhat circular. **2.** a strap or band that goes around, as the band that goes beneath the body of a horse to hold a saddle in place. *v.* **3.** to go around or encircle. **4.** to hold or fasten with a girth.

GIS *n.* See geographic information system.

glacial *adj.* (Latin *glacialis*, frozen) **1.** pertaining to glaciers or the results of glaciation. **2.** affected by very large ice masses. **3.** very cold conditions. **4.** happening very slowly (like the movement of a glacier). **5.** disdainful and unsympathetic, as a glacial stare.

glacial drift *n.* the assemblage of both sorted and unsorted earthy material transported by a glacier and deposited either by the ice or by meltwater from the glacier; a combination of glacial outwash and glacial till.

glacial meal *n.* See rock flour.

glacial outwash *n.* sorted and stratified earthy material deposited by meltwater flowing away from a glacier, including outwash plains, valley trains, kames, eskers, and deltas. Also called glaciofluvial deposits.

glacial spillway *n.* a valley produced by glacial meltwater.

glacial till *n.* unsorted earthy material deposited either beneath a glacier or left behind as a glacier melts; a mixture of particles ranging from clay size to large boulders. Also called ground moraine or till.

glaciation *n.* (Latin *glaciatus*, made into ice + *atio*, action) the process of glaciers covering and modifying a landscape.

glacier *n.* (Latin *glacies*, ice) a large mass of ice and snow that flows slowly outward from an area of accumulation to an area of attrition where it melts. Glaciers may be classified as either mountain glaciers (sometimes called valley glaciers) or continental glaciers (also called ice sheets). Glaciers need to be about 50 ft (15 m) thick to be able to flow very slowly under their own weight, and continental glaciers may be or have been miles thick. Glaciers now cover about 10% of the Earth's land surface, but as much as 30% was covered during the glacial maxima of the Pleistocene.

glaciology *n.* (Latin *glacies*, ice + *logia*, word) the geological science that studies the nature, action, effects, and distribution of glaciers. Also called glacial geology.

glacis *n.* (French *glacier*, to slip or slide) **1.** a bare level area at the foot of a fortification leading to a steep, plastered wall with a fortified wall above it (remains of such are often present in a tell). An approaching enemy crossing the glacis is exposed to gunfire from the fortification. **2.** a smooth, gently sloping surface at the foot of a hill, mountain, other promontory, or former high place that has been eroded away (similar to a pediment but without implying a specific genesis).

glade *n.* (Anglo-Saxon *glaed*, bright) **1.** an open area in a forest with few or no trees that is commonly caused by either excessive wetness or very shallow soil over bedrock and usually covered with grass but may include areas of bare rock. **2.** an everglade.

gladiolus *n.* (Latin *gladiolus*, small sword) **1.** any of many plants (*Gladiolus* sp., family Iridaceae) native to Africa that have long, blade-shaped leaves and spike flowers and are often grown in flower gardens. **2.** the spike flowers that grow in a variety of bright colors. Sometimes spelled gladiola or called glads.

Gladioluses in bloom. The lower blossoms open first (like the spike on the right).

glanders *n.* (Latin *glandulae*, swollen glands) a disease of horses that is caused by bacillus bacteria (*Malleomyces mallei*) and is communicable to humans. It is marked by a pussy inflammation of mucous membranes and skin nodules that turn to deep ulcers (some may penetrate cartilage and bone). See farcy.

glasshouse *n.* See greenhouse.

glauconite *n.* (Greek *glaukon*, bluish green + *ite*, mineral) an iron-rich, greenish mica. It aggregates so strongly that it behaves physically like sand but chemically like clay.

glean *v.* (Latin *glennare*, to gather) **1.** to gather, usually slowly and laboriously, especially to harvest grain that was missed by the first group of reapers. **2.** to learn or make new discoveries slowly, bit by bit.

glebe *n.* (Latin *gleba*, clod) **1.** soil; a lump of earth or a field. **2.** a tract of land owned by an English parish church.

glei or **gley** *adj.* or *n.* (Serbo-Croatian *glej*, clay) reducing conditions and the accompanying bluish- or greenish-gray colors that occur in saturated soils or soil layers.

gleization or **gleysation** *n.* (Serbo-Croatian *glej*, clay + Latin *atio*, action) soil formation under reducing conditions. A concept of soil formation in wetlands that was used in some older systems of soil classification. Gleization resulted in a structureless bluish or greenish gray horizon. See wetlands.

Glinka, K. D. *n.* an early Russian soil scientist (1867–1927) who was leader of the Dokuchaev school of pedology and author of a compilation in German of Russian soil science, *Die Typen der Bodenbildung* (1914), that C. F. Marbut translated into English (published in 1927) as the background for his system of soil classification.

gloaming *n.* (Anglo-Saxon *glomung*, twilight) dusk, the evening time when daylight is fading.

global positioning system (gps) *n.* a system that uses distance readings from three or more satellites to establish the specific location of its user.

global warming *n.* a gradual increase in the overall average temperature of the Earth as a result of human influence, primarily through the greenhouse effect from a buildup of certain gases (especially carbon dioxide from fossil fuels). The forecast temperature increase is enough to shift climatic

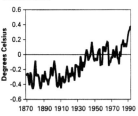

Global surface temperature variations.

Note: Global surface temperature is relative to a 1950-1979 reference period mean.

Average annual global surface temperature 1870–1990

belts and crop production zones of the Earth and to cause glacial ice melting to raise sea level and flood many large cities.

globin *n.* (Latin *globus*, round body + *in*, in) the protein component of hemoglobin; a chain of amino acids that wraps around a heme group and forms a pocket where either oxygen or carbon dioxide can enter and become attached to the iron in the heme group.

glochidium *n.* (Greek *glōchidion*, little arrow) **1.** barbed hairlike growth, as the bristles on certain cacti or the spines on certain ferns. **2.** the parasitic larval stage of certain freshwater mussels (family Unionidae) that attach to the gills of fish.

glossina *n.* the tsetse fly (*Glossina* sp., especially *G. mortisans*), which transmits African sleeping sickness to people and animals in the humid lowlands of Africa. See trypanosomiasis.

glowworm *n.* a luminescent larva or wingless female of any of about 1000 species of beetles whose flying males are known as fireflies.

glucan *n.* (Greek *glykys*, sweet + *an*, belonging to) a polymer of about 2000 to 6000 glucose units chained together through β-1,4 linkages. About 30 to 40 glucan chains held together by hydrogen bonds and overlapping at random form a typical fiber of cellulose.

glucoamylase *n.* (Greek *glykys*, sweet + *amylon*, starch + *ase*, enzyme) an enzyme produced by fungi (especially *Aspergillis* sp.) that breaks down starches and dextrins into glucose and is used industrially to convert corn starch into corn syrup.

glucose *n.* (Greek *glykys*, sweet + *ose*, sugar) the most abundant hexose sugar ($C_6H_{12}O_6$), one of the first products of photosynthesis. Glucose units chain together to form starch and cellulose and combine with fructose to form sucrose; when fermented with yeast, ethyl alcohol (ethanol) can be produced. See ethanol.

glucoside *n.* (Greek *glykys*, sweet + *ide*, chemical) any of a group of organic compounds that yield glucose when hydrolized.

glucosinolate *n.* any of a group of chemicals produced by certain plants (many of *Brassica* sp.) that have potential allelopathic effects for inhibiting seed germination (especially that of monocot grasses).

glume *n.* (Latin *gluma*, husk) a bract, usually one of a pair, that covers a seed and is characteristic of the inflorescence of grasses, sedges, and some other plants. It becomes chaff when the grain is threshed.

glutamate *n.* (Latin *gluten*, glue + *amate*, salt of amic acid) a salt or ester of glutamic acid. It is commonly used as a flavor enhancer, especially monosodium glutamate.

glutamic acid *n.* an amino acid ($C_5H_9NO_4$) that is a major constituent of proteins and used as a flavor enhancer.

gluten *n.* (Latin *gluten*, glue) a protein that is abundant in wheat and present in other grains. Gluten binds flour together when bread is baked and is composed mostly of gliadin (which contributes elasticity) and glutenin (which adds strength).

glyceraldehyde *n.* (Greek *glykys*, sweet + *aldehyde*, dehydrogenated alcohol) a 3-carbon sugar ($C_3H_6O_3$). An early step in photosynthesis produces glyceraldehyde-3-phosphate (glyceraldehyde with a phos-phate group attached to its third carbon) as a precursor to hexose sugars.

glycerol or **glycerin** (Greek *glykeros*, sweet) *n.* a triple alcohol ($C_3H_8O_3$); a 3-carbon chain with one hydroxyl group attached to each carbon, forming an odorless, colorless, syrupy liquid. Glycerol with fatty acids linked to (reacted with) each of its hydroxyl groups constitutes fats and oils (triacylglycerols).

glycogen *n.* (Greek *glykys*, sweet + *gen*, produced) a polysaccharide composed of glucose isomers ($C_6H_{10}O_5$)$_n$ that is stored in human and animal livers and in bacterial tissues as an energy reserve that can be changed into glucose as required by the body. Also called animal starch.

glycophyte *n.* (Greek *glykys*, sweet + *phyt*, plant) a plant that has a low salt tolerance; one that cannot grow in a saline soil. Opposite of halophyte.

glycoside *n.* (Greek *glykys*, sweet + *ide*, chemical) any of many sugar derivatives that can be hydrolyzed to produce a sugar and one or more other organic compounds (e.g., an alcohol).

glyoxysome *n.* (Greek *glykys*, sweet + *oxy*, oxygen + *soma*, body) a specialized vesicle found in the cells of certain plants that contains the enzymes needed for the glyoxylate cycle to convert stored lipids into carbohydrates. It is usually found in seeds that contain significant quantities of lipids.

GMO *n.* a genetically modified organism; a plant or animal with one or more genes transferred from an unrelated species to improve some characteristic or produce a desired product.

gnarled *adj.* **1.** bent, twisted, and/or knotted, as the growth of some trees in harsh habitats. **2.** worn and weatherbeaten.

gnat *n.* (Anglo-Saxon *gnaet*, gnat) a small two-winged insect (*Culex* sp.) that annoys, bites, or stings.

gnatcatcher *n.* any of several small American warblers (*Polioptila* sp., family Muscicapidae) with slender bills, bluish-gray backs, and white fronts that catch insects in flight.

gneiss *n.* (German *gneis*, gneiss) a metamorphic rock with coarse grains and a banded appearance. Its mineralogy is usually dominated by feldspars, quartz, and mica, much like that of an intrusive igneous rock of similar chemical composition.

gnu *n.* (French *gnou*, gnu) a large African antelope that is either silvery gray (*Connochaetes taurinus*) or black (*C. gnou*) with a head and body resembling those of an ox and a mane and tail more like those of a horse. Also called wildebeest. See antelope.

goad *n.* (Anglo-Saxon *gad*, spearhead) **1.** a prod; e.g., a pointed stick or an electric prod used to drive livestock (especially cattle). **2.** any driving impulse or stimulus to action.

goat *n.* (Anglo-Saxon *gat*, goat) **1.** a small, horned, ruminant mammal (*Capra* sp., family Bovidae) known for its agility, for its willingness to eat almost anything (even brush), and for butting other animals or humans with its head. It grows wild in some mountainous areas. Domesticated goats (*C. hircus*) are used to produce milk, meat, hair or wool, and leather. Male goats are bucks and are generally bearded, females are does or nannies, and the young are kids. **2.** a person who is the target of a joke or is blamed for a mistake (a scapegoat). **3.** a lecherous or cantankerous man (usually called an old goat).

goat fever *n.* a disease of goats and sheep caused by eating agave (*Agave lophantha*, family Amaryllidaceae) that causes them to become sensitized to bright light and produces jaundice, swelling of the face and head, and lesions in the kidney and liver.

goatfish *n.* red mullet (family Mullidae). See surmullet.

goat milk *n.* very rich, nutritious milk obtained from female (nanny) goats that is noted for its very small fat globules that make it naturally homogenized and for being readily digested even by some people who are allergic to cow milk. It is used as milk (especially for infants or others who cannot tolerate cow's milk) or to make cheese.

goethite or **göthite** *n.* a yellowish, reddish, or brownish mineral ($Fe_2O_3 \cdot H_2O$) that is a common constituent of rust. It is named after Wolfgang von Goethe (1749–1832), a German poet and dramatist.

goiter *n.* (French *goitre*, goiter) an enlargement of the thyroid gland producing a swelling in the front of the neck and caused by insufficient iodine in the diets of humans or animals. Seafoods and vegetation grown near the seashore are rich in iodine. The use of iodized salt has greatly reduced the incidence of goiter in humans.

gold (Au) *n.* (Anglo-Saxon *gold*, gold) element 79, atomic weight 196.96654; a well-known heavy metal with very low reactivity and a beautiful golden yellow. The most malleable and ductile of all metals, gold occurs naturally as free metal in veins with quartz or pyrite and also in tellurides. It is widely used in coins, jewelry, dentistry, decoration, etc.

golden nematode *n.* a yellowish nematode (*Heterodera rostochiensis*, family Heteroderidae) that is a very serious potato parasite, native to Europe and introduced in the U.S.A. It causes potato sickness that delays sprouting and stunts both the plants and the tubers of Irish potatoes and also attacks other plants of the nightshade family.

goldenseal *n.* an herb (*Hydrastis canadensis*, family Ranunculaceae) native to North America with showy leaves, red fruit, and a rootstock that is poisonous to livestock but has been used to make medicinal tonics for people. Also called yellow root, orange root, or turmeric root.

goldfinch *n.* (Anglo-Saxon *goldfinc*, goldfinch) a small European songbird (*Carduelis carduelis*) with yellow markings on its wings and crimson on its face. **2.** any of several related American songbirds (*Spinus* sp.) whose males have mostly yellow feathers during the summer.

goldfish *n.* a small fish (*Carassius auratus*, family Cyprinidae), usually yellow or orange and 2 to 4 in. (5 to 10 cm) long, that Chinese breeders developed from carp and that is widely used in aquaria and home fish tanks.

gold thread *n.* an herb (*Coptis* sp., family Ranunculaceae) with yellow rhizomes that have been used to make tonics in folk medicine.

Golgi body *n.* an area of endoplasmic reticulum in an animal or eukaryotic plant cell that serves as a site for a biosynthetic process (e.g., combining protein and sugar to form a glycoprotein). It is named after Camillo Golgi (1843–1926), an Italian histologist.

gomuti *n.* (Malay *gumuti*, gomuti) a Malayan palm (*Arenga pinnata* = *A. saccharifera*, family Arecaceae) with feathery fronds and a sweet sap used to make crude sugar and a palm wine (arrack).

gonad *n.* (Greek *gonad*, seed) an organ in animals or humans that produces reproductive cells; an ovary or testis.

gonadotropin *n.* (Greek *gonad*, seed + *tropos*, turning + *in*, pertaining to) any of several hormones produced in the pituitary gland or placenta that stimulate the development of the gonads.

gonidium *n.* (Greek *gonad*, seed + *idion*, little) **1.** an asexual reproductive spore of algae that is produced inside a case rather than free; an endospore. **2.** any of the chlorophyll-bearing algae that occur in lichens.

gonopore *n.* (Greek *gonos*, seed + *poros*, passage) in insects, the external opening through which gametes (eggs or sprem) are released.

goober *n.* (Kimbundu *nguba*, peanut) peanut.

goose (pl. **Geese**) *n.* (Anglo-Saxon *gos*, goose) **1.** a wild or domestic plump, web-footed aquatic bird (*Anser* sp., *Branta* sp., or *Chen* sp., family Anatidae) with a long neck. It is larger than a duck but smaller than a swan. **2.** any of several domestic breeds (*A. domesticus*). **3.** a mature female goose, as distinguished from a male gander or an immature gosling. **4.** goose flesh used as food. **5.** a foolish person, especially one that is the butt of a joke. *v.* **6.** to prod into action, especially by poking a person in the buttocks or by briefly pressing the accelerator of a vehicle.

gooseberry *n.* **1.** a low-growing prickly shrub (*Ribes* sp., family Saxifragaceae) that bears a small sour berry. Gooseberry is an alternate host of white pine blister rust. **2.** the small, sour, greenish berry produced by the plant and used to make pies, jellies, jams, and preserves. **3.** any of several similar shrubs.

goosefoot *adj.* **1.** any of a family of dicotyledonous plants (family Chenopodiaceae), including beets and spinach. *n.* **2.** any of several mostly weedy plants (*Chenopodium* sp., family Chenopodiaceae) native to tropical America, including wormseed goosefoot (*C. ambrosioides*), an herb that contains ascaridole that has been used to kill intestinal worms but is also poisonous to livestock.

gopher *n.* (Amerindian *magofer*, gopher) **1.** a burrowing rodent (*Thomomys* sp., family Geomyidae) with a sturdy body, cheek pouches, and a short bushy tail. Gophers live in colonies, produce extensive tunnel systems that may damage lawns and gardens, and eat earthworms, roots, and stems of grasses. Also called pocket gopher. **2.** a ground squirrel. **3.** a gopher tortoise. **4.** a resident of Minnesota (the state nickname). **5.** a person who runs errands (often gofer).

gopher turtle *n.* a high-domed land tortoise (*Gopherus polyphemus*, family Testudinidae), native to the southern coastal states of the U.S.A., that escapes hot weather by making deep burrows in dry sandy soils.

gorge *n.* (French *gorge*, throat) **1.** a deep, narrow valley with steep, rocky sides; a narrow canyon. **2.** the throat. **3.** a very large meal, or the contents of the stomach. **4.** a large mass that obstructs a passage, as an ice gorge that acts as a dam. **5.** a primitive fishhook made of bone or horn. **6.** a feeling of disgust. *v.* **7.** to feed or eat excessively.

This narrow valley is a gorge.

gorilla *n.* (Greek *gorilla*, gorilla) **1.** the largest of the great apes *Gorilla gorilla*) and a native of equatorial Africa that is typically shy, intelligent, and vegetarian. Males grow to a height of about 6 ft (1.8 m) and a weight of about 500 lb (225 kg). **2.** a large man with brutish strength and actions. **3.** a criminal; a thug.

goshawk *n.* (Anglo-Saxon *goshafoc*, goose hawk) any of several hawks (e.g., *Accipiter gentilis*, family Accipitridae) with short, rounded wings that prey on small animals and birds such as squirrels and grouse and are sometimes used in falconry.

gosling *n.* (Anglo-Saxon *gosling*, little goose) **1.** a young goose. **2.** a foolish person who is young, inexperienced, and overly bold.

gossypol *n.* a yellow phenolic substance ($C_{30}H_{30}O_8$) found in cottonseed that is toxic to most nonruminant animals and humans but is inactivated by heat. Gossypol lowers sperm count and has been used experimentally as a male contraceptive.

gout *n.* (Latin *gutta*, a drop) **1.** a human disease caused by an excess of uric acid ($C_5H_4N_4O_3$) in the blood, forming crystals in small joints that cause pain and swelling, especially in the big toe. **2.** a visceral disease of chickens caused by crystals of sodium urate. **3.** drops of blood, especially when clotted.

GPS *n.* See global positioning system.

graben *n.* (German *graben*, trough or ditch) a long, narrow trough in the Earth's crust with parallel normal faults on its sides and the bottom having fallen relative to the nearby landscape. A corresponding rising structure is called a horst.

gradation *n.* (Latin *gradatio*, a gradual change) **1.** a gradual change, either as a continuum or involving a series of many steps. **2.** a step in such a series. **3.** a geologic process that gradually reduces elevation differences by erosion and deposition, leading toward a level surface.

grade crossing *n.* an intersection between two routes (roads, railroads, or paths) at the same level. Also called a level crossing.**gradient** *n.* (Latin *gradiens*, walking or going) **1.** the rate of ascent or descent of a surface or a stream, road, or other feature, usually expressed in degrees or percentage (units of elevation change per 100 units of horizontal distance). **2.** any rate of change with respect to distance, as a temperature or pressure gradient (in the direction that gives the maximum rate of change).

grafting *n.* (Latin *graphium*, stylus) **1.** the joining of part of two plants by inserting a scion, shoot, or bud from one plant in a cut made in the stem or a branch of another, usually to either produce more than one kind of fruit or color of flower on a single plant or to provide a hardy rootstock for the asexual reproduction of a plant with a high quality of fruit or appearance. **2.** to attach living tissue from one part of the body of a person or animal to another or from one body to another.

grain *n.* (Latin *granum*, seed) **1.** the edible seed of a grass; e.g., corn, wheat, rice, barley, rye, sorghum, oats, millet, and teff. See small grain, cereal. **2.** the crop plants that produce grain, especially the small grains. **3.** any small hard particle, such as a grain of sand or a grain of gunpowder. **4.** a small unit of weight in the avoirdupois system (originally the weight of a grain of wheat), equal to 1/7000 lb (0.0648 g). **5.** a small unit of weight equal to ¼ carat (50 mg), used for pearls and sometimes diamonds. **6.** a tiny amount of anything, as a grain of truth. **7.** the pattern of fibers forming hard and soft lines in a piece of wood or meat, or a similar pattern in some other material (cloth, metal, stone, etc.).

grain alcohol *n.* ethyl alcohol (C_2H_5OH), especially that made from grain.

grain aphid *n.* a tiny insect (*Macrosiphum avenae*, family Aphididae) no larger than ¼ in. (6 mm) long that feeds on the sap of grain crops and transmits several viruses that cause plant diseases. Also called English grain aphid.

grain bin *n.* a structure used to store grain, either a separate building or a portion of a larger building.

grain drier *n.* a structure equipped to reduce the moisture content of grain so it will store without heating or spoiling, usually by circulating air (either heated or unheated) through it.

grain drill *n.* an implement used to plant small grain or other seeds in closely spaced rows at a suitable depth for emergence.

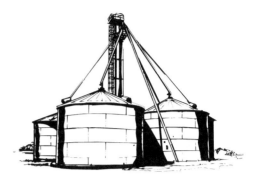

Bins like these often store grain on large farms. Note: Entering a grain bin can be hazardous, especially when grain is being added or removed. In the U.S.A., about 30 lives per year are lost by suffocation of persons buried in grain.

grain dust *n.* dust accumulations from soiled and broken kernels of grain, especially in grain elevators and other grain-handling structures. Such dust is a hazard because it is combustible and often explosive.

grain elevator *n.* a building used for storing grain and equipped with elevators and other machinery to move the grain in and out. Most grain elevators are commercial buildings associated with scales and offices for weighing, buying, and selling grain.

grain fed *adj.* readied for market by feeding with grain rather than pasture or hay, especially cattle but also other livestock.

Grain Inspection, Packers and Stockyards Administration (GIPSA) *n.* an agency formed in the U.S. Department of Agriculture in 1994 by a merger of the Federal Grain Inspection Service and the Packers and Stockyards Program to facilitate agricultural trade and assure the quality of agricultural products imported to or exported from the United States.

grain sorghum *n.* a warm-season crop (*Sorghum bicolor*, family Poaceae) that withstands heat and moisture stress better than corn. It is grown extensively in the Great Plains and southwestern U.S.A. for its grain (used mostly as livestock feed) rather than being used as a forage sorghum or made into syrup.

gram *n.* (Latin *gramma*, a small weight) **1.** (**gramme** in French; abbr. **g** or **gm**) the basic unit of weight in the metric system; the weight of 1 cm^3 of distilled water at 4°C (15.432 grains). **2.** a legume used as fodder (especially in India), especially the chickpea (*Cicer arietinum*) or mung bean (*Vigna radiata*).

grama or **grama grass** *n.* any of several grasses (*Bouteloua* sp., family Poaceae), native to America, that grow extensively in the Great Plains and western states and are warm-season producers of high-quality forage on range and pasture land. Two of the best-known species are blue grama (*B. gracilis*) and side-oats grama (*B. curtipendula*).

Gramineae *n*. (Latin *gramineus*, of grass) the older name for plants of family Poaceae.

Gram-negative (Gram–) bacteria *n*. bacteria with cell walls that resist absorption of gentian violet dye because the peptidoglycan in the cell wall is covered by another membrane. Gentian violet is decolorized by acetone or alcohol treatment. Counterstaining with safranine gives them a pink appearance. Examples are *Escherichia coli*, *Pseudomonas aeruginosa*, *Salmonella*, *Vibrio*, and *Pasteurella*.

Gram-positive (Gram +) bacteria *n*. bacteria with cell walls composed of a thick layer of peptidoglycan that retains a gentian violet dye even when rinsed with alcohol or acetone. Examples are *Staphylococci*, *Streptococci*, and *Bacillus anthracis*.

Gram stain *n*. a staining method devised in 1884 by a Danish physician, Hans C. J. Gram, to aid in the identification of bacteria by classifying them as Gram positive or Gram negative. Also called Gram's method.

granary *n*. (Latin *granarium*, granary) **1.** a storage building for threshed grain. **2.** a region that produces large amounts of grain.

Grange *n*. See Patrons of Husbandry.

granite *n*. (Italian *granito*, grainy) **1.** a coarse-grained igneous rock composed mostly of feldspars, some quartz, and smaller amounts of mica and dark minerals such as hornblende or augite. **2.** something that is hard and durable.

granoculture *n*. (Latin *granum*, grain + *cultura*, cultivation) agriculture based on the cultivation of grains, especially in the early evolution of agriculture. Three centers of origin of granoculture are recognized: the fertile crescent from Israel through Turkey and Iraq to Iran, an area in northwestern China, and an area in the southern portion of the Mexican Plateau. See vegeculture.

granular *adj*. (Latin *granulum*, a grain) **1.** grainy; composed of grains, containing grains, or having a grainy appearance. Sometimes called granulose or granulous. **2.** a type of soil structure composed of small soil aggregates with roughly spherical shape, classified by diameter as fine (1 to 2 mm), medium (2 to 5 mm), or coarse (5 to 10 mm).

granular pesticide *n*. a formulation of a pesticide applied to granules of an inert carrier such as clay or ground corn cobs.

granulometric *adj*. (Latin *granulum*, a grain + *metricus*, metric) pertaining to particle size.

granulosis *n*. (Latin *granulum*, a grain + Greek *osis*, condition) a disease of insects caused by a virus (*Bergoldia* sp.) and characterized by minute granules in infected cells.

granville wilt *n*. a bacterial disease (*Xanthomonas solancearum*) that attacks tobacco, peanuts, potatoes, tomatoes, eggplant, peppers, several common weeds, etc., causing plant roots to decay and the plants to wilt and die. Stalks show dark brown threadlike streaks in the woody tissue. First observed in North Carolina on tobacco about 1900, it destroyed 20% to 50% of the crop during its peak years from 1920 to 1940. Losses have been reduced by using crop rotation and resistant varieties.

grape *n*. (French *grape*, a cluster of grapes) **1.** a small round berry with smooth skin and typically purple, red, or green color that grows in clusters on a woody vine. It can be eaten raw, eaten as dried raisins, or made into jelly, juice, or wine. **2.** a climbing woody vine (*Vitis* sp., family Vitaceae) that bears clusters of grapes. **3.** a dark purple.

grapefruit *n*. **1.** a large citrus fruit with a yellow rind and an edible but somewhat sour yellow or pink pulpy flesh, so named because it grows in clusters. It is often eaten raw for breakfast or made into juice. **2.** the tropical tree (*Citrus paradisi*, family Rutaceae) that bears the fruit.

graphite *n*. (Greek *graphein*, to write or draw) a form of carbon crystallized in the hexagonal system that cleaves readily into very thin sheets. Dark gray with a soft, smooth feel, it is a good electrical conductor, is highly resistant to heat, and is used to make pencils, electrodes, lubricants, crucibles, and moderator rods for nuclear reactors.

GRAS *n*. generally recognized as safe; a political/scientific acronym used by the U.S. Food and Drug Administration for substances not known to constitute a health hazard.

grass *n*. (Anglo-Saxon *graes*, grass) **1.** a mono-cotyledonous plant that has a hollow stem with solid joints and a leaf at each joint consisting of a sheath that wraps around the stem and holds an elongated blade with parallel venation. Grass flowers are minute, are borne on tiny branchlets, are usually crowded together, and are wind pollinated. The grass family (Poaceae) includes grain crops (corn, wheat, barley, etc.), various forage and lawn grasses, sugarcane, bamboo, various weeds, etc. **2.** grass plants collectively, or even all nonwoody plants collectively. **3.** land that is covered with grass (often grassland). **4.** (slang) marijuana. *v*. **5.** to plant or produce grass (usually for forage) on an area. **6.** to put livestock in a grassy area for forage.

grasscutter *n.* a field rodent native to much of Africa whose flesh is considered a delicacy and is sold in the meat markets of several nations.

grassed waterway *n.* a drainage channel stabilized by environmentally adapted, erosion-resistant grasses, including natural low areas, terrace outlets, irrigation water outlets, road ditches, etc. Also called vegetated waterway. Some popular grasses for grassed waterways are Bermuda grass, Italian ryegrass, Kentucky bluegrass, redtop, reed canarygrass, smooth bromegrass, tall fescue, and western wheatgrass.

grass egg *n.* an egg with an olive-colored yolk; usually regarded as low quality or not acceptable in the market.

grass fattened *adj.* refers usually to cattle fattened on grass forage in contrast to those fed grain or high-protein feeds. Meat from a grass-fattened animal usually contains less fat than that from a grain-fed animal.

grasshopper *n.* (Anglo-Saxon *gaershoppe*, grass-hopper) any of about 5000 species of jumping insects (family Acridadae) that often occur in droves that seriously damage or destroy tender vegetation. See locust.

grassland *n.* (Anglo-Saxon *graes*, grass + *land*, land) **1.** an area of natural prairie vegetated mostly with perennial grasses and associated forbs, characteristic of subhumid, semiarid, and some arid regions before human intervention. **2.** an area seeded to grasses and/or legumes for forage production.

grass-seed nematode *n.* a nematode (*Anguina agrostis*) that produces a purple-to-black gall on seeds of Chewings fescue and bentgrasses grown in the Pacific Northwest. The galls are poisonous to livestock.

grass tetany *n.* a disease of cattle or sheep caused by a deficiency of magnesium or of magnesium and calcium in the forage and in the blood. It can be controlled by adding magnesium to the soil as indicated by a soil test. Also called grass staggers or hypomagnesemia.

graupel *n.* (German *graupein*, to sleet) a porous ice mass between 2 and 5 mm in diameter falling as precipitation; a soft hail or snow pellet that has a random shape rather than that of a snowflake. See sleet, hail.

gravel envelope *n.* a layer (envelope) of gravel or crushed stone placed around a tile line to facilitate entrance of seepage water into a tile line or seeping of effluent from a septic drain line into the soil.

gravitational constant (g) *n.* the rate of acceleration caused by gravity; equal to 32 ft/sec^2 (980 cm/sec^2) under frictionless conditions at the Earth's surface.

gravitational water *n.* soil water that temporarily occupies aeration pore space as it moves downward into drier soil below, generally considered to be water that is held by a soil water potential of less than ⅓ atm. Note: Gravity is a factor, but the principal driving force is actually the high soil moisture tension of the underlying dry soil.

gravity *n.* (Latin *gravitas*, weight) **1.** the force of attraction that exists between objects because of their masses. **2.** heaviness, as center of gravity. **3.** density, as specific gravity. **4.** seriousness or solemnity, as the gravity of a situation.

gravity feed *n.* movement under the influence of gravity alone (without any pump or auger), as a fuel system using gravity feed or the gravity feed of grain out of a bin.

gray (Gy) *n.* the unit for absorbed dose of radiation in the SI system; the amount of ionizing radiation absorbed when a body is exposed to a dose of 1 joule of energy per kilogram of body mass. It is named for Louis Harold Gray (1905–1965), an English radiobiologist.

grayling *n.* (Anglo-Saxon *graeg*, gray + *ling*, small or young) **1.** a game fish (*Thymallus* sp.) about 16 to 18 in. (40 to 45 cm) long that resembles a trout with a large, brightly colored dorsal fin and that is typically found in clear, rapid streams. **2.** any of several butterflies (family Satyridae) with spotted gray or brown wings.

gray speck *n.* a disease of oats caused by manganese deficiency. First reported in the Netherlands in 1909, it causes chlorotic and grayish brown spots on leaf blades (usually appearing first on the basal half of the middle leaf), stunts the plants, and reduces yield. It can be corrected by applying manganese sulfate or another manganese source to either the soil or the plant foliage.

graywacke *n.* (German *grauwacke*, gray stone) a rock similar to sandstone with an unusually wide range of particle sizes from clay size to pebbles. Also called lithic sandstone or wacke.

gray water *n.* wastewater from kitchen sinks and bathtubs, in contrast to black water wastes from toilets. See black water.

graze *v.* (Anglo-Saxon *grasian*, to eat grass) **1.** to eat forage from a pasture, as sheep graze in the meadow. **2.** to put livestock in a pasture for forage. **3.** to touch marginally, with or without minor injury.

grazer *n.* (Anglo-Saxon *grasian*, to eat grass + *er*, one that) an animal that grazes, such as a cow or sheep in a pasture.

grazier *n.* one who manages livestock on grazing land.

grazing land *n.* land used for pasturing domestic livestock, usually without reseeding or fertilization, including rangeland, rain-fed pastures, and irrigated pastures. About 22.4% of the Earth's land surface is used as grazing land, roughly twice as much as the area used as cropland. It is mostly land that is too arid or stony to be used for cropland or intensive pasture.

grazing method *n.* a management practice chosen by a grazing manager; e.g., alternate or rotational grazing whereby two or more pastures are grazed in succession, continuous grazing whereby a pasture is grazed throughout the season, or deferred grazing whereby the forage is permitted to grow for a time with no grazing.

greasewood *n.* a woody shrub (*Sarcobatus vermiculatus*, family Chenopodiaceae) that grows in alkaline soils in arid areas of western U.S.A.

Great Basin *n.* an area of about 210,000 mi^2 (544,000 km^2) between the Sierra Nevada and the Wasatch Mountains of western U.S.A. (most of Nevada plus parts of adjoining states) that has no outlet to the sea. The climate is mostly arid and semi-arid. The streams of the Great Basin flow into lakes (e.g., Great Salt Lake), and the water evaporates.

great circle *n.* any line representing the intersection of the surface of a sphere and a plane passing through the center of the sphere. Long-distance plane flights prefer a great circle route because it is the shortest distance across the surface between any two points on the surface.

Great Lakes *n.* the five large freshwater lakes that empty into the Saint Lawrence River: Lake Michigan in northern U.S.A. and Lakes Superior, Huron, Erie, and Ontario between the U.S.A. and Canada.

Great Plains *n.* the semiarid region of western U.S.A. and Canada east of the Rocky Mountains.

Great Plains Conservation Program *n.* a special program established in 1956 for the semiarid region extending from the Canadian border of North Dakota to Texas and from eastern Kansas to the Rocky Mountain foothills. It is administered by the Soil Conservation Service to conserve soil and water in this drought-prone region. This program became part of the Environmental Quality Incentives Program (EQIP) in 1996.

Great Salt Lake *n.* the largest saltwater lake in the U.S.A., ranging in size from 1000 to 2400 mi^2 (2600 to 6200 km^2), with a maximum depth of about 60 ft (18 m). Located in northern Utah with no outlet, it has become much more saline than seawater. It is bordered by a large area of salt flats that was once covered by water.

green *n.* (Anglo-Saxon *grene*, grow) **1.** one of the seven colors of the rainbow (wavelength between 500 and 570 nm); a blend of blue and yellow. **2.** a patch or area of grass, as a golf green. **3.** the symbol of anything environmentally friendly. **4.** money. **5.** a green light; the signal to go. *adj.* **6.** being green, as green foliage or green clothing. **7.** inexperienced, as a person still learning how to do a task. **8.** uncured, as green lumber or green cheese. **9.** a flavor defect of milk.

green algae *n.* any of at least 7500 species of photosynthetic eukaryotes (division Chlorophyta) that are mostly aquatic, though some grow in soil, on tree trunks, rocks, snow, etc., and some are components of lichens. It is closely related to higher plants and shares the same kinds of chlorophyll (a and b) and similar energy storage (starch) but lacks a vascular system.

greenbelt *n.* **1.** an area around a city restricted in use to woodland, parks, or farming for environmental reasons such as esthetics and pollution control. **2.** a strip of irrigated land planted to prevent encroachment by an adjacent desert.

greenbug *n.* a pale green aphid (*Toxoptera graminum*, family Aphidae) that damages rice, wheat and other small grains, corn, sorghum, alfalfa, etc., and helps to spread sugarcane mosaic disease.

green chop *n.* livestock feed consisting of freshly chopped forage used to feed animals in confinement in lieu of pasturing the animals.

greenery *n.* **1.** green plants. **2.** cuttings from green plants (especially evergreen trees) used for decoration. **3.** a greenhouse or other place where green plants are grown.

greenheart *n.* **1.** any of several tropical evergreen trees, especially bebeeru (*Ocotea = Nectandra rodiei*, family Lauraceae), a native of South America. Its bark produces bebeerine, a valuable alkaloid used to combat malaria. **2.** the dense, hard, durable, greenish wood of the tree, often used to build wharves, bridges, or ships.

greenhouse *n.* a structure with walls made of clear glass or clear plastic, usually heated during cold weather and used to grow out-of-season plants, to start sensitive plants such as tomatoes for later transplanting outdoors, or to provide controlled growing conditions for research. Also called glasshouse or hothouse (if heated).

greenhouse effect *n.* the primary cause of the gradual warming of the Earth's environment. The effect of a buildup of carbon dioxide and/or other trace gases that allows light from the sun's rays to heat the Earth but reduces the loss of heat (long-wave radiation) back into the atmosphere.

greenhouse gases *n.* gases whose buildup in the atmosphere contributes to the greenhouse effect that causes global warming. The most important greenhouse gases and their estimated percentage contribution to the total effect are carbon dioxide (50%), methane (20%), chlorofluorocarbons (15%), ozone (10%), and nitrous oxide and water vapor (5%).

greenhouse whitefly *n.* a tiny sucking insect (*Trialeurodes vaporariorum*, family Aleyrodidae) that attacks tomatoes, cucumbers, and many other plants, especially in greenhouses, causing stunting, chlorosis, wilting, and death.

green June beetle *n.* an insect pest (*Cotinus nitida*, family Scarabaeidae) that feeds on the foliage of a number of trees and plants as a beetle and whose larvae damage plant roots in lawns, gardens, and fields.

green lacewing *n.* an important environmentally friendly predator (*Chrysopa californica*, family Chrysopidae) on the citrus mealybug (*Pseudococcus citri*). See aphis lion.

green manure *n.* a forage or other crop grown and plowed under or disked in while immature to enrich the soil by adding humus and improving soil structure.

green onion *n.* See scallion.

Greenpeace *n.* an international environmental organization founded in Canada in 1969 that is noted for confrontational measures taken to interfere with actions it opposes, such as commercial whaling and sealing and the disposal of toxic wastes.

green revolution *n.* **1.** a marked increase in rice and wheat yields in the 1960s and 1970s as a result of new varieties that are highly responsive to fertilizer. The term was coined by the international news media after the release of rice variety I-R-8 by Norman Borlaug and others of the International Rice Research Institute in the Philippines. **2.** advances in controlling environmental hazards for scientific wildlife management.

greens *n.* **1.** potherbs; edible green leaves of wild and domestic plants. Wild plants most commonly used for greens are dandelion, lamb's-quarter, and dock. Commonly used tame plants are spinach, turnips, chard, and kale. **2.** greenery; green foliage used in a floral arrangement or other decoration. **3.** the area adjacent to the holes on a golf course. Such greens are specially designed for good drainage and a smooth surface, and the grass is kept very short.

green snake *n.* a harmless, slender, green snake (*Opheodrys* sp.) native to North America.

green thumb *n.* exceptional ability for growing plants, especially in a garden or indoors.

Greenwich mean time *n.* the time on the world's prime meridian as established by an international conference in 1884; that of a line passing north–south through Greenwich, a borough of London and the site of Britain's Royal Greenwich Observatory. See international date line.

gregarious *adj.* (Latin *gregarius*, belonging to a flock) **1.** sociable; enjoying life with others. **2.** flocking together, as a herd of animals. **3.** growing in open clusters; not matted or overcrowded. **4.** pertaining to a crowd or flock.

Gregorian calendar *n.* a corrected version of the Julian Calendar introduced by Pope Gregory XIII in 1582 and adopted in most countries. It provides for a normal year of 365 days and leap years of 366 days when the year is evenly divisible by 4 except that century leap years must be evenly divisible by 400. The result is an average year with 365.2425 days. See Julian Calendar.

gribble *n.* a small marine crustacean (*Limnoria* sp., order Isopoda) that damages wood submerged in water by boring into it.

grist *n.* (Anglo-Saxon *grist*, a grinding) **1.** grain that has been or is to be ground. **2.** something that a person can use to gain an advantage, such as accurate information.

grits *n.* (Anglo-Saxon *grytt*, grits) coarsely ground hominy (dehulled corn) that is often boiled and served as a hot breakfast cereal in southern U.S.A. Also called hominy grits.

groin *n.* (Anglo-Saxon *grynde*, low place) **1.** the lower abdomen, adjoining the thigh. **2.** an architectural structure formed of intersecting curved pieces, as the arches supporting a vaulted ceiling. **3.** jetties extending out from a shoreline to reduce beach erosion.

grosbeak *n.* (French *grosbec*, large beak) any of several large finches (family Emberizidae). These seed-eating birds with bright plumage and wide, short conical beaks are most common in woodlands; e.g., the black-headed grosbeak (*Pheucticus melanocephalus*), the blue grosbeak (*Guiraca caerulea*), the evening grosbeak (*Coccothraustes vespertinus*), the pine grosbeak (*Pinicola enucleator*), and the rose-breasted grosbeak (*Pheucticus ludovicianus*).

gross primary productivity *n.* the rate of solar energy conversion and biomass production in a given ecosystem or site.

ground *n.* (Anglo-Saxon *grund*, ground) **1.** soil; the surface layer of the dryland portion of the Earth. **2.** support for one's position, as being on firm ground. **3.** an electrical connection to the soil. *adj.* **4.** located near or pertaining to the soil, as ground level. **5.** operating at the land surface, as a ground attack. **6.** hitting the ground, as a ground ball in baseball. **7.** cut into small pieces by a grinder. **8.** abradcd, roughcncd, smoothcd, or shapcd by grinding, as a finely ground lens. *v.* **9.** to give something a good foundation or someone a good understanding. **10.** to establish an electrical connection with the soil. **11.** to run a ship into shallow water where its hull hits the bottom. **12.** to prevent a pilot or aircraft from flying. **13.** to punish a person (esp. a teenager) by restricting activities. **13.** the past tense and past participle of grind. (Also see grounds.)

ground beetle *n.* any of a large number of beetles (family Carabidae) that live on the ground, often under rubbish or stones, and prey on other insects, especially caterpillars.

A ground beetle

groundbreaking *n.* **1.** plowing or otherwise tilling a new field for the first time. **2.** the act or ceremony marking the beginning of the construction of a new building. **3.** developing new techniques or pioneering a new endeavor or field of study.

ground cover *n.* low-growing vegetation used for ornamental purposes and/or to control erosion; usually perennial plants not of the grass family.

ground hemlock *n.* a low-growing evergreen shrub (*Taxus canadensis*, family Taxaceae) with flat needles and red berry-like fruit. It is toxic to livestock. Also called American yew or Canada yew.

groundhog or **ground hog** *n.* a woodchuck. See prairie dog.

ground ice *n.* a lens of clear or nearly clear ice within a frozen soil. Ground ice is the main cause of frost heaving. It is produced at the frost line as capillary action moves water to it from unfrozen soil below.

ground moraine *n.* unsorted earthy material deposited across a land surface by glacial ice, either by being beneath the ice sheet as it advanced or by the melting of a stagnant or receding ice sheet. Also called till or glacial till. See end moraine, lateral moraine, terminal moraine, glacial outwash.

groundnut or **ground nut** *n.* **1.** a twining plant (*Apios tuberosa*, family Fabaceae) with an edible tuber and fragrant brown flowers. **2.** any of several other plants with edible parts that grow underground; e.g., peanuts. **3.** the edible tuber or other underground part of such a plant.

grounds *n.* **1.** a tract of land around a home or other building or land designated for a particular purpose, as picnic grounds. **2.** the basis for believing or doing something, as grounds for action. **3.** dregs, as coffee grounds.

groundsel *n.* a weed (*Senecio* sp., especially *S. vulgaris*, family Asteraceae) that bears small yellow flowers and contains an alkaloid that is poisonous to livestock.

ground squirrel *n.* any of several burrowing rodents of the squirrel family (especially *Spermophilus* sp., family Sciuridae), including gophers, chipmunks, etc.

groundswell or **ground swell** *n.* **1.** a giant rise of water level that travels for long distances across the sea. It is caused by a windstorm or an earthquake. **2.** a wave of popular opinion in support of a person or a cause.

ground truth *n.* identities and characteristics of objects and materials established by on-site observations to supplement remote sensing images such as aerial photographs.

groundwater or **ground water** *n.* water that is loosely held in the pore space of soil and rocks, especially that in continuous passages in the saturated zone so it may flow or seep from one place to another. Groundwater supplies wells and springs. Much of it enters streams and keeps them flowing even during periods with little or no rain. Pollution of groundwater by agricultural and industrial pollutants and by leaking underground storage tanks is a major environmental concern. Also known as phreatic water. See aquifer, vadose water.

grouse *n.* **1.** a plump, ground-dwelling, medium-sized game bird (subfamily Tetraoninae) with a short bill, feathered legs, and mottled feathers; e.g., the ruffed grouse (*Bonasa umbellus*). **2.** a complaint. *v.* **3.** to complain and grumble.

grove *n.* (Anglo-Saxon *graf*, grove) **1.** a small wooded area, usually easy to walk through because it has little or no undergrowth. **2.** an orchard, especially a small citrus orchard; e.g., an orange grove.

growing degree days *n.* the sum of the number of degrees that the average air temperature exceeds a specified base temperature for each day during a growth period. The base temperature is chosen as the minimum growth temperature for the crop or other plant species under consideration; e.g., oats will begin growth at about 40°F (4.5°C), and corn will begin growth at about 50°F (10°C). A maximum temperature is also used as a cutoff when temperatures are too warm. A record of growing degree days is useful for predicting when plants will pass through various growth stages such as germination, blooming, and ripening and when pesticide applications may be needed to control developing weeds, insects, or plant diseases.

growing season *n.* the time interval when plants (or a specific kind of plant) may grow in a particular place; usually the number of days between the average date of the last killing frost in the spring and that of the first killing frost in the fall.

growth factor *n.* a substance that influences the growth of an organism, especially a protein that influences the development and maintenance of cells and tissues.

growth hormone *n.* **1.** a growth factor produced by the pituitary gland that promotes normal growth. **2.** any substance that promotes the growth of any specific organism.

growth ring *n.* a double layer of new tissue (a soft layer of rapid growth and a hard layer of slow growth) representing one year's growth, seen as a ring in the cross section of a stem, branch, or root; e.g., in a tree trunk. Also called annual ring or tree ring.

growth stage *n.* any of a succession of periods in the development of a plant or animal. Each growth stage marks a change in the dominant developmental processes occurring in the organism; e.g., growth stages for an annual plant include germination, vegetative growth, flowering, and seed production (and may be divided into several shorter periods).

grub *n.* (German *grubilon*, to dig) **1.** the larva of certain beetles and other insects; usually a fat-bodied, often slimy or wooly, slow-moving wormlike organism. **2.** (slang) food. *v.* **3.** to clear land of roots and other obstacles in preparation for planting a crop. **4.** to dig or do other menial labor.

grubbing hoe or **grub hoe** *n.* a sturdy hand tool used for digging out tree roots; a mattock.

grubby *adj.* **1.** infested with grubs, especially cattle or sheep infested with botfly larvae. **2.** dirty and unkempt; wearing clothing that is soiled and wrinkled.

grumous *adj.* (Latin *grumus*, hillock) **1.** clustered into granules or clots (e.g., clotted blood). **2.** stabilized, as the soil structure produced by endogeic earthworms in their casts.

grumusol *n.* an older name for Vertisol; a soil with a high content of swelling clay that cracks open during dry seasons and swells and mixes during wet seasons.

grunion *n.* (Spanish *grunion*, grunter) a small fish (*Leuresthes tenuis*) that spawns on sandy beaches along the coast of southern California.

guaiacum *n.* (Spanish *guayacán*, guaiacum) **1.** a tropical South American ornamental tree or shrub (*Guaiacum officinale* and *G. sanctum*, family Zygophyllaceae) with pinnate leaves and blue flowers. **2.** the very hard, dense, dark greenish brown wood of such trees or shrubs that is used for carving and to make furniture, mallets, bearings, etc. Also known as lignum vitae.

guanaco *n.* (Quecha *wanaku*, guanaco) a wild, woolly, reddish brown ruminant (*Lama guanicoe*, family Camelidae) native to the Andes of South America. It is related to the vicuna, is a distant relative of the camel, and is believed to be the source of domesticated alpacas and llamas.

guano *n.* (Spanish *guano*, fertilizer or dung) **1.** an accumulation of bird droppings, especially those of seabirds on islands off the coast of Peru, that is used as fertilizer. **2.** an accumulation of bat droppings or other similar manure or substance (e.g., ground fish).

guayule *n.* (Spanish *guayule*, tree gum) **1.** a wild shrub (*Parthenium argentatum*, family Asteraceae) that grows in semiarid Texas and Mexico and yields a form of rubber. **2.** the rubber made from this shrub.

guinea fowl *n.* a plump gallinaceous fowl (family Numididae) about the size of a chicken with spotted gray plumage and a fleshy horn on each side of the head, especially the domesticated species (*Numida meleagris*). Native of Africa, they make excellent "watchdogs." Also called guinea or guinea hen. The young are called keet.

guinea pig *n.* **1.** a small, fat rodent (*Cavia* sp.) with a short tail or tailless that is often kept as a household pet (especially *C. porcellus*) or used in medical research. **2.** any person or animal used in a trial or experiment.

guinea worm *n.* an African roundworm (*Dracunculus medinensis*) common in warm climates that infests humans and other mammals as a subcutaneous parasite and causes painful abscesses.

gulch *n.* a steep, narrow canyon or ravine.

gulf *n.* **1.** an area of ocean largely enclosed by land; e.g., the Gulf of Mexico bordered by southern U.S.A., eastern Mexico, and a line of islands. **2.** a broad, deep, open space; an abyss or chasm. **3.** a large divergence, real or figurative, as the gulf between what one says and what one does or a gulf between opposing proposals.

Gulf Coast *n.* **1.** the southern U.S. coast from the southern tip of Texas to the southern tip of Florida. **2.** any coastal area along a gulf.

Gulf Coast tick *n.* a serious insect pest (*Amblyomma maculatum*, family Ixodidae) that sucks blood from people and animals, especially from around the ears of cattle, sheep, and goats.

Gulf Stream *n.* a warm ocean current flowing from the Gulf of Mexico around Florida and along the eastern coast of the U.S.A. and Canada to where it feeds the North Atlantic Current flowing toward Europe.

gulfweed *n.* a coarse olive-brown seaweed (*Sargasum bacifferum*) with air vessels that cause it to float that is common in the Gulf Stream and the Sargasso Sea.

gull *n.* (Welch *gwylan*, gull) a long-winged, wide-tailed, web-footed aquatic bird (especially *Larus* sp., family Laridae) that is typically white but with many species having gray or black markings on the upper wings and back. Marine species are often called seagulls.

gullet *n.* (French *goulet*, throat) **1.** the esophagus or some other similar structure. **2.** the pharynx or throat. **3.** a water channel; e.g., a gully or ravine.

gully *n.* an erosion channel that is too big and steep-banked to be crossed and smoothed with ordinary farm equipment. Gullies are classified as small, medium (3 to 15 ft or 1 to 5 m deep), or large; as active or inactive (vegetated); and as u- or v-shaped (the steep sides of the u shape commonly form in loess). See ephemeral gully, rill.

gully erosion *n.* **1.** cutting of a landscape by one or more gullies, especially where the gullies interfere with the use of the nongullied land between them. **2.** the soil and stone fragments carried away from an area by such erosion.

A gullied landscape

gullyhead *n.* the upper end of a gully; typically an area of active erosion where water falls into the gully and undercuts the soil, causing the gully to grow toward the source of the water.

gullying *n.* the process or effect of gully erosion.

gum *n.* (Latin *gummi*, gum) **1.** an amorphous plant exudate that hardens when dry but dissolves or softens in water. **2.** other plant secretions, such as resins. **3.** chewing gum. **4.** certain adhesives. **5.** a gum tree or its wood. **6.** the firm tissue surrounding the teeth of a person or animal. *v.* **7.** to chew with the gums (as a person who has no teeth). **8.** to stick together with gum as the adhesive. **9.** to exude a gummy material. **10.** to clog or become clogged with a gummy substance. **11.** to cause trouble or prevent something from happening, as to gum up the works.

gumbo *n.* (French *gombo*, gumbo) **1.** a chicken or seafood soup thickened with okra pods. **2.** okra. **3.** a soil that is hard to work and becomes very sticky when wet.

gumbo till *n.* a compact layer of glacial till.

gummite *n.* (Latin *gummi*, gum + *ite*, mineral) a yellowish or reddish brown mineral with a gumlike appearance that contains oxides of uranium, thorium, and lead.

gummosis *n.* (Latin *gummosis*, gummosis) secretion of tree sap that dries to a gummy mass as a result of mechanical injury or damage by insects or fungi.

gum tree *n.* any tree that produces an exudate that dries to a gummy substance; e.g., black gum, sweet gum, or eucalyptus.

gunnysack *n.* a brown fabric bag made from coarse jute. It is commonly used for sacking feed or seed grains, produce such as potatoes, etc. Also called burlap bag.

Gunter's chain *n.* a steel measuring device named after Edmund Gunter (1581–1626) that was used for surveying land in early U.S. history. It consisted of 100 links each 7.92 in. long, for a total length of 4 rods (66 ft or 20.13 m). This odd size was used because it comes to exactly 80 chains per mile, and an area 10 chains long by 1 chain wide is exactly 1 acre. Also called pole chain.

gust *n.* (Icelandic *gustr*, a gust) **1.** a sudden brief air movement; a burst of wind. **2.** a sudden burst of rain, fire, smoke, or noise. **3.** an outburst of emotion (laughter, anger, etc.).

gut *n.* (Anglo-Saxon *guttas*, guts) **1.** (or guts) the digestive tract, especially the intestines. **2.** strong cord made from animal intestines and used for stringing musical instruments or tennis rackets or for medical sutures. **3.** (guts) the internal organs of a person or animal. **4.** (guts) the internal working parts of a machine. **5.** (guts) the basic idea of a story. **6.** (guts) fortitude; determination. *v.* **7.** to disembowel or eviscerate. **8.** to destroy the interior of a building or take out the working parts of a machine. **9.** to remove vital parts from a story or from an organization. *adj.* **10.** internal or instinctive, as a gut feeling. **11.** fundamental or basic, as a gut issue.

guttapercha *n.* (Malay *getah*, gum + *perca*, tree) **1.** the milky white sap of certain Malayan trees (*Palaquium gutta* and other species of *Palaquium* and *Payena*, family Sapotaceae). **2.** a tough rubbery gum obtained from the sap; used in dentistry, golf balls, and electrical insulators.

guttation *n.* (Latin *gutta*, a drop + *atio*, action) the release of water droplets from the veins of plant leaves; often seen in the early morning but reabsorbed as the air warms up.

Gy *n.* gray, a measure of absorbed radiation. See gray.

gymnosperm *n.* (Greek *Gymnos*, naked + *sperma*, a seed) a plant whose seeds are not enclosed in an ovary (fruit); e.g., a pine, fir, or other conifer, cycad, or ginkgo. See angiosperm.

gynandromorph *n.* (Greek *gyne*, a woman + *andros*, male + *morphe*, form) an individual animal or human with part male and part female morphology. See hermaphrodite.

gynoecium *n.* (Greek *gynaikeion*, female place) the female part of a flower. See pistil.

gypsum *n.* (Latin *gypsum*, chalk or gypsum) hydrated calcium sulfate ($CaSO_4 \cdot 2H_2O$) occurring naturally in both massive and fibrous forms and as alabaster and selenite. It is used to make gypsum board and plaster of Paris and to reclaim high-sodium soils.

gypsum block *n.* See Bouyoucos block.

gypsy moth *n.* a European moth (*Porthetria* = *Lymantria dispar*, family Lymantriidae) that came to the U.S.A. about 1870. Its larvae (caterpillars) are highly destructive defoliating insects that strip the leaves from large areas of both evergreen and deciduous trees, including many fruit trees. Note: The worm sometimes found inside an apple is the larva of the gypsy moth.

H

H *n.* **1.** the chemical symbol for hydrogen. **2.** henry. See henry.

Haber-Bosch process *n.* the process used for the production of synthetic ammonia (NH_3). Nitrogen from the air and hydrogen (usually from either natural gas or obtained by electrolysis of water) are subjected to intense pressure (between 200 and 1000 atm) at temperatures up to 1200°C in the presence of a catalyst (commonly magnetite, Fe_3O_4, plus additives) that causes them to react and form ammonia.

habit *n.* (Latin *habitus*, appearance or dress) **1.** a practice that a person has repeated so many times that it has become almost automatic. **2.** a personal style or custom of doing something, such as shaking hands with people one meets or always putting things back where they belong. **3.** an addiction, as the habit of smoking. **4.** an attitude or action that has become a personal characteristic, as a habit of criticizing, or a habit of coughing frequently. **5.** the general appearance of a plant, animal, or person. **6.** the kind of costume one wears, as a nun's habit. **7.** the way a plant grows, as a twining habit.

habitant *n.* (Latin *habitans*, inhabiting) **1.** an inhabitant. **2.** a plant, animal, or person that grows or resides in a location for many years. **3.** or **habitan:** a French-speaking farmer in Quebec or Louisiana.

habitat *n.* (Latin *habitat*, it inhabits) **1.** the usual kind of environment of a plant, animal, or person. See environment, ecology, niche. **2.** a place that has an appropriate environment for a plant, animal, or person.

habituation *n.* (Latin *habituat*, conditioned + *tion*, action) **1.** the process of adjusting to an environment or situation. **2.** the condition of having adjusted. **3.** the process of becoming or the fact of being addicted to a drug. **4.** increased tolerance and decreased response to a type of stimulus that has been often repeated.

hachure *n.* (French *hachure*, cut up condition) any of a series of short, parallel lines used on a map or drawing, usually to indicate an elevated area, a shaded area, or a slope.

hack *n.* (Anglo-Saxon *haccian*, cut to pieces) **1.** a person (e.g. a politician or writer) who will do anything for payment. **2.** a horse, carriage, or taxicab for hire. **3.** a rack for drying food or to hold fodder for livestock to eat. **4.** a cut or gash, an axe or other tool that makes such cuts, or the action of making such cuts. *v.* **5.** to cut into, creating a gash or notch. **6.** to cough, usually raspingly and spasmodically **7.** to make available for hire. **8.** to do new things with a computer, especially inconsiderate or illegal action via a computer network.

hackberry *n.* (Danish *haeggebaer*, hagberry) **1.** a tree or shrub (*Celtis* sp., family Ulmaceae) of the elm family; used as shade trees, in shelterbelts, and for making furniture, boxes, and baskets. Also called sugarberry or nettle tree. **2.** either the wood or the small, round, purplish edible fruit of the tree.

hadal *adj.* (Greek *Haidēs*, Hades + Latin *alis*, of) of or in the deepest part of the ocean; below about 20,000 ft (6000 m).

haddock *n.* a bottom-dwelling, carnivorous, edible fish (*Melanogrammus aeglefinus*, family Gadidae) of the North Atlantic coastal regions of Europe and North America.

haemonchosis *n.* infection by a large stomach roundworm (*Haemonchus contortus*); a very serious disease causing anemia, weakness, and weight loss in sheep and cattle, especially in young animals. Affected animals often exhibit a swelling of the throat known as bottlejaw.

hafnium (Hf) *n.* element 72, atomic weight 178.49; a toxic heavy metal with a valence of +4 that usually accompanies zirconium in minerals. It is used in control rods for nuclear reactors and in some gas-filled and incandescent lamps.

hail *n.* (Anglo-Saxon *haegel*, hail) **1.** frozen precipitation in the form of balls or clumps of ice produced by accretion during thunderstorms. Updrafts lift the hailstones to high elevations with freezing temperatures, then let them fall to a level where it is raining before they are caught again in another updraft. The cycle may be repeated many times before the hailstones finally fall to the ground. Hail occurs in warm seasons because warm temperatures are required to produce strong updrafts. See graupel, sleet. **2.** a cluster or shower of anything, such as a hail of bullets. *v.* **3.** to precipitate in the form of hail. **4.** to greet or welcome someone, especially as an indication of honor. **5.** to call to someone, as to hail a taxi.

hailstone *n.* (Anglo-Saxon *hagolstän*, hailstone) an individual ball or clump of ice in a hail storm.

hairball *n.* a wad of hair formed in the stomach of a cat or cow caused by the animal licking its fur. See bezoar, egagropilus, stomach ball.

hairlessness *n.* **1.** a lack of hair for any reason. **2.** a disease of calves, pigs, or lambs with a severe iodine deficiency that causes bare or thin areas in the hair or wool. **3.** a lethal genetic factor in some Holstein cattle that causes calves to be born with very little hair.

hairy vetch *n.* a viny annual legume (*Vicia villosa*, family Fabaceae) with pinnate leaves, purple flowers, and small round seeds that shatter readily when their pods are dry; used as a forage crop in northern U.S.A. or as a winter cover and green manure crop in southern U.S.A.

Haleakala volcano *n.* a large dormant volcano in Haleakala National Park, Maui, Hawaii. It is 10,023 ft (3055 m) high and has a crater of 19 mi^2 (50 km^2).

half-life *n.* **1.** the time required for half of the atoms of a radioactive element to undergo decay; e.g., the half-life of cobalt 60 is 5.3 years, that of carbon 14 is 5,730 years, and that of uranium 235 is 710,000,000 years. Note: after 1 half-life, half of the atoms remain; after 2 half-lives, one-fourth of them remain; after 3 half-lives, one-eighth, etc. **2.** the time required for a chemical reaction to consume half of the reactant present; e.g., for a pollutant to lose half of its effect on the environment. The half-life of DDT in the environment is 15 years. **3.** the time required for the body to eliminate half of the medication in a dose.

half-moon *n.* **1.** a moon stage halfway between full moon and new moon; at first quarter or last quarter, when half of its face is lit. **2.** anything shaped like the lighted part of a half-moon.

half-sib *n.* **1.** related through only one parent; a half brother or half sister. **2.** plants related through one parent only.

halieutics *n.* (Latin *halieutica*, pertaining to fishing) the study of fishing.

haline *adj.* (Greek *halos*, salt + *inos*, like) saline; dominated by ocean salt.

halite *n.* (Greek *halos*, salt + *ite*, mineral) crystalline sodium chloride (NaCl); rock salt.

halloysite *n.* a layer silicate clay mineral of the kaolinite family [1 : 1 structures with the idealized formula Al$_4$Si$_4$O$_{10}$(OH)$_8$] with a layer of water molecules between the aluminosilicate layers that gives it a layer spacing of 1.0 nm. Named after Jean-Baptiste-Julian Omalius d'Halloy (1783–1875), a Belgian geologist.

hallucinogen *n.* (Latin *hallucinatio*, wandering of the mind + *gen*, cause) a psychedelic drug; a drug that distorts the senses; an organic compound that causes hallucinations. Examples: marijuana and LSD.

halo *n.* (Latin *halos*, circle around the sun or moon) **1.** a band of light or a golden band over or around the head (or sometimes around the entire image) in the depiction of an angel or other holy being. **2.** a circle of light around the sun or moon produced by the refraction of light by ice particles in Earth's atmosphere. **3.** a similar circle around any object.

haloclastism *n.* (Greek *halos*, salt + *klastos*, broken + *isma*, condition) disintegration of rock caused by salt crystallization.

halogen *n.* (Greek *halos*, salt + *gen*, origin) any of the five chemically related nonmetallic elements that form Group 7 of the Periodic Table: fluorine, chlorine, bromine, iodine, and astatine. The halogens occur in compounds as anions with a single negative charge; e.g., Cl$^-$.

halogeton *n.* (Greek *halos*, salt + *geitōn*, neighbor) a coarse annual herb (*Halogeton glomeratus*, family Chenopodiaceae) from Siberia that is now a noxious weed in many rangeland areas of western U.S.A. Halogeton is toxic to livestock because it has a high oxalate content.

halomorphic soil *n.* a soil with characteristics that reflect high salinity and/or high alkalinity. It is a term used in the 1938 U.S. system of soil classification as an intrazonal soil suborder composed of the Solonchak, Solonetz, and Soloth great soil groups.

halophile *n.* (Greek *halos*, salt + *philos*, beloved) a plant or animal that requires a salty environment.

halophyte *n.* (Greek *halos*, salt + *phyton*, plant) a plant capable of growing in saline or alkaline soil. Opposite of glycophyte.

halter *n.* (Anglo-Saxon *haelfter*, halter) **1.** a person or animal who halts or limps. **2.** a rope or strap that wraps around and over the head of an animal so it can be held in place or led. **3.** a garment that ties around the neck and waist but leaves the arms and back exposed; worn by a woman or girl. **4.** a hangman's noose.

ham *n.* (Anglo-Saxon *hamm*, bend of the knee) **1.** the upper leg (thigh and buttock) of an animal or person, especially that of a hog. **2.** the meat from the upper leg of a hog. **3.** an entertainer who overacts; one who exaggerates expressions and motions. **4.** a licensed amateur radio operator.

hamadryad *n.* See king cobra.

hamburger helper *n.* a food product made from soybeans and used to extend or substitute for hamburger.

hammock *n.* (Spanish *hamaca*, hammock) **1.** a type of bed consisting of a rope net or a canvas stretched between two supports; often used outdoors in warm weather. **2.** a covering of certain caterpillars. **3.** an area of fertile land on the U.S. Gulf Coast with wetland forest vegetation.

hamster *n.* (German *hamster*, hamster) a short-tailed burrowing rodent (*Mesocricetus auratus* and related species, family Cricetidae) of European origin. Hamsters are often used in nutrition research or kept as pets.

hand *n.* (Anglo-Saxon *hand*, hand) **1.** the part of a human or primate arm from the wrist outward, including the wrist, palm, fingers, and thumb. **2.** the terminal part of a limb of some other animal or bird. **3.** a pointer, as a hand on a clock. **4.** an employee who does manual labor. **5.** an experienced person, as an old hand at such work. **6.** a designated degree of skill, as a poor hand or a master's hand. **7.** a position of control, as being in his hand(s). **8.** the active agent, as by his hand. **9.** aid or assistance, as give me a hand. **10.** style of writing. **11.** a round of applause. **12.** agreement or promise, as to give one's hand in marriage. **13.** a bunch, as a hand of bananas, a bundle of tobacco leaves, or a hand of cards. **14.** a unit of measurement equal to 4 in. (10.2 cm) used to designate the height of a horse. *v.* **15.** to give or deliver, as to hand it to him. *adj.* **16.** manually operated, as a hand tool.

handpick *v.* **1.** to collect by hand labor, as handpick corn or handpick insect pests. **2.** to choose personally, as handpick an assistant.

handweed *v.* to use hand labor (e.g., with a hoe) rather than a herbicide to kill weeds in a crop.

hank *n.* (Icelandic *honk*, hank, coil, or skein) **1.** a loop used to fasten something in place, such a gate or a sail. **2.** a skein, especially one holding a standard length of yarn; e.g., cotton or silk yarn = 840 yd (768 m), worsted yarn = 560 yd (512 m). **3.** a coil or bundle, as a hank of hair.

hantavirus pulmonary syndrome (HPS) *n.* an often fatal viral disease that is transmitted only by rodent bites, body fluids, or droppings. First recognized in 1993, it causes high fever, muscular aches, and fluid accumulation in lungs.

haploid *adj.* (Greek *haplos*, single + *eidos*, form) having only one set of chromosomes (in contrast to diploid with two sets), e.g., the drone bee, male ant, or the reproductive cells of most organisms. See diploid, tetraploid, polyploid.

harbor *n.* (Anglo-Saxon *herebeorg*, lodgings) **1.** an inlet along a coastline where a ship can enter and be protected from storms and currents. **2.** a place of refuge; a shelter. *v.* **3.** to provide shelter or a habitat, e.g., for insects and/or diseases. **4.** to conceal or hide, as to harbor fugitives. **5.** to hold or have in the mind, as to harbor rebellious ideas or suspicions.

hardening *n.* **1.** the process of becoming hard. **2.** or **hardener:** a substance that makes another material hard. **3.** a preconditioning process for plants (e.g., subjecting them to a period of cool temperatures and/or reduced water availability) to enable them to withstand environmental stress such as adverse temperature or moisture extremes, e.g., to prepare plants that were started in a greenhouse for transplanting in the field.

hardiness *n.* the relative ability of a plant or animal to tolerate adverse environmental conditions; e.g., resistance to diseases, insects, wet soil, dry soil, hot weather, and/or cold weather.

hardness scale *n.* See Mohs' scale.

hardpan *n.* a dense soil horizon that restricts root development and water infiltration; e.g., a duripan, fragipan, or plow sole.

hard rock *n.* **1.** a consolidated rock; especially an igneous rock (e.g., granite or basalt) or a hard metamorphic rock (e.g., gneiss or quartzite). **2.** rock-and-roll music; music with a loud, strong beat.

hard seed *n.* seeds with a covering that resists water absorption and thus prevents germination and growth. See scarify.

hardware disease *n.* an animal disease resulting from eating pieces of metal (e.g., nails or wire) that lodge in the stomach, where they may pierce the wall and cause abscesses, peritonitis, and sometimes death. It affects mostly cattle and other ruminants, usually occurring in the reticulum (second stomach). See bezoar.

hard water *n.* water that contains an excess of dissolved salts such as calcium and magnesium chlorides, sulfates, and carbonates. Hard water (e.g., that with more than 60 ppm calcium + magnesium) reacts with soap and forms a scum. See soft water.

hard wheat *n.* **1.** any wheat variety that produces reddish-colored, hard, flinty kernels. Typically, it is grown in drier areas than soft wheat. See soft wheat, durum wheat. **2.** the grain produced by such wheat varieties. Hard wheat is used mostly to make bread flour.

hardwood *n.* **1.** wood that is dense and hard. **2.** any angiosperm (broad-leaved; seeds enclosed in an ovary) tree species whose wood has true vessels, in contrast to gymnosperms (pines, firs, and other evergreens). **3.** the wood of an angiosperm (even if it is soft). See softwood.

hardy *adj.* (French *hardir*, to harden) **1.** able to withstand environmental stress, especially cold weather (plants are often called winter hardy). **2.** able to endure fatigue, exposure, and other forms of stress. **3.** bold, brave. **4.** brash or foolhardy.

hare *n.* (Anglo-Saxon *hara*, a hare) **1.** any of several gnawing mammals (family Leporidae, order Lagomorpha) with soft fur, long ears, large hind legs, and a short tail. Hares are related to rabbits but are usually larger than rabbits. **2.** the player or piece that is chased in certain games, e.g., hare and hounds. **3.** a constellation directly beneath Orion.

harlequin bug *n.* a black and red stinkbug (*Murgantia histrionica*, family Pentatomidae). It is the most important insect pest of the cabbage family, especially in southern U.S.A., but also attacks many other plants. Its nymphs suck the sap from leaves and stems. Also called fire bug, collard bug, or calico bug.

harmattan *n.* (Spanish *harmatán*, hot, dry wind) a hot, dry wind blowing from the African interior toward the west coast and Atlantic Ocean. Harmattans generate clouds of dust during the period from November to March.

harmonic *adj.* (Latin *harmonicus*, musical) **1.** pertaining to harmony; tones, frequencies, colors, patterns, or behaviors that are pleasing when combined (consonant). **2.** pertaining to frequencies (e.g., of sound) that harmonize because they are small-integer multiples of the same base frequency. **3.** varying cyclically according to a sine or cosine function.

harmony *n.* (Latin *harmonia*, a joining) **1.** agreement; harmonious relations. **2.** a pleasing, orderly arrangement of components, as a blending of harmonious tones, colors, or patterns.

harness *n.* (French *harness*, equipment or baggage) **1.** the assemblage of collar, straps, bands, etc., that form the gear of a work horse or other draft animal. **2.** any similar gear attaching a load, support, or restraint to an animal or person; e.g., a parachute harness. *v.* **3.** to fit or equip with a harness. **4.** to gain control and put to use, as to harness the power of wind.

harrow *n.* (Anglo-Saxon *hearge*, harrow) **1.** a toothed implement used to break clods, smooth the soil surface, and kill weeds prior to planting a crop. *v.* **2.** to work the soil with a harrow. **3.** to cause great distress; to disturb the mind and emotions; to cause a harrowing experience.

harsh *adj.* (German *harsch*, harsh) severe or unpleasant; difficult conditions for life or work, as a harsh climate.

Hartig net *n.* a network of fungal hyphae in the intercellular space between cortical cells of a root of a tree, shrub, or other plant that forms an ectomycorrhiza.

harvester ant *n.* an ant (*Pogonomyrmex* sp., family Formicidae) that clears all the vegetation from an area adjacent to its nest and collects seeds for food; common in southwestern U.S.A.

harvest moon *n.* the full or nearly full moon occurring nearest the autumnal equinox (September 22 in the northern hemisphere or March 22 in the southern) because it furnishes light for harvesting crops after sunset. See hunter's moon.

hashish *n.* (Arabic *hashīsh*, dry vegetation) **1.** a concentrated, resinous form of marijuana. **2.** the flowers and leaves of the hemp plant that is the source of marijuana.

Hatch Act *n.* the federal law passed in 1887 that allotted money for the land-grant colleges to establish agricultural experiment stations.

haunch *n.* (French *hanche*, haunch) an animal's hindquarter; the fleshy part of the upper leg and adjacent part of the body.

haw *interj.* (Anglo-Saxon *hāwian*, to beware) **1.** a command for work horses or other draft animals to turn left. See gee, whoa. *n.* **2.** hawthorn or its fruit. **3.** a break or pause in one's speech.

hawk *n.* (Anglo-Saxon *hafoc*, hawk) **1.** a bird of prey (especially *Accipiter* sp. and *Astur* sp., family Accipitridae) with rounded wings, a long tail, curved talons, and a short, hooked beak. **2.** most other large, soaring birds of prey (e.g., falcons, buzzards, harriers, kites, ospreys, etc., excluding eagles and vultures). **3.** a person who advocates strong military preparedness and the use of military measures in international relations (opposite of dove). **4.** a person who sells goods aggressively and noisily. *v.* **5.** to hunt in flight like a hawk, or to hunt with trained hawks. **6.** to peddle loudly in public. **7.** to publicize orally; to spread news or rumors.

Hawk

hawthorn *n.* (Anglo-Saxon *haguthorn*, hawthorn) a thorny shrub (*Crataegus* sp., family Rosaceae) with beautiful white early spring flowers and small applelike red fruit; the state flower of Missouri. Hawthorn is often planted as an ornamental or as a hedge. Also called haw, thorn, or thorn apple.

hawthorn rust *n.* a disease caused by a fungus (*Gymnosporangium globosum*, family Pucciniaceae). Its aecial stage produces lesions in the fall on leaves of hawthorn, apple, or pear trees, and its telial stage produces wedge-shaped dark orange teliospores in the spring on red cedar and related species (*Juniperus* sp.). Also known as cedar-hawthorn rust.

hay *n.* (Anglo-Saxon *hieg,* hay) leafy plant material cut, dried, and (usually) stored for use as forage; commonly alfalfa, clover, or grasses, either singly or in mixtures.

haycock *n.* a small stack of hay drying in a field.

haylage *n.* forage cut and partially dried (to about 45% moisture), chopped, and put in a silo to be stored and fed like silage. See green chop.

hayloft or **haymow** *n.* a hay storage area in the upper part of a barn.

hayseed *n.* **1.** seed for grasses and legumes to be used for hay, especially when shaken from hay (hence often including weed seed). **2.** slang for a rustic farmer or other person living in a remote area (usually used in a derogatory sense).

hazard *n.* (Arabic *az-zahr,* the die) **1.** randomness, chance, or a game of chance. **2.** danger; a possible cause of accident or injury (sometimes classified as toxic, causing an immediate health hazard; carcinogenic, causing a delayed health hazard; ignitable; or explosive). **3.** an obstacle on a golf course, e.g., a sand trap or a body of water.

hazardous substance *n.* any material whose release in a significant amount would pose a threat to human health and/or the environment, especially those listed in the Clean Water Act, the Clean Air Act, the Toxic Substances Control Act, or other pertinent legislation. Typical hazardous substances are toxic, ignitable, explosive, or chemically reactive.

hazardous waste *n.* any waste material with chemical, physical, or biological characteristics that may damage either human health or the natural environment. Waste materials may be listed as hazardous under the Resource Conservation and Recovery Act if specific tests show them to be excessively toxic, ignitable, corrosive, or reactive.

hazardous waste site *n.* any site known to hold hazardous wastes that could injure the health and well-being of humans and animals, often either the site of or a discard area from an old factory or industrial area. The Environmental Protection Agency lists such sites in the U.S.A. and oversees their remediation. See Superfund.

haze *n.* (Anglo-Saxon *hasu,* ashen or dusky) **1.** an atmospheric condition such as water droplets, dust, or smoke that diffuses light and reduces visibility. **2.** mental confusion; vagueness of understanding. *v.* **3.** to produce haziness or become hazy. **4.** to harass or abuse with humiliating tricks and ridicule, often as part of an initiation.

haze coefficient *n.* a measure of the interference in visibility caused by haze in the atmosphere.

hazel *n.* (Anglo-Saxon *haesel,* hazel shrub) **1.** a shrub or small tree (*Corylus* sp., family Betulaceae) with toothed ovate leaves and edible reddish brown nuts. **2.** the wood of a hazel tree. **3.** a light reddish brown color like that of a hazelnut.

hazelnut *n.* (Anglo-Saxon *haeselhnutu,* hazelnut) a small rounded reddish brown nut produced by a hazel tree. Also called filbert.

head louse *n.* a tiny wingless insect (*Pediculus humanus capitis,* family Pediculidae) that sucks blood from the scalp of human heads and attaches its eggs (nits) to hairs. It causes itching in infested areas and can lead to infection.

head smut *n.* a fungal (*Sphacelotheca reiliana*) disease of corn or sorghum—especially in western U.S.A., Asia, Africa, Australia, and eastern Europe—characterized by large reddish brown to black smut balls replacing the tassel and ear of corn or the panicle of sorghum. The head smut races that attack corn are separate from those that attack sorghum and do not affect the other species.

headwaters *n.* the small streams that join together to form a river; sometimes defined as streams with average annual flow less than 5 ft^3/sec (0.14 m^3/sec).

health *n.* (Anglo-Saxon *haelth,* health) **1.** the general physical and mental condition of a person or organism. Good health means having vigor and being free from disease and pain. **2.** the status or condition of a business, organization, or nation, such as good economic health.

healthy *adj.* (Anglo-Saxon *haelth,* health + *y,* characterized by) **1.** having good health. **2.** healthful; helping to produce or maintain good health, as a healthy diet or healthy exercise. **3.** characteristic or indicative of good health, as a healthy appearance. **4.** vigorous; sound; as a healthy organization or business. **5.** fairly large, as a healthy stream of water.

heartseases *n.* See pansy.

heartwood *n.* the inner part of a tree trunk or branch, usually dead cells that are darker in color than the surrounding sapwood. Also called duramen.

heartworm *n.* a parasitic nematode (*Dirofilaria immitis*) transmitted by a mosquito that infests the heart and pulmonary arteries of a dog or other canid or cat.

heat *n.* (Anglo-Saxon *haetu,* heat) **1.** the energy of motion of molecules. **2.** added warmth; a source of energy that increases molecular motion as indicated by higher temperature. **3.** the degree of hotness or the rate of supply of such energy, as moderate heat. **4.** a source of heat, as a furnace or other heating system. **5.** warm feeling, passion. **6.** sexual receptiveness in a female animal. See estrus. *v.* **7.** to make warmer or hotter. **8.** to become warm or hot by fermentation, as grain or hay stored with too much moisture. **9.** to arouse and excite emotionally.

heat capacity *n.* the amount of heat required to raise the temperature of a body or object 1°C. See specific heat.

heath *n.* (Anglo-Saxon *haeth*, heath) **1.** open uncultivated land covered with low shrubs, especially in the British Isles. **2.** the low-growing, shrubby vegetation common on such land, especially heather (*Erica* sp., family Ericaceae).

heather *n.* (Anglo-Saxon *haedre*, heather) any of several low-growing evergreen shrubs (*Erica* sp. or *Calluna* sp., especially *C. vulgaris* family Ericaceae), common in England and Scotland, that have tiny pink or purple bell-shaped flowers and scalelike leaves. It is used as winter forage for sheep and cattle and for making brushes and brooms.

heat index *n.* an indicator of how hot the weather actually feels to a person based on the interaction of temperature and humidity. High humidity reduces the cooling effect of evaporation and thus makes the temperature feel hotter.

Heat Index

Relative Humidity	Actual Air Temperature (°F)					
	70	80	90	100	110	120
%	Apparent Temperature (°F) (How hot it feels)					
0	64	73	83	91	99	107
20	66	77	87	99	112	130
40	68	79	93	110	137	
60	70	82	100	132		
80	71	86	113			
100	72	91				

heat island *n.* an elevated temperature above urban areas. Usually a dome-shaped area extending upward from the tops of the buildings may be 5° to 7°C warmer than the air at ground level during night and early morning hours, but the differential may disappear near midday.

heat stroke or **heatstroke** *n.* debility of people and animals caused by exposure to high temperatures, especially sunstroke.

heat-tolerant *adj.* able to tolerate warmer temperatures than most other species or individuals; e.g., cacti, Brahman cattle, people who live in the tropics.

heat wave *n.* a period of unusually warm or hot weather at a particular location, usually lasting for several days.

heaves *n.* (Anglo-Saxon *hebban*, to heave) **1.** a respiratory disease of horses characterized by heaving flanks, difficult breathing, and double

exhaling. It is most common in older horses after heavy exercise. Also called broken wind. **2.** severe vomiting with a heaving contraction of the abdominal muscles.

heavy metal *n.* **1.** a metallic element with high atomic weight (usually more than 50). Many heavy metals are useful in industry; some are micronutrients for plants and/or animals (e.g., copper, iron, manganese, and zinc), but others are toxic to life (e.g., cadmium, chromium, lead, mercury, plutonium). **2.** a type of loud rock music with a strong beat.

heavy soil *n.* a soil that is high in clay and sticky when wet, so called because of the heavy draft required to pull a plow or other implement through it. Note: The dry density of such soil is usually less than that of a sandy soil, but the clay soil holds more water, so its wet weight is usually more than that of a sandy soil. See light soil.

heavy water *n.* **1.** deuterium oxide (D_2O). Its molecular weight (20) being heavier than that of H_2O raises its freezing point to 3.8°C and its boiling point to 101.4°C. Heavy water is used as an energy moderator in some nuclear reactors. **2.** water with a molecular weight greater than 18 because it contains deuterium, tritium, and/or oxygen with atomic weights heavier than 16.

hectare (ha) *n.* (Greek *hekaton*, hundred + Latin *are*, area) a unit of surface measure in the SI (metric) system that is equal to 100 ares (10,000 m² or 2.471 ac).

hedgehog *n.* **1.** a small (about 9 in. or 23 cm long) short-legged, insect-eating, nocturnal mammal (*Erinaceus europaeus* and related species, family Erinaceidae) with spiny hairs on its sides and back; native to Europe and Asia. **2.** a similar Eurasian mammal with soft hair instead of spines. **3.** any of several similar animals with spines, such as the porcupine. **4.** a leguminous plant (certain *Medicago* sp., family Fabaceae) with spiny seed pods. **5.** a dredging machine with buckets or spades attached around the perimeter of a large digging wheel. **6.** a defensive military device inspired by the hedgehog to obstruct the passage of ships, tanks, or other vehicles.

hedgerow *n.* (Anglo-Saxon *heggerewe*, hedgerow) a line of bushes, shrubs, or small trees planted close together to serve as a border or barrier.

heel fly *n.* a serious pest (*Hypoderma lineatum* and *H. bovis*, family Hypodermatidae, the common and northern U.S. species, respectively) of cattle. The flies

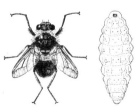

Heel fly and its larva

live only a few days during warm spring weather; they mate and lay their eggs, mostly on the hair of the lower legs of cattle, agitating the animals even to the point of causing a stampede. The eggs hatch a few days later as larvae that are called either cattle grubs or heel grubs. See cattle grub, warble.

heel-in *v.* to store seedling trees or other plants in a trench and cover the roots with moist soil until they are planted.

heliocentric *adj.* (Greek *hēlios*, sun + *kentrikos*, centered) centered around the sun, as a heliocentric view of the universe, or measurement of distances from the center of the sun.

heliophilous *adj.* (Greek *hēlios*, sun + *philos*, loving) adapted to full sunlight; e.g., sunflowers.

heliophobic *adj.* (Greek *hēlios*, sun + *phobos*, fear) adapted to growing in the shade; e.g., creeping red fescuegrass.

heliophyte *n.* (Greek *hēlios*, sun + *phyton*, plant) a plant that does best in bright sunlight (including some that even require intense light for survival). Such plants typically grow in exposed hot, dry habitats and have small vertical leaves or other adaptations that conserve water but are not photosynthetically efficient.

heliotropism *n.* (Greek *hēlios*, sun + *tropos*, turning + *isma*, condition) the characteristic of certain plants to turn in response to the sun. **1.** positive heliotropism (eutropism) means turning toward the sun. **2.** negative heliotropism (diaheliotropism) is turning away from the sun. **3.** paraheliotropism means turning leaves parallel to the light rays, thus minimizing the amount of light absorbed.

helium (He) *n.* (Greek *helios*, the sun + Latin *ium*, element) element 2, atomic weight 4.0026; a colorless, odorless inert gas with the lowest density of any gas other than hydrogen and the lowest boiling point (4.2 K) of any substance. Helium is commonly obtained from natural gas; it is used for buoyancy in balloons, as a coolant for work at very low temperatures, and as a vehicle in general anesthetics for surgery, and it is blended with oxygen as a breathing mixture for deep-sea diving.

helix *n.* (Greek *helix*, a spiral) **1.** a spiral, especially the line traced around the surface of a rotating cylinder or cone by a marker moving progressively along its length, as the thread of a bolt or screw. **2.** the rim of cartilage around the outer ear. **3.** any of a group of spiral-shelled land mollusks (*Helix* sp.), especially the common edible European snail *H. pomatia*.

hellebore *n.* (Greek *helleboros*, hellebore) **1.** any of a group of plants (*Helleborus* sp., family Ranuncu-

laceae) related to buttercups. The European variety (*H. niger*) has poisonous rhizomes. See Christmas rose. **2.** any plant (*Veratrum* sp., family Liliaceae) that has poisonous rhizomes. Also called false hellebore. **3.** a poisonous or medicinal substance obtained from any of these plants, especially the powdered root of white hellebore (*V. album*) applied to plants to kill caterpillars and lice.

hellgrammite *n.* the aquatic larva of dobsonfly (*Corydalus cornutus*, family Corydalidae); a predator on aquatic insects and a popular live bait for fishing.

helminth *n.* (Greek *helmins*, worm) any of several kinds of parasitic worms, such as lungworms, roundworms, flatworms, tapeworms, and hookworms.

helminthology *n.* (Greek *helmins*, worm + *logos*, word) scientific study of the nature and action of worms.

helminthosporium blight *n.* a fungus (*Helminthosporium* sp., family Dermatiaceae) that attacks many different plants of the grass family (Poaceae), especially small grains and corn, causing seedling blight, leaf spots, or root rot. For example, *H. avenae* causes leaf blotches and root rot of oats; *H. bromi* causes brown spot on smooth bromegrass; *H. caronum*, *H. Maydis*, and *H. turcicum* produce spots on corn leaves and other grasses; *H. oryzae* produces yellow or brown spots on rice leaves; *H. sativum* causes leaf spots on barley, wheat, and several grasses. See corn leaf blight.

helolac *n.* (Greek *helos*, marsh + Sanskrit *lac*, mark or sign) a shallow lake covered by aquatic vegetation such as water lilies or water hyacinth.

helotism *n.* (Greek *Helot*, slaves from Helos + *isma*, condition) **1.** a caste system wherein a minority group is constantly dominated and degraded by the majority. **2.** a system among certain ants wherein a dominant species makes a subordinate species do the work necessary to sustain both species.

hematite *n.* (Greek *haimatitēs*, like blood) a very common iron oxide (Fe_2O_3), the most abundant iron ore. It is often an accessory mineral in igneous rocks but is much more concentrated in certain sedimentary rocks. Hematite also is common in brown coatings on subsoil peds and commonly occurs as steely gray to black crystals or as an earthy reddish brown mass.

hematophagous *adj.* (Greek *haimat*, of blood + *phagos*, eating) feeding on blood, as a female mosquito or a vampire bat.

hematozoon (pl. **hematozoa**) *n.* (Greek *haimat*, of blood + *zōion*, animal) any parasitic animal living in the blood of an animal or human.

hemicellulose *n.* (Greek *hēmi*, half + Latin *cellula*, little cell + French *ose*, sugar) includes several different materials, such as xyloglucans (mixtures of glucose and xylose plus small amounts of other five- and six-carbon sugars), xylan (a pure polymer of xylose, a five-carbon sugar) in dicotyledon plants, and arabinoxylans (xylose with various side chains) in monocots and legumes.

hemiepiphyte *n.* (Greek *hēmi*, half + *epi*, outer + *phyton*, plant) a semiepiphyte; a plant that is epiphytic for only part of its lifetime; e.g., the banyan tree of India.

hemimetabolic or **hemimetabolous** *adj.* (Greek *hēmi*, half + *metabolikos*, metabolic) characterized by incomplete metamorphism; applicable to certain insects that pass through some but not all of the usual metamorphic states; e.g., the mayfly goes through only three stages—egg, nymph, and adult.

hemiparasite *n.* (Greek *hemi*, half + *parasitos*, one who eats at another's table) **1.** a plant, such as mistletoe, that is a parasite on trees but that can carry on photosynthesis as a nonparasite. **2.** a facultative parasite; any organism that can live either independently or as a parasite.

hemispherical scale *n.* a very serious fungal pest (*Saissetia coffeae*, family Coccidae) on avocado, citrus, and several greenhouse plants. It reproduces asexually.

hemlock *n.* (Anglo-Saxon *hymlic*, hemlock) **1.** a poisonous plant (*Conium maculatum*, family Apiaceae) that contains the deadly alkaloid coniine ($C_8H_{17}N$). Hemlock has white flowers, finely divided leaves, and purple spots on its stems. Although native to Europe, it is widely distributed in North America. **2.** a poisonous beverage made from the hemlock plant. **3.** any of several similar plants, especially water hemlock (*Cicuta occidentalis*, family Apiaceae) and related species. **4.** any of several evergreen trees (*Tsuga* sp., family Pinaceae) with short needles and small cones; the bark is used in tanning leather. Also called hemlock spruce. **5.** the soft light wood from any of these trees; used in construction and for making paper.

hemlock poisoning *n.* poisoning of animals that browse on hemlock plants; characterized by nervous tremors, rapid pulse, weakness and unsteady gait, dilated pupils, salivation, digestive problems, and difficulty breathing.

hemophiliac *n.* (Latin *hemo*, blood + *philos*, loving + *acus*, having characteristics of) an animal or a person affected by hemophilia. Also called a bleeder because even a minor cut or injury can lead to excessive bleeding.

hemorrhage *n.* (Latin *haemorrhagia*, violent bleeding) **1.** any escape or discharge of large quantities of blood. **2.** a rapid loss of assets. *v.* **3.** to lose blood rapidly. **4.** to lose large amounts of cash or other assets rapidly.

hemorrhagic septicemia *n.* an acute infection in cattle, hogs, chickens, etc., caused by bacteria (*Pasteurella multocida*) and characterized by pneumonia and bleeding in subcutaneous tissues and internal organs. It is commonly associated with stress such as during shipping or crowding in a stockyard. Also called pasteurellosis, shipping fever.

hemp *n.* (Anglo-Saxon *henep*, hemp) **1.** a tall (up to 16 ft or 5 m) annual herb (*Cannabis sativa*, family Cannabinaceae) grown for its strong fiber used in making rope and to produce marijuana from its flowers and leaves. Some plants are monoecious (male and female flowers on the same plant), and some are dioecious (flowers of only one sex on a plant). Also called marijuana, locoweed. See marijuana. **2.** fibers from the plant, or rope made from the fibers. **3.** similar fibers from other plants.

hemp sesbania *n.* a semitropical annual herb (*Sesbania exaltata*, family Fabaceae) with yellow flowers. It is used as a cover and green-manure crop, but it can be a troublesome weed if it escapes.

hen *n.* (Anglo-Saxon *henn*, hen) **1.** a mature female chicken or other fowl. **2.** a female lobster or similar crustacean. **3.** (slang) a woman, especially one who gossips.

henbane *n.* (Anglo-Saxon *henn*, hen + *bana*, striking) a very poisonous plant (*Hyoscyamus niger*, family Solonaceae), especially to fowls; native of Eurasia. It has greenish yellow flowers and hairy foliage and contains a narcotic that is used in medicine.

henna *n.* (Arabic *hinnā*, henna) **1.** an Asian shrub or small tree (*Lawsonia inermis*, family Lythraceae) that grows in warm regions. Its white or red flowers smell like roses. **2.** a reddish brown dye made from dried, ground leaves of the shrub or tree and used to tint hair auburn or to draw decorative patterns on women's hands and feet. **3.** a reddish brown color.

henry (H) *n.* the unit for inductance in the SI system; the inductance in a closed circuit where a current varying at a rate of 1 ampere per second produces an electromotive force of 1 volt.

hepatitis *n.* (Greek *hēpatitis*, liver inflammation) a common viral disease of the liver that causes liver enlargement, jaundice, and fever. Hepatitis A (infectious hepatitis), commonly caused by fecal contamination in drinking water or by unsafe sexual practices, is normally mild and curable. Hepatitis B (serum hepatitis), transmitted through blood or other body fluids, is slower developing, more persistent, more serious, and sometimes fatal. Hepatitis C is caused by a retrovirus transmitted mostly through blood. Hepatitis D affects only persons who have had hepatitis B. Hepatitis E, transmitted mostly through contaminated water, has caused fatalities of pregnant women.

heptachlor *n.* a chlorinated hydrocarbon insecticide ($C_{10}H_7Cl_7$) banned for use on food products since 1978 and for other uses, such as seed treatment, since 1983; similar to chlordane.

herb *n.* (Latin *herba*, herb) **1.** a nonwoody flowering plant whose stem dies to the ground each year (an annual herb if its roots also die or a perennial if the roots survive for several years). **2.** plant material (seeds whole or ground, ground leaves, etc.) used in seasoning food or for medical purposes; e.g., anise, basil, dill, sage.

herbaceous *adj.* (Latin *herbaceus*, like grass) **1.** nonwoody; used to describe vascular plants or plant material. **2.** characteristic of an herb, especially having the color, shape, and texture of leaves or other foliage.

herbage *n.* (French *herbage*, herbage) succulent parts of plants, especially when eaten by animals; leaves, stems, flowers, etc.

herbarium *n.* (Latin *herbarium*, herbarium) **1.** a collection of dried or otherwise preserved plants that are appropriately mounted for display and labeled with scientific and common names. **2.** a room or building where such plants are displayed.

herbicide *n.* (Latin *herba*, herb + *cida*, killer) a chemical used to kill plants (especially weeds). A residual or preemergence herbicide acts in the soil and kills new seedlings; a postemergence herbicide is applied to the plants and may be either selective (to kill only certain types of plants) or nonselective. The action of the herbicide may be either contact (nonmobile) or systemic (translocated throughout the plant). Herbicides can be classified chemically as chlorinated aliphatic acids, triazines, urea compounds, etc. Also called defoliant.

herbicide resistance *n.* the ability of a plant either to resist absorbing or to tolerate the presence of a herbicide. Such characteristics can be a very favorable feature of crop plants, but they can be a serious problem in weeds. Herbicide resistance tends to increase with succeeding generations because the resistant plants are more likely to survive and reproduce.

herbivore *n.* (Latin *herba*, herb + *vorus*, eating) an animal that sustains itself by eating plant material rather than other animal flesh; an animal that converts plant biomass into animal biomass; e.g., cattle, sheep, goats, horses, deer. See carnivore, omnivore.

herd *n.* (Anglo-Saxon *heorde*, herd) **1.** a coherent group of animals, either domestic or wild. **2.** (slang, often disparaging) a large group of people. *v.* **3.** to drive or guide and care for a group of animals. **4.** to gather a group of animals into a confined area or to form a herd.

herder *n.* (Anglo-Saxon *heorde*, herd + *er*, one who does) one who tends a herd, especially a flock of sheep or goats; e.g., a shepherd.

herdsman *n.* (Anglo-Saxon *hyrdemann*, herdsman) one who tends a herd, especially of cattle; e.g., a cowboy.

heritable *adj.* (French *hérit*, inherit + *able*, able) **1.** able to be passed from one generation to the next. **2.** controlled by genetics rather than environmental influences.

hermaphrodite *n.* (Greek *hermaphroditos*, hermaphrodite) a plant, animal, or person having both male and female reproductive organs on the same individual; e.g., oak trees, pine trees, earthworms. See monoecious, gynandromorph.

hermetic *adj.* (Latin *hermeticus*, pertaining to Hermes) **1.** airtight; sealed to keep air from getting in or out; e.g., canned food. **2.** isolated from outside influence.

heron *n.* (French *hairon*, heron) any of a large number of wading birds (family Ardeidae) with long legs and neck; a long, straight, pointed bill; a short tail; and broad wings. Most herons feed on aquatic animals in shallow water.

herpes *n.* (Greek *herpēs*, creeping) any of a number of viral infections in humans and many animals that causes blisters on the skin or mucous membranes. Some forms are intensified by emotional disturbances.

herpetology *n.* (Greek *herpeton*, reptile + *logos*, word) the branch of zoology that studies snakes and other reptiles and amphibians.

hertz (Hz) *n.* the international unit for frequency; named after Heinrich Rudolph Hertz (1857–1894), a German physicist. One hertz equals one cycle per second.

Hessian fly *n.* a tiny fly (*Phytophaga destructor*, family Cecidomyiidae) whose larvae are very destructive on certain grasses, especially wheat. Each female lays hundreds of eggs on the leaves of young plants; the eggs produce small red maggots that make their way down the leaf, inside the sheath, and eat the tender stem tissues. Injury can be reduced by planting Hessian fly–resistant varieties or by planting late to avoid the major population levels of Hessian fly. See fly-free date.

heterocyst *n.* (Greek *heteros*, other + *kystis*, bag or pouch) a nonphotosynthetic cell that provides an anaerobic environment for nitrogen fixation in a filament of cyanobacteria (blue-green algae).

heteroecious *adj.* (Greek *hetero*, different + *oikos*, house) requiring more than one plant host for a parasite to complete a life cycle. Example: cedar-apple rust requires a cedar or juniper tree for one stage of the rust and an apple tree for another stage.

heterogametic *adj.* (Greek *hetero*, other + *gamos*, reproduction + *ikos*, of) having more than one type of gamete, as a male producing some sperm with an X chromosome and some with a Y chromosome.

heterogamy *n.* (Greek *hetero*, other + *gamos*, reproduction) sexual reproduction that involves fusion of two types of gametes, as sperm and eggs or pollen and ovules.

heterogeneous *adj.* (Greek *heterogenēs*, of different kinds) **1.** different; foreign; incongruous. **2.** of mixed composition; not homogeneous. **3.** present in more than one state, as a mixture of ice and water.

heterogenesis *n.* (Greek *hetero*, other + *genesis*, origin) alternation between sexual and asexual reproduction in successive generations (the sexual and asexual generations commonly differ in appearance). Also called xenogenesis.

heterophyte *n.* (Greek *hetero*, other + *phyton*, plant) a plant that obtains its nutrition from living or dead plants or animals; a parasite or a saprophyte.

heterosis *n.* (Greek *heterosis*, alteration) hybrid vigor; increased rate of growth, size, yield, or other characteristic in the offspring of unlike parents; e.g., a mule, hybrid corn.

heterotroph or **heterotrophic organism** *n.* (Greek *hetero*, other + *trophos*, one who eats) an organism that depends on organic compounds from other organisms for its nutrition, including consumers such as humans and animals and decomposers such as bacteria and fungi. (In contrast to an autotroph.)

heterozygote *n.* (Greek *hetero*, one of two that differ from each other + *zygōtos*, a fertilized egg) a hybrid plant or animal that has different alleles in gene pairs that control one or more heritable characteristics. Its offspring inherit the different alleles randomly.

heterozygous *adj.* (Greek *heteros*, one of two that differ from each other + *zygos*, of the egg) **1.** having different alleles as gene pairs for one or more heritable characteristics. **2.** having to do with a heterozygote.

hexamitiasis *n.* an infectious disease caused by parasitic protozoa (*Hexamita meleagridis*) affecting the intestines of fowls, especially young turkeys.

hexose *n.* (Greek *hex*, six + French *ose*, sugar) a six-carbon sugar ($C_6H_{12}O_6$) that can have either a chain or ring form. Glucose, 1 of the 16 possible aldehyde forms, is the most abundant monosaccharide, and fructose is the most abundant of the eight possible ketone forms. A linkage of glucose and fructose produces sucrose, the most abundant disaccharide.

Hg *n.* (Latin *hydrargyrum*, liquid silver) the chemical symbol for mercury.

hibernaculum *n.* (Latin *hibernaculum*, winter residence) **1.** a winter den where an animal such as a bear hibernates. **2.** a structure built by a larva for winter protection; e.g., the bag of a bagworm. **3.** an overwinter enclosure for a plant reproductive organ; e.g., a bulb.

hibernation *n.* (Latin *hibernare*, to pass the winter + *atio*, action) spending the winter in a dormant state, as bears, snakes, and moles do. The body temperature is low, and metabolic activity is very slow; the animal survives on stored fat. See estivation.

hibiscus *n.* (Greek *hibiskos*, marshmallow) **1.** a woody plant (*Hibiscus rosa-sinensis*, family Malvaceae) with large, showy blossoms; the state flower of Hawaii. **2.** any of about 250 related species (*Hibiscus* sp.) of tropical shrubs, trees, and other plants with large, showy blossoms. Several species are used as ornamentals, some are edible (e.g., okra, *H. esculentus*), and some are used as fiber crops (e.g., *H. cannabinus*).

hickory *n.* (American Indian *pohickory*, hickory tree) **1.** a North American tree (*Carya* sp., family Juglandaceae) with large compound leaves divided into five or more pointed leaflets, greenish blossoms, and smooth-shelled nuts. The nuts of some species are edible (e.g., pecans, *C. illinoensis*). **2.** the hard, tough wood of certain of these species, commonly used for tool handles, fuel wood, furniture, flooring, boxes, and crates. **3.** a cane or switch made of such wood. **4.** a nut from any of these trees. Also called hickory nut.

hidden hunger *n.* a nutrient deficiency that reduces yield but is not severe enough to produce visible symptoms. Although usually used to describe crop plants, the concept could also apply to people and animals with inadequate supplies of essential vitamins and minerals.

hiemal *adj.* (Latin *hiemalis*, of winter) like winter; pertaining to winter.

highbush blueberry *n.* a bushy shrub (*Vaccinium corymbosum*, family Ericaceae) that produces edible round, bluish black berries (the source of most blueberries in the market); native of northeastern U.S.A.

high-density polyethylene *n.* a low-permeability polymer of ethylene used to make sheet plastic (e.g., for landfill liners) and containers such as milk jugs. It produces toxic fumes when burned.

high-level radioactive waste *n.* radioactive waste material that emits more than 100 nanocuries per gram; material that requires perpetual storage in isolation; e.g., spent nuclear fuel rods and much of the waste from the production of nuclear weapons.

high-lime soil *n.* a soil that contains free calcium carbonate (usually identified by effervescence with dilute hydrochloric acid).

high pressure *n.* **1.** an anticyclone; a mass of air that has higher atmospheric pressure than the surrounding air; the wind blows clockwise around a center of high pressure in the northern hemisphere or counter-clockwise in the southern hemisphere. High pressure usually indicates that fair weather is approaching. *adj.* **2.** using persuasive argument, as a high-pressure salesman. **3.** operating under high fluid pressure, as a high-pressure irrigation system or a high-pressure lubrication system.

Hilgard, Eugene Woldemar *n.* an early American soil scientist (1875–1905) who noted that soil regions correspond with vegetative regions and climatic regions. He made the first county soil maps and founded the College of Agriculture and the University of California Agricultural Experiment Station.

hill *n.* (Anglo-Saxon *hyll*, hill) **1.** a topographic feature higher than the surrounding landscape but smaller than a mountain. **2.** a mound built by ants or other animal life. **3.** a mound of soil around a plant or small group of plants, or the plants themselves, as a hill of beans. Note: A hill of beans is sometimes used to represent a small amount or something of little value.

hill country *n.* an area of rolling topography; a series of hills or low mountains separated by valleys.

hillock *n.* (Anglo-Saxon *hyll*, hill + *ock*, small) a small hill or large mound.

hinny *n.* (Latin *hinnus*, hinny) the offspring of a female donkey sired by a male horse. See mule.

hippiatrics or **hippiatry** *n.* (Greek *hippos*, horse + *iatreia*, healing) the branch of veterinary medicine specializing in diseases of horses.

hippopotamus *n.* (Greek *hippos*, horse + *potamos*, river) a large vegetarian mammal (*Hippopotamus amphibius*) of African origin that spends much time in the water. It has a thick body, short legs, and a broad head and is hairless except for the end of its tail.

hircine *adj.* (Latin *hircinus*, of a goat) pertaining to a goat, especially smelling like a goat.

Hiroshima *n.* the Japanese seaport destroyed on August 6, 1945, by a U.S. atomic bomb, the first ever used in warfare. See Nagasaki.

hirsute *adj.* (Latin *hirsutus*, rough or bristly) **1.** hairy or bristly. **2.** having stiff feathers.

histamine *n.* (Greek *histos*, tissue + *amine*, ammonia chemical) beta-imidazolyethyl-amine ($C_5H_9N_3$); an amine that occurs in all plant and animal tissues. It stimulates gastric secretions and dilates capillaries. The release of histamine in allergic reactions lowers blood pressure and causes nasal irritation, sneezing, and eye watering.

histic epipedon *n.* a surface soil layer high in organic matter that characterizes Histosols in Soil Taxonomy. Its thickness ranges from 20 to 60 cm (8 to 24 in.), and the organic carbon content must be at least 12% to 18% (12% + 0.1 × % clay up to 60% clay).

histology *n.* (Greek *histos*, tissue + *logos*, word) **1.** the structure of plant, animal, and human tissues, especially on a microscopic scale. **2.** the branch of biology concerned with a study of the structure of organic tissues.

histone *n.* (Greek *histos*, tissue + *one*, chemical) any of five types of protein that contain numerous positively charged amino acids (arginine and lysine) that bond to negatively charged sites on the DNA molecule and help fold the long chain of DNA into a condensed chromosome. See chromosome, nucleosome.

Histosol *n.* (Greek *histos*, tissue + French *sol*, soil) a soil with a histic epipedon; a member of the soil order in Soil Taxonomy that includes peats and mucks (organic soils). Histosols represent about 1% or 2% of the soils in the U.S.A. and the world. See histic epipedon, organic soil.

hive *n.* (Anglo-Saxon *hyf*, hive) **1.** a home for honeybees. Also called a beehive. **2.** the colony of bees that inhabit one beehive. **3.** a very busy place and/or the participants who make it busy, as a beehive of activity. **4.** a skin eruption caused by an insect bite or an allergy; usually called hives or urticaria. *v.* **4.** to gather or enter into a hive. **5.** to live together in a busy place. **6.** to build a reserve of goods for the future, as bees hive honey.

hives *n.* a skin eruption that causes itching and burning on the skin of humans or animals. Causes include insect bites, food sensitivity, cold, drugs, or other environmental stress. Also called urticaria.

hoar frost *n.* delicate white ice crystals formed by condensation of moisture from the air when the dew point (frost point) is below freezing. Also called rime, frost.

hoe *n.* (French *houe*, hoe) **1.** a hand tool for working the soil, consisting of a handle and a transverse blade with variable shape according to its use for light weeding or deeper cultivation in sandy or clayey soils, etc. U.S. hoes typically have long handles, but people in developing nations often prefer short handles (users must bend down, but they work faster and more effectively). **2.** a similar hand tool used for other purposes, such as mixing mortar for brickwork. *v.* **3.** to use a hoe; e.g., to hoe weeds in a garden.

Hoar frost on a shrub

hoecake *n.* unleavened bread made of cornmeal (or flour) and water, formed into a patty, and baked (in colonial times in southern U.S.A.) on the blade of a hoe held over an open fire.

hoe culture *n.* growing crops with hoes used as the primary tillage implements as well as for weeding (a common practice in developing countries where labor is abundant and machinery is expensive and scarce).

hog *n.* (Anglo-Saxon *hogg*, swine) **1.** a pig; a domesticated swine, especially one weighing 120 lb (54 kg) or more and being prepared for market. **2.** any of several hoofed omnivorous mammals (family Suidae), including wild boars and warthogs as well as domesticated swine. **3.** a greedy or filthy person. **4.** (slang) a large motorcycle. *v.* **5.** to take as much as possible (much more than one's fair share).

hogback *n.* **1.** something with an arched back like that of a hog. **2.** a narrow ridge with steeply sloping sides, usually formed of tilted rock strata that are resistant to erosion.

hog cholera *n.* an acute, highly contagious, often fatal viral (*Tortor suis*) disease of swine that causes fever, diarrhea, loss of appetite, emaciation, and damage to internal organs. It often devastated herds in the late 19th and early 20th centuries prior to the development of an anti-hog-cholera serum at Ames, Iowa, about 1907.

hogget *n.* (Anglo-Saxon *hogg*, swine + French *ette*, small) **1.** a sheep about a year old that has not yet been sheared, especially in Australia and New Zealand. **2.** the fleece of such a sheep. **3.** (British) a young boar between one and two years old.

hog off *v.* to harvest a grain crop by turning the hogs in to eat it when the grain is nearly ripe (often because it is a poor crop not worth harvesting otherwise).

hog louse *n.* the only louse species that attacks hogs; a blood-sucking louse (*Haematopinus adventicius*, family Haematopinidae) that is the largest of all lice (about ¼ in. or 6 mm long and almost as wide).

hogshead *n.* **1.** a volume of liquid, especially one equal to 63 U.S. gal (52.5 imperial gal or 238 L). **2.** a large cask, especially one with a volume of 100 to 140 U.S. gal (378 to 530 L).

hog-wallow land *n.* nearly flat land with small basins and mounds measuring only a few yards (or meters) across and up to about 3 ft (1 m) high; characteristic of black clay Vertisols in Texas.

holding pen *n.* a pen used to hold sheep or other livestock for a short period of time.

holding pond *n.* a pond made with an earthen dam to hold runoff until sediment can settle or other pollutants can be either treated or removed. See lagoon.

holly *n.* (Anglo-Saxon *holegn*, holly) any of a number of trees and shrubs (*Ilex* sp., family Aquifoliaceae) with red berries and glossy leaves that have several points. **2.** the foliage and berries from a holly plant, often used for Christmas decorations.

Holly with berries

holmium (Ho) *n.* (Latin *Holmia*, Stockholm + *ium*, element) element 67, atomic weight 164.93; a soft, malleable rare earth metal with a bright silver luster and unusual magnetic properties.

holocaust *n.* (Latin *holocaustum*, burnt whole) **1.** devastation; great destruction, especially by fire. **2.** reckless destruction of life. **3.** a religious offering or sacrifice that is completely burned. **4. The Holocaust:** the systematic killing of more than 6 million European Jews by the German Nazis before and during World War II.

Holocene *n.* (Greek *holos*, whole + *kainos*, recent) the geologic period since the last major glaciation, a period of approximately 10,000 years. Also known as the Recent period.

holophytic *adj.* (Greek *holos,* whole + *phyton*, plant + *ic*, of) autotrophic; able to nourish itself by synthesizing organic compounds from carbon dioxide (CO_2) and water (H_2O); e.g., chlorophyll-bearing plants and chemoautotrophs. See autotrophic nutrition, photosynthesis.

holotype *n.* (Greek *holos*, whole + *typos*, blow) the single specimen of a plant or animal described as the representative of a new species in taxonomy.

holozoic *adj.* (Greek *holos*, whole + *zōion*, an animal) heterotrophic; nutritionally dependent upon organic foods; e.g., animals and humans.

homegrown *adj.* produced locally, as crops or livestock that are grown on one's own property, in the local vicinity, or in the same state or nation where they are offered for sale and consumption.

homeostasis *n.* (Greek *homoios*, unchanging + *stasis*, standing) a tendency for internal constancy in an organism even though the external environment changes; e.g., constant body temperature, turgor pressure, pH, redox potential.

homeothermic *adj.* (Greek *homos*, same + *thermē*, heat + *ic*, of) homeothermal; warm-blooded; maintaining a constant body temperature; e.g., humans and other mammals. See poikilothermic, endotherm.

homestead *n.* (Anglo-Saxon *hamstede*, homestead) **1.** a dwelling place, especially a farm/ranch home and associated buildings and grounds; often such a place that has remained in the same family through generations. **2.** a dwelling that is occupied by the family that owns it and is legally exempted from seizure in bankruptcy. **3.** a unit of public land acquired under the U.S. Homestead Act of 1862 and subsequent acts. *v.* **4.** to acquire land by homesteading.

homesteading *n.* (Anglo-Saxon *hamstede*, homestead + *ing*, action) the process of acquiring a unit of undeveloped federal (or sometimes state) land under the regulations of the U.S. Homestead Act of 1862.

Homestead Act *n.* a law enacted by the U.S. government in 1862 (following the Civil War) to help jobless and displaced people secure land. A homesteader could claim a tract of government (or sometimes state) land, usually 160 acres (65 ha), by registering the tract, building a house on it, and living there for a minimum of 5 years.

hominid *n.* any modern or ancestral human or closely related primate (*Homo* sp. and *Australopithecus* sp., family Hominidae).

hominize *v.* (Latin *hominid*, human + *izein*, making) **1.** to alter the environment on Earth to suit human needs. **2.** to advance the evolutionary development of humankind.

hominoid *n.* (Latin *hominid*, human + *oides*, similar to) any member of a superfamily of primates (Hominoidea) that includes all hominids and manlike apes and their ancestors. Scientists announced the discovery of hominoid remains that are 4.4 million years old in 1994.

hominy *n.* (Algonquin Indian *auhuminea*, parched corn) coarse maize that has been dehulled and boiled; a southern U.S.A. delicacy originally dehulled by soaking kernels of corn in lye from wood ashes. Ground hominy is known as grits.

homogametic *adj.* (Greek *homos*, same + *gamete*, wife + *ic*, like) having only one type of reproductive gamete, as ova that contain one X chromosome in each egg, in contrast to heterogametic sperm that contain either an X or a Y chromosome.

homogeneous *adj.* (Greek *homos*, same + *genēs*, kind) **1.** uniform; having the same composition or character throughout, as a homogeneous population. See heterogeneous. **2.** alike in essential characteristics. **3.** mathematically comparable, as all component terms having the same units.

homogenized *adj.* (Greek *homos*, same + *genēs*, kind + *ized*, made) made uniform by mixing or other treatment; e.g., homogenized milk has been passed through small openings that break the fat globules into such small pieces that no visible cream separation occurs after 48 hours of refrigerator storage. See goat milk.

homoiotherm *n.* (Greek *homoios*, like + *thermē*, heat) a warm-blooded animal.

homoiothermic *adj.* (Greek *homoios*, like + *thermē*, heat + *ic*, having the nature of) having an internal mechanism to maintain a constant body temperature, as in humans, other mammals, and birds. Also called homothermal, homothermic, warm-blooded. See poikilothermic.

Homo sapiens *n.* the human species (family Hominidae). Modern humans are designated *Homo sapiens sapiens*. See hominid.

homosphere *n.* (Greek *homos*, same + *sphaira*, ball) the atmosphere of the Earth extending from ground level to a height of about 53 mi (85 km); turbulent mixing maintains an essentially uniform composition throughout this zone. The composition of the nonmixed zone (heterosphere) above it varies with elevation.

homozygote *n.* (Greek *homos*, same + *zygōtos*, yoked) a plant, animal, or human with identical alleles for the gene or genes that control a particular characteristic; an organism that will breed true to type for the designated characteristic(s).

honey *n.* (Anglo-Saxon *hunig*, honey) **1.** a natural sweet gathered from the nectar of flowers by honeybees, partially digested by the bees, stored in a comb, and dehydrated to a thick supersaturated liquid that will crystallize if left undisturbed for an extended time. It keeps well because it is an acid solution (pH 3.5 4) with a high osmotic concentration; it contains mostly fructose and glucose along with plant coloring, enzymes, and pollen grains. Honey is used as a sweetener on various foods and in baking, as a salve, and with many medicines to make them palatable. **2.** anything sweet, either literally or figuratively, as a honey of a deal. **3.** sweetheart; an affectionate title, especially for a close friend or spouse. *v.* **4.** to flavor with honey. **5.** to speak affectionately; to flatter or coax someone.

honeybee *n.* (Anglo-Saxon *hunig*, honey + *beo*, bee) a communal insect (*Apis mellifera*, family Apidae) that builds combs and fills them with honey; commonly domesticated and kept in hives. Each colony has three kinds of bees: one queen bee—a female that lays eggs and rules the hive by chemical messages known as pheromones; worker bees—sterile females that gather nectar to make honey; and drone bees—male bees whose only function is to fertilize the queen. Honeybees are the principal pollinators of more than 50 agricultural crops.

honeybee, Africanized *n.* an aggressive strain of honeybee that stings in swarms, sometimes killing an animal, and displaces feral bees. It migrated from Africa to Brazil in 1957, reached Texas in 1990, and has since moved northward and westward. Also called killer bee.

honeycomb *n.* (Anglo-Saxon *hunig*, honey + *comb*, toothed strip) **1.** a waxy material created by bees to make combs composed of hexagonal cells that are either brood cells, where the queen deposits her eggs, or honey cells, which the workers fill with honey. *adj.* **2.** having the appearance of a honeycomb, as a honeycomb pattern.

honeydew *n.* (Anglo-Saxon *hunig*, honey + *deaw*, dew) **1.** a sweet liquid that exudes from the leaves of certain plants when the weather is hot. **2.** a sweet liquid secreted by a group of aphids, scale insects, and leafhoppers (Homoptera). Certain ants (*Lasius allenus americanus*) protect the aphids (*Aphis maidiradicis*) so they can feed on the aphid's honeydew. See ant cow. **3.** a type of winter melon (*Cucumis melo inodorus*, family Cucurbitaceae) with a pale greenish rind and a tasty pale green interior.

honeylocust or **honey locust** *n.* a large thorny tree

Honeylocust leaves and pod

(*Gleditsia triacanthos*, family Fabaceae) having pinnate leaves and seed pods 12 to 18 in. (30 to 45 cm) long and 1 in. (3 cm) or more wide; native to east-central U.S.A. Livestock and wildlife eat the pods; the wood is used for fence posts, furniture, railroad ties, etc.

honey plant *n.* any plant used by bees as a source of nectar, especially one that gives honey a distinctive flavor. The principal honey plants in the U.S.A. are alfalfa, aster, buckwheat, citrus, clovers, goldenrod, fireweed, cotton, locust, mesquite, sweetclover, and sumac.

honeysuckle *n.* (Anglo-Saxon *hunigsuce*, honeysuckle) **1.** any of several bushy or climbing plants (*Lonicera* sp., family Caprifoliaceae) with small, fragrant pink, yellow, or white tubular blossoms. **2.** any of several similar plants.

hoof (pl. **hooves**) *n.* (Anglo-Saxon *hof*, hoof) **1.** a horny covering that protects the foot of a horse, donkey, or other similar animal. **2.** the entire foot of such an animal. **3.** a cloven hoof; the horny covering that protects the toes of certain animal, such as cows and pigs. **4.** (slang) a human foot.

hoof-and-mouth disease *n.* See foot-and-mouth disease.

hookworm *n.* (Dutch *hoek*, hook + Anglo-Saxon *wyrm*, worm) a small parasitic roundworm (family Ancylostomatidae) that has teeth or hooks around its mouth; most common in the tropics or in mines or tunnels with warm, moist conditions. It infests the small intestines of animals and humans, causing digestive disturbances, internal bleeding, diarrhea, and anemia; the disease is called hookworm disease, miners' anemia, tunnel anemia, ancylostomiasis, or ankylostomiasis.

hoop house *n.* a temporary structure placed over high-value crops in a field to provide a greenhouse effect, especially at night (it may be removed during the day). It is made of either wire or plastic semicircular hoops covered with either plastic or cloth.

Hopkins bioclimatic law *n.* See bioclimatic law, Hopkins.

hops (pl.) *n.* (Dutch *hoppe*, hop) **1.** a climbing vine (*Humulus lupulus*, family Moraceae) with separate male and female flowers. New vines growing each year from a perennial crown reach lengths of 15 to 25 ft (4.5 to 7.5 m). Hops are grown extensively with irrigation in Washington, Oregon, Idaho, and California. **2.** the ripened and dried cones from the female flowers that supply resins and essential oils used to flavor beer and ale and to make certain medicines.

horehound *n.* (Anglo-Saxon *harhune*, horehound) an aromatic perennial herb (*Marrubuim vulgare*, family Lamiaceae); a small bush with wrinkled leaves and whorls of small white flowers. Its juice is used in tea and for flavoring sugar candy; supposedly, horehound is good for colds.

horizon *n.* (Latin *horizon*, boundary circle) **1.** the skyline; the juncture where the sky appears to meet the surface of the Earth or sea. **2.** the outer boundary of one's perception, knowledge, or interest. **3.** a soil layer produced by soil-forming processes that have given it distinct characteristics (e.g., color, texture, structure, pH) different than those in the layers above and below it.

horizonation *n.* (Latin *horizon*, boundary circle + *ation*, condition) **1.** the process of soil formation producing distinct horizons in a soil. **2.** the identification of the horizons in a soil and designation of their nomenclature. The major soil horizons are designated O for organic, A for mineral surface horizons, E for bleached eluvial horizons, B for illuvial horizons, C for weathered rock that is not yet true soil, and R for hard bedrock.

hormone *n.* (Greek *hormōn*, stimulant) **1.** a chemical substance (e.g., insulin, growth hormone) produced in a gland or organ in the body of a person, animal, or insect and transported in the blood to another organ, where it either starts or stops a specific biochemical reaction. **2.** a similar substance (e.g., an auxin or gibberellin) produced in one part of a plant to control growth or other activity in another part. Also called phytohormone. **3.** a synthetic substance used medically to simulate the effect of a natural hormone.

horn *n.* (Anglo-Saxon *horn*, horn) **1.** a hollow bony growth, commonly curved and pointed, protruding from each side of the upper part of the head of certain ungulate mammals, including most cattle, sheep, goats, and antelopes. **2.** a similar growth in the middle of the snout of a rhinoceros or the tusk of a male narwhal. **3.** an antler. **4.** any similarly shaped projection from the head of any animal. **5.** the bony substance that constitutes such a growth, or any of several other similar substances. **6.** a wind instrument made either from a hollow animal horn or from brass or other suitable material. **7.** a sound producer, as those used in loudspeakers or as a warning device for a motor vehicle. **8.** a post at the front of a saddle where a rope may be held. **9.** the pointed end of an anvil. **10.** either point of a crescent, especially when pointing upward. **11.** a mountain peak with a pyramidal shape, especially one shaped by glaciation. **12.** either alternative of a dilemma. **13.** (slang) telephone.

hornbean *n.* See ironwood.

horned lark *n.* a small bird (*Eremophila alpestris*) that has a small tuft of feathers on each side of its head, is mostly brown on the back with black tail feathers, and is marked with yellow and black on the face and chest. Usually seen in large open fields or spaces, it is the only lark native to America.

horned lizard *n.* any of several lizards (*Phrynosoma* sp., family Iguanidae) resembling iguanas, with hornlike spines on the head and spiny scales covering a flattened body; native to arid parts of southwestern U.S.A. and Mexico. Also called horned toad.

hornet *n.* (Anglo-Saxon *hyrnet*, hornet) any of several large, yellow and black, stinging wasps (family Vespidae) that build papery nests with a honeycomb form.

horn fly *n.* a blood-sucking fly (*Siphona = Haemotobia irritans*, family Muscidae) that feeds primarily on cattle during warm weather, often causing a 10% to 20% drop in milk production or a weight loss in beef cattle. Horn flies lay their eggs in fresh cattle dung, where the maggots live for a few days until they pupate and become flies that then seek the nearest cow. Thousands of horn flies may live on a single cow during a warm, humid summer.

hornworm *n.* (Anglo-Saxon *horn*, horn + *wyrm*, worm) the larva of any of several hawk moths, so named for a hornlike growth at the rear of the abdomen; includes the tobacco hornworm (*Manduca sexta*, family Sphingidae) and the tomato hornworm (*M. quinquemaculata*), large green worms that eat the leaves of tobacco and tomato, respectively.

horse *n.* (Anglo-Saxon *hors*, horse) **1.** a large herbivorous mammal (*Equus caballus*, family Equidae) with long hair in a flowing mane and tail, four legs, and solid hooves. Horses have been domesticated since ancient times; some breeds are used for riding or racing, others for draft animals. Mature horses are called stallions (males) or mares (females); young horses are colts. See also filly, gelding. **2.** any member of the horse family (Equidae), including donkeys, zebras, etc. **3.** anything one rides on, as a rocking horse. **4.** a support or frame, as a saw horse or a clothes horse. **5.** a piece of exercise equipment, as a pommel horse or a vaulting horse. **6.** a rock mass surrounded by ore.

horsefly *n.* (Anglo-Saxon *hors*, horse + *fleogan*, fly) a troublesome fly (*Tabanus* sp., family Tabanidae) that attacks horses, deer, cattle, dogs, humans, etc. The female sucks blood and can carry anthrax or other diseases. Although similar to a deerfly, a horsefly is larger (about 2 cm long—almost an inch); also known as gadfly. Horseflies lay their eggs on plants or objects where the larvae can drop into the water and live there until they pupate. See horseguard.

horseguard or **horse guard** *n.* a black and yellow sand wasp (*Bembix carolina*, family Sphecidae) that buzzes loudly and captures horseflies and botflies to feed them to its larvae in its nest in the sand. It is common around horses and cattle in southern U.S.A.

horse latitudes *n.* either of the two climatic belts of warm, dry weather with high barometric pressure and little wind located about 30° to 35° North or South Latitude. So called because becalmed sailing ships carrying horses to America had to throw their horses into the sea when their supply of freshwater ran low.

horsepower *n.* (Anglo-Saxon *hors*, horse + French *poueir*, to be able) **1.** a unit of power in the foot-pound-second system equal to 550 foot-pounds per second (745.7 watts); used to express the power of an engine or motor, and sometimes of an animal or other power source. **2.** (slang) the resources needed to accomplish a task, as he doesn't have enough horsepower to do that.

horseradish *n.* **1.** a garden plant (*Armoracia rusticana*, family Brassicaceae) with small white flowers and a pungent fleshy root. **2.** a condiment prepared from the ground root of the plant; often mixed with mustard.

horseshoe *n.* (Anglo-Saxon *hors*, horse + *sceoh*, shoe) **1.** a U-shaped piece of metal nailed to a horse's hoof to protect it from injury. **2.** something with the shape of a U, especially a natural feature such as a curved part in a stream channel.

horseshoe crab *n.* a large arthropod with an armored body that includes a U-shaped front section as much as 22 in. (55 cm) wide with four pairs of legs and a roughly triangular rear section. *Limulus polyphemus* (family Limulidae) is found along the Atlantic and Gulf coasts of North America, and three related species are found along the Southeast Asia and East Indies coasts. Also called king crab.

horsetail *n.* **1.** an evergreen rushlike plant (*Equisetum hyemale = arvense*, family Equisetaceae) that has hollow jointed stems and grows in wet places. It is toxic to domestic animals, especially horses and sheep. See equisetum. **2.** giant horsetail (*E. telmateia*); a jointed plant found in wet meadows; native to northwestern U.S.A. Also called ivory horsetail.

horst *n.* (German *horst*, thicket or aerie) an elongated ridge with a normal fault on each side. A corresponding sunken structure is called a graben.

Horst (top) and graben (bottom)

horticulture *n.* (Latin *hortus*, garden + *cultura*, culture) the science that deals with growing vegetables, fruits, flowers, and other ornamental plants in gardens, lawns, nurseries, and commercial enterprises.

host *n.* (Latin *host*, host, guest, or enemy) **1.** a large number of individuals; a multitude. **2.** a person who receives and entertains guests in a home or elsewhere. **3.** a moderator or other lead person for a radio or television program. **4.** a city or organization that provides a meeting place and associated services for a professional meeting, sporting event, or other gathering. **5.** a living animal or plant that serves as a parasite's source of nutrition and shelter. **6.** a genetically modified organism, usually a bacterium, carrying a transplanted gene from another organism, e.g., to produce a hormone or other product related to the other organism. *v.* **7.** to act as a host; to receive guests or sponsor an event.

hotbed *n.* **1.** a heated bed of growth media, usually under an enclosure or in a greenhouse, used to grow plants out of season. **2.** an environment of intense activity, often of a radical nature; e.g., a hotbed of religion or of politics.

hot cap *n.* a temporary covering to protect a plant from frost in early spring, often a paper sack, a paper or cardboard cone, or a plastic jug with the bottom removed.

hothouse *n.* **1.** a heated greenhouse. **2.** a room kept at a high temperature; a drying room, e.g., for pottery. **3.** a place of intense activity, as a political hothouse.

hot manure *n.* fresh manure from horses or poultry; manure that generates considerable heat when it decomposes. See cold manure.

housefly or **house fly** *n.* (Anglo-Saxon *hus*, house + *fleogan*, fly) a household pest (*Musca domestica*, family Muscidae) of worldwide distribution that bothers both humans and animals; a medium-sized gray-striped fly that breeds in manure and decaying garbage and transmits such diseases as typhoid fever.

household waste *n.* garbage, sewage, and other waste from a private residence or other nonindustrial site (sometimes including similar waste from business districts).

house sparrow *n.* a small European finch (*Passer domesticus*, family Passeridae) with a streaked brown back, a gray breast, and in males, a black bib. It was intentionally introduced into the U.S.A. circa 1850 to destroy insects and caterpillars, only to become a pest that prefers grain, buds, etc. Also called English sparrow.

hovel *n.* **1.** a shed with an open side, especially one used to shelter cattle, produce, or machinery. **2.** a hut or other small, humble dwelling. **3.** any cluttered, dirty dwelling.

howl *n.* (Dutch *huilen*, howl) **1.** a prolonged, mournful cry produced by a dog or wolf. **2.** a similar sound made by some other animal or a person, often indicating pain or anger. **3.** a similar sound produced by the wind or a machine. *v.* **4.** to utter or produce a howling sound.

huckleberry *n.* **1.** a small shrub (*Gaylussacia baccata* = common or black huckleberry, *G. brachysera* = box huckleberry, or *G. dumosa* = dwarf huckleberry, family Ericaceae) that grows in well-drained acid soils, especially in mountainous areas. Also called lowbush blueberry. **2.** the small, round, edible, dark blue berries produced by such shrubs and often gathered from wild plants for home use.

hue *n.* (Anglo-Saxon *heow*, color or appearance) **1.** a spectral color, as red, blue, green, or yellow. See Munsell color. **2.** a particular shade or tint of a color, or a range of such tints, as pale yellowish hues or dark green hues. **3.** a loud outcry, especially that of a group of pursuers or objectors, used in the phrase hue and cry.

human *adj.* (Latin *humanus*, human) **1.** pertaining to people; having the nature of a person. **2.** consisting of people. **3.** the social concerns of people, as human affairs. **4.** humane.

humectant *n.* (Latin *humectans*, to moisten) a substance used to retain moisture even in a drying environment; e.g., glycerol used as a moisture retainer in food or as a component of skin lotions.

humic acid *n.* dark-colored soil humus that can be extracted from soil with an alkaline solution and precipitated when the solution is acidified to pH 2 or below.

humic substance *n.* any of a variety of black or brown organic materials produced in soils by decomposition of plant and animal residues, including fulvic acid, humic acid, and humin.

humid *adj.* (Latin *humidus*, moist) **1.** containing much water vapor, as humid air. **2.** receiving much precipitation (enough to have humid air and leached soils), as a humid area (one with annual precipitation more than about 15 in. or 38 cm in a cold climate such as Alaska, more than about 30 in. or 75 cm in central U.S.A., or more than about 60 in. or 150 cm in a tropical setting such as Hawaii).

humidify *v.* (Latin *humidus*, moist + *ificare*, to make) to add water vapor to the air; to make air more humid.

humidity *n.* (Latin *humiditas*, humidity) **1.** dampness; moist or wet conditions. **2.** relative humidity.

3. humid air that combines with warm temperatures to cause discomfort (stuffiness).

humification *n.* (Latin *humus*, earth or ground + *ificare*, to make + *atio*, action) the process that converts plant and animal residues into humus; a combination of biological (especially microbiological) and chemical processes.

humin *n.* any of the components of soil humus that are insoluble in a dilute alkaline solution. Also called ulmin. See humic substance.

hummingbird *n.* any of several small birds (family Trochilidae) with a long slender bill whose rapidly beating wings produce a humming sound. Hummingbirds can hover to obtain nectar from a flower and fly backward to exit from the flower. Most species are quite colorful.

Hummingbird

hummock *n.* **1.** a small mound, often built up around a clump of vegetation in a swamp or in an area with drifting sand or silt; sometimes produced by humans mounding an area of soil to grow desired plants. **2.** a densely wooded area with its ground level a little above that of an adjacent swamp. **3.** a ridge on a glacier.

hummocky *adj.* covered with hummocks; an undulating land surface.

hummus, hommos, or **hummous** *n.* (Arabic *hummus*, chickpea) a paste or dip used in Middle Eastern cookery. It is made from chickpeas ground with garlic, sesame seeds, oil, and lemon juice and is usually served with pita bread.

humus *n.* (Latin *humus*, earth or ground) **1.** nonliving organic materials remaining in soil after most of the original plant and animal residues have decomposed and are no longer recognizable. **2.** soil organic matter.

hundred weight (cwt) *n.* **1.** 100 lb (45.4 kg) = 1/20 of a short ton. **2.** 112 lb (50.8 kg) = 1/20 of a long ton (the original use and still common in England).

hunger *n.* (Anglo-Saxon *hungor*, hunger) **1.** the craving of a person or animal for food; a need that progresses from appetite to great discomfort, weakness, and starvation. **2.** a driving urge to obtain or achieve something, as a hunger for winning or for righteousness. *v.* **3.** to have hunger. **4.** to desire strongly and seek actively, e.g., to hunger for power.

hunter *n.* **1.** a person who hunts wild game animals or birds for either food or sport. **2.** one who searches for anything, as a treasure hunter. **3.** an animal trained for use in hunting, especially a horse or dog. **4.** an animal that hunts its own prey, as a cat or spider. **5. The Hunter**; the constellation Orion.

hunter's moon *n.* the first full moon after the harvest moon (providing light for hunting after crops are harvested).

hunting and gathering *n.* the mode of life based on hunting wild animals, fishing, and gathering food, such as edible berries and shoots from wild plants; living off the land.

hurricane *n.* (Spanish *huracán*, hurricane) a large, violent cyclone that develops over tropical waters with winds of 74 or more mph (120 or more km/h). See Beaufort wind scale. It is called a typhoon if it occurs in the China Sea or near the Philippines or a cyclone if it occurs near India or Australia. (Tornadoes are smaller and may be spun off by a hurricane or typhoon.)

husbandry *n.* (Anglo-Saxon *husbonda*, husband + *ry*, practice) **1.** the art or science of care and management of livestock and/or (especially in older usage) field crops; often used with a designator, as animal husbandry or dairy husbandry. **2.** careful management of resources to avoid waste; frugality.

husk *n.* (Anglo-Saxon *hosu*, pod or husk) **1.** the natural covering around certain seeds, especially that around an ear of corn, also called a shuck. **2.** a dry exterior that protects the interior of anything. *v.* **3.** to remove husks; to shuck (especially corn).

hutch *n.* (Anglo-Saxon *hwicce*, chest) **1.** a pen or cage for a small animal; e.g., a rabbit hutch. **2.** a storage cabinet or shelves, usually open in front and located over a cupboard or desk, as a china hutch or a computer hutch. **3.** a hut.

hyacinth *n.* (Latin *hyacinthus*, blue larkspur) **1.** a bulbous plant (*Hyacinthus orientalis*, family Liliaceae) that produces long, narrow leaves and fragrant, colorful (white, yellow, red, blue, or purple) flowers on spikes. **2.** any of several related plants, including the water hyacinth (*Eichhornia crassipes* or *E. azurea*, family Pontederiaceae). **3.** certain gem stones, especially in ancient times, either a blue gem (sapphire) or a reddish orange gem (zircon, garnet, etc.). **4.** a purplish blue color.

hyaline *adj.* or *n.* (Greek *hyalinos*, glassy) **1.** transparent or glassy, as a clear sky or a smooth sea. **2.** degenerate tissue that has become transparent or translucent. **3.** amorphous; lacking crystal structure.

hybrid *n.* (Latin *hybrida*, an offspring of a tame sow and a wild boar) **1.** offspring of mixed parentage; either animals or plants produced by crossing different breeds, varieties, or species, especially with human intervention. **2.** a person whose heritage combines two contrasting cultures. **3.** anything produced as an unusual combination of two contrasting components. **4.** a word whose formative parts come from two languages, as television (Greek and Latin). *adj.* **5.** coming from two different sources. **6.** composite; having incongruous parts.

hybrid corn *n.* corn produced from hybrid seed. The seed is produced by crossing inbred lines and is used to increase yields through hybrid vigor.

hybrid vigor *n.* heterosis; increased size, strength, yield, or other characteristics associated with a hybrid animal or plant as compared with its parents; e.g., a mule or hybrid corn.

hydatid *n.* (Greek *hydatis*, watery vesicle) a watery cyst formed by a tapeworm larva. See *Echinococcus*.

hydra *n.* (Latin *hydra*, water serpent) **1.** a freshwater polyp (family Hydridae) with a round body and a ring of tentacles surrounding its mouth. **2.** a problem that keeps presenting new difficulties even as some parts are solved (based on Greek mythology).

hydrate *n.* (Greek *hydōr*, water + *ate*, chemical) **1.** a chemical compound that contains water molecules; e.g., gypsum ($CaSO_4 \cdot 2H_2O$). *v.* **2.** to add water of hydration; e.g., to change anhydrite ($CaSO_4$) into gypsum.

hydration *n.* (Greek *hydōr*, water + *atio*, action) attraction to water molecules, either to form a hydrated compound (e.g., gypsum, $CaSO_4 \cdot 2H_2O$) or to form hydrated ions in solution (e.g., hydrated Na^+ and Cl^- ions from NaCl, and similar ions from most other soluble salts). See hydrolysis.

hydraulic conductivity *n.* the constant in Darcy's law representing the permeability of a porous medium (e.g., soil) to water flow under saturated conditions.

hydraulic ram *n.* a device that uses the energy of flowing water to lift a small portion of the water to a higher elevation. In the illustration, the hydraulic ram allows water to escape through valve A until the velocity becomes great enough to overcome the spring pressure and close the valve; the inflow momentum then opens valve B and forces water into

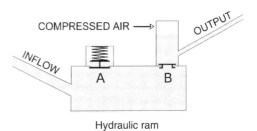

Hydraulic ram

the output chamber until the velocity slows; the slower velocity allows valve B to close and valve A to reopen, and the cycle repeats.

hydraulics *n.* (Greek *hydraulis*, a water organ) the scientific study of how water and other liquids flow.

hydric soil *n.* any soil that is saturated with water for long enough periods to have anaerobic conditions influencing the growth of plants. Since 1994, U.S. law requires hydric soils to be classified as wetlands; adding new artificial drainage is prohibited on such land.

hydrilla *n.* an aggressive weed (*Hydrilla verticillata*, family Hydrocharitaceae) native to Asia, Africa, and Australia that grows underwater, clogging streams by its massive growth and producing tubers that can remain viable for years. The use of herbicides is limited by environmental contamination problems. The use of fish or insects to control hydrilla is being researched.

hydrocarbon *n.* an organic compound composed of hydrogen and carbon, including aliphatic (chain) structures such as methane, ethane, propane (CH_4, C_2H_6, C_3H_8), etc., and aromatic (ring) structures such as benzene (C_6H_6). Methane and other lightweight hydrocarbons are gaseous, most middleweight hydrocarbons are liquids, and the large, complex molecules form solids.

hydrocyanic acid *n.* a solution of hydrogen cyanide (HCN) in water; a toxic compound produced by several species of plants (e.g., sorghum) under environmental stress. It reacts with bases to form cyanides, many of which are highly toxic. Also called prussic acid.

hydrodynamics *n.* (Greek *hydōr*, water + *dynamikos*, dynamic) the study of the dynamics of liquids, especially the physics of flowing liquids.

hydroelectric *adj.* (Greek *hydōr*, water + *ēlektron*, amber + *ikos*, of) pertaining to the generation of electricity from water power.

hydrogen (H) *n.* (French *hydrogène*, hydrogen) element 1, atomic weight 1.00797, the lightest of all known elements and the principal component of stars. Hydrogen forms a colorless, odorless, flammable gas (H_2) that produces water (H_2O) when burned; it bonds readily to carbon and to many other elements. It is an essential element for all known forms of life.

hydrogenate *v.* (French *hydrogène*, hydrogen + *ate*, to treat with) to react with hydrogen to reduce the number of double bonds in an organic compound; e.g., to modify petroleum products or to increase the saturation of oils (e.g., cottonseed or soybean oil), often turning them into solids and making them more resistant to oxidation and rancidity.

hydrogen bomb *n.* an extremely powerful explosive device based on the principle of nuclear fusion, whereby the heavy isotopes of hydrogen (deuterium and tritium) are fused under intense heat and pressure to form helium and release energy.

hydrogen bond *n.* a weak bond whereby a hydrogen atom that is strongly bonded to another atom on one side is weakly attracted to an exposed pair of electrons on a nearby atom on its opposite side.

hydrogen sulfide (H_2S) *n.* a toxic gas commonly formed during the decomposition of organic materials that contain sulfur; an acute irritant that reacts with bases to form sulfides. It is used in metallurgy, in the production of certain chemicals, and for certain chemical tests. Also known as rotten egg gas.

hydrogeology *n.* (Greek *hydōr*, water + *geō*, earth + *logos*, word) the study of groundwater in geologic strata.

hydrolase *n.* (Greek *hydōr*, water + *ase*, enzyme) any enzyme that catalyzes the hydrolysis of an organic compound; e.g., urease causes urea to react with water and form ammonium carbonate. See enzyme.

hydrologic cycle *n.* the natural water cycle, including water transpiring from plants and evaporating from both sea and land, forming clouds, falling again as rain, snow, etc. (with a net transfer from sea to land), and returning to the sea in streams and as groundwater seepage.

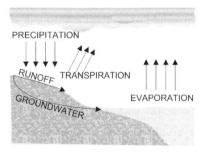

Hydrologic cycle

hydrologic process *n.* any process that involves water movement; e.g., infiltration, runoff, rill and gully erosion, stream flow.

hydrology *n.* (Greek *hydōr*, water + *logos*, word) the science of water distribution and movement, often divided into specialties, as groundwater hydrology, surface hydrology, wetland hydrology, etc.

hydrolysis *n.* (Greek *hydōr*, water + *lysis*, loosening) reaction of a compound with hydrogen (H^+) and/or hydroxyl (OH^-) ions of water to form new ions or compounds; e.g., the replacement of cations in a mineral structure with hydrogen ions from water. See hydration.

hydrometer *n.* (Greek *hydōr*, water + *metron*, measure) a device consisting of a bulb with an elongated stem used to measure the density of a liquid. Enough weight is placed in the bulb to submerge the bulb but leave part of the stem above the surface; graduations on the stem indicate the density of the liquid.

hydrometer analysis *n.* a means of determining the silt and clay content of a soil or other sample by measuring the density of a suspension with a hydrometer after allowing enough time for the sand to settle out and again after the silt has settled out. See particle-size analysis, pipette analysis, sieve analysis.

hydromorphic *adj.* (Greek *hydōr*, water + *morphē*, form + *ikos*, of) having characteristics associated with wetness; developed in a wetland environment, as hydromorphic plants (reeds, sedges, etc.) or a hydromorphic soil.

hydromorphic soil *n.* a wetland soil; any of seven great soil groups of the 1938 U.S. system of soil classification that included soils with poor drainage (e.g., soil formed in a bog or swamp).

hydrophilic *adj.* (Greek *hydōr*, water + *philos*, loving + *ikos*, of) strongly attracted to water; capable of uniting with or absorbing water even in a relative dry environment; a characteristic of gelatins, vegetable gums, algin, pectin, collagen, silica gel, calcium chloride and certain other chemicals, and soil colloids such as fine clay and humus.

hydrophobia *n.* (Greek *hydōr*, water + *phobos*, fear) **1.** an abnormal fear of water. **2.** rabies (from one of its symptoms, the inability to swallow water).

hydrophobic *adj.* (Greek *hydōr*, water + *phobos*, fear + *ikos*, of) water repellent; resistant to absorbing water; a characteristic of most oils, fats, waxes, and resins.

hydrophyte *n.* (Greek *hydor*, water + *phyton*, a plant) a plant that grows in water or saturated soil. Hydrophytes may be:
- free-floating; e.g., algae
- submerged-floating with roots in the soil at the bottom of the water; e.g., water lilies
- emersed (not floating) with roots growing in the mud; e.g., cattails, wetland rice.

hydroponics *n.* (Greek *hydōr*, water + *ponos*, labor) the growing of plants in water without any soil. The essential nutrients are dissolved in the water.

hydrosphere *n.* (Greek *hydōr*, water + *sphaira*, ball or globe) **1.** the water in the Earth's oceans, lakes, streams, etc., sometimes construed to include the groundwater in saturated soil and rock strata, thus making the hydrosphere nearly continuous around the Earth. **2.** as above plus the water vapor in the atmosphere.

hydrostatic *adj.* (Greek *hydōr*, water + *statikos*, stationary) dealing with pressure and equilibria in liquids.

hydrothermal deposit *n.* a mineral deposit formed by the action of hot water and/or gases; usually occurs as a vein in a joint, crack, or fault in older rock with a contrasting appearance and composition.

hydrothermal vent *n.* an opening where hot mineral water and gases escape, often on the floor of the sea.

hydrotropism *n.* (Greek *hydōr*, water + *tropos*, turning + *ismos*, action) **1.** the tendency of an organism to grow toward water; e.g., willow tree roots that grow into drainage tiles and plug them. **2.** the influence of water on the growth or movement of an organism.

hydroxide clay *n.* a hydrated oxide of mostly iron or aluminum (hydrated Fe_2O_3 or Al_2O_3) that forms clay-size particles, mostly without a definite crystalline structure. It commonly is present in small amounts as ped coatings in many soils (especially subsoils) and as the dominant component in certain highly weathered soils.

hyena *n.* (Latin *hyaena*, hyena) a carnivore (family Hyaenidae) the size of a large dog with a hunched back; a native of Africa and southern Asia. Hyenas commonly travel in large packs, hunting mostly at night or scavenging; they are known for their laughing cry.

hyetograph *n.* (Greek *hyetos*, rain + *graphein*, to write) a chart showing rainfall distribution for a period of time (especially a year) or across a geographic area.

hyetography *n.* (Greek *hyetos*, rain + *graphia*, writing) the branch of meteorology that deals with annual rainfall distribution over a geographic area.

hygiene *n.* (French *hygiène*, health rules) **1.** the science of maintaining good health and avoiding the spread of disease. **2.** cleanliness and other practices that help maintain good health and esthetics.

hygrometer *n.* (Greek *hygros*, wet + *metron*, measure) any instrument that measures the water content of the air. See psychrometer, relative humidity.

hygrophyte *n.* (Greek *hygros*, wet + *phyton*, plant) **1.** a plant that requires a very humid atmosphere. **2.** a hydrophyte.

hygroscope *n.* (Greek *hygros*, wet + *skopos*, watcher) an instrument for determining the approximate relative humidity of the air.

hygroscopic *adj.* (Greek *hygros*, wet + *skopos*, watcher + *ikos*, like) tending to absorb moisture from the atmosphere.

hygroscopic coefficient *n.* the water content of an air-dry soil, expressed as a percentage of the oven-dry weight; for precise measurements, the water content of a soil equilibrated with air at 25°C and 98% relative humidity.

hygroscopic water *n.* the water held in an air-dry soil that is driven off by oven-drying.

hyla *n.* (Greek *hylē*, wood) any of a large genus of tree frogs (*Hyla* sp.); also called spring peepers because their voices are loudest in spring.

hylophagous *adj.* (Greek *hylophagos*, eating wood) feeding on young shoots of trees, as certain insects.

hymenoptera *n.* (Greek *hymen*, membrane + *pteron*, wing) an order of insects that typically have a biting or sucking mouth and either no wings or four membranous wings, including ants, bees, wasps, ichneumon flies, and sawflies.

hyoscyamine *n.* (Latin *hyoscyamus*, henbane + *ine*, chemical) a very poisonous alkaloid ($C_{17}H_{23}NO_3$) produced by henbane and other plants of the nightshade family. Used in medicine as a sedative, analgesic, and antispasmodic.

hyperparasite *n.* (Greek *hyper*, above + *parasitos*, a parasite) a parasite that eats other parasites. A parasite that preys on a hyperparasite is called a second degree hyperparasite.

hyperpnea *n.* (Greek *hyper*, over + *pnoiē*, breathing) abnormally heavy breathing, either labored or very deep. See polypnea.

hypersaline *n.* (Greek *hyper*, over + Latin *salinus*, salty) excessively salty.

hyperthermia *n.* (Greek *hyper*, over + *thermē*, heat) fever; an excessively warm body temperature.

hyperthermic *adj.* (Greek *hyper*, over + *thermē*, heat + *ikos*, like) having abnormally high temperature; used in Soil Taxonomy to name soil subgroups that have mean annual temperatures above 22°C (71.6°F).

hypertrophy *n.* (Greek *hyper*, over + *trophia*, nutrition) excessive enlargement of an organ or other part of an organism caused by oversized cells.

hypha *n.* (pl. **hyphae**) (Greek *hyphē*, a weaving) a filament in the mycelium of a fungus. The hyphae constitute the main body of a fungus.

hypocalcemia *n.* (Greek *hypo*, under or low + Latin *calx*, lime + *emia*, blood condition) too little calcium in the blood. It causes milk fever in cows and weak bones in people and animals.

hypodermic flow *n.* flow through cracks, wormholes, and other passages large enough that capillary forces are not a significant factor. Sometimes called preferential flow.

hypolimnion *n.* (Greek *hypo*, below + *limnē*, lake + *ion*, small) the noncirculating lower layer of cold water below the thermocline in a thermally stratified lake, usually deficient in oxygen. See profundal zone, epilimnion, monimolimnion.

hypomagnesemia *n.* a deficiency of magnesium in the blood. Also called grass tetany or grass staggers in cattle.

hyponasty *n.* (Greek *hypo*, below + *nastos*, close + *y*, condition) an upward curve in the growth of a leaf caused by slower growth on the upper surface than on the lower side. See epinasty.

hypoxia *n.* (Greek *hypo*, below + *oxy*, oxygen + *ia*, condition) oxygen levels too low to support animal life (e.g., fish and shrimp) in deep water, especially if caused by human activity. It is a common condition in water bodies enriched with nitrogen and all other essential nutrients to the extent that life forms such as marine phytoplankton grow excessively and deplete the supply of dissolved oxygen; e.g., an area of up to 7000 mi^2 (18,000 km^2) in the Gulf of Mexico affected by nutrients carried in by the Mississippi River.

hypsometer *n.* (Greek *hypsos*, height + *metron*, measure). **1.** an instrument for measuring heights (e.g., of trees) by triangulation. **2.** a device for determining elevation above sea level by measuring changes in air pressure or boiling point. See barometer.

hysteresis *n.* (Greek *hysterēsis*, a shortage) **1.** a lagging behind; a delayed response, especially to a change in a magnetic field. **2.** a difference based on history, especially a difference in the water content of a soil at a particular moisture tension depending on whether the soil is drying from a wet condition (small pores tend to remain full) or being moistened from a dry condition (small pores tend to remain empty). In the illustration, hysteresis is shown by the drying and wetting curves for a representative soil (the curves are similar in form for other soils).

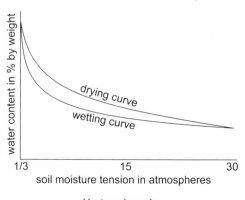

Hysteresis cycle

Hz *n.* hertz. See hertz.

I

I *n.* **1.** the 9th letter in the English alphabet, derived from the Greek *iota*. **2.** the symbol for 1 in Roman numerals. **3.** the chemical symbol for iodine. **4.** the symbol for electric current (normally expressed in amperes) or for moment of inertia. *pron.* **5.** the nominative singular pronoun for the person speaking (pl. = we).

IBPGR *n.* International Board for Plant Genetic Resources, an international agricultural research center founded in 1974 and headquartered in Rome. It focuses on conserving gene pools of current and potential crops and forages.

ICARDA *n.* International Center for Agricultural Research in the Dry Areas, an international agricultural research center founded in 1977 with headquarters in Aleppo, Syria. It emphasizes improved farming systems for northern Africa and western Asia. Research covers wheat, barley, chickpea, lentils, pasture legumes, and small ruminants.

ice *n.* (Anglo-Saxon *is*, ice) **1.** the solid form of water. Ice floats in water because it expands about 9% when it freezes, from a density of 1.0 g/cc for water to 0.9166 g/cc for ice. **2.** any substance that resembles ice; e.g., dry ice = frozen carbon dioxide. **3.** slang for diamonds. *v.* **4.** to apply icing to a cake.

ice age *n.* **1.** the Pleistocene glacial epoch, a period from about 1 million years ago to about 10,000 years ago that included four major glacial stages. **2.** a glacial epoch at any time in geologic history.

iceberg *n.* (Dutch *ijsberg*, ice mountain) a large mass of ice broken off from a glacier and floating in the sea; a detached floe.

ice core *n.* a long core of ice extracted from a glacier for analysis. Impurities trapped in the ice give clues to past climate and vegetation. Air entrapped in the ice can show the degree of pollution that existed at the time the ice was formed.

ice crystal *n.* a small piece of frozen ice. Ice crystallizes in the hexagonal system, sometimes forming intricate patterns such as snowflakes. Cirrus clouds are formed of ice crystals at high elevations.

ice floe *n.* See floe.

icehouse *n.* a building used to store winter ice for use through the summer; commonly used before refrigerators were invented.

ice jam *n.* **1.** a mass of jumbled chunks of ice blocking a narrow place in a river or large stream, most common after a spring thaw followed by a period of freezing. Ice jams can cause flooding, block navigation, and put intense pressure on a bridge or other structure that blocks the stream flow. **2.** a jumbled mass of ice blocks driven against the coastline of a lake by strong prevailing winds.

ice pack *n.* **1.** a large expanse of floating ice broken into pieces that have refrozen together; e.g., the jumbled mass of ice that covers much of the Arctic Ocean. Also called pack ice. **2.** an ice jam. **3.** a bag (usually rubber or plastic) containing chipped ice used to reduce swelling and ease pain following an injury.

Ice Patrol *n.* See International Ice Patrol.

ice plant *n.* a low-growing, thick-leaved, succulent plant (*Mesembryanthemum* sp., especially *M. crystallinum* or *M. edule*) whose leaves have a shiny, icelike appearance. Native to southern Africa and the Mediterranean region, it is widely used as ground cover in southern California.

ichneumon fly *n.* a wasplike fly of family Ichneumonidae. Its larvae are parasitic on the larvae of other insects, especially moths and butterflies.

ichthyology *n.* (Greek *ichthys*, a fish + *logos*, word) a branch of zoology dealing with the classification, form, structure, habits, and use of fishes.

icicle *n.* (Anglo-Saxon *īsgicel*, icicle) a hanging piece of tapered ice formed by gradual freezing of water that seeps down its length and drips from its point.

ICLARM *n.* See International Center for Living Aquatic Resources Management.

ICRAF *n.* See International Centre for Research in Agroforestry.

ICRISAT *n.* See International Crops Research Institute for the Semi-Arid Tropics.

icterus *n.* (Greek *ikteros*, jaundice) **1.** jaundice. **2.** a yellowing of plant leaves that is caused by cold, wetness, or other stress.

identical twins *n.* two siblings that have the same genotype because they developed from a single fertilized ovum. Identical twins normally resemble each other closely in both appearance and behavior. Nonidentical twins (developed from separate fertilized ova) are called fraternal twins or dizygotic twins.

IFM *n.* integrated farm management, the use of a variety of management practices to control crop pests with a minimum of pesticides and expense.

IFPRI *n.* See International Food Policy Research Institute.

igloo *n.* (Inuit *igdlu*, snow house) a dome-shaped hut made of packed blocks of snow for shelter from cold weather.

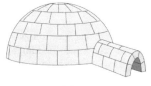

Igloo

igneous *adj.* (Latin *igneus*, of fire) **1.** produced under conditions with enough heat to cause melting followed by cooling and solidification; e.g., igneous rocks. **2.** fiery; having something to do with fire. **3.** or **igneus** on fire or flame colored (as some flowers).

igneous rock *n.* a rock formed by the cooling of magma. Igneous rocks are considered extrusive if the magma flowed out on the surface as lava (e.g., basalt) or intrusive if the magma cooled slowly beneath the surface (e.g., granite).

ignescent *adj.* (Latin *ignescens*, taking fire) **1.** emitting fiery sparks, as when flint or chert is struck with steel. **2.** igniting in flame.

ignis fatuus *n.* (Latin *ignis*, fire + *fatuus*, foolish) **1.** a phosphorescent light that sometimes flickers over marshy places that is believed to result from spontaneous combustion of marsh gas (methane = CH_4). Also called will-o'-the-wisp. **2.** something that deludes or misleads; a false hope.

iguana *n.* (Arawak *iwana*, iguana) a tree-dwelling New World lizard (*Iguana* sp., especially *I. iguana*) with a sturdy body and a scaly crest along its back and tail. Iguanas grow up to 7 ft (2 m) long and may weigh as much as 30 lb (14 kg). Most are green, but chameleons that change color according to temperature or surroundings are included in the iguana family.

IIMI *n.* See International Irrigation Management Institute.

IITA *n.* See International Institute for Tropical Agriculture.

illite *n.* a 2:1 nonexpanding layer silicate intermediate between muscovite and smectite. Illite is the most common clay mineral in recent marine sediments and in clayey rocks and is common in soils that have not been intensely weathered. It has a moderate cation-exchange capacity (averaging about 300 mmol of charge per kilogram).

illuvial *adj.* (Latin *il*, in + *luvi*, wash + *alis*, pertaining to) moved into a soil layer; e.g., colloidal clay formed in the A and/or E horizon but moved into the B horizon. The B horizon is illuviated, and the A and E horizons are eluviated. See eluvial.

illuviation *n.* (Latin *il*, in + *luvi*, wash + *atio*, action) the process of forming an illuvial soil horizon by coagulating, filtering, or precipitating material eluviated from overlying horizons by percolating water. See cutan, eluviation.

imago *n.* (Latin *imago*, a likeness) **1.** an insect's adult stage. **2.** a subconscious idealized image of a loved person carried over from childhood into adult life; e.g., a loving mother or a teacher.

imbibe *v.* (Latin *imbibere*, to drink in) **1.** to drink, especially alcoholic beverages. **2.** to absorb water or other liquid. **3.** to absorb anything, as light or ideas.

immigrant *n.* (Latin *immigrans*, moving into) a plant, animal, or person that came from elsewhere and moved into the area they now occupy.

immigration *n.* (Latin *immigrare*, to move into + *tion*, action) the process of moving into a new area, usually as a permanent resident. The departure process is called emigration. See emigration.

immobilization *n.* (Latin *immobilis*, not mobile + *izatio*, making) **1.** the process or act of preventing something from moving; e.g., the immobilization of a broken arm by placing it in a cast. **2.** the process of using a mobile plant nutrient to form an immobile organic molecule in microbial or plant tissue; e.g., the production of protein by using nitrate nitrogen.

immune *adj.* (Latin *immunis*, exempt) **1.** having protection against a disease because one has inherited resistance, has had the disease, or has been inoculated against it. **2.** being exempt from prosecution because of diplomatic status. **3.** able to resist something, as immune to temptation.

immune system *n.* the internal defense system of a person or animal that causes antibodies to be produced when the body is threatened by a disease.

immunity *n.* (Latin *immunitas*, exemption) **1.** resistance against an infection or disease. **2.** exemption from a responsibility or freedom from a hazard, as exemption from a tax or immunity from prosecution.

imp *n.* (Anglo-Saxon *impa*, a shoot or graft) **1.** offspring; either a child or a plant shoot. **2.** a small demon. **3.** a mischievous child. *v.* **4.** to graft a scion onto a plant. **5.** to repair an injury by filling out or splicing onto the injured part.

impala *n.* (Zulu *impala*, impala) a reddish brown African antelope (*Aepyceros melampus*) about 2½ to 3 ft (80 to 90 cm) tall at the shoulders with long, twisted, ringed horns. See antelope.

imperfect *adj.* (Latin *imperfectus*, unfinished) **1.** having defects or weaknesses. **2.** incomplete; lacking some parts. **3.** reproducing asexually and having no known sexual stage, as certain fungi. *n.* **4.** a past-tense verb indicating a continuing action; e.g., it was going on.

imperfect flower *n.* a unisexual flower having either stamens or pistil but not both (the other part is in another flower). See perfect flower, incomplete flower.

impermeable *adj.* (Latin *impermeabilis*, not permeable) preventing the passage of a liquid or gas.

impervious *adj.* (Latin *impervius*, not pervious) **1.** resistant to penetration; impermeable, as impervious to water. **2.** unaffected; not influenced, as impervious to reason.

impetuous *adj.* (Latin *impetuosus*, impetuous) **1.** impulsive; acting rashly, suddenly, often emotionally. **2.** acting with violent force, as the wind in a tornado.

impetus *n.* (Latin *impetere*, to rush upon) a driving force; the power causing something to happen; the momentum of a moving mass or process.

impinge *v.* (Latin *im*, to put in + *pangere*, to drive in) **1.** to infringe; to enter into an area occupied by another object or being, as to impinge on someone's rights. **2.** to fall upon or strike, as light impinging on an object. **3.** to impress, as an idea that impinges on someone's imagination.

impoundment *n.* (Anglo-Saxon *im*, in + *pund*, enclosure + Latin *mentum*, condition) **1.** a body of water held in place by a dam, dike, or other barrier. **2.** holding of something in confinement, as the impoundment of an automobile.

impoverish *v.* (Latin *im*, to put in + *pauper*, poor + *iscere*, tending to) **1.** to make someone poor; to reduce someone to poverty. **2.** to deplete or exhaust a resource, as to impoverish a soil by removing crops without adding fertilizer.

impregnate *v.* (Latin *impraegnare*, to make pregnant) **1.** to make a female pregnant. **2.** to saturate with something, as to impregnate a substance with disinfectant.

imprinting *n.* (Latin *imprimere*, to impress + *ing*, action) quickly forming a close bond for a long-lasting relationship, as that between a Seeing Eye dog and its master.

impulse *n.* (Latin *impulsus*, sudden pressure) **1.** a sudden driving force, usually brief but commonly repeated. **2.** a sudden decision to act in a certain way, often somewhat emotional. **3.** mathematically, the product of force multiplied by time. An impulse produces an equivalent change in momentum.

inane *adj.* (Latin *inanis*, empty) **1.** empty. **2.** nonsensical; silly. *n.* **3.** emptiness, especially the void of outer space.

inanimate *adj.* (Latin *in*, not + *anima*, life) **1.** without life; not a living being, as an inanimate object. **2.** dull, sluggish, lacking volition.

inanition *n.* (Latin *inanire*, to make empty + *tion*, action) **1.** exhaustion resulting from starvation. **2.** lack of ambition; lethargy.

inarticulate *adj.* (Latin *inarticulatus*, not jointed) **1.** incapable of speaking clearly, as an inarticulate speaker. **2.** not intelligible, as the baby made inarticulate sounds. **3.** mute; unable to speak. **4.** not jointed; not divided into segments.

inborn *adj.* (Anglo-Saxon *inboren*, native) innate; already present at birth.

inbred *adj.* (Anglo-Saxon *in*, within + *bredan*, to nourish) **1.** inborn; innate. **2.** bred within a family or closely related group.

inbreeding *n.* (Anglo-Saxon *in*, within + *brēdan*, to nourish + *ing*, action) repeated breeding within closely related genetic stock. Inbreeding often reveals genetic problems, especially in animals, but it is useful in developing pure lines that can be crossed for hybridization as in hybrid corn.

Inca *n.* (Quecha *inca*, prince) **1.** a member of the royal family in the pre-Columbian civilization that controlled much of South America from its capitol at what is now Cuzco, Peru. The Incan empire was conquered by Francisco Pizarro and his Spanish army in the 1530s, as he took advantage of a civil war among the Incas. **2.** any member of this advanced, wealthy pre-Columbian civilization that built great cities and connected them with roads. **3.** a modern descendant of these people. See Aztec, Maya.

incendiary *adj.* (Latin *incendiarius*, setting on fire) **1.** designed for lighting fires, as incendiary bombs. **2.** intended to incite emotional action, as an incendiary speech.

incentive payment *n.* payment to a producer from a government agency or other external source to promote production of a particular commodity.

Inceptisol *n.* (Latin *inceptum*, beginning + *solum*, soil) an order in Soil Taxonomy composed of soils in an early stage of development whose profiles show horizons with differences in color and organic matter content but not in clay eluviation. Inceptisols occur in humid climates and occupy about 10% of the land area in the U.S.A. and about 9% of the world's land area.

inch *n.* (Latin *uncia*, one-twelfth) **1.** a unit of linear measure equal to one-twelfth of a foot in U.S. customary units, or precisely 2.540005 cm as now used by the U.S. Coast and Geodetic Survey. **2.** a very small part out of a much larger total, as he won by an inch. *v.* **3.** to move in small increments, as to inch along in bad conditions.

inchoate *adj.* (Latin *inchoatus*, begun) in the beginning stage; started but not yet fully developed.

inchworm *n.* See cankerworm.

incident *n.* (Latin *incidens*, falling upon) **1.** an event or occurrence, especially a small one of little importance. **2.** one of several successive events in a story. **3.** an associated event, often with little relevance to the main sequence or story. **4.** a minor event that could have serious consequences. *adj.* **5.** likely to occur; a natural part of life, as obstacles incident on the way to an accomplishment. **6.** subordinate to the main goal or consideration. **7.** falling directly on something, as incident light.

incineration *n.* (Latin *incineratus*, burned + *tion*, action) burning to ashes under controlled conditions. Combustible wastes are often incinerated to reduce their volume and destroy disease vectors; the incinerator must have safeguards to avoid air pollution.

incipient *adj.* (Latin *incipient*, taking in hand) in a very early stage; barely beginning.

incipient gully *n.* See ephemeral gully.

incision *n.* (Latin *incisio*, cut) **1.** a cut or the act of cutting, especially the cut made by a surgeon to begin an operation. **2.** channel cutting in a landscape, especially gully erosion on an otherwise relatively smooth topography. **3.** keenness; detailed understanding.

inclement *adj.* (Latin *in*, not + *clemens*, lenient) **1.** harsh, stormy, disagreeable (weather). **2.** not merciful; unforgiving.

incomplete flower *n.* a flower that lacks one or more of the four parts of a complete flower: pistil, stamens, petals, and sepals. See complete flower, imperfect flower.

increaser *n.* (Latin *increscere*, to grow) a plant species that increases its population under moderate grazing while other plants decrease.

incubate *v.* (Latin *incubare*, to sit on) **1.** to sit on eggs so they will hatch. **2.** to hold in a warm environment to make eggs hatch, seeds sprout, plants or animals grow, etc. **3.** to develop an idea, as to incubate a scheme.

incumbent *n.* (Latin *incumbens*, leaning on) **1.** the present occupant of an office or position. *adj.* **2.** current, as the incumbent senator. **3.** required, as it is incumbent on him to do this. **4.** resting upon, as one rock stratum incumbent on another.

incursion *n.* (Latin *incursio*, running against) **1.** a damaging inroad. **2.** a hostile raid; an invasion. **3.** an entry with negative effect, as an incursion of seawater into an aquifer.

indehiscent *adj.* (Latin *in*, not + *dehiscens*, opening) not dehiscent; not splitting open at maturity, mostly fruits that bear individual seeds rather than groups of seeds in a pod.

indeterminate growth *n.* unrestricted growth, as in plants with vegetative apical meristems that continue branching indefinitely. Such plants produce flowers and fruit during an extended period rather than all at once. See determinate growth.

index of aridity *n.* a measure of the aridity of a location based on precipitation and temperature:

$$I = P/(T + 10)$$

where
I = index of aridity,
P = average annual precipitation in cm,
T = mean annual temperature in °C.

Indian *n.* **1.** a native of India or of the East Indies. **2.** an American aborigine, sometimes called an American Indian, an Amerindian, or an Amerind; the preferred term now is Native American. *adj.* **3.** an object or cultural practice related to Indian people.

Indian corn *n.* **1.** maize (*Zea mays*, family Poaceae), commonly called corn in the U.S.A. See maize. **2.** a variety of maize with variegated kernels.

indiangrass or **Indian grass** *n.* a tall, coarse grass (*Chrysopogon nutans*, family Poaceae) native to the southern Great Plains of the U.S.A.

Indian land *n.* **1.** land held in trust by the U.S. government for an Indian tribe or an individual Native American. **2.** any land belonging to an Indian tribe or a Native American. **3.** during the settlement period, any land still under Indian control, usually the land west of that occupied by the U.S. government and European settlers.

Indian mallow *n.* See velvetleaf.

Indian meal moth *n.* See meal moth.

Indian paintbrush *n.* a semiparasitic plant (*Castilleja* sp., family Scrophulariaceae) with brightly colored brachts that is common in arid and semiarid parts of western U.S.A.

Indian pony *n.* a horse descended from horses brought to America by Spaniards and other Europeans and either lost or released. The Indians captured and rode these horses. The Nez Percé tribe used them to develop the Appaloosa breed, which has a distinctive spotted pattern.

Indian ricegrass *n.* a tufted perennial bunchgrass (*Oryzopsis hymenoides*, family Poaceae) native to western U.S.A. A good forage grass whose seeds were used by the Indians to make meal, flour, and bread, it is drought resistant and grows best on sandy soils.

Indian summer *n.* a warm period after the first autumn frosts, usually in late September or early October. It is especially pleasant because frost has killed many insect pests.

Indian tobacco *n.* a common weed (*Lobelia inflata*, family Lobeliaceae) in western U.S.A. Its leaves can be dried and used like tobacco, but they are toxic to livestock, causing exhaustion, stupor, dilated pupils, convulsions, coma, and death. It is also the source of certain medicines used as emetics and purgatives.

indicator *n.* (Latin *indicator*, one who points) **1.** someone or something that points to or is a sign of a value or condition; e.g., the pointer on the dial of an instrument. **2.** a chemical that changes color with pH or some other property; used to measure the pH or other property or to show the endpoint of a titration. **3.** a species of plant or animal whose presence is evidence of a particular environmental condition; e.g., cheatgrass indicates a degraded condition of rangeland, and vultures circling overhead indicate a dead animal.

indicator plant *n.* a plant species whose presence indicates certain soil conditions; e.g., wetness, salinity, or acidity.

indifferent *adj.* (Latin *indifferens*, not caring) **1.** having or showing no interest. **2.** unbiased, impartial, especially on religious or political topics. **3.** only moderate in amount or fair in quality, as an indifferent performance. **4.** neutral in some quality, as a day-neutral plant.

indigenous *adj.* (Latin *indigenus*, native) **1.** native to the area where it occurs. **2.** native to a designated location, as maize is indigenous to America. **3.** characteristic of, as instincts indigenous to a particular type of animal.

indigo *n.* (Spanish *indigo*, blue dye) **1.** a hairy plant (*Indigofera* sp., especially *I. tinctoria*, family Poaceae) that bears clusters of red or purple flowers. **2.** a blue dye ($C_{16}H_{10}N_2O_2$) derived from the plant or produced synthetically. *adj.* **3.** a blue like that produced by the dye.

Indium (In) *n.* (Latin *indicum*, blue dye + *ium*, element) element 49, atomic weight 114.82; a very soft metal with a brilliant silvery luster that is chemically much like gallium and aluminum. It is used in alloys, semiconductors, mirrors, etc.

indoor climate *n.* the environment inside a building or room, including the temperature, lighting, humidity, noise level, and degree of air pollution.

induction *n.* (Latin *inductio*, act of inducing) **1.** the process of persuasion or leading to a particular action. **2.** formal placement in an office. **3.** a reasoning process that leads to a logical but not forced conclusion. **4.** a conclusion reached by the process of induction. **5.** the generation of an electric current and/or magnetism resulting from either movement of an electric conductor through a magnetic field or a change in its magnetic environment (e.g., by an electric current being turned on or off). **6.** a biochemical response to an environmental influence; e.g., enzyme synthesis stimulated by a buildup of substrate, or differentiation of a part of an embryo in response to the influence of another part.

induration *n.* (Latin *induratio*, hardening) **1.** the process of making or becoming hard. See lithification. **2.** the product or state resulting from such a process, as an indurated soil horizon or rock layer. **3.** an abnormal hardening of a body part (an organ, tissue, or joint).

industrial waste *n.* waste from any factory or related facility, often including chemicals, metals, and other materials not normally significant in household waste.

inert *adj.* (Latin *iners*, inert) **1.** unable to move or cause movement, as inert matter. **2.** chemically nonreactive, as atmospheric nitrogen or a noble gas. **3.** having no medical effect, as the inert ingredients in a pill.

inert gas *adj.* a nonreactive gas, especially a noble gas, i.e., helium, neon, argon, krypton, xenon, or radon.

inertia *n.* (Latin *inertia*, lack of skill) **1.** the property of matter that tends to resist changes in its velocity of motion. **2.** a similar property of a force such as electricity. **3.** sluggishness, inactivity.

inexhaustible *adj.* (Latin *inexhaustus*, not exhausted + *ibilis*, able) in unlimited supply; not likely to be depleted, as an inexhaustible supply (but many resources that were once thought to be inexhaustible have been seriously depleted, e.g., land, water supplies, and petroleum).

infection *n.* (Latin *infectio*, dyeing) invasion by a pathogen and the resulting reaction of the body tissue.

infectious *adj.* (Latin *infectio*, dyeing + *ius*, having) caused by or capable of causing an infection, as a source of pathogenic bacteria, fungi, protozoa, or virus.

inferoflux *n.* subterranean flow of water beneath a surface stream.

infertile *adj.* (Latin *infertilis*, not fertile) **1.** incapable of reproducing. **2.** not fertilized, as an infertile egg. **3.** barren, unproductive, as infertile soil.

infest *v.* (Latin *infestare*, to assail) to attack, annoy, or inhabit in large numbers, generally in a troublesome manner, as lice infest a chicken or mice infest a building.

infestation *n.* (Latin *infestatio*, infestation) **1.** the process of becoming infested or the condition that results therefrom. **2.** an invasion by a pest, as an infestation of termites.

infiltration *n.* (Latin *in*, in + *filtratus*, filtered + *tion*, action) **1.** the process of a fluid entering into or passing through a solid; e.g., water infiltrating into soil. **2.** any process of entering into or passing through, as spies infiltrating a country or an organization.

infiltration rate *n.* the maximum rate that water can enter soil or other material. The National Cooperative Soil Survey classifies the infiltration rate of water into soil as

- very low   less than 0.1 in. (0.25 cm) per hour
- low   0.1–0.5 in. (0.25–1.25 cm) per hour
- medium   0.5–1.0 in. (1.25–2.5 cm) per hour
- high   faster than 1.0 in. (2.5 cm) per hour.

infiltrometer *n.* (Latin *in*, in + *filtratus*, filtered + Greek *metron*, measure) equipment designed to measure the infiltration rate of water into soil; e.g., a ring infiltrometer that uses two concentric metal bands pressed into the soil. Both are filled with water, and the infiltration rate is taken as the rate the water level drops in the inner ring.

infinitesimal *adj.* (Latin *infinitesimus*, infinitely small + *alis*, like) approaching zero in size; too small to be measured.

inflammable *adj.* (Latin *inflammabilis*, combustible) an old term for flammable; combustible. The use of the word inflammable is now discouraged to avoid confusion,

inflammation *n.* (French *inflammatio*, setting on fire) **1.** the act of igniting a fire or of inciting strong emotions. **2.** a state of emotional agitation. **3.** a swollen, usually reddened condition of injured or diseased human or animal tissue.

inflorescence *n.* (Latin *inflorescentia*, blossoming) **1.** the part of a plant that blossoms. **2.** the opening of flowers on a plant. **3.** the arrangement of blossoms on a plant stem.

inflow *n.* (Anglo-Saxon *in*, in + *flowan*, flow) something that flows into an area, as an inflow of water into a pond or an inflow of sewage into a stream.

influent *n.* (Latin *influens*, flowing in) **1.** water flowing into a stream from a tributary. **2.** anything flowing into a larger body.

influenza *n.* (Latin *influentia*, flowing in) **1.** an acute viral disease of the respiratory tract of humans or animals that often causes an epidemic. Infection usually causes fever, catarrhal inflammation of the nasal passages and throat, headaches, and weakness. These symptoms may last from a few days to two weeks; the weakness may last for some time beyond that. Relapses are common. The causal virus has many strains and mutates frequently. Influenza is commonly called flu.

influx *n.* (Latin *influxus*, inflow) **1.** inflow. **2.** the entry point where a stream flows into a larger stream, lake, or sea. **3.** the arrival of a large number of immigrant people or things.

infrared *n.* (Latin infra, *below* + Anglo-Saxon *read*, red) electromagnetic radiation beyond the red end of the visible spectrum but shorter than radio waves; radiation with a wavelength between 800 nm and 1 mm. Near-infrared is that portion of the infrared spectrum that can be recorded on photographic film. Thermal infrared is that portion that is sensed as radiant heat; its heating effect can penetrate flesh.

infrastructure *n.* (Latin *infra*, below + *structura*, structure) **1.** the underlying social and economic framework of a society. **2.** the buildings and the transportation, power, and communications systems of a geographic area. **3.** a nation's military installations.

infringe *v.* (Latin *infringere*, to break or weaken) to encroach or trespass in a restricted area or right; e.g., to infringe on someone's privacy or to violate a patent.

ingest *v.* (Latin *ingestus*, poured into) to eat or absorb food, drink, or drugs into the body.

ingesta *n.* (Latin *ingesta*, taken in) food or other substances that enter a body, especially through the stomach.

ingestion *n.* (Latin *ingestus*, poured into + *tion*, action) the process of eating, drinking, or absorbing food and other materials.

ingluvies *n.* (Latin *ingluvies*, gizzard) **1.** the crop (craw) of a bird. **2.** the first stomach (rumen) of a cow or other ruminant animal.

inhabitant *n.* (Latin *inhabitant*, dweller) a person or animal that lives in a designated place (house, city, state, etc.), especially as a permanent resident.

inhalable *adj.* (Latin *inhalare*, to breathe + *abilis*, able) capable of being inhaled; e.g., a vapor or particulate matter suspended in the air.

inhere *v.* (Latin *inhaerere*, to stick) to belong as an inseparable quality.

inherent *adj.* (Latin *inhaerens*, sticking) innate, inseparable, impossible to remove; e.g., the inherent right to life, liberty, and the pursuit of happiness.

inherit *v.* (Latin *inhereditare*, to make heir) **1.** to receive property, title, position, or other possessions as an heir from one's progenitors, predecessors, or other benefactor. **2.** to have certain characteristics transmitted by hereditary factors.

inheritance *n.* (French *enheritance*, inheritance) **1.** possessions that pass from a person to a successor or heir, especially at death. **2.** genetic characteristics transmitted from parents to offspring. **3.** the legacy one has received from preceding generations or predecessors in a position.

inhibitor *n.* (Latin *inhibitio*, a holding back) **1.** one who holds back the action of another. **2.** a compound that slows or stops a chemical reaction, especially one that reduces the activity of an enzyme, usually by binding to the enzyme either at its active site or at a nearby site that influences the active site.

injection well *n.* a well used to insert water into the ground to recharge an aquifer, dispose of wastes, or supply water for solution mining a substance such as salt.

inky cap *n.* a poisonous mushroom (*Coprinus* sp., especially *C. atramentarius*) whose gills liquefy into an inky liquid as they mature.

inland *adj.* (Anglo-Saxon *inland*, in land) within a country; away from the ocean.

inland waters *n.* surface water surrounded by land, including rivers, lakes, reservoirs, and ponds over 2 ac (0.8 ha) in size.

Inland Waterways Corporation *n.* an entity created in 1924 by the U.S. Congress to coordinate water and rail transportation in the U.S.A. It was a part of the Department of Commerce from 1939 to 1953 and then was sold to a private corporation.

inlet *n.* **1.** a small bay or other indentation in a shoreline, especially one that is long and narrow. **2.** an entryway.

innate *adj.* (Latin *innatus*, inborn) **1.** inborn; a natural characteristic of a person or animal. **2.** an inherent characteristic of something, as an innate defect in a theory. **3.** originating from intellect rather than experience.

innocuous *adj.* (Latin *innocuus*, harmless) **1.** harmless; not likely to cause injury or ill effect; not nocuous. **2.** unlikely to offend.

inoculant *n.* (Latin *inoculatus*, grafted + *ant*, agent) inoculum.

inoculation *n.* (Latin *inoculatio*, inoculation) **1.** administration of a vaccine, antibody, or antigen to generate immunity to a disease. **2.** treatment of seed, soil, or other medium with a beneficial microbe to cause infection of seedlings; e.g., treating legume seed with a compatible strain of *Rhizobium* for nitrogen fixation.

inoculum *n.* (Latin *inoculum*, inoculum) a microbial culture, vaccine, or other substance used for inoculation. Also called inoculant.

inorganic *adj.* (Latin *in*, not + *organicus*, of carbon) **1.** not organic; matter that is not a component or product of living entities. **2.** composed of chemical compounds that do not contain organic carbon. Organic compounds have carbon structures and covalent bonds with hydrogen and often with oxygen, nitrogen, phosphorus, and sulfur. Most inorganic materials have cations with ionic bonds to oxygen, sulfur, or some other anion. Note: Certain carbon compounds (carbonates, cyanides, and cyanates) are inorganic.

inosilicate *n.* a silicate mineral with a chain structure, either single chains such as the pyroxenes or double chains such as the amphiboles. The chains align themselves in parallel bundles, producing crystals with a fibrous structure.

inquiline *n.* (Latin *inquilinus*, a lodger) an insect or other animal that lives in an abode that belongs to another, usually without harming the host animal or insect; e.g., an insect that lives in a gall produced by a gall insect.

insect *n.* (Latin *insectum*, cut into or divided) a 6-legged, air-breathing animal (class Insecta of phylum Arthropoda) with a 3-segment body. Some insects are useful to humans, and some are harmful (see insect vector). Examples are bees, beetles, flies, and mosquitoes.

insecticide *n.* (Latin *insectum*, cut into or divided + *cida*, killer) a chemical used to kill insects by either contact or ingestion.

insecticide resistance *n.* the ability of an insect populations to tolerate an insecticide (insect populations tend to become resistant to insecticides used against them). Housefly resistance to DDT was noted as early as 1947, and more than a hundred other types of resistance have been identified since then.

insectifuge *n.* (Latin *insectum*, cut into or divided + *fugia*, put to flight) a preparation or thing used to repel or drive away insects; e.g., a chemical repellant or a strong fan.

insectivore *n.* (Latin *insectum*, cut into or divided + *vore*, eating) **1.** an animal that habitually eats insects; e.g., bats and many birds. **2.** a mammal of the order Insectivora, including hedgehogs, moles, and shrews. **3.** a plant that traps insects and absorbs them; e.g., the pitcher plant and roundleaf sundew.

insect metamorphosis *n.* some insects undergo "complete" four-stage metamorphosis consisting of egg, larva, pupa, and adult stages, whereas others undergo "incomplete" three-stage metamorphosis including egg, nymph, and adult stages.

insect resistance *n.* the ability of certain plants to resist attack by certain insects; e.g., a muskmellon variety that resists aphids, or sweet-corn resistance to corn earworm.

insect vector *n.* an insect that helps spread a disease by carrying virus, bacteria, spores, or mycelia that can infect plants, animals, or humans.

insemination *n.* (Latin *inseminatus*, implanting + *tion*, action) impregnation of a female either by coitus or by artificial placement of semen in the vagina.

insidious *adj.* (Latin *insidiosus*, deceitful) **1.** intended to deceive; sly and treacherous. **2.** hardly noticeable but with serious consequences, as an insidious disease.

insipid *adj.* (Latin *insipid*, not sapid) **1.** lacking distinctive or interesting flavor; e.g., bland food. **2.** uninteresting; nonstimulating; e.g., an insipid personality.

in situ *adj.* (Latin *in situ*, in place) in its natural place; not moved; e.g., soil formed in place from the local bedrock. See ex situ.

insolation *n.* (Latin *insolatio*, exposure to the sun) **1.** solar radiation received at the surface of the Earth (or on some other surface of interest). Such heat may be stored or it may be reradiated into the atmosphere at longer wavelengths than the incoming radiation, e.g., light energy reradiated as heat energy. **2.** the intensity of solar radiation striking a surface. **3.** exposure to the sun's rays for treatment of a disease. **4.** sunstroke.

instability *n.* (Latin *instabilitas*, lack of stability) **1.** the condition of being not stable; subject to being upset or changed in some other undesirable way. **2.** change-ability; lack of determination to follow a chosen course.

instar *n.* (Latin *instare*, to stand near) any of the forms that occur between molts of an insect or other arthropod.

instinct *n.* (Latin *instinctus*, impulse) inherent understanding or ability; an inborn response to a certain type of stimulus. Instincts are controlled by a more primitive part of the brain than that involved in learned reasoning processes.

instrumentation *n.* (Latin *instrumentum*, instrument + *atio*, action) **1.** equipment designed to supply data for a user; e.g., a pH meter in a laboratory or a speedometer in an automobile. **2.** arrangement of a musical composition for an instrument or a group of instruments.

instrument weather *n.* weather conditions that cause such poor visibility that aircraft must fly by instruments.

insulation *n.* (Latin *insulatus*, isolate + *tion*, action) **1.** material that does not conduct or is a very poor conductor, especially of heat or electricity. **2.** the process of installing insulation.

insulator *n.* (Latin *insulatus*, isolate + *or*, one that) an object or material that insulates, especially a glass or ceramic support for an electric wire.

insulin *n.* (Latin *insulin*, island) **1.** a pancreatic hormone that is essential for the metabolism of glucose. A deficiency of insulin causes diabetes. **2.** a commercial product used for the treatment of diabetes. It is obtained from sheep pancreas or, recently, from genetically altered bacteria.

intangible *adj.* (Latin *intangibilis*, not tangible) **1.** not material; something that cannot be touched; e.g., popularity or goodwill. **2.** vague, unclear in the mind. **3.** not subject to monetary valuation.

integral *adj.* (Latin *integralis*, part of) **1.** an essential part of an entity. **2.** having a number of parts that combine to form the whole entity. **3.** complete; whole, as an integral number (not a fraction). **4.** a symbol in calculus for the procedure used to find the sum of all the parts indicated by a function (e.g., the area of a figure).

integrated pest management (IPM) *n.* an ecological approach to pest control that avoids excessive use of chemicals by rotating crops, using pest-resistant varieties, natural predators, growth hormones, and/or mechanical control when possible. A chemical pesti-cide is considered a last resort to be used only after scouting for pests has shown that the economic threshold has been exceeded.

integration *n.* (Latin *integratio*, renewal) **1.** the process of combining parts so they work together harmoni-ously; blending together. **2.** the mixing of different races or other groups. **3.** forming a consistent person-ality from a variety of inputs. **4.** the mathematical procedure used in calculus to evaluate an integral.

integrity *n.* (Latin *integritas*, wholeness) **1.** complete-ness; not lacking any parts. **2.** soundness; not injured or likely to break. **3.** honesty, reliability; the quality of uncompromising moral character.

integument *n.* (Latin *integumentum*, a covering) a natural covering, e.g., the skin of an animal, the shell of an egg, a seed coat, an orange rind, or the bark of a tree.

intensity *n.* (Latin *intens*, stretched out + *itas*, condition) **1.** concentration; focus; brightness of light or comparable quality of other forms of energy. **2.** the magnitude of force applied per unit area. **3.** the chroma (degree of saturation) of a color. **4.** degree of excitement or depth of feeling.

interaction *n.* (Latin *inter*, among + *actio*, action) **1.** acting reciprocally, each having influence on the other. **2.** acting together to give a different effect than any of the components produces alone, as toxicity produced by chemical interaction.

interbedded *adj.* having alternating layers of different composition, as interbedded limestone and shale.

intercalary *adj.* (Latin *intercalarius*, interposed) inserted; added; e.g., February 29 is an intercalary day inserted in a leap year, and a leap second is an intercalary second inserted to adjust time to the movement of the Earth around the sun.

intercardinal *adj.* (Latin *inter*, between + *cardinalis*, principal) between cardinal points or other references; e.g., northeast, southeast, southwest, and northwest are intercardinal points.

intercellular *adj.* (Latin *inter*, between + *cellula*, small cell) located in the space between or among cells.

interception *n.* (Latin *interceptio*, act of intercepting) **1.** catching or seizing something that is moving before it reaches its normal destination; e.g., diverting the flow of runoff water with a berm or ditch, retention of rainfall or snow on foliage, or reduction in sunlight intensity on a cloudy day. **2.** intersection, as the point where two lines cross. **3.** intrusive listening, as intercepted communication.

intercropping *n.* growing more than one crop in an area by planting one or more other species between the rows or plants of the primary crop; e.g., pumpkins or melons planted in a corn patch or rice grown between banana plants. Also called mixed intercropping. See multiple cropping.

interface *n.* (Latin *inter*, between + *facies*, surfaces) **1.** the contact surface between two bodies or phases; e.g., where a gas meets a liquid or solid, where a liquid meets another liquid or a solid, or where two solids meet. **2.** the connection that links two bodies, machines, or organizations. *v.* **3.** to work together; to mesh.

interference *n.* (Latin *inter*, between + *ferire*, to strike + *entia*, condition) **1.** opposition; blocking an action. **2.** the result of merging two beams of equal wavelength of radiant energy. Constructive interference results in an additive effect if the waves are in phase, whereas destructive interference results in a subtractive effect if they are out of phase. **3.** incoherence resulting from obtrusive energy, as noise interfering with speech or static interfering with radio signals.

interferometer *n.* (Latin *inter*, between + *ferire*, to strike + *metrum*, measure) a device that measures wavelengths of light by splitting a beam, reflecting it so one branch must travel farther, then recombining the two branches and determining their interference patterns. Such instruments are used to measure indices of refraction, astronomical distances, the sizes of giant stars, and the angular distance between double stars.

interferon *n.* an antiviral protein produced by cells in response to a viral infection.

interflow *n.* (Anglo-Saxon *under*, among + *flowan*, to flow) water that infiltrates into the soil and then moves laterally through the water table until it reappears in a seep spot, a spring, a stream, or some other body of water.

interfluve *n.* (Latin *inter*, between + *fluvius*, river) the upland area between adjacent streams.

intergelisol *n.* a layer of soil that freezes and does not melt for one or a few years and is akin to permafrost layers that remain frozen indefinitely. It underlies soil that freezes and thaws every year. Also called pereletok.

interglacial period *n.* the time interval between glaciations, especially one of those between the four major periods of Pleistocene glaciation, known in the U.S.A. as the Aftonian, Yarmouth, and Sangamon periods. Buried soils that developed during those periods are known as paleosols.

interlunar *adj.* (Latin *inter*, between + *luna*, moon) the period between the old moon and the new moon when the moon appears too close to the sun to be seen.

interment *n.* (Latin *inter*, among + *mentum*, condition) burial.

intermingle *v.* to mix, mingle, and associate with each other and yet retain the distinct identities of individual items or beings.

intermittent *adj.* (Latin *intermittent*, leaving off) alternating periods of occurrence and nonoccurrence of a phenomenon; e.g., a pain or a sound.

intermittent stream *n.* a stream that has surface water present more than 30 days but not more than 11 months of a year. See ephemeral stream, perennial stream.

intermountain or **intermontane** *adj.* (Latin *inter*, between + *montana*, mountain) located in a valley, basin, or other relatively low area between mountains or mountain ranges.

Intermountain Region *n.* **1.** the area in southwestern U.S.A. between the Sierra Nevada Mountains in eastern California and the Rocky Mountains of Montana, Wyoming, Colorado, and New Mexico. Also known as the Great Basin, this area contains many basins with rivers that do not reach the ocean and many small mountain ranges, including the area known as the Basin and Range Province. Much of the lowland area is arid. **2.** similar areas located between mountain ranges in Mexico, Australia, and other nations.

internal black spot *n.* an ailment of garden beets (*Beta vulgaris*) caused by boron deficiency, often as a result of high soil pH. The first symptoms are stunted, narrow, reddish leaves developing into a cluster of dead leaves at the crown. Black, corky masses of dead cells develop inside the beet roots, making them unmarketable. Also called heart rot.

internal browning of cauliflower *n.* an ailment of cauliflower (*Brassica oleracea botrytis*) caused by boron deficiency. Its effects include poor growth, yellowing of terminal growth, death of terminal buds, and small watery areas in the stem and main branches within the head.

internal cork of apples *n.* an ailment of apples (*Malus sylvestris*) caused by boron deficiency. Its effects include poor growth, yellowing of terminal growth, death of terminal buds, brittle twigs, vascular discoloration, and dead corky spots inside the fruit.

internal drainage *n.* the ability of a soil or other material to allow water to pass through and drain from it; the permeability of the material to water.

international agricultural research centers *n.* There are 23 international agricultural research centers and affiliates; 18 are in the CGIAR system (Consultative Group on International Agricultural Research). See CGIAR, CIAT, CIFOR, CIMMYT, CIP, IBPGR, ICARDA, ICLARM, ICRAF, ICRISAT, IFPRI, IIMI, IITA, and IRRI.

International agricultural research centers

International Center for Living Aquatic Resources Management (ICLARM) *n.* the world fish center, an international food and environmental organization with headquarters in Manila, the Philippines. It works to improve the productivity, management, and conservation of aquatic resources, especially to benefit users and consumers in developing countries in ways that will provide them with food without destroying the environment.

International Centre for Research in Agroforestry (ICRAF) *n.* an international agricultural research center founded in 1977 with headquarters in Nairobi, Kenya. It emphasizes the initiation and support of research on integrating trees in land-use systems in developing countries.

International Crops Research Institute for the Semi-Arid Tropics (ICRISAT) *n.* an international agricultural research center founded in 1972 with headquarters in Andhra Pradesh, India. It focuses on crop improvement and cropping systems. Research covers sorghum, millet, chickpea, pigeon pea, and groundnut.

international date line *n.* an arbitrary line drawn through the Pacific Ocean approximately along the 180th meridian (halfway around the world from the prime meridian passing through Greenwich, England). The area westward from this line to the midnight line as it moves around the Earth is considered a new calendar day, whereas the area westward from the

midnight line to the international date line is still the day before. See Greenwich mean time.

International Food Policy Research Institute (IFPRI) *n.* an international agricultural research center founded in 1975 with headquarters in Washington, D.C. It focuses on food policy and socioeconomic research related to agricultural development and provides policy research and institution building assistance to developing countries.

International Ice Patrol *n.* an organization established in 1914 using U.S. Coast Guard vessels to track icebergs in the North Atlantic and provide warnings to help ships avoid colliding with them after the Titanic struck an iceberg and sank April 15, 1912, drowning 1513 people.

International Institute for Tropical Agriculture (IITA) *n.* an international agricultural research center founded in 1967 with headquarters in Ibaden, Nigeria. It focuses on farming systems, crop improvement, and land management in humid and subhumid tropics.

International Irrigation Management Institute (IIMI) *n.* an international agricultural research center founded in 1984 with headquarters in Colombo, Sri Lanka. It focuses on the performance of irrigation in developing countries. Research covers institutional conditions for managing irrigation systems and facilities, management of water resources, and irrigation support to farmers.

International Rice Research Institute (IRRI) *n.* the international center for the study of rice production founded in 1960 by the Rockefeller and Ford Foundations. It is located at Los Baños on Luzon in the Philippines.

International Service for National Agricultural Research (ISNAR) *n.* an international organization established to assist individual nations in the development and strengthening of national agricultural research systems.

international system of units *n.* See *Système Internationale d'Unités.*

interpolate *v.* (Latin *interpolatus*, polished up) **1.** to estimate a value between established data points. See extrapolate. **2.** to insert new material in a text, especially deceptively. **3.** to insert additional (usually extraneous) things or parts among those already in place.

interrupted stream *n.* a stream that flows for a distance and then disappears into its gravel bed or an underground channel. It is called a lost river if it does not reappear downstream in the same valley.

intersection *n.* (Latin *intersectio*, crossing) **1.** a place where two or more streets or roads cross or junction.

2. a place where two or more bodies overlap. **3.** the process when two or more bodies meet. **4.** the set of elements that two or more larger mathematical sets have in common.

intersex *n.* (Latin *inter*, between + *sexus*, sex) an individual with abnormal sexual characteristics that are between those of normal males and females of its species.

interspecific competition *n.* competition for available resources involving individuals of different species.

interspecific cross *n.* a cross between two different species.

interstate waters *n.* waters that cross more than one state or form part of state or international boundaries; e.g., coastal waters, the Great Lakes, and the Mississippi River.

interstitial *adj.* (Latin *interstitium*, to put between + *ial*, of) in the space between particles or parts of a body; e.g., between sand grains in soil or between adjoining tissues in a body.

interstock or **interstem** *n.* a piece of plant grafted between a scion and rootstock, usually either to form a trunk or to enable the grafting of otherwise incompatible stock.

intertidal *adj.* (Latin *inter*, between + Anglo-Saxon *tid*, time + *alis*, of) in the zone between high and low tides on a coastal area.

intertropical *adj.* (Latin *inter*, between + *tropicus*, tropics + *alis*, of) in the band on both sides of the Earth's equator between the Tropic of Cancer and the Tropic of Capricorn.

intestine *n.* (Latin *intestinus*, internal) the digestive tract between the stomach and the anus, including the small intestine (small in diameter but long and coiled), where the body extracts much of the nutrition from its food, and the large intestine, where the remaining waste is dried and concentrated before expulsion.

intolerant *adj.* (Latin *intolerans*, impatient) **1.** unwilling to accept or respect others whose character, beliefs, opinions, or manners are different than one's own, especially in racial, religious, or political matters. **2.** unable to endure an adverse circumstance; e.g., a plant that is killed by frost.

intracellular *adj.* (Latin *intra*, within + *cellula*, small cell) inside a cell or a coherent group of cells.

Intracoastal Waterway *n.* a protected waterway for small craft that extends from Norfolk, Virginia, south to Key West, Florida, and west from there to Brownsville, Texas. It includes 2666 miles (4300 km) of canals and natural bays and channels that are sheltered from the open sea.

intraspecific competition *n.* competition for available resources involving individuals of the same species.

intrazonal *adj.* (Latin *intra*, within + *zona*, belt + *alis*, of) within a zone; a designation used in the 1938 U.S. system of soil classification for soils with developed profiles that reflect a dominant influence of either topography (especially wetness) or unique parent material. See zonal soil, azonal.

intrinsic *adj.* (Latin *intrinsecus*, inward) naturally part of something; inherent rather than acquired.

introgression *n.* (Latin *introgressus*, gone in + *sion*, action) the transfer of genes from one plant or animal species to another.

intrusive *adj.* (Latin *intrusus*, pushed in + *iv*, function) **1.** disturbing; entering where unwanted. **2.** forced entrance. **3.** injected rock material that separates layers of preexisting rock.

Inuit *n.* (Inuit *inuit*, people) a native inhabitant of eastern Siberia, Alaska, northern Canada, or Greenland. The term Inuit is preferred to Eskimo. These people have lived in this cold climate for more than 2000 years. The Inuit are more Asiatic in appearance than the Native Americans to their south.

inukshuk *n.* (Inuit *inukshuk*, resembling a person) a stone cairn or figure made by nomadic Inuit people to serve as a route marker and/or historical reminder that is now used as a symbol of the Inuit people and their heritage.

Inukshuk

inundate *v.* (Latin *inundatus*, flooded) **1.** to flood; to cover with water. **2.** to overwhelm with an excess of something, as inundated by letters.

invasion *n.* (Latin *invasion*, invasion) **1.** an entrance that overwhelms the native population of an area, often to take possession of the area. **2.** the establishment of a plant or animal species in an area it has not previously occupied.

invasive plant or **invasive weed** *n.* any weed species that invades a new area. Certain invasive weeds are either serious threats or significant added costs to agricultural crop and livestock production; e.g., Canada thistle, field bindweed, and velvetleaf on cropland; kudzu on cropland and woodland; cheatgrass and medusahead on western rangeland; and alligatorweed, hydrilla, and water hyacinth in waterways.

inversion *n.* (Latin *inversio*, turned the wrong way) **1.** a reversal of the usual sequence or direction; e.g., c-b-a instead of a-b-c. **2.** an unusual turning, as a foot that twists inward. **3.** a temperature inversion; i.e., a layer of warm air above a layer of cool air. **4.** an overturned sequence of rock layers with younger rocks beneath older rocks. **5.** the process of converting direct current electricity into alternating current. **6.** a mathematical reversal of dependent and independent variables. **7.** an exchange of musical tones with a normally lower tone raised to a higher position, or vice versa.

invertebrate *adj.* or *n.* (Latin *invertebratus*, not vertebrate) not having a spinal column, as an invertebrate animal.

inverted siphon *n.* a pipeline or other closed passageway that is high at both ends and low in the middle; e.g., a buried pipeline used to carry irrigation water from one high point to another.

invert sugar *n.* a mixture of glucose and fructose formed by hydrolyzing sucrose; so called because the optical rotation of polarized light is reversed in direction when sucrose molecules are split into glucose and fructose. The enzyme involved is called invertase or sucrase.

in vitro *adj.* (Latin *in vitro*, in glass) separated from the living organism or its natural environment; occurring under controlled conditions; e.g., cells grown in a test tube or petri dish.

in vivo *adj.* (Latin *in vivo*, in a living being) occurring in a natural environment inside the body of a living animal or human.

iodine (I) *n.* (Greek *iōdēs*, rust colored or violet colored + *ine*, element) **1.** element 53, atomic weight 126.90; a halogen that forms anions with a charge of −1. Iodine is an essential element for humans and animals but not for plants. Iodine deficiency causes goiter, whose incidence has been greatly reduced by the use of iodized salt containing a small amount of potassium iodide. Silver iodide (AgI) is light sensitive and is used in photography. Commercial sources of iodine include the sodium nitrate deposits in Chile, natural brines in Michigan, seaweed, and seawater. **2.** tincture of iodine, a 3% solution of iodine in alcohol used as a topical antiseptic. It is toxic if swallowed.

iodine value *n.* a measure of the degree of unsaturation (double bonds in the carbon chain) of a fat. It is obtained by titration with an iodine solution. Animal fats tend to have lower iodine values (higher saturation) than vegetable oils.

iomoth *n.* (Greek *Io*, mythological daughter of Inachus + Anglo-Saxon *moththe*, moth) a large North American moth (*Automeris io*) marked with a large eyelike spot on each hind wing. Its larvae have hairs that cause a skin rash.

ion *n.* (Greek *ion*, going) an atom or group of atoms that is either positively charged (a cation such as Ca^{2+} or NH_4^+) or negatively charged (an anion such as Cl^- or SO_4^{2-}). Soluble salts form ions when they are dissolved in water; e.g., NaCl ionizes to Na^+ and Cl^-. The ionic charge is developed by the transfer of the single electron sodium (Na) has in its outer shell to the chlorine (Cl), where it completes an outer shell of 8 electrons.

ion exchange *n.* the replacement of the ions adsorbed on a charged surface with others of like charge; e.g., the replacement of sodium ions (Na^+) on the negatively charged resin in a water softener with cations such as calcium, magnesium, and iron (Ca^{2+}, Mg^{2+}, and Fe^{2+}) from hard water, and the reversal of this exchange when the softener is recharged by passing a concentrated salt (NaCl) solution through it. Similarly, a positively charged resin can be used to exchange anions.

ionic *adj.* (Greek *ion*, going + *ikos*, like) composed of ions. Most mineral substances are composed of ions held together by ionic bonds (electrostatic attraction).

ionic strength *n.* the sum of the concentrations of all ions present in a solution. Ionic strength influences protein solubility and behavior, colligative properties, etc.

ionium (Io) *n.* a radioactive isotope of thorium with a mass number of 230 and a half-life of 8×10^4 years that is formed naturally by the decomposition of uranium.

ionization *n.* (Greek *ion*, going + *ization*, process) **1.** the gain or loss of one or more electrons that converts a neutral atom or cluster of atoms into a charged ion. **2.** the separation of charged ions from a neutral molecule or mineral mass that normally occurs when a salt dissolves in water.

ionosphere *n.* (Greek *ion*, going + *sphaira*, ball) a series of atmospheric layers with a concentration of electrons and anions, designated as D, E, and F layers. The D layer is present only in daytime, from about 30 to 55 mi (50 to 90 km) above the Earth's surface; the E layer is from about 60 to 70 mi (100 to 115 km); the F layer is a single layer at night but divides into two parts in the daytime, with F1 beginning at about 140 mi (225 km) and F2 from about 200 to 260 mi (320 to 420 km). Other doughnut-shaped charged layers centered about 2000 and 9000 mi (3200 and 14,500 km) above the Earth's surface are known as the Van Allen belts. The composition of the ionosphere varies with elevation, solar radiation, and other factors. Radio waves of various frequencies are reflected by different levels of the ionosphere.

iota *n.* **1.** the 9th letter of the Greek alphabet (I, ι). **2.** a very small amount, as in not one iota was left.

ipecac *n.* (Portuguese *ipecacuanha*, ipecac) **1.** a small, creeping, perennial, tropical shrub (*Cephaelis ipeca-cuanha* or *C. acuminata*) with small drooping flowers that is native to South America. **2.** the dried and powdered roots and rhizomes of the shrub that are used as an emetic.

IPM *n*. See integrated pest management.

IR-8 *n*. a variety of rice developed and released in the late 1960s by the International Rice Research Institute. It is day neutral, high yielding, and broadly adapted.

iridescent *adj*. (Greek *iris*, a rainbow + Latin *escens*, becoming) exhibiting a display of various colors, like a rainbow.

iridium *n*. (Greek *iris*, a rainbow + Latin *ium*, a metallic element) element 77, atomic weight 192.22; a relatively rare, heavy, transitional metal of the platinum group. It is significant as an indicator of meteorites and meteoric dust; e.g., there is a notable concentration of iridium in the thin layer of dust that marks the end of the Cretaceous Period and the extinction of the dinosaurs.

Irish moss *n*. carrageen (*Chondrus crispus*), a red algae seaweed common on rocky Atlantic seashores of North America and northern Europe. It is the source of carrageenan used in food products and medicines. See carrageen.

Irish potato *n*. the white potato (*Solanum tuberosum*, family Solanaceae), native to South America but so important in Ireland that two years of crop failure in the 1840s caused the Irish Potato Famine that killed more than a million people and resulted in thousands migrating to the U.S.A. The fungal disease responsible was *Phytophthora infestans*. See potato.

iron (Fe) *n*. **1.** (Anglo-Saxon *iron* = Latin *ferrum*, iron) element 26, atomic weight 55.847; an essential micronutrient for plants, animals, and humans. Iron is an unusually abundant element for its weight, but much of it is concentrated in the Earth's core, and much of the remainder is tied up tightly in mineral structures. Iron ores are abundant and used to produce iron and steel products. Iron deficiencies are common in alkaline soils. They cause chlorosis in plants and anemia in animals and humans. See iron chlorosis, anemia, heme. **2.** any of several kinds of tools made of iron or iron alloys, including irons for pressing clothes, for hitting golf balls, for shooting (guns), for lifting weights, etc. *adj*. **3.** made of iron. **4.** strong like iron. **5.** gray, like the color of iron.

Iron Age *n*. the period between about 1000 B.C. and 100 A.D. (following the Bronze Age) when the use of iron developed and expanded into a dominant factor in human culture.

iron chlorosis *n*. a yellowing of foliage, especially between the veins of younger leaves, caused by iron deficiency. It is most common in alkaline soils. When produced by excessive liming, it is called lime-induced chlorosis.

iron deficiency *n*. See anemia.

ironstone *n*. (Anglo-Saxon *iron*, iron + *stān*, stone) **1.** hardened plinthite, formerly called laterite. Plinthite contains positively charged hydrous oxides of iron and aluminum along with negatively charged clay minerals (mostly kaolinite). Drying causes these materials to bond irreversibly into ironstone. Soft plinthite can be cut into blocks and dried in the sun to form durable bricks. **2.** any rock material that is high enough in iron to be used as iron ore. **3.** a type of dense, white earthenware pottery.

ironweed *n*. (Anglo-Saxon *iron*, iron + *weod*, weed) any of several perennial plants (*Vernonia* sp., family Asteraceae) with hard stems and clusters of tubular flowers with purple to rose colors that turn to rust color.

ironwood *n*. (Anglo-Saxon *iron*, iron + *wudu*, wood) **1.** any of several trees, especially the hornbeams (*Carpinus* sp., family Betulaceae) that produce very hard, dense wood. **2.** the wood of such a tree.

irradiance *n*. (Latin *ir*, in + *radians*, radiating + *antia*, condition) the intensity of solar radiation actually received per unit of time on an actual land surface; the insolation minus the radiation absorbed by the atmosphere and as distributed by the angle of incidence.

irradiation *n*. (Latin *ir*, in + *radiatio*, radiation) **1.** illumination or other form of electromagnetic energy. **2.** exposure to some form of electromagnetic energy. **3.** medical treatment by X rays or ultraviolet rays. **4.** dispersion; radiating out in many directions. **5.** the newest form of food preservation. Using X rays, electron beams, or gamma rays from radioactive cobalt or cesium to kill microbes and insects without causing high temperatures. There is little or no effect on flavor and nutrition. Irradiation greatly increases the shelf life of many food products while reducing the hazard of spoilage and disease.

IRRI *n*. See International Rice Research Institute.

irrigation *n*. (Latin *irrigatus*, bringing water + *tion*, action) **1.** an artificial supply of water supplied through ditches, pipes, or other means to increase plant growth, especially of field crops. About 16% of the world's cropland is irrigated. **2.** the application of a stream of water or other fluid to a body part for medical purposes.

irrigation methods *n*. irrigation water is applied by sprinkler irrigation, surface irrigation (basin, border, corrugation, furrow, or wild flooding irrigation), subirrigation (also called subsurface irrigation), or drip irrigation (also called trickle irrigation).

irrigation system *n*. the canals, ditches, pipes, and other equipment used to deliver water to irrigated land. Most surface and subsurface irrigation uses canals and ditches. Sprinkler irrigation uses pipelines, and drip irrigation uses pipelines and plastic tubing.

irrigation water *n.* water for irrigation comes from surface water, groundwater, or wastewater. Surface water may be diverted or pumped from a stream, or surface runoff may be caught and stored in a reservoir until needed. Groundwater is usually obtained by pumping from a well and is subject to depletion by overuse (see qanat for an exception). Wastewater is available only in certain locations and commonly is of poor quality.

isinglass *n.* (Dutch *huysenblase*, sturgeon bladder) **1.** a sheet of muscovite mica, especially when used as a window in a stove. **2.** translucent or transparent gelatin obtained from air bladders of sturgeon and certain other fish that is used in jams and jellies.

island *n.* (Anglo-Saxon *igland*, island) **1.** an area of land that is completely surrounded by water and is too small to be a continent. **2.** an isolated body of something surrounded by something else, as a counter in the middle of a room, a clump of trees surrounded by prairie, or a cluster of cells or body of tissue notably different from the surrounding tissue.

ISNAR *n.* See International Service for National Agricultural Research.

isobar *n.* (Greek *isos*, equal + *baros*, weight) **1.** a line connecting points of equal barometric pressure on a weather map at a specified time. **2.** one of two or more forms of atoms of different elements having the same atomic weight (one has more protons, the other has more neutrons); e.g., carbon and nitrogen have isobars with an atomic mass of 14.

isoelectric point *n.* the pH that results in a molecule having equal positive and negative electrical charges; e.g., a protein having equal numbers of ionized carboxyl and amino groups and, therefore, minimum solubility in water.

isogonic *adj.* (Greek *isos*, equal + *gōnia*, angle) **1.** having all angles equal; e.g., an equilateral triangle, a rectangle, or a hexagon. **2.** showing equal magnetic declination; e.g., an isogonic line on a map. See agonic line.

isohel *n.* (Greek *isos*, equal + *hēlios*, sun) a line connecting points on a weather map representing places that receive the same number of hours of sunshine each day.

isohyet *n.* (Greek *isos*, equal + *hyetos*, rain) a line on a weather map connecting points of equal precipitation.

isolate *v.* (Italian *isolare*, to isolate) **1.** to separate a thing or group from others; to place apart; to prevent contact with others. **2.** to prepare a pure substance or organism,

usually for study or to evaluate its properties. *n.* **3.** a being or thing that has been separated and set apart.

isoline *n.* (Greek *isos*, equal + Latin *linea*, line) a line that represents a constant value of a specified item; e.g., a contour line on a topographic map or an isobar on a weather map. Also called isogram or isopleth.

isomer *n.* (Greek *isos*, equal + *meros*, part) **1.** one of two or more chemical entities that have the same elemental composition but differ in the spatial arrangement of their component atoms or ions; e.g., butane (C_4H_{10}) can have a simple chain of four carbons, or it can have three carbons attached to a central carbon. Also, many molecules are designated as left-hand or right-hand isomers because they have a mirror-image counterpart (optical isomers rotate polarized light in opposite directions); e.g., most of the amino acids. Many biological processes either produce or require a specific isomer. **2.** one of two or more nuclei configurations with the same proton and neutron composition but differing in energy states and radioactive properties.

isomerase *n.* (Greek *isos*, equal + *meros*, part + *ase*, enzyme) any enzyme that rearranges a molecular structure to convert an organic chemical into one of its isomers; e.g., glucose isomerase converts glucose to fructose. See enzyme, isomer.

isometric *adj.* (Greek *isos*, equal + *metron*, measure) **1.** having two or more equal dimensions. **2.** having a crystal form that has three equivalent axes perpendicular to one another. **3.** projected in a manner to represent a three-dimensional object in a drawing, usually with the third dimension drawn at a 30° angle from the horizontal. **4. isometric exercise** involves tensing a muscle against resistance strong enough to prevent movement (in contrast to isotonic exercise).

isomorphic or **isomorphous** *adj.* (Greek *isos*, equal + *morphē*, form) **1.** having the same form; e.g., a mineral whose chemical composition can vary within certain limits without changing its internal structure that controls the shape of its crystals. **2.** similar appearance of two or more organisms of different species or races.

isomorphic substitution *n.* the replacement of a component by another without changing the overall form; e.g., certain mineral structures apply to a family of minerals that have different octahedral cations (Al^{+3}, Fe^{+2}, Mg^{+2}, Cu^{+2}, Zn^{+2}, etc.); others permit Al^{+3} to substitute for some of the Si^{+4} that typically occupies tetrahedra. Isomorphic substitution can produce cation exchange capacity when the substituting cation carries less charge than the cation it replaces. Also called isomorphous substitution. See cation exchange.

isoneph *n.* (Greek *isos*, equal + *nephos*, cloud) a line on a weather map connecting points of equal cloud cover.

isoprene *n.* a 5-carbon organic molecule (C_5H_8) that serves as the building block of terpenes and sterols.

isostasy *n.* (Greek *isos*, equal + *stasis*, stationary) **1.** equal in pressure at all points. **2.** equal pressure at depth in the Earth's crust; the principle that causes rocks of relatively low density to form mountains.

isostatic adjustment *n.* a shift in the Earth's crust to equalize pressure at depth, usually as a result of erosion having changed the previous balance.

isotherm *n.* (Greek *isos*, equal + *thermē*, heat) **1.** a line on a weather map connecting points of equal temperature. **2.** a curve on a pressure–volume graph representing conditions of constant temperature.

isothermal *adj.* (Greek *isos*, equal + *thermē*, heat + *alis*, of) occurring with no change in temperature.

isotonic *adj.* (Greek *isos*, equal + *tonos*, accent) **1.** equal in osmotic pressure. **2.** having the same osmotic concentration as blood; e.g., a 0.9% solution of sodium chloride. **3.** contraction of muscles that overcomes any resisting force (in contrast to isometric contraction).

isotope *n.* (Greek *isos*, equal + *topos*, place) one of two or more forms of an element having different atomic weights resulting from different numbers of neutrons in their nuclei. The number of neutrons affects the stability of the nucleus, so one isotope may be much more plentiful and another may be more radioactive than its counterpart.

issue *n.* (French *issue*, passage out) **1.** the act of sending out or distributing. **2.** a flowing out or discharge, as water issues from an opening or pus issues from a sore. **3.** one of a series, as today's issue of a newspaper (either one copy or the entire printing). **4.** a matter that is in dispute, especially one being emphasized as being important to the public, as a political issue. **5.** the results of an undertaking. **6.** offspring or progeny. **7.** an outlet or point of egress. *v.* **8.** to send out; to put into circulation; to publish. **9.** to discharge or emit, as to issue radiation or a signal. **10.** to supply, as to issue a uniform or to distribute food. **11.** to go forth. **12.** to be born or produced, as a crop issues from the land.

isthmus *n.* (Greek *isthmos*, neck of land) **1.** a narrow neck of land connecting two larger land areas; e.g., the Isthmus of Panama connecting North and South America. **2.** a narrow strip of tissue joining two larger parts or a narrow passageway connecting two body cavities.

itai-itai *n.* (Japanese *itai-itai*, ouch-ouch) a painful, debilitating bone disease caused by excessive cadmium in food and drink.

itch *n.* (Anglo-Saxon *giccan*, to itch) **1.** a skin irritation that causes an impulse to scratch it. **2.** a desire to do something. *v.* **3.** to feel an itch.

itch mite *n.* a parasitic mite (*Sarcoptes scabiei*, family Sarcoptidae) that burrows into the skin to lay its eggs and causes blisters and intense itching, sometimes known as the 7-year itch, scabies in humans, or mange in animals.

ivy *n.* (Anglo-Saxon *ifig*, ivy) **1.** a climbing vine (*Hedera helix*) with a long woody stem and shiny evergreen leaves that is often seen adorning buildings and walls; native to Eurasia and northern Africa. Also called English ivy. **2.** any of several similar viny plants that attach themselves to walls by means of aerial roots. Note: This includes poison ivy, which causes dermatitis.

Izaak Walton *n.* English ironmonger and biographer (1593–1683), best known for authoring a philosophical book about fishing, *The Compleat Angler* (1653), often considered the "Bible" of fishermen.

Izaak Walton League *n.* an association founded in 1922 to promote conservation of natural resources and encourage their use for recreational activities such as fishing.

J

J or **j** *n.* the symbol for joule in the metric system.

jackal *n.* (Turkish *chagāl*, jackal) **1.** a wild dog (*Canis aureus* and related species, family Canidae) native to southern Europe and northern Africa. Jackals are smaller than wolves and live as hunters and scavengers that travel in packs, often at night. **2.** an individual that does menial, often dishonest, tasks for someone else.

jackass *n.* **1.** a male donkey. **2.** a foolish person.

jackfruit *n.* **1.** a large tropical tree (*Artocarpus heterophyllus*, family Moraceae) with glossy leaves. **2.** the large (watermelon-size) edible, yellow, knobby fruit of the tree. **3.** the fine-grained, yellowish wood of the tree.

jack pine *n.* a short-lived, scrubby, evergreen tree (*Pinus banksiana*, family Pinaceae) native to New England, the states bordering the Great Lakes, and Canada. Jack pine has a wide environmental adaptation and grows on nearly sterile sand as well as on rich soils. It also has great genetic diversity, so seed for regeneration should be collected as close as possible to the area to be planted. It is used for pulpwood, box lumber, and fuel wood.

jackrabbit *n.* **1.** a large hare (*Lepus* sp., family Leporidae) with large ears and long back legs; native to western U.S. rangelands. It feeds on grain crops and the bark of citrus trees. *adj.* **2.** like a jackrabbit, especially in terms of sudden jumping movements, as a jackrabbit start.

jaeger *n.* (German *jäger*, hunter) an aggressive sea bird (*Stercorarius* sp., family *Stercorarius*) that steals food from gulls and terns.

jagsiekte or **jaagsiekte** *n.* (Dutch *jagen*, go quickly + *ziekte*, sickness) a contagious viral disease affecting mostly range sheep (plus some goats and guinea pigs), causing labored breathing, weakness, and death. First identified in South Africa, it was reported in northwestern U.S.A. in 1922.

jaguar *n.* (Tupi *jaguara*, jaguar) the largest American cat (*Panthera onca*, family Felidae); native to tropical regions in Central and South America. It typically is 6 to 7 ft (about 2 m) long and has a yellowish brown coat and black, open spots.

jalap *n.* (Spanish *jalapa*, purgative) **1.** a viny plant (*Exogonium purga*, family Convolvul-aceae) of eastern Mexico with tuberous roots. **2.** a powder prepared from the tubers of the plant for use as a cathartic.

Jaguar

jalapeño *n.* **1.** a pepper plant (*Capsicum annuum*, family Solonaceae) that produces very pungent green or red peppers. Named after Jalapa, Mexico. **2.** the fruit from the plant, used as a spicy seasoning, especially in Mexican cooking.

jalousie *n.* (French *jalousie*, jealousy) a window shutter with adjustable horizontal slats for regulating light and air coming into a room.

jam *n.* **1.** a mass of objects packed tightly together, as a log jam. **2.** the process of packing objects so tightly together that they cannot move. **3.** a type of preserved fruit, crushed and sweetened but not filtered; used as a sandwich spread or a sweetener for baking. **4.** an embarrassing situation; a predicament. *v.* **5.** to force objects tightly together, often into an enclosed space. **6.** to hurt or injure by squeezing too tightly, as to jam one's hand between 2 blocks. **7.** to block by overcrowding. **8.** to press violently, as to jam one's foot on the pedal. **9.** to interfere with radio signals by broadcasting a strong conflicting signal. **10.** to be stopped by a blockage.

Jamestown weed *n.* See jimsonweed.

Japan clover *n.* a drought-resistant annual legume (*Lespedeza striata*, family Fabaceae) used for hay, pasture, and soil improvement; native to Asia.

Japanese barberry *n.* a beautiful and hardy shrub (*Berberis thunbergii*, family Berberidaceae) that is resistant to wheat rust; native to Japan. It often is grown as a decorative border shrub because of its yellow flowers, red berries, and red fall foliage.

Japanese beetle *n.* a plump, shiny, brown and green insect pest (*Popillia japonica*, family Scarabaeidae) about ½ in. (1.2 cm) long that feeds on the leaves of hundreds of different plants, damaging leaves, blossoms, and fruits. Common in Japan, it first was seen in the U.S.A. in New Jersey in 1916 and now is widespread in eastern U.S.A. The beetle (adult) is seen only in summer; its grubs (larvae) feed on plant roots.

Japanese rose *n.* See multiflora rose.

japygid *n.* a wingless insect (family Japygidae) that eats many species of soil arthropods.

jarovization *n.* See vernalization.

jasmine or **jasmin** *n.* (Persian *yāsamīn*, jasmine) **1.** any of several tropical or subtropical plants (*Jasminum* sp., family Oleaceae) with fragrant yellow, red, or white blossoms. The fragrance is used in perfumes and teas. **2.** any of several similar plants with fragrant blossoms.

jaundice *n.* (French *jaunisse*, yellow skin) **1.** a yellowish pigmentation of the skin of humans and animals caused by deposits of bile pigments; often first noted in the whites of the eyes. Also called icterus. **2.** a prejudiced viewpoint with judgment distorted by resentment or envy. *v.* **3.** to incite prejudice and resentment.

Java black rot *n.* a fungal (*Riplodia tubericola*) disease of stored sweet potato tubers. It causes a dry rot that turns the decayed tissues black and hard.

jay *n.* (Latin *gaius*, jay) any of several medium-to-large omnivorous passerine birds with a gregarious nature; most often found in forests; e.g., blue jay, Steller's jay.

jejune *adj.* (Latin *jejunus*, meager, dry, empty) **1.** lacking nutrition; barren. **2.** lacking maturity; childish; empty phrases. **3.** insignificant; lacking meaning and/or interest.

jellyfish *n.* **1.** any of several marine animals (class Hydrozoa or Scyphozoa) with an umbrella-shaped, jellylike body and tentacles with stinging hairs. Also called medusa. **2.** a person who yields easily; a weakling.

jennet *n.* (French *genet*, Zenetic horse) **1.** a female donkey. Also called a jenny. **2.** a breed of small Spanish horses.

jenny *n.* **1.** a jennet. **2.** a female wren. **3.** a multi-spindled spinning machine.

jerky *n.* **1.** narrow strips of meat (usually beef) dried in the sun. *adj.* **2.** spasmodic; irregular short sudden movements. **3.** silly, ridiculous, and/or inconsiderate.

Jerusalem artichoke *n.* an annual plant (*Helianthus tuberosus*, family Asteraceae) that resembles a sunflower and has edible, crooked, potato-like tubers that are high in fructose ($C_6H_{12}O_6$), a sugar recommended for diabetics. The plant is so aggressive it can become a weed pest. Also called Canada potato.

jet lag *n.* a temporary imbalance in the circadian rhythm associated with high-speed travel across several time zones that disrupts a traveler's sleeping and eating cycles.

jet stream *n.* a meandering west-to-east (except for a summer jet stream from southeast Asia to Africa) air current that moves through the upper troposphere, typically about 6 to 9 miles (10 to 15 km) above the Earth's surface at a speed of 100 to 250 mph (150 to 400 km/h).

jetty *n.* (French *jetée*, projection) **1.** a barrier made of con-crete, stones, or wood that extends into the sea or a lake to protect a harbor. **2.** a pier or the piling or other structure that supports a pier. **3.** a barrier that extends into a stream or body of water to prevent bank erosion by diverting a current. **4.** an overhang where an upper story of a building extends beyond the wall of the floor below.

jicama *n.* (Nahuatl *xicama*, jicama) **1.** a twining tropical vine (*Pachyrhizus erosus*, family Fabaceae). See *Pachyrhizus*. **2.** the large, turnip-shaped, tuberous root of the plant with a fibrous brown skin and a white fleshy interior that may be eaten raw, cooked, or marinated in fruit juices. Also called Mexican potato, Mexican turnip, sakula, xiquima, or yam bean.

jig *n.* **1.** a very lively dance. **2.** the prancing of a horse. **3.** a bouncy or jerky movement to attract attention, as for a fish lure. **4.** a clamp or other device to hold work in place for a machine tool to cut or shape it. **5.** an apparatus used to separate coal or ore from reject material. **6.** a system for dying cloth by taking it from a roller, pulling it through a vat, and wrapping it on another roller. *v.* **7.** to play or dance to a lively tune or pace. **8.** to cause to move in a bouncy manner, as a fishing lure. **9.** to place material in a holder so it can be worked.

jimmies *n.* **1.** a disease of sheep caused by eating bulbs of the cloak fern (*Notholaena sinuata*, family Polypodiaceae). Affected animals tremble, breathe rapidly, and walk stiffly with their backs arched. **2.** a trademark name for small, colored, rod-shaped confections (usually containing chocolate) often sprinkled on ice cream, cake, or pastry.

jimmyweed *n.* a perennial herb (*Aplopappus heterophyllus*, family Asteraceae) that grows in southwestern U.S.A. and contains tremetol, a toxin that causes trembles in livestock and milksickness in humans. Also called rayless goldenrod.

jimsonweed *n.* an annual weed (*Datura stramonium*, family Solanaceae) with tubular flowers and prickly fruit. Its leaves contain alkaloids, hyoscyamine and scopolamine, that are very poisonous to people and animals, causing nausea, thirst, blindness, convulsions, and sometimes death; used in medicine to cure gallstones. Also called Jamestown weed, thorn apple.

Job's-tears *n.* **1.** an Asian grass (*Coix lacrima-jobi*, family Poaceae) that bears edible grains wrapped in hard, beadlike bracts. **2.** the hard, nearly spherical structure formed by the bracts, used as beads.

joey *n.* (Austral *joey*, joey) a young animal in Australia, especially a young kangaroo.

Johne's disease *n.* a chronic infectious disease of cattle, sheep, deer, and goats named after Heinrich A. Johne (a German pathologist). The bacterium *Mycobacterium paratuberculosis* causes chronic diarrhea that often leads to death; it is transmitted orally and can survive for a year or more in feces, soil, or water.

johnsongrass *n.* a tall (up to 6 ft or 1.8 m) perennial sorghum (*Sorghum halepense*, family Poaceae) that spreads by rhizomes; native to the Mediterranean area. It is a highly productive forage crop in southern U.S.A. but becomes a difficult weed to control in cropped fields. Named after William Johnson, a U.S. agriculturalist who began planting it in 1840.

joint *n.* (French *joint*, joint) **1.** either a flexible or rigid connection between two parts of a body or object; e.g., a knee or elbow of a person, a node in a plant stem, or the junction of two pieces of wood in a chair. **2.** a fracture plane in a rock mass (usually one of a group with parallel orientation). **3.** a marijuana cigarette. **4.** a disreputable estab-lishment of public entertainment. *adj.* **5.** combined; shared, as joint custody of children or property or a joint effort.

jointworm *n.* (French *joint*, joint + Anglo-Saxon *wyrm*, worm) a small grub; the larva of a tiny wasp (*Harmolita tritici*, family Eurytomidae). The wasp lays eggs in early spring in the succulent stems of wheat; the larvae produce gall-like swellings— usually just above the second or third joint of the plant—that cause lodging and reduce the yield.

jojoba *n.* (Spanish *jojoba*, jojoba) a deciduous shrub (*Simmondsia chinensis*) native to desert areas of southwestern U.S.A. and northern Mexico; used as a hedge. It bears edible fruit and seeds that yield a valuable oil used as a replacement for whale oil in cosmetics and high-pressure lubricants.

jonquil *n.* (French *jonquille*, little junco) a narcissus (*Narcissus jonquilla*, family Amaryllidaceae) with long, narrow leaves and fragrant yellow or white flowers; a perennial ornamental flower that regrows from bulbs each year.

joule (J or **j)** *n.* **1.** a unit of energy equal to the work done when a force of 1 newton acts through a distance of 1 meter; equal to 10^7 ergs, 0.735 ft-lb, 0.239 cal, or 0.000952 Btu. **2.** a unit of electrical energy equal to the work done when a current of 1 ampere is passed through a resistance of 1 ohm for 1 second.

Judas goat *n.* a goat trained to lead sheep to the killing pens in a slaughterhouse; named after Judas Iscariot, the biblical traitor who betrayed Jesus.

Judas tree *n.* a purple-flowered Eurasian tree (*Cercis siliquastrum*, family Fabaceae) believed to be the kind of tree Judas Iscariot used to hang himself after betraying Jesus. **2.** other similar trees, such as the redbud.

judicious *adj.* (Latin *judiciosus*, prudent) sound in principle; based on good judgment. **2.** wise in the use of resources; not wasteful.

jughead *n.* **1.** a contrary horse not amenable to training. **2.** a foolish, noncooperative person.

juglone *n.* (Latin *juglans*, walnut) a toxic crystalline substance ($C_{10}H_6O_3$) in black walnut, butternut, and pecan nuts, hulls, and leaves; an antibiotic with antihemorrhagic and sedative properties; toxic to most garden crops.

jujube *n.* (Latin *jujuba*, jujube tree) **1.** a tree (*Ziziphus jujuba* and related species, family Rhamnaceae) of warm climates of the Old World with oval leaves and clusters of green or white flowers. **2.** the small, brown, oval, edible fruit from the tree; a drupe. Also called Chinese date.

Julian calendar *n.* a solar calendar introduced by Julius Caesar in 46 B.C.; it had a 12-month year of 365 days and a leap year of 366 days every 4 years. The resulting average year of 365.25 days was slightly too long, so a 10-day correction was made when it was replaced by the Gregorian calendar in 1582. See Gregorian calendar.

Julian day *n.* **1.** the number of the day counted consecutively from January 1, 4713 B.C. **2. modified Julian day** the Julian day minus 2,400,000.5 (equivalent to counting from noon November 5, 1857).

jumping bean *n.* a triangular seed from any of several shrubs (*Sebastiana* sp. and/or *Sapium* sp., family Euphorbiaceae) that contains the larva of a small moth (*Cydia saltitans*). Movements of the larva cause the seed to jump, jerk, and roll. Sometimes called Mexican jumping bean.

jumping gene *n.* a gene that moves from one site to another on the same chromosome, on another chro-mosome of the same organism, or that even transfers to another organism. Barbara McClintock (1902–1992) was awarded the Nobel Prize for Medicine in 1983 for her work in the 1940s and 1950s that demonstrated jumping genes in plants. This dis-covery helped explain some mutations and many anomalies, including the development of some forms of resistance to pesticides and antibiotics. Also called transposon. See transposon, gene.

jumping spider *n.* a type of spider that is used in China to control insect pests in rice fields. See wolf spider.

juncture *n.* (Latin *junctura*, juncture) **1.** a joining together. **2.** the line or point marking the junction of two bodies. **3.** a particular time, usually a crucial time for an event or process or a time of special interest to the parties involved. **4.** a crisis; a very serious state of affairs. **5.** the transition between two sounds or between a sound and silence in the enunciation of words.

June bug or **June beetle** *n.* **1.** a large brown beetle (*Phyllophaga portoricensis*, family Scarabaeidae) of northern U.S.A. that eats leaves of several tree species and resembles a Japanese beetle; so called because it is seen flying after about June 1. The larvae are white, fat grubs that feed on roots of grasses and many garden vegetables and are a favorite food of moles. Sometimes called May beetle. See white grub. **2.** a similar large green beetle (*Cotinis nitida*) of southern U.S.A. Also called figeater.

June bug

jungle *n.* (Hindi *jangal*, wooded area) **1.** a wild, dense growth of trees, shrubs, and vines in a warm, damp climate. See tropical rainforest. **2.** the land area covered by jungle vegetation. **3.** a jumbled mass of miscellaneous objects. **4.** a dangerous place where violence may occur, as the city streets were a jungle. **5.** a confusing situation, as a jungle of rules.

juniper *n.* (Latin *juniperus*, juniper) any of several evergreen trees or shrubs (*Juniperus* sp., family Cupressaceae) with scaly leaves and small (¼ to ½ in.

or 6 to 12 mm across) berrylike cones that contain only one to four seeds. Junipers are used as ornamentals and for fuelwood and fenceposts.

Jupiter *n.* (Latin *Juppiter*, Jupiter) **1.** the largest planet in the solar system, a dense gaseous body located in the middle of the sequence between Mars and the asteroid belt on the inside and Saturn on the outside. **2.** the Roman god of thunder and the skies, and the chief of the Roman deities.

jute *n.* (Hindi *jūt*, matted hair) **1.** a tropical herbaceous plant (*Corchorus capsularis* or *C. olitorius*, family Tiliaceae) grown mostly in India, China, and Bangladesh. **2.** the strong, coarse, brown fiber obtained from jute plants and used to make burlap, gunny sacks, and other coarse fabrics.

juvenile *n.* (Latin *juvenilis*, young) **1.** a young person, animal, or plant. **2.** gas or water coming to the surface for the first time. See fossil water. *adj.* **3.** pertaining to or suitable for young persons or animals. **4.** immature; acting like a young person or animal, as juvenile behavior.

juvenile hormone *n.* a hormone secreted by insects that regulates growth and prevents the insects from molting to their adult form. Synthetic juvenile hormones can sometimes be used to control insects by keeping them in an immature stage and thus preventing reproduction. See integrated pest management.

juvenility *n.* (Latin *juvenilis*, young + *itat*, condition) the growth period preceding maturity. Flowering cannot be induced in juvenile plants, but cuttings from juvenile shoots root more readily than those from mature shoots.

K

K *n.* **1.** chemical symbol for the element potassium; from Latin *kalium.* **2.** symbol for temperature in degrees Kelvin. **3. k** kilo (1000), as in 1 kg = 1000 g or 1 km = 1000 m. **4.** computer language symbol for $2^{10} = 1024$.

K-A age or **K-A date** *n.* the radioactive age of a rock as determined from the ratio of potassium 40 (^{40}K) to argon 40 (^{40}A) present in the rock. The age is based on a half-life of 1.3 billion years for the decay of ^{40}K to ^{40}A.

kaavie *n.* in Scotland, a heavy snowfall.

kafir or **kafir corn** *n.* (Arabic *kāfir,* grain sorghum) a tropical African variety of grain sorghum (*Sorghum bicolor* var. *caffrorum*) with short-jointed, sturdy, leafy stalks; suitable for semiarid regions such as southern U.S. Great Plains; used for grain and forage.

kainite *n.* (Greek *kainos,* recent + *ite,* mineral) a mineral ($MgSO_4 \cdot KCl \cdot 3H_2O$) used as a source of fertilizer for supplying magnesium and potassium.

kaki *n.* (Japanese *kaki,* kaki) Japanese (or Chinese) persim-mon, an Asiatic tree (*Diospyros kaki,* family Ebenaceae), a source of hard wood with straight grain. Also called date plum. **2.** the edible fruit of the tree.

kala-azar *n.* (Hindi *kālāāzār,* black disease) a tropical parasitic disease of the liver and spleen caused by protozoa (*Leishmania donovani*) transmitted by sand flies (*Phlebotomus* sp.). It causes fever, anemia, enlarged spleen, and death.

Kalahari Desert *n.* a desert area of 360,000 mi^2 (930,000 km^2) in southern Africa, mostly in Botswana.

kale *n.* (Scottish *kale,* cole) an annual or biennial, cool-season, leafy vegetable (*Brassica oleracea* var. *acephala,* family Brassicaceae) related to broccoli, cabbage, cauliflower, mustard, etc.

kalmia *n.* See lambkill kalmia, laurel.

kame *n.* (Anglo-Saxon *camb,* comb) a knob or short ridge composed of stratified sand and gravel deposited by glacial meltwater. It is commonly associated with low areas called kettles in kame and kettle topography.

kanat *n.* See qanat.

kandic horizon *n.* a diagnostic subsurface horizon in Soil Taxonomy. Clay content is at least 4% to 8%

more than that of the overlying epipedon, but clay coatings characteristic of argillic horizons are not required. Most of the clay is kaolinitic, and the cation exchange capacity is relatively low.

kangaroo *n.* (Austral *ganurru,* kangaroo) a marsupial (family Macropodidae) with a relatively small head, short front limbs, powerful rear legs that it uses for leaping, and a thick tail that helps it balance. It is a native of Australia and adjacent islands. Mothers carry their young in a marsupium (pouch).

kankar or **kunkur** *n.* **1.** calcium carbonate ($CaCO_3$) nodules formed usually in alluvium of semiarid areas, especially in India in Vertisols. **2.** limestone rock containing such nodular concretions, especially in India. **3.** precipitated calcium carbonate coatings on pebbles in porous sediments, especially in India. **4.** a residual calcium carbonate deposit in the U.S.A.; e.g., a deposit of caliche.

kaolin *n.* (Chinese *kaoling,* high ridge; the name of a clay hill in southeastern China) any of a family of 1:1 layer silicate clay minerals (kaolinite, halloysite, nacrite, and dickite) common in highly weathered soils and sedimentary rocks that all have the same formula [$Al_4Si_4O_{10}(OH)_8$] but with differences in the alignment of the layers. Kaolin has a relatively low cation exchange capacity (about 80 mmol/kg or 8 meq/100 g), a layer spacing of 0.72 nm, and low shrink–swell potential. It is used to make fine chinaware, as a filler in textiles, paper, and certain medicines, and as an absorbent in the alimentary tract of animals. Also called China clay, white clay, or bolus alba.

kaolinite *n.* (Chinese *kaoling,* high ridge + *ite,* mineral) **1.** the form of kaolin with all of the layers aligned. **2.** any of the kaolin family of silicate clay minerals, including kaolinite and its poorly crystallized polymorphs, nacrite and dickite, its magnesium equivalent, antigorite (with Mg$_3$ substituted for Al$_2$ in the formula), and halloysite. See kaolin.

kapok *n.* (Malay *kapoq,* kapok) **1.** a large spiny tropical tree (*Ceiba pentandra,* family Bombacaceae) native to America. Also called silk-cotton tree. **2.** a fibrous material obtained from the fruit of the tree. It is used as filler for life preservers and as acoustic insulation and is very resistant to wetting.

karez or **kareze** *n.* qanat, a name used in Pakistan.

karroo or **karoo** *n.* (Hottentot *karusa,* dry, hard) about 100,000 mi^2 (260,000 km^2) of high, semiarid tableland in South Africa with highly weathered red clay soils where grass grows only during the wet summer season.

karst *n.* a type of topography characterized by sinkholes and streams that disappear from the surface and flow through underground solution channels and caves in limestone, dolomite, or gypsum bedrock. It is named after a humid area with old limestone rock north of Trieste in the northwestern part of the former Yugoslavia and northeastern Italy and is now used for similar topography around the world. See thermokarst.

kat *n.* See qat.

katabatic *adj.* (Greek *katabatikos*, going down) going down; descending, as a local gravity wind, a cool breeze that moves down a valley or a slope, typically as a result of radiation cooling in the evening. See anabatic.

katabolism *n.* See catabolism.

katamorphic *adj.* (Greek *kata*, down or against + *morphe*, form + Latin *ic*, of) near Earth's surface; occurring in rock or soil at or near the surface.

katamorphism *n.* (Greek *kata*, down or against + *morphē*, form + *isma*, action) the result of natural geologic processes changing complex minerals to simpler, less dense minerals by oxidation, hydration, and solution in the katamorphic zone. Also spelled catamorphism.

katarobic zone *n.* a water environment in which much of the oxygen is used in the decomposition of organic materials but enough oxygen remains to support aerobic life.

katydid *n.* any of several large green insects (order Orthoptera, especially family Tettigonidae), related to grasshoppers and crickets, that dwell mostly in trees. Named for the shrill mating sound the males make by rubbing their forewings together.

kava *n.* (Maori *kawa*, bitter) **1.** a plant (*Piper methysticum*, family Piperaceae) whose roots are a source of a narcotic and intoxicating drink. **2.** the drink produced from the plant roots.

kcal, K Cal, Kcal or **Cal** *n.* kilogram calorie, a unit of heat or energy commonly used to evaluate the energy value of food or the energy output of an organism; the amount of heat required to warm 1 kg of water 1°C (1.8°F), now defined as 4184 joules, equivalent to 1000 small calories or 3.97 Btu.

keet *n.* a young guinea fowl (*Numida meleagris*) named for the sound of its call.

kelp *n.* **1.** any of many large brown alga (family Laminariaceae) commonly known as seaweed and harvested for use as feed and as a source of algin, iodine, and potassium carbonate. Kelp grows in cool waters of both the Pacific and Atlantic Oceans. See giant kelp. **2.** a mass of such seaweed, either floating or washed up on a beach. See Sargasso Sea. **3.** ashes from burnt kelp, used as a source of iodine.

Kelvin scale (K) *n.* the SI standard temperature scale, named after Lord Kelvin (William Thomson, 1824–1907), an English physicist and inventor, that is measured from absolute zero with degrees matching those of the Celsius scale (equal to 1.8°F). Under 1 atm pressure, water freezes at 273.15 K and boils at 373.15 K.

kenaf *n.* (Persian *kanab*, hemp) **1.** a giant tropical grass (*Hibiscus cannabinus*, family Malvaceae) that can be grown from Kansas southward and is used as a high protein (30% in the leaves) livestock feed and as a fiber crop for newsprint. Also called Indian hemp or bastard jute. **2.** the fiber from the bark of the plant.

Kenaf

Kentucky bluegrass *n.* a cool-season perennial grass (*Poa pratensis*, family Poaceae) widely grown as a turf grass and for pasture.

Kentucky coffee tree *n.* a deciduous North American tree (*Gymnocladus dioica*, family Fabaceae) with large brown seed pods that hang on the tree over winter. The seeds were formerly used for making coffee.

keratin *n.* (Greek *keratos*, horn) a durable insoluble protein that is the principal component of hair, nails, horns, hooves, claws, and feathers.

keratitis *n.* (Greek *keras*, cornea + *itis*, inflammation) inflammation of the cornea caused by foreign bodies, wounds, or other irritants that causes tear formation, encrustations, and redness and affects humans and many species of animals. Keratitis, conjunctivitis, and other eye diseases are sometimes called pinkeye.

kernel *n.* (Anglo-Saxon *cyrnel*, little corn seed) **1.** a seed, as a kernel of corn, wheat, or other grain crop, usually covered by husks or integuments during growth. **2.** the usually edible soft interior of a nut or of the pit of a stone fruit. **3.** a small hardened place in the skin. **4.** the essence of an idea, as the kernel of an argument or a kernel of truth.

kernite *n.* hydrated crystalline sodium borate ($Na_2B_4O_7 \cdot 4H_2O$), the principal source of boron. It is named after Kern County, California, where some of it is mined.

kerogen *n.* (Greek *kēros*, wax + *genesis*, becoming) a fossilized organic substance in oil shale that, upon heating, yields shale oil.

kerosene *n.* (Greek *kēros*, wax + *enus*, hydrocarbon) a petroleum product with a maximum distillation temperature of about 400°F (200°C) and a minimum flash point of 100°F (38°C); used in space heaters, water heaters, cook stoves, and wick lamps.

kestrel *n.* (French *quercerelle*, kestrel) a small falcon (*Falco sparverius* of North America or *F. tinnunculus* of Europe) that has a rust-colored back and is able to hover in the air as it hunts (mostly for insects).

ketone *n.* (German *keton*, ketone) an organic compound with an oxygen double bonded to a carbon that is not at the end of the chain. See aldehyde.

ketose *n.* (German *keton*, ketone + *ose*, sugar) a ketone sugar, e.g., fructose.

kettle *n.* (Latin *catillus*, a small container or pot) **1.** a container used to heat water, usually with a handle and a spout for pouring the hot water; e.g., a teakettle. **2.** a pot or pan used to heat water or cook foods. **3.** a basin in glaciated areas formed by the melting of a block of ice that was buried and left behind as the glacier receded; commonly associated with kames in kame and kettle topography. **4.** a kettledrum, a drum shaped like half of a sphere covered with a taut skin or parchment.

key species *n.* a species chosen as an indicator to represent the assemblage of species in an area, especially regarding environmental conditions. Populations of certain bird species are often used as key species.

kg *n.* kilogram.

khamsin *n.* (Arabic *khamsīn*, fifty) a hot, southerly wind that blows across Egypt from the Sahara Desert for about 50 days each spring from late March to early May.

khapra beetle *n.* a tiny beetle (*Trogoderma granurium*, family Dermestidae) that is a serious pest of stored grain.

khat *n.* **1.** (Arabic *qāt*, khat) an evergreen shrub (*Catha edulis*, family Celastraceae) native to Africa and Arabia. Also spelled kat or qat. **2.** the leaves of the shrub chewed fresh as a stimulant or made into a tea.

kiang *n.* (Tibetan *kyang*, wild ass) a wild ass (*Equus hemionus*, subspecies kiang) of the Asian mountains and high plateaus of Tibet and Mongolia.

kibbutz *n.* (pl. **kibbutzim**) (Hebrew *qibbūs*, gathering) a collective community settlement in Israel (usually agricultural).

kibbutznik *n.* (Hebrew *qibbūs*, gathering + *nik*, person) a member of a kibbutz.

kid *n.* (Norse *kith*, kid) **1.** a young goat. **2.** the meat from a young goat. **3.** kidskin leather. **4.** (slang) a child. *v.* **5.** to joke or tease.

kidney *n.* **1.** a bean-shaped glandular organ that the body of a human or other vertebrate uses to remove liquid and soluble wastes from the blood for secretion as urine. **2.** a similar organ in invertebrate animals. **3.** the meat from animal kidneys.

kidney bean *n.* **1.** a legume seed shaped like a kidney (commonly about a half-inch or 1.2 cm long, a quarter-inch wide, and slightly curved), especially such a seed that is reddish brown. **2.** a plant that produces such seed (certain varieties of *Phaseolus vulgaris*, family Fabaceae).

kidney stone *n.* a calculus (pl. calculi) in a kidney; a hardened mineral deposit that crystallizes in the kidney from calcium carbonate, urates, phosphates, or cystine.

kidney vetch *n.* a leguminous herb (*Anthyllis vulneraria*, family Fabaceae), native to the Old World, that was formerly used to treat kidney diseases. Also called woundwort.

kidskin *adj.* (Icelandic *kith*, kid + *skinn*, skin) **1.** leather made from the skin of a young goat. **2.** an object made from such leather, especially a kidskin glove or shoe.

kieselguhr *n.* (German *kiesel*, flint + *guhr*, an earthy sediment) a German name for diatomaceous earth.

kieserite *n.* hydrous magnesium sulfate ($MgSO_4 \cdot H_2O$) named after a German scientist, D. G. Kieser (1779–1862). It occurs in saline residues and is used as a magnesium fertilizer and also as a mild laxative.

Kilauea *n.* (Hawaiian *kilauea*, spewing) the largest active volcano in the world; a shield volcano with a crater 8 mi (13 km) in circumference surrounding a fire pit at an elevation of 3646 ft (1111 m) on the slopes of Mauna Loa Mountain (13,679 ft or 4169 m high) on the island of Hawaii. It is located in Hawaii Volcanoes National Park.

Kilimanjaro, Mount *n.* a mountain in Tanzania composed of three extinct volcanoes. The center peak (Kibo) reaches 19,340 ft (5895 m) and is the highest point in Africa. Although near the equator, the peak is snowcapped the year around.

killer bee *n.* See honeybee, Africanized.

killer whale *n.* the orca, the largest species of dolphin (*Orcinus orca*). It is a predator about 30 ft (9 m) long that kills and eats large fish, penguins, porpoises, sea lions, and seals and even attacks whales that are much larger than itself. Also called toothed whale, but orca is the preferred name.

killing frost *n.* a temperature that drops below freezing long enough to kill sensitive plants. See hardy.

kiln dried *adj.* dried in a kiln (a drying oven or furnace); e.g., kiln-dried lumber.

kilocalorie *n.* See kcal.

kilogram (kg) *n.* (French *kilo*, 1000 + Greek *gramma*, weight) the base unit of mass in the metric system of weights and measures. It equals 1000 g or 2.2046 lb.

kilohertz (kHz) *n.* the newer term for kilocycle, a frequency of 1000 cycles per second.

kilometer (km) *n.* (French *kilo*, 1000 + Greek *metron*, a measure) a metric unit of distance equal to 1000 m, 3281 ft, or 0.6214 mi.

kilowatt (kW or **kw)** *n.* a unit of electrical power equal to 1000 W or approximately 1.34 horsepower.

kilowatt-hour (kWh or **kwh)** *n.* the usual unit for marketing electricity, equivalent to a current of 1000 W maintained for 1 hr.

kinesics *n.* (Greek *kinēsis*, responsive movement) body language; the study of nonverbal communication based on facial expressions, gestures, posture, and body movements. Also called body language.

kingbird *n.* (Anglo-Saxon *cyng*, king + *bridd*, young bird) any of several aggressive American flycatchers (*Tyrannus* sp.) with brown backs and white or yellow breasts that frequent rural areas and grow to a length of about 7 in. (18 cm). The most common species are the eastern kingbird (*T. tyrannus*), with a black head, white breast, and white fringe across the end of its tail (also called bee martin), and the western kingbird (*T. verticalis*), with a yellow breast and a narrow white fringe on each side of its tail.

king cobra *n.* a large, poisonous snake (*Ophiophagus hannah*, family Elapidae) that preys mostly on other snakes in both forests and fields. Native of the Philippines and southeast Asia, typical adults are 7 to 13 ft (2 to 4 m) long, but some grow to 18 ft (5.5 m). Also called hamadryad.

king crab *n.* **1.** a large, edible crab (*Paralithodes camtschatica*) common in the Pacific coastal waters of Alaska, Japan, and Siberia. **2.** horseshoe crab.

kingdom *n.* (Anglo-Saxon *cyng*, king + *dom*, dominion) **1.** a nation ruled by a king or queen. **2.** the highest category in the classification of living things; any of five such kingdoms, including the animal, plant, fungus, protist, and moneran kingdoms. **3.** one of the three broad classes of natural things, the animal, vegetable, or mineral kingdom.

kingfisher *n.* (Anglo-Saxon *cyng*, king + *fiscere*, fisher) any of several birds (family Alcedinidae, order Coraciiformes) with large heads, ragged crests, and short tails. They dive into the water to catch fish with their long, sharp beaks.

king snake *n.* a large, spotted, nonpoisonous snake (*Lampropeltis* sp.) that eats rodents and other snakes.

Kirtland's warbler *n.* an endangered songbird (*Dendroica kirtlandii*) that nests on the ground but only among stands of jack pine in north-central Michigan. It has a gray back and a yellow breast.

kissing bug *n.* any of several assassin bugs (*Triatoma* sp., family Reduviidae) that attack mostly insects but will also bite humans and other mammals, especially around the face and lips. Their bite is toxic and very painful. Also called conenose.

kitchen midden *n.* a mound or other deposit containing refuse, shells, bones, etc., such as that discarded from a kitchen. It is a common indicator of a former habitation, especially one that represents a prehistoric site.

kite *n.* (Anglo-Saxon *cyta*, a bird of the kite family) **1.** a lightweight frame covered with paper or other thin material designed to fly in the wind at the end of a long string. **2.** any of several graceful, medium-sized American hawks (family Milvinx) that can fly swiftly, soar, or hover while hunting before swooping down to catch their prey. **3.** a bad check or other fraudulent monetary document. **4.** a greedy person who defrauds others. *v.* **5.** to fly with a rapid gliding motion like a kite. **6.** to write a bad check, as to kite a check.

kiwi *n.* (Maori *kiwi*, kiwi) **1.** a flightless bird (*Apteryx* sp., family Apterygidae) with a long bill and no tail. A native of New Zealand about the size of a chicken, it feeds mostly on insects and worms. **2.** a woody Chinese vine (*Actinidia chinensis*, family Actinidiaceae) with edible fruit. **3.** the edible fruit from the vine, about the size of a small lemon with a fuzzy brown skin and a green interior. Also called Chinese gooseberry.

Kjeldahl *n.* a process for measuring the nitrogen content of organic materials by reducing the nitrogen to ammonia and measuring the ammonia by a distillation procedure. It was developed in 1883 by J. G. C. Kjeldahl, a Danish chemist.

klebsiella *n.* any of a genus of schizomycetes that are pathogenic on humans, including a species (*K. pneumoniae*) that causes pneumonia. They are rod-shaped, gram-negative bacteria (family Enterobacteriaceae) that closely resemble *Aerobacter aerogenes*.

klendusity *n.* the ability of a plant to avoid infection with a disease because it has a thick cuticle or other protection that forms a barrier to microbes.

kleptoparasite *n.* (Greek *kleptēs*, thief + *parasitos*, one who eats at another's table) a parasite that steals food from its host species, as happens among ants and bees.

klinotaxis *n.* (Greek *klinein*, to lean + *taxis*, arrangement) movement of an organism (e.g., a motile microbe or an insect) toward or away from a stimulus such as light, heat, or other variable in the environment. See chemotaxis.

Knapp, Seaman Asahel *n.* a farmer, a livestock breeder, a banker, a publisher, a professor, and a president of Iowa State College, but the public knows Dr. Knapp (1833–1911) best as the founder of the Agricultural Extension Service in the United States. In 1904, the USDA Bureau of Plant Industry gave him the assignment of stopping the boll weevil. He organized on-farm demonstrations to change the date of planting and rotate crops to starve the boll weevil. In the process, he established county agents, home demonstration agents, and the boys' and girls' 4-H clubs in the Agricultural Extension Service.

knob and kettle terrain *n.* hummocky topography with relatively high relief produced by glacial action where blocks of stagnant ice became detached from a receding glacier. Kettles (depressions) were left when the ice melted. The knobs are kames formed from sand and gravel that washed from the surface of the glacier and filled crevices and cavities in the ice.

knot *n.* (Anglo-Saxon *cnotta*, a knot) **1.** a place where two strings, cords, or ropes are tied together or where a cord is wrapped or tangled around itself. **2.** any kind of real or figurative bond; e.g., a marriage bond. **3.** a close-packed group or cluster of people, animals, or things. **4.** a hard lump in a muscle or other tissue. **5.** a cross-grained mass in a board or log representing the base of a tree branch (sometimes described as a loose knot or a solid knot). **6.** a unit of speed used at sea; 1 nautical mile per hour (6080.27 ft, 1853 m, or 1.15 statute miles per hour), originally measured by throwing a log overboard and counting the passage of knots in an attached cord. *v.* **7.** to tie together with a knot. **8.** to become tangled and form a knot.

knotgrass *n.* (Anglo-Saxon *cnotta*, a knot + *graes*, grass) **1.** any of several weedy grasses (*Polygonum aviculare* and related species, family Polygonaceae) with narrow leaves, slender stems, and small axillary blossoms. **2.** a creeping grass (*Paspalum distichum*, family Poaceae) that grows in wet areas in southern U.S.A. Also called knotweed.

knothole *n.* (Anglo-Saxon *cnotta*, a knot + *hol*, hole) a hole in a board where a loose knot has fallen out.

koala *n.* (Austal *kūlla*, koala) an Australian marsupial (*Phascolarctos cinereus*) that dwells in trees and resembles a gray teddy bear about 2 ft (0.6 m) tall.

kochia *n.* an annual herbaceous weed (*Kochia scoparia*, family Chenopodiaceae) with some forage value; named after W. D. J. Koch, a German botanist. It is a densely branched plant with foliage that turns purplish red in the fall. Also called summer cypress or bassia.

kola *n.* or *adj.* See cola, kola nut.

kola nut *n.* the bitter brown seed of the cola tree (*Cola acuminata*). Kola nuts are about the size of chestnuts, high in caffeine, and the source of the cola extract used in soft drinks. Also called cola nut.

Komodo dragon *n.* the largest living lizard (*Varanus komodoensis*), a flesh-eating native of Indonesian jungles that grows to about 10 ft (3 m) long; named for Komodo Island, Indonesia.

konimeter *n.* (Greek *konia*, dust + *metron*, measure) a device for measuring the amount of dust in the atmosphere. A measured volume of air is blown against a glass surface coated to hold dust particles so they can be counted under a microscope.

Köppen climate classification *n.* the first widely used system for classifying world climates that coincided well with the world patterns of vegetation and soils (published in 1900), and the basis for most subsequent systems. Named after Wladimir Köppen, it designates five major types of climate:
A tropical moist climates
B dry climates
C moist midlatitude climates with mild winters
D moist midlatitude climates with severe winters
E polar climates.
Several subdivisions are made for detail, and a 6th designation, H for highland climates, is used in mountainous areas where changes in elevation cause local climatic differences that are impractical to show on a map.

Köppen-Supan line *n.* an isotherm representing a mean temperature of 50°F (10°C) for the warmest month of the year. The areas toward the poles from this line will not support tree growth and are designated as polar climates.

Köppen, Wladimir *n.* a Russian-German climatologist (1846–1940) who published (in 1900) the first widely used classification system for world climates that coincided well with the Earth's vegetation and soil patterns.

kosher *adj.* (Hebrew *kāshēr*, fit, proper) **1.** fit to eat because the animal was killed and/or food prepared according to Jewish dietary laws. **2.** preparing or dealing with kosher food, as a kosher kitchen.

3. (slang) proper and acceptable, as activity that is kosher.

krait *n.* (Hindi *karait*, a venomous snake) any of several highly venomous nocturnal snakes (*Bungarus* sp.), native of India and southeast Asia, with broad black-and-white or black-and-yellow bands.

Krakatoa or **Krakatau** *n.* a volcanic island in Indonesia that blew apart in a violent volcanic explosion on August 26 and 27, 1883, creating a tsunami 120 ft (36 m) high that killed 36,000 people. Dust from the eruptions obscured the sun for 2½ days and drifted around the world several times, causing spectacular red sunsets for a year.

krill *n.* (Norwegian *kril*, small fish) small, shrimplike marine crustaceans (order Euphausiacea, especially *Euphasia* sp.) that eat small marine plants and animals and are the principal food of baleen whales, penguins, seals, squids, and certain seabirds.

krotovina or **crotovina** *n.* (Russian *krotovin*, mole-hill) contrasting soil material that represents an animal burrow filled with material from another horizon or from precipitation of calcium carbonate ($CaCO_3$) or gypsum ($CaSO_4 \cdot 2H_2O$).

krummholz *n.* (German *krumm*, crooked + *holz*, wood) shrubs or trees with stunted and distorted growth in either a polar or a high mountain ecological environment, commonly in shallow soil over permafrost. Also called elfinwood. See tundra.

krypton (Kr) *n.* (Greek *kryptos*, hidden) element number 36, atomic weight 83.8; an inert noble gas used in fluorescent and incandescent lamps.

kudzu *n.* (Japanese *kuzu*, kudzu) a fast-growing, herbaceous to woody vine (*Pueraria lobata*, family Fabaceae) native to China and Japan. Kudzu was promoted in the 1930s as an erosion-control plant for southern U.S.A., but it was too aggressive. It controls erosion, and its leaves are excellent forage for livestock, but the plants climb trees and sometimes smother them.

Kudzu growing like a blanket over roadside trees

kwashiorkor *n.* (Kwa language of coastal Ghana *kwashioko*, red child) a nutritional deficiency disease that occurs mostly in southern Africa and afflicts young children who receive inadequate protein because their diet is based mostly on maize. It is characterized by swollen hands, feet, and face and black patches on knees and elbows. Also called nutritional dystrophy. See marasmus, pellagra for related diseases.

L

L *n.* **1.** liter. **2.** lambert, the SI (metric or centimeter-gram-second) unit for brightness.

labdanum *n.* (Latin *ladanum*, mastic) a dark resin produced by some plants, such as rockrose (*Cistus* sp., family Cistaceae), and used to make flavorings, perfumes, and insecticides.

Labiatae *n.* the older name for plants of family Lamiaceae. See Lamiaceae.

labile *adj.* (Latin *labilis*, liable to slip or fall) **1.** unstable; likely to change. **2.** unreliable; subject to error or lapse. **3.** grains in a rock that weather easily; e.g., calcite ($CaCO_3$). **4.** a nutrient form that can readily exchange with a dissolved ion and thus become available to plants; e.g., nitrogen (N) in chemical fertilizers and in animal manures (usually determined by measuring exchange with a radioactive isotope of the same ion). **5.** changeable mood in a person.

Labrador Current *n.* a cold ocean current that flows south along the Labrador and Newfoundland coasts and turns east after it meets the warm Gulf Stream feeding the North Atlantic Current. It carries icebergs into North Atlantic shipping lanes. See Gulf Stream.

lac *n.* (Hindi *lākh*, a hundred thousand) **1.** the principal ingredient of shellac; a resin secreted by a tiny female scale insect (*Luccifer lacca*, family Lacciferidae) and deposited on certain southeastern Asian trees (*Acacia, Butea, Zizyphus, Schleichera,* and *Ficus religiosa*), where the insect feeds on leaves and twigs. It is used for making lacquers, dyes, and sealing wax. **2.** a hundred thousand in India, especially in counting large sums of money. **3.** French for lake.

laccolith *n.* (Greek *lakkos*, a cistern + *lithos*, stone) a large pluton that pushes the overlying rock layers up and produces a dome when it is intruded between the layers of older rock. See pluton.

lacewing *n.* a beneficial insect (*Chrysopa* sp. and others of order Neuroptera) with delicate transparent wings. Lacewing larvae, known as aphis lions, feed on aphids, thrips, mites, and young corn earworms, etc.

lachrymator or **lacrimator** *n.* (Latin *lacrima*, a tear + *ator*, producer) tear gas; a vapor that irritates the eyes.

lacquer *n.* (French *lacre*, lac) **1.** a resin and/or cellulose ester that forms a clear coating used to protect wood and other surfaces. Pigments may be added to change the color. **2.** such a material dissolved in a fast-drying volatile solvent (usually a mixture of esters, aromatic hydrocarbons, etc., that can be thinned with acetone) so it can be painted on a surface. **3.** any resinous varnish suitable for producing a hard, lustrous finish on wood, especially varnish obtained from the lacquer tree. *v.* **4.** to apply lacquer to a surface (usually by spraying).

lacquer tree *n.* a tree of eastern Asia (*Rhus verniciflua*, family Anacardiaceae) with a toxic latex exudate that once was widely used for making lacquer (a dark lacquer now known as Oriental or Chinese lacquer).

lactase *n.* (latin *lactis*, milk + *ase*, enzyme) an enzyme that splits lactose ($C_{12}H_{22}O_{11}$) into glucose ($C_6H_{12}O_6$) and galactose ($C_6H_{12}O_6$). It is produced by certain yeasts and is present in the intestinal juices of mammals.

lactate *n.* (Latin *lactare*, to secrete milk) **1.** a salt or ester of lactic acid. *v.* **2.** to produce milk, especially for suckling young.

lactic acid *n.* an organic acid ($CH_3CHOHCOOH$) formed by the fermentation of lactose (or of sucrose by certain microbes). Lactic acid is a syrupy liquid that provides the tangy flavor and custardlike consistency of yogurt. It accumulates in tired muscles.

lactobacillus *n.* (Latin *lactis*, milk + *bacillum*, little stick) anaerobic bacteria (*Lactobacillus* sp., e.g., *L. acidophilus*) that can ferment milk carbohydrates into lactic acid. It is used to make yogurt and related milk products.

lactoferrin *n.* (Latin *lactis*, milk + *ferrum*, iron + *in*, chemical) an iron-binding protein that occurs in milk, bile, saliva, tears, and the specific granules of neutrophil leukocytes. It functions in iron transport and can interfere with iron metabolism of bacteria.

lactose *n.* (Latin *lactis*, milk + *ose*, sugar) milk sugar, a reducing disaccharide ($C_{12}H_{22}O_{11}$) found in milk, composed of linked galactose and glucose units. It is used in pharmaceuticals, infant foods, and bakery products.

lactose intolerance *n.* inability to tolerate lactose (and therefore milk products), usually results from a lactase deficiency in the intestines; typically causes gastroenteritis. The U.S. Department of Agriculture estimated the percentage of lactose intolerance in U.S. population groups is African-American, 45%–81%; Asian-American, 65%–100%; Mexican-American, 47%–74%; Native American, 50%–75%; and North European, 6%–25%.

lacustrine *adj.* (Latin *lacus*, a lake + *inus*, characteristic of) **1.** of or pertaining to a lake. **2.** living or growing in a lake. **3.** formed by or in a lake, as a lacustrine terrace or a deposit of lacustrine clay.

lady beetle, ladybird, or **ladybug** *n.* any of several small, round beetles (family Coccinellidae) that are beneficial to farmers and home gardeners. Most species are predators that eat aphids, scale insects, spider mites, insect eggs, and larvae of many pests. An adult lady beetle can devour as many as 56 aphids in a day, and its larva can eat half this number.

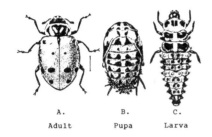

A.   B.   C.
Adult   Pupa   Larva

Adult, pupa, and larva of the convergent lady beetle (the most common ladybug)

LAFTA *n.* See Latin American Free Trade Association.

lag gravel *n.* **1.** desert pavement; a layer of gravel covering the surface of a desert where wind has removed the finer particles. **2.** a layer of gravel covering a stream bed or other surface where water has washed away the finer particles.

lagomorph *n.* (Greek *lagōs*, a hare + *morphē*, form) any of several herbivorous mammals (order Lagomorpha), including rabbits, hares, and pikas. Lagomorphs are similar to rodents but have two pairs of upper incisor teeth.

lagoon *n.* (Italian *laguna*, shallow lake) **1.** a long, narrow body of shallow water along a coast; separated from the sea by a sandy ridge. **2.** a shallow pond or lake that connects to a larger body of water. **3.** the body of water surrounded by the coral reef of an atoll. **4.** a surface impoundment; a body of water used to store and treat organic wastes. The soil beneath a lagoon must have a very slow permeability to prevent seepage from polluting the groundwater. Such lagoons are aerobic if they are aerated (usually by some kind of pump) or anaerobic if they are not.

lahar *n.* (Javanese *lahar*, lava) a mudflow or debris flow in the loose material deposited on the side of a volcano by a large eruption.

LAI *n.* See leaf-area index.

LAIA *n.* See Latin American Integration Association.

lair *n.* (Anglo-Saxon *leger*, a bed or place of rest) **1.** the resting place or den of a wild animal. **2.** a secret base of illegal operations; a hideout; e.g., a pirate's lair.

lake *n.* (Latin *lacus*, basin or lake) **1.** a body of freshwater in a natural depression fed by a spring or stream (excluding flooded rice fields, excavated ponds, etc.). **2.** a body of salt water surrounded by land. **3.** a large impoundment behind a dam. **4.** a similar body of some other liquid, as a lake of oil.

Lake Baikal or **Baykal** *n.* (Russian *baikal*, rich lake) a large lake in southern Siberia, the deepest lake in the world. At 5,315 ft (1620 m) deep, about 390 mi (630 km) long, and up to 50 mi (80 km) wide, it holds about one-fifth of the freshwater in the world.

lake effect *n.* or **lake-effect** *adj.* the climatic and other environmental effects of a large body of water. For example, the Great Lakes temper the weather in adjacent areas such as the state of Michigan, extending the growing season and reducing the hazard of frost, thus making it more favorable for fruit production.

lake-effect snow *n.* heavy snowstorms that occur downwind from a large lake in late fall or early winter while the lake is still warm.

Lake Superior *n.* the largest of the Great Lakes and one of the largest freshwater lakes in the world, bordered on the south by Michigan's Upper Peninsula and Wisconsin and on the north by Minnesota and Ontario, Canada. It is 350 mi (565 km) long from west to east and up to 160 mi (260 km) wide, with an area of 31,700 mi^2 (82,100 km^2) and a maximum depth of 1333 ft (406 m). See Lake Baikal.

Lake Titicaca *n.* the lake at the highest elevation (12,508 ft or 3812 m) of any large lake in the world, occupying 3200 mi^2 (8290 km^2) between Bolivia and Peru. It is a tourist attraction known for clear water, reed boats, and Incan ruins and terraces.

Lamarck, Jean Baptiste de Monet *n.* a French naturalist (1744–1829) who coined the word *biologie* (biology), published important work on the classification of plants and animals, and helped establish the Museum of Natural History in Paris but is best known for some of his theories that have been refuted. See Lamarckism.

Lamarckism *n.* an evolutionary proposal based on two theories proposed by Jean Baptiste de Monet Lamarck, who believed that life could be generated spontaneously and that characteristics acquired from the environment can be inherited. Both of these theories have been refuted.

lamb *n.* (Anglo-Saxon *lamb*, lamb) **1.** a young sheep of either sex less than a year old. **2.** meat from a young sheep. See mutton. **3.** a gentle, innocent person, especially one who is young. **4.** a person who is easily misled or cheated, like leading a lamb to the slaughter.

lambent *adj.* (Latin *lambens*, licking or lapping) **1.** flickering lightly across as surface, as lambent tongues of fire. **2.** twinkling, as lambent starlight. **3.** light, playful, and brilliant, as lambent wit is clever but not unkind. **4.** glowing gently; radiant.

lambert (L) *n.* the unit of brightness in the International System of Units (metric system); the brightness of a perfectly diffusing surface that emits or reflects 1 lumen of light per square centimeter, equal to 0.32 candles per square centimeter. Named after Johann Heinrich Lambert (1728–1777), a German mathematician.

lambert crazyweed *n.* a poisonous plant (*Oxytropis lambertii*, family Fabaceae) that grows on rangeland in the U.S. Great Plains and in Mexico. Whether green or dried, the foliage of this plant is toxic to livestock, causing listlessness, loss of appetite, shagginess, and death. Also called white locoweed.

lambkill kalmia *n.* an evergreen shrub (*Kalmia angustifolia*, family Ericaceae) that grows in swamps and wet pastures; native to the eastern U.S.A., Canada, and Cuba. Its leaves and crimson flowers are poisonous to lambs and other young animals. Also called sheep laurel.

lambsquarters or **lamb's-quarters** *n.* an annual weed (*Chenopodium album*, family Chenopodiaceae) with pale leaves and dense clusters of small green flowers; native to Europe, but a nuisance that can grow up to 6 ft (1.8 m) tall in humid regions of the U.S.A. It sometimes is used as a potherb (boiled green). Also called pigweed.

Common lambsquarters

Lamiaceae *n.* the family name for aromatic herbaceous plants with square stems, simple leaves, and clusters of two-lipped blossoms, including basil, catnip, lavender, marjoram, various mints, etc. Formerly called family Labiatae.

laminar flow *n.* smooth, layered flow within a fluid moving at relatively slow velocity; an orderly, nonmixing flow that moves in parallel streamlines, even if it must go around obstacles, in contrast to the turbulent, mixing flow characteristic of higher velocities. See Reynolds' number, turbulence.

lammergeier *n.* (German *lämmer*, lambs + *geier*, vulture) a large predatory Eurasian vulture (*Gypaetus barbatus*, family Vulturidae) with a wingspan of about 10 ft (3m) that is seen mostly in the high mountains of Europe, Asia, and Africa. Also called bearded vulture because it has a tuft of feathers below its bill.

lamprey *n.* (Latin *lampetra*, a rock licker) any of a group of eel-like cyclostomes (family Petromyzonidae) from 6 to 40 in. (15 to 100 cm) long; some are parasitic. The oldest form of vertebrate, the lamprey has a skeleton of cartilage and jawless, toothed, sucking mouthparts. Parasitic species, e.g., the sea lamprey (*Petromyzon marinus*), attach themselves to fish and suck their blood, scarring or killing the host.

land *n.* (Anglo-Saxon *land*, land) **1.** the solid part of the Earth's surface that is in contact with the atmosphere rather than covered by a body of water. See lithosphere. **2.** a specific nation, property, or other part of the Earth's land area, as the land of America, the land of the Vikings, or a small parcel of land. **3.** a rural area rather than an urban setting, as a distinction between the land and the city. **4.** one of the three principal factors of production: land, labor, and capital; a composite of soil, rock, mineral deposits, climate, water supply, location, vegetative cover, and buildings. **5.** a strip across a field that is tilled as a unit. **6.** a surface between two ridges or furrows, as a land in a rifle barrel or on a millstone. **7.** a situation or domain, as the land of the living. *v.* **8.** to move from air or water to solid ground, as to land an airplane or to land a fish that one has caught. **9.** to obtain, as to land a job. **10.** to place in a situation, as doing that could land him in jail.

land capability class *n.* a rating assigned by the Natural Resources Conservation Service in their system of classifying land based on the most intensive agricultural use it is capable of sustaining without loss of long-term soil productivity. The classes are designated from I (suitable for any use) to VIII (restricted to recreation, wildlife, watershed protection, and esthetic uses). Land capability classes II through VIII are usually divided into subclasses that designate the type of limitation as erosion (e), wetness (w), soil (s), or climate (c), and into numbered units (e.g., IIe1 or IIIs2) that are locally defined. Formerly called land-use capability class.

land capability map *n.* a map of a farm or other designated land area showing the land capability classes of the various soils therein, usually colored to show the ratings as green = I, yellow = II, red = III, blue = IV, dark green or white = V, orange = VI, brown = VII, and purple = VIII.

land drainage *n.* **1.** the natural flow of runoff water as it accumulates in low areas, forms small streams, enters creeks and rivers, and eventually reaches the sea. **2.** the artificial removal of excess water from the soil to improve plant growth or to increase soil strength for supporting buildings and roads. See soil drainage, wetland.

landfill *n.* (Anglo-Saxon *land*, land + *fyllan*, fill) **1.** an area where a low place has been filled with soil or other material to make it more useful, e.g., as a base for a road or a building or to improve the natural drainage of the area. **2.** a system of burying wastes and covering them with soil. See sanitary landfill. **3.** waste or other material that is buried in a landfill. *v.* **4.** to make an area more useful by adding material that raises its elevation. **5.** to place material in a landfill.

landform *n.* (Anglo-Saxon *land*, land + Latin *forma*, form) **1.** the shape of the land surface. **2.** a geomorphic feature of the Earth's surface; e.g., a mountain, hill, valley, plain, plateau, etc.

land-grant college *n.* originally, any public institution of higher learning that was supported financially by the sale of public lands. Now, specifically, the colleges and universities established in each state under the Morrill Act of 1862 to teach "agriculture and the mechanic arts" (engineering); 17 more were established from 1866 to 1912 especially for African-Americans (called Negroes at the time) in states where segregation was practiced. In 1994, 29 Native American tribal colleges were added to the family of land-grant colleges (many of them now called universities), making a total of 105.

landlocked *adj.* **1.** surrounded or nearly surrounded by land, as a landlocked bay or a landlocked country that has no seaport. **2.** in a lake or other body of water shut off from the sea, as landlocked salmon. **3.** having no public access; property that cannot be reached without crossing someone else's private property.

landrace *n.* an early form of a crop species; one that was derived from a wild population.

land reclamation *n.* the process and practices used to make land suitable for cultivation, including (where needed) soil drainage, irrigation, removal of stones, trees, structures, and debris, and correction of chemical problems such as acidity, alkalinity, or salinity.

LANDSAT *n.* (short for land satellite) **1.** any of several U.S. satellites used for mapping and gathering natural resource data from around the world. *adj.* **2.** obtained by means of a LANDSAT satellite; e.g., a map or an aerial photograph.

landscape *n.* (Anglo-Saxon *landsceap*, landscape) **1.** an area of land that can be seen at a single view with a certain combination of physical, biological, and anthropological physical elements and dynamics that interact coherently; e.g., an urban landscape or a picturesque landscape. *v.* **2.** to beautify a specific area

by shaping the land surface and planting grass, shrubs, and trees. *adj.* **3.** wider than it is tall, as a landscape photo or landscape orientation on a page (as opposed to portrait orientation).

landscape architect *n.* a professional person who plans and guides the land shaping and the placement of structures and plants around buildings and in parks or other areas to make them convenient and pleasing for human use.

landscape ecology *n.* the interaction of the different parts, processes, and populations in a patchy landscape.

landside *n.* the part of a plow that presses against the vertical wall of the furrow and keeps the plow from shifting sideways as the share lifts and turns the soil.

landslide *n.* (Anglo-Saxon *land*, land + *slidan*, to slide) **1.** a sudden movement of a mass of earth, rock, and vegetation sliding down a slope, often with catastrophic effects. **2.** the mass of material deposited by a landslide. **3.** the landform produced by a landslide. **4.** an overwhelming victory in an election.

land survey *n.* **1.** the marking of locations and boundaries on the land, especially the public land survey authorized in 1785 by the U.S. Continental Congress to lay out townships 6 mi square, each containing 36 sections. See metes and bounds. **2.** the preparation of maps showing landscape features in their proper positions and often indicating the elevations of reference points.

land tenure *n.* the system of land ownership, whether it belongs to a king or lord as in the feudal system, is owned collectively as in communism, is held by large landholders as in a plantation system, or belongs to individuals and families under private enterprise.

land use *n.* a classification of land according to its main function; e.g., the land in the U.S.A. can be classified as:
- Forestland: 746 million acres (302 million ha)
- Grassland: 633 million acres (256 million ha)
- Cropland: 458 million acres (185 million ha)
- Desert, etc.: 292 million acres (118 million ha)
- Cities, etc.: 142 million acres (57 million ha)

land-use capability *n.* See land capability class.

langbeinite *n* a mineral [$K_2Mg_2(SO_4)_3$] named after A. Langbein, the German scientist who identified it in the 19th century; used as a fertilizer that provides 18% potassium (K) (22% K_2O equivalent), 10.8% magnesium (Mg), and 22% sulfur (S).

La Niña *n.* (Spanish *la niña*, little girl) a cold current in the equatorial part of the eastern and central Pacific Ocean. It strengthens the westerly flow of the trade winds and has the opposite climatic effects of *El Niño.*

lanolin *n.* (Latin *lana*, wool + *olein*, oil) a purified form of wool wax; a fatty substance derived from sheep wool. It is used as an emollient in cosmetics, soaps, and ointments.

lantana *n.* (Italian *lantana*, wayfaring tree) any of a group of mostly tropical flowering shrubs (*Lantana* sp., family Verbenaceae). *L. camara* is sometimes cultivated for its small aromatic flowers that grow in dense yellow and orange or blue and violet spikes; it became a thorny pest in Hawaii, so seed flies from Mexico were imported to control it. See seed fly.

lantern fish *n.* a small fish (family Myctophidae) that has phosphorescent light organs along each side. Lantern fish live in the deep sea but come to the surface at night.

lanthanum (La) *n.* (Greek *lanthanein*, to lie hidden) element 57, atomic weight 138.9055; a soft, malleable, ductile rare earth metal that oxidizes rapidly in air and also reacts with water. It is used in carbon lights, lighter flints, optical glass, alloys, etc.

lapillus (pl. **lapilli**) *n.* (Latin *lapillus*, little stone) a piece of pyroclastic material about an inch (2 or 3 cm) across that is ejected by a volcano. See cinder, tephra.

lapse rate *n.* the rate of decline in temperature with increasing altitude above the Earth's surface. The air temperature decreases at an average lapse rate of 3.6°F per 1000 ft (6.5°C per km) in the troposphere (up to an altitude of about 7 mi, or 11 km).

larch *n.* (Latin *larix*, larch) **1.** a coniferous tree (*Larix* sp.) sometimes described as a deciduous evergreen because it turns brown and sheds its needles in the fall. Also called tamarack. **2.** tough, durable wood obtained from the tree and used for construction, flooring, shipbuilding, etc.

lard *n.* (Latin *lardum*, pork fat) processed pork fat, especially abdominal fat that has been rendered (melted), bleached, deodorized, and sometimes hydrogenated; used as shortening for cooking, but much of that use has been replaced by vegetable products.

larder *n.* (French *larder*, pantry) **1.** a reserve supply of food. **2.** a pantry; a room where food is stored.

largemouth bass *n.* a very popular freshwater game fish (*Micropterus salmoides*) with a protruding lower jaw and a dark stripe along each side of its body. Largemouth bass reach lengths up to 30 in. (75 cm) or more and weights up to 15 lb (7 kg).

lark *n.* (Anglo-Saxon *lauerce*, lark) **1.** any of several migratory songbirds (family Alaudidae); native to the Old World. **2.** the horned lark.

larkspur *n.* any of several annual or perennial herbs (*Delphinium* sp. and *Consolida* sp., family Ranunculaceae) with a spur-shaped flower. It is a very popular plant for flower gardens, but it contains a toxin that is very poisonous to all classes of livestock and to people.

larva (pl. **larvae**) *n.* (Latin *larva*, a ghost or skeleton) **1.** the early-development stage of many insects, sometimes called caterpillars, grubs, or maggots. Larvae hatch from eggs, feed vociferously as they develop, and then pass into the pupal (inactive) stage before becoming adult insects. **2.** the immature form of certain animals; e.g., a tadpole is the larva of a frog.

larvicide *n.* (Latin *larva*, a ghost or skeleton + *cida*, killer) a chemical used to kill larvae.

late blight *n.* **1.** a potato disease caused by a fungus (*Phytophthora infestans*) that infests both the foliage and the tubers, causing the plants to wither. It nearly wiped out the Irish potato crop in 1845 and 1846, leading to widespread famine. **2.** similar diseases that affect tomatoes (*Phytophthora infestans*, but a different strain than the one that causes potato blight), celery (*Septoria apii* var. *graveolentis*), and certain other plants, causing leaves and stems to darken and wither.

latent *adj.* (Latin *latens*, hiding) **1.** potential; present but hidden or not expressed, as latent ability. **2.** inactive or dormant, as a latent phase of a disease. **3.** psychologically dormant, as a latent emotion. **4.** undeveloped plant parts, as latent buds that have not yet emerged to be seen.

latent heat *n.* the heat absorbed or released without any change in temperature during a phase change, such as when a substance thaws, freezes, evaporates, or condenses.

latent period *n.* **1.** the time interval between exposure to an infectious organism or a carcinogen and clinical appearance of the disease. **2.** the interval between a stimulus and the response.

lateral *adj.* (Latin *lateralis*, of the side) **1.** of the side or directed to or from a side, as a lateral movement or a lateral view. **2.** roughly equivalent, as a lateral job change is neither a promotion nor a demotion. *n.* **3.** a horizontal extension, as a horizontal mine shaft or a horizontal tree branch or root.

lateral moraine *n.* a moraine that formed along the side of a glacier.

laterite *n.* (Latin *later*, brick + *ite*, rock) **1.** an older term for plinthite; a highly weathered soil material with a high iron content and red and yellow colors that forms under humid tropical conditions. It is soft and easily cut while it remains moist, but it hardens irreversibly into a rocklike mass called ironstone if it is allowed to dry. **2. Laterite** an older name for some of the highly weathered tropical soils now classified as Oxisols.

Cutting building blocks from laterite

laterization *n.* (Latin *later*, brick + *izatio*, formation) the soil-forming process involving long, intense weathering that produces red soils high in laterite; an emphasis in some of the older systems of soil classification such as that outlined in *Soils and Men*, the USDA Yearbook of Agriculture for 1938.

latex *n.* (Latin *latex*, juice) a suspension of proteins and resins forming a milky liquid that can be tapped or squeezed from certain plants, such as hevea trees, milkweed, and poppies. Latex coagulates into rubber when exposed to air; latex from hevea trees (*Hevea braziliensis*, family Euphorbiaceae) is used to make natural rubber.

lathyrism *n.* a selenium-toxicity disease that can cause skeletal deformities and rupture of the aorta in animals (especially horses) or loss of nerve function leading to leg paralysis in humans by interfering with lysine linkages in collagen. It is caused by eating seeds of chickpea (*Lathyrus sativus*), annual sweet pea (*L. odoratus*), or related species that are high in selenium. Also called neurolathyrism.

Latin America Free Trade Association *n.* a trade agreement ratified in 1961 by Argentina, Brazil, Chile, Colombia, Ecuador, Mexico, Paraguay, Peru, and Uruguay and reformed in 1980 as the Latin American Integration Association.

Latin American Integration Association *n.* an international trade agreement signed in 1980 by Argentina, Bolivia, Brazil, Chile, Colombia, Ecuador, Mexico, Paraguay, Peru, Uruguay, and Venezuela, with organizational headquarters in Montevideo, Uruguay. Its functions have been supplemented and partly replaced by more regional organizations such as Mercosur. See Mercosur, Latin American Free Trade Association.

latitude *n.* (Latin *latitudo*, breadth) **1.** the angular distance north or south of the equator of a position on the Earth measured in degrees; the equator is at zero latitude, the North Pole at 90° north latitude, and the South Pole at 90° south latitude. It is used along with longitude to define locations. **2.** freedom, often in terms of permissible range, as a dictatorship may permit only a small latitude of expression. **3.** the range of brightness that can be shown as a satisfactory image on a particular photographic material.

latrine *n.* (Latin *latrina*, bath or privy) **1.** an outdoor toilet, often a pit in the ground. **2.** a toilet for a camp, barracks, or other place where there is a large group of people. See outhouse, privy.

laurel *n.* (Latin *laurus*, bay tree) **1.** any of a group of evergreen shrubs or trees (*Laurus* sp., especially *L. nobilis*, family Lauraceae) with aromatic glossy green leaves and black berries; native to northern Africa and southern Europe. Laurel is grown as an ornamental and is used to make laurel wreaths for crowning winners of athletic events. Also called bay, bay tree, or sweet bay. **2.** any of several similar shrubs and trees (mostly of family Ericaceae), including cherry laurel (*Prunus laurocerasus*), spurge laurel (*Daphne laureola*), mountain laurel (*Kalmia latifolia*), and great laurel (*Rhododendron maximum*).

Laurentian ice sheet *n.* a sheet of glacial ice that, prior to 10,000 years ago, extended from the Laurentian Highlands in the southeastern part of Quebec, Canada, over North America east of the Rocky Mountains from the Arctic Ocean south as far as the Ohio River, New York City, St. Louis, and Kansas City.

lava *n.* (Latin *labes*, sliding down) **1.** molten rock; e.g., the material that flows from a volcanic vent. **2.** rock formed by solidification of lava. The most common types are basalt, rhyolite, and andesite. See magma.

lava flow *n.* **1.** a mass of molten rock flowing from a volcano or other vent. **2.** a layer of rock formed by cooling and solidification of such a flow of molten rock. See magma.

lavender *n.* (Latin *lavendula*, lavender plant) **1.** a perennial mint (*Lavandula angustifolia* and related species, family Lamiaceae) with spikes of fragrant, pale purplish flowers. It is used for a pleasant fragrance in perfumes, toilet waters, and cosmetics.

2. dried flowers, leaves, etc., of the plant used to fill sachets. **3.** a pale purplish color like that of the flowers.

lawn *n.* (French *lande*, glade) an area of grass around a house or other building, normally well tended and kept mowed to provide an even surface.

law of gravity *n.* **1.** all particles of matter attract each other with a force equal to the product of their respective masses divided by the square of the distance between their centers of mass. See Newton's law of gravitation. **2.** "What goes up must come down" (a common saying).

law of the minimum *n.* a principle expressed in 1862 by Justus von Liebig (1803–1873), a German chemist, indicating that plant growth is limited by whichever growth factor is present in the least adequate amount. Liebig thought that growth would be proportional to the supply of the limiting factor, but later work showed that the returns tend to diminish as the supply increases and that more than one factor may be limiting at the same time.

layer silicate *n.* a phyllosilicate; a silicate mineral that has silica tetrahedra linked together in sheets that are parts of distinct layers in the mineral structure. It includes 2:1 structures (with an octahedral sheet sandwiched between two tetrahedral sheets), such as the micas (e.g., biotite and muscovite) and smectite clay minerals, and 1:1 structures (with one octahedral sheet and one tetrahedral sheet bonded together) such as kaolinite.

lay of the land *n.* **1.** topography or terrain; the appearance of visible land features. **2.** (figurative) the general situation or state of affairs, especially factors that influence what can or cannot be accomplished.

LC$_{50}$ *n.* (lethal concentration, 50%) the concentration of a chemical in air (inhalation toxicity) or water (aquatic toxicity) that will kill 50% (or some other indicated percentage) of the target organisms in a specific test situation.

LD$_{50}$ *n.* (lethal dose, 50%) the amount of a pesticide or other chemical that will kill 50% (or some other indicated percentage) of the target species, usually expressed on the basis of milligrams of chemical per kilogram of body weight.

lea *n.* (Anglo-Saxon *leah*, meadow) **1.** grassland; a pasture, meadow, or grassy field. **2.** tillable land used for pasture or hay for a few years in a rotation with crops. **3.** fallow.

leaching *n.* (Anglo-Saxon *leccan*, to wet) lixiviation; the removal of materials from a soil or a soil horizon by dissolving them in percolating water. See eluviation.

lead (Pb) *n.* (Anglo-Saxon *lædan* = Latin *plumbum*, lead) **1.** element 82, atomic weight 207.2; a heavy, soft, malleable metal that forms many compounds, all of them toxic to animals and humans. Before it was banned for such uses, much lead entered the environment as an ingredient in house paint (replaced since 1978 in the U.S.A. by titanium oxide and other pigments) and in leaded gasoline. Leaded gasoline was phased out in the U.S.A. from 1976 to 1980; in 1991, the EPA made lead illegal for use as pellets in shotgun ammunition. Lead still has many uses, e.g., in car batteries, radiation shielding, and weights and as an alloy with other metals. **2.** a plummet with a line attached for sounding depths at sea or in a lake. **3.** bullets, as in a hail of lead. **4.** a grooved strip of lead used to join pieces and form lines in stained glass windows. **5.** a graphite cylinder in or for a pencil, often called pencil lead.

lead plant *n.* a small shrub (*Amorpha canescens*, family Fabaceae) with grayish leaves and twigs; native to the U.S.A. Some early miners thought that lead plants indicated the presence of lead ore. Known also as false indigo.

lead poisoning *n.* **1.** plumbism; a toxic condition caused by excessive lead in the body, causing anemia, brain and nerve problems, slow growth, convulsions, etc. **2.** being shot (with a lead bullet).

lead tree *n.* any of several tropical trees or shrubs (*Leucaena* sp., family Fabaceae) with pinnate leaves and white flowers.

leaf (pl. **leaves**) *n.* (Anglo-Saxon *leaf*, leaf) **1.** a flat organ growing from a stem or twig of a plant, usually green from chlorophyll that uses light energy to produce carbohydrates by photosynthesis. **2.** a sheet of paper, as a leaf in a book. **3.** a thin sheet of metal; e.g., gold leaf. **4.** a piece used to lengthen a table, either a loose or a hinged board. **5.** one of several long, thin pieces of metal that form a leaf spring. *v.* **6.** to leaf out; to put forth new leaves, especially in the spring. **7.** to leaf through; e.g., to quickly or casually turn the pages of a book or to leaf through a stack of paper.

leaf-area index *n.* a measure of the density of vegetative cover expressed as the ratio of the total surface area of the leaves present divided by the area of the underlying ground surface.

leaf blight *n.* **1.** a symptom of plant disease, usually showing as discolored and/or necrotic spots on the leaves of a particular plant species. **2.** a plant disease (usually fungal) that causes such symptoms.

leaf burn *n.* brown discoloration and/or necrosis of the tip and edges of a leaf or leaves caused by drought, plant disease, chemical damage, or injury.

leaf-cutting ant *n.* any of several mostly tropical American ants (*Atta* sp., family Formicidae) that cut pieces of green leaves and flower petals that they carry back to their nest to feed a fungus. The ants grow the fungus in underground chambers and use it for food.

leaflet *n.* (Anglo-Saxon *leaf*, leaf + French *let*, little) **1.** a small leaf, especially one just beginning to develop. **2.** a single part of a compound leaf. **3.** a small publication of printed matter, often a single folded sheet used for publicity.

leaf mold *n.* decaying leaves and associated material suitable for incorporation into soil to improve structure and fertility; e.g., composted leaves.

leaf roll *n.* abnormal rolling of leaves caused by disease (e.g., rhizoctonia disease of potatoes) or inadequate supply of water.

leaf spot *n.* discolored and/or necrotic spots, usually caused by leaf blight.

leafy spurge *n.* a troublesome weed (*Euphorbia esula*, family Euphorbiaceae) in the Great Plains of the U.S.A. that came from Europe without its natural insect and disease enemies. It produces latex that can be used to make rubber or gasoline.

league *n.* **1.** (French *legue*, a Gallic mile) a unit of distance that varied with time and place, commonly equal to about 3 mi (4.8 km). **2.** (French *ligue*, alliance) a group of nations, other organizations, or persons that have agreed to work together for a common purpose or for mutual protection; e.g., a bowling league or a league of nations. **3.** a level of skill or accomplishment, as they are not even in the same league with you.

League of Nations *n.* an international organization created by the Treaty of Versailles in 1919 to promote world peace following World War I. It was disbanded and replaced by the United Nations in 1946 following World War II.

leather *n.* (Anglo-Saxon *lether*, leather) **1.** processed animal skin that has had the hair and nonskin material removed to form rawhide that is then tanned and softened to make it supple. **2.** objects made from leather. *adj.* **3.** made of leather.

leaven *n.* (Latin *levamen*, something that lightens) **1.** an ingredient such as yeast or baking powder that ferments sugar and causes dough to rise and become lighter in preparation for baking. **2.** a piece of fermented dough kept for use with the next batch of batter. **3.** anything that has a lightening influence, as the leaven of laughter. *v.* **4.** to cause fermentation and rising of dough. **5.** to permeate slowly through a mass or group, causing a gradual change.

lectin *n.* (Latin *lectus*, selecting + *in*, chemical) a protein that can identify and bind to a specific sugar sequence. Some have toxic properties, some prevent plants from self-pollinating, and some affect animal life processes and health.

lee *n.* (Anglo-Saxon *hleo*, shelter) **1.** partial shelter, especially from wind and rain, as the lee of a large rock. **2.** the downwind (leeward) side, as the lee of a mountain. **3.** (usually **lees**) sediment from a liquid; e.g., wine dregs.

leech *n.* (Anglo-Saxon *læce*, a leech or physician) **1.** a flattened annelid worm (class Hirudinea) that lives in wet places. Most species are bloodsuckers; they formerly were used by physicians to bleed patients in attempts to cure many human diseases. Leech contains hirudin, a blood thinner that is now produced synthetically. **2.** a person who lives as a parasite on the resources of another person and gives little or nothing in return. **3.** (archaic) a physician. *v.* **4.** to use leeches to withdraw blood. **5.** to cling to someone and use their resources.

leek *n.* (Anglo-Saxon *leac*, leek) a biennial plant (*Allium ampeloprasum*, family Amaryllidaceae) native to the Middle East and Mediterranean area; the national emblem of Wales. A relative of onions and garlic, its bulb and leaves are used as a vegetable, especially in European soups and stews.

lee side *n.* the downwind side of a mountain, hill, structure, or other obstacle. The lee side is at least partially sheltered from the wind and/or is warmed by catabatic warming. Also called leeward side.

leeward *adj.* facing away from the source of the prevailing wind (as opposed to windward).

left bank *n.* the stream bank that is on your left when you face downstream.

leghemoglobin *n.* (Latin *legumen*, legume + *hemo*, blood + *globus*, sphere + *in*, chemical) a red pigment similar to hemoglobin produced in nodules containing rhizobia on legume roots. Leghemoglobin transports oxygen and maintains the low oxygen supply that rhizobia require for nitrogen fixation.

legume *n.* (Latin *legumen*, legume) **1.** any of a very large and valuable family of plants (family Fabaceae, formerly Leguminosae) that bear their seeds in pods with seams along both sides, including alfalfa, clovers, peas, beans, locust trees, and many others. Most legumes have a symbiotic relationship with nitrogen-fixing bacteria (*Rhizobium* sp. and *Bradyrhizobium* sp.) that grow in nodules on their roots. This supplies nitrogen to the host legume and to the crops that follow. **2.** a table vegetable from a legume plant; e.g., peas or beans. **3.** a pod from a legume plant.

Leguminosae *n.* the older name for legume plants; now family Fabaceae.

leishmaniasis *n.* a disease of people and animals caused by infection with microscopic, flagellate, protozoan parasites (*Leishmania* sp.) transmitted by bloodsucking sand flies (*Phlebotomus* sp.); named after William Boog Leishman (1865–1926), a Scottish bacteriologist. American or Brazilian leishmaniasis, caused by *L. braziliensis*, is of common occurrence in Central and South America; it is characterized by ulcers in the mucous membranes of the nose and throat that can lead to considerable tissue damage. Canine leishmaniasis, caused by *L. infantum*, affects dogs and children in the Mediterranean region. Cutaneous or dermal leishmaniasis, caused by *L. tropica*, is endemic in Asia, Africa, Mediterranean countries, and parts of South America; it causes skin tubercles, scabs, and ulcers. Visceral leishmaniasis (also known as kala-azar) is usually fatal; it is caused by *L. donovani* that infest the epithelial cells of the spleen and liver, causing fever, anemia, wasting away, and dropsy. It occurs in the Mediterranean area, Asia, and Brazil.

lemming *n.* (Norwegian *læmingi*, lemming) a small, mouselike rodent (*Lemmus* sp. and similar genera) of northern regions. The common lemming (*L. lemmus*) is very prolific, forming large hordes that sometimes mass migrate from the mountains to the seas and occasionally drown themselves.

lemon *n.* (French *limon*, lemon) **1.** a small, spiny, subtropical citrus tree (*Citrus limon*, family Rutaceae). **2.** the yellow-skinned fruit of the tree with its sour pulp that provides juice rich in vitamin C. It is used as a seasoning and to make lemonade and mixed drinks. **3.** a yellowish color similar to that of the fruit. **4.** (slang) a defective product that requires frequent repairs, especially an automobile. *adj.* **5.** resembling a lemon in color, taste, or character.

lemongrass or **lemon grass** *n.* lemon-scented grass (*Cymbopogon citratus* and related species, family Poaceae) native to India and Sri Lanka. Its aromatic oil is used in flavorings, perfumes, and medicines.

lemur *n.* (Latin *lemures*, specters) a small, monkeylike mammal (*Lemur* sp. and related genera, family Lemuridae) native to Madagascar and East Indian islands; a nocturnal tree-dweller with large eyes and soft fur.

lentic or **lenitic** *adj.* (Latin *lentus*, slow, smooth) living in or relative to quiet water such as that in a lake or pond (as opposed to lotic, in flowing water).

lentil *n.* (Latin *lentis*, lentil) **1.** a bushy, annual cool-season plant (*Lens culinaris*, family Fabaceae) with pinnately compound leaves similar to those of vetch, small, white to light purple blossoms, and small pods containing two or three seeds. It was cultivated at least as early as the Bronze Age. **2.** the small, flattened, biconvex seeds of the plant; used as food, especially in soups and salads.

leopard *n.* (Latin *leopardus*, leopard)**1.** a large member of the cat family (*Panthera pardus*, family Felidae), about 5 ft (1.5 m) long plus tail, with a tawny coat spotted with black markings; native to Africa and southeastern Asia. Also called panther. **2.** any of several other similar members of the cat family. **3.** the pelt of such an animal.

Leopold, Aldo *n.* an early naturalist (1887–1948) and Forest Service director of fish and game projects in southwestern U.S.A.; a founder of the Wilderness Society (1935) and generally considered the founder of applied ecology in the U.S.A. He expressed clearly his philosophy of the relationship between humans and the environment in his well-known book, *A Sand County Almanac* (1949).

lepidopteron or **lepidopter** *n.* (Greek *lepido*, scale + *pteron*, wing or feather) any of a large group of insects (order Lepidoptera) that as adults have four membranous wings covered with tiny scalelike feathers, often brightly colored, including butterflies, moths, and skippers.

leptandra *n.* (Greek *leptos*, slender + *aner*, anther) the rhizome of *Veronica virginica*; the source of leptandrin.

leptandrin *n.* (Greek *leptos*, slender + *aner*, anther + *in*, chemical) a bitter glycoside that acts as a violent emetic and cathartic. See Culver's physic.

lepton *n.* (Greek *leptos*, small) a subatomic particle with a spin of $\pm\frac{1}{2}$ that does not interact strongly with other particles or nuclei; believed to have no subunits. It includes the electron, electron-neutrino, muon, mu-neutrino, tau lepton, and tau-neutrino. See neutrino.

leptospirosis *n.* (Greek *leptos*, slender + *speira*, coil + *osis*, condition) infection in the blood of people, horses, cattle, dogs, swine, and rodents caused by a schizomycetes microorganism (*Leptospira interrogans*, family Trepo-nemataceae) transmitted by contact or from contaminated water. The principal serotypes can cause fever, muscle pain, hepatitis, nephritis, meningitis, or febrile disease. Also known as swamp fever.

lesion *n.* (French *lesion*, injury) **1.** an injury, wound, or diseased tissue. **2.** an abnormal structural change or growth in a body. **3.** legal damage resulting from failure to fulfill a contractual agreement.

lethal *adj.* (Latin *lethalis*, deadly) **1.** capable of causing death; e.g., a toxic substance or a lethal injection. **2.** extremely harmful; fatal; e.g., a lethal dose of a chemical. See LD_{50}.

lettuce *n.* (Latin *lactuca*, lettuce) **1.** a plant with succulent, edible leaves (*Lactuca sativa*, family Asteraceae), either wrapped around each other to form a round head or as individual leaves. **2.** the leaves of the plant, commonly used to make salad. **3.** any plant of the same genus (*Lactuca* sp.). **4.** (slang) paper money, especially U.S. currency.

leucaena *n.* (Greek *leukain*, to turn white) any of several tropical evergreen trees and shrubs (*Leucaena* sp., family Fabaceae); e.g., the lead tree.

leucine *n.* (Greek *leukos*, white + *ine*, chemical) a water-soluble amino acid ($C_6H_{13}NO_2$) essential for human and animal nutrition. It is produced by pancreatic enzymes hydrolyzing proteins during digestion, by putrefaction of nitrogenous organic materials, and by synthesis.

leucocyte or **leukocyte** *n.* (Greek *leukos*, white + *kytos*, a hollow vessel) **1.** any of several types of amorphous, colorless cell masses that can move like an amoeba, especially the white blood cells that combat disease. **2.** a similar cell in the lymph. **3.** a pus corpuscle.

leucoplast *n.* (Greek *leukos*, white + *plastos*, protoplasm) an organelle in a plant cell where starch granules are stored, usually in a root or tuber.

leukopenia *n.* (Greek *leukos*, white + *penia*, poverty) an abnormally low concentration of leukocytes in the blood (counts below 5000 in humans).

levee *n.* (Latin *levare*, to raise) **1.** an embankment along a river, either natural or constructed to prevent a river from flooding. Many levees are built to protect cities from floods. **2.** a reception, especially one held by a royal person in the early morning shortly after rising. **3.** an early afternoon British court assembly for men only.

level *n.* (Latin *libella*, a small set of scales) **1.** a tool for testing whether a surface is precisely horizontal or vertical, usually indicated by an air bubble in a glass tube with a very slight curvature. Also called spirit level. **2.** a survey instrument used to measure deviations from a horizontal plane. **3.** a field tool used to smooth the land surface; also called a leveler. **4.** an elevation, as mean sea level or a high-water level. **5.** a horizontal surface, either real or imagined. **6.** a degree of status, accomplishment, or intensity, as at the professional level, an average level of skill, or a loud level of sound. *adj.* **7.** horizontal. **8.** smooth; flat. **9.** at a specified elevation, as level full. **10.** equal. **11.** sensible, not easily disturbed, as level headed. **12.** unchanging, as a level tone. *v.* **13.** to adjust to a horizontal plane. **14.** to smooth a surface. **15.** to equalize. **16.** to demolish, as to level a building or a city. **17.** to aim, as to level a gun.

level of saturation *n.* the surface of a water table.

levorotatory *adj.* (Latin *laevus*, left + *rotatorius*, turning) causing polarized light to rotate to the left (counterclockwise), as certain crystals or solutions of certain chemicals.

levulose *n.* (Latin *laevus*, left + *ose*, sugar) the levorotatory form of fructose ($C_6H_{12}O_6$); the sugar in honey that does not crystallize but readily darkens when honey is overheated. Most diabetics can tolerate levulose.

lice (sing. **louse**) *n.* See louse.

lichen *n.* (Greek *leichēn*, lichen) a symbiotic growth involving a fungus (class Ascomycetes or, rarely, class Basidiomycetes) growing with an alga (*Cladonia* sp.) imbedded in the fungal hyphae. The algae carry on photosynthesis like all green plants, and some of them fix atmospheric nitrogen (N_2). The fungus supplies water and nutrients for the alga and provides protection. Lichens are able to grow on previously bare rock surfaces and make the environment more favorable for mosses, ferns, and later, for higher plants. There are an estimated 18,000 kinds of lichens, many of them living in harsh conditions (dry, cold, hot, underwater, etc.); they are classified into three types: crustose (lichens that form a tight crust), foliose (have leaflike growth), and fruticose (are bushy or have branched stalks). See usnea.

licorice *n.* (Latin *liquiritia*, licorice) **1.** a stoloniferous perennial plant (*Glycyrrhiza glabra* and related species, family Fabaceae) native to southwestern Asia and the Mediterranean region. **2.** an extract from the roots of the plant; the source of the common licorice flavor used in candy, tobacco, and medicine. *adj.* **3.** having the flavor of licorice, as licorice candy.

Liebig, Justus von *n.* a German chemist (1803–1873); author of *Chemistry and Its Application to Agriculture and Physiology* (1840). Liebeg established that plants obtain their carbon from the carbon dioxide in the air rather than from humus and that they obtain mineral elements from the soil; he is well known for his law of the minimum. See law of the minimum.

life (pl. **lives**) *n.* (Anglo-Saxon *lif*, life) **1.** vitality; the difference between an inanimate or dead object and a plant, animal, or microbe that can grow and reproduce. **2.** life span; the duration of the life of an organism from its birth or sprouting until death. **3.** a certain part or the remainder of one's life span, as middle life or a life sentence. **4.** all living things, as life on Earth. **5.** a specific class of living things, as plant life or animal life. **6.** the duration of usefulness of a machine or object, as the life of an automobile. **7.** a quality or manner of living, as a hard life or a military life. **8.** the prevailing condition of human existence, as life is like that. **9.** the essence of something, as the life of a democracy or the life of the party. **10.** activity, as a city that is full of life. **11.** action, resiliency, as a ball with a lot of life.

life-and-death or **life-or-death** *adj.* of life-determining importance; having a possible outcome of death for one or more participants, as a life-and-death struggle.

life blood *n.* **1.** the blood of a person or animal, especially when it is being lost at a rate that threatens the individual's life. **2.** a vital, animating component, as agriculture is the life blood of the nation.

life cycle *n.* the sequence of forms that characterize the life of an organism from the early development of an individual to the corresponding stage in the next generation.

life expectancy *n.* **1.** the remaining period of time that an individual is likely to live. **2.** the projected period of usefulness of a machine, reservoir, or other thing.

life science *n.* any scientific discipline that deals with living organisms and life processes; e.g., biology, botany, ecology, zoology.

ligase *n.* (Latin *ligare*, to tie or bind + *ase*, enzyme) any enzyme that catalyzes a reaction that combines two organic structures into one and hydrolyzes ATP in a coupled reaction; e.g., DNA polymerase I (also called Pol I) helps replicate DNA strands. Also called synthetase.

light *n.* (Anglo-Saxon *leoht* or *liht*, light) **1.** electromagnetic radiation that can be seen by the human eye, including wavelengths in the range from 380 nm (violet) to 760 nm (red). **2.** similar radiation beyond the range of human vision, as ultraviolet light or infrared light. **3.** a source of such radiation, as a candle or a window, often designated as sunlight, starlight, electric light, etc. **4.** day (as opposed to night), as at first light (dawn). **5.** a flame, as a light that ignites a fire. **6.** a source of intelligence or inspiration, as those ideas shed light on the subject. **7.** public awareness, as something brought to light by an investigation. **8.** an expression of recognition or understanding, as the light in her eyes. **9.** a good example, as a light for others to follow. **10.** viewpoint or attitude, as seeing things in a cheerful light. *v.* **11.** to ignite or switch on, as to light a fire or turn on a light. **12.** to provide illumination, as to light the room. **13.** to show the way, as to light a path. **14.** to alight; to get down, as from a vehicle or a high place. **15.** to land or settle somewhere, as the bird will light on its perch. *adj.* **16.** having little weight; e.g., as light as a feather. **17.** of low density, as balsa wood is light. **18.** thin, as a light fabric. **19.** small in amount, as light trading. **20.** exerting little pressure, as a light touch. **21.** easy and entertaining rather than serious, as light reading. **22.** low in something that is considered undesirable, as light soda has few calories. **23.** porous, spongy; e.g., light bread or cake. **24.** nimble, as light on her feet. **25.** carefree, as a light heart. **26.** not serious, as he made light of the problem. **27.** small in size or scale, as light industry or light artillery.

light air *n.* in meteorology, a wind speed of 1 to 3 mph (about 1.6 to 5 km/h or 0.44 to 1.3 m/sec). See Beaufort wind scale.

light breeze *n.* in meteorology, a wind speed of 4 to 7 mph (about 6 to 12 km/h or 1.7 to 3.5 m/sec). See Beaufort wind scale.

light compensation point *n.* the light intensity needed for a plant to produce photosynthate by photosynthesis at the same rate that its respiratory processes consume photosynthate.

light intensity *n.* the total energy content of a light beam or other light source crossing or impacting an area; commonly expressed in langleys (calories per square centimeter), watts (Joules per second), or einsteins (6×10^{23} photons).

lightning *n.* a visible flash of light caused by an electrical discharge either between clouds or between a cloud and the Earth; characteristic of thunderstorms. It often is associated with cumulonimbus clouds.

lightning arrestor *n.* an electrical device that protects electronic equipment by providing an alternate path (usually via a spark gap) to ground for lightning or other electric surges in a power source or antenna.

lightning bug *n.* a firefly; any of several species of luminous insects (family Lampyridae and some of family Elateridae) that produce light by the oxidation of luciferin, catalyzed by the enzyme luciferinase in the presence of adenosine triphosphate (ATP) and magnesium ions (Mg^{2+}). The light intensity and duration serve as specific signals to identify the male lightning bug to a female of the same species and for the female to signal her receptivity. See firefly, luciferin.

lightning rod *n.* **1.** a grounded metal rod placed on a roof ridge or other high point of a structure to protect the structure from lightning. **2.** an unpopular person or thing that attracts hostile feelings and attention, thereby diverting them from others.

lightning speed *n.* very fast, like the speed of a lightning bolt flashing across the sky.

light quality *n.* the relative amounts of various wavelengths (colors) contained in a light beam or emission.

light reaction *n.* any reaction that requires the presence of light, especially those that are part of the process of photosynthesis; specifically, the reactions that use light energy to produce ATP and NADPH.

light soil *n.* soil that is easy to till; sandy soil. Note: The actual dry density of such soil is usually more than that of a "heavy" soil. See heavy soil.

light year *n.* the distance that light travels in a year; about 5.88 trillion mi (9.46 trillion km).

lignification *n.* (Latin *lignum*, wood + *ficatio*, making) making or becoming like wood. Many plant stems undergo lignification as they mature.

lignin *n.* (Latin *lignum*, wood) a complex organic material formed by polymerization of coniferyl alcohol, sinapyl alcohol, and *p*-hydroxycinnamyl alcohol; a component of plant cell walls, especially in wood and older cells that stop growing and rigidify. Lignin has low digestibility because it is highly resistant to both chemical and enzymatic attack.

lignite *n.* (Latin *lignum*, wood + *ite*, mineral) soft dark brown coal with characteristics midway between those of peat and subbituminous coal. See peat, coal.

lignum vitae *n.* either of two tropical American tree species (*Guaiacum* sp.) whose wood is very dense, durable, beautiful, and hard. See guaiacum.

ligule *n.* (Latin *ligula*, spoon or strap) **1.** a membrane that grows from the base of the leaf blade upward around the stem of most grasses and some other plants. **2.** a strap-shaped corolla in certain flowers, especially those of composite plants.

lilac *n.* (Arabic *līlak*, lilac) **1.** any of several shrubs or small trees (*Syringa vulgaris* and related species, family Oleaceae) that produce large clusters of tiny flowers in the spring; commonly used as an ornamental shrub. **2.** the flower clusters of such plants, commonly light lavender but ranging from white to crimson. **3.** a lavender color similar to that of the flowers.

lily (pl. **lilies**) *n.* (Latin *lilium*, lily) **1.** any of a group of plants (*Lilium* sp., family Liliaceae) that grow from a bulb and produce tall stems, slender leaves arranged either in whorls or alternately, and bell- or trumpet-shaped flowers (often in clusters). **2.** any of several other plants with similar bell-shaped flowers; e.g., the calla lily, pond lily, or water lily. **3.** a blossom or bulb from any of these plants.

Easter lily

lily-of-the-valley *n.* a small, perennial, shade-loving plant (*Convallaria majalis*, family Liliaceae) grown for its racemes of beautiful white bell-shaped flowers and fragrance.

lima bean *n.* (named for Lima, Peru) **1.** a bean plant (*Phaseolis limensis*, family Fabaceae) with large pods; native to tropical America. **2.** the large, flat, near-white, edible beans produced by the plant.

limberneck *n.* a fatal disease that causes fowls (especially chickens and ducks) to become weak in their necks, wings, and legs; the result of eating feed containing toxins produced by bacteria (*Clostridium botulinum*).

lime *n.* (Anglo-Saxon *lim*, lime) **1.** calcium oxide (CaO), sometimes called burnt lime or quicklime; used in paper and steel production and in some mortars. It reacts with water to form slaked lime [Ca(OH)$_2$]; the slaked lime reacts with carbon dioxide to form calcium carbonate (CaCO$_3$) or with silicone dioxide (SiO$_2$) to form calcium silicate, which serve as binding agents when the mortar dries. **2.** finely ground limestone (dominantly CaCO$_3$, with or without some MgCO$_3$), also known as aglime; used to raise the pH of acid soil or of acid water. See liming material. **3.** a small citrus tree (*Citrus aurantifolia*, family Rutaceae) native to Asia. **4.** the acidic fruit of the tree; similar to a lemon but with a green skin. It is used for flavoring drinks and some foods. *v.* **5.** to apply liming material to a field to raise the pH of the soil or to a body of water to reduce its acidity. **6.** to apply liming material to manure or other excrement so it will decompose faster and produce less odor. **7.** to soak hides in a lime solution to loosen the hair. **8.** to whitewash; to paint with a slurry of lime in water.

lime kiln *n.* a furnace (often an electric furnace) used for heating shells, chalk, or limestone (CaCO$_3$) to a high enough temperature to drive off carbon dioxide and make quicklime (CaO), especially for use in making mortar.

lime scrubbing or **limestone scrubbing** *n.* flue gas desulfurization; a process whereby powdered limestone slurries are used to scrub sulfur compounds from flue gas.

limestone *n.* (Anglo-Saxon *lim*, lime + *stān*, stone) a sedimentary rock composed mostly of calcium carbonate (CaCO$_3$) formed by precipitation, mostly in shallow water, triggered by biochemical processes but also including evaporation (e.g., in a cave or from a geyser) and other chemical processes. Limestone that contains a significant component of dolomite (CaCO$_3$·MgCO$_3$) is called dolomitic limestone. Impure limestones grade into calcareous shales and sandstones.

Limicolae *n.* any of several wading birds, including plovers (families Charadriidae), sandpipers (family Scolopacidae), and curlews (*Numenius* sp.).

limicoline *adj.* (Latin *limus*, slime + *colere*, to inhabit) living at the shoreline; relating especially to wading birds (Limicolae).

liming material *n.* a material that can be used to neutralize the acidity of an acid soil, acid waste, or acid body of water; most commonly ground limestone, chalk, or shells. The main chemical component is usually calcium carbonate, but any carbonate, oxide, hydroxide, or silicate of calcium or magnesium present in a liming material can contribute to acid neutralization. Note: Many other materials (e.g., sodium carbonate, Na_2CO_3) can also neutralize acids, but calcium and magnesium ions are the most desirable cations in soil materials.

limiting factor *n.* **1.** the factor that is in least adequate supply for the growth of a plant (e.g., water, available nitrogen, heat, etc.) according to Liebig's law of the minimum. **2.** the environmental factor that limits the population of a species in a particular area. **3.** the rate-limiting (slowest) step in a series of actions or reactions needed to produce an end product.

limnetic *adj.* (Greek *limnē*, marsh + *etikos*, relating to) **1.** pertaining to the open waters of a pond or lake. **2.** living on open water; e.g., vegetation floating on a pond.

limnology *n.* (Greek *limnē*, marsh or lake + *logos*, word) the branch of hydrology that studies freshwater aquatic environments and associated life, including physical, chemical, biological, and other environmental relationships of fresh (not salty) water.

limonite *n.* (Greek *leimōn*, meadow + *ite*, mineral) amorphous ferric oxide with variable hydration ($Fe_2O_3 \cdot xH_2O$) giving it colors ranging from yellow to dark brown; formed by biogenic precipitation in wet places such as bogs. It is used as an iron ore. See goethite.

limose *adj.* relating to or growing in mud.

lindane *n.* benzene hexachloride ($C_6H_6Cl_6$); a long-lasting chlorinated insecticide named after T. von der Linden, the Dutch chemist who first isolated it in 1912. It also is used to kill the mites that cause scabies. See benzene hexachloride.

linden *n.* (Anglo-Saxon *linden*, linden) **1.** any of several large fast-growing trees (*Tilia* sp., family Tiliaceae) with heart-shaped leaves and fragrant yellowish white blossoms; used for shade and ornamental trees. American linden (*T. americana*) is also called basswood. **2.** the soft, lightweight, white wood from the tree; used as boxwood and for furniture and building construction.

linen *adj.* (Anglo-Saxon *linen*, made of flax) **1.** made from flax; e.g., linen thread woven into linen fabric and used to make linen tablecloths, sheets, etc.

n. **2.** (often **linens**) tablecloths, sheets, shirts, etc., made of linen or similar items made of cotton or other fabric. **3.** fine stationery made of linen or with a linen texture.

link *n.* (Danish *lænkia*, chain) **1.** a connecting piece; e.g., one of the pieces of a chain. **2.** a unit of length; e.g., each of the 100 links in a surveyor's (Gunters) chain is 7.92 in. (20.1 cm) long, and each link in an engineer's chain is 1 ft (30.48 cm) long. **3.** a figurative connection, as a link with the past or a link in a communication system. **4.** a sausage that is part of a string of sausages. *v.* **5.** to connect two or more objects, especially to form a series.

Linnaean *adj.* pertaining or according to the binomial system developed by Carolus Linnaeus, using Latin genus and species names for plants and animals.

Linnaeus, Carolus *n.* the latinized name of Carl von Linn (1707–1778), a Swedish botanist who developed the binomial system used to identify plants and animals by Latin genus and species names. For example, in the Linnaean system, white oak = *Quercus alba*.

linoleic acid *n.* (Latin *linum*, flax + *oleum*, oil) an unsaturated fatty acid ($C_{18}H_{32}O_2$) found in linseed oil and other drying oils.

linseed *n.* (Anglo-Saxon *lin*, flax + *sæd*, seed) flax (*Linum usitatissimum*) seed, especially when made into linseed oil and linseed cake.

linseed oil *n.* a drying oil extracted from flax seed and used to make paint, varnish, ink, linoleum, soap, etc. The remaining material, called linseed cake, is often used in cattle feed.

lion *n.* (Greek *leōn*, lion) a large, usually tawny yellow, carnivorous cat (*Panthera leo*) native to Africa and southwestern Asia. The adult male carries a large mane and is sometimes called the king of the beasts. The lion is the national symbol of Great Britain. **2.** any other large wild cat; e.g., the cougar. **3.** a bold, strong, influential person.

A male lion

lipase *n.* (Greek *lipos*, fat + *ase*, enzyme) an enzyme that splits lipids into components such as fatty acids and glycerin. It is produced by organs such as the liver and pancreas for use in the digestive system and also is found in certain plants.

lipid *n.* (Greek *lipos*, fat) any of a group of fat-related substances that are important components of all plant, animal, and human cells; organic compounds that are insoluble in water but soluble in various organic solvents, especially fatty acids and their derivatives (e.g., waxes, steroids, and steroid esters).

lipophilic *adj.* (Greek *lipos*, fat + *philos*, loving) able to bond strongly to lipids.

liquid *n.* **1.** (Latin *liquidus*, fluid) a fluid with a definite volume but able to fit any shape; its molecules flow freely from one arrangement to another but have enough cohesive force to hold a definite volume; not rigid like a solid and not expansive as a gas. **2.** a consonant with a flowing sound, as l, m, n, or r. *adj.* **3.** able to flow but not expand. **4.** composed of liquids, as a liquid diet. **5.** smooth and flowing, as graceful movement or sounds such as those of l and r, of certain languages, or of certain person's voices. **6.** clear and bright, as liquid eyes. **7.** readily available, as liquid assets can be converted into cash.

liquid fertilizer *n.* fertilizer applied to either soil or plants in liquid form, often through an irrigation system or a sprayer. Note: All essential plant nutrients are available in forms that are soluble in water. See fertigation, foliar fertilization.

liquid limit *n.* the moisture content of a soil when it is barely able to flow like a liquid. See Atterberg limits.

liquid measure *n.* any of the units of measure typically used for liquids; e.g., liters (L) and derivatives thereof in the SI (metric) system. In the U.S. version of the foot-pound-second system, 8 liquid drams = 1 liquid ounce (1 liquid ounce = 29.57 ml); 4 liquid ounces = 1 gill; 4 gills = 1 pint; 2 pints = 1 quart (0.946 L); 4 quarts = 1 gallon (231 in.^{3} or 3.785 L).

liquid nitrogen *n.* liquefied nitrogen. At atmospheric pressure, liquid nitrogen boils at a temperature of 77°K (–196°C). It is used to preserve materials (including sperm and embryos) at very low temperatures, to study the effects of very low temperatures on living plant and animal tissue (in cryobiology), to perform certain forms of brain surgery or to freeze off skin growths (in cryosurgery), and to cool irons for freeze-branding animals, etc.

liquor *n.* (Latin *liquor*, liquid or juice) **1.** a distilled alcoholic beverage; e.g., whiskey or brandy. **2.** a broth or juice, especially one used for or produced by cooking. **3.** a solution used for pharmaceutical purposes or industrial arts, especially a highly concentrated solution.

LISA *n.* See low input sustainable agriculture.

lister *n.* a type of plow that throws soil to both sides and leaves a large open furrow; usually made like two moldboard plows back-to-back. It is used to make drainage channels and raised beds or for lister planting. Also called a lister plow.

listeriosis *n.* an animal and occasionally human disease caused by bacteria (*Listeria monocytogenes*). Symptoms include mononucleosis, loss of motor control, paralysis, and fever. Named after Joseph Lister.

Lister, Joseph *n.* an English surgeon (1827–1912) considered the founder of modern antiseptic surgery.

lister planting *n.* the placement of seed in moist soil at the bottom of lister furrows.

listing *v.* **1.** tilling the soil with a lister, usually on the contour to conserve water, conserve soil, or to allow a crop to be planted in moist soil in the bottom of the furrow. **2.** making a list of things. *n.* **3.** the list or catalog so compiled. **4.** an item on such a list.

liter (**L**) *n.* (Greek *litra*, pound) the basic unit of volume in the metric system, originally equal to the volume of 1000 g of water at 4°C; redefined in 1964 to equal 1 cubic decimeter (1000 cm^3 or 1.0567 U.S. liquid quarts).

lithiasis *n.* (Greek *lithiasis*, stone formation) calculosis; formation of calculi (stones) in one or more passages or glands (bladder, gall bladder, pancreas, etc.) of the animal or human body.

lithic *adj.* (Greek *lithikos*, of stone) **1.** stony; made of stone or having the nature of stone. **2.** containing stone fragments, as sedimentary or volcanic rock that contains fragments of older rocks. **3.** occurring as a stony mass (calculus) in the body. See lithiasis. **4.** containing or having to do with lithium.

lithification *n.* (Greek *lithos*, stone + Latin *ficatio*, making) induration; the compacting, cementation, and other processes that convert a fresh deposit of sediment into an indurated (hardened) rock. Depending on the environment, the time may be a few years, centuries, or millenia.

lithium (**Li**) *n.* (Greek *lithos*, stone + Latin *ium*, metallic element) element 3, atomic weight 6.94; the lightest of all metals, silver-white in color; highly reactive, forming Li$^+$ ions. It is used in metallurgy to harden lead and aluminum alloys, in ceramics, and in thermonuclear explosions. Lithium carbonate is used to treat persons with mania or depression.

lithology *n.* (Greek *lithos*, rock + *logos*, word) **1.** a scientific study of rock formation and the physical and chemical characteristics of rocks. **2.** the medical science that deals with understanding and treating lithiasis.

lithometeor *n.* (Greek *lithos*, rock + *meteōron*, something in the air) a suspension of dry particulate matter in the atmosphere, such as dust, smoke, or haze. See particulates.

lithophyte *n.* (Greek *lithos*, rock + *phyton*, plant) **1.** plant life that grows on rock; e.g., a lichen. **2.** an organism such as coral whose skeletal material produces a stony formation.

lithosere *n.* (Greek *lithos*, rock + Latin *sere*, series) a succession of plant communities that begins with bare rock.

lithosphere *n.* (Greek *lithos*, rock + *sphaira*, sphere) **1.** the solid portion of the Earth as distinguished from the atmosphere and the hydrosphere. Also called petrosphere. **2.** the Earth's crust to a depth of about 50 mi (80 km); the part that is affected by changes such as volcanic activity, erosion, and sedimentation. See atmosphere, hydrosphere, biosphere.

lithotroph *n.* See autotroph.

litter *n.* (Latin *lectaria*, litter) **1.** discarded trash, especially that discarded in public areas. **2.** several offspring born together, as a litter of kittens, puppies, or pigs, etc. **3.** a layer of loose organic debris on the forest floor; also known as duff. **4.** straw, hay, or other plant material used as bedding for animals or poultry (and often subsequently incorporated with their manure) or as mulch in a garden. **5.** a stretcher; canvas or other fabric stretched between two long supports to carry an injured or sick person. *v.* **6.** to carelessly discard trash or scatter objects. **7.** to give birth to a litter of offspring. **8.** to place plant material as bedding or mulch.

litter control *n.* efforts to reduce litter in public areas, either by penalties (e.g., fines for discarding litter) or by cleanup (e.g., adopt-a-highway programs).

little black ant *n.* a small ant (*Monomorium minimum*, family Formicidae) that bothers people at picnics, in kitchens, and in other settings. The ants feed on sweets, fruits, vegetables, meats, etc.

Little Dipper *n.* a group of seven stars that form the outline of a dipper in the Ursa Minor (Little Bear) constellation. The tip of the handle is Polaris, the North Star. See Ursa Minor.

little fire ant *n.* a small ant (*Wasmannia auropunctata*, family Formicidae) with a vicious sting. These ants commonly inhabit citrus orchards and make harvesting by hand difficult.

little-leaf *n.* **1.** a disease of shortleaf (*Pinus echinata*) and loblolly (*P. taeda*) pines that severely retards growth, turns the needles yellow and reduces their size and number, and after some years kills the trees; caused by a parasitic fungus (*Phytophthora cinnamomi*) that attacks the tree roots. It occurs mostly in southeastern U.S.A. **2.** small, crinkled, yellowish leaves on fruit trees (apples and stone fruits) and grape vines caused by zinc deficiency, nematodes, root rot, or unknown factors.

littoral *adj.* (Latin *litoralis*, of the shore) **1.** relating to, living in, or occurring in the shallow water in the region along a seashore, especially the zone between the high and low water lines, and sometimes including the adjacent land. **2.** a similar region along the shore of a lake out to the maximum depth of plant rooting, including the organisms that live there. See riparian habitat.

live cave *n.* a cave where the formations are still growing, as indicated by the presence of moisture.

live oak *n.* **1.** a slow-growing, broad, evergreen oak (*Quercus virginiana*, family Fagaceae) native to the southern U.S.A. in coastal areas near the Gulf of Mexico. It is used as a shade tree and was formerly used in shipbuilding because its wood is durable for that use. **2. interior live oak** a related evergreen oak (*Q. wislizen*) native to California and Mexico that occurs in the foothills of inland mountain ranges and offers good browse for cattle and goats and good soil protection.

liver fluke *n.* a trematode worm parasite (*Fasciola hepatica*, *Clonorchis sinesis*, and similar species) that attacks the livers of sheep, cattle, goats, horses, and swine, obstructing bile passages and causing the liver to swell and degenerate. Occasionally also occurring in human liver. Several snail species (*Lymnaea* sp.) and some fish are hosts.

livestock *n.* (Anglo-Saxon *lif*, life + *stocc*, a stick) domestic animals raised on a farm or ranch for food production (e.g., cattle, hogs, sheep) or for work (e.g., horses, mules, donkeys).

living fossil *n.* a relict species; an organism such as the ginkgo tree (*Ginkgo biloba*) or the coelacanth fish (*Latimeria chalumnae*) that has survived for ages in spite of environmental change. These are the only living species of these types of organisms—other related species exist only as fossils.

living mulch *n.* a cover crop; a living plant cover that protects soil from erosion between crops (e.g., during the winter) and/or between crop plants that do not cover the entire surface (e.g., in an orchard or vineyard).

lixiviate *v.* (Latin *lixivium*, lye + *atus*, action) to leach; to dissolve and carry away water-soluble substances, especially by water percolating through soil.

lizard *n.* (French *lezard*, lizard) **1.** any of thousands of scaly reptiles (suborder Sauria, order Squamata, class Reptilia) with slender bodies and long tails. Most have four legs; e.g., the chameleon, gecko, horned lizard, and iguana. They are cold-blooded animals that cannot live in areas with cold winters. **2.** any other reptile that resembles a lizard, including alligators, crocodiles, and dinosaurs. *n. or adj.* **3.** leather made from lizard skins, or objects made from such leather (shoes, purses, etc.).

llama *n.* (Quecha *llama*, llama) a mule-sized ruminant animal (*Lama glama*) with a long wooly coat.Llamas and alpacas are believed to have been domesticated and developed from the guanaco; they are a close relative of the vicuna and a more distant relative of the camel. Llamas are surefooted and tolerant of high altitudes; they are used as pack animals in mountainous areas and also as wool producers.

A herd of llamas

llano *n.* (Spanish *llano*, plain) an extensive semiarid, fairly level plain with grass/shrub vegetation in western Texas, Mexico, and South America.

Llano Estacado *n.* (Spanish *Llano Estacado*, staked plain) the High Plains of western Texas and southeastern New Mexico, a cattle-raising region representing the southern extension of the Great Plains.

L-layer *n.* a surface layer of undecomposed leaves in a forest.

lm *n.* lumen. See lumen.

LMO *n.* a living modified organism intended for consumption as a food, feed, or raw material for processing rather than for seed (especially in international trade). See GMO.

loam *n.* (Anglo-Saxon *lam*, loam) soil material that exhibits the properties of sand, silt, and clay almost equally. See soil texture.

loam texture *n.* soil material that contains between 23% and 52% sand, 28% and 50% silt, and 7% and 27% clay, as determined by pipet or other mechanical analysis (according to the U.S. Department of Agriculture definition).

loamy *adj.* **1.** having any soil texture that includes loam in its name; loam, clay loam, sandy loam, sandy clay loam, silt loam, or silty clay loam. **2.** describing a soil family that has sandy clay loam, loam, or finer

texture up to 35% clay and contains less than 35% stones in the upper subsoil. **3.** having good fertility and friable soil structure; soil that is naturally productive and easy to work.

loblolly pine *n.* **1.** a large evergreen tree (*Pinus taeda*, family Pinaceae) of southeastern U.S.A. with bundles of three slender pale green needles about 6 to 9 in. (15 to 23 cm) long and deeply fissured reddish brown bark. **2.** wood from the tree; used for lumber and pulpwood.

lobo *n.* (Spanish *lobo*, wolf) the large gray timber wolf (*Canis lupus*, family Canidae) of western U.S.A., the largest living member of the dog family.

lobster *n.* (Anglo-Saxon *loppestre*, spidery creature) **1.** a marine crustacean (*Homarus* sp. and related genera, family Homaridae) with two claws for front appendages, one larger than the other. **2.** some other similar crustacean, as certain crayfish (especially the spiny lobster, *Palinurus vulgaris*). **3.** the edible meat of such a creature.

lobster caterpillar *n.* the larva of the lobster moth, so named for its legs with long segments.

lobster moth *n.* the European bombycid moth (family Bombycidae, order Lepidoptera).

lock *n.* (Anglo-Saxon *loc*, a fastener) **1.** a device for latching or securing a door, gate, case, cabinet, or other closure. It usually requires a key or a combination to open the lock. **2.** a gun lock; the firing mechanism of a gun, or a trigger lock that prevents the gun from firing. **3.** a system of gates and ponds that can raise or lower ships in a canal so they can cross a neck of land. **4.** an air lock; a sealable compartment where air can be pumped in or out, e.g., for the entrance to a space vehicle. **5.** a strong hold, as a wrestler with a lock on an opponent. **6.** a figurative hold, as he has a lock on the nomination. **7.** a jam, as the logs have formed a lock in the river. *v.* **8.** to fasten something securely. **9.** to hold together, as to lock one's jaws, lock arms, lock horns, or lock the gears. **10.** to lodge, as to lock logs in a river or pieces in a passage.

locoweed *n.* (Spanish *loco*, insane + Anglo-Saxon *weod*, weed) any of several leguminous weeds (*Astragalus* sp. or *Oxytropis* sp., family Fabaceae) of western U.S.A. and Mexico with dense clusters of small flowers resembling those of peas. The plants accumulate selenium preferentially if it is available in the soil of an arid region; excess selenium causes blind staggers in cattle and horses. Also called crazyweed.

locus *n.* (Latin *locus*, place) **1.** a location, especially one that serves as a focal point for an activity. **2.** the site of a gene on a chromosome. **3.** a point, line, or plane that satisfies a set of mathematical conditions, or a set of points and/or lines that satisfies a set of equations.

locust *n.* (Latin *locusta*, grasshopper) **1.** any of several leguminous trees, especially black locust (*Robinia pseudoacacia*, family Fabaceae) and honeylocust (*Gleditsi triacanthos*) and related species with sharp spines, pinnate leaves, fragrant white flowers, and pods; sometimes used as an ornamental tree or in shelterbelts. **2.** the hard, durable, yellowish wood of the tree (especially the black locust); often used for fenceposts, poles, furniture, construction, etc. **3.** any of several thousand species of grasshoppers (family Locustidae = Acrididae). Locusts sometimes occur in swarms of millions that devour any tender vegetation in their path, especially when the weather is hot and dry. **4.** any of several cicadas (family Cicadidae); e.g., the 17-year locust.

Leaves and pods of honeylocust (left) and black locust (right)

locust borer *n.* a beetle (*Megacyllene robiniae*, family Cerambycidae) whose larvae bore into the sapwood and heartwood of the black locust tree, producing swollen areas, cracked bark, and both exposed and hidden tunnels.

lode *n.* (Anglo-Saxon *lad*, way or course) **1.** a vein containing metallic ore filling a fissure in a rock. **2.** any deposit of valuable ore in a well-defined layer or fissure filling. **3.** a rich source of ore or (figuratively) of anything of value, sometimes called a mother lode if it is unusually large and valuable.

lodestone or **loadstone** *n.* (Anglo-Saxon *lad*, way or course + *stān*, stone) a natural magnet; either magnetite (Fe_3O_4) or maghemite (Fe_2O_3). See ferromagnetic.

lodge *n.* (French *loge*, hut) **1.** a crude shelter, e.g., one made of tree branches. **2.** a Native American structure or tent used as a residence or a meeting place. **3.** a temporary residence; e.g., a hunting lodge or a building used for retreats and related activities, or a secondary residence such as a caretaker's lodge. **4.** the main building of a camping area or a resort hotel. **5.** a fraternal organization, as the Masonic Lodge, or its members or meeting place. **6.** the structure built in a pond by a beaver for use as a den. *v.* **7.** to house, usually temporarily. **8.** to lock in place, as a bone may lodge in one's throat or logs lodge in a river and form a logjam. **9.** to fall down or beat down, as a grain crop may lodge because of heavy rain and

wind, too much nitrogen, or damage by insects or disease.

lodgepole pine *n.* a Rocky Mountain pine (*Pinus contorta*, var. *latifolia*, family Pinaceae) that produces long, slender trunks used for poles, posts, fuel, lumber, pulpwood, etc. It has two needles about 2 in. (5 cm) long in each cluster. The state tree of Wyoming. Also called Jack pine, knotty pine.

lodging *n.* **1.** the flattening of the crop in an area, often under the impact of heavy rain and strong wind. Excess nitrogen may cause cereal grains to grow so tall and produce such heavy heads that they are susceptible to lodging. A lodged crop may still mature, but it dries out slowly and is more difficult to harvest than a standing crop. **2.** a place to stay, especially on a temporary basis; a room with a bed and other amenities.

Lodging in a field of oats

loess *n.* (German, *löss*, loose) mostly silt-size particles (0.002 to 0.05 mm in diameter) of mineral matter deposited by the wind. Most loess is calcareous, has a narrow range of particle size, and is highly subject to water erosion, especially to gully formation. It has a tendency to stand better in nearly vertical banks than on moderate slopes. Loess covers the landscape like a blanket that gets gradually thinner with distance from the source. Much of the world's loess was deposited during the Pleistocene epoch when meltwater from glaciers deposited sediment on broad open floodplains, and the temperature differences near the glaciers produced strong winds to carry the silt particles for long distances.

☐ Loess

Loess deposits in the U.S.A.

log *n.* (Danish *laag*, felled tree) **1.** a section of a tree trunk (or sometimes of a large branch) suitable to be sawed into lumber or other wood products. **2.** a record of a trip or of an activity such as drilling a well, filming a motion picture, or maintenance work on an aircraft. **3.** an object that can be thrown overboard to measure the speed of a ship as indicated by the rate of passage of knots in a cord attached to the log. See knot. **4.** a logarithm; the exponent (power) needed to make a base number (usually 10 or e, the base of natural logarithms) equal a specified value. *v.* **5.** to harvest trees from a forest. **6.** to make a record of an activity (keep a log).

logjam *n.* **1.** a tangle of logs lodged in a river or other large stream. **2.** an overload on a system, as Congress sometimes has a logjam of bills.

loin disease of cattle *n.* a form of botulism caused by cattle eating rotten meat containing toxins produced by bacteria (*Clostridium botulinum*), often because a phosphorus deficiency has caused them to have a depraved appetite.

lone star tick *n.* a bloodsucking tick (*Amblyomma americanum*, family Ixodidae) with a white spot on the back of adults. It infests most animals, both wild and domestic, and people and serves as a vector for diseases such as spotted fever.

long-day plant *n.* a plant that blooms during the spring when the photoperiod (daylight hours) becomes longer than their minimum requirement (actually, the critical feature is that the dark period becomes shorter than a maximum length). See photoperiodism.

longhorn *n.* **1.** See Texas longhorn. **2.** a breed of English beef cattle with long horns, now nearly extinct. **3.** a long-horned beetle. **4.** (slang) a Texan.

long-horned beetle *n.* any of several beetles (family Cerambycidae) that typically have long antennas and bright body colors. The larvae bore into the wood of both living and dead trees.

longitude *n.* (Latin *longitudo*, length) angular distance on the surface of the Earth measured east or west from the prime meridian passing through Greenwich, England, up to 180 degrees at or near the International Date Line; used along with latitude to define locations.

long-range forecast *n.* a weather prediction that extends more than a week, sometimes called an extended forecast (e.g., for the next 6 to 10 days) or an outlook (e.g., for 30 days or 90 days).

loofah or **loofa** *n.* (Arabic *lūfah*, loofah) **1.** any of several Old World tropical vines (*Luffa* sp.) that bears a fruit with a fibrous sponge-like interior. **2.** the interior part of the fruit, dried and used as a sponge.

loon *n.* **1.** any of several large aquatic birds (*Gavia* sp., family Gaviidae) with short tails, webbed feet, and a weird, laughing call. Loons swim and dive to catch fish or crustaceans; they live in the water except when they come ashore to breed. **2.** a person with crazy ideas.

lophotrichous *adj.* (Greek *lophos*, tuft + *trichos*, hair) having a tuft of hairlike appendages, as lophotrichous flagellation of certain bacteria with a cluster of flagella. See flagellum.

loris *n.* (French *loris*, a clown) any of several small, slender, tree-dwelling Asiatic lower primates (family Lorisidae) with large eyes, nocturnal habits, no tails, and slow movements; native to southern India and Sri Lanka (*Loris gracilis*) and to southeastern Asia (*Nycticebus* sp.).

lost river *n.* a disappearing river. It may flow through a sinkhole into an underground channel in karst topography, into an ice fissure in cryokarst topography, into an underground passage in lava rock, etc., or it may evaporate in extremely arid conditions.

lotic *adj.* (Latin *lot*, washed + *ic*, pertaining to) **1.** flowing, as the water in creeks or rivers. **2.** living in flowing waters (either plant or animal life).

lotus *n.* (Latin *lotus*, lotus) **1.** a plant in Greek legends (probably a jujube or elm) whose fruit was supposed to cause contentedness and loss of memory. **2.** a tropical water lily (*Nymphæa* sp., family Nymphæaceae), especially the blue African or Indian lotus or the white Egyptian lotus. **3.** a group of shrubs and herbs (*Lotus* sp., family Fabaceae) with pinnate leaves and yellow, purple, or white flowers like those of peas; e.g., birdsfoot trefoil.

Louisiana Purchase *n.* a land purchase that more than doubled the size of the U.S.A. at the time; the area from the Mississippi River westward to the Rocky Mountains and from the Gulf of Mexico northward to the Canadian border. It was purchased from Napoleon Bonaparte of France by President Thomas Jefferson for the U.S.A. in 1803 for $15 million and was explored by the Lewis and Clark expedition (1804–1806).

louse (pl. **lice**) *n.* (Anglo-Saxon *lus*, louse) **1.** any of several small, parasitic, wingless insects (order Anoplura) with flat bodies and either sucking or biting mouthparts. Lice live in the hair or on the skin of humans and other mammals and either bite or suck blood. They include the body or clothes louse (*Pediculus humanus corporis*), head louse (*Pediculus humanus capitis*), chicken louse (*Dermangal linae*), crab louse (*Pthirus pubis*), goatlouse (*Linognathus stenopis*), horse louse (*Trichodectes pilosus*), and sucking louse (order Anoplura). **2.** any of several other small parasitic insects, arachnids, or crustaceans that infest animals or plants. **3.** (slang) a contemptible person who has abused a trust or otherwise mistreated another.

love apple *n.* the tomato or a tropical plant (*Solanum aculeatissimum*, family Solanaceae) that produces similar red fruit.

love bird *n.* **1.** any of several small parrots (especially *Agapornis* sp.) that show strong affection for their mates and are often kept as pets. **2.** a person who shows deep affection for a companion, especially one of a happily married couple that is a pair of lovebirds.

Love Canal *n.* a site in the city of Niagara Falls, New York, where an abandoned canal was used as a disposal site by the Hooker Chemical and Plastics Co. in the 1920s and 1930s. Many homes and a school were built on the site in the 1950s. Wet years raised the water table, and many people became ill in the 1970s; the water was tested and found to be contaminated with more than 200 hazardous chemicals. The school was closed, many families had to be relocated, and several million dollars worth of lawsuits were filed against the Hooker Company.

love vine *n.* any of several species of dodder (*Cuscuta* sp., family Convolvulaceae); a pernicious, parasitic, seed-bearing, leafless weed that grows on other plants and often strangles them. Also called strangle weed. See dodder.

low clouds *n.* clouds below 6500 ft (2000 m), including stratus, stratocumulus, and nimbostratus.

low input sustainable agriculture (LISA) *n.* a system of agriculture that endeavors to sustain long-term productivity through soil-conserving cropping systems, the return of crop residues to the soil, manure applications, integrated pest management, and other techniques that minimize the use of inputs such as chemical pesticides, mineral fertilizers, and antibiotics.

lowland *n.* land such as a valley floor that is at a lower elevation than nearby areas (often called highlands). Lowlands tend to be more level than highlands and often have a high water table. See bottomland.

lubricant *n.* (Latin *lubricus*, slippery + *ant*, thing) a substance used to reduce friction between moving parts, especially those involving a bearing surface. Lubricants include various oils and greases, mostly with a petroleum-product base.

LUCA *n.* last universal common ancestor; theorized to have been a single-celled microbe that lived about 4 billion years ago prior to branching into separate evolutionary lines.

lucerne *n.* (Latin *lucerna*, lamp) alfalfa (*Medicago sativa*, family Fabaceae). See alfalfa.

luciferase *n.* (Latin *lucis* light, + *fer*, to bear + *ase*, an enzyme) an enzyme that produces light by catalyzing the oxidation of luciferin in bioluminescent organisms. See lightning bug, firefly, luciferin.

luciferin *n.* (Latin *lucis*, light + *fer*, to bear + *in*, chemical) a chemical in bioluminescent organisms that produces an almost heatless bluish green light when oxidized by ATP with luciferase as a catalytic enzyme. See luciferase, firefly, luminescent.

lucuma *n.* an Andean tree (*Pouteria lucuma*) that produces starchy sweet fruit that is suitable as a basic staple and can be dried and stored for years.

lumber *n.* **1.** wood products such as boards or planks. **2.** a miscellany of mostly useless stored items. *v.* **3.** to cut trees and saw them into lumber. **4.** to become or be stored away as useless. **5.** to clutter an area or impede movement with useless items. **6.** to move ponderously and often clumsily, as when overloaded; e.g., a truck lumbering down the road or an old man lumbering along.

lumbricus *n.* (Latin *lumbricus*, worm) any of several types of worms (*Lumbricus* sp., family Lumbricidae), including earthworms, intestinal worms, etc. See earthworm.

lumen (lm) *n.* (Latin *lumen*, light) **1.** the base unit for evaluating light output, equal to the light flux passing through an area of a unit solid angle from a point source with an intensity of 1 international candle. **2.** the space inside the walls of a plant cell. **3.** the channel or cavity inside a tubular organ in an animal body; e.g., the interior of a duct, an artery, or an intestine.

lumen-hour (lm-hr) *n.* the amount of light represented by a flux of 1 lumen lasting for 1 hr.

luminescent *adj.* (Latin *lumen*, light + *escentis*, being) **1.** producing light with little or no heat as a result of chemical, biological, or electrical processes. **2.** glowing in the dark, as phosphorescent paint or a lightning bug. See lightning bug, firefly, luciferase.

lunar day *n.* See tidal day.

lunar eclipse *n.* the passage of the moon through the umbra (shadow) of the Earth, an event that darkens the full moon. See eclipse.

lunar month *n.* a synodic month; the average period between new moons, equal to 29 days, 12 hr, 44 min, and 2.8 sec. See lunar year, sidereal month.

lunar soil *n.* See moon soil.

lunar tide *n.* a tide caused by the gravity of the moon, occurring twice each 24 hr and 50 min, especially on seacoasts but also to a lesser extent on the coasts of large lakes.

lunar year *n.* 12 lunar months, equal to 354.37 days, or about 11 days less than a solar year. See lunar month, solar year.

lunette *n.* (French *lunette*, little moon) **1.** any crescent-shaped formation, object, or space; e.g., a decorative crescent in or on a piece of wood or a crescent-shaped ridge composed of a salty wind deposit forming part of the rim around a depression (Australian usage). **2.** a room framed by an archway. **3.** a ring (especially one shaped like a half-moon) used to tow a vehicle.

lung *n.* (Anglo-Saxon *lungen*, lung) **1.** either of the two spongelike respiratory organs in the thorax of a human or animal. Lungs oxygenate the blood, dispose of carbon dioxide from the blood, and exchange air with the atmosphere by expanding and contracting. **2.** a balloonlike sac that performs a similar function in the body of a reptile or amphibian. **3.** a book lung; the folded structure that certain large spiders use for respiration. **4.** an iron lung; a machine that encloses the body of a person (except for the head) and induces respiration by alternately increasing and decreasing the air pressure surrounding the body; formerly used in the treatment of poliomyelitis.

lungworm *n.* (Anglo-Saxon *lungen*, lung + *wyrm*, worm) **1.** any nematode worm (family Metastrongylidae, phylum Nematoda) parasitic in the lungs of humans or animals. **2.** a similar nematode (*Rhabdias* sp.) that infests the lungs of amphibians or reptiles.

lungwort *n.* (Anglo-Saxon *lungen*, lung + *wyrt*, plant) **1.** a common garden flower (*Pulmonaris officinalis* or *Mertensia* sp., family Boraginaceae) with blue flowers and large spotted leaves. Some lungworts have been used for treating lung disease. **2.** a lichen (*Sticta pulmonaria*) that grows on tree trunks in alpine areas.

lupine *n.* **1.** (Latin *lupinus*, lupine) a leguminous plant (*Lupinus* sp., family Fabaceae) sometimes grown in flower gardens; produces tall spikes of blue, pink, or white blossoms. About 200 species are native to western U.S.A. Some species are used for forage and some produce edible seeds (e.g., European lupine, *L. albus*), but others are toxic (e.g., Washington lupine, *L. polyphyllus*);.seeds of the white lupine (*L. albus*) are edible after they have been detoxified. The bluebonnet (*L. subcarnosis*) is

A yellow lupine with fingerlike leaves, a spike flower, and a detached flower and pod.

the state flower of Texas. *adj.* **2.** (Latin *lupus*, wolf + *inus*, like) like a wolf. **3.** hungry; predatory; wild.

lure *n.* (French *leurre*, bait) **1.** an object that attracts, especially a decoy or artificial bait or a device to recall a falcon to its roost. **2.** an attraction, e.g., of an activity, as the lure of the circus. *v.* **3.** to attract or entice.

lurk *v.* (Swedish *lurka*, to lurk) **1.** to lie in wait or move furtively in concealment, ready to observe or attack a prey. **2.** to be hidden, as an evil intent lurks in his heart.

lush *adj.* (French *lasche*, loose) **1.** luxuriant; prolific, tender, and succulent, as lush vegetation. **2.** covered by lush vegetation, as a lush meadow. **3.** luxurious, as lush surroundings. *n.* **4.** liquor. **5.** a person who frequently is intoxicated. *v.* **6.** to drink liquor.

lutein *n.* See xanthophyll.

lutetium (Lu) *n.* (Latin *Lutetia*, Paris + *ium*, element) element 71, atomic weight 174.967; a radioactive rare earth metal that is associated in small amounts with yttrium and difficult to purify.

lux (lx) *n.* the unit for illuminance in the SI system; the illumination produced by a flux of 1 lumen falling perpendicularly on a surface 1 meter square; equal to 0.0929 foot-candle.

luxuriant *adj.* (Latin *luxurians*, thick or immoderate) **1.** lush; abundant and vigorous, as luxuriant vegetation. **2.** extravagant; rich appearing; excessive; as a luxuriant setting.

luxury consumption *n.* plant use of nutrients that are in excess of the amounts needed for the amount of yield produced, especially more than is needed to produce maximum growth. Luxury consumption is most common with nitrogen and potassium, but it can occur with any nutrient that is available in excessive amounts. Concentrations of a nutrient that are high enough to cause luxury consumption are also likely to cause pollution.

lyase *n.* (Greek *lysis*, loosening or releasing + *ase*, an enzyme) any enzyme that catalyzes a breakdown reaction, releasing an organic radical and forming a double bond in the remaining structure; e.g., lysozyme degrades bacterial cell walls, leading to lysis of bacterial cells. See enzyme, lysis.

lycopod *n.* (Greek *lykos*, wolf + *podion*, little foot) an evergreen moss (*Lycopodium clavatum*, *L. saururus*, and similar species, family Lycopodiaceae) with either erect or creeping growth that has small scaly leaves and produces spores at the tips of stems or in leaf axils. Its spores are used as an absorbent powder, as a covering for pills, and as a size standard in microscopy. Also called club moss.

lye *n.* (Anglo-Saxon *leáh,* lye) **1.** a caustic liquid formerly obtained by leaching wood ashes with water and used for making soap, cleaning and processing fruit, processing leather, etc. **2.** a concentrated highly alkaline solution of either sodium hydroxide (NaOH) or potassium hydroxide (KOH).

lymph *n.* (Latin *lympha,* water) **1.** a transparent watery fluid similar to blood plasma found in the lymphatic vessels of human and animal bodies and containing transparent lymphocyte cells that can be seen under a microscope. **2.** any clear watery fluid similar to true lymph, especially a clear liquid excreted from inflamed tissues.

lynchet *n.* a terraced field on a steeply sloping hillside.

lynx *n.* (Greek *lynx,* lynx) **1.** any of several types of wildcat (*Lynx* sp.) of the Northern Hemisphere having long, tufted ears, a tuft of hair on each side of the face, long legs, and a short tail. The principal North American species are the Canada lynx (*L. lynx*) with grayish brown fur marked with white and the bobcat (*L. rufus*) with a brownish coat marked with black spots. **2.** the fur of a lynx. **3.** a northern constellation near the Big Dipper.

lyophilic *adj.* (Greek *lyein,* to dissolve + *philos, loving + ic,* of) hydrophilic; having a strong affinity for a solute, usually water, wherein the lyophilic material is dispersed as a colloid.

lyosome *n.* (Greek *lyein,* to dissolve + *soma,* body) a spherical vesicle produced by a Golgi body and bounded by a membrane in the cytoplasm of an animal cell; the site of hydrolytic enzymes that digest foreign material and recycle waste products.

lysimeter *n.* (Greek *lysis,* dissolution + *metron,* measure) a container installed around a volume of undisturbed soil with a chamber beneath it where percolating water can be caught for analysis to study the leaching process in soil.

lysin *n.* (Greek *lysis,* dissolution) an antibody that dissolves or disintegrates cells. Bacteriolysin dissolves bacterial cells; hemolysin dissolves red blood cells.

lysis *n.* (Greek *lysis,* dissolution) **1.** the process whereby lysins break down cell walls. **2.** the gradual regression of a disease accompanying a return to good health.

lysozyme *n.* (Greek *lysis,* dissolution + *enzymos,* leavened) a natural enzyme in tears, saliva, white blood cells, egg white, and that acts as a mild antiseptic in some plants, destroying certain bacteria (e.g., *Micrococcus lysodeikticus, Escherichia coli,* and *Salmonella typhosa*).

M

maar *n.* (Latin *mare*, sea) a circular volcanic crater with a flat bottom formed by an ash eruption; commonly holding a lake.

MAB *n.* man and the biosphere, an international program of the U.N. Educational, Scientific, and Cultural Organization established in 1971 to focus on people and their needs, especially regarding human relations with the environment and the rational management and conservation of natural resources.

macabre *adj.* (French *macabre*, dance of death) gruesome; suggesting the horror of death and decay.

macadam *n.* pavement made of stone and asphalt; named after J. L. McAdam (1756–1836), the Scottish engineer credited with inventing it (though it had been used long before on roads built by the Incas in South America). Also called blacktop.

macadamia *n.* **1.** an Australian tree (*Macadamia ternifolia* and related species, family Proteaceae) with shiny green leaves up to 1 ft (30 cm) long in whorls of three or four and clusters of pink flowers; named after John Macadam (?–1865), an Australian chemist. **2.** the edible nuts of the tree with their thick leathery shells about the size of golf balls. Macadamia nuts have become a major product of Hawaii.

mace *n.* (French *mace*, a club) **1.** a weapon with a heavy metal head (often with spikes) at the end of a handle; formerly used in hand combat to break armor. **2.** an East Indian spice made from the dried outer covering of the nut from a nutmeg tree (*Myristica fragrans*, family Myristicaceae).

Mach or **Mach number** *n.* a system of evaluating the speed of an aircraft or other object moving at high speed by comparing it with the speed of sound through the same medium; e.g., Mach 2 is twice the speed of sound. It is named after Ernst Mach (1838–1916), an Austrian physicist.

machete *n.* (Spanish *machete*, short sword) **1.** a large, heavy knife used for cutting sugarcane, to clear a path through dense brush, or as a weapon, especially in Central and South America. **2.** a fish (*Elops affinis*) found in the Pacific Ocean.

mackerel sky *n.* a group of altocumulus or cirrocumulus clouds with a rippled appearance resembling the scales of a fish; an indicator of fair weather.

macrobe *n.* (Greek *makros*, long + *bios*, life) any living organism larger than a microbe; a plant or animal large enough to be seen by the unaided eye.

macrobiotics *n.* (Greek *macrobiotos*, having long life) **1.** the science that studies longevity. **2.** a philosophical program of Asian origin that seeks to prolong life by emphasizing harmony with nature (balancing yin and yang) and promoting a diet based on whole grains and vegetables with some seafood and fruit.

macrocosm *n.* (Greek *makros*, large + *kosmos*, universe) **1.** the entire universe considered as a whole. See microcosm. **2.** an overall view of a topic, as the macrocosm of war.

macrography *n.* (Greek *makros*, large + *graphein*, to draw) examination of an object with the unaided eye rather than with a microscope. See micrography.

macrohabitat *n.* (Greek *makros*, large + Latin *habitat*, a dwelling) an extensive area that encompasses a large variety of ecological niches and supports several kinds of plants and animals.

macronutrient *n.* (Greek *makros*, large + Latin *nutriens*, nourishing) any nutrient that plants and animals require in relatively large amounts. These are carbon, oxygen, hydrogen, nitrogen, phosphorus, potassium, calcium, magnesium, and sulfur for plants, and all of these plus sodium and chlorine for animals and humans. See essential element, micronutrient, primary nutrient, secondary nutrient.

macrophyte *n.* (Greek *makros*, large + *phyton*, plant) a plant that is large enough to be visible to the unaided eye, especially a marine plant.

macroplankton (Greek *makros*, large + *planktos*, wandering) free-floating aquatic organisms that are large enough to be seen with the unaided eye (about 1 mm long or larger).

macropore *n.* (Greek *makros*, large + *poros*, passage) a large enough passage between soil particles (or other media) for water to flow through with little hindrance; practically, any pore visible to the unaided eye.

macroscopic *adj.* (Greek *makros*, large + *skopos*, aim) visible to the unaided human eye without the use of a microscope.

mad cow disease *n.* bovine spongiform encephalopathy (BSE), a deadly neurological disease affecting the brains of mature cattle that was first identified in 1985 in England. The disease transforms and attacks at least 18 other species of animals. A related disease, Creutzfeldt-Jacob disease (CJD) affects humans.

made land *n.* a designation used on soil maps for an area covered with soil formed of material placed there by people, either to fill a low area or dispose of waste; e.g., an old sanitary landfill or a reclaimed strip mine.

mad itch *n.* See Aujeszky's disease.

mad tom *n.* any of several small freshwater North American catfish (*Noturus* sp.) that have poisonous pectoral spines.

maelstrom *n.* (Dutch *malen*, to grind or whirl + *stroom*, a stream) **1.** a famous dangerous whirlpool off the northwestern coast of Norway. **2.** any whirlpool that is large and violent. **3.** a tumultuous state of affairs, as the maelstrom of a controversial political campaign.

maggot *n.* (Danish *maddik*, maggot) the soft-bodied, legless larva of a fly (order Diptera), commonly found in decaying organic wastes or diseased tissues.

magma *n.* (Greek *magma*, kneaded) molten rock material from within the Earth. Magma forms granite and other intrusive (coarse grained) igneous rocks when it cools slowly beneath the surface. It forms lava when it reaches the surface through a volcano or vent. The lava cools relatively quickly and forms basalt or other extrusive (fine grained) igneous rocks.

magnesia *n.* (Greek *Magnēsie*, of Magnesia, a district in Thessaly, Greece) magnesium oxide (MgO), a tasteless white solid used as a soil amendment or feed supplement to overcome grass tetany in cattle. It is also used in laxatives and in medicines to neutralize stomach acidity.

magnesian limestone *n.* See dolomitic limestone.

magnesite *n.* (Greek *Magnēsie*, of Magnesia, a district in Thessaly, Greece + *ite*, mineral) magnesium carbonate ($MgCO_3$) that usually is formed by carbonaceous waters reacting with crystalline rocks high in magnesium or by magnesium-rich waters reacting with limestone. It is used in the manufacture of magnesium oxide (MgO) or magnesium metal. Also called giobertite.

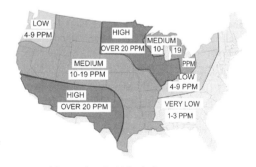

Magnesium in U.S. drainage waters

magnesium *n.* (Greek *Magnēsia*, a district in Thessaly, Greece + *ium*, element) element 12, atomic weight 24.312; a lightweight metal that burns readily. It is an essential macronutrient for plants as a component of chlorophyll and for animals as a cofactor for several enzymes and is used in lightweight alloys, fireworks, flashbulbs, batteries, etc.

magnet *n.* (Latin *magnetis*, magnet) **1.** a piece of iron, steel, or certain minerals (e.g., magnetite) that attracts iron objects and either attracts or repels other magnets, depending on their orientation. The magnetic property may be resident in the material (a permanent magnet) or induced by an electric coil (an electromagnet). **2.** a body that contains enough magnetic material to exhibit magnetic properties, as the Earth is a magnet. **3.** any place, person, or thing that attracts, as the circus is a magnet for children.

magnetic declination *n.* the angle between lines pointing to magnetic north and true north at a particular location and time. Also called magnetic variance. See agonic line, magnetic pole.

magnetic north *n.* the direction from a point of observation toward the location of the north magnetic pole of the Earth.

magnetic pole *n.* **1.** either of two focal points for the magnetic field surrounding any magnet, commonly designated as a north pole (that will point toward the Earth's magnetic north pole if used as a compass) and a south pole. The north pole of any magnet will attract the south pole and repel the north pole of another magnet. **2.** the magnetic poles of the Earth; the two locations on the Earth's surface where a compass needle will point toward the center of the Earth. Anywhere else the compass needle orients parallel to the great circle line passing through the location and the two poles (unless deflected by a local magnet). The north and south magnetic poles both move about 0.1 or 0.2 degrees per year (in a northwesterly direction in recent times). In 2000 A.D., the magnetic north pole was located near the Canadian coast on the Arctic Ocean near 79°N latitude and 103°W longitude and the magnetic south pole was located off the Antarctic coast south of Australia near 64°S latitude and 133°E longitude.

magnetic storm *n.* a worldwide disturbance in the magnetic field of the Earth caused by an eruption of charged particles from the sun.

magnetite *n.* (Latin *magnetis*, magnet + *ite*, mineral) a black iron oxide mineral (Fe_3O_4) with strong natural magnetic properties; an important commercial iron ore with about 65% iron content by weight. Magnetite deposits (e.g., those in Iron County, Michigan) can make a magnetic compass unreliable. Also called lodestone or loadstone.

magnitude of an earthquake *n.* See Richter scale.

magnolia *n.* **1.** any of a number of trees and shrubs (*Magnolia* sp., family Magnoliaceae) with large, fragrant, white, pink, or purple blossoms; named after Pierre Magnol (1638–1715), a French botanist. Some species are used as shade trees or ornamentals, and some of the wood is used to make furniture, boxes, etc. Southern magnolia (*M. grandiflora*) is the state tree and state flower of Louisiana and Mississippi. **2.** a blossom from such a tree or shrub.

magpie *n.* **1.** a bird (*Pica* sp.) with a black hood, white markings on the wings and breast, a tail that is longer than its body (total length up to 18 in. or 45 cm), and a raucous call. It is native to Eurasia and the Rocky Mountain area (the black-billed magpie, *P. pica*) or California (the yellow-billed magpie, *P. nuttalli*). **2.** any of several similar birds with black-and-white markings. **3.** a person who chatters noisily.

maguey *n.* See century plant.

mahonia *n.* any of several evergreen shrubs (*Mahonia* sp., family Berberidaceae) that bear grapes, including Oregon grape (*M. aquifolium*). It is named after Bernard McMahon (1775–1816), a U.S. botanist of Irish birth. See Oregon grape.

maidan *n.* (Hindi *maidān*, open space) an open space set aside for public use; e.g., as a public park, parade ground, or open marketplace, especially in India.

maiden *n.* (Anglo-Saxon *maegden*, little maid) **1.** an unmarried young woman. **2.** an unbred female animal. **3.** a 1-year-old, single-stemmed fruit tree seedling used for budding or grafting. *adj.* **4.** unmarried; or the name of a woman before she was married (maiden name). **5.** unbred; or having not yet won a race, as a maiden horse. **6.** first or initiating, as a ship's maiden voyage or a new representative's maiden speech.

maiden flight *n.* the flight of a queen bee when she mates with several drones before returning to the hive to lay eggs.

maidenhair tree *n.* See ginkgo.

maize *n.* (Spanish *maíz*, corn) **1.** corn (*Zea mays*, family Poaceae). See corn. **2.** a light yellow similar to that of ripe corn grain.

malachite *n.* (Greek *malachē*, mallow + *ite*, mineral) a green carbonate mineral [$Cu_2CO_3(OH)_2$], a secondary mineral formed by reaction of water with primary copper minerals such as cuprite. It is used as highly polished green stones for making jewelry and as a source of copper.

malachite green *n.* **1.** a brilliant green, analine dye. **2.** the green of the mineral malachite or of the dye.

malacology *n.* (French *malacologie*, study of mollusks) the science that studies mollusks such as shellfish and snails.

malamute *n.* (Inupiaq *malimiut*, a group of native Alaskans) **1.** a large, strong dog with long, thick gray or black and white hair, a breed developed in for use as sled dogs. **2. Malamute** a native Alaskan of the Inupiaq group from the Kotzebue Sound region in western Alaska.

malaria *n.* (French *malaria*, bad air) an infectious, debilitating, human disease caused by a protozoan (*Plasmodium* sp., family Plasmodidae) that destroys red blood cells and causes chills, fever, and sweating. Malaria is transmitted by the bite of an infected female mosquito (*Anopheles* sp.). It occurs mostly in lowlands in the humid tropics. See quinine.

malathion *n.* an organophosphate insecticide ($C_{10}H_{19}O_6PS_2$) that is relatively safe for humans and other mammals and that degrades relatively quickly.

male *adj.* or *n.* (Latin *masculus*, male) **1.** a person or animal that produces sperm cells (male gametes) that can fertilize the eggs of females of the same species. **2.** a plant or the part of a plant (stamen) that produces pollen. **3.** masculine; often associated with strength and boldness. **4.** made to enter an opening, as a bolt is a male part and the nut it screws into is a female part.

male fern *n.* a bright green fern (*Dryopteris filix-mas*), native to Europe and northeastern North America, whose rhizomes and stalks yield an oleoresin used to expel tapeworms.

malignant *adj.* (Latin *malignans*, wicked) **1.** having an evil intent; causing harm deliberately. **2.** very damaging; tending to go from bad to worse. **3.** likely to cause death; e.g., cancer.

malignant edema *n.* an acute, infectious disease caused by bacteria (*Clostridium septicum*, family Bacillaceae) that results in fluid retention (edema) and can infect people and all domestic animals. The edema spreads rapidly, destroying tissue and producing gas.

mallard *n.* (French *malart*, mallard) one of the most common and widespread wild ducks (*Anas platyhynchos*, family Anatidae), believed to be an ancestor of domestic ducks.

mallow *n.* (Latin *malva*, mallow) **1.** any of several plants (*Malva* sp., family Malvaceae) with lobed leaves. **2.** any of many other plants, shrubs, and trees of the Malvaceae family, including cotton, okra, hollyhock, and several other ornamental plants with large, showy blossoms.

malnutrition *n.* (Latin *mal*, bad + *nutritio*, a feeding) either too little or unbalanced food or feed, including starvation, undernourishment, unbalanced diet, and various deficiencies. It commonly causes weight loss and sometimes deformation. Also sometimes construed to include overnutrition.

malt *n.* (Anglo-Saxon *mealt*, malt) **1.** a mass of grain (usually barley) that has been soaked to make it sprout and then dried in a kiln and used to make beer or ale. **2.** an alcoholic beverage made from malt. **3.** malted milk, a beverage made by dissolving powdered malt in milk. *v.* **4.** to prepare malt from barley or other grain. **5.** to add malt as a flavoring.

Malta fever *n.* See brucellosis.

Malthus, Thomas Robert *n.* a British political economist (1766–1834) known for having expounded the Malthusian principle in 1798 in his *Essay on the Principle of Population.*

Malthusian principle *n.* a theory named after Thomas Robert Malthus, stating that mass starvation is inevitable at some future time because population increases geometrically (1, 2, 4, 8, 16, etc.) while food supply increases arithmetically (1, 2, 3, 4, 5, etc.).

mamba *n.* (Zulu *imamba*, mamba) any of four long, slender tree snakes (*Dendroaspis* sp.) of tropical Africa whose bite is often fatal, including the black mamba (*D. polyepsis*), which is typically about 10 ft (3 m) long and hunts small mammals in woods and scrublands and is one of the deadliest known snakes, and three green species that are mostly tree dwellers, including the common green mamba (*D. viridis*), the eastern green mamba (*D. angusticeps*), and Jameson's mamba (*D. jamesoni*), which often catch their prey by falling onto it from a higher tree branch.

mammal *n.* (Latin *mammalis*, of the breast) any warm-blooded vertebrate (class Mammalia), including humans. The females produce milk for their young.

mammary gland *n.* one of two or more organs of a female mammal capable of producing milk for young to suckle; e.g., the breasts of a woman or the udder of a cow.

Mammoth Cave *n.* a live and growing cave system in Kentucky with 300 mi (500 km) of explored passageways formed by carbonic acid (H_2CO_3) in water dissolving passages through limestone over millions of years. Mammoth Cave National Park with an area of 79 mi^2 (204 km^2) has been established in the vicinity. See karst.

Man and the Biosphere (MAB) *n.* an international environmental program established by the U.N. Educational, Scientific, and Cultural Organization (UNESCO) in 1971 to establish ecosystem reserves and to develop scientific knowledge relating to the rational management and conservation of natural resources. By 2000, they had databases for the vascular plants and vertebrate animal species in more than 660 protected areas in 97 nations.

manatee *n.* (Spanish *manatí*, manatee) any of several large, slow-moving, aquatic plant-eating mammals (*Trichechus* sp.) with two front flippers and a broad tail. It is an endangered species found in shallow waters along the coasts and nearby rivers of the Caribbean Sea and Gulf of Mexico, Brazil, and Africa. Formerly heavily hunted for their meat, oil, and hides, manatees are now protected in most places but are still threatened by polluted waters, boat propellers, gill nets, etc.

manchineel *n.* (Spanish *manzanilla*, little apple) **1.** a tropical American tree or shrub (*Hippomane mancinella*, family Euphorbiaceae) with poisonous yellowish green fruit shaped like small apples and a milky sap that causes skin blisters on contact. **2.** the wood of the tree.

mandible *n.* (Latin *mandibula*, a jaw) **1.** the lower jaw of a vertebrate animal. See maxilla. **2.** either of a pair of jaws of an insect or other arthropod. **3.** either jaw of a cephalopod or other animal with a beak.

mandrake *n.* (Latin *mandragora*, mandrake) **1.** an herb (*Mandragora officinarum* and related species, family Solonaceae), native to southern Europe, with a short stem and purple or white flowers. **2.** the thick and often forked roots of this plant that were once thought to be magical because they form figures that sometimes resemble a man. The roots of this and other plants of the same family contain a poisonous alkaloid, hyoscyamine ($C_{17}H_{23}NO_3$), used medically for its narcotic and emetic effects. **3.** the May apple.

manganese *n.* (French *manganèse*, Magnesia, a district in Thessaly, Greece) element 25, atomic weight 54.938; a heavy metal that is an essential micronutrient for plants and animals. It is commonly present in small amounts in iron minerals and forms a few minerals of its own such as pyrolusite (MnO_2), manganite ($MnOOH$), and rhodochrosite ($MnCO_3$). It is used with iron to toughen alloy steels and also to form alloys with aluminum and copper.

mange *n.* (French *mangeue*, itch) any of several contagious skin diseases of cattle (e.g., *Psoroptes bovis*, family Psoroptidae), horses, dogs, several other animals, and sometimes humans, caused by parasitic mites (family Sarcoptidae) and characterized by intense itching, loss of hair, and lesions that eventually spread over the entire body. Also called psoroptic scab, scabies, barn itch, or cattle mange.

mange mite *n.* any of several species of mites (*Sarcoptes scabiei, Chorioptes* sp., *Demodex* sp., *Notoedres* sp., and *Psoroptes* sp., family Sarcoptidae) that are vectors for mange. Also called itch mite.

mangium *n.* a fast-growing tropical forest leguminous tree that fosters a lush ground cover of understory rainforest vegetation beneath its canopy and supplies degraded land with nitrogen. It is now being used for reforestation in Indonesia and other parts of Asia.

mango *n.* (Portuguese *manga*, mango) **1.** a tropical evergreen tree (*Mangifera indica*, family Anacardiaceae) of Asiatic origin whose fruit has been greatly improved by centuries of plant breeding. **2.** the edible fruit from the tree; a stone fruit with an oval shape, a thick yellowish red rind, and a juicy pulp. **3.** any of several large hummingbirds (*Anthracothorax* sp.).

mangrove *n.* (Portuguese *mangue*, mangrove) a woody plant that grows in swampy areas along ocean shores of Florida and tropical America. Its vinelike adventitious roots extend downward from the trunk and branches and form a thick, tangled mass that helps prevent coastal erosion. The principal species, American (red) mangrove (*Rhizophora mangle*, family Rhizophoraceae), grows to a height of 40 ft (12 m). Another significant species is the black mangrove (*Avicennia germinans*, family Verbenaceae), whose flowers are a good source of nectar for honeybees in the Florida Everglades.

Manila hemp *n.* fibers from the abaca plant (*Musa textilis*, family Musaceae) that are used to make rope, fabrics, paper, etc.

manioc *n.* See cassava.

manna *n.* (Aramaic *mannā*, manna) **1.** the food supplied miraculously to the Israelites wandering in the wilderness, described in Exodus 16 as covering the ground like hoarfrost with small particles like white coriander seed that tasted like wafers made with honey. **2.** anything badly needed that comes from an unanticipated source. **3.** a sweet, gummy juice containing mannitol from any of several plants, insects, and sponges.

mannitol *n.* a polyhydric alcohol [$C_6H_8(OH)_6$], named after manna, that occurs in flowering ash (*Fraxinus ornus*), certain sponges, and other sources. It is used as a sweetener and to keep foods moist and medically as a diuretic to reduce intracranial and intraocular pressure.

mano *n.* (Spanish *mano*, hand) a small handheld stone used to grind grain (especially corn) in a shallow depression in a larger stone.

manometer *n.* (Greek *manos*, loose + *metron*, measure) an instrument for measuring the pressure of liquids or gases; originally a U-shaped tube partly filled with liquid that indicated pressure by the level of the liquid.

manta *n.* (Spanish *manta*, blanket) **1.** a cloak or mantle (especially in Spanish-speaking nations). **2.** a blanket used to cover the load on the back of a packhorse or mule. **3.** any of several large tropical ray fish (*Manta* sp., family Mobulidae), including those known as manta ray, devil ray, or devilfish.

mantis or **mantid** *n.* (Greek *mantis*, a kind of locust) any of several very beneficial insects (order Mantidae) with a long body and long, slender legs. They prey on harmful insects (aphids, beetles, caterpillars, flies, wasps, etc.). Also called praying mantis. See Chinese mantis, praying mantis.

mantle *n.* (French *mantel*, cloak) **1.** a loose, sleeveless outer garment worn like a cloak. **2.** anything that covers or conceals, either literally, as a cloth mantle or a soil mantle or, figuratively, as the mantle of darkness or a mantle of pretense. **3.** a shelf that covers and projects from an object such as a fireplace. **4.** that part of the Earth immediately beneath the crust down to the core at a depth of about 1800 mi (2900 km). See Mohorovičić discontinuity. **5.** a noncombustible hood that covers a flame and produces a bright light.

manufactured fertilizer *n.* a fertilizer material with a guaranteed composition produced at a factory by blending two or more single-compound fertilizer materials and forming the mixture into granules. It is usually sold in bags with the guaranteed analysis printed on the bag.

manufactured gas *n.* methane (CH_4) produced by gasification of coal. Also called synthetic gas or coal gas.

manure *n.* (French *manouvrer*, to cultivate) **1.** an organic material used as fertilizer to grow crops or ornamental plants, especially partially decomposed animal excrement with or without added straw or other plant material. Note: In many developing countries, animal excrement is dried and used as fuel for cooking. **2.** other materials used as fertilizer; e.g., an immature crop tilled into the soil to add organic matter is called green manure. *v.* **3.** to add manure to soil or other growth media to improve fertility and physical condition. Note: Livestock on farms and ranches in the U.S.A. produce an estimated 1.5 billion tons of excrement each year. About half of it is discharged on ranges and pastures by grazing animals, where it helps to fertilize the forage. The remaining half, mostly from cattle and hogs, is best used as fertilizer for crops and gardens and should be properly applied at rates consistent with the nutrient needs of the plants. Otherwise it may constitute a pollution hazard. See compost.

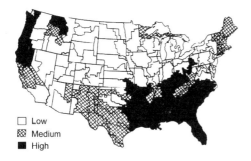

□ Low
▨ Medium
■ High

Potential for pollution of surface water
by animal wastes

map *n.* (Latin *mappa*, map) **1.** a chart; a schematic representation of a surface and of selected features located thereon, as a geographic map, road map, contour map, soil map, etc. **2.** a similar representation of a portion of the sky as seen from Earth (or any other designated position). **3.** any surface that shows features that can be interpreted like a map, as an aerial photograph may serve as a map. **4.** (slang) an expressive face of a person or animal. *v.* **5.** to make a map of an area. **6.** to plan, as to map out a course of action.

maple *n.* (Anglo-Saxon *mapel*, maple tree) **1.** any of about 200 species of deciduous trees and shrubs (*Acer* sp., family Aceraceae). Many are used as ornamentals and for lumber, maple syrup, etc. Maples typically have thick foliage of medium to large leaves with either three or five lobes and produce clusters of small flowers and pairs of winged nutlet seeds. **2.** the hard wood of such a tree or the lumber produced from it, often used to make furniture, plywood, flooring, etc. *adj.* **3.** having the flavor of maple syrup.

maple sugar *n.* a tan sugar product made by evaporating maple sap or maple syrup to dryness. Sucrose is the dominant component of maple sugar, but it also contains other solids that contribute to its flavor and color.

maple syrup *n.* **1.** a sweetener made by collecting sap from sugar maple (*Acer saccharum*) or several other maple species during the freeze–thaw period of early spring and boiling it down to the consistency of syrup. **2.** a commercial product based on the syrup from maple trees but often with cane sugar and/or other sweeteners added.

map unit *n.* any of the areas separated on a map by boundary lines, symbols, and/or patterns. Formerly called mapping unit. See soil map unit.

marama bean *n.* a legume used for food in the Kalahari and other sandy semidesert regions of southern Africa. Its seeds are high in protein (37%) and oil (33%) and have a rich, nutty flavor. It also produces large edible tubers, often consumed as young tubers weighing about 1 kg (2.2 lb) that are baked, boiled, or roasted and eaten as a vegetable dish.

marasmus *n.* (Greek *marasmos*, wasting away) **1.** progressive emaciation, especially of infants or young children, that is caused by too little food and unbalanced nutrition. Also called infantile atrophy, pedatrophy, marasmus infantilis, or marasmus lactantium. See kwashiorkor. **2. enzootic marasmus** a similar condition in domestic animals, especially that caused by too little copper and/or cobalt. Known as salt sickness in Florida, bush sickness in New Zealand, and pine or pining in Scotland.

marble *n.* (Anglo-Saxon *marmel*, marble stone) **1.** a metamorphic rock formed from limestone subjected to metamorphism by heat and pressure. Marble is composed largely of calcium carbonate but may be dolomitic and typically has enough impurities to form decorative patterns. **2.** a statue or other sculpture made from such stone. **3.** plastic or other material imprinted with a pattern that looks like marble stone. **4.** a small sphere made of glass, plastic, or stone, often used in games. *adj.* **5.** made of marble. **6.** having a variegated pattern that resembles marble in color, pattern, etc. **7.** to form or imprint with a pattern like that of marble.

Marbut, Curtis F. *n.* an American geologist and soil chemist (1863–1935) who gained the reputation of being the world's leading soil scientist for his work on soil surveying and soil classification. He was the Scientist in Charge of Soil Survey for the USDA Bureau of Soils from 1910 to 1935.

marc *n.* (French *marche*, treading) **1.** solid residue from processing fruits, sugarcane, and sugar beets, especially from trampling or squeezing grapes to extract juice. Many of these residues make good livestock feed. **2.** brandy made by distilling marc residues from grapes. **3.** residues left from extractions using a solvent; e.g., from processing vegetable matter to obtain medicines.

marcescent *adj.* (Latin *marcescens*, withering) withering but remaining in place, as withered leaves hanging on a plant.

marcot *adj.* starting a new plant by wrapping rooting media around a vine or branch. Also called air-layering.

mare *n.* **1.** (Anglo-Saxon *mere*, female horse) a mature female horse, donkey, or other equine animal. **2.** (Latin *mare*, sea; pl. *maria*) a sea. **3.** a large, dark-toned plain on the surface of the moon or Mars.

Marek's disease *n.* a poultry diseases named after Jozsef Marek (1868–1952), a Hungarian veterinarian who described the disease in 1907. Caused by a herpes virus and spread by either a tick (*Argas persicus*) or the darkling beetle (family Tenebrionidae), symptoms include cancer and paralysis of a limb or the neck. Chicks can be vaccinated against Marek's disease by injection through the eggshell three days before hatching (after about 18 days of incubation).

margarine *n.* (French *margarine*, a salt of margaric acid) a semisolid product similar to butter. It is made with vegetable oils (especially soybean or coconut oil, sometimes with animal fats added), emulsified with water or milk, and used like butter as a spread and for cooking. Also called oleomargarine or oleo.

margarite *n.* (Latin *margarita*, a pearl) **1.** a natural pearl composed of layers of mostly calcium carbonate deposited around a grain of sand or other irritant by a mollusk (e.g., species of family Margaritiferidae are important sources of pearls). **2.** a calcium–aluminum silicate mineral [$CaAl_2(AlSi_3O_{10})(OH)_2$] that commonly forms scales with a pearly luster. **3.** small globules like a row of beads in glassy lava rocks.

margin *n.* (Latin *marginis*, border) **1.** the outer limit or border of a solid object or surface. **2.** the unprinted area around the outside of each page in a book or other printed page or around the outside of a postage stamp or other sticker. **3.** a small extra amount beyond what is required, as a margin for error. **4.** an amount that must be deposited with a broker before a purchase can be made that could lead to either a gain or a loss. **5.** the difference between the selling price and the acquisition cost, as a profit margin. **6.** the difference between the value of collateral pledged as security and the loan secured by it. **7.** a difference in the number of votes, as he won by a margin of 50 votes or the measure passed by a margin of 5 votes. **8.** the added cost of producing one more of something (marginal cost) or the added income from selling it (marginal return).

Mariana Trench *n.* a trench in the Pacific Ocean south and west of the Mariana Islands, including the Challenger Deep that reaches 36,201 ft (11,034 m) below sea level; the deepest known depression in any ocean.

mariculture *n.* (Latin *mare*, sea + *cultura*, culture) the production of crops in seawater; e.g., the culture of algae to be fed to crabs grown for the commercial market. Also called aquaculture, marine culture, sea farming, or ocean farming. See kelp.

marigold *n.* **1.** a composite flower, usually with golden blossoms but ranging from yellow to red. **2.** a short, bushy plant that produces these flowers (*Tagetes erecta* and related species, family Asteraceae) and its strongly scented foliage. The roots contain an oil that repels root parasites. **3.** any of several similar plants, such as the pot marigold (*Calendula officinalis*, family Asteraceae).

marijuana or **marihuana** *n.* (Spanish *mariguana*, marijuana) **1.** wild hemp (*Cannabis sativa*, family Moraceae). See hemp. **2.** an illegal narcotic derived from the dried leaves and female flowers of the plant; known chemically as delta-9-tetrahydrocannabinol (THC). Also called ganja, bhang, bang, or hashish, etc.

marine *adj.* (Latin *marinus*, of the sea) **1.** in, of, or from the sea (or ocean), as a marine environment or a marine deposit. **2.** having to do with maritime affairs such as shipping by sea or navigating the oceans. **3.** designed for use at sea, as marine equipment. **4.** seagoing ships. **5.** trained to serve at sea, as a member of the merchant marine or of the Marine Corps.

maritime influence *n.* a moderating influence affecting the weather of land near an ocean or other large body of water.

marjoram *n.* (Latin *majorana*, marjoram) a perennial, aromatic herb (*Origanum* sp. or *Majorana* sp., family Lamiaceae) native to North Africa and Asia; especially sweet marjoram (*M. hortensis* = *O. majorana*), whose oval grayish green leaves are used for flavoring foods.

marl *n.* (French *marle*, marl) **1.** a soft sedimentary deposit composed mostly of calcium carbonate ($CaCO_3$), especially in the form of shell fragments with some clay; often used as an impure lime to reduce soil acidity. **2.** a soft, earthy material. *v.* **3.** to lime a field with marl. **4.** to wrap a rope with marline (a two-fiber cord), forming a knot with each wrap.

marmoset *n.* (French *marmouset*, marmoset) a small monkey (*Callithrix* sp., *Leontocebus* sp., etc.), native to South and Central America, with soft bright-colored fur and a long tail. Also called squirrel monkey.

marmot *n.* (French *marmotte*, mutter) **1.** a large, fat-bodied, burrowing squirrel (*Marmota* sp.). Also called a woodchuck or groundhog. **2.** a related animal such as a prairie dog.

Mars *n.* (Latin *Mars*, god of war) **1.** the 4th planet from the sun, sometimes called the red planet, a relatively small planet (7th largest in the solar system) located between Earth and the asteroid belt. **2.** the Roman god of war.

marsh *n.* (Anglo-Saxon *merisc*, marsh) wetland dominated by sedges and grasses with little accumulation of peat or muck; often treeless. It may be freshwater or seawater and tidal or nontidal. See wetland.

marsh gas *n.* impure methane (CH_4) produced by anaerobic decomposition of organic matter in a wet area. Also called swamp gas.

marsh grass *n.* any grass that commonly grows in marshes, especially cordgrass (*Spartina* sp.).

marsupial *n.* or *adj.* (Latin *marsupium*, a pouch) a nonplacental mammal (order Marsupialia), including bandicoots, kangaroos, opossums, and wombats. The females of most species have a marsupium (pouch) for carrying and nursing their young.

mash *n.* (German *meisch*, mash) **1.** a soft mass; e.g., a grain mixture soaked in water for livestock feed or for making beer. *v.* **2.** to soften grain by soaking it in water. **3.** to crush with high pressure; e.g., to mash a bug by stepping on it.

mass *n.* (Latin *massa*, mass) **1.** any body of solid, coherent matter, especially a large body. **2.** a collection of similar particles considered as a whole, as a mass of sand. **3.** a large group, as a mass of people. **4.** the size or magnitude of an object or assemblage. **5.** the majority of a certain group. **6.** the physical magnitude of an object defined as its weight divided by the gravitational force where the weight was measured. **7.** a paste used to fill capsules or make pills in a pharmacy. *adj.* **8.** affecting or reaching many people, as mass unemployment, mass education, or mass media. **9.** done on a large scale, as mass destruction. **10.** influenced by the presence of large numbers, as something that happens by mass action. *v.* **11.** to assemble into a group.

mass flow *n.* movement as part of a convective flow, especially the transport of ions dissolved in soil water to plant roots as they absorb water for plant use.

massive *adj.* (French *massif*, large) **1.** bulky and heavy. **2.** large and heavy in appearance, as a massive head. **3.** on a large scale, as massive unemployment. **4.** not hollow; solid throughout. **5.** lacking visible structure; having no obvious internal surfaces; e.g., massive soil or a massive rock.

mass movement *n.* bulk movement of unsorted soil and/or rock material down a slope with gravity as a driving force, either slowly as by soil creep and solifluction, more rapidly as by mudflows and slumps, or very rapidly as by landslides. Water may act as a lubricant but not as the transporting agent.

mass selection *n.* crop improvement by the traditional method of selecting superior individuals from a large population and using their seed for planting the next crop.

mass wasting *n.* the combination of all forms of mass movement, including the effects of all processes that move soil and rock material downslope with gravity as a driving force. See backwearing, downwearing.

mast *n.* (Anglo-Saxon *maest*, mast) **1.** seeds, acorns, and nuts eaten by domestic and wild animals from oak, beech, chestnut, various pines, and other species of trees and shrubs, especially where such seeds have accumulated on the forest floor. **2.** a tall spar rising vertically from the keel or deck of a vessel to support a sail. **3.** any upright pole.

mastitis *n.* (Greek *mastos*, breast + *itis*, inflammation) an inflammation of the mammary gland (udder or breast) of dairy cattle, sheep, goats, and other animals and humans that is caused by many species of bacteria. Symptoms include swollen mammary glands, reduced milk production, abnormal milk, and fibrosis. Also called garget or bad quarter.

mate *n.* (German *mate*, companion) **1.** spouse, a husband or wife. **2.** an animal's reproductive partner. **3.** either one of a pair, as these two shoes are mates. **4.** a partner or comrade, as a classmate or a roommate. **5.** an officer on a ship, as the first mate or a boatswain's mate. *v.* **6.** to match pairs; e.g., to mate stockings. **7.** to marry or to pair male and female animals for reproduction.

maté *n.* (Quechua *mati*, calabash) **1.** a South American holly (*Ilex paraguariensis*, family Aquifoliaceae). Also called yerba maté. **2.** the dried leaves of this holly tree. **3.** the strong tealike hot drink made from the dried leaves, usually brewed in and sipped from a calabash; popular in South America.

A maté gourd and a bombilla used to sip the maté

mating flight *n.* the flight of a new honeybee queen when she is ready to mate. She flies high in the air and mates with from one to ten drones on one or more mating flights.

matriclinous, matroclinous, or **matroclinal** *adj.* (Latin *mater*, mother + German *klinein*, to bend + *osus*, full of) inheriting traits mostly from the mother, as opposed to patroclinous.

matrix *n.* (Latin *matrix*, source) **1.** internal substance or structure. **2.** the fine-grained component that surrounds larger crystals or fragments in certain rocks, in concrete, in crystalline alloys, etc. **3.** point of origin, as the womb, the matrix of a tooth, or the matrix of civilization. **4.** a 2-dimensional mathematical array of numbers, functions, or symbols that is manipulated (added, multiplied, etc.) according to a set of rules called matrix algebra. **5.** a 2-dimensional array for linguistic elements whose presence or absence is indicated by + or − signs for each item. **6.** a master mold that can be used to make many reproductions of a recording.

matter *n.* (Latin *materia*, material) **1.** substance; solid material. Physical objects are composed of matter. **2.** any or all solid, liquid, and gaseous material, as opposed to qualities, ideas, actions, and other nonsubstances. **3.** a situation or affair, as that was a trivial matter. **4.** an amount, as a matter of 10 dollars or a matter of a few hours. **5.** information or understanding, as subject matter, or matter for deep thinking. **6.** written or printed material, as reading matter. **7.** pus and/or other discharge, as from a wound. **8.** a legal allegation. **9.** a decision to be made, as a matter of importance or of life and death. *v.* **10.** to be important. **11.** to produce pus.

mattock *n.* (Anglo-Saxon *mattuc*, mattock) a digging instrument that resembles a pickax, but one or both blades are broad rather than pointed.

maturation *n.* (Latin *maturus*, ripe + *ation*, action) the process of maturing or ripening.

mature *adj.* (Latin *maturus*, ripe) **1.** fully developed; the relatively stable period between the rapid growth stage of youth and the waning stage of old age; e.g., a mature person, plant, animal, idea, plan, or soil. For example, a mature landscape has a fully integrated drainage system and relatively high relief. **2.** ripe, as a mature crop is ready for harvest. **3.** at the end of its term, as a mature bond. **4.** intended for adults, as for mature audiences only. *v.* **5.** to reach full development. **6.** to ripen. **7.** to become due.

maturity *n.* (Latin *maturitas*, of full age) **1.** the condition of being full grown, fully developed, or ripe. **2.** the time when a financial matter becomes due and payable.

Mauna Loa *n.* an active volcano in Hawaii; height 13,680 ft (4170 m).

mauve swallowtail *n.* a beautiful butterfly that is cultured in Papua New Guinea. Each sells for about $50.

maverick *n.* **1.** an unbranded calf, cow, or steer on open range, especially one separated from its mother; named after Samuel A. Maverick (1803–1870), a pioneer Texan who refused to brand his cattle. **2.** a strong-willed, independent person; a lone dissenter.

mavericker *n.* a person who rustles cattle by putting his/her brand on any unbranded cattle he/she finds.

maxilla *n.* (Latin *maxilla*, little jaw) **1.** the upper jaw of a vertebrate animal. **2.** an accessory jaw or appendage located just behind the mandible of an insect, crab, or other arthropod.

maximum sustainable harvest *n.* the maximum annual harvest of any renewable resource such as a forest or grass that can be sustained over the long term. For example, the maximum sustainable harvest of rangeland forage is about 50% of annual growth. Taking more than that will weaken the plants and reduce future growth.

Maya *n.* (Spanish *Maya*, Maya) **1.** a member of a pre-Columbian civilization that inhabited the Yucatán Peninsula (Guatemala and southern Mexico) and attained a high degree of development from about 300 to 900 A.D. (an older civilization than that of the Aztecs and Incas). The Mayans were conquered by a Spanish army led by Francisco de Montejo in 1542. **2.** a modern descendant of these people. **3.** the language of the Mayan people. See Aztec, Inca.

mayapple *n.* **1.** a rhizomatous plant (*Podophyllum peltatum*, family Berberidaceae), native to North America, with elongated leaves and large, individual, cup-shaped flowers. Its roots, rhizomes, leaves, and seeds are poisonous. Also called mandrake. **2.** the lemon-shaped, yellow, edible fruit of the plant.

May bug or **May beetle** *n.* See June bug.

mayfly or **May fly** *n.* a fragile insect (order Ephemeroptera) with large membranous front wings and smaller ones behind. It develops from an aquatic nymph and lives as an adult for only about two days.

mayweed *n.* (Anglo-Saxon *magothe*, mayweed) a foul-smelling weed (*Anthemis cotula*, family Asteraceae) with a composite flower that has a yellow disk and white rays. Also called stinking camomile.

McClintock, Barbara *n.* an American geneticist (1902–1992), winner of the 1983 Nobel Prize for Physiology or Medicine for her 1940s and 1950s work on transposons (jumping genes) in plants. See jumping gene, transposon.

McKinley, Mount *n.* a mountain peak in south central Alaska in the middle of the Alaska Range. At 20,320 ft (6194 m), it is the highest point in North America; named for William McKinley (1843–1901), 25th President of the U.S.A. (1897–1901).

M discontinuity *n.* See Mohorovičić discontinuity.

meadow *n.* (Anglo-Saxon *mædwe*, meadow) **1.** an area naturally vegetated with mostly grasses and forbs suitable for hay or pasture and having very few or no trees; usually a valley floor or other relatively flat area. **2.** a field set aside for hay or pasture, either long term or as part of a crop rotation.

meadowlark *n.* **1.** a North American songbird about the size of a robin, including the eastern meadowlark (*Sturnella magma*, family Icteridae) and the western meadowlark (*S. neglecta*), both with brown backs marked with black and white and yellow breasts marked with a bold black V. **2.** any of several similar birds commonly seen in meadows.

meadow mouse or **meadow vole** *n.* any of several short-tailed, plant-eating rodents (*Microtus* sp. and related genera, family Cricetidae) that are larger than a house mouse but smaller than a rat. Meadow mice commonly live in burrows in fields and meadows of temperate regions. Also called field mouse or field vole.

meadow nematode *n.* a wormlike nematode (*Pratylenchus* sp., family Tylenchoidae) that infests the roots of many tree species such as apple, cherry, and walnut, and of herbaceous plants such as alfalfa, tobacco, and cotton. It causes brown root rot, sloughing off of root tissue, and proliferation of fine roots near the soil surface.

meal *n.* **1.** (Anglo-Saxon *mæl*, time or repast) the food served at one time, especially to a family or other gathering as breakfast, lunch, dinner, or supper. **2.** (Anglo-Saxon *melu*, meal), an edible grain that has been coarsely ground; e.g., cornmeal. **3.** any coarsely ground substance; e.g., nutmeal.

meal moth *n.* a lepidopterous insect (*Asopia farinalis*) whose larvae commonly infest all kinds of stored grain and seed and cause large losses. They typically spin silken threads that cover the surface kernels and mat them together. Also called Indian meal moth.

mealworm *n.* either the dark mealworm (the larva of the darkling beetle, *Tenebrio obscurus*, family Tenebrionidae) or the yellow mealworm (the larva of *T. molitor*). These larvae feed on grains, especially those stored in grain bins, and survive in dry conditions by making metabolic water from carbohydrates. See darkling beetle.

meander *n.* (Greek *maiandros*, a winding course, as that of the Menderes River) **1.** one of a series of looping turns a river makes as it flows. Sometimes a meander loop is cut off and the old channel forms an oxbow lake. **2.** a winding or circuitous pathway or travel route. **3.** a journey that follows a meandering route. **4.** a form of decoration with interlacing lines sometimes used on vases or in architecture. *v.* **5.** to follow a meandering course. **6.** to wander slowly, often aimlessly.

meandered shore *n.* the banks of a large stream or the shore of a lake that has shifted over time because of natural forces such as a stream meandering across its bottomland. Property lines defined according to such boundaries shift with the shoreline unless a specific position or date is noted in the legal description.

meandering stream *n.* a stream that makes looping turns and crosses back and forth across much or all of the width of its floodplain.

mean distance *n.* **1.** the average of the longest and shortest distances of a planet or satellite from the star or other body it orbits around. **2.** some other average distance.

mean sea level *n.* the average level of the water surface between high and low tides, commonly used as a reference level for surveying elevations. The 1929 sea-level data have been used by the National Geodetic Survey as the mean sea level for mapmaking.

measurable precipitation *n.* any amount of precipitation received that can be measured in a rain gauge (more than 0.005 in. or 0.125 mm). Any less is recorded as a trace (T).

meat *n.* (Anglo-Saxon *mete*, meat) **1.** animal flesh used as food, especially that of mammals. The flesh of birds and fish is called meat by some people but not by others (e.g., fish and chicken are not meat to Hindus). **2.** the edible part of a nut or fruit. **3.** any solid food, as distinguished from liquid, as meat and drink. **4.** the gist or main part of something, as the meat of a story. **5.** a specialty or something one enjoys, as racing was his meat.

mechanical aeration *n.* the use of a motor-driven impeller to stir the contents of a lagoon and thus add oxygen so aerobic bacteria can help purify the wastes. See lagoon, aerobic, anaerobic.

mechanical analysis *n.* determination of the percentages of sand, silt, clay, and sometimes gravel in a soil sample by sieving and/or sedimentation; now called particle-size analysis. See hydrometer analysis, pipette analysis, sieve analysis.

mechanical control *n.* methods of pest control (especially weed control) that depend on tillage tools such as cultivators and hoes.

mechanical practice *n.* a tillage practice, barrier, dam, terrace, or other structure used to control soil erosion. See vegetative practice.

mechanoreceptor *n.* a sense organ that responds to such mechanical stimuli as vibration, tension, or pressure (e.g., an eardrum or a cat's whiskers).

medic or **medick** *n.* (Latin *medica*, Median grass) **1.** any of several leguminous plants (*Medicago* sp., family Fabaceae) with trifoliate leaves, especially purple medic (*M. sativa*). Medics are used as forage crops. **2.** (slang) short for a medical doctor, especially one in military service.

Mediterranean climate *n.* an environment where most of the precipitation occurs during the cooler part of the year, and the summers are hot and dry; the opposite of a monsoon climate.

Mediterranean flour moth *n.* a small gray moth (*Anagasta kuehniella*, family Pyralidae) whose larvae feed on stored wheat flour, wheat grain, buckwheat flour, cornmeal, corn grain, and cottonseed meal.

Mediterranean fruit fly or **medfly** *n.* a small black-and-white banded fly (*Ceratitis capitata*, family Tephritidae) that is a very serious pest on citrus and other succulent fruits and vegetables, mostly in warm regions. It lays eggs that hatch into maggots inside the fruit. A serious infestation in Florida in 1929 was successfully eradicated in 1930 by means of an embargo coupled with vapor-heat and cold-treatment processes.

medusahead *n.* an annual weed (*Taeniatherum caput-medusae*, family Poaceae) that threatens much of the rangeland of western U.S.A. It was introduced from Eurasia in the late 1800s and has since spread over more than 1 million ac (400,000 ha) of rangeland. Medusahead is an aggressive competitor with a high silica content in its foliage that gives it a very low palatability to livestock and a slow decomposition rate. It seldom grows more than 1 ft (30 cm) high, but its residues accumulate as a dense thatch that resists erosion and causes a high fire danger. Also called medusahead wildrye.

megatherm *n.* (Greek *megas*, large + *thermē*, heat) **1.** any plant that requires a warm climate, plenty of water, and a long growing season; e.g., palm trees and bananas. **2.** any organism that lives in a hot, moist environment.

megaton *n.* **1.** one million tons. **2.** the explosive power of 1 million tons of TNT, the unit used to evaluate the blast effect of a nuclear weapon.

meiosis *n.* (Greek *meiōsis*, making smaller) the process of cell division that produces gametes with a single set of chromosomes instead of the double (diploid) set present in most cells. See mitosis.

melaleuca *n.* (Greek *melanos*, black + *leukós*, white) any of 100 or more shrubs and small trees (*Melaleuca* sp., family Myrtaceae) from Australia and Malaysia with black trunks and white branches; now considered as weeds in the Florida Everglades because they spread rapidly and their flaky bark burns readily. One species produces oil of cajuput that is used in medicines. Also called honey myrtle, river tea tree, weeping tea tree, or bottlebrush.

melanic epipedon *n.* a thick, dark epipedon high in allophane that gives it a low bulk density and a high capacity to adsorb anions.

melanin *n.* (Greek *melanos*, black) any of a large number of dark brown or black pigments in eyes, skin, hair, feathers, and soil.

melanism *n.* (Greek *melanos*, black + *isma*, condition) **1.** dark color in the skin, hair, eyes, etc., of an individual or race. **2.** abnormally dark coloration in all or part of the body of a person, animal, or bird (the opposite of albinism).

melanization *n.* (Greek *melanos*, black + *izatio*, becoming) **1.** the darkening of skin or other animal tissue by accumulation of melanin granules. See melanoma. **2.** the darkening of a soil as melanins accumulate from the use of abundant compost, manure, straw, or other organic materials.

melanocyte *n.* (Greek *melanos*, black + *kytos*, cell) a cell that produces and/or contains melanin.

melilot *n.* (Greek *melilōtos*, clover) an Old World legume (*Melilotus* sp.) with 20 or more annual or biennial species of clover and related legumes with white or yellow flowers; commonly used as forage plants in hay or pasture. See sweet clover.

meliorate *v.* (Latin *meliorare*, to make better) to ameliorate, improve, and make better; e.g., to create a more favorable environment.

melissophobia *n.* (Greek *melissa*, bee + *phobia*, great fear) apiphobia, an abnormal fear of bees and/or bee stings.

melissotherapy *n.* (Greek *melissa*, bee + *therapeia*, healing) apiotherapy, the use of bee venom for treatment of illness.

melittology *n.* (Greek *melissa*, bee + *logos*, word) apiology, the scientific study of bees.

meloid *n.* a beetle of family Meloidae; a blister beetle or other closely related insect. See blister beetle.

meltwater *n.* water from melting snow and ice. Meltwater produces more intense runoff than rainfall in some areas.

membrane *n.* (Latin *membrana*, a thin skin) **1.** a thin, pliable layer, especially of plant or animal tissue, that covers an organ or lines a body cavity. **2.** the thin, flexible covering around a cell, just inside the more rigid outer covering if the cell has a cell wall, or a similar flexible covering around an internal part of a cell.

Mendel, Gregor Johann *n.* an Austrian monk and botanist (1822–1884) who published the principles of plant genetics in 1866 based on his study of peas. His work was lost for a time but was rediscovered in 1900 and forms the basis for classical genetics.

Mendel's law or **Mendelian law** *n.* a statement of scientific facts of inheritance expressed by Gregor Mendel in 1866 on the basis of segregation and dominance. For example, when Mendel crossed tall peas with short peas, the progeny were all tall because each plant had one gene from each parent

(Ts) and the gene for tallness (T) was dominant over the recessive gene for shortness (s). When these progeny were crossed with each other, the first generation (F1) segregated according to the ratio of one homozygous plant that was true-breeding for tallness (TT) to two plants that were heterozygous for tallness and shortness (Ts), and one homozygous plant that was true-breeding for shortness (ss). This fit with the statistical probability of offspring inheriting either T or s from each parent. Mendel found that this 1:2:1 ratio also applied to other characteristics that were controlled by a single gene and that each of these characteristics was inherited independently of the others.

meniscus *n.* (Greek *mēniskos*, crescent or little moon) **1.** a curved interface between a liquid and a gas enclosed in a tube or other container. The surface extends upward toward the gas at the edge and hangs down in the middle if the liquid wets the container wall or the reverse if it does not wet the wall. **2.** a lens that is convex on one side, concave on the other, and thicker in the middle than on the sides. **3.** any crescent shape or crescent-shaped body. **4.** a disk of cartilage that separates the bones in a joint.

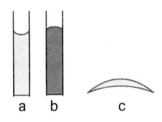

a. The meniscus of water
b. The meniscus of mercury
c. A meniscus lens

mensuration *n.* (Latin *mensuratio*, measurement) **1.** the branch of mathematics dealing with the determination of length, area, and volume. **2.** the process of measuring. For example, forest mensuration is the process of measuring the growth, development, and volume of individual trees and/or of the total stand of trees in a designated area and of the amount of lumber or other product that can be made from it.

mephitic *adj.* (Latin *mephiticus*, poisonous) **1.** foul smelling; having an offensive odor. **2.** poisonous, as a noxious gas.

mephitis *n.* (Latin *mephitis*, poison) **1.** a poisonous gas escaping from a vent in the ground. **2.** any strong offensive or poisonous odor.

mercaptan *n.* (Latin *mercurium captans*, seizing mercury) any of a large group of organic compounds with a strong offensive odor related to one or more sulfhydryl (-SH) attached to a carbon structure; similar in structure to alcohols but with SH groups instead of OH groups.

Mercator projection *n.* a cylindrical system of map drawing named for Gerhardus Mercator (1512–1594), a Flemish cartographer. The equator is represented as a straight line with longitudinal lines drawn perpendicular to it. Latitude lines are parallel to the equator and spaced increasingly far apart with increasing distance from the equator in an effort to maintain appropriate shapes for the areas shown. Such a projection accurately represents areas along the equator but progressively enlarges areas as their distance from the equator increases (e.g., Greenland looks larger than South America on a Mercator projection).

mercury (Hg) *n.* element 80, atomic weight 200.59; a heavy metal that is a silvery liquid in its elemental form and sometimes occurs naturally as small liquid globules but more commonly as cinnabar (HgS) or other combined forms. Like many other heavy metals, mercury is very toxic when inhaled, swallowed, or absorbed through the skin. It has been used in thermometers, electric switches, tooth fillings, medicines, etc., but its use is being reduced to avoid mercurialism (mercury toxicity). Also called quicksilver. See Minamata disease.

Mercury *n.* (Latin *Mercurius*, Mercury) **1.** the planet closest to the sun; the smallest planet in the Solar system (excluding Pluto). **2.** the Roman god of commerce, supposedly the messenger of the gods.

meridian *n.* (Latin *meridianus*, of noon) **1.** the highest point, as the position of the noonday sun, the middle period of one's life, or the apex of culture, power, prosperity, or splendor. **2.** a north–south line along the surface of the Earth joining the North and South Poles. See prime meridian. **3.** a similar line projected into the heavens.

meristem *n.* (Greek *meristos*, divisible) an area of plant cells that divide to form new cells, as the region near the tip of a root or a shoot (much of the growth in older parts of the plant is by cell enlargement rather than cell division).

meromictic *adj.* (Greek *meros*, part + Latin *mixtus*, mixing) mixing partly but not completely; descriptive of a lake whose upper part (the mixolimnion) circulates and mixes down to a certain depth (the chemocline boundary). The layer below the chemocline (the monimolimnion) does not mix because it is saline and therefore denser than the overlying freshwater. The monimolimnion water is generally anaerobic. See amictic lake.

meropidan *n.* (Greek *merops*, bee eater + *idēs*, family) a multicolored bird (*Merops* sp. and related genera, family Meropidae), native to the Old World, that eats bees.

mesa *n.* (Spanish *mesa*, table) high, nearly level tableland that drops off steeply on all sides; a small plateau, typically with a shallow layer of soil over bedrock. Mesas are common in arid regions such as southwestern U.S.A.

mescal *n.* (Spanish *mezcal*, agave) **1.** any of more than 300 species of agave (family Agavaceae), a desert plant with long, thick leaves growing outward like oversized blades of grass. Also called century plant. **2.** a distilled alcoholic liquor made from the cooked, fermented heads and hearts of the pulque agave. Called pulque if fermented but not distilled. **3.** a spineless cactus (*Lophophora williamsii* or *L. diffusa*, family Cactaceae), native to Texas and northern Mexico, that produces mescal buttons at the ends of its stems; the buttons are chewed for the narcotic effect of the mescaline they contain. Also called peyote.

mescal bean *n.* an evergreen shrub or small tree (*Sophora secundiflora*, family Fabaceae), native to southwestern U.S.A. and northern Mexico, with fragrant blue flowers and pods containing bright red seeds that are poisonous to cattle, sheep, and goats.

mescaline *n.* (Spanish *mezcal*, agave + *ine*, alkaloid) an alkaloid ($C_{11}H_{17}NO_3$) contained in mescal buttons. Chewing these buttons can give a hallucinogenic euphoria with delusions of colors and music and can cause serious permanent psychological disturbances. See mescal.

mesh *n.* (Anglo-Saxon *masc*, mesh) **1.** an opening in a screen or net, commonly described as the number of openings per inch; e.g., a 10-mesh screen has 10 openings per linear inch (4 openings per centimeter). **2.** a screen, net, or fabric with numerous openings, usually of uniform size and spacing. **3.** the cords, wires, or other material that form such a screen, net, or fabric. **4.** the particles that can pass through a mesh of a specified number of openings per inch (often specified as upper and lower size limits, as between 4 and 8 mesh). **5.** a network of lines or passageways. **6.** the way the teeth of mating gears meet each other. *v.* **7.** to enmesh; to catch, as with a net. **8.** to engage, as when the teeth of two gears meet so one drives the other. **9.** to match or agree, as their plans must mesh. *adj.* **10.** having openings like mesh, as mesh stockings or a mesh bag.

mesic *adj.* (Greek *mesos*, middle) **1.** having a moderate water requirement; e.g., mesic plants. **2.** supplying a moderate amount of water, as a mesic habitat. **3.** having moderate temperatures (between frigid and thermic). Mesic is used in Soil Taxonomy to name soil subgroups that have mean annual temperatures between 8° and 15°C (46.4° to 59°F).

mesocarp *n.* (Greek *mesos*, middle + *karpos*, fruit) the middle layer of a pericarp, as the fleshy part of a stone fruit.

mesoderm *n.* (Greek *mesos*, middle + *derma*, skin) the middle layer of an embryo; the part that becomes skeletal, muscular, vascular, and connective tissue. Also called mesoblast.

mesodont *adj.* (Greek *mesos*, middle + *odontos*, tooth) **1.** having medium-sized teeth. **2.** belonging to or resembling a certain class of extinct animals (*Mesodonta* sp.).

mesopause *n.* (Greek *mesos*, middle + *pausis*, a stop) a level in the atmosphere about 50 mi (85 km) above the surface of the Earth where the temperature is at a minimum (about –130°F or –90°C); the boundary between the mesosphere (below), where the temperature decreases with increasing height, and the thermosphere (above), where the temperature increases with increasing height. See atmosphere.

mesopelagic *adj.* (Greek *mesos*, middle + *pelagos*, the sea) pertaining to or living in seawater at a depth between 600 and 3000 ft (180 to 900 m).

mesophile *n.* (Greek *mesos*, middle + *philos*, loving) a parasite or pathogenic bacterium that grows at moderate temperatures, especially one that grows best at normal human body temperature, 37°C (98.6°F).

mesophyll *n.* (Greek *mesos*, middle + *phyllon*, leaf) internal leaf tissue (green parenchyma).

mesophyte *n.* (Greek *mesos*, middle + *phyton*, plant) a plant that grows in moist soil; less water-loving than a hydrophyte, which grows in water, but needing more water than a xerophyte, which grows in a dry habitat.

mesosphere *n.* (Greek *mesos*, middle + *sphaira*, sphere) the atmosphere above the stratosphere and below the thermosphere; between about 20 and 50 mi (32 to 85 km) above the Earth's surface, where temperatures decrease with height from about 50°F (10°C) to –130°F (–90°C). See atmosphere.

mesquite *n.* (Nahuatl *mizquitl*, mesquite) **1.** a spiny tree or shrub (*Prosopis juliflora* and related species, family Fabaceae) that grows in thickets in semiarid climates such as southwestern U.S.A. and bears flowers that attract bees and pods rich in sugar that are eaten by cattle. Seeds in the manure spread the mesquite. Also called algarroba. **2.** the wood of the tree, often used for cooking. **3.** any of several similar trees or bushes.

meta *n.* (Greek *meta*, along with) **1.** a column or post at either end of a Roman racetrack. *adj.* **2.** occupying two positions on a benzene ring that have one carbon between (e.g., attached to carbons 1 and 3). **3.** a combining form indicating an acid or salt that is the least hydrated of a series; e.g., metaphosphoric acid (HPO_3) as distinguished from ortho-phosphoric acid (H_3PO_4) and pyrophosphoric acid ($H_4P_2O_7$). **4.** a prefix meaning with, after, or beyond.

metabiosis *n.* (Greek *meta*, along with + *biosis*, way of life) a form of symbiosis of organisms wherein one organism depends on the other to make the environment suitable for it to live. Example: An alga provides a suitable environment for a fungus in lichens.

metabolic *adj.* (Greek *metabolos*, changeable + *ikos*, having to do with) **1.** pertaining to metabolism, as metabolic processes. **2.** pertaining to metamorphosis.

metabolism *n.* (Greek *metabolos*, changeable + *isma*, condition) the assembly of processes needed for life and growth of plants, animals, and people, including building up (anabolic) and tearing down (catabolic) processes that take place within all living cells.

metabolite *n.* (Greek *metabolos*, changeable + *ite*, product) any substance produced by metabolism; e.g., sugars are common metabolites.

metal *n.* (Latin *metallum*, metal) **1.** a chemical element that tends to form cations in salts or in solution. **2.** the pure form of such an element; typically an opaque solid that is ductile and able to conduct electricity. **3.** an alloy formed by fusing a mixture of such elements (and sometimes with some nonmetal additives). *adj.* made of metal.

metal fume fever *n.* an acute influenzalike occupational disorder caused by the inhalation of finely divided dust or fumes of metals such as magnesium or zinc or their oxides. Symptoms include fever and muscular pain.

metamorphic *adj.* (Greek *meta*, change + *morphos*, form + *ikos*, having to do with) **1.** causing or produced by metamorphism, as metamorphic activity or metamorphic rock. **2.** pertaining to a change in form of an organism, as metamorphosis.

metamorphic rock *n.* a rock that originally formed as either a sedimentary or igneous rock but later was subjected to enough heat and pressure to change it mineralogically without remelting; e.g., gneiss, schist, slate, marble, or quartzite.

metamorphism *n.* (Greek *meta*, change + *morphos*, form + *isma*, condition) a change in mineralogical, structural, and/or textural composition of rocks under heat, pressure, and/or chemical stress. For example, limestone changes into marble, granite into gneiss, shale into slate, etc. The metamorphic process takes place deep within the Earth, well below the zone of weathering, and the rocks are exposed later by erosion.

metamorphosis *n.* (Greek *meta*, change + *morphos*, form + *osis*, condition) **1.** a marked change in form; a transformation, especially that of an insect as it goes from one stage to another (e.g., egg to larva to pupa to adult). **2.** any considerable change in the form of an organism, a structure, or an organization.

metaphosphate *n.* (Greek *meta*, change + *phos*, light + *ate*, salt) a salt or ester of metaphosphoric acid. See polyphosphate.

metaphosphoric acid *n.* a dehydrated form (HPO_3) of phosphoric acid. Its molecules commonly link to form polyphosphoric acid.

metastasis *n.* (Greek *meta*, change + *stasis*, standing) **1.** the spreading of pathogens or cancer cells from one site in the body to another site or sites, often via the blood or lymphatic system. **2.** the condition resulting from such spreading of a disease. **3.** metabolism. **4.** an abrupt change of topic in a discussion. **5.** a change in the position or orbit of a subatomic particle; e.g., the excitation of an electron to a higher energy level.

metayer *n.* (Latin *medietas*, the mean) a tenant farmer who tills the land for half of the produce. The landlord receives the other half for having supplied the land and equipment.

meteor *n.* (Greek *meteōron*, a meteor) **1.** a body from space that enters the Earth's atmosphere and is heated to a luminescent temperature so it can be seen as a light streaking across the sky. Most meteors are small and are completely vaporized; those that reach the Earth's surface are called meteorites. Also called a shooting star. **2.** any person, thing, or process that moves or happens very rapidly (with the speed of a meteor). **3.** originally anything that fell from the sky or was seen in the air, including dust, hail, snow, etc.

meteoric water *n.* **1.** water that falls from the sky as rain, snow, hail, etc. Note: This includes water from meteors; most small meteors are thought to contain ice from disintegrated comets. **2.** groundwater that fell as precipitation and percolated through soil and/or cracks until it reached the water table.

meteorite *n.* (Greek *meteōron*, a meteor + *ite*, mineral solid) a mass that has fallen to Earth from space; a meteor that reaches the Earth's surface.

meteoroid *n.* (Greek *meteōron*, a meteor + *eidēs*, like) a small solid body traveling through space (e.g., fragments from a comet or an asteroid). Meteoroids become meteors when they enter the Earth's atmosphere.

meteorological satellite *n.* See weather satellite.

meteorology *n.* (Greek *meteōron*, a meteor + *logos*, word) the science that studies the atmosphere and the causes and effects of weather as a major component of the environment. Meteorology deals with weather at a specific time and place, whereas climatology deals with atmospheric conditions over large areas and extended periods of time.

meter or **metre (m)** *n.* (Greek *metron*, measure) **1.** the basic unit of length or distance in the metric system. It was originally defined as 1 ten-millionth of the distance along a meridian from the Earth's equator to a pole, redefined as the distance between two lines on a platinum-iridium bar stored in Paris (1889 to 1960) and then as 1,650,763.73 wavelengths of orange-red light from krypton 86 (1960 to 1983). Since 1983, it is the distance light travels in 1 sec in a vacuum divided by 299,762,458; equal to 39.37 in., 3.281 ft, or 1.094 yd. **2.** an instrument that measures, especially one that does so automatically, as a gas meter, water meter, or electric meter. **3.** the rhythm followed in music or verse. *v.* **4.** to measure. **5.** to control the flow of a fluid.

metes and bounds *n.* a system of defining the boundaries of a property or other piece of land by the distance and direction between specified markers (e.g., trees, large stones, or other identifiable objects). It was commonly used in Europe and in the U.S.A. before the rectangular system of land surveys was established in 1785. See land survey.

methane *n.* a colorless, odorless, very flammable, hydrocarbon gas (CH_4) produced naturally by the anaerobic decomposition of organic matter (e.g., in saturated soil and in digestive systems of animals). Methane is the chief component of natural gas used as a fuel and of explosive firedamp in mines. It is also a greenhouse gas that contributes to global warming. Known also as marsh gas.

methanogen *n.* any anaerobic microorganism that can use organic substances and/or carbon dioxide as carbon sources to produce methane gas (CH_4). They exist in swamp mud, in paddy rice fields, in cow stomachs, and in other anaerobic environments under plentiful organic matter, high moisture content, and anaerobic conditions. Six genera of methanogen bacteria have been identified: *Methanobacterium* sp., *Methanobrevibacter* sp., *Methanococcus* sp., *Methanogenium* sp., *Methanomicrobium* sp., and *Methanosarcina* sp.

methanol *n.* methyl alcohol, the simplest alcohol (CH_3OH) and a highly toxic, flammable, volatile liquid that is miscible with water and other alcohols. It is produced by destructive distillation of wood, partial oxidation of methane, or reaction of carbon monoxide with hydrogen and is used as a solvent, as a fuel, as an antifreeze, to denature ethyl alcohol, and in the synthesis of formaldehyde. Also called wood alcohol.

methanotroph *n.* any bacterium that can use methane (CH_4) as a sole carbon source, oxidizing some methane to carbon dioxide for energy generation and using some to produce organic chemicals. Methanotrophs can also use other 1-carbon molecules such as methanol (CH_3OH) as a carbon source.

methemoglobinemia *n.* **1.** a genetic disease that causes iron in hemoglobin to be oxidized from ferrous (Fe^{2+}) to ferric (Fe^{3+}), producing methemoglobin and reducing the oxygen-transporting capacity of the blood. The blood turns brown and the skin turns blue from lack of oxygen. **2.** a similar effect caused by excess nitrite bonding to hemoglobin in infants less than six months old, before their digestive tracts contain enough acid to kill the bacteria that can reduce nitrate (NO_3^-) to nitrite (NO_2^-). Note: Public health standards commonly limit nitrate nitrogen in drinking water to 10 ppm (10 mg N/L = 44 mg NO_3/L) or less to avoid methemoglobinemia in infants.

methyl alcohol *n.* See methanol.

methyl bromide *n.* a colorless, very toxic gas (CH_3Br) used as a fumigant to kill insects and as a soil sterilant that was banned for use beyond the year 2001 because bromine damages the ozone layer.

metolachlor *n.* a liquid herbicide ($C_{15}H_{22}ClNO_2$) that controls annual grasses and some broad-leaved weeds by inhibiting their germination in a number of field crops (corn, soybeans, peanuts, grain sorghum, cotton, safflower, sugar beets, sunflowers, etc.), vegetable crops (potatoes, pod crops, etc.), and woody ornamentals. Its half-life in soil averages about 90 days, but its vapor may degrade in the atmosphere within a few hours.

metric system *n.* (Greek *metron*, a measure) the standard system of measurement in scientific work and for all use in most of the world outside the U.S.A. Basic units include the meter (39.37 in.; 1 in. = 2.54 cm), the gram (0.0022046 lb; 453.6 g = 1 lb), and the liter (1.057 qt or 61.025 in.^{3}; 1 qt = 0.946 L). See SI, SI base units, SI-derived units, *Systém Internationale d'Unités*.

metric ton *n.* a unit of weight equal to 1000 kg (2204.62 lb). Also written tonne. See ton.

Mexican bean beetle *n.* a brown, spotted ladybird beetle (*Epilachna varivestis*, family Coccinellidae) whose larvae are a serious pest feeding on soybeans, kidney beans, and cowpeas.

Mexican blindness *n.* See onchocerciasis.

Mexican fruit fly *n.* a brightly colored fly (*Anastrepha ludens*, family Tephritidae) that is larger than a housefly and is the principal citrus-infesting fruit fly native to Mexico. Its larvae attack citrus, mangoes, plums, peaches, pears, etc. It destroys much fruit in Mexico and as far south as Panama and infests fruit in Texas during the winter but disappears by late spring.

Mexican jumping bean *n.* See jumping bean.

Mexican milkweed *n.* a perennial herb (*Asclepias mexicana*, family Asclepiadaceae) with a milky juice. Native to North America from Mexico to Washington state, its leaves are poisonous to sheep and poultry. Also called narrow-leaved milkweed.

Mexican onyx *n.* a form of translucent calcite ($CaCO_3$) that typically has brown and white bands and is often carved to make ornaments and decorative pieces. It resembles the quartz form of onyx but is much softer. See onyx.

Mexican vanilla *n.* a tropical climbing vine (*Vanilla planifolia*, family Orchidaceae) with long, fleshy, climbing vines, oblong leaves, and yellow flowers. Native to Mexico, its seed pods are gathered green, fermented, and dried to produce a volatile oil called vanillin, which is extracted with alcohol to make the vanilla flavoring used in foods.

mho *n.* a unit of electrical conductivity equal to the reciprocal of an ohm and to a siemen, the corresponding SI metric unit. Also called reciprocal ohm.

mica *n.* (Latin *mica*, morsel or particle) any of several 2:1 layer silicate minerals that can be divided into very thin, tough sheets that are generally shiny, flexible, good electrical and heat insulators, and often transparent or translucent; e.g., biotite $[K(Mg,Fe)_3(AlSi_3)O_{10}(OH)_2]$ (also called black mica) or muscovite $[KAl_2(AlSi_3)O_{10}(OH)_2]$ (also called white mica or isinglass).

micaceous *adj.* (Latin *mica*, morsel or particle + *aceus*, resembling or made of) **1.** containing a large component of mica, as micaceous rock or micaceous sediments. **2.** pertaining to mica.

micelle *n.* (Latin *mica*, morsel or particle + *elle*, small) a colloidal particle that carries many electrical charges; in soil, either a small clay or a humus particle.

microbe or **microorganism** *n.* (Greek *mikros*, small + *bios*, life) any microscopic living organism, formerly classified as either microphytes (algae, fungi, and bacteria; also called microflora) or microzoa (protozoa and tiny nematodes and insects; also called microfauna). Newer classifications based on RNA sequencing divide the old bacterial class plus certain algae (prokaryotes) into archaea and bacteria. All other microbes plus all macrobes are now classified as eukaryotes.

microbial pesticide *n.* a microbe that can be used to control insects or disease-producing bacteria, fungi, viruses, or protozoa.

microbicide *n.* (Greek *mikros*, small + *bios*, life + Latin *cida*, killer) a chemical or mixture of chemicals that kills microbes.

microbiologist *n.* (Greek *mikros*, small + French *biologie*, life study + *iste*, person) a career scientist who specializes in microbes, such as protozoa, algae, fungi, bacteria, viruses, and rickettsiae.

microbivorous *adj.* (Greek *mikros*, small + *bios*, life + Latin *vorus*, feeding) feeding on microbes; e.g., protozoa feed on bacteria.

microclimate *n.* (Greek *mikros*, small + *klima*, zone) the climatic conditions of a small area in the immediate vicinity of a plant or animal, or even that of an individual leaf or other part of an organism; e.g., a small frost pocket or the area where lichens grow on the north side of a tree. The climate of a somewhat larger area may be called a local climate.

microcline *n.* See potassium feldspar.

microcosm *n.* (Latin *microcosmus*, small world) **1.** a miniature world. See macrocosm. **2.** humans and/or the Earth taken as representing the universe in miniature. **3.** a community or environment as a model of the world or other large unit, especially one with a complex, diverse nature.

microelement *n.* See micronutrient.

microenvironment *n.* (Greek *mikros*, small + French *environnement*, environment) a microhabitat; the microclimate and other environmental conditions of a small area in the immediate vicinity of a plant or animal. It may be a small protected area or an area influenced by the decaying remains of an animal or tree, etc.

microevolution *n.* (Greek *mikros*, small + Latin *evolutio*, unrolling) mutations and other changes in genetic structure that result in minor hereditary changes in a species. Microevolution coupled with natural selection often leads to better adaptation to an environment, and such adaptations can have either positive or negative effects from a human viewpoint. For example, they can produce desirable species or they can produce insects and weeds that are resistant to pesticides and disease organisms that are resistant to antibiotics.

microfauna *n.* (Greek *mikros*, small + Latin *fauna*, animal life) microscopic animal life; e.g., protozoa or amoebas and microscopic insects and nematodes. See microbe.

microflora *n.* (Greek *mikros*, small + *flora*, plant life) microscopic plant life; e.g., microscopic forms of algae. Bacteria and fungi were also included in older systems of classification but are now considered to be separate kingdoms rather than part of the plant kingdom. See microbe.

microhabitat *n.* See microenvironment.

microirrigation *n.* (Greek *mikros*, small + Latin *irrigatus*, bringing water) irrigation that minimizes water requirement by placing the water precisely where it is needed, usually by using a drip irrigation system.

micrometeorology *n.* (Greek *mikros*, small + *meteoron*, a thing in the air + *logos*, word) the climate of a small area, usually that of a thin stratum of air at or near ground level.

micron (**μ**) or **micrometer** (**μm**) *n.* (Greek *mikros*, small, minute) a unit of length equal to 1 millionth of a meter (10^{-6} m; there are 25,400 microns in 1 in.).

micronutrient *n.* (Greek *mikros*, small + Latin *nutriens*, nourishing) a vitamin or mineral nutrient that is required in very small amounts by plants, animals, and humans. Micronutrients required by plants are boron, chlorine, copper, iron, manganese, molybdenum, nickel, and zinc. Nutrients required by animals and humans include chromium, cobalt, fluorine, iodine, selenium, silicon, and sodium and all those required by plants except boron (sodium and chlorine are macronutrients for animals). Sometimes called minor element or trace element, but micronutrient is the preferred term. See essential element, macronutrient, cobalt.

microorganism *n.* See microbe.

microrelief *n.* (Greek *mikros*, small + French *relever*, to lift up again) small irregularities in a land surface; either small depressions (that may hold water after a rain) or small rises (e.g., an anthill or a molehill).

microsatellite *n.* (Greek *mikros*, small + Latin *satellitis*, attendant) a repeat segment of DNA that is nonfunctional and highly variable, constituting a reservoir of genetic variability and useful for tracking past genetic changes.

microseism *n.* (Greek *mikros*, small + *seismos*, shaking) a weak shaking of the Earth's crust that produces a Richter scale reading less than 1; probably caused by a minor earthquake or possibly by a large storm at sea.

microsite *n.* (Greek *mikros*, small + Latin *situs*, position) a small area with distinctly different conditions than the surrounding area; e.g., the shaded area beneath a tree in a pasture.

microsymbiont *n.* (Greek *mikros*, small + *symbiountos*, living together) a microbial member of a symbiotic relationship; e.g., *Rhizobium* bacteria when associated with a legume.

microtherm *n.* (Greek *mikros*, small + *thermē*, heat) a plant that can survive and grow with only a small amount of heat (low temperatures and/or short summers); e.g., willows and lichens.

microwave *n.* (Greek *mikros*, small + Anglo-Saxon *wafian*, to fluctuate) **1.** a high-frequency form of electromagnetic wave energy with a wavelength between 1 mm and 30 cm. **2.** an oven that uses microwaves to generate heat in food or other materials placed inside it. *v.* **3.** to cook or heat something in a microwave oven.

midden *n.* (Danish *mødding*, dung heap) **1.** refuse such as a pile of kitchen garbage or a pile of manure. **2.** a mound of earthworm casts or scat (excrement) from wild animals or insects. **3.** a kitchen midden.

midge *n.* (Anglo-Saxon *mycg*, midge) **1.** a small, very annoying, gnatlike insect (especially one of family Chironomidae) that often flies in swarms near wetlands. **2.** a tiny person.

midget *n.* (Anglo-Saxon *mycg*, midge + *et*, small) **1.** a tiny person with normal physical proportions (a small, deformed person is a dwarf). **2.** anything that is much smaller than usual for its kind. *adj.* **3.** much smaller than usual. **4.** miniaturized, as a small replica or model.

midnight sun *n.* the sun as seen a little above the horizon at midnight of days near the summer solstice in either the Arctic or Antarctic region.

migrant *n.* or *adj.* (Latin *migrans*, moving) a plant, insect, animal, or person that changes location frequently, often in response to changes in environmental factors; e.g., tumbleweeds, monarch butterflies, lemmings, and gypsies.

migration *n.* (Latin *migratio*, mover) **1.** movement of a large group of people or animals to a new location. **2.** movement of a mass of atoms or ions; e.g., by diffusion.

mildew *n.* (Anglo-Saxon *mildeáw*, honeydew) **1.** a fungal growth that produces a white or gray covering or discoloration as it grows on living or dead organic materials, especially in a damp environment. **2.** any plant disease resulting from the growth of such a fungus; e.g., powdery mildew. *v.* **3.** to affect or be affected with mildew.

mile *n.* (Latin *milia*, 1000 paces) **1.** a land distance equal to 5280 ft (1609.35 m), 80 Gunter's chains, or

320 rods. Also called a statute mile. **2.** a nautical mile, 6080.2 ft (1853.25 m) in former U.S. usage (also called a sea mile or geographic mile) or the slightly shorter international nautical mile, 1852 m, used both at sea and in the air. **3.** any of several other units of distance used in various countries during different historical periods. **4.** any considerable distance, as he missed by a mile.

milk *n.* (Anglo-Saxon *meolc*, milk) **1.** the white, bluish white, or light-yellowish liquid secreted by the mammary glands of female mammals to nourish their young. **2.** this liquid drawn from animals (especially cows, sometimes goats, or occasionally other animals) and used as a beverage or to make milk products such as butter, cheese, yogurt, or ice cream. **3.** any similar liquid from some other source; e.g., coconut milk. *v.* **4.** to draw or extract milk from a female mammal or other source. **5.** to exploit or draw out some benefit or information, often excessively or illicitly, as to milk a benefactor or a source.

milk cow *n.* a cow that is bred and kept primarily for milk production rather than as a meat animal.

milk fever *n.* **1.** a mild fever resulting from an infection and associated with the beginning of lactation following childbirth. **2.** a fever caused by consumption of contaminated milk. **3.** an acute disorder affecting a dairy cow after calving, causing low blood calcium and paralysis (also called parturient paralysis or parturient hypocalcemia) or paralysis accompanied by below normal body temperature (also called cerebral anemia).

milk intolerance *n.* See lactose intolerance.

milk irradiator *n.* a lamp that produces ultraviolet light for converting ergosterol ($C_{28}H_{43}OH$) into vitamin D_2 (ergocalciferol) and is used to prevent or cure rickets.

milk sickness *n.* **1.** a toxic condition that causes inactivity, leg trembling, constipation, rapid breathing, nausea, and ultimately collapse after cattle eat white snakeroot (*Eupatorium rugosum*, family Asteracea), a tall, white-flowered, perennial herb, or jimmyweed (*Aplopappus heterophyllus*, family Asteraceae). **2.** digestive problems and severe tremors in humans that consume milk, milk products, or meat from cattle affected by milk sickness.

milk vetch *n.* any of several plants (*Astragalus* sp., especially *A. glycyphyllos*, family Fabaceae) that are believed to increase milk production in goats.

milkweed *n.* **1.** any of several perennial plants (*Asclepius syriaca* and related species, family Asclepiadaceae) that produce a milky juice when cut or squeezed. Their stems are thick and 2 to 5 ft (0.6 to 1.5 m) tall with large leaves 4 to 8 in. (10 to 20 cm) long and clusters of small, bell-like flowers.

Some species (e.g., *A. eriocarpa* and *A. tuberosa*) are planted for their showy flowers, ornamental seed pods, and attraction for monarch butterflies. They can cause milkweed poisoning in most animals that eat their foliage. Also called silkweed. **2.** any of several other plants that produce a milky juice; e.g., certain spurges.

milky disease *n.* a disease that kills Japanese beetle larvae and white grubs; it causes their blood to turn milky white as bacterial spores proliferate (*Bacillus popilliae* and *B. lentimorbus*). Milky disease spores can be mixed with talc and used as a pesticide by dusting it on areas where the grubs are working in the soil. The bacteria become established in the soil and spread to adjoining areas, thus controlling the Japanese beetle.

Milky Way *n.* the spiral galaxy that contains our solar system. Its central disk shows as a milky band across the night sky.

mill *n.* (Latin *molina*, a mill) **1.** a factory where a product is manufactured or processed, such as a sawmill, paper mill, steel mill, or textile mill. **2.** a building where grain is ground into flour. **3.** a machine that grinds material into small grains, as a coffee mill, or one that squeezes out juice, as a cider mill. **4.** a machine that shapes material as it passes through rollers or past cutters. **5.** an institution or organization that does things routinely without regard to quality, as a diploma mill or divorce mill. **6.** a group that often spreads information that may not be true, as a rumor mill. **7.** one-thousandth, as 0.001 dollar (a 10th of a cent), often used as a unit for tax rates. *v.* **8.** to do the grinding, cutting, shaping, or squeezing work of a mill. **9.** to move slowly and aimlessly in a small area, as a crowd or herd mills around.

millet *n.* (Latin *milium*, millet) any of several small-grained cereals (family Poaceae) used either as human food or animal fodder, especially in areas with short growing seasons that are cut off by either a hot, dry summer or by cold weather, including finger millet (*Eleusine coracana*), foxtail millet (*Setaria italica*), Japanese millet (*Echinochloa frumentacea*), kodo millet (*Paspalum scrobiculatum*), pearl millet (*Pennisetum typhoideum*, also called cattail millet), and proso millet (*Panicum miliaceum*).

millimho (mmho) *n.* a unit of electrical conductivity commonly used to evaluate the salinity of soil and equal to 0.001 mho, 0.001 siemen, or the reciprocal of 1000 ohm. See mho, saline soil.

millipede *n.* (Latin *millipeda*, thousand feet) any of various arthropods (class Diplopoda) with elongated bodies composed of a head with chewing mouthparts and a pair of antennae, followed by many segments with a pair of legs on each segment and a hard

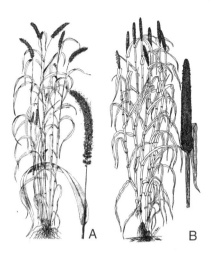

Foxtail millet (left) and pearl millet (right)

covering over each pair of segments. Most live in warm climates and damp environments such as soil or rotting wood and eat dead leaves or decaying organic materials, but a few species damage living plants by feeding on their roots.

millstone *n.* (Latin *molina*, a mill + Anglo-Saxon *stān*, stone) **1.** a large round stone used in mated pairs for grinding grain into flour. Water-powered mills using such stones were common in Colonial times and are still used in some places. Also called buhrstone. **2.** anything that pulverizes material. **3.** a heavy burden.

milo *n.* a type of grain sorghum (*Sorghum vulgare* var. *subglabrescens*, family Poaceae), native to tropical Africa, with yellowish seeds. It is grown in the southern Great Plains of the U.S.A. and in other parts of the world with similar climate, mostly for livestock feed. Also called milo maize.

Milorganite *n.* granular, dried sewage sludge bagged and sold for use as an organic fertilizer; named after its source, Milwaukee, Wisconsin.

milpa *n.* (Nahuatl *milpan*, milpa) **1.** a form of slash-and-burn agriculture used in the forests of Mexico and Central America. An area is cleared, cropped for a few seasons (especially to corn), and then abandoned to forest regrowth while other areas are cropped. **2.** a corn (maize) plant.

mimesis *n.* (Greek *mimēsis*, imitation) **1.** mimicking the words and/or actions of another person. **2.** imitation of one organism by another. **3.** simulation of the symptoms of a disease by another.

Minamata disease *n.* mercury poisoning that causes neurological disorders, tremors, numbness, anemia, birth deformities, etc. It is named for Minamata, Japan, where seafood contaminated with mercury

from wastes discharged by a plastic factory killed 43 people and damaged the health of thousands in the 1950s. Note: The limit for mercury in fish for human consumption has been set by the EPA at 1 ppm Hg.

mine *n.* (French *mine*, mine) **1.** a site where material (coal, iron ore, diamonds, etc.) is removed from a natural deposit. It is a surface mine if the material is obtained by stripping off any overburden and working in the open air. It is a shaft mine if one or more shafts and/or tunnels are used to penetrate beneath the surface. **2.** a source; e.g., the library is a mine of information. **3.** an explosive device that is typically buried in the soil to kill or injure an enemy or destroy an enemy vehicle. **4.** an explosive device that is anchored underwater to destroy a passing ship. *v.* **5.** to excavate for the purpose of mining. **6.** to seek as much information as possible from a source. **7.** to bury explosive devices in the soil or place them underwater. **8.** to use resources with little regard for future needs, as to mine the soil or the forest of an area or nation. *pronoun* **9.** my property; a possessive pronoun meaning this belongs to me.

minefield *n.* a dangerous area where explosive mines have been placed in either land or water.

mineral *n.* (Latin *minerale*, mineral) **1.** a homogeneous solid substance that forms through natural physical and chemical (but not biochemical) processes as a component of a rock mass. Each mineral has a consistent and distinct set of physical and chemical properties and crystal structure. The chemical composition of most minerals can vary somewhat from one location to another by isomorphic substitution but must fall within certain natural or arbitrary limits. See isomorphic substitution. **2.** ore; material obtained by mining. **3.** a nonliving substance, neither animal nor plant in origin. **4.** an element that is required for good nutrition; e.g., calcium, iron, magnesium, potassium, zinc, copper, or manganese (often supplied along with vitamins as a dietary supplement). *adj.* **5.** composed entirely or largely of minerals. **6.** inorganic; having the nature of minerals rather than of living forms, as mineral matter. **7.** containing a higher than usual amount of minerals, as mineral water.

mineral deposit *n.* a rock layer or zone that contains a high concentration of a certain mineral or minerals, especially those that are of value for mining.

mineralization *n.* (Latin *minerale*, mineral + *izatio*, making) **1.** the decomposition of organic materials into simple inorganic ions or molecules. **2.** fossilization, the process whereby mineral materials such as calcium carbonate or silicon dioxide replace organic constituents in the dead body of an organism (e.g., producing petrified wood). See fossilization.

mineralogy *n.* (Latin *minerale*, mineral + *logia*, description) the scientific study of the formation, occurrence, properties, and classification of minerals.

mineral resource *n.* a mineral deposit suitable for mining.

mine tailings *n.* waste materials from hard-rock mining, commonly left in piles with ponds intermingled. Acid conditions and heavy-metal toxicities are common in both the spoil piles and the ponds. Stabilization by vegetation is essential but very difficult because of the hostile environment. Smoothing the surface and adding lime to correct acidity, organic matter to loosen the material, and fertilizer usually help to enhance plant growth. Applications of sewage sludge, sewage effluent, and/or manure have proven helpful on such materials.

minimum tillage *n.* **1.** the use of tillage only when it is essential for growing the crop or crops designated for the land in question. A drill or planter may be the only tillage tool needed when chemicals are used to control weeds and other pests. Crop residues are left on the soil surface where they protect against erosion, and the soil structure and living organisms (worms, etc.) are disturbed as little as possible. **2.** (loosely) reduced tillage that avoids the use of a plow. See conservation tillage, conventional tillage, reduced tillage, no-till.

mining *n.* (French *mine*, a mine + Anglo-Saxon *ing*, doing) **1.** the process of removing material (coal, iron ore, diamonds, etc.) from an area where there is a natural deposit. **2.** the industry that extracts material from natural deposits. **3.** the process of placing explosive mines in soil or water.

mink *n.* (Swedish *menk*, mink) **1.** a semiaquatic mammal (*Mustela* sp.) that resembles a large weasel; native to cool climates in the Northern Hemisphere. **2.** the thick, lustrous brown fur of this animal, especially when used to make a fur coat.

mint *n.* (Latin *mentha*, mint) **1.** an aromatic herb (*Mentha* sp., family Lamiaceae) with characteristic square stems, simple leaves, and clusters of 2-lipped blossoms; e.g., basil, catnip, coleus, lavender, marjoram, oregano, peppermint, rosemary, sage, spearmint, and thyme. It is commonly used for flavoring food, chewing gum, candy, or medicine. **2.** a candy with a mint flavor. **3.** a place where money is made by government authority. **4.** a factory that produces ornaments or other products by printing or stamping. **5.** a very large amount, as a mint of money. *adj.* **6.** having a mint flavor, as a mint candy. **7.** in perfect condition, just as it came from the mint. *v.* **8.** to make coins at a mint. **9.** to invent or fabricate, as to mint new words.

mirage *n.* (Latin *mirari*, to wonder at) a false image; an illusion caused by atmospheric conditions that produce a reflection or refraction so that something appears to be where it is not; e.g., an oasis in the desert or a pool of water on a hot road (commonly a reflection of the sky).

mire *n.* (Icelandic *myrr*, a bog) **1.** muddy land; a small area that is marshy or swampy. See wetlands. *v.* **2.** to sink into the mud. **3.** to become stuck and unable to move.

miscible *adj.* (Latin *miscere*, to mix + *ibilis*, able) able to mix or be mixed, as alcohol and water are miscible but oil and water are not.

mist *n.* (Anglo-Saxon *mist*, mist or darkness) **1.** a cloudy condition that is less dense than fog (fog has smaller droplets but more of them). **2.** a hazy condition caused by smoke or dust. **3.** a spray of fine droplets (typically between 40 and 500 µm in diameter; droplets less than 40 µm are fog). **4.** a coating of fine droplets deposited on a surface by either a spray or condensation. **5.** a coating that blurs the vision, as a mist of tears. *v.* **6.** to spray a mist into the air or onto a surface.

mistletoe *n.* (Anglo-Saxon *misteltan*, mistletoe twig) **1.** any of about 20 genera of plants (family Loranthaceae) that are true seed-bearing but parasitic on other plants. **2.** Old World mistletoe (*Viscum album*) with leathery evergreen leaves, tiny yellow blossoms, and white berries that are toxic if eaten; parasitic on apple trees, hawthorns, poplars, willows, etc. **3.** Christmas mistletoe (*Phoradendron flavescens*), a woody plant that deforms the branches of various deciduous trees that it parasitizes in eastern U.S.A. Its branches, with shiny green leaves and preferably bearing red berries (even though the berries are toxic if eaten), are used as a Christmas decoration (traditionally, anyone may kiss a person who stands under the mistletoe). **4.** eastern mistletoe (*Phoradendron serotinum* and related species), a leafy plant with toxic white berries that grows on many different hardwoods in eastern U.S.A. **5.** dwarf mistletoe (*Arceuthobium* sp.), a plant with no leaves that is parasitic on conifers.

mistletoe cactus *n.* a leafless, epiphytic, tropical American cactus (*Rhipsalis baccifera*).

mistral *n.* (French *mistral*, master wind) a cold, dry wind that blows from the Mediterranean Sea northward into France and nearby parts of Europe.

mite *n.* (Anglo-Saxon *mite*, mite) **1.** any of a multitude of tiny arachnids (family Acarae, subclass Acarina). Some live on stored foods or decaying organic matter, some are serious parasites on leaves of almost all plants, and some are pests on livestock and humans. Plant mites commonly avoid chemical pesticides (miticides) because they live on the underside of leaves and suck plant juices. Dust mites are no larger than a printed period, live on human skin flakes, cause 50% to 80% of the asthma in people, and are resistant to all common pesticides. See miticide. **2.** a small coin of very little value; a very small sum of money, especially when that is all a person can afford; e.g., the widow's mite. **3.** a very tiny object or creature.

mithun *n.* a species of noncattle bovine animals of tropical Asia.

miticide *n.* (Anglo-Saxon *mite*, mite + Latin *cida*, killer) a pesticide for killing mites and ticks. Also called acaracide.

mitigate *n.* (Latin *mitigare*, to make mild) to alleviate or moderate a stressful condition, making it less severe.

mitigation *n.* (Latin *mitigatio*, calming) **1.** action to reduce negative effects; e.g., mitigation of damage caused by pollutants. **2.** reduction of punishment, as mitigation of a sentence. **3.** appeasing, soothing, or quieting, as mitigation of anger.

mitochondrion (pl. **mitochondria**) *n.* (Greek *mitos*, thread + *chondrion*, small grain) a subcellular organelle where respiratory processes produce energy in a plant or animal cell. Mitochondria resemble bacteria in size, shape, content of DNA, etc.

mitomycin *n.* (Latin *mitos*, a thread + *mycin*, antibiotic) any of three highly toxic, chemically related antibiotics ($C_{15}H_{18}N_4O_5$) produced by a soil bacterium (*Streptomyces caespitosus*) that inhibit DNA synthesis and are used to treat tumors or bacterial infections when other treatments have failed.

mitosis *n.* (Latin *mitos*, a thread + *osis*, action) normal cell division whereby each new cell receives a full diploid set of chromosomes. See meiosis.

Mitscherlich, E. A. *n.* a German chemist who published a mathematical expression in 1909 for growth that follows a law of diminishing returns; that is, the growth response per additional unit of a limiting element diminishes as the supply increases. Mitscherlich expressed this as

$$dy/dx = (A - y)C$$

where *dy/dx* is the increase in yield per additional unit of input *x*; *A* is the maximum possible yield; *y* is the yield at the actual input level; and *C* is a proportionality constant. See baule unit.

mixed cropping *n.* growing more than one crop, either as a crop rotation (and often simultaneously in different fields) or as intercropping.

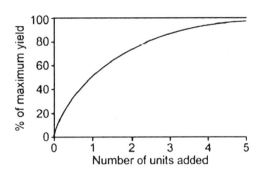

A curve illustrating the Mitscherlich equation

mixed fertilizer *n.* a fertilizer material prepared by physically mixing two or more single-compound fertilizers, either a bulk blend fertilizer wherein the components are simply mixed or a manufactured fertilizer that has the components fused into granules.

mixed intercropping *n.* See intercropping.

mixohaline *adj.* containing between 0.5 and 30 parts of ocean salts per thousand parts of water (0.5‰ to 30‰). See brackish water, mixosaline.

mixolimnion *n.* (Latin *miscere*, to mix + Greek *limnē*, pool) the part that mixes in a lake whose upper layer mixes but the lower layer (monimolimnion) is stagnant saline water. The interface between the layers is called a chemocline. See meromictic lake, epilimnion.

mixosaline *adj.* containing between 0.5 and 30 parts of land-derived salts per thousand parts of water (0.5‰ to 30‰). See mixohaline, salinity.

mobilization *n.* (Latin *mobilis*, movable + *izatio*, making) **1.** the process of making something mobile; e.g., the mobilization of plant nutrients as organic matter decays. **2.** the activation of a military unit.

moderate breeze *n.* a wind velocity of 13 to 18 mph (5.8 to 8 m/sec). See Beaufort wind scale.

moderate gale *n.* a wind velocity of 32 to 38 mph (14 to 17 m/sec). See Beaufort wind scale.

Mohave Desert *n.* See Mojave Desert.

Mohorovičić discontinuity (Moho) *n.* the contact between the Earth's crust and the underlying mantle, as marked by an abrupt increase in the velocity of earthquake waves. It occurs at a depth ranging from about 3 mi (5 km) under much of the ocean to about 40 mi (65 km) under some mountain ranges. It is named for Andrija Mohorovičić (1857–1936), a Croatian geologist who described it. Also called M discontinuity.

Mohs' scale *n.* an empirical scale of mineral hardness based on the ability of a harder mineral to scratch a softer mineral; a sequence of 10 minerals proposed by Friedrich Mohs (1773–1839), a German mineralogist who numbered them from 1 (softest) to 10 (hardest): 1 = talc, 2 = gypsum, 3 = calcite, 4 = fluorite, 5 = apatite, 6 = orthoclase, 7 = quartz, 8 = topaz, 9 = corundum, and 10 = diamond.

moist *adj.* (French *moiste*, moist) **1.** damp; slightly or moderately wet; containing some water. **2.** having high humidity, as moist air.

moisture *n.* (French *moisteur*, moisture) **1.** water, usually in a distributed form, as soil moisture, moisture in the air, or droplets of condensed moisture. **2.** a small amount of water or other liquid; enough to moisten a material or surface.

moisture equivalent *n.* the moisture content of a soil sample 1 cm thick that has been saturated with water and then centrifuged at 1000 times gravity for 30 min. This procedure was devised to approximate field capacity but is no longer much used.

moisture index *n.* a means of evaluating climate that was developed in 1955 by C. W. Thornthwaite (1899–1963) and is based on the difference between precipitation and potential evapotranspiration. Humid climates have high moisture index values, whereas arid climates have negative values. See P/E index.

moisture meter *n.* **1.** a device that measures the electrical resistance of a gypsum block buried in soil to indicate the relative moisture content of the soil, usually scaled from 0% to 100% of the available water-holding capacity of the soil. It was designed by Dr. George J. Bouyoucos, Michigan State University. See Bouyoucos block. **2.** a neutron probe that can be lowered into an aluminum access tube to measure the water content of a soil by radiating neutrons into the soil and counting how many neutrons are deflected back to its detector in a given time (the hydrogen ions in water deflect the course of electrons). **3.** a device similar to either of the preceding used to measure the moisture content of some other material (grain, wood, coal, etc.).

moisture regime *n.* the pattern of water availability throughout the year (especially during the growing season). For example, an aquic soil moisture regime means the soil is so wet that it has reducing conditions, an aridic soil moisture regime means the soil is dry most of the time, a udic soil moisture regime means the soil is usually moist but not saturated, a ustic soil moisture regime is between aridic and udic, and a xeric soil moisture regime has cool, moist winters and warm, dry summers.

moisture tension *n.* the force that must be overcome to remove water from soil or other media; now expressed in megapascals. Most older data were expressed in atmospheres or bars (1 megapascal =

9.9 atm or 10 bars). Also called suction or moisture potential.

moit *n.* (Anglo-Saxon *mot*, speck) a bur, seed, or other foreign particle found in wool.

Mojave Desert *n.* an arid area of about 15,000 mi^2 (39,000 km^2) in the Great Basin area in southeastern California. Also spelled Mohave Desert.

mol *n.* mole.

mold or **mould** *n.* (French *molle*, mold) **1.** mellow soil that is high in organic matter. **2.** a fuzzy surface covering, usually off-white or light gray, composed mostly of fungal mycelium, a common occurrence when organic materials are stored in a moist environment. **3.** a fungus that produces such a growth. **4.** a hollow form used to hold a liquid until it solidifies to form a desired product; e.g., a frozen dessert or a plaster figure. **5.** a distinctive shape or form, often one used as a pattern or model. **6.** the form produced by casting something in a mold or by pressing something against a surface with a distinctive shape. *v.* **7.** to become covered with mold. **8.** to shape something, as to mold clay into a figure.

moldboard *n.* (French *molle*, mold + *bord*, side of a ship) **1.** a curved surface on a moldboard plow; the part that causes the soil to curl and turn upside down as it rises from a new furrow and falls into the previous furrow. **2.** a board used to hold poured concrete and shape its surface.

moldboard plow *n.* a tillage implement that uses a share to cut a furrow in the soil and a moldboard to invert the soil as it falls into the previous furrow. Most moldboard plows have several such units (bottoms) that turn their furrows simultaneously.

Reversible 3-bottom moldboard plow

mole *n.* (German *mol*, mound maker) **1.** one gram-molecular weight of a compound or element (abbr. mol). **2.** any of several small mammals (family Talpidae) that burrow through surface soil and eat insects such as grubworms and earthworms. Their burrows form small ridges that roughen the surface of a lawn. **3.** a torpedo-shaped object that is pulled through wet soil to create a drainage channel. Depending on the soil, such channels may last for several years. **4.** a large machine used to bore tunnels through soil or rock. **5.** a spy within the ranks of an organization being spied upon. **6.** a small, dark blemish on the skin of a person or animal. Also called a nevus.

molecular biology *n.* the biological science that studies the molecular structure of DNA, RNA, and other large molecules involved in inheritance and cell functions. See genome.

molecule *n.* (Latin *moles*, mass + *cula*, small) the smallest group of atoms that can exist naturally as a particle of an element or compound. The atoms in the molecule may be identical, as in N_2 or H_2, or different, as in H_2O and CO_2. Molecules range from small groups of two or three atoms as in these examples to large, complex structures containing many thousands of atoms.

mollic epipedon *n.* a thick, dark surface horizon that characterizes Mollisols in Soil Taxonomy. The soil must contain at least 0.6% organic carbon and have a base saturation of at least 50% for a thickness of at least 10 cm (4 in.) over bedrock or one-third of the solum up to 25 cm (10 in.) in a deep soil.

A mollisol soil profile

Mollisol *n.* (Latin *mollis*, soft + *solum*, soil) a soil order in Soil Taxonomy that includes any soil with a mollic epipedon. Mollisols typically develop under grass vegetation in a semiarid to subhumid climate and are usually quite fertile. They occupy 22% of the U.S. and 9% of the world soil area.

mollusc *n.* See mollusk.

molluscicide *n.* (Latin *molluscus*, soft + *cida*, killer) a pesticide used to control mollusks, especially slugs and snails.

mollusk *n.* (Latin *molluscus*, soft) any invertebrate of phylum Mollusca, including snails, slugs, oysters, clams, mussels, squids, octopi, etc. Also written mollusc.

molt or **moult** *v.* (Latin *mutare*, to change) to shed hair, outer skin, horns, or feathers and replace with new growth, as in birds and reptiles.

molybdenosis *n.* (Latin *molybdaena*, galena + Greek *osis*, condition) poisoning caused by eating forage or other food high in molybdenum (generally more than 10 ppm Mo) or drinking water containing excess molybdenum; especially hazardous for cattle and other ruminants. Excess molybdenum in an animal's diet decreases the absorption of copper, causing diarrhea and lighter hair, especially around the eyes (see teart). Natural areas with soils high in molybdenum have been identified in several western U.S. states (California, Colorado, Idaho, Montana, Nevada, and Washington) and in Florida. Sites where industrial wastes have caused molybdenosis have been identified in Missouri, North Dakota, Pennsylvania, and Texas.

molybdenum (Mo) *n.* (Latin *molybdaena*, galena) element 42, atomic weight 95.94; a transition metal used in alloys to harden steel. It is a micronutrient for plants and animals, needed for the reduction of nitrate nitrogen to the amine form. Unlike other micronutrients, molybdenum is more readily available in neutral to alkaline soils than in acid soils. Legume seeds are often treated with molybdenum because the *Rhizobium* bacteria that fix atmospheric nitrogen (N_2) require adequate molybdenum.

monad *n.* (Latin *monas*, unit) **1.** any simple single-celled organism, especially certain amoebas (*Monad* sp.) with three flagella. **2.** an ion with a single unit of charge, or an element that forms such an ion. Similar ions with two or three units of charge are called dyads and triads, respectively. **3.** a single entity, especially one that is considered to be an indestructible microcosm of the universe according to the metaphysics of Leibniz.

monadnock *n.* (Algonquian *monadnock*, isolated mountain) a remnant hill or mountain of resistant rock rising above an eroded plain, composed of rock that has resisted erosion. It is named after Mount Monadnock in southwestern New Hampshire.

monarch butterfly *n.* a large, orange, migratory North American butterfly (*Danaus plexippus*, family Danaidae) with black-and-white markings that ranges north as far as Hudson Bay in the summer and south to California, Texas, and Florida in winter. Birds avoid eating it because of its foul taste and odor. Its larvae feed on milkweed.

monensin *n.* (Latin *cinnamonensis*, the microbe that produces monensin) a broad-spectrum antibiotic ($C_{36}H_{62}O_{11}$) obtained from a soil actinomycete (*Streptomyces cinnamonensis*). It is often used as a feed additive for beef cattle in a feedlot.

Monera *n.* (Greek *monērēs*, single) a taxonomic kingdom for prokaryotic organisms that were formerly classified as part of the plant kingdom, mostly bacteria and cyanobacteria (formerly called blue-green algae). These organisms have relatively simple cell structures and reproduce by asexual budding.

moneywort *n.* (Latin *moneta*, mint + Anglo-Saxon *wyrt*, plant) a creeping evergreen plant (*Lysimachia* *nummularia*, family Primulaceae) with rounded leaves and individual yellow flowers. It reproduces by stolons, is often grown in rock gardens, and is sometimes used as an herb for healing wounds. Also called creeping Charlie or creeping Jenny.

mongoose *n.* (Marathi *mangūs*, mongoose) a slender carnivorous mammal (*Herpestes edwardsi* and related species, family Viveridae) about 20 in. (0.5 m) long, that eats rodents, birds, eggs, and snakes. Noted for killing cobras and other venomous snakes, its quick movements and thick fur prevents the snake venom from penetrating its skin. It is a native of Africa, Asia, and southern Europe.

mongrel *n.* or *adj.* (Anglo-Saxon *mengan*, to mix) **1.** a dog of unidentified, mixed breeds. **2.** any animal or plant produced by crossbreeding different breeds or varieties. **3.** any incongruous cross of different things.

monimolimnion *n.* (Latin *minimus*, least + Greek *limnē*, pool) the stagnant lower layer of a permanently stratified lake whose upper layer (mixolimnion) mixes, but whose lower layer does not mix and is generally anaerobic. See meromictic.

monitor *n.* (Latin *monitor*, one who prompts or warns) **1.** a person who observes and supervises, e.g., while students take a test. **2.** one who admonishes persons whose conduct is not proper. **3.** something that gives warning when a problem occurs. **4.** a tube or screen for viewing a television signal or computer output. **5.** a hose mounting capable of directing a stream of water in any direction for use in hydraulic mining or in fighting a fire. **6.** a large flesh-eating lizard (family Varanidae) that supposedly gives a warning when crocodiles are present; native to Africa, southern Asia, and Australia. **7.** a rotating toolholder. *v.* **8.** to observe and supervise. **9.** to watch closely and keep a record of one's observations; e.g., to keep track of the degree of pollution in a stream or lake. **10.** to receive and check a television transmission.

monitoring well *n.* a well drilled near a hazardous waste facility or deposit to detect any pollution that may occur in the groundwater.

monkey *n.* (Spanish *mono*, monkey) any of several small primate mammals (order Primate, excluding humans, apes, and other large species) with long tails. Monkeys are mostly tropical forest dwellers native to areas of Africa, Asia, and South America and are mostly vegetarian, though some may eat eggs or insects.

monkshood *n.* a poisonous, hardy, perennial herb (*Aconitum napellus* and related species, family Ranunculaceae), native to northern Europe, with blue, white, or yellow flowers that include a hood-shaped sepal. Dried monkshood leaves and roots contain aconite, a very poisonous alkaloid that is used medically in certain ointments, liniments, diuretics, etc. Also called aconite, friar's-cap, or wolfsbane.

monocotyledon or **monocot** *n.* (Greek *monos*, one + *kotylēdōn*, a cavity) a plant with only one cotyledon (seed leaf); e.g., a plant of the grass family such as corn, wheat, bluegrass, etc. Monocots have parallel venation in contrast to the netted or reticulate venation of dicots. See cotyledon, dicotyledon.

monoculture *n.* (Greek *monos*, one + Latin *cultura*, cultivation) growing the same crop in the same field year after year. This system enables the user to specialize, but it generally encourages a buildup of diseases and insects and may cause the use of more pesticides.

monoecious *adj.* (Greek *monos*, one + *oikos*, house) having separate staminate (male) and pistillate (female) flowers, both on the same plant, as in corn, cucumbers, maples, and oaks. See dioecious.

monoestrous or **monestrous** *adj.* (Greek *monos*, one + *oistros*, frenzy) having only one estrous (heat) period a year; e.g., a female deer or dog.

monogastric *adj.* (Greek *monos*, one + *gastēr*, stomach + *ikos*, of) having a single stomach; e.g., people or swine.

monomictic lake *n.* a lake whose surface and deep waters mix once during the year, either in the summer for a cold monomictic lake or in the winter for a warm monomictic lake. See amictic lake, dimictic lake, polymictic lake.

monophagia *n.* (Greek *monos*, one + *phagein*, to eat + *ia*, condition or state) **1.** eating only one meal per day (common among animals). **2.** eating or craving only one kind of food.

monosaccharide *n.* (Greek *monos*, one + *sakkharon*, sugar + *ide*, chemical) a simple sugar. See disaccharide, sugar.

monosodium glutamate (MSG) *n.* a flavor-enhancing soluble substance ($C_5H_8NO_4Na$), usually used in the monohydrate form, that is derived from vegetable protein and often used in food and tobacco. Some people are allergic to it.

monsoon *n.* (Arabic *mawsim*, season) **1.** a seasonal wind system that blows from the cooler surrounding oceans across southern Asia toward central China where warm air rises above the land during the summer and then reverses its direction in winter, blowing from central China outward toward these oceans. **2.** any wind system that reverses directions every summer and winter, especially one involving a large landmass and a large area of ocean.

monsoon low *n.* any low-pressure center that tends to develop over a continental landmass in the summer or over a large oceanic area in winter.

monsoon season *n.* the season of the summer monsoon crossing India and nearby lands. The summer monsoon gives India 80% of its annual precipitation during June, July, and August.

montane *adj.* (Latin *montanus*, pertaining to a mountain) growing on or pertaining to a mountain in the cool, humid zone just below the timberline; typically an area dominated by evergreen trees.

month *n.* (Anglo-Saxon *monath*, moon) **1.** the time period required for the moon to complete one orbit around the Earth, especially the time from one new moon to the next; a period of 29.531 days (the lunar or synodic month). **2.** the time required for the moon to return to the same position relative to the stars, a period of 27.3217 days (the sidereal month). **3.** the time required for the moon to return to the same position on the ecliptic (the apparent path of the sun as seen from Earth), a period of 27.2122 days (the nodical or draconic month). **4.** the average length of a lunar month relative to the vernal equinox, a period of 27.3216 days (the tropical month). **5.** the period between successive perigees of the moon, 27.555 days (the anomalistic month). **6.** one-twelfth of the solar year, 30.437 days (the solar month). **7.** any of the 12 periods commonly used to subdivide a year, each containing 28, 29, 30, or 31 days (a calendar month). **8.** the period from any date in one month to the corresponding date in the next month, or an arbitrary period of either 4 weeks or 30 days.

monticule *n.* (Latin *monticulus*, small mountain) **1.** a secondary cone on the side of a volcano. **2.** any small mountain or large hill.

montmorillonite *n.* (French *Montmorillon*, a church in France + *ite*, mineral) **1.** an expanding-lattice silicate clay mineral of the smectite family with tetrahedral layers on both sides of each octahedral layer and a variable composition. An idealized formula for it is $(Al_{1.8}Mg_{0.2})(Si_{3.9}Al_{0.1})O_{10}(OH)_2 \cdot xH_2O$ + enough exchangeable cations (especially Ca^{2+}) to balance the charge. The lattice spacing varies with water content from 0.96 to 2.14 nm, and it has a relatively high cation exchange capacity averaging about 800 mmol of charge per kilogram (80 meq/100 g in the older notation). **2.** formerly any smectite mineral.

Montreal protocol *n.* an international agreement in force since 1989 that aimed to freeze chlorofluorocarbons at 1986 levels and then to reduce their production and use gradually.

moon *n.* (Anglo-Saxon *mona*, moon) **1.** the natural satellite with a diameter of 2160 mi (3476 km; 27% of the diameter of the Earth) that orbits the Earth at an average distance of 238,900 mi (384,400 km). It is a solid body with no atmosphere and no liquids, and its light is reflected from the sun. **2.** a similar body orbiting any other planet. **3.** a month. See blue moon. **4.** a designated phase of the moon, as a waxing moon goes from new moon to crescent moon to half moon to full moon with more and more of its face being lit by the sun, and a waning moon goes the other way. **5.** an object with a round or a crescent shape that resembles the lighted part of either the full or the crescent moon. *v.* **6.** to mope or wander listlessly or nostalgically.

moonarian *n.* (Anglo-Saxon *mona*, moon + *arian*, person) a person who uses the phases of the moon as a guide for planting crops; e.g., one who plants aboveground crops such as tomatoes in the light of the moon and belowground crops like potatoes in the dark of the moon. See book farmer.

moon gravity *n.* the gravitational pull of the moon. Gravity on the surface of the moon is only about one-sixth (17%) of that at the Earth's surface. Thus, a person weighing 120 lb on Earth would weigh only 20 lb on the moon. Even so, the moon is close enough to Earth that moon gravity is a stronger factor than that of the sun for causing tides on Earth.

moonquake *n.* a series of seismic vibrations in the moon similar to an earthquake.

moon rock *n.* stones and powdered rock material from the moon (48.5 lb of it) were first brought back to Earth by U.S. astronauts Edwin E. Aldrin, Jr. and Michael Collins in July 1969. Minerals identified included plagioclase and pyroxene, both common in Earth's volcanic rocks.

moor *n.* (Anglo-Saxon *mor*, wasteland) **1.** an area of flat, poorly drained, open land, often covered with heath. See heath, wetland. **2.** an area reserved for hunting wild game in Britain. *v.* **3.** to tie a ship or other craft in place with cables or other attachments. **4.** to tie anything securely in a place.

moorland *n.* (Anglo-Saxon *mor*, wasteland + *land*, land) an area that is mostly moor but with inclusions of other types of soil and vegetation.

moose *n.* (Algonquian *mos*, moose) a large, hoofed mammal (*Alces alces*, family Cervidae) native to forest and swamp areas of northern North America and Europe. The males (bulls) stand as tall as 7.7 ft (2.35 m) at the shoulder, weigh as much as 1800 lb (over 800 kg), and carry large antlers with broad flat surfaces. The females (cows) are about ¾ as large as the bulls. Called elk in Europe.

mor *n.* (Danish *mor*, humus) a layer of rotted organic debris covering a mineral soil in a forest. See mull.

moraine *n.* (French *moraine*, moraine) **1.** an area of irregular hills or mounds representing the border of a former glacier. It is composed of unsorted materials with particles sizes ranging from clay to boulders. Sometimes called lateral moraine at the side of the glacier and either terminal or end moraine where the glacier stopped moving forward. **2.** glacial till, a layer of unsorted material deposited by a glacier as it moved across an area. Also called ground moraine or till.

morass *n.* (Dutch *moeras*, a marsh) **1.** an area of low-lying wetland such as a marsh or bog. **2.** an entangling predicament; a troublesome situation that is difficult to escape, especially in a political or social sense.

morel *n.* (French *morille*, morel) **1.** the edible fruiting body of a small mushroom (*Morchella* sp., especially *M. esculenta*, family Helvellaceae), one of the most easily identified of the edible fungi. It resembles a brown sponge on a short stalk with a rounded, reticulated top and grows mostly in the spring under trees in a humid environment. It is the state mushroom of Minnesota. **2.** morelle; the black nightshade (*Solanum nigrum*, family Solonaceae) or a related species.

mores (sing. **mos**) *n. pl.* (Latin *mores*, customs) **1.** the commonly accepted customs and moral standards of a particular social group; folkways generally accepted without question by members of the group. **2.** groups of organisms that inhabit the same ecological environment and have the same reproductive season.

moringa *n.* a fast-growing woody species that produces pods like giant green beans with an asparagus flavor as well as leaflets that can be boiled and eaten like spinach.

Mormon cricket *n.* a large wingless katydid (*Anabrus simplex*, family Tettigoniidae) of intermountain and far-western U.S. areas, so named for large hordes of them that were consuming the crops of Mormon settlers in Utah until the Mormon crickets were eaten by flocks of California seagulls (*Larus californicus*) in 1848. Crickets feed on at least 250 species of crop and range plants, and their numbers may build up rapidly.

morning glory *n.* **1.** any of several twining vines (*Ipomoea* sp., family Convolvulaceae); considered a weed in fields and gardens but sometimes grown on a trellis for its beautiful white to pink bell-shaped flowers that open mostly in the morning. **2.** any of several similar plants; e.g., field bindweed (*Convolvulus* sp., family Convolvulaceae).

morphallaxis *n.* (Greek *morphē*, form + *allaxis*, exchange) the ability to regenerate a new body part when one has been cut off or destroyed; e.g., an earthworm can regenerate itself after being cut in two.

morphogenesis *n.* (Greek *morphe*, form + *genesis*, origin) shaping of the land surface by erosional and depositional processes.

Morning glory

morphogenic process *n.* any of the natural processes that shape land surfaces, including erosion and deposition by wind, water, and ice, landslides and other forms of mass movement, uplift, and displacements that accompany earthquakes.

morphography *n.* (Greek *morphe*, form + *graphē*, writing) qualitative description of landforms.

morphology *n.* (Greek *morphe*, form + *logos*, word) **1.** the shape and/or structure of a living organism, object, or organization. **2.** the science that is concerned with the form and structure of plants, animals, soils, or other entities. See phenotype. **3.** the composition, form, pronunciation, and inflection of the words in a language.

morphometry *n.* (Greek *morphe*, form + *metron*, measure) measurement and quantitative description of the surface shapes of objects, organisms, landforms, etc. (e.g., elevations, slope gradients, curvatures, and spacings).

Morrill Act *n.* **1.** a U.S. law authored by Vermont Representative Justin S. Morrill (1810–1898) and passed in 1862 to provide each state 30,000 ac (about 12,000 ha) per congressman to be used to establish land-grant colleges that would teach agriculture and the mechanic arts. **2.** a second U.S. law authored by Vermont Senator Justin S. Morrill and passed in 1890 to provide annual funding for the land-grant colleges and to add a second land-grant college for African-American students in states that practiced segregation. Both Morrill Acts continue in effect with modifications by subsequent amendments. See land-grant college.

mortal *n.* (Latin *mortalis*, subject to death) **1.** one who is subject to death; especially a human being. *adj.* **2.** subject to death; having a life that will end. **3.** human. **4.** bitter and deadly, as a mortal enemy or mortal combat. **5.** likely to cause death, as a mortal wound. **6.** dire, as a mortal fear.

mortality *n.* (Latin *mortalitas*, mortal condition) **1.** the condition of being subject to death. **2.** death on a large scale, as that caused by a war, an epidemic, or another disaster. **3.** the death rate or statistical probability of death as shown by data or in a mortality table for humans, for a specific group (e.g., infant mortality), or for a specified organism (e.g., baby chick mortality) or losses attributed to a certain cause (e.g., plants killed by damping off or some other disease).

mosaic *n.* or *adj.* (French *mosaique*, artistic) **1.** a decoration made with small tiles of various colors fit together to form a figure or pattern on a surface. **2.** the process of gluing tiles to a surface to form such a figure or pattern. **3.** an assemblage of aerial photos cut and fit together to form a single picture of a large area of land.

mosquito *n.* (Spanish *mosquito*, little fly) any of more than 2000 species of small flying insects (family Culicidae, order Diptera) that are annoying pests and disease vectors (e.g., *Aedes aegypti* carries yellow fever and dengue fever; *Anopheles quadrimaculatus* and *A. freeborni* carry malaria). Female mosquitoes use a proboscis to pierce the skin and feed on the blood of people and animals. Male mosquitoes live on flower nectar and plant juices.

Three species of mosquitoes. Each one is ~1 cm long.

mosquito larva *n.* an immature state of any species of mosquito. Mosquitoes lay their eggs in small bodies of stagnant water, and their larvae grow there until they pupate. One method of mosquito control is either to eliminate stagnant water or to cover it with a thin layer of oil that prevents the larvae from breathing.

moss rose *n.* See portulaca.

most probable number (MPN) *n.* an estimation of the population of a type of organism based on a statistical interpretation of data regarding its growth or lack of growth in culture media prepared with various dilutions of the source material; e.g., an estimation of the number of bacteria in 1 mL of a suspension.

mote *n.* (Anglo-Saxon *mot*, speck) a tiny particle or spot, especially a dust particle.

moth *n.* (Anglo-Saxon *moththe*, moth) any of a large number of adult insects (order Lepidoptera) with large wings relative to their body size. Moths are distinguished from butterflies by their feathery antennae and by fluttering about mostly in the evening or at night. Most are not as brightly colored as a typical butterfly. See clothes moth, butterfly.

mothball *n.* **1.** a ball made of naphthalene or camphor for use as a moth repellant to protect woolen items such as clothing or blankets from being eaten by the larvae of clothes moths. *v.* **2.** to inactivate and put in storage, as to mothball a ship.

mother-of-pearl *n.* a hard iridescent layer on the inner surface of the shells of certain mollusks, especially in oyster shells. It is composed of very thin alternating layers of calcium carbonate ($CaCO_3$) and hornlike material and is used to make buttons and other ornaments. Also called nacre.

motile *adj.* (Latin *motus*, moved + *ilis*, ability) capable of movement; e.g., flagellate bacteria are motile.

motility *n.* (Latin *motus*, moving + *ility*, condition of) the ability to move spontaneously; the possession of locomotive faculties.

motte or **mott** *n.* (Latin *matta*, cover) a small grove of trees growing in a slight depression or other favorable local environment on a prairie, especially in Texas.

mottle *n.* **1.** a blotch, streak, or spot that differs in color from its surroundings. Mottles on plant leaves can be either normal coloration or signs of disease. Rust mottles in soil are usually a sign of intermittent wetness, and gray mottles indicate more permanent reducing conditions as in wetland soils. Mottled teeth can result from excess fluorine in drinking water.

mound *n.* (Anglo-Saxon *mund*, hand or protection) **1.** a raised, usually rounded, surface, especially one composed of or covered with soil. It may be tiny (as a mound of soil pushed up by a mole or placed to induce a branch to root), small (as a mound of soil placed to protect a plant from freezing or to cover a grave), medium (e.g., an earthen fortification), or large (as a rounded hill). **2.** a pile of anything, as a mound of papers or a mound of garbage. *v.* **3.** to push soil into a pile in the form of a mound.

Mound Builders *n.* prehistoric Americans who built burial mounds and other earthen mounds (some conical and some linear; others in the shapes of animals, sometimes called effigy mounds) in the central part of what is now the U.S.A.

mount *n.* (Latin *montis*, a hill) **1.** a high place; a hill or mountain. **2.** a support where something can be held in place, as a telescope mount. **3.** a riding horse. **4.** the act of getting on a horse or some other animal or machine that can be ridden. *v.* **5.** to go up or get on. **6.** to place on a support. **7.** to place something where it can be observed without damage. **8.** to prepare to impregnate a female, as a bull mounts a cow.

mountain *n.* (Latin *montana*, a mountain) **1.** a landform larger than a hill; one that stands high above its surroundings, usually at least 2000 ft (600 m) above nearby lowlands. Major mountain chains are produced by shifts in the Earth's crust that cause folding, faulting, intrusion of a granitic core, and uplift of a major area. Other mountains are formed by large earthquake faults and/or volcanic action. Mountains are gradually worn down by erosion. **2.** a large amount of anything, as a mountain of mail.

mountain ash *n.* any of several species of small trees (*Sorbus* sp., family Rosaceae) that bear white flowers and bright red berries that birds can eat.

mountain bluebird *n.* a slender bluebird (*Sialia currucoides*) about 6 in. (15 cm) long, native to the Rocky Mountains. Adult males have a bright blue color, including a blue breast rather than the red breast of other bluebirds. Females and young birds have bluish gray plumage and gray breasts.

mountain goat *n.* a long-haired white antelope (*Oreamnos americanus*), native to the Rocky Mountains, that resembles a goat about 5 ft (1.5 m) long and 3 ft (1 m) tall with short black horns. Also called Rocky Mountain goat.

mountain laurel *n.* an evergreen shrub (*Kalmia latifolia*, family Ericaceae) of eastern North America and the state flower of Connecticut and Pennsylvania. It bears pink and white flowers and has poisonous, shiny, leathery leaves. Also called calico bush.

mountain lion *n.* See cougar.

mountain range *n.* a series of mountains that form a continuous line of peaks and saddles. Large mountain systems have several parallel mountain ranges with major valleys between the ranges.

mountain sheep *n.* **1.** the bighorn sheep, also called Rocky Mountain sheep. **2.** any wild sheep in a mountainous region.

mountain sickness *n.* an ailment caused by a deficiency of oxygen in the rarefied air at high elevation. Fortunately, people can become acclimatized to the rarefied air. Also called altitude sickness.

mountain wind or **mountain breeze** *n.* an evening wind that blows down from a mountain because the higher areas cool earlier than those below. See valley wind.

mourning dove *n.* a North American dove (*Zenaida macroura*, family Columbidae) with a long, pointed tail and a mournful cooing voice. It is gray, about 10 in. (25 cm) long, found in urban areas and woodlands in all 48 conterminous states of the U.S.A., and is hunted as a game bird in 31 states but protected in others. Sometimes called turtledove.

mouse deer *n.* a tiny deer (about 1 ft or 30 cm high) that weighs only about 2 lb (1 kg) and is considered a delicacy in Malaysia.

mouthbreeder *n.* any of several genera of fish (especially *Haplochromis* sp. and *Tilapia* sp.) that carry their eggs and/or their young in their mouth.

MPN *n.* See most probable number.

MTBE *n.* methyl tertiary butyl ether, an additive used to oxygenate gasoline and make it burn cleaner. Unfortunately, it can also mix with water and become a water pollutant.

muck *n.* (Icelandic *myki*, cow dung) **1.** organic soil developed from decaying plants in a wetland that has been drained and oxygenated. See peat, wetland, Histosol. **2.** moist farmyard manure. **3.** decaying plant materials. **4.** mud, slime, or filth. **5.** wet mine spoil. **6.** valueless material that must be removed to reach good ore in a mine. **7.** derogatory remarks, especially in political campaigns. **8.** chaos and confusion. *v.* **9.** to move or work in muck, as to muck around. **10.** to fertilize with manure or decaying organic materials. **11.** to make muddy or dirty. **12.** to clean mucky premises, as to muck out a barn.

mucor *n.* (Latin *mucor*, mold) a common and very destructive genus of fungi (*Mucor* sp., family Mucoraceae). A few species are animal parasites, whereas others cause disease in squash and pumpkins, sweet potatoes, or Irish potatoes, and another produces common bread mold.

mucous *adj.* (Latin *mucosus*, slimy) **1.** like mucus, consisting of mucus, or pertaining to mucus. **2.** containing, covered by, or secreting mucus, as mucous membranes.

mucus *n.* (Latin *mucus*, snot) slimy, viscous material containing water, soluble salts, and carbohydrates secreted by glands in the mucous membranes lining the nasal passages, esophagus, and other body cavities, plus dead cells and leukocytes (especially when there is an infection). It serves to lubricate and protect these surfaces.

mud *n.* (German *mudde*, wet) **1.** soft, wet, sticky soil. **2.** (slang) strong, bitter coffee. **3.** a wet soil and chemical mixture used to lubricate the bit while drilling a large-diameter, deep well. See drilling mud. **4.** derogatory statements and other negative assertions, as slinging mud in a political campaign. *v.* **5.** to coat with mud (e.g., to seal the walls of a log cabin). **6.** to make water turbid, as by stirring up sediment.

mud dauber *n.* a small wasp (*Sceliphron caementarium* and related species, family Sphecidae) with a dark brown body, yellow markings, and yellow legs. Mud daubers kill spiders, other insects, and insect larvae and store them in mud cells as food for their own larvae. Also called common mud dauber or mud wasp.

mud fence *n.* a ridge made from mud and allowed to dry, usually topped with rocks or brush to protect it from raindrop splash (useful in arid regions).

mud flat *n.* a nearly level area that floods frequently and is covered with mud rather than vegetation, as a tidal flat or a dry lake bed.

mudflow *n.* (German *mudde*, wet + *flouwen*, to flow) **1.** a landform produced by mass movement where saturated soil became fluid and flowed down a sloping area onto a bottom. Some mudflows have been large enough to bury buildings and block roads. (Similar flows containing more than 50% gravel and stone fragments are called debris flows.) **2.** the process that produces such a landform.

muggy *adj.* (Icelandic *mugga*, drizzly and misty) oppressively warm and humid, as muggy weather.

Muir, John *n.* an early naturalist (1838–1914) born in Scotland who crusaded after the Civil War to save wilderness areas, especially in western U.S.A. He founded the Sierra Club and induced President Theodore Roosevelt to greatly expand the national park system.

mulch *n.* (German *molsch*, soft, rotting) **1.** a layer of plant residues or other organic materials or a plastic sheet that protects the soil from erosion, holds moisture in the soil for plant growth, and inhibits weed growth. *v.* **2.** to provide such a protective cover for the soil in a field or garden.

mulch tillage *n.* tillage that leaves plant residues on the soil surface; e.g., use of a chisel implement or one with blades that cut beneath the surface without inverting the soil. See conservation tillage, reduced tillage.

mule *n.* (Latin *mulus*, mule) **1.** a cross between a horse and a donkey, usually with a horse as the mother (see hinny). Most mules are sterile and are shaped like a horse with long ears and small feet. They are stubborn, strong, surefooted, and valued as work animals. **2.** a sterile hybrid plant, animal, or bird. **3.** an obstinate person. **4.** a person employed by a smuggler to carry contraband (especially drugs). **5.** a smalllocomotive used to move goods, vehicles, or ships at a local site.

mule deer *n.* a large deer (*Odocoileus hemionus*) with large ears, a gray coat, and a black tip on its tail. It is native to western U.S.A., Canada, and Mexico.

muley *n.* or *adj.* (Welsh *moel*, hornless + *ey*, being) a cow or deer that does not grow horns.

mulish *adj.* (Latin *mula*, mule + *ish*, like) very stubborn, like a mule.

mull *n.* (Anglo-Saxon *myl*, dust) **1.** humic material mixed into the upper part of a forested mineral soil, thus forming a darkened A horizon that is usually quite thin. See mor. **2.** dust, dirt, or rubbish. *v.* **3.** to grind or pulverize into dust, dirt, or rubbish. **4.** to ponder or consider carefully; e.g., to mull it over in one's mind.

mullet *n.* (Latin *mullus*, red mullet) **1.** any of several edible fish (family Mugilidae) of both salt water and freshwater that have nearly cylindrical bodies, small mouths, spiny fins, and mostly gray or reddish brown colors, especially a striped mullet (*Mullus auratus*). **2.** a surmullet (goatfish). **3.** a sucker (especially *Moxostoma* sp.).

multiflora rose *n.* a very aggressive climbing or trailing rose (*Rosa multiflora*, family Rosaceae) with clusters of white flowers. It was native to Japan and Korea and was introduced into the U.S.A. for erosion control, living fences, and wildlife. It has been so aggressive that it now is considered a noxious thorny weed. The seeds are spread by animals and birds that eat the rose hips (fruits). Also called Japanese rose.

multiple cropping *n.* growing two or more crops in the same field in the same year; either sequential cropping with one crop following another (also called double cropping) or intercropping with two or more crops intermixed and growing simultaneously. See crop rotation for different crops in different years.

mung bean *n.* **1.** an Asian bean (*Vigna radiata*, family Fabaceae) grown for its edible seeds and pods used as vegetables and its young sprouts used in salads and oriental dishes. **2.** the seed of such plants.

Munsell color *n.* a system developed by A. H. Munsell and adopted in the 1940s for describing soil colors by matching them to samples that quantify three variables: hue (the spectral color, as red, yellow, green, blue, or purple), value (brightness expressed as the square root of the percent light

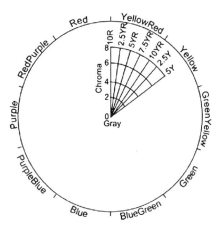

Munsell hues and chromas that are common in soil

reflected), and chroma (color intensity from 0 = gray to 20 = pure color). Colors are noted by a hue designation followed by a fraction representing value/chroma; e.g., 10YR 4/2.

murine *adj.* (Latin *muris*, mouse + *ine*, pertaining to) relating to mice and/or rats (family Muridae).

murine typhus *n.* a mild but long-lasting form of typhus caused by rickettsia (*Rickettsia typhi*). It is a common disease near seacoasts in all parts of the world. It affects rats and mice (family Muridae) and is transmitted to people via fleas (*Xenopsylla cheopis*). Also called rat typhus.

murky *adj.* (Anglo-Saxon *myrce*, dark) cloudy; gloomy; obscure; as a murky storm or a murky solution.

murrain *n.* (French *morine*, a plague) **1.** an infectious disease of cattle; e.g., foot-and-mouth disease. **2.** an infectious disease or plague that affects various animals or plants.

muscarine *n.* (Latin *muscarius*, of flies) a highly toxic alkaloid ($C_8H_{19}NO_3$) found in red mushrooms (*Amanita muscaria*) and in rotting fish or other rotting flesh.

Muscatine melon *n.* an improved cultivar of large, sweet-tasting cantaloupe (*Cucumis melo cantalupensis*, family Cucurbitaceae) named after a city in eastern Iowa.

muscle *n.* (Latin *musculus*, muscle) **1.** an animal body part composed of fibrous tissue that can contract forcibly to move a limb or other part, including such things as the heart muscle. Muscle becomes meat when cooked for a meal. **2.** strength; power; ability to do work or cause something to happen, either literally or figuratively, as that organization has a lot of muscle.

muscovite *n.* white mica, a semitransparent layer silicate [idealized formula, $KAl_2(AlSi_3)O_{10}(OH)_2$] that occurs in granitic, metamorphic, and sedimentary rocks in sizes ranging from tiny flakes to pegmatite crystals several inches across. It is used as an electrical insulator and to make heat-resistant windows. Also called isinglass.

mushroom *n.* (Latin *mussirionis*, mushroom) **1.** the edible fruiting body of a fungus (family Agaricaceae, order Agaricales); e.g., a cultivated species (often *Agaricus campestris*) or the popular morel (*Morchella* sp.), as distinguished from similar toxic bodies called toadstools. **2.** any fleshy fungal fruiting body (including the toadstools, puffballs, etc.). **3.** anything that grows rapidly and produces a shape similar to a mushroom; e.g., the cloud produced by a nuclear explosion. *adj.* **4.** containing mushrooms or pieces thereof, as a mushroom omelet. **5.** shaped like a mushroom. **6.** growing rapidly like mushrooms, as a mushroom town. *v.* **7.** to grow rapidly. **8.** to flatten into a mushroom shape, as when a bullet mushrooms.

mushroom root rot *n.* a serious fungal disease (*Armillaria mellea* or *Armillariella mellea*, family Agaricaceae) that infects the roots and lower trunk of many fruit and forest trees, especially of oak species. Affected trees are stunted and become progressively more chlorotic and wilted until they die after 1 to 10 years or more (depending greatly on the tree species). The disease is identifiable by white mycelial plaques or fans that form in or under the tree bark. Also called oak root fungus disease, shoestring root rot, or armillaria root rot.

musk *n.* (Latin *muscus*, musk) **1.** a greasy secretion with a strong odor produced in a sac near the navel of the male musk deer (*Moschus moschiferus*), used as a base in perfumes. **2.** a similar substance produced by other animals such as the civit, muskrat, or otter. **3.** a synthetic substitute for the natural musk. **4.** the odor of musk. **5.** a plant that produces a musky fragrance; e.g., musk plant (*Mimulus moschatus*) or musky heron's bill (*Erodium moschatum*).

muskeg *n.* (Cree *maskeek*, swamp) poorly drained land in northern North America with a mat of living vegetation (mostly sphagnum moss and sedges) overlying acid peat that ranges in thickness from a few inches (several centimeters) to several feet (a few meters). Some muskegs have a few small black spruce or tamarack trees, and some are used for raising cranberries.

muskmelon *n.* **1.** a trailing vine (*Cucumis melo reticulatus*) that bears edible melon fruit. **2.** the round or oval fruit of such vines with a usually ribbed tan or greenish gray rind and an orange, light green, or white edible flesh with a sweet flavor and an internal cavity containing seeds. Also called cantaloupe.

muskrat *n.* (Algonquian *musquash*, muskrat) **1.** a large aquatic rodent (*Ondatra zibethica*) with a musky odor, whose body is stocky and about 1 ft (30 cm) long, with a tail nearly as long as the body. It behaves like a small beaver, chewing down small trees and building dams to form ponds. **2.** the thick, light brown fur of this animal, used to make or trim coats, hats, etc.

musk thistle *n.* either of two weedy species of thistle (*Carduus nutans* and *C. theormeri*, family Asteraceae) that spread rapidly in pastures and are difficult to control with chemicals. Biological control by the musk-thistle head weevil (*Rhinocyllus conicus*) is showing some success.

mussel *n.* (Latin *musculus*, small mouse or mussel) a small bivalve mollusk (especially family Mytilidae in salt water or family Unionidae in freshwater). Mussel shells are sometimes made into buttons or other ornaments. The saltwater forms of mussel are edible. See zebra mussel.

mustard *n.* (French *moustarde*, mustard) **1.** any of a large number of herbaceous plants (family Brassicaceae, especially *Brassica* sp.) with alternate leaves, clusters of four-petaled flowers, pungent juice, and several types of edible vegetables, including broccoli, cabbage, cauliflower, cress, radishes, etc. Some species become weeds when they escape; e.g., black mustard (*B. nigra*) and white mustard (*Sinapis alba*). **2.** a condiment made from powdered seeds of certain species of mustard, usually used in the form of a yellow paste but also used medically in plasters and poultices as a counterirritant. **3.** a strong yellow like that of the condiment.

mustard plaster *n.* a paste containing powdered mustard that is spread on a cloth and applied to the skin as a counterirritant.

mutagen *n.* (Latin *mutare*, to change + *gen*, agent) any environmental agent or condition that causes mutation in the genetic material of a cell.

mutagenesis *n.* (Latin *mutare*, to change + *genesis*, creation) the process of inducing a mutation, including the application of certain chemicals for this purpose.

mutant *n.* (Latin *mutans*, changing) **1.** an organism with new characteristics produced by mutation. *adj.* **2.** changing or causing change, especially as a form of evolution.

mutation *n.* (Latin *mutatio*, change) an abrupt change in the inheritable genetic makeup of a plant, animal, or human; a revision of the genome by either an environmental cause (e.g., radiation or chemical influence) or a genetic accident.

mutton *n.* (French *mouton*, mutton) sheep meat, especially that from a sheep more than a year old. See lamb.

mutualism *n.* (Latin *mutualis*, reciprocal + *isma*, condition) **1.** a biological interaction of two organisms with benefit to both; e.g., pollination of flowers by bees as the bees obtain nectar to make honey, or *Rhizobium* bacteria supplying a legume plant with nitrogen and the plant supplying the bacteria with carbohydrate. See symbiosis. **2.** an ethical or sociological principle of people working as a group for group benefits rather than as individuals for individual benefits.

mycelium (pl. **mycelia**) *n.* (Greek *mykēs*, fungus + *ēlos*, nail or wart + *ium*, object) **1.** a threadlike tube of fungal vegetative material. **2.** a branching, interconnected (often tangled) mass of such threadlike fungal tubes.

mycology *n.* (Greek *mykēs*, fungus + *logos*, word) the scientific study of fungi.

mycorrhiza (pl. **mycorrhizae**) *n.* (Greek *mykēs*, fungus + *rhiza*, root) a beneficial, symbiotic combination of mycorrhizal fungi and roots on trees, shrubs, and most herbaceous plants, either ectomycorrhizae with mantles of mycelia around the roots of conifers, beeches, oaks, etc., or endomycorrhizae (also called arbuscular mycorrhizae, AM, or vesicular arbuscular mycorrhizae, VAM) with mycelia penetrating the root cells of most other plants and extending out into the soil much like long, slender root hairs. Mycorrhizae help the plants absorb water and nutrients and are especially important for absorbing phosphorus in low-fertility soils.

mycotoxin *n.* (Greek, *mykēs*, fungus + *toxikon*, a poison) a toxic proteinlike substance produced by certain fungi (especially *Aspergillus* sp. and *Fusarium* sp.). The poisoning of people and animals by mycotoxins in food has been a serious problem for more than 5000 years. Rapid drying and the use of ammonia on grains are two practical treatments for mycotoxins.

myiasis *n.* (Greek *myia*, fly + *asis*, condition) a disease caused by either maggots or adult flies; e.g., a skin infestation where fly larvae grow in the skin of an animal or person. See foot maggot, screwworm.

myna *n.* (Hindi *mainā*, myna) a tropical bird (family Sturnidae, especially *Acridotheres* sp. and *Gracula* sp.) native to India and certain islands in the Pacific. Some species can be kept as pets and trained to talk much like parrots; e.g., the hill myna (*G. religiosa*), a bird that resembles a starling with glossy black plumage and yellow neck wattles. Also spelled mynah, mina, or minah.

myrmecology *n.* (Greek *myrmex*, ant + *logos*, word) the branch of entomology that studies ants.

myrmecophile *n.* (Greek *myrmēx*, ant + *philos*, loving) a beetle that shares the nest of an ant colony.

myrrh *n.* (Latin *myrrha*, myrrh or bitter) **1.** a fragrant, gummy resin exuded by certain shrubs (*Myrrhis* sp.) used to make medicines, incense, and perfume. **2.** any of several spiny shrubs (*Myrrhis* sp., especially *M. odorata*, family Apiaceae) with few leaves, small green blossoms, and small oval fruit; native to Arabia and eastern Africa.

myrtle *n.* (Latin *myrtus*, myrtle) **1.** any of several pleasant-smelling plants (*Myrtus* sp., family Myrtaceae) with evergreen leaves, small white or pink blossoms, and dark purple, fragrant berries. **2.** any of several similar plants; e.g., periwinkle or California laurel.

N

N *n.* (Greek *nu*, n) **1.** the chemical symbol for nitrogen. **2.** north. **3.** newton.

Na *n.* the chemical symbol for sodium, from the Latin *natrium*.

nacre *n.* (Arabic *naqqārah*, drum) mother-of-pearl.

nacreous *adj.* (Arabic *naqqārah*, drum + Latin *osus*, having) **1.** iridescent, like mother-of-pearl. **2.** pertaining to or resembling mother-of-pearl.

nacreous cloud *n.* a rare high cloud that appears iridescent with a soft pearly luster; seen mostly in high latitudes when the sun is just below the horizon.

nadir *n.* (Arabic *nazīr*, opposite) **1.** a point in the heavens that would be directly overhead on the opposite side of the Earth (opposite from the zenith; straight down from the observer). **2.** the lowest point in adversity, despair, or mental depression.

NAFTA *n.* See North American Free Trade Agreement.

nag *n.* (Dutch *negge*, a small horse) **1.** a horse that isn't worth much; often old and decrepit. **2.** (slang) any racehorse; sometimes any horse. **3.** a small horse. **4.** a person who nags, or an instance of nagging. *v.* **5.** to annoy with persistent demands and complaints.

nagana *n.* (Zulu *unakane*, nagana) a disease of horses, other animals, and humans that is common in equatorial Africa; caused by a pathogen (*Trypanosoma brucei*) transmitted by the tsetse fly (*Glossina* sp.). See trypanosomiasis.

Nagasaki *n.* a Japanese seaport; the target of the second atomic bomb used in warfare, August 9, 1945. Japan sued for peace the next day and surrendered to the U.S.A. August 15. See Hiroshima.

naiad *n.* (Greek *Naias*, a water nymph) **1.** a Greek mythological creature supposed to preside over streams, springs, and fountains. **2.** a juvenile mayfly, dragonfly, or damselfly. **3.** an aquatic plant (*Najas* sp., family Najadaceae) with narrow leaves and solitary flowers. **4.** an expert female swimmer.

naked flower *n.* a blossom with no petals or sepals.

naked seed *n.* a seed that is not enclosed in an ovary.

ñandu *n.* (Spanish *ñandu*, ostrich) the South American ostrich (*Rhea* sp.). See ostrich.

nannoplankton or **nanoplankton** *n.* (Greek *nanos*, dwarf + *planktos*, drifting) the tiniest form of plankton; extremely small aquatic plants and animals that can pass through fine-mesh plankton nets. See plankton.

naphtha *n.* (Latin *naphtha*, naphtha) a flammable oily liquid obtained by fractional distillation of crude oil (e.g. at temperatures from 80°C to 110°C); used as a solvent and as a fuel. **2.** petroleum. **3.** any of several similar flammable liquids distilled from other organic materials (e.g., coal, coal tar, wood).

naphthalene *n.* (Latin *naphtha*, naphtha + *al* for alcohol + *ene*, hydrocarbon) a coal tar product ($C_{10}H_8$) that forms white crystals; used as a soil fumigant, as a moth repellant, and in the production of certain dyes. See mothball.

nappe *n.* (Latin *nappa*, napkin) **1.** a continuous surface; e.g., a cone has two nappes that meet at a point. **2.** a sheet of water flowing over a weir, the spillway of a dam, or other structure. **3.** a large nonconforming layer of rock that has been thrust horizontally across other rocks in a fault zone or in a recumbent fold.

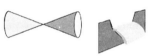

Two nappes of a cone (left), and a nappe of water (right)

Narcissus *n.* (Greek *narkissos*, narcotic) **1.** a youth in Greek mythology who was so much in love with his own image in a pool of water that he was transformed into a flower. **2.** any of the flowers that bear his name (*Narcissus* sp., family Amaryl-lidaceae); bulbous flowering plants with smooth leaves and cup-and-saucer-shaped white, yellow, or orange blossoms (daffodils, jonquils, poet's narcissus, etc.).

narcotic *n.* (Greek *narkōtikos*, deadening) **1.** an agent that causes numbness, stupor, or loss of consciousness; common narcotics include opium and derivatives such as codeine and morphine and such synthetic drugs as methadone; used to relieve pain, sedate, or induce sleep. **2.** a person addicted to the use of narcotics. **3.** anything that dulls the senses and induces sleep; e.g., a dull lecture can be a narcotic. *adj* **4.** causing sleep or stupor. **5.** having to do with narcotics and their use or misuse.

narrows *n.* a narrow strip of water between two islands or between an island and a mainland, especially the strip between Staten Island and Long Island, New York.

narwhal *n.* (Danish *narhval*, narwhal) an arctic whale (*Monodon monoceros*) prized for oil and ivory. The males grow a prominent spiraling tusk from the upper jaw.

NASA *n.* See National Aeronautics and Space Administration.

nascent *adj.* (Latin *nascens*, being born) **1.** beginning to develop; in the early stages of existence. **2.** free, as when an element is released from a chemical compound and has not yet reacted into a new compound. Also called nascent state.

nasturtium *n.* (Latin *nasturtium*, turning the nose) **1.** any of several herbaceous plants (*Tropaeolum* sp., family Tropaeolaceae) with a pungent odor; commonly grown for their showy orange, yellow, or red blossoms or for their fruit. **2.** a blossom from such a plant. **3.** a type of water cress (family Cruciferae) that grows in wetlands and has pinnate leaves and white or yellow blossoms; e.g., *Nasturtium officinale*, the common water cress found in springs and small streams.

natality *n.* (Latin *natalis*, birth + *itas*, condition) birth rate.

National Aeronautics and Space Administration (NASA) *n.* a U.S. government agency established in 1958 to promote and coordinate U.S. space programs.

National Agricultural Center and Hall of Fame *n.* a 172-acre complex including a museum and hall of fame located at Bonner Springs, Kansas. Although chartered by the U.S. Congress on August 31, 1960, to honor American farmers and the history, science, and technology of farming, it is privately supported and receives no government funds. Formerly named the Agricultural Hall of Fame.

National Air Monitoring System (NAMS) *n.* a network of about 1500 stations in the U.S.A. that monitor particulate matter in the air. Many stations also monitor other pollutants covered by the national ambient air quality standards.

national ambient air quality standards *n.* standards for maximum permissible levels of air pollutants established or proposed by the U.S. Environmental Protection Agency in furtherance of the U.S. Clean Air Act. There are standards for carbon monoxide (CO), nitrogen dioxide (NO_2), ozone (O_3), sulfur dioxide (SO_2), lead (Pb), particulates less than 10 microns in diameter, and particulates less than 2.5 microns in diameter.

National Association of Conservation Districts (NACD) *n.* an organization founded in 1946 as the National Association of Soil Conservation Districts to coordinate the nongovernmental activities of soil conservation districts. The initial emphasis on soil conservation has since broadened to include water conservation and other environmental concerns, so the word soil was dropped from the name.

National Audubon Society *n.* a society founded in 1905 to promote natural ecosystems with an emphasis on birds and other wildlife. Named after John James Audubon (1785–1851), renowned as a naturalist for his paintings and descriptions of North American birds.

National Cattlemen's Association *n.* an association of cattle producers formed in 1977 through a merger of previous organizations dating back as far as 1898; a nonprofit organization that communicates on behalf of the U.S. beef cattle industry.

National Cooperative Soil Survey *n.* the program for making and publishing soil surveys in the U.S.A. under the leadership of the Natural Resources Conservation Service in cooperation with the land-grant colleges and other public agencies and private organizations.

National Farmers Organization (NFO) *n.* an organization founded in 1955 to serve as a bargaining agent to secure higher prices for farm products sold by its members.

national forest *n.* an area of land administered by the Forest Service of the U.S. Department of Agriculture for multiple uses, including timber production, wildlife management, grazing, outdoor recreation, and watershed protection.

national grassland *n.* an area of land covered mostly with grass and/or shrubs and administered, along with the national forests, by the Forest Service of the U.S. Department of Agriculture for use as grazing land and for wildlife management, outdoor recreation, and watershed protection.

national monument *n.* an area set aside for preservation and public use because of its archaeological, scientific, or esthetic interest; operated by the National Park Service (usually smaller than a national park).

National Oceanic and Atmospheric Administration *n.* an agency in the U.S. Department of Commerce that was formed in 1970 as a reorganization of earlier programs tracing back as far as 1807. The agency is responsible for geographic, atmospheric, oceanic, and meteorological information; it compiles and disseminates data gathered on land and sea and in the air, plus that obtained by satellite imagery. See National Weather Service.

national park *n.* an area of special historical or scenic interest set aside by the U.S. government for preservation and public use. It is administered by the National Park Service with the purpose of preserving both plant and animal life along with its scenery for public enjoyment in perpetuity. The first, largest, and most famous national park is Yellowstone National Park, created in 1872.

National Park Service *n.* an agency in the U.S. Department of the Interior; created in 1916. It is responsible for administering national parks, national monuments, national historical sites, and national recreational areas.

National Response Team (NRT) *n.* a team organization involving 11 government agencies under the leadership of the Environmental Protection Agency to develop methods of responding to discharges or releases of hazardous materials (e.g., petroleum or anhydrous ammonia) and to train, equip, and coordinate response teams.

National Weather Service (NWS) *n.* an agency in the National Oceanic and Atmospheric Administration of the U.S. Department of Commerce. NWS is responsible for keeping weather records, making both short-range and long-range weather forecasts, and issuing warnings of incoming severe weather. It provides information to meteorologists who work for television stations and other local users.

National Wildlife Refuge System (NWRS) *n.* a network of more than 92 million acres of land and water in more than 500 wildlife refuges distributed across the U.S.A. from tropical oceanic islands to northern Alaska. It is managed by the U.S. Fish and Wildlife Service of the Department of the Interior to preserve as much habitat and species diversity as possible. The first wildlife refuge was Pelican Island in Florida, established in 1903.

native *adj.* (Latin *nativus*, inborn, natural) **1.** innate; natural rather than acquired, as native talent. **2.** of one's place of birth, as a native land. **3.** original or first, as one's native language. **4.** natural and undisturbed by humans, as an environment in its native state. **5.** relating to the inhabitants of an area, especially the aboriginal inhabitants, as native customs. **6.** occurring naturally, as a deposit of native gold. *n.* **7.** a person born in a particular place, as an American native, especially a descendant of the indigenous people. **8.** a plant or animal that is indigenous to a designated place.

native American *n.* **1.** a person born in the U.S.A. **2. Native American** a descendant of the people who inhabited any part of North or South America prior to its discovery by Columbus and the subsequent European colonization, except that the Inuit (Eskimos) of the far north are often considered a separate group. Previously called American Indian; also called Amerind or Amerindian.

natric horizon *n.* a diagnostic subsurface horizon in Soil Taxonomy that qualifies as an argillic horizon, has more than 15% exchangeable sodium on its cation exchange sites, and has columnar or prismatic soil structure.

natural *adj.* (Latin *naturalis*, pertaining to nature) **1.** pertaining to or dealing with nature. **2.** existing, occurring, or produced in a natural environment with no human influence. **3.** not changed artificially; e.g., natural food produced without chemical additives and with minimum processing, or a natural landscape without human structures or cultivation. **4.** having to do with observable physical things rather than spiritual concepts or fictitious items. **5.** innate rather than acquired, as natural ability. **6.** realistic, as a natural likeness. **7.** ordinary or customary, as a natural reaction. **8.** free of pretense, as natural behavior. **9.** simple, as a natural number rather than an imaginary number or a derived number such as a logarithm. **10.** neither sharp nor flat (as a musical note).

natural area *n.* an area that has not been modified by human influence; e.g., an area of virgin forest or of ungrazed rangeland.

natural bridge *n.* a place where water and/or wind has cut through a divide, leaving a high passageway crossing over an opening below. Some such sites have been made into national monuments (e.g., Natural Bridges National Monument and Rainbow Bridge National Monument, both in Utah) or other tourist attractions (e.g., Natural Bridge of Virginia and Ayres Natural Bridge in Wyoming). At least one is formed of a large petrified tree trunk.

This petrified tree trunk makes a natural bridge.

natural control *n.* reducing the population of a pest to an acceptable level through parasites, predators, disease, heat, cold, rain, drought, or other environmental factors rather than by using chemical pesticides.

natural disaster *n.* any extremely damaging environmental event such as a tornado, hurricane, tidal wave, flood, avalanche, landslide, or earthquake, especially one accompanied by loss of lives.

natural enemy *n.* a predator or other organism that kills individuals of a species. Natural enemies help to keep populations balanced and are sometimes introduced for biological control of pests.

natural gas *n.* **1.** a combustible mixture of gases trapped in geologic deposits, usually produced above a petroleum deposit by anaerobic decomposition of buried organic materials. Natural gas typically is composed of about 80% methane (CH_4) gas and smaller amounts of other combustibles such as ethane (C_2H_6), propane (C_3H_8), and butane (C_4H_{10}) and noncombustibles such as nitrogen (N_2), helium (He), carbon dioxide (CO_2), and sulfur compounds. **2.** a purified form of the above consisting almost entirely of methane gas. It is piped long distances and used as a fuel for heating, cooking, producing electricity, etc., and as a raw material for making ammonia, carbon black, and many other products.

natural habitat *n.* the environment where an animal or plant occurs naturally.

natural history *n.* the study of natural objects, organisms, and events, including ecology, geology, botany, zoology, anthropology, etc., especially in the early development of these sciences.

naturalist *n.* (Latin *naturalis*, pertaining to nature + *ista*, practitioner) **1.** one who carefully observes and studies nature; e.g., a botanist or zoologist. **2.** one who believes in adhering to nature and minimizing human changes.

naturalized *adj.* **1.** introduced into and established in a new place or environment, as a naturalized plant or animal. **2.** accepted as a nonnative citizen; a citizen who was born in another nation.

natural levee *n.* a raised bank along a stream formed by the deposition of sediment (usually mostly sand) when the stream floods.

natural resource *n.* any useful thing in nature, such as coal, oil, minerals, water, air, arable soil, vegetation, and wildlife.

Natural Resources Conservation Service (NRCS) *n.* a federal agency in the U.S. Department of Agriculture responsible for providing leadership in soil and water conservation and related topics throughout the U.S.A. and, often, internationally. It provides free technical assistance relative to land use to individuals and organizations. This agency was called the Soil Conservation Service from 1935 to 1995.

natural science *n.* any science that deals with observable things in nature; e.g., biology, chemistry, geology, physics, and soil science as distinguished from theoretical topics such as mathematics and philosophy.

natural selection *n.* the adaptive process whereby species change gradually from one generation to the next as individuals that are well adapted to their environment thrive and outnumber less successful individuals. See Darwin, endangered species, survival of the fittest.

nature *n.* (Latin *natura*, conditions at birth) **1.** the collection of living and nonliving things that exist independent from or in spite of human influence. **2.** the scenery or landscape appearance when not obstructed by human structures or other changes. **3.** the inherent characteristics of an organism or thing. **4.** inborn disposition, as human nature; the inherited tendencies and instincts of a person or animal. **5.** the combination of all forces at work in the universe.

nature study *n.* direct observation of plant and animal life, especially in a natural setting and in a popular rather than a technical manner.

nature trail *n.* a footpath through a forest, wilderness, garden, or other suitable area maintained for people to observe and study plants and animals in their natural environment, usually accompanied by a guide book and/or placards along the path.

Nautical Almanac *n.* a book published annually by the U.S. Naval Observatory and the Nautical Almanac Office, Royal Greenwich Observatory. It contains astronomical data for the sun, moon, and various stars and other information useful for marine and celestial navigation.

nautical mile *n.* a unit of distance commonly defined as 1 minute of arc on a great circle route around the Earth, specified equal to 1852 m (6076.11549 ft or 1.15 mi) in 1929 by the International Hydrographic Bureau. Also called a sea mile.

navigable *adj.* (Latin *navigabilis*, suitable for sailing) **1.** wide, deep, and open enough for ship traffic, as a navigable river. **2.** able to be steered, as a navigable ship, aircraft, or missile.

Neanderthal *n.* or *adj.* **1.** a very robust human species (*Homo sapiens neanderthalensis*) that lived in Europe and part of Asia during the late Pleistocene Period, originating about 300,000 years ago and becoming extinct about 10,000 years ago. Named for a German valley (Neander Tal), where the remains of Neanderthals were first identified. **2.** a person who is extremely backward, out-of-date, and often reactionary.

neap tide *n.* an ocean tide that occurs midway between spring tides and has the smallest range between high and low tide for a particular site. Neap tide happens once every 14 or 15 days, when the moon is at first or last quarter—the times when a line from the moon to the Earth forms a right angle with a line from the Earth to the sun. See spring tide.

neat's-foot oil *n.* a pale yellow oil obtained by boiling the feet and shinbones of cattle; used mostly to soften and preserve leather.

neck *n.* (Anglo-Saxon *hnecca*, neck) **1.** the part of a human or animal that connects the head to the rest of the body. **2.** the part of a shirt or other garment that goes around the neck of a person. **3.** a narrow part connecting two larger parts of an object, as the neck of a violin or the neck of a leg bone. **4.** a narrow strip of land bordered by water (e.g., an isthmus or a cape), contrasting material, or distinctly different topography on both sides. **5.** a narrow strip of water (e.g., a strait) bordered by land on both sides. **6.** the smooth part between the threads and the head of some bolts.

neck ail *n.* an ailment that causes emaciation and listlessness in cattle; a result of cobalt deficiency.

necrosis *n.* (Greek *nekrōsis*, death) death and/or decay of part of a plant, animal, or human.

nectar *n.* (Greek *nektar*, the drink of the gods) **1.** the beverage that was supposed to make a Greek god immortal. **2.** the sweet secretion produced by most flowers and used by bees to make honey. **3.** any delicious beverage, especially one based on fruit juice.

needlefish *n.* any of several saltwater fish (family Belonidae) with long, sharp beaks and pointed teeth. See gar.

neem *n.* (Hindi *nim*, neem) **1.** the neem tree; a tall tropical tree (*Azadirachta indica*, family Meliaceae) that grows in most of India and in nearby tropical areas. Its leaves are used in poultices, and its juices are used to treat fevers and diseases. Also called margosa. **2.** an extract from the leaves and seeds of the tree used to kill tapeworms and insects; a natural insecticide that paralyzes the digestive system of an insect but is harmless to birds and animals.

nekton *n.* (Greek *nēkton*, swimming) all of the aquatic organisms that swim (not just float) in either freshwater or seawater, from tiny insects, amphibians, and fish to whales.

nematocide or **nematicide** *n.* (Greek *nēmatos*, thread + Latin *cida*, killer) a chemical meant to kill nematodes.

nematode *n.* (Latin *nematoda*, nematode) a nonsegmented, threadlike worm (phylum Nematoda), including parasites on nearly all species of animals and plants of the world. Some are called hookworms, pinworms, roundworms, or threadworms.

nematology *n.* (Greek *nēmatos*, thread + *logos*, word) the branch of zoology that studies nematodes, usually including their effects on plant and animal life.

nemoral *adj.* (Latin *nemoralis*, woods) pertaining to a grove of trees or other wooded area.

neocolonial *adj.* (Greek *neos*, new + Latin *colonia*, colony + *alis*, pertaining to) **1.** pertaining to a newly established colony of people. **2.** having to do with the entrance of new plant cover into an area, especially a previously barren area. **3.** the status of a new kind of animal population in an area.

neo-Darwinism *n.* a modified form of Darwin's theory of evolution, using modern understanding of genetics to explain how natural selection works in the evolution of plants and animals. Named after Charles Darwin (1809–1892). See Darwinism, survival of the fittest.

neodymium (Nd) *n.* (Greek *neos*, new + *didymos*, twin) element 60, atomic weight 144.24; a bright, silvery rare-earth metal that quickly tarnishes in air; used in lasers and for coloring glass and enamels.

neolithic *adj.* (Greek *neos*, new + *lithikos*, pertaining to stone) pertaining to the latter part of the Stone Age (from about 8000 to 3500 years ago); of the time when humans developed polished stone tools and weapons, textiles, and pottery and began raising cattle and cultivating crops.

neomycin *n.* (Greek *neos*, new + *mykēs*, fungus + *in*, chemical) an antibiotic ($C_{23}H_{46}N_6O_{13}$) obtained from cultures of actinomycetes (*Streptomyces fradiae*). It is effective against a broad range of common bacteria, including *Salmonella* sp., *Shigella* sp., and *Aerobacter aerogenes*, and is used especially to treat infections of the skin and eye.

neon (Ne) *n.* (Greek *neon*, new) element 10, atomic weight 20.1797; a noble gas that does not form any known compounds. It is obtained by distillation of liquid air and is used in electric signs (neon lights).

neonate *n.* (Latin *neonatus*, newborn) a newborn human or animal, especially one less than four weeks old.

neoplasia *n.* (Greek *neos*, new + *plasis*, molding) **1.** growth of a neoplasm. **2.** growth of any new tissue.

neoplasm *n.* (Greek *neos*, new + *plasma*, formed) new growth of abnormal tissue; e.g., a tumor.

neotenin *n.* (Latin *neotenia*, new stretching) a juvenile hormone that controls growth of an insect larva.

neoteny *n.* (Latin *neotenia*, new stretching) **1.** production of offspring by larvae or other juvenile forms of an insect (the adult phase is eliminated). Also called pedogenesis. **2.** arrested development that causes adults to retain features from earlier stages of their evolutionary development.

nepheloid *adj.* (Greek *nepheloeidēs*, a cloud form) cloudy; e.g., nepheloid water or nepheloid urine.

nepheloid layer *n.* cloudy water above the sea floor near a land mass where a boundary current keeps fine particles of clay and organic matter in suspension.

nephology *n.* (Greek *nephos*, cloud + *logos*, word) the branch of meteorology that deals with clouds and their relationships to weather and climate.

nephometer *n.* (Greek *nephos*, cloud + *metron*, measure) an instrument for measuring the percentage of cloud cover.

Neptune *n.* (Latin *Neptunus*, Neptune) **1.** one of the outermost planets, located beyond Uranus but inside most of the orbit of Pluto. **2.** the Roman god of the sea.

nerite *n.* (Greek *nēritēs*, a kind of sea snail) a mollusk (*Nerita* sp.) found mostly in tropical climates.

neritic zone *n.* the shallow water part of an ocean or lake from the low-tide line to a depth of about 100 fathoms (600 ft or 180 m); the zone penetrated by enough sunlight for green plants to carry on photosynthesis.

nerve *n.* (Latin *nervus*, nerve) **1.** a neuron; either a sensory (afferent) nerve that detects pain, sound, odor, or taste and transmits a signal accordingly or a motor (efferent) nerve that transmits an action impulse. **2.** a sinew or tendon, as to strain every nerve (the original meaning but not much used now). **3.** courage, strength, and determination, as the nerve to keep going. **4.** emotional control. **5.** audacity, gall, brazenness.

nervous system *n.* the network of nerves and related parts that control the sensations and actions of an organism, including the brain, spinal cord, nerves, and ganglia.

nesosilicate *n.* (Greek *nêsos*, island + Latin *silicis*, flint + *atus*, having) a mineral composed of independent silicon tetrahedra (SiO_4) with enough octahedral cations between the tetrahedra to balance the charge; e.g., olivene [$(Mg, Fe)_2SiO_4$]. Sometimes called orthosilicate.

nester *n.* **1.** a squatter; a person who occupies land not his own, either legally or illegally, especially a farmer or homesteader who settled on what was previously grazing land in western U.S.A. during pioneer times. **2.** a bird that nests.

net duty of water *n.* the amount of irrigation water actually delivered to a field (excluding losses from the storage and canal systems).

net precipitation *n.* average annual precipitation minus the evaporation from an open water surface such as a lake; used as an estimate of the amount of leachate that might pass through materials deposited in a waste disposal site.

net primary productivity *n.* the amount of biomass produced in excess of the energy requirements of the primary producers; the amount available to support other living beings. Herbivores consume part of it; the rest goes to decomposers.

net reproductive rate (R_o) *n.* the capacity of an animal population to increase its numbers. An R_o of 2 indicates that its population can double in one generation.

nettle or **nettles** *n.* (Anglo-Saxon *netele*, nettle) **1.** any of several plants (*Urtica* sp., family Urticaceae) with small spines that scratch and sting, including European stinging nettle (*U. dioica*) and dog or burning nettle (*U. urens*). **2.** any of several other similar plants. *v.* **3.** to annoy, bother, or irritate. **4.** to cause a stinging sensation like that of the plant spines.

nettle tree *n.* **1.** a tree (*Celtis* sp., family Ulmaceae) that resembles an elm tree, including the American hackberry (*C. occidentalis*) and the European honeyberry (*C. australis*) **2.** the Australian nettle tree (*Urtica gigas*) that attains heights up to 70 ft (21 m).

neuron *n.* (Greek *neuron*, nerve) a nerve cell; an elongated cordlike cell that transmits nerve impulses. An enlarged part at one end contains the cell nucleus and several branching dendrites; the other end branches into axon terminals.

neuropteran *n.* or *adj.* (Latin *neuroptera*, nerve winged) a carnivorous insect (family Neuroptera) with four membranous wings, including the lacewings, ant lions, dobsonflies, alderflies, fishflies, snakeflies, spongillaflies, and mantispids. Larvae of the spongillaflies feed on freshwater sponges; the others feed on other insects.

neutralism *n.* (Latin *neutralis*, neuter + *isma*, doctrine) **1.** a policy of not taking sides in disputes, especially in international conflicts. **2.** advocacy of such a policy. **3.** coexistence of two or more species that have no effect on each other. See ecological interaction.

neutrino *n.* (Latin *neuter*, neither + *ine*, chemical substance) an uncharged lepton with little or no mass (a subatomic particle that is very difficult to detect). See lepton.

neutron *n.* (Latin *neutral*, neuter + *on*, elementary particle) an uncharged component of the nucleus of an atom. It has a spin of $\pm\frac{1}{2}$ and a mass of 1.675×10^{-24} g or 1.0087 atomic mass units.

neutron bomb *n.* a type of nuclear bomb that would release mostly neutrons, thereby killing people and animals but minimizing property damage and residual radioactivity.

neutron probe *n.* a device used to measure water content in soil and other media. A unit containing a neutron source and a detector is introduced through an aluminum access tube (typically 2 in. or 5 cm in diameter); the hydrogen in water deflects the neutrons and increases the frequency of neutrons returning to the detector.

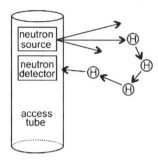

Neutron probe (schematic) and a possible neutron path

Newcastle disease *n.* a contagious disease of poultry and wild birds caused by a virus (*Tortor furens*); characterized by pneumonia and nervous disorder. It may be transmitted to humans as a mild conjictivitis. Sale of poultry from diseased areas is prohibited. Also called avian pneumoencephalitis or California flu.

new moon *n.* the time of the lunar month when the moon passes most directly between the Earth and the sun so only the dark side of the moon is seen from Earth. Note: A solar eclipse occurs at the time of the new moon when the sun-moon-Earth alignment is exact. See eclipse, full moon, moonarian.

newt *n.* (Anglo-Saxon *efete*, newt) **1.** a colorful salamander (especially *Triturus* sp. and *Notophthalmus* sp., family Salamandridae); an amphibious lizard no more than 6 or 7 in. (maximum 15 to 17 cm) long with a slender body and flattened tail that is called a newt when it lives in water or an eft when it lives on land. It commonly occurs in North America, Europe, and northern Asia. **2.** any of several other salamanders.

newton (N) *n.* a unit of force in the SI (metric) system equal to that required to accelerate a mass of 1 kilogram by a speed of 1 meter per second per second; $1 \text{ N} = 1 \text{ kg m s}^{-2}$. Named for Sir Isaac Newton.

Newton, Sir Isaac *n.* an English philosopher and mathematician (1642–1727) known for his system of calculus and for developing Newtonian physics, including the law of gravitation and three laws of motion.

Newton's law of gravitation *n.* every particle of matter attracts every other particle of matter with a force whose magnitude is proportional to the product of their respective masses and inversely proportional to the square of the distance between their respective centers of mass:

$$F = (\text{G} \times m_1 \times m_2)/d^2$$

where F = the attractive force, G = the gravitational constant (6.670×10^{-8} dyne-cm^2/g^2), m_1 and m_2 are the two masses, and d is the distance between their centers of mass. Also called the law of gravity.

Newton's laws of motion *n.* three principles expressed by Sir Isaac Newton and used as the basis for classical mechanics: (1) A particle remains at rest or moves with constant speed in a straight line as long as it is not subject to external forces. (2) The acceleration of a particle is directly proportional to the resultant of all external forces acting on the particle and is inversely proportional to its mass. (3) When two particles interact, the forces they exert on each other are equal in magnitude but opposite in direction; for every action there is an equal and opposite reaction.

Newton's rings *n.* concentric rings showing a rainbow of color (or of shades if all colors are not present in the light) where light is refracted by passing through a transparent curved surface; e.g., on the surface of a soap bubble.

niacin *n.* nicotinic acid ($C_6H_5NO_2$); a water-soluble vitamin of the B complex (vitamin B_3) that helps metabolize food to provide energy. Niacin is commonly added to flour as a nutritional supplement and is used to prevent or treat pellagra. Foods naturally high in niacin include yeast, liver, peanuts, fish, and lean meats. See pellagra.

niaye *n.* a freshwater swamp with drainage blocked by coastal dunes.

niche *n.* (Latin *nidus*, a nest) **1.** a specific set of environmental conditions especially favorable for a particular species of plant or animal. See biotope. **2.** a job or opportunity that is especially suitable for a particular person or organization, as his niche in society or a niche in the market. **3.** a recess in a wall where an object (or objects) can be displayed. **4.** a defect, cranny, or depression in a surface that is mostly smooth.

niche amplitude or **niche breadth** *n.* an organism's tolerable range in one of the environmental characteristics of its niche. An organism with a narrow niche amplitude is called a specialist; one that tolerates a wide range of conditions is called a generalist.

niche diversity *n.* variations that permit two or more organisms to occupy what appears to be the same niche without being eliminated by competition; e.g., plants with different rooting depths, trees with different canopy height and shade tolerance, birds with different feeding habits or nesting sites, predatory animals that hunt different prey, etc.

nick *n.* (Norse *hnykla*, to wrinkle) **1.** a small cut, notch, or chip indenting the edge or surface of an object. **2.** a small wound that indents the skin. **3.** a small groove that guides the placement of type for printing. **4.** a break in one of the two strands in a DNA or RNA molecule.

nickel (Ni) *n.* (German *Nickel*, Satan) **1.** element 28, atomic weight 58.7; a hard, ductile element with a silvery appearance. It is used to make batteries and corrosion-resistant alloys and coatings. **2.** a micro-nutrient added to the list of essential elements for plants in 1992 when it was shown that nickel is required for iron absorption and for the breakdown of urea. However, it is more likely to be present in high enough levels to be toxic to plant growth than it is to be deficient. **3.** a coin worth 5 cents in either U.S. or Canadian money.

nicotine *n.* (French *nicotiane*, tobacco plant) a toxic alkaloid ($C_{10}H_{14}N_2$) derived mostly from the leaves of tobacco; one of the hazardous, addictive materials found in tobacco smoke. Readily absorbed through the lungs, skin, or gut, it acts as a short-term stimulant and long-term depressant and is a leading cause of lung cancer. Nicotine has been used as a greenhouse fumigant and as an insecticide.

nidicolous *adj.* (Latin *nidus*, nest + *colous*, inhabiting) remaining in the nest after hatching, as a young bird not yet able to fly, in contrast to nidifugous.

nidifugous *adj.* (Latin *nidus*, nest + *fugous*, taking flight) leaving the nest shortly after hatching, in contrast to nidicolous.

nidus *n.* (Latin *nidus*, nest) **1.** a nest, especially one where insects or spiders lay their eggs. **2.** a place where a disease organism can infect a body. **3.** a place where spores or seeds can develop.

night-blooming (nocturnal) plants *n.* plants that bloom only at night. Examples: night jasmine (*Nyctanthes arbor tristis*), evening primrose (*Oenothera biennis*), cacti of genera *Hylocereus*, *Nyctocereus*, *Peniocereus*, and *Selenicereus*.

night crawler *n.* a large earthworm (family Lumbricidae) often seen crawling when the soil is moist near dawn or after a rain and highly prized as fish bait. See earthworm.

nightshade *n.* (Anglo-Saxon *nihtscada*, nightshade) **1.** any of several herbs, vines, or shrubs (*Solanum* sp., family Solonaceae) with star-shaped white or purplish flowers, orange berries, and poisonous sap; e.g., bittersweet nightshade (*S. dulcamara*), black nightshade (*S. nigrum*), ball nightshade, silverleaf nightshade (*S. elaegnifolium*), or horse nettle (*S. carolinense*). **2.** a family of plants (Solonaceae) that includes all of the above plus potatoes (*S. tuberosum*), eggplant (*S. melongena*), henbane (*S. niger*), Jerusalem cherry (*S. pseudocapsicum*), tomatoes (*Lycopersicon* sp.), tobacco (*Nicotiana* sp.), petunias (*Petunia* sp.), deadly nightshade or belladonna (*Atropa bella*), etc.

night soil *n.* human feces and urine used as a fertilizer. Although highly regarded in olden times and still used in some developing countries, especially in Asia, its use is prohibited for health reasons in most developed nations.

Nile crocodile *n.* a large crocodile (*Crocodylus niloticus*, family Crocodylidae) that grows to a length of about 20 ft (6 m) and is ferocious enough to attack livestock and people. Once common in all of Africa except in the Sahara Desert and the Mediterranean coast, it is now rare in many areas.

nimbostratus cloud *n.* a low, dark, uniform layer of clouds that commonly precipitates rain, snow, or sleet but has no thunder or lightning. Also called rain cloud.

nimbus *n.* (Latin *nimbus*, rain cloud) **1.** a combining form meaning rain cloud, especially one that produces violent rain, as in nimbostratus or cumulonimbus. **2.** an aura around a person or thing, as a bright disk, cloud, or halo around a head or image in a painting. **3. Nimbus** any of several polar-orbiting satellites launched by the U.S.A. for meteorological and environmental research.

niobium (Nb) *n.* (Greek *Niobē*, a Greek goddess + *ium*, element) element 41, atomic weight 92.90638; a soft, ductile, shiny white metal used in alloys to improve strength, in special arc-welding rods, and as a superconductor. Also called columbium (Cb).

nipa *n.* (Malay *nipah*, nipa) a large palm tree (*Nipa frutescens*) native to India, Australia, and the Philippines. Its sap is used to make an alcoholic drink, and its leaves are excellent for use as thatch on roofs of houses and for making baskets, etc.

nit *n.* (Anglo-Saxon *hnitu*, nit) **1.** the egg of a parasitic insect, such as a louse, often attached to a hair or thread. **2.** the young that hatch from such an egg. **3.** a unit of luminosity; 1 candela/m^2.

niter or **nitre** *n.* (Latin *nitrum*, niter) potassium nitrate (KNO_3), a highly soluble salt that forms naturally in soils and accumulates in certain hot, dry areas by evaporation at the soil surface (often accompanied by sodium nitrate, known as soda niter or Chile saltpeter). It is used in making gunpowder, fertilizer, and preservatives. Also called saltpeter.

niton *n.* (Latin *nitere*, to shine + *on*, gas) an early name for radon; element 86, a heavy, radioactive noble gas.

nitrate (NO₃) *n.* (Latin *nitrum*, niter + *ate*, salt) **1.** a salt of nitric acid; nitrogen surrounded by three oxygens, forming an ion with one negative charge that balances the positive charge of a cation to form a salt (e.g., niter, KNO_3). Almost all such salts are highly soluble in water; nitrate ions are absorbed by plant roots as a readily available form of nitrogen. Excess nitrates in soil are subject to leaching and accumulation in groundwater and surface water. See eutrophic, methemoglobinemia. **2.** an ester of nitric acid; an organic compound formed by reacting a nitrate group with an alcohol.

nitrate toxicity *n.* actually nitrite toxicity that can result from nitrates being reduced to nitrites. Public health standards usually limit nitrates in drinking water to 10 ppm nitrogen (10 mg N/L = 44 mg NO_3/L). Certain vegetables (kale, spinach, celery, beets, radishes, etc.) and forage plants (sudangrass, immature cereal grains, silage corn, Kentucky bluegrass, tall fescuegrass, bromegrass, etc.) tend to accumulate relatively high nitrate concentrations. See methemoglobinemia.

nitric *adj.* (Latin *nitrum*, native soda + *icus*, of) **1.** pertaining to nitrogen, especially to niter. **2.** a more oxidized form of nitrogen than nitrous, especially the pentavalent form (N^{5+} or NO_3^-).

nitric acid (HNO₃) *n.* the acid that reacts with bases to form nitrates; a slightly yellowish, fuming acid that is soluble in water and has strong oxidative powers. It is highly corrosive and caustic, and its fumes can be suffocating.

nitric oxide (NO) *n.* a colorless, poisonous gas formed by the partial oxidation of nitrogen at high temperatures and pressures, as in an internal combustion engine, or by microbial activity. It contributes to smog and can be further oxidized to nitrogen dioxide and nitrates.

nitrification *n.* (French *nitrifier*, to form nitrates + *ic*, containing + *atio*, action) **1.** the oxidation of ammonium (NH_4^+) salts to nitrites (NO_2^-) followed by the oxidation of nitrites to nitrates (NO_3^-). In soil, this process is carried out by bacteria (mostly by *Nitrosomonas* sp. and *Nitrosococcus* sp. for the first step and *Nitrobacter* sp. for the second). See nitrogen cycle. **2.** the process of forming nitrogen compounds, especially nitrates. **3.** the process of accumulating nitrates.

nitrifier *n.* (French *nitrifier*, to form nitrates) something that converts other nitrogen compounds (esp. ammonia or ammonium compounds) to nitrates, usually a bacterium.

nitrile *n.* (Latin *nitrum*, niter + *ilis*, like) an organic compound that has a nitrogen triple-bonded to a carbon.

nitrilotriacetic acid *n.* a combustible white crystalline chemical, $N(CH_2COOH)_3$, used as a replacement for phosphorus compounds in detergents to reduce the eutrophication hazard to streams and bodies of water.

nitrite (NO₂) *n.* (Greek *nitron*, native soda + *ite*, partly oxidized salt) **1.** a salt of nitrous acid; an incompletely oxidized form of nitrogen; an anion with one negative charge (NO_2^-). **2.** an ester formed by the reaction of nitrous acid with an alcohol (and releasing a molecule of water).

Nitrobacter *n.* a genus of bacteria (family Nitrobacteraceae) that oxidize nitrites (a harmful form of nitrogen) to nitrates (NO_3^-) (a less harmful/extremely beneficial form) in the environment.

nitrogen (N) *n.* (Greek *nitron*, native soda + *genēs*, producing) element 7, atomic weight 14.0067. Nitrogen composes about 78% of the Earth's air (by volume, excluding water; 75.6% by weight) in the form of dinitrogen (N_2), a colorless, odorless, tasteless inert gas. Nitrogen is an essential macronutrient for plants, animals, and people and is used to make ammonia, nitric acid, explosives, fertilizers, dyes, etc. See nitrogen cycle.

nitrogenase *n.* (Greek *nitron*, native soda + *genēs*, producing + *ase*, enzyme) an enzyme system that converts dinitrogen (N_2) into combined nitrogen in the form of amines by using adenosine triphosphate (ATP) from a host plant as an energy source to reduce nitrogen and hydrogen simultaneously; includes dinitrogenase (a protein that contains iron and molybdenum) and dinitrogenase reductase (an iron protein). The two component enzymes (and some alternative forms known as alternative nitrogenases) require an anaerobic environment because they are inactivated by oxygen. Nitrogenase is able to break the triple bonds of several molecules, including those of dinitrogen ($N{\equiv}N$) and acetylene ($HC{\equiv}CH$).

nitrogen cycle *n.* the complex cycling of nitrogen among its various reservoirs, from the atmosphere to the soil-plant system by nitrogen fixation and through a myriad of organic and inorganic compounds involving numerous subcycles. Many of the processes are microbial in nature.

nitrogen dioxide (NO₂) *n.* a highly toxic gas with a reddish brown color and a pungent odor. An air pollutant produced by internal combustion engines and a major component of photochemical smog, it attacks the lining of the lungs of humans and animals and can cause death. Also called nitrogen peroxide.

nitrogen fixation *n.* **1.** the conversion of atmospheric nitrogen (N_2) to organic compounds such as amines ($-NH_2$) by either free-living bacteria (*Azotobacter*

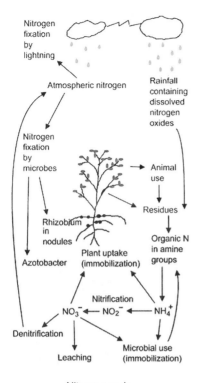

Nitrogen cycle

sp., *Clostridium* sp., etc.) or symbiotic bacteria (*Rhizobium* sp. and *Bradyrhizobium* sp.), actinomycetes (*Frankia* sp.) or blue-green algae (Cyanobacteria). **2.** conversion of atmospheric nitrogen to oxides of nitrogen by lightning or other processes. **3.** synthesis of nitrogen compounds (especially ammonia) from atmospheric nitrogen.

nitrogen-free extract *n.* the remaining material of undetermined nature after determining moisture, crude protein, ash, crude fiber, and ether extract (crude fat) in animal feed analyses.

nitrogenous *adj.* (Greek *nitron*, native soda + *genes*, producing + *ous*, chemical) containing nitrogen as a part of its chemical composition.

nitrogenous waste *n.* waste that contains nitrogen compounds, especially wastewater that contains oxidizable dissolved inorganic nitrogen such as ammonium and nitrite compounds.

nitrogen oxide (NO$_x$) *n.* any of the gases composed of nitrogen and oxygen, including nitrous oxide (N$_2$O), nitric oxide (NO), nitrogen dioxide (NO$_2$), and the less common forms, NO$_3$, N$_2$O$_3$, N$_2$O$_4$, and N$_2$O$_5$. Nitric oxide and nitrogen dioxide are considered to be serious air pollutants.

nitroglycerin *n.* (Greek *nitron*, niter + *glykeros*, sweet) a pale yellow, oily liquid [C$_3$H$_5$(NO$_3$)$_3$] that explodes violently on concussion or exposure to heat. It is made by reacting glycerol with nitric acid, is used to make dynamite and rocket propellants, and is a heart stimulant medicine.

nitrosamine *n.* (Latin *nitrosus*, full of natron + *amine*, ammonia chemical) an organic compound with a nitroso group (nitrogen double-bonded to oxygen) attached to a nitrogen that is part of an organic structure (R$_2$NN=O). Nitrosamines form in meat by conversion of nitrites, and most of them are carcinogenic.

nitrous *adj.* (Latin *nitrosus*, full of natron) **1.** pertaining to nitrogen that is less oxidized than the corresponding nitric form; e.g., nitrous oxide (N$_2$O) versus nitric oxide (NO). **2.** containing nitrogen, especially trivalent nitrogen as in nitrous acid (HNO$_2$).

nitrous oxide (N$_2$O) *n.* the least oxidized form of nitrogen oxide (NO$_x$), a colorless gas with a sweet taste and smell used as an anesthetic in dental work and surgery and as a propellant in aerosol cans. Also called laughing gas.

nival *adj.* (Latin *nivalis*, of snow) anything related to, living in, or growing in or under snow.

niveous *adj.* (Latin *nivis*, snow + *osus*, full of) very white; resembling snow.

noble gas *n.* a nonreactive gas; one of the elements in group 8a or 0 of the periodic table: helium, neon, argon, krypton, xenon, or radon. These elements are nonreactive because their atoms have a full outer shell of electrons.

nocturnal *adj.* (Latin *nocturnalis*, of the night) **1.** active at night, as an animal or person that sleeps during the day and is active at night. **2.** opening blooms at night, as a nocturnal plant. **3.** occurring at night, as a nocturnal event. See diurnal.

nocuous *adj.* (Latin *nocuus*, harmful) likely to cause injury or negative effects; harmful, poisonous, or noxious; opposite of innocuous.

nodding pogonia *n.* a rare North American orchid (*Triphora trianthophora*), native to eastern U.S.A., that produces nodding flower buds and white or pink blossoms in groups of three.

node *n.* (Latin *nodus*, knot) **1.** a knot, swelling, joint, or other point of concentration on a string, stem, or line. **2.** the knoblike swelling on a plant stem where a leaf or branch emerges. **3.** a central point with ties to several outlying points, as a node of operations. **4.** a stationary point in a standing wave. **5.** a point where a satellite crosses the orbital plane of its planet or a planet or other celestial body crosses the solar ecliptic.

nodical month *n.* a lunar month of 27.2122 days based on the repeat time for the moon's crossing of the ecliptic plane of the Earth's path around the sun, especially the time from its passage from one ascending node to the next ascending node. Also called dracontic (sometimes spelled draconic) month.

nodulation *n.* (Latin *nodulus*, little knot + *atio*, action) **1.** the process of forming nodules **2.** the presence of nodules.

nodule *n.* (Latin *nodulus*, little knot) **1.** a tubercle where bacteria (*Rhizobium* sp.) live on legume roots and fix atmospheric nitrogen. **2.** a small node in a plant stem or root. **3.** a small mass or lump; e.g., a concretion of calcium carbonate ($CaCO_3$), iron oxide (e.g., Fe_2O_3), or manganese dioxide (MnO_2) developed around a point or line in a soil. **4.** a small, hard, somewhat rounded mass that can be detected by touch, such as a corn on a toe.

nodulin *n.* (Latin *nodulus*, little knot + *in*, protein) a protein that has either a nodule-specific or a nodule-amplified biochemical function in nitrogen-fixing nodules.

noise abatement zone *n.* a designated area where the noise level is legally limited by a local law or ordinance.

noise control *n.* efforts to limit noise pollution by reducing the emission of noise (e.g., by use of a muffler on an engine) or by installing noise barriers (e.g., a line of trees).

nomad *n.* (Greek *nomas*, pasturing flocks) **1.** a person who customarily moves from place to place (usually as a member of a group) in search of food, water, shelter, and forage for grazing livestock. See transhumance. **2.** any person who moves frequently.

nonconforming use *n.* land use that violates a local zoning ordinance; usually something that predates the ordinance and is allowed under a grandfather clause.

nonfat dry milk *n.* milk solids obtained by pasteurizing the milk, removing the fat, and drying the milk under a partial vacuum until the water content is less than 5% and the fat content less than 1.5%. Also called powdered milk.

noninfectious chlorosis *n.* yellowing of plant leaves resulting from an abiotic (nonpathogenic) environmental condition such as waterlogged soil, strongly acid or alkaline soil, or a deficiency of nitrogen, potassium, calcium, magnesium, sulfur, or a micronutrient.

noninversive tillage *n.* tillage with a chisel or other implement that loosens the soil without turning it and with minimal mixing.

nonnodulating *adj.* failing to produce nodules on the roots of a plant species that is normally expected to have them; e.g., nonnodulating soybeans.

nonnutritive sweetener *n.* a synthetic, low-calorie substitute for sugar. Examples: sodium or calcium salts of saccharin ($C_7H_5NO_3S$) that are 200 to 700 times as sweet as sucrose (the usual table sugar) and cyclamates that are 30 times as sweet as sucrose. Note: The use of cyclamates in the U.S.A. has been banned by the EPA since 1970 because they have carcinogenic effects.

nonpersistent *adj.* (Latin *non*, not + *persistens*, persisting) of short duration rather than long-lasting; e.g., a pesticide that breaks down to nonpolluting substances in a few days.

nonpoint source *n.* a source of pollution that comes from a broad area, such as runoff from fields, parking lots and streets, surface mining, etc., rather than from a point source, such as a smokestack or a sewage pipe.

nonpolar solvent *n.* a solvent such as benzene, chloroform, gasoline, and other hydrocarbons whose molecules are not polarized (as distinguished from molecules like water and alcohol that have positive and negative poles). Nonpolar solvents are generally able to dissolve nonpolar substances like fats and oils but are not good solvents for ionic substances such as salts that are highly soluble in water.

nonrenewable *adj.* (Latin *non*, not + *re*, again + *novus*, new + *abilis*, able to) **1.** not replaceable; produced only over very long periods of time (geologic ages). For example, resources such as coal and other products of mining, petroleum, and soil, in contrast to renewable products like field crops, trees, and water power. **2.** limited to the stated period without the right of further extension, as a nonrenewable lease.

nonselective *adj.* (Latin *non*, not + *selectus*, gathered apart + *ivus*, tendency) not choosing or selecting; applicable to a broad spectrum; e.g., a herbicide that kills many different weeds or an insecticide that kills a wide variety of insects.

nonsweating *adj.* not able to perspire, usually because the species (e.g., cattle, chickens, dogs, sheep, and swine) does not have sweat glands.

nonsymbiotic nitrogen fixation *n.* conversion of atmospheric nitrogen (N_2) to a fixed form such as an amine ($-NH_2$) by free-living microbes, including *Azotobacter* sp. (aerobic soil bacteria), *Clostridium* sp. (anaerobic soil bacteria), *Beijerinckia* sp. (aerobic bacteria in surface residues or soil), several facultative anaerobes under anaerobic conditions (e.g., *Klebsiella* sp., *Enterobacter* sp., and *Achromobacter* sp.), and blue-green algae (Cyanobacteria).

nontarget *adj.* (French *non*, not + *targette*, small shield) not the object or goal of a particular action or attack; e.g., a good pesticide controls the target species without harming nontarget organisms.

nontoxic *adj.* (Latin *non*, not + *toxikon*, poison) not poisonous; will not cause illness or death of plants and animals.

nontronite *n.* a form of smectite that has most of its aluminum replaced by iron (Fe^{3+}). See smectite.

noria *n.* (Arabic *nā ūrah*, noria) a waterwheel with buckets attached to its perimeter; used to lift water from a stream, especially for use in irrigation in Spain and the Orient.

normal *adj.* (Latin *normalis*, made square) **1.** resulting from natural circumstances; e.g., geologic erosion can be called normal erosion. **2.** common, ordinary, likely to occur in everyday life; a normal occurrence. **3.** close to average; within the usual range, as normal rainfall or a person with normal intelligence. **4.** healthy; free of disease. **5.** perpendicular, as a vertical line rising from a horizontal surface. **6.** having a chemical concentration of 1 equivalent weight per liter of solution. **7.** unbranched, as a straight-chain aliphatic hydrocarbon.

normal soil *n.* a term used by C. F. Marbut in his system of soil classification to describe a soil with a profile in equilibrium with its ambient environment; i.e., a soil that reflects the influence of the climate and vegetation of the area rather than extreme youth, wetness, or other local factor. The term was replaced with zonal soil in the 1938 U.S. system of soil classification to avoid the implication that other soils are abnormal; neither of these terms is used in the current system of Soil Taxonomy.

north *n.* (Anglo-Saxon *north*, north) **1.** a cardinal direction; the direction faced by a person with the rising sun to the right or the setting sun to the left; the opposite of south. See agonic line for the relationship between true north and compass north. **2.** a region or area that is located in a northerly direction from a reference location; e.g., the far north. *adj.* **3.** located in a northerly direction, as the north side of a property. **4.** coming from a northerly direction, as a north wind.

North American Free Trade Agreement (NAFTA) *n.* an agreement to establish free trade among the U.S.A., Canada, and Mexico that began January 1, 1994, with tariffs on various items being eliminated over a period of 10 years. It includes detailed requirements for protecting the environment, especially in the treatment of effluents.

norther *n.* a sudden cold wind from the north, especially one that reaches Texas or the Gulf Coast.

north magnetic pole *n.* the Earth's magnetic pole that is closest to the North Pole. See magnetic pole, agonic line.

North Pole *n.* **1.** the northernmost point on Earth; the north end of the Earth's axis of rotation, located at 90° north latitude. **2.** a point in the heavens about 1° from the North Star, representing the northward extension of the Earth's axis of rotation.

North Star *n.* Polaris; the end star of the handle of the Little Dipper (Ursa Minor), located within 1° of due north of the Earth. The North Star is often located by using the two outer stars of the bowl of the Big Dipper (Ursa Major) as pointers.

north temperate zone *n.* that part of the Earth between the tropics and the Arctic Zone; i.e., between 23°27' north latitude (the Tropic of Cancer) and 66°33' north latitude (the Arctic Circle).

Norway rat *n.* the common rat (*Rattus norvegicus*) with a grayish brown body, long scaly tail, and light-colored underparts; a native of Asia but now found worldwide. It is a prolific, omnivorous animal that can swim, burrow through soil, and eat its way through wooden walls. The Norway rat will attack larger animals, including humans, is very destructive to the environment, and is a vector for many diseases. Also called barn rat, brown rat, sewer rat, or wharf rat. See black rat.

Norway spruce *n.* a hardy evergreen tree (*Picea abies*, family Pinaceae) with shiny green needles that is one of the most widely grown ornamental evergreen trees. There are many varieties, ranging from dwarf varieties only 1 to 2 ft (30 to 60 cm) tall to large trees up to 150 ft (45 m) tall.

nosema disease *n.* an intestinal disorder of honey-bees caused by protozoa (*Nosema apis*, family Nosematidae).

nose slope *n.* the area around the end of a hill or ridge where contour lines wrap around from one side to the other. A nose slope is relatively dry because the diverging slope lines cause water to spread rather than to concentrate.

notch *n.* (French *oche*, notch) **1.** an indentation, often either rectangular or V-shaped, cut into an edge or surface. **2.** such a mark made on a stick as a record or tally or in the ear of an animal to identify ownership. **3.** a deep, narrow valley between mountains. **4.** an opening for water to flow over a dam or spillway. **5.** an undercut used to control the direction a tree will fall when the main cut is made.

no-till, notill, or **no-tillage** *n.* or *adj.* the agricultural system of planting a new crop directly through the residues of the previous crop without plowing, disking, or other tillage. Also called zero tillage. See conservation tillage, minimum tillage.

noxious *adj.* (Latin *noxius*, harmful) **1.** harmful, injurious, and/or poisonous; e.g., noxious fumes such as carbon monoxide (CO) and hydrogen sulfide (H_2S). **2.** morally degrading or corrupting, as a noxious plan. **3.** an aggressive weed that is difficult to control; states compile lists of noxious weeds.

NPK *n.* nitrogen, phosphorus, and potassium—the three elements that must be present in guaranteed amounts for a fertilizer to be called a complete fertilizer.

NRGIS *n.* Natural Resources Geographic Information System. See geographic information system.

nuance *n.* (Latin *nubes*, a cloud + *antia*, condition) **1.** a barely detectable difference in shade or color. **2.** a very small difference in sound, taste, smell, or feel. **3.** a subtle difference in meaning or perception.

nuclear energy *n.* energy released by fission (splitting) of a heavy nucleus such as uranium 235 or plutonium 239 or by fusion (union) of deuterium or tritium to form helium. Fission is used in atomic reactors for generating electricity, powering submarines and other ships, and in atomic bombs. Fusion produces the energy released by the sun and other stars and is used in the hydrogen bomb.

nucleation *n.* (Latin *nucleatus*, having a kernel + *tion*, action) formation of a nucleus; e.g., the nucleation of a raindrop provides a center for water to condense.

nuclear power plant *n.* a facility that converts atomic energy into electricity by the fission of heavy atoms (either uranium 235 or plutonium 239) to make steam that drives turbines.

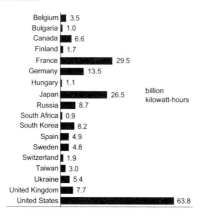

Electricity generated by nuclear energy, December 1999

nuclear waste *n.* radioactive materials produced as by-products of nuclear energy, usually classified as either low-level or high-level wastes. Low-level wastes are so slightly radioactive that they can be safely disposed in a properly designed sanitary landfill. However, high-level wastes (spent fuel, other materials that have been in or near the reactor, and part of the structure itself after it has been decommissioned) constitute an enduring problem that has not yet been solved.

nuclear winter *n.* a prediction by some scientists that multiple nuclear detonations in a nuclear war would raise enough dust to blot out the sun for a long enough time to drastically cool the Earth.

nucleosome *n.* (Latin *nucleus*, a small kernel + Greek *sōma*, body) a beadlike unit on the DNA molecule of a chromosome. It contains two molecules of each of four types of histone wrapped around a protein core. See chromosome, histone.

nucleus *n.* (Latin *nucleus*, a small kernel) **1.** a central core. A group or mass is arranged around its nucleus in an organized manner. **2.** a body in the protoplasm of plant, animal, and human cells containing chromosomes that carry the genes of inheritance. **3.** the central part of an atom containing the protons and neutrons that constitute almost all of the mass of the atom. Electrons orbit around the nucleus. **4.** a particle (often of dust) that serves as a center for water to condense and form a raindrop. **5.** a dense core in a variety of structures, such as a benzene ring in certain organic compounds or a mass of ice in the head of a comet. **6.** a beginning concept, as the nucleus of a plan or an idea. **7.** the content at the beginning, as the nucleus of a collection.

nudation *n.* (Latin *nudatio*, making bare) the process of making something bare or becoming bare. Example: toxic fumes can kill vegetation and cause nudation of the soil.

nugget *n.* (Anglo-Saxon *nug*, lump + *et*, small) **1.** a small lump of something, especially of gold or other precious metal. **2.** any small item of great value or importance, as a nugget of wisdom. **3.** the fused metal in a spot weld.

nuisance *n.* (Latin *nocere*, to annoy + *antia*, action) **1.** an annoyance or inconvenience, especially one that is repetitive. **2.** a person, animal, or thing that causes such an annoyance; a pest. **3.** a legal violation of certain standards or rights; e.g., erosion that causes sedimentation on someone else's property may be declared a nuisance in many jurisdictions.

nullah *n.* (Hindi *nālā*, brook or ravine) a gully or an intermittent stream in India or Pakistan.

nutant *adj.* (Latin *nutans*, nodding) drooping; having the top bent down, as a plant that is stressed by dry weather, insects, or disease.

nutation *n.* (Latin *nutatio*, nodding) **1.** nodding one's head, especially when drowsy. **2.** drooping or twisting of a plant caused by differential growth rates or by stress of dry weather, insects, or disease. **3.** the rotation of a plant during the day in response to the changing position of the sun, as the action of a sunflower head. **4.** the small 18.6-yr cyclic variation in the precession of the Earth's axis and equinoxes resulting from the combined influence of the gravity of the sun and moon. **4.** a similar variation in the axis of a spinning gyroscope.

nuthatch *n.* any of several small birds (family Sittidae) 3 to 6 in. (8 to 15 cm) long with relatively large heads and short tails, with bluish gray backs, black or brown caps, and white or red breasts, depending on species. Nuthatches eat insects from the bark of trees.

nutmeg *n.* (French *nois muguete*, nutmeg) a large, tropical, evergreen tree (*Myristica fragrans*, family Myristicaceae) native to the East Indies. Its fruit is a drupe that resembles a small pear containing a hard oval nut. **2.** a spice made by grating nuts from the tree. **3.** a similar spice made from other trees or shrubs of *Myristica* or other genera.

nut pine *n.* a pine tree (*Pinus* sp., family Pinaceae) that produces edible nutlike seeds; e.g., certain varieties of piñon pine (*P. monophylla* or *P. edulis*).

nutria *n.* (Spanish *nutria*, nutria) **1.** an aquatic South American rodent (*Myocastor coypus*) that resembles a small beaver. Also called coypu. **2.** the soft, short, brown fur of the animal, often used to make fur coats.

nutrient *n.* or *adj.* (Latin *nutriens*, nourishing) **1.** something that provides or conveys nourishment. **2.** an essential element for plants, animals, or people.

nutrient flux rate *n.* the amount of a nutrient that enters or leaves a nutrient pool per unit of time.

nutrient pool *n.* the supply of nutrients available for use by a growing plant. Also called available nutrient pool.

nutriment *n.* (Latin *nutrire*, to nourish + *mentum*, means) food; any substance that serves a living organism as a source of energy and/or promotes growth.

nutrition *n.* (Latin *nutritio*, a feeding) **1.** the bodily processes needed to ingest food, assimilate it, and use it for energy and growth. **2.** the nutriment that serves such purposes. **3.** the science and art that studies these processes and the nutrients needed by organisms, especially in humans.

nutritional anemia *n.* a deficiency in either the quantity or the quality of the blood resulting from inadequate intake of one or more essential elements (e.g., inadequate iron to make hemoglobin).

nutritional dystrophy *n.* See kwashiorkor.

nutsedge *n.* either of 2 Old World perennial, grasslike sedges, yellow nutsedge (*Cyperus esculentus*, family Cyperaceae) and purple nutsedge (*C. rotundus*), that grow up to about 3 ft (0.9 m) tall and have small aromatic nutlike tubers. Nutsedges spread by seeds and creeping tendrils and are weeds in fields and gardens in most of the U.S.A. and elsewhere.

nux vomica *n.* **1.** a tropical tree (*Strychnos nux-vomica*, family Loganiaceae) of the East Indies whose orangelike fruit contains poisonous alkaloids such as brucine ($C_{23}H_{26}N_2O_4 \cdot 4H_2O$) and strychnine ($C_{21}H_{22}N_2O_2$). **2.** the seeds contained in the fruit of the tree. **3.** strychnine; a medicine extracted from the seeds of the tree for use as a heart stimulant.

nyctitropism *n.* (Greek *nyktos*, night + *tropos*, turning + *isma*, action) turning of leaves or flowers of plants such as clovers into a nighttime position that is different than their daytime position.

O *n.* the chemical symbol for oxygen.

oak *n.* (Anglo-Saxon *ac*, oak) **1.** any of a large group of slow-growing, mostly deciduous trees and shrubs (*Quercus* sp., family Fagaceae) whose leaves turn to various shades of red, yellow, and brown in autumn. Some oaks live for hundreds of years and reach heights of 100 to 150 ft (30 to 45 m). Oaks bear nuts called acorns that are commonly eaten by forest animals and sometimes by humans. The bark of oak trees contains tannin that is used for tanning leather, etc. **2.** the hard, durable wood of an oak tree; commonly used to make furniture and in construction, especially for decorative trim. *adj.* **3. oak** or **oaken** made of oak, as an oak chair or an oaken bucket.

oak wilt *n.* a serious fungal disease (*Ceratocystis fagacearum*, also known in its imperfect stage as *Chalara quercina*, family Ophiostomataceae) that plugs water passages and causes oak leaves to wilt, discolor, and fall prematurely. Oak wilt usually begins in the upper part of the tree and the outer ends of the branches.

OAS *n.* See Organization of American States.

oasis *n.* (Greek *oasis*, a fertile spot) **1.** a spot in the desert where enough water is available to grow trees and other vegetation, usually with a spring or well that provides water for animals and people. **2.** a refuge; a quiet place in the midst of much activity and turmoil, as the library provides an oasis for students.

oat or usually **oats** *n.* (Anglo-Saxon *ate*, oat) **1.** a small-grain crop (*Avena sativa*, family Poaceae) grown mostly in the temperate zone for its edible seed or for annual forage. **2.** the seed from the crop, commonly fed to horses and other animals and also used in human food such as oatmeal (rolled oats). **3.** any of several other related grasses (*Avena* sp.) grown for seed or forage; some species are considered weeds (e.g., wild oats, *A. fatua*).

oatmeal *n.* (Anglo-Saxon *ate*, oat + *melu*, meal) **1.** oats that have been crushed or ground to form meal. **2.** such oats mixed with water or milk, cooked to a pasty consistency, and eaten as a breakfast cereal. Also called porridge.

OAU *n.* See Organization of African Unity.

obese *adj.* (Latin *obesus*, to eat away) excessively overweight; very fat.

oblate *adj.* (Latin *oblatus*, devoted) **1.** dedicated to a monastic life. **2.** flattened at the poles. An oblate spheroid such as the Earth has a larger circumference around the equator than it has around the poles.

obligate *v.* (Latin *obligatus*, bound) **1.** to bind; to legally or morally require that one do a certain thing. **2.** to commit or pledge funds or other resources to a specified purpose. *adj.* **3.** legally or morally constrained or required to act a certain way. **4.** able to live only in a certain way, as an obligate parasite or an obligate anaerobe (certain bacteria that cannot live in the presence of oxygen). Humans are obligate aerobes because they must have air to breathe. See facultative.

obliquity *n.* (Latin *obliquitas*, being oblique) **1.** an oblique condition. **2.** immorality; deviation from accepted conduct or thinking. **3.** deviation from parallelness or perpendicularity. **4.** the angle between the planes of a planet's equator and its orbit or between its axis and the perpendicular to its orbital plane. The Earth has an obliquity of 23.5° that varies from 22.5° to 24.5° over a 41,000-year cycle.

obscuration *n.* (Latin *obscuration*, a darkening) **1.** the act of making something dark, indistinct, or unclear. **2.** the condition of being darkened, as the obscuration of the moon during an eclipse or the obscuration of the sky by a heavy fog.

obsidian *n.* (Latin *Obsidianus*, stone of Obsius) black or nearly black, translucent volcanic glass with conchoidal fracture and a composition similar to that of granite.

obstruction *n.* (Latin *obstructio*, a blocking up) **1.** something that blocks or impedes flow, movement, or progress; an obstacle. **2.** an act of blockage or the condition of being blocked. **3.** an action that delays the activity of a person, group, or deliberative body; e.g., a filibuster.

oca *n.* (Quecha *oqa*, oca) an exceptionally hardy root crop (*Oxalis tuberosa*) grown by the Incas and still produced in the Andes for its edible tubers. There are many varieties; their various sugar content levels in the tubers yield sweet to sour yet pleasant flavors. They are grown as yams in New Zealand. Also called wood sorrel. Sometimes spelled oka.

occlude *v.* (Latin *occludere*, to close up) **1.** to block or close a passage or opening. **2.** to incorporate something notably different, as air bubbles occluded in glass or in rocks. **3.** to meet and fit closely, as her upper and lower teeth occlude properly. **4.** to form an occluded weather front.

occluded front *n.* a combined warm and cold front, typically formed when a cold front overtakes a warm front and pushes the warm air aloft, often producing heavy showers.

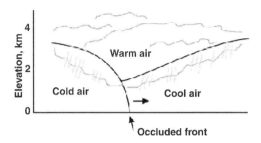

An occluded front moving from left to right, causing warm air to rise and produce showers

occult precipitation *n.* precipitation that is usually not measured; e.g., dew, rime, fog.

Occupational Safety and Health Administration (OSHA) *n.* an agency established in 1970 as a division of the U.S. Department of Labor to establish and enforce occupational safety and health standards.

ocean *n.* (Latin *oceanus*, ocean) **1.** the body of salt water that covers 70.8% of the Earth's surface to an average depth of 12,234 ft (3729 m). Ocean water contains an average of 3.5% salt, mostly sodium chloride (NaCl) but including many other salts, with calcium carbonate ($CaCO_3$) near saturation. **2.** a large subdivision of this body, especially the Antarctic, Arctic, Atlantic, Indian, and Pacific oceans. **3.** a large quantity or expanse of anything, especially something abstract, as an ocean of love. *adj.* **4.** oceanic; relating to, occurring in, or sailing on the ocean, as ocean depths, ocean fish, or ocean liner.

ocean depths *n.* the average depth of Earth's oceans is 12,234 ft (3729 m). The deepest water in the Pacific Ocean is 35,840 ft (10,924 m); Atlantic Ocean, 28,232 ft (8605 m); Indian Ocean, 23,376 ft (7125 m); Arctic Ocean, 17,881 ft (5450 m); Mediterranean Sea, 16,896 ft (5150 m). (Data from the U.S. Department of Defense.)

ocean dumping *n.* the practice of disposing of wastes in the ocean (banned under the Ocean Dumping Ban Act of 1988).

oceanic climate *n.* a climate with temperatures moderated by the presence of a nearby ocean.

oceanodromous *adj.* (Latin *oceanus*, ocean + Greek *dromos*, running) relating to fish that migrate to salt water; e.g., salmon.

oceanography or **oceanology** *n.* (Latin *oceanus*, ocean + *graphia*, writing or drawing or + *logos*, word) the geography of oceans. *Oceanography* is usually used for a scientific study of oceans, and *oceanology* is the practical application of such study.

ocelot *n.* (Nahuatl *ocelotl*, jaguar) a nocturnal wildcat (*Felis pardalis* or *Leopardus pardalis*) with black spots on a yellow or gray coat; common in Texas and in Central and South America. Rodents are a favorite food.

ocher or **ochre** *n.* (Greek *ōchra*, yellow ocher) a mixture of iron oxides and other associated earthy materials ranging in color from yellowish brown limonite ($2Fe_2O_3 \cdot 3H_2O$) to reddish brown hematite (Fe_2O_3). Ochers are used as pigments in such things as paints and sometimes are used as iron ores.

ochric epipedon *n.* an epipedon that is too thin, too light in color, or too low in organic carbon to qualify as a mollic epipedon in Soil Taxonomy.

octahedral *adj.* (Greek *oktaedros*, eight-sided) having eight surfaces in the form of an octahedron.

octahedron *n.* (Greek *oktaedron*, eight-sided) **1.** a solid figure with eight plane surfaces, shaped like two pyramids placed base to base. **2.** an arrangement of six atoms or ions, each centered where a point would be on a solid octahedron. See tetrahedron.

octane rating *n.* a system for evaluating the antiknock characteristics of gasoline and similar fuels; the percentage of isooctane in a mixture with normal heptane that has antiknock characteristics comparable to that of the fuel being tested.

octopus *n.* (Greek *oktōpous*, eight-footed) **1.** any octopod (*Octopus* sp.) with a soft saclike body, a large head with a mouth on the underside, and eight tentacles covered with suckers; a coastal-water, bottom-dwelling, oceanic cephalopod whose tentacles may span from a few feet up to 16 ft (4.8 m).

odonate *n.* (Greek *odontos*, tooth) any of several large, aquatic, predatory insects (order Odonata) with two pairs of membranous wings; e.g., a damselfly or a dragonfly.

oestrus *n.* See estrus.

offal *n.* (German *abfall*, offal) **1.** waste parts (especially entrails) from a butchered animal. **2.** rubbish; solid or liquid waste; anything discarded because it is considered useless.

Office of Solid Waste and Emergency Response (OSWER) *n.* an administrative unit in the Environmental Protection Agency that is responsible for overseeing the management of hazardous wastes and has teams that respond to spills of hazardous chemicals.

off-season *n.* **1.** an unfavorable part of the year; a time when plants cannot grow, birds may leave for a warmer climate, animals may hibernate, etc. **2.** a time of slow business activity or other form of poor performance.

offset disk *n.* a tillage implement with two gangs of large disks mounted in tandem so one gang moves soil to the right and the other to the left; used for either primary or secondary tillage.

offshore *adj. or adv.* **1.** in or on the water near the shore, as an offshore oil well, or they anchored the boat offshore. **2.** moving away from the land, as an offshore wind, or they pushed the boat offshore. **3.** in another country, as they relocated offshore.

Ogalala *n.* **1.** a Native American member of the Teton tribe of Sioux of the Black Hills Region of South Dakota. Also spelled Oglala. **2.** this tribe of Native Americans. **3.** the language spoken by these people. **4.** the largest aquifer of fossil groundwater in the U.S.A., underlying some 174,000 mi^2 (450,000 km^2) in eight states, from South Dakota to Texas. Also spelled Oglala or Ogallala.

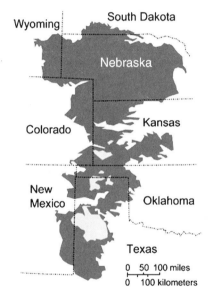

The Ogalala aquifer underlies most of Nebraska and parts of 7 other states. The water table has dropped 50 ft (15 m) or more in the lighter shaded areas.

ohm (Ω) *n.* the unit for electrical resistance in the SI system; the resistance in a circuit that passes a current of 1 ampere with an electromotive force of 1 volt. Named after George S. Ohm (1787–1854), a German physicist.

O horizon *n.* a usually thin surface soil horizon dominated by organic matter; e.g., the litter layer at the surface of an undisturbed soil, especially forest soil, or a peat or muck layer in a wetland soil. It sometimes is subdivided into undecomposed and partially decomposed plant parts.

oil *n.* (Latin *oleum*, oil). **1.** a triacylglycerol that is liquid at room temperature and miscible with organic liquids such as ether or alcohol, but not with water; a triple ester composed of three fatty acids linked to glycerol. It is used as a lubricant, fuel, cooking oil, etc. Also called a triglyceride. **2.** any liquid that looks, feels, and behaves like the liquid triacylglycerols. **3.** petroleum.

oil cake *n.* a mass of seeds (e.g., linseed, cottonseed, or soybean seed) or other plant parts remaining after having been compressed to squeeze out oil, generally used as feed for livestock.

oil meal *n.* oil cake that has been ground into granules for use as a livestock feed.

oil of turpentine *n.* a flammable essential oil, with a pungent odor and taste, obtained from coniferous trees by distillation of oleoresins. It is used in paints, varnishes, and medicines.

oil paint *n.* paint made with a drying oil (e.g., oil of turpentine) as a base, as distinguished from latex paint with a water base.

oil palm *n.* a tall tree (*Elaeis guineensis*) native to tropical Africa; the source of two kinds of oil— edible palm oil, obtained from the fleshy coating around the fruit, and nonedible palm kernel oil, obtained from the hard kernels. The palm fronds are used to make brooms and mats, and the bark is made into baskets. See palm oil, palm kernel oil.

oil shale *n.* shale rock impregnated with bitumens called kerogen. Shale oil is extracted from it by destructive distillation. It constitutes a large reserve of oil that is more expensive to extract than that from oil wells. Large deposits of oil-bearing shale exist in Utah, Colorado, and Wyoming.

oil skimmer *n.* a device used to remove oil from the surface of a body of water.

oil slick *n.* an area of water with a smooth surface because it is covered by a layer of oil floating on the water.

oil spill *n.* a discharge of petroleum into a body of water from an oil tanker, an offshore oil well, or other facility. The oil slick produced by a large oil spill presents a hazard to marine life, seabirds, beaches, and the environment in general.

oil well *n.* a well drilled through underground rocks to reach a petroleum deposit. Natural gas is often trapped above the petroleum and sometimes forces the oil to spout from the well as a gusher.

ojo *n.* (Spanish *ojo*, eye) a small pond, lake, or spring in southwestern U.S.A., especially a hot spring (*ojo caliente*).

An oil well pump with a storage tank behind it

okapi *n.* (Bambuba *okapi*, okapi) an animal (*Okapia johnstoni*) native to Africa. It is closely related to the giraffe but is smaller (about 5 ft or 1.5 m at the shoulders) and has a much shorter neck than the giraffe.

okra *n.* **1.** an annual garden crop (*Abelmoschus esculentus*, family Malvaceae) of tropical origin; a bush that produces edible pods. **2.** the sticky green pods of the plant harvested at an immature stage for use in stews, soups, etc. **3.** a food made with these pods; also known as gumbo.

old age *n.* **1.** the latter part of a life or existence, commonly a period of declining vigor and productivity. For humans, it is often assumed to be 65 years of age or older. **2.** the geologic stage of a topographic surface that has been eroded down to a nearly featureless plain. **3.** the status of a soil that has been weathered and leached until most of its weatherable minerals are gone, its available nutrient supply has diminished, and its productivity has declined. See mature.

Old Faithful *n.* the most widely known geyser in Yellowstone National Park in Wyoming. Its eruptions are quite regular, occurring about 65 min apart, sending a column of hot water and steam to a height of 116 to 175 ft (35 to 53m), and lasting about 4 min.

Old Faithful geyser

old growth *n.* **1.** a stand of trees in a forest that has never been cut. **2.** a stand of mature trees.

Old Wives' summer *n.* a period of warm, sunny weather following a colder period in the fall; the European equivalent of Indian summer in the U.S.A.

Old World *n.* or *adj.* the Eastern Hemisphere, especially the parts of Europe, Asia, and Africa near the Mediterranean Sea.

oleander *n.* (French *oléandre*, oleander) an evergreen shrub or small tree (*Nerium oleander*, family Apocynaceae) with narrow leaves and very fragrant white, rose, or purple flowers. All parts of it are poisonous. Although native to the Orient and southern Europe, it is widely planted (often as a hedge) in southern U.S.A. and other areas with warm climates.

oleic acid *n.* (Latin *oleum*, oil) a colorless, odorless, oily liquid ($C_{17}H_{33}COOH$) with one double bond in its chain. It is present in the form of an ester in most animal and vegetable fats and oils. Oleic acid is used to make soap and cosmetics.

olfactory *adj.* (Latin *olfactus*, smelly) of, relating to, or contributing to the sense of smell; e.g., olfactory nerves detect odors.

oligomer *n.* (Greek *oligos*, little + *méros*, part) a protein composed of two or more identical subunits called protomers; e.g., hemoglobin is an oligomer with two identical subunits (also called a dimer).

oligotroph *n.* (Greek *oligos*, little + *trophē*, nourishment) a plant that can grow where nutrient concentrations are low; e.g., sphagnum moss (*Sphagnum* sp.).

oligotrophic *adj.* (Greek *oligos*, little + *trophia*, nutrition) low nutrient content in water (especially in a lake) that therefore supports little vegetative growth but stays clear and maintains a high level of dissolved oxygen. See dystrophic, eutrophic.

olive *n.* (Latin *oliva*, olive) **1.** a semitropical evergreen tree (*Olea europaea*, family Oleaceae) widely grown for its fruit and wood, often in orchards around the Mediterranean and in other warm areas. **2.** the drupe fruit of such a tree, commonly black, green, or brown. Olives are widely used as a relish or seasoning or as a source of oil for cooking and other uses. **3.** the wood of such a tree, often used for ornamental work. **4.** the dull yellowish green color of the unripe fruit or foliage of the tree. *adj.* **5.** having a color resembling the unripe fruit or foliage. **6.** made from or pertaining to the wood, foliage, or fruit of the tree.

omasum *n.* (Latin *omasum*, paunch) the third of the four stomachs of a ruminant (e.g., a cow), where ingested feeds are ground to a fine consistency. Also called manyplies or psalterium. See abomasum, reticulum, rumen.

ombrogenous *adj.* (Greek *ombros*, rain + *genos*, produced by) kept wet by rainfall, as a peat deposit on a slope rather than in a depression.

ombrograph *n.* (Greek *ombros*, rain + *graphē*, drawing) a recording rain gauge.

ombrometer *n.* (Greek *ombros*, rain + *metron*, measure) a pluviometer; a rain gauge.

omnivore *n.* (Latin *omnis*, all + *vorare*, to eat) an animal or human that eats both plant and animal products. See carnivore, herbivore.

onager *n.* (Latin *onager*, wild ass) **1.** a wild ass (*Equus hemionus*) native to desert areas of central and southwestern Asia. **2.** a catapult used for hurling stones in ancient and medieval times.

onchocerciasis *n.* (Greek *onkos*, barb + *kerkos*, tail + *iasis*, condition) river blindness; a condition that affects about 20 million people plus millions of animals, mostly in tropical areas of Africa and Central and South America. It is caused by filarial worms (*Onchocerca volvulus* attacks humans, *O. cervicalis* attacks horses and mules, and *O. gibsoni* attacks cattle and zebras, superfamily Filarioidea), a type of nematode parasite transmitted by flies (family Simuliidae) that reproduce in flowing river water. The disease causes an itchy rash, nodules under the skin, eye lesions, and sometimes elephantiasis. The disease is so feared that it prevents use of fertile land in many river valleys, especially in western Africa. Other names are onchocercosis, Robles' disease, blinding filarial disease, mal morado, volvulosis, African river blindness, or Mexican blindness.

oncornavirus or **oncovirus** *n.* (Greek *onkos*, mass, tumor + RNA + Latin *virus*, poison) any RNA virus that causes tumors in humans, animals, and/or birds.

onion *n.* (French *oignon*, onion) **1.** a biennial herbaceous plant (*Allium cepa*, family Amaryllidaceae) with long, tubular leaves and an edible bulbous root with multiple spherical layers inside the bulb. **2.** the bulb (and sometimes lower leaves) of the plant. It has a pungent odor and taste and is used to season meats, stews, and other foods. **3.** the flavor, odor, seasoning, or other derivative of onions. *adj.* **4.** of or related to onions, as onion flavor, onion salt, or onion flakes.

onshore *adj.* (German *an*, on + *schore*, point of division) **1.** moving from water to land, as an onshore breeze. **2.** located or occurring on land (usually near the shore), as an onshore lighthouse or an onshore activity for sailors.

ontogeny or **ontogenesis** *n.* (Greek *ontos*, being + *genesis*, generation) the life cycle (origin and development) of an individual organism. See phylogeny.

onyx *n.* (Greek *onyx*, nail) **1.** a form of agate (quartz, SiO_2) with alternate light and dark bands; often used to carve cameos. **2.** a fingernail or toenail. **3.** an accumulation of pus in the anterior part of the eye. **4.** Mexican onyx, also called onyx marble or calcitic alabaster, a form of calcite ($CaCO_3$) that resembles onyx but is much softer than the quartz form. See Mexican onyx.

oocyte *n.* (Greek *ōion*, egg + *kytos*, hollow) an immature egg cell of a human or animal.

oogonium *n.* (Greek *ōion*, egg + *gonos*, offspring + Latin *ium*, thing) **1.** the large cell that constitutes the female reproductive organ in thallophytic plants; the cell that produces eggs. **2.** any cell that produces oocytes.

oolite *n.* (French *oölithe*, egg stone) a form of limestone (dominantly $CaCO_3$) with a granular structure that resembles fish eggs, including roestone with distinct, well-rounded granules and pisolite or peastone with large granules about the size and shape of peas. Also called egg stone.

oology *n.* (Greek *ōion*, egg + *logos*, word) the study of birds' eggs; a branch of ornithology.

ooze *n.* (Anglo-Saxon *wase*, mud) **1.** soft, spongy mud or slime. **2.** a soft, boggy area. **3.** a deposit on the ocean floor composed mostly of calcareous or siliceous shells of tiny organisms (mostly foraminifers) plus silt particles that settle from the air to the water and down to the bottom. *v.* **4.** to flow slowly through openings (e.g., between toes). **5.** to exude liquid. **6.** to express or exhibit, as to ooze charm. **7.** to disappear, as to ooze away.

opacity *n.* (Latin *opacitas*, darkness or obscurity) **1.** the condition of being opaque; not transparent or nontranslucent. **2.** the degree of darkness; e.g., the proportion of the incident light that is absorbed as it passes through a medium. Opacity of air and water is sometimes used as a measure of air pollution and water contamination. **3.** dull mentality. **4.** obscurity in a statement.

opacus *adj.* (Latin *opacus*, shaded) sufficiently opaque to obscure the sun or moon; densely cloudy of any cloud type, especially altocumulus, altostratus, stratocumulus, stratus, cumulus, and cumulonimbus.

opal *n.* (Latin *opalus*, opal) **1.** amorphous silica (SiO_2) that contains some water of hydration, pore space, and small amounts of impurities that give it any of a variety of colors, often with streaks or patterns. It is used as an abrasive, as insulation, and as a filter material. **2.** an iridescent form of this material used as a semiprecious gem stone or cut into thin sections to make ornaments.

opalized wood *n.* petrified wood. See chalcedony, petrify.

opaque *adj.* (Latin *opacus*, shaded) **1.** impenetrable to light; neither transparent nor translucent. **2.** dull;

dark; not reflecting much light. **3.** impenetrable to heat, electricity, or sound. **4.** obscure; difficult to understand. **5.** not intelligent; dull; stupid. *n.* **6.** a substance that is opaque, especially coloring matter used to darken parts of a photographic negative.

open formula *n.* a statement attached to a container identifying the ingredients and their percentages in a mixed feed or fertilizer.

open-pit mining *n.* mining accomplished by stripping the overburden from the deposit so the coal or mineral deposit can be excavated in the open air, as contrasted to a shaft mine.

open pollination *n.* random pollination of the plants of a given species resulting from natural dispersal of pollen among many plants. Open pollination promotes genetic mixing and plant diversity.

open system *n.* an assembly of materials and/or organisms that receives energy and/or substances from outside and emits energy and/or substances to the outside. Normally the inputs and outputs are balanced.

open weather *n.* pleasing atmospheric conditions; mild temperatures without fog or severe storms.

ophidian *adj.* (Greek *ophis*, snake + Latin *ian*, belonging to) **1.** pertaining to snakes (suborder Serpentes). *n.* **2.** a snake.

ophiology *n.* (Greek *ophis*, snake + *logos*, word) the science that studies snakes (suborder Serpentes).

opium *n.* (Greek *opion*, poppy juice) **1.** a very strong narcotic made by drying the sap from the pods of opium poppies; a mixture of codeine, morphine, and other alkaloids. It is used medically to relieve pain, diarrhea, and coughing. **2.** either a raw or a processed form of the opium poppy that is smoked as an illicit drug. **3.** anything that numbs the senses and/or quiets the emotions.

opium poppy *n.* a plant (*Papaver somniferum*, family Papaveraceae) with grayish green leaves and red, white, or purple flowers; the source of opium. It is sometimes grown as an ornamental.

opossum *n.* (Algonquian, *opossum*, white beast) **1.** a carnivorous, nocturnal, arboreal, marsupial (*Didelphis marsupialis*, also called *D. virginiana*, family Didelphidae), native to eastern U.S.A., that has a long snout and a long prehensile tail resembling that of a rat. It eats dead animals and birds and stays motionless when it senses danger. Also called possum. **2.** any of about 75 related species (mostly in 11 genera of family Didelphidae) native to other parts of the world.

opportunity cost *n.* the profit that could have been made if one had chosen a different way; e.g., the profit that could have been made by planting a crop where a grass filter strip was planted to protect a body of water from siltation and eutrophication.

optimum *n.* or *adj.* (Latin *optimus*, best) the most favorable condition; e.g., the highest point on a curve or the condition (temperature, light level, humidity, etc.) that produces the fastest growth.

opuntia *n.* (Greek *Opus*, a city of ancient Greece) any of a large genus of cacti (*Opuntia* sp.) with red, purple, or yellow blossoms and berries that are either fleshy or dry; e.g., the prickly pear cactus that bears Barbary figs.

orange *n.* (French *orenge*, orange) **1.** an evergreen tree (*Citrus* sp., especially *C. aurantium* or *C. sinensis*, family Rutaceae), native to India and southern China, with white blossoms, ovate pointed leaves, and greenish brown bark. **2.** the edible citrus fruit of such a tree, with a nearly spherical shape, a yellowish red rind, and a segmented, fleshy interior. The fruit of *C. aurantium* is quite bitter, but that of *C. sinensis* is much sweeter and commonly is eaten as fruit or squeezed to make a beverage, especially with breakfast. **3.** the yellowish red color of the fruit rind, with a wavelength between 590 and 610 nm. **4.** the flavor of the orange fruit (often a synthetic simulation).

orangutan *n.* (Malay *oran*, man + *utan*, forest) a large anthropoid ape (*Pongo pygmaeus*) that grows to a height of about 5 ft (1.5 m). It is native to Borneo and Sumatra and has long arms and a body covered with reddish brown hair.

orb *n.* (Latin *orbis*, a circle) **1.** a spherical object; e.g., a star, planet, moon, or globe. **2.** an eyeball. **3.** a circular form or pathway. **4.** a combination of a sphere and a cross representing royal power, especially British royal power.

orbit *n.* (Latin *orbita*, course or circuit) **1.** the usually elliptical pathway of one celestial body around another, as the moon around the Earth or the Earth around the Sun. **2.** a similar pathway of a satellite around the Earth or another celestial body. **3.** the path of an electron around the nucleus of an atom. **4.** the skin or other tissue surrounding the eye of a bird or insect. **5.** the ordinary series of routine activities of a person. **6.** the range of influence of a powerful nation, referring to other nations in its orbit. *v.* **7.** to travel in an orbit around something.

orbital *n.* (Latin *orbitalis*, orbiting) **1.** a wave function that describes the probability of an electron being at various locations as it orbits around the nucleus of an atom. **2.** an electron with a particular energy level and type of orbit. *adj.* **3.** pertaining to an orbit.

orca *n.* (Latin *orca*, whale) the killer whale (*Orcinus orca*), the largest dolphin of the oceans, about 30 ft (9 m) in length. It typically is found in schools in coastal waters. Orcas eat fish, penguins, porpoises, sea lions, and seals and will attack whales much larger than themselves. Also called toothed whale, but orca is the preferred name.

orchard *n.* (Anglo-Saxon *ort*, herb + *geard*, garden) **1.** an area of land used to grow trees that produce edible fruit or nuts. **2.** the trees grown on such land.

orchardgrass *n.* a tall, coarse, long-lived perennial bunchgrass (*Dactylis glomerata*, family Poaceae) native to Europe. It is used for early spring and summer pasture and for erosion control. Called cocksfoot in Europe.

Orchardgrass

orchid *n.* (Latin *orchis*, plant with hanging roots) **1.** any of a large family of tropical or temperate epiphytic flowering plants (family Orchidaceae) that typically have bright, showy blossoms. Orchids often are grown as ornamental plants. **2.** a blossom from such a plant. **3.** a light purple color.

ordeal bean *n.* See calabar bean.

ordeal tree *n.* **1.** a tree (*Tanghinia* or *Cerbera venenata*) with poisonous seeds in a plumlike fruit. In Madagascar, persons suspected of a crime are forced to eat the seeds as a test of guilt or innocence; a lance dipped in juice extracted from the seeds is used to put criminals to death. **2.** any of several other trees whose seeds or leaves have been similarly used.

ordure *adj.* (Latin *horridus*, frightful) dung, excrement, filth, manure.

ore *n.* (Anglo-Saxon *ora*, unwrought metal) **1.** a mineral or native-metal rock that is mined as a source of metal. **2.** a similar deposit that is mined as a source of a nonmetal such as sulfur.

oregano *n.* (Latin *origanum*, an aromatic herb) an aromatic herb (*Origanum vulgare*, family Lamiaceae) whose fragrant leaves are used in cooking as a pot herb. Also called pot marjoram.

Oregon grape *n.* **1.** a hardy evergreen shrub with spiny leaves (*Mahonia aquifolium*, family Berberidaceae) that grows to a height of about 3 ft (0.9 m); dainty yellow blossoms appear in late winter or very early spring followed by edible blue berries that resemble small grapes. Also called mahonia. **2.** the blossoms and/or fruit of this shrub; the state flower of Oregon.

organan *n.* an illuvial coating that is dominantly organic in composition on the surfaces of soil peds; an organic cutan.

organelle *n.* (Latin *organella*, small organ) a small body surrounded by a membrane and suspended in the cytoplasm of a eukaryotic cell outside the nucleus. Each type of organelle has a specific biochemical function. See plastid.

organic *adj.* (Greek *organikos*, like an instrument) **1.** relating to an organ, to living tissue, or to a living organism. **2.** inherent; a component part of an organism. **3.** having a complex structure comparable to that of a living organism. **4.** containing carbon and hydrogen (and often oxygen, nitrogen, phosphorus, and/or sulfur) with atoms linked together by covalent bonding, often in large, complex structures. Although originally thought to be limited to natural compounds, many organic compounds have been made synthetically; e.g., urea $[CO(NH_2)_2]$.

organic carbon *n.* carbon atoms that are linked together by covalent bonds to form organic compounds. Average or assumed factors are sometimes used to convert organic carbon data to or from organic matter data.

organic farming *n.* farming without the use of synthetic chemicals. Organic fertilizers such as manure or compost may be used, but mineral fertilizers, synthetic pesticides, growth hormones, and mineral feed additives for livestock are excluded.

organic fertilizer *n.* any fertilizer composed of organic compounds of plant or animal origin (often accompanied by naturally associated mineral compounds); e.g., manure, green manure, or compost.

organic food *n.* food produced by organic farming techniques.

organic gardening *n.* gardening without the use of synthetic chemicals. See organic farming.

organic load *n.* the amount of organic matter that enters a body of water in a specified time period, especially that added by human activities. Decomposing such material generates a biological oxygen demand.

organic matter *n.* **1.** any material composed of organic compounds, especially plant and animal residues. **2.** the organic component of soil or rock material, often expressed as a percentage of the dry weight.

organic nitrogen *n.* nitrogen that is covalently bonded in organic compounds. Organic nitrogen in soil or water must be mineralized before the nitrogen can be used to support plant growth.

organic phosphorus *n.* phosphorus that is covalently bonded in organic compounds. Organic phosphorus must be mineralized before it can be used to support plant growth.

organic soil *n.* **1.** a soil that has a greater total thickness of organic soil horizons than of mineral soil horizons in its profile; a soil classified as a Histosol in Soil Taxonomy. **2.** an organic soil horizon, a soil layer that contains at least 12% to 18% organic carbon (12% + 0.1 × % clay up to 60% clay). Organic soil materials are called muck if they have been oxidized or peat if they have not been oxidized.

organic solvent *n.* an organic liquid suitable for dissolving organic compounds such as fats and oils. It is usually a hydrocarbon such as benzene or toluene or a carbon–hydrogen–oxygen compound such as acetone or an alcohol.

organic sulfur *n.* sulfur that is covalently bonded in organic compounds. Organic sulfur must be mineralized before it can be used to support plant growth.

organism *n.* (Greek *organikos*, like an instrument + *isma*, state or condition) **1.** any living thing—a plant, animal, microbe, or person. **2.** any complex entity with properties that depend on the way its parts work together, as a governmental organism.

organ meat *n.* the edible organs from an animal or bird, especially the liver, heart, and kidney. Organ meats of birds are often called giblets.

organochlorine *n.* See chlorinated hydrocarbon.

organogen *n.* (Greek *organon*, organ + *genes*, kind) a chemical element that is a common component of organic substances. For example: carbon, hydrogen, oxygen, nitrogen, phosphorus, sulfur, chlorine.

organogenesis *n.* (Greek *organon*, organ + *genesis*, development) the growth and development processes of organs.

organogenic *adj.* (Greek *organon*, organ + *genic*, producing) pertaining to or produced in an organ.

organoleptic *adj.* (Greek *organon*, organ + *lēptikos*, accepting) either causing or receiving a sensory perception by sight, feel, smell, taste, and/or sound.

organology *n.* (Greek *organon*, organ + *logos*, word) the study and body of knowledge regarding the bodily organs.

organomercurial *adj.* or *n.* (Greek *organon*, organ + Latin *mercurialis*, containing mercury) a complex of an organic molecule and mercury; e.g., methyl mercury. Such compounds are generally toxic to humans and animals, especially in their effects on the nervous system.

organometallic *adj.* (Greek *organon*, organ + *metallikos*, containing metal) a combination of any metallic ion with an organic structure.

organophosphate *n.* (Greek *organon*, organ + *phōs*, light + *ate*, salt) any of several organic insecticides that contain phosphorus as a component in their molecular structure; e.g., diazinon, malathion, parathion. Organophosphates are highly toxic, relatively nonpersistent, widely used insecticides that are also hazardous to humans and animals. Affected humans experience weakness, dizziness, and headache that can lead to convulsions, paralysis, and coma in serious cases. Related materials were developed as nerve gases during World War II.

organotropic *adj.* (Greek *organon*, organ + *tropos*, turning toward) attracted to and tending to concentrate in a certain bodily organ or tissue.

oriental poppy *n.* an Asiatic poppy (*Papaver orientale* or *P. bracteatum*, family Papaveraceae) with showy scarlet, pink, or white blossoms and bristly stems and leaves. It is grown as an ornamental.

orientation *n.* (French *orienter*, to orient + *atio*, action) **1.** the determination or standardization of direction relative to cardinal or compass points, a landscape, or a map. **2.** guidance to help newcomers adjust to new surroundings or activities. **3.** the relative positions of various objects in three-dimensional space; e.g., the structural arrangement of the atoms in an organic molecule. **4.** the homing instinct of certain animals and insects.

origin *n.* (Latin *originis*, rising) **1.** beginning; source. **2.** starting or zero point, as the origin of a line or of a journey. **3.** lineage; ancestry.

oriole *n.* (Latin *aureolus*, golden) a colorful arboreal bird of the Old World (family Oriolidae) or of the New World (*Icterus* sp., family Icteridae) with pointed beaks and barred wings. Adult males have bright yellow or orange (or some red) breasts, bellies, and rumps and black hoods, backs, and tails; females and young are more drab than the males. The Baltimore oriole (*I. galbula*) is also called firebird or northern oriole.

A male Baltimore oriole

ornamental *adj.* (Latin *ornamentum*, equipment or adornment) **1.** valued as an adornment or embellishment rather than for utilitarian purposes. *n.* **2.** a plant, shrub, or tree used for decorative effect.

ornithology *n.* (Latin *ornithologia*, bird study) the scientific study of birds.

ornithosis *n.* (Greek *ornithos*, bird + *osis*, condition) psittacosis, a contagious disease of birds caused by a virus (*Chlamydia psittaci*). It is transmittable to humans as a respiratory infection called psittacosis.

orogeny or **orogenesis** *n.* (Greek *oros*, mountain + *genesis*, formation) mountain building; the intrusion of granitic magma combined with uplift, thrusting, folding, and faulting processes that form mountain chains.

orographic *adj.* (Greek *oros*, mountain + *graphikos*, drawing or writing) influenced by mountains. Clouds and rain formed in an air mass as it rises to pass over a mountain range produce orographic clouds and orographic precipitation on the windward side. A rain shadow occurs where the air descends on the lee side.

orography or **orology** *n.* (Greek *oros*, mountain + *graphia*, writing or *logos*, word) the scientific study of mountains.

orometer *n.* (Greek *oros*, mountain + *metron*, measure) an aneroid barometer with a scale that indicates elevation above mean sea level.

orophyte *n.* (Greek *oros*, mountain + *phyton*, plant) a plant that grows at high elevations, under subalpine conditions on mountains.

ort (usually **orts**) *n.* (German *ort*, early) **1.** a morsel of food remaining after a meal. **2.** rejected feed left by animals fed *ad libitum*. **3.** a scrap of refuse.

ortho *adj.* (Greek *orthos*, straight, upright) **1.** occupying or related to two adjacent positions on a benzene ring (as opposed to meta and para). **2.** a combining form in chemical names of acids and salts meaning most hydrated or highest water content; e.g., orthophosphoric acid (H_3PO_4) as opposed to pyrophosphoric acid ($H_4P_2O_7$) or metaphosphoric

acid (HPO_3). **3.** a combining form meaning straight, upright, or vertical. **4.** a combining form meaning right angle, as in orthorhombic. **5.** a combining form meaning proper or correct, as in orthography.

orthoclase *n.* (Greek *orthos*, straight + *klasis*, fracture) a tektosilicate mineral of the feldspar family with an ideal formula of $KAlSi_3O_8$; hardness 6 on Mohs' scale. Orthoclase is a very abundant mineral that is orange-pink to white. See potassium feldspar.

orthogenesis *n.* (Greek *orthos*, straight + *genesis*, formation) **1.** determinative evolution, the theory that the evolution of a species is a continual, non-branching progression toward a predetermined end controlled by genetic rather than by environmental factors. **2.** a sociological theory that any culture or society follows a developmental course similar to that of others.

orthophosphate *n.* (Greek *orthos*, straight + *phōs*, light + *ate*, salt) a salt, ester, or other product of a reaction with orthophosphoric acid (H_3PO_4) that contains its trivalent phosphate group (PO_4^{3-}).

os *n.* (Latin *os*, bone) **1.** a bone (pl. **ossa**). **2.** an opening or entrance; e.g., a mouth (pl. **ora**). **3.** an esker (pl. **osar**).

osage orange *n.* a small thorny tree (*Maclura pomifera*, family Moraceae) that is dioecious (has male and female flowers on different plants) and has pointed leaves 2 to 5 in. (5 to 13 cm) long and round, yellowish fruit 4 to 5 in. (10 to 13 cm) in diameter; native to south-central U.S.A. It is used as hedges, and the wood makes long-lasting fenceposts.

OSHA *n.* See Occupational Safety and Health Administration.

osmeterium *n.* a gland on many caterpillars that secretes a foul-smelling substance that wards off many predatory wasps and flies.

osmics *n.* (Greek *osmē*, smell + *ikos*, knowledge) the science that studies the sense of smell and the olfactory nerves used in smelling.

osmium (Os) *n.* (Greek *osmē*, smell + Latin *ium*, element) element 76, atomic weight 190.2; an extremely hard and dense, lustrous bluish white metal that is brittle even at high temperatures. Its tetroxide is highly toxic to lungs, skin, and eyes. Osmium is used to harden alloys.

osmolarity *n.* the molarity of an ideal solution that would have the same osmotic pressure as a test solution (typically expressed in osmoles or milliosmoles for biological fluids with complex compositions). The body fluids of mammals generally have an osmolarity of about 300 milliosmoles.

osmosis *n.* (Greek *ōsmos*, impulse + *osis*, condition) **1.** the passage of water through a semipermeable membrane from a dilute to a more concentrated solution; e.g., water passes from a soil solution of low solute concentration through the walls of roots into cell sap, where solutes are more concentrated. See dialysis, exosmosis, osmotic pressure. **2.** a slow absorption or mixing process with little or no obvious effort, as learning by osmosis.

osmosis, reverse *n.* See reverse osmosis, exosmosis.

osmotic pressure *n.* the force that must be applied to a solution to prevent it from being diluted by water passing through a semipermeable membrane from a pure water source on the opposite side of the membrane. The osmotic pressure (OP) in atmospheres in an ideal system can be calculated from the equation $OP = MRT$, where M represents the total molarity of all molecules and ions in solution, R is the ideal gas constant (0.0821 liter atmospheres per mole degree), and T is the temperature in degrees Kelvin.

osprey *n.* (Latin *ossifraga,* bone breaker) a long-winged hawk (*Pandion haliaetus,* family Accipitridae) with a wing span of about 54 in. (1.4 m). Its plumage is dark brown above and mostly white below with conspicuous black wrist markings. Its habitat is along seacoasts, lakes, and rivers; it eats only fish, sometimes plunging beneath the water to catch its prey. Also called fish hawk or water eagle.

ossification *n.* (Latin *os*, bone + *ficatio*, making) **1.** the hardening of soft tissue into bone or a bony mass, either natural or abnormal. **2.** a hard, bony condition or formation produced by the process of ossification. **3.** the process of becoming unduly rigid in behavior, beliefs, or habits.

osteoporosis *n.* (Greek *osteon*, bone + *poros*, passage + *osis*, condition) a condition wherein low bone density and excessive porosity make human or animal bones easy to fracture. Wrist, hip, and leg fractures are common; extremely painful collapsed vertebrae can occur. Loss of bone density is a natural consequence of aging, especially in postmenopausal women, and is more severe when there is little exercise and/or a deficiency of vitamin D, calcium, or phosphorus; genetics and liver diseases can also be significant factors. A similar condition in poultry is called rickets.

ostrich *n.* (French *ostrusce*, ostrich) **1.** a large bird (*Struthio* sp., family *Struthionidae*) with a long neck and very long legs with two toes on each foot; a fast runner (up to 40 mph or 65 km/h) but unable to fly. The African ostrich (*S. camelus*) is the largest of all living birds, growing to 6 to 8 ft (1.8 to 2.4 m) tall. Ostriches eat grain and grass; they are sometimes grown for meat. Males are black with white markings on the wings and tail; females are mostly brown. Ostrich feathers are used for millinery and decoration. **2.** any of three similar but smaller (up to 4 ft or 1.2 m tall) brown and white South American birds (*Rhea americana*, *R. darwini*, or *R. macrorhyncha*, family Struthionidae) with three toes on each foot. Also called rhea or ñandu.

ostrich fern *n.* a tall fern (*Onoclea Struthiopteris*) that grows in circular clumps in swampy alluvium in North America and Europe.

OSWER *n.* See Office of Solid Waste and Emergency Response.

ounce *n.* (Latin *uncia*, one twelfth) **1.** a unit of weight equal to either 1/12 of a pound troy (31.103 g) or 1/16 of a pound avoirdupois (28.349 g). **2.** a fluid ounce; a liquid measure equal to 1/8 of a cup (29.573 mL or 1.8047 in.^{3}) in the U.S.A. or 1/20 of an imperial pint (28.413 mL or 1.7339 in.^{3}) in Great Britain. **3.** a small amount, often figuratively, as an ounce of goodness.

outback *n.* (Anglo-Saxon *ut*, out + *baec*, back) **1.** or **Outback** the sparsely populated arid land in the interior of Australia. Sometimes called the bush. **2.** any remote area with a sparse population that has little contact with more populated areas.

outfall *n.* (Anglo-Saxon *ut*, out + *feallan*, to fall) a point of discharge, as the place where a drain tile, sewer pipe, or river empties into an open or larger body of water.

outhouse *n.* (Anglo-Saxon *ut*, out + *hus*, house) **1.** a separate building that serves as a privy. Also called a latrine. **2.** any smaller building separate from a main building or house.

output *n.* (Anglo-Saxon *ut*, out + *putung*, impelling) **1.** productivity, the rate or amount of product coming from something that is growing or a factory that is producing. **2.** the act or process of production. **3.** the liquid and gaseous waste produced by a living organism (sweat, urine, and exhalant). **4.** the amount of power generated by a motor, engine, or other source. **5.** the printed matter or electronic media produced by a computer from information it receives as input.

outwash *n.* (Anglo-Saxon *ut*, out + *waescan*, to wash) stratified sand and gravel deposited by meltwater from a glacier.

outwash plain *n.* a smooth landscape with sandy (sometimes gravelly) soil formed in front of a terminal moraine in alluvial material deposited by floodwaters from a melting glacier.

oven-dry *adj.* dried in an oven until there is no more weight loss; e.g., soil that has been dried for 24 hr (or longer for large samples and soils high in clay) at 105°C or plant material that has been dried at 70°C for 48 hr or longer.

overburden *n.* (Anglo-Saxon *ofer*, above + *byrthen*, load) **1.** soil and rock materials lying on top of a minable deposit (e.g., coal or ore), especially one that is being, has been, or might be mined by open-pit mining. **2.** any load that is too heavy for the person, animal, machine, or structure that must support or carry it. *v.* **3.** to overload someone or something, either with excess weight or excess work or other stress.

overcast *adj.* (Anglo-Saxon *ofer*, above + *casten*, to throw) **1.** covered with clouds; less than 5% clear sky. **2.** dark and gloomy. **3.** sewn with a stitch that wraps around the edge. **4.** thrown beyond the intended target, as an overcast fishing line.

overconsolidated *adj.* (Anglo-Saxon *ofer*, over + Latin *consolidatus*, made solid) a deposit that has been compacted by a heavy overburden that has since disappeared; e.g., by a glacier that melted away or a mass of material that has been eroded away.

overfish *v.* (Anglo-Saxon *ofer*, over + *fiscian*, to fish) to deplete the fish population in a body of water by catching too many. See Pacific salmon.

overflow *v.* (Anglo-Saxon *ofer*, above + *flowan*, to flow) **1.** to flood; to cover a floodplain or other area near a stream. **2.** to spill out the top of a container when its capacity is exceeded, as a bucket overflows. **3.** to have an excess, more than enough for one's needs or more than expected, as a heart may overflow with gratitude. **4.** to expand into a new area, as the population of a crowded city overflows into new developments. *n.* **5.** the part that exceeds capacity and overflows. **6.** the assemblage that exceeds capacity or need, as an overflow of applicants for a job. **7.** an outlet for an excess to escape, as an overflow pipe on a reservoir.

overgraze *v.* (Anglo-Saxon *ofer*, above + *grasian*, to graze) to cause livestock to consume too much of the forage of a pasture or range area, resulting in loss of the most palatable species, reduced plant cover, and excessive soil erosion. A common rule to avoid overgrazing is to use half and leave half.

overgrown *adj.* (Anglo-Saxon *ofer*, above + *growan*, to grow) **1.** covered with a thick stand of weeds or other uncontrolled vegetation. **2.** grown too large, as he has overgrown his clothes.

overland *adj.* (Anglo-Saxon *ofer*, above + *land*, land) **1.** on land, as to travel overland rather than by sea or air. **2.** crossing land, as overland flow of water.

overpopulation *n.* (Anglo-Saxon *ofer*, above + Latin *populatio*, population) a density of plants, animals, or people in excess of environmental capacity; a cause of poor living conditions and high mortality rates.

overproduction *n.* (Anglo-Saxon *ofer*, above + *productio*, production) production of more than can be sold profitably or more than can be used beneficially; a common cause of low prices.

overshot *adj.* (Anglo-Saxon *ofer*, above + *sceot*, shot) **1.** hitting above or beyond the target, as an overshot arrow. **2.** extending over and beyond, as an overshot jaw is an upper jaw that extends beyond the lower jaw. **3.** supplied with water from above, as an overshot water wheel (often used to power a mill or a machine).

Overshot waterwheel

oversowing or **overseeding** *n.* the practice of seeding one crop into another that is still growing; e.g., seeding a legume hay crop into a crop of oats that is about to begin forming heads. This allows the oats to develop with minimal competition and provides the legume with an optimal light intensity for seedling growth while the oats are present and for forage production after the oats are harvested.

overstory *n.* the tallest trees in a multilayered forest; the trees that form a canopy over everything else.

overturn *v.* (Anglo-Saxon *ofer*, above + Latin *tornare*, to turn) **1.** to tip over or turn upside down. **2.** to defeat, as to overturn a government or overturn someone's plans. **3.** to invert, as the layers of temperature-stratified water in a lake, especially in late fall when the upper layer of water becomes colder than the deeper water.

overyield *n.* an increase in crop yield resulting from intercropping as compared with the production of the component crops in monoculture on an equivalent site.

ovicide *n.* (Latin *ovum*, egg + *cida*, killer) a pesticide that kills insect eggs.

ovine *adj.* (Latin *ovinus*, like sheep) pertaining to or characteristic of sheep.

oviparous *adj.* (Latin *oviparus*, egg bearing) producing eggs that hatch outside the body of the female; e.g., eggs of birds, snails, and most reptiles and fishes. See viviparous.

ovipositor *n.* (Latin *ovum*, egg + *positus*, placed + *or*, agent) **1.** an organ used to deposit eggs from the rear of the abdomen of certain female insects. **2.** a similar organ in certain fish.

ovule *n.* (Latin *ovulum*, little egg) **1.** an immature ovum. **2.** a plant part that contains the female germ cell. **3.** any small structure that resembles an egg.

ovum (pl. **ova**) *n.* (Latin *ovum*, egg) **1.** the female reproductive cell of an animal or plant. **2.** an architectural ornament shaped like an egg.

owl *n.* (Latin *ulula*, owl) **1.** a nocturnal bird of prey (order Strigiformes) with a large head, large eyes, and a short neck, including the barn owls (family Tytonidae) and other owls (family Strigidae). Owls hoot noisily, but they fly quietly as they hunt for small mammals such as gophers, moles, ground squirrels, chipmunks, mice, and rats **2.** a person with a solemn appearance, especially one who stays up late at night, often called a night owl.

An owl

ox (pl. **oxen**) *n.* (Anglo-Saxon *oxa*, ox) **1.** a castrated adult male bovine, often used as a draft animal. See steer. **2.** any bovine (*Bos* sp.). **3.** a man who is physically strong, clumsy, and stupid.

oxalic acid *n.* a poisonous, crystalline, organic acid ($(COOH)_2 \cdot 2H_2O$) occurring in sheep sorrel (*Rumex acetosella*) and many other plant leaves and in lichens. It causes oxalic acid poisoning in sheep and cattle and is a strong weathering agent that can dissolve minerals in hard rocks.

oxbow *n.* (Anglo-Saxon *oxa*, ox + *boga*, bow, as in bow and arrow) **1.** a U-shaped wooden piece that extends downward from a yoke to encircle the neck of an ox. **2.** a U-shaped section of a river or stream that has meandered and left only a narrow neck of land. The stream may cut across the narrow neck, leaving the old channel as an oxbow lake. **3.** the land that is nearly encircled by an oxbow channel of a river.

oxcart *n.* (Anglo-Saxon *oxa*, ox + *craet*, a little car) a cart pulled by an ox or by oxen.

oxic horizon *n.* a diagnostic subsurface horizon in Soil Taxonomy that characterizes Oxisols. Oxic horizons have accumulated iron and aluminum oxides, kaolinite clay, and other resistant minerals from intense weathering in a tropical environment.

oxidation *n.* (Greek *oxys*, sour + Latin *atio*, action) **1.** the process whereby a substance reacts with oxygen; e.g., combustion or rusting. **2.** the oxidized material resulting from such reactions; e.g., tarnish or rust. **3.** technically defined as the loss of electrons from an atom or ion, thus increasing its positive charge or reducing its negative charge; the opposite of reduction.

oxidation pond *n.* a sewage lagoon, a shallow body of water that is aerated so aerobic bacteria can decompose organic wastes. See lagoon.

oxide mineral *n.* any of a group of amorphous compounds that accumulate in soil because they are highly resistant to weathering, especially the oxides of iron (Fe_2O_3), aluminum (Al_2O_3), and silicon (SiO_2), often in hydrated forms such as limonite and bauxite. Most soils have some oxide minerals present in coatings on soil peds, especially in the subsoil. Some highly weathered tropical soils (Oxisols) are composed mostly of oxide minerals.

oxidorectase *n.* any enzyme that catalyzes an oxidation-reduction process; e.g., alcohol dehydrogenase catalyzes the removal of hydrogen from alcohol, thus converting the alcohol to an aldehyde. See enzyme.

Oxisol *n.* (French *oxide*, oxide + *sol*, soil) a soil that contains an oxic horizon or has a continuous layer of plinthite within 1 ft (30 cm) of the soil surface; a highly weathered soil that forms in a humid, tropical environment. Oxisols are a soil order in Soil Taxonomy; they occupy less than 1% of the U.S.A. but 8% of all world soils.

oxpecker *n.* either of two African starlings (*Buphagus africanus* or *B. erythrorhyncus*) that eat parasitic ticks on the backs of domestic and wild animals.

oxygen (O) *n.* (Greek *oxys*, sour + *genes*, produced) **1.** element 8, atomic weight 15.9994 (set at 16.0 and used as the reference to establish other atomic weights until the reference was changed to carbon 12 in 1961); a reactive element that can combine with most other elements. Oxygen forms large ions that represent about 60% of all the ions and more than 90% of the volume of all ions in the crust of the Earth. Oxygen is an essential constituent of all vegetable, animal, and human tissues. **2.** a colorless, tasteless, odorless gas (O_2) that makes up about 21% by volume of the ambient air (about 23% by weight), excluding water vapor, and is the component needed for respiration and combustion.

oxygenation *n.* (Greek *oxys*, sour + *genes*, produced + *atio*, action) **1.** the addition of dissolved molecular oxygen, as oxygenation of the blood or the release of oxygen into water by aquatic plants as a byproduct of photosynthesis. **2.** aeration, as when water is aerated or a lagoon is aerated. **3.** reaction with oxygen.

oxyhemoglobin *n.* (Greek *oxys*, sour + *haimatos*, heme + Latin *globus*, globe) a red complex of oxygen and hemoglobin formed when blood is oxygenated. The bond is weak, so the oxygen is readily released where it is needed. See carboxyhemoglobin, methemoglobinemia.

oyster *n.* (Latin *ostrea*, oyster) **1.** an edible marine mollusk (*Ostrea* sp., family Ostreidae) that grows in seawater attached to an object; its shell is a bivalve with a flat top valve and a cupped bottom valve. Oyster shells are about 95% calcium carbonate ($CaCO_3$); they are ground for use as agricultural lime and as a calcium supplement in diets and feeds. **2.** any of several similar bivalve mollusks, such as the pearl oyster, scallop, etc. **3.** either of the two dark-colored tenderloin muscles of a fowl located in the hollow of the side bone in front of the hip bone. **4.** a quiet person who is good at keeping secrets. **5.** a source of easy money and/or pleasure, as the world is his oyster.

ozone *n.* (Greek *ozein*, to smell) a molecular form of oxygen (O_3) that occurs as an unstable, highly reactive, blue gas with a pungent odor. Ozone is used to disinfect air, water, and sewage effluent. In the troposphere (lower atmosphere), ozone concentrations above 2 parts per million can irritate eyes and lungs and become a damaging constituent of smog and acid rain. Ozone accumulates in the stratosphere and serves as a filter for ultraviolet light that would otherwise cause skin cancer, cataracts, and genetic mutations.

ozone hole *n.* an area above the Antarctic where the concentration of ozone in the stratosphere is seriously depleted each year, with the hole becoming larger and longer-lasting as the years go by. It is believed to result from certain chemicals, including CFCs and nitrogen dioxide (NO_2), that have been released into the atmosphere. Smaller, less-persistent holes have been found in the stratosphere over other parts of the world.

P

P *n.* **1.** the chemical symbol for phosphorus. **2.** the parental generation in a discussion of genetic inheritance.

Pa *n.* **1.** the chemical symbol for the element protactinium. **2.** pascal.

pabulum *n.* (Latin *pabulum*, food) **1.** any nourishing food for either animal or plant life. **2.** food for the mind, especially greatly simplified, as for infants; pablum.

paca *n.* a large, white-spotted rodent (*Agouti paca*) about 2½ ft (0.75 m) long with almost no tail. It lives in the forests of Central America and the northern part of South America, and its flesh is considered a delicacy.

pace *n.* (Latin *passus*, step) **1.** a step; half a stride. It is often used as a unit for measuring distance and commonly equal to 30, 36, or 40 in. (75, 90, or 100 cm). **2.** a rate of activity, especially the speed of walking or of working to produce standard objects. **3.** the gait of a horse, especially when both feet on one side move together. *v.* **4.** to walk, often aimlessly and nervously, as to pace the floor while waiting. **5.** to measure distance by counting paces. **6.** to determine the speed, as to set the pace in a marathon.

pachyderm *n.* (Greek *pachys*, thick + *derma*, skin) **1.** an animal with a thick skin, especially an elephant but also a hippopotamus or a rhinoceros. **2.** a person who generally ignores criticism, ridicule, etc. (often called thick-skinned).

pachyote *adj.* (Greek *pachys*, thick + *ōtos*, ear) **1.** thick-eared. *n.* **2.** a bat with thick ears (e.g., *Pachyotus* sp.).

pachypterous *adj.* (Greek *pachys*, thick + *pteron*, wing) **1.** thick-winged (as a bat or certain insects). **2.** thick-finned (as certain fish).

Pachyrhizus *n.* any of three or more species of tropical vines (*P. ahipa, P. erosus,* and *P. tuberosus,* family Fabaceae, plus other species names that are not widely accepted) up to 20 ft (6 m) long, with tuberous roots, whose pods grow 6 to 8 in. (15 to 20 cm) long (including jicama). Its pods and seeds contain rotenone but can be eaten after thorough cooking (sometimes called yam bean or potato bean). See jicama.

pacific *adj.* (Latin *pacific*, peacemaking) **1.** calm, peaceable, not boisterous or warlike. **2.** encouraging peace and calmness. **3.** at peace, the world is pacific when there are no wars. **4.** calm and smooth, as a lake or river with pacific water.

Pacific *n.* **1.** the Pacific Ocean, the world's largest ocean, about 70 million mi^2 or 181 million km^2, west of North and South America and east of Asia and Australia. *adj.* **2.** bordering or near the Pacific Ocean, as the Pacific coast.

Pacific Coast tick *n.* a tick (*Dermacentor occidentalis,* family Ixodidae) that is a vector of Rocky Mountain spotted fever.

Pacific Coast tick

Pacific Northwest *n.* Washington, Oregon, and Idaho.

Pacific salmon *n.* any of several species of salmon (*Onchorhynchus* sp., especially the Chinook salmon, *O. tshawytscha*) that occur along the Pacific Coast from California to Alaska and that provided tens or hundreds of millions of pounds of salmon per year in the late 19th century while it was being overfished but only a few million pounds per year in the late 20th century.

pack *n.* (Flemmish *pac*, pack) **1.** a bundle of things prepared for transport, especially by a pack animal or on a person's back (in a backpack). **2.** a standard package containing a usual number of cans, bottles, or other units, as a 6-pack of soda. **3.** the amount of something that is harvested and packed in a season or a year, as the annual pack of peaches. **4.** a group, often in a derogatory sense, as a pack of fools or a pack of lies. **5.** a band of certain animals (especially predators), such as a wolf pack. **6.** a deck of playing cards. **7.** an ice pack, either a jumbled mass of chunks of ice blocking a stream or a wrapped bundle of ice used to relieve pain and swelling. *v.* **8.** to prepare a bundle of things or to place garments and other items in a suitcase or other suitable container for travel or storage. **9.** to prepare meat, fruit, or other food products for shipment or storage. **10.** to carry, either personally (e.g., to pack a gun) or in one's entourage (e.g., to pack a lot of luggage or freight). **11.** to become compacted, as a snowpack. **12.** to press tightly into a recess to form a seal around a shaft or in a pocket.

pack animal *n.* a beast of burden; commonly a donkey, mule, horse, camel, llama, or alpaca used to carry heavy loads, often over rough terrain.

Packers and Stockyards Programs *n.* a function of the Grain Inspection, Packers and Stockyards Administration (GIPSA) that facilitates the marketing of livestock, poultry, meat, and related agricultural products and promotes fair and competitive trading practices for the overall benefit of consumers and American agriculture. GIPSA was formed in 1994 as a merger of the Federal Grain Inspection Service and the Packers and Stockyards Programs.

pack ice *n.* See ice pack.

packinghouse *n.* a building where meats or fruit and vegetable products are processed, canned, frozen, and/or packed in preparation for shipping to wholesale or retail outlets.

pack rat *n.* **1.** a large rat (*Neotoma cinerea*) with a bushy tail that collects miscellaneous things in its nest. Also called a trade rat because it often leaves another item in exchange for one it takes. **2.** a person who imitates a pack rat by collecting many things in a clutter.

pad *n.* (Dutch *pad*, path) **1.** a footpath. **2.** a soft sound of footsteps (usually made by an animal walking quietly). **3.** a cushion on the underside of an animal's foot. **4.** a type of cushion or saddle without a tree for riding on the back of an animal. **5.** a cushion used for protecting items or for comfortable seating, etc. **6.** a piece of protective equipment worn when playing certain games, such as American football. **7.** living quarters or the bed in such quarters. **8.** a small tablet of writing paper, often used for making notes. **9.** an ink pad, used for inking a rubber stamp. **10.** a floating plant leaf, as a lily pad. **11.** a launch pad, where rockets are launched. *v.* **12.** to walk, especially steadily and quietly. **13.** to cushion or fill with cushioning material. **14.** to add extra verbiage or items, usually unneeded items or illicit charges.

paddlefish *n.* a large fish (*Polyodon spathula*) with hard, bony scales and an elongated flat paddlelike snout. It is found in the Mississippi River and its larger tributaries. Also called spoonbill.

paddock *n.* (Anglo-Saxon *pearroc*, enclosure) **1.** a small area near a barn or stable, usually fenced or otherwise enclosed. It is used for pasture or for exercising animals, especially horses. **2.** a small enclosure where horses are saddled and mounted in preparation for a race. **3.** (in Australia) any fenced field or pasture. **4.** (especially in Scotland) a frog or toad.

paddy *n.* (Malay *padi*, husked rice) **1.** a flooded field where rice is grown. The water depth should be carefully controlled at about 4 in. (10 cm) depth. **2.** rice, especially while still in its husks.

paddy rice *n.* **1.** rice growing in flooded fields. **2.** harvested rice that is still in its husks.

padouk or **padauk** *n.* (Burmese *padauk*, padouk) **1.** any of several tropical Asian or African trees (*Pterocarpus indicus* and related species, family Fabaceae) with broadly winged seedpods. **2.** the hard wood from such trees, typically black with reddish stripes or mottles, that is used for making decorative cabinetwork, paneling, furniture, etc. Also called amboyna.

paedomorphism *n.* (Greek *paido*, child + *morphē*, form + *ismus*, condition) retention of infantile or juvenile characteristics in an adult mammal. Also called pedomorphism.

paha *n.* a low, elongated ridge in northeastern Iowa consisting of a core of glacial till covered by a loess mantle. The core is believed to be a remnant left when the surrounding landscape was eroded severely, leaving an area known as the Iowan erosion surface.

painted bunting *n.* a small bird of southern U.S.A. and northern Mexico (*Passerina ciris*). The male is brilliantly colored with a blue head, yellow back, red underparts and rump, and black tail and wings with colored markings. The female is yellow, green, and gray.

painter's colic *n.* constipation and intense intestinal pain caused by exposure to paint containing lead. Also called lead colic. Note: The use of lead in paint has been banned in the U.S.A. since 1978; other countries have also banned it at various dates.

palaeotemperature *n.* See paleotemperature.

palatability *n.* (Latin *palatum*, roof of the mouth + *abilis*, worthy of + *itat*, condition) **1.** the quality of being palatable. **2.** the degree of palatability that a food or idea has. Palatability varies with both the consumer (type of animal, habits, degree of hunger, etc.) and the environment (temperature, disturbances such as precipitation, noise, presence of enemies, etc.) as well as the nature of the food (chemical composition, texture, age, condition, etc.).

palatable *adj.* (Latin *palatum*, roof of the mouth + *abilis*, worthy of) **1.** pleasing or acceptable to the taste; appetizing; easy to eat. **2.** acceptable or agreeable to science, logic, and/or the desires and feelings.

palate *n.* (Latin *palatum*, roof of the mouth) **1.** the upper surface of the mouth, including the bony front part (hard palate) and the muscular back part (soft palate). **2.** sense of taste, as in good food is pleasing to the palate. **3.** aesthetic taste.

Palearctic or **Paleoarctic** *n.* (Greek *palaios*, ancient + *arktikos*, bear) the biogeographic region comprising Europe, part of northern Africa, most of the Arabian peninsula, and Asia north of the Himalayan Mountains.

paleethnology or **paleoethnology** *n.* (Greek *palaios*, ancient + *ethnos*, people + *logos*, word) the study of prehistoric human races.

paleoagriculture *n.* (Greek *palaios*, ancient + Latin *ager*, field + *cultura*, culture) ancient agriculture; the prehistoric systems used to grow crops and tend livestock.

paleoanthropology *n.* (Greek *palaios*, ancient + *anthrōpos*, human + *logos*, word) the scientific study of fossils and other indicators of the origins of early humans (*Homo sapiens*) and humanlike creatures that preceded them.

Paleoarctic *n.* See palearctic.

paleobiology *n.* (Greek *palaios*, ancient + *bios*, life + *logos*, word) the study of life forms based on fossils, especially with regard to the origin and evolution of plant and animal species.

paleobotany or **palaeobotany** *n.* (Greek *palaios*, ancient + *botanē*, plant) the study of plant fossils. Also called paleophytology.

paleoclimate *n.* (Greek *palaios*, ancient + *klima*, region or zone) the climate of past geologic time, including the warm periods (global warming) and the ice ages (global cooling), as evidenced by fossils, isotopic ratios of elements, etc.

Paleocene *n.* (Greek *palaios*, ancient + *kainos*, recent) the earliest epoch of the Tertiary Period of the Cenozoic Era (about 65 to 55 million years ago), when mammals proliferated and became dominant; preceding the Eocene Epoch.

paleoenvironment *n.* (Greek *palaios*, ancient + French *environnement*, surroundings) the biological, chemical, and physical conditions, circumstances, and influences that formed the environment of an ancient organism.

paleogeography *n.* (Greek *palaios*, ancient + *geōgraphia*, geography) the study that endeavors to reconstruct the geographic features of the Earth at past geologic times.

paleogeology *n.* (Greek *palaios*, ancient + *gē*, Earth + *logos*, word) the science that endeavors to reconstruct the geologic conditions that existed at specified times in the geologic history of Earth.

paleolimnology *n.* (Greek *palaios*, ancient + *limnē*, marsh + *logos*, word) the study of sediments and fossils to characterize ancient lakes.

Paleolithic or **paleolithic** *n.* or *adj.* (Greek *palaios*, ancient + *lithos*, stone) **1.** the period when humans began using stone tools; the early part of the Stone Age. **2.** the late Pliocene and most of the Pleistocene epochs, from about 2 million years ago to about 10,000 years ago.

paleomagnetism *n.* (Greek *palaios*, ancient + *Magnētis*, stone of Magnesia) **1.** the alignment of magnetic crystals of iron and nickel minerals in rocks formed in ancient times. **2.** the study of such alignments as indicators of the wandering and reversals of the magnetic poles of the Earth. The study of the reversals is also called magneto-stratigraphy.

paleontology *n.* (Greek *palaios*, ancient + Latin *ontologia*, nature of being) the study of fossils to decipher prehistoric forms of life.

paleopathology *n.* (Greek *palaios*, ancient + *pathos*, suffering + *logos*, word) the study of diseases and other medical conditions of prehistoric humans based on mummies, skeletal remains, and artifacts.

paleosol or **palaeosol** *n.* (Greek *palaios*, ancient + French *sol*, soil) a soil that exhibits morphological properties indicating that it formed in a distinctly different environment. It is considered a relict if it has persisted on the landscape without adapting to a changing environment; a buried soil if it has been covered, perhaps by a glacial deposit, a loess deposit, or a lava flow; or an exhumed paleosol if it was buried but has been reexposed by erosion.

A glacial deposit (top half) overlying a paleosol

paleotemperature *n.* (Greek *palaios*, ancient + Latin *temperatura*, temperature) the prevailing temperature of an ancient body of water at the time of deposition of sedimentary rock. Paleotemperatures interpreted from fossils and the ^{18}O isotope contents of organic carbonates indicate that the equatorial oceans were 5° to 6°C warmer at one time than they are now.

paleophytology *n.* (Greek *palaios*, ancient + *phyton*, plant + *logos*, word) paleobotany.

Paleozoic or **Palaeozoic** *adj.* (Greek *palaios*, ancient + *zōē*, life + *ikos*, pertaining to) the geologic period when most forms of life developed and fishes, insects, reptiles and giant ferns were the dominant forms of life. Large peat deposits were formed that have since become coal. The Paleozoic is dated at 570 million years ago to 325 million years ago and is divided into seven periods, with the Cambrian being the oldest followed by the Ordovician, Silurian, Devonian, Mississippian, Pennsylvanian, and lastly the Permian.

palladium (Pd) *n.* (Greek *Pallas*, goddess of wisdom + *ium*, element) element 46, atomic weight 106.42; a very malleable steel-white metal that does not tarnish in air. The lowest-melting and least-dense member of the platinum group, it absorbs up to 900 times its volume of hydrogen and is a good catalyst for hydrogenation and dehydrogenation reactions, It is also used in alloys for dentistry, watchmaking, surgical instruments, and electrical contacts and jewelry and is added to gold to make white gold.

palm *n.* (Latin *palma*, hand or palm tree) **1.** the concave inner surface of the hand between the fingers and the wrist. **2.** the corresponding part of an animal's front foot. **3.** the part of a glove that covers the palm of a hand. **4.** a linear measure based on either the width of a hand, generally 3 to 4 in. (8 to 10 cm) or its length, generally 7 to 10 in. (18 to 25 cm). **5.** a broad, flat surface on an antler of a moose or deer. **6.** a tropical or subtropical mono-cotyledonous plant (family Arecaceae). Most are tall trees with unbranched stems topped with a crown of large leaves with many deep clefts. Some produce coconuts, dates, or other useful fruit; some are used as a source of wood; and some are sources of useful resins, fibers, oils, or drugs. **7.** a symbol of victory and/or honor.

Palmae *n.* the older name for plants of family Arecaceae. See palm.

palmate *adj.* (Latin *palmatus*, like a hand) **1.** shaped like a hand with four or more separate fingers, especially certain clusters of veins, lobes, or leaflets of certain trees (sometimes called palmatisect if the divisions reach all the way to the base of the leaflets). See pinnate. **2.** web-footed, as many waterbirds or amphibians.

palmatisect *adj.* See palmate.

palmetto *n.* (Spanish *palmito*, a small palm tree) any of several small palm trees with fan-shaped leaves native to the West Indies and southern U.S.A.; e.g., the cabbage palmetto (*Sabal palmetto*, family Arecaceae) of the South Atlantic and Gulf Coasts with thick, leathery, fan-shaped, evergreen leaves 4 to 7 ft (1.2 to 2.1 m) long that are used to make baskets, brooms, hats, mats, and thatch. An unbranched trunk that provides durable wood for pilings, docks, and poles and fibers for brushes and whisk brooms and is used as an ornamental. It is the state tree of Florida and South Carolina.

palmito *n.* (Spanish *palmito*, a small palm tree) **1.** palmetto. **2.** the edible white heart of a palm tree (where the leaves attach).

palm kernel oil *n.* oil obtained from the hard-kernel seeds of the oil palm tree, a nonedible oil used to make lubricants, soaps, paints, and varnishes. See palm oil.

palm leaf *n.* a frond (leaf) of a palm tree, especially of the fan palmetto, that is used to make fans, hats, mats, or thatch.

palm oil *n.* an edible oil obtained from the fleshy fruit surrounding the seeds of the oil palm tree. It is used to make cooking oil, margarine, confections, soaps, paints, and pharmaceuticals. See palm kernel oil.

palp or **palpus** (pl. **palpi**) *n.* (Latin *palpus*, a caress) a taste sensor and/or feeler attached to the mouth or front of the head of certain insects, crustaceans, or worms.

palpate *v.* (Latin *palpare*, to touch) **1.** to examine by touch. *adj.* **2.** having one or more palp.

palpitate *v.* (Latin *palpitare*, to pulsate) **1.** to beat rapidly, as a heart affected by overexertion, strong emotion, or disease. **2.** to tremble, shake, or quiver.

palpus *n.* See palp.

paludal *adj.* (Latin *paludis*, marsh + *alis*, pertaining to) **1.** swampy; pertaining to a marsh or swamp. **2.** produced by or developed in a marsh or swamp, as certain diseases.

paludinal or **palustrine** *adj.* **1.** living in a marsh. **2.** paludal.

pampas *n.* (Quechua *pampas*, boundless flat plain) **1.** a large grassland plain, especially an area in Argentina about 1000 mi (1600 km) long north to south with the *pampa humeda* (humid pampa) in the east with its highly fertile, poorly drained soils. It is bordered on the east by the Paraná River and the Atlantic Coast. The *pampa seca* (dry pampa) to the west is bordered on the west by the foothills of the Andes. **2.** a similar but smaller area in the valley floors between mountain ranges in Chile. **3.** a Native American from the region of the pampas in Argentina.

pampas grass *n.* the dominant bunchgrass (*Cortaderia selloana*, family Poaceae) in the pampas of Argentina. The leaves grow up to 8 ft (2.4 m) tall, and the feathery white flower plumes reach as high as 12 ft (3.6 m).

pampero *n.* (Spanish, *pampero*, of the pampas) a strong, cold, west or southwest wind that blows from the Andes across the pampas, especially if it reaches Buenos Aires and the Rio de la Plata.

pan *n.* (Anglo-Saxon *panne*, a broad dish) **1.** a shallow metal dish with an upward sloping edge used for cooking, baking, etc. **2.** a similar dish used for other purposes such as a weighing pan for a balance. **3.** the contents or capacity of such a dish, as a pan of beans. **4.** a container where valuable metals are separated from other materials in ore. **5.** a depression in the ground, either natural or artificial; e.g., one used to allow saline water to evaporate and leave salt. **6.** a soil layer sufficiently hard to restrict root growth; e.g., a hardpan, fragipan, or plow pan. **7.** bread (in Spanish). **8.** the rear part of a whale's lower jaw. *v.* **9.** to use a pan to wash lighter materials from a mixture and retain gold or other heavy metal. **10.** to rotate a camera to follow action. **11.** to criticize someone's actions or work, either orally or in writing.

PAN *n.* See peroxyacetylnitrate.

panache *n.* (French *pennache*, little wing) **1.** an ornamental plume or bunch of feathers adorning a helmet or cap. **2.** a flamboyant manner; elegant and dashing. **3.** a scopula, an upright tuft of hair on the foot of a spider or certain bees.

panda *n.* (Nepalese *panda*, panda) **1.** the giant panda (*Ailuropodo melanoleuca*), a black-and-white bearlike animal about 6 ft (1.8 m) long and 2 ft (0.6 m) high at the shoulder, native to Nepal but now found mostly in bamboo forests of central China or in zoos. **2.** the lesser panda (*Ailurus fulgens*), a reddish brown animal with head and body about 2 ft (0.6 m) long and a raccoonlike tail about 1.5 ft (0.45 m) long, native to the Himalayan Mountains and northern India. Pandas eat bamboo, fruit, and insects. Also called wah.

pandemic *adj.* (Greek *pan*, all + *dēmos*, people + *ikos*, of) occurring over an entire area such as the world or a nation; said of a widespread disease or attitude, as a pandemic fear of war. See epidemic.

Pangaea or **Pangea** *n.* (Greek *pan*, all + *gē*, Earth) the supercontinent believed to have contained all the landmass of Earth from about 300 million years ago to about 200 million years ago (when it began splitting into separate continents). The term was coined in 1912 by Alfred Wegener (1880–1930), a German meteorologist.

pangolin *n.* (Malay *pengguling*, pangolin) any of several mammals of order Pholidota (e.g., *Mavis javanica*) whose bodies are covered with broad, overlapping, horny plates. It feeds mostly on ants and termites, and rolls into a ball when threatened. Also called scaly anteater.

panic grass *n.* (Latin *panicum*, a millet) any of many species of grass (*Panicum* sp., family Poaceae) including several millets (e.g., *P. miliaceum*), switchgrass (*P. virgatum*), fall panicum (*P. dichotomiflorum*), etc. Many have edible grain and/or are used as forage.

panicle *n.* (Latin *panicula*, a plant tuft) a loose cluster of flowers; a compound raceme.

panleukopenia *n.* (Greek *pan*, all + *leukós*, white + Latin *penuria*, poverty) a disease of cats that causes abnormalities in white blood cells, inactivity,

diarrhea, vomiting, and refusal to eat. It is caused by a virus (*Tortor felis*). Also called cat distemper, cat plague, or feline enteritis.

panmixia *n.* (Greek *pan*, all + *mixis*, mixing) random mating within a heterogeneous population.

panning *n.* **1.** searching or mining for gold or other heavy metals in loose sediments by swirling the loose material in a pan of water to skim off the light material while retaining the denser material in the pan. **2.** rotating a camera to follow action. **3.** criticizing someone's actions or work, either orally or in writing.

panorama *n.* (Greek *pan*, all + *horama*, view or sight) **1.** an unobstructed view of a wide area, especially one that is very scenic and impressive. **2.** a pictorial representation of such a scene, often presented as a series of photographs taken by panning. **3.** a chronological sequence of scenes or events, as a panorama of national history. **4.** a comprehensive coverage of a topic.

pansy *n.* (French *pensée*, a thought or fancy) **1.** a small, popular ornamental plant (*Viola tricolor hortensis*, family Violaceae). Also called heartsease. **2.** the flowers from such plants. Many different colors are common. **3.** (offensive slang) a weak, cowardly, effeminate, and/or homosexual man.

panther *n.* (Latin *panthera*, panther) **1.** a cougar (*Felis concolor*). Also called puma. See cougar. **2.** a leopard (*Panthera pardus*) or any black leopard. **3.** a jaguar (*Panthera onca*).

panzoism *n.* (Greek *pan*, all + *zōē*, life + *ism*, principle) a combination of all factors that contribute vitality and energy to life.

panzootic *adj.* (Greek *pan*, all + *zōion*, animal + *ikos*, of) widespread among animals; pandemic.

pap *n.* (Dutch *pap*, pap) **1.** soft food (often semiliquid) for infants or invalids. **2.** a nipple or teat of a female mammal. **3.** any soft, pulpy mass. **4.** political favors; patronage. **5.** an idea, suggestion, or publication with little real value.

papain *n.* (Spanish *papaya*, papaya fruit + *in*, chemical) **1.** a proteolitic enzyme found in the latex of green papaya fruit. **2.** a refined form of this enzyme used to tenderize meat by catalyzing the hydrolysis of proteins to amino acids and polysaccharides. It is also used medically to digest nonliving tissues without harming living tissues; e.g., to promote healing of skin lesions. Urea and/or other materials are used with it to increase its activity.

papaw or **pawpaw** *n.* (Spanish *papaya*, papaya fruit) **1.** a small tree (*Asimina triloba*, family Annonaceae) with large oblong leaves and purplish flowers that grows in central and southern U.S.A. **2.** its edible banana-shaped fruit filled with many seeds. **3.** papaya.

papaya *n.* (Spanish *papaya*, fruit of the papaw) **1.** a tropical American shrub or small tree (*Carica papaya*, family Caricaceae). Also called papaw. **2.** the large yellow fruit of the tree, shaped like a melon and eaten either raw or cooked. **3.** the juice from the fruit.

paper birch *n.* a medium to large North American tree (*Betula papyrifera*, family Betulaceae) with white or light-gray bark that can be peeled off in sheets like thick paper. It is native to Canada, Alaska, and the northern tier of states in eastern U.S.A. and is used as an ornamental tree and to make toothpicks, spools, and other turned items. The bark was used by Native Americans to make canoes. Also called white birch, canoe birch, or silver birch.

paper chromatography *n.* a method of chemical analysis that uses strips of paper and a liquid that will wet the paper. The material to be analyzed is impregnated into a band near one end of a strip of paper and allowed to dry. The end of the paper is then dipped into the water or other liquid. As the liquid wicks up through the paper, it dissolves the chemicals in the sample at different rates and moves them upward along the strip at different rates, thus forming bands on the paper. Each band can then be analyzed separately. See chromatography.

paper hornet or **paper wasp** *n.* any of several social insects (family Vespidae, especially *Vespra crabro*), including yellow jackets and hornets, that build nests from chewed wood pulp. There are three castes: queens, workers, and males. The queens and workers have a very serious sting.

paprika *n.* (Greek *peperi*, pepper) **1.** the fruit from a type of pepper plant (*Capsicum annuum* and others, family Solanaceae). **2.** a mild, red condiment made by grinding the dried fruit.

papyrus *n.* (Greek *papyros*, reed) **1.** a tall aquatic sedge (*Cyperus papyrus*, family Cyperaceae) native to the Nile Valley in Egypt and elsewhere. Called bulrushes in the Bible. **2.** a paperlike material made from these plants by cutting the pith into thin strips, placing crossed layers of these strips together, pressing the water out, and allowing them to dry. It was used in ancient times by Egyptians, Greeks, Romans, and others to make an early type of paper. **3.** a document or manuscript written on papyrus.

PAR *n.* See photosynthetically active radiation.

parabiosis *n.* (Greek *para*, side by side + *biōsis*, mode of life) **1.** an ecological situation wherein two or more species intermix colonies in harmony; e.g., mixed species of ants or birds or a mixture of cattle, sheep, and goats. **2.** an exchange of blood between two individual animals, either naturally or experimentally. **3.** a temporary loss of function of a nerve cell.

para cress *n.* a composite-flower plant (*Spilanthes acmella*, family Asteraceae) of tropical areas in America and Asia that is the source of a powerful mosquito larvicide and formerly was used as a toothache remedy.

paradichlorbenzene *n.* an aromatic chemical ($C_6H_4Cl_2$) used to protect woolen articles from moths, to fumigate combs in beehives, and to fumigate soil.

paradise *n.* (Greek *paradeisos*, a park or pleasure ground) **1.** a heavenly place, either the final abode of the righteous or a place for righteous souls awaiting resurrection. **2.** a very beautiful, delightful place.

paraheliotropism *n.* (Greek *para*, parallel + *hēlios*, sun + *tropos*, turning) the turning of plant leaves to parallel the direct rays of the sun and thus reduce the amount of light intercepted. See heliotropism.

parallax *n.* (Greek *parallaxis*, wandering change) **1.** a deviation between where an object is and where it seems to be from the position of the eye of an observer. **2.** the angle of such a deviation; e.g., in astronomical observations, the angle between the line of sight of an observer and a line from the center of the Earth to the same object. **3.** the difference in the view of an object as seen from two different positions; e.g., simultaneously by the two eyes of a person; the difference that permits stereoscopic vision. **4.** the difference in the view photographed by a camera and that seen through its separate viewfinder.

parallelism *n.* (Greek *parellēlos*, side by side + *ismos*, condition) **1.** the condition of being parallel, of being equidistant at all positions. **2.** similarity in direction, tendency, appearance, sound, meaning, or other characteristics. **3.** parallel evolutionary development of species that are isolated from each other. Also called parallel evolution. **4.** the use of a series of similar words (in sound or meaning), phrases, or clauses (a common feature of Hebrew writing). **5.** simultaneous processing of more than one task in a computer through the use of separate units of memory. **6.** the teaching that mind and body are distinct and not causally related but that they nevertheless coordinate and develop in harmony with each other.

paralysis *n.* (Greek *paralysis*, to loosen on one side) **1.** a loss of voluntary control over part or all of the body, usually caused by an illness or disease that attacks the nervous system, often affecting one side (e.g., one arm) but not the other. **2.** a disease that causes symptoms of paralysis, especially palsy. **3.** a stoppage of all productive activity at a place or of a particular kind, as paralysis of the bus system.

paramagnetic *adj.* (Greek *para*, beside + *Magnētis*, stone of Magnesia + *ikos*, like) slightly magnetic; weakly attracted to an iron magnet. All iron-bearing minerals are included plus some others (e.g., beryl). See ferromagnetic, diamagnetic.

paramo *n.* (Spanish *páramo*, barren plain) a mostly barren, windswept plain at very high elevation near the equator; especially a plateau in the Andes of South America. See Altiplano.

paraquat *n.* a popular nonselective contact herbicide, dimethyl dipyridilium ($C_{12}H_{14}N_2 \cdot 2CH_3SO_4$). Paraquat residues adhere strongly to soil particles until degraded by soil organisms; new seedlings can grow in its presence. Paraquat can be absorbed through the skin and cause skin irritation. It is much more toxic to humans and animals than are most herbicides; large doses cause lung damage, and swallowing a teaspoonful can be fatal.

parasite *n.* (Greek *parasitos*, one who eats at another's table) **1.** an organism that lives on a host organism and uses the host as a source of nourishment but usually does not kill the host. **2.** a person who takes advantage of another (or others) for food, shelter, labor, or other advantage but gives little or nothing in return.

parasiticide *n.* or *adj.* (Greek *parasitos*, parasite + Latin *cida*, killer) a chemical that kills parasites.

parasitism *n.* (Greek *parasitos*, parasite + *isma*, action) **1.** the relationship between a host and a parasite: the parasite benefits at the expense of the host. See symbiosis. **2.** the mode of life of a parasite. **3.** a diseased condition caused by parasites.

parasitize *v.* (Greek *parasitos*, parasite + *izein*, to subject to) to infest and live as a parasite in or on a host.

parasitoid *n.* (Greek *parasitos*, parasite + *eidēs*, like) the parasitic larvae of an insect such as a wasp (e.g., *Trichogramma* sp.) that lays its eggs in a caterpillar. The host usually dies after the larvae mature. Some parasitoids are used for biological control of certain pests.

parasitology *n.* (Greek *parasitos*, parasite + *logos*, word) the study of parasites and parasitism (a branch of biology).

parathion *n.* an early, popular organophosphate insecticide ($C_{10}H_{14}O_5NPS$) that binds tightly to clay and humus and is highly toxic to humans and animals through inhibition of nerve functions.

parchment *n.* (Latin *pergamina*, paper of Pergamus) **1.** a thin, flexible, durable writing material made from the skin of an animal, usually a sheep or goat. **2.** a document written on parchment. **3.** a thick paper that resembles parchment made from animal skin.

pare *v.* (Latin *parare*, to prepare) **1.** to remove the peeling or other covering, such as that of a fruit or vegetable. **2.** to reduce gradually, usually as much as possible, as to pare expenses to the bone.

pareira or **pareira brava** *n.* (Portuguese *parreira brava*, wild vine) the root of a South American plant (*Chondrodendron tomentosum*) and the source of curare, a poison used medically and by certain South American aborigines on their arrows.

parenchyma *n.* (Greek *parenchyma*, something poured in beside) **1.** functional organ tissue, as distinguished from connective or supporting tissue. **2.** relatively soft tissue in the stems or fruit of plants, consisting of thin-walled cells with a rounded shape. **3.** soft tissue in the bodies of certain invertebrates such as protozoa or flatworms. **4.** soft, growing tissue in a tumor.

parent material *n.* unconsolidated mineral matter that may develop into soil; e.g., a layer of unconsolidated material between a soil solum and the underlying bedrock.

Paris green *n.* a poisonous green powder that was an early homemade insecticide composed of a mixture of sodium arsenite, copper sulfate, and acetic acid.

park *n.* (French *parc*, park) **1.** an area designated for public use for relaxation, recreation, etc., usually by a local, state, or federal government, that is commonly equipped with picnic tables, playgrounds, and walkways. **2.** a stadium or other enclosed area used for sports and other public events; e.g., a baseball park. **3.** a scenic area of grassland, woods, lakes, etc., surrounding a large estate. **4.** (British) a game preserve. **5.** a large open valley in the midst of mountains, especially in western U.S.A. **6.** an area used by a military unit to assemble vehicles and equipment. **7.** a setting that locks an automobile transmission. *v.* **8.** to place a vehicle where it can be left for a period of time. **9.** to similarly place any item, as to park your coat over there. **10.** to establish a satellite in a stable orbit.

parr *n.* **1.** a young salmon that has not yet entered salt water; marked with dark bars on its sides. **2.** certain other young fish; e.g., young codfish.

parrot fever *n.* See psittacosis.

par-sec *n.* a measure of astronomical distance that gives a parallax reading of 1 arc-second annually as the Earth moves from one side of the sun to the other. It is equal to 3.26 light years or 19.2 trillion mi (3.09×10^{13} km).

Parshall flume *n.* a device for measuring water flow in an open ditch with a minimum of head loss. See weir.

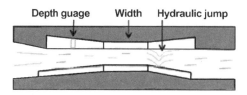

Depth guage   Width   Hydraulic jump

Water flowing through a Parshall flume regains most of its head loss at the hydraulic jump. Its water delivery is approximately:

$Q = 4.0 \ WH^{1.55}\text{ft}^{0.45}/\text{sec}$ or $Q = 2.34 \ WH^{1.55}\text{m}^{0.45}/\text{sec}$

where W is the width of the middle section and H is the head at the depth guage. All units for the first equation are in feet, and those for the second equation are in meters.

parthenocarpy *n.* (Greek *parthenos*, virgin + *karpos*, fruit) development of fruit without fertilization of the ovary.

parthenogenesis *n.* (Greek *parthenos*, virgin + *genesis*, birth) development of an egg or a fruit (a mature ovary) without fertilization. Examples are drone bees, aphids, and certain other insects, algae, bananas, etc., depending partly on environmental factors.

particle *n.* (Latin *particula*, little part) **1.** a small individual piece or fragment, as a dust particle. **2.** a very tiny amount, as a particle of faith. **3.** a small part in a document. **4.** one of the tiniest components of matter, so small that its size is negligible.

particle density *n.* weight per unit volume of solids (excluding the volume of any pore space in the material). The particle density of soil is often assumed to be 2.65 g/cm^3. Formerly called specific gravity. See bulk density.

particle-size analysis *n.* determination of the percentages of sand, silt, and clay in a soil sample, usually by a combination of sieve analysis for the sand sizes and sedimentation rates for the smaller particles. These percentages of soil separates are used to place the sample in a textural class such as sandy loam or silty clay. Formerly called mechanical analysis. See hydrometer analysis, pipette analysis, sieve analysis.

particulates *n.* finely divided liquid or solid particles suspended in the air, especially those that are pollutants. Examples are smoke, dust, spray droplets, and other small particles or droplets.

partridge *n.* (Latin *perdix*, partridge) any of several plump, mostly ground-dwelling game birds; e.g., the common European partridge (*Perdix cinerea*) or any of several North American game birds, including the ruffed grouse (*Bonasa umbellis*) and the bobwhite (*Colinus* sp.).

parts per million (ppm) *n.* a unit for designating small concentrations that is numerically the same as milligrams per liter (mg/L), milligrams per kilogram, or grams per metric ton. Note: Still smaller concentrations are sometimes designated as parts per billion or parts per trillion.

parturition *n.* (Latin *parturitio*, in labor) the process of giving birth to young. Also called birthing, calving, farrowing, lambing, etc., according to species.

pascal (Pa) *n.* the SI (metric) unit of pressure, defined as a force of 1 newton/m^2, that is named after Blaise Pascal (1623–1662), a French philosopher and mathematician. Often used as megapascals (MPa) for a larger unit; 1 MPa equals 1 million Pa or approximately 10 bars, 9.9 atm, 10.2 kg/cm^2, or 145 lb/in^2.

passenger pigeon *n.* a prime example of a formerly abundant bird species that became extinct by human intervention. A migratory pigeon (*Ectopistes migratorius*) that was killed for its meat until the last of the species died in 1914 in a zoo in Cincinnati, Ohio.

passerine *adj.* (Latin *passerinus*, like a sparrow) belonging to the group of small to medium-sized songbirds (order Passeriformes) with feet adapted for perching (more than half of all birds).

passive permafrost *n.* permafrost in a setting where the average annual temperature is above 0°C. It formed in an earlier, colder climate. If it thaws, it will not refreeze under present conditions. Also called fossil permafrost.

passive smoking *n.* the inhalation of tobacco smoke in an enclosed area where others are smoking.

pasteurellosis *n.* See hemorrhagic septicemia.

pasteurization *n.* a process for killing pathogenic microbes in food products by heating the product to 140° to 148°F (60° to 65°C) for 30 min or to 161°F (72°C) for 15 sec (or some combination in between); named after Louis Pasteur.

Pasteur, Louis *n.* a French chemist and microbiologist (1822–1895). He was the first person to demonstrate that microbes can cause disease, and he therefore is considered the father of microbiology. He developed vaccines for anthrax, rabies, and chicken cholera.

pastoral *adj.* (Latin *pastoralis*, like a shepherd) **1.** of the activities and way of life of a shepherd. **2.** rural; generally considered as a charmingly serene setting with pastures, fields, and often woods intermixed. **3.** ministerial, as the religious services and counsel offered by a pastor or other member of the clergy.

pasture *n.* (Latin *pastura*, feeding area) **1.** an area of mostly grass and herbs used for grazing livestock, especially where the forage has been improved by seeding and fertilizing. Note: Larger areas of mostly unimproved forage are called range or rangeland. **2.** a specific area containing forage for livestock, usually bordered by a fence or other obstacle. *v.* **3.** to allow livestock to graze in an area.

pataua *n.* a palm tree native to the Amazon basin that bears large bunches of fruit that produce a high-quality cooking oil similar to olive oil. It is marketed in Colombia but not much elsewhere.

patchiness *n.* (German *plakke*, patch + *nis*, quality or state) the intermixing of contrasting ecological environments; e.g., agricultural fields, pastures, and woodlands; vegetation types reflecting different soil and topography; or the results of natural but spotty disturbances such as fire, storm damage, or landslides.

patchy *adj.* (German *plakke*, patch + *ig*, characterized by) **1.** spotty; having an appearance like a worn garment that has been patched. **2.** few and scattered; representing only a small fraction of the total area.

path *n.* (Anglo-Saxon *paeth*, path) **1.** a place for humans and/or animals to walk (or for bicycles to travel) from one location to another, either made smooth for that purpose or worn by the footsteps of many people or animals. See road, trail. **2.** a place to walk through a garden, park, or other area to be visited. **3.** the line an object follows, as the path of a baseball or of a meteor. **4.** a line of conduct or activity that one follows, as the path of righteousness. **5.** a list of operations that a computer executes or references.

pathogen *n.* (Greek *pathos*, disease + *genēs*, born) anything that can cause disease, especially a living microorganism or virus capable of causing disease.

pathogenesis *n.* (Greek *pathos*, disease + *genesis*, origin) the developmental process of a disease, especially the changes that occur at the cellular level.

pathology *n.* (Greek *pathos*, disease + *logos*, word) **1.** the scientific study of origin, cause, and development of diseases. **2.** an instance of disease; a deviation from normal health.

pathway *n.* (Anglo-Saxon *paeth*, path + *weg*, way) **1.** a path. **2.** a sequence of biochemical reactions, usually involving several enzymes, that produces a specific protein or other organic substance.

Patrons of Husbandry *n.* a fraternal association of farmers organized in 1867 by Oliver Hudson Kelly as an advocacy group for the welfare of farmers and rural Americans in general. It is involved in community service and legislative lobbying and commonly called the Grange. See Farm Bureau, Farmers Union.

patterned ground *n.* a land surface that shows a repetitive set of circles, polygons, stripes, or other forms marked by high and low points, cracks, or stone lines. It is commonly formed in areas with permafrost or other strong frost action, often with vegetative differences accompanying the soil pattern.

paucity *n.* (Latin *paucitas*, fewness) **1.** fewness; very small numerically. **2.** deficiency; scarceness.

Pb *n.* the chemical symbol for lead (an abbreviation of Latin *plumbum*, lead). See lead.

PCB *n.* See polychlorinated biphenyl.

PCP *n.* See pentachlorophenol.

pea *n.* (Latin *pisum*, pea) **1.** any of several cool-season plants (*Pisum* sp., family Fabaceae) with white or pink flowers and green seedpods; e.g., field pea (*P. arvense*) and English, green, or garden pea (*P. sativum*). **2.** the round seeds, especially of the garden pea, used as a vegetable. **3.** any of several related leguminous plants; e.g., Cowpea (*Vigna sinensis*), chickpea (*Cicer arietinum*), pigeon pea (*Cajanus cajan*), commonly used as forage crops and/or cover crops. **4.** something small and spherical that resembles a pea seed.

pea aphid *n.* a green aphid (*Macrosiphum pisi*, family Aphidae) that can develop with or without wings. They suck the sap from the leaves, stems, blossoms, and pods of pea plants, killing or stunting the plants and spreading viral diseases in fields and gardens.

Peace Corps *n.* a U.S. government organization founded in 1961 under President John F. Kennedy as a means for U.S. citizen volunteers to serve in developing countries, especially as teachers and agriculturalists.

peach *n.* (French *pêche*, peach) **1.** a small fruit tree (*Prunus persica*, family Rosaceae) native to China. It was domesticated long ago and is often grown in orchards in temperate climates. The trees have glossy green, lance-shaped leaves and pink or white blossoms. **2.** the rounded, juicy, sweet, edible fruit of the tree with a single pit as its core covered by a yellowish flesh and an orange-yellow fuzzy skin. **3.** an orange-yellow similar to that of the fruit. **4.** (slang) a person or thing that is beautiful and/or admired. *adj.* **5.** made of peaches, as a peach pie or cobbler.

peach palm *n.* (Spanish *pejibaye* or Brazilian *pupunha*, peach palm) a tropical palm that produces highly nutritious chestnutlike fruit and very tasty palmitos.

peacock *n.* (Latin *pavo*, peacock + *cok*, a male bird) **1.** a male peafowl (*Pavo cristatus*) known for its bright iridescent colors, long tail that can spread into a fan shape with multicolored eyelike spots, and crest. **2.** a vain person, as one who struts like a peacock.

peafowl *n.* either a peacock or a peahen (*Pavo* sp.), a mostly ground-dwelling bird native to India and southeastern Asia.

peak *n.* (Norse *pike*, peaked summit) **1.** a mountain or large hill with a pointed top. **2.** the top of such a mountain or hill. **3.** the highest point of anything, as the peak of a career or the peak output of a factory.

peanut *n.* (Latin *pisum*, pea + *nux*, nut) **1.** a tropical or semitropical trailing plant (*Arachis hypogaea*, family Fabaceae) native to Bolivia and northeastern Argentina, with self-pollinated yellow flowers that produce pegs that grow downward into the soil and produce pods underground. **2.** the elongated light-colored pods, each usually containing two nuts but with up to six nuts in Valencia types. **3.** the high-protein (about 25%), edible nuts produced in the underground pods of the plant. **4.** (slang, usually plural) a small amount of money, as that's just peanuts.

pear *n.* (Latin *pirum*, pear) **1.** a fruit tree (*Pyrus communis*, family Rosaceae) with round to oval, leathery leaves that is native to western Asia or nearby Europe. **2.** the edible fruit of the tree with its rounded base and a neck on the stem end, usually picked green. When ripe, it has a yellow skin and sweet white flesh with small seeds in its core. **3.** certain related species; e.g., Oriental pear (*P. pyrifolia*), callery pear (*P. calleryana*), sand pear (*P. serotina*), and snow pear (*P. nivalis*).

pearl *n.* (Latin *perna*, sea mussel) **1.** a lustrous, smooth, spherical body formed inside the shell of a mollusk (e.g., an oyster such as *Meleagrina margaritifera*) by the deposition of thin layers of mostly calcium carbonate ($CaCO_3$) crystallized as aragonite or calcite with a grain of sand or other irritant at its core. Mostly white with a slight bluish tinge, it is highly valued as a gem, especially for necklaces. See margarite. **2.** a cultured pearl; a similar stone resulting from an irritant implanted in a mollusk by human intervention. **3.** a simulated pearl made of plastic or other similar material; commonly used in costume jewelry. **4.** a very precious thing, person, or concept, as a pearl of wisdom. **5.** a white with a slight bluish tinge.

peasant *n.* (French *paisant*, country dweller) **1.** a farmer or farmworker, especially in Europe. **2.** a person of low rank; uneducated and unsophisticated. *adj.* **3.** of a peasant, the clothing of a peasant, or the way of life of a peasant.

pea shrub or **pea tree** *n.* a small leguminous tree or shrub (*Caragana* sp., family Fabaceae) native to central Asia. It is used as an ornamental for its showy yellow flowers and in windbreaks because of its hardiness.

peat *n.* (Latin *peta*, piece of turf) **1.** organic material accumulated in a bog or other wet area under reducing conditions so it is sufficiently preserved to identify its type of plant material; e.g., fibrous peat, sphagnum moss peat, and woody peat. Peat is used in some horticultural soil mixtures, and dried peat is sometimes used as a fuel. See muck, Histosol. **2.** sedimentary peat, a gelatinous material that represents an accumulation of fecal material from organisms living in a body of water. Also called coprogenous earth.

peat bog *n.* a wetland composed mostly or entirely of peat; generally a former pond or lake that has been filled with peat over a long interval of time.

peat moss *n.* **1.** sphagnum moss peat, a very lightweight, low-fertility organic peat that can hold a lot of water. It is often used in soil mixes for potted plants to provide aeration and water storage. **2.** any other peat composed mostly of moss; often used for mulching flowers.

peat soil *n.* a wetland soil composed mostly of peat; generally somewhat acid. It is most common in cool glaciated areas with broad plains, such as Alaska, Minnesota, and Canada. Oxidation following drainage changes peat into muck. See peat, muck soil.

pea tree *n.* See pea shrub.

pea weevil *n.* one of the most serious pests on peas. A beetle (*Bruchus pisorum*, family Bruchidae) that lays its eggs on green pea pods. Its larvae bore into the pea seeds and pupate there.

pebble *n.* (Anglo-Saxon *papol-stan*, pebblestone) **1.** a small stone, especially one worn round and smooth in a stream of water. **2.** an individual piece of gravel, between 2 and 75 mm in diameter, usually somewhat rounded. Pebbles may be subdivided into fine pebbles (2 to 5 mm in diameter), medium pebbles (5 to 20 mm), and coarse pebbles (20 to 75 mm). **3.** a transparent mineral crystal (often quartz) or a lens made from such a crystal.

pebrine *n.* a protozoan (*Nosema bombycis*, order Microsporidia) and the cause of a disease of silkworms that was epidemic in France and Italy in the 19th century.

peccary *n.* (Spanish *pecarí*, peccary) any of several 2-toed American mammals (*Tayassu* sp.) with tusks and musk glands; related to pigs.

pecking order *n.* **1.** a social hierarchy in a flock of birds, such as chickens. Each bird pecks any subordinate bird. **2.** a similar hierarchy among other animals or people wherein either individuals or groups are ranked from highest to lowest; e.g., a caste system.

pecten *n.* (Latin *pecten*, a comb or a shellfish) **1.** a scallop, a type of marine bivalve mollusk (*Pecten* sp., family Pectinidae) with a fan-shaped shell. **2.** a comblike body part, especially a folded membrane over the eyes of birds, fish, and reptiles.

pectin *n.* (Greek *pēktos*, congealing) a water-soluble colloidal carbohydrate occurring in most fruits. The most soluble component of cell walls (easily extracted with hot water), it is composed mainly of linear chains of galacturonic acid (a hexose sugar) plus some rhamnose, etc. The chains are linked by hydrogen bonding and by calcium ions (Ca^{2+}) reacted with carboxyl groups. It is used in food, especially as a congealing agent in jams and jellies, and in certain cosmetics and pharmaceuticals.

pectization *n.* (Greek *pēktos*, congealing + *izein*, to cause + Lation *atio*, process) the process of congealing into a jelly or other gel.

ped *n.* (Greek *pedon*, soil) a naturally occurring unit of soil structure; a natural aggregation of soil particles; e.g., a soil granule, plate, block, or prism formed as a result of the physical, chemical, and biological environment in the soil. See clod.

pedalfer *n.* (Greek *pedon*, soil + Al and Fe, the chemical symbols for aluminum and iron) a soil that accumulates aluminum and iron as it forms, typically a soil that forms in a humid climate and is well leached (a designation that was used in some of the older systems of soil classification but not in the current systems). See pedocal.

pedate *adj.* (Latin *pedatus*, like a foot) **1.** having one or more feet. **2.** similar to a foot in shape or function. **3.** arranged as a group of leaves or leaflets that fan out from a common point of attachment.

pedicel *n.* (Latin *pedicellus*, a small foot) **1.** a small stalk that bears a single flower of a cluster on a peduncle. **2.** a small, short footstalk on a plant. **3.** a footstalk that holds certain lower-animal life forms in place.

pedicle *n.* (Latin *pediculus*, a small foot) **1.** a small stalk that connects the cephalothorax to the abdomen in certain arachnids.

pedicular *adj.* (Latin *pedicularis*, of lice) having to do with lice or infestation with lice.

pediculate *n.* (Latin *pediculatus*, having a foot) **1.** a type of fish (family Pediculati) whose pectoral fins are attached to a stem that resembles a peduncle. *adj.* **2.** having to do with such fish. **3.** on a pedicel.

pediculosis *n.* (Latin *pedis*, louse + *osis*, condition) infestation with lice (family Pediculidae); a lousy environment. Also called phthiriasis.

pedigree *n.* (Latin *pedi*, foot + *grus*, crane) **1.** a list of ancestry, especially one that proves a superior lineage or that an animal belongs to a particular breed. **2.** the derivation of anything, as the pedigree of a word.

pediment *n.* (Latin *pedi*, foot + *mentum*, state or condition) **1.** the triangular part at the end of a building beneath the roof, especially where it sits above a portico. **2.** a curved or otherwise different version of this part of a building. **3.** a landform with a smooth surface and a gentle slope that develops at the foot of a steeper slope and gradually enlarges as the steeper slope erodes back into the hill. Sometimes called a footslope. See peneplain, piedmont.

pedocal *n.* (Greek *pedon*, soil + *cal*, short for calcium) a soil that accumulates calcium as it forms, typically a soil of an arid or semiarid climate (a designation used in some older systems of soil classification but not in the present systems). A high level of calcium usually indicates a pH above neutral and a relatively high soil fertility. See pedalfer.

pedogenesis *n.* (Greek *pedon*, soil + *genesis*, formation) **1.** the process of soil formation, including the formation of soil horizons by weathering, eluviation, illuviation, and accumulation of organic materials. **2.** the production of offspring by larvae or other juvenile form. Also called neoteny.

pedogenic *adj.* (Greek *pedon* soil + *genēs*, kind + *ikos*, of) related to the natural processes of soil formation, as distinguished from characteristics of rock materials that are not yet soil and from characteristics caused by human influence.

pedologist *n.* (Greek *pedon*, soil + *logos*, word + *istēs*, person) a person who studies, describes, and classifies soils; a soil scientist.

pedology *n.* (Greek *pedon*, soil + *logos*, word) **1.** soil science; the study of soil as a natural entity for its own sake rather than for its use as a medium for plant growth. See edaphology. **2.** the scientific study of the nature, behavior, and development of children. See pediatrics.

pedomorphism *n.* See paedomorphism.

pedon *n.* (Greek *pedon*, soil) the smallest three-dimensional unit that can represent a soil. It is large enough to show all important features of the soil and to serve as the root zone for a medium-sized plant such as corn and is arbitrarily taken as 1 m^2 unless special features require it to be larger (up to 10 m^2 maximum). See polypedon.

pedosphere *n.* (Greek *pedon*, soil + *sphaira*, ball) the soil layer that mantles the Earth.

pedotubule *n.* (Greek *pedon*, soil + Latin *tubulus*, small tube) a tubular concretion of cemented soil material, commonly formed in or along a root channel, likely as a result of chemical changes caused by a root that is now gone.

peduncle *n.* (Latin *ped*, foot + *unculus*, small stem) **1.** a specialized stem that supports either a simple or a compound flower. See pedicel. **2.** a stalk that supports a fungal fruiting body. **3.** a stalklike structure in an organ of an animal; e.g., certain structures in the brain or a pedicle in certain arachnids.

peepul or **pipal** *n.* See bo tree.

pegmatite *n.* (Greek *pēgma*, framework + *ite*, rock) a rock with very large crystals. Usually a granitic or other high-silica material (mostly potassium feldspar and quartz) formed as a hydrothermal deposit in a vein or dike.

P/E index *n.* a complex factor developed in 1931 by C. W. Thornthwaite (1899–1963) to evaluate climate based on the ratio of precipitation to potential evaporation. See moisture index.

pelagic *adj.* (Latin *pelagicus*, of the sea) of the surface waters of the open sea, as distinguished from coastal areas and from benthic waters; e.g., pelagic fish live mostly in the upper part of the open sea.

pelican *n.* (Greek *pelekan*, pelican) any of several large, aquatic, web-footed birds (family Pelicanidae) that have a very large bill with a pouch that enables them to swallow large fish.

pelite *n.* (Greek *pēlos*, clay + *ite*, member) shale, mudstone, or other clayey rock.

pelleted *n.* **1.** formed into spherical or tubular pellets, as pelleted feed or pelleted fertilizer. **2.** mixed with other material and formed into pellets, as pelleted seed.

pellicle *n.* (Latin *pellicula*, little skin) **1.** a membrane or other skinlike surface on an organ or other part of an organism. **2.** a thin scum on the surface of a body of water. **3.** a thin coating on a photographic emulsion. **4.** a partially reflective surface on certain mirrors.

pellicular water *n.* a thin film of capillary water held by soil or rock particles. See vadose water.

pellote *n.* See peyote.

pellucid *adj.* (Latin *pellucidus*, shining through) **1.** transparent or translucent; admitting maximum passage of light; e.g., pellucid water. **2.** clearly expressed; easy to understand, as pellucid writing.

pelophyte *n.* (Greek *pēlos*, clay + *phyton*, plant) a plant that grows in clayey soil.

pelt *n.* (Latin *pellis*, skin) **1.** an animal skin, usually one that has been removed from the animal but not yet tanned. **2.** an animal skin used as a garment. *v.* **3.** to strike with thrown objects, as to pelt with stones. **4.** to strike repeatedly, as hail pelts a crop.

pemmican or **pemican** *n.* (Cree *pemikkân*, fat meat) **1.** a food product prepared by Native Americans that consists of dried meat pounded to a powder, mixed with fat and dried fruit or berries, and formed into small loaves or cakes. **2.** a similar mixture of dried beef, suet, and raisins carried by explorers.

peneplain *n.* (Latin *pene*, almost + *planus*, flat) almost a plain; a term proposed in the 1890s by W. M. Davis (1850–1934) to describe a nearly featureless eroded plain representing the end result of landscape development according to his theory of downwearing. See backwearing, downwearing, mass wasting, pediment.

penestable *adj.* (Latin *pene*, almost + *stabilis*, standing) in a steady state; almost in equilibrium; e.g., a land surface that does not change noticeably over a long period.

penetrance *n.* (Latin *penetratus*, penetrated + *antia*, quality) the frequency of expression of a specific genotype in a given environment; the percentage of the individuals that show the trait in a population.

penguin *n.* (Welsh *pen*, head + *gwyn*, white) **1.** a large, flightless bird (*Aptenodytes* sp., family Spheniscidae) of cold southern regions that walks upright and swims gracefully. It can be more than 3 ft (1 m) tall and weigh as much as 80 lb (36 kg). **2.** the great auk.

penicillin *n.* the first antibiotic ($C_9H_{11}N_2O_4SR$), or actually a group of similar compounds, developed in 1928 by Alexander Fleming (1881–1955) from the molds of an ascomycete (*Penicillium notatum*) that grows naturally on cheese and decaying fruits and a soil microbe (*P. chrysogenum*).

penicillium *n.* (Latin *penicillum*, brush) any of several fungi (*Penicillium* sp.) found on cheese and other foods, in soils, and elsewhere, whose fruiting bodies branch and spread like the straws of a broom or the bones of a hand. Some species are a source of penicillin, some are used to make cheese, and some are parasitic on animals or humans.

pentachlorophenol (PCP) *n.* (C_6Cl_5OH) a chemical used to preserve wood against termites and as a fungicide and disinfectant. It is toxic to humans and animals.

pentose *n.* (Greek *pente*, five + *ose*, sugar) a 5-carbon sugar ($C_5H_{10}O_5$) that can have either a chain or ring form. L-arabinose, D-ribose, and D-xylose (all aldehydes) are the most common of the 12 possible pentoses (4 ketones and 8 aldehydes).

penumbra *n.* (Latin *pene*, almost + *umbra*, shade) **1.** the partial shadow occurring in an area when a partial eclipse takes place. **2.** a marginal, shaded area that is not completely dark. **3.** (figuratively) an action or suggestion that is questionable but not completely wrong.

peon *n.* (Spanish *peón*, day laborer) **1.** an unskilled migratory worker who works by the day, especially on a farm. **2.** an indentured servant working to pay a debt. **3.** a foot soldier or a native police officer in India. **4.** a messenger, attendant, or other servant. **5.** any low-ranking person.

pepo *n.* (Latin *pepo*, large melon) **1.** any large gourd with a hard rind, a fleshy interior, and many seeds; e.g., a melon, squash, cucumber, or gourd. **2.** dried ripe pumpkin seeds (*Cucurbita pepo*) eaten as an anthelmintic (an agent to kill internal worms) or as a diuretic.

pepper *n.* (Latin *piper*, pepper) **1.** any of several tropical shrubs (*Piper* sp., family Piperaceae), especially black pepper (*P. nigrum*) used to make the common condiment of the same name. **2.** the garden pepper (*Capsicum annuum*, family Solanaceae), commonly grown in gardens and picked green for use in salads and other foods. Also called bell pepper or sweet pepper. **3.** any of several varieties of hot peppers (also *C. annuum* and related species); e.g., cayenne pepper, chili pepper, jalapeño pepper, and Tabasco pepper. **4.** a condiment made by drying and crushing the berries from either black pepper (for the condiment commonly used along with salt) or certain hot peppers; e.g., cayenne pepper or chili pepper.

peppercorn *n.* (Latin *piper*, pepper + *cornu*, horn) the dried berry of black pepper (*Piper nigrum*).

peppermint *n.* (Latin *piper*, pepper + *mentha*, mint) **1.** a perennial mint (*Mentha piperita*, family Lamiaceae) that is a sterile hybrid believed to be a natural cross between an aquatic mint (*M. aquatica*) and spearmint (*M. spicata*). It grows well in wet soils and spreads by underground stolons. Its green or dried leaves are used in soups, sauces, and to flavor meats. See mint. **2.** an aromatic flavor obtained from the plant as an oil; used to flavor chewing gum, toothpaste, mouthwash, medicines, candy, etc.

peppermint oil *n.* the oil obtained by steam distillation of partially dried peppermint plant material (hay); used to flavor various foods and other products.

peptide *n.* (Greek *peptos*, digested + *ide*, binary compound) a chain of two or more (up to about 50) linked amino acids that can be a component of protein (a polypeptide).

peptide bond *n.* the linkage that ties amino acids into long chains that is produced by reaction of the amino group of one amino acid with the carboxyl group of another, accompanied by the release of a molecule of water. Multiple peptide bonds link groups of amino acids into polypeptide chains and protein molecules.

The two amino acids on the left lose a water molecule when they are linked by a peptide bond, as on the right. Note: C = carbon, H = hydrogen, N = nitrogen, O = oxygen, and R = any organic radical.

peptidoglycan *n.* a polymer of sugars and amino acids that is the main component of the cell walls of bacteria, especially of Gram-positive bacteria.

P/E ratio *n.* See precipitation-evaporation ratio.

percent slope *n.* the slope gradient expressed as a percentage; 100 times the elevation difference divided by the horizontal distance (or sometimes the sloping distance).

perch *n.* (French *perche*, perch) **1.** any of several species of freshwater game fish (*Perca* sp. and related genera) with sharp spines and powerful dorsal fins, especially the common European perch (*P. fluviatilis*) or the American yellow perch (*P. flavescens*). **2.** a support that a bird can grasp with its feet for a resting place, especially a horizontal rod, wire, branch, etc. **3.** an elevated seat (often small and not very secure) that provides a broad view. **4.** a pole that connects the front and rear running gear of certain carriages. **5.** an obsolete measure of length equal to a rod (16.5 ft, 5.5 yd, or about 5 m), or of area equal to a square rod, or of a volume of stone (usually 24.75 ft^3 or about 0.7 m^3).

perched water table *n.* a zone in soil or rock that is underlain by a layer of low permeability, causing it to be temporarily saturated with an aerated zone between it and the permanent water table below.

percolation *n.* (Latin *percolatio*, filtering) **1.** downward flow, usually at a slow or moderate rate, especially that of water through the pores of a soil or other porous medium that is saturated or nearly saturated. **2.** extraction by leaching, as coffee brewing in a percolator or when a pharmacist extracts a soluble drug from solid material.

percolation rate *n.* the rate that the water level drops in a hole in the soil, a test commonly used to determine suitability of a soil as a site for a septic tank drain field (a hole is dug, filled with water, and allowed to drain once before being refilled for the measurement).

peregrine *adj.* (Latin *peregrinus*, foreign) **1.** foreign; coming from somewhere else. **2.** wandering, migratory.

peregrine falcon *n.* a widely distributed large falcon (*Falco peregrinus*) that is noted for swift flight and often trained by falconers for hunting and demonstrations. See falcon.

pereletok *n.* See intergelisol.

perennate *v.* (Latin *perennatus*, to continue for a long time) to survive for years, as a perennial plant.

perennating organ *n.* an underground absorptive organ that helps a plant survive over winter.

perennial *adj.* (Latin *per*, through + *annus*, year) **1.** lasting throughout the year, as a perennial stream. **2.** living more than two years, as a perennial plant. See annual, biennial. **3.** continuing year after year or recurring every year, as a perennial pest. *n.* **4.** a plant that lives more than two years, either as an evergreen that retains its leaves or as a deciduous plant that regrows them each year.

perennial stream *n.* a stream that has surface water flowing all the time, or at least more than 11 months of the year. See intermittent stream.

perfect flower *n.* a flower that has both male (stamen) and female (pistil) parts. See complete flower, imperfect flower.

perfoliate *adj.* (Latin *per*, through + *folium*, a leaf) piercing a leaf, as in plants whose leaves grow around the stem.

pergelic *adj.* (Latin *per*, very + *gelat*, frozen + *ic*, of) permanently frozen; used in Soil Taxonomy to name subgroups of soils that have permafrost. See permafrost.

pergelisol *n.* See permafrost.

perhumid *adj.* (Latin *per*, very + *humidus*, moist) very wet. The wettest climate in the C. W. Thornthwaite classification was called perhumid or superhumid.

pericarp *n.* (Greek *peri*, around + *karpos*, fruit) the ovary wall of a plant, commonly composed of an exocarp (rind of a fruit or shell of a nut), mesocarp (middle part, often fleshy), and endocarp (the hard seed of a stone fruit or the juicy part of a citrus fruit).

perigee *n.* (Greek *perigeion*, near Earth) the nearest point in the orbit of a moon or other satellite around a larger body. See apogee, apsis.

periglacial *adj.* (Greek *peri*, around + Latin *glacialis*, icy) around the outer edge of a glacier.

perihelion *n.* (Greek *peri*, near + *hēlios*, the sun) **1.** the point in its orbit when Earth is closest to the sun, on or about January 3 of each year. **2.** the point in the orbit of any planet or comet when it is closest to the sun. See aphelion, perigee.

perilla *n.* an annual mint plant (*Perilla frutescens* and related species, family Lamiaceae) native to Asia. Its seeds are the source of perilla oil.

perilla oil *n.* a light yellow oil extracted from perilla seeds that is used as a cooking oil and to make paint, varnish, ink, and artificial leather.

perimeter *n.* (Greek *peri*, around + *metron*, measure) **1.** the outer limit of a designated area or figure, often marked by a line or a fence. **2.** the distance around something; the length of a line marking its perimeter. *adj.* **3.** the outer part of something, as one's perimeter vision.

period *n.* (Latin *periodus*, period) **1.** an interval of time that is meaningful in some way, as a period of prosperity, a period of change, or a period of illness. **2.** an interval of time designated for a purpose, as the class comes in second period. **3.** the time between two repetitions of an event, as a comet that has a period of 70 years. **4.** a female's time of menstruation. **5.** an end mark, as the period at the end of a sentence. **6.** a subdivision of a geologic era. **7.** the duration of a complete cycle of some action or oscillation (the inverse of frequency). **8.** a division in a musical composition, typically containing either 8 or 16 measures.

periodical cicada or **periodical locust** *n.* a type of locust (*Magicicada septendecim*, family Cicadidae) of eastern U.S.A., usually about 2 in. (5 cm) long and green, with red and black markings. Its 17year life cycle (in northern U.S.A.; 13 years in southern U.S.A.) consists mostly of a juvenile (nymph) stage spent underground. When it emerges from the ground, it soon becomes a winged adult that lives for only a few weeks, during which the males pierce the air with their shrill song, and the females cause great damage to trees by laying hundreds of eggs apiece in twigs. The nymphs hatch, fall to the ground, and burrow to where they can suck sap from a tree root until time to emerge again. Also called 17-year cicada (or locust).

periphyton *n.* (Greek *peri*, around + *phyton*, plant) a community of microscopic plants and animals such as algae and small crustaceans that live attached to the surface of objects at the bottom of a body of fresh water.

peritrichous *adj.* (Greek *peri*, around + *trichos*, hair) **1.** having hairlike appendages around the perimeter of a body, as peritrichous flagellation of certain bacteria. See flagellum. **2.** having cilia around the mouth of certain protozoa.

periwinkle *n.* (Latin *pervinca*, periwinkle) **1.** any of several small plants (*Vinca* sp., family Apocynaceae) of European origin with shiny dark green leaves and blue, white, or pink 5-petaled blossoms; often grown as ground cover. **2.** any of several small saltwater snails (family Littorinidae) with thick globular shells. **3.** the empty shell left by such a snail.

perlite *n.* (French *perle*, pearl + *ite*, rock) a type of volcanic glass that fractures into small, pearly, spherical fragments, mostly of rhyolitic composition. It is used as insulation and as a component of plant growth media for potted plants.

permafrost *n.* (Latin *permanent*, remaining + Anglo-Saxon *frost*, frozen) permanently frozen soil and/or rock layers ranging from a few inches (several centimeters) to hundreds of feet (more than 100 m) thick, occupying about 20% to 25% of the world's land area (where the average annual temperature is below freezing). A soil layer above the permafrost freezes and thaws annually. The layer below the permafrost is thawed by the internal heat of the Earth. Also called pergelisol.

permafrost table *n.* the lower limit of the soil that thaws above a layer of permafrost.

permanent cover *n.* vegetation that covers the soil continuously, year after year.

permanent pasture *n.* land that is used for grazing livestock and has a perennial cover of grasses and/or legumes. It is not used for cultivated crops. Note: Similar use applies to permanent grass, permanent hay, and permanent meadow.

permanent wilting point *n.* **1.** the condition of a wilted plant that has lost too much water to recover even if placed in a saturated atmosphere. **2.** the water content of the soil around the roots of such a plant, commonly assumed to have a soil matric potential of minus 1.5 MPa (15 atm).

permeability *n.* (Latin *permeabilis*, permeating + *abilitas*, ability) **1.** the ability to permit a fluid to pass through. **2.** the rate of such passage through a membrane or through a porous material such as soil, commonly expressed in distance per unit time (e.g., inches or centimeters per hour). Permeability restricted to one-dimensional flow is the same as hydraulic conductivity. It becomes a percolation rate when outward flow is included. **3.** the corresponding ability to transmit magnetic lines of force, often expressed as a fraction of the intensity transmitted through a vacuum.

permeable *adj.* (Latin *permeabilis*, permeating) porous; able to permit a fluid to pass through.

permeant *adj.* (Latin *permeant* pervading) **1.** able to penetrate, spread throughout, and/or pass through. *n.* **2.** an animal or person that moves readily from one place to another. **3.** a material that passes through; a gas or liquid that enters or passes through a porous medium.

Permian extinction *n.* the most complete mass extinction known from the fossil record. About half of all animal species and 95% of marine species (e.g., trilobites) were wiped out at the end of the Permian Period, about 248 million years ago, possibly as a result of intense volcanic activity.

perosis *n.* a disease of chicks and other young fowl that is caused by insufficient manganese and choline in the diet. Symptoms include bone deformation and swollen hocks, often with a bluish green cast to the skin of the hocks. In severe cases, the Achilles tendon slips out of place in the hock joint. Also called hock disease or slipped tendon.

peroxisome *n.* (Latin *per*, very + Greek *oxys*, sour + *sōma*, body) a specialized vesicle in animal and plant cells where amino acids and fatty acids are degraded. It contains antioxidants to prevent harm from hydrogen peroxide and free radicals released by the degradation process.

peroxyacetylnitrate (PAN) *n.* any of a group of organic compounds ($RCONO_2$ where R represents a hydrocarbon group) present in photochemical smog; toxic to plants and an eye irritant.

persimmon *n.* (Algonquian *pessemins*, dried fruit) **1.** any of several trees (*Diospyros virginiana*, *D. kaki*, and related species, family Ebenaceae) that produce dark, heavy wood that resembles ebony. **2.** the edible, plumlike fruit of these trees, about 3 in. (8 cm) in diameter, high in vitamins A and C, sour and astringent when green but sweet when ripe. It is eaten fresh or used to make pudding or jam.

persistence *n.* (Latin *persistere*, to stand firm + *entia*, quality) **1.** steadfastness; continuing in spite of difficulties. **2.** lasting a long time. **3.** the length of time that something lasts or that its effects endure; e.g., the persistence of a pesticide.

perspiration *n.* (Latin *perspirare*, to breathe through + *tion*, action) **1.** the process of perspiring; a human or animal producing sweat, commonly as a result of strong exertion, a warm environment, or psychological pressure. **2.** human or animal sweat, the fluid exuded from sweat glands that consists mostly of water with dissolved salts and urea.

perturbation *n.* (Latin *perturbare*, to cause confusion + *tion*, action) **1.** a disturbance; anything that causes a rearrangement of items, often causing agitation, disorder, and anxiety; e.g., an earthquake. **2.** a small deviation from normal; e.g., the presence of a planet causes a perturbation in the movement of a star.

pervasive *adj.* (Latin *pervadere*, to go through + *ivus*, tendency) the tendency to spread and permeate, as a strong odor or an aggressive species.

pest *n.* (Latin *pestis*, plague) **1.** an insect, animal, person, or other organism that annoys or bothers a person or animal. **2.** any organism that annoys or injures humans, animals, or economic plants. **3.** a plaque or pestilence.

pesticemia *n.* (Latin *pestis*, plaque + Greek *haima*, blood + *ia*, disease) **1.** the presence of bubonic plague bacteria (*Pasteurella pestis*, also called *Yersinia pestis*) in the blood. **2.** septicemia (blood poisoning).

pesticide *n.* (Latin *pestis*, plague + *cida*, killer) any substance made or used to kill a pest; e.g., herbicides, insecticides, etc.

pesticide drift *n.* movement of a pesticide through the atmosphere into an adjoining area, especially when there is a significant wind velocity. Damage to a neighbor's vegetation may result. Such damage has led to lawsuits.

pesticide management plan (PMP) *n.* a designation of safeguards to be required for pesticide handlers to avoid environmental pollution through improper use of pesticides. The EPA has assigned primary responsibility to the states for the development of pesticide management plans to deal with agricultural chemicals.

pesticide persistence *n.* the length of time that a pesticide remains effective and/or has a negative effect on either target or other organisms.

pesticide pollution *n.* the contamination of food, air, water, soil, or other media by pesticides or their residues or decomposition products.

pesticide residue *n.* the active pesticide or its hazardous decomposition products that remain on plant material after a pesticide has been applied to the plants either directly or indirectly, especially that remaining on food when it is marketed or eaten.

pesticide resistance *n.* the ability of an organism to resist the effects of a pesticide. Crops, livestock, and other nontarget organisms should have high resistance to pesticides applied to them. Target organisms often build up resistance by natural selection when the same pesticide is used repeatedly.

pesticide tolerance level *n.* the maximum concentration of pesticide residue allowable on food or feed. This value must be established by the EPA before the pesticide can be applied to a food or feed crop that is intended to be sold, distributed, or consumed in the U.S.A.

pestilence *n.* (Latin *pestilentia*, full of plague) **1.** an epidemic of any virulent contagious or infectious disease. **2.** a disease that causes such an epidemic.

pestology *n.* (Latin *pestis*, plague + *logos*, word) a scientific study of pests.

petiole *n.* (Latin *petiolus*, leafstalk) **1.** the stalk that connects a leaf to a stem. **2.** a peduncle, the slender part that connects an insect's abdomen to its midsection.

petrifaction or **petrification** *n.* (Latin *petra*, stone + *factio*, making) **1.** the process of changing plant or animal tissue into stone. **2.** a fossil produced by this process.

petrify *v.* (Latin *petra*, stone + *ficare*, to make) **1.** to change plant or animal tissue to stone by the grad-ual replacement of their cellular material with mineral matter, mostly carbonates and/or silica. See opal. **2.** to harden and rigidify, as to petrify one's attitude. **3.** to paralyze with fear or other strong emotion.

Cross-section of petrified wood

petroglyph or **petrograph** *n.* (Greek *petra*, stone + *glyphē*, carving or *graphos*, drawing) prehistoric picture writing drawn, painted, or carved on rock faces.

petrochemical *n.* **1.** any material obtained or made from petroleum or natural gas, including gasoline, fuel oil, lubricants, plastics, soaps, solvents, fertilizers, explosives, etc. *adj.* **2.** having to do with the various petrochemical products and the process of manufacturing them.

petrogenesis *n.* (Greek *petra*, stone + *genesis*, origin) **1.** the branch of geology that studies the origin and formation of rocks. **2.** the processes involved in the formation of rocks.

petrography *n.* (Greek *petra*, stone + *graphos*, written) the branch of geology that describes and classifies rocks.

petroleum *n.* (Greek *petra*, stone + Latin *oleum*, oil) crude oil, an oily mixture of hydrocarbons formed by the breakdown of ancient plant (and possibly animal) material buried in the Earth's crust. It is used to make fuels, lubricants, plastics, and other petrochemicals.

petrology *n.* (Greek *petra*, stone + *logos*, word) rock science, a combination of petrography and petrogenesis.

petrosphere *n.* (Greek *petra*, stone + *sphaira*, ball) See lithosphere.

peyote *n.* (Nahuatl *peyotl*, caterpillar) a small spineless cactus (*Lophophora williamsii*) with a dome shape; native to southwestern U.S.A. and northern Mexico. The flower heads (buttons) contain the hallucinogenic alkaloid mescaline ($C_{11}H_{17}O_2N$) used by Native Americans in certain religious rituals. Also called mescal, peyotl, or pellote.

pH *n.* (French *pouvoir hydrogène*, power of hydrogen) the negative logarithm of the hydrogen ion concentration of a water solution, a system proposed by S. L. Sorensen (1869–1939) in 1909. At 24°C, a pH of 7.0 is neutral ($H^+ = OH^- = 1 \times 10^{-7}$ mol/L); below 7.0 is acid (contains more H^+ than OH^-) and above 7.0 is alkaline (contains more OH^- than H^+).

The approximate pH values of some common substances

Hydrochloric acid, 1.0 N	0.1
Gastric juice	1.3
Lemons	2.3
Cider	3.0
Soft drinks	3.0
Oranges	3.5
Carbonic acid (saturated)	3.8
Coffee	5.0
Fresh cow's milk	6.4
Human saliva	7.0
Human blood	7.4
Agricultural lime ($CaCO_3$)	8.3
Sodium bicarbonate, 0.1 N	8.4
Ammonium hydroxide, 1.0 N	11.6
Sodium hydroxide, 1.0 N	14.0

phage *n.* See bacteriophage.

phagocyte *n.* (Greek *phagein*, to eat + *kytos*, cell or hollow) a cell that ingests bacteria, foreign particles, or cell remains; e.g., white blood cells that combat bacteria that cause infectious diseases.

phagotroph *n.* (Greek *phagein*, to eat + *trophē*, nutrients) any organism that ingests solid food for nourishment, including protozoa, animals, and humans, in contrast to autotrophs such as green plants that use photosynthesis to manufacture carbohydrates.

phalarope *n.* any of three species of small shorebirds (red phalarope, *Phalaropus fulicarius*, Wilson's phalarope, *Steganopus tricolor*, and northern phalarope, *Lobipes lobatus*, family Phalaropodidae) that nest in Arctic regions and migrate long distances to winter in South America, Asia, southern Africa, and certain islands in the Pacific. They are similar to sandpipers with long, slender necks and thin bills but with lobate toes.

phalloidin or **phallotoxin** *n.* a heat-stable, poisonous, glucoside ($C_{35}H_{48}N_8O_{11}S$) from the death cap mushroom (*Amanita phalloides*); used as a stain for dead cells.

phanerogam *n.* (Greek *phaneros*, visible + *gamos*, marriage) a plant that has visible reproductive parts; a plant that produces flowers and/or seeds, as opposed to a cryptogam.

phanerophyte *n.* (Greek *phaneros*, visible + *phyte*, plant) a perennial tree, shrub, vine, or herb whose overwintering buds are exposed above the ground surface; usually at least 10 in. (25 cm) above the ground surface for the new shoots of trees and shrubs.

pharaoh ant *n.* a small, yellowish-red ant (*Monomorium pharaonis*) native to Europe that has become a worldwide pest, infesting homes and eating human food.

phase *n.* (Latin *phasis*, stage) **1.** any of a series of states or conditions that something passes through sequentially; e.g., a phase of life or a phase of a disease. **2.** any of the changing appearances of an animal as it adapts to warm and cold seasons. **3.** any stage in the appearance of the moon as its lighted part waxes and wanes during a month, or any corresponding stage in the appearance of some other celestial body. **4.** the position in the cycle of anything cyclic, such as a light wave, a sound wave, or alternating electric current. **5.** any of the three forms of matter: solid, liquid, or gas. **6.** any component of a mixture that is distinguishable from the rest. **7.** a subdivision of a soil series made for utilitarian purposes but not for soil classification; e.g., any designation of surface texture, stone content, slope, erosion, salinity, etc., added to the name of the soil series. *v.* **8.** to synchronize, as to put two electric generators in phase with each other. **9.** to do something in stages, as to phase in a new system and phase out the old.

phaseolin *n.* a phytoalexin ($C_{20}H_{18}O_4$) with antifungal properties that is produced by kidney beans (*Phaseolus vulgaris*) and related species in response to stress.

pheasant *n.* (Greek *phasianos*, bird of the Phasis River) **1.** any of several relatively large, gallinaceous game birds (*Phasianus* sp., family Phasianidae) with bright colors and long tails that are Eurasian in origin but are now widely distributed. See ring-necked pheasant. **2.** any of several somewhat similar birds, such as the ruffed grouse.

phenocopy *n.* (Greek *phaino*, showing + Latin *copia*, copy) a plant or animal affected by environmental conditions that cause its phenotype to mimic a form usually produced by a different genotype; appearing like a mutation but not hereditary.

phenocryst *n.* (Greek *phaino*, showing + *krystallos*, crystal) a large, conspicuous crystal in a porphyritic rock.

phenol *n.* (French *phen*, from benzene + *ol*, alcohol) **1.** carbolic acid (C_6H_5OH), an organic compound with a benzene ring structure and an attached hydroxyl group; a very poisonous disinfectant. **2.** an organic compound containing one or more ring structures with attached hydroxyl group(s). Phenols are synthesized by plants but not by animals. They serve as building blocks for aromatic amino acids and other complex molecules.

phenology *n.* (Greek *phainomenon*, phenomenon + *logos*, word) the study of the responses of plants, animals, and people to periodic (especially seasonal) changes in their environment.

phenotype *n.* (Greek *phainomenon*, phenomenon + *typos*, impression) **1.** the observable characteristics of an organism, as opposed to its genetic characteristics. See genotype. **2.** the general appearance of a plant, animal, or human as a result of its heredity as modified by its environment.

phenylphenol *n.* either of two isomers of a toxic, combustible, double-ring compound ($C_6H_5C_6H_4OH$) used as a fungicide and in making dyes and rubber.

pheromone *n.* (Greek *pherein*, to carry + *hormone*, stimulating) a substance secreted by an organism to cause a specific reaction in another organism of the same species, as a sex attractant, an alarm signal, a trail marker, etc. Synthetic pheromones serve as bait or decoys to control insects.

This plastic strip contains codling moth sex pheromones.

phloem *n.* (Greek *phloos*, bark) generally the outer layer of a plant. The assemblage of sieve tubes, fibers, parenchyma, and companion cells that transport food such as glucose from the photosynthetic parts where it is produced to other parts of the plant. See xylem.

phon *n.* (Greek *phōnē*, a sound) a measure of loudness equal in decibels to an apparently equally loud sound with a pure tone of 1000 cycles/sec.

phonics *n.* (French *phonique*, of sound) **1.** acoustics, the quality of sound transmission in an environment. **2.** a method of teaching students to read, enunciate, and spell by sounding the individual letters rather than seeing the word as a single unit.

phoresy *n.* (Greek *phoresis*, carrying) a kind of commensalism wherein one organism transports another organism, especially among arthropods and certain fish.

phorology *n.* (Greek *phoros*, carrying + *logos*, word) the study of carriers of diseases.

phosphate *n.* (Greek *phōs*, light + *ate*, salt) **1.** a salt of phosphoric acid (H_3PO_4); a compound that contains PO_4^{3-} ions. Phosphates are important constituents of bones, teeth, and certain rocks. See phosphorus. **2.** a fertilizer that provides available phosphorus for plant growth. **3.** phosphorus pentoxide (P_2O_5), especially as a means of expressing the available phosphorus content of a fertilizer or the amount of phosphorus needed by a crop.

phosphate rock *n.* rock composed mostly of the mineral apatite [$Ca_5(PO_4)_3$(F, Cl, or OH)], of either sedimentary or igneous origin. It is used as a raw material for making phosphorus-containing products, especially phosphorus fertilizer, and sometimes used directly as a powdered material to fertilize acid soil. Also called rock phosphate. Morocco has about half of the known deposits of phosphate rock. The U.S.A. has significant deposits in Florida, North Carolina, Utah, Wyoming, and Idaho. Also called rock phosphate, especially when powdered for use as a phosphorus fertilizer.

phospholipid *n.* (Greek *phōsphoros*, light + *lipos*, fat) a lipid consisting of glycerol linked to fatty acids on its first two carbons and to a phosphate group on the third carbon (usually with another group linked through the phosphate to the glycerol). Phospholipids are important intermediaries in many biochemical processes. They are hydrophobic on the fatty-acid end and hydrophyllic on the phosphate end, which makes them important building blocks of membranes and other interfaces between water and lipid environments.

phosphor *n.* (Greek *phōsphoros*, light) a substance that emits light (glows) when subjected to certain types of radiation.

phosphorescence *n.* (Greek *phōsphoros*, light + Latin *essentia*, essence) **1.** the white light produced by the slow spontaneous combustion of yellow elemental phosphorus in contact with air. **2.** the light emitted by a phosphor when subjected to radiation. **3.** the ability to produce light without combustion or heat, as exhibited by certain minerals, insects, and fungi, and to continue to glow for some time after excitation.

phosphorism *n.* (Greek *phōsphoros*, phosphorus + *isma*, condition) a diseased condition caused by chronic exposure to excessive amounts of phosphorus.

phosphorite *n.* (Greek *phōsphoros*, bearing light + *ite*, rock) a sedimentary rock with a high enough content of phosphorus to be used to make fertilizer.

phosphorus (P) *n.* (Greek *phōsphoros*, bearing light) element 15, atomic weight 30.97376; a highly reactive element that is essential to plants, animals, and humans as a component of protein, bones, and teeth and of the adenosine phosphates (ADP and ATP) involved in energy transfer. After nitrogen, it is the most commonly deficient nutrient, but it is plentiful in meat, eggs, and whole-grain cereals. Present in rocks mostly as apatite, it is used to make phosphate fertilizers, pesticides, cleaning agents, matches, electronic parts, etc. See phosphate.

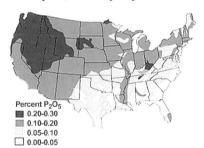

Percent P_2O_5
- ■ 0.20–0.30
- ▨ 0.10–0.20
- ▩ 0.05–0.10
- ☐ 0.00–0.05

Phosphorus content (as P_2O_5) of soils in the U.S.A.

photic *adj.* (Greek *phōtos*, light + *ikos*, of) **1.** related to light. **2.** responsive to light, as a plant that turns in response to light. **3.** able to produce light, as phosphorescent organisms.

photics *n.* (Greek *phōtos*, light + *ics*, principles) the science that studies light.

photic zone *n.* the upper part of a body of water; the part where there is enough light for plants to carry on photosynthesis.

photoautroph *n.* (Greek *phōtos*, light + *autos*, self + *trophē*, food) an organism that uses light energy to manufacture its own carbohydrates from carbon dioxide and water; e.g., green plants and algae.

photochemical smog *n.* a noxious mixture of irritating noncombustible gases, including oxides of nitrogen (NO_x), sulfur dioxide (SO_2), ozone (O_3), peroxyacetylnitrate (PAN), etc., that irritates eyes and causes respiratory distress. Concerns about photochemical smog led to the passage of legislation to limit emissions of its component gases by automobiles and other sources.

photokinesis *n.* (Greek *phōtos*, light + *kinēsis*, movement) movement of an organism in response to light, as when insects are attracted to light or a plant turns toward a light source.

photometer *n.* (Greek *phōtos*, light + *metron*, measure) an instrument for measuring light intensity and quality.

photon *n.* (Greek *phōtos*, light + *on*, a subatomic particle) a quantum of light or other electromagnetic energy that has no mass or charge but does have momentum and exhibits characteristics of both wave and particle motion. Also called light quantum. Symbolized by γ.

photoperiod *n.* (Greek *phōtos*, light + *periodos*, time interval) the daylight period, commonly expressed in hours per day.

photoperiodism *n.* (Greek *phōtos*, light + *periodos*, time interval + *ismos*, action) response of an organism to the duration of light and darkness in each 24-hour period. Research by W. W. Garner and H. A. Allard in 1920 divided plants into three groups: (1) short-day plants that bloom after the dark period exceeds their critical minimum, (2) long-day plants that flower after the dark period becomes sufficiently short, and (3) day-neutral plants. Photoperiodism has also been shown to control bird migrations and breeding activities of certain birds and mammals. See phytochrome, critical light period.

photophygous *adj.* growing best (or at least well) in the shade.

photoreceptor *n.* (Greek *phōtos*, light + Latin *receptor*, receiver) a plant or animal cell or group of cells that senses light; especially the rods and cones in an eye.

photorespiration *n.* (Greek *phōtos*, light + Latin *respiratio*, breathing out) a form of light-dependent respiration that is a part of the photosynthetic cycle of C_3 plants and reduces the photosynthetic efficiency of C_3 plants in comparison with C_4 plants that have a pathway that avoids photorespiration. It wastes energy by using oxygen in place of carbon dioxide in the dark reactions of photosynthesis.

photosynthate *n.* (Greek *phōtos*, light + *synthesis*, putting together + *ate*, chemical) the simple sugars produced by photosynthesis.

photosynthesis *n.* (Greek *phōtos*, light + *synthesis*, putting together) any of three pathways catalyzed by chlorophyll that permit plants to use light energy to produce carbohydrates from carbon dioxide and water (CO_2 and H_2O):

$$6\ CO_2 + 12\ H_2O + \text{light energy} = C_6H_{12}O_6 + 6\ H_2O + 6\ O_2$$

Note: Water is shown on both sides of the equation because radioactive tracers show that the oxygen produced comes from the water, not from the carbon dioxide.

In C3 photosynthesis, the first stable product of the dark reaction is a 3-carbon carbohydrate. In C4 photosynthesis, 4-carbon compounds are formed first and then split to produce 3-carbon carbohydrates. In CAM (crassulacean acid metabolism) photosynthesis, malate is formed at night and serves as a source of carbon dioxide for daytime photosynthesis. The C3 pathway works well where daylight temperatures are cool and is used by most cool-season plants, including wheat, rice, beans, squash, tomatoes, etc. The C4 pathway has an advantage where moisture stress accompanies warmer temperatures and is used by warm-season plants such as corn, sorghum, sugarcane, etc. The CAM pathway is advantageous in hot, dry environments such as deserts and is used by cacti and many succulent plants.

photosynthetically active radiation (PAR) *n.* light energy that is absorbed by chlorophyll and used in photosynthesis; light with wavelengths between 400 and 760 nm, with the bands between 400 and 470 nm (violet and blue) and between 650 and 680 nm (orange-red) being most effective.

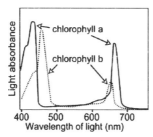

Light absorbed by chlorophyll (PAR wavelengths)

photosynthetic rate *n.* the rate of conversion of carbon dioxide (CO_2) into carbohydrates by a plant (as it carries on photosynthesis, usually expressed as net photosynthesis (gross photosynthesis minus respiration) in units of mg CO_2 fixed per dm^2 of leaf surface per hour.

phototonic *adj.* (Greek *phōtos*, light + *tonos*, stretching) responsive to stimulation by light, especially when irritated by light.

phototaxis *n.* (Greek *phōtos*, light + *taxis*, arrangement) movement of a plant or animal either toward or away from light; e.g., a plant growing toward a light source, a moth attracted to a light, or a bat that returns to its cave as daylight approaches.

phototoxic *adj.* (Greek *phōtos*, light + *toxikon*, poison) **1.** causing skin to be subject to damage when exposed to light, especially ultraviolet light; e.g., the effects of sunburn. **2.** pertaining to or caused by a substance that causes such sensitivity to light; e.g., certain medications and cosmetics.

phototroph *n.* (Greek *phōtos*, light + *trophē*, nourishment) an organism that uses light as its main source of energy; e.g., a green plant. See photosynthesis, autotroph.

phototropism *n.* (Greek *phōtos*, light + *tropos*, a turning + *ismos*, action) growth of plants in response to light, either toward or away from the light source.

phragmites *n.* (Greek *phragmítēs*, growing in hedges) any of several wetland reeds with large plumelike panicles (*Phragmites* sp.). They grow up to 20 ft (6 m) tall in ditches and other wet areas and are valuable in lagoons to help purify water contaminated with effluent from animal waste. The stems are used to make mats, screens, and arrows.

phreatic *adj.* (Greek *phreatos*, a well) **1.** pertaining to groundwater in the saturated zone; the source of water for springs and wells. Phreatic water is also known as groundwater. **2.** pertaining to steam produced from groundwater, as in an explosive volcano or in geysers such as those in Yellowstone National Park.

phreatophyte or **phreatotype** *n.* Greek *phreatos*, a well + *phyton*, plant or *typos*, mark) a plant that grows where its roots can reach phreatic water; e.g., near a pond, a stream, or a spring. See phragmites, cattail, water lily.

phrynin *n.* a poison found in the skin and secretions of various toads that is similar in composition to digitalin.

phthiriasis *n.* (Greek *phtheir*, louse + *iasis*, condition) infestation with lice, especially crab lice. Also called pediculosis.

phthisis *n.* (Greek *phthisis*, decay) **1.** a general decay and wasting away of the body or a body part. **2.** a disease such as pulmonary tuberculosis that causes a wasting away.

phycology *n.* (Greek *phykos*, seaweed + *logos*, word) the botanical study of algae. Also called algology.

phyllosilicate *n.* (Greek *phyllon*, leaf + Latin *silex*, flint) layer silicate, a mica or clay mineral with either a 1:1 (one tetrahedral to one octahedral sheet; e.g., kaolinite) or a 2:1 (two tetrahedral sheets with an octahedral sheet between them; e.g., smectite or mica) layered structure. Interlayer cations in the 2:1 structures and hydrogen bonds in the 1:1 structures are much weaker than the bonding within the layers.

phyllosphere *n.* (Greek *phyllon*, leaf + *sphaira*, ball) the microenvironment surrounding a plant leaf. High microbial populations are present on leaf surfaces in some humid environments, including bacteria that fix nitrogen or produce gums or slimes.

phylogeny *n.* (Greek *phylon*, race or lineage + *genesis*, origin) **1.** the evolutionary line of descent of any species (or higher class) of plant or animal, sometimes shown as a family tree. Also called phylogenesis. See ontogeny. **2.** the evolutionary development of a part of an organism; e.g., that of an organ. **3.** the historical development of a group of organisms; e.g., that of a race or tribe. **4.** the historical development of nonliving things such as a language or customs.

phylum *n.* (Greek *phylon*, race) the highest (most general) classes in the taxonomic classification within either the plant or animal kingdom. The hierarchy is phylum, class, order, family, genus, and species.

physical property *n.* a characteristic that can be detected and evaluated by sight or feel; e.g., size, shape, color, hardness, density, and porosity.

physiogeny *n.* (Greek *physis*, nature + *genesis*, origin) the branch of ontogeny that studies the development of organic functions in the body of an individual.

physiognomy *n.* (Greek *physis*, nature + *gnōmōn*, one who knows) **1.** the physical features and expression of the face, especially as used to evaluate character and mood. **2.** the practice of judging character and other personal qualities from facial and other bodily features. **3.** outward appearance; apparent characteristics of an individual or other entity.

physiological ecology *n.* the study of how an individual organism lives in and adapts to the physical and chemical characteristics of its environment, including environmental limitations on where the organism can live.

physiological zero *n.* the minimum temperature for a biological process to function; e.g., for seeds to germinate or roots to grow. The physiological zero for many cool-season plants is about 40°F (4°C), and for warm-season plants about 50°F (10°C).

physiology *n.* (Greek *physis*, nature + *logos*, word) the science that studies the organs and other parts of living organisms and the vital processes associated with each part.

physostigmine *n.* a poisonous alkaloid ($C_{15}H_{21}N_3O_2$) contained in the Calabar bean (*Physostigma venenosum*). It depresses motor functions and is used to relieve spasms, as an expectorant, and to stimulate constipated bowels. Large doses can cause death by respiratory arrest.

phytoalexin *n.* (Greek *phyton*, plant + *alexein*, to ward off) a compound produced by a plant that accumulates at the site of a microbial infection and resists development of disease.

phytochrome *n.* (Greek *phyton*, plant + *chrōma*, color) the plant pigment that regulates photoperiodism in plants. Phytochrome has one form that absorbs red light (peak at 660 nm) and another that absorbs far-red light (peak at 730 nm). Light rapidly converts the red-absorbing form to the far-red-absorbing form, whereas darkness causes a gradual reversion to the red-absorbing form. Dominance of the red-absorbing form causes short-day plants to bloom, whereas the far-red-absorbing form causes long-day plants to bloom. See photoperiodism.

phytoclimate *n.* (Greek *phyton*, plant + *klima*, zone) the microclimate produced by a plant or a community of plants in a particular location.

phytocoenosis *n.* (Greek *phyton*, plant + *koinos*, common + *osis*, condition) the plant population as a whole in a particular environment.

phytoecology *n.* (Greek *phyton*, plant + *oikos*, house + *logos*, word) See plant ecology.

phytogenesis *n.* (Greek *phyton*, plant + *genesis*, origin) the origin and evolutionary development of plants.

phytography *n.* (Latin *phytographia*, phytography) the botanical science that describes plants.

phytol *n.* (Greek *phyton*, plant + *ol*, alcohol) a water-repelling alcohol ($C_{20}H_{40}O$) that occurs in the side chain of chlorophyll molecules.

phytolite *n.* (Greek *phyton*, plant + *lithos*, stone) a fossilized plant or plant part.

phytolith *n.* (Greek *phyton*, plant + *lithos*, stone) a rock composed mostly of plant parts; e.g., coal, peat, and some limestones.

phytotomy *n.* (Greek *phyton*, plant + *temnein*, to cut) the dissection of plants, especially to study their anatomy.

phyton *n.* (Greek *phyton*, plant) a propagule; any small part of a plant (stem, root, leaf, or other part) that can grow into a new plant.

phytopathology *n.* (Greek *phyton*, plant + *pathos*, disease + *logos*, word) See plant pathology.

phytophagous or **phytophagan** *adj.* (Greek *phyton*, plant + *phagos*, eating) herbivorous; a plant eater, especially an insect that eats plants.

phytophthora *n.* (Greek *phyton*, plant + *phthora*, destruction) any of several parasitic fungi

(*Phytophthora* sp., family Phthiaceae) that cause damping off, foot rot, and other plant diseases. See damping-off, foot rot. Some of the species and some crops they attack are listed in the following table:

P. cactorum	strawberry fruit
P. capsici	melons
P. cinnamomi	avocado
P. citrophthora	citrus fruit
P. drechsleri	melons
P. fragariae	strawberries
P. infestans	potatoes
P. megasperma	soybeans
P. palmivora	coconut
P. parasitica	melons
P. phaseoli	lima beans
P. sojae	soybeans

phytoplankton *n.* (Greek *phyton*, plant + *planktos*, drifting) tiny plants that drift with water currents in aquatic environments and are eaten by fish. See plankton, zooplankton.

phytoremediation *n.* (Greek *phyton*, plant + Latin *remediatus*, remedial + *atio*, action) any form of bioremediation that uses plants to remove pollutants from water or soil. For example, some plants are able to use their metabolic processes to destroy organic pollutants such as pesticides or solvents, whereas other plants absorb heavy metals and concentrate them in plant tissues that can then be harvested and removed.

phytostabilization *n.* (Greek *phyton*, plant + Latin *stabilis*, firm + *izatio*, making) protection of a landform by a thick permanent plant cover that resists erosion.

phytostasy *n.* (Greek *phyton*, plant + *stasis*, standing state) landscape stability produced by good vegetative cover.

phytotoxic *adj.* (Greek *phyton*, plant + *toxikon*, toxic) **1.** poisonous to plants. **2.** inhibiting the growth of plants. **3.** pertaining to phytotoxins.

phytotoxin *n.* (Greek *phyton*, plant + *toxikon*, toxic + *in*, chemical) a toxic chemical produced by a plant.

phytotron *n.* (Greek *phyton*, plant + *tron*, experimental chamber) a large room with environmental controls for light, temperature, and humidity; used for growing plants, often for research purposes. Also called biotron.

piassava *n.* (Portuguese *piaçába*, piassava) **1.** coarse fibers that grow at the base of the petioles of certain palm trees; used for stuffing and cordage. **2.** a palm tree that furnishes such fibers (especially *Attalea funifera* of Brazil or *Leopoldinia piassaba* of oases in the Sahara Desert).

pica *n.* (Latin *pica*, magpie) a strong desire of a person or animal under stress to eat something that is not food, such as chalk or clay. See bone chewing, geophagia, parorexia, depraved appetite.

pickax or **pickaxe** *n.* (French *picquois*, pickax) a digging tool with a long point on one side and a chisel-shaped head on the other; used for loosening hard soil and similar materials. Also called mattock or pick.

pickle *n.* (German *pekel*, brine) **1.** a brine, acid, and/or spicy solution used to preserve and flavor food. **2.** a cucumber or other vegetable that has been preserved by soaking in such a solution. **3.** a solution (usually acid) used to remove oxide scale or other impurities from metal surfaces. **4.** (informal) an awkward, embarrassing situation. *v.* **5.** to soak food in a pickle solution to preserve it. **6.** to dip a metallic object in an acid solution to clean it. **7.** to give wood an aged appearance by painting it and then removing or bleaching part of the paint.

picocurie *n.* a measure of radioactivity equal to one trillionth (10^{-12}) of a curie, or the quantity of a material that has 2.2 radioactive particle disintegrations per minute.

piedmont *n.* (Latin *pedis*, foot + *montis*, mountain) a depositional plain that is formed from the coalescing sediments of many streams, and that slopes gently away from the foot of a mountain or mountain range. See bajada.

piezometer *n.* (Greek *piezein*, to press + *metron*, measure) **1.** any instrument used for measuring pressure in a fluid. **2.** an instrument used to measure the compressibility of a fluid. **3.** an observation well used to measure the water pressure of underground water at a particular location.

pig *n.* (Anglo-Saxon *picga*, pig's food) **1.** a young hog (domestic swine, *Sus scrofe*) weighing up to 120 or 130 lb (54 or 59 kg). **2.** any domestic or wild hog. **3.** (slang) a gluttonous, selfish, or filthy person. **4.** (derogatory) a law officer. **5.** a container made of lead for the storage or transportation of radioactive materials. **6.** a device used to clean the interiors of pipes. **7.** a mass of iron or other metal poured into a sand mold and solidified. *v.* **8.** to farrow; to give birth to baby pigs. **9.** to eat too much or too fast (often called pig out). **10.** to pour metal into sand molds.

pig deer *n.* a large animal (*Babirussa babirussa*) of the East Indies that resembles a pig and has large, curved tusks.

pigeon *n.* (Latin *pipion*, squab) **1.** any of many birds (family Columbidae) with small head, compact body, and short legs, living in most parts of the world. A pigeon is of the same family as a dove, but generally somewhat larger than a dove and has a square tail rather than a pointed one. **2.** (slang) a person who is easily cheated or deceived.

pigeon hawk *n.* any of several small hawks, especially the merlin (*Falco columbarius*, family Falconidae), a small falcon with a dark back, brown and white streaks below, barred tail, and long pointed wings that flies low over open ground to catch pigeons, mice, and insects.

pig iron *n.* iron that has been smelted in a blast furnace and poured into pigs to be further processed into various types of iron and steel.

pigment *n.* (Latin *pigmentum*, coloring) **1.** coloring matter that can be mixed with oil, water, or other carrier to make paint or ink. **2.** natural coloring matter in the tissues and cells of plants or animals or products derived from them.

pigmentation *n.* (Latin *pigmentatus*, colored + *atio*, condition) **1.** coloration, especially that in the skin of a person or animal or the tissue of a plant. **2.** the process of adding color.

pigpen *n.* **1.** a pigsty, an enclosure where swine are kept. **2.** a room or area that is dirty and messy.

pigsty *n.* See pigpen.

pigweed *n.* **1.** any of several annual, coarse, weedy plants (*Chenopodium album* and related species, family Chenopodiaceae) of temperate regions that grow 2 to 6 ft (60 to 180 cm) tall and produce spikes of small green flowers. *Chenopodium album* is also called lambsquarters. **2.** an annual weed (*Amaranthus retroflexus*, family Amaranthaceae) with red roots that is possibly native to the Great Plains and is widespread in gardens, fields, and waste areas. It grows 2 to 6 ft (60 to 180 cm) tall and produces small green flowers in dense panicles at the top of the plant and in upper leaf axils. It has been shown to have some allelopathic effect on beans. Also called redroot pigweed.

Redroot pigweed

pika *n.* (Siberian *peeka*, pika) a small brown to gray mammal (*Ochotona* sp.) that resembles a rabbit with no tail. It is found in mountainous areas of western North America, eastern Europe, and Asia. Also called rat hare or calling hare. See lagomorph.

pike *n.* (French *pique*, pike or pickax) **1.** a large (up to 4.5 ft or 1.4 m long), slender, freshwater game fish (*Esox* sp., especially *E.lucius*) with a long snout, sharp teeth, and a projecting lower jaw; found in North America, Europe, and northern Asia. **2.** any of several similar fish; e.g., garfish or walleye. **3.** a spearlike weapon once used by foot soldiers. **4.** a pickax or other pointed tool or weapon. **5.** a hill or mountain with a sharp peak. **6.** a toll road, the tollbooth, or the toll paid for using the road. *v.* **7.** to kill with a pike. **8.** to hurry, especially when leaving.

pillbug *n.* any of several small segmented insects with many legs (*Armadillidium* sp., especially *A. vulgare*, and *Oniscus* sp.) that is able to roll into a ball that resembles a pill, lives in damp places, and is a pest on flowers and shrubs, especially in greenhouses. Also called sowbug.

pimpernel *n.* (Latin *piperinus*, pimpernel) a low, spreading, ornamental plant (*Anagallis* sp., especially *A. arvensis*, family Primulaceae) with small scarlet, blue, or white flowers that is sometimes called poor man's weatherglass because the blossoms close when rain approaches.

pine *n.* (Latin *pinus*, pine tree) **1.** any of several coniferous trees (*Pinus* sp., family Pinaceae) with clusters of evergreen needles for leaves, ranging in size from bushes to trees as tall as 300 ft (90 m). Several species are used to make lumber, turpentine, tar, and pitch; e.g., western yellow or ponderosa pine (*P. ponderosa*), white pine (*P. strobus* or *monticola*), red pine (*P. resinosa*), and Scotch pine (*P. sylvestris*). **2.** any of several similar conifers in related genera. **3.** wood from any of these trees, which is typically light colored, soft, and easy to use for construction.

pineal gland *n.* a small body with a flattened cone shape that grows out from the epithalamus in the brains of humans and animals and secretes the hormone melatonin. It helps regulate biorhythms and sexual maturity. Extracts from pineal glands of slaughtered meat animals are used in human medicine. Also called pineal body.

pineapple *n.* **1.** a low-growing perennial plant (*Ananas comosus*, family Bromeliaceae) with swordlike leaves 2 to 4 ft (0.6 to 1.2 m) long. Pineapple production requires a frost-free environment and a well-drained soil with moderate to high rainfall. Hawaii and the West Indies are important producing areas. **2.** the fruit from the plant with an external appearance like that of a large pine cone but with a fleshy light-yellow interior that is made into pineapple slices or juice.

pine barren *n.* a tract of poor land with sandy or peaty soil that is mostly covered with pine trees; e.g., certain low-lying areas along the Atlantic and Gulf Coasts of the U.S.A.

pinecone *n.* the seed-bearing structure of a pine tree. See Coulter pine.

pinegrass *n.* (Latin *pinus*, pine tree + Anglo-Saxon *graes*, grass) a tall, perennial bunchgrass (*Calamagrostis rubescens*, family Poaceae) that grows in open areas intermixed with ponderosa and lodgepole pine forests in western U.S.A.

pineland *n.* a forested area dominated by pine trees.

pine mouse *n.* any of several voles (*Pitymys* sp.) with small ears and short tails that are common in forested areas. Also called pine vole.

pine needle *n.* a long, slender leaf from any pine tree. Needles typically occur in clusters of 2, 3, or 5, depending on species. They stay green on the tree but turn brown on the ground, where they accumulate in a litter layer that decomposes very slowly and produces an acid leachate.

pine nut *n.* the edible seed from certain species of pine cones; e.g., stone pine (*Pinus pinea* of southern Europe and *P. cembra* of the Swiss Alps) and piñon pine (*P. cembroides* of southwestern U.S.A.). Roasting removes the turpentine flavor common to some species.

pine oil *n.* an oil obtained from pine and fir trees that is similar to turpentine and is used in varnishes.

pinery *n.* (Latin *pinus*, pine tree + *aria*, place) **1.** a pine forest or a grove of pine trees. **2.** a pineapple plantation or a hothouse where pineapples are grown.

pinetum *n.* (Latin *pinus*, pine tree + *etum*, grove) an arboretum or grove of pine trees or other conifers that is kept primarily for aesthetic and/or educational purposes rather than as a forest.

pingo *n.* (Eskimo *pinguq*, conical hill) a conical hill with an ice core produced by frost heaving on a tundra plain underlain by permafrost. It is commonly 100 ft (30 m) high in Arctic regions.

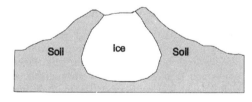

Cross section of a pingo

pink bollworm *n.* the pinkish larva of a small brown moth (*Pectinophora gossypiella*) native to Asia. It is very destructive as it feeds on the seeds in cotton bolls.

pinnate *adj.* (Latin *pinnatus*, like a feather) **1.** shaped like a feather. **2.** composed of several leaflets arranged on two sides of a common stalk.

pin oak *n.* a large deciduous tree (*Quercus palustris*, family Fagaceae) with deeply lobed leaves that is native to east-central U.S.A. and often used as an ornamental but needs acid soil to avoid iron deficiency. Its wood is used for fuel, charcoal, flooring, furniture, etc.

piñon pine *n.* a small tree (*Pinus edulis*, family Pinaceae) of the Rocky Mountains in southwestern U.S.A. and northern Mexico, with needles in clusters of two each or sometimes three. The seeds are eaten by people, and it is the state tree of New Mexico. Also called pinyon pine, nut pine, or Mexican pine. See pine nut.

pint *n.* (French *pinte*, pint) **1.** a measure of liquid volume. In the U.S.A., it is equal to 16 fluid oz, 2 cups, or ½ qt (about 29 in.^{3} or 473 mL). In Great Britain and Canada, the imperial pint is equal to 568.25 mL. **2.** a measure of dry volume of powders or granular materials that can flow. It is equal to ½ dry quart (about 33.6 in.^{3} or 550 mL) in the U.S.A.

pinworm *n.* a parasitic nematode (family Oxyuridae) that lives mostly in the large intestine. Its eggs are passed in the feces and may become attached to the anal region and cause severe itching. Examples are the human pinworm (*Enterobius vermicularis*), especially common in children; the equine pinworm (*Oxyuris equi*); and the sheep and goat pinworm (*Skrjabinema ovis*). **2.** the cecum worm (*Heterakis gallinae*) of poultry that develops in the intestines, concentrates in the cecum, and is a vector for blackhead disease of poultry.

pinyon pine *n.* See piñon pine.

pioneer plant *n.* a plant that establishes itself in the first stage of a new succession in a disturbed area. Also called pioneer species.

pipette analysis or **pipet analysis** *n.* a means of determining the silt and clay content of a soil or other sample by withdrawing a measured volume with a pipette after allowing enough time for the sand to settle out and again after the silt has settled out. The samples are dried, weighed, and multiplied by a volume factor to calculate the contents of silt + clay and of clay alone. See particle-size analysis, hydrometer analysis, sieve analysis.

piping *n.* **1.** a set of pipes. **2.** the music made by certain instruments with pipes, or the action of playing such an instrument. **3.** a shrill sound; e.g., the shriek of a child. **4.** lines of icing used to decorate cakes. **5.** a pipelike fold in the fabric of a garment, upholstery, or other product. **6.** an erosion process where water flows through a crack or other opening and produces an underground channel. Sometimes the overlying soil will collapse into it and produce a pothole. See karst, thermokarst. *adj.* **6.** delivering a substance through a set of pipes. **7.** playing peaceful music on an instrument with pipes. **8.** decorating a cake with icing squeezed out of a tube.

piranha *n.* (Tupi *piranha*, toothed fish) a small, voracious, freshwater fish (*Serrasalmus* sp.) with sharp teeth that is native to tropical areas of South America. Piranhas eat other fish and will attack large mammals such as cattle or humans, rapidly consuming their flesh. Also called caribe.

piroplasmosis *n.* See babesiosis.

pisciculture *n.* (Latin *piscis*, fish + *cultura*, culture) artificial breeding, rearing, and transplanting of fish. See aquaculture.

pistachio *n.* (Italian *pistacchio*, pistachio) **1.** a small, drought-tolerant Eurasian tree (*Pistacia vera*, family Anacardiaceae) grown for its nuts. **2.** the edible greenish nuts from the tree. **3.** the flavor of the nuts. **4.** a yellowish green similar to the color of the nuts.

pistil *n.* (Latin *pistillum*, pistil) the female (seed-bearing) organ of a flowering plant, including the stigma (where pollen is received), style (the connection between the stigma and the ovary), and ovary (where the egg cell is produced). Some flowers have more than one set of these parts in a pistil. Also called gynoecium. See stamen.

pitcher plant *n.* any of several plants that grow in bogs, trap insects, and digest them in liquid held in leaves shaped like pitchers. They include North American species (*Sarracenia* sp., family Sarraceniaceae and *Darlingtonia californica*), with the best-known species (*S. purpurea*) having red or green leaves and a large dull red flower that nods in a breeze; South American species (*Heliamphora* sp.); and Eurasian species (*Nepenthes* sp., family Nepenthaceae).

pith *n.* (Anglo-Saxon *pitha*, pith) **1.** soft, spongy, cellular tissue that fills the stems of certain plants. **2.** a soft core in other tissues such as bones, spinal cord, or feathers. **3.** the essential part of a topic; the gist. **4.** power, strength, force, energy, vigor, importance.

Pitot tube *n.* a curved tube, named after Henri Pitot (1695–1771), that is used to measure the velocity of a fluid such as the current in a stream or the speed of an airplane. A static tube is added to the system if the fluid is enclosed and under pressure (as in a pipe), and the difference in fluid level in the two tubes indicates the velocity. A pressure gauge may be used to obtain the reading instead of measuring the fluid level.

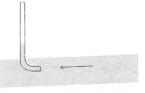

The water level in a pitot tube indicates water velocity.

pituitary *adj.* (Latin *pituitarius*, secreting phlegm) **1.** pertaining to the pituitary gland, to its secretions, or to an extract from it. **2.** pertaining to a form of gigantism believed to be caused by excessive pituitary secretions that involves large bones and abnormally long arms and legs.

pituitary gland *n.* a small gland with two lobes (anterior and posterior) located near the base of the brain; the master endocrine gland with secretions that control hormonal functions throughout the body to regulate growth and metabolism.

pit viper *n.* any of several poisonous snakes (family Crotalidae) that have a prominent heat-sensing pit below each eye, including the copperhead, rattlesnake, water moccasin, and fer-de-lance.

placenta *n.* (Latin *placenta*, a cake) **1.** the membranous tissue in the uterus that surrounds a developing fetus and connects to the umbilical cord to transfer nourishment from the mother to the fetus and waste from the fetus to the mother. Also called afterbirth. **2.** the corresponding structure in the ovary of a flowering plant; the part of the ovary that produces ovules (or sporangia in ferns).

placer *n.* (Spanish *placer*, sandbank) **1.** an alluvial or glacial deposit containing concentrations of valuable heavy minerals such as gold, platinum, or rutile. **2.** a site where placer mining is or has been practiced.

placer mining *n.* mining precious metals (especially gold) from a placer by hydraulic means, including panning (swirling a pan so water washes the gangue over the side), hydraulic mining, and sluices.

plage *n.* (Latin *plagia*, shore) **1.** a sandy beach, especially one near a resort. **2.** a bright spot near a sunspot in the sun's chromosphere.

plaggen epipedon *n.* (German *plaggen*, sod) an epipedon thickened to at least 50 cm (20 in.) by multiple additions of manure, bedding, and/or sod over long periods that usually contains artifacts, sand lenses, and spade marks.

plagioclase *n.* (Greek *plagios*, slanting + *klasis*, fracture) a very abundant tektosilicate named for its slanting cleavage surfaces. It is a mineral of the feldspar family with a range of composition from albite ($NaAlSi_3O_8$) to anorthite ($CaAl_2Si_2O_8$). Typical colors range from bluish white to bluish black.

plague *n.* (Latin *plaga*, pestilence) **1.** any very deadly, contagious, epidemic disease, especially that known as the bubonic plague, caused by a toxin produced by bacteria (*Yersinia pestis*) and transmitted by a rat flea. **2.** any widespread influence with a bad effect, especially when interpreted as punishment from God, as a plague of locusts or of war. *v.* **3.** to harass, annoy, or torment. **4.** to cause illness and death, especially as an epidemic. **5.** to afflict with any kind of evil.

plain *n.* (Latin *planus*, flat) an extensive area of land without large hills or other prominent features; an expanse of nearly level, gently sloping, undulating, or even rolling land, generally at a lower elevation than nearby hilly or mountainous areas; e.g., the Great Plains of west-central U.S.A.

planar *adj.* (Latin *planus*, flat + *aris*, like) **1.** flat or level; like a plane. **2.** pertaining to or characteristic of a plane.

planate *adj.* (Latin *planus*, flat + *atus*, having) being flat; having a flat surface.

planation *n.* (Latin *planus*, flat + *atio*, action) the leveling of a land surface by erosion, especially where a drifting river channel produces a large area of nearly level land.

Planck, Max *n.* the German physicist (1858–1947) who proposed the quantum theory in 1900 to explain the behavior of light and won the Nobel prize for physics in 1918. This work became the basis for quantum mechanics, a new approach to theoretical physics.

Planck's constant (*h*) *n.* the constant, named for Max Planck, that relates the energy content (*E*) to the frequency of electromagnetic radiation (*v*): $E = hv$; $h = 6.626 \times 10^{-34}$ joule-sec.

plane *n.* (Latin *planus*, flat) **1.** a flat surface; smooth and without curvature. A horizontal plane has uniform elevation. **2.** a level of position, status, or development. **3.** short for airplane. **4.** a carpenter's tool with a sharp blade that shapes and smoothes the surface of wood by shaving off a thin layer. **5.** a trowel used to smooth the surface of sand, clay, wet concrete, etc., especially in a brick mold. *v.* **6.** to shape and smooth a surface, as with a plane. **7.** to erode a land surface, as by the action of a river.

plane survey *n.* **1.** a map made without any correction for the curvature of the Earth. **2.** the process of making such a map. Note: The errors involved are negligible for small areas.

planet *n.* (Latin *planeta*, wandering stars) **1.** any of the nine major bodies orbiting the sun: Mercury, Venus, Earth, Mars, Jupiter, Saturn, Uranus, Neptune, and Pluto. Note: Pluto is sometimes excluded, and asteroids in the belt between Mars and Jupiter are sometimes classed as minor planets. **2.** a corresponding body orbiting some other star.

plane table *n.* **1.** a drawing board mounted on a tripod for use in making maps. Usually an alidade setting on the board is used to measure distances and show the direction to the location of an assistant with a stadia rod. *v.* **2.** to use such equipment to make a map.

planetarium *n.* (Latin *planetarius*, of a planet) **1.** a model representing the solar system, usually capable of demonstrating the movement of the planets as they orbit the sun. **2.** a similar model showing the positions and movements of the planets relative to the sun, moon, and stars, often on the interior surface of a dome. **3.** the structure that contains such a model.

plank *n.* (Latin *planca*, board) **1.** a thick piece of lumber that can be used to support considerable weight. Lumber less than 1 in. (2.5 cm) thick is usually called a board, whereas a thicker piece is called a plank, especially if it is also more than 6 in. (15 cm) wide. **2.** a support where one can stand or cling. **3.** a statement in a political platform. *v.* **4.** to cover or make a level surface with planks. **5.** to cook and serve on a board, as broiled fish or steak.

plankton *n.* (Greek *planktos*, drifting) minute plants and animals that float in water and are eaten by fish; mostly microscopic algae and protozoa. See benthos, nekton, phytoplankton, zooplankton.

plant *n.* (Latin *planta*, a plant) **1.** any of a great many multicellular life forms (kingdom Plantae) generally characterized by the presence of chlorophyll (usually indicated by green color), by lack of locomotion, and usually by absorption of nutrients through roots rather than ingestion of solid food. See chlorophyll, photosynthesis. **2.** a small vegetative life form with a nonwoody stem (excluding trees and shrubs). **3.** a seedling or a cutting that can be used to grow a new plant. **4.** the structures, machinery, and other physical parts of a factory or business, as a manufacturing plant. **5.** an object placed where it will deceive someone, as the evidence was a plant (it was false). **6.** a spy; a person secretly placed in an organization to obtain information. **7.** an idea expressed briefly at a point (e.g., in a theatrical production) for reference in light of later developments. *v.* **8.** to place seedlings or cuttings in soil or other growth media. **9.** to seed an area; e.g., to plant corn in a field. **10.** to suggest and nurture an idea or philosophy in the mind of another person, as to plant a love for music in one's child. **11.** to introduce an animal species to an area; e.g., a new breed of cattle. **12.** to stock a body of water with fish. **13.** to place something relatively permanently, as to plant a post in the ground or to plant a police officer on a street corner. **14.** to place something emphatically, as to plant himself in the doorway or to plant a kiss on her cheek. **15.** to release deceptive information, as to plant a story that will mislead the police, a criminal, or the voters.

planta *n.* (Latin *planta*, sole of the foot) **1.** the bottom surface of the foot. **2.** a bone in the back of the body of a bird or of certain insects.

plantain *n.* (French *plantain*, plane tree) **1.** a large tropical plant (*Musa paradisiaca*, family Musaceae) with a crown of broad leaves. It is closely related to banana plants but without the purple spots on its stem. **2.** the fruit from such plants, which is larger and more angular than a banana and usually is cooked before eating. **3.** any of a family of weeds, especially common or broadleaf plantain (*Plantago major*, family Plantaginaceae) with large leaves growing close to the ground and small flowers in slender spikes. See buckhorn plantain.

plantation *n.* (Latin *planta*, plant + *atio*, place) **1.** a large estate located in a warm climate, usually with many workers growing labor-intensive crops such as coffee, cotton, sugarcane, or tobacco. **2.** an area covered with trees that have been planted in a systematic pattern (often square or rectangular). **3.** a newly established colony.

plant breeding *n.* the process of applying genetic principles to develop new cultivars by inbreeding, hybridization, and selection of superior types. See biotechnology, genetic engineering.

plant bug *n.* any of a large number of bugs (family Miridae, order Capsidae) that damage the leaves of plants by sucking juice from them. Most are brightly colored insects with sucking mouth parts and two pairs of wings.

plant ecology *n.* the science that studies the relationships between plants and their environment. Also called bionomics or phytoecology.

planter *n.* **1.** a person who places living plants in soil or other media where they can grow. **2.** a machine that places seeds in the ground (usually large seeds such as corn that can be placed individually, as distinguished from a drill that places smaller seeds closer together). **3.** a decorative container used for ornamental plants. **4.** the owner of a plantation, or the manager.

plant hardiness zone *n.* an area of relatively uniform climate, especially with regard to winter temperatures that affect the survival of plants.

Plant hardiness zones in the U.S.A. based on average annual minimum temperature

planting *n.* **1.** the process of placing plants in the ground where they can grow, either as seeds or as transplants. **2.** an area of land where such plants are

growing. **3.** the beginning of something, as the planting of an idea, the establishment of a movement in a new location, or the placement of the first row of stones in a foundation.

planting season *n.* the appropriate part of the year for planting a desired crop; the interval when it is likely to become established and have enough time remaining to mature or reach some stage appropriate for the intended use.

plant kingdom *n.* the total of all plant species. Living things were divided into two kingdoms (plant and animal) until the 1960s but are now divided into five kingdoms: plants, animals, monera (bacteria), protista (protozoa), and fungi.

plant materials center *n.* a station managed by the U.S. Department of Agriculture Natural Resources Conservation Service to collect and maintain a wide variety of plant material as a resource for plant breeding and for conservation of natural resources. There is a national center at Beltsville, Maryland, and 25 other centers are widely distributed throughout the U.S.A.

plant nutrient *n.* an element that plants need for growth and development; either a macronutrient needed in large amounts (carbon, hydrogen, and oxygen from carbon dioxide and water plus nitrogen, phosphorus, potassium, calcium, magnesium, and sulfur obtained mostly from the soil) or a micronutrient needed in small amounts (boron, chlorine, copper, iron, manganese, molybdenum, nickel, and zinc).

plant pathology *n.* the science that studies plant diseases and methods of controlling them. Also called phytopathology.

plant physiology *n.* the science that studies the response of plants to environmental factors such as temperature, light, water, and plant nutrient supplies.

plash *n.* (Dutch *plasch*, puddle) **1.** a shallow pool of water; a puddle. *v.* **2.** to form a hedge by intertwining the branches of bushes.

plasma *n.* (Greek *plasma*, something molded) **1.** the clear, watery component of bodily fluids such as blood, lymph, or milk. **2.** the cytoplasm in a cell. **3.** whey. **4.** an ionized gas that contains both electrons and positively charged particles. **5.** a pale green form of chalcedony (quartz).

plasma membrane *n.* the semipermeable membrane that surrounds and protects the cytoplasm of a cell. It is composed primarily of a bilayer of phospholipid and protein and is considered the outer boundary of the cell even though many cells have a cell wall surrounding it.

plasmid *n.* (Greek *plasma*, something molded + *is*, origin) a small segment of DNA that is not part of a chromosome but floats freely in the cytoplasm of a bacterial or yeast cell, can move from one cell to another, and is capable of reproducing itself. It is used to transfer genetic material from one cell to another in recombinant DNA procedures.

plasmodium *n.* (Greek *plasma*, molded figure + *eidos*, appearance + *ium*, thing) **1.** a mass of protoplasm formed by the fusion of many cells into a multinucleate body; e.g., of slime mold (class Myxomycetes). **2.** a parasitic protozoan (*Plasmodium* sp., family Plasmodiidae) that attacks the red blood cells of mammals, birds, and lizards. *Plasmodium malariae*, *P. falciparum*, *P. ovale*, and *P. vivax* cause malaria in humans. The organism is transmitted by the bite of certain mosquitoes (*Anopheles* sp.).

plasmology *n.* (Greek *plasma*, molded form + *logos*, word) the study of the smallest components of living cells.

plasmolysis *n.* (Greek *plasma*, molded form + *lysis*, breakdown) shrinkage of the protoplasm in plant cells, occurring when water is lost faster than it is absorbed.

plaster of Paris *n.* powdered calcium sulfate hemihydrate ($CaSO_4 \cdot \frac{1}{2}H_2O$). When mixed with water, it soon hardens into gypsum ($CaSO_4 \cdot 2H_2O$). It is used in making plaster casts, walls, and objects.

plastic *adj.* (Greek *plastikos*, able to be molded) **1.** able to be molded and hold its new shape. **2.** able to undergo metabolic change. **3.** formed by molding. **4.** easily influenced, as a plastic personality. **5.** synthetic; false; as a plastic society. **6.** made of plastic. *n.* **7.** any of a wide variety of either synthetic or natural polymers, resins, cellulose derivatives, etc., that can be produced as a soft, moldable material that will harden and retain the shape it has been given. Plastic is used to make multitudes of modern objects, films, and other products. **8.** a credit card used instead of money.

plasticity *n.* (Greek *plastikos*, able to be molded + Latin *itas*, condition) the extent of plastic conditions exhibited by a material. For example, a slightly plastic soil can be molded into short flat or round segments if its water content is within a narrow range, whereas a very plastic soil can be molded into much longer segments if its water content is within a relatively broad range.

plasticity index *n.* the percentage of water held in a soil at the liquid limit minus that at the plastic limit. Also called plasticity number.

plastic limit *n.* the percentage water content of a soil when it is barely plastic. See Atterberg limits.

plastid *n.* (Greek *plastis*, creator) an organelle that performs a specific function within a plant cell, such as manufacturing food (chloroplasts), storing starch grains (leucoplasts), and producing colors (chromoplasts).

plastron *n.* (Italian *piastra*, thin metal plate) **1.** the underside of a turtle shell. See carapace, tortoise. **2.** a metal plate worn over the front of the body under a coat of mail. **3.** a padded front piece worn as protection while fencing. **4.** a starched shirt front or a dickey trimming the front of a dress.

plat *n.* (French *plat*, flat) **1.** a small plot of land. **2.** a plan or map showing the division of an area of land into plots, as a city plat. **3.** a braid of hair. *v.* **4.** to plan and/or map a set of plots for a piece of land. **5.** to braid hair.

plateau *n.* (French *platel*, small flat area) **1.** an area of relatively level upland that rises considerably above the land on one or more sides and is commonly cut by deep gullies or canyons. **2.** a relatively level period in the development of an individual or the progress of a project, as he reached a plateau and made little progress for a time. *v.* **3.** to come into a state of little or no progress.

plate count *n.* the number of bacterial or other microbial colonies that grow on an agar plate when it is incubated. It is used as an estimate of microbial population in water, soil, or other material (usually performed with several different dilutions of the material to obtain a reasonable number of colonies).

plate tectonics *n.* the theory that the Earth's lithosphere is composed of a number of rigid plates that gradually shift locations, producing earthquakes (especially around their margins), volcanoes, and mountains. The six major plates are the African, American, Antarctic, Eurasian, Indian, and Pacific plates, which are subdivided into a large number of minor plates.

platinum *n.* (Latin *platina*, silvery) element 78, atomic weight 195.08; a ductile, silvery-white, heavy metal that resists acids and oxidation. It is a precious metal that is used in jewelry, electronic devices, petroleum refining, etc.

platinum black *n.* a finely divided form of platinum that can absorb hydrogen, oxygen, and other ions. It is used to catalyze chemical reactions and to make hydrogen electrodes.

platy *adj.* (Greek *platys*, broad) **1.** divided or divisible into broad, thin layers. Shale and slate rocks are typically platy, and certain soil structures are described as platy. **2.** platyfish.

platyfish *n.* (Greek *platys*, broad + Anglo-Saxon *fisc*, fish) any of several small, mostly gray to yellow, freshwater fish (*Xiphophorus variatus* of Mexico and related species) that are popular for home aquariums.

platyhelminth *n*. (Greek *platys*, broad + *helminthos*, worm) any of thousands of species of flatworms (phylum Platyhelminthes) with a soft, generally flattened body. It is a primitive life form with bilateral symmetry (left and right sides). The term includes both parasitic and nonparasitic (free-living) types and flukes.

platypus *n*. (Greek *platys*, broad + *pous*, foot) a small aquatic animal (*Ornithorhyncus anatinus*) native to Australia and Tasmania. It is an egg-laying mammal with webbed feet, a tail like a beaver, and a bill like a duck. Also called duckbill.

playa *n*. (Spanish *playa*, beach) **1.** a desert basin in southwestern U.S.A. that becomes an intermittent salty lake immediately after a rain but dries out to a dry, barren surface between rains. **2.** (Spanish) a sandy beach.

plinthite *n*. (Greek *plinthos*, brick + *ite*, mineral) a highly weathered soil material with a high clay, high iron, and low humus content, commonly marked with dark red mottles. It develops on old land surfaces in the wet/dry tropics and hardens irreversibly to ironstone when exposed to dry heat. Pieces cut from plinthite and dried can be used as bricks for home building. See ironstone, laterite.

plot *n*. (Anglo-Saxon *plot*, secret plan) **1.** a small area of land marked for a particular purpose, as a building plot or a garden plot. **2.** a diagram or map of such an area. **3.** a conspiracy; a secret plan or scheme, especially one that is illegal or against the interests of another person or group. **4.** the underlying plan for a story; an outline of highlights. *v*. **5.** to make a plan for development of an area, or to mark such a plan on the land at the site. **6.** to develop a plan for an activity or a story. **7.** to scheme a way to gain an advantage over another person or group.

plover *n*. (French *plovier*, rainbird) **1.** any of several small to medium-sized shorebirds (family Charadriidae) of North America with short tails, pointed wings, and slender bills that are not as long as those of many other shore birds. Plovers have mostly gray or brown plumage marked with white. **2.** any of several other related shorebirds with longer bills and necks; e.g., the sandpipers.

plow or (mostly British) **plough** *n*. (German *pflug*, plow) **1.** a farm implement used for primary tillage. The first plows were made of wood and pulled by animals.

A wooden plow, as currently used in some developing nations. It has a metal point, but the rest is made of wood.

They cut the soil and moved it sideways a short distance. Modern chisel plows have a similar action but cut several furrows at once. **2.** a moldboard or disk plow that cuts, lifts, and inverts the soil, usually several furrows at once. **3.** the part of the implement that cuts and moves the soil. **4.** a snowplow used to remove snow from roads and other areas. **5.** a tool used by carpenters to cut a groove in wood. **6. Plow** Ursa Major (the Big Dipper). *v*. **7.** to use a plow to till soil. **8.** to use a snowplow to remove snow or a carpenter's plow to cut a groove. **9.** to force one's way, as to plow through a crowd, or a ship plows through the water. **10.** to do something slowly and laboriously, as to plow through the work. **11.** to invest, especially heavily, as to plow a lot of money into a project or to plow the profits back into a company.

plow layer *n*. the layer of soil that is moved by a plow, usually the top 6 to 10 in. (15 to 25 cm) of soil. Also called furrow slice.

plow pan or **plow sole** *n*. a soil layer that has been compacted by the downward pressure at the bottom of a plow (especially a moldboard plow) as it lifts and turns the tilled layer. Such a layer typically occurs at a depth of 6 to 10 in. (15 to 25 cm) and may impede water movement and root growth. Also called a tillage pan.

plow planting *n*. an early system of reduced tillage that mounted a planter behind a plow so that the plowing and planting were done at the same time with no secondary tillage. See wheel-track planting.

plowshare *n*. (German *pflug*, plow + Anglo-Saxon *scear*, cutter) the part of a moldboard plow that cuts the soil (usually removable so it can be sharpened or replaced). Also called share.

plum *n*. (Anglo-Saxon *plume*, plum) **1.** any of about 100 species of fruit trees (*Prunus* sp., family Rosaceae), mostly native to northern temperate regions. **2.** the fruit of the tree; a smooth drupe commonly about 2 in. (5 cm) in diameter with a smooth purple, red, or yellow skin. A dried plum is called a prune. **3.** a deep purple. **4.** a choice position or thing, as this job is a plum.

plumage *n*. (Latin *pluma*, small feather + *aticum*, belonging to) **1.** the assemblage of feathers that cover a bird. **2.** feathers in general.

plumbism *n*. (Latin *plumbum*, lead + *isma*, condition) chronic lead poisoning. Symptoms include loss of appetite, insomnia, constipation, dizziness, and a blue line on the margin of the gums. Also called saturnism. The EPA has specified 15 parts per billion as the maximum "safe" limit for lead in drinking water.

plum curculio *n*. See curculio.

plume *n.* (Latin *pluma*, small feather) **1.** a feather, especially a long, showy feather. **2.** a prize; an emblem or token of accomplishment. **3.** an emission from a smokestack or other point source that forms a visible or measurable streak through air, water, or other medium, often with a feathery appearance; e.g., a plume of smoke. **4.** an area of measurable and potentially harmful radiation leaking from a damaged reactor.

Pluto *n.* (Greek *Ploutōn*, Pluto) **1.** the 9th planet from the sun (small enough that some authorities do not consider it a planet). Discovered in 1930, it is usually the outermost planet, but its elliptical orbit carries it inside the orbit of Neptune. **2.** the mythological Greek and Roman god of the underworld.

plutology *n.* (Greek *ploutos*, wealth or *Ploutōn*, Pluto + *logos*, word) **1.** the science that studies wealth, especially from the viewpoint of politics and economics. **2.** the science that studies the interior of the Earth.

pluton *n.* (Greek *Ploutōn*, Pluto) a mass of coarse-grained rock formed by slow cooling of magma within the crust of the Earth or other celestial body.

plutonic *adj.* (Greek *Ploutōn*, Pluto + *ikos*, like) intrusive igneous, as a pluton is a plutonic rock formed by plutonic activity, either as small bodies such as dikes and sills or as large bodies such as laccoliths and batholiths.

plutonium (Pu) *n.* (Greek *Ploutōn*, Pluto + *ium*, element) element 94, atomic weight 244; a trace element in uranium deposits, but mostly synthesized by bombarding uranium with deuterium (heavy hydrogen) nuclei. It is used in atomic bombs. Plutonium 239, a bone-seeking poison, is one of the most hazardous substances known.

pluvial *adj.* (Latin *pluvialis*, pertaining to rain) **1.** pertaining to or caused by rain, especially sustained intensive rain. **2.** formed by rain and the associated runoff; e.g., an alluvial deposit.

pluviograph *n.* (Latin *pluvia*, rain + Greek *graphē*, drawing) a recording rain gauge. Also called ombrograph.

pluviometer *n.* (Latin *pluvia*, rain + *metrum*, measure) a rain gauge. Also called ombrometer.

PMP *n.* See pesticide management plan.

pneumatolysis *n.* (Greek *pneumatos*, breath + *lysis*, decomposition) the alteration of rocks into ores or other minerals by reaction with vapors released from a nearby mass of magma.

pneumatophore *n.* (Greek *pneumatos*, breath + *phoros*, bearing) a specialized structure that develops on the roots of certain woody plants growing in swamps; e.g., bald cypress (*Taxodium distichum*), tupelo (*Nyssa sylvatica*), and mangroves (principal genera: *Rhizophora*, *Avicennia*, *Conocarpus*, and *Laguncularia*). Pneumatophores grow upward through the water to reach the air and facilitate root respiration. Also called stilt roots or root knees.

pneumoconiosis *n.* (Greek *pneumōn*, lung + *konis*, dust + *osis*, condition) any chronic lung disease caused by inhaling dust particles that embed in the lungs, causing fibrosis and loss of lung function, including aluminosis, anthracosis, asbestosis, black lung, and silicosis. Also called pneumonoconiosis.

pneumonia *n.* (Greek *pneumōn*, lung + *ia*, disease) an infectious inflammation of the lungs, often affecting a person or animal that has been weakened by some other disease. It is aggravated by overcrowding, poor sanitation, chilling, and inadequate ventilation. Symptoms include loss of appetite, listlessness, fever, and wheezing.

pneumonic plague *n.* plague that seriously affects the lungs and causes expectoration of large quantities of sputum loaded with the organisms that cause the disease.

Poa *n.* (Greek *poa*, grass) a genus of grasses (family Poaceae) that includes the bluegrasses, meadow grass, wire grass, etc. Mostly native to the northern hemisphere, it is widely used for forage, lawn grasses, etc.

Poaceae *n.* the name adopted in 1972 for plants of the grass family. Formerly called family Graminae. See grass.

poach *v.* (French *pochier*, to pocket) **1.** to kill or capture game animals or fish illegally, either out of season or in a prohibited area. **2.** to cook in a simmering liquid, especially to cook the intact contents of an egg in water until the white coagulates. **3.** to trample an area of wetland, puddling and roughening its surface, or to become puddled, soggy, and roughened by trampling. **4.** to mix water and soil into a uniform paste. **5.** to pierce, stab, or poke. **6.** to cheat; e.g., by hitting a ball when it is the partner's turn to hit it.

pocket *n.* (French *poquette*, little pouch) **1.** a pouch in a garment; used for carrying small items. **2.** a separate pouch or enclosure used for carrying or storing items. **3.** a recess in a wall or piece of furniture; used for storage or to enclose moving parts. **4.** an area that differs from its surroundings, physically and/or socially, as a pocket of poverty. **5.** a low area surrounded by mountains. **6.** a depression where cold air accumulates and frost may occur even when surrounding areas are above freezing. **7.** an isolated mass of ore or other unique geologic deposit. **8.** an atmospheric condition such as a downdraft that causes an aircraft to drop suddenly.

9. a sac in the body of an animal. *adj.* 10. kept in a pocket, as a pocket handkerchief or pocket change. 11. a small item, often folded for storage, as a pocketknife. *v.* 12. to place an object in a pocket. 13. to surround or enclose, as to pocket an animal. 14. to suppress and conceal, as to pocket one's pride. 15. to take or steal.

pocket mouse *n.* any of several species of long-tailed burrowing rodents (*Perognathus* sp.) native to southwestern U.S.A. and northern Mexico that carry items in cheek pouches.

pocosin or **pocoson** *n.* (Algonquin *pocosin*, dismal) a depressional swampy wetland with mixed forest and grass vegetation in the coastal plains of eastern U.S.A.

pod *n.* **1.** a fruiting body developed from a carpel that splits open at maturity to release the seeds produced within it; e.g., that of a legume such as a pea or bean plant. Also called a seedpod. **2.** a mass of insect eggs or the case that holds them. **3.** a container shaped like a pod, as a cocoon or a pouch. **4.** a group, such as a pod of whales or seals or a small flock of birds. **5.** a streamlined protective enclosure, such as an engine pod on the wing of an airplane. **6.** a section or compartment such as a pod where gauges are located in an automobile.

podocarpus *n.* (Greek *podos*, foot + *karpos*, fruit) any of several tropical or subtropical coniferous evergreen trees (*Podocarpus* sp.), especially *P. macrophyllus* used as an ornamental.

pododerm *n.* (Greek *podos*, foot + *derma*, skin) the skin inside the horny part of the hoof of an animal.

pododermatitis *n.* (Greek *podos*, foot + *derma*, skin + *itis*, inflammation) an infectious bacterial (*Spherophorus necrophorus*) disease that affects the feet of sheep, goats, and cattle, causing lameness by eroding horny tissue and inflaming soft tissues.

Podzol or **Podsol** *n.* (Russian *pod*, under + *zol*, ash) a class of intensely leached acidic soils in some older systems of soil classification, characterized by a light gray layer at a shallow depth in the topsoil, often underlain by dark and/or bright reddish brown layers. Podzols form in humid temperate climates, especially under coniferous vegetation. Classified as Spodosol in Soil Taxonomy.

podzolization *n.* (Russian *pod*, under + *zol*, ash + *izatio*, formation) a soil-forming process characterized by leaching of bases, resulting in the formation of acid soils, especially Podzols (Spodosols in current Soil Taxonomy). See calcification.

poi *n.* (Hawaiian *poi*, poi) a Hawaiian food prepared by baking, grinding or pounding, moistening, and fermenting roots of the taro plant (*Colocasia esculenta*, family Araceae). It is commonly served at luaus and is useful as a cereal substitute for infants allergic to milk.

poikilothermic *adj.* (Greek *poikilos*, various + *thermē*, heat + *ikos*, of) cold-blooded; relying on the ambient environment to maintain a suitable body temperature. Examples are snakes, lizards, and fish. See ectotherm, homeothermic.

poikilothymia *n.* (Greek *poikilos*, various + *thymos*, spirit + *ia*, disease) a mental condition involving unusual variations in mood.

poinsettia *n.* an ornamental plant (*Euphorbia pulcherrima*, family Euphorbiaceae) with brilliant red bracts (leaves) surrounding yellow flowers. Native to Mexico and Central America, it is commonly used for Christmas decorations. It is named for J. R. Poinsett (1799–1851), a U.S. Ambassador to Mexico.

point source *n.* a localized source of pollution; e.g., waste emitted through a pipe or smoke from a smokestack. See nonpoint source.

poison *n.* (Latin *potio*, poisonous drink) **1.** any chemical agent that kills or damages the health of a plant, animal, or human; e.g., arsenic compounds, various cyanides, carbon monoxide, and mercury compounds. **2.** detrimental words or actions, as the poison of slander or of tyranny. *v.* **3.** to kill or injure by means of a chemical agent. **4.** to administer poison or add it to food, drink, or air. **5.** to have a strong negative influence, as to poison someone's mind or to poison a discussion.

poison control center *n.* any of hundreds of locations in the U.S.A. that give advice for treatment of persons or animals that have been poisoned. Many are located at medical centers and are accessible at any time of day.

poison dogwood *n.* See poison sumac.

poison hemlock *n.* See hemlock.

poison ivy *n.* a vine or shrub (*Rhus radicans*, family Anacardiaceae) that contains urushiol, a toxic chemical that causes a skin rash on sensitive persons. Common in eastern U.S.A., it is characterized by pointed leaves that grow in groups of three.

poison oak *n.* either of two woody vines or shrubs (*Rhus toxicodendron* of eastern U.S.A. or *R. diversiloba* of western U.S.A., family Anacardiaceae) with pointed leaves in groups of three similar to those of poison ivy. Poison oak contains urushiol and causes a skin rash similar to that of poison ivy on sensitive persons. Sometimes called poison ivy.

poison sumac *n.* a shrub or small tree (*Rhus vernix*, family Anacardiaceae) with pinnate leaves that grows in swamps throughout eastern U.S.A. It contains urushiol, a toxic chemical that causes dermatitis similar to that of poison ivy in sensitive persons. Also called poison dogwood.

poisonvetch *n.* any of several plants (*Astragalus* sp., family Fabaceae) of the mountains, foothills, and valleys of the Rocky Mountain states that cause weakness, nausea, and difficulty breathing in cattle and sheep.

polar *adj.* (Latin *polaris*, polar) **1.** pertaining to the North Pole or the South Pole or the surrounding area. **2.** pertaining to any other pole, such as that of a magnet. **3.** divided into opposing groups or characteristics, as polar personalities. **4.** capable of ionizing; soluble salts are polar. **5.** asymmetrical in charge distribution, as water molecules are polar, so water is a polar solvent. See nonpolar solvent. **6.** pivotal; a central concept or a guiding principle.

polar bear *n.* a large bear (*Ursus maritimus*) with white fur; native to the Arctic Region.

polar cap *n.* **1.** the ice cap at either the North Pole or the South Pole. **2.** a similar-appearing area around each pole of the planet Mars.

polar circle *n.* the circle delineating the part of the Earth that experiences continual sunlight at the summer solstice and continual darkness at the winter solstice; either the Arctic circle, located 23° 27′ from the North Pole (66° 33′ north latitude), or the Antarctic Circle (similarly located in the Southern Hemisphere).

polar climate *n.* a climate where the average temperature during the warmest month is below 50°F (10°C). It is typical of the coastal areas around the North Sea and of Greenland and Antarctica.

polar desert *n.* the area where the average temperature is below freezing every month of the year. Water is perpetually frozen, so there is no plant growth. Glacial ice accumulates to great depths even though there is little precipitation, generally less than 4 in. (10 cm) per year. Also called Arctic desert.

polar front *n.* the shifting frontal zone that occurs in either northern or southern latitudes where polar air masses meet tropical air masses, often causing heavy precipitation. See subpolar low.

polar ice cap climate *n.* the climatic type in the polar desert.

Polaris *n.* (Latin *polaris*, polar) the North Star, located within 1° of due north of the Earth. Polaris is the end star of the handle of the Little Dipper and is often located by using the two outer stars of the bowl of the Big Dipper as pointers.

polar orbit *n.* the orbital path of a satellite that passes over the poles of the Earth or some other celestial body, thus permitting its cameras or sensors to cover the entire area of the planet.

polder *n.* (Dutch *polder*, pool) an area reclaimed from the sea, especially those areas in the Netherlands where dikes were built to isolate areas in the Zuider Zee. These areas were then drained and converted to cropland and other uses. About half of the people in the Netherlands live on polders below sea level. See Zuider Zee.

pole bean *n.* any variety of climbing beans that need support, as by a set of poles or on a trellis.

polecat *n.* (French *poule*, chicken + *chat*, cat) **1.** a European weasel (*Mustela putorius* and related species) with blackish fur. It ejects a foul-smelling liquid when attacked or disturbed. Called a ferret when domesticated. **2.** any of several skunks native to North America.

pole chain *n.* Gunter's chain.

pollard *n.* (German *polle*, hair of the head + *ard*, characterized by) **1.** a naturally hornless animal; an ox, sheep, stag, etc., that lacks horns. **2.** a tree with a dense growth of branches because its crown was pruned severely (coppiced).

polled *adj.* (German *polle*, hair of the head) **1.** hornless, especially naturally hornless. **2.** shorn; hairless. **3.** having had its branches cut off; coppiced.

pollen *n.* (Latin *pollen*, fine flour) the male cells on the stamens of flowering plants, typically a yellowish powdery mass that may be distributed by the wind or by insects such as bees that collect it to make honey. It causes hay fever in sensitive persons. See allergen.

pollen basket *n.* a corbiculum.

pollen count *n.* a measure of the amount of pollen (especially that of ragweed) in the air, usually the number in a cubic yard (0.76 m^3) of air.

poll evil *n.* a swelling on the head of a horse or mule produced by severe injury.

pollination *n.* (Latin *pollen*, fine flour + *atio*, action) the first step in fertilization of a flower; the transfer of pollen (male gametes) from the anthers to the pistils (female structures). In nature, pollination takes place mostly by insects passing from one flower to another, but sometimes wind is the agent.

pollutant *n.* (Latin *pollutio*, pollution + *antem*, agent) any substance that contaminates or degrades the environment; e.g., a waste product or escaped chemical that makes the air, water, soil, or other part of the environment less suitable for the life of people, animals, or plants.

pollute *v.* (Latin *polluere*, to soil) **1.** to contaminate; to make impure or unclean; to degrade, especially some part of the environment, as air, water, or soil. **2.** to corrupt; to make immoral. **3.** to defile or make religiously unclean.

polluter *n.* one who pollutes, corrupts, or defiles.

pollution *n.* (Latin *pollutio*, pollution) **1.** the process of introducing pollutants into the environment. **2.** the condition of an environment that contains pollutants.

polonium (Po) *n.* (Latin *Polonia*, Poland + *ium*, element) element 84, atomic weight 209; an unstable radioactive metal (half-life 138.39 days) formed by the decay of bismuth (^{210}Bi), It is a low-melting, volatile metal that is highly toxic.

polyandrous *adj.* (Greek *polyandros*, having several husbands) **1.** married to several husbands. **2.** mating with more than one male (among certain species of animals). **3.** having more than 20 stamens in a single flower.

polyandry *n.* (Greek *polyandria*, many men) **1.** the practice of a woman having more than one husband at the same time. **2.** the mating of a female animal with more than one male. **3.** the condition of having more than 20 stamens in a single flower.

polychlorinated biphenyl (PCB) *n.* polychlorinated biphenyl; any of several synthetic hydrocarbons containing two benzene rings and more than two chlorine atoms attached to each molecule; formerly used as plasticizers and to keep electrical transformers from overheating. Their production in the U.S.A. was banned in 1979 because they cause skin diseases, birth defects, and cancer. They are resistant to degradation, are a very persistent hazard to wildlife, and are subject to bioaccumulation, but some are still in use.

polyclimax *n.* (Greek *poly*, many + *klimax*, ladder) an ecological community that maintains itself across or in spite of variations in environmental factors such as soils, topography, vegetation, fire, and interactions with other species.

polyculture *n.* (Greek *poly*, many + Latin *cultura*, tillage) multiple cropping, the practice of growing two or more crops simultaneously in the same area; e.g., the traditional corn, bean, and squash combination used in Central America and some other warm climates.

polygamous *adj.* (Greek *poly*, many + *gamos*, marrying) having more than one mate during a designated period.

polygonal *adj.* (Greek *polygōnos*, many angled) a pattern of lines that surround areas with several angles and sides (usually more than four). Soils often form polygonal patterns either by shrinkage as they dry or by frost action.

A polygonal pattern of soil cracks

polygonatum *n.* (Greek *poly*, many + *gonatos*, knee) a plant (*Polygonatum biflorum*, family Liliaceae) with a thick rootstock, a crooked stem marked with scars and curves where previous stems have broken away, greenish yellow flowers, and red or blue berries. When eaten in small doses, it is a tonic and purgative. In large doses, it is a cardiac poison. Also called Solomon's seal.

polygyny *n.* (Greek *polygynia*, having many wives) **1.** the condition of a man married to more than one woman at the same time. **2.** the breeding of several female animals by the same male. **3.** the presence of two or more functioning queens in an insect colony at the same time. **4.** the presence of multiple female parts (pistils and/or styles) in a single flower.

polyhedrose *n.* (Greek *polyedros*, having many bases + Latin *osus*, full of) a disease of insects caused by a virus (*Borrelina* sp.) characterized by polyhedral inclusions in the nuclei of the infected cells. Electron microscopy shows bundles of viruses (usually rod shaped and about 40 × 300 μm in size) in the polyhedra.

polymictic lake *n.* (Greek *poly*, many + Latin *mixtus*, mixing) a lake that mixes its surface and deep water either constantly or most of the time; common at high altitudes and in equatorial climates. See amictic lake, dimictic lake, monomictic lake.

polymorph *n.* (Greek *polymorphos*, many forms) **1.** one of two or more adult forms of the same organism, as different castes of ants or a worker bee and a queen bee. **2.** any of the various forms of a mineral that exhibits polymorphism.

polymorphism *n.* (Greek *polymorphos*, many forms + *isma*, condition) **1.** the occurrence of two or more forms of the same kind of plant, animal, or thing; e.g., different colors or shapes, often in response to environmental conditions. **2.** the occurrence of more than one genetically controlled phenotype of an organism; e.g., eye color or blood type. **3.** the existence of two or more minerals with the same chemical composition, as diamonds and graphite (both pure carbon but of different crystal structure), or high and low quartz, cristobalite, and tridymite (all SiO_2).

polynya or **polyna** *n.* (Russian *polynya*, open area) an area of open water in the Arctic ice cap, believed to be formed by a combination of wind and water currents enlarging fissures in the ice.

polyp *n.* (Latin *polypus*, many feet or nasal tumor) **1.** any of several small aquatic animals with a tubelike body and a set of tentacles around the mouth area; e.g., the hydra and sea anemone. **2.** an octopus. **3.** an abnormal growth on a mucous membrane of the nasal passage, bladder, intestine, etc.

polyped *n.* (Greek *poly*, many + *ped*, footed) an organism or object that has many feet; e.g., many kinds of insects or a multilegged table.

polypedon *n.* (Greek *poly*, many + *pedon*, soil) a group of contiguous pedons of the same kind of soil. Soil map units on detailed soil maps approximate polypedons as closely as it is practical to map them. See pedon, soil map unit.

polypeptide *n.* (Greek *poly*, many + *peptos*, digested + *ide*, binary compound) a group of amino acids linked by a straight chain of peptide bonds; a building block of protein that can have a molecular weight up to about 10,000.

polyphagia *n.* (Greek *poly*, many + *phagein*, to eat) **1.** excessive appetite (can be a sign of diabetes); eating too much. **2.** eating many different kinds of food, as certain animals. See geophagia, omnivore.

polyphagous *adj.* (Greek *polyphagos*, eating too much) **1.** consuming a wide variety of foods, as a polyphagous animal. **2.** feeding on a wide variety of living hosts, as a polyphagous aphid.

polyphenol *n.* (Greek *poly*, many + French *phén*, from benzene + *ol*, alcohol) an organic structure that contains multiple phenolic groups (ring structures with attached hydroxyls); e.g., lignin and tannin.

polyphosphate *n.* (Greek *poly*, many + *phōs*, light + *ate*, salt) a salt of polyphosphoric acid. Also called metaphosphate. Sodium polyphosphate is used as a water softener and as a dispersing agent for mechanical analysis of soils.

polyphosphoric acid *n.* either a ring or a chain of phosphoric acid groups $[(HPO_3)_n]$. Also called metaphosphoric acid.

polyploid *n.* (Greek *poly*, many + *ploos*, fold + *eidos*, form) an organism (usually a plant) that has more than two sets of chromosomes; e.g., a tetraploid produced by doubling the number of chromosomes that a normal diploid plant has.

polypnoea or **polypnea** *n.* (Greek *poly*, many + *pnoia*, breathing) (hyperpnea) rapid breathing or panting. See hyperpnea.

polypody *n.* (Latin *polypodium*, many feet) any of several ferns (*Polypodium* sp., family Polypodiaceae) that have creeping rootstalks and can cling to a surface; e.g., wall fern.

polysaccharide *n.* (Greek *poly*, many + *sakcharon*, sugar) a long chain or network of sugar molecules linked together, forming pectin, starch, cellulose, glycogen, etc.

polysaprobic *adj.* (Greek *poly*, many + *sapros*, rotten + *bios*, life + *ikos*, of) anaerobic because of pollution with organic wastes; e.g., the condition of a grossly polluted zone in a lake or stream.

polystyrene *n.* or *adj.* (Greek *poly*, many + *styrene*, an aromatic liquid) a polymer of styrene ($C_6H_5CHCH_2$). It is used to make very lightweight foamed plastic that is molded into cups and other containers and made into flakes or beads that are used in growth media and packing materials. It is commonly called by the brand name of Styrofoam.

polyunsaturated *adj.* having several double bonds and therefore containing less hydrogen than the corresponding saturated molecule. Unsaturated fatty acids have lower melting points than the corresponding saturated forms and are less likely to form cholesterol deposits.

polyurethane *n.* (Greek *poly*, many + French *uréthanne*, ester of carbamic acid) any of a group of polymers made by a condensation reaction of OH radicals and NCO groups. Polyurethane is used to make various forms of plastic, including protective coatings, adhesives, and lightweight foamed plastic for insulation or filler.

polyuria *n.* (Greek *poly*, many + *ouron*, urine) excessive passage of urine, often a symptom of diabetes.

polyvinyl chloride (PVC) *n.* a polymer ($CH_2=CHCl)_n$ produced from vinyl chloride. It can be softened by heating, shaped or cast in various forms, and hardened by cooling. It is commonly used in food packaging, pipes, gutters, and electrical insulation for wires and cables. Products made from it are smooth, shiny, impermeable to liquids and gases, chemically inert, and very stable. Finely powdered PVC is a health hazard similar to powdered asbestos. Incineration of PVC produces hydrochloric acid that is noxious and a contributor to acid rain unless it is caught by a scrubber.

pomace *n.* (Latin *pomacium*, cider) **1.** the pulpy residue that remains when apples are pressed to make cider. **2.** any similar crushed pulpy residue.

pome *n.* (Latin *pomum*, apple) a fruit that contains a core with several seeds in it; the fruit of any tree of the apple family, including apples, pears, and quince.

pomegranate *n.* (Latin *pomum*, apple + *granatum*, grained) **1.** a small fruit tree (*Punica granatum*, family Punicaceae) with glossy green leaves, large scarlet, white, or yellowish flowers, and red-gold spherical fruit. **2.** the edible red-gold spherical fruit that is filled with many seeds enclosed in a juicy red pulp that has an acid taste.

pomelo *n.* (Dutch *pompelmoes*, shaddock) **1.** shaddock. **2.** grapefruit.

pomology *n.* (Latin *pomum*, apple + *logos*, word) the horticultural study of fruits and fruit culture.

pond *n.* (Anglo-Saxon *pound*, enclosure) **1.** a small body of water (smaller than a lake) collected in a low place, often one that dries out when there is no rain. **2.** an artificial water reservoir formed by building a dam or digging into the landscape or both and filled either with water from a stream or by runoff. It is often used to store water for livestock or to allow sediment to settle from runoff water. See lagoon. *v.* **3.** to accumulate water in a low place, as on bottomland, behind a barrier, or on a flat part of a roof.

ponderosa pine *n.* **1.** a large pine tree (*Pinus ponderosa*, family Pinaceae) that is abundant in western U.S.A. and Canada. It has needles mostly in clusters of three and light reddish brown cones, and its dark brown bark can be quite thick and deeply furrowed. It is an important timber tree and the state tree of Montana. Also called western yellow pine. **2.** the pale yellowish wood of this tree. It is a soft wood used for construction, crates, furniture, timbers, railroad ties, fuel, etc. Also called yellow pine.

pond scum *n.* a green mass, usually composed mostly of algae, growing on the surface of a pond.

poodle *n.* (German *pudel*, to splash) a breed of standard-sized (more than 15 in. or 38 cm high at the shoulders), miniature (10 to 15 in. or 25 to 38 cm high), or toy-sized (less than 10 in. or 25 cm high at the shoulders) dogs with white or black curly hair, often trimmed to special patterns of long and short hair.

pool *n.* (Anglo-Saxon *pol*, pool) **1.** a puddle; a small body of ponded water or other liquid; e.g., a pool of blood. **2.** a calm part of a stream where the water is deep and quiet. **3.** a swimming pool. **4.** a group of people who regularly ride together in an automobile, as to and from work. **5.** a game played on a flat table with 15 colored balls to be knocked into pockets by means of a cue stick and a white cue ball. Also called pocket billiards. **6.** a collection of resources, often of money representing combined resources either to be used for a common purpose or being placed as wagers on the outcome of an event. **7.** a group of people who agree to control the supply and price of a marketable product. **8.** a group of people who are available for work, as a labor pool or a typing pool. *v.* **9.** to form a pool. **10.** to combine the resources of several people or organizations.

popcorn *n.* **1.** any of several varieties of corn (*Zea mays*, family Poaceae) whose kernels explode into a puffy white mass when heated sufficiently. These varieties generally have relatively small ears and small pointed kernels. **2.** the popped kernels, eaten as a snack, often salted, and sometimes made into caramel corn by adding caramelized sugar that sweetens and aggregates the popped kernels into larger masses.

poppy *n.* (Latin *papaver*, poppy) **1.** a flowering plant (*Papaver* sp., family Papaveraceae) with bright blossoms that is grown as an ornamental, as a source of poppy seeds, or as a source of opium, and also occurs as a weed in many fields. See opium poppy. **2.** any of several similar plants of the same family, such as the California poppy (*Eschscholzia californica*) or the prickly poppy (*Argemone mexicana*). **3.** a flower from such a plant or a paper or plastic imitation thereof. **4.** an orange-red typical of such flowers. Also called poppy red.

A poppy growing in a field

poppy seed *n.* seed from poppy plants, especially that from *Papaver orientalis*. A small, dark, nearly spherical seed used in cooking, often sprinkled on the surface of bread or rolls before baking, and commonly included in bird seed.

population *n.* (Latin *populatio*, population) **1.** the number of plants, animals, or people of a single species (or other type of group) in a given area at a specific time, expressed either as a total (e.g., the number of inhabitants in a city) or as a density (e.g., the number of corn plants per acre). **2.** all of the people or other species in an area, considered collectively. **3.** the total group considered in a study, especially those represented in a statistical analysis. **4.** the process of establishing a population in an area.

population ecology *n.* the study of factors that influence population size, growth, and stability, especially in regard to the capacity of a particular environment to support it.

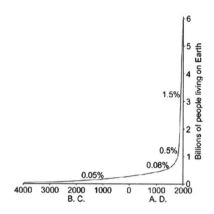

Approximate human population of the Earth. Average annual population growth is shown beside the curve.

population growth *n.* the rate of growth of a population, commonly expressed as an annual percentage or as a time required to double the population. It is sometimes called a population explosion when it appears to be very rapid and/or likely to exceed the carrying capacity.

porcupine *n.* (Latin *porcus*, pig + *spina*, spine or thorn) a gnawing animal that moves slowly but is protected by its sharp quills, including the Old World porcupine (*Hystrix crystata*, family Hystricidae) with quills up to 1 ft (30 cm) long, the porcupine of eastern North America (*Erethizon dorsatus*, family Erethizonidae), and the porcupine of western North America (*E. Epixanthus*). Both of the latter have short quills with barbed points that readily stick in flesh and detach from the porcupine.

pore *n.* (Greek *poros*, passage) *n.* **1.** a tiny opening and/or passageway that permits movement and/or storage of fluids such as air and water (especially those microscopic in size or barely visible), as the pores in the skin of an animal or the leaves of a plant. **2.** the tiny spaces between particles of soil or rock.

pore-size distribution *n.* the volume of pore space representing designated size ranges of pore space in a medium such as soil. The following size ranges are used to evaluate soil pores:

Coarse macropores	>5000 μm
Medium macropores	2000–5000 μm
Fine macropores	1000–2000 μm
Very fine macropores	75–1000 μm
Mesopores	30–75 μm
Micropores	5–30 μm
Ultramicropores	0.1–5 μm
Cryptopores	<0.1 μm

pore space *n.* the nonsolid portion of a soil or other material; voids; space occupied by fluids such as air and water, commonly expressed as a percentage. Also called porosity. Soil pore space is often calculated as:

$$\% \text{ pore space} = 100 \times (1 - \text{b.d.}/\text{p.d.})$$

where b.d. = bulk density and p.d. = particle density (p.d. is usually assumed to be 2.65 g/cm^3).

pork *n.* (Latin *porcus*, pig) **1.** flesh of swine. **2.** government projects authorized for political favors rather than for their own worth.

porosity *n.* (Latin *porositas*, porosity) **1.** the percentage of the volume of any material that is not solid; the voids that hold and/or permit passage of fluids such as air and water. **2.** pore space.

porous *adj.* (Greek *poros*, passage + *osus*, having) **1.** having many pores. **2.** permeable to fluids; allowing air, water, etc., to pass through. **3.** allowing almost anything to pass through, as the defense was porous.

porphyry *n.* (Greek *porphyros*, purple) **1.** a hard rock with large red and white feldspar crystals embedded in a purplish matrix, as found in certain areas in Egypt. **2.** other igneous rocks with large phenocrysts embedded in a finer-textured matrix.

porpoise *n.* (Latin *porcus*, pig + *piscis*, fish) **1.** a small (typically 5 to 8 ft or 1.5 to 2.4 m long), friendly cetacean (*Phocoena* sp.) with a pale belly and a black back with a triangular fin in the middle. **2.** a dolphin or other small member of the cetacean (whale) family.

porridge *n.* (Latin *porrata*, leek broth) a cereal (especially rolled or ground oats) mixed with water or milk and cooked to a pasty consistency; often served for breakfast. See oatmeal.

port *n.* (Latin *portus*, harbor or haven) **1.** a location where ships can dock to load and unload cargo and provisions. **2.** an entry point for people and/or cargo to arrive and depart from a nation or other political entity, usually with customs officers present; often called a port of entry. **3.** a haven that offers refuge from storms. **4.** an opening in the side of a ship or machine. **5.** the left side of a ship or other vessel (when facing forward), opposite the starboard side. **6.** a place on a computer where a peripheral device can be attached. **7.** a very sweet wine of a type that originated in Portugal.

portland cement *n.* a product that is usually mixed with sand, gravel, and water to make concrete. The portland cement adheres to stone fragments and hardens by absorption of water. It is produced by mixing pulverized limestone and shale, heating it in a furnace until it fuses into a clinker, and grinding it to a very fine powder. Developed in 1824 and patented by Joseph Aspdin, an English bricklayer, it is named for its similarity to stone found on the island of Portland.

portulaca *n.* (Latin *portulaca*, purslane) an herbaceous plant (*Portulaca* sp., family Portulacaceae) with fleshy leaves and showy blossoms of various colors used as an ornamental (especially *P. grandiflora*). Some varieties (e.g., *P. oleracea*) are used as herbs in salads and other foods. Also called rose moss, moss rose, or purslane.

possum *n.* See opossum.

post *n.* (Latin *postis*, placed) **1.** an upright piece of wood, metal, or other material placed to support a fence, sign, gate, building, etc. **2.** anything similar to such a post in form or use. **3.** a place marked by a post; e.g., a starting point for a survey or race. **4.** the outer lane of a racetrack. **5.** a place where a soldier is stationed as a lookout or where a body of troops is stationed. **6.** a unit of troops stationed at a post, or a unit of an organization of veterans. **7.** a position, job, or workplace. **8.** a trading post or a post office. *v.* **9.** to place a notice on a post as a form of publicity. **10.** to publicize or announce, as to post a reward. **11.** to put up warning signs such as no trespassing or no hunting. **12.** to station someone at a post. **13.** to mail, as to post a letter. **14.** to enter an accounting item in a ledger or other official record. *adj.* **15.** after or since.

post oak *n.* any of several varieties of American oak trees (especially *Quercus stellata*) used to make posts.

postpartum *adj.* (Latin *post partum*, after bringing forth) pertaining to the period following the birth of a child.

posy *n.* (French *poesie*, poetry) **1.** originally a poem or inscription expressing a sentiment. **2.** a flower or bouquet.

pot *n.* (Dutch *pot*, pot) **1.** a container used for cooking food, usually fairly deep as compared wiyh a pan, and commonly made of metal, glass, or ceramic and having one or two handles. **2.** the food cooked in such a container, as a pot of stew. **3.** such a container used for other purposes, as a firepot in a stove or furnace or a chamber pot formerly used as a night toilet. **4.** a flowerpot, a container used for growing plants, usually with one or more drainage holes in the bottom. **5.** an accumulation of a considerable amount of something, as a pot of money. **6.** (slang) a pot belly. *v.* **7.** to plant or transplant into a pot. **8.** to capture or shoot game or to win a prize.

potable *adj.* (Latin *potabilis*, drinkable) drinkable; safe for drinking by humans. The EPA has set standards for water to qualify as potable. It must be free from objectionable odors and taste and contain no higher than the following concentrations of the designated elements:

- nitrate nitrogen (NO_3–N)   10 ppm
- fluoride (F)   4 ppm
- barium (Ba)   1 ppm
- arsenic (As)   0.05 ppm
- chromium (Cr_6)   0.05 ppm
- lead (Pb)   0.05 ppm
- silver (Ag)   0.05 ppm
- selenium (Se)   0.01 ppm
- mercury (Hg)   0.002 ppm

Note: 1 ppm is equal to 1 mg/L.

potamology *n.* (Greek *potamos*, river + *logos*, word) the science that studies rivers.

potamometer *n.* (Greek *potamos*, river + *metron*, measure) a device that measures the velocity of the current in a river or other body of water.

potamoplankton *n.* (Greek *potamos*, river + *planktos*, drifting) tiny organisms that float in river water. See plankton.

potash *n.* (Dutch *potasch*, pot ashes) **1.** impure potassium carbonate (K_2CO_3) such as that obtained by leaching ashes with water and then evaporating the water. **2.** potassium hydroxide (KOH). **3.** potassium oxide (K_2O), the form used for calculating the percentage of potassium in a fertilizer or for expressing the amount of potassium needed on a soil. **4.** (loosely) potassium in any form.

potassium (K) *n.* (Dutch *potasch*, pot ashes + *ium*, element) element 19, atomic weight 39.0983; a highly reactive monovalent metal that is an essential macronutrient for plants and animals. It is highly soluble and mobile as an ion and important for ionic balance. The K symbol comes from its Latin name, kalium.

potassium-argon dating *n.* a means of estimating the age of a rock by the proportions of potassium and argon contained in it. It is based on the half-life of ^{40}K as it decays into ^{40}A, 1,265,000,000 years.

potassium chloride *n.* the most abundant salt of potassium (KCl) and the principal potassium fertilizer, a white, crystalline mineral (often with pink shades or streaks from impurities) containing about 50% K and 60% K_2O equivalent. It occurs in evaporite deposits, usually mixed with sodium chloride (NaCl). The largest known deposit is in southern Saskatchewan, Canada. The U.S.A. has deposits in California, Utah, and New Mexico. Also called muriate of potash.

potassium cyanide *n.* a very poisonous compound (KCN) used as a fumigant, an insecticide, in the extraction of gold and silver from their ores, and in photography.

potassium feldspar *n.* any feldspar that contains potassium as an important component (tektosilicates with idealized formula $KAlSi_3O_8$ but usually containing some substitution of Na and Ba for K). Orthoclase is the most abundant, occurring as a major component of most igneous and metamorphic rocks. Sanidine is its high-temperature polymorph, occurring in volcanic rocks. Microcline is a more ordered triclinic form, whereas the other two are monoclinic. Adularia is an intermediate form found in low-temperature hydrothermal deposits.

potassium nitrate *n.* a white crystalline salt (KNO_3) used to pickle meat, to make matches and gunpowder, and as a fertilizer. It is especially good in starter fertilizer because it has a high analysis (13-0-44) and no extra ions that would increase the salt index. Also called saltpeter or niter.

potassium sulfate *n.* a white crystalline salt (K_2SO_4) used as a fertilizer (0-0-50 plus 18% sulfur), especially on crops such as potatoes and tobacco that are sensitive to chlorine and on soils with sulfur deficiencies.

potato *n.* (Spanish *patata* from *batata*, sweet potato) **1.** an annual plant (*Solanum tuberosum*, family Solanaceae) that produces edible underground tubers. It is native to the Andes of Bolivia and Peru in South America and known to have been cultivated by the Incas as early as 200 B.C. It is often called white potato or Irish potato to distinguish it from sweet potato. **2.** the somewhat rounded but variable shaped, white-fleshed, brown-, tan-, or red-skinned tubers produced by the plant and used as a high-energy starchy food. Potatoes became so popular after they were introduced in Ireland that the Irish potato famine resulted when potato rot wiped out the crop.

potato dry rot *n.* a fungal disease (*Fusarium sambucinum*) common in stored potatoes that is usually treated with the chemical thiabendazole (TBZ).

potato rot *n.* a downy mildew disease caused by a fungus (*Phytophthora infestans*) that can kill potato plants within two weeks, especially if the weather is humid. Dead spots on the leaves and discolored foliage make the plants look blighted. Potato rot wiped out the potato crop and caused the Irish potato famine of 1846 to 1848, causing a mass migration to the U.S.A.

potato scab *n.* a soft-rot disease caused by a soil-borne actinomycete (*Streptomyces scabies*) that infects potato tubers.

potency *n.* (Latin *potentia*, having power) **1.** the power to produce a desired effect; e.g., the potency of a medicine to cure or heal or the potency of an engine to power a machine. **2.** the ability of a male of any plant, animal, or human species to fertilize female germ cells of the same species. **3.** the ability of an embryo to develop into a viable plant or animal.

potherb *n.* (Dutch *pot*, pot + Latin *herba*, herb) any plant leaves cooked in a pot for food; e.g., chard or spinach, or as seasoning; e.g., thyme. See greens.

pothole *n.* (Dutch *pot*, pot + German *hohl*, hollow) **1.** a pit or deep hole in soil or rock. **2.** a hole worn in the bed of a stream by the grinding action of stones driven by the water. **3.** a hole in the pavement of a road. Also called a chuckhole.

potsherd *n.* (Dutch *pot*, pot + Anglo-Saxon *sceard*, fragment) a fragment of broken pottery, especially one found in an excavation of an archaeological site. Such fragments were used much like scratch paper is now used and often have ancient writing on them. Also called sherd or shard.

potter wasp *n.* any of several predaceous wasps (especially *Eumenes fraterna*) that build a mud nest shaped like a pot. Also called mud dauber.

pottery *n.* (French *poterie*, from pot) **1.** earthenware containers such as pots, bowls, and vases made of clay that is shaped and then hardened in a furnace. Pottery is often covered with a glaze and/or decorated by painting. Also called ceramics. The designs and patterns used on ancient pottery provide important clues in archaeological investigations. See potsherd. **2.** a place or business that produces pottery. **3.** the pottery trade.

potting soil *n.* any rooting medium used to grow plants in pots or other containers. It is usually a mixture of sandy or loamy topsoil with enough coarse sand, peat, perlite, and/or vermiculite to make it porous enough to drain well and provide good aeration.

poultry *n.* (French *pouleterie*, small fowl) domesticated fowls kept for the production of meat, eggs, feathers, etc., including chickens, ducks, geese, guineas, and turkeys.

pound *n.* (Latin *pondo*, pound) **1.** a unit of weight or mass with various magnitudes depending on the several systems that use it; e.g., 1 lb avoirdupois weighs 16 oz (7000 grains or 453.592 g), but 1 lb apothecary or troy weighs 12 oz (5760 grains or 373.242 g). **2.** the monetary unit of any of several nations (each with its own value), including the United Kingdom, Ireland, Egypt, Sudan, Syria, Lebanon, and Cyprus. **3.** an enclosure or structure where animals are kept or sheltered, especially stray animals. **4.** a trap or enclosure for animals or fish. **5.** an enclosure or structure where something is kept; e.g., impounded automobiles. *v.* **6.** to strike repeatedly, as with fists, a hammer, artillery and/or bombs, etc. **7.** to pulverize, as to pound wheat into flour. **8.** to run with heavy steps, as to pound down the track. **9.** to play loudly, as to pound a piano and pound out a tune.

poverty adjustment *n.* a term coined in 1936 by P. Macy to represent the range in concentration of a nutrient in plant tissue from the minimum percentage for life and growth up to a critical percentage beyond which there is no more increase in yield. He used the term luxury consumption for any further increase in concentration beyond that point.

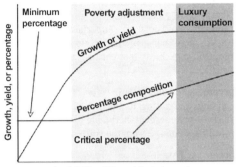

The central (gray) part represents poverty adjustment.

poverty grass *n.* any of several grasses that grow in poor soil; e.g., poverty oatgrass (*Danthonia spicata*) or churchmouse threeawn grass (*Aristida dichotoma*).

powdery mildew *n.* **1.** a parasitic fungus (an ascomycete of order Erysiphales) that covers plant parts (especially the underside of leaves) with mycelia, making the surface appear to have been powdered. It attacks apples, cherries, grapes, peaches, strawberries, barley, oats, wheat, beans, clovers, peas, and many other food and ornamental plants. **2.** a disease caused by the fungus, causing yellowing and death of foliage.

power line *n.* a thick wire (or usually a set of two or three wires) used to carry electricity from one place to another, usually supported on poles or towers, but sometimes buried.

pox *n.* (Anglo-Saxon *poc*, pustule) **1.** any of several diseases that cause skin eruptions; e.g., smallpox or chicken pox. **2.** any of several diseases that cause similar skin eruptions in animals. For example, cowpox affects cattle (and exposure to it may induce immunity to smallpox in humans), swinepox affects swine, sheep pox (also called ovine pox) affects sheep, and equine pox affects horses. **3.** a disease of sweet potatoes caused by a fungus (*Streptomyces ipomoea*) that produces pitlike lesions on the roots. **4.** a curse, used as an interjection to express dislike and resentment; e.g., a pox on you for what you have done!

ppb *n.* parts per billion.

ppm *n.* See parts per million.

prairie *n.* (French *praerie*, meadow) an area of natural grassland, usually with rich, dark soil on relatively smooth or gently rolling topography and a semiarid to subhumid climate. It may have scattered trees and shrubs, especially along streams and on valley slopes. Large areas of prairie occur in central U.S.A. and Canada and in comparable areas in other parts of the world. Smaller areas occur in grassed areas surrounded by forest in more humid climates and in relatively humid sites of drier climates.

prairie chicken *n.* either of two gallinaceous game birds of prairie areas in western U.S.A. with mixed black, brown, and white plumage: the greater prairie chicken (*Tympanuchus cupido*) and the lesser prairie chicken (*T. pallidicinctus*).

prairie dog *n.* any of several burrowing rodents (*Cynomys* sp.) that live in colonies on the prairies of central and western U.S.A. Their burrows are considered a nuisance by cattlemen because a horse can step in them and break its leg. See woodchuck.

praseodymium (Pr) *n.* (Greek *prasios*, green + *didymos*, twin + *ium*, element) element 59, atomic weight 140.90765; a soft, malleable, ductile, silvery rare earth metal that gradually tarnishes and spalls when exposed to air. It is used for carbon arc lights, to color glasses and enamels yellow, and to make high-refractory glass.

prawn *n.* any of several edible crustaceans (*Palaemon serratus* and related species) of the shrimp family, an ocean shellfish with a thin, leathery shell covered with reddish brown dots.

praying mantid or **mantis** *n.* any of several long, slender, predatory insects (especially *Mantis religiosa*, family Mantidae) that waits in ambush to capture an insect with its front legs, thus consuming many harmful insects (e.g., grasshoppers).

Praying mantid

precious metal *n.* a metal that is highly valued because it is relatively rare and can be made into beautiful objects that are resistant to tarnish by oxidation and hydrolysis, etc., including gold, silver, platinum, palladium, rhodium, iridium, ruthenium, and osmium. These are heavy metals that are good conductors of heat and electricity and are used to make jewelry, electronics, and other items.

precious stone *n.* a gem stone; any of several minerals that can be cut and polished to make beautiful ornaments for jewelry and other items. They include diamonds, rubies, sapphires, and emeralds; pearls usually are also included, and several other minerals are considered semiprecious stones.

precipice *n.* (Latin *praecipitium*, steep place) **1.** a bluff or cliff with a nearly vertical or overhanging face. **2.** a very dangerous situation, as when nations are on the precipice of war.

precipitation *n.* (Latin *praecipitatio*, a falling down) **1.** the process of falling or casting down, especially when sudden and/or rapid. **2.** great haste, especially with tumult and inadequate preparation. **3.** causing to begin, as the sudden precipitation of a crisis. **4.** the formation and settling of an insoluble substance from a solution, usually caused by the addition of another chemical or a change in temperature. **5.** the settling of particulate matter from the air. **6.** any form of moisture falling to Earth, including rain, snow, sleet, hail, and dew.

precipitation-evaporation index (P/E index) *n.* the annual sum of the P/E ratios for a given locale according to Thornthwaite's system for classifying climates, which is used to distinguish five humidity provinces as follows:

Map letter	Humidity province	Vegetation	P–E index
A	Wet	Rainforest	>127
B	Humid	Forest	64–127
C	Subhumid	Grassland	32–63
D	Semiarid	Steppe	16–31
E	Arid	Desert	<16

precipitation-evaporation ratio (P/E ratio) *n.* the monthly precipitation divided by the potential evaporation at a given site. It is used as an indication of the effectiveness of precipitation when calculating the P/E index in Thornthwaite's system for classifying climates.

precipitator *n.* (Latin *praecipitare*, to cast down + *or*, person or thing) **1.** someone or something that triggers or incites an action. **2.** a chemical or other agent that causes a precipitate to form. **3.** a mechanical, electrical, or chemical device that collects particulates; e.g., from the smoke in a chimney.

precision agriculture *n.* the application of carefully controlled rates of fertilizer and pesticides to meet the needs of growing crops without using excesses that would damage the environment and increase costs. It involves the use of detailed soil tests and scouting to determine when and where pest levels require treatment.

precursor *n.* (Latin *prae*, before + *cursor*, runner or courier) **1.** a person, procedure, or thing that goes before another. **2.** a harbinger, something that signals the approach of someone or something; e.g., the initial appearance of a robin is often considered a precursor of spring. **3.** a chemical that is synthesized and then used to produce another, either naturally or artificially. See provitamin. **4.** a cell or tissue that produces a more mature or specialized form.

predacious insect *n.* an insect that feeds on and kills other insects. Some benefit humanity by controlling harmful insects (e.g., the dragonflies, damselflies, aphis lions, praying mantids, and lady beetles).

predation *n.* (Latin *praedatio*, plundering) killing and eating organisms of another species.

predator *n.* (Latin *praedator*, plunderer) an animal that lives as a carnivore, killing and consuming other animals. Predators are important links in the food chain, and many predators are important for pest control; e.g., birds that eat insects or snakes that eat mice.

preemergent *adj.* (Latin *prae*, before + *emergens*, arising) pertaining to the period before emergence; e.g., a preemergent herbicide is applied before the seedlings emerge from the soil.

preen *n.* (Anglo-Saxon *preon*, clasp) **1.** a pin or brooch used to adorn one's clothing. *v.* **2.** to clean and arrange; a bird preens its feathers with its beak. **3.** to dress neatly, especially with adornments. **4.** to exhibit vanity and self-satisfaction.

preferential flow *n.* rapid flow through large passages such as cracks or worm channels. Preferential flow moves pollutants and other dissolved or suspended material down to the water table or tile lines much more rapidly than water flow through the soil mass. Also called hypodermic flow.

prehensile *adj.* (French *préhensile*, taking) seizing, grasping, holding, as the prehensile tail of a monkey.

premunition *n.* (Latin *praemunitio*, advance provision) **1.** preparation such as advance fortification or development of a logical argument to defend one's position. **2.** a balanced condition between a host and an infectious agent whereby the host is able to resist further infection but unable to eliminate the existing infection.

prescribed burn *n.* **1.** a planned fire, one that is set by people as part of their land management; e.g., to improve rangeland pasture or to prevent forest from invading grassland. Normally, fire control provisions are included to limit the fire to a specified area. **2.** a controlled forest fire used to destroy residues, surplus brush, and unwanted species and to reduce the hazard of an uncontrolled fire. Also called controlled burning.

preservation *n.* (Latin *praeservatio*, preserving) protection against destruction; maintenance of the status quo; prevention of injury or material change. Preservation may prevent use, whereas conservation attempts to allow use while protecting from damage. See conservation.

pressure *n.* (Latin *pressura*, pressure) **1.** force applied to a surface, quantified as force per unit area (e.g., pounds per square inch, kilograms per square centimeter, atmospheres, bars, or pascals). **2.** coercion or other urging to persuade one to take a certain action. **3.** external conditions that tend to cause a certain result or a problem; e.g., the pressure of everyday life. *v.* **4.** to exert strong influence toward a certain choice or action.

prevailing wind *n.* the usual wind direction at a particular location.

prevailing winds *n.* the broad wind patterns that characterize large sections of the Earth, as controlled by the rotation of the Earth combined with large cells of air circulating between the equator and the poles.

prey *n.* (Latin *praeda*, booty) **1.** an animal that is hunted, especially one that may be caught and eaten by a carnivore. See predator. **2.** goods (and sometimes captives) taken as plunder, as in a war or a raid. **3.** a victim of a swindler or other high-pressure salesperson. **4.** the action of seizing prey, as by a bird of prey. *v.* **5.** to hunt, kill, and/or eat animals for food. **6.** to plunder and rob. **7.** to victimize by overcharging or swindling, as to prey upon the elderly.

prickly pear *n.* **1.** any of several cacti (*Opuntia* sp.) with spiny, flattened, green, rounded lobes linked into branched stems, yellow to red blossoms, and fruit filled with many seeds. See sabra cactus. **2.** the prickly fruit of such cacti (edible in most species). Also called Barbary fig.

Prickly pear cactus forming a living fence

prill *n.* **1.** a type of European flatfish (*Scophthalmus rhombus*). Also called brill. **2.** the highest quality of ore from a mine. **3.** a globule of metal formed during an assay. **4.** pellets or granules formed by spraying droplets of a liquid that solidify as they cool in the air; e.g., prills of fertilizer.

primary color *n.* **1.** red, yellow, or blue. These can be mixed to produce other colors such as orange, green,

purple, and brown. **2.** red, green, or blue monochromatic light that can be added to each other to produce other colors.

primary consumer *n.* an animal that eats plants. Animals that eat other animals are secondary consumers.

primary feather *n.* any of the large feathers on the outer part of a bird's wing. Also called flight feathers.

primary mineral *n.* a mineral that formed in a rock when the rock was formed. A mineral formed later by weathering processes is called a secondary mineral.

primary nutrient *n.* any of the three most frequently applied fertilizer nutrients. Guaranteed analyses for these three nutrients are included in the fertilizer grade printed on a bag of fertilizer as nitrogen (%N), phosphorus ($\%P_2O_5$ equivalent), and potassium ($\%K_2O$ equivalent).

primary producer *n.* an organism at the base of the food chain, either a green plant that uses energy from sunlight to convert carbon dioxide and water into carbohydrates or a decomposer (a microbe, worm, or insect) that lives on the remains of dead organisms. See food chain, trophic level.

primary production *n.* the photosynthetic use of light energy to produce plant biomass.

primary root *n.* the first root produced by a seedling; usually a root that grows straight down and becomes the center of the root system. It is known as a taproot in plants that have one large main root.

primary standard *n.* **1.** a chemical standard that is used as a beginning point for quantitative analyses; e.g., a pure reactive metal or chemical that can be weighed precisely and then titrated to determine the strength of the titrating solution. **2.** a limit set by the EPA for the maximum permissible concentration of a pollutant in either air (sulfur dioxide, ozone, carbon monoxide, nitrogen dioxide, lead, and particulate matter) or drinking water (20 different materials are included) to protect human health. Secondary standards are set for uses such as irrigation water that do not affect human health.

primary succession *n.* the stages of an ecosystem that develops in a previously uninhabited area; e.g., on bare rock, on a glacial deposit, or on a recently eroded surface.

primary treatment *n.* the first steps in wastewater treatment. Screens and sedimentation tanks are used to remove materials that will float or settle. See sewage treatment.

primate (Latin *primatis*, of the first) *n.* **1.** a high-ranking official of certain churches. **2.** any of a group of omnivorous mammals (order Primates), including humans, apes, monkeys, and prosimians; a mammal that can interact in a complex society and can manipulate tools with its hands.

primatology *n.* (Latin *primatis*, of the first + *logos*, word) the branch of zoological science that studies primates, including early hominids.

primavera *n.* (Spanish *primavera*, spring) **1.** the spring season (Spanish and Italian). **2.** a bignonia tree (*Cybistax donnel-smithii*, family Bignoniaceae) native to central America. **3.** the hard, yellowish wood of the tree; used for making furniture. Also called white mahogany. *adj.* **4.** fresh vegetables lightly cooked in cream sauce, in a springtime Italian style.

prime agricultural land or **prime farmland** *n.* the land with the most productive and least erodible soils. The land best suited for raising the common food and fiber crops. See unique agricultural land.

prime meridian *n.* the meridian that passes through Greenwich, England. The reference used to measure east and west longitude around the Earth.

primeval *adj.* (Latin *primaevus*, first) pertaining to the first periods in the development of the world, before there were any humans.

primitive *adj.* (Latin *primitivus*, first of its kind) **1.** the earliest form, especially a life form that existed during an early geologic age or humans and human activities during an early period of civilization. **2.** characteristic of an early period; crude and simple.

primitive area *n.* undeveloped land reserved for wildlife and recreational use without any roads or buildings. Also called wilderness area.

primrose *n.* (French *primerole*, a flower) **1.** any of several plants (*Primula* sp., family Onagraceae) with tubelike corollas that have five lobes and a variety of colors. **2.** the blossoms produced by these plants. **3.** the yellow characteristic of some of these flowers. **4.** the primrose family (Onagraceae or Primulaceae) of plants, a group of dicots with blossoms at the ends of leafless stems. See evening primrose.

prion *n.* (Greek *priōn*, saw) **1.** an aquatic bird (*Pachyptila* sp.) of the petrel family that is native to the oceans of the Southern Hemisphere and characterized by serrated edges on its bill. **2.** a brain disease agent; a variant form of a glycoprotein that normally occurs in nerve membranes; a causal agent for scrapie in sheep and goats, for mad cow disease, and for similar conditions causing a spongelike structure in the brain tissue of mink, deer, cats, and humans.

prior appropriations *n.* a legal principle that assigns water rights to those who have established a history of beneficial use of water. In the U.S.A., prior appropriations is used in Minnesota and all states to its west, whereas eastern states use the riparian rights principle, and a few states (California, Oklahoma, North Dakota, and South Dakota) use a combination of the two. Also called appropriative right. See riparian right.

prismatic structure *n.* an arrangement of soil peds that are taller than they are wide and bounded by irregular vertical cracks that form a polygonal pattern, usually with a flat top surface (columnar prismatic structure has a rounded top) and often with an indistinct lower boundary. See blocky structure, granular.

Subsoils with moderate prismatic structure in a subhumid region (left) and strong prismatic structure in an arid region (right)

pristine *adj.* (Latin *pristinus*, pure state) remaining in a pure or original state, uncorrupted by civilization; primitive.

privet *n.* any of several deciduous or evergreen shrubs (*Ligustrum vulgare* and related species, family Oleaceae) with small white blossoms; commonly grown for hedges because they can withstand severe trimming.

proboscis *n.* (Greek *proboskis*, trunk of an elephant) **1.** an elephant's trunk. **2.** the long flexible snout of certain other animals; e.g., that of a tapir. **3.** the long beak of certain insects that is used for piercing and sucking blood or plant sap. **4.** certain elongated sensory or defensive organs of leeches and worms. **5.** (slang) a long human nose.

procambium *n.* (Latin *pro*, before + *cambium*, an exchange) meristematic tissue that produces vascular bundles.

procaryote *n.* prokaryote.

procreate *v.* (Latin *procreare*, to reproduce) **1.** to reproduce; to give birth to young. **2.** to create; to produce something new.

procryptic *adj.* (Greek *pro*, from protective + *kryptikos*, hidden) concealed from predators; e.g., a chameleon that matches its color to its surroundings.

procumbent *adj.* (Latin *pro*, for + *incumbens*, leaning upon) lying down; trailing; having stems that extend across the ground surface but do not root at the joints; e.g., a squash vine.

produce *v.* (Latin *producere*, to bring forth) **1.** to bring forth; to show or reveal, as to produce evidence. **2.** to bear fruit or give birth to offspring. **3.** to cause, as to produce results. **4.** to manufacture, as to produce automobiles. **5.** to yield, as to produce a profit. **6.** to present to the public, as to produce a play or a motion picture. **7.** to create something of value. *n.* **8.** the yield or result from something that produces; e.g., fresh produce from a garden.

producer *n.* (Latin *producere*, to bring forth) **1.** an organism (usually a green plant) that uses light energy to produce biomass from carbon dioxide, water, and plant nutrients. **2.** a person who generates or grows a product. **3.** a person who manages the development and directs the presentation of a play, a motion picture, or a radio or television program. **4.** equipment that generates producer gas.

producer gas *n.* a mixture of nitrogen and carbon monoxide produced by burning coal with a restricted supply of air. It is used as an industrial fuel and to manufacture ammonia.

productivity *n.* (Latin *producere*, to bring forth + *activus*, action) **1.** productiveness; active generation of a product. **2.** yield or rate of production.

proenzyme *n.* (Greek *pro*, before + *en*, in + *zyme*, leaven) any of several proteins that are converted into enzymes by reaction with acids or other enzymes.

proestrus *n.* (Latin *pro*, before + *oestrus*, gadfly, frenzy) the interval immediately preceding estrus.

profundal zone *n.* the bottom part of a deep lake. The part of a freshwater ecosystem where there is not enough light penetration for photosynthesis. See abyss, hypolimnion.

profuse *adj.* (Latin *profusus*, lavish) **1.** occurring or growing in great abundance. **2.** yielding or giving generously or excessively, as profuse flowering. **3.** spending or giving extravagantly.

progenitor *n.* (Latin *progenitor*, family founder) **1.** a direct ancestor. **2.** the originator of a family, model, concept, or technique.

progeny *n.* (Latin *progenies*, offspring) **1.** a direct descendant. A person, animal, or plant that descends from an indicated progenitor. **3.** all of the descendants of one progenitor, collectively. **4.** an outcome that results from something that happened before, as the progeny of a disaster.

progesterone *n.* (Latin *pro*, before + *gestare*, to bear + *sterol*, from cholesterol + *one*, ketone) a female sex hormone ($C_{21}H_{30}O_2$) that causes the uterus to prepare for and nurture a fertilized ovum. Synthetic versions are used to correct deficiencies and make pregnancy possible or in contraceptives to prevent ovulation.

prokaryote or **procaryote** *n.* (Greek *pro*, before + *karyōtis*, an organism) an organism such as a bacterium or a blue-green alga that has prokaryotic cells and reproduces by fission. See eukaryote.

prokaryotic cell *n.* a cell that has only one chromosome, no clearly defined nucleus, and no internal membrane-bounded structures; e.g., a bacterial or blue-green algal cell. See cell, eukaryotic cell.

prolific *adj.* (Latin *prolificus*, fertile) **1.** producing a large number of offspring. **2.** producing a large quantity, as a prolific harvest or a prolific writer. **3.** a time of unusually abundant production, as a prolific year.

promethium (Pm) *n.* (Greek *Promētheus*, a Greek god + *ium*, element) element 61, atomic weight 145; an unstable radioactive rare earth metal (half-life, 17.7 years for its most stable isotope, ^{145}Pm).

prominence *n.* (Latin *prominentia*, projection) **1.** a conspicuous high point that projects above or beyond its surroundings; e.g., a large butte or nose. **2.** a notable solar eruption; a flamelike tongue of dense gases. **3.** conspicuousness; high visibility.

A prominence (conspicuous high point)

promiscuous *adj.* (Latin *promiscuus*, mixing) **1.** unorganized mixing of components or entities, as a promiscuous mass or a promiscuous crowd. **2.** not discriminating, especially having sexual relations with several partners. **3.** able to nodulate (a plant; e.g., *Phaseolus vulgaris*) with any of several strains of *Bradyrhizobium*, *Rhizobium*, or *Frankia* or to be infected by various strains of other microbes such as

VAM fungi (most mycorrhizal associations are promiscuous). Promiscuous plants are less likely to respond to inoculation than those with specific symbiotic partners.

promontory *n.* (Latin *promontorium*, a forward mountain) **1.** a high point, especially one that projects outward over water or a low area. **2.** a prominent anatomical part.

promycelium *n.* (Greek *pro*, forward + *mykēs*, fungus + *ēlos*, nail or wart + *ium*, object) a short filament of mycelium that bears sporidia.

prong *n.* (Anglo-Saxon *pronge*, pointed instrument or pain) **1.** a tine, as a long, pointed part of a fork. **2.** a pointed projection, as a prong of an antler. **3.** a fork or branch, as a division of a stream. **4.** a sturdy projection used to hold something, as a prong that grips a gem in a piece of jewelry.

pronghorn *n.* a species of hollow-horned, ruminant antelope (*Antilocapra americana*) about 3 ft (0.9 m) tall at the shoulder that is native to western U.S.A. and Mexico, inhabiting the Great Plains in summer and Rocky Mountains in winter.

propagate *v.* (Latin *propagatus*, pegged or enlarged) **1.** to reproduce or cause to reproduce by either sexual or asexual means. **2.** to spread from one individual to another, as a disease propagates. **3.** to pass on to succeeding generations as an inherited characteristic propagates. **4.** to move outward from a source, as sound propagates.

propagule or **propagulum** *n.* (Latin *propagulum*, small shoot) **1.** a disseminule, a small part of a plant or microbe that is able to reproduce the species; e.g., a seed, spore, bud, fragment of fungal hyphae, or plant cutting. **2.** the minimum number of individuals of a species necessary to colonize a habitable island successfully.

propane *n.* a highly flammable hydrocarbon (C_3H_8) that boils at −44°C. A gas obtained from petroleum or natural gas and used as a liquefied petroleum fuel and as a raw material to synthesize certain organic products.

propylene glycol *n.* a colorless, viscous liquid ($C_3H_8O_2$) that can be mixed with water for use as an antifreeze. See ethylene glycol.

prosimian *n.* (Latin *pro*, before + *simia*, ape) any of the lower primates (order Primate, suborder Strepsirhini), including lemurs, lorises, and tarsiers, species that developed earlier and are less evolved than the higher primates (monkeys, apes, and humans).

proso millet *n.* an annual grass (*Panicum miliaceum*, family Poaceae) grown for its seed that is used to feed livestock, poultry, and wild birds or is sometimes used to make porridge or flatbread,

especially in Europe; native to Asia. Also called broom corn millet, broom millet, or hog millet.

protactinium (Pa) *n.* (Greek *prōtos*, first + *aktinos*, ray or beam + *ium*, element) element 91, atomic weight 231.0359; a radioactive heavy metal with very high melting (2840°F or 1560°C) and boiling points. Previously called protoactinium.

protean *adj.* (Greek *Proteus*, a minor sea god) **1.** extremely variable and changeable in form; e.g., an amoeba. **2.** highly versatile, as an actor that can play many different roles. **3.** of Proteus, a mythological sea god that was supposedly able to change form and prophesy.

protein *n.* (Greek *prōtos*, first + *inos*, pertaining to) a polymer of amino acids joined by peptide linkages into one or more straight-chain polypeptides totaling approximately 100 to 3000 amino acid units (with or without additional components such as complex carbohydrates, polycyclic compounds, or metal ions) giving a molecular weight of about 10,000 to 300,000. Proteins serve as structural components of cells, enzymes, etc. Each protein molecule has a distinctive shape that enables it to perform a specific function and to recognize or be recognized by other molecules (e.g., antigens).

proteinase *n.* (Greek *prōtos*, first + *inos*, pertaining to + *ase*, enzyme) an enzyme that catalyzes the cleavage of large protein molecules into smaller polymers.

proteolytic *adj.* (Greek *prōtos*, first + *lysis*, loosening or dissolving) tending to break down proteins and flesh that contains protein.

protist *n.* (Greek *prōtistos*, the very first) any single-celled eukaryotic organism (kingdom Protista), including algae, yeasts, protozoa, etc.

protocooperation *n.* (Greek *prōtos*, first + Latin *cooperatio*, working together) an optional interaction between two organisms. Symbiosis that is beneficial to both organisms when it occurs but is not essential for the survival of either organism.

protomer *n.* (Greek *prōtos*, first + *meros*, part) either or any of two or more identical subunits in a protein; e.g., hemoglobin is a dimer containing two large, complex protomers. The combined parts are called an oligomer.

protoplasm *n.* (German *protoplasma*, first formed) the essential living material in all living cells of microbes, plants, animals, and people; a colloidal system including water, proteins, lipids, carbohydrates, and salts that constitute the cytoplasm and nucleus.

protozoan (pl. **protozoans** or **protozoa**) *n.* (Latin *protozoa*, protozoa) a single-celled, generally aquatic, nonphotosynthetic, eukaryotic organism of kingdom Protista (formerly classified in kingdom Animalia) that is capable of locomotion and can exist as an individual or as part of a colony. It is generally microscopic in size but is larger than a bacterium and has a changeable external form. It can engulf and consume bacteria for food. See amoeba.

provender *n.* (Latin *praebenda*, state support to a private person) **1.** dry food or feed such as hay and grain fed to livestock. **2.** food.

province *n.* (Latin *provincia*, official charge) **1.** an administrative division, territory, or colony; e.g., a territory outside Rome that was part of the Roman Empire. **2.** an administrative division of certain countries (e.g., Canada), especially the parts away from the capitol and other large cities. **3.** an ecological area smaller than a region wherein the organisms and habitat are in harmony. **4.** a geologic area where a particular mineral can be mined. **5.** a field or branch of learning, as the province of chemistry. **6.** the sphere of authority associated with a certain position, office, or training as the province of the County Assessor or of a teacher.

provincial *adj.* (Latin *provincialis*, pertaining to an official charge) **1.** local; pertaining to a particular province, as a provincial government. **2.** characteristic of the customs, manners, or styles of a particular province, especially if rustic and rigid or unsophisticated. **3.** narrow-minded; not considering matters outside of one's province, as a provincial treatment of a topic. *n.* **4.** a person who lives in or comes from a province. **5.** an unsophisticated or narrow-minded person. **6.** an ecclesiastical official who presides over an order in a province.

provitamin *n.* (Latin *pro*, before + *vita*, life + *amine*, made of ammonia) a precursor of a vitamin. For example, carotene is a provitamin that is changed to vitamin A by the liver of a human or animal, and calciferol is a provitamin that is changed to vitamin D_2 by ultraviolet radiation.

Proxima Centauri *n.* the nearest star to our solar system (4.2 light years away); a red dwarf star that is part of a triple-star system called Alpa and Proxima Centauri, the brightest object in the constellation Centaurus and the third brightest star in the heavens as seen from Earth.

proximal *adj.* (Latin *proximus*, nearest) nearby; near the center or near the point of attachment; opposite of distal.

prune *n.* (Latin *prunum*, plum) **1.** a dried plum, or a plum suitable for drying. **2.** (slang) a person who behaves in a silly or contrary manner. *v.* **3.** to trim away excess material, such as long or unwanted branches of a vine or tree. **4.** to remove or dispose of any excess thing or amount.

pruning *n.* **1.** the process of removing excess branches, items, or things. Trees, shrubs, bushes, and vines are often pruned to limit their size, control their shape, control fruiting, or remove weak or dead parts. **2.** the material that a falcon preens from its feathers.

pruning hook *n.* a tool with a long handle and either a curved blade or shears with one curved and one straight blade. It is used to prune small branches from trees and other plants.

Psamment *n.* (Greek *psammos*, sand + *ent*, from recent) a sandy Entisol in Soil Taxonomy; a droughty soil with little or no profile differentiation and with sand or loamy sand texture, at least throughout the subsoil.

psammophyte *n.* (Greek *psammos*, sand + *phyton*, plant) a plant that grows with its roots in sand or very sandy soil.

pseudoaquatic *adj.* (Greek *pseudēs*, false + Latin *aquaticus*, of water) indigenous to wetlands but not truly aquatic.

pseudomonas *n.* (Latin *pseudomonas*, false monad) any of several rod-shaped bacteria (*Pseudomonas* sp.). Many species are common in soil. Some cause root rot of various plants, whereas others cause blights or spots on plant leaves.

pseudomorph *n.* (Greek *pseudēs*, false + *morphē*, form) **1.** anything with a false form that makes it deceptive and hard to identify or classify. **2.** a mineral that has the crystal shape of another mineral that it has replaced by recrystallization.

pseudoparenchyma *n.* (Greek *pseudēs*, false + *parenchyma*, something poured in beside) a compact mass of intertwined fungal hyphae that resembles plant tissue.

pseudopod or **pseudopodium** *n.* (Latin *pseudopodium*, false little foot) **1.** a retractable temporary projection of protoplasm that extends from the body of certain protozoa to serve for locomotion or for ingestion of food. **2.** the caudal (rear) end of a rotifer. **3.** a stalklike pedicel that supports sporangia on certain mosses.

pseudorabies *n.* (Greek *pseudēs*, false + Latin *rabies*, rage or madness) a highly contagious, deadly disease of swine, cattle, dogs, cats, rats, and most other mammals, including humans, that is caused by a herpes virus. Symptoms resemble those of rabies. See Aujeszky's disease.

pseudoscorpion *n.* (Greek *pseudēs*, false + *skorpios*, scorpion) any of several small arachnids (order Chelonethida) with large maxillary palpi that serve as nipping claws. They resemble miniature scorpions, but lack the arching tail, and prey on many small insects.

pseudotuberculosis *n.* (Greek *pseudēs*, false + Latin *tuberculosis*, a tuberlike swelling disease) an acute disease caused by any organism other than *Mycobacterium tuberculosis* (especially *Yersinia pseudotuberculosis*) that forms nodules similar to those produced by tuberculosis. It affects rodents, birds, humans, etc., and is sometimes fatal.

pseudovary *n.* (Greek *pseudēs*, false + Latin *ovarium*, egg producer) the structure in certain viviparous aphids that produces pseudova. It lacks certain female reproductive parts that are present in true ovaries.

pseudovum *n.* (pl. **pseudova**) (Greek *pseudēs*, false + Latin *ovum*, egg) an egglike body that produces the young of viviparous aphids without impregnation by sperm.

psilocybin *n.* (Greek *psilós*, bare + *kybē*, head + *in*, chemical) a hallucinogen ($C_{12}H_{17}N_2O_4P$) isolated from mushrooms (*Psilocybe mexicana*).

psittacosis *n.* (Greek *psittakos*, parrot + *osis*, disease) a viral (*Chlamydia psittaci*) disease of parrots and other birds, including pigeons, parakeets, ducks, turkeys, pheasants, and probably chickens. Also called ornithosis or parrot fever. It can be transmitted to humans by infected birds and cause pneumonia, hepatitis, myocarditis, delirium, and/or coma.

psychrometer *n.* (Greek *psychros*, cold + *metron*, measure) an instrument that measures relative humidity by comparing the readings of a wet-bulb thermometer (one that is cooled to the dew point by evaporation) to that of a dry-bulb thermometer and reading the appropriate value from a table. Also called hygrometer.

psyllium *n.* **1.** a branching annual herb (*Plantago psyllium*), native to the eastern Mediterranean, with hairy, sticky, linear leaves up to about 3 in. (8 cm) long and dense spikes of small flowers. Also called fleawort. **2.** the seeds of the plant, used as a mild laxative that swells into a gelatinous mass when wet.

ptarmigan *n.* (Scotch *tarmachan*, ptarmigan) any of several grouses (*Lagopus* sp.) with feathered feet; native to mountainous and cold northern regions. Adults are about 10 to 13 in. (25 to 33 cm) long, plump bodied, and have mostly brown summer colors that change to white in winter.

pterodactyl or **pterosaur** *n.* (Latin *pterodactylus*, winged finger) any of several winged reptiles (*Pterodactylus* sp.) with wingspans up to 20 ft (6 m) that became extinct near the end of the Mesozoic Era. Their wings were sheets of skin stretched along each side of the body from a very long finger of each front limb to the corresponding hind limb.

pteropod *n.* (Greek *pteron*, wing + *pod*, foot) any of several genera of mollusks (group Pteropoda) with a winglike lobe extending from each foot that is useful for swimming. They either lack a shell or have a very thin, fragile shell.

ptomaine *n.* (Greek *ptōma*, corpse + *ine*, chemical) any foul-smelling alkaloid amine produced by bacterial putrefaction of proteins in improperly refrigerated foods such as fish and chicken. Some of them are toxins that cause ptomaine poisoning.

pubescence *n.* (Latin *pubescens*, becoming hairy) **1.** a covering of fine, short hairs that insulates a leaf or other plant or animal tissue. **2.** the attainment of sexual maturity.

public domain *n.* **1.** belonging to the public rather than to an individual; e.g., an invention whose patent has expired, a literary work whose copyright has expired, or a work that was never so protected. **2.** government-owned land that is accessible to the public; e.g., a national forest.

public land *n.* land that is held by the government for the people, especially such land that is open to settlers.

public sector *n.* an activity, affair, or property that belongs to everyone and is under governmental control rather than that of an individual.

puddle *n.* (German *pudel*, puddle) **1.** a shallow pool of water, especially stagnant rainwater on the ground. **2.** a small stagnant pool of any liquid. **3.** a mixture of water thickened with clay, soil, or other fine material. *v.* **4.** to form puddles of water or other liquid. **5.** to mix soil or other fine material with water. **6.** to work wet soil into a structureless mass. **7.** to work molten iron into a plastic mass called a puddle ball that is then passed through puddle rolls that shape it into a puddle bar.

pueblo *n.* (Spanish *pueblo*, town or people) **1.** a communal village composed of a group of flat-roofed stone or adobe structures that house a group of families. **2.** any Native American village or any village or town in Spanish-speaking areas of America. **3.** the group of Native Americans that lives in a pueblo. **4.** a town or township in the Philippines.

puff adder *n.* **1.** a large African viper (*Bitis arietans*) that hisses and inflates its body when disturbed. **2.** a related, similar viper (*B. inornata*) native to southern Africa.

puffball *n.* **1.** a round, fungal fruiting body with a papery brown skin filled with white flesh and brown spores. Puffballs resemble mushrooms and emit a cloud of spores when broken open. **2.** any of several fungi (*Lycoperdon* sp.) that produce puffball fruiting bodies.

puffer fish *n.* **1.** a small bony marine fish (family Tetraodontidae) that can blow itself up like a balloon by swallowing water or air. Some species contain tetrodotoxin. **2.** other similar species.

puffin *n.* any of several stocky marine diving birds (*Fratercula* sp. and *Lunda* sp.) about 11 to 12 in. (28 to 30 cm) long with short necks and large, triangular, grooved beaks with red markings on the beaks and legs; black or dark brown on the back but white on the front except for the tufted puffin (*F. cirrhata*) whose front is darker, especially in winter.

pug *n.* **1.** a breed of small, stocky short-haired dogs that carry their tail in a curl. Pugs have a smooth black or silver coat and a wrinkled face with a very short nose. **2.** wet clay; e.g., that prepared for packing into a brick mold. **3.** an animal footprint, especially that of a game animal. **4.** (slang) a boxer or other fistfighter. *v.* **5.** to make a paste or plastic ball of clay and water, or to use such material to fill holes. **6.** to coat a floor or wall with mortar to so it will dry smooth and deaden sound. **7.** to track an animal by following its footprints.

pulp *n.* (Latin *pulpa*, flesh) **1.** the fleshy part between the rind and the seed of any fruit. **2.** the soft interior of a plant stem. **3.** the sensitive inner part of a tooth where the blood circulates and the nerve is located. **4.** any soft, somewhat coherent mass. For example, flesh that has been injured by pounding or wood that has been chemically treated and pounded to separate its fibers so it can be made into paper. **5.** a magazine or book printed on poor-quality paper. **6.** pulverized ore mixed with water in a mining operation.

pulque *n.* (Nahuatl *pulque*, pulque) a fermented alcoholic beverage made from the sap of certain species of agave, especially in Mexico. The sap may also be distilled to make mescal or tequila.

pulse *n.* (Latin *pulsus*, a beat) **1.** the rhythmic increase in blood pressure in an artery caused by the beating of the heart. **2.** the rate of the heartbeat, especially as felt where an artery is near the surface, as in the wrist. **3.** a large-seeded legume used for food; e.g., a pea or bean. **4.** a single brief buildup or burst of energy, radiation, or other kind of output (it may or may not be repeated). **5.** any stroke, swing, vibration, or undulation, either as a single event or as one of a rhythmic series. **6.** a marked increase in population of a particular species in an area followed by a corresponding decline.

Normal pulse rates of humans and domestic animals in pulses (heart beats) per minute	
Humans	50–100
Cats	110–300
Cattle	60–70
Chickens	200–400
Dogs	70–120
Goats	70–80
Horses	32–44
Sheep	70–80
Swine	60–80

puma *n.* See cougar.

pumice *n.* (Latin *pumex*, pumice) a porous, lightweight, light-colored volcanic glass with a composition similar to that of rhyolite; commonly ejected from a volcano in chunks a few centimeters across. It breaks easily into tiny particles that are used as an abrasive. See scoria, tephra.

pungent *adj.* (Latin *pungens*, pricking) **1.** strongly acrid, as a pungent taste or smell. **2.** sharp and penetrating; painful to the nose or palate. **3.** very clever and expressive; stimulating, as a pungent speech. **4.** ending with a sharp, hard point, as the spine at the end of some leaves.

punk *n.* **1.** a substance that will smolder without flaming, often made into a stick to light fuses. **2.** dry wood that has decayed into a soft mass suitable for tinder. **3.** a similar spongy mass of fungal origin. **4.** (slang) a person of little skill or value, especially a young male. **5.** a young hoodlum or male prostitute. *adj.* **6.** of poor quality; rotten or decayed.

pup *n.* (Latin *pupa*, doll) **1.** a young dog (puppy), fox, wolf, rat, seal, whale, etc. **2.** a small plant growing as an offshoot from a larger plant. **3.** to whelp, to give birth to a litter of pups.

puppy *n.* (French *poupée*, doll or puppet) a young dog, especially one less than one year old.

pure line *n.* a breed of animals or strain of organisms that has been inbred selectively to make it nearly uniform genetically.

purine *n.* (Latin *purus*, pure + *uricum*, uric acid) a colorless organic compound ($C_5H_4N_4$) with a double-ring structure that is a substrate in the formation of nucleic acids, caffeine, theobromine, theophylline, etc. It combines with adenine to form adenosine in ADP and ATP.

purslane *n.* See portulaca.

purtenance *n.* (French *partenence*, viscera) the internal organs of an animal, including the heart, liver, and lungs.

pussy willow *n.* a common small willow (*Salix discolor* and related species, family Salicaceae) with small, fuzzy, gray catkins arranged alternately on stems. Native to North America, it grows naturally in wet areas and is grown as an ornamental or to provide branches for decorations.

pustule *n.* (Latin *pustula*, pimple or blister) **1.** a bump on the skin; caused by an infection and an accumulation of pus. **2.** any pimple, blister, or other bump on the skin of an animal or the surface of a plant; caused by infection or injury.

putrefaction *n.* (Latin *putrefactio*, making putrid) **1.** microbial decomposition of organic materials, generally accompanied by a disagreeable odor; the process of rotting. **2.** the state or condition of rotted organic materials.

putrescible *adj.* (Latin *putrescere*, to rot + *ibilis*, able) likely to rot, as an organic substance that rots quickly and produces unpleasant odors.

putrid *adj.* (Latin *putridus*, rotten) **1.** decomposing; rotting; breaking down and producing foul odors. **2.** accompanying or produced by putrefaction. **3.** smelling like rotten flesh. **4.** of very poor quality; rotten. **5.** corrupt and evil, as a putrid government.

Pygmy *n.* (Latin *pygmaeus*, Pygmy) any of several modern, historical, or legendary African and Asian races of human dwarfs.

pyrethrin *n.* (Greek *pyrethron*, feverfew) a natural insecticide that is one of the safest available, suited for indoor use but not stable enough for outdoor use. Either of two liquid esters ($C_{21}H_{28}O_3$) or ($C_{22}H_{28}O_5$) derived from chrysanthemums and related flowers (especially *Chrysanthemum cinerariaefolium* and *C. coccineum*). Pyrethrin stuns insects quickly by paralyzing the nervous system but does not always kill them.

Pyrgota fly *n.* a parasitic fly that lays an egg in the body of a May beetle (or June beetle) in flight. The larva develops within a few days, gradually consuming the body of the beetle. The stricken beetle burrows into the soil and dies, whereas the larva pupates and remains in the dead body of the beetle until it emerges as an adult fly the next spring.

pyrheliometer *n.* (Greek *pyros*, fire + *helios*, sun + *metron*, measure) an instrument that measures the intensity of solar radiation.

pyric *n.* (French *pyrique*, pyric) related to, caused by, or associated with burning; e.g., the influence of fire on vegetation, soil, and other parts of the environment.

pyrimidine *n.* (German *pyrimidin*, pyrimidine) a colorless organic compound ($C_4H_4N_2$) with a ring structure that serves as a building block for organic bases (especially cytosine, thymine, and uracil) that are constituents of nucleic acids.

pyrite *n.* (Greek *pyritēs*, firey) iron disulfide (FeS_2), a brassy yellow mineral with a metallic luster that resembles gold but is harder than gold. It is sometimes burned to produce sulfuric acid. Also called pyrites, iron pyrite, or fool's gold.

pyrobitumen *n.* (Latin *pyro*, fire + *bitūmen*, asphalt) any dark, solid, naturally occurring hydrocarbon, including peat, coal, and black shale.

pyroclastic *adj.* (Greek *pyro*, fire + *klastos*, broken in pieces) formed from volcanic debris that is either blown through the air or flows across the ground as a fiery cloud of loose ash.

pyrolysis *n.* (Greek *pyro*, fire + *lysis*, loosing) **1.** chemical decomposition induced by heat; e.g., the decomposition of calcium carbonate into calcium oxide and carbon dioxide:

$$CaCO_3 + heat \rightarrow CaO + CO_2$$

2. decomposition of organic substances by heat in a low-oxygen environment; e.g., to make charcoal from wood, bones, or coal or to produce methane and other fuel gases from organic wastes.

pyrometer *n.* (Greek *pyro*, fire + *metron*, measure) an instrument for measuring high temperatures based on reading electromagnetic radiation or changes in electrical conductivity.

pyrophoric *adj.* (Greek *pyro*, fire + *phoros*, bearing + *ikos*, like) a chemical that self-ignites at room temperature; e.g., elemental phosphorus (P) or lithium hydride (LiH).

pyrophyte *n.* (Greek *pyro*, fire + *phyton*, plant) a plant species more capable than most of surviving a fire; e.g., longleaf pine, jack pine, or lodgepole pine.

pyrosphere *n.* (Greek *pyro*, fire + *sphaira*, ball) the zone of the Earth below the lithosphere. Some parts are hot enough to melt and form magma. Also called barysphere, centrosphere, or magmosphere.

pyroxine *n.* (Greek *pyro*, fire + *xenos*, alien) an inosilicate mineral with a single-chain structure that forms crystals with a fibrous texture having 87° angles between faces of the fibers; e.g., augite, diopside, and enstatite. Compare amphibole.

python *n.* (Greek *pythōn*, python) any of several large nonpoisonous snakes (*Python* sp., subfamily Pythoninae) of Asia, Africa, and Australia that wrap around their prey and crush it; a boa constrictor up to 20 ft (6 m) long.

Q

qanat *n.* a tunnel with just enough slope to transport underground water to the surface by gravity flow for irrigation and other uses. Also spelled kanat, ghanat, karez, or kareze.

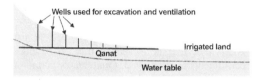

A qanat carries underground water to the surface.

qat *n.* (Arabic *qät*, qat) an evergreen shrub (*Catha edulis*) grown in Africa and Arabia, especially in Yemen, whose leaves are chewed as a stimulant and used to prepare a beverage that has narcotic properties similar to those of coca. Also spelled kat or khat.

quack grass *n.* a weedy perennial grass (*Agropyron repens*, family Poaceae) of European origin that spreads rapidly by rhizomes. It has value for erosion control and as forage but is a weed that spreads aggressively, it is hard to kill, and both growing quack grass and its residues have allelopathic effects on many crop plants (e.g., alfalfa, clover, and barley). Also known as couch grass.

quad *n.* **1.** a quadrat, a filler used to leave a blank space in a typeset line. **2.** short for quadrangle, quadraphonic sound, or quadruplet. **3.** a unit of energy equal to 1 quadrillion (10^{15}) Btu; the amount produced by approximately 1 trillion (10^{12}) cubic feet of natural gas, 182 million barrels of oil, 42 million tons of coal, or 293 billion kilowatt hours of electricity. *adj.* **4.** short for quadraphonic or quadruple; e.g., quad occupancy provides quarters for four persons.

quadrat *n.* (Latin *quadratus*, squared) **1.** a piece of metal used in typesetting to leave a blank space in a line equal in length to an en-pica (–) or 1, 2, or 3 em-pica (—). **2.** a small research plot used for an ecological study of plant and animal life in a microenvironment, often 1 m^2 in area.

quadrennial *adj.* (Latin *quadriennium*, four years) **1.** occurring once every four years. **2.** lasting four years. *n.* **3.** an event that occurs once every four years; e.g., a celebration.

quadriplegia *n.* (Latin *quattuor*, four + *plēgia*, paralysis) paralysis of all four limbs (often including the entire body from the neck down).

quadrumane *n.* (Latin *quadrumanus*, four-handed) an animal such as a monkey whose four feet are all adapted for use as hands that can grip.

quadruped *n.* (Latin *quadrupedis*, four-footed) an animal (especially a mammal) that walks on four feet.

quagmire *n.* **1.** a marsh, bog, or swamp; wetland that will not support much weight and that shakes when one walks on it. **2.** a situation that makes escape difficult, as a swampy morass or a financial predicament.

quail *n.* (French *caille*, quail) **1.** any of several small, upland, gallinaceous game birds (*Coturnix* sp.). **2.** any of several similar species, especially the bobwhite (*Colinus* sp.) *v.* **3.** to show fear and draw back. **4.** to subdue; to cause another to draw back in fear.

quake *v.* (German *quackeln*, to shake) **1.** to shake or tremble from cold, anger, fear, etc. **2.** to shake from an external force, such as an earthquake. *n.* **3.** an episode of shaking, especially an earthquake.

quaking aspen *n.* a widely distributed, short-lived deciduous tree (*Populus tremuloides*, family Salicaceae), native to northern U.S.A. and Canada, named for its roundish leaves that shake in a slight breeze, showing both their shiny green top surface and their dull green underside. The bark of young trees is smooth and white to yellowish green but becomes darker and furrowed with age. The wood, which is soft and of poor quality, is used mostly for pulpwood, boxes, excelsior, and matchsticks. Also called trembling poplar, golden aspen, or mountain aspen.

quantum (pl. **quanta**) *n.* (Latin *quantus*, quantity) **1.** amount; how much or how many. **2.** a specific amount or portion. **3.** a bulk amount, especially a large quantity. **4.** the smallest unit of energy that can be emitted, equal to Planck's constant (h) multiplied by the radiation frequency. **5.** the smallest unit of some other physical phenomenon, such as angular momentum. *adj.* **6.** large and sudden, as a quantum change in output.

quantum evolution *n.* a situation where rapid environmental change requires organisms to adapt or perish.

quantum jump *n.* **1.** a change in some part of a system (as envisioned by quantum mechanics) from one energy level to another; e.g., the movement of an electron from one orbital position to another, accompanied by the absorption or release of a specific wavelength of light or other electromagnetic energy. **2.** any sudden significant change (usually an increase in value or status).

quantum mechanics *n.* a system of theoretical physics, based on the quantum theory, that describes the behavior of particles and waves at the atomic scale.

quantum number *n.* a value equal to an integer (or an integer minus 0.5) that evaluates a discrete state in a quantum mechanics system; e.g., the energy level of an electron in an atom.

quantum theory *n.* a proposal made in 1900 by Max Planck, a German physicist, to explain the behavior of light by assuming that it is emitted in very small discrete units (quanta) that cannot be subdivided. The theory was expanded to include both the emission and absorption of all forms of energy. It became the basis for modern theoretical physics known as quantum mechanics.

quarantine *n.* (Italian *quarantina*, 40 days) **1.** a 40-day period of mourning and reprieve for a widow under English law. Her husband's creditors could not evict her or take property from her during this time. **2.** a waiting period (originally 40 days) for a vessel to be isolated in port if it is suspected of carrying a contagious disease. **3.** the place where such a ship waits at anchor. **4.** a system used by a government to prevent entry of diseased plants, animals, or people at their ports or frontiers. **5.** the place where such measures are enforced or the enforcement agency. **6.** isolation and restriction of persons, plants, or animals suspected of having a contagious disease. **7.** the place of isolation where such persons, plants, or animals are kept; e.g., a special section in a hospital. **8.** isolation of an individual or group imposed as punishment for social, political, or economic reasons. *v.* **9.** to turn away or send one or more persons, plants, or animals into isolation for health, political, or social reasons.

quarry *n.* (Latin *quadrāria*, where stone is squared) **1.** a pit where stone is mined. **2.** a source that supplies something abundantly. **3.** anything being hunted, especially a small animal hunted by a hawk. *v.* **4.** to mine stone from a quarry. **5.** to dig a large pit to be used as a quarry.

quart *n.* (Latin *quartus*, quarter) **1.** a liquid measure equal to a quarter of a gallon, 57.749 in.^{3} (0.946 L) in the U.S.A., or 69.355 in.^{3} (1.1365 L) in Canada and Great Britain. **2.** a dry measure equal to an eighth of a peck, or 67.201 in.^{3} (1.101 L). **3.** a container with a 1-quart capacity.

quarter section *n.* one-fourth of a section (square mile) of land; 160 acres. Usually a half-mile square described as the NE, NW, SW, of SE quarter of a specified section.

quartz *n.* (German *quarz*, quartz) a very abundant mineral, silicon dioxide (SiO_2) crystallized in the hexagonal system. It is an essential component of all acid igneous rocks and is also abundant in sandstone, quartzite, and many other rocks and in soils. It has a hardness of 7 on Mohs' scale and is either colorless and transparent or made milky or colored by impurities. Some colored varieties are semiprecious stones. Quartz crystals are piezoelectric, making them useful in strain gauges and for controlling the frequencies of radio transmitters.

quartzite *n.* (German *quarz*, quartz + *ite*, mineral or rock) a metamorphic rock formed by subjecting sandstone to enough heat and pressure to recrystallize its minerals without melting them.

quaternary *adj.* (Latin *quaternarius*, four at a time) **1.** a set of four or in sets of four. **2.** having four constituents in an alloy. **3. Quaternary** pertaining to the latter part of the Cenozoic Era, from the beginning of the Pleistocene up to the present time; e.g., Quaternary rock formations. *n.* **4.** the number four or a set of four. **5.** the **Quaternary** Period of geologic time.

quaver *v.* (a combination of quake and waver) **1.** to shake or tremble, as tree leaves that quaver in the breeze. **2.** to vary rapidly in pitch, as a voice that quavers with fear or that trills in singing. *n.* **3.** a variable tone produced by quavering.

queen *n.* (Anglo-Saxon *cwen*, wife or woman) **1.** either the wife of a king or a female monarch ruling a kingdom. **2.** a woman who excels in her field, as a beauty queen or a drag queen. **3.** a reproductive female ant, bee, termite, or wasp; typically much larger than the females workers that do not lay eggs. **4.** the most powerful piece in a game of chess, able to move forward, backward, sideways, or diagonally as far as the line is open. **5.** a playing card with a representation of a queen (normally ranking below the king and above the jack).

Queen Anne's lace *n.* an annual or biennial weed (*Daucus carota*, family Apiaceae), native to Europe and Asia, with an attractive broad umbel of white blossoms suitable for use in bouquets. It imparts an objectionable odor to milk if eaten by cows. Also known as wild carrot.

queen cell *n.* an atypically large cell where a queen bee develops, usually at the bottom of a comb.

queen fish *n.* an edible, silvery and bluish fish (*Seriphus politus*) that inhabits shallow waters off the coast of California.

queen pheromone *n.* a substance secreted by a queen bee, queen ant, queen termite, or queen wasp. Its odor tells others in the hive or colony that a queen is present; without this odor, the workers will begin work to produce a new queen.

quelea *n.* (Latin *qualea*, quail) a sparrowlike African finch (*Quelea quelea*) that flies in large flocks that can strip grain fields of sorghum, rice, millet, etc., bare. Entrepreneurs in Zimbabwe have found that they can attract the flocks to roost at night in blocks of tall grass. They can then capture thousands of the birds and export them to Hong Kong as tiny poultry.

quench *v.* (Anglo-Saxon *cwencan*, to quench) **1.** to extinguish, as a fire. **2.** to repress, as to quench one's enthusiasm. **3.** to fulfill or satisfy, as to quench a thirst or to quench the capacity of a soil to fix a fertilizer element (especially phosphorus as phosphate). **4.** to cool suddenly, e.g., by immersing hot metal in water.

quetsch *n.* (German *quetsche*, plum) a variety of plum, or a dry, white, unaged brandy made from such plums.

quetzal *n.* (Nahuatl *quetzalli*, quetzal tail feathers) **1.** any of several large, beautiful birds (*Pharomachrus* sp.) with golden green back feathers and scarlet feathers in front, a crest, and a long tail, especially the resplendent quetzal (*P. mocino*), Guatemala's national bird. Sacred to the Aztecs, it is now an endangered species. **2.** the monetary unit of Guatemala.

quick *adj.* (Anglo-Saxon *cwicu*, living) **1.** alive. **2.** done rapidly and/or promptly, as a quick response. **3.** brief; completed within a short time, as a quick snack. **4.** able to comprehend rapidly, as a quick eye, a quick mind, or a quick student. **5.** moving or able to move suddenly and swiftly, as a quick flip of the hand, or a chipmunk is quick. **6.** responding without delay, as a quick worker or a quick wit. **7.** easily provoked, as a quick temper. **8.** sharp or sudden, as a quick bend in the road or a quick change of plans. *n.* **9.** living people, used especially as the quick and the dead. **10.** soft, tender flesh, especially that under the fingernails and toenails. **11.** the core of one's being, as sarcasm that cuts to the quick. **12.** a line of living plants (or a plant in such a line), especially in Britain for hawthorn shrubs forming a hedge.

quick clay *n.* fine-textured material that is so weak when wet that it cannot support a load and may flow even across a land surface with only 1% slope. Also called sensitive clay. See thixotropy.

quicken *v.* **1.** to cause to move more rapidly, as to quicken the pace. **2.** to arouse, as to quicken the imagination. **3.** to bring back to life, as rain will quicken dormant vegetation. **4.** to show signs of life, usually by movement of a fetus in the womb.

quicksand *n.* a condition that exists when upward-flowing water separates sand particles enough to keep them from interlocking. Quicksand will yield under any load, but buoyant objects can float in it (it will not pull them in).

quicklime *n.* lime (CaO). Also called burnt lime, caustic lime, or calcium oxide.

quick test *n.* a chemical test (usually for nitrogen, phosphorus, potassium, or pH) of either plant tissue or soil that can be performed in a few minutes to check the adequacy of soil fertility. It is usually used in the field, especially to check the adequacy of fertilizer already applied or to check problem spots.

quid *n.* (Anglo-Saxon *cudu*, chewed food) **1.** cud. **2.** a chewable portion of something that is not to be swallowed, especially a cut of chewing tobacco. **3.** (British slang) a pound sterling.

quince *n.* (Greek *kydōnion*, apple of Cydonia) **1.** either of two species of small trees (*Cydonia oblonga or C. sinensis*, family Rosaceae), native of Asia, that bear beautiful white or pink blossoms and edible fruit. **2.** the edible pear- or apple-shaped pome fruit with a sour, astringent taste produced by the tree. Up to 3 in. (8 cm) in diameter with green skin that turns yellow at maturity, it is used mostly for stewing or to make jams and jellies.

quinhydrone *n.* a dark green solid ($C_{12}H_{10}O_4$) that is slightly soluble in water and can be used along with a platinum wire as a quinhydrone electrode for measuring the pH of a solution. It is especially useful in tropical conditions where other pH meters are likely to fail by corrosion.

quinoa *n.* (Quecha *kinoa*, quinoa) a nutritious grain crop (*Chenopodium quinoa*) of the Incas that is still grown in the Andes, an annual, broad-leaved herb that produces clusters of small white or pink seeds that contain 12% to 19% protein with a favorable balance of amino acids (high in lysine and methionine), sold in some health-food stores and vegetarian restaurants. Also spelled quinua.

quintal *n.* (Arabic *qintār*, 100 lb) **1.** a metric unit of weight equal to 100 kg (220.46 lb avoirdupois) that is used to express yields in quintals per hectare. **2.** a hundredweight; 100 lb in the U.S.A. or 112 lb in Great Britain.

quoin *n.* (French *coin*, corner) **1.** an external corner of a building. **2.** a construction stone used at such a corner; a cornerstone. **3.** a tapered piece of wood or metal used as a wedge; e.g., to secure printer's type in a chase or as a block to prevent a barrel from rolling.

quokka *n.* (Nyungar *kwaka*, quokka) a small wallaby (*Setonix brachyurus*) native to swampy areas of southwestern Australia and nearby islands.

R

R *n.* **1.** the usual symbol for the universal gas constant. **2.** the usual symbol for a radical in a chemical structure. **3.** a symbol used to indicate electrical resistance. **4.** the rainfall erosivity factor in the universal soil-loss equation. See rainfall erosivity. **5.** restricted; a rating for motion pictures indicating that children under the age of 17 are excluded unless accompanied by an adult.

Ra *n.* **1.** the chemical symbol for the element radium. **2.** the ancient Egyptian name for their sun god. **3.** the name Thor Heyerdahl gave to a large reed boat he used in 1970 to demonstrate how ancient seafarers might have crossed the Atlantic Ocean.

rabbit *n.* (French *rabotte*, rabbit) **1.** a burrowing rodent (*Lepus coniculus*, family Leporidae) with soft fur, long ears, large front teeth, and a short tail; known for its ability to make long hops and to run fast. **2.** any of several other species of the hare genus (*Lepus*), usually distinguished from those called hares by being smaller, blind, and hairless when born. **3.** the fur of such animals, often used to make coats.

rabbitbrush *n.* any of several shrubs (*Chrysothamnus* sp., family Asteraceae) with gray branches and yellow flowers. It is native to arid regions in western U.S.A. and Mexico and usually larger than associated sagebrush; named for the shelter it gives to rabbits.

rabbit fever *n.* See tularemia.

rabbit food *n.* vegetables prepared to be eaten raw, either as carrot sticks, celery sticks, etc., or in salads.

rabbit hutch *n.* a cage or box that is raised off the floor and used for raising rabbits.

rabies *n.* (Latin *rabies*, rage) an RNA viral disease of the rhabdovirus group transmitted by the bite of an infected animal, usually a wolf, dog, cat, bat, skunk, fox, raccoon, or woodchuck. It affects the brain and nervous system and is usually fatal unless a series of prophylactic treatments is administered. All animal pets should be vaccinated against rabies. Also called hydrophobia or canine madness.

raccoon *n.* (Algonquian *aroughcun*, raccoon) **1.** a tree-climbing nocturnal carnivore (*Procyon lotor*), native to North and Central America, with a black marking across its face that looks like a mask, a pointed snout, and a bushy, ringed tail about 1 ft (30 cm) long in addition to its body, which is about 2 ft (60 cm) long. **2.** the valuable, thick, brownish gray fur of the animal; well known as coonskin caps. **3.** any of several related species found on islands off the coast of Central America.

race *n.* (Anglo-Saxon *raes*, a rush) **1.** a subdivision of a species. **2.** all people, as the human race, or a group of people with some notable shared characteristics, such as national origin, language, or skin color. **3.** a group of plants or animals that are distinguished from others of their species by some notable characteristic. **4.** a contest to see who, which animal, or which machine can go the fastest (reach a destination in the least time). *v.* **5.** to participate in a contest of speed. **6.** to work rapidly, as to race against time. **7.** to cause to run fast, as to race an animal or an engine.

raceme *n.* (Latin *racemus*, cluster of grapes) a type of inflorescence with blossoms borne on short pedicels attached at short intervals along a rachis (stem); e.g., lily of the valley.

rachis *n.* (Greek *rhachis*, spine or ridge) **1.** the central stem of a raceme. **2.** the stemlike part that supports the leaflets of a compound leaf. **3.** the central shaft of a feather; the part that supports the web. **4.** the spinal column of an animal or person.

rack *n.* (Dutch *rek*, framework) **1.** a structure of bars, wires, pegs, etc., used to hold articles, as a clothes rack or a luggage rack. **2.** a framework on a wagon or truck used to hold a load of hay, straw, or other produce, as a hayrack or a grain rack. **3.** a framework for arranging items such as type for printing or the balls for a game of pool. **4.** a toothed bar that moves when the pinion turns in a rack and pinion. **5.** the antlers of a deer or other animal. **6.** the neck portion of the carcass of a lamb, pig, or calf. **7.** a bed or bunk. **8.** an instrument of torture that pulls on a victim's limbs. **9.** violent strain, anguish, or torment. **10.** ruin; destruction (also spelled wrack). **11.** a single-foot gait of a horse. **12.** a group of clouds drifting across the sky. *v.* **13.** to torment, torture, or otherwise cause acute distress to a person; e.g., by stretching them on a rack. **14.** to exert as much physical force as possible. **15.** to concentrate on a problem or try to remember something, as to rack one's brains. **16.** to put things (e.g., pool balls) in a rack.

rad *n.* (acronym for radiation absorbed dose) a unit of energy of ionizing radiation, equal to 100 erg absorbed per gram of medium or 0.01 joule per kilogram of irradiated material; a replacement for the older term, roentgen.

radar *n.* (acronym for radio detection and ranging) **1.** a device that uses electromagnetic energy to detect and locate objects such as clouds, precipitation, aircraft, hills, buildings, vehicles, bedrock, or water tables. **2.** the electromagnetic energy used by such a device.

radar mile *n.* the time required for a radar pulse to travel from a radar to a target 1 mi away and return; 10.75 μsec for a land mile or 12.355 μsec for a nautical mile.

radiant energy *n.* heat, light, or other electromagnetic energy transmitted across space or through a medium as electro-magnetic waves rather than by conductance or convection.

Radiata *n.* (Latin *radiatus*, radiating) a grouping once used for animals that have approximate radial symmetry, including jellyfish, sea anemones, corals, starfish, sea cucumbers, etc.

radiate *v.* (Latin *radiare*, to shine) **1.** to emit rays of light, heat, or other energy. **2.** to extend outward from a center. **3.** to show energy and happiness, as people that radiate good cheer.

radiation *n.* (Latin *radiatio*, radiation) **1.** the emission of electromagnetic energy (light, heat, etc.) from a source, its transmission through a medium or across space, and its absorption by one or more receiving bodies. **2.** the energy so emitted or transferred. Also called radiant energy. **3.** energy emitted as gamma rays, X rays, neutrons, α and β particles, etc. **4.** the use of radioactive materials to treat disease (e.g., radiation treatment for cancer) or to sterilize food so it will keep longer (also called irradiation). **5.** radial placement of parts.

Sources of radiation exposure to people in the U.S.A. in 1986

Natural background (cosmic rays and radioactive elements in soil and rocks)	67.5%
Medical	30.7%
Fallout	0.6%
Occupational	0.5%
Nuclear energy	0.2%
Miscellaneous	0.5%

radiation cooling *n.* loss of radiant energy from plants, soil, or other objects, causing the temperature of the object to become cooler than the air temperature, especially on a cloudless night.

radiation fog *n.* fog that develops at night when the soil temperature is reduced by radiation cooling and the air temperature cools to the dew point (the temperature of water condensation). Also called ground fog.

radiation sickness *n.* illness caused by exposure to ionizing radiation. Depending on severity, it may include nausea, vomiting, diarrhea, headache, falling teeth and hair, cataracts, cancer, reduction in red and white blood cells, and sterilization, or even result in death.

radical *n.* (Latin *radicalis*, having roots) **1.** a root part of something. **2.** a group of atoms that form a chemical unit, either an ion such as the ammonium radical (NH_4^+) or a branch of an organic molecule (especially one that can be changed to form other members of a family of compounds); e.g., RNH_2, where R could represent CH_3, C_2H_5, C_3H_7, etc. **3.** a mathematical quantity that serves as the root of another quantity. **4.** a person who advocates an extreme position on an issue, often with a strong, uncompromising attitude. *adj.* **5.** of fundamental importance; going to the root of the matter. **6.** extreme and often thorough and permeating, as a radical change. **7.** favoring an extreme position on an economic, political, religious, or social issue. **8.** inherent, as a radical defect in character. **9.** growing from the root or the base of the stem of a plant.

radicle *n.* (Latin *radicula*, small root) **1.** the first root of a new plant; the lower part of the developing embryo that develops into the first root. Also called caudicle. **2.** any of the tiniest branches of a nerve or vein in an animal body.

radioactive *adj.* (Latin *radius*, ray + *activus*, active) able to emit radiation from the nuclei of atoms as α particles (helium nuclei), β particles (electrons), or γ rays (electromagnetic radiation).

radioactive dating *n.* the process of measuring the age of materials by the relative proportions of a radioactive isotope and its decay product present in a material; e.g., radiocarbon dating, potassium-argon dating, or uranium-lead dating. Also called radiometric dating.

radioactive element *n.* an element that has one or more isotopes that spontaneously emit charged particles from the nuclei of its atoms and thus change into another element. All isotopes of heavy elements such as uranium are radioactive, whereas lighter elements have both stable and radioactive isotopes (e.g., ^{12}C is stable but ^{14}C is radioactive). Many radioactive isotopes that can be created artificially decay so rapidly that they are not found in nature. See half-life.

radioactive tracer *n.* an isotope of an element that undergoes radioactive decay at a moderate rate (fast enough to be measured, but slow enough to last through an experiment) and can be used as a label to follow the course of chemical reactions and processes that usually involve a more stable form of the element; e.g., ^{14}C, ^{22}Na, ^{36}S, ^{45}Ca, and ^{131}I. Also called radiotracer.

radioactive waste *n.* radioactive by-products of nuclear processes, especially those resulting from the operation of a nuclear reactor or from the production of fuel for such reactors; also including spent radioactive materials from military, medical, and research applications. It is often divided into high-level radioactive waste that is highly radioactive and extremely hazardous to handle and low-level radioactive waste that poses a much lower hazard.

radio astronomy *n.* the branch of astronomy that studies the universe by using a radio telescope to detect radio waves rather than light waves.

radiocarbon *n.* **1.** a radioactive isotope of carbon, ^{14}C, with a half-life of 5730 years. **2.** any isotope of carbon other than the stable isotopes, ^{12}C or ^{13}C; i.e., ^{10}C, ^{11}C, ^{14}C, ^{15}C, or ^{16}C.

radiocarbon dating *n.* a technique for measuring the age of carbon-bearing material derived from plants or animals within the last 70,000 years. The system is based on measuring the radioactivity of ^{14}C in the object studied for comparison to the radioactivity of ^{14}C in the atmosphere. (Living plants and animals constantly absorb ^{14}C from the atmosphere or from their food, but dead plants and animals do not replenish their supply.) The half-life of ^{14}C is 5730 ± 30 years.

radioecology *n.* (Latin *radius*, ray + Greek *oikos*, house + *logos*, word) the study of the effects of radiation on plants, animals, and people, especially those effects resulting from radiation and radioisotopes that represent human sources rather than natural sources.

radioisotope *n.* (Latin *radius*, ray + Greek *isos*, equal + *topos*, place) a radionuclide, especially one that is produced artificially.

radionuclide *n.* (Latin *radius*, ray + *nucleo*, nucleus + Greek *eidos*, shape) any radioactive nucleus; a radioactive isotope of any element.

radiosonde *n.* (French *radio*, radio + *sonde*, depth sounder) an instrument carried aloft by a balloon to collect and transmit meteorological data (e.g., temperature, pressure, and humidity).

radio telescope *n.* an antenna that gathers radio waves and focuses them on a receiver; usually either a large parabolic dish or a dipolar antenna used to gather radio waves from celestial sources.

radio wave *n.* a form of electromagnetic radiation with a wavelength between 1 mm and 30,000 m (a frequency between 10 kHz and 300,000 MHz).

radium (Ra) *n.* (Latin *radius*, ray + *ium*, element) element 88, atomic weight 226 for its most stable isotope; a rare radioactive metal with a half-life of 1622 years. It occurs mostly as a component of uranium ore and is used to provide radiation for cancer treatment.

radon (Rn) *n.* element 86, atomic weight 222 for its most stable isotope, half-life 3.8 days; a colorless, odorless, naturally occurring, dense noble gas that was called niton shortly after it was discovered. It is chemically inert but poses a radiation hazard, especially as a cause of lung cancer. Produced by radioactive decay of radium in soil and rocks, it tends to accumulate in low areas that are poorly ventilated, especially in the basements of homes.

radurization *n.* treatment of food with sufficient low-level radiation to enhance its keeping quality by reducing the numbers of viable microorganisms that cause spoilage.

raffia *n.* (Malagasy *rofia*, raffia) **1.** a hard palm-leaf fiber used for making mats, baskets, hats, etc., and for cord to tie plants and other objects. **2.** raffia palm (*Raphia farinifera* and related species), a palm with pinnate leaves up to 65 ft (20 m) long (the longest leaves of any plant) that yield raffia fiber; native to Madagascar and tropical Africa.

raffinose *n.* (French *raffiner*, to refine + *ose*, sugar) a nonsweet trisaccharide ($C_{18}H_{32}O_{16}\cdot5H_2O$) that is common in sugar beets, cottonseed, and other sources. It hydrolyzes into fructose, galactose, and glucose and becomes sweet when it hydrolyzes.

raft *n.*(Norse *raptr*, log) **1.** a platform made of logs, boards and barrels, rubber sheets and tubes, or other materials designed to float on the water as a means of transport. **2.** an accumulation of fallen trees, logs, or other debris that block a stream and impede navigation. **3.** a mass of insect (e.g., cockroach) eggs that stick together. **4.** a large number or amount of something. **5.** a large slab of reinforced concrete used to spread the weight of a building over a large area of unstable soil. *v.* **5.** to travel by raft or move something on a raft. **6.** to tie together logs or other materials to make a raft.

raft ice *n.* a jumbled mass of blocks of ice that impedes the water flow in a stream.

ragweed *n.* (short for ragged weed) an annual herbaceous plant (*Ambrosia artemisifolia* and related species, family Asteraceae) native to North America and of widespread occurrence, some (e.g., giant ragweed, *A. trifida*) growing as tall as 8 ft (2.4 m). Ragweed pollen causes millions of people to suffer from hay fever.

Ragweed plant parts

rail *n.* (French *raale*, rail) **1.** any of several small wading birds (family Rallidae) with short wings and tail, large feet, a long pointed beak, and a harsh cry. Rails live in marshes, grasslands, and forests in much of the world. **2.** a horizontal bar used as a support (e.g., a handhold) or a barrier (e.g., a rail fence). **3.** a fence marking either the inner or the outer boundary of a racetrack. **4.** either of the two parallel steel bars that form a train track. **5.** railroad, as to travel by rail. *v.* **6.** to place rails in designated places. **7.** to complain bitterly, denouncing someone or something, as to rail against a policy or against one's fate.

rail fence *n.* a fence made of wooden rails either laid in a zigzag line so they support one another where they cross or supported by crossed stakes. Rail fences were popular when pioneers were clearing trees to open new fields.

rain *n.* (Anglo-Saxon *regn*, rain) **1.** water that condenses in the atmosphere and falls to land as liquid drops from about 0.02 to 0.3 in. (0.5 to 7 mm) in diameter (smaller droplets are called drizzle). **2.** rainfall, the event when such drops fall. **3.** something that is repeated many times in a short to moderate interval, as a rain of blows or a rain of criticism. *v.* **4.** to fall from the air, as it may rain on our picnic. **5.** to fall in fragments, as when ashes rain from the sky. **6.** to bestow repeatedly, as to rain favors on a favorite grandchild. **7.** to hit repeatedly, as to rain bullets against the tank's armor.

rainbird *n.* (Anglo-Saxon *regn*, rain + *bridd*, bird) any of several birds whose cries and/or actions are considered as indicators of oncoming rain; e.g., the rain crow, the black-billed cuckoo (*Coccyzus erythropthalmus*), or the yellow-billed cuckoo (*C. americanus*).

rainbow *n.* (Anglo-Saxon *regnboga*, rainbow) a colorful arc in the sky produced by the refraction, reflection, and dispersion of light rays (usually those of the sun) passing through droplets of water. The colors are in the same sequence as those produced by a prism: violet, indigo, blue, green, yellow, orange, and red. Sometimes there are two such arcs: an inner primary rainbow and an outer secondary rainbow.

rainbow trout *n.* a cool-water game fish (*Salmo gairdnerii*) native to western U.S.A. and Canada and marked by a strip of bright rainbow colors on each side.

raindrop erosion *n.* detachment of soil particles by the impact of raindrops on bare soil, often accompanied by a net downwind or downslope movement of soil. It is a major source of loose soil particles for both wind and water erosion. Also called splash erosion.

rainfall *n.* (Anglo-Saxon *regn*, rain + *feallan*, to fall) **1.** a rain event; a time when rain is falling. **2.** the amount of precipitation, usually in either inches or millimeters, received during a year or other specified time in a particular place, including drizzle and solid forms (snow, sleet, and hail) expressed on the basis of their water content when melted. Annual rainfall ranges from nearly zero in certain desert areas to 463.4 in. (11,770 mm) at Tutunendo, Colombia, in South America.

rainfall erosivity *n.* the ability of rainfall to cause erosion. In the universal soil-loss equation (and others derived from it), rainfall erosivity is evaluated in the R factor as a product of the total kinetic energy of each rainstorm multiplied by its maximum 30-min intensity (R is also called EI_{30}). Rainfall erosivity data are shown on maps as iso-erodent lines.

rainfall intensity *n.* the rate of rainfall; the amount of rain that falls during a specified time, as a 1-min intensity or a 30-min intensity.

rainfall interception *n.* rainfall that never reaches the ground because it is caught on leaves, branches, and stems of plants or on structures or other objects.

rainfall simulator *n.* a machine that produces artificial rainfall either in the field or in a laboratory by means of either sprinklers or drippers.

rainfed or **rain-fed** *adj.* dependent on natural precipitation, especially for the growing of farm crops (as distinguished from an irrigated crop).

rain forest or **rainforest** *n.* a dense forest in high-rainfall, tropical areas, especially in areas where the annual rainfall is 100 in. (2540 mm) or more in Brazil, the Philippines, and the African Congo. Usually there are three levels of vegetation: low-growing plants and shrubs, short to moderately tall trees, and tall trees. Also called jungle or selva.

rain gauge *n.* a device for measuring rainfall, usually with some amplification to increase the precision for reading small amounts. A standard U.S. Weather Service rain gauge has a funnel-shaped top 8 in. (20.3 cm) in diameter that drains into a tube with a cross section one-tenth as large. Also called a pluviometer or udometer.

rain shadow *n.* the area of relatively low rainfall on the lee side of a mountain range where the prevailing winds descend after having crossed the mountains; e.g., the Great Plains (eastern) side of the Rocky Mountains, where the climate is much drier than on the western side. The descending air has low relative humidity because it is heated by catabatic warming.

rainstorm *n.* (Anglo-Saxon *regn*, rain + *storm*, storm) a rainfall event, especially one with heavy rain accompanied by wind, either as it occurs in a particular location or as it moves across an area.

raised bed *n.* an area where the land surface has been raised for agricultural use. Such beds provide deeper soil with higher fertility, better drainage, reduced frost damage, and better pollution control than the surrounding land. For example, the Incas built about 300,000 ac (122,000 ha) of raised beds around Lake Titicaca between 1000 B.C. and 400 A.D. and used them mostly for potato and quinoa production; they are about 1 m high, 4 to 10 m wide, and 10 to 100 m long.

rake *n.* (Anglo-Saxon *raca*, rake) **1.** a long-handled implement with tines at one end; used to gather loose grass, hay, leaves, etc., or to smooth the surface of loose soil. **2.** a machine designed to gather loose hay into rows or piles. **3.** any implement designed to gather small, loose items. **4.** a vertical or inclined vein of rock or ore that cuts across rock strata. **5.** a steep incline. **6.** an acute angle between a cutting edge and a line perpendicular to the plane it cuts. **7.** a dissolute, licentious person, especially a man who seeks to impose himself on women. *v.* **8.** to gather loose grass, leaves, or other material with a rake, or to smooth loose soil with a rake. **9.** to gather abundantly, as to rake in a lot of money. **10.** to search, as to rake through a pile of debris, or a gaze that raked across the room. **11.** to criticize severely, as to rake him over the coals. **12.** to remove excess mortar from the joints of freshly laid bricks or blocks (usually leaving a shallow depression at the mortar line). **13.** to slope steeply. **14.** to hunt, as a hawk flying after game, or a dog with its nose near the ground.

ram *n.* (Anglo-Saxon *ramm*, ram) **1.** a male sheep capable of breeding. **2.** a plunger or piston that pushes through a cylinder or channel. **3.** a type of pump. See hydraulic ram. **4.** a heavy log or other device used to batter down a door or wall. **5.** a sturdy projection on the front of a warship used to batter another ship or a fortification. **6. Ram** the constellation Aries, or its zodiacal sign. *v.* **7.** to run against something, striking it with great force. **8.** to push into place with great force.

ramble *v.* **1.** to roam or wander slowly, often idly and aimlessly. **2.** to meander, like a stream with many curves rambling across a countryside. **3.** to grow without a predictable pattern, as a vine that rambles over the rubble. **4.** to talk or write at length, expressing random ideas about various topics.

ramification *n.* (Latin *ramus*, branch + *ficatio*, making) **1.** the branching of a tree or other plant. **2.** a product or effect of branching, as a new branch. **3.** the effect or consequence of an action, as that ruling has many ramifications.

ramify *v.* (Latin *ramus*, branch + *ficare*, to make) to divide into branches or subdivisions.

rammed earth *n.* a soil mixture that has been tightly compressed in a form for use as a construction material.

rampant *adj.* (French *rampant*, climbing) **1.** growing profusely, as rampant plant growth. **2.** spreading rapidly; unchecked. **3.** running wild; uncontrolled; as a rampant beast or rampant inflation. **4.** standing on its rear legs with its front legs reaching forward (a 4-footed animal). **5.** an architectural arch or vault with the support on one side higher than that on the other side.

ranch *n.* (Spanish *rancho*, ranch) **1.** an agricultural operation with its main emphasis being to raise cattle on rangeland. **2.** a farm in western U.S.A., sometimes designated to indicate its principal crop, as a cattle ranch, a mink ranch, or a wheat ranch. See dude ranch.

rancid *adj.* (Latin *rancidus*, stinking) spoiled, rotten, putrid, stinking.

range *n.* (French *renge*, row) **1.** the extent or interval between two limits, as a range of prices, the difference between the largest and smallest values in a set of results, or the interval between the highest and lowest notes in a piece of music or that a singer can reach. **2.** the distance something reaches, as the range of a projectile or of one's vision, or the distance a vehicle can travel without refueling. **3.** a target area, as a rifle range or a missile range. **4.** a line of mountains, hills, or buildings. **5.** the area covered by a particular animal or type of animal, sometimes designated seasonally, as winter range or summer range. **6.** the area where a particular type of plant occurs naturally. **7.** rangeland. **8.** a row of townships in the public land survey of the U.S.A. **9.** a kitchen stove. *v.* **10.** to arrange items or beings in a formation. **11.** to pass over or through an area, as cattle range across a pasture.

range cattle *n.* cattle raised on rangeland.

range condition *n.* the state of the vegetation and soil on rangeland, ranging from excellent (a full stand of native vegetation) through good and fair to poor (a greatly reduced stand dominated by invader species; erosion may be excessive).

range indicator *n.* a plant whose presence or absence serves as an indication of range condition.

rangeland *n.* a large area of generally untilled open land, especially grassland used for grazing livestock or left for the benefit of wild animals. Also called range.

Rankine scale *n.* a temperature scale named after John R. Rankine (1820–1872), a Scottish engineer, that measures temperature from absolute zero in degrees that match those of the Fahrenheit scale.

Under 1 atm of pressure, the freezing point of water is 491.69°R and the boiling point of water is 671.69°R. See Kelvin scale.

rapacious *adj.* (Latin *rapax*, greedy + *osus*, full of) **1.** greedy; inclined to take things by force. **2.** committing extortion. **3.** predatory; capturing living animals for food.

rape *n.* **1.** (Latin *rapum*, turnip) a plant (either *Brassica napus* or *B. campestris*, family Brassicaceae) closely related to mustard and used as forage for sheep, swine, etc., and as a source of rape oil. **2.** (Latin *rapere*, to take by force) an illegal act of sexual intercourse forced on a woman. **3.** any forced act of sexual intercourse. **4.** an act of plundering and despoiling. **5.** (Latin *raspa*, to scrape together) the residue (skins and stems) remaining after the juice is squeezed from grapes. *v.* **6.** to force someone to have sexual intercourse against their will. **7.** to take by force; to plunder or destroy.

raptor *n.* (Latin *raptor*, one who seizes) a predatory bird (e.g., a hawk or an eagle) that seizes live prey.

rare earth *n.* any of the lanthanide series of metallic elements, numbers 57 through 71 of the periodic table, or one of their oxides. These elements have similar properties and occurrence because their differences occur in internal electron orbits rather than in the valence shell. Several of them are used in alloys and as coloring agents for glass and ceramics.

rare species *n.* a plant or animal species with only a small world population but that is not presently considered endangered or threatened. See endangered species, threatened species.

rarefaction *n.* (Latin *rarefactio*, making thin) **1.** a decrease in density and pressure in air. **2.** a process that creates such a condition.

rasorial *adj.* (Latin *rasus*, scraping + *ialis*, like) **1.** tending to scratch the ground to find food; gallinaceous, like a chicken. **2.** pertaining to the foot of a bird that scratches.

raspberry *n.* **1.** any of several thorny vines or shrubs (*Rubus* sp., family Rosaceae). **2.** the small, knobby, juicy dark blue, purple, black, red, or yellow berries born by the vines and often used to make jam or pie. **3.** a dark reddish purple characteristic of the berries. **4.** a heckler's sign of derision.

rat *n.* (Anglo-Saxon *raet*, rat) **1.** either the Norway rat (also called brown rat, gray rat, barn rat, sewer rat, or wharf rat) or the black rat (also called roof rat, climbing rat, or gray rat) (*Rattus norvegicus* or *R. rattus.*, family Muridae). A rat is an aggressive, gnawing rodent that resembles a large mouse and has learned to live near people in buildings, ships, etc., and is often feared because it can gnaw or burrow its way into enclosed areas, make a mess, consume and contaminate food, and transmit several serious diseases (e.g., Bubonic plague, rabies, and typhus). **2.** any of several related rodents (*Rattus* sp.) that are most common in the East Indies that generally avoid people and are much less offensive. **3.** (loosely) any of several similar but unrelated animals such as muskrats, rat kangaroos, and rat opossums. **4.** a derogatory term for a person who reveals secrets and/or betrays a friend or associate. *v.* **5.** to hunt for and kill rats, especially with a trained dog. **6.** to betray a friend or associate and/or reveal secrets.

rata *n.* (Maori *rata*, rata) **1.** a large tree (*Metrosideros robusta*) native to the forests of New Zealand. **2.** the wood of the tree, used by the Maori to make canoe paddles, war clubs, etc.

rat-bite fever *n.* either of two infectious diseases caused by bacteria (*Streptobacillus moniliformis* or *Spirillum minus*) transmitted by the bite of a rat. Symptoms include fever, muscular pain, and a bluish red rash.

ratel *n.* (Dutch *rateldas*, honeycomb badger) a quadruped carnivore (*Mellivora capensis*), native to Africa and India, that resembles a badger with a body about 2 ft (60 cm) long.

ratite *adj.* (Latin *ratitus*, ratite) any of several flightless birds (formerly classified in group Ratitae), including emus, ostriches, tinamous, etc., that have flat breastbones without the central ridge characteristic of birds that fly.

ratoon *n.* (Spanish *retoño*, sprout again) **1.** a shoot growing from the base of a plant that has been cut down, especially sugarcane that has been harvested. Also spelled rattoon. *v.* **2.** to grow new shoots after having been cut down.

ratooning *n.* the harvesting of several crops from one planting by cutting off the plant near the soil and encouraging new sprouts to grow from the base or root. Plants suitable for ratooning when conditions are favorable include banana, bamboo, cotton, sugarcane, and sorghums. Similar management of tree species such as willow is called coppice. See coppice forest.

rat snake *n.* any of several large nonpoisonous snakes (*Elaphe* sp.) that feed on rats, birds, etc., and often enter houses. Also called house snake.

rattail cactus *n.* a Mexican cactus (*Aporocactus flagelliformis*) with crimson flowers and slender cylindrical stems that are used for weaving unusual designs.

rattan *n.* (Malay *rotan*, rattan) **1.** any of several climbing palms (*Calamus* sp. and *Rhapis* sp.) with long, round stems that are jointed but unbranched.

2. tough, durable, flexible strips from the stems of such palms used for weaving (e.g., caned chair seats) and wickerwork. **3.** a cane made of such material for use as a walking stick.

rat terrier *n.* any of several breeds of small dogs developed and trained to catch rats.

rattlesnake *n.* any of several venomous American snakes (*Crotalus* sp. and *Sistrurus* sp., family Crotalidae). Adults range from about 2 to 8 ft (60 to 240 cm) long and are characterized by a triangular head, vertical pupils, and a segmented horny rattle on the tail that they shake as a warning signal. It is typically gray or brown with either a diamond pattern or diagonal stripes across the body.

rattlesnake fern *n.* any of several grape ferns (*Botrychium virginianium*) native to America that have clusters of sporangia that resemble the rattles of a rattlesnake.

rattlesnake root *n.* any of several composite plants (*Prenanthes* sp.) whose roots and tubers have served as sources for snakebite remedies.

rattlewort *n.* a leguminous plant (*Crotalaria sagittalis*, family Fabaceae) with an inflated pod that rattles when its enclosed seeds are dry.

rauwolfia *n.* **1.** a poisonous tropical shrub or tree (*Rauwolfia serpentina* and related species, family Apocynaceae) of India and southeast Asia; named for Leonhard Rauwolf, a 16th-century German botanist. **2.** an extract from the powdered root of the shrub or tree; the source of tranquilizing alkaloid drugs such as reserpine.

raven *n.* (Danish *ravn*, raven) **1.** any of several large black birds (*Corvus corax* and related species) that fly in flocks, grow to more than 20 in. (50 cm) long, and have lustrous plumage and a loud, harsh call. See crow for a smaller black bird. **2.** the constellation Corvus. **3.** a lustrous black, as raven hair. *v.* **4.** to devour voraciously. **5.** to plunder; to seize spoil or prey by force.

ravenous *adj.* (French *ravineus*, ravenous) **1.** extremely hungry; famished. **2.** intensely predaceous, as a ravenous lion stalking its prey. **3.** eager for some type of gratification, as ravenous for praise.

ravine *n.* (French *ravine*, violent rush) **1.** a large gully or deep, narrow valley eroded by running water. **2.** a torrent of water that can erode such a valley.

rawhide *n.* animal skin that has had the hair and nonskin material removed but has not been processed into leather. Rawhide is sometimes made into a rope or a whip.

raw milk *n.* fresh, unpasteurized milk. Grade A raw milk must be cooled immediately and maintained at a temperature of 50°F (10°C) or below, must contain no detectable pesticide residues, and cannot exceed 100,000 bacteria per milliliter. See pasteurization.

raw sewage *n.* untreated wastewater from septic tanks and municipal sewers.

ray *n.* (French *rai*, ray) **1.** a thin beam of light. **2.** a line radiating from a center, either uniform, as a radius, or widening with distance; e.g., to represent a beam of light. **3.** a beam of particles radiated in a straight line. **4.** a half-line; a straight line that extends outward from a point of origin. **5.** any one of a group of parts that extend outward from a center; e.g., an arm of a starfish. **6.** a radiating petal of a circular flower blossom. **7.** a line that radiates from a lunar crater. **8.** an incident of mental inspiration, as a ray of enlightenment. **9.** a small amount, especially where there seemingly was none, as a ray of hope. **10.** (Latin *raia*, a ray) any of several fishes (class Elasmobranchii, order Batoidei) with flattened bodies, a slender whiplike tail, and large pectoral fins extending laterally from the sides and head; adapted for life on the sea bottom.

rayless goldenrod *n.* See jimmyweed.

razorback *n.* **1.** a half-wild swine with a lean body, ridged back, and long legs that is common in many parts of southern U.S.A. See feral. **2.** a finback whale (*Balaenoptera* sp.); any of several baleen whales with a prominent dorsal fin.

RDF *n.* See refuse-derived fuel.

real estate or **real property** *n.* nonmovable property that is owned, including land, mineral rights, growing crops, bodies of water, and structures that are fixed in place (as distinguished from personal property such as livestock, vehicles, and portable items).

reap *n.* (Anglo-Saxon *ripan*, to gather) **1.** to cut down or to cut down and gather a ripe crop (especially a grain crop). **2.** to receive a reward or recompense, as he reaps what he deserves.

reaper *n.* (Anglo-Saxon *ripan*, to gather + *er*, agent or machine) **1.** a person who reaps. Note: The Grim Reaper is a personification of death. **2.** a machine that harvests standing grain. Also called a combine.

recalcitrant material *n.* any material that resists decomposition for a long time.

recalescence *n.* (Latin *re*, again + *calescence*, growing hot) a sudden glowing caused by the release of the latent heat of transformation when a hot metal cools.

recapitulate *v.* (Latin *recapitulare*, to repeat main points) **1.** to summarize; to make a brief review. **2.** to pass through successive ancestral stages of development. Note: Recapitulation theory states that ontogeny recapitulates phylogeny, meaning that a developing organism passes through successive stages of ancestral development.

recessional moraine *n.* a glacial deposit, usually in the form of a line of low hills, formed during a minor readvance or a temporary halt in the retreat of a waning glacier.

recessiveness *n.* (Latin *recessus*, receded + *ivus*, tendency + Anglo-Saxon *nes*, state) a characteristic of genes that are not expressed when a corresponding dominant gene is present; the recessive gene is expressed only if matched with an allele of like nature. For example, blue eyes are recessive to brown eyes. See dominance.

recharge *n.* (French *recharge*, reload) **1.** to load, fill, or charge again, as to recharge a battery with electricity. **2.** to attack again with another charge. **3.** to revitalize, as relaxing for a while can recharge a person. **4.** the process of renewing the supply of groundwater, either from natural rainfall or by pumping or diverting water to where it can soak down to the aquifer.

reciprocal crosses *n.* the crossing of two plants or strains in both directions. One cross is made with one plant or strain as the female parent fertilized by the other and another cross is made with the roles reversed.

reciprocal recurrent selection *n.* plant breeding using two separate populations improved by recurrent selection that are crossed to produce seed for a crop or for further plant breeding, especially for crops such as corn, sugar beets, or sunflowers that are commonly produced as hybrids.

reclamation of mined land *n.* the process of restoring a mined site to its original use or to an approved alternative use. It usually involves smoothing the topography, replacing topsoil, loosening compacted materials, adding lime and fertilizer as needed, and establishing crops or permanent vegetation.

recognition protein *n.* a plasma membrane component that identifies a specific molecule if it is present in the environment. Also called a receptor.

recombinant DNA *n.* genetically engineered DNA prepared by inserting genes from another source into a chromosome. The inserted genes can cause the resulting progeny to express new characteristics such as higher production, higher concentrations of desired components, resistance to pests, and/or production of specific materials such as human insulin or artificial interferon by bacteria such as *Escherichia coli*, etc.

recommended daily allowance (RDA) *n.* guidelines established by the U.S. Food and Drug Administration for the daily dietary needs of people for nutrients such as vitamins, minerals, and calories.

reconnaissance *n.* (French *reconoissance*, recognition) **1.** a spotty survey of an area to make a generalized map of the soils, geology, range condition, or other features, often preliminary to a detailed map to be made later. **2.** gathering of information, often of a military nature behind enemy lines.

reconstituted *adj.* (Latin *re*, again + *constitutus*, set up) **1.** made liquid again, as when water is added to a dehydrated product; e.g., reconstituted milk or orange juice. **2.** restored to an original form or composition.

recovered resources *n.* useful material or energy derived from surplus or waste materials. See recycle.

recovery zone *n.* a section of a stream where bacteria break down pollutants and restore the natural environment, usually a flowing, aerated stream. See riffle.

recreation *n.* (Latin *recreation*, restoration or recovery) **1.** an activity that refreshes the mind or body, especially following hard work. **2.** a leisure activity such as playing a game, walking, swimming, fishing, boating, hunting, or taking photographs.

recurrent selection *n.* a process of plant breeding using repeated cycles of selection to increase the frequency of desired alleles in a crop while still maintaining genetic variability for further selection in future generations, including mass selection whereby seed is gathered each year from selected individual plants in a field. Other forms include various degrees of outcrossing and/or inbreeding in the selection process.

recycle *v.* (Latin *re*, again + *cyclus*, cycle or wheel) **1.** to use again; to process waste products so they can be used again either for the same use or for another, as cooling water may be recycled for cooling again or for watering an area of grass, or aluminum cans and glass bottles may be reprocessed to make new cans or bottles. **2.** to adapt to a new use, as recycling an old factory to make a new product. **3.** to repeat a process, as when one recycles the drier because one cycle does not dry the clothes enough.

red algae *n.* any of several unicellular algae (phylum Rhodophyta) of worldwide distribution; found mostly in marine environments, but some species grow in lakes or on snow. The chlorophyll of the algae is masked by red protein pigments called phycocyanin and phycoerythrin. Red algae grows in filamentous to leafy shrublike masses wherever light penetrates to the ocean floor or to the floor of an estuary.

redbud *n.* **1.** a common name for a small leguminous American tree (*Cercis canadensis*, family Fabaceae) that has small, light pink to purple blossoms in early spring, is used as an ornamental, and is sometimes called Judas tree. **2.** any of several other similar trees.

red cedar or **redcedar** *n.* any of several evergreen trees with scalelike leaves and reddish wood that has a low density and a spicy odor and taste, including eastern red cedar (*Juniperus virginiana,* family Cupressaceae), Rocky Mountain red cedar (*Juniperus scopulorum*) (also called Rocky Mountain juniper), southern red cedar (*Juniperus silicicola*), and western red cedar (*Thuja plicata,* family Cupressaceae).

red clover *n.* a forage clover (*Trifolium pratense,* family Fabaceae) with round, purplish red flower heads (the state flower of Vermont).

red eye *n.* **1.** any of several fishes (e.g., the rudd or the rock bass) whose eyes are red. **2.** red-colored eyes in photographs caused by reflection of light from the retina of wide-open eyes when a flash is used on the camera.

red fescue *n.* a fine-leaved cool-season grass (*Festuca rubra,* family Poaceae) used extensively as a lawn grass in the northern two-thirds of the U.S.A. and southern Canada.

redfin *n.* any of several freshwater minnows that have orange or red lower fins, especially a shiner (*Notropis umbratilis*).

red fir *n.* **1.** an evergreen tree (*Abies magnifica,* family Pinaceae), native to western U.S.A., with deeply fissured dark red bark. Also called California red fir. **2.** the light-colored wood of such trees, widely used for construction, crates, pulpwood, etc. Also called white fir.

red fox *n.* a common fox (*Vulpes vulpes*) with orange to reddish brown fur.

red maple *n.* a deciduous tree (*Acer rubrum,* family Aceraceae), native to eastern North America, that grows in moist soil, produces red blossoms, and has notched leaves that turn red, brown, or orange in the fall.

red mite *n.* any of several spider mites (family Tetranychidae) that feed on plants and may severely damage field crops, greenhouse plants, and orchard trees.

red mulberry *n.* a deciduous tree (*Morus rubra,* family Moraceae), native to North America, that produces red or purplish red edible fruit.

red oak *n.* **1.** any of several North American deciduous hardwood trees (e.g., *Quercus rubra, Q. falcata,* or *Q. borealis,* family Fagaceae) that reproduce from acorns. **2.** the hard wood of such trees, used as lumber (especially for flooring), timbers, railroad ties, charcoal, etc.

red ocher *n.* a natural red pigment obtained from certain soils or geologic deposits that contain hematite (Fe_2O_3).

redolent *adj.* (Latin *redolens,* having a scent) **1.** pleasant smelling; fragrant. **2.** having the scent of something, as redolent of garlic or of flowers.

red osier *n.* **1.** a North American dogwood tree (*Cornus sericea* or *C. stolonifera,* family Cornaceae) with red branches and twigs and white fruit. Also called red-osier dogwood. **2.** any similar willow with flexible red branches that can be woven to make baskets or wickerwork.

redox *adj.* (short for reduction–oxidation) **1.** the potential for reduction or oxidation in an environment, as measured by an appropriate electrode. **2.** transfer of electrons from atoms being oxidized to atoms being reduced in oxidation-reduction (redox) reactions.

redoximorphic feature *n.* any soil property that indicates reducing conditions associated with wetness; typically the presence of gleying, rust mottles, concretions, or other forms of reduced iron and manganese compounds.

red pepper *n.* **1.** a perennial plant (*Capsicum annuum longum,* family Solonaceae), native to central America, that is grown as an annual in temperate climates. **2.** a very spicy condiment made from the ground red or yellow pods and seeds of such plants and used in many spicy sauces. Also called cayenne pepper. **3.** the ripe form of the green, bell, or sweet pepper (*Capsicum annuum grossum*) that is used as a vegetable.

red pine *n.* **1.** an evergreen tree (*Pinus resinosa,* family Pinaceae), native to northeastern U.S.A. and southeastern Canada, with reddish brown bark and with two long, slender needles in each cluster; sometimes used as an ornamental and shade tree. **2.** the wood of such trees, widely used in building construction, for making boxes and crates, as pulpwood, etc.

redroot *n.* any of several plants with red roots, including New Jersey tea (*Ceanothus americanus*), stoneweed (*Lithospermum tinctorium*), a bloodwort (*Lacnanthes tinctoria*), and pigweed (*Amaranthus retroflexus*).

redroot amaranth *n.* an annual herbaceous weed (*Amaranthus retroflexus,* family Amaranthaceae) native to tropical America and introduced elsewhere; an erect plant with a red root that commonly grows in cultivated fields. Also called pigweed, redroot pigweed, or rough pigweed.

red rot *n.* any of several fungal diseases that produce a reddish discoloration in the pith and may break out into lesions and cause early death of young sugar plants, leading to reduced stands and low yields; also affects sorgho and certain trees.

red seaweed *n.* any red marine alga (especially *Polysiphonia* sp.). The thallus of most species has many branches.

red snow *n.* snow with a red surface from either airborne red dust or from red algae growing on it (common in Arctic and Alpine regions).

red spider *n.* See spider mite.

red squill *n.* a sea onion plant (*Urginea maritima*) with a red bulb; used to make rat poison.

red squirrel *n.* a North American tree squirrel (*Tamiasciurus hudsonicus*, family Sciuridae) with a reddish coat.

red-tailed hawk *n.* a common North American soaring hawk (*Buteo jamaicensis*, family Accipitridae) that nests in woodlands but feeds in open country. Its tail is red above and pink beneath, but its body color is variable, generally brown above with much white beneath.

red tide *n.* a brownish-red discoloration in marine waters that is caused by populations of microscopic flagellate protozoa (especially *Gymnodinium* sp. and *Gonyaulax* sp.) that produce a potent neurotoxin that accumulates in the flesh of shellfish. Eating such shellfish can produce paralytic shellfish poisoning in people, animals, and birds.

redtop *n.* any of several perennial bent grasses (especially *Agrostis alba*, family Poaceae) with red panicles that grow in cool humid regions on a wide variety of soil conditions and spread slowly by rootstocks, forming a coarse turf. Redtop is used for pasture, hay, and erosion control.

reduced tillage *n.* any tillage system that involves less tillage than is conventionally used for the crop being grown. It generally eliminates plowing and/or one or more subsequent tillage operations. See conventional tillage, conservation tillage, minimum tillage, no-till, plow planting, strip tillage, wheel-track planting.

reducing sugar *n.* a sugar whose double-bonded oxygen can be oxidized and therefore bring about the reduction of another reactant. A reducing sugar must be in the chain form because the ring form involves a linkage to the reactive oxygen. The aldehyde sugars produce carboxyl groups (organic acids) when oxidized. The ketone sugars are less reactive than the aldehyde sugars and produce a variety of compounds when they do react.

reduction *n.* (Latin *reductio*, bringing back) **1.** a decrease in the amount, size, number, or value of something. **2.** a decrease in degree, intensity, rank, or prestige. **3.** the process of decreasing an amount, size, intensity, etc. **4.** the removal of oxygen or other negatively charged ions from a chemical compound.

5. the addition of electrons to an atom or ion (reducing its positive charge or increasing its negative charge). See oxidation.

reduviid *n.* any of a family of insects (family Reduviidae) that attack humans, other mammals, and other insects, including conenose, kissing bugs, etc. Some species transmit Chagas' disease. See Chagas' disease.

red wine *n.* wine with a reddish color produced by pigments from the skins of the grapes used to make it.

red-winged blackbird *n.* an abundant North American bird (*Agelaius phoeniceus*) of marshes and fields. Males are dominantly black with a notable patch of red trimmed with yellow near the bend of each wing (may be covered when wings are folded). Females and young are dark brown with streaks and may or may not have red wing patches.

redwood *n.* **1.** a very large coniferous tree (*Sequoia sempervirens*, family Pinaceae), native to Pacific coastal areas of California and southern Oregon, with small scalelike and needlelike leaves and small reddish-brown cones about 1 in. (2.5 cm) long. Redwoods are the tallest tree species in the world, the largest specimen being about 30 ft (9 m) in diameter, 350 ft (106.7 m) tall, and over 2000 years old. See sequoia. **2.** any of several trees with reddish wood that grow in various parts of the world, including Indian dyewood (*Pierocarpus santalinus*), Turkish redwood (*Cornumascula* sp.), Jamaican redwood (*Gordonia haematoxylon*), and Bahamanian redwood (*Ceanothus colubrinus*). **3.** the reddish wood of any of these trees. California redwood is noted for its light density, straight grain, and resistance to rot.

reed *n.* (Anglo-Saxon *hreod*, reed) **1.** a grassy plant with a tall, straight stalk, especially one (*Phragmites* sp., *Arundo* sp., or *Ammophila* sp., family Poaceae) that grows in marshy areas or standing water, especially the common reed (*P. communis*). **2.** the jointed, hollow stem of such a plant; used to make thatched roofs, woven baskets, arrows, etc. **3.** a musical instrument made from a section of hollow stem from such a plant. **4.** a thin flexible strip of flexible material (reed stem, wood, metal, plastic, etc.) used to produce sound in certain musical instruments (e.g., a clarinet or an oboe). *v.* **5.** to thatch a roof or decorate with reeds.

reed canarygrass *n.* a perennial cool-season forage grass (*Phalaris arundinacea*, family Poaceae), native to Europe and widely grown in

Reed canarygrass

wet soils, that is used as a forage grass and for erosion control.

reef *n.* (Norse *rif*, a rib) **1.** an offshore ridge of rock or sand with its upper surface near water level; e.g., a coral reef. **2.** a lode or vein of ore in a mine. **3.** a variety of sponge. **4.** part of a sail that has been rolled up or folded so it will not catch the wind. *v.* **5.** to reduce the exposed area of a sail by tying up a rolled or folded section. **6.** to reduce the length of a mast or bowsprit by sliding sections together.

reek *n.* (Anglo-Saxon *rec*, reek) **1.** a strong undesirable odor; a stench. **2.** a vapor, fume, or smoke. *v.* **3.** to emit a strong offensive odor. **4.** to emit steam or smoke. **5.** to be permeated with anything unpleasant or offensive.

reflectance or **reflectivity** *n.* (Latin *reflectere*, to bend + *antia*, action or *ivus*, tending to) the fraction of light or other radiant energy falling upon a surface that is redirected either back toward the source or in a new direction; the part that is not absorbed by the surface and does not pass through the material whose surface it impacts.

refluent *adj.* (Latin *refluens*, flowing back) flowing back, as the waters from a wave or from the tide.

reforestation *n.* (Latin *re*, again + *forestis*, unenclosed woods + *atio*, act of making) establishment of a new stand of trees on land that was once forested but lost its trees by logging, fire, or other cause. See afforestation, forestation.

refraction *n.* (Latin *refractio*, bending) **1.** a bend in the path of light or other radiant energy as it passes obliquely from one medium into another where it moves at a different velocity; the process that enables a lens (as in the eye, a camera, a microscope, or a telescope) to focus rays of light or other energy and form an image on a receptive surface. **2.** the amount of change in the apparent position of an object caused by the passage of light or other detectable energy through an intervening medium such as the light from a star passing through Earth's atmosphere or the image of a fish in water as seen from outside the water.

refrigeration *n.* (Latin *refrigeratio*, making cool) **1.** the process of cooling or of maintaining a low temperature. **2.** the equipment used for this process.

refuge *n.* (Latin *refugium*, a retreat or refuge) **1.** protection from enemies or shelter from distress. **2.** a place that offers protection or shelter; e.g., a tract of land set aside for wildlife protection.

refuse *n.* (French *refus*, rejection) **1.** things and materials that are discarded as trash, rubbish, or garbage. *adj.* **2.** rejected and discarded.

refuse-derived fuel (RDF) *n.* fuel derived from garbage or other refuse, usually in a resource recovery plant by shredding and air classification to sort the low-density combustible materials from high-density noncombustible materials. The RDF is then mixed with coal and used in a power plant to generate electricity.

reg *n.* **1.** aeolian deflation pavement composed of a surface layer of polished stones left behind where the wind has blown away the finer material. Also called desert pavement. **2.** short for regulation.

regelation *n.* (Latin *re*, again + *gelatio*, freezing) the fusion of blocks of ice by melting under pressure and refreezing.

regeneration *n.* (Latin *regeneratio*, bringing forth again) **1.** growth of a new organ or tissue to replace a lost or damaged organ or tissue; e.g., a starfish will grow a new arm if one is cut off. **2.** an electronic feedback process that amplifies a signal by feeding output energy back into the input circuit. **3.** a spiritual revival; religious rebirth. **4.** any action that renews the form or spirit of a being or organism.

region *n.* (French *region*, boundary or area) **1.** an area or district that is considered or discussed as a unit for political, social, economic, ecological, or other purposes. **2.** a part of the Earth's surface, usually a large area with a particular type of climate or other feature, as the Arctic region. **3.** a section of the universe, as a galactic region. **4.** an area of interest or field of study or expertise, as the region of logic. **5.** a political subdivision of a city or territory. **6.** a part of an organism, as the shoulder region.

registered seed *n.* the seed produced by plants grown from either foundation seed or breeder seed and handled in a manner that preserves its genetic identity and purity. Registered seed must be certified by an agency designated for that purpose. See foundation seed, certified seed.

regolith *n.* (Greek *rhēgos*, blanket + *lithos*, stone) the layer of soil and/or unconsolidated rock material from the soil surface down to solid bedrock.

Regosol *n.* an azonal soil without any genetic soil horizons according to the system of soil classification presented in *Soils and Men*, the USDA Yearbook of Agriculture for 1938; a young soil developing in a soft mineral deposit such as loess or glacial drift (classified as an Entisol or an Inceptisol in Soil Taxonomy).

regression *n.* (Latin *regressio*, going back) **1.** reversion; return to an earlier form or place. **2.** retrogression; change from a more highly developed to a simpler life form; reversal of the succession from a highly developed to a simple type. **3.** return to less appropriate, more immature behavior. **4.** decreased expression of disease symptoms; e.g., reduction in the size of a tumor. **5.** regression analysis, a statistical method for using the value(s) of one or more independent variables to predict or estimate the value of a dependent variable; normally expressed as the equation of a straight line. **6.** an increase in land area along a coast by either uplift of the land or a lowering of sea level (e.g., by formation of more glaciers).

regrowth *n.* plant growth following either a dormant period (e.g., in the spring after winter dormancy or in the fall after a dry summer) or after removal of much of the foliage by grazing or by harvesting hay or silage.

regur *n.* (Hindi *regur*, regur) a fertile, dark, calcareous soil high in swelling clay, especially such soils on the Deccan Plateau of India; now classified in the Vertisol order of Soil Taxonomy.

rein (usually pl. **reins**) *n.* (French *rêne*, rein) **1.** either of the two leather straps fastened to each side of the bit in the mouth of a horse or other animal so the rider or driver can guide the animal. **2.** any of certain other leather straps in the harness of an animal. **3.** any type of restraint used to control action. **4.** (figurative) the means of governing, as the reins of government. *v.* **5.** to guide or hold back an animal by pulling on one or both reins, or to respond to such guidance. **6.** to curtail the action of a person, animal, or process, as to rein him in.

reindeer *n.* (Swedish *rendjur*, reindeer) **1.** the caribou (*Rangifer caribou*) of the far-northern part of North America, either wild or domesticated as a source of milk, meat, and hides. **2.** any of several other large deer of northern regions, with large, curved, branched antlers that are renewed annually.

rejuvenate *v.* (Latin *re*, again + *juvenis*, young + *atus*, to cause to become) **1.** to restore youthful vigor and appearance. **2.** to rehabilitate, renew, or enhance the environment; e.g., to reactivate erosional processes by uplift of an old landscape.

relapsing fever *n.* any of several infectious diseases of worldwide occurrence caused by schizomycetes (*Borrelia* sp., family Treponemataceae) that are spread by ticks and body lice to rodents and people. Symptoms include 5- to 7-day periods of fever, chills, muscular pains, and headache alternating with fever-free relapses that last several days.

relative humidity *n.* the amount of moisture in the air as a percentage of the amount the air would hold at saturation at the same temperature (usually determined by comparing the temperature readings of wet- and dry-bulb thermometers). See dew point.

relict *n.* (Latin *relictus*, left or forsaken) **1.** a species of organism that survives as a relic of earlier times even though the environment and other species have changed; e.g., the ginkgo tree or the coelacanth fish. Sometimes called a living fossil. **2.** a widow.

relief *n.* (French *relief*, rise in elevation) **1.** topographic form; elevation difference (may be quantified as the difference in elevation from the highest to the lowest points in a given area). **2.** variations in thickness or surface height on any object, especially on a carved, sculpted, or molded surface. Small differences are called bas-relief, and larger differences are high relief. **3.** a projecting part in a sculpture or an apparent projection in a drawing or painting; a third dimension, either real or perceived. **4.** something that eases pain or reduces discomfort or anxiety. **5.** a pleasant change; e.g., relief of monotony. **6.** assistance given in time of need or danger. **7.** release from duty, as when a replacement arrives.

rem *n.* (acronym for roentgen equivalent in man) the quantity of any ionizing radiation that has the same biological effectiveness as 1 rad of X rays: rem = rad × relative biological effectiveness. The comparable SI unit is the sievert: 1 sievert = 100 rem.

remediation *n.* (Latin *remediatus*, correction) **1.** the process of correcting a fault. **2.** a purification process used to cleanse polluted air, water, or soil.

remote sensing *n.* the detection and recording of electromagnetic radiation reflected from surfaces or emitted by materials; a no-contact, nondestructive means of obtaining information.

render *v.* (French *rendre*, to give back) **1.** to deliver, give, or submit, as to render an account or a bill or to render thanks or obedience. **2.** to surrender, as to render up the fort. **3.** to return, as to render good for evil. **4.** to make or cause to be, as the potion will render him helpless. **5.** to depict, as the artist will render a good likeness. **6.** to perform, as to render (recite) a poem or to render (sing) a duet. **7.** to revise the wording or translate, as to render a passage in a different style or in another language. **8.** to melt and purify, as to render the fat from a carcass. **9.** to process, as to render carcasses for market. **10.** to cover a brick wall or other surface with a first coat of plaster.

renewable resource *n.* a resource that is replenished on a short-term basis (e.g., annually), such as field crops, wood, or water power, as opposed to a nonrenewable resource such as coal or oil.

rennet *n.* (German *rinnen*, to curdle) **1.** the membrane lining the abomasum (4th stomach) of a calf or other young ruminant. **2.** an extract from the abomasum lining that contains rennin. **3.** a processed product from the abomasum lining that is used to curdle milk, as in the making of cheese.

rennin *n.* (German *rinnen*, to curdle + *in*, chemical) an enzyme that coagulates milk. It is the active agent of rennet obtained from the abomasum (4th stomach) of young ruminants and used to make cheese and other curdled dairy products. Also called chymosin.

reovirus *n.* any of a group of large viruses (family Reoviridae) that contain double-stranded RNA and are associated with respiratory and gastrointestinal diseases in animals and people.

repand *adj.* (Latin *repandus*, bent backward) **1.** slightly wavy. **2.** having a wavy border; e.g., the leaves of common nightshade.

repellent *n.* (Latin *repellens*, driving back) **1.** a substance that repels, as an insect repellent. **2.** a substance applied to a garment or fabric to make it repel water, stain, moths, mildew, etc. **3.** a medication that reduces swelling. *adj.* **4.** causing revulsion; driving away. **5.** resistant to wetting by water or to attack by moths, mildew, etc.

replication *n.* (Latin *replicatio*, folding back) **1.** a response or reply, especially in response to a previous response. **2.** a copy. **3.** one of two or more repetitions of an experiment, usually optimized for statistical analysis of the results. **4.** the genetic process whereby DNA makes a copy of itself.

reproduction *n.* (Latin *re*, again + *productio*, production) **1.** the act or process of making copies or duplicates. **2.** a copy or duplicate of an original, as the vase is a reproduction. **3.** the biological process that creates new individuals of a species, either sexually or asexually.

reptile *n.* (Latin *reptilis*, a creeping animal) **1.** any cold-blooded vertebrate animal (class Reptilia), includeing snakes, turtles, lizards, crocodiles, alligators, etc. **2.** any animal that crawls or creeps. **3.** a derogatory term for a mean, despicable person.

rescuegrass *n.* a short-lived perennial grass (*Bromus unioloides*, family Poaceae), native to Argentina, that has been adapted to humid areas with mild winters and is used for forage and erosion control.

reservoir *n.* (French *réservoir*, storehouse) **1.** a place for storing resources held in reserve, especially where large amounts are stored. **2.** an impoundment for holding water; e.g., for municipal use, irrigation, flood control, or production of electricity. **3.** a receptacle that holds a fluid; e.g., an inkwell. **4.** a sac or cavity in the body of an animal or plant where liquid accumulates. **5.** a large supply of anything; e.g., of infectious agents.

residence time *n.* the length of time that a given molecule remains in a particular environment; e.g., the time that an air pollutant remains in the atmosphere.

resident *n.* (Latin *residens*, dwelling) **1.** a person, animal, or bird whose home is in the area; not a visitor or transient. **2.** a professional person assigned to work (and often live) in a specific place, as a resident (physician or nurse) in a hospital or a diplomat who is resident in a foreign locale. *adj.* **3.** living in the place designated. **4.** remaining in a particular area; not migratory. **5.** intrinsic; an inherent characteristic.

residual soil *n.* soil formed from residuum and underlain by bedrock of the same type that weathered to form the soil, as distinguished from soil formed in alluvial or colluvial material that has been moved.

residuum *n.* (Latin *residuum*, leftover material) **1.** weathered rock material that remains in place as the rock disintegrates. See alluvium, colluvium, regolith, saprolite. **2.** residue that remains after something has been used, burned, or evaporated.

resilient *adj.* (Latin *resiliens*, leaping back) returning or able to return to a previous state following a disturbance.

resin *n.* (Latin *resina*, resin or gum) **1.** any of a group of amorphous solid or semisolid organic materials that dissolve in organic solvents such as alcohol or ether but not in water. They generally contain a mixture of aromatic acids and esters and do not conduct electricity. Resins are either obtained from plants or synthesized and are used to make medicines, varnishes, and plastics. **2.** a substance such as rosin prepared by distillation of resin from certain species of pine trees.

resistance *n.* (Latin *resistantia*, resistance) **1.** opposition; something that slows or prevents an action. **2.** the ability to resist, especially the ability of a plant, animal, or human to ward off injury from a disease, a pesticide, or any other unfavorable environmental factor. **3.** electrical resistance; the energy that is converted into heat when an electric current flows through a material; usually measured in ohms = voltage/current. **4.** a material or object that has a specified strength of electrical resistance.

resistant *adj.* (Latin *resistens*, resisting) having resistance; able to stand against a force or ward off a disease.

resonance *n.* (Latin *resonantia*, echoing) **1.** the condition or quality of vibrating or otherwise responding at the same rate as an input signal. **2.** amplification or prolongation of sound or other wave energy by bodies that respond to an input and vibrate at the same frequency. **3.** the condition of an electrical circuit that maximizes current flow by minimizing or eliminating resistance. **4.** resonance structures; molecular structures that include two or more alternate sets of electron positions (usually double-bond positions) that are equivalent and equally probable; e.g., the structure of a benzene ring, of an ozone molecule, or of a nitrate ion. **5.** the sound produced by tapping the exterior of an air chamber; e.g., for medical diagnosis.

resource *n.* (French *ressource*, rising anew) **1.** a supply of something that meets a need. **2.** a source of information or wisdom; e.g., a library or a wise person. **3.** (usually pl.) the abilities of a person or organization. **4.** (pl.) wealth; money, property, personnel, and other assets.

resource recovery plant *n.* a facility that uses shredders, magnets, sieves, and air classification to sort garbage and other refuse so that metals can be recycled, combustible materials can be used to generate electricity, and only the minimum amount of residue will be sent to a landfill.

respiration *n.* (Latin *respiratio*, breathing out) **1.** the process of breathing; inhaling and exhaling air. **2.** the combination of physical and chemical processes that people, animals, and plants use to oxidize foods (especially glucose and other carbohydrates) to produce energy. Aerobic oxidation of glucose may be generalized as

$$C_6H_{12}O_6 + 6\ O_2 \rightarrow 6\ CO_2 + 6\ H_2O + \text{energy}$$

Anaerobic respiration is less efficient than aerobic respiration because the products are fermented (only partially oxidized). Fermentation of glucose by yeast may be generalized as

$$C_6H_{12}O_6 \rightarrow 2\ CO_2 + 2\ C_2H_5OH + \text{energy}$$

response *n.* (Latin *respons*, an answer) **1.** an answer or reply; something done or said to answer an action, statement, order, request, or question. **2.** an action or physiological change in an organism that results from an external condition, usually an environmental effect or the action of another organism.

restoration *n.* (Latin *restauratio*, act of restoring) **1.** renewal; bringing back to an earlier, better, or more natural state or condition; e.g., restoration of a polluted or damaged environment. **2.** the return of something that was lost or stolen. **3.** an object (e.g., a structure or a painting) that has been restored, or its restored condition. **4.** religious salvation; the redemption of humanity from sin and reconciliation with God.

restricted-use pesticide *n.* a pesticide that is considered hazardous enough to people or the environment that government regulations permit its use only by a certified applicator.

restriction enzyme *n.* any of hundreds of enzymes that can cut fragments from strands of DNA, thus facilitating chromosome mapping and the transfer of genetic material from one organism to another. Some restriction enzymes cut DNA at specific locations, whereas others cut at random.

resurgence *n.* (Latin *resurgens*, rising again) coming back to life or vigor; resurrection or recovery, as a resurgence of growth after winter dormancy.

resuscitation *n.* (Latin *resuscitacio*, reviving) the act of reviving a person or animal from apparent death; artificial renewal of heartbeat and respiration.

ret *v.* (Dutch *reten*, to wet) to soak flax, hemp, wood, or other fibrous material in water to partially rot and soften the material so the fibers will separate from the other tissue.

retention basin *n.* a reservoir that is used to control stormwater runoff. Part of the water is held as a permanent lake, and the rest is released slowly enough to prevent downstream flooding. See detention dam.

reticular *adj.* (Latin *reticularis*, netlike) **1.** having a netlike form or pattern, usually intricate and entangled. **2.** pertaining to the reticulum of a ruminant.

reticulum *n.* (Latin *reticulum*, little net) **1.** a network structure or pattern, especially one with a fine mesh; e.g., the structure in certain cells, nervous tissues, or plant leaves. **2.** the second of the four stomachs of a cow or other ruminant, between the rumen and the omasum. **3.** a small southern constellation located between Dorado and Hydrus.

retrorsine *n.* a poisonous alkaloid ($C_{18}H_{25}O_6N$) from the groundsel plant (*Senecio retrorsus*) that can cause cirrhosis of the liver in horses or cattle that eat the plants.

retrovirus *n.* (Latin *retro*, backward + *virus*, slime or poison) any of a family of single-stranded DNA viruses that can reverse the normal transcription of DNA to RNA and incorporate themselves into the DNA of host cells, thus causing the host to reproduce the retrovirus; includes the AIDS virus and certain cancer-causing viruses.

retting *n.* **1.** the process of soaking flax or other fibrous material in water so its fibers can be separated from the surrounding woody tissue. **2.** the place where such work is done. Also called a rettery.

revegetate *v.* (Latin *re*, again + *vegetare*, to enliven) **1.** to grow plants again in places that once had vegetation; e.g., on eroded areas, construction sites, or mine spoils. **2.** to grow again, as perennial plants revegetate after winter dieback and dormancy.

reverse osmosis *n.* a water purification process that uses pressure to force water through a semipermeable membrane whose pores are too small for dissolved salts and other contaminants to pass through. The salts are carried away by permitting some water to escape from the upstream side of the membrane. Small reverse-osmosis units are used in homes, larger ones in laboratories, and much larger ones for wastewater treatment or to desalinize seawater.

reverse transcriptase *n.* an enzyme produced by a retrovirus that reverses the usual synthesis of RNA from DNA and instead synthesizes DNA from viral RNA. It is used in genetic engineering to transfer genetic material from one species to another.

revetment *n.* (French *revêtement*, protective cover) **1.** a layer of masonry, stones, or other material that protects a surface (e.g., an embankment) from erosion. **2.** a retaining wall. **3.** a decorative surface covering a wall or structure.

Reynolds' number *n.* a dimensionless parameter relating the velocity, density, and thickness of a fluid stream to its viscosity. Developed in 1883 by Osborne Reynolds (1842–1912), an English physicist, it is used to predict the borderline velocity between laminar and turbulent flow in a fluid.

R factor *n.* genetic material in certain bacteria that can conjugate with that of other bacteria and thus transmit heritable characteristics such as resistance to antibiotics from one bacterium to another.

rhabdovirus *n.* (Greek *rhabdos*, rod or wand + Latin *virus*, slime or poison) any of a group of rod-shaped plant and animal viruses (family Rhabdoviridae) that contain RNA; includes the rabies virus and viruses that cause vesicular stomachitis.

rhamnose *n.* (Greek *rhamnos*, thornbush) a dextrotatory deoxyhexose sugar ($C_6H_{12}O_5$); an important component of polysaccharides in the cell walls of plants. Also called deoxymannose or isodulcitol.

rhea *n.* any of three species of large, flightless South American birds (*Rhea americana*, *R. darwini*, or *R. macrorhyncha*) that resemble a small ostrich but have three toes instead of two and have feathers on their head and neck. Also called ostrich or ñandu.

rhenium (Re) *n.* (Latin *Rhenus*, Rhine + *ium*, element) element 75, atomic weight 186.207; a lustrous silvery white ductile metal with very high density and melting point. It is used in superconductors; for electrical contacts, themocouples, and filaments for mass spectrography; and as a catalyst.

Rh factor *n.* (named after rhesus monkey) a substance on the surface of red blood cells of humans and monkeys that are Rh positive. It induces a strong antibody response in individuals that are Rh negative (lacking that substance). Blood transfusions or bearing an Rh-positive child can cause an Rh-negative mother to produce antibodies that will necessitate an immediate complete blood transfusion, or even an intrauterine blood transfusion, for any subsequent Rh-positive babies she may bear.

rhinoceros *n.* (Latin *rhinoceros*, nose-horned) any of several large, plant-eating mammals (family Rhinocerotidae), native to Africa and India, with thick skins and a single large horn (or two horns, one in front of the other, on some species) growing from the snout. Adults are typically 5 to 6 ft (1.5 to 1.8 m) tall and up to 10 ft (3 m) long.

rhinoceros beetle *n.* any of several large beetles (*Dynastes* sp. and some others, family Scarabaeidae) with one or more horns on their head or forebody. They are found in tropical areas and are a pest on coconuts. Also called dynastid.

rhinoestrus *n.* any of several flies (*Rhinoestrus* sp., family Oestridae) whose larvae invade the nasal passages of horses and the eyes of humans in Asia, Africa, and Europe.

rhipicentor *n.* a Mexican tick (*Rhipicentor* or *Ixodes bicornis*) that infests cougars, causes a high fever in adults, and may be lethal to a child.

rhipicephalus *n.* (Greek *rhipis*, fan + *kephalē*, head) a cattle tick (*Rhipicephalus* sp.) that causes bovine anaplasmosis, East Coast fever, typhus, Rocky Mountain spotted fever, and related diseases in cattle, horses, sheep, and dogs.

rhizobacteria *n.* (Greek *rhiza*, root + *baktērion*, little staff) bacteria that live in the rhizosphere.

Soybean roots with rhizobium nodules (round objects) where nitrogen is fixed

rhizobia *n.* plural of rhizobium; used generically for the various classes of symbiotic nitrogen-fixing bacteria, including *Azorhizobium*, *Bradyrhizobium*, *Photorhizobium*, and *Rhizobium* sp.

Rhizobium n. (Greek *rhiza*, root + *bios*, life) the symbiotic bacteria (*Rhizobium* sp.) that form nodules on legume roots and fix atmospheric nitrogen (N_2), converting it into amines or other combined forms. Six species of *Rhizobium* and *Bradyrhizobium* infect different families of legumes:

B. japonicum	soybean
R. leguminosarum	peas, sweet peas, and vetches
R. lupini	lupines and trefoils
R. meliloti	alfalfa, medics, and sweet clover
R. phaseoli	beans
R. trifolii	clovers

rhizoctonia n. (Greek *rhiza*, root + *ktonos*, murder) any of several soil fungi (*Rhizoctonia* sp.). Some species cause plant diseases such as damping off of seedlings, root rot, rotting of tubers, root and stem cankers, and foliage blights that affect a wide variety of plants, including clovers, corn, cotton, grasses, potatoes, etc.

rhizome n. (Latin *rhizoma*, taking root) a stem that grows horizontally beneath or on the soil surface and produces new shoots and roots at nodes; e.g., in bamboo, bluegrass, bracken, Bermuda grass, horsetail, and iris. The new shoots may develop into separate plants as the internode connections die. See soboles, stolon.

rhizomorph n. (Greek *rhiza*, root + *morphē*, form) fungal filaments that tie a fungal mass together. One rhizomorph mass of honey mushroom (*Armillaria ostoyae*) extending through an area of 2200 ac (890 ha) in the Malheur National Forest in eastern Oregon to a depth of about 3 ft (0.9 m) has been shown to be genetically uniform—the largest living organism ever identified.

rhizophagous adj. (Greek *rhiza*, root + *phagos*, eating) root-eating.

rhizophore n. (Greek *rhiza*, root + *phoros*, bearing) a stem that grows downward and does not produce leaves but may reach the ground and produce roots and a new plant; e.g., in certain club mosses (*Seleginella* sp.), raspberries, or forsythia shrubs.

rhizosphere n. (Greek *rhiza*, root + *sphaira*, ball) a term proposed by Hiltner in 1904 for the zone in and around plant roots where the presence of roots and of root exudates influence soil characteristics; e.g., microbial activity is generally much higher in the rhizosphere than elsewhere.

rhizotron n. (Greek *rhiza*, root + *tron*, chamber) an underground chamber used to study growing roots by viewing them through a glass wall in contact with soil.

rhodium (Rh) n. (Greek *rhodon*, rose + *ium*, element) element 45, atomic weight 102.9055; a durable, hard, silvery white metal that is present in small amounts in platinum ores. It is used as a hardener in platinum and palladium alloys, in electrical contacts and spark plugs, and in optical instruments.

rhododendron n. (Greek *rhodon*, rose + *dendron*, tree) any of about 800 species of evergreen shrubs and trees (*Rhododendron* sp., family Ericaceae) with large, showy pink, purple, or white blossoms and oval or oblong leaves. Several are grown as ornamental rhododendrons and azaleas.

rhodotoxin n. (Greek *rhodon*, rose + *toxikon*, poison) any of several poisonous substances present in the flowers and leaves of various species of *Rhododendron*, *Kalmia*, and *Leucothoe*. Honey made by bees that visit these flowers may also be toxic to humans.

rhubarb n. (Latin *rheubarbarum*, foreign rhubarb) any of several very hardy, cool-season, perennial herbs (*Rheum* sp., family Polygonaceae), especially the variety (*R. rhabarbarum*) grown for its long petioles that are used to make pies and sauces (the leaves are large and should not be eaten because they contain toxic levels of oxalic acid [$COOH]_2$), or Chinese rhubarb (*R. officinale* or *R. palmatum*), whose rhizomes contain chemicals used medicinally as laxatives, astringents, and tonics to treat indigestion, diarrhea, and hemorrhoids.

rhus n. (Greek *rhous*, sumac) any of several deciduous shrubs and vines (*Rhus* sp., family Anacardiaceae), including poison ivy, poison oak, and poison sumac. Many people are sensitive to the urushiol chemicals found in these plants and develop a severe skin rash if they contact the leaves, stems, or roots. The leaves of these plants occur in groups of three, so a safety rule is "leaves of three, leave it be."

rhyolite n. (Greek *rhyax*, stream of lava + *lithos*, stone) a volcanic rock with a high content of silica, usually pinkish; equivalent to granite in composition, but coming from lava. See basalt.

ribes n. (Arabic *ribas*, rhubarb) any of several shrubs (*Ribes* sp., family Saxifragaceae) native to cool, humid, temperate climates of Europe and America, including currant and gooseberry plants. Growth of these plants is banned in some parts of northwestern U.S.A. because they are alternate hosts of the white pine blister rust.

ribgrass or **ribwort** n. See buckhorn plantain.

riboflavin n. (German *ribose*, ribose + Latin *flavus*, yellow) a growth-promoting, orange-yellow, crystalline compound ($C_{17}H_{20}N_4O_6$) that occurs naturally in fresh meats, egg yolks, leafy vegetables, and milk. Synthetic forms are often added to enriched flour or included in vitamin pills as part of the B-complex vitamins. Also called vitamin B_2, vitamin G, or lactoflavin.

ribonucleic acid (RNA) *n.* the transcription material formed as a complementary molecule from DNA. The composition of RNA is the same as that of DNA except that uracil and ribose are substituted for the thymine and deoxyribose of DNA. RNA is composed of single-stranded molecules that receive hereditary coding from DNA to control growth and life processes in human, animal, and plant cells. Many viruses transmit hereditary information via RNA without involving DNA.

ribose *n.* (German *ribose*, ribose) a water-soluble pentose sugar ($C_5H_{10}O_5$) that occurs in riboflavin, nucleotides, and nucleic acid. It combines with adenine and phosphate groups to form ADP and ATP.

ribosomal RNA *n.* a large RNA molecule; the part of a ribosome that carries the genetic code for synthesizing a protein.

ribosome *n.* (German *ribose*, ribose + Greek *sōma*, body) a cell structure that synthesizes a protein; composed of RNA and protein.

rice *n.* (Italian *riso*, rice) **1.** a small grain plant (*Oryza sativa*, family Poaceae), native to southeast Asia, that grows 2 to 6 ft (0.6 to 1.8 m) tall and can produce several tillers from one seed, each producing an open panicle at the top of an upright stem. Paddy rice is grown in fields flooded under 2 to 4 in. (5 to 10 cm) of water (oxygen absorbed above the water level is transported internally to the roots), whereas upland rice is grown as a rainfed crop, especially in tropical and subtropical environments. **2.** the grain produced by the plant; a starchy food that is the principal food of more than half of the world's population. Increased rice yields were a major factor in the Green Revolution of the 1960s and 1970s.

ricegrass *n.* (Italian *riso*, rice + Anglo-Saxon *graes*, grass) a native perennial bunchgrass (*Oryzopsis hymenoides*, family Poaceae) of wide occurrence on dry sandy soils in western U.S.A. The plants grow 1 to 2 ft (30 to 60 cm) tall with slender leaves almost as long as the stalks and produce black, plump, milletlike seeds that are tipped with short awns. Native Americans used them to make meal, flour, and bread. Also called Indian ricegrass.

rice paddy *n.* a field where rice is grown under flooded conditions.

Richter scale *n.* a logarithmic scale for indicating the intensity of an earthquake; named after Charles Francis Richter (1900–1985), a U.S. seismologist. The numbers range from 1 to 10, with each number representing 10 times the intensity of the next lower number. The highest earthquake intensity ever measured was 8.9 on March 2, 1933, in Japan; 2990 persons were killed.

ricin *n.* (Latin *ricinus*, castor-oil plant) a poisonous albumin contained in castor beans (*Ricinus communis*, family Euphorbiaceae) that coagulates red blood corpuscles and causes an antitoxin to be produced. Ricin is so toxic that eating several seeds can kill a person.

rickettsia *n.* any of a group of rod-shaped to spherical microbes (*Rickettsia* sp., family Rickettsiaceae) with characteristics between those of virus and bacteria, including organisms that cause typhus and Rocky Mountain spotted fever. Named after H. T. Ricketts (1871–1910), a U.S. pathologist, the microbes live in the gut of lice, fleas, and ticks and are transmitted from rodents, rabbits, dogs, etc., to other animals and to people through insect bites.

ridge *n.* (Scottish *rigging*, roof of a house) **1.** a long, narrow hilltop or a line of hills or mountains. **2.** any long, narrow elevation; e.g., the back of a cow, horse, or other four-footed animal. **3.** a horizontal line where two sloping surfaces meet; e.g., a roof peak. **4.** a raised strip on a surface; e.g., on cloth. **5.** an elongated area of high pressure on a weather map. *v.* **6.** to shape something into a ridge or ridges, as to ridge the soil in a field.

ridge planting *n.* a system of planting row crops on small ridges. Usually the ridges are maintained in the same place year after year. It provides well-drained soil in the ridges and allows residues to accumulate in the furrows to control erosion and reduce weed growth.

riffle *v.* (French *riffler*, to scratch) **1.** to produce a rippled effect. **2.** to hold the edge of a book, magazine, or other sheaf of papers and let them pass quickly through the fingers one at a time. **3.** to shuffle a deck of cards by hand. *n.* **4.** a shallow place in a stream that forms ripples where the water flows over a bed of stones. This helps to oxygenate the water. **5.** a series of slats or other low blocks that cross the bottom of a trough or sluice used as a fish ladder or in gold mining, etc. **6.** the action of shuffling cards by bending and releasing them from two stacks with one's hands.

rift *n.* (Danish *rift*, cleavage) **1.** an open crack, cleft, or fissure. **2.** a geologic fault zone. **3.** a division, as an argument can cause a rift between two people, groups, or nations. *v.* **4.** to split open or divide into two parts.

rift valley *n.* **1.** a graben; e.g., the Great Rift Valley that extends from Jordan in southwestern Asia about 4000 mi (6400 km) to Mozambique in southern Africa, with a typical width of 30 to 40 mi (50 to 65 km), including the lowest dry land on Earth, bordering the Dead Sea. **2.** a deep-sea chasm through the midst of the midoceanic ridge that runs about 40,000 mi (64,000 km) through the North Atlantic, South Atlantic, Indian, and South Pacific Oceans where the ocean floor is spreading and magma flows are common.

Rift Valley fever *n.* a contagious viral disease of people and animals in southern and eastern Africa that is transmitted by insects (especially mosquitoes), causing headaches, fever, aching muscles, poor vision (even blindness), and anorexia.

right bank *n.* the bank on a person's right when one faces downstream.

rigor *n.* (Latin *rigor*, stiffness) **1.** strictness; adherence to exact meaning of laws, rules, or principles. **2.** severity of living conditions; e.g., difficulties caused by climate, weather, or war. **3.** stiffness; a rigid condition with muscles locked and unable to respond to stimuli.

rigor mortis *n.* body stiffness that develops progressively within a few hours after death, as muscle proteins coagulate.

rile *v.* (variation of roil) **1.** to irritate or make angry. **2.** to roil water; to stir up settled material; to make water muddy.

rill *n.* (German *rille*, rivulet) a small erosion channel that follows tillage lines. It is small enough to be crossed and filled by normal tillage equipment. See ephemeral gully, gully.

rill erosion *n.* loss of soil by repeated formation and smoothing of rills. The long-term effect is similar to that of sheet erosion.

rime *n.* See hoarfrost.

rimrock *n.* a line of rock, usually with a vertical surface that forms a cliff, as around a mesa, plateau, or basin.

rincon *n.* (Spanish *rincón*, an inner corner) **1.** the inside of a corner where two or three walls or surfaces meet. **2.** a hidden or semiconcealed place, as lost in a rincon. **3.** a narrow valley between two hills or cut into the border of a mesa.

rind *n.* (German *rinde*, rind) a peel; a firm outer covering on fruit (e.g., a watermelon rind or orange rind) or certain cheeses or meats (e.g., bacon rind).

rinderpest *n.* (German *rinder*, cattle + *pest*, pestilence) an acute and usually fatal disease of ruminants such as cattle and sheep that is caused by a virus (*Morbillivirus*) that spreads through contact, can be controlled by a vaccine, and is not transmitted to humans. Symptoms include fever, diarrhea, inflammation of mucous membranes, and intestinal lesions. Also called cattle plague.

Ringelmann chart *n.* a chart with various shades of gray produced by black grid lines covering 20%, 40%, 60%, and 80% of a white surface that can be compared with a plume of smoke to estimate its density on a scale from 0 for clear white to 5 for solid black.

ringnecked pheasant *n.* a popular, large, gallinaceous game bird (*Phasianua colchicus*) that frequents open woods, hedgerows, and cornfields. The male is colorful with a white ring around its neck, a red patch around its eye, black-and-white markings on brown plumage, and a long tail that gives him a total length of up to 3 ft (0.9 m). The female has various shades of brown but lacks the brighter colors. The bird is native to Asia and was introduced into other areas, including North America. Also called Chinese pheasant.

Ring-necked pheasant

ring-tailed *adj.* **1.** having rings of alternating colors on the tail; e.g., a raccoon is ring-tailed. **2.** having a tail coiled into a ring; e.g., a capuchin monkey (*Cebus capucinus*) or certain Australian marsupials (*Pseudocheirus* sp., family Plalangeridae).

ringworm *n.* any of several contagious skin diseases caused by parasitic fungi (*Microsporum audouini* in people or *Trichophyton* sp. in animals) that form an itchy reddish circular area that gradually enlarges into a circle of eruptions. Also called tinea.

riparian *adj.* (Latin *riparius*, of a riverbank) associated with or adjacent to the bank of a stream or lake, especially the land subject to flooding.

riparian buffer *n.* a wooded filter strip that affords protection from erosion and reduces pollution of an adjacent body of water; usually suitable for wildlife use.

riparian habitat *n.* the area adjacent to the bank of a river, lake, or pond; a place for unique vegetation to grow that provides a living area for certain birds, animals, and people.

riparian right *n.* a legal principle based on old English law that associates the right to use water from a stream or lake or to catch fish in the water with ownership of land that borders the water. Riparian rights are commonly used in eastern U.S.A. (Wisconsin, Iowa, Missouri, Arkansas, Louisiana, and states to their east) but are replaced by the prior appropriations doctrine in most of western U.S.A.

rip current *n.* an undertow; subsurface flow of water away from the land, cycling with surface water that flows toward the land.

ripe *adj.* (Anglo-Saxon *ripe*, ripe) **1.** fully mature; ready for harvest, as ripe grain or fruit, or ready for slaughter, as ripe meat animals. **2.** ready for eating or drinking, as ripe cheese or ripe wine. **3.** fully developed, as ripe judgment comes in one's ripe years. **4.** appropriate, as when the time is ripe. **5.** ready to be lanced or drained, as a ripe abscess. **6.** ready for use, as a ripe idea or ripe plans.

ripple *n.* **1.** a small undulation or wrinkle on a surface, especially on the surface of water. See riffle. **2.** a small disturbance, as the notice didn't even cause a ripple in the activities. **3.** a toothed device used to comb seeds or capsules from flax or hemp. *v.* **4.** to flow with undulations on the surface, or to cause such flow, as the wind ripples the water. **5.** to have or make undulations or wrinkles on a solid surface. **6.** to fluctuate in loudness, as the music ripples or the brook makes a rippling sound.

ripple mark *n.* one of a series of surface undulations caused by wind or water flow, either in a loose surface such as sand or preserved from the past by the cementation of such a surface into sandstone or siltstone.

riprap *n.* **1.** a layer of loose stones placed over a surface (e.g., a streambank or the face of a dam) to reduce erosion. Where the water flow is rapid, riprap may need to be a graded filter with sand next to the soil covered by gravel, small stones, and larger stones or blocks. The pieces in each layer are too large to pass through the openings in the layer above it, and those in the top layer are too heavy for the water to move. **2.** an irregular wall of stones, sometimes used as the foundation for a building or structure.

riptide *n.* a strong, narrow return current in the opposite direction from other currents, especially one flowing outward from a coast and fed by incoming water on both sides.

risk assessment *n.* evaluation of the probability that damage or loss will occur and of its likely magnitude if a particular action is taken. It may be applied to health hazards associated with the use of pesticides or other chemicals, to possible losses by erosion or theft if specified designs and procedures are followed, or to physical danger inherent in various activities or situations.

risk management *n.* the procedures involved in assessing and responding appropriately to risks that one faces (e.g., as a business owner) or that threaten public well-being (e.g., pollution problems). It may involve regulations regarding the use of hazardous substances, the inclusion of safety considerations in the design of buildings and structures, and specification of appropriate ways to perform certain activities.

ritual *n.* (Latin *ritualis*, ritual) a standardized procedure for performing a certain activity, as a religious ritual or the mating ritual of a given species of animal or bird.

rive *v.* (Danish *rive*, to tear) **1.** to tear or break apart, as to rive meat from a bone or weathering rives granite into sand. **2.** to split, as to rive shakes from a block of wood. **3.** to severely stress or break, as to rive one's heart or feelings.

river *n.* (French *rivière*, river) **1.** a large stream of water flowing in its own natural course and emptying into the ocean, a lake, or another river. **2.** a large flow of something other than water, as a river of ice, lava, tears, or words.

riverbank *n.* soil and rock materials (usually forming a ridge or levee) that have a sloping surface bordering a river, both above and below the waterline. See levee.

river basin *n.* the land area that supplies water to a river and its tributaries. Also called watershed, catchment area, or drainage basin.

riverbed *n.* the layer of sand, gravel, stones, and rock beneath a river. Water may flow through the porous materials in a riverbed whether or not visible water is flowing in the channel.

riverine *adj.* (French *rivière*, river + *ine*, pertaining to) **1.** associated with or pertaining to a river; e.g., a dam across a river produces a riverine lake, or a riverine ecosystem occupies an area near a river. **2.** riparian.

river system *n.* the main river and all of its tributaries.

rivulet *n.* (Latin *rivulus*, small stream) a small stream of water, either intermittent or perennial. Also called runnel or streamlet.

R layer *n.* the underlying solid bedrock beneath a soil; material that is too hard to excavate with hand tools.

RNA *n.* See ribonucleic acid.

road *n.* (Anglo-Saxon *rade*, a place to ride) **1.** a travel route; a long, smoothed passage that leads from one place to another, usually connecting with other roads, for travel by vehicles of any kind. See path, trail. **2.** a highway. **3.** a railroad. **4.** a mine tunnel. **5.** a course of action, as the road to fame or to success.

roadrunner *n.* either of two species of birds, native to western U.S.A., Mexico, and Central America, that run fast, often along roads, but seldom fly, especially the greater roadrunner (*Geococcyx californianus*), a crested bird with rounded wings, long legs and tail, and mostly black-and-white plumage. Roadrunners eat mostly lizards, snakes, and insects.

roaring sand *n.* a dry desert sand dune that makes a soft roaring sound as it is moved by the wind.

robber bee *n.* a honeybee that steals honey from another hive.

robber fly *n.* any of a large group of large flies (family Asilidae) up to 1.5 in. (3.8 cm) long that prey on other insects. They attack bees, dragonflies, grasshoppers, wasps, etc., in flight, and their larvae feed on larvae of other insects.

robin *n.* (French *robin*, little Robert) **1.** a widespread, well-known American thrush (*Turdus migratorius*, family Muscicapidae). Adults are about 8 in. (20 cm) long and have a reddish breast. The breasts of the young are orange with black spots. Robins feed on insects and earthworms. **2.** a smaller European bird (*Erithacus rubecula*) with a reddish breast. **3.** any of several other similar birds of European and tropical origin, including some that have gray breasts.

robusta coffee *n.* **1.** a small coffee tree (*Caffea canephora*, family Rubiaceae) with fragrant flowers and red fruit that is native to tropical areas in western Africa. **2.** seed from this tree. **3.** coffee made from the seed.

rock *n.* (Latin *rocca*, rock) **1.** hard bedrock; the consolidated mineral matter that underlies soil, water, and other loose material at variable depths. **2.** any large mass of mineral matter, either consolidated or unconsolidated, that has not been transformed into soil, including alluvial, glacial, and wind deposits as well as hard bedrock. **3.** a boulder, stone, or pebble; any fragment of consolidated rock more than 2 mm in diameter. **4.** anything similar to solid rock, especially something used as a foundation for building, either literally or figuratively, as founded on the rock of hard work. **5.** (slang) a diamond or other gemstone. *v.* **6.** to swing or roll back and forth, especially in a soothing manner, as to rock a baby to sleep. **7.** to sway, as the wind rocked the house.

rock barnacle *n.* a barnacle (especially *Balanus balanoides*) that attaches itself to rocks near the shore.

rockfall *n.* the breaking away and falling of a large stone or group of stones from a high place, often leaving a cliff or escarpment above and a scree slope below.

rock flour *n.* powdered rock material, especially that ground from bedrock by a glacier. Also called glacial meal.

rock fragment *n.* a pebble or stone; any piece of stony material that is detached from bedrock and is more than 2 mm in diameter.

rock garden *n.* a garden in either naturally rocky or stony ground or, more often, with stones placed for a decorative effect, often in a raised area with plants growing in soil between the stones. Rock gardens are often watered with either drip or sprinkler irrigation.

rockiness *n.* degree of exposure of bedrock at the surface or so near the surface that the soil cannot be tilled or cropped. Soil survey classifies rockiness on a scale from 0 to 5:

0 No bedrock exposed or so little that it does not significantly affect tillage.

1 Enough bedrock exposed to cause some problems but not to prevent intertilled crops (commonly 2% to 10% of the surface area).

2 Enough bedrock exposed to make intertilled crops impractical (commonly 10% to 25% of the surface area).

3 Enough rock outcrops to severely limit the use of machinery (commonly 25% to 50% of the surface area).

4 Enough rock outcrop to limit land use to untilled pasture, forest, or wildlife (commonly 50% to 90% of the surface).

5 Over 90% of the surface area is exposed bedrock or soil too shallow for any use.

Good soil in the field, but class 4 rockiness behind it

rock outcrop *n.* **1.** a place where bedrock is exposed at the soil surface or above the surrounding soil. **2.** an area large enough to delineate on a soil map as a miscellaneous land type that is dominated by exposed bedrock (usually class 5 rockiness).

rock phosphate *n.* **1.** finely ground phosphate rock, especially when used as a phosphorus fertilizer (suitable for use on acid soils but too insoluble to be of much benefit if the soil is neutral or alkaline). **2.** phosphate rock.

rock salt *n.* **1.** a sedimentary deposit composed mostly of common salt (impure NaCl), usually formed by evaporation of seawater or of saline lake water; e.g., the salt flats around Great Salt Lake. **2.** salt taken from such a deposit and not completely purified, often coarse in texture or still in large fragments.

rock weathering *n.* the formation of soil or of parent material that may become soil by the disintegration of rock into small fragments and chemical decomposition of the minerals accompanied by the formation of secondary minerals.

Rocky Mountain goat *n.* a long-haired wild goat (*Oreamnos americanus*) of the Rocky Mountain region in western North America. It is covered with long white hair, and its horns are black, curved, and about 10 in. (25 cm) long.

Rocky Mountain spotted fever *n.* a disease of humans, domestic animals, and many wild animals that is caused by a microorganism (*Rickettsia rickettsii*) that causes high fever, muscular pain, bone pain, headaches, and a spotty red skin eruption. It is transmitted from rabbits, rodents, and other animals to people by various ticks, including the Rocky Mountain wood tick, the Pacific Coast tick, lone star tick, brown dog tick, American dog tick, and rabbit tick, among others.

Rocky Mountain wood tick *n.* a hard tick (*Dermacentor andersoni*, family Ixodidae) that occurs in western U.S.A. and transmits Rocky Mountain spotted fever, Q fever, encephalomyelitis, and anaplasmosis. See *Dermacentor*.

rod *n.* (Anglo-Saxon *rodd*, a beam or staff) **1.** a long, straight piece of wood, metal, or other material used as a walking stick, measuring stick, club, fishing pole, etc. **2.** a long, straight stem of a plant, either still growing or severed from the plant. **3.** a linear measure equal to 16.5 ft (5.5 yd or 5.029 m). **4.** a stick used to whip a person or animal as punishment. **5.** a stick marked with a scale and numbers for measuring elevations. See stadia rod. **6.** an elongated cell shaped like a tiny rod, as rod-shaped bacteria or the light-detecting rods in eyes.

rodent *n.* (Latin *rodentia*, rodent) a gnawing mammal (family Rodentia), including mice, rats, squirrels, beavers, etc.

rodenticide *n.* (Latin *rodentia*, rodent + *cida*, killer) any substance used to kill rodents.

rodeo *n.* (Spanish *rodeo*, a place for gathering cattle) **1.** a gathering and counting of range cattle. **2.** an enclosure to hold such cattle. **3.** a public exhibition of and/or competition in cowboy skills such as riding horses or bulls and roping calves. **4.** some other contest with similar organization, as a bicycle rodeo for children. *v.* **5.** to participate in such an event.

roe *n.* (Danish *rogn, roe*) **1.** spawn; the mass of eggs of a female fish, also called hard roe. **2.** milt; the mass of sperm of a male fish, also called soft roe. **3.** the spawn of certain crustaceans. **4.** a pattern of light and dark shades in some wood, especially mahogany. **5.** a roe deer (*Capreolus capreolus*), a small, agile deer native to Europe. The male has 3-point antlers and is also called a roebuck.

roentgen *n.* a unit formerly used to measure radiation; named after Wilhelm Konrad Roentgen (1845–1923), a German physicist. It is equal to the quantity of ionizing radiation that will produce 1 electrostatic unit of electricity in 1 cm^3 of dry air at 0°C (32°F) and standard atmospheric pressure. A roentgen is equal to 2.58×10^{-4} coulomb per kilogram of air. See rad.

Roentgen ray *n.* X ray.

rogue *n.* or *adj.* (Latin *rogare*, to beg) **1.** an inferior plant; one with undesirable characteristics that differ markedly from the norm. **2.** a vicious and unpredictable person or animal, usually a solitary individual; e.g., a rogue elephant. **3.** a vagabond, scoundrel, or unreliable person. *v.* **4.** to remove inferior and undesirable plants from a seed plot or field. **5.** to behave as a rogue.

roil *v.* (Latin *robigo*, rust) **1.** to stir up sediment; to make water or other liquid turbid or cloudy. **2.** to agitate and vex, or irritate a person or an animal.

rolled oats *n.* oats processed by crushing between rollers rather than grinding. The cooked product is less pasty and more palatable than that produced by grinding.

ronnel *n.* a chemical ($C_8H_8Cl_3O_3PS$) formerly used as a residual insecticide to protect livestock from insects. It was especially useful against flies and cockroaches, but its registration was canceled by the EPA in 1991.

roof *n.* (Anglo-Saxon *hrof*, roof) **1.** the upper surface of a building. **2.** (figurative) a home, as a roof over one's head. **3.** the upper part of a cavity, as the roof of the mouth or the roof of a cave. **4.** a very high place, as the Himalayan Mountains are the roof of the world.

roof garden *n.* a garden located on top of a flat-roofed building, sometimes as an added attraction for a restaurant or other business (perhaps located on the top floor rather than the actual roof).

roof rat *n.* a black rat (*Rattus rattus alexandrinus*) that is a common pest in the upper floors of buildings in warm climates.

rook *n.* (Anglo-Saxon *hroc*, crow) **1.** a gregarious, black, European crow (*Corvus frugilegus*) that often nests in trees near buildings. **2.** a cheater, especially in gambling. **3.** a chess piece, also called castle. *v.* **4.** to cheat or swindle. **5.** to squat or sit, as a hen on eggs.

rookery *n.* (Anglo-Saxon *hroc*, crow + *ery*, place) **1.** a place where a large number of wild rooks (crows) congregate, especially a breeding place. **2.** a similar place for other gregarious birds or animals, such as penguins or seals. **3.** a dilapidated tenement house or district, especially one that is overcrowded.

Roosevelt, Theodore *n.* an American statesman who was also a cowboy, soldier, naturalist, and explorer (1858–1919); the 26th president of the U.S.A. (1901–1909). Many of the national parks and forests were established during his administration.

roost *n.* (Anglo-Saxon *hrost*, roost) **1.** a perch where a bird (wild or domestic) sleeps or rests. **2.** a place with many such perches. **3.** a place for a person to lodge or rest, especially on a short-term basis. **4.** the place where one lives or spends a lot of time, as she rules the roost. *v.* **5.** to sit, sleep, or rest on a perch. **6.** to settle down or stay for the night. **7.** to boomerang, as he will regret it when his actions come home to roost.

rooster *n.* (Anglo-Saxon *hrost*, roost + *er*, doer) **1.** a mature male chicken or other male gallinaceous bird. Also called a cock. **2.** any bird that rests or sleeps on a roost. **3.** (informal) a cocky person, especially a small man who acts and speaks boldly.

root *n.* (Anglo-Saxon *rote*, root) **1.** the part of a plant that normally grows downward into the soil and serves to anchor the plant, absorb water and nutrients (through root hairs), and often store energy reserves. **2.** any underground part of a plant; e.g., a rhizome or a tuber. **3.** a part of a plant that serves for anchorage, as the part of a vine that clings to a wall. **4.** the base of a hair, nail, nerve, or tooth; the embedded part. **5.** the basic source or fundamental part of something, as the root of a problem. **6.** a number that produces another number when multiplied by itself a given number of times, as 2 is the square root of 4, the cube root of 8, and the 4th root of 16. **7.** a morpheme used to form a word and its derivatives. *v.* **8.** to begin producing roots. **9.** to become established in a place. **10.** to dig, as a pig roots with its snout. **11.** to pull up or remove completely, as to root out weeds or to root out crime.

root aphid *n.* any of a large number of insects (family Aphidae) that suck sap from and may produce galls on the roots of many species of plants. These aphids, also called ant cows, often inhabit mud shelters that ants make for them at the base of plants. The aphids produce a honeydew that the ants use as food.

root cap *n.* a covering of loose cells on the tip of a root that protects the meristematic tissue as the root grows through the soil.

root cellar *n.* a storage shelter either partly or entirely underground in well-drained soil, used to store vegetables, especially root crops such as turnips and beets.

root grafting *n.* **1.** the grafting of a scion of a desired cultivar to a root of another, usually more hardy, cultivar. **2.** a plant produced by such grafting. **3.** the natural intergrowth of the roots of two or more plants (usually trees).

root hair *n.* a short-lived tubular extension from an epidermal root cell near the tip of a growing root. Root hairs and mycorrhizae are the active sites where water and plant nutrients are absorbed.

root interception *n.* the absorption of nutrient ions that are directly in the path of a growing root so that the ions do not require transport by either mass flow or diffusion.

root knot *n.* a disease that attacks many trees, shrubs, crop plants, and ornamentals. It is caused by a nematode (*Meloidogyne* sp.) that produces galls on the roots and stunts plant growth. Also called eelworm disease.

root pressure *n.* osmotic pressure resulting from the absorption of water by root cells. Enough pressure to raise a column of water 200 ft (60 m) has been measured in tomato roots.

root pruning *n.* mechanical severing of roots by passage of a cultivator or other machine, by trimming as a ball of soil is removed with a plant being transplanted, or deliberately to slow the growth of a plant.

root rot *n.* decay of root tissue, usually by the growth of bacteria or fungi, often at a site that has been damaged, e.g., by the action of nematodes or other parasites.

rootstock *n.* **1.** the plant that provides the root system for a grafted plant. Also called stock or understock. **2.** a rhizome.

rosarian *n.* (Latin *rosarium*, rose garden) a specialist in roses; a person who appreciates, grows, and/or studies roses.

rosary pea *n.* a woody tropical vine (*Abrus precatorius*, family Fabaceae) with poisonous scarlet-and-black seeds that are used as beads and roots that are chewed like licorice. Also called crab's eyes or Indian licorice.

rose *n.* (Latin *rosa*, rose) **1.** a rosebush; any of a large group of either wild or cultivated shrubs (*Rosa* sp., family Rosaceae) with sharp stickers, pinnate leaves, and showy blossoms. **2.** a blossom from such a plant, usually red, pink, yellow, or white. The rose (any rose) was declared the national flower by the U.S. Congress in 1986. **3.** a red like that of certain roses. **4.** a similar color in the cheek of a person. **5.** a pattern of repeating loops that shows symmetry around a central point.

rose acacia *n.* a low-growing tree (*Robinia hispida*, family Fabaceae), native to southeastern U.S.A., that produces clusters of large rose-colored flowers in early spring and is often planted for erosion control. Also called bristly locust.

rose aphid *n.* a dark green aphid (*Macrosiphum rosae*) that is parasitic on roses and related plants.

rose apple *n.* **1.** an oriental tree (*Eugenia = Syzygium jambos*, family Myrtaceae) with pale green flowers and yellowish oval fruit. Also called jambos. **2.** any of several similar, closely related species. **3.** the rose-scented fruit of such a tree; used for making jams and jellies.

rosebush *n.* See rose.

rose chafer *n.* a tan, long-legged, slender beetle (*Macrodactylus subspinosus*, family Scarabaeidae) that feeds on roses, grapes, cherries, etc. Young chickens that eat these beetles become very ill and may die. Also called rose beetle.

rose hip *n.* the ripe, berrylike, fleshy stem end of a rose (especially a wild rose) where a flower forms, usually red and rich in vitamin C. It is eaten by wildlife and used to make tea and jelly. Also called hip.

rosemary *n.* (Latin *ros*, dew + *marinus*, marine) **1.** an evergreen shrub (*Rosmarinus officinalis*, family Lamiaceae), native to the Mediterranean region, with bell-shaped pale blue flowers that is traditionally a symbol of faithful remembrance. **2.** the narrow, leathery, aromatic leaves of the plant that are used in cooking and to make perfumes and medicines.

rose moss *n.* See portulaca.

rosette *n.* (French *rosette*, little rose) **1.** an ornament made by tying ribbons or strings into a rose-shaped bow. **2.** a design, arrangement, or formation that resembles a rose. **3.** a cluster of leaves, petals, or other parts arranged in a circle. **4.** an ornamental piece around the head of a nail or screw. **5.** a symptom of several diseases that cause short internodes and a clustering of leaves. **6.** an animal marking with a roselike form, such as one of the compound spots on a leopard.

A rosette of leaves on an iron-deficient pin oak tree

rosewood *n.* **1.** a valuable hard, reddish wood marked with black streaks, sometimes with a rose scent, that is used to make furniture and cabinets. **2.** any of several tropical trees (*Dalbergia* sp., especially *D. nigra*, family Fabaceae) that produce such wood.

rot *v.* (Dutch *rotten*, to rot) **1.** to slowly decay and decompose, to spoil, or to cause decay or spoilage. **2.** to become corrupt; to decay morally. **3.** to ret flax or hemp. *n.* **4.** the process of decaying; putrefaction. **5.** the softened, often spongy, material produced by rotting, sometimes accompanied by a strong odor. **6.** corruption; decay of moral and social standards. **7.** a disease that causes tissues to rot. *interj.* **8.** nonsense! an expression of strong disagreement.

rotation *n.* (Latin *rotatio*, a rolling or turning) **1.** the action of revolving around an axis. **2.** one complete turn around an axis, as the rotation of the Earth around its axis (one day) or the rotation of the Earth around the sun (one year). **3.** regular alternation, as a crop rotation or a rotation of officials.

rotation grazing *n.* grazing one pasture or area of rangeland at a time, usually out of a group of three or more. Each area recuperates during the time it is not grazed. Production is usually increased by greater plant growth and more uniform grazing in comparison to continuous grazing. The system is particularly effective on western rangeland when combined with deferred grazing so that each area has a chance to mature seed before being grazed at least once every three years. The combination is called either rotational deferred grazing or deferred rotation grazing. See deferred grazing.

rotavirus *n.* (Latin *rota*, wheel + *virus*, slime or poison) any of a group of double-stranded RNA viruses (*Rotavirus* sp., family Reoviridae) that have the shape of a spoked wheel under a microscope.

Rotavirus is the leading cause of acute gastroenteritis in human infants and is also responsible for diarrhea in calves, pigs, and mice. It is often spread by fecal contamination in food.

rotenone (Japanese *roten*, derris + Greek *ōnē*, chemical) *n.* an organic insecticide and scabicide ($C_{23}H_{22}O_6$) extracted from the root of derris (*Derris elliptica*, family Fabaceae), grown mainly in Malaysia and Peru. It is nontoxic to plants, mildly toxic to humans (especially when inhaled rather than eaten), moderately toxic to animals, and very toxic to fish and swine. It is used in insecticides, to kill or paralyze fish, and to treat scabies, chiggers, and head lice.

rotifer *n.* (Latin *rotifera*, wheel carrier) any of a large number of complex microscopic invertebrate animals (phylum Rotifera) that live mostly in freshwater (but some in salt water) and have one or more rings of cilia that look like wheels on the front part of their bodies.

rotten *adj.* (Norse *rotinn*, rotting) **1.** partly decomposed; decaying; often putrid and foul-smelling. **2.** easy to break apart, as rotten wood or rotten rock. **3.** morally corrupt. **4.** bad; unsatisfactory; as a rotten day, rotten weather, or rotten work. **5.** contemptible, nasty, dishonest; e.g., a rotten liar or a rotten trick.

roughage *n.* (Anglo-Saxon *ruh*, rough + *age*, aggregate) **1.** coarse food, especially fodder for livestock, usually bulky and high in fiber. **2.** any coarse, rough material.

roundworm *n.* **1.** an ascarid; a nematode (*Ascaris* sp., especially *A. lumbricoides*) that is parasitic in human intestines, causing diarrhea and colic. **2.** a nematode that is parasitic in animal intestines; e.g., the common roundworm (*Parascaris equorum*), a large parasite, 5 to 15 in. (13 to 38 cm) long and about the diameter of a pencil, that lives in the intestines of young horses. Heavy infestations are serious or even fatal.

rowen *n.* (French *rewain*, regain) regrowth; the second cutting of hay from a field in a season.

royal jelly *n.* a viscous milky white secretion produced by young honeybees that contains all of the essential amino acids, sugars, etc., and is rich in B vitamins. It is fed to all larvae for the first three days and then only to larvae chosen to develop into queens. It is also used in making face cream and certain tonics.

rubber *n.* **1.** a cream- to amber-colored, soft, solid material formed by polymerization [e.g., $(C_5H_8)_n$] as the latex (milky juice) obtained from rubber trees is dried. **2.** a similar material obtained from guayule (*Parthenium argentatum*), a desert shrub of western North America. **3.** a highly elastic material made by adding sulfur to this material and heating it to produce cross-links between the polymer chains for use in making a wide variety of rubber objects. See vulcanization. **4.** synthetic rubber, a similar substance produced synthetically by polymerization of styrene (C_8H_8) and butadiene (C_4H_6), mostly obtained from petroleum. Also called styrene-butadiene rubber. **5.** a pencil eraser made from rubber. **6.** (informal) a tire or set of tires. **7.** a low overshoe. **8.** a person who smoothes or polishes something by rubbing, or a tool used for this purpose. **9.** a series of three (or five) games of bridge, whist, or backgammon, or the decisive game in such a series.

rubber tree *n.* a type of tall tree (*Hevea brasiliensis*, family Euphorbiaceae) that grows wild in the tropical rain forest of Brazil. Commercial plantings in Malaysia and Indonesia are the principal source of natural rubber in the world.

rubbish *n.* (Latin *rubbosa*, rubble) **1.** solid waste; material that is discarded as not worth keeping. **2.** rubble. **3.** a worthless idea.

rubble *n.* (French *robel*, rubble) **1.** pieces of buildings and other property left from the destruction of one or more buildings, either naturally, e.g., by a tornado, earthquake, or volcano, or by an explosion or a wrecking operation. **2.** the crumbling upper part of weathered rock. **3.** masonry built of rough pieces of stone.

rubidium (Rb) *n.* (Latin *rubidus*, red + *ium*, element) element 37, atomic weight 85.4678; an alkali metal with chemistry similar to potassium. Generally present in small amounts in potassium minerals and ores, it is a silvery metallic element that ignites spontaneously in air and reacts violently with water, liberating and igniting hydrogen. It alloys with mercury, gold, cesium, sodium, and potassium; gives a yellowish violet color to a flame; and is used to make special glass and in photocells and vacuum tubes.

rudd *n.* (Anglo-Saxon *rudu*, red) a freshwater European fish (*Scardinius erythrophthalmus*, family Cyprinidae) with a small head and a deep body with an olive back, red ventral and tail fins, and red markings on its yellowish sides and belly. See red eye.

ruderal *n.* or *adj.* (Latin *ruderalis*, growing in rubble) a plant that can colonize open environments and grow well on disturbed soils such as roadsides or trashy areas; includes aggressive species that can produce a lot of seed during a short life span; e.g., most annual crop plants and annual weed species.

rudiment *n.* (Latin *rudimentum*, initial stage) **1.** the beginning or initial stage of something. **2.** an organ or part in its embryonic or vestigial stage. **3.** a basic principal or fundamental of a subject for study, as the rudiments of art.

rudimentary *adj.* (Latin *rudimentum*, initial stage + *arius*, pertaining to) in the earliest stage of development; embryonic or vestigial.

ruff *n.* **1.** a bulky, pleated collar made of muslin or lace, etc., worn by men and women in the 16th and 17th centuries. **2.** a similar circular part such as a protrusion of feathers or hairs on the neck of a bird or animal. **3.** a male sandpiper (*Philomachus pugnax*) that has a ruff on its neck during mating season; native to Europe and Asia. The female is called a reeve.

ruffed grouse *n.* a stocky game bird (*Bonasa umbellus*) that grows to about 14 in. (35 cm) long and resides in clearings of open woods. It has brownish plumage with white-and-black markings, either gray or red bands across its tail, and a black band across the broad tip of the tail. The male has a display pattern with its tail spread and ruffs of black feathers on each side of its neck to attract females. Also called pheasant in southern U.S.A., partridge in northern U.S.A., and birch partridge in Canada.

rumen *n.* (Latin *rumen*, gullet) the first of the four stomachs of a cow or other ruminant. Also called paunch or farding bag.

ruminant *n.* (Latin *ruminans*, cud chewer) any four-footed, hoofed mammal with even toes (suborder Ruminantia) that can regurgitate and rechew its food (cud), including cattle, buffalo, goats, deer, sheep, camels, antelopes, and giraffes. A ruminant has four stomachs: rumen, reticulum, omasum, and abomasum.

runnel *n.* (Anglo-Saxon *rynel*, little runner) **1.** a rivulet or small brook. **2.** a narrow channel for water. **3.** a troughlike depression formed on a beach by the action of waves or tides.

runner *n.* **1.** one who runs, especially a person, animal, or machine that races. **2.** a messenger or one who runs errands or collects bills. **3.** a person who operates a machine. **4.** a narrow surface that supports a sled or skate as it slides across snow or ice. **5.** a channel for ball bearings or other moving parts. **6.** a stolon, a stem that runs across the ground and forms new roots at one or more nodes. **7.** a plant that spreads by producing runners (stolons); e.g., strawberries, scarlet runners, or runner beans. **8.** a bird that runs, as the roadrunner.

runoff *n.* water that flows over the land surface and away from the area where it fell as rain, seeped from a spring, or was applied; often picking up pollutants from the land surface. Also called surface runoff.

runoff agriculture or **runoff farming** *n.* a cropping system that uses a dike system to divert runoff water from intermittent streams onto adjoining fields on the floodplain each time it rains. In desert areas, catchment areas 20 to 30 times as large as the fields are cleared so the soil will puddle to increase runoff.

run-on *n.* **1.** water that flows onto an area; runoff from higher ground. *adj.* **2.** continuing from a previous line or thought, as a run-on sentence.

runt *n.* **1.** an unusually small animal or plant. **2.** the smallest (and often weakest) animal in a litter, especially in a litter of pigs or puppies. **3.** (derogatory) a small person.

runway *n.* **1.** a paved strip for airplanes to take off and land. **2.** a parking strip, loading area, or merging lane for automobiles. **3.** a path used by wild animals. **4.** a channel, track, or bed followed by a moving part. **5.** the bed of a stream.

RUSLE *n.* revised universal soil loss equation, an updated version of the USLE used for predicting soil loss by water erosion.

Russian thistle *n.* an annual weed (*Salsola kali*, family Chenopodiaceae) with narrow spiny leaves that is troublesome in semiarid areas of western U.S.A. The tops break off from dead plants and blow across the countryside as tumbleweeds, spreading their seeds. Also called Russian tumbleweed.

rust *n.* (Anglo-Saxon *rust*, red) **1.** oxidized iron (Fe_2O_3), often hydrated, with an orange or reddish color. **2.** the process that oxidizes iron (Fe) to FeO, to $Fe(OH)_2$, and to Fe_2O_3. **3.** a reddish stain colored by rust; e.g., on a concrete surface. **4.** any of the many fungal diseases that produce reddish or brownish spots on the leaves and/or stems of plants. **5.** a reddish brown. *v.* **6.** to oxidize and deteriorate (an object that contains iron).

rustler *n.* **1.** a cattle thief. **2.** an energetic person who gets a lot done.

rust mottle *n.* a spot with a rusty color on a background of a different color, a common feature in

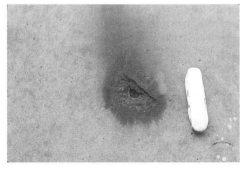

A rust mottle on a concrete surface.
The pocketknife is about 3 in. (8 cm) long.

the part of a soil that is saturated for extended periods but aerated at other times.

rut *n.* **1.** (French *route*, route) a channel cut into the ground by the passage of a wheeled vehicle or worn by the passage of many vehicles. **2.** a habitual way of doing things that is difficult to change. **3.** the time or the condition when a male deer, goat, sheep, camel, etc., becomes sexually excited, corresponding to estrus in a female.

ruthenium (Ru) *n.* (Latin *Ruthenia*, Russia + *ium*, element) element 44, atomic weight 101.07; a member of the platinum group of metals that occurs as a component of platinum ores. A hard, white metal that does not tarnish at room temperature, it is resistant to hot and cold acids and aqua regia but oxidizes explosively if potassium chlorate is added to such solutions. It is used to harden platinum and palladium. Its tetroxide is very toxic and may explode.

R-value *n.* a rating of the effectiveness of a material such as insulation to impede the flow of heat. Higher R-values indicate more resistance to the passage of heat.

rye *n.* (Anglo-Saxon *ryge*, rye) **1.** a hardy cereal grain (*Secale cereale*, family Poaceae) that resembles wheat and will intercross with wheat to some extent (see triticale). It is a widely grown crop, especially where the crop needs to be more winter hardy or drought resistant than wheat. Rye suppresses weed growth through allelopathic chemicals that it releases from either growing plants or plant residues. It is used for bread, animal feed, and alcoholic beverages. **2.** any of several related species (*Secale* sp.) that are not used commercially, partly because their seeds shatter more than those of *S. cereale*.

S

S *n.* **1.** sulfur. **2.** south. **3.** second. **4.** siemens.

sabadilla *n.* (Spanish *cebadilla*, little barley) **1.** a plant (*Schoenocaulon officinale*, family Liliaceae) native to Mexico and Central America that produces long grasslike leaves, long racemes covered with blossoms, and brown seeds that contain several alkaloids, especially cevadine and veratrine. **2.** a natural insecticide extracted from the seed of the plant; formerly used medically as an emetic and purgative, but extremely irritating to mucous membranes and can cause death by paralyzing the heart and respiratory system.

sable *n.* (Danish *sabel*, sable) **1.** a small carnivorous mammal (*Mustela zibellina*, family Mustelidae) native to cold climates in Europe and Asia. **2.** a similar American mammal (*M. americana*). Also called a marten. **3.** the glossy dark brown fur from such animals, or a coat made from this fur. **4.** a very dark brown or black.

sabra cactus *n.* (Hebrew *sabra*, prickly pear) a variety of prickly pear cactus (*Opuntia ficus-indica*, family Cactaceae) cultivated for its edible fruit and for use as a fence. Also called Indian fig or prickly pear.

sacbrood *n.* a very infectious disease caused by a filterable virus that infects the larvae of honeybees, making them shrivel and die.

saccharide *n.* (Latin *saccharum*, sugar + *ide*, chemical) sugar; a carbohydrate with a sweet taste. Monosaccharides include aldehydes and ketones with 3-, 4-, 5-, 6-, or 7-carbon chains with one hydroxyl group attached to each carbon except the one with the double-bonded oxygen (the most common monosaccharides are 6-carbon sugars, especially glucose and fructose, both $C_6H_{12}O_6$). Disaccharides are composed of two monosaccharides linked together (the most common one is sucrose, $C_{12}H_{22}O_{11}$). Polysaccharides have many monosaccharide units linked together (starch and cellulose are important examples).

saccharification *n.* (Latin *saccharum*, sugar + *ificatio*, making) the conversion of polysaccharides into sugar.

saccharin *n.* (Latin *saccharum*, sugar) a strong sweetening agent ($C_7H_5O_3NS$) that is about 500 times as sweet as sucrose but has no calories; used as a sugar substitute for diabetics, etc., and in pharmaceuticals.

saccharomyces *n.* (Greek *sakcharon*, sugar + *mykēs*, mushroom) a fungus (*Saccharomyces* sp.) that resembles yeast but lacks asci. It can ferment sugars into alcohols, is commonly present in milk as a lactose fermenter, and is used in the fermentation of wine and beer (*S. cerevisi*), in animal feed supplements, and in pharmaceuticals. Some species cause animal diseases. Formerly called *Torula* sp.

saddle *n.* (Anglo-Saxon *sadol*, saddle) **1.** a padded leather seat strapped to a horse or other animal so a rider can sit on its back. **2.** a similar seat mounted on a bicycle or other machine. **3.** the low part of the back of an animal where a rider may sit with or without a saddle. **4.** the padded section of a harness in the saddle area of a draft animal. **5.** a cut of meat including part of the backbone and both loins from a sheep or deer. **6.** a ridge with a peak or summit at each end. **7.** the covering on the ridge of a roof. **8.** the threshold piece in a doorway. **9.** any part or formation that has the general shape of a saddle. *v.* **10.** to strap a saddle on an animal. **11.** to burden someone with a difficult task, a lot of work, or a costly obligation.

saddle horse *n.* a horse used for riding; usually a smaller, faster horse than a draft horse.

saddle sore *n.* **1.** a sore place on the back of an animal caused by a poorly fit saddle. **2.** a sore spot on a rider caused by a saddle. *adj.* **3.** stiff and sore from horseback riding.

safety factor *n.* **1.** the amount of strength or capacity that is not likely to be required, often expressed as a ratio of the amount available divided by the amount expected to be needed. **2.** a comparison of the amount of something (e.g., a pesticide) that probably could be tolerated to the amount specified as a legal limit, as a safety factor of 2 to 1 or allowing a safety factor of 100%.

safflower *n.* (French *saffleur*, safflower) **1.** false saffron, an annual composite plant (*Carthamus tinctorius*, family Asteraceae), native to southern Asia, with finely toothed leaf margins and large reddish orange blossoms much like those of a thistle. **2.** the dried florets of the plant used to make medicines and dyes. **3.** the medicine or dye stuff made from the flowers.

safflower oil *n.* an oil obtained from the seeds of the safflower plant. It is very high in oleic and/or linoleic fatty acids and is used in cooking, to make margarine, or as an industrial oil.

saffron *n.* (French *safran*, saffron) **1.** a crocus (*Crocus sativus*, family Iridaceae) with large purple blossoms. **2.** a spice and natural orange-coloring agent made from dried stigmas of the female flowers that is used in many countries to flavor and color cooked rice and other foods.

sage *n.* (Latin *salvia*, safe) **1.** a perennial herb (*Salvia officinalis*, family Lamiaceae), native to the Mediterranean region, whose thick grayish green leaves are popular for seasoning foods, especially dressings and strong meats, and are used in medicines. **2.** any related mint plant or shrub (*Salvia* sp.). **3.** sagebrush. **4.** a person, especially an elderly man, whose wisdom, experience, and calm judgment are recognized and respected.

sagebrush *n.* any of several woody shrubs (*Artemisia* sp., family Asteraceae) with small, aromatic grayish green leaves, especially big sagebrush (*A. tridentata*), that grows abundantly in arid regions such as western U.S.A., especially in overgrazed rangeland.

sage grouse *n.* a large grouse (*Centrocercus urophasianus*) that is common where sagebrush grows in western U.S.A. It is marked with white borders on its mostly brown back and tail feathers, a pointed tail, and a black area on its lower belly. The male has a white breast and a black throat with a V-shaped white band running up toward the eyes.

sage hen *n.* a female sage grouse.

sage thrasher *n.* a short-tailed thrasher (*Oreoscoptes montanus*) that is common in sagebrush habitats of western U.S.A. It is marked with a dark gray back and tail, white bars on the wings, and black and white streaks on the front.

sago *n.* (Malay *sagu*, sago) a starchy food obtained from the pith inside the stem of various sago palms; used to make pudding.

sago palm *n.* any of several species of palm trees (*Metroxylon* sp. and *Caryota* sp., especially *M. rumphii*, family Palmaceae), native to Indonesia, that produce sago.

saguaro *n.* (Spanish *saguaro*, saguaro) a giant cactus (*Carnegiea = Cereus gigantea*, family Cactaceae) that grows in southwestern U.S.A. and Mexico. It bears an edible fruit used in the preparation of sweetmeats. Also called giant cactus or sahuaro.

Saguaro is a giant cactus.

Sahara desert *n.* (Arabic *sahārā*, deserts) the largest desert in the world, extending across northern Africa and covering 3,500,000 mi² (9,000,000 km²).

Sahel *n.* the arid area along the southern border of the Sahara Desert from Senegal to Chad.

sailfish *n.* **1.** any of several large fish (*Istiophorus* sp., family Istiophoridae) with a large dorsal fin that resembles a sail and a swordlike protrusion from the upper jaw. **2.** the basking shark (*Selache maximus*) that swims at the surface with its dorsal fin above water.

sainfoin *n.* (French *sainfoin*, wholesome hay) a perennial forage legume (*Onobrychis viciifolia*, family Fabaceae) that does not cause cattle to bloat, as do most legumes. See bloat.

salamander *n.* (Latin *salamandra*, salamander) **1.** a small amphibian (order Caudata) with a long tail and soft, moist skin that burrows extensively in sandy soils of southern U.S.A. The skin of some species is poisonous. See newt. **2.** a mythical reptile presented as being able to live in fire. **3.** a person who tolerates and enjoys great heat. **4.** any of several items associated with fire, as an iron poker, a metal plate or grill for cooking pastry, a large mass of iron that accumulates in the bottom of a blast furnace, or certain types of gas ovens.

salar *v.* (Spanish *salar*, to salt) a salt flat such as that around Great Salt Lake in Utah or the sodium nitrate deposits in Chile.

salic horizon *n.* a soil horizon at least 15 cm (6 in.) thick that is enriched in soluble salts (e.g., NaCl) to a concentration of at least 2% salt by weight and a minimum value of 60 for the product of thickness in centimeters multiplied by percent salt.

salicylic acid *n.* 2-hydroxybenzoic acid (C_6H_4OH-COOH), an organic acid that is slightly soluble in water and is produced by many plants. It is commonly obtained from oil of wintergreen or synthesized for use in making aspirin or as a preservative or fungicide.

saliferous *adj.* (Latin *sal*, salt + *ferous*, bearing) **1.** containing a high concentration of salt; e.g., an evaporite deposit. **2.** producing a saline solution, as saline seeps are saliferous.

saline *adj.* (Latin *salinus*, salty) **1.** resembling or containing salt, especially sodium chloride (NaCl). **2.** pertaining to any chemical salt, especially one that is highly soluble in water. *n.* **3.** a sterile, isotonic sodium chloride solution used to add liquid intravenously or to dilute medications. **4.** any salty solution. **5.** a salt lick or a salt deposit such as that produced by a saline seep.

saline seep *n.* a place where salt-bearing underground water reaches the soil surface and produces salt

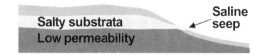

A landscape that produces a saline seep

incrustations and other forms of salty environment, e.g., in the Great Plains of the U.S.A.

saline-sodic soil *n.* a soil with both the high salt content of a saline soil and the high percentage of exchangeable sodium of a sodic soil. Its appearance and behavior are like those of a saline soil unless the salts are leached from it. To prevent it from becoming sodic, it should always be treated with a soil amendment before leaching. An older name is saline-alkali soil. See saline soil, sodic soil.

saline soil *n.* a soil with high salt content, technically one whose saturated extract has an electrical conductivity of 4 dS/m (4 mmhos/cm) or more at 25°C. This much salinity makes a white salt crust on the soil when it is dry and reduces the growth of most plants (some sensitive plants are affected at half this much salinity, and some will tolerate up to twice this much).

saline soil reclamation *n.* treatment of a soil to remedy excess salinity. This usually involves improving soil drainage by installing ditches or tile lines and applying enough irrigation water to leach out the excess salts. The soil should be tested for sodium content before leaching. If it is also sodic, then enough gypsum or other soil amendment should be applied before leaching to keep it from becoming a sodic soil. See sodic soil reclamation.

saline water reclamation *n.* The removal of salts from brackish or saline water by ion exchange, crystallization, distillation, evaporation, or reverse osmosis.

salinity *n.* (Latin *salinus*, salty + *itas*, condition) the total soluble salts in solution (usually mostly sodium chloride, NaCl). It is generally measured by electrical conductivity with a Wheatstone bridge and expressed in parts per million (ppm), parts per thousand (‰), or percent (%). Some typical salinity values are:

- Potable well water (maximum)   300–500 ppm
- Limit for irrigation (maximum)   ~1 000 ppm
- Typical seawater   35 000 ppm

salinization *n.* (Latin *salinus*, salty + *izatio*, making) the accumulation of sufficient salts in a soil to make it saline. It is typically caused by water moving upward through soil and evaporating from the soil surface, especially where irrigation water is applied either to the affected soil or to nearby soils at a higher elevation. It is common in the rooting media used for potted plants as well as in soils of arid regions.

salinometer *n.* (Latin *salinus*, salty + *metrum*, measure) an instrument for measuring the salt content of water; e.g., an induction salinometer that measures voltage as an indicator of salt content.

sallee *n.* any of a group of acacias native to Australia.

sallow *n.* (Anglo-Saxon *sealh*, willow or *salu*, yellowish) **1.** any of several shrubby Old World willows (especially *Salix atrocinerea* and *S. caprea*, family Salicaceae) with large catkins in the spring before the leaves appear. **2.** a twig from such a plant. **3.** sallee. *adj.* **4.** having a pale, sickly, yellowish skin color.

salmon *n.* (Latin *salmonis*, leaper) **1.** a large fish (family Salmonidae) that spawns in freshwater and then dies. The young salmon swim downstream to the ocean where they live for about 6 or 7 years and then swim back to the area of their origin to spawn, including Atlantic salmon (*Salmo* sp.) that live in the North Atlantic (except for some that are landlocked in large lakes such as Lake Pend Oreille and Lake Coeur d'Alene in Idaho) and Pacific salmon (*Oncorhynchus* sp.) that live along the northern Pacific coast of North America and Asia. See chinook salmon. **2.** the flesh of such fish used as food. **3.** a yellowish pink like that of the flesh of the fish.

salmonella *n.* any of several species of gram-negative, rod-shaped bacteria (*Salmonella* sp., family Enterobacteriaceae) named after Daniel E. Salmon (?–1914), an American veterinarian. It is a cause of diarrhea in animals and humans (especially young children, the elderly, and other persons with weak immune systems) that spreads through foods such as raw eggs, raw meat, (especially poultry), animal feed, or feces. Ten species are pathogenic to humans and to most warm-blooded animals. *Salmonella enteritidis* is a widely distributed parasite of rodents and chickens (often infecting their eggs) that occurs in several serotypes and can produce very serious diarrhea in humans. *Salmonella typhimurium* infects pigs, and even a pig that has been cured with antibiotics can be a carrier of the disease.

salt *n.* (Latin *sal*, salt) **1.** crystalline sodium chloride (NaCl), especially used as a flavoring agent or preservative. **2.** a chemical compound that ionizes in solution into a metallic cation (e.g., Na^+, K^+, and Ca^{2+}) or a complex cation (e.g., NH_4^+) and an anion such as a halide (F^-, Cl^-, etc.), sulfide (S^{2-}), or a complex anion containing oxygen (e.g., NO_3^- and SO_4^{2-}). The reaction of an acid with a base produces a salt plus water. **3.** something that adds wit, interest, or liveliness, as the salt of a story. **4.** reservations, as take it with a grain of salt. *v.* **5.** to season with salt. **6.** to provide salt, as for cattle. **7.** to treat with salt, as to salt the streets to melt snow. **8.** to create a false impression by placing valuable gems, minerals, or rich ore in a mine or vein where someone will see them and be deceived, as to salt a mine. *adj.* **9.** salty, having a salt taste. **10.** preserved with salt, as salt pork. **11.** containing saline water, as a salt marsh.

Salts in ocean water	% of total
Sodium chloride (NaCl)	77.8
Magnesium chloride ($MgCl_2$)	10.9
Magnesium sulfate	4.7
Calcium sulfate	3.6
Potassium sulfate	2.5
Calcium carbonate	0.3
Magnesium bromide	0.2

saltation *n.* (Latin *saltatio*, dancing) **1.** a jumping or dancing movement. **2.** an abrupt change in a movement or process. **3.** the jumping movement of sand particles (mostly those 0.05 to 0.5 mm in diameter) impacted by wind or flowing water. Saltation is the most significant initiator of wind erosion because the saltating particles knock loose particles of a wide range of sizes that otherwise would not move. See avalanching. **4.** a single mutation that drastically alters the phenotype (appearance) of a plant or animal.

The path of a saltating sand particle

salt block *n.* a solid block of salt suitable for placement in a large pasture or on rangeland to supply salt to livestock. Salt blocks can be used to induce cattle to graze remote areas that otherwise might not be used.

saltbush *n.* any of a group of shrubs and herbs (*Atriplex* sp., family Chenopodiaceae) with grayish green leaves and tiny flowers that grow in saline and alkaline soils of arid regions in western North America and Australia. Animals browse on some of them for their salty taste.

salt effect *n.* an effect such as plasmolysis resulting from the osmotic effect of a material (especially fertilizer) in water, causing dehydration.

salt flat *n.* **1.** a large area of level land covered with a salt deposit formed by precipitation in an intermittent or former shallow lake and/or by evaporation of rising ground water. **2.** any similar salt-covered area large enough to be shown as a miscellaneous land type on a soil map.

salt grass or **saltgrass** *n.* **1.** a salt-tolerant grass (*Distichlis stricta*, family Poaceae) that is widely distributed on saline soils of western U.S.A. Its forage value is fair. Also called inland salt grass or desert salt grass. **2.** a related salt-tolerant grass (*D. spicata*) that grows along seashores and in marshy areas. Also called seashore salt grass or spike grass. **3.** any of several other salt-tolerant grasses.

salt haze *n.* a cloudiness in the atmosphere caused by tiny salt particles left by the evaporation of sea spray in the air.

salt index *n.* a measure of the osmotic effect of a particular fertilizer in relation to an equal weight of sodium nitrate ($NaNO_3$) taken as a value of 100. The higher the salt index is, the lower the amount of the fertilizer that can be safely applied as foliar fertilizer or as starter fertilizer.

salting out *n.* a process of precipitation. For example, salting out may occur in a liquid fertilizer if the concentrations are too high, or dissolved protein can be caused to salt out by increasing the ionic strength of a solution beyond its optimum for solubility.

salt lick *n.* **1.** a place where salt occurs naturally (e.g., a saline seep) and animals may lick it. **2.** a site where salt is placed to benefit livestock and/or cause them to graze a pasture more uniformly.

salt marsh *n.* a wetland with brackish or saline water, either a tidal marsh or a low area in an arid or semiarid region.

salt spray *n.* droplets of salty water thrown into the air by breaking waves, especially when carried inland by the wind. The salt accumulation can cause plant leaves to burn and even drop.

salt-tolerant crop *n.* any crop that can grow satisfactorily in a saline soil. Some of the most salt-tolerant crops are Bermuda grass, tall wheatgrass, barley, sugar beets, and cotton. See halophyte, glycophyte.

salt water *n.* **1.** water that contains a high concentration of salt, especially seawater. *adj.* **2.** pertaining to or living in salt water, as saltwater fish.

saltwater intrusion *n.* the entrance of seawater into an aquifer that normally carries fresh water; e.g., when freshwater wells near a seacoast are pumped heavily enough to lower the water table below sea level.

salvia *n.* (Latin *salvia*, sage) an ornamental plant (*Salvia* sp., family Lamiaceae), especially the scarlet sage with its brilliant red blossoms.

salvinia *n.* any of a large group of aquatic ferns (*Salvinia* sp., family Salviniaceae), including some (e.g., *S. molesta*) that cause problems by their rapid massive growth under tropical conditions in Asia, Africa, and in the Pacific islands. Named for Antonio

Maria Salvini (1653-1729), a Greek professor. A weevil (*Cyrtobagous salviniae*) has been used to help control such water ferns.

samarium (Sm) *n*. (German *samarskit*, a mineral + *ium*, element) element 62, atomic weight 150.36; a rare earth metal with a bright silver luster that occurs in several minerals along with other rare earth metals, It is used in carbon-arc lights, in permanent magnets, and in lasers.

San Andreas Fault *n*. the zone of contact between the North American plate and the Pacific plate roughly paralleling the California coast. It is a major source of earthquakes in southern California as the Pacific plate slides northwest relative to the North American plate.

sanctuary *n*. (Latin *sanctuarium*, holy place) **1.** a temple or other holy place, especially the holy of holies in a house of worship. **2.** a place of refuge; a place where hunted persons or wildlife are protected. **3.** immunity from arrest or injury while protected in such a place.

sand *n*. (Anglo-Saxon *sand*, sand) **1.** mineral particles that are smaller than gravel (usually less than 2 mm in diameter) and larger than silt and clay. The International Soil Science system defines sand as having a diameter range from 2.0 to 0.02 mm, whereas the U.S. Department of Agriculture system classifies sand as being from 2.0 to 0.05 mm in diameter with subdivisions as:

Sand separate	Size (mm)
Very coarse sand	1–2
Coarse sand	0.5–1.0
Medium sand	0.25–0.5
Fine sand	0.1–0.25
Very fine sand	0.05–0.1

2. the sand in an hourglass, representing the sands (particles) of time. **3.** a tract of land with very sandy soil; e.g., a beach. *v*. **4.** to smooth or polish with sandpaper. **5.** to cover, fill, or mix with sand.

sandalwood *n*. (Latin *santalium*, sandal + Anglo-Saxon *wudu*, wood) **1.** an Asian evergreen tree (especially *Santalum album*, family Santalaceae) with ovate leaves and yellow blossoms that turn red. **2.** any of several related or similar trees, including red sandalwood (*Pterocarpus santalinus*, family Fabaceae). **3.** the fragrant heartwood of such trees, burned for incense or used to make ornamental carvings.

sandbur *n*. (Anglo-Saxon *sand*, sand + *burre*, bur) **1.** any of a group of small, weedy annual grasses (*Cenchrus* sp., family Poaceae) that grow in sandy soils. Their seeds are enclosed in prickly burs that get caught in animal fur or clothing and cause pain to bare feet. **2.** any of several other plants with similar characteristics (e.g., *Solanum rostratum*, family Solonaceae).

sand dune *n*. See dune.

sand-dune stabilization *n*. the process of establishing vegetation that will hold dune sand in place. It usually involves reducing wind velocity (e.g., with snow fence), starting on the upwind side of the dune area, and planting drought-tolerant, salt-resistant grasses or shrubs (transplanting plants—seeds can blow away) such as Bermuda grass, salt grass, switchgrass, and alkaligrass.

sand fly *n*. any of several biting, bloodsucking flies (families Psychodidae, Heleidae, and Simuliidae) of tropical regions that transmit several diseases.

sandhill crane *n*. a large gray (brown when immature) wading bird (*Grus canadensis*, family Gruidae) with a pointed bill, long legs and neck, and wide wingspan (6⅔ ft or 2 m). It feeds on small rodents, frogs, and insects and is often seen in large flocks. See whooping crane.

sandpiper *n*. any of a large number of coastal, marsh, or upland birds (family Scolopacidae) with long legs and long, slender bills that are named for a sandy habitat and a piping call.

sandstone *n*. (Anglo-Saxon *sand*, sand + *stān*, stone) a sedimentary rock formed mostly of sand-sized particles that have been subjected to either chemical cementation (by silica, iron oxide, and/or calcium carbonate) or to enough heat and pressure to fuse them together but not enough to change them into a metamorphic rock.

sandstorm *n*. (Anglo-Saxon *sand*, sand + *styrman*, storm) a strong wind carrying so much sand that visibility is markedly reduced. See dust storm.

sand-wireworm *n*. a worm (*Horistonotus uhlerii*, family Elateridae) that infests the roots of many field crops such as potatoes, cotton, and corn.

sandworm *n*. any of several marine worms (*Nereis* sp. and *Arenicole* sp.) that live in coastal areas and are used for fish bait.

sanidine *n*. See potassium feldspar.

sanitary *adj*. (Latin *sanitas*, health + *aris*, like) **1.** clean; hygienic; in a condition that promotes health. **2.** dealing with health, cleanliness, and conditions that promote good health.

sanitary landfill *n*. a burial site for nonhazardous solid waste materials. Wastes are compacted and covered with a layer of soil the same day that they are placed in the landfill to keep from providing breeding grounds for insects and rodents and to avoid spread of disease, fires, and other environmental hazards. The landfill should be above the groundwater, lined with fine clay, have a mounded clay cap, and have one or more monitoring wells.

sanitary sewer *n.* a sewer that carries wastewater from homes, businesses, and/or factories. See sewage treatment, sewage.

sanitation *n.* (Latin *sanitas*, health + *atio*, action) **1.** the study and use of clean and healthful practices. **2.** the application of practices that produce a more healthful environment, such as drainage, ventilation, and water purification. **3.** provisions for safe waste disposal, such as sewer systems and solid waste-disposal services.

San Jose scale *n.* a small insect (*Aspidiotus perniciosus*, family Coccidae) of widespread occurrence that sucks sap from the leaves, stems, and fruits of many kinds of fruit trees and shrubs.

Santa Ana *n.* (Spanish *Santa Ana*, Saint Anne) a warm, dry summer wind that blows down the mountains toward the California or Chilean coast late in the day when the land is much warmer than the ocean. Also called sundowner.

sap *n.* (Anglo-Saxon *saep*, sap) **1.** water-based fluid that circulates in a plant, especially that in trees. **2.** any liquid that functions in the life processes of a living organism. **3.** the energy and vitality of a living organism. **4.** a fool; a person who is easily duped. **5.** a narrow trench dug to attack or defend a military stronghold. *v.* **6.** to drain sap from, as to sap a maple tree. **7.** to drain strength from, as to sap one's energy or resolve. **8.** to dig or use narrow trenches to approach a military stronghold.

sapling *n.* (Anglo-Saxon *saep*, sap + *ling*, young) **1.** a young tree, especially one of a commercial timber species that has a diameter between 1 and 5 in. (2.5 to 13 cm) at breast height. **2.** a young person.

saponin *n.* (French *saponine*, soap) any of several glycosides found in many forage legumes, consisting of a polar sugar linked to either a steroid or other similar polycyclic compound. Saponin, which has a bitter taste and strong detergent properties, can reduce serum cholesterol in the body.

saprobe *n.* saprophyte.

saprolite *n.* (Greek *sapros*, rotten + *lithos*, stone) disintegrated and decomposed rock material that retains the original rock structure. Also called saprolith.

sapropel *n.* (Greek *sapros*, rotten + *pēlos*, mud) a deposit of mostly organic materials accumulated on the bottom of a shallow anaerobic lake, swamp, or seabed.

saprophyte or **saprobe** *n.* (Greek *sapros*, rotten + *phyton*, plant) any organism such as a fungus or bacterium that lives on dead, dying, or decaying organic matter from either plants or animals. Also called saprophage. See parasite.

saprophytic *adj.* (Greek *sapros*, rotten + *phyton*, plant + *ikos*, like) **1.** living on decaying organic matter. **2.** pertaining to saprophytes. Also called saprobic.

saprozoic *adj.* (Greek *saprós*, rotten + *zōē*, life + *ikos*, like) **1.** feeding on decaying organic matter; e.g., maggots or dung beetles. **2.** nourished on decaying organic material by absorption of the organic and inorganic decay products; e.g., protozoans and some fungi. **3.** an animal parasite such as a tapeworm.

sapsucker *n.* (Anglo-Saxon *saep*, sap + *súcan*, suck + *er*, one who) any of several small American woodpeckers (*Sphyrapicus* sp.) with mostly black and white plumage and either a yellow belly (*S. varius* or *S. thyroideus*) or a red breast and head (*S. ruber*) that drill holes in trees to drink the sap and eat insects they find there.

sapwood *n.* (Anglo-Saxon *saep*, sap + *wudu*, wood) the relatively soft, new cells of trees between the cambium layer and heartwood. Also called alburnum. See heartwood.

sarcobiont *n.* (Greek *sarx*, flesh + *bioun*, to live) a flesh-eating organism.

sarcoptic mange *n.* See mange mite.

Sargasso Sea *n.* a large area of water that is relatively warm and calm and has a clockwise movement located in the North Atlantic Ocean, southeast of the Gulf Stream. It is largely covered by a brown alga called gulfweed or kelp (*Sargassum* sp.) and an accumulation of floating solid wastes.

sarsaparilla *n.* (Spanish *zarzaparrilla*, little parra bush) any of several tropical vines (*Smilax* sp., family Liliaceae) with either climbing or trailing habit, alternate leaves, and flowers on stalks that radiate from the center. **2.** a root extract from such plants used for flavoring beverages such as root beer and to treat psoriasis.

sassafras *n.* (Spanish *sasafras*, sassafras) **1.** a medium to large tree (*Sassafras albidum*, family Lauraceae), with deeply furrowed reddish brown bark and a notable aroma, that grows in fencerows in much of the eastern half of the U.S.A. and is used as a shade tree and to make fence posts. The bark of the roots has been used to make tea and as a source of sassafras oil. **2.** any of several related trees (*Sassafras* sp.) with yellow blossoms and bluish fruit.

sassafras oil *n.* a yellow to orange aromatic oil obtained from sassafras roots and root bark; used to perfume soap and make food flavorings and medicines.

sassafras tea *n.* tea made from the dried bark of the roots of sassafras trees that is used medically as a blood thinner, pain reliever, diuretic, and stimulant but is also suspected as a cause of cancer.

satellitism *n.* (Latin *satellitis*, attendant + *ismus*, action) stimulation of growth by an unrelated species. For example, certain species of bacteria grow more vigorously near colonies of other unrelated species that release a stimulating metabolite.

saturated air *n.* air holding the maximum amount of water vapor possible without causing condensation at its ambient temperature.

saturated fat *n.* a fat or oil whose molecules have no double bonds. Each carbon atom is fully hydrogenated (most common in animal fat).

saturation point *n.* **1.** a physical limit; the maximum amount of a substance that will dissolve or evaporate or that can be placed in a particular area, etc. **2.** an abstract limit, as the saturation point of one's mental capacity. **3.** the minimum light intensity that fully stimulates the photosynthetic pigments to their maximum photosynthetic rate.

Saturn *n.* (Latin *Saturnus*, Saturn) **1.** the second largest planet in the solar system, located between Jupiter and Uranus. Also called the ringed planet. **2.** the Roman god of agriculture.

saturnid *n.* (Latin *saturniidae*, saturnid) any of several brightly colored, large silkworm moths (family Saturnidae) with hairy bodies, broad wings, and small heads.

saturnism *n.* (Latin *saturnus*, lead + *ismos*, condition) lead poisoning. See plumbism.

sauerkraut *n.* (German *sauer*, sour + *kraut*, greens) a food made by anaerobic fermentation of chopped cabbage with about 2.5% salt by weight. It contains about 1% lactic acid ($C_3H_6O_3$) and 0.3% acetic acid (CH_3COOH).

sault *n.* (Latin *saltus*, a leap) a waterfall or rapids in a stream.

sausage *n.* (French *saulsage*, sausage) **1.** ground meat (especially pork but sometimes veal or beef), usually highly spiced, either in bulk (usually made into patties for cooking) or enclosed in a tubular casing as hot dogs or bologna. **2.** the fruit of the tropical sausage tree. **3.** any other object with an elongated tubular shape similar to a hot dog; e.g., a sausage-shaped observation balloon.

sausage tree *n.* a tropical African tree (*Kigelia pinnata*, family Bignoniaceae) with whorls of pinnately compound leaves, large scarlet bell-shaped blossoms, and gourdlike fruit on stalks more than 40 in. (1 m) long.

savanna or **savannah** *n.* (Spanish *sabana*, savanna) a tropical or subtropical ecosystem with a nearly continuous herbaceous (mostly grass) cover coexisting with scattered trees and/or shrubs in an area with strongly seasonal rainfall. Sometimes called a glade. See veldt.

savanna sparrow *n.* any of several sparrows (*Passerculus sandwichensis* and similar species), common in grassy areas of North America, whose plumage is mostly light brown with white streaks on the breast and a notch in the tail.

savin or **savine** *n.* (Anglo-Saxon *safine*, savin) **1.** an evergreen shrub (*Juniperus sabina*, family Cupressaceae) whose fresh tops yield an acrid volatile oil used in folk medicine for rheumatism and intestinal worms and for making perfume. **2.** the drugs made from the volatile oil extracted from dried tops of these shrubs.

savory *adj.* (French *savouré*, tasted) **1.** appetizing; pleasant to taste or smell. **2.** morally pleasing; respectable. **3.** spicy or salty but not sweet; e.g., a relish. **4.** any of several annual or perennial herbs (*Satureja* sp., family Lamiaceae) whose flowers and leaves are used to flavor food, especially summer savory (*S. hortensis*) and winter savory (*S. montana*).

sawfish *n.* (Anglo-Saxon *seax*, knife + *fisc*, fish) a large ray (*Pristis antiquorum* and related species) that grows 12 to 18 ft (3.6 to 5.4 m) long, with a long snout and with sharp teethlike spines on each side, that lives along coasts and nearby portions of rivers in tropical settings.

sawfly *n.* (Anglo-Saxon *seax*, knife + *fleoge*, fly) any of a large number of hymenopterous insects (family Tenthredinidae) whose females have a sawlike ovipositor used to deposit eggs in the tissues of host plants.

sawlog *n.* a tree trunk that is large enough to be sawed into lumber; generally at least 8 ft (2.4 m) long and 6 in. (15 cm) in diameter inside the bark (larger standards are common locally or for certain species).

sawmill *n.* (Anglo-Saxon *seax*, knife + *mylen*, grinder) a structure equipped with machines to saw logs into boards, planks, etc.

saw palmetto *n.* a small palm tree (*Serenoa repens*, family Arecaceae), native to southern U.S.A., that has spiny teeth on its leafstalks.

sawtimber *n.* a stand of live trees that have potential for sawlogs.

sawtooth *n.* (Anglo-Saxon *seax*, knife + *toth*, tooth) **1.** a tooth in a saw. *adj.* **2.** having a profile with a series of points that resemble the teeth of a saw, as a sawtooth mountain range.

saxicoline or **saxicolous** *adj.* (Latin *saxum*, rock + *cola*, dweller) living among rocks; e.g., plants in a rock garden.

saxifragous *adj.* (Latin *saxifragus*, stone breaking) **1.** growing in cracks in rocks, especially a group of mostly perennial plants (*Saxifraga* sp., family Saxifragaceae) found mostly in mountainous areas. **2.** stone dissolving; breaking or dissolving of calculi, especially in the bladder.

saxitoxin *n.* (Latin *saxum*, stone + Greek *toxikon*, poison) a potent neurotoxin ($C_{10}H_{17}N_7O_4$) that accumulates in shellfish that feed on dinoflagellates (*Gonyaulax catanella*) that produce the toxin. Eating such shellfish causes food poisoning in humans. See red tide.

scab *n.* (Norse *skabb*, scar or itch) **1.** a crusty covering the body produces to cover a wound while the skin heals. **2.** mange; a skin disease that afflicts sheep and other animals. **3.** a fungal disease that attacks plants, or a roughened spot produced by the disease. **4.** a person who is detested, especially one who takes the place of a striking worker.

scabies *n.* (Latin *scabies*, the itch) a contagious skin disease caused by itch mites (*Sarcoptes scabiei* on humans or *Psoroptes ovis* on cattle) that deposit their eggs under the skin and cause intense itching.

Cattle hair on the fence is an indication of scabies.

scabland *n.* (Norse *skabb*, scar + Anglo-Saxon *land*, land) an area covered mostly by barren rock (often a lava flow) with little soil and scant vegetation.

scald *v.* (Latin *excaldare*, to wash in hot water) **1.** to heat to almost the boiling point of water [100°C (212°F)]; e.g., to scald milk. **2.** to immerse in or subject to hot water, as to scald tomatoes. **3.** to cause a serious burn by contact with hot water or steam. **4.** a serious burn of plant leaves or bark caused by intense sunlight or heat. **5.** a plant disease that causes

blanching or browning and withering of leaves, either by a fungus that attacks cranberries or by excessive sunlight or improper storage conditions.

scale *n.* (Latin *scala*, ladder) **1.** a ladder, stairway, or other means of ascent. **2.** a measuring stick or tape with markings that show linear distances, or such markings placed on a surface. **3.** the ratio between distance on a map, photograph, or diagram and the corresponding distance on the surface being represented, commonly expressed as a ratio (e.g., 1:20,000, or 1 to 10) or as two measurements (e.g., 4 in./mi, ½ in. = 1 ft, or 5 cm = 1 m). **4.** a grouping system that has a series of steps, as a temperature scale, wage scale, or musical scale. **5.** an instrument used to weigh objects. **6.** one of the thin horny plates that cover the body of certain fishes and reptiles. **7.** any small, thin piece that peels off the skin of a person or animal, the surface of a plant, or the surface of a rock, piece of metal, or painted object. **8.** a precipitated incrustation that forms in pipes, boilers, kettles, etc. **9.** a rudimentary leaf that is part of the cover of a bud or other plant part. *v.* **10.** to climb a steep surface, as to scale a cliff. **11.** to measure size or weight or to estimate the amount of lumber that can be cut from a log or tree. **12.** to remove the scales from a fish, reptile, or surface. **13.** to peel off in scales. **14.** to change the size of a thing or of its representation, as to scale down a project or to scale a drawing. *adj.* **15.** made proportionally larger or smaller, as a scale model or a scale drawing.

scale insect *n.* any of a wide variety of tiny insects (superfamily Coccoidea) that suck sap from plants and whose females cover themselves with a waxy, scalelike secretion. See San Jose scale.

scallion *n.* (French *escalogne*, scallion) a green onion with a long, thick stem and very little enlargement for its bulb (e.g., *Allium ascalonicum*, family Amaryllidaceae). Also called green onion.

scallop *n.* (French *escalope*, shellfish) **1.** any of several bivalve mollusks (*Argopecten* sp. and related genera) that move by flapping their shells to pump water. **2.** the main muscle of such mollusks, used as a seafood. **3.** a small dish made from the shell of such a mollusk or shaped to represent it. **4.** a curved shape that is applied repeatedly around the edge of a tablecloth or other covering or surface. *v.* **5.** to cut such curved shapes into the edge of a covering or surface. **6.** to bake in a milk sauce, usually with bread crumbs and seasoning, as to scallop potatoes.

scalp *n.* (Icelandic *skalpr*, leather sheath) **1.** the skin that is typically covered with hair on the top, back, and sides of the head. **2.** a trophy, especially part or all of the scalp of an enemy. **3.** the corresponding part of a dog, wolf, or whale. *v.* **4.** to remove the scalp from a person or animal. **5.** to remove the vegetation and part or all of the soil from an area, either in preparation for planting trees or other plants or to use the soil elsewhere. **6.** to separate the dust, chaff, or other worthless parts from the flour or other product of a mill. **7.** to cheat or extort; e.g., by selling tickets at an exorbitant price. **8.** to buy and sell frequently in an effort to make a quick profit.

scandium (Sc) *n.* (Latin *Scandia*, Scandinavia + *ium*, element) element 21, atomic weight 44.95591; a very light, relatively soft, silvery white metal that tarnishes to a yellowish or pinkish cast. It occurs in small amounts in many different minerals, reacts readily with various acids, and is used in high-intensity lights and as a tracer in refinery crackers for crude oil.

scar *n.* (Latin *eschara*, scar) **1.** a mark left on the skin where a wound or sore has healed. **2.** a blemish that looks like a scar on any surface. **3.** psychological damage, as a mental scar. **4.** a blemish where a mold did not fill properly. **5.** a parrot fish (family Scaridae). **6.** (mostly British) a protruding rock, or a steep, rocky face. *v.* **7.** to damage a surface, leaving a scar.

scarab *n.* (Latin *scarabaeus*, altered) **1.** the winged black dung beetle (*Scarabaeus sacer*). Its image was worn as a charm or used as a seal in ancient Egypt. **2.** any of a large family of beetles (family Scarabaeidae) with mostly stout bodies and bright colors, including June bugs, dung beetles, and others.

scarify *v.* (Latin *scarifare*, to scratch) **1.** to scratch or abrade a surface, as to scarify the skin for a vaccination or to scarify hard seed so it will absorb water and germinate. **2.** to treat with concentrated sulfuric acid for 30 min and then rinse with distilled water as a seed treatment. **3.** to break the hard crust on a puddled soil so seedlings can emerge. **4.** to break up a road surface as an initial step in its repair.

scarp *n.* (Italian *scarpa*, slope) **1.** a steep slope or cliff, especially one formed by vertical displacement along a fault zone; an escarpment. **2.** a steep slope built as part of the defense system around a fortification, especially that on the inner side of a ditch or moat. *v.* **3.** to make a steep slope.

scatology *n.* (Greek *skatos*, dung + *logos*, word) **1.** the science that identifies species of insects, birds, or other animals by the feces they leave behind. **2.** an obsession with fecal matter and processes, especially in obscene writing.

scavenger *n.* (French *scawage*, inspection) **1.** an animal, bird, or insect that feeds on carcasses of dead animals; e.g., a hyena, a vulture, or a scavenger beetle. **2.** a person that sorts through trash to try to salvage something of value; a form of recycling.

3. a chemical or process that removes pollutants from a fluid or a mixture. Rain and snow are scavengers when they remove dust and other pollutants from the air.

scavenger well *n.* a well located near the coast to intercept salt water that might otherwise intrude into freshwater aquifers and wells.

scenery *n.* (Latin *scenarius*, of scenes) **1.** the appearance of a landscape, especially one that is attractive, as we have a nice view of the beautiful scenery. **2.** a stage setting, especially one that represents a landscape.

schist *n.* (Greek *schistos*, divided) a metamorphic rock with a fine-layered structure, usually with a high mica content giving it a flaky appearance.

schistosomiasis *n.* (Greek *schistos*, divided + *sōma*, body + *iasis*, disease) snail fever. See bilharziasis.

schizogenesis *n.* (Greek *schizo*, split + *genesis*, origin) reproduction by fission, as that of bacteria.

schizomycete *n.* (Greek *schizo*, split + *mykēs*, fungus) any microbe (class Schizomycetes) that reproduces by fission, including bacteria and fission fungi.

school section *n.* land designated for the establishment and support of public schools. In 1785, as part of the survey of the Northwest Territory, the U.S. government set aside section 16 of each township (there are 36 sections of 640 ac each in a township) for the support of public schools. Section 36 was set aside later for the same purpose, with the provision that alternate tracts in units of 40 ac would be substituted if section 16 or 36 was known to contain minable minerals.

scientific name *n.* a binomial Latin name (genus and species) assigned to an organism according to the system developed by Carl von Linné (1707–1778), a Swedish physician better known as Linnaeus.

scintillation *adj.* (Latin *scintillatio*, sparking) **1.** a spark or the action of producing a spark. **2.** sparkling or twinkling, as the scintillation of a star. **3.** any shimmering of an image caused by temperature or other variations in the atmosphere. **4.** a pulse or flash of light produced by a photon of energy; e.g., the dots of light that form an image on a cathode-ray tube such as a television set, a computer monitor, or a radar display.

scintillation counter *n.* a device that measures radiation (e.g., from radioactive materials) by counting the flashes of light produced in a scintillator material.

scion *n.* (French *cion*, a cutting) a detached woody shoot or twig containing buds to be grafted to a different plant or rootstock. The scion usually contains two or three buds, although it may contain more. When the scion is only a single bud, the form of grafting is called budding.

sciophyte *n.* (Greek *skio*, shadow + *phyton*, plant) a plant that grows in and/or does well in the shade.

sclerotinia *n.* (Latin *sclerotinia*, hard fungi) fungi (*Sclerotinia* sp.) that form sclerotia. Many species cause injury to lettuce, clovers, fruits, alfalfa, and kidney beans.

sclerotium (pl. **sclerotia**) *n.* (Latin *sclerotium*, hard mass) a hard, dry, irregularly shaped mass of fungal protoplasm that constitutes a resting structure formed asexually by certain slime molds as a means of surviving over winter or during unfavorable periods.

scoliid wasp *n.* any of several brightly colored predaceous wasps (*Scolia* sp., family Scoliidae) that lay eggs on the larvae of the Asiatic and Oriental beetles. Upon hatching, the young wasps kill the beetle larvae.

scoparius *n.* (Latin *scopa*, thin branch + *arius*, connected to) a legume (*Cytisus scoparius*, family Fabaceae) with a broomlike growth habit. The foliage contains scoparin and the alkaloid sparteine. Leaves of scoparius are sometimes smoked for their euphoric properties. Also called broom.

scopula *n.* (Latin *scopula*, a little broom) a dense tuft of hair on the feet of certain spiders (used in building webs) or on the lower legs of certain bees.

scoria *n.* (Greek *skōria*, refuse) **1.** porous, cinderlike, basaltic lava rock. **2.** slag from ore that has been smelted. **3.** the residue from a burned out coal bed.

scorpion *n.* (Latin *scorpio*, scorpion) an eight-legged arachnid (order Scorpionida) whose tail curves forward and can inject a poison into an insect, animal, or person. Scorpions live mostly on insects.

Scotch pine *n.* a large coniferous pine tree (*Pinus sylvestris*, family Pinaceae) with spreading branches. Its needles grow 2 to 3 in. (5 to 8 cm) long, are two in a cluster, and are a bluish green. Native to Europe, it is now grown widely throughout temperate climates and used in plantations, shelterbelts, and as an ornamental (the most popular Christmas tree species in the U.S.A.).

scotophil *adj.* (Greek *skotos*, darkness + *philia*, affinity) growing best in darkness, as a sciophyte.

scotophobin *n.* (Greek *skotos*, darkness + *phobos*, fear + *in*, pertaining to) a peptide ($C_{62}H_{97}N_{23}O_{26}$) found in the brains of rats and mice that have been conditioned to fear darkness. When injected into normal rodents, it induces fear of darkness.

scour *v.* (Latin *excurare*, to clean or care for) **1.** to rub, often hard and with an abrasive, to remove dirt, grease, stain, etc. **2.** to remove loose material, as a stream scours its channel. **3.** to purge, as to scour an animal or to scour a nation of spies. **4.** to eliminate debris and impurities. **5.** to corrode a surface. **6.** to polish a surface. **7.** to not stick, as this soil scours well from a plowshare. *n.* **8.** the act of scouring. **9.** a material that scours; e.g., sand. **10.** a place that has been scoured, as a streambed. **11.** the erosive scouring action of water.

scouring *adj.* the erosive action of running water, flowing air, or glacial ice.

scours *n.* infectious animal diarrhea, especially in calves, lambs, baby pigs, and foals only a few days old, that causes diarrhea, fever, and pneumonia. It is the most devastating killer of beef and dairy cattle in the U.S.A., claiming nearly 10% of the 40 million calves born each year, and is best controlled by providing a clean living area with adequate ventilation and proper nutrition. Also called common scours, white scours, or dysentery.

scouting *n.* See crop scout, integrated pest management.

scrapie *n.* an infectious brain disease of sheep that is usually fatal and can be spread when sheep leave wool on objects by scraping against them to relieve itch. It also causes them to grind their teeth and twitch their heads.

scree *n.* (Norse *skritha*, landslide) loose rock debris on a steep slope (talus) at the base of a cliff or other rocky area.

screen analysis *n.* sieve analysis. See particle-size analysis, sieve analysis.

screenings *n.* foreign, fragmented, or undersize material removed by passing grain or other product over a screen. Grain screenings are sometimes used as livestock feed.

screwworm *n.* the larva of a fly (*Cochliomyia hominivorax*, family Calliphoridae) that is a very serious pest infesting wounds and nasal drainage of livestock and people. It was eliminated from the U.S.A. and Mexico by releasing large numbers of sterilized males, thus causing the females to lay sterile eggs. See myiasis.

scrub *n.* (Danish *skrub*, brushwood) **1.** small trees or shrubs, usually of inferior quality. **2.** an area covered by such trees or shrubs. **3.** a runt or inferior animal. **4.** anything that is undersized or inferior in quality. *adj.* **5.** scrubby. *v.* **6.** to clean by rubbing with a brush or cloth soaked in water. **7.** to remove foreign matter from smoke or exhaust fumes by a system of filters and/or chemical agents.

scrubby *adj.* small or inferior. **6.** covered with scrub trees or shrubs.

scrub typhus *n.* an infectious disease that affects rodents and people, especially in Japan and the East Indies, that is caused by a rickettsia (*Ricksettia tsutsugamushi*), which is transmitted by harvest mites and chiggers. Symptoms include headache, fever, apathy, and weakness.

scud *v.* (Anglo-Saxon *scudan*, to hurry) **1.** to move fast or quickly. **2.** to sail in a high wind with little sail set. **3.** to miss widely with an arrow. **4.** to cleanse hair and dirt from a hide. *n.* **4.** a quick movement. **5.** a cloud or fine mist driven by a strong wind. **6.** the hair and dirt removed from a hide.

sculpt *v.* (Latin *sculpere*, to carve) **1.** to carve, mold, or otherwise shape a sculpture. **2.** to shape or be shaped by erosion, as the wind sculpts a rock.

scum *n.* (Dutch *schum*, scum) **1.** a thin layer of algae, fat, oil, or other foreign matter on the surface of a body of water; e.g., pond scum. **2.** a similar layer on the surface of some other liquid. **3.** the refuse or other material that forms such a layer; e.g., scoria in a furnace. **4.** a person with an evil nature or of the lowest class, or such persons collectively, as the scum of the Earth. *v.* **5.** to form scum. **6.** to remove scum from a surface.

scurf *n.* (Norse *scurfa*, crust) **1.** dandruff; thin scales that flake loose from the epidermis. **2.** any scaly incrustation or flaking on a surface.

scurfpea *n.* any of several bushy herbaceous plants (*Psoralca* sp., family Fabaceae) that grow in midwestern and southwestern U.S. rangelands, prairies, and open forests, including some that are toxic to livestock; e.g., Arabian scurfpea (*P. bituminosa*), silverleaf scurfpea (*P. argophylla*), and slimflower scurfpea (*P. tenuiflora*). Also called scurfy psoralea.

scythe *n.* (Anglo-Saxon *sithe*, scythe) **1.** a cutting instrument with a long, arcing blade and a long, crooked handle with handholds for both hands that is used to cut grass or ripe grain near ground level without stooping. **2.** to harvest with a scythe.

Se *n.* the chemical symbol for selenium.

sea *n.* (Anglo-Saxon *sae*, sea) **1.** a large body of salt water surrounded by land. **2.** a part of an ocean. **3.** the oceans in general. **4.** the rising, falling, rolling action of ocean waves, or the lack thereof, as a rough sea or a calm sea. **5.** a vast expanse or amount of anything, as a sea of faces or a sea of debt. **6.** the way of life associated with the sea.

sea anemone *n.* any of several sedentary marine animals (phylum Coelenterata) with tentacles surrounding a mouth at the top of a columnar-shaped body that expands and contracts.

sea breeze *n.* a breeze that blows from sea to land, especially on a warm day. A land breeze blowing from the land to the sea is more likely at night (the land warms and cools faster than the sea; whichever side is warmer causes the air to rise and the wind to blow toward it).

seacoast *n.* the land next to a sea or ocean. It normally has a view of the sea and has its climate and other characteristics influenced by the sea.

sea duck *n.* any of several diving ducks (family Anitidae) that frequent salt water, especially the American eider.

sea eagle *n.* any of several large eagles (*Haliaetus* sp.) that feed on fish.

sea elephant *n.* See elephant seal.

seafood *n.* any fish or shellfish that is taken from the sea and eaten by humans. Seafood is a good source of iodine and may lower the risk of heart disease.

sea horse *n.* any of several small fish (*Hippocampus* sp., family Syngnathidae) that swim in an upright position. Their bodies are covered with bony plates, and their head is bent at an angle like that of a horse.

seal *n.* (Anglo-Saxon *seolh*, seal) **1.** any of several large sea mammals, including the true seals (family Phocidae) and the eared seals (family Otariidae). A seal has a plump, smooth body, a doglike head, and four webbed feet. They move with a jumping motion on land, are excellent swimmers in cool or cold sea waters, and eat mostly fish. **2.** sealskin; the fur of certain seals, or leather made from the skin. **3.** a dark grayish brown. **4.** an emblem, figure, stamp, or other mark used to hold something closed or as a mark of authenticity, e.g., on an official document. **5.** a tight closure, especially one that keeps air and water from passing in or out of a container or enclosure. **6.** a stamp used for decoration. **7.** confirmation or acceptance, as a seal of approval. *v.* **8.** to close with a seal. **9.** to hunt seals.

sea lamprey *n.* an eel-like external parasite (*Petromyzon marinus*, family Petromyzontidae) that kills fish by attaching itself to the body and sucking the blood. Although primarily a saltwater creature, it now spawns in the freshwater Great Lakes.

sea level *n.* the mean water level halfway between high tide and low tide, taken as the zero reference level for land elevations and ocean depths.

sea lion *n.* a large-eared seal, either Steller's sea lion (*Eumetopias jubatus*) of the northern Pacific or the California sea lion (*Zalophus californicus*).

sea mile *n.* See nautical mile.

seashell or **sea shell** *n.* the shell (external skeleton) of any marine mollusk. Ground seashells contain up to 99% calcium carbonate and are a good source of calcium for laying hens, farm animals, and humans.

season *n.* (French *saison*, due time) **1.** an appropriate or customary time, as winter is topcoat season, the growing season here is from May to October, or it's breeding season now for these animals. **2.** any of the four seasons, spring, fall, summer, or winter. **3.** the part of the year when a certain condition prevails, as a wet season or a dry season. **4.** a time for celebration, as Christmas season or the holiday season. *v.* **5.** to flavor food with salt and/or spices. **6.** to add interest or relish, as to season a conversation with wit. **7.** to allow time for something to ripen or mature. **8.** to make less severe, as to season one's reaction with discretion.

seasonal *adj.* (French *saison*, due time + *al*, pertaining to) varying with the season of the year, as the rainfall here is seasonal, or the fresh produce is seasonal.

sea spray *n.* tiny droplets of salt water thrown up by breaking waves and carried inland by the wind.

sea turtle *n.* any of several large reptiles (families Cheloniidae and Dermochelyidae) whose bodies are enclosed in dome-shaped shells. They can pull their limbs and head inside the shell for protection, and their limbs form paddles more suitable for swimming than for walking. They are widespread in tropical and subtropical seas but either threatened or endangered by ocean-water pollution.

sea urchin *n.* **1.** any of a group of echinoderms (class Echinoidea) with round or globular calcareous shells formed of fused plates covered with projecting spines. **2.** a small evergreen tree or tall shrub (*Hakea laurina*), native to Australia, with narrow leaves and dense clusters of crimson flowers.

seawall *n.* a sturdy wall built along the shore to prevent breaking waves from eroding the land.

seaweed *n.* **1.** a marine alga, especially kelp (e.g., *Fucus* sp. and *Sargassum* sp.). **2.** any plant that grows in seawater, either free floating or anchored to the seafloor. Several seaweeds are useful to make gelatin (*Gelidium* sp.) or other food products; e.g., brown seaweed (*Laminaria* sp. and *Undaria* sp.), grown in abundance in China and Japan, and some red seaweed (*Chondhus crispus*). Some are used to make fertilizer, paper, or furniture stuffing.

sebkha or **sebka** *n.* a salt flat formed in a closed depression where water accumulates in an arid area.

secobarbital *n.* an odorless barbiturate ($C_{12}H_{18}N_2O_3$) used as a sedative, anticonvulsant, and hypnotic.

second (**s**) *n.* (Latin *secundus*, second) **1.** a short interval of time, now defined as the duration of 9,192,631,770 cycles of radiation in a cesium atom in its transition state. There are 60 sec in 1 min and 3600 sec in 1 hr. **2.** any brief interval of time. **3.** an angular measure equal to 1/60th of a minute or 1/3600 degree. **4.** an alternate, substitute, or assistant, as his second will do that. **5.** second gear in a transmission, as shift into second now. *adj.* **6.** number two in a series. **7.** one of every two, as every second day. **8.** not the highest quality, as second best. **9.** the person addressed, as the second person in grammatical construction.

secondary nutrient *n.* a plant macronutrient (calcium, magnesium, or sulfur) that is less often needed as fertilizer than are the primary nutrients (nitrogen, phosphorus, and potassium). The secondary nutrients are often incidental components of primary nutrient fertilizers and are used as soil amendments in relatively large amounts. See essential element, macronutrient, primary nutrient.

secondary succession *n.* the redevelopment sequence of an ecosystem on a site that has lost its previous vegetation by a disturbance such as fire, flooding, wind damage, or overgrazing.

secondary treatment *n.* See sewage treatment.

second bottom *n.* a former floodplain that is now a low terrace and no longer subject to flooding. See bottomland, floodplain.

secondhand smoke *n.* cigarette, cigar, or pipe smoke inhaled from air in a room where someone else is smoking. The EPA concluded in 1993 that secondhand tobacco smoke is a human lung carcinogen that is responsible for about 3000 deaths each year in the U.S.A.

secretion *n.* (Latin *secretio*, sifted) **1.** the action of releasing and discharging a substance from a cell either to perform a specific function or to be excreted as waste. **2.** such a substance released and discharged by a plant or animal. See exocytosis.

secretory vesicle *n.* a structure produced by a Golgi body to transport its biosynthetic product to the cell membrane for discharge outside the cell.

section *n.* (Latin *sectio*, a cutting) **1.** a subdivision of something; a distinct part, e.g., of a piece of furniture, a fishing rod, a publication, or a legal code. **2.** any one of the 36 sections of land in a township; an area 1 mi (1.609 km) square comprising 640 ac (259 ha), except where adjustments had to be made to fit the survey to the landscape. This system is used in most of the U.S.A. west of Ohio. **3.** the action of making an incision, cut, or separation into parts. **4.** a view of a layer within an object or organism, as a cross section or a thin section to be viewed through a microscope. **5.** a segment of a fruit such as an orange or grapefruit. *v.* **6.** to divide or cut something into sections. **7.** to show sections, e.g., by shading or marking patterns on parts of a drawing.

sector *n.* (Latin *sector*, cutter) **1.** part of a circular area bounded by two radii; e.g., like a piece of pie. **2.** a mechanical device with scales for making proportional drawings or for measuring angles; e.g., an astronomical instrument used to measure the angle between a star and the zenith.

sedentary *adj.* (Latin *sedentarius*, sitting) **1.** pertaining to or done in a sitting position, as a sedentary occupation. **2.** not very active; sitting and resting most of the time. **3.** not migratory; staying in or attached to one spot; e.g., a barnacle. **4.** residual; formed in place, as a residual soil.

sedge *n.* (Anglo-Saxon *secg*, sedge) any of many wetland plants (*Carex* sp., family Cyperaceae) that are similar to grasses but with nonjointed stems.

sediment *n.* (Latin *sedimentum*, settling of dregs) **1.** solids that have settled from suspension in a liquid such as sand, silt, or clay (popularly known as silt). **2.** organic or mineral solids that have been or are being transported and deposited by air, water, wind, or glaciers. See Stokes' law.

sedimentary peat *n.* See colloidal peat.

sedimentary rock *n.* rock formed by sedimentation, with or without chemical cementation; e.g., limestone, shale, siltstone, sandstone, or loess.

sedimentation *n.* (Latin *sedimentum*, settling of dregs + *atio*, action) **1.** the process of deposition and accumulation of materials transported by water or air.

sedimentation rate *n.* the rate of settling and accumulation of alluvium in a water environment or of windblown materials such as loess.

sedimentology *n.* (Latin *sedimentum*, settling of dregs + *logia*, description) the geologic science that deals with sedimentation and sedimentary rocks.

sediment yield *n.* **1.** the amount of soil and/or rock material in an area that is detached and moved by erosion. **2.** the amount of such material that is carried into the streams that drain the area. **3.** the amount of such material that is carried past a downstream measuring point. **4.** the dry weight of sediment per unit volume or per unit of total weight of the water-sediment mixture.

sediment yield rate *n.* the sediment yield per unit area occurring in a unit time, often expressed in tons/ac-yr or metric tons/ha-yr.

sedum *n.* (Latin *sedum*, houseleek) any of a group of plants (*Sedum* sp., family Crassulaceae) that grow on rocks and walls. Sedums are mostly perennials with fleshy stems and leaves and white, yellow, or pink blossoms.

seed *n.* (Anglo-Saxon *saed*, seed) **1.** an embryonic plant in a resting state; a mature ovule with a supply of stored food and the structures needed to initiate growth of a new plant. **2.** any similar plant part suitable for seeding, including true seeds, dried fruits containing one or more seeds, bulbs, and tubers. **3.** a quantity of seeds, as seed for next year's crop. **4.** sperm or semen. **5.** descendants, as Abraham's seed. **6.** a small beginning, as the seed of a revolution. *v.* **7.** to plant seeds. **8.** to remove seeds, as from fruit. **9.** to scatter silver iodide in a cloud in an endeavor to cause rain to fall. **10.** to place something where it may start a process, as to seed crystals in a supersaturated solution. **11.** to rank an athlete or team for a certain position in a contest.

seed bank *n.* **1.** a place for long-term storage of seeds, especially as a means of preserving germ plasm. Also called gene bank or germ-plasm bank. **2.** the total amount of seeds present in a soil.

seedbed (Anglo-Saxon *saed*, seed + *bed*, a plot of ground) **1.** a strip or area where the soil has been prepared for seeding, commonly by tillage that kills any existing vegetation, loosens the soil, and recompacts it to the desired density for seed to sprout and seedlings to grow. **2.** a plot of ground where seedlings will be grown prior to transplanting. **3.** a bed of soil covered with glass that serves as a miniature greenhouse for seedling growth. **4.** the place where an idea or movement originates.

seed-borne disease *n.* any plant disease that may be transmitted from one generation to the next via the seed (e.g., *Diplodia* sp. and *Fusarium* sp.). See seed treatment.

seed coat *n.* the integument covering a seed; a hard outer surface that protects the interior.

seed corn maggot *n.* the pale yellow, legless maggot of a fly (*Hylemya platura*, family Anthomyiidae) that was introduced from Europe into the U.S.A. and is now widespread. The flies emerge in early spring and lay eggs in the soil about the time corn is planted. The developing maggots prevent seedling growth by boring into the sprouting seeds of corn, beans, peas, and many other plants.

A seed corn maggot is about 3/8 in. (1 cm) long.

seed cotton *n.* cotton as picked from the boll with seed and lint attached, not yet ginned.

seed fly *n.* a fly (*Agromyza lantanae*, family Agromyzidae) imported from Mexico for the biological control of lantana shrubs (*Lantana camara*, family Verbenaceae). It eats the berries of the shrubs, thus preventing birds from spreading the seeds.

seed label *n.* a tag or label required by federal and/or state seed laws, generally requiring seed lots to be labeled with information such as net weight, % pure seed, % weed seed, % other crop seed, % inert matter, % hard seed, % viable seed, date of germination test, and name and address of the supplier.

seedling *n.* (Anglo-Saxon *saed*, seed + *ling*, little) **1.** a plant grown from a seed rather than from a bud or other plant part. **2.** a young plant, especially one to be transplanted. **3.** a small tree, usually less than 3 ft (0.9 m) tall.

seedling blight *n.* a disease that attacks germinating seeds and young seedlings, often caused by fungi (*Fusarium* sp., *Diplodia* sp., *Gibberella* sp., etc.). It may cause serious damage to crops, garden plants, and greenhouse plants.

seedpod *n.* See pod.

seed testing *n.* a laboratory analysis of seed quality, including purity, freedom from weed seed, germination, etc. Each U.S. state operates a laboratory that tests seeds for sale, and most other nations do likewise.

seed tick *n.* a small, young tick that resembles a seed.

seed treatment *n.* **1.** application to seed of a chemical (especially a fungicide and/or insecticide) in a dust, slurry, liquid, or vapor form to prevent the growth of microbes that cause disease. Mercury compounds are often used for this purpose, and the treated seed cannot be used for animal or human food. **2.** some other process used for the same purpose, such as heating in water. **3.** inoculation; application of nitrogen-fixing bacteria (*Rhizobium* sp.) to legume seed.

seep *n.* (Anglo-Saxon *sipian*, soak) **1.** a seep spot. **2.** the water or petroleum oozing from such a place. *v.* **3.** to ooze or flow slowly through or from a porous material. **4.** to gradually spread or diffuse, as fog seeps across a landscape or a new method seeps through an organization.

seep spot *n.* a small wet area where water or petroleum is oozing from underground. See saline seep.

segment *n.* (Latin *segmentum*, cut part) **1.** a distinct part of a body, especially of an arthropod; a section or division. **2.** a part cut off from a body or from a figure in a drawing, as a segment of a circle. **3.** a part of something, as a segment of a computer program or of a magazine article. *v.* **4.** to break, cut, or divide into segments.

seiche *n.* (French *seiche*, seiche) a wave that oscillates like a tide moving back and forth across a lake, bay, or gulf for a few hours after having been initiated by an earthquake, strong wind, or a marked change in atmospheric pressure. See wind tide.

seine *n.* (Latin *sagena*, seine) **1.** a large net fitted with weights at the bottom and floats at the top. It is either placed across the path of fish or towed through the water to catch fish commercially. *v.* **2.** to use a seine to catch fish.

seism *n.* (Greek *seismos*, shaking) an earthquake.

seismic *adj.* (Greek *seismos*, shaking + *ikos*, of) pertaining to earthquakes or other vibrations in the Earth.

seismogram *n.* (Greek *seismos*, shaking + *gramma*, drawing) a graph of an earthquake as recorded by a seismograph.

seismograph *n.* (Greek *seismos*, shaking + *graphos*, drawer) an instrument that measures and records vibrations in the Earth, as those caused by an earthquake. See seismometer.

seismology *n.* (Greek *seismos*, shaking + *logos*, word) a scientific study of earthquakes. Also called seismography.

seismometer *n.* (Greek *seismos*, shaking + *metron*, measure) a seismograph that accurately indicates the direction and intensity of each movement of the ground during an earthquake, usually by recording the apparent movement of a heavy object that is suspended on springs that enable it to remain stationary while its supports move.

selective *adj.* (Latin *selectus*, gathered apart + *ivus*, tendency) **1.** choosing or acting according to some characteristic or criterion. **2.** affecting some but not others, as a selective herbicide that kills target species but not crop species.

selenium (Se) *n.* (Greek *selēnē*, moon + *ium*, element) element 34, atomic weight 78.96; similar to sulfur but heavier. It is a constituent of glutathione peroxidase, essential for domestic animals (see white-muscle disease) and assumed so for humans, but excessive amounts cause selenosis. It is used to make light sensors because the electrical resistance of selenium crystals varies with light intensity.

selenosis *n.* (Greek *selēnē*, moon + *osis*, condition) a disease caused by selenium poisoning. Certain plants such as mustard (*Brassica* sp.), milkvetch (*Astragalus* sp.), woody aster (*Machaeranthera* sp.), and prince's plume (*Stanleya* sp.) accumulate selenium and, on high-selenium soils in semiarid regions (e.g., north-central Great Plains and the San Joaquin Valley in California), become toxic to livestock because selenium replaces sulfur in amino acids, thus altering the proteins formed with those amino acids and making them inactive. Also called alkali disease.

self-pollination *n.* the fertilization of a plant ovum by pollen from the same plant. Also called selfing when done by a plant breeder.

self-purification *n.* the natural process whereby a flowing stream decomposes organic pollutants with the aid of naturally occurring aerobic bacteria.

self-replicating *adj.* **1.** making exact copies of itself, as a strand of DNA or a virus. **2.** reproducing asexually; e.g., bacteria and some other organisms.

self-sterile *adj.* not able to fertilize itself.

selva *n.* (Latin *silva*, forest) a dense tropical rain forest, usually composed of three levels: scattered very tall trees that emerge above the rest, a nearly continuous canopy dominated by broad-leaved trees that remain green all year, and an understory of trees and brush; e.g., in the Amazon basin of South America.

Three levels of vegetation in a selva

semiarid *adj.* (Latin *semi*, half + *aridus*, to be dry) between arid and subhumid; receiving enough precipitation to grow short bunchgrasses or other vegetation that provides partial cover but is generally dry enough to cause a dormant period in the summer; commonly receiving between 10 and 20 in. (25 to 50 cm) of precipitation annually in temperate regions, less than that in cold regions, and up to about 50 in. (125 cm) in warm climates. See steppe.

senecio *n.* (Latin *senecio*, old) any of a large, diverse group of trees, shrubs, vines, and herbs (*Senecio* sp., family Asteraceae) with alternate or basal leaves, and usually with yellow ray flowers, including many weeds; e.g., golden ragwort (*S. aureus*), a plant that is poisonous to livestock.

senescence *n.* (Latin *senescentia*, growing old) the process of aging that is applicable to all organisms (plants, animals, and people) and to certain other entities, such as lakes or landscapes.

senility *n.* (Latin *senilis*, old + *itas*, condition) **1.** a state of physical and mental deterioration associated with old age in humans and animals. **2.** the end of a geologic erosion cycle when the topography approaches a base-level plain.

sensible temperature *n.* the temperature a person feels, including the effects of radiant energy, wind speed, and relative humidity as well as the actual temperature. See heat index, wind chill index.

sensor *n.* (Latin *sensus*, felt + *or*, thing) **1.** a detection device that responds to light, temperature, movement, or some other environmental factor. **2.** a radiation detector for a specific spectral band (a narrow range of frequencies or wavelengths). **3.** a sense organ; e.g., an eye, nose, or tongue.

sepal *n.* (Modification of Latin *petalum*, petal) any one of the green leaves that cover a bud or surround the base of an open blossom.

sepedon *n.* (Greek *sepedon*, rottenness) putrefaction; a rotten, septic condition.

sepsis *n.* (Greek *sēpsis*, decay) poisoning by products of putrefaction, either pathogenic microbes or their toxins, especially septicemia.

septage *n.* (Latin *septicus*, rotting + *age*, product) **1.** the sludge that accumulates in a septic tank or cesspool; equivalent to sewage sludge. **2.** such material pumped from a septic tank to clean the tank. About 6 billion gallons (23 billion liters) of septage are produced each year in the U.S.A.

septic *adj.* (Greek *sēptikos*, rotting) **1.** rotten, putrid; infective; mostly anaerobic and foul smelling. See antiseptic, septic tank. **2.** not sterile; infected or contaminated with microbes. **3.** produced by or filled with putrefaction.

septic tank *n.* a tank for anaerobic decomposition and the settling of solids from the effluent of one or more homes. Excess liquid runs into a drain field, where it seeps into the soil. The sludge containing the solids is pumped out periodically and either taken to a sewage plant or spread on the soil, where it can decompose harmlessly.

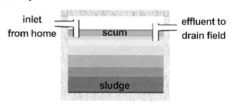
A cross section of a septic tank.
Usually, the entire tank is buried in the ground.

septum *n.* (Latin *septum*, enclosure) a partition; a membrane, wall, or other structure that divides a plant or animal part into two cavities.

sequence analysis or **sequencing** *n.* determination of the sequence of amino acids forming the genetic code on a chromosome or a section thereof.

sequential cropping *n.* growing two or even three crops one after the other on the same land in a one-year period; e.g., growing rice during the wet season and cassava during the dry season in a tropical area, or growing winter wheat followed by soybeans in southeastern U.S.A. Also called double cropping when two crops are grown during a year.

sequester *v.* (Latin *sequestrare*, to remove or lay aside) **1.** to segregate or set apart, as to sequester a jury. **2.** to take possession of; to claim, as to sequester property. **3.** to take one kind of thing out of a mixture. **4.** to chelate; to form a stable, water-soluble complex that prevents metallic ions (e.g., Al^{3+}, Ca^{2+}, Cu^{2+}, Fe^{2+}, Fe^{3+}, Mg^{2+}, Mn^{2+}, and Zn^{2+}) from circulating in solution and precipitating in insoluble compounds. Natural sequestrants and the ions they sequester include vitamin B_{12} (Co^{2+}), chlorophyll (Mg^{2+}), and hemoglobin (Fe^{2+}). Sequestrants are used in micronutrient fertilizers and to prevent rancidity in fats and oils. See chelate, EDTA.

sequestrant *n.* (Latin *sequestrare*, to lay aside + *antem*, one that) a chelating agent; an organic chemical that sequesters divalent (or higher-charged) cations. See chelate, EDTA.

sequoia *n.* (Cherokee *sikwo ya*, big tree) **1.** a very large coniferous tree (*Sequoia giganteum*, family Taxodiaceae) named after a Cherokee Indian scholar, Sequoya (1770?–1843). Sequoias grow in small areas near the California coast. The largest one has a circumference of 83 ft 2 in. (25.35 m) and a height of 275 ft (83.8 m). Also called big tree or giant sequoia. **2.** a redwood (*Sequoia sempervirens*).

sere *n.* (Latin *series*, order) **1.** the entire succession of stages that an ecosystem passes through from bare soil (e.g., in recovering from a severe disturbance) to climax vegetation. Each change in the sere is called a seral stage. *adj.* **2.** seared; dried out; withered.

serotinous *adj.* (Latin *serotinus*, coming late) **1.** bearing flowers later in the season than most species. **2.** remaining attached to the plant for many months after maturity, as the seeds in certain pinecones.

serotype *n.* (Latin *serum*, whey + *typus*, the mark of a blow) **1.** a group of individuals or cells that have the same antigens. **2.** the set of antigens that distinguishes such a group from other similar groups.

serrate *adj.* (Latin *serratus*, like a saw) **1.** notched around the margin like a saw, as a serrate leaf. Also called serrated or sawtooth. **2.** grooved on the edge, like some coins. *v.* **3.** to make notches around the edge.

serum *n.* (Latin *serum*, whey) **1.** blood serum; the clear or yellowish liquid left when blood clots that is used as an antitoxin to transfer immunity from a disease. **2.** any watery fluid obtained from the body of an animal. **3.** a similar watery fluid obtained from a plant. **4.** whey.

serval *n.* (Portuguese *cerval*, deer wolf) a nocturnal African cat (*Felis serval*) about 2 ft (60 cm) tall at the shoulders, with long legs and a tawny coat with black spots.

sesame *n.* (Greek *sēsamē*, sesame plant) **1.** a drought-tolerant, tropical, herbaceous plant (*Sesamum indicum*, family Pedaliaceae), native to India, that has alternate leaves and solitary yellow or pink axillary flowers. It produces oil and seeds used in cooking. **2.** the small, flat, oval seeds used to flavor bread, crackers, and other foods.

sesquioxide *n.* (Latin *sesqui*, more by a half + French *oxide*, oxide) a molecular compound with 1.5 oxygen ions for each cation; aluminum and ferric iron oxides (Al_2O_3 and Fe_2O_3). Sesquioxides have very low solubility in water and high resistance to weathering. They are very common in soils as coatings on sand and silt grains and as clay-sized minerals, especially in highly weathered soils. Ferric oxide is the principal source of reddish colors in soils, becoming orange when hydrated and brown when mixed with gray or black minerals and organic matter. Also called oxide minerals.

sessile *adj.* (Latin *sessilis*, low enough to sit on) **1.** attached directly, without any stem, as a leaf with no petiole or a blossom with no pedicel. **2.** permanently attached and unable to move about, as a barnacle or coral.

seston *n.* solid material suspended or floating in water, including living and dead bodies of plants and animals, other organic debris, and mineral matter.

settler *n.* (Anglo-Saxon *setl*, a seat + Latin *er*, a person) **1.** a person who settles, especially in a previously unsettled area. **2.** an animal or thing that settles.

seventeen-year locust *n.* See periodical cicada.

sewage *n.* (French *seuwiere*, a drain + *age*, product) wastes that flow through sewers. Liquid and accompanying solid wastes that humans discharge from homes, businesses, and factories into municipal sewers. Also called sullage. See septage.

sewage effluent *n.* the effluent from a municipal sewage system. The watery remnant after sewage has been treated to remove solid and oxidizable wastes.

sewage lagoon *n.* a body of water used to treat organic wastes, e.g., from a household or a rest stop. Aerobic bacteria decompose organic compounds in the upper layers, and anaerobes degrade the material that settles to the bottom. Lagoon effluent is usually discharged to a stream.

sewage sludge *n.* the semisolid material that settles out of sewage, is filtered from sewage, or is skimmed from the top of sewage; a waste product that requires disposal. It is sometimes incinerated, deposited in a landfill, or dumped in the ocean, but better spread on land (fields with nonfood crops, pastures, forests, mine spoils, etc.) where it serves as a source of plant nutrients and organic matter with an estimated value of $5 per ton.

sewage treatment *n.* the processing of sewage in a sewage treatment plant to make it safe to discharge the effluent into a stream, usually involving primary and secondary treatment, and sometimes tertiary treatment. Primary treatment involves holding sewage in a pond or tank to remove materials that either float or settle out of quiet water. Secondary treatment involves the use of bacteria to decompose organic materials in either an activated sludge tank or a trickling filter. Tertiary treatment involves removal of dissolved ions (especially plant nutrients such as nitrogen and phosphorus) either by chemical treatment or, preferably, by applying the effluent to pasture or other land growing nonfood crops.

sewer *n.* (French *seuwiere*, a drain) **1.** a sanitary sewer, a system of underground pipes used to carry liquid wastes from a city or town to a sewage treatment plant. **2.** a storm sewer, a system of underground pipes used to carry runoff water from streets and adjoining areas in a city or town, usually emptying directly into a stream.

sewerage *n.* (French *seuwiere*, a drain + *age*, product) **1.** sewage. **2.** the entire sewage collection system for a city or other entity. **3.** the process of operating a sewage system.

sewer gas *n.* a mixture of gases produced in a sewer by decomposition of transported wastes, typically including methane (CH_4), hydrogen sulfide (H_2S), and carbon dioxide (CO_2). Sewer gas can be toxic and/or explosive.

sex *n.* (Latin *sexus*, dividing) **1.** reproductive gender. Individual animals and persons are of either the male or female sex or both (see hermaphrodite). Individual plants commonly have both male and female parts. **2.** all of the characteristic parts, functions, and behavior associated with being either male or female.

3. the attraction between two individuals that exists because one is male and the other female. **4.** coitus; sexual intercourse.

sex chromosome *n.* either of the pair of chromosomes that determines the sex of an individual. In humans and other mammals, the female sex chromosomes are XX and the male XY. In birds, the male is designated either ZZ or XX and the female ZW or XY.

sex hormone *n.* any of several steroid chemicals that affect the development of male or female characteristics in an individual.

sex-linked *adj.* associated with one sex but not the other. It applies to characteristics controlled by genes located on the sex chromosomes.

sex pheromone *n.* a chemical produced by either a male or a female to attract an individual or individuals of the other sex of the same species. Some sex pheromones are now made synthetically and used to bait traps in integrated pest management. See pheromone.

sextant *n.* (Latin *sextans*, one-sixth) an instrument designed to determine latitude and longitude by measuring the angle between the position of a heavenly body (sun, moon, or star) and a horizontal line. It is named for its adjustable arm moving through one-sixth of a circle (with a mirror mechanism permitting measurement of angles from 0° to 120°).

sexually transmitted disease (STD) *n.* See venereal disease.

sexual reproduction *n.* any reproductive process involving interaction of genetic material from male and female parents; e.g., the union of a sperm with an egg of an animal species or the union of pollen with an ovule in a plant species.

shad *n.* (Anglo-Saxon *sceadda*, a kind of fish) **1.** any of several saltwater fish (*Alosa* sp.) that resemble a herring but have a deeper body and spawn in fresh water. The common shad (*A. sapidissima*) is a valuable food fish in Europe and along the Atlantic Coast of North America. **2.** any of several other similar species.

shaddock *n.* **1.** a citrus tree (*Citrus maxima*, family Rutaceae) named after Captain Shaddock, who carried it from the East to the West Indies in 1696. **2.** the very large yellow to orange fruit of the tree, similar to a grapefruit. Also called pomelo or pompelmous.

shade garden *n.* a garden composed of shade-tolerant plants grown in a shaded area.

shade house *n.* a greenhouse with low light levels that is used to grow plants that need shade.

shade-tolerant *adj*. able to grow and develop in dense shade, as under a forest canopy; e.g., sugar maple and Norway spruce trees, many ferns, ivies, anemones, columbine, lungwort, primrose, and violets.

shade tree *n*. **1.** a tree grown to provide shade, either to grow shade-loving plants in its shade or to provide a shady area for people. **2.** any tree species commonly used for this purpose.

shaft mine *n*. an underground mine where materials are extracted through a network of shafts and tunnels. See strip mine.

shag *n*. (Anglo-Saxon *sceacga*, coarse hair) **1.** wool or hair that is heavily matted. **2.** a heavy cloth or a carpet with a rough nap. **3.** a tangled mass. **4.** an uneven hairdo. **5.** a type of tobacco cut into shreds. **6.** a dance step involving hops, first on one foot and then on the other. **7.** a small cormorant (*Phalacrocorax aristotelis*, family Phalacrocoracidae) that frequents European coastal areas. *adj*. **8.** shaggy; having an irregular or matted surface. *v*. **9.** to roughen or make hairy. **10.** to dance the shag. **11.** to hurry or pursue, especially to chase a baseball and throw it back.

shagbark *n*. (Anglo-Saxon *sceacga*, coarse hair + *barc*, bark) **1.** a North American hickory tree (*Carya ovata*, family Juglandaceae) with rough gray bark that peels off in long strips. Also called shagbark hickory. **2.** the wood from such trees. **3.** the nuts from such trees.

shakes *n*. (Anglo-Saxon *sceacan*, to shake) **1.** a common name for any disease that causes uncontrollable trembling; e.g., hatter's shakes caused by mercury absorbed while making hats and spelter shakes caused by metal fumes affecting brass-foundry workers, both causing intermittent fever and chills along with shaking. **2.** a short time, as I'll be there in a couple of shakes or in two shakes of a lamb's tail.

shale *n*. (German *schale*, a thin layer) a sedimentary rock composed mostly of thin layers of clay. Moderate pressure causes the clay particles to bond together into a rock.

shale oil *n*. crude oil extracted from the kerogen in carbonaceous shale; e.g., in rock layers in Colorado, Utah, and Wyoming. The oil reserves in shale oil are large (about the same as in petroleum reserves), but they are costly to extract. See kerogen.

shallot *n*. (French *eschalotte*, a kind of onion) **1.** a garden plant (*Allium cepa* = *A. ascalonicum*, family Amaryllidaceae) similar to an onion with a divided bulb and similar to garlic but milder. **2.** the bulb of the plant used as a seasoning in cooking. **3.** a small green onion.

shallow *adj*. (Anglo-Saxon *sceald*, shallow) **1.** having little depth, as shallow water or a shallow trench. **2.** occurring not far below the surface, as a shallow layer. **3.** not very intellectual, as a shallow mind. **4.** not very loud; weak, as a shallow wail. **5.** not very full, as shallow breathing.

shallows *n*. (Anglo-Saxon *scealw*, to dry out) **1.** a shallow part in a stream. **2.** a shoal; a sandbar that is sometimes exposed and sometimes covered with shallow water.

shambles *n*. (Anglo-Saxon *scamol*, bench) **1.** a place of killing and/or destruction, as the invaders left the city in shambles. **2.** a cluttered, disorderly place, as his desk was a shambles.

shamrock *n*. (Irish *seamrog*, little clover) **1.** any of several trifoliate plants, especially the small yellow-flowered clover that is the national emblem of Ireland (*Trifolium procumbens*, family Fabaceae). **2.** a cluster of three leaflets plucked from such a plant.

shard *n*. (Anglo-Saxon *sceard*, shard) a piece of a broken pot or other ceramic item. It was often used in ancient times for short notes, like scratch paper is used now. Also called potsherd or sherd.

sharecropper *n*. a tenant who crops an assigned tract of someone else's land and receives a share of the crops grown as remuneration. The landowner usually supplies the tenant a house and all or most of the necessary farming equipment, seed, and fertilizer and closely supervises the tenant. Also called cropper.

shark *n*. (German *schurke*, scoundrel) **1.** any of several mostly large marine fish (families Carchariidae, Lamnidae, Carcharhinidae, Squalidae, and several others). Sharks are mostly fish eaters, but some will attack humans. **2.** a person who commonly cheats, tricks, or takes advantage of others, as a card shark or a loan shark. **3.** a person who is unusually talented in a specialty.

sheaf *n*. (Anglo-Saxon *scheaf*, bundle) **1.** a small bundle of wheat, rye, or other cereal, often tied around the middle with a few stems of the crop. **2.** any bundle, stack, or other group, as a sheaf of papers or a sheaf of arrows (especially 24 arrows in a quiver).

shearing *n*. (Anglo-Saxon *scear*, cutter + *ing*, act) **1.** the act or process of cutting, especially with scissors; e.g., shearing sheep means cutting off their wool. **2.** the material cut off, as a shearing from the sheep. **3.** a shearling. **4.** the harvesting process of cutting stalks of grain. **5.** making vertical cuts to loosen coal at the exposed surface of a seam.

shearling *n*. (Anglo-Saxon *scear*, cutter + *ling*, small) **1.** a 1-year-old sheep that has been sheared once.

2. the wool from such a sheep. **3.** a tanned sheepskin with the wool still on it from such a sheep.

sheath *n.* (Anglo-Saxon *sceath*, sheath) **1.** a case to enclose a knife blade or other sharp cutter. **2.** a covering around the petiole of a leaf, the stem below a leaf, the wings of an insect, or the outside of a muscle.

sheep (pl. **sheep**) *n.* (Anglo-Saxon *scēp*, sheep) **1.** any of several ruminant mammals (*Ovis* sp., family Bovidae), especially those domesticated from *O. aries* and used to produce wool, mutton, and sheepskin leather. **2.** sheepskin; leather made from the skin of these animals. **3.** a meek person who follows wherever he is led.

sheep dog *n.* a dog trained to protect and herd sheep, especially a collie or certain other large, gentle dogs.

sheep laurel *n.* See lambkill kalmia.

sheep loco *n.* a plant (*Astragalus nothoxys*, family Fabaceae) found in southern Arizona that is toxic to sheep in both the green and dry stages. It causes dullness, difficulty walking, loss of appetite, weight loss, and shagginess. Also called locoweed.

sheep pox *n.* **1.** a highly infectious disease of sheep that is spread by both direct and indirect contact, forming dark-red eruptions that turn into blisters, especially on the bare or thin wool parts of a sheep. It is caused by a filterable virus closely related to the vaccinia virus that infects humans. **2.** ovinia (*Ecthyma contagiosum*), a viral infection in sheep similar to smallpox that can also affect humans. Also called ovine smallpox.

sheep scab mite *n.* a tiny white mite (*Psoroptes equi* var. *ovis*, family Psoroptictae) that embeds itself in the skin of sheep and causes psoroptic mange, the most common mange of sheep. The scab mite causes severe itching as it feeds, resulting in wool loss as the sheep rubs to alleviate the itch.

sheepskin *n.* **1.** the pelt of a sheep, especially one that has been processed with the wool on it, e.g., one used as a rug or blanket with protrusions for the neck and legs of the sheep. **2.** leather or other materials made from the skin of a sheep. **3.** (colloquial) a diploma.

sheet erosion *n.* water erosion that removes a thin layer from an exposed land surface; removal by sheet wash of soil particles detached by the beating action of raindrops. It generally makes the soil sandier and less productive by removing mostly fine particles.

sheet wash *n.* **1.** a thin layer of water accumulating on and flowing across the land surface, generally agitated by falling raindrops. Also called sheet flow in contrast to channel flow. **2.** sheet erosion, especially as an obvious, severe expression.

shelf *n.* (Anglo-Saxon *scylfe*, shelf) **1.** a flat surface inside a cabinet or extending from a wall where objects may be placed for display or storage. **2.** the objects placed on a shelf, or its capacity to hold objects. **3.** a ledge along a cliff. **4.** a nearly level surface in a body of water, as a sandbar, a shallow bottom area, or a continental shelf. **5.** a nearly level bedrock surface covered by sediments or other rock.

shell *n.* (Anglo-Saxon *scell*, shell) **1.** a hard outer covering, especially that protecting certain animals, eggs, or fruits, as a clamshell, an eggshell, or a coconut shell. **2.** any hard surface protecting a hollow interior. **3.** the material used to make such a protective surface. **4.** the framework and outer covering of a structure. **5.** a cartridge or bullet. **6.** a long, lightweight rowboat used for racing. **7.** a reserved attitude that reveals little of one's personality. **8.** any of the energy levels of electrons orbiting around the nucleus of an atom. *v.* **9.** to remove from a shell or pod, as to shell peas. **10.** to shell corn, removing the kernels from the ear. **11.** to fall from the exterior, as when a surface sloughs or peels. **12.** to bombard with big guns, as the ships will shell the enemy position.

shellfish *n.* (Anglo-Saxon *scilfisc*, shellfish) an aquatic animal with a shell, including mollusks (clams, oysters, scallops, and abalone) and crustaceans (crabs, lobsters, and shrimp) that live in the bottoms of bays, sounds, tidal rivers, and other sheltered areas. Some shellfish are mobile and others are sedentary, some are edible, some carry diseases such as typhoid fever, and some are useful indicators of environmental conditions.

shelterbelt *n.* a long line of trees and shrubs placed to protect a field from wind erosion. Most shelterbelts have a line of tall trees in the middle with shorter trees and shrubs on both sides, for a total of either five or seven rows. See windbreak.

shelterwood cutting *n.* a technique of harvesting timber whereby only the mature marketable trees are removed. Enough trees always remain to protect the soil and water environment.

shepherd *n.* (Anglo-Saxon *sceap*, sheep + *hyrde*, herd or protection) **1.** a sheep herder, a person who guides and guards a flock of sheep. **2.** a minister or pastor. *v.* **3.** to watch over a flock or a congregation.

sherd *n.* a shard.

shield volcano *n.* a low-lying volcano composed of basic lava.

shifting agriculture or **shifting cultivation** *n.* a rotation of a few years of cultivated crops alternating with forest practiced on close to a third of the arable land in the world and used to support about 300 million people, mostly in tropical countries. Trees are cut and burned (but some stumps may be left to sprout), a farm crop is planted for two or three years until yields decline, and then the field is allowed to revert to trees for ten or more years to rejuvenate the soil. Little or no lime or fertilizer is used. Also called slash-and-burn agriculture or swidden and by many country names, including taungya in Burma, chena in Ceylon, kumri in India, shamba in Kenya, ladang in Maylaysia, kaingin in the Philippines, and parcelero in Puerto Rico.

An area being cleared for shifting cultivation

shigellosis *n.* bacillary dysentery, a disease caused by any of many species of bacteria (*Shigella* sp., family Enterobacteriaceae) that cause intestinal pain, diarrhea, and blood in the stool in people and monkeys. These bacteria are very abundant in feces, septage, sewage sludge, and sewage effluent, especially in tropical conditions.

shoal *n.* (Anglo-Saxon *sceald*, shallow) **1.** a shallow area in a stream, lake, or sea. **2.** a sandbar, especially one that is exposed when the water is low. **3.** a large group; a crowd, a throng, or a school of fish. *v.* **4.** to become shallow. **5.** to gather or move about as a crowd or as a large school of fish.

shock *n.* (French *choc*, a shock) **1.** a sudden impact or violent blow that disturbs a physical equilibrium or process, e.g., breathing may stop or a heartbeat may be restarted. **2.** a mental disturbance caused by a severe emotional problem. **3.** an electric shock caused by a sudden passage of an electric current through the body. **4.** a medical condition caused by any of the above producing a circulatory disorder, weakness, rapid pulse and low blood pressure, and often unconsciousness. **5.** a neatly arranged stack of hay or group of bundles of hay or grain to be picked up later, often after a period of drying. **6.** a mass of long, thick hair, often shaggy. *v.* **7.** to produce any of the above conditions of shock. **8.** to cause surprise and disgust by a statement or action. **9.** to stack hay or grain in shocks.

shoot *n.* (Anglo-Saxon *sceotan*, shoot) **1.** a young, tender stem or branch of a plant. **2.** a sudden burst of growth or speed. **3.** a shot from a gun or an arrow from a bow or an event where guns or arrows are fired. **4.** a missile launching. **5.** a picture-taking session. **6.** a small ore vein or mining tunnel that branches from a larger vein or tunnel. *v.* **7.** to produce new branches or other growth or to grow rapidly. **8.** to move suddenly, especially in a straight line. **9.** to fire a weapon. **10.** to injure, kill, damage, or destroy by firing one or more projectiles. **11.** to do something suddenly, as to shoot a smile or to shoot a light. **12.** to attempt to score a goal in certain games, as to shoot a basket. **13.** to push quickly into place, as to shoot a door bolt into its open or locking position. **14.** to mark or be marked with a contrasting color, as a blue sky shot with white clouds or a red fabric shot with yellow. **15.** to measure, as to shoot the angle of a star with a sextant or to shoot an elevation with a transit. **16.** to take a picture or group of pictures. **17.** to inject, as to shoot a dose of insulin.

short-day plant *n.* a plant that matures and bears flowers and fruit when the days are becoming shorter (actually triggered by the nights becoming longer). See long-day plant, day-neutral plant, photoperiodism.

shortgrass *n.* (Anglo-Saxon *scort*, short + *graes*, grass) any of several bunchgrasses that grow up to about 1½ ft (0.45 m) tall on rangeland in semiarid climates, e.g., on the western Great Plains, including buffalo grass (*Buchloe dactyloides*), grama (*Bouteloua* sp.), bluegrasses (*Poa* sp.), cheat grass (*Bromus tectorum*), etc.

shrew *n.* (Anglo-Saxon *screawa*, shrew) **1.** any of several small animals (*Sorex* sp. and related genera, family Soricidae) that resemble mice with soft brown fur and long, sharp snouts. Shrews eat insects and earthworms. **2.** a noisy, complaining, bad-tempered woman.

shrike *n.* (Anglo-Saxon *scric*, shrike) **1.** any of several robin-sized birds (*Lanius* sp., family Laniidae) with short hooked bills and dominantly white or gray colors and black markings that pursue insects, small birds, and rodents. **2.** any of several other similar birds.

shrimp *n.* (Anglo-Saxon *scrimman*, to shrink) **1.** any of several small marine crustaceans with long tails (mostly *Crangon* sp., especially *C. vulgaris*, family Crangonidae). Many are used for food. **2.** a small or unimportant person. *v.* **3.** to catch or fish for shrimp. *adj.* **4.** made with shrimp, as a shrimp salad.

shrinkage *n.* (Anglo-Saxon *scrincan*, to shrivel up + *age*, related to) **1.** the process of shriveling; a reduction in volume, number, and/or value. **2.** loss caused by breakage, spillage, spoilage, misplacement, shoplifting, and/or robbery. **3.** the amount of such reduction or loss; e.g., the percentage of weight lost by livestock during the marketing process or the loss in weight of grain as it dries in storage. **4.** the reduction in soil volume when a wet soil dries, often resulting in the opening of cracks.

shrink-swell potential *n.* the changes in the volume of a soil as it goes from wet to dry or dry to wet. For example, Vertisols, which have high shrink-swell potentials that cause wide cracks to open during dry periods, often shift enough to disrupt building foundations and road pavement.

shrub *n.* (Anglo-Saxon *scrybb*, brushwood) a perennial woody plant smaller than a tree and usually having several stems branching from near ground level.

Si *n.* silicon.

SI *n.* the international system of units (from French *Le Système Internationale d'Unités*). See SI base units, SI-derived units, metric system.

sib *n.* (Anglo-Saxon *sibb*, sib) **1.** a blood relative; a kinsman, especially a sibling. **2.** the descendants of one ancestor. *adj.* **3.** related by birth. **4.** from the same plant, either a full sib with the same male and female parentage or a half-sib that shares either the ovule or the pollen parentage.

SI base units *n.* individually defined units in the international (SI) system of units. There are seven SI base units:
m = meter (length)
kg = kilogram (weight or mass)
s = second (time)
A = amp = ampere (electric current)
K = Kelvin (temperature)
Cd = candela (luminosity)
mol = mole (molecular amount)
See SI-derived units, metric system, *Système Internationale d'Unités*.

Siberian peashrub *n.* See caragana.

sibling *n.* (Anglo-Saxon *sibb*, sib + *ling*, young) **1.** a brother or sister, especially one who is not a twin. **2.** a plant with the same parentage (the same genetic heritage). *adj.* **3.** pertaining to siblings, as sibling support or sibling rivalry. See sib.

sick building syndrome *n.* a condition wherein more than 20% of the occupants of a building report a common illness that is thought to be building related (possibly caused by drafts, poor ventilation, mold, vapors from building materials, etc.). Common symptoms include eye, ear, nose, and throat irritations, coughing, hoarseness, headaches, fatigue, nausea, dizziness, etc.

sickle *n.* (Latin *secula*, a cutter) **1.** a semicircular cutting blade with a short handle; a hand tool used for cutting grass, weeds, or ripe grain. **2.** a line of triangular blades that moves back and forth in the cutter bar of a mowing machine or harvester.

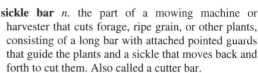
A hand sickle. The blade diameter is about 1 ft (30 cm). The blade is sharp on the inside.

sickle bar *n.* the part of a mowing machine or harvester that cuts forage, ripe grain, or other plants, consisting of a long bar with attached pointed guards that guide the plants and a sickle that moves back and forth to cut them. Also called a cutter bar.

sickle feather *n.* either of a pair of long curved feathers in the middle of the tail of a rooster.

side dressing *n.* **1.** the application of a band of fertilizer a few inches from one side of a crop row, either on the soil surface or injected below the surface. **2.** the fertilizer so applied.

sidereal *adj.* (Latin *sidereus*, belonging to the stars + *alis*, of) **1.** pertaining to the stars. **2.** measured according to the stars, as sidereal time.

sidereal day *n.* the time it takes for the Earth to rotate once relative to the stars (a solar day is about 4 min longer than a sidereal day because it includes the Earth's orbit around the sun).

sidereal month *n.* the average time for the moon to orbit around the Earth relative to the stars: 27 days, 7 hr, 43 min, and 11.47 sec of mean solar time (the lunar month is about 2 days and 5 hr longer because it includes the Earth's orbit around the sun).

sidereal year *n.* one revolution of the Earth around the sun, returning to a comparable alignment with the stars, equal to 365.2564 mean solar days (the calendar approximates the solar year based on the time from an equinox one year to the same equinox the next year: 365.2422 mean solar days). See solar year.

SI-derived units *n.* units that are formed as a combination of other more basic units. There are 17 derived units in the SI system:
N = newton (force)
Pa = pascal (pressure, stress)
J = j = joule (energy)
W = watt (power)
C = coulomb (electric charge)
V = v = volt (electric potential difference)
Ω = ohm (electric resistance)
S = siemens (electric conductance)
F = farad (electric capacitance)

Wb = weber (magnetic flux)
H = henry (inductance)
T = tesla (magnetic flux density)
lm = lumen (luminous flux)
lx = lux (illuminance)
Hz = hertz (frequency)
Bq = becquerel (activity of radioactive source)
Gy = gray (absorbed dose of radiation)

See SI base units.

siderophore *n.* (Greek *sidēros*, iron + *phoros*, bearing) either of two types of small proteins (catecholate and hydroxamate) with a high affinity for ferric iron (Fe^{3+}). It is produced by certain microbes.

side slope *n.* **1.** the slope of an embankment such as the side of a highway or of a drainage ditch. It is usually expressed as the ratio of the horizontal distance to the vertical distance, up or down. For example, 2:1 means for every 2 ft horizontal, the vertical rise or fall is 1 ft (or other units such as 2 m for every 1 m). **2.** the side of a hill or ridge. See nose slope.

siemens (S) *n.* the unit of electrical conductance in the SI system; equal to 1 amp per volt. It is named for Ernst Werner von Siemens (1816–1892), a German inventor and electrical engineer and an early developer of the telegraph and the electric locomotive.

sienna *n.* (Italian *Siena*, a city in Italy) **1.** a yellowish brown earthy material with a high iron content; used as a pigment. **2.** the yellowish brown of this material. **3.** burnt sienna, a reddish brown pigment produced by burning raw sienna.

sierra *n.* (Spanish *sierra*, saw) **1.** a range of hills or mountains with peaks that resemble a saw blade. **2.** any of several saltwater game and food fish (*Scomberomborus* sp.) that resemble mackerel.

Sierra Club *n.* an activist organization founded in 1892 by John Muir to advocate the preservation of wilderness areas.

sieva bean *n.* a lima bean (*Phaseolus lunatus*, family Fabaceae) with small, light-colored seeds that is native to tropical America. Also called butter bean.

sieve *n.* (Anglo-Saxon *sife*, sieve) **1.** a container with either a perforated or a wire-mesh bottom with opening sizes defined on the basis of the number of openings (or wires) per inch (2.54 cm) that is used to separate solid materials by particle size, for sieve analyses, or to filter solid materials from liquids. **2.** something with holes that will not hold water, as it leaks like a sieve. **3.** a person who tells secrets. *v.* **4.** to pass materials through a sieve.

sieve analysis *n.* a particle-size analysis made by passing soil or other material through a nest of sieves to separate it into various sand sizes. Usually the fines (silt and clay) pass through the finest sieve in the nest and are analyzed by either hydrometer or pipette analysis. See particle-size analysis.

sievert (Sv) *n.* the SI unit for the effect of radiation on the body of an organism. It is named for Rolf Maximilian Sievert (1896–1966), a Swedish radiologist. See rem.

silage *n.* (Spanish *silo*, fruit cellar or silo + *age*, related to) chopped forage preserved in a succulent condition (generally with a water content of 65% to 70%) by partial fermentation in an airtight structure. Anaerobic bacteria produce lactic acid and acetic acid in the silage until microbial activity ceases at a pH between 3.6 and 4.2. The acidic silage retains its nutritive value for a long time and can be fed to livestock at any time throughout the year. Forages commonly used for silage include corn, forage sorghums, alfalfa, and cereal grains. Also called ensilage. See haylage. *Caution:* Entering a silo is hazardous because the oxygen supply is depleted.

silage additives *n.* materials added at the time of ensiling to increase the feeding value and/or reduce spoilage of silage, including urea, ammonium bicarbonate, ammonium sulfate, anhydrous ammonia, molasses, grains, enzymes, bacteria (e.g., *Lactobacillus buchneri*), and/or vitamins.

silica *n.* (Latin *silex*, flint) silicon dioxide (SiO_2) in the form of quartz, flint, agate, opal, etc., that is ground into sand-sized particles for use as an abrasive or used to make glass, ceramics, etc.

silicate *n.* (Latin *silex*, flint + *ate*, salt) a mineral with silica tetrahedra (Si^{4+} surrounded by four O^{2-}) as major components of its structure. Silicates, which constitute more than 90% of the Earth's crust, are classified according to the linkage of their tetrahedra into nesosilicates (unlinked), sorosilicates (pairs), cyclosilicates (rings), inosilicates (chains), phyllosilicates (sheets), and tektosilicates (three-dimensional structures).

silicate clay *n.* a clay-sized soil mineral with a layer structure composed mostly of silicon and oxygen. The principal silicate clays are kaolinite, smectite (montmorillonite), illite, vermiculite, and chlorite.

silicatosis *n.* See silicosis.

siliceous *adj.* (Latin *siliceus*, of flint) **1.** resembling and/or containing a high content of silica. **2.** growing in soil that has a high content of silica.

silicified wood *n.* petrified wood. Fossilized wood formed by the gradual replacement of fibers and other structures with silica (SiO_2), usually with the wood structure still visible.

silicon (Si) *n.* (Latin *silicis*, flint) element 14, atomic weight 28.0855, second only to oxygen in abundance in the Earth's crust (20% of all atoms there). It typically forms tiny cations with +4 charge (Si^{4+}) that fit into the space inside an oxygen tetrahedron, the basic structural unit of silicate minerals. Silicon is used in steelmaking, alloys, and semiconductors.

silicon tetrafluoride *n.* a colorless, very pungent gas (SiF_4) evolved in the acidulation of rock phosphate to make superphosphate fertilizer and used to manufacture fluosilicic acid.

silicosis *n.* (Latin *silex*, flint + *osis*, disease) a lung disease resembling pneumonia caused by inhalation of dust containing silicon dioxide (SiO_2). Also called silicatosis, a form of pneumoconiosis.

silk *n.* (Latin *sericus*, Chinese fabric) **1.** the lustrous, slender fibers obtained from the cocoons of silkworms. **2.** thread or fabric made from such fibers. **3.** any silklike filament, as the long slender tube that carries a pollen grain to the egg cell so that corn kernels will develop, or the fibers from a milkweed pod. **4.** a parachute, as they hit the silk. *adj.* **5.** made of silk, as a silk shirt.

silk oak *n.* any of several tropical evergreen trees (*Grevillea* sp., family Proteaceae), especially *G. robusta*, grown as an ornamental tree in Florida and California for its fernlike leaves and large orange or yellow flowers. It is planted extensively in India to shade tea gardens.

silk tree *n.* a short, flat-topped, deciduous, Asian mimosa tree (*Albizia julibrissin*, family Fabaceae) with dark green fernlike leaves and clusters of fragrant pink stamens up to an inch (2.5 cm) across. Silk trees are hardy enough to tolerate frost but are killed by prolonged periods below 0°F (–18°C).

silkworm *n.* (Anglo-Saxon *seolcwyrm*, silkworm) **1.** a large caterpillar about 3 in. (7.5 cm) long that wraps itself with a long silken thread to form a cocoon that is unwound to produce the silk of commerce; the larva of a moth (*Bombyx mori*, family Bombycidae) that feeds on the leaves of mulberry (*Morus alba* var. *multicaulis*, family Moraceae). It is native to Asia. **2.** the larva of any of several similar moths (family Saturniidae) that spin cocoons.

sill *n.* (Anglo-Saxon *syll*, base or foundation) **1.** a stone or timber placed horizontally to serve as the foundation of a wall or building. **2.** the horizontal frame piece at the bottom of a window opening or doorway. **3.** a thin pluton intruded as a lens between layers of older (usually sedimentary) rock.

silo *n.* (Spanish *silo*, fruit cellar or silo) **1.** a tall, usually cylindrical, airtight structure where fodder or forage can be stored to make silage. It is normally filled from the top and emptied from the bottom. **2.** a trench silo or pit silo; a trench, pit, or other structure that can be filled with fodder or forage and sealed so it will make silage. See trench silo. **3.** an underground structure to store a long-range ballistic missile.

silt *n.* (Norwegian *sylt*, salt marsh) **1.** any earthy sediment composed of small particles deposited by running water, glacial ice, or wind. **2.** a soil separate consisting of mineral particles that are larger than clay but smaller than sand; particles between 0.002 and 0.05 mm in diameter by U.S. Department of Agriculture standards, between 0.002 and 0.02 mm by International Soil Science Society standards, or between 0.005 and 0.08 mm according to the American Association of State Highway Officials. **3.** a soil texture class containing at least 80% silt and less than 12% clay. *v.* **4.** to deposit silt or receive a silt deposit, especially when it fills an area, as the lake silted full or the channel silted in.

siltation *n.* (Norwegian *sylt*, salt marsh + Latin *atio*, action) the settling of silt particles (often including some clay and/or sand) on the bottom of a body of quiet water. Also called silting.

silt loam *n.* a soil texture class; in the U.S. Department of Agriculture system, it contains at least 50% silt and less than 27% clay, excluding the part that qualifies as a silt texture. See silt, soil texture triangle.

silt pond *n.* a pond built for the purpose of holding water long enough for silt to settle out. Usually the water is fairly shallow, and the pond fills with sediment before many years.

siltstone *n.* (Norwegian *sylt*, salt marsh + Anglo-Saxon *stān*, stone) a fine-grained sedimentary rock formed mostly of silt plus enough clay to bond it into a rock mass by having been subjected to moderate pressure.

silva *n.* (Latin *silva*, woodland) **1.** the trees that grow in a particular forest. **2.** a book about the trees of a particular area.

silver (Ag) *n.* (Gothic *silubr*, silver) **1.** element 47, atomic weight 107.8682; a metallic element that is stable either as a shiny metal or as a cation (+1 or +2) in various compounds, including light-sensitive chlorides, bromides, iodides, nitrates, etc., that are used in photography. **2.** the shiny light-gray metal, especially as used in coins, jewelry, tableware, mirrors, or electrical conductors. **3.** metallic coins in general. **4.** the color of silver. *adj.* **5.** containing, made of, or resembling silver. **6.** based on silver, as a silver standard for currency. **7.** 25th, as a silver anniversary. **8.** eloquent, as having a silver tongue.

silverfish *n.* (Anglo-Saxon *seolfer*, silver + *fisc*, fish) **1.** any of several species of fish that have a silvery color; e.g., a silvery goldfish (*Carassius auratus*). **2.** a small wingless insect (*Lepisma saccharina*, family Lepismatidae) with silvery scales, long feelers, and a bristly tail that lives in dark places and feeds on paper products, destroying books by eating the starchy glue.

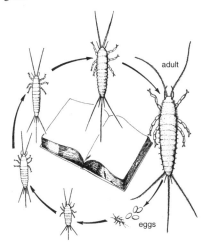

The life cycle of silverfish insects

silver iodide *n.* a chemical (AgI) used in powdered form to seed clouds with crystals that can act as nuclei for raindrops (with erratic results). Silver iodide is also a light-sensitive compound used in photography.

silvex *n.* a selective herbicide ($C_9H_7O_3Cl_3$) used to kill woody and aquatic plants and for weed control in paddy rice.

silvicide *n.* (Latin *silva*, woodland + *cida*, killer) a chemical pesticide intended to kill trees, shrubs, and other woody plants.

silvics *n.* (Latin *silva*, woodland + *icus*, study) the science that studies trees and their environment. See silviculture.

silviculture *n.* (Latin *silva*, woodland + *cultura*, cultivation) forestry; woodland management; applied silvics. See silvics.

simian *n.* (Latin *simia*, an ape) **1.** an ape or monkey. **2.** *adj.* of or about an ape or monkey.

simple sugar *n.* any monosaccharide; a chain of 3, 4, 5, 6, or 7 carbons with attached hydrogens and hydroxyls and one oxygen that is either double-bonded to one carbon or linked to two carbons in a ring ($C_nH_{2n}O_n$). A sweet food that supplies energy stored by photosynthesis. See sugar.

simuliid *n.* any of about 600 species of flies (*Simulium* sp., family Simuliidae) variously known as black-flies, buffalo gnats, and turkey gnats. The females bite people and animals.

singing sand *n.* loose sand that makes a crunching sound when a person walks on it. Also called sounding sand.

single cross *n.* or *adj.* **1.** the process of crossing two inbred lines. **2.** the progeny from such a cross, as single-cross hybrid corn.

sink *n.* (Anglo-Saxon *sincan*, to sink) **1.** a basin, usually set in a counter and equipped with a water supply and a drain, as a kitchen sink or a lavatory. **2.** a low place; e.g., a depression in the land surface. **3.** a storage place; e.g., productive soil and living organisms are carbon sinks, a cesspool is a sink for pollutants, or a large mass of metal may act as a heat sink. **4.** a sinkhole. **5.** a drain or sewer. **6.** a process that removes environmental pollutants; e.g., a process that removes pollutants from the atmosphere. *v.* **7.** to move slowly to a lower level, especially to become entirely or mostly submerged in a body of water or other liquid. **8.** to settle, as to sink into a chair or as a building with inadequate foundation will sink into the soil. **9.** to pass out of sight, as the sun sinks in the west. **10.** to penetrate, as a stone sinks into the mud or as an idea sinks into the mind. **11.** to occupy a depression, as eyes that sink deep into the eye sockets. **12.** to lose value, as the price sinks. **13.** to become depressed, as when one's mood sinks. **14.** to pass gradually, as to sink into sleep or despair. **15.** to weaken, especially when near death. **16.** to become lower in musical pitch. **17.** to subside, as a sound sinks in intensity. **18.** to dig, as to sink a well. **19.** to destroy, as to sink a ship. **20.** to invest, as to sink a lot of money into a project.

sinkhole *n.* (Anglo-Saxon *sincan*, to sink + *hol*, hollow) a depression in the landscape that connects to an underground passage, usually produced by solution of limestone bedrock or melting of ice. See karst, thermokarst.

sinter *n.* (German *sinter*, dross) **1.** a crust formed by evaporation of water leaving a deposit of calcium carbonate or silica, as around the geysers and springs in Yellowstone National Park. **2.** dross formed when iron is heated to near the melting point. *v.* **3.** to fuse iron particles by heating them.

sire *n.* (French *sire*, master) **1.** a male parent, especially of a quadruped such as a horse. **2.** a title of respect, now usually reserved for a king. *v.* **3.** to father, especially when a stallion sires a colt.

siriasis *n.* (Latin *siriasis*, scorching) sunstroke.

sirocco *n.* (Italian *scirocco*, hot wind) **1.** a hot wind blowing north or northwest from the Sahara Desert. It is a hot, dry, dusty wind in northern Africa but becomes a hot, humid wind as it crosses the Mediterranean Sea and reaches Italy and other parts of southern Europe. **2.** any hot wind, especially a dry wind blowing toward a low pressure center.

sisal *n.* (Maya *sisal*, cold waters) **1.** a type of fiber obtained from the leaves of any of several species of agave named for a former seaport of Yucatán. The sisal fibers are mostly sheathing cells that enclose the vascular bundles of leaves that are several years old and leaning to a nearly horizontal position. They are used to make rope, rugs, etc. **2.** the agave plants that produce these fibers (*Agave* sp., family Agavaceae, especially *A. sisalana*), native to the Yucatán area in Central America that is now grown mainly in Brazil and eastern Africa.

site index *n.* a measure of the suitability of a site for its dominant tree species, normally expressed as the height in feet the trees reach or are projected to reach at a specified age (usually 25, 50, or 100 years); e.g., the height of 50-year-old dominant and codominant sugar maples at a given site.

sitotropism *n.* (Greek *sitos*, food + *tropos*, a turning + *isma*, condition) the effect of food on living cells, either attraction or repulsion.

skate *n.* (Norse *skata*, ray fish) **1.** any of several large rays (*Raja* sp.) up to 8 ft (2.4 m) across with a pointed snout and a slender tail. **2.** a blade used for sliding across a surface, as an ice skate or a skid on a lifeboat. **3.** a roller skate that uses small wheels for movement similar to ice skating on nonslippery surfaces. **4.** a person; either a good skate or a contemptible cheapskate. **5.** an old horse; a nag. *v.* **6.** to glide across a surface on skates.

skeleton *n.* (Greek *skeletos*, dried up) **1.** the bony structure that forms the framework of a body, especially that of a human or other vertebrate. **2.** the framework of other types of animals, including both endoskeletons and exoskeletons. **3.** a comparable framework of a leaf, building, ship, or other structure. **4.** an extremely lean or emaciated person or animal, as you are so thin you are nothing but a skeleton. **5.** a machine with protective covers removed, leaving nothing but essential or structural parts. *adj.* **6.** skeletal. **7.** minimal, as a skeleton crew.

skim *v.* (French *escumer*, to scum) **1.** to remove something that floats on top, as oil on water, or cream from milk. **2.** to form a thin layer on a surface, as ice skims a pond. **3.** to look through quickly, as to skim a book. **4.** to glide quickly across, as a skater skims across the ice.

skink *n.* (Latin *scincus*, lizard) any of several small lizards (family Scincidae) with scaly bodies that live either on the ground or in trees in both the Old World and the New World.

skunk *n.* (Algonquian *sekakwa*, skunk) **1.** a carnivorous quadruped (*Mephitis* sp., family Mustelidae) about the size of a cat. It usually has black fur marked by white stripes down its back and is known for the strong offensive smell of the glandular secretion that it sprays when molested. **2.** an offensive person. *v.* **3.** to beat badly, especially to prevent the other person or team from scoring.

skunk cabbage *n.* **1.** a low, broad-leaved plant (*Symplocarpus foetidus*, family Araceae) with thick roots, wide leaves, and a strong disagreeable odor. Native to North America, it grows in moist places. **2.** a similar plant (*Lysichiton americanum*) native to western North America.

sky *n.* (Norse *sky*, cloud) **1.** the atmosphere, especially the upper part, often described as blue sky, clear sky, cloudy sky, etc. **2.** the expanse that one sees when looking outward from the Earth. **3.** the heavens.

skylark *n.* (Norse *sky*, cloud + *laevirki*, lark) an Old World migratory songbird (*Alauda arvensis*, family Alaudidae), with speckled brown plumage, that is known for its melodious song.

sky pond *n.* (dew pond) a pond built near the top of a hill or mountain in a very high rainfall and low-evaporation area, e.g., in eastern India where even a small watershed can fill the pond.

SLAR *n.* side-looking airborne radar, which is a means of obtaining aerial images beneath cloud cover by taking them from the side rather than overhead.

slaked lime *n.* calcium hydroxide [$Ca(OH)_2$] formed by hydration of burnt lime (CaO), as when lime is used in mortar. The slaked lime crystallizes and bonds to mineral matter.

slash *n.* (French *escleche*, a portion) **1.** brush, tree limbs, etc., that are cut but not removed when a forested area is logged or cleared. **2.** a tract of marshy land overgrown with shrubs and trees. **3.** a virgule (/) or other mark or cut made by a single stroke. **4.** a stroke with a pen, pencil, knife, or sword that makes such a mark or cut. **5.** a drastic reduction in prices. *v.* **6.** to make a stroke that marks or cuts. **7.** to reduce prices drastically. **8.** to criticize severely; to attack savagely.

slash-and-burn *adj.* cut and burn; a type of agriculture that cuts and burns the trees and brush in a tropical forest in preparation for raising crops for a few years. See shifting agriculture.

slash pine *n.* **1.** a large evergreen tree (*Pinus elliottii*, family Pinaceae), native to swampy areas in southeastern U.S.A., with needles 8 to 12 in. (20 to 30 cm) long, mostly in clusters of three, that is a source of lumber, turpentine, and rosin. **2.** the hard, durable lumber from such trees used in general construction and shipbuilding. Also called Cuban pine, swamp pine, or pitch pine.

slate *n.* (French *esclat*, a splinter) **1.** a metamorphic rock with a fine, smooth-layered structure formed by compression of shale into a denser, harder mass. **2.** a thin piece of such rock used as a surface for writing with chalk or as a roofing tile. **3.** a bluish gray characteristic of slate. **4.** a list of candidates representing a political party in an election.

slave-making ant *n.* (*Formica sanguinea*) a European ant species that raids the nest of other ants and carries off the pupae to obtain workers for their own colony. See Amazon ant.

sleepygrass *n.* a coarse bunchgrass (*Stipa robusta*, family Poaceae) that grows 3 to 6 ft (0.9 to 1.8 m) tall in semiarid areas of southwestern U.S.A. and Mexico. Its forage quality is poor, and it is said to produce sleepiness in horses, sheep, and cattle that eat it.

sleet *n.* (Anglo-Saxon *sliete*, hail) **1.** precipitation in the form of small ice pellets (generally 5 mm or less in diameter) formed of raindrops that freeze as they fall. See hail, graupel. **2.** (British) snow and rain falling together. *v.* **3.** to fall as sleet.

slickenside *n.* (Anglo-Saxon *slician*, to make smooth + *sīde*, side) **1.** a polished and striated ped surface resulting from soil peds pressed together and sliding past each other during the wetting and drying of a Vertisol. **2.** a rock surface smoothed and marked with striations by the movement of an adjacent rock mass.

slick spot *n.* a small, level, barren spot of nearly impervious sodic soil that is slippery when wet. See sodic soil.

slime *n.* (Danish *sliim*, mucus) **1.** a viscous organic secretion from either an animal or a plant; e.g., mucus. **2.** any stringy viscous substance, especially one that smells bad. **3.** thin, stringy mud. **4.** any moist, gummy substance with a displeasing nature. **5.** a person who does despicable things. *v.* **6.** to coat with slime. **7.** to remove slime from a surface.

sling psychrometer *n.* a device with two thermometers mounted on it. The bulb of one thermometer is covered with a wet wick, and the device is whirled through the air, causing water to evaporate from the wick and cool its temperature to the dew point. The relative humidity of the air is read from a table according to the temperatures recorded by the wet-bulb and dry-bulb thermometers.

slob *n.* (Irish *slab*, mud) **1.** mud or muddy land. **2.** any wet, muddy surface. **3.** a person who is clumsy and sloppy.

sloe *n.* (Anglo-Saxon *sla*, plum) **1.** a shrub or small tree (*Prunus spinosa*, family Rosaceae) with black bark, finely serrated leaves, and white blossoms. **2.** the small, bluish black, bitter plums that grow on the plant. **3.** any of several related plants (*Prunus* sp.).

slop *n.* (Anglo-Saxon *sloppe*, slime) **1.** slush; watery snow or thin mud. **2.** liquid waste. **3.** swill; food scraps in water and/or milk used to feed animals. **4.** an unkempt, careless person. **5.** clothing, bedding, etc., supplied to a sailor on a ship. **6.** a loose outer garment such as a frock used as a cloak or nightgown. *v.* **7.** to splash liquid from a container or from a puddle. **8.** to splash or spill liquid on something. **9.** to drink noisily and greedily. **10.** to feed slop to animals, especially pigs.

slope *n.* (Anglo-Saxon *aslopen*, slipped away) **1.** an inclination; an oblique surface or line that is neither horizontal nor vertical. **2.** slope gradient. Slopes are uniform if the gradient is constant, convex if the outer surface becomes steeper at lower elevations, or concave if the outer surface becomes flatter at lower elevations. (Concave and convex also apply to contour lines. See contour curvature.) *v.* **3.** to have a slope. **4.** to move along a slope.

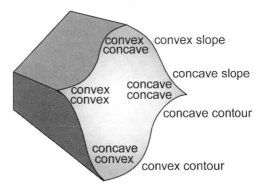

Both slope lines and contour lines
can be either concave or convex.

slope gradient *n.* the rate of deviation up or down of a surface from the horizontal (or sometimes from the vertical), expressed in percent, degrees, or a ratio of horizontal distance to vertical distance of the surface in its steepest direction. Also called gradient.

sloth *n.* (Anglo-Saxon *slaewth*, slow one) **1.** either of two species of mammals (*Bradypus tridactylus* of South America or *B. didactylus* of the West Indies) about 2 ft and 1 ft (60 cm and 30 cm) long, respectively, that dwell in trees and move very slowly, often hanging beneath a branch. **2.** laziness;

lack of desire to work or move. **3.** slowness, especially when it delays others.

slough *n.* (Anglo-Saxon *sloh*, slough) **1.** a small marsh. **2.** a backwater area; an old river channel that has become a marsh. **3.** a deep mud hole, as in a road.

slow-release nitrogen fertilizer *n.* a fertilizer that supplies available nitrogen over a period of weeks rather than all at once, because it has either low solubility (e.g., urea formaldehyde or trimethylene-urea) or a resistant coating (e.g., sulfur-coated urea); also known as a controlled-release nitrogen fertilizer. Note: Organic materials that require decomposition are inherently slow-release fertilizers.

sludge *n.* (German *slote*, mud) **1.** ooze; a semisolid residue composed of materials that settle to the bottom of a body of water or other liquid, e.g., in a silt pond or a sewage lagoon. See sewage sludge. **2.** a deposit of semisolid residue that collects in the bottom of a boiler, hot-water tank, or other similar container.

sludgeworm *n.* (German *slote*, mud + *wurm*, worm) a small freshwater worm (*Tubifex tubifex*) that is commonly found in muddy bottoms of lakes, ponds, and streams and often inhabits sewage sludge.

slug *n.* (Scandinavian *sluggje*, a heavy, slow person) **1.** a small, soft, slimy gastropod (family Limnacidae) that resembles a snail without a shell and is a serious pest of leafy garden crops. **2.** any of several caterpillars that resemble the gastropod slugs. **3.** an animal, person, or vehicle that moves very slowly. **4.** a metal disk that resembles a coin; e.g., those punched out of an electrical box. **5.** a piece of lead or other metal used as a bullet. **6.** a block of crude metal, especially as it comes from a furnace. **7.** a filler strip used to leave a blank space in printing, or a line of type from a Linotype machine. **8.** a unit of mass (approximately 32.2 lb or 14.6 kg) that accelerates at the rate of 1 ft-sec^{-2} when acted upon by a force of 1 lb. **9.** a shot of liquor taken in one gulp. **10.** a heavy blow. *v.* **11.** to make corrections for printing by replacing an entire line of type as set by a Linotype machine. **12.** to use pieces of metal to partially fill a joint being welded. **13.** to strike with a heavy blow, especially with the fist. **14.** to hit a baseball a long distance. **15.** to keep going slowly ahead in spite of resistance or obstacles, e.g., through mud or snow.

sluice *n.* (Latin *exclusa*, water barrier) **1.** an artificial channel for conducting water or other liquids, often fitted with a sluice gate at the upper end to control the flow. **2.** the water held back by the sluice gate. **3.** any device used to regulate water flow in or out of a container. **4.** a channel where a flow of water is used to transport solid objects such as logs. **5.** a sluice box. *v.* **6.** to allow water to pass through a sluice. **7.** to drain a wet area through a sluice. **8.** to clean an area by flushing water quickly across it. **9.** to transport logs or other objects through a sluice. **10.** to use a sluice box to process ore from a mine.

sluice box *n.* a sloping channel, usually made of wood with slats or grooves across the bottom, used to catch dense particles (e.g., pieces of gold) as ore is washed through it by a stream of water.

sluice gate *n.* a barrier used to control the flow of water entering a sluice.

sluiceway *n.* a sluice, especially one with a sluice gate at the entrance.

slump *n.* (German *slumpen*, accidental) **1.** a sudden collapse, sinking, or fall, especially of a heavy body or a person, animal, or mass of soil and/or rock. **2.** a period of poor or ineffective performance of a person, a group, or an athletic team. **3.** a period of low spirits; a downhearted feeling. **4.** a sharp decline in business activity and/or prices. *v.* **5.** to collapse or fall suddenly. **6.** to sink into mud or snow or through the ice on a body of water. **7.** to decrease (business activity) or decline (prices). **8.** to sit with the hips forward and the back rounded. **9.** to walk in a leaning position with rounded back.

slurry *n.* or *adj.* **1.** a mixture of water or other liquid carrier with a heavy load of solid particles ranging from powder to sand dispersed in it, often requiring agitation to maintain dispersion, that is used as a means of spreading fertilizer, transporting coal through a pipeline, transporting wastes, etc. **2.** a similar natural mixture, as watery mud at the bottom of a pond or lake.

small grain *n.* a grass crop with edible seeds smaller than those of corn and sorghum; wheat, rice, barley, oats, and rye. See cereal, grain.

smallpox *n.* (Anglo-Saxon *smael*, small + *poc*, pox) an acute infectious disease of humans caused by the variola virus (also called poxvirus). Declared extinct in 1979 following a worldwide vaccination program, smallpox produced vomiting, prolonged fever, and pustules that commonly left scars and sometimes resulted in blindness, pneumonia, and up to 40% mortality. Also called variola. See pox.

smaze *n.* (Anglo-Saxon *smoca*, smoke + *hasu*, ashen or dusky) a dry mixture of smoke and haze in the atmosphere.

smectite *n.* (Greek *smēktos*, smeared + *ite*, mineral) **1.** an expanding-lattice silicate clay mineral with tetrahedral layers on both sides of each octahedral layer [with a variable composition; an idealized formula is $(Al_{1.7}Mg_{0.3})Si_4O_{10}(OH)_2 \cdot xH_2O$ + enough exchangeable cations (especially Ca^{2+}) to balance the charge]. Its lattice spacing varies with water content from 0.96 to 2.14 nm, and it has a relatively high cation-exchange capacity averaging about 800 mmol of charge per kilogram. Formerly called mont-morillonite. **2.** a family of silicate clay minerals, including smectite, montmorillonite, and their close relatives (hectorite, nontronite, saponite, and sauco-nite).

smelt *v.* (German *smelten*, to melt) **1.** to extract a metal from its ore by melting the ore. *n.* **2.** any of several small, silvery, food fish (*Osmerus* sp., family Osmeridae) that spawn in tidal waters.

Smith-Lever Act *n.* the federal law passed in 1914 that provided funds for establishing agricultural extension in the U.S.A.

smog *n.* (Anglo-Saxon *smoca*, smoke + Danish *fog*, spray) **1.** an irritating, visibility-limiting mixture of smoke and other air pollutants with fog trapped near ground level by temperature inversion. **2.** photo-chemical smog. See ozone, peroxyacetylnitrate, photochemical smog.

smoke *n.* (Anglo-Saxon *smoca*, smoke) an aerosol produced by incomplete combustion of organic substances that releases solid particles of carbon, silica, metal oxides, etc., and gases such as sulfur dioxide and oxides of nitrogen. Smoke can kill vegetation and injure the lungs.

smoke-cured *adj.* a meat product such as certain bacon, ham, sausage, and bologna that has been heated in a smokehouse to an internal temperature of 65.6° to 76.7°F (18.7° to 24.8°C) to impart a smoke flavor and improve the storage quality. Hardwood such as hickory is generally used for smoke curing.

smolder or **smoulder** *v.* **1.** to burn slowly, producing heat and smoke but no flame. **2.** to hold a subdued negative feeling, as smoldering with inward resentment. **3.** to display such a negative inward feeling, as a smoldering glare. *n.* **4.** dense smoke produced by combustion without flame, often of wet material. **5.** slow combustion without flame.

smooth bromegrass *n.* an awnless perennial sod-forming grass (*Bromus inermis*, family Poaceae), native to Europe, that grows as tall as 5 ft (1.5 m), produces palatable hay or pasture, and is used extensively to seed pastures and rangeland.

smother *v.* (Anglo-Saxon *smorian*, to suffocate) **1.** to kill or nearly kill by cutting off the air supply of a person or animal. **2.** to extinguish a fire by depriving it of oxygen. **3.** to prevent plant growth by covering with a layer of plastic, mulch, soil, ice, etc. **4.** to cover one food with another, as to smother a steak with mushrooms. **5.** to repress feelings, as to smother one's emotions. **6.** to die of suffocation, as to smother to death. *n.* **7.** a dense covering, as a smother of smoke, spray, or thick fog. **8.** an overload, as a smother of paperwork.

smother crop *n.* a dense, fast-growing crop planted to suppress weeds by intense competition for light, water, and nutrients.

smudge *n.* (Dutch *smotsen*, to besmirch) **1.** thick, stifling smoke. **2.** a smoky fire, e.g., that from a smudge pot. **3.** a dark mark, usually smeared. *v.* **4.** to produce a thick smoke; e.g., to prevent or reduce frost damage or to drive away insects. **5.** to make a smeared mark. **6.** to be marked with a smudge, as shoes that smudge easily.

smudge pot *n.* a pot that holds oil and a wick to produce a smoky fire intended to reduce frost damage of fruits and vegetables.

smudging *n.* the use of smudge pots or fires made with wet straw, manure, garbage, old tires, or other material that will produce thick smoke to drive away insects or reduce frost damage to sensitive fruits or vegetables. Unfortunately, the smoke produced is an air pollutant.

smut *n.* (German *smutt*, dirt) **1.** any of several diseases of wheat, rye, oats, barley, sorghum, and corn caused by fungi (of class Basidiomycetes, order Ustilaginales) that infect seedlings, grow with the plant, and cause it to produce masses of black spores instead of seed. Seed smuts are also called bunt. **2.** any of many other similar fungal diseases that attack hundreds of grass species, affecting the seeds, flowers, leaves, and/or stems. **3.** the fungi that attack the seeds (*Tilletia* sp.), flowers (*Sorosporium* sp., *Sphacelotheca* sp., and *Ustilago* sp.), leaves (*Entyloma* sp., *Urocystis* sp., and *Ustilago* sp.), and stems (*Ustilago* sp.) of grasses. **4.** a particle of dark, sooty matter, or a smudge made by such fungi. **5.** indecent (especially sexually oriented) language or printed material.

Sn *n.* the chemical symbol for tin.

snail *n.* (Anglo-Saxon *snegel*, snail) **1.** any of several land or aquatic mollusks (*Helix* sp., family Helicidae, class Gastropoda) that feed mostly at night and eat many kinds of vegetation. Each snail lives in its own

Land snails look like blossoms on these plants.

coiled shell and moves slowly on a single muscular foot. See slug. **2.** a slow-moving or lazy person. **3.** a spiral-shaped cam.

snake *n.* (Anglo-Saxon *snaca*, snake) any of a large group of cold-blooded vertebrate animals (suborder Serpentes) with flexible elongated bodies covered by scaly skin that is shed periodically. Snakes have no legs but move by twisting their bodies, and they have keen senses for sight and smell but no hearing. Most people fear them because some snakes are poisonous (e.g., rattlesnakes) or strong enough to strangle victims (e.g., pythons), but most kill and eat small animals such as mice. See poikilothermic.

snakeroot *n.* (Anglo-Saxon *snaca*, snake + *rote*, root) **1.** any of several plants reputed to cure the effects of snakebite, including Seneca snakeroot (*Polygala senega*, family Polygonaceae) and white snakeroot (*Eupatorium rugosum*, family Asteraceae), among others. White snakeroot foliage contains tremetol, a toxin that causes milk sickness. **2.** the root of such a plant.

snare *n.* (Norse *snara*, a snare) **1.** a trap for small animals, usually made with a noose that is pulled tight when a trigger is bumped. **2.** a temptation to enter into a risky venture. **3.** a wire noose used in surgery to excise tumors or other growths. *v.* **4.** to trap with a snare. **5.** to trick someone into doing something.

snipe *n.* (Norse *snipa*, beak) **1.** any of a group of birds (*Gallinago* sp. and *Limnocryptes* sp., family Scolopacidae), with long, slender bills, that inhabit marshy areas. The common snipe (*G. gallinago*) is about 9 in. (23 cm) long with a brown body marked with white stripes, a white belly, and an orange tail. **2.** any of several related sandpipers. **3.** a shot from a concealed position. *v.* **4.** to go snipe hunting. **5.** to shoot from a concealed position. **6.** to make critical comments remotely or anonymously.

snout *n.* (German *schnauze*, snout) **1.** muzzle; the front part of an animal's head, including the nose and mouth. **2.** anything that resembles the snout of an animal; e.g., the mouth of a snout beetle. **3.** a person's nose, especially if it is long. **4.** a long pouring spout.

snout beetle *n.* any of a group of beetles (family Curculionidae) whose head extends forward and includes biting mouthparts that enable it to chew deep holes in plant tissue, where it lays its eggs. Grain weevils are part of this family. Also called curculio beetle.

snow *n.* (Anglo-Saxon *snaw*, snow) **1.** frozen water in the form of snowflakes that form in the upper atmosphere, fall gently, and accumulate like a white blanket on the landscape. On the average, the water content of 10 in. of snow is equal to 1 in. of rain.

2. an episode of snowfall. **3.** the bright white of fresh snow. **4.** extraneous white spots or streaks on a television screen. *v.* **5.** to fall as snow. **6.** to form a deposit of snow, as it snowed 4 in. (10 cm). **7.** to persuade by deceptive flattery. **8.** to bury or overwhelm with snow or (figuratively) with work (usually snow under with work).

snowbank *n.* (Anglo-Saxon *snaw*, snow + *banki*, bench) a deep pile of snow, often with a steep face, as a snowdrift, a pile pushed up by a snowplow, or a mound made with a shovel.

snow bunting *n.* a chunky songbird (*Plectrophenax nivalis*) about 6 in. (15 cm) long that is commonly seen in the tundra and sometimes on beaches, dunes, or grassy areas. It is identified by white wing patches and generally white, yellow, and tan coloration with black markings. Also called snowflake.

snow cover *n.* **1.** a layer of snow on the ground. It serves as an insulating blanket that reduces frost penetration. **2.** the thickness of the snow layer. **3.** the percentage of the area covered with snow.

snow devil *n.* a column of snow rotating in an air current as it falls. See dust devil.

snowdrift *n.* (Anglo-Saxon *snaw*, snow + *drifan*, to drive) a pile of snow deposited by wind, often with a steep face similar to a sand dune.

snowdrop *n.* (Anglo-Saxon *snaw*, snow + *dropa*, drop) any of several bulbous flowering plants (*Galanthus* sp., family Amaryllidaceae), especially *G. nivalis* with its drooping white blossoms marked with green.

snowfall *n.* (Anglo-Saxon *snaw*, snow + *feallan*, to fall) **1.** a period when snow is falling. **2.** the amount of snow that falls, as a depth of accumulation.

snow fence *n.* **1.** fencing material composed of slats about 2 in. (5 cm) wide and 4 ft (1.2 m) long held together with wires. It is used to form a wind barrier or a fence. **2.** a wind barrier that reduces wind velocity and causes a snowdrift to form on the downwind side rather than elsewhere. Either a wooden-slat fence, a line of shrubs or trees, or another obstacle that slows the wind.

snowflake *n.* (Anglo-Saxon *snaw*, snow + *flacor*, flying) **1.** a single feathery crystal of frozen water. Snowflakes form as they fall and occur in multitudes of shapes with hexagonal symmetry. **2.** a snow bunting. **3.** a bulbous plant (*Leucojum* sp., family Liliaceae) with drooping white blossoms that is of European origin.

snowline *n.* (Anglo-Saxon *snaw*, snow + *line*, a cord) the line marking the lowest elevation on a mountain that is covered with snow throughout the year. It is strongly influenced by altitude, latitude, and the direction the slope faces (aspect).

snow mold *n.* a disease of grasses, grains, and alfalfa caused by fungi (*Typhula* sp., *Calonectria* sp., and *Pithium* sp.) that grow and spread beneath snow cover. When the snow melts, dead plants with white mold can be seen in circular patches, usually 1 to 12 in. (2.5 to 30 cm) or more in diameter. See eastern snow mold.

snowpack *n.* (Anglo-Saxon *snaw*, snow + *pakke*, pack) the accumulation of snow on the ground during a winter, especially in mountainous areas.

snowquake or **snow tremor** *n.* (Anglo-Saxon *snaw*, snow + *cwacian*, to shake) a sudden collapse of one or more layers of snow, accompanied by an explosive sound and often producing an avalanche.

soakage *n.* (Anglo-Saxon *socian*, to soak + French *age*, result) **1.** the act of soaking in a liquid. **2.** the condition of being soaked. **3.** the amount of moisture a substance will absorb from the air or soak up from a liquid. For example, the fire hazard in a forest is related to the amount of soakage held in the litter layer.

soap *n.* (Anglo-Saxon *sape*, soap) **1.** a cleansing agent made by treating natural fats or oils with an alkali (usually potassium hydroxide or sodium hydroxide). Synthetic detergents have largely replaced soaps because the soaps react with divalent cations (esp. Ca^{2+}) of hard water and form precipitates. (See detergent.) **2.** any metallic salt of a fatty acid. *v.* **3.** to wash something with soapy water. **4.** to coat or mark a surface with soap.

soboles *n.* (Latin *sub*, under + *olere*, to grow) **1.** a creeping stem, usually underground, that produces new plants at short intervals, similar to a rhizome or stolon but usually associated with a woody plant, e.g., sumac (*Rhus* sp.) or lilac (*Syringa* sp.). **2.** a sucker growing near ground level from a shrub or tree.

social dominance *n.* a group situation wherein one or more aggressive individuals (persons or animals) assert leadership such as that exhibited in the pecking order in chickens, the dominant stallion in a group of horses, or natural leaders in human society.

social insects *n.* bees, wasps, ants, termites, and other insects that live together, usually subdivided into groups that perform specific functions. See honeybee.

sociobiology *n.* (Latin *socius*, companion + French *biologie*, life study) the study of the social habits of animals, including survival tactics, reproduction, behavior, ecological relationships, etc.

sociophagous *adj.* (Latin *socius*, companion + Greek *phagos*, eating) living at the expense of others; preying on one or more social groups.

sod *n.* (Dutch *sode*, turf) **1.** a thin layer of soil with a dense mat of grass growing in it. Also called sward or turf. **2.** an area of land covered with growing grass. *adj.* **3.** made of sod, as a sod house. *v.* **4.** to cover with sod, as to sod the new lawn.

sodbound *adj.* (Dutch *sode*, turf + Anglo-Saxon *bindan*, to bind) a stand of grass that is doing poorly because its dense mat of roots is no longer able to obtain enough air, water, or nutrients or because there is an insect or disease problem.

sodbuster *n.* (Dutch *sode*, turf + Anglo-Saxon *berstan*, to break) one who tills an area that has long been in grass to convert it into cropland.

sodbuster law *n.* a provision in the 1985 U.S. Food Security Act that requires erosion protection when highly erodible land is cropped. Also called the conservation compliance provision.

sodden *adj.* (Anglo-Saxon *sodden*, seethed) **1.** saturated with water or other liquid, as a sodden area in a field. **2.** heavy and excessively moist, as sodden bread or cake. **3.** looking soaked and bloated, as a sodden expression on his face, especially when caused by drunkenness. **4.** dull and listless, as sodden spirits.

sod house *n.* a house or farm building with walls made of sod cut from native grasslands, as was done by some of the pioneers who settled the plains of central U.S.A.

sodic soil *n.* a soil with 15% or more of its cation-exchange capacity occupied by exchangeable sodium. The high sodium content causes a high pH (8.5 or above), dissolves organic matter (giving the soil water and soil surface a dark color and the common name black alkali), and disperses the clay in the soil. The dispersed clay plugs soil pores and drastically reduces soil permeability. Little if anything will grow on such soil, which typically occurs in small level spots a few feet or meters across that are often called slick spots.

sodic soil reclamation *n.* treatment of a sodic soil to remedy adverse soil conditions. This usually involves application of large amounts of gypsum or other soil amendment to neutralize the sodium, mixing it physically with the soil, installing a drainage system if needed, and leaching out sodium salts. It is slow, expensive, and usually not economical.

sodium (Na) *n.* (Latin *soda*, burnt plant + *ium*, element) element 11, atomic weight 22.98977; a highly reactive alkali metal that forms monovalent cations. It is an essential nutrient for animals and humans but present in such abundance that it is often in excess (causing water retention and contributing to high blood pressure) and hardly ever deficient. In the form of table salt (NaCl), it is commonly added to food to enhance flavor and is sometimes used as a preservative.

sodium arsenate *n.* a poisonous water-soluble chemical ($Na_3AsO_4 \cdot 12H_2O$) used as a germicide and insecticide and in printing and dyeing.

sodium arsenite *n.* a poisonous water-soluble chemical ($NaAsO_2$) used as an herbicide, insecticide, and germicide and for cleaning hides.

sodium benzoate *n.* a water-soluble chemical ($NaC_7H_5O_2$) used as a food preservative, as a fungicide, as a test for liver function, and in the manufacture of pharmaceuticals.

sodium chlorate *n.* a strong oxidant ($NaClO_3$) used in the manufacture of explosives, in bleaching agents, and as an herbicide.

sodium chloride *n.* halite ($NaCl$); the principal solute in seawater and the dominant component in many evaporite deposits in the form of large cubical crystals. It is used to season food, to supplement fodder in livestock feed, to preserve food, and to melt snow and ice. Also called common salt or table salt (usually as iodized salt with potassium iodide, KI, added).

sodium fluoride *n.* a water-soluble chemical (NaF) often added to toothpaste and to public drinking water supplies at a rate of 0.7 to 1.0 part fluorine (F) per million to control tooth decay (high levels would be toxic). It is also used to control cockroaches and to control chewing lice on poultry and other animals. See fluorapatite.

sodium hypochlorite *n.* an oxidant (NaClO) that decomposes in hot water. Used in household bleaches, disinfectants, and fungicides and to purify water, it is sold under several trade names; e.g., Clorox™ is 5.25% sodium hypochlorite.

sodium molybdate *n.* a molybdenum salt ($Na_2MoO_4 \cdot 2H_2O$) used as a micronutrient fertilizer and sometimes dusted on legume seed to promote nitrogen fixation by *Rhizobium* bacteria.

sodium nitrate *n.* a water-soluble salt ($NaNO_3$) that occurs naturally in a large evaporite deposit in Chile. It was the first important mineral nitrogen fertilizer but was largely supplanted by higher-analysis nitrogen fertilizers. It is also used in explosives and to cure meat and enhance its color. Also called nitrate of soda or Chilean saltpeter.

sodium selenate *n.* a poisonous, water-soluble chemical ($Na_2SeO_4 \cdot 10H_2O$) used as an insecticide, especially by adding it to water that plant roots will absorb, thus making the plant sap toxic to parasites such as aphids and red spider mites.

sodokosis *n.* (Japanese *so*, rat + *doku*, poison + Greek *osis*, condition) rat-bite fever. Either of two infectious diseases (one caused by *Spirillum minus*, the other by *Streptobacillus moniliformis*) that cause pain and fever and are transmitted by the bite of a rodent, especially a rat.

sod planting *n.* **1.** seeding directly into sod that has either matured (as an annual crop that has been harvested) or treated with an herbicide that kills or stunts growth (e.g., of a pasture or hay crop in a rotation). **2.** seeding a cool-season crop such as rye or white clover into an established stand of warm-season grass that goes dormant in the fall. **3.** laying strips of sod for a new lawn, roadside cover, or other area where permanent grass cover needs to be established quickly. Also called sodding. **4.** planting cuttings such as stolons or rhizomes to establish certain grasses (e.g., Bermuda grass).

soft rock *n.* **1.** a sedimentary rock, especially those that are unconsolidated (e.g., loess, alluvium, or glacial till) or weakly consolidated (e.g., shale). **2.** a type of music that emphasizes arrangement and lyrics more than the beat that characterizes hard rock music.

soft water *n.* **1.** water that naturally has a low concentration of solutes, especially divalent cations (e.g., less than 60 ppm $Ca^{2+} + Mg^{2+}$), that would react with soap and form a precipitate. **2.** water that has been passed through a water softener to replace divalent cations with sodium ions (Na^+). Also called softened water. Note: Distilled water, deionized water, and reverse-osmosis water contain very little solute and are very soft.

soft wheat *n.* **1.** any wheat variety that produces soft, starchy kernels. See hard wheat, durum wheat. **2.** the grain produced by such wheat varieties. Soft red wheat is typically used to make cake and pastry flour. Soft white wheat is used mostly for bread, breakfast cereals, and pastries.

softwood *n.* **1.** any wood that is relatively soft, low in density, and easy to cut. **2.** any tree species that produces such wood. **3.** any coniferous tree (gymnosperm); e.g., a pine, spruce, fir, or tamarack. See gymnosperm, hardwood, angiosperm.

soil *n.* (Latin *solum*, ground) **1.** the highly variable, unconsolidated mixture of mineral and organic matter that supports plant life at the surface of the Earth. **2.** the natural product formed at the surface of the Earth by the action of climate and living organisms on weathered rock, as influenced by topography and time. **3.** any unconsolidated earthy material. Also called regolith. **4.** an environment where something may develop, as despair provides the soil for crime. *v.* **5.** to make unclean or dirty; to smudge or stain. **6.** to tarnish or damage figuratively, as to soil one's reputation.

soil aeration *n.* the exchange of gases between soil air and the atmosphere above the soil by either diffusion or mass movement resulting from water moving in and out of the soil or mechanical action such as tillage or animals digging in the soil.

soilage *n.* (Latin *solum*, ground + *age*, related to) **1.** green forage crops grown in the field and cut and hauled to feed confined livestock. **2.** the process or condition of being soiled.

soil aggregate *n.* a group or clump of soil particles held together more strongly than they are attracted to other aggregates or particles; commonly occurring as components of larger structural units such as blocks, plates, or prisms.

soil air *n.* See soil atmosphere.

soil amendment *n.* a material added to the soil to alter its physical or chemical characteristics (factors other than its supply of plant nutrients); e.g., lime, animal manure, compost, sewage sludge, sand, or clay. Soil amendments are generally applied in relatively large amounts (thousands of pounds per acre or thousands of kilograms per hectare). See fertilizer.

soil association *n.* a soil map unit used on generalized soil maps. The units are groups of soil series that occur together and collectively occupy extensive areas.

soil atmosphere *n.* the air contained in a soil profile; the part of the pore space that is not filled with water. Respiration by plant roots and soil organisms depletes the soil air of oxygen, increases its carbon dioxide content, and may make the soil anaerobic if there is inadequate aeration (generally associated with less than 10% air by volume in the soil).

soil biology *n.* the science that studies life in the soil, including soil microbes and larger forms such as insects, nematodes, and earthworms, and processes such as nitrogen fixation that are associated with soil organisms.

soilborne *adj.* transmitted via soil, as certain pathogenic organisms (e.g., nematodes and viruses) that survive cold or dry periods in the soil and/or may be transported in soil that sticks to machinery, shoes, etc., and thus infect plants the next year or in a new area.

soil bulk density *n.* See bulk density.

soil burning *n.* burning the soil is a traditional method of increasing the availability of phosphorus in Vertisols in Ethiopia. An area of grassland is broken into large clods with a village plow, small mounds of sod are built, manure and/or wood is added, and a fire is built in the center of each pile. After smoldering for a few days, the field is smoothed, and a crop, perhaps of sorghum, is planted.

soil category *n.* Soil Taxonomy has six categorical levels: order, suborder, great group, subgroup, family, and series. See Soil Taxonomy.

soil class *n.* **1.** a subdivision of any soil series based on the texture of the surface horizon. **2.** any group of soils classified together in Soil Taxonomy or some other soil classification system. **3.** any of the eight classes in the land-use capability classification system.

soil classification *n.* systematic arrangement of soils into classes at various categorical levels on the basis of defined soil characteristics; e.g., Soil Taxonomy used for the National Cooperative Soil Survey in the U.S.A.

soil colloid *n.* either a mineral or an organic soil component composed of solid particles (mostly silicate clay minerals, iron and aluminum oxides, and humus) so finely divided that it has a very large surface area per gram. The individual particles are so small, less than about 1 µm across, that they will not settle out if they are individually suspended in water. In the soil, these particles are chemically active ion-exchange sites and are usually clumped together in soil aggregates.

soil color *n.* a visible soil property that can be evaluated by comparing a piece of soil to the color chips in a Munsell soil-color book. The Munsell system classifies color on the basis of its hue (in soil, mostly gradations of yellow and red), value (brightness or darkness ranging from 0 for black to 10 for white), and chroma (purity of color ranging from 0 for white, gray, or black to about 8 for soil or 20 for an absolutely pure color). These three variables are designated by a symbol; e.g., 10YR 5/3 is a brown soil with a 10 yellow-red hue, a value of 5, and a chroma of 3.

soil complex *n.* a mixture of two or more soil series that are so intermixed that they are combined even on a detailed soil map.

soil conditioner *n.* a material added to the soil to improve its structure, especially to make it more granular on a long-term basis. Large applications of organic materials such as manure will do this and also will increase soil fertility. Synthetic soil conditioners made of organic polymers that resist decomposition (the first was Krilium, introduced by Monsanto Chemical Company in 1952) can stabilize soil structure without affecting soil fertility, thus allowing soil structure and fertility to be controlled separately, especially for experimental purposes (such soil conditioners are generally too expensive for farm use).

soil conservation *n.* **1.** any combination of procedures that allows soil to be used without excessive loss by erosion or other damage to its productive capacity. **2.** the movement and organizations that foster the use of soil-conserving practices.

soil conservation district *n.* See conservation district.

Soil Conservation Service *n.* the agency in the U.S. Department of Agriculture with the responsibility of assisting land users in applying soil-conserving practices from 1935, when it replaced the Soil Erosion Service (organized in 1933), until it was converted into the Natural Resources Conservation Service in 1995.

soil consistence *n.* the tendency of soil particles to cohere together or fall apart. Soil consistence is described in dry soil as loose, soft, slightly hard, hard, very hard, or extremely hard. In the moist state, the consistence is described as loose, very friable, friable, firm, very firm, or extremely firm. In wet soil, the consistence is described as nonplastic, slightly plastic, plastic, or very plastic and as nonsticky, slightly sticky, sticky, or very sticky.

soil core *n.* a cylindrical soil sample extracted with either a hand probe or a hydraulic probe mounted on a pickup or tractor.

soil creep *n.* gradual downslope movement of soil under the influence of gravity, freezing and thawing, and animal traffic, occurring mostly when the soil is wet.

soil degradation *n.* deterioration in soil quality, either chemical (e.g., becoming very acid, alkaline, saline, or low in fertility) or physical (e.g., loss of soil structure and permeability or formation of a hard crust).

soil development *n.* **1.** the combination of physical, chemical, and biological processes that convert rock material and organic residues into soil and gradually change it into an older soil. **2.** the changes that occur in the material as it becomes soil and as the soil ages; e.g., the formation of soil structure in a young soil, the accumulation of clay in a mature subsoil, or the depletion of fertility in an old soil.

soil drainage *n.* **1.** the tendency of a soil either to accumulate or to release water, especially when there is an excessive supply of water. External soil drainage is usually described in terms of runoff classes as ponded, very slow, slow, medium, rapid, or very rapid. Internal soil drainage classes used in soil survey in the U.S.A. are excessively drained, somewhat excessively drained, well drained, moderately well drained, somewhat poorly drained, poorly drained, and very poorly drained. **2.** the removal of excess water from soil by a system of ditches or tile drainage. See land drainage.

soil erosion *n.* the removal of soil by water, wind, and sometimes ice, especially that carried out of the field or vicinity, including much that becomes an air or water pollutant and settles as dust or sediment in places where it is not wanted.

Soil Erosion Service *n.* a temporary public works program established in the U.S. Department of the Interior in 1933 to work on controlling soil erosion on both public and private lands that was replaced by the Soil Conservation Service in the U.S. Department of Agriculture in 1935.

soil fabric *n.* the spatial arrangement of soil particles and pore space, including size, shape, and orientation of the components.

soil fertility *n.* the ability of soil to supply plant nutrients for use by growing plants, involving both the amount present and the availability of each essential nutrient. See soil productivity.

soil formation *n.* the processes involved in changing weathered rock material and organic residues into soil; the action of climate and living organisms (plants, animals, and microbes) on parent materials over time, as conditioned by topography. Also called soil genesis.

soil fraction *n.* any of the solid components of a soil; usually divided into sand, silt, clay, and organic fractions.

soil fumigant *n.* a volatile chemical (e.g., ethylene dibromide, $C_2H_4Br_2$) used to kill soil pests (usually nematodes or disease agents).

soil genesis *n.* See soil formation.

soil horizon *n.* a layer that is approximately parallel to the soil surface and can be distinguished from the horizons above and below by its color, texture, structure, density, consistency, pH, living organisms present, etc.

soil-improving crop *n.* a crop that has a positive effect on future crops by improving either the fertility (e.g., nitrogen fixation by a legume) or the physical condition (e.g., root penetration opening new pores and added organic matter stabilizing soil structure).

soil map *n.* a map showing delineations and symbols to indicate the kinds of soils present in an area, commonly published by the Natural Resources Conservation Service and cooperating agencies on an aerial photo base as a part of a soil survey report. Soil maps may be detailed (mapping units are mostly subdivisions of soil series that have been mapped throughout the area), reconnaissance (more generalized maps made by a partial survey), or soil association maps (very generalized to show large areas).

soil map unit *n.* **1.** a soil or group of soils that occur in identifiable areas on a landscape, usually defined in terms of soil series, surface texture, and slope gradient, often with a designation of severity of past erosion and sometimes indicating other factors such as position, wetness, or alkalinity. **2.** an area delineated on a soil map and marked with a symbol to indicate the kind of soil present. Formerly called soil mapping unit. See map unit, polypedon.

soil mechanics *n.* the branch of civil engineering that deals with soil as either a foundation or a construction material for roads and structures, considering properties such as compression and shear strengths and permeability.

soil microbe *n.* any microscopic soil organism, including bacteria, actinomycetes, fungi, algae, and protozoa.

soil microbiology *n.* the science that studies soil microbes and processes such as nitrogen fixation and decomposition of organic matter that are associated with soil microbes. It is sometimes broadened and called soil biology to include visible insects, nematodes, earthworms, etc.

soil mineralogy *n.* the science that studies the composition of soil minerals and the weathering processes that change rock minerals into soil minerals such as silicate clays.

soil moisture profile *n.* **1.** the distribution of water in a soil profile. **2.** a curve plotted to indicate the soil moisture content throughout the soil profile.

soil moisture potential *n.* See soil water potential.

soil moisture regime *n.* an evaluation of the degree of wetness or dryness normally associated with a particular soil, described as aquic (wet enough to cause reducing conditions), aridic (too dry for plant growth through more than half of the growing season), torric (aridic but used in warmer climates), udic (humid but not aquic), ustic (between aridic and udic), or xeric (moist during the cool season but dry during summer).

soil moisture tension *n.* the attractive force between soil water and soil particles that is now expressed in pascals by soil scientists but was formerly expressed in bars, atmospheres, or other units of negative pressure. Also called capillary potential, matric potential, or suction (negative pressure).

soil monolith *n.* a mounted soil profile for display or study; a vertical section taken from a soil bank or pit and mounted on a board or other support, often with an accompanying description of each soil horizon.

soil morphology *n.* the physical characteristics of a soil and its constituent horizons, usually described for a representative pedon plus a defined range of important characteristics. See pedon, soil horizon.

soil order *n.* the highest (most generalized) category in Soil Taxonomy. Lower categories are suborders, great groups, subgroups, families, and series. Twelve soil orders are identified in Soil Taxonomy:

Soil order	% of U.S. surface area
Alfisols	14
Andisols	2
Aridisols	8
Entisols	12
Gelisols	9
Histosols	2
Inceptisols	10
Mollisols	21
Oxisols (0.02%)	0
Spodosols	3
Ultisols	9
Vertisols	2
Water, ice, bare rock	8

soil organic matter *n.* the organic fraction of any soil, especially the soil humus; usually including any organic materials that remain after the soil is passed through a sieve with 2-mm openings.

soil particle *n.* an individual grain of sand, silt, or clay, or a fragment of organic matter.

soil phase *n.* a subdivision of any class in a soil classification system such as Soil Taxonomy, based on a property that affects land use but is not defined in the classification system, especially a subdivision of a soil series on the basis of slope or texture of the surface horizon.

soil physics *n.* the science that studies the physical properties of soil, especially texture, structure, water relations, and water flow.

soil porosity *n.* **1.** the pore space in the soil; the space occupied by water and air. **2.** the percentage of the soil volume occupied by pore space, sometimes also including pore size distribution and how well the pores are connected to each other.

soil productivity *n.* the amount of crop produced per unit area (e.g., bushels per acre, tons per acre, or quintals per hectare). Soil productivity depends on the crop, soil fertility, environmental factors, pest problems, and management.

soil profile *n.* a vertical section through the horizons of a soil as viewed in an embankment, pit, or soil monolith.

soil reaction *n.* the degree of acidity or alkalinity of a soil, normally expressed in pH values for each horizon. See pH. The following terminology is sometimes used to describe soil acidity and alkalinity:

Extremely acid	Below 4.5
Very strongly acid	4.5–5.0
Strongly acid	5.1–5.5
Medium acid	5.6–6.0
Slightly acid	6.1–6.5
Neutral	6.6–7.3
Mildly alkaline	7.4–7.8
Moderately alkaline	7.9–8.4
Strongly alkaline	8.5–9.0
Very strongly alkaline	9.1 and higher

soil reclamation *n.* any remedial process applied to a soil to make it suitable for cropping, especially draining wet land and correcting extremely acid or alkaline conditions. See reclamation of mined land, saline soil reclamation, sodic soil reclamation.

soil salinity *n.* **1.** the concentration of soluble salts in the soil solution. See saline soil. **2.** the soil condition when excess soluble salts are present.

soil science *n.* **1.** pedology. **2.** any curriculum that studies soil; e.g., soil genesis, soil classification, soil chemistry, soil physics, soil fertility, and soil biology.

soil separate *n.* any of the divisions of mineral soil particles commonly made on the basis of size (diameter). The U.S. Department of Agriculture uses the following size limits:

Soil separate	Mineral soil particle size (mm)
Very coarse sand	1–2
Coarse sand	0.5–1
Medium sand	0.25–0.5
Fine sand	0.1–0.25
Very fine sand	0.05–0.1
Silt	0.002–0.05
Clay	<0.002

soil series *n.* **1.** a class at the lowest (most specific) categorical level in Soil Taxonomy and several other systems of soil classification. **2.** a group of polypedons that are classified together because their sequence of soil horizons and the properties of those horizons all fall within defined ranges. Moving to higher categories in Soil Taxonomy, soil series are grouped into families, subgroups, great groups, suborders, and orders.

soil slip *n.* a short downslope movement of a mass of soil, usually a much smaller mass and shorter movement than a landslide.

soil solution *n.* the liquid phase in a soil, composed of water and dissolved materials.

A soil slip that moved about 5 ft (1.5 m). The lines above the slip are road cuts.

soil survey *n.* **1.** the process of examining soils in the field and making a soil map. **2.** the program that makes such surveys, describes and classifies the soils, and publishes soil survey reports. See National Cooperative Soil Survey.

Soil Taxonomy *n.* the U.S. Department of Agriculture system of soil classification since 1975 (when it officially replaced the system published in *Soils and Men*, the 1938 USDA Yearbook of Agriculture), including 12 soil orders, about 50 suborders, 200 great groups, 1000 subgroups, 4500 families, and more than 12,000 soil series.

soil testing *n.* **1.** using chemical tests to measure soil pH, the availability of phosphorus and potassium, and sometimes the availability of nitrogen, magnesium, sulfur, and/or some of the micronutrients as a basis for making lime and fertilizer recommendations. **2.** any chemical, physical, or biological procedure used to evaluate the suitability of a soil for supporting plant growth. **3.** any procedure that evaluates the suitability of a soil for use as a foundation or construction material. See soil mechanics.

soil texture *n.* the overall physical effect of the proportions of the various soil separates present in a soil material. Soil textures are named by the U.S. Department of Agriculture with various combinations of the terms sand, silt, clay, and loam, sometimes with size designations added to sandy soils; e.g., fine sandy loam. Either a soil texture triangle or a written definition is used to determine the soil texture name from the results of a mechanical analysis.

soil texture triangle *n.* a diagram used to determine soil texture from the percentages of sand, silt, and clay.

soil type *n.* **1.** a combination of soil series and surface texture; the lowest categorical level in the 1938 system of soil classification (replaced by Soil Taxonomy in 1975). **2.** kind of soil (current usage).

soil water potential *n.* the combined effect of gravitational, capillary, and osmotic forces acting on soil water and reducing its free energy; the force that plant roots must overcome when they absorb water. Also called soil moisture potential.

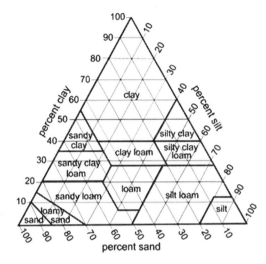

The U.S. Department of Agriculture uses this soil texture triangle to name soil textures.

sol *n.* **1.** a colloidal system composed of solid particles dispersed in a liquid. **2.** the French word for soil, often used as an ending for soil names or terms. **3.** a musical note; the 5th tone on a scale. **4.** a monetary unit in Peru and formerly in France.

solanaceous *adj.* (Latin *solanum*, nightshade + *aceus*, like) belonging to the nightshade family of plants (*Solanum* sp. and related genera, family Solanaceae), a large family with wide distribution, including several food plants and some species that are weedy and/or poisonous:

Common name	*Solanum* species
Eggplant	*S. melongena*
Potatoes	*S. tuberosum*
Tomatoes	*Lycopersicon esculentum*
Bittersweet	*S. dulcamara*
Black nightshade	*S. nigrum*
Horse nettle	*S. carolinense*
Jerusalem cherry	*S. pseudocapsicum*
Nipplefruit	*S. mammosum*

solanine *n.* (Latin *solanum*, nightshade + *inus*, like) a green, bitter, toxic alkaloid ($C_{45}H_{73}O_{15}N$) that develops in solanaceous plants, e.g., on the outer part of potato tubers when the tuber is exposed to the sun while growing.

solar *adj.* (Latin *solaris*, of the sun) **1.** pertaining to or coming from the sun, as solar mass, solar activity, or solar radiation. **2.** using solar energy, as a solar building or solar industry. **3.** according to the sun, as solar time.

solar cell *n.* a photovoltaic cell that uses a semiconductor (made of silicon, gallium arsenide, gallium antimonide, etc.) to convert sunlight into electricity. It is a clean, safe, silent means of producing electricity that is currently used, because of high cost and limited efficiency, mostly in remote locations and in spacecraft.

solar constant *n.* the average amount of energy from the sun impacting 1 cm^2 of the Earth's outer atmosphere on a straight line toward the center of the Earth, 2 cal/min ± 2%.

solar cooker *n.* a solar-powered cooking device; usually one that focuses the sun's rays on a container to boil water or cook food.

solar day *n.* the 24-hr day marked by the position of the sun as the Earth rotates. It is about 3.94 min longer than the sidereal day marked according to the position of a star because the orbit of the Earth around the sun causes the solar year to have one less day than the sidereal year.

solar energy *n.* energy derived from the sun's radiation, often by using sunlight either to heat water or to produce electricity in solar cells.

solarium *n.* (Latin *solaris*, of the sun + *ium*, place) a room or glass-enclosed porch exposed to the sun for the enjoyment of people and pets or for growing plants. It is sometimes used in homes, resort hotels, or convalescence areas of hospitals.

solarization *n.* (Latin *solaris*, solar + *izatio*, process) **1.** conversion to the use of solar energy for heat or electricity. **2.** covering moist soil with a plastic film for a month or two to cause weed seed to germinate and die and to reduce or eliminate pathogen populations before planting (also likely to reduce favorable microbes and increase the need for inoculation with symbionts). **3.** exposure of a photographic print to light during development, causing a partial reversal of light and dark tones.

solar month *n.* one-twelfth of a solar year, totaling 30.43685 days or 30 days, 10 hr, 29 min, and 3.8 sec. See solar year.

solar radiation *n.* light, heat, and other energy radiated from the sun.

solar system *n.* the sun and all the planets, asteroids, comets, etc., that orbit around it.

solar wind *n.* a high-speed stream of ionized particles (mostly electrons and protons) emitted from the sun. See stellar wind.

solar year *n.* the time required for the Earth to make a complete revolution around the sun (e.g., from an equinox one year to the corresponding equinox the next year), equal to 365.2422 days or 365 days, 5 hr, 48 min, and 45.51 sec. Also called astronomical year or tropical year. See sidereal year.

Solenopsis *n.* a genus of stinging ants (*Solenopsis* sp.), including the fire ants that sting animals and people. *Solenopsis geminata* is common in southern U.S.A., and *S. saevissimi richteri* is an aggressive South American species that has migrated to the U.S.A.

solid solution *n.* **1.** a homogeneous mixture of components in certain solids; e.g., colored glass or certain alloys. **2.** isomorphic substitution; the replacement of part or all of certain ions or molecules in a substance with others that are different but fit the same space.

solid waste *n.* unwanted material that will not flow as a fluid but must be carried in a vehicle, including debris, garbage, industrial wastes, demolition materials, mining residues, agricultural wastes, semisolid materials such as sewage sludge, and even containers filled with liquid or gaseous materials.

solifluction *n.* (Latin *solum*, soil + *fluctio*, flowing) slow downslope movement of soil materials or detritus, often associated with saturated soil overlying a frozen layer. Also called soil creep or creep erosion.

solonetz *n.* an intrazonal great soil group in the 1938 U.S. system of soil classification; a sodic soil, normally shallow but having a strongly developed argillic horizon resulting from the eluviation of dispersed clay.

soloth *n.* an intrazonal great soil group in the 1938 U.S. system of soil classification with a clayey subsoil and strong profile development; a former solonetz soil that has been reclaimed by natural leaching over a long period. See solonetz, sodic soil reclamation.

solstice *n.* (Latin *solstitium*, the sun standing) either of two times a year when the position of the noonday sun is at its greatest distance from the equatorial plane. The summer solstice in the Northern Hemisphere (winter solstice in the Southern Hemisphere) occurs on or about June 21. The winter solstice in the Northern Hemisphere (summer solstice in the Southern Hemisphere) occurs on or about December 21. See equinox.

solum *n.* (Latin *solum*, ground) the true soil, including whatever A, B, and E horizons are present.

solute *n.* (Latin *solutus*, dissolved) **1.** material dissolved in a solution. *adj.* **2.** dissolved; held in solution. **3.** soluble. **4.** free or separate, as a solute tendril on a plant.

soma *n.* (Greek *sōma*, body) **1.** the body of an animal or plant, excluding its germ cells. **2.** the head, trunk, and tail of an animal, excluding the limbs.

somatropin or **somatotropin** *n.* (Greek *sōma*, body + *trophē*, nutrition + *in*, chemical) a growth hormone produced by the pituitary gland or synthesized. Human somatropin has been used to treat pituitary dwarfism. Bovine somatropin (BST) has increased milk production up to 15% when fed to dairy cows, and no adverse effects have been found in people who drank the milk or ate the meat of such animals.

sombric horizon *n.* a diagnostic subsurface horizon in Soil Taxonomy. It is darkened by illuvial humus, has a base saturation of less than 50%, and occurs in well-drained soils at high elevations in humid tropical or subtropical regions.

songbird *n.* (Anglo-Saxon *sang*, song + *bridd*, bird) **1.** a bird that makes musical sounds. **2.** (slang) a woman that sings well.

sonic *adj.* (Latin *sonus*, sound + *icus*, like) **1.** dealing with or pertaining to sound. **2.** at the speed of sound (about 1115 ft/sec, 760 mph, 1220 km/hr, or 339 m/sec at sea level through air at room temperature or about 1087 ft/sec, 741 mph, 1193 km/hr, or 331 m/sec at 0°C in dry air).

sonication *n.* (Latin *sonus*, sound + *icus*, like + *atio*, action) the use of high-frequency sound waves to disrupt bacterial or viral cells or to disperse other substances.

sonic boom *n.* the shock wave and accompanying loud noise created by an aircraft flying faster than Mach I (the speed of sound). See sonic.

sonifaction *n.* (Latin *sonus*, sound + *factio*, making) making sound; e.g., stridulation, when certain insects make noise by rubbing their legs together.

soot *n.* (Anglo-Saxon *sot*, soot) dust-sized carbon particles formed by the incomplete combustion of organic materials. Soot accumulates on surfaces that contact hot combustion gases.

sorb *n.* (Latin *sorbum*, serviceberry) **1.** a small European tree (*Sorbus domestica*, family Rosaceae). **2.** the small, sour fruit of this tree. Also called sorb apple. *v.* **3.** to attract and hold, either next to a surface (adsorb) or internally (absorb).

sorghum *n.* (Italian *sorgo*, sorghum) any of several grasses (*Sorghum* sp., family Poaceae), native to southeastern Africa, that is used to produce grain, fodder, sugar, syrup, and broomstraws. It is related to corn but more drought resistant and produces large, loose heads at the top of the plant and ends of branches rather than ears like corn. Under certain conditions, cattle, sheep, and goats may be poisoned by dhurrin ($C_{14}H_{17}NO_7$), a glucoside by-product of hydrocyanic acid (HCN), found in young sorghum leaves.

sorgo *n.* (Italian *sorgo*, sorghum) a sorghum plant or crop (*Sorghum* sp.) with sweet juicy sap. It is grown for sugar, syrup, or fodder. Also called sweet sorghum.

sorosilicate *n.* (Greek *sōros*, heap + Latin *silex*, flint + *ate*, salt) a silicate mineral with pairs of linked tetrahedra (Si_2O_7) as its basic structural unit; the least abundant class of silicate minerals; e.g., hemimorphite [$Zn_4Si_2O_7(OH)_2 \cdot H_2O$].

sorption *n.* a process that attracts and holds either a liquid or a gas, either internally (absorption) or next to a surface (adsorption); e.g., the sorption of water by a sponge. See absorption, adsorption.

sorrel *n.* (French *sorel*, a little brown) **1.** a light reddish brown. **2.** a horse with a sorrel color, often with a pale mane and tail. **3.** any of several plants (*Rumex* sp., family Polygonaceae; or *Oxalis* sp., family Oxalidaceae) with edible sour leaves that are used in salads and sauces. The tree sorrel (*R. lunaria*) grows to the height of a small tree. *adj.* **4.** having a sorrel color.

sorrel tree *n.* a North American tree (*Oxydendrum arboreaum*, family Ericaceae) with sour-tasting leaves and drooping white blossoms. Also called sourwood.

sound *n.* (Latin *sonus*, sound) **1.** a vibration that stimulates the sense of hearing by fluctuations in air pressure on the eardrums. Most people can hear such vibrations at frequencies between 8 and 20,000 vibrations per second with an intensity between 0 and 120 decibels (dB). See decibel. **2.** a distinctive form of such vibrations, as the sound of music or the sound of an engine. **3.** a voice or noise that can be heard. **4.** an effect that information has on a person, as she didn't like the sound of that. **5.** a narrow water passage, as that between an island and the nearby mainland. **6.** an inlet, as Puget Sound. *v.* **7.** to produce or emit sound; e.g., to sound a bugle or to sound an alarm. **8.** to pronounce, as to sound all the letters in a word. **9.** to give an impression by what one says, as it sounds like you are serious about that. **10.** to examine by tapping sounds, as to sound a patient's chest. **11.** to measure the depth of water. **12.** to search for more information, as to sound out the speaker. **13.** to plunge underwater, as the whale may sound at any moment. *adj.* **14.** firm, solid, as a sound footing. **15.** free of defect or disease, as sound health or a sound mind. **16.** financially secure, as a sound investment. **17.** reliable, as sound advice. **18.** good, as sound moral values or sound sleep.

sound barrier *n.* a hypothetical barrier to flight because sound waves no longer influence the air ahead of an aircraft after it reaches the speed of sound and atmospheric friction increases greatly at that speed. It is more feasible to fly at either a faster or a slower speed than at or near the speed of sound. Also called sonic barrier or transonic barrier. See sonic boom.

sounding sand *n.* See singing sand.

sound pollution *n.* offensive noise; environmental sound that is annoying and/or harmful. Also called noise pollution.

sour *adj.* (Anglo-Saxon *sur*, sour) **1.** one of the four basic tastes (sour, sweet, salt, and bitter). **2.** having an acid taste; tart. **3.** made acid by fermentation. **4.** distasteful; unpleasant, as a sour deal or a sour attitude. **5.** off-pitch, as a sour musical note. *v.* **6.** to become sour; to ferment. **7.** to become unpleasant or unfavorable, as their relations may sour with time. **8.** to become acid, as nonpasteurized milk can sour or intense leaching causes soil to sour.

source *n.* (French *source*, source) **1.** place of origin; e.g., a pollutant may have a point source or a nonpoint source. **2.** a person, place, or thing that produces or supplies something, as milk is a source of calcium. **3.** an authority, as a source of information. **4.** a spring or fountain of water.

source-water protection area *n.* the area around a well that provides recharge water that is likely to reach the well within a specified period (e.g., 2 years or 10 years).

sour clover *n.* a winter annual plant (*Melilotus indica*, family Fabaceae) with yellow blossoms that is a close relative of the sweet clovers and used as a improving crop in the southern tier of U.S. states.

southern blight *n.* a fungal disease (*Sclero rolfsii*) that attacks the base of the stem many plants (lespedeza, lupines, peanuts, peppers, soybeans, tomatoes, etc.), often covering the lower stem with white fruiting bodies, killing many plants in southeastern and gulf-coast states of the U.S.A., especially in hot weather.

southern house mosquito *n.* a bloodsucking insect (*Culex quinquefasciatus*, family Culicidae) that transmits filariasis (elephantiasis) and encephalitis (inflammation of the brain).

South Pole *n.* **1.** the southernmost point on Earth, located at 90° south latitude; the southern end of the Earth's axis of rotation. **2.** a point in the heavens representing the southern extension of the Earth's axis of rotation. The longest limb of the Southern Cross points toward the South Pole in the heavens from a distance of about 30°. See North Pole.

sow *n.* (Danish *sow*, sow) **1.** an adult female pig, especially one that is bearing or has borne a litter. **2.** an adult female of various other species, such as a bear. **3.** a large mass of solidified pig iron, or the channel through which the molten iron flows.

sow *v.* (Anglo-Saxon *sawan*, seed) **1.** to plant seeds where they can grow. **2.** to scatter seed widely. **3.** to spread an idea or feeling, as to sow discord or unrest. **4.** to scatter anything across a surface.

SO$_x$ *n.* a combination symbol for whatever oxides of sulfur may be present (mostly SO$_2$ and SO$_3$).

soy *n.* (Japanese *soy*, soy) **1.** soy sauce. **2.** soybean plants or seeds.

soya *n.* soybean.

soybean *n.* (Japanese *soy*, soy + Anglo-Saxon *bean*, bean) **1.** an annual legume (*Glycine max*, family Fabaceae); a bushy plant with trifoliate leaves and purple to pink blossoms that has about the same climatic adaptation and growing season as corn. It is widely grown for its seed because it can produce more protein per acre than any other plant or animal source and has a rich content of essential amino acids, including lysine, leucine, and isoleucine. **2.** the seed of these plants, used to make high-protein meat substitutes, soy flour, animal feed, soybean oil and meal, soybean ink, etc.

soybean ink *n.* any ink made from soybeans; generally more biodegradable than petroleum-based ink and now competitive in quality and price. Also called soy ink.

soybean oil *n.* oil pressed or extracted by solvent from soybeans. It is used as a cooking oil a, solvent, and an ingredient in paint, soap, candles, ink, etc. Also called soy oil.

soybean meal *n.* the material left after oil is extracted from soybeans. It is used to make soy flour, meat extenders or substitutes, and protein concentrates for animal feed. Also called soybean oil meal.

soy sauce *n.* a salty, dark, fermented sauce prepared from soybeans that is used to season Asian cuisine.

spadix *n.* (Latin *spadix*, a torn-off palm bough) a fleshy spike that bears tiny flowers, usually covered by a spathe.

Spanish moss *n.* an epiphyte (*Tillandsia* sp., family Bromeliaceae) that is common in tropical areas, including southern U.S.A. It hangs in long strips from tree branches in shaded areas beneath a canopy, has narrow grayish stems with tiny leaves and flowers, and is a good indicator of air pollution. Also called Florida moss.

sparse *adj.* (Latin *sparsus*, scattered) **1.** widely distributed with open space between; scattered. **2.** spaced apart; not thick, as sparse hair. **3.** meager, as sparse rations. **4.** positioned irregularly; neither opposite nor alternate, as sparse branches, leaves, or peduncles.

spathe *n.* (Latin *spatha*, flat blade) either one or two large bracts that enclose a flower cluster, especially a spadix.

spatulate *adj.* (Latin *spatulatus*, like a blade) **1.** shaped like a spatula or small spoon. **2.** narrow at the base but widening into a broad, rounded end, as a spatulate leaf.

spawn *v.* (French *espandre*, to expand) **1.** to reproduce (especially fish, amphibians, crustaceans, or mollusks) by depositing eggs, sperm, or young. **2.** to bring forth something, especially a great abundance of something that has little value. **3.** to plant mycelia of mushrooms. *n.* **4.** a mass of eggs or young produced by fish, amphibians, crustaceans, or mollusks. **5.** an overabundance of offspring or of some product that has little value. **6.** fungal mycelia, especially of mushrooms being cultured.

spearmint *n.* **1.** a perennial aromatic herb (*Mentha spicata*, family Lamiaceae) with square stems, lance-shaped leaves, and purple blossoms. **2.** an oil obtained from the plant and used as a flavoring for foods and chewing gum. **3.** green or dried leaves from the plant used in tea, soup, sauces, and for meat flavoring; also used to flavor puddings and gelatin desserts.

specialist *n.* (Latin *specialis*, of a given species + *ista*, person) **1.** a person who studies or works on a particular subject rather than a broad area. **2.** a medical doctor that tends to a particular type of disease or problem, in contrast to a general practitioner. **3.** a plant or animal species that requires environmental conditions within a narrow range of tolerance.

specialty crop *n.* a crop grown by a relatively small number of producers, often concentrated in one area or a few specific areas.

specialty fertilizer *n.* a fertilizer intended for a specific use, e.g., for fertilizing roses or houseplants.

speciation *n.* (Latin *species*, form or kind + *atio*, making of) the development of a new species of plant or animal.

species *n.* (Latin *species*, form or kind) **1.** a distinct kind or type of anything, as a species of humor or a species of logic. **2.** a group of plants, animals, or microbes with a high degree of similarity that can interbreed among themselves. A species is a subdivision of a genus, designated by the second word in a binomial Latin name (customarily italicized); e.g., *Quercus alba* (white oak), with *alba* being the species name and *Quercus* the genus. Some commonly used species names and their English equivalents are

- *alba*: white
- *annua*: yearly, annual
- *arida*: dry
- *bicolor*: two-colored
- *borealis*: northern
- *circulare*: round
- *edulis*: edible
- *indurata*: hardened, hard
- *occidentale*: western
- *orientalis*: eastern
- *perennans*: perennial
- *vulgaris*: common

species evenness *n.* the degree of variability in the spatial distribution of the various species present in a community or ecosystem.

species richness *n.* the number and variety of species of plants and animals present in a given environment.

specific adsorption *n.* holding of ions in positions on the surface of solids (e.g., soil particles) where they fit precisely and cannot be replaced by ions of another element.

specific gravity *n.* the ratio of the weight (mass) of a given volume of a substance to that of an equal volume of water.

specific heat *n.* the amount of heat required to raise the temperature of 1 g of a substance 1°C. Specific heat is sometimes described as a ratio comparing the amount of heat required to raise the temperature of a substance 1°C to that required to raise the temperature of the same amount of water 1°C. Water has one of the highest specific heats known, equal to 1.0 cal/g-degree or 4.18 joules/g-degree.

spectral signature *n.* the assemblage of wavelengths and intensities of electromagnetic radiation emitted by a body. It is useful for identifying celestial bodies, brain activity, diseased organs, etc.

spectrometry *n.* (Latin *spectrum*, appearance or image + *metrum*, measure) the use of a spectrometer to study the various wavelengths of light or other electromagnetic energy and measure their intensities coming from a given source.

spectrum *n.* (Latin *spectrum*, appearance or image) **1.** the series of visible colors sequenced by wavelength, e.g., by a prism or in a rainbow, including violet, blue, green, yellow, orange, and red. **2.** the visible light sequence plus extensions into the ultraviolet and infrared wavelengths. **3.** radio waves

ranging in wavelength from 3 cm to 30,000 m. **4.** a range of ideas or philosophies; e.g., the political spectrum. **5.** any array of entities arranged in sequence according to the magnitude of some property shared by all.

speed of light *n.* light travels at a speed of 299,792,458 m/sec or 186,300 mi/sec.

spent nuclear fuel *n.* the rods containing uranium or plutonium that have been used to fuel a reactor for about three years and can no longer sustain the efficient continuous chain reaction required in the reactor. It is a form of high-level nuclear waste that constitutes a serious disposal problem.

sperm *n.* (Latin *sperma*, seed) one (or many) mature male gamete (reproductive cell) produced in the testicles and carried in the semen. Also called spermatozoon.

spermatheca *n.* (Latin *sperma*, seed + *theca*, case) the receptacle in a female insect (e.g., a queen bee) that receives and stores the sperm of the male insect.

spermatophyte *n.* (Greek *spermatos*, seed + *phyton*, plant) a plant that bears seeds; a vascular plant; e.g., a flowering plant or conifer.

spermophile *n.* (Greek *sperma*, seed + *philos*, loving) any of several burrowing rodents (*Spermophilus* sp., family Sciuridae), including chipmunks, groud squirrels, and woodchucks. Some of them do serious damage to crops, and stepping in one of their burrows can cause a horse or other large animal to break its leg.

sphagnum *n.* (Greek *sphagnos*, a moss) a very low density moss (*Sphagnum* sp., family Sphagnaceae) that grows in freshwater lakes with its roots in the water. It is used in potting mixes for growing plants, as packing to keep plants moist during storage or shipping, and as a dressing for wounds. Also called sphagnum moss or peat moss.

spice *n.* (French *espice*, spice) **1.** any of a large number of substances of plant origin with a strong flavor or odor suitable for seasoning food; e.g., cinnamon, nutmeg, and pepper. **2.** all such substances as a group. **3.** a flavor or fragrance produced by such a substance. **4.** something that flavors a conversation, as a spice of humor. **5.** a small amount of something, as a mystery spiced with a few clues.

spider *n.* (Danish *spinder*, spinner) any of several predaceous arachnids (order Arenea) that have eight legs and spin webs. Most are helpful to humans because they catch many flies and mosquitoes, but some are poisonous; e.g., the brown recluse spider (*Loxosceles reclusa*) and the black widow spider (*Latrodectus mactans*).

spider mite *n.* any of several small mites (family Tetranychidae) that spin webs. Several are serious pests in gardens and orchards. Also called red spider.

spider monkey *n.* any of several monkeys (*Ateles* sp.) native to tropical America that swing easily through trees with their long, slender bodies and limbs and prehensile tail.

spike *n.* (German *spīker*, nail) **1.** a large, sturdy nail; a nail more than 3 in. (7.5 cm) long. **2.** any similar long, slender metal fastener; e.g., a railroad spike or a tooth in a spike-tooth harrow. **3.** one of a group of bluntly pointed projections on the bottom of a shoe. **4.** an abrupt increase followed by an equally sharp decrease, as an electrical spike or a spike in a curve. **5.** an unbranched antler on the head of a young deer. **6.** a flower stalk with several or many blossoms along its length. **7.** an ear of grain with several or many seeds attached along a stalk. Also called a head.

spile *n.* (German *spile*, splinter or peg) **1.** a wooden plug, especially one used as a spigot. **2.** a spout used to drain sap from a sugar maple. **3.** a tube or wooden box passing through the bank of an irrigation ditch with a gate on the outer end to control water flow. Sometimes called a lath box. **4.** a stake or timber driven into the ground to serve as a piling for an abutment or wharf. **5.** a timber used to case a mine shaft.

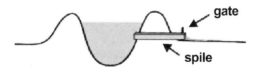

A spile through a ditch bank controls irrigation water.

spirochete *n.* (Latin *spirochaeta*, coiled hair) any of several motile bacteria (family Spirochaetaceae) with spiral shapes. Some are free-living, some are saprophytic or parasitic, and some are pathogenic, causing syphilis, relapsing fever, and other diseases in people and animals.

spirochetosis *n.* (Latin *spirochaeta*, coiled hair + *osis*, condition) a septicemic disease caused by a spirochete; e.g., rheumatism in human joints (spirochetosis arthritica) caused by a spirochete (*Spirochaeta forans*) and a disease of fowls (*Borrelia anserina*) spread by the fowl tick (*Argas persicus*). Also called spirillosis.

spirogyra *n.* (Greek *speira*, coil + *gyros*, ring) any of several widely distributed filamentous algae (*Spirogyra* sp.) that form bright green, slimy masses in freshwater. Their chlorophyll forms bands that spiral to the right.

spit *n.* (Anglo-Saxon *spitu*, a spit) **1.** a narrow strip of land extending into open water. **2.** a long shoal extending outward from the shore. **3.** a long, pointed metal bar used to impale meat for broiling or roasting it over coals or open fire. **4.** any of a variety of bars, rods, or pins used for several purposes. **5.** (slang) saliva. **6.** the act of ejecting saliva or other material from the mouth. **7.** light rainfall or snowfall, as a spit of snow. *v.* **8.** to impale on a metal bar for cooking, as to spit a roast. **9.** to expectorate; to eject saliva or other material from the mouth. **10.** to rain or snow lightly. **11.** to make a popping or hissing noise, as when a fire spits or a cat spits. **12.** (figurative) to express hatred or contempt, as he spits on them.

spitting rock *n.* a rock that breaks off (exfoliates) small pieces with a cracking sound when weather changes cause it to heat up, cool down, or freeze.

splash erosion *n.* detachment and jumping movement of soil particles caused by the impact of raindrops. It is the primary process of detaching soil particles that can then be carried away in sheet erosion or picked up later by wind erosion. Also called raindrop erosion.

spodic horizon *n.* a diagnostic subsurface horizon in Soil Taxonomy used to identify Spodosols. Characterized by illuvial organic matter and iron and aluminum oxides, it is evidenced by bright orange and brown, especially near the middle of the horizon (the upper part may be black), often with wavy or irregular depth boundaries.

Spodosol *n.* (Greek *spodos*, wood ash + French *sol*, soil) a soil order in Soil Taxonomy for mineral soils that have highly leached surface horizons and spodic subsurface horizons. Spodosols form under forest vegetation in humid environments with acid leaching, mostly in sandy materials, and represent about 5% of U.S. and 4% of world soils.

A spodic soil profile with a spodic horizon from 1 ft (30 cm) down to about 3 ft (90 cm)

spoil *n.* (French *espoille*, spoil) **1.** waste, residue, and debris left from mining, excavating, quarrying, etc., often in the form of spoil banks or spoil piles that are difficult to vegetate and may be a source of pollutants such as acid drainage water as well as sediment. **2.** an object of plunder, as the spoils of war. **3.** prizes, as the spoils from a party. **4.** a product rejected as defective. *v.* **5.** to injure or ruin and make worthless; e.g., bacteria cause food to spoil. **6.** to diminish or destroy the quality of an event, as a quarrel spoils a meeting or rain spoils a picnic. **7.** to harm someone's character, as to spoil a child by lack of discipline. **8.** to defeat an enemy and plunder enemy belongings.

spoilage *n.* (French *espoille*, spoil + *age*, result of) **1.** material that has spoiled, especially by decaying and rotting. **2.** the amount or percentage of such material.

spontaneous combustion *n.* self-ignition of a fire without external heat; e.g., white phosphorus in air or damp hay in a barn.

spoonbill *n.* (Anglo-Saxon *spon*, spoon + *bile*, beak) **1.** any of several large wading birds (family Plataleidae) with a spoonlike tip on a long, flat bill, especially the roseate spoonbill (*Ajaia ajaja*) with its mostly pink plumage and legs. **2.** any of several other birds with a similar bill; e.g., the shoveler duck (*Anas* sp.). **3.** the paddlefish.

spoor *n.* (Anglo-Saxon *spor*, spoor) **1.** the track or trail of a person or animal, especially one being hunted. *v.* **2.** to follow the footprints and other traces of passage left by an animal or person.

sporangium (pl. **sporangia**) *n.* (Greek *spora*, seed + *angeion*, vessel) a spore case where spores are formed, e.g., by a fungus or a fern.

spore *n.* (Greek *spora*, seed) **1.** a small reproductive body of certain bacteria, fungi, ferns, protozoans, etc., consisting of either one or several cells with a protective covering that enables them to rest during unfavorable conditions (e.g., cold or dry) and produce a new individual when conditions become favorable. **2.** any small cell, seed, or organism capable of producing a new individual.

sporicide *n.* (Latin *spora*, seed + *cida*, killer) a substance or combination of substances used to kill spores.

sporidium (pl. **sporidia**) *n.* (Latin *spora*, seed + *ium*, structure) a small spore or a case filled with small spores. See promycelium.

sporocarp *n.* (Greek *spora*, seed + *karpos*, fruit) a fertilized multicellular fruiting body where spores develop in red algae, lichens, and certain fungi.

sporophyte *n.* (Greek *spora*, seed + *phyton*, plant) a diploid plant in its diploid stage. Most seed-producing plants have a long period as sporophytes that alternates with a brief haploid stage as gametophytes. More primitive plants such as ferns

have alternate generations of these two stages. Still more primitive mosses live mostly as gametophytes and only briefly as dependent sporophytes.

sporozoan *n.* (Latin *sporozoa*, spore animal) any of a large number (class Sporozoa) of parasitic protozoans that alternate sexual and asexual generations, often in different hosts, including those that cause malaria and Texas fever. Formerly called coccidia.

sport (biological) *n.* an organism that exhibits a notable difference from preceding generations; a mutant.

spotted owl *n.* a rare and endangered owl (*Strix occidentalis*, family Tytonidae), a western counterpart of the barred owl. It nests in tree cavities in old-growth coniferous forests along the West Coast of the U.S.A.

sprawl *v.* (Anglo-Saxon *spreawlian*, to move convulsively) **1.** to spread one's limbs widely, as a cat sprawls on the floor or a tree sprawls across a lawn. **2.** to crawl awkwardly. **3.** to spread out farther than necessary. *n.* **4.** the condition of spreading widely, as urban sprawl takes cropland out of production.

sprig *n.* (Dutch *sprik*, dry twig) **1.** a small shoot, twig, or small branch from a plant with attached leaves and (often) blossoms. **2.** an ornament having a form similar to that taken from a plant. **3.** a small triangular piece of metal used to hold a pane of glass in its sash. **4.** a small headless nail; a brad with no head. **5.** a young man. **6.** an offspring or young representative of a family or institution. *v.* **7.** to cut a sprig or sprigs from a tree, bush, or other plant. **8.** to form sprigs into an ornament. **9.** to use small nails or metal pieces to hold something in place.

spring *n.* (Anglo-Saxon *springan*, spring) **1.** a source of water flowing out of the ground. **2.** a source of anything, as a spring of inspiration. **3.** the season of the year between winter and summer; the period when temperatures are warming. **4.** the time between the vernal equinox and the summer solstice (from March 21 or 22 to June 21 or 22 in the Northern Hemisphere or from September 22 or 23 to December 21 or 22 in the Southern Hemisphere). **5.** a quick jump or bounce, as a spring in his step. **6.** elasticity, as the pole has a lot of spring. **7.** an elastic object; e.g., a wire coiled so it will compress, stretch, or twist under load and exert a force that tends to restore its normal shape. *v.* **8.** to leap, bounce, or come forth quickly. **9.** to react suddenly, as to spring to attention or a trap springs shut. **10.** to come into being quickly, as a crop springs up in the field or new developments spring up in the suburbs. **11.** to have a specified origin, as to spring from noble ancestry. **12.** to bend under load. **13.** to break loose, as a flap springs free in a storm. **14.** to extend upward, as a spire or arch springs toward the heavens. **15.** to open a hole or crack, as to spring a leak. **16.** to do something unexpected, as to spring a trick on a friend. *adj.* **17.** occurring in or pertaining to the spring season, as spring rains or spring flowers. **18.** using one or more springs, as having a spring-mounted assembly or operating by spring pressure.

springbok *n.* (Dutch *springen*, to spring + *boc*, male goat) a South-African gazelle (*Antidorcas marsupialis*) that is known for springing into the air when startled. It is about 5 ft (1.5 m) long and half that tall, is brown on top and white underneath, and has unbranched, curved antlers

spring crop *n.* **1.** a crop that is planted in the spring (especially one that might be planted in the fall), as spring barley, spring rye, or spring wheat. **2.** a crop that is harvested in the spring, as a spring cutting of alfalfa.

spring tide *n.* the time of highest and lowest tide at any particular site. Spring tides occur twice during each lunar month, coinciding with or immediately following a full moon or a new moon, because the sun, moon, and Earth are aligned at those times. See neap tide.

springwood or **spring wood** *n.* the relatively soft and usually lighter-colored layer of wood made in the spring (especially early spring) when rapid cell growth in woody plants produces large cells with thin walls. See summerwood.

sprinkler irrigation *n.* any watering system that sprays pressurized water through nozzles over a designated area. The lines and nozzles may be fixed in place, portable, or self-propelled (e.g., a center-pivot system). See surface irrigation, drip irrigation, subirrigation.

sprout *v.* (Anglo-Saxon *sprutan*, to sprout) **1.** to initiate new growth; to produce seedlings, buds, or shoots. **2.** to grow rapidly, especially while young. *n.* **3.** a new seedling, bud, or shoot. **4.** new growth from a root, bulb, tuber, or other underground part. **5.** a part that is young and growing rapidly. **6.** a young person.

spruce *n.* (French *Pruce*, Prussia) any of several evergreen trees (*Picea* sp., family Pinaceae) that commonly have their longest branches near ground level with branches becoming shorter with height up to a pointed top. They have short, sharp needles (leaves) attached individually all around twigs, hanging cones, and berrylike fruit.

spruce aphid *n.* a green insect (*Aphis abictina*, family Aphidae) that is very destructive to Sitka spruce (*Picea sitchensis*) in Washington and Oregon.

Engelmann spruce (*P. engelmanni*, left) and blue spruce (*P. pungens*, right) twigs and cones

spruce budworm *n.* a very destructive moth (*Choristoneura fumiferana*, family Tortricidae) whose larvae feed on terminal shoots of trees in northern U.S.A. and Canada.

spruce sawfly *n.* a sawfly (*Neodiprion abietis*, family Diprionidae) whose larvae defoliate spruce and pine trees. Sawflies did serious damage in the 1930s, especially in southern Canada, before they were controlled by importing spruce sawfly parasites from Europe.

spruce spider mite *n.* a black mite (*Oligonychus ununguis*, family Tetranychidae) that sucks sap from spruce trees and spins webs between the needles. Young spruce trees may die within a year, and older trees may die later.

spur *n.* (Anglo-Saxon *spura*, spur) **1.** a device worn on the heel of a riding boot to prick the side of a horse to make it move faster. **2.** any action that goads a person or animal to act faster. **3.** a spinelike growth on the leg or wing of a bird, especially one on the leg of a fighting cock, or a similar metal device attached to the leg. **4.** a sharp point attached to the inside of a boot for use in climbing wooden poles or trees. **5.** a short branch or shoot projecting from the trunk of a tree. **6.** a ridge that branches from the main part of a mountain chain. **6.** a tubelike structure extending from the petals or sepals of certain blossoms such as columbine or larkspur. **7.** a dike extending from the bank of a stream to protect against bank erosion. Also called a spur dike or jetty. **8.** a branch from an ore vein in a mine. **9.** a sidetrack on a railroad used for switching or to allow another train to pass. *v.* **10.** to goad; to urge an animal or person to move or to go faster. **11.** to equip with spurs, as to spur a pair of boots. **12.** to attack or injure with a spur.

spurge *n.* (French *espurger*, to cleanse) **1.** any of several plants (*Euphorbia* sp., family Euphorbiaceae) that have milky sap and whose blossoms often have showy bracts but lack petals and sepals. **2.** any of a large number of plants of the spurge family (Euphorbiaceae), including cassava, poinsettia, spurge, etc. Some spurges are sources of castor oil, tung oil, rubber, and natural insecticides.

squab *n.* (Swedish *sqvabb*, loose flesh) **1.** a young pigeon that cannot yet fly. **2.** a person with a short, stout build. **3.** a soft, thick cushion or couch.

squall *n.* (Norse *skvala*, shriek) **1.** a sudden, violent windstorm, usually brief and often carrying rain, snow, or sleet. **2.** trouble; a brief period of violence or of loud debate. **3.** loud crying, as a baby squalls. *v.* **4.** to produce a squall. **5.** to cry and shriek loudly.

squall line *n.* a long line (often hundreds of miles or kilometers) of severe showers and thunderstorms, usually with strong winds ahead of an advancing cold front.

squash *n.* (Algonquian *asquash*, squash) **1.** a viny crop plant (*Cucurbita maxima*, *C. mixta*, *C. moschata*, or *C. pepo*, family Cucurbitaceae) whose large leaves suppress weeds both by their shading effect and by an allelopathic effect of chemicals washed from their leaves into the soil by rainwater. **2.** the edible gourds produced by such plant. **3.** the flesh of such gourds eaten as a vegetable. **4.** a soft, pulpy mass. **5.** either of two games for two or four people, played with rackets and a ball that is bounced off the walls of a room (squash rackets) or played on a court similar to tennis (squash tennis). *v.* **6.** to crush into a soft mass, as to squash a bug. **7.** to destroy, as to squash someone's ambition. **8.** to silence, as to squash a speaker. **9.** to force one's way into a tight place or through a crowd.

squash bug *n.* a dark brown wingless nymph (*Anasa tristis*, family Coreidae) that sucks the sap from leaves of plants of the gourd family such as squash and pumpkin. Small plants may die.

squash vine borer *n.* the larva of a clear-winged moth (*Melittia cucurbita*, family Aegeriidae). The larvae bore into the stems of squash, pumpkin, cucumbers, and other gourd plants, entering near ground level and eating tunnels through the stems into the leaves.

squawroot *n.* a leafless North American plant (*Conopholis americana*, family Orobanchaceae) that grows in parasitic clusters on the roots of oaks and other trees.

squid *n.* any of several cephalopods with 10 arms (*Loligo* sp., *Ommastrephes* sp., and *Architeuthis* sp.). Squids dwell in the open sea, some near the surface and others at great depth. Small squids (e.g., *L. pealeii*) are about 8 to 12 in. (20 to 30 cm) long, whereas giant squids (*Architeuthis* sp.) have bodies up to 20 ft (6 m) long, have tentacles up to 35 ft (20 m) long, and weigh up to nearly 1000 lb (450 kg).

squill *n.* (Latin *squilla*, squill) **1.** the sea onion (*Urginia martima*, family Liliaceae), native to southern Europe and northern Africa, that is related to onions and hyacinths. **2.** the bulb of such plants, especially when sliced, dried, and used medically as an expectorant. **3.** any of several related plants with blue, white, or purple blossoms.

squirrel *n.* (Latin *sciurus*, shadow tailed) **1.** any of several tree-dwelling animals (*Sciurus* sp., family Sciuridae) with bushy tails. **2.** any of several other members of the Sciuridae family, including chipmunks, flying squirrels, ground squirrels, and woodchucks. **3.** the pelt, fur, or meat of such animals. *v.* **4.** to store or hide food, money, or other resources for later use (e.g., in winter).

squirrel corn *n.* an herb (*Dicentra canadensis*, family Fumariaceae) native to eastern U.S.A. that has finely dissected leaves, racemes with clusters of cream-colored, heart-shaped flowers, and yellow grainlike buds that look like kernels of corn on its roots. The foliage is toxic to cattle.

stability *n.* (Latin *stabilitas*, stable quality) **1.** having a stable condition. **2.** reliability; firmness; ability to resist change. **3.** resistance to chemical reactions. **4.** consistency; sound character, as emotional stability. **5.** the ability to resist disruption. See aggregate stability. **6.** the ability to return to a former steady state after a disturbance; e.g., the ability of an ecosystem to restore itself after a fire or the ability of an aircraft to resume stable flight after a disruption.

stabilization *n.* (Latin *stabilis*, stable + *izatio*, process or result) **1.** the process of making something stable, as stabilization of an embankment. **2.** the condition of being stabilized. **3.** processing that makes waste material acceptable for release into the environment; e.g., stabilization of sewage by the action of aerobic bacteria.

stabilization pond *n.* a body of water designed to receive organic wastes and supply oxygen for aerobic bacteria to transform the undesirable substances into environmentally safe products. See lagoon.

stabilized grade *n.* a construction slope or channel gradient where neither erosion nor sedimentation occurs.

stable *adj.* (Latin *stabilis*, stable) **1.** firm; steady; able to resist change, destruction or deterioration. **2.** likely to last for a long time. **3.** mentally sound; reliable. **4.** not undergoing radioactive decay, as a stable isotope. **5.** remaining in one place; a stable air mass (e.g., one trapped by a temperature inversion) holds, rather than dissipates, air pollutants. *n.* **6.** a building where horses or other livestock are housed and fed; especially one with separate stalls for the animals. **7.** the animals housed in such a building. **8.** a group of people who work together. **9.** a collection of the products of an industry, as a stable of new automobiles. *v.* **10.** to house in a stable.

stable fly *n.* a common two-winged fly (*Stomoxys calcitrans*) with biting mouthparts that annoys both animals and people, especially around stables and other areas where manure accumulates. It can transmit diseases such as anthrax, tetanus, and infectious anemia. Also called biting housefly or leg sticker.

stag *n.* (Anglo-Saxon *stagga*, male bird) **1.** a full-grown male of the deer family. **2.** any of several other adult male animals. **3.** a castrated boar or bull. **4.** a man who attends a social gathering without a woman companion or a gathering for men only.

stage *n.* (Latin *status*, stood) **1.** a step or phase in a process of development; the form or condition of an organism or thing at a given time. **2.** an elevated platform for a speaker, ceremony, theater, or other presentation. **3.** the setting for an action or activity. **4.** a stagecoach or a stopping place for a stagecoach. **5.** a group of sedimentary rocks that represent a given geologic age. **6.** the platform on a microscope that supports a slide or other specimen to be viewed. **7.** a section of a rocket containing an engine or cluster of engines that work at the same time. *v.* **8.** to present or exhibit, as to stage a theatrical event or to stage a demonstration. **9.** to plan, prepare, and conduct, as to stage a major attack in a war.

staggerbush *n.* a deciduous shrub (*Lyonia mariana*, family Ericaceae), native to eastern U.S.A., with white or pink flowers and poisonous leaves that can cause sheep to convulse and froth at the mouth.

stagnant *adj.* (Latin *stagnans*, standing still) **1.** not flowing or moving, as stagnant water. **2.** fouled by pollutants accumulated while not moving; e.g., air pollution can be a problem when a temperature inversion causes air to stagnate. **3.** not growing or changing, as a stagnant economy or a stagnant mind.

stagnum *n.* (Latin *stagnum*, pool or swamp) a small body of water with no outlet. See kettle.

stalactite *n.* **1.** (Latin *stalactites*, dripping) an evaporite deposit (mostly calcium carbonate) hanging like an icicle from the roof of a cavern. Note: As a memory aid, a stalactite must hang on tight, whereas a stalagmite might grow up to meet a stalactite and thus form a column. **2.** any similar hanging formation; e.g., those formed in lava tunnels by dripping of hot, fluid lava.

stalagmite *n.* (Latin *stalagmites*, result of dripping) an evaporite deposit that grows upward from the floor of a cavern, usually from calcium carbonate dissolved in water dripping from a stalactite.

stalk *n.* (Danish *stilk*, a stalk) **1.** the stem of a plant; an upright part that supports the leaves, flowers, and fruit, as a cornstalk. **2.** any elongated supporting part

Stalactites, stalactites, and columns in a cavern

on a plant or insect, as a petiole or a peduncle. **3.** any similar part on an object, as the stalk of a pipe or the stalk that holds controls for lights and windshield wipers in an automobile. **4.** an ornamental design that resembles the stalk of a plant. **5.** a slow, stiff, haughty marching step. **6.** the act of hunting or observing an animal or enemy while avoiding detection. *v.* **7.** to walk or march in a slow, stiff, haughty manner. **8.** to follow or observe and try not to be seen or heard.

stalk borer *n.* the larva of certain moths (e.g., *Gortyna nitela* and *Papaipema nebris*) that burrow into the stalks of many garden plants (e.g., tomatoes, strawberries, raspberries, and asters). See corn borer.

stalk rot *n.* a widespread form of several plant diseases that weaken stalks and cause much lodging in corn, small grains, and many other crops. Similar diseases are called stem rot.

stall *n.* (Anglo-Saxon *steall*, place or station) **1.** a compartment for an individual animal (e.g., a horse) in a stable, barn, or other building. **2.** a booth or stand used to conduct business, often at a temporary location. **3.** an individual seat or position, especially one inside a small enclosure, in a church, theater, place of business, or factory. **4.** an enclosed space used for a particular purpose, as a shower stall. **5.** a loss of lift causing an airplane to fall, especially after a slow climb. **6.** an opening between pillars in a mine. *v.* **7.** to place an animal in a stall. **8.** to delay, often while waiting for someone to arrive or something to happen. **9.** to come to a halt, as from an overload. **10.** to lose control and start falling from inadequate flying speed.

stamen *n.* (Latin *stamen*, warp threads) the male reproductive part of a flower. It is composed of a filament that supports a pollen-bearing anther. See pistil.

standing crop *n.* **1.** a crop standing in a field, especially one that is ready for harvest. **2.** the total plant biomass present in an ecosystem at a given time.

stannum *n.* (Latin *stannum*, an alloy of silver and lead) tin.

staphylococcus *n.* (Greek *staphylē*, bunch of grapes + *kokkos*, berry) any of about 20 species of gram-positive, spherical bacteria (*Staphylococcus* sp., family Micrococcaceae) that tend to clump together like bunches of grapes and are commonly found in milk and other dairy products or free-living in water. A few species are pathogenic and are the most common cause of illnesses such as pneumonia and septicemia (often called staph infection). See food poisoning.

starboard *n.* or *adj.* (Anglo-Saxon *steorbord*, steering board) the right-hand side of a ship or aircraft when facing forward (where the steering board used to be mounted), as opposed to the port or left side.

starch *n.* (German *stärken*, to stiffen) **1.** the principal carbohydrate stored in seeds and tubers of plants; a polysaccharide composed of glucose units ($C_6H_{10}O_5$) joined into long chains called amylose (with α-1,4 linkages) and branched chains called amylopectin (with α-1,6 linkages at the branch points, typically after about 25 α-1,4 linkages). The long amylose chains roll into a helix, forming a tube that can enclose other molecules (e.g., iodine that thus forms a blue complex with amylose). **2.** a commercial product containing this material for use in laundering to stiffen fabrics and make them easier to clean in the next washing. **3.** stiffness, as he is so full of starch that he doesn't know how to relax. *v.* **4.** to add starch to laundry water to stiffen clothing or other articles. **5.** to make one's actions or a setting very formal.

starfish *n.* (Anglo-Saxon *steorra*, star + *fisc*, fish) any echinoderm (class Asteroidea) with five or more arms forming a star-shaped body with a hard, spiny outer covering and a mouth in the center of the underside.

starling *n.* (Anglo-Saxon *staerling*, starling) a short-tailed bird (*Sturnus vulgaris* and related species, family Sturnidae) about 6 in. (15 cm) long with iridescent brownish black plumage and a yellow bill in the spring and summer but with its winter plumage heavily speckled with white. Starlings came from Europe, but their range now includes most of the area east of the Rocky Mountains. They are often considered pests, partly because they often assemble in large communal roosts and leave messy droppings in city parks, suburbs, and farms. Their food is varied, but they seem to prefer suet, dead animals, and sites with scattered food scraps.

starter *n.* (Anglo-Saxon *styrtan*, to start + *er*, person or thing) **1.** a person, animal, or thing that initiates an action, either as a participant or as an activator, e.g., by giving a signal or closing a switch. **2.** a specific bacterial culture used to inoculate milk or cream to make such products as cheese and yogurt. **3.** the first food supplied to newborn animals or humans. **4.** a sample of something (e.g., medicine) to be used until a regular supply can be obtained.

starter fertilizer *n.* a relatively small amount of fertilizer placed near or with the seed (often placed by the planter in a band about 2 in., or 5 cm, to one side and 2 in. below the seed) when a crop is planted. It is intended to accelerate seedling growth.

star thistle *n.* (Anglo-Saxon *steorra*, star + *thistel*, thistle) an annual weed (*Centaurea calcitrapa* and *C. solstitialis*, family Aetheopappus) with beautiful yellow flowers that is poisonous to horses only. Introduced from Greece about 1850 and now growing in 23 states of the U.S.A., it is particularly troublesome on 8 million ac (3 million ha) of annual grasslands in California. Control efforts include introduction of a 1/4-in. (6 mm)-long weevil (*Larinus curtus*) and intensive grazing by sheep and goats.

static *adj.* (Greek *statikos*, standing) **1.** stationary; not moving. **2.** not changing, as a static condition. **3.** staying traditional in social activities. **4.** pertaining to electricity in the atmosphere, especially as it interferes with radio signals. **5.** nonmoving weight, as a static load or static pressure. **6.** not growing or shrinking, as a static population. *n.* **7.** atmospheric electricity. **8.** crackling sounds or other interference with radio or television signals caused by atmospheric electricity. **9.** trouble, resistance, or disagreement, as they gave me a lot of static when I wanted to leave.

stationary front *n.* a nonmoving boundary between a cold air mass and a warm air mass. Winds blow parallel to the front in one direction on one side and in the opposite direction on the other side.

steady state *n.* a constant condition in a system resulting from any processes that would cause change being offset by other compensating processes; e.g., inputs equal outputs and heat gain equals heat loss. See dynamic equilibrium.

steckling or **stechling** *n.* **1.** a root of a biennial crop such as a carrot, turnip, or beet that may be dug up, stored over winter, and replanted for seed production the following year. **2.** a transplanted seedling or a cutting.

steep *adj.* (Anglo-Saxon *steap*, steep or lofty) **1.** having a large slope gradient; deviating greatly from horizontal, as a steep hill or steep stairway. **2.** very high, as steep prices. **3.** lofty; at a high elevation. **4.** difficult to believe, as a steep claim. **5.** a liquid used to soak material to be worked (e.g., reeds to be woven) or food to be cooked. *v.* **6.** to soak in hot water. **7.** to wet thoroughly with any liquid. **8.** to imbue or indoctrinate, as to steep in the tradition of the group, or a story steeped in mystery.

steer *n.* (Anglo-Saxon *steor*, a steer) **1.** a male bovine castrated before maturity so it will make better beef. See ox. **2.** a suggestion, as they gave her a good steer. *v.* **3.** to control or guide, as to steer an automobile. **4.** to oversee and direct, as to steer an organization. **5.** to follow a course, as to steer a straight line.

stellar atmosphere *n.* the gas surrounding a star. The gas is typically composed of about 90% hydrogen and 9% helium.

stellar wind *n.* ionized gases (mostly electrons and protons) flowing away from a star at high velocity like a wind that travels through space. This phenomenon associated with the sun is called the solar wind.

Steller's jay *n.* a crested jay (*Cyanocitta stelleri*) about 11 in. (28 cm) long, with mostly blue plumage with brown and black markings and a prominent dark crest. It is common in the coniferous forests of western U.S.A.

stem *n.* (Anglo-Saxon *stemn*, tree trunk) **1.** the main aboveground stalk of a plant; the central part that supports branches, leaves, blossoms, and fruit; e.g., a tree trunk or the corresponding part of any rooted plant. **2.** a stalk that supports any part of a plant, including petioles, peduncles, etc. **3.** a stemlike part of any object, as the stem of a goblet or a winding stem in a watch. **4.** the forward part of a ship, especially the piece that forms the prow where the two sides meet. **5.** the initial part of a word; the part that carries the basic meaning and can have endings such as –s, –ed, –ing, or –ly attached to alter its usage.

stem cutting *n.* any part of a stem used to propagate a new plant asexually; e.g., sections of sugarcane stalk with three or four eyes are used to establish a new crop of sugarcane. Softwood or green stem cuttings are succulent pieces cut while the plant is in leaf. Hardwood stem cuttings are taken from a dormant plant.

stem-end rot *n.* a disease that causes various kinds of fruit to discolor, shrivel, and decay, especially near the stem. It is caused by any of several fungi (*Diplodia* sp. and *Phomopsis* sp.).

stemflow *n.* (Anglo-Saxon *stemn*, tree trunk + *flowan*, to flow) rainwater or other precipitation that flows gently down the trunks of trees and the leaves and stems of other plants and seeps into the soil.

stem rot *n.* **1.** any of several fungal diseases that attack the stems of many plants; e.g., fusarium wilt. Similar diseases in crop plants are often called stalk rot. **2.** the decay caused by such diseases.

stem rust *n.* **1.** any of several fungal diseases that produce red pustules followed by black spores on the stems of wheat and other grass plants. **2.** the condition produced by such diseases.

stenohaline *adj.* (Greek *stenos*, narrow or close + *hals*, salt + *inos*, like) capable of living only in water within a narrow range of salinity.

stenopetalous *adj.* (Greek *stenos*, narrow + *petalos*, spread out) having narrow petals.

stenophagous or **stenophagic** *adj.* (Greek *stenos*, narrow or close + *phagos*, eating) feeding only on one kind or a very limited variety of food. See euryphagous.

stenotherm *n.* (Greek *stenos*, narrow or close + *thermē*, heat) an organism that can live only in a narrow range of temperatures. See eurytherm, niche.

stenovalent *adj.* (Greek *stenos*, narrow or close + *valentia*, worth) able to live only in a specific niche; requiring habitat with a narrow range of characteristics.

steppe *n.* (Russian *step*, great plain) an extensive plain covered with mostly short bunchgrasses and small shrubs; e.g., much of the semiarid plains in Russia and the similar area east of the Rocky Mountains in North America.

sterilization *n.* (Latin *sterilis*, barren + *izatio*, making) **1.** the process of making or being made unable to reproduce; e.g., the removal or disruption of part of the reproductive system (e.g., the use of chemicals or radiation to sterilize insects or other pests so they cannot reproduce). **2.** the destruction or killing of all living organisms, especially of microbes in a material or on a surface. See pasteurization, disinfectant.

sterile *adj.* (Latin *sterilis*, barren) **1.** having no living organisms, especially no microbial life present. **2.** infertile; unable to reproduce. **3.** unsuitable to serve as a growth medium. **4.** not producing any results or ideas.

sterilize *v.* (Latin *sterilis*, barren + *izare*, to cause) **1.** to render incapable of reproduction; e.g., the surgical removal of reproductive parts or treatment with chemicals or radiation. **2.** to remove or kill all living organisms in a medium or on a surface; e.g., by treatment with heat, chemicals, or radiation. **3.** to treat soil with a chemical, e.g., sodium chlorate ($NaClO_3$), so nothing will grow. **4.** to delete secret information from a document or report.

sternutator *n.* (Latin *sternutatorius*, a cause of sneezing) a substance in the environment (e.g., pollen or a chemical) that causes sneezing and/or coughing.

Stewart's disease *n.* a bacterial wilt (*Bacterium stewartii*) that infects the vascular system of corn and is most damaging to early yellow sweet corn and inbred lines but also affects dent corn and popcorn throughout eastern U.S.A. It is transmitted through wounds on the foliage made by flea beetles eating the leaves. Also called bacterial wilt or Stewart's wilt.

stickiness *n.* (Anglo-Saxon *stecan*, to stick + *nes*, state) the tendency to adhere to other objects. The stickiness of wet soil is related to clay content and is evaluated as a feature of soil consistency. Soils may be nonsticky, slightly sticky, sticky, or very sticky.

sticktight *n.* any of several weedy plants (*Bidens* sp. and others, family Asteraceae) whose seeds have barbs that cause them to adhere to clothing or fur. Also called beggarticks.

sticktight flea *n.* an insect pest (*Echidnophaga gallinacea*, family Pulicidae) that attaches itself in clusters to nonfeathered parts of the head of a chicken or other fowl and is also troublesome to dogs and cats. Found mostly in southeastern U.S.A., it weakens the host by sucking blood and is difficult to remove.

stiff-lamb disease *n.* an ailment associated with deficiencies of vitamin E and selenium affecting young lambs 1 to 10 weeks old. Symptoms include weakness, stiffness, staggering, and difficulty nursing. Also called white-muscle disease.

stilbestrol *n.* See diethylstilbestrol.

stillage *n.* (Latin *stillare*, to drop + *age*, pertaining to) **1.** the residue from grain used to make alcohol and as a livestock feed supplement. **2.** a low platform used to keep goods off the floor and to make them easier to pick up with a forklift in a warehouse or factory.

stilling basin *n.* **1.** a concrete-lined basin holding a pool of water at the base of a dam, spillway, or drop structure that dissipates the energy and reduces the erosiveness of falling water. **2.** a widened area in a channel protected by a concrete lining. It is used to dissipate energy and slow the flow downstream from the outlet of a pipeline or tunnel. See stilling pool.

stilling pool *n.* a pool of water that dissipates flow energy in a basin excavated in soil or rock. See stilling basin.

stilt *n.* (German *stilte*, pole) **1.** either of two poles or other supports used to hold the feet some distance above the ground while walking. **2.** any of a group of supports driven into the ground to hold a structure above the land or water surface. **3.** a three-point support used to hold a ceramic item in an oven while it is being fired. **4.** any of several black-and-white wading birds (especially *Himantopus nexucabys* and *Cladorhynchus leucocephalus*) with long, slender pink legs and long black bills. **5.** (British) a plow handle. **6.** (slang) a crutch.

stingray or **sting ray** *n.* any of the large rays of tropical ocean waters (family Dasyatidae, especially *Dasyatis centroura*), so named because their tails have bony spines projecting on both sides and serve as whips that can inflict serious wounds.

stink *v.* (Anglo-Saxon *stincan*, to emit an odor) **1.** to emit a foul, offensive odor. **2.** to be offensive, as that program stinks. **3.** to be inferior, as the workmanship in that project stinks. *n.* **4.** a foul odor, usually one that is quite strong. **5.** an expression of strong disagreement, as he raised a stink about what they did.

stinkbug *n.* **1.** any of several hemipterous insects (family Pentatomidae) with broad shield-shaped bodies. Stinkbugs are pests that suck sap from plants and emit a foul odor when disturbed; e.g., the green stinkbug (*Acrosternum hilare*), southern green stinkbug (*Nezara viridula*), and tropical stinkbugs such as *Antiteuchus tripterus*. See conchuela. **2.** any of several other insects that emit offensive odors.

A stinkbug

stinkdamp *n.* (German *stinken*, to emit an odor + *dampf*, vapor) hydrogen sulfide (H_2S) that accumulates in a mine from decomposition of pyrite (FeS_2). It is a colorless, toxic, flammable gas that smells like rotten eggs.

stinkgrass *n.* (Anglo-Saxon *stincan*, to emit an odor + *graes*, grass) a weedy annual grass (*Eragrostis cilianensis*, family Poaceae) that emits a disagreeable odor and can be toxic to horses if consumed in large amounts over a period of time. It is found in fields and waste areas throughout much of the U.S.A., Mexico, and Argentina.

stinkquartz *n.* (German *stinken*, to emit an odor + *quarz*, quartz) a variety of quartz that gives off an offensive odor when struck sharply.

stinkstone *n.* (German *stinkstein*, stinkstone) a bituminous limestone rich in organic phosphate that gives off an offensive odor when struck or rubbed.

stipe *n.* (Latin *stipes*, tree trunk) **1.** a stem or other support for a plant part; e.g., a petiole of a frond on a fern or the stalk of a mushroom. **2.** a stipes; a stalklike part in the body of an insect or crustacean.

stirofos *n.* tetrachlorvinphos ($C_{10}H_9Cl_4O_4P$), a chemical used as a feed additive for cattle, horses, or swine to control fecal flies. It is also used to control insects in stored feed products.

Stokes' law *n.* a mathematical expression for the settling rate of spherical particles of coarse clay, silt, and fine sand in water. Named after George G. Stokes (1819–1903), it is used as the official basis for both the pipette and hydrometer methods of mechanical analysis. In its simplest form,

$$v = kD^2$$

where v = settling velocity (usually in cm/min or cm/hr), k = a constant that can be calculated from water viscosity and the density difference between the particle and water, and D = the particle diameter in mm (actually the diameter of a sphere that would settle at the same rate); e.g., k = 6000 mm^{2}cm/min for particles with a density of 2.65 g/cm^3 settling in water at 25°C to give v in cm/min.

stolon *n.* (Latin *stolonis*, branch or shoot) an aboveground horizontal stem capable of forming a new plant vegetatively, such as a runner on a strawberry plant. Also called a flagellum. See rhizome, soboles.

stoloniferous *adj.* (Latin *stolonis*, branch or shoot + *ferous*, bearing) having or producing stolons, as a strawberry is a stoloniferous plant.

stoma (pl. **stomata**) *n.* (Greek *stoma*, mouth) **1.** any of the minute openings in the surface of a leaf or other plant part where atmospheric gases are exchanged, especially the intake of carbon dioxide and the release of oxygen during photosynthesis. **2.** any small pore, orifice, or opening on a free surface of animals or people; e.g., a skin pore or the mouth of a tiny organism. **3.** an artificial opening into a body cavity or between body cavities; e.g., a stoma used to sample the contents of a stomach.

stomach *n.* (Greek *stomachos*, opening) **1.** the large digestive cavity in vertebrates (or collective group of as many as four digestive cavities in ruminants), between the esophagus and the intestines, where most of the early stages of digestion occur. **2.** a corresponding digestive cavity in invertebrates. **3.** appetite or desire for food, as a hungry stomach. **4.** character, inclination, or ability to accept, as "I don't have the stomach to do that." **5.** (loosely) the abdomen, as that animal has a large stomach. *v.* **6.** to be able to eat something. **7.** to bear or accept or to be able to do so.

stomach ball *n.* an accumulation of hair, dirt, or other indigestible material in the stomach of an animal that can only be removed by surgery. See bezoar, hair ball, egagropilus.

stomach worm *n.* a parasitic nematode (*Haemonchus contortus* and *H. placei*, family Trichostrongylidae) that is the largest of the stomach worms, resembling a coarse hair from 1 to 1.5 in. (2.5 to 3.8 cm) long. It lives in the stomachs of sheep, goats, and cattle, especially in warm, humid climates such as southeastern U.S.A., and can affect humans. It causes anemia, weakness, paleness, swollen jaws, and dry, hard feces. Also called twisted stomach worm or wireworm.

stomate (pl. **stomata**) *n.* (Greek *stoma*, mouth) a stoma, especially a small opening in the epidermis of a leaf bounded by guard cells that control its size and thereby control the exchange of gases from within the leaf and the outside atmosphere.

stone *n.* (Anglo-Saxon *stān*, stone) **1.** a rock fragment larger than a pebble and smaller than a boulder. **2.** defined for soil survey as a rounded rock fragment between 10 and 24 in. (25 to 60 cm) in diameter or a flat rock fragment between 15 and 24 in. (38 to 60 cm) long. **3.** the mineral matter that constitutes rocks and rock fragments. **4.** a piece of mineral matter cut from bedrock for a specific purpose, as a building stone or a paving stone. **5.** a gemstone. **6.** any of several units of weight, especially the British unit equal to 14 lb (6.4 kg). **7.** a hard seed such as the pit in a peach or prune. **8.** a concretion such as a kidney stone or a gallstone. **9.** a tombstone used to mark a grave. **10.** a cut piece of either natural rock material or a synthetic substitute used as a whetstone to sharpen knives, a grindstone to shape metal, or a millstone to grind flour or similar material. **11.** a hailstone. **12.** a marker used to play certain games such as dominoes or checkers. *adj.* **13.** made of stone, as stoneware or a stone wall. *v.* **14.** to throw stones at a person or animal; either to drive away or to kill by throwing stones. **15.** to pave, line, or otherwise cover with stones. **16.** to smooth, polish, or sharpen by rubbing on a stone. **17.** to remove the pits from fruit.

Stone Age *n.* the stage in human development when human tools and weapons consisted mostly of pounding stones and cutting stones. The Stone Age is subdivided into the Paleolithic, Mesolithic, and Neolithic Periods now estimated to have begun about 2.6 million years ago and ended with the beginning of the Bronze Age between 6000 B.C. and 2500 B.C., depending on locality.

stone crab *n.* an edible crab (*Menippe mercenaria*) found along rocky shores of southern U.S.A. and Caribbean nations.

stonefish *n.* (Anglo-Saxon *stān*, stone + *fisc*, fish) a tropical scorpion fish (*Synanceja verrucosa*) with spiny dorsal fins that discharge a deadly toxin.

stone fly *n.* (Anglo-Saxon *stān*, stone + *fleoge*, fly) any of several flying insects (order Plecoptera) with soft, flattened bodies. Their nymphs live under stones in and along streams and are a major source of food for bass, trout, and various other fish. The nymphs are prized for use as fish bait and are popular models for flies made for fly-fishing.

stone fruit *n.* any fruit with a hard endocarp at its center, as an apricot, cherry, peach, or plum. Also called a drupe.

stone land *n.* an area of land that has some type of stone (e.g., granite or marble) that has economic value for quarrying.

stone line *n.* a soil layer with a concentration of gravel, cobblestones, or stones that shows in banks or cuts as a line. It is generally a remnant of a former soil surface where stony material was concentrated by erosion and later covered by alluvium, colluvium, glacial till, or loess.

stone wall *n.* a wall surrounding a field built of stones cleared from the field. Such walls are notable landscape features in some glaciated areas.

Stone walls surround many fields in northeastern U.S.A.

stonewort *n.* (Anglo-Saxon *stān*, stone + *wyrt*, plant) any of several jointed aquatic green algae (*Chara* sp., order Charophyceae) with whorled branches that grow rooted to the bottom of shallow ponds with such high calcium contents that the plants are usually encrusted with calcium carbonate ($CaCO_3$).

stoniness *adj.* (Anglo-Saxon *stān*, stone + *nes*, state or quality) the proportion of soil surface or soil material that is composed of stones. Soil Taxonomy defines six classes of stoniness:

- Class 0: Stones do not interfere with tillage and generally cover less than 0.01% of the surface.
- Class 1: Stones cause some interference but not enough to prevent growing intertilled crops and generally cover 0.01% to 0.1% of the surface.
- Class 2: Stones make intertilled crops impractical (but the area can be used for hay or pasture) and generally cover 0.1% to 3% of the surface.
- Class 3: Stones make the use of machinery impractical except for hand tools (but the area can be used for pasture) and generally cover 3% to 15% of the surface.
- Class 4: Stones make all use of machinery impractical (but the area may have some use for grazing or forestry) and generally cover 15% to

90% of the surface. The soil is often classified as stony land and is not placed in a soil series.
- Class 5: The land is essentially paved with stones that occupy more than 90% of the surface.

stony land *n.* land that has too many loose stones on the surface to be used as cropland. Commonly more than 15% of the soil surface is covered by stones (class 4 or 5 stoniness).

stool *n.* (Anglo-Saxon *stol*, seat) **1.** a flat seat on either a pedestal or legs with no back support or arms, sometimes described by use or form as a milk stool or a three-legged stool, etc. **2.** a low support used for standing, kneeling, or as a footstool to rest one's feet while seated nearby. **3.** a toilet, privy, bowel movement, or fecal material. **4.** a cluster of stems produced from a single seed, as in barley or wheat with a low seeding rate. **5.** the freshly cut stump of a tree that produces new shoots, or the group of shoots produced by such a stump. **6.** the base of a plant that produces new stems or shoots each year. **7.** a decoy used by hunters to attract ducks or other birds. **8.** a seat of authority, as the bishop's seat in certain churches or the ruler's seat in certain African nations. *v.* **9.** to send out new stems or shoots from the crown or stump of a plant. **10.** to have a bowel movement. **11.** to be an informer, as a stool pigeon.

stork *n.* (Anglo-Saxon *storc*, stork) any of several large wading birds (*Mycteria americana* in the U.S.A. and *Ciconia* sp. in Europe, family Ciconiidae) with mostly white bodies and long legs and bill. Storks nest in colonies in trees and feed on fish, reptiles, and amphibians. See wood stork.

storm *n.* (Anglo-Saxon *storm*, storm) **1.** any strong atmospheric disturbance, usually including strong and/or gusty winds that are commonly accompanied by rain, snow, hail, or sleet and sometimes by thunder and lightning. **2.** a very strong atmospheric disturbance with winds blowing 64 to 72 mph (103 to 116 km/hr). See Beaufort wind scale. **3.** a large social or political commotion. **4.** a strong expression of sentiment, as a storm of applause or a storm of criticism. **5.** a concentration of missiles, as a storm of bullets. **6.** a sudden military attack on a fortified position. *v.* **7.** to blow, rain, snow, and/or sleet heavily. **8.** to express rage or strong dissatisfaction, usually loudly and actively. **9.** to attack with sudden military action. *adj.* **10.** showing the effects of a storm, as storm damage. **11.** offering protection from the weather, as a storm window or a storm shelter.

storm cellar *n.* **1.** an underground structure built for protection against violent storms (e.g., tornadoes) or warfare. **2.** an underground structure used to store garden produce, either fresh or canned, for the winter.

storm sewer *n.* a sewer that carries runoff water, usually emptying directly into a stream without treatment.

storm surge *n.* the rise in water level caused by a strong wind (e.g., a hurricane) blowing toward an oceanic coastline, sometimes augmented by a low-pressure center near the coast. Also called wind tide.

stover *n.* (French *estover*, provisions) mature stalks of corn, sorghum, or other cereal crop after the grain has been harvested. Stover is often used as roughage for livestock feed. See straw.

strain *n.* (Anglo-Saxon *streon*, lineage) **1.** any group of organisms that are similar to each other because they have a common lineage; sometimes called a race. **2.** a relatively uniform artificial group of plants or animals produced by human selection. **3.** a unique characteristic, especially one that is inherited, as an athletic strain or a strain of insanity. **4.** a strong force or pressure either exerted or experienced. **5.** deformation of a body or structure resulting from an applied force, often expressed in displacement distance per unit of force. **6.** muscular damage caused by excess load or effort. **7.** physical or mental stress, as that caused by uncertainty, too much work, or expectations that are too high. **8.** the sound of music or other melody, as the strains of an orchestra. **9.** the tone or style of written or oral expression, as a somber strain. *v.* **10.** to exert a strong effort. **11.** to sensitize as much as possible, as to strain one's hearing. **12.** to apply as much stress as possible to something, as to strain a rope. **13.** to injure by excess effort, as to strain one's muscles. **14.** to extend to or beyond the normal usage, as to strain the meaning of a word. **15.** to filter solids from a liquid. **16.** to seep, as water strains from a crack.

strangler fig *n.* See banyan tree.

strangles *n.* (Latin *strangulare*, to choke) equine distemper caused by bacteria (especially *Streptococcus equi*). Symptoms include fever, weakness, mucus, and swelling with a strangling effect in horses and some other animals. It can be fatal, especially to colts. Also called distemper or colt distemper.

stratification *n.* (Latin *stratificatio*, covering or spreading) **1.** the formation of layers. **2.** the condition of having layers, as stratification in rocks. **3.** social hierarchy, as a society with stratification according to rank or caste.

stratified *adj.* (Latin *stratificare*, stratified) **1.** having distinct layers, as a stratified rock or a stratified atmosphere. **2.** arranged in grades or ranks, as a stratified social order.

stratigraphic sequence *n.* the sequence of rock layers as deposited through time, usually listed from the bottom (oldest) to the top (youngest), either at a specific site or as a composite of many sites.

stratigraphy *n.* (Latin *stratum*, a cover + *graphia*, writing) **1.** the sequence of rock strata occurring at a particular site or in a particular area. **2.** the geologic science that studies rock strata, including the form, distribution, composition, and properties of layers of rocks and the environmental condition when each was formed.

stratopause *n.* (Latin *stratus*, cover + *pausa*, stopping) the atmospheric boundary between the stratosphere, where the air temperature increases with altitude up to about 28°F (–2°C) at the stratopause, and the mesosphere, where the air temperature decreases with altitude. It is located at an altitude of about 31 mi (50 km), depending on latitude and season.

stratosphere *n.* (Latin *stratus*, cover + *sphaera*, sphere) that part of the atmosphere above the troposphere and below the mesosphere, from about 6 mi (10 km) high at the North and South Poles and 12 mi (20 km) high at the equator up to about 31 mi (50 km) high. About 97% of atmospheric ozone occurs in the stratosphere, where high clouds form. Temperature is nearly constant through the lower third of the stratosphere and then increases with altitude up to the stratopause. See atmosphere, tropopause, troposphere, stratopause.

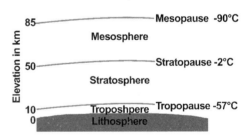

The statosphere is above the troposphere and below the mesosphere.

stratum (pl. **strata**) *n.* (Latin *stratum*, a cover) **1.** a layer, especially where it is one of many parallel layers (strata), e.g., those common in sedimentary rock. **2.** any division that may be perceived as a layer or level, as low, medium, and high strata of vegetation in a tropical forest or a statement with several strata of meanings. **3.** a layer of water in the ocean or air in the atmosphere; e.g., the stratosphere. **4.** a social level, as the lowest stratum in the nation.

stratus cloud *n.* **1.** a sheetlike or layered cloud that covers a large area, often the entire sky visible from one place; typically gray, at low elevation, and dry or accompanied by light drizzle. **2.** a combining form for cirrostratus (high stratus), altostratus (middle stratus), stratocumulus (lumpy and layered), or nimbostratus (wet-looking and stormy) clouds.

straw *n.* (Anglo-Saxon *streaw*, straw) **1.** a hollow stem from a cereal (especially barley, oats, rye, and wheat) that has been threshed (especially the part cut loose from the stubble) that is often used for bedding, fodder, etc. See stover. **2.** a drinking straw; a tube similar to a hollow cereal stem but made of paper, plastic, or glass for sucking up a beverage. **3.** something of very little value. **4.** a pale yellow.

strawberry *n.* (Anglo-Saxon *streaw*, straw + *berie*, berry) **1.** a low-growing perennial plant (*Fragaria* sp., family Rosaceae) with trifoliate leaves, white blossoms, and long slender runners that produce new plants several inches from the parent plant. See stolon. **2.** the red conical-shaped fruit (actually a fleshy receptacle bearing achenes) of such plants that is eaten as a fruit dessert or as a topping on other desserts or is made into salad, jam, or preserves.

straw itch mite *n.* a small mite (*Pyemotes ventricosus*, family Pyemotidae) that normally lives in straw and eats the larvae of insects but that can bite people and cause a hivelike eruption over a person's body. See hive, mite.

streak *n.* (Anglo-Saxon *strica*, a line or stroke) **1.** a long, thin mark. **2.** a layer, vein, or stratum, as a streak of quartz in a rock or a streak of fat in meat. **3.** a series of events during an interval of time, as a streak of good luck or a losing streak. **4.** a tendency, as a person with a nervous streak. **5.** the color produced by rubbing a mineral on a rough white ceramic surface as a help in identification. **6.** an elongated lesion on a stem or leaf of a plant, or a disease that produces such lesions. *v.* **7.** to mark with streaks, as beams of light streak a surface. **8.** to mark with streaks of a different shade or color, as to streak one's hair. **9.** to spread on a glass slide for examination under a microscope. **10.** to run or work rapidly. **11.** to flash across, as lightning streaks across the sky.

stream *n.* (Anglo-Saxon *stream*, stream) **1.** a flow of water in a channel, as a river, brook, creek, or rivulet. **2.** a current of water within an ocean or other large water body, as the Gulf Stream. **3.** a flowing mass of any liquid, as a stream of lava. **4.** a flowing mass of air or other gas, as a stream of steam. **5.** a beam of light. **6.** any succession of things, as a stream of words, a stream of events, or a waste stream. *v.* **7.** to flow as a coherent body or as a series of bodies following the same general path, as tears stream

down her cheeks. **8.** to flow as light streams through an opening in the clouds. **10.** to wave, as a flag streams in the breeze. **11.** to hang loosely, as long hair streams down her back.

streambank *n.* (Anglo-Saxon *stream*, stream + *banke*, table) either of the two borders of a stream, usually with a steep slope facing the stream and often rising into a levee that slopes down to the land level away from the stream.

streambank erosion *n.* erosion caused by a stream impacting against its bank, especially the downstream part of the outer bank of a curve in the channel. The USDA–Natural Resources Conservation Service estimates that there are about 300,000 mi (500,000 km) of streambanks in the U.S.A. in need of erosion control. Also called bank erosion.

streambed *n.* (Anglo-Saxon *stream*, stream + *bed*, base) the land surface beneath a stream, mostly sand, gravel, and stones that are moved periodically by the stream plus some areas where bedrock is exposed and being eroded.

stream gauge *n.* a current meter, weir, or other measuring device used to determine the volume and velocity of water flow in a stream.

streamlet *n.* (Anglo-Saxon *stream*, stream + *let*, small) a small stream. Also called a rivulet or runnel.

stream load *n.* the total weight of sediment and dissolved material being carried by a stream (usually a segment of a stream) at a given time.

streptococcus *n.* (Greek *streptos*, twisted + *kokkos*, berry) any of a large number of gram-positive, facultative anaerobic bacteria (*Streptococcus* sp., family Streptococceae) with a spherical shape and commonly occurring in pairs or chains, including three groups: (1) human and animal pathogens that may cause septic sore throat, scarlet fever, erysipelas, etc.; (2) forms that occur as normal flora in the upper respiratory and intestinal tract; and (3) forms associated with the souring of milk and used in making buttermilk and cheese.

strid *n.* a ravine, a narrow stream course between steep rock walls.

stridulation *n.* (Latin *stridulus*, creaking + *atio*, action) making a shrill sound by rubbing the legs or other body parts together, as the chirp of a cricket.

striga *n.* (Latin *striga*, stiff bristle) **1.** a stiff bristle on a plant. **2.** witchweed.

strip cropping *n.* **1.** contour strip cropping; growing strips of erosion-resistant plants on the contour to intercept runoff and erosion that may occur on the intervening cropped strips. The erosion-resistant strips and the cropped strips may be rotated from one year to the next. **2.** buffer strip cropping; contour strip cropping with permanent vegetation in erosion-control strips that may be variable in width to compensate for variable slopes. **3.** field strip cropping; growing erosion-resistant plants in strips parallel to a field boundary that crosses the general slope of the land with cropped strips between them. **4.** wind strip cropping; growing tall plants in strips across the path of the prevailing erosive winds to protect intervening cropped strips from wind erosion.

Contour strip cropping

strip mine *n.* an open mine where the overlying layers are removed so the layer containing ore or minable minerals can be excavated. Many old strip mines were left as pits and piles of rubble, but current regulations require miners to restore the area after mining is completed so it will be at least as productive as it was before mining.

strip tillage *n.* a system of reduced tillage that uses a single machine to loosen narrow strips of soil, plant a row crop, usually place fertilizer near the row, and compact the soil over the seeds. Erosion control is enhanced by leaving crop residues in place on the area between the tilled strips.

strobila *n.* (Greek *strobilos*, a twisted thing) **1.** a stage in the development of certain larvae when their bodies twist into a form like that of a pinecone as they subdivide; e.g., the larvae of certain jellyfish (class Scyphozoa). **2.** the segmented body of a mature tapeworm.

strobilation *n.* (Greek *strobilos*, a twisted thing + Latin *atio*, action) an asexual form of reproduction whereby segments of an adult body separate to form new individuals, as in tapeworms and certain jellyfish.

stroma *n.* (Greek *strōma*, mattress) **1.** the framework or matrix of an organ or cell, as the supporting structure of a leaf or of a red blood cell or the matrix of a chloroplast. **2.** a mass of fungal tissue that supports the fruiting bodies.

strongyle *n.* (Greek *strongylos*, round) any of a large group of nematodes (family Strongyloidea), including hookworms and related intestinal parasites of people, horses, cattle, sheep, pigs, goats, rabbits, rats, etc.

strontium (Sr) *n.* element 38, atomic weight 85.62; a metallic element that behaves much like calcium except that it weighs more than twice as much as calcium; named after Strontian parish in Scotland. A heavy form released by atomic bombs, ^{90}Sr, has a radioactive half-life of 28 years and is hazardous to life because it can replace calcium in bones. Strontium is used to produce red in fireworks, flares, and tracer bullets.

structural stability *n.* See aggregate stability.

structure *n.* (Latin *structura*, structure) **1.** a building or other entity composed of parts that are linked together. **2.** the arrangement of parts within an entity. **3.** a complex system considered as a whole. **4.** the way a business, program, written or spoken work, etc., is organized. **5.** the position, slope, and other physical features of the geologic strata in an area. **6.** the arrangement of the crystals of the various minerals present in a rock. **7.** the arrangement of soil particles in peds to form various types (granular, angular blocky, subangular blocky, platy, prismatic, or columnar) and grades (weak, moderate, or strong) of soil structure. **8.** the arrangement of atoms in a molecule, especially in an organic molecule; e.g., the chair or boat structure of a sugar molecule or the double-helix structure of DNA. **9.** the organization and interaction of social elements in a community or ecosystem. **10.** the organization of linguistic units in a language. *v.* **11.** to organize or arrange the components of any physical or conceptual entity.

strychnine *n.* (Greek *strychnos*, nightshade) **1.** an extremely poisonous alkaloid ($C_{21}H_{22}N_2O_2$) obtained from any of several related plants (*Strychnos* sp.) that excites all parts of the central nervous system. Its principal use has been in rat poison to kill rodents. Small doses have been used as a tonic and stimulant for persons and animals to increase the acuity of hearing, seeing, feeling, tasting, and smelling. **2.** the tree (*Strychnos nux-vomica*, family Loganiaceae) whose seeds are the principal source of the poisonous alkaloid. Native to India, it has small yellowish white blossoms and clusters of berrylike fruit.

stubble *n.* (French *stuble*, a small appendage at the base of a leaf) **1.** the bottom parts of plant stems left standing in a field after harvest, especially the remains of small grains or other plants with high population and small stems (as distinguished from stalks of corn and stumps of trees). **2.** the short stubs of plants left by grazing animals. **3.** any short growth remaining after having been trimmed; e.g., a short growth of whiskers.

stubble mulch *n.* the stubble from a previous crop left entirely or mostly in place to protect the soil until time to plant the next crop, which is sometimes planted through the standing stubble.

stud *n.* (Anglo-Saxon *studu*, prop or pillar) **1.** a small ornamental object fastened on a surface. **2.** a button-like object used as an ornament and/or fastener on clothing. **3.** an upright piece in the framework of a wall, as a 2 × 4 stud. **4.** a pin, lug, or other protruding part used to align parts in a machine. **5.** a stud bolt that is threaded at both ends so it can screw into a hole and protrude as a stud. **6.** any of a number of metal pieces embedded in the rubber of a tire to improve traction on ice. **7.** a type of earring used in pierced ears. **8.** a male animal kept for breeding, especially a stallion (also called a stud horse and said to be at stud). **9.** a stable or other establishment where horses are kept for breeding. **10.** a group of horses kept by an owner for racing or hunting. **11.** a bull, ram, other male animal, or a group of male animals, kept for breeding. **12.** (slang) a virile young man, especially one who is sexually promiscuous.

stumpage *n.* (Norse *stumpr*, end + *age*, condition) **1.** the standing timber on a piece of land. **2.** the value of such standing timber. **3.** a tax on timber when it is cut.

stunt *v.* (Anglo-Saxon *stunt*, blunt or stupid) **1.** to reduce growth and development of a plant, animal, or person to less than the normal range. **2.** to display unusual skill or dexterity; to do a trick or perform a difficult and/or dangerous maneuver. *n.* **3.** a halting or slowing of normal growth and development. **4.** a plant, animal, or person whose growth and development have been slowed or stopped too soon by inadequate nutrition, disease, an unfavorable environment, or a genetic factor. **5.** any disease that causes dwarfing or stunting of plants. **6.** a display of skill, dexterity, and/or bravado. **7.** a trick.

stunting *n.* (Anglo-Saxon *stunt*, blunt or stupid + *ing*, action of) reduced growth resulting from inadequate nutrition, disease, an unfavorable environment, or a genetic factor.

stupp *n.* toxic residue produced when mercury is smelted, consisting of a sooty black mixture of mercury, mercuric oxide, soot, hydrocarbons, dust, etc.

sturgeon *n.* (French *esturgeon*, sturgeon) any of several large fishes (*Acipenser* sp. and related genera, family Acipenseridae) with long bodies lined with five rows of bony plates. They are sought for their flesh, their eggs (caviar), and their swim bladders (a source of isinglass gelatin). Most species live in the ocean but swim up a river to spawn. Some beluga sturgeons in the Caspian and Black Seas (*A. huso* or *Huso huso*) grow to as large as 2866 lb (1300 kg).

The largest North American species is the white sturgeon (*A. Transmontanus*), weighing up to 1800 lb (820 kg). The common sturgeon (*A. sturio*) is found along the west coast of Europe or (*A. oxyrhynchus*) along the east coast of North America.

subalpine *adj.* (Latin *sub*, below + *alpinus*, of the Alps or other high mountains) pertaining to the area just below the tree line; the upper zone of tree and shrub growth on a high mountain.

subaquatic *adj.* (Latin *sub*, below + *aquaticus*, of the water) partly aquatic; growing or occurring partly under water and partly on land. See amphibian, emergent vegetation.

subaqueous *adj.* (Latin *sub*, below + *aqueus*, of water) occurring, located, or performed underwater.

subarctic *adj.* (Latin *sub*, below + *arcticus*, northern) pertaining to the area or climatic conditions immediately south of the Arctic Circle, or to other areas with similar climate and living conditions. See subpolar climate.

subartesian water *n.* underground water confined in a porous formation under sufficient natural pressure to rise above the water table but not enough pressure to flow to the soil surface.

subclimax *adj.* (Latin *sub*, below + *climax*, ladder) pertaining to a developmental state of an ecological community that has not reached the expected stable climax; e.g., an area that is recovering from a forest or range fire.

subgelisol *n.* (Latin *sub*, below + *gelatus*, frozen + *solum*, soil) a zone of unfrozen soil or rock material beneath permafrost.

subhumid *adj.* (Latin *sub*, below + *humidus*, moist) receiving less precipitation than a humid region but more than a semiarid region, between about 20 and 30 in. (50 to 75 cm) per year in central U.S.A., about half that much in central Canada, and up to twice that much in the tropics. A subhumid region typically supports a stand of tall, thick grass vegetation but not a forest.

subirrigation *n.* (Latin *sub*, below + *irrigatus*, bringing water + *tion*, action) **1.** the practice of supplying irrigation water through ditches or pipelines into a porous underground layer. Relatively few areas meet the required conditions of having smooth topography with a suitable porous layer underlain by a layer that is impermeable enough to restrict downward flow and occurring in a climate humid enough to keep the soil from becoming saline. Also called subsurface irrigation. **2.** a similar practice of supplying water from underneath for plants grown in pots or other containers.

submarine *n.* (Latin *sub*, below + *marinus*, the sea) **1.** a ship that can operate and navigate while submerged. **2.** a plant or animal that lives beneath the surface of the sea. **3.** a hero sandwich. *adj.* **4.** occurring or living underwater, as a submarine rock formation, submarine warfare, or a submarine plant or animal. *v.* **5.** to operate a submarine, especially in warfare.

submarine spring *n.* a freshwater spring that emerges from an offshore opening in the seafloor.

subpolar climate *n.* a climate such as that in parts of Alaska and Canada where the mean air temperature is below freezing more than half of the year but may reach temperatures of 80°F (27°F) or higher during the long days of the short summer. Also called subarctic climate, taiga climate, or boreal climate.

subpolar low *n.* a low-pressure zone located between 50° and 70° north or south latitude where cold air from around the North or South Pole meets mild air from the temperate region, rises, and produces stormy conditions and a relatively humid climate. See polar front.

subsere *n.* (Latin *sub*, below + *serere*, to join) a secondary series of ecological communities that begins after the original succession has been interrupted, e.g., when an area is converted to cropland and later abandoned, after a fire, or when an area that has been overgrazed is allowed to recover. See sere.

subsidence *n.* (Latin *subsidere*, to settle + *entia*, action) **1.** a gradual lowering of the land surface caused by (1) underlying limestone rock being dissolved (see karst), (2) melting of blocks of ice (see permafrost), (3) pumping of oil or water, (4) dehydration when a soil (especially an organic soil) is drained, or (5) decomposition of an organic soil by oxidation following drainage (see eremacausis). **2.** sinking of an air mass as it is cooled by radiant heat loss.

subsistence agriculture *n.* farming to raise food to be consumed by the farm family, usually with little surplus or profit beyond the family livelihood.

subsoil *n.* (Latin *sub*, below + *solum*, soil) the lower part of the soil. The part below the topsoil or plow layer, often but not always equated with the B horizon.

subspecies *n.* (Latin *sub*, below + *species*, kind) a subdivision of a plant or animal species, usually based on geographic distribution.

substrate *n.* (Latin *sub*, below + *stratum*, covering or layer) **1.** a layer of material that underlies and supports the main part of something. **2.** the material that an enzyme acts upon and converts to a product. **3.** a medium for culturing bacteria or other microbes. **4.** a base used for fabricating a group of electrical circuits.

substratum (pl. **substrata**) *n.* (Latin *sub*, below + *stratum*, covering or layer) underlying layers, especially any unconsolidated layers and the rock layers beneath a soil.

subsurface *adj.* (Latin *sub*, below + *superficies*, surface) beneath the surface, as underwater or in the subsoil or substratum.

subsurface irrigation *n.* See subirrigation.

subterranean *adj.* (Latin *subterraneus*, under earth) **1.** located, existing, or occurring underground; e.g., a subterranean cave or mine. **2.** hidden, done in secret, operating from concealment, as smuggling is a subterranean activity. *n.* **3.** a person, thing, or activity that is underground or secret. Also called a subterrane.

succession *n.* (Latin *successio*, a following) **1.** a sequence of individuals, things, or events that follow one after another (each one succeeding the one before). **2.** the process or designated order whereby one individual or event succeeds another. **3.** the process whereby one plant community replaces another until the climax vegetation is reached.

successional mosaic *n.* a patchwork of adjacent plant communities in different stages of succession.

succulent *adj.* (Latin *succulentus*, juicy) **1.** filled with juice (sap). **2.** composed of fleshy, juicy parts. **3.** having desirable characteristics. **4.** mentally stimulating. *n.* **5.** a fleshy plant; e.g., a cactus or sedum.

succus *n.* (Latin *sucus*, juice) any fluid secreted by or extracted from a plant, insect, or other living tissue; e.g., succus entericus, intestinal juice secreted by glands in the walls of the intestine.

sucker *n.* (Anglo-Saxon *sucan*, to suck + *er*, person or thing) **1.** a person or animal that sucks, especially a baby pig or whale. **2.** a part that serves for sucking. **3.** any of several different fishes that suck (mostly family Catostomidae). **4.** a rootlike organ of certain parasitic plants. **5.** a younger shoot that comes from the root or stem of a plant. **6.** a piece of candy that dissolves slowly when sucked, usually mounted on a stick or other support. **7.** a person or animal that can be cheated easily.

suckle *v.* **1.** to supply or take milk from the breast or udder. **2.** to nourish.

suckling *n.* (Anglo-Saxon *sucan*, to suck + *ling*, young thing) an infant or young animal that suckles; one that has not yet been weaned.

sucrose *n.* (French *sucre*, sugar) a disaccharide ($C_{12}H_{22}O_{11}$) that is the principal commercial sugar. It is produced mostly from sugarcane or sugar beets and is composed of a glucose molecule linked to a fructose molecule (with a molecule of water released). The linkage locks both the glucose and the fructose molecules in their ring (hemiacetal) forms, thus making sucrose a nonreducing sugar.

Sudan grass or **Sudangrass** *n.* an annual sorghum (*Sorghum vulgare*, var. *sudanense*, family Poaceae), native to Africa, that is grown for summer pasture, hay, and silage.

suffrutescent *adj.* (Latin *suffrutescens*, moderately shrublike) somewhat or partly woody or shrublike; e.g., a carnation, a perennial plant with a woody base and herbaceous, annual growth above. See suffruticose.

suffruticose *adj.* (Latin *suffruticosus*, moderately shrubby) woody below but herbaceous above; e.g., trailing arbutus (*Ericacea repens*, family Ericaceae).

sugar *n.* (Arabic *sukkar*, sugar) a water-soluble carbohydrate with a sweet taste, having at least a three-carbon chain with either an aldehyde (-CHO) group on an end carbon or a ketone (double-bonded oxygen) on an inner carbon. All other carbons in the molecule have an alcohol (-OH) group and enough hydrogens to satisfy the remaining bond positions. The simple sugars (monosaccharides) all have the empirical formula $C_nH_{2n}O_n$ and are called trioses, tetroses, pentoses, hexoses, or heptoses for those with 3-, 4-, 5-, 6-, or 7-carbon chains, respectively. Disaccharides are formed by linkage of two monosaccharide molecules through an oxygen, and a water molecule is released in the process. Similar linkage of many sugar molecules produces starch or cellulose. Common monosaccharides such as glucose, dextrose, and fructose occur in honey, grape sugar, potato sugar, and starch sugar. Common disaccharides such as sucrose, lactose, and maltose occur in sugarcane, sugar beets, sugar maple sap, palm tree sap, and milk sugar.

sugar beet or **sugarbeet** *n.* a large biennial beet (*Beta vulgaris*, family Chenopodiaceae) grown extensively for its sugar content. Beet sugar is the source of about 40% of the world's commercial sugar. The beet leaves and the pulp remaining after sugar extraction are used as cattle fodder.

sugarberry *n.* See hackberry.

sugarcane or **sugar cane** *n.* a tall grass (*Saccharum officinarum*, family Poaceae) with stout, jointed stalks 1 to 2 in. (2 to 5 cm) in diameter that grow 10 to 26 ft (3 to 8 m) tall. It is grown in tropical and semitropical regions for its sugar content. Cane sugar constitutes about 60% of the world's commercial sugar.

sugar maple *n.* **1.** any of several large maple trees (especially *Acer saccharum*, family Aceraceae) with short trunks, rounded crowns, long branches, pointed-lobe leaves, and sweet sap. It is the state tree of New York. Sugar maples are used for lumber and to produce maple sugar and maple syrup and are popular as ornamental trees in northeastern U.S.A. **2.** the hard wood from the tree; used for making furniture, flooring, handles, plywood, etc.

Sugar maple leaf and seed pods

sugar weather *n.* weather suitable for tapping the sap of sugar maples; the late winter period several weeks before the trees leaf out when daytime temperatures are above freezing but night temperatures go below freezing.

suicide plant *n.* a plant (*Phyllanthus engleri*, family Euphorbiaceae), native to northern Rhodesia, that causes instant death when its bark or root is smoked.

sulfur or **sulphur (S)** *n.* (Latin *sulphur*, sulfur) element 16, atomic weight 32.066; an essential element for plants, animals, and people. A component of certain amino acids (cystine, cysteine, and methionine) and of two vitamins (thiamine and biotin), it reacts chemically much like oxygen as a divalent anion or forms covalent bonds. It forms sulfide and sulfate minerals, yellow sulfur deposits, gaseous hydrogen sulfide (H_2S) and sulfur dioxide (SO_2), and many organic compounds.

sulfur dioxide *n.* a colorless gas (SO_2) with an offensive odor; an air pollutant that dissolves in water and produces sulfurous acid (H_2SO_3), a component of acid rain. It is formed by combustion of coal, petroleum products, and many other organic materials. Emission controls have gradually reduced its concentrations in air in the U.S.A. from a high of about 16 parts per billion in 1976 to about half that in 2000.

sulfur oxides (SO_x) *n.* a mixture of sulfur dioxide (SO_2) and sulfur trioxide (SO_3).

sullage *n.* (French *souiller*, to soil) **1.** sewage. **2.** muddy sediment deposited by still or slow-moving water. **3.** the oxide coating that forms on molten metal. Also called scoria.

sultry *adj.* (Anglo-Saxon *sweltan*, to die or faint) **1.** oppressively hot and humid, as the weather just before a summer rain. **2.** emitting enough heat to cause sultry conditions, as a sultry sun. **3.** done under sultry conditions, as sultry work. **4.** acting or appearing in a manner that arouses sexual passion, as sultry movements or sultry eyes.

summer fallow *n.* See fallow.

summer savory *n.* See savory.

summer solstice *n.* the time when the sun reaches its greatest declination north or south of the equator, June 21 or 22 in the Northern Hemisphere or December 21 or 22 in the Southern Hemisphere. See Tropic of Cancer, Tropic of Capricorn.

summerwood or **summer wood** *n.* the lines of harder, usually darker wood produced by a tree during the summer in contrast to the more rapid springwood growth produced earlier. Summerwood forms the tree rings that are counted to determine the age of a tree.

summit *n.* (Latin *summum*, highest) **1.** the highest part of a hill, object, or route. **2.** one's greatest attainment, as the summit of a career. **3.** a high-level diplomatic meeting.

sumpweed *n.* (German *sump*, a swamp + *wēd*, weed) a group of herbaceous plants (*Iva* sp., family Asteraceae) that grow in wet places in the Rocky Mountain states and westward. Their pollen is a cause of hay fever.

sun *n.* (Anglo-Saxon *sunne*, sun) **1.** the star at the center of our solar system, about 93 million mi (150 million km) from Earth. The source of light and heat on Earth (produced by thermonuclear reactions in the sun's interior that convert hydrogen to helium and release energy), it has a diameter of 864,000 mi (1.4 million km) and a mass about 330,000 times that of the Earth. **2.** sunlight; the light, heat, and other energy provided by the sun. **3.** the position of the sun in the sky or the length of time that the sun can be seen at a particular date and place. **4.** a site that receives such sunlight, as a place in the sun. **5.** any star that has planets around it. **6.** any incandescent body that provides light and heat; e.g., a sunlamp. **7.** a representation of the sun, as in artwork or ornamentation. *v.* **8.** to expose oneself to the rays of the sun. **9.** to place something where it will receive the rays of the sun, as to sun the clothes so they will dry.

sunburn *n.* (Anglo-Saxon *sunne*, sun + *baernan*, to kindle) **1.** injury to plants, plant products, animals, or people that is caused by too much exposure to the ultraviolet light of intense sunlight; e.g., sunburned Irish potatoes turn green and develop toxic solanine, and sunburned skin on people or animals becomes inflamed, turns red, and later peels off its surface layer. Each year, an estimated 300,000 people get skin cancer from excessive exposure to the sun. *v.* **2.** to be injured by excessive exposure to sunlight.

sun-cured *adj.* preserved by being dried by the heat of the sun's rays; e.g., sun-cured hay, fruit, fish, and meat.

sundial *n.* (Anglo-Saxon *sunne*, sun + Latin *dialis*, daily) an instrument that indicates apparent local time by the position of the shadow cast by a gnomon (pointer) on a horizontal dial.

sundowner *n.* See Santa Ana.

sunfish *n.* (Anglo-Saxon *sunne*, sun + *fisc*, fish) **1.** a sluggish ocean fish (*Mola mola*, family Molidae) with a short, thick, bony body, a small mouth, long dorsal and anal fins, and almost no tail. **2.** any of a number of other fishes of the family Molidae. **3.** any of several brightly colored, North American, freshwater fish (*Lepomis* sp.) with deep, compressed bodies, including bluegills, crappies, and pumpkinseeds. **4.** any of the large jellyfishes.

sunflower *n.* (Latin *flos solis*, flower of the sun) **1.** any of several plants (*Helianthus* sp., family Asteraceae); e.g., the common sunflower (*H. annus*), a native of the U.S.A. that grows 3 to 20 ft (1 to 6 m) tall and produces showy, round, yellow flowers 6 to 15 in. (15 to 38 cm) across that rotate during the day to face in the direction of the sun. It is the state flower of Kansas. *adj.* **2.** pertaining to or coming from sunflowers, as sunflower seeds much used to feed birds, sunflower oil widely used for cooking, and sunflower meal used as a feed supplement.

sunscald *n.* (Anglo-Saxon *sunne*, sun + French *escalder*, to wash in hot water) sunburn injury to a woody plant caused by exposure to sunlight, generally accompanied by dehydration in hot weather or alternating with freezing temperatures in winter.

sunspot *n.* (Anglo-Saxon *sunne*, sun + German *spot*, speck) a dark area in the photosphere of the sun where the surface temperature is lowered, presumably by magnetic storms that disrupt radio signals on Earth at the same time. Sunspots appear in groups and vary in abundance on about an 11-year cycle that seems to influence the weather on Earth.

superficial *adj.* (Latin *superficialis*, of the surface) **1.** on, affecting, or of the surface; not penetrating. **2.** the uppermost geologic stratum. (See surficial.) **3.** not very profound; easy to discern but possibly incomplete or faulty. **4.** based on outward appearance only, as a superficial resemblance. **5.** areal extent.

Superfund *n.* a fund established in 1980 that is administered by the U.S. Environmental Protection Agency to remediate hazardous waste sites where the responsible polluter is unknown or cannot be required to perform the remediation.

supernodulation *n.* (Latin *super*, above + *nodulus*, little knot + *atio*, action) the formation of an unusually large number of nodules on the roots of a legume and/or the formation of nodules even in the presence of abundant nitrate. It is evidently a result of a mutation that negates the regulation of nodulation by an inhibitor normally formed in the leaves and translocated to the roots.

superphosphate *n.* (Latin *super*, above + Greek *phos*, light + *ate*, salt) phosphorus fertilizer that has been treated with acid to make it more soluble and available for plant use. Single superphosphate is produced by treating apatite [(Ca$_5$(PO$_4$)$_3$F)] that is between 0% and 4% available P$_2$O$_5$ equivalent with sulfuric acid (H$_2$SO$_4$) to form a mixture of monocalcium phosphate [Ca(H$_2$PO$_4$)$_2$] and gypsum (hydrated CaSO$_4$) that is about 20% available P$_2$O$_5$ equivalent, also known as ordinary or simple superphosphate. Triple superphosphate is produced by treating apatite with phosphoric acid (H$_3$PO$_4$) to form relatively pure monocalcium phosphate with up to 46% available P$_2$O$_5$ equivalent, also called treble or concentrated superphosphate.

supersaturation *n.* (Latin *super*, above + *saturatio*, saturation) a concentration of dissolved material that exceeds the normal saturation level, usually the result of cooling that reduces the solubility in the absence of nuclei that would trigger the formation of liquid or solid particles. For example, a supersaturated atmosphere can produce rain or snow, and a supersaturated solution such as honey can crystallize and harden.

supersonic *adj.* (Latin *super*, above + *sonus*, sound + *icus*, of) **1.** faster than the speed of sound in the same fluid at the ambient temperature. Faster than 1129 ft/sec (770 mph, 344 m/sec, or 1239 km/hr) in air at 70°F or 21°C, 1088 ft/sec at 0°C; or 1165 ft/sec at 40°C. **2.** traveling or capable of traveling at supersonic speed, as a supersonic aircraft. **3.** ultrasonic; vibrating too fast to be heard by the human ear; faster than about 20,000 Hz.

supplemental irrigation *n.* irrigation that is used only when needed in an area with enough natural precipitation to provide most of the water needed by the crop being grown.

supraglacial *adj.* (Latin *supra*, above + *glacialis*, frozen) carried on and deposited from the upper surface of a glacier (and therefore less consolidated than material deposited beneath the ice).

suprapermafrost *n.* (Latin *supra*, above + *perma-nens*, remaining + Anglo-Saxon *frost*, frozen) the soil layer that freezes and thaws annually in a cold climate where the soils are underlain by permafrost. Also known as the active layer. See frost zone.

surcle or **surculus** *n.* (Latin *surculus*, twig or branch) a small leafy shoot or twig; a sucker.

surculose *adj.* (Latin *surculosus*, having suckers) producing leafy shoots (suckers) that grow from the base of a plant.

surface area *n.* the total exposed area of an object or mass. The surface area of a given mass is inversely proportional to the size of its particles (assuming the different sizes have similar shapes). For example, the surface area of a 1-cm cube is 6 cm^2, but that same mass has a total surface area of 60 cm^2 if it is cut into 0.1-cm cubes (1000 sand-size particles), 600 cm^2 if cut into 0.01-cm cubes (silt size), and 6000 cm^2 if cut into 0.001-cm cubes (coarse clay size). The clay probably has even more surface area because its particles tend to be platy rather than cubic or spherical.

surface fire *n.* a fire that burns the litter, trash, grass, and low brush in a wooded area but not the treetops. It is generally much slower moving and less destructive than a crown fire.

surface impoundment *n.* See lagoon.

surface irrigation *n.* the artificial application of water to the soil surface, usually with its flow guided by either channels (corrugation or furrow irrigation) or ridges (flood irrigation between borders or in basins). See sprinkler irrigation, drip irrigation, subirrigation.

surface mining *n.* See strip mine.

surface runoff *n.* See runoff.

surfactant *n.* (short for surface-active agent) a wetting agent. For example, a detergent that reduces the surface tension of water, enabling it to mix with other (e.g., oily) liquids and to wet surfaces that would otherwise resist wetting; surfactants are added to pesticide sprays to make them spread and stick to the target species.

surficial *adj.* (Latin *superficialis*, on the surface) occurring on or pertaining to the surface, especially that of land; e.g., soil is a surficial material.

surge *n.* (Latin *surgere*, to rise up) **1.** an unusually large incoming wave or group of waves, as a surge of seawater or a surge of electricity. **2.** the swelling movement of such an incoming wave. **3.** any dramatic increase, as a surge of activity triggered by a surge of adrenalin. **4.** irregular pressure changes and flow, as water surging through a conduit or splashing violently in a container. *v.* **5.** to come in as a large wave or waves, as the sea surges against the shore. **6.** to rise and fall with the waves, especially a ship in a rough sea. **7.** to move by a series of strong pulses. **8.** to slacken or slip, as a rope tied to a capstan.

surge irrigation *n.* a method of furrow irrigation that turns the irrigation water on and off (usually back and forth between two sets of rows) to cause the water to flow in surges that reach the bottom end sooner than continuous flow with smaller streams would, thus improving the uniformity of irrigation.

suricate *n.* (Dutch *surikat*, macaque) a small South African burrowing carnivore (*Suricata suricatta*) with a gray coat marked with dark bands, especially on the back. It is related to the mongoose and civet but smaller than most domestic cats and is useful for killing rats and mice.

surmullet *n.* (French *surmulet*, red mullet) any of several fish (family Mullidae), especially the red mullet (*Mullus barbatus*), a saltwater food fish about 12 in. (30 cm) long that is common in the Mediterranean Sea. Surmullet is characterized by its reddish brown color and two barbels attached to the lower lip.

surra *n.* (Marathi *sura*, heavy breathing) a highly infectious disease caused by a parasitic protozoan (*Trypanosoma evansi*, family Trypanosomatidae) carried by the gadfly and horsefly (family Tabanidae). It affects elephants, horses, mules, camels, and most other domestic animals. Symptoms include fever, tiny red spots on mucous surfaces, edema, anemia, and emaciation. It can be fatal. See trypanosomiasis.

survival of the fittest *n.* a 19th-century interpretation of Darwin's concept of evolution by natural selection, suggesting that the best-fit and best-adapted individuals survive and reproduce the species, whereas the unfit are eliminated in the struggles of life. See natural selection.

suspended load *n.* that part of the total load of a stream that is carried in suspension by the flowing water, as distinguished from the material in solution (dissolved load) and that rolled and bounced along the bottom (bed load or traction load).

suspended water *n.* See vadose water.

suspension *n.* (Latin *suspensio*, hanging up) **1.** an overhead support that allows objects to hang down from it. **2.** the act of hanging something in a suspended position. **3.** the springs, etc., that support the body of an automobile on its undercarriage. **4.** the state of a system that has solid particles distributed through a moving fluid. **5.** the mixture of a liquid with suspended solid particles. **6.** an interval of barring from office or privilege, holding back from payment, or other delay in an activity. **7.** a dissonant delay in certain musical tones. **8.** the condition of suspense created by deliberately withholding critical information or conclusions from an interested party.

sustainability *n.* (Latin *sustinere*, to uphold + *abilitas*, capacity or ability) the potential for being continued without being depleted.

sustainable agriculture *n.* an agricultural system adapted to a particular area so its plant and animal production does not decline over time and is reasonably stable over normal fluctuations of weather and other uncontrollable factors. It should satisfy human needs for food and fiber, protect natural resources and environmental quality, make efficient use of nonrenewable resources, and sustain the economic viability and quality of life of farm operations and of society as a whole. See alternative agriculture.

sustainable development *n.* development that meets present needs in ways that will not make it more difficult for future generations to meet their own needs.

swag *n.* (Norwegian *svagga*, to sway) **1.** a drape, garland, or other ornamentation that hangs down in the middle between two points of support. **2.** an ornamental wreath or other item made of boughs, flowers, and/or fruit. **3.** a swale. **4.** an unsteady, lurching movement. **5.** money, other valuables or belongings, especially when stolen. *v.* **6.** to hang down loosely. **7.** to sway and move unsteadily. **8.** (Australia) to travel with a small bundle of belongings.

swale *n.* (Norse *svalr*, cool) **1.** a low-lying narrow strip of land where water flows slowly when there is runoff. **2.** a shallow hollow or depression, especially one in a marshy area. **3.** a cool, shady place.

swallet *n.* (Anglo-Saxon *swelgan*, swallow + *et*, little) **1.** an opening where a stream goes underground. **2.** an underground stream. **3.** underground water flowing into a mine.

swallow *n.* (Anglo-Saxon *swalewe*, swallow) **1.** any of several small, swift-flying birds (family Hirundinidae) with pointed wings and short, wide bills that catch insects as they fly. Several species (e.g., the barn swallow, *Hirundo rustica*) have deeply forked tails. They are noted for migrating 1000 mi (1600 km) or more. **2.** any of several similar but not closely related birds; e.g., the chimney swift (*Chaetura pelagica*, family Apodidae) or the swallow shrike (*Artamus fuscus*). **3.** a bite of food or a gulp of water that can be taken at one time. **4.** the act of passing food or drink from the mouth into the esophagus and stomach. **5.** the esophagus or gullet. **6.** a channel to hold a rope in a pulley. *v.* **7.** to ingest, especially to pass food or drink from the mouth through the esophagus and into the stomach. **8.** to move the throat as though eating or drinking, as he swallowed hard when she said that. **9.** to cause to disappear from view, as the sea swallows a diving submarine. **10.** to subdue, as she had to swallow her pride. **11.** to accept, as the people had to swallow the new prices. **12.** to take back, as he will have to swallow those words.

swallowtail *n.* (Anglo-Saxon *swalewe*, swallow + *taegel*, tail) **1.** a deeply forked tail, like that of many swallows. **2.** any of several butterflies (*Papilio* sp.) with back wings that extend rearward like a forked tail. **3.** anything that has a forked shape like that of a swallow's tail; e.g., certain dove-tailed joints, some topcoats, some fortifications, etc.

swamp *n.* (Danish *svamp*, sponge) **1.** a waterlogged area with spongy soil covered mostly with water-loving trees and shrubs. Also called swampland. See bog, marsh, wetland. **2.** a low place that collects water in a coal bed. *v.* **3.** to sink slowly, as into swampy ground. **4.** to work in a swamp, e.g., to build a road through it or to clear the ground for a logging operation. See swampbuster law. **5.** to flood with water, as the heavy load caused the boat to swamp. **6.** to overwhelm, as all those bills can swamp a new business, or too many details can swamp a student.

swampbuster law *n.* provisions in the 1985 Food Security Act and subsequent legislation that forbid or restrict the drainage of wetlands.

swamper *n.* (Danish *svamp*, sponge + *er*, person) **1.** a person who works in, lives in, or studies swamps. **2.** a person who does chores or runs errands; a menial laborer. **3.** a woodsman who trims the branches from logs and clears the ground for a logging operation.

swamp sumac *n.* See poison sumac.

swan *n.* (German *schwan*, swan) **1.** any of several large, graceful, long-necked, aquatic birds (*Cygnus* sp.) most species have all white plumage when mature. See cygnet. **2.** a person who sings sweetly or has unusual beauty, purity, or excellence. **3.** the constellation Cygnus. *v.* **4.** declare surprise, as "I swan, I never expected to see that happen."

sward *n.* (Anglo-Saxon *sweard*, skin or rind) **1.** an area of close-growing grass that is kept mowed. **2.** any area covered with grass. **3.** the soil layer that is held together by grass roots. Also called sod or turf. **4.** any rind or hard skin, as that on bacon. *v.* **5.** to establish a grass cover on an area.

swarm *n.* (Anglo-Saxon *swearm*, tumult) **1.** a large cluster of insects flying together, especially a group of honeybees being led by a queen to establish a new colony, or a colony so established. **2.** any group that is in active motion; e.g., a swarm of people around a celebrity or a swarm of bacteria around a decaying root. *v.* **3.** to fly together as a compact group, as bees or termites swarm. **4.** to assemble and move in a large group, as a crowd swarms. **5.** to fill or be filled, as sailors swarm the marketplace when a ship docks, or the playground swarms with children. **6.** to climb a tree or pole by wrapping the arms and legs around it and moving them upward by increments. Also called shinny.

swarm spore *n.* a zoospore.

swash *n.* (Swedish *svasska*, to make a washing noise) **1.** a body of fast-flowing water, especially one that is flowing through a break in a sandbank. **2.** the surge of water from an ocean wave as it strikes the shore. **3.** the sound made by such waves. **4.** the splashing of water washing over a sandbar. **5.** an ornamental flourish attached to some cursive or italic letters in certain styles or fonts. **6.** a swagger or a person who swaggers or blusters. *v.* **7.** to splash noisily, as waves dashing over or against a bank or a pier. **8.** to move quickly and violently. **9.** to swagger, bluster, and/or brag.

sweetbread *n.* **1.** stomach sweetbread, the pancreas gland of an animal (especially a calf or lamb) used as food. **2.** neck or throat sweetbread, the thymus gland similarly used as food.

sweet clover or **sweetclover** *n.* any of about 20 species of mostly biennial, and some annual, plants (*Melilotus* sp., family Fabaceae) native to the Mediterranean region. White sweet clover (*M. alba*) and yellow sweet clover (*M. officinalis*) are important forage and soil-improving crops in the U.S.A. and Canada; both are named for the color of their blossoms. They need a near neutral or alkaline soil (pH 6.0 to 8.5) and 17 in. (43 cm) or more of precipitation or irrigation water and are used as forage crops but must not be cut or grazed below 1 ft (30 cm) in height. They contain coumarin (dicumarol) that accumulates if the forage spoils, can become toxic to livestock, is a component in warfarin rat poison, and is useful as a blood thinner.

sweet corn *n.* a variety of corn (*Zea mays* var. *saccharata*, family Poaceae), with a relatively high sugar and low starch content, that is used as a vegetable either on or off the cob.

sweet gum *n.* a large, fast-growing tree (*Liquidamber styraciflua*, family Altingiaceae), native to south-eastern U.S.A., with 5-pointed shiny leaves, spiny ball-shaped fruit, and fragrant juice; used as an ornamental tree. **2.** the hard, reddish brown wood from the tree; used to make furniture and veneer. **3.** the amber gum exuded by the tree; used in perfumes and drugs.

sweet pea *n.* an annual climbing vine (*Lathyrus odoratus*, family Fabaceae), native to India, with alternate pinnately compound leaves, tendrils, and sweet-scented multicolored blossoms up to 2 in. (5 cm) across, that is used as an ornamental.

sweet pepper *n.* **1.** a variety of pepper (*Capsicum annum*, family Solanaceae) that produces larger, milder, sweeter fruit than the hot peppers; an annual plant often grown in gardens. **2.** the fruit of these plants with a squarish lobed shape, usually picked when it is bright green but can be left until it turns bright red. It is used as a vegetable, often served with other vegetables to add color and flavor. Also called green pepper or bell pepper.

sweet potato *n.* **1.** a warm-season perennial plant that is grown as an annual (*Ipomea batatas*, family Convolvulaceae) that requires about 175 freeze-free days to mature. **2.** the fleshy yellow to orange-brown swollen roots produced by the plant and used as a vegetable. See yam.

sweet-potato weevil *n.* a beetle (*Cylas formicarius elegantulus*, family Curculionidae) with a long, slender body and six legs that is the most destructive insect pest on sweet potatoes.

Sweet potato weevil

sweetsop *n.* **1.** a tropical American tree or shrub (*Annona squamosa*, family Annonaceae). **2.** the sweet pulpy fruit with a thin, tuberculate rind produced by such trees or shrubs. Also called sugar apple.

sweet sorghum *n.* See sorgo.

sweet william or **sweet William** *n.* a perennial flowering plant (*Dianthus barbatus*, family Caryophyllaceae) with clusters of small blossoms with various colors.

swellhead *n.* **1.** a disease that causes a watery discharge from the nostrils and swelling of the sinuses under the eyes of turkeys, leading to loss of weight and making the turkeys unmarketable. Also called sinusitis. **2.** a disease of sheep and goats grazing on little-leaf horsebrush (*Tetradymia glabrata*) and related species on rangeland from southern Idaho and Oregon to western Utah and eastern California, especially in the early spring. The toxic component affects the liver and causes edema that shows mostly as a swollen head about a day after the forage is eaten. Also called bighead.

swidden *n.* See shifting cultivation.

swill *n.* (Anglo-Saxon *swilian*, swill) **1.** a mixture of water and kitchen refuse, usually fed to swine. **2.** kitchen refuse; garbage. **3.** any watery refuse; slop. **4.** a large swig of liquor. *v.* **5.** to feed swill to swine or other animals. **6.** to drink eagerly. **7.** to clean by flooding with water.

swine *n.* (Anglo-Saxon *swin*, swine) **1.** any of several omnivorous mammals (family Suidae) with cloven hooves and thick hides covered with sparse, coarse hair; commonly known as pigs or hogs. Adult males are boars and adult females are sows. Domestic types are raised for meat (pork, though eating pork is forbidden in certain religions), and wild types are often hunted. **2.** a contemptible, brutish, coarse person.

Swiss chard *n.* See chard.

switchgrass *n.* a vigorous, perennial, warm-season, sod-forming grass (*Panicum virgatum*, family Poaceae) with an open, branching inflorescence. Native to the U.S.A., it grows 3 to 5 ft (0.9 to 1.5 m) tall and provides good summer forage and good erosion control that is especially useful where atrazine resistance is needed, e.g., in grassed waterways in a cornfield.

syconium *n.* (Greek *sykon*, fig + Latin *ium*, plant) a multiple fruit, e.g., that of a fig tree.

sylvan *adj.* (Latin *sylvanus*, forest) **1.** related to or characterized by trees; wooded. **2.** made of boughs or other tree parts. *n.* **3.** a person who lives in a forest or wooded area. **4.** a mythical spirit of the woods. Also spelled silvan.

symbiont *n.* (Greek *symbiont*, living together) any organism that lives with another kind of organism in a symbiotic relationship. See symbiosis.

symbiosis *n.* (Greek *symbiōsis*, living together) any relationship between different organisms that live close enough together to influence each other, especially mutualism (when the influence is mutually beneficial), but also including amensalism, commensalism, parasitism, and synnecrosis.

symbiotic nitrogen fixation *n.* **1.** a mutualism relationship between bacteria (*Rhizobium* sp. and *Bradyrhizobium* sp.) and most leguminous plants (family Fabaceae). The plant supplies carbohydrate and a place for the bacteria to live (in nodules on the plant roots). The bacteria fix atmospheric nitrogen (N_2) and release some of it in combined forms (mostly amine, $-NH_2$) that the plant can use. Some of this combined nitrogen escapes into the soil and can be used by other plants, either the same season or the next season. Measured amounts of symbiotic nitrogen fixation commonly range from 20 to 200 lb/ac (22 to 225 kg/ha) per year, with occasional measurements up to twice that much. **2.** a similar relationship between certain trees and shrubs (e.g., alder trees, *Alnus* sp.) and an actinomycetes (*Frankia* sp.).

sympatric *adj.* (Greek *syn*, together + *patēr*, father + *ikos*, of) pertaining to or occurring in the same geographic area; e.g., different species of organisms that share the same territory.

symplastic growth *n.* coordinated growth of two tissues so that their cell walls maintain contact.

synchronous *adj.* (Latin *synchronus*, together in time) **1.** occurring simultaneously; taking place at the same time. **2.** progressing at the same rate and time; synchronized. **3.** having the same frequency and matching in phase. **4.** having uniform time intervals.

5. geostationary, as a satellite that orbits the Earth once a day and remains above the same point at all times.

syncline *n.* (Greek *syn*, together + *klinein*, to lean) the junction point or line for two downward sloping lines or surfaces, especially the central axis of a trough (sunken area) in rock strata; the opposite of an anticline. Synclines and anticlines alternate in folded rock strata.

synecology *n.* (Greek *syn*, together + *oikos*, house + *logos*, word) the study of groups or communities of organisms in relation to their environment. See ecology, autoecology.

synergism *n.* (Latin *synergismus*, working together) **1.** the increased effect achieved when two or more entities work together and produce a greater total effect than the sum of the results their separate efforts would accomplish, as opposed to antagonism. **2.** the combined action of two or more agents that increase each other's effectiveness; e.g., two drugs that work together to combat a disease.

synnecrosis *n.* (Greek *syn*, together + *nekrōsis*, death) a symbiotic relationship that is detrimental to both symbionts.

synodic month *n.* **1.** a lunar month from one new moon to the next. **2.** the average time between successive new or full moons: 29.531 days, or 29 days, 12 hr, and 44 min.

synoptic *adj.* (Greek *synoptikos*, together as a whole) **1.** a general overview; a summary or compendium. **2.** extending across a broad area (e.g., a continent), as the meteorological events occurring at any particular time.

synthetic variety *n.* a cultivar produced by human intervention; the product of cross-pollination of parent sources chosen as carriers of certain desirable traits followed by growing several generations and selecting plants that exhibit the desired traits for use as seed sources.

syntrophism *n.* (Greek *syn*, together + *trophikos*, pertaining to food + *ismos*, action) a mutualistic biological relationship between organisms of two different species that depend on each other for nutrition; e.g., the mutualistic relationship between *Rhizobium lupini* and a lupine or trefoil.

syringa *n.* (Greek *syringos*, pipe) any of several shrubs or trees (*Syringa* sp., family Oleaceae), including the common lilac (*S. vulgaris*). Some species are also called mock orange. A wild species with white blossoms is the state flower of Idaho.

Syringa in bloom

Système Internationale d'Unités (SI) *n.* the modern international system of measuring units and nomenclature that is recognized and used throughout most of the world, commonly called the metric system; its basic units are:

Meter	Length
Kilogram	Mass/weight
Second	Time
Kelvin	Temperature
Ampere	Electrical current
Candela	Luminous intensity
Mole	Amount of a substance

systemic *adj.* (Latin *systema*, system + *icus*, of) **1.** throughout a physiological system in a body; e.g., throughout the nervous system or the circulatory system of an animal or plant. **2.** affecting or pertaining to an entire body or system, especially in a pathological or other negative sense.

systemic pesticide *n.* a chemical absorbed by either the roots or the leaves and carried throughout the plant. Either a herbicide that kills the plant or an insecticide that makes the plant parts toxic to insect pests.

T

T *n.* **1.** the chemical symbol for the tritium isotope of hydrogen. **2.** tesla, a unit of magnetic induction. **3.** temperature. **4.** *T* value for soil loss. See tolerable soil loss.

Ta *n.* chemical symbol for the element tantalum.

tabanid *n.* (Latin *tabanus*, gadfly) any of a large family of bloodsucking dipterous flies (family Tabinidae), including deerflies and horseflies. The female horsefly (*Tabanus* sp.) sucks blood from most mammals, including humans. Also known as gadfly. Tabanids can transmit anthrax and trypanosomes. See anthrax, trypanosomiasis.

tableland *n.* (French *table*, table + Anglo-Saxon *land*, land) a large area of nearly level land at a relatively high elevation, often bordered by steep slopes or cliffs; a plateau or mesa.

tachina fly *n.* (Latin *tachinos*, swift thing) any of a large group of dipterous insects (family Tachinidae) whose larvae parasitize several species of caterpillars, beetles, and other insects. For example, *Madremyia saundersii* lays its eggs on cabbageworms (*Pieris rapae*) and cabbage loopers (*Trichoplusia ni*) so its larvae will parasitize the pest species. Also called tachinid fly. See parasite.

taconite *n.* a low-grade iron ore found in rocks around Lake Superior, named for the Tacon mountain range. It contains at least 25% iron (Fe) in the form of hematite (Fe_2O_3) and enough magnetite (Fe_3O_4) to make it magnetic.

tactile *adj.* (Latin *tactilis*, touchable) **1.** having the sense of touch. **2.** detectable by the sense of touch; subject to being felt or evaluated by touching.

taction *n.* (Latin *tactio*, touch) **1.** making contact by touching. **2.** the sense of touch; perception by touch.

taenia or **tenia** *n.* (Latin *taenia*, ribbon or tape) **1.** a headband (ancient Greek). **2.** a fillet in Doric architecture. **3.** an elongated, ribbonlike structure in muscle, nerve tissue, or the brain. **4.** a long, narrow bandage. **5.** any of a group of large tapeworms (family Taeniidae, especially *Taenia* sp.) that infest bobcats, dogs, foxes, other mammals, and humans.

taeniacide or **teniacide** *n.* (Latin *taenia*, ribbon or tape + *cida*, killer) a chemical that kills tapeworms (class Cestoda) that are parasites in vertebrate intestines.

tagua *n.* **1.** a slow-growing tropical palm tree with large, graceful fronds; native to South America. **2.** the nut produced in clusters by the palm tree. These nuts resemble a chicken egg in size and shape and go through 3 stages: (1) they fill with a liquid similar to coconut milk (2) the liquid congeals, forming a sweet, edible gelatin, (3) the gelatin hardens into a white substance that resembles ivory from elephant tusks and can be carved and polished like ivory.

taguan *n.* (East Indian *taguan*, flying squirrel) a large flying squirrel (*Pteromys petuarista*) that has a body about 2 ft (60 cm) long and uses folds of skin between its front and back legs plus its long bushy tail to glide from tree to tree; native to the East Indies.

taiga *n.* (Russian *taiga*, taiga) subarctic coniferous forest (especially fir, spruce, and larch) in a broad band below the tundra in northern U.S.A., southern Canada, northern Europe, and northern Asia; intermixed with muskeg bogs.

tailings *n.* (Anglo-Saxon *taegel*, to tear off + *ing*, product of) fine-particle refuse from distilling, milling, mining, etc. Reclamation of mine tailings commonly is difficult because they are often strongly acid and contain toxic heavy metals.

tailrace *n.* (Anglo-Saxon *taegl*, tail or end + *raes*, swift movement) an artificial channel for carrying water away from a mill, mining operation, turbine, waterwheel, etc.

tail water *n.* **1.** water that flows from the bottom ends of the rows in furrow or border irrigation. It can sometimes be used to irrigate lower-lying land. **2.** water downstream from a dam. **3.** water that is passing or has passed through a tailrace.

taint *n.* (French *teint*, tint) **1.** a small amount of contaminant or a trace of infection. **2.** a suggestion or likelihood of dishonor. **3.** a color, shade, or tint. *v.* **4.** to contaminate slightly; e.g., to taint milk with the flavor of wild onions. **5.** to infect with a harmful organism. **6.** to tarnish a reputation. **7.** to dye with a new color or tint.

take-all *n.* a root rot disease caused by a fungus (*Ophiobolus graminis*) that blackens and decays the lower stems of wheat, rye, barley, and oats, stunting the plants and markedly reducing yield.

talik *n.* (Russian *talik*, talik) either a permanent or a seasonal layer of unfrozen ground with permafrost beneath it and either permafrost or seasonally frozen soil above it. Also called tabetisol.

tall buttercup *n.* See crowfoot.

tall grass *n.* **1.** any of several tall grasses characteristic of areas such as that in central U.S.A. immediately west of the Mississippi River, especially little bluestem, big bluestem, indiangrass, and switchgrass. *adj.* **2. tallgrass** characterized by tall grasses, such as tallgrass prairie, as distinguished from the shortgrass prairie of drier climates.

tallow *n.* (Anglo-Saxon *taelg*, a color) **1.** the relatively solid, nearly colorless animal fat stripped from carcasses of sheep, cattle, etc., or melted from the meat for use in making lubricants, soaps, candles, etc. **2.** vegetable tallow; any plant material similar in nature and use to animal tallow.

talus *n.* (Latin *talutium*, a gold-bearing slope) **1.** a sloping surface formed of rock fragments (scree) that have fallen from a cliff. **2.** any slope, especially a sloping wall that widens at its base as part of a fortification.

talweg *n.* See thalweg.

tamarisk *n.* (Latin *tamariscus*, tamarisk) any of several evergreen tree or shrub species (*Tamarix* sp.) native to the Mediterranean area and southern Asia. The twigs are grayish green, wiry, and jointed and have a tiny scale leaf at each joint; flowers are feathery clusters of tiny pink blossoms. The rapidly growing trees tolerate salinity; they are used in windbreaks and as ornamentals.

tanager *n.* (Tupi *tangara*, tanager) any of several brightly colored red and/or yellow forest birds that have black and white wing markings (*Piranga* sp. and *Spindalis zena*, family Emberizidae) and are about the size of a robin. Tanagers include the scarlet tanager (*P. olivacea*) of northeastern U.S.A. and eastern Canada, the summer tanager (*P. rubra*) of southern U.S.A., the western tanager (*P. ludoviciana*) of western U.S.A. and Canada, the hepatic tanager of southwestern U.S.A. and Mexico, and the stripe-headed tanager (*S. zena*) of the West Indies and southern Florida.

tanbark *n.* the bark of an oak or any of several other trees with a high tannin content and tan color. Such bark is shredded for use as a source of tannin for tanning leather; the remnants are used to cover circus arenas, racetracks, exhibit hall floors, etc.

tankage *n.* (Portuguese *tanque*, tank or pond + French *age*, related to) **1.** storage in tanks, or the fee charged for such storage. **2.** the capacity of a tank or group of tanks. **3.** the residue left as a byproduct of rendering fat from carcasses by the meatpacking industry. It is used in some feeds for livestock, and organic gardeners use it as a fertilizer that averages 9% total nitrogen (N), 4.4% total phosphorus (P), and 1.3% total potassium (K).

tannic acid *n.* a yellowish astringent chemical ($C_{14}H_{10}O_9$) derived from the bark of oak and hemlock trees. It is used to tan animal hides and, medically, to stop bleeding. See tannin, tanbark.

tannin *n.* (Latin *tannum*, oak bark) any of a number of polyphenols (including tannic acid) that are able to precipitate proteins from solution by bonding with collagen and other proteins, thus reducing the palatability and digestibility of foods, making wood more durable, etc. Named for their use in tanning leather.

tanning *n.* **1.** a major part of the process of converting animal skin into leather. After the hair and flesh are removed and the skin has been washed, soaked, and pickled, it is tanned by treatment with tannins and/or chromium (often also involving an alkaline and or glucose treatment) that makes the leather resistant to decay and shrinkage. **2.** darkening of the skin caused by exposure to ultraviolet rays from the sun or sun lamps. **3.** (slang) a spanking.

tan oak *n.* a leathery-leaved evergreen tree (*Lithocarpus densiflorus*) that produces tanbark used as a source of tannin and to cover circus arenas, exhibit hall floors, etc.; native to California and Oregon.

tansy *n.* (Greek *athanasia*, immortality) an Old World plant (*Tanacetum vulgare* and related species, family Asteraceae) with a strong scent, serrated leaves, and small tubular yellow flowers. Now a common weed in the U.S.A., it has some medicinal and herbal uses but also has a toxic effect that can cause paralysis, convulsions, and even death.

tansy mustard *n.* (*Descurainia pinnata*, family Brassicaceae) **1.** an annual herb with bitter-tasting pinnate leaves similar to those of tansy. When moisture is favorable, it grows on arid and semiarid ranges of western U.S.A. When eaten while in bloom, it gives a "paralyzed tongue" to cattle. **2.** a plant (*Sisymbrium canescens*, family Brassicaceae) with leaves similar to those of tansy.

tantalum (Ta) *n.* (Greek *Tantalos*, a Greek god + *ium*, element) element 73, atomic weight 180.9479; a heavy metal of the vanadium family that forms hard, rustproof, heat-resistant alloys with other metals.

tapeworm *n.* (Anglo-Saxon *taeppe*, strip + *wyrm*, worm) any of several species of parasitic, hermaphroditic, segmented, flat intestinal worms (phylum Platyhelminthes) that have no alimentary canal of their own but infest that of humans and other vertebrate animals, usually entering through the mouth in infected food. Most tapeworms have different species of hosts for their larval and adult stages.

taphonomy *n.* (Greek *táph*, grave + *nomia*, knowledge of) **1.** the process and conditions of fossilization. **2.** the scientific study of the environmental conditions that favor fossilization of plant and animal remains.

tapioca *n.* (Tupi *tipioca*, juice) a starchy food made from the root of cassava (*Manihot esculenta*), widely grown in tropical climates, especially on poor soils. It is used in puddings and as a thickener. See cassava.

tapir *n.* (Tupi *tapira*, tapir) any of several large hoofcd animals (*Tapirus* sp., family Tapiridae) with three toes on their back feet and four in front, and a stout build similar to swine; native to Central and South America and the Malayan peninsula. It is related to the rhinoceros but is smaller.

taproot *n.* (Anglo-Saxon *taeppe*, tap + *rote*, root) **1.** a large central root that extends downward from a radicle and produces smaller branch roots. *adj.* **2.** having such a root; e.g., a plant with a taproot system, as distinguished from a fibrous root system composed of many fine roots.

tarantula *n.* (Italian *tarantola*, Taranto, a city in southern Italy + *ula*, little) **1.** any of several large, hairy, venomous spiders (family Theraphosidae) whose bite is painful; native to southwestern U.S.A. **2.** any of several other similar related spiders. **3.** a large European wolf spider (*Lycosa tarantula*) that was formerly thought to cause tarantism (a crazed impulse to dance) when it bit someone.

tar ball *n.* a ball of solidified tar that resists decomposition. Formed from an oil spill or natural seepage of petroleum into the sea, it often washes onto a beach and damages the environment.

tare *n.* (Arabic *tarha*, reject) **1.** the empty weight, as the weight of a container to be deducted from the total to obtain the net weight. **2.** any of several species of vetch (*Vicia* sp.), especially common vetch (*V. sativa*). **3.** a noxious weed. The biblical tare is believed to have been darnel (*Lolium temulentem*). *v.* **4.** to weigh an empty container, or to allow for its weight, e.g., by deducting it from a total, by resetting the zero point, or by balancing with a similar container.

target value *n.* a quantitative goal of a program; e.g., to keep nitrates below 10 ppm in drinking water.

tarn *n.* (Icelandic *tjörn*, a tarn) a small lake in a mountainous region, especially one formed in a cirque.

tarnish *v.* (French *ternir*, to make dull) **1.** to make or become dull; to lose luster or become discolored. **2.** to damage or sully, as to tarnish a reputation. *n.* **3.** the oxide or other material that dulls or discolors a tarnished surface, or the surface itself. **4.** any stain, blemish, or discoloration.

tarnished plant bug *n.* an insect (*Lygus lineolaris*, family Muridae) that infests many fruit and garden plants, causing dead areas, distorted growth, and stunting, and releases a toxin that may deform the fruit.

Tarnished plant bug

taro *n.* (Polynesian *taro*, taro) **1.** a perennial herb (*Colocasia esculenta*, family Araceae) native to southeast Asia; grown in warm climates as an ornamental with decorative foliage and as a root crop. **2.** the large root tubers produced by the plant, used as a starchy food.

tarpan *n.* (Russian *tarpan*, tarpan) a small wild horse with a tan coat and black mane and tail; native to southern Russia. It became extinct in the early 20th century but has been simulated by selective breeding.

tarpon *n.* (Dutch *tarpoen*, tarpon) a large game fish (*Megalops atlantica*) with silvery scales that inhabits warm waters in the western part of the Atlantic Ocean. It grows up to 8 ft (2.4 m) long and weighs up to 200 lb (90 kg). Also called silverfish.

tarragon *n.* (French *targon*, tarragon) **1.** a bushy Old World perennial plant (*Artemisia dracunculus*, family Asteraceae) with long, narrow, aromatic leaves. **2.** the leaves of the French variety used as an herb for seasoning meats and vegetables. This variety does not produce seeds, so it is propagated vegetatively.

tarsier *n.* (French *tarsier*, tarsier) one of the lesser primates (*Tarsius* sp., family Tarsidae), a tiny mammal with big eyes and a long tail. It is a nocturnal tree dweller native to Indonesia and the Philippines.

Tasmanian devil *n.* a fierce nocturnal marsupial (*Sarcophilus harrisii*) that lives in open forest and eats carrion or preys on sheep and other animals; native to Tasmania. Its head and body are about 28 in. (0.7 m) long, and its tail is about 10 in. (25 cm) long; it has a black coat marked with white bands on its chest and rump.

tassel *n.* (French *tassel*, knob, knot, or button) **1.** the pollen-producing male part at the top of a corn plant. **2.** the hanging head or blossom at the top of some plants. **3.** a hanging ornament, often made with a clump of strings. *v.* **4.** to put forth tassels. **5.** to adorn with one or more tassels.

taste *n.* (French *taster*, to handle or taste) **1.** one of the five senses; the sensation of sweet, sour, bitter, or salty detected by the taste buds in the mouth; a component of flavor along with smell and texture.

2. a small sample of a food or other substance taken to test its flavor, or the characteristic of such a sample that is determined in the mouth. **3.** a characteristic that can be determined by the sense of taste, as a sweet taste. **4.** a preference that one likes to satisfy or express, as a taste for adventure. **5.** a small amount of anything, as a taste of politics or a taste of vengeance. **6.** the ability to evaluate, as a taste for good music. **7.** a lingering impression, as that experience left a bad taste in her mouth. *v.* **8.** to test something, originally to test by feel, but now mostly by taste in the mouth. **9.** to detect a taste sensation, as I taste salt in it. **10.** to have a new experience, as he will taste the adventure of skiing.

tautonym *n.* (Greek *tauto*, same + *onyma*, name) a Latin name that uses the same word for both genus and species; e.g., the black rat is *Rattus rattus*. This indicates a type species, but new names like this are no longer approved.

taw *v.* (Anglo-Saxon *tawian*, to prepare) **1.** to prepare a raw material for further processing, especially to treat an animal skin with alum, salt, etc., as part of the process of making it into leather. *n.* **2.** a fancy marble used for shooting in a game of marbles. **3.** the boundary circle or line in a game of marbles.

taxidermy *n.* (Greek *taxis*, order + *dermis*, skin) the art of preserving a dead animal in lifelike form, including stuffing and mounting.

taxis *n.* (pl. **taxes**) (Greek *taxis*, order) **1.** ordering or arrangement of items, as in a physical science. **2.** movement of a motile organism toward or away from a light or other stimulus. **3.** returning of a displaced body part to its proper place (e.g., a protruding hernia) without cutting any tissue.

taxon *n.* (Greek *taxis*, order + *on*, unit) a particular class of organisms at any categorical level, as a family or a genus.

taxonomy *n.* (Greek *taxis*, order + *nomia*, law) **1.** the science that deals with the methods and principles of classifying objects or living things. **2.** a system of classification; e.g., plant taxonomy or soil taxonomy.

TDN or **t.d.n.** *n.* See total digestible nutrients.

tea *n.* (Malay *teh*, tea) **1.** a small, long-lived, perennial shrub (*Camellia sinensis*, family Theaceae) grown in humid, semitropical countries. See silk oak. **2.** dried leaves from the plant; used to make a beverage by steeping the leaves in hot water. **3.** the beverage made from such tea leaves, often sweetened with sugar and consumed either hot or cold. **4.** any of several other usually hot beverages made with plant leaves; e.g., various herbal teas, mint tea, or maté.

teak *n.* (Portuguese *teca*, teak) **1.** a tall evergreen tree (*Tectona grandis*, family Verbenaceae) of south-eastern Asia. **2.** the very hard, durable, yellowish brown, resinous wood from the tree; used to make furniture and to build ships.

teal *n.* (Dutch *taling*, teal) **1.** any of several small ducks (*Anas* sp., family Anatidae) that frequent freshwater lakes, ponds, and marshes worldwide. These surface feeders have a mostly vegetarian diet. **2.** a dark grayish blue color that is characteristic of the wings and/or backs of many of these birds.

teart *n.* **1.** a soil or plant that contains an unusually high concentration of molybdenum. **2.** a form of molybdenosis, a disease of cattle on certain pastures in England caused by excessive intake of molybdenum, e.g., by being pastured on a teart soil. Symptoms include weakness and diarrhea.

teasel *n.* (Anglo-Saxon *taesel*, tease) **1.** any of several thistlelike plants (*Dipsacus sylvestris* and related species, family Dipsacaceae) with prickly leaves and bristly yellow or purple blossoms. **2.** the dried, bristly flower head of such plants; used to raise the nap on cloth. **3.** a mechanical device used to raise the nap on cloth. *v.* **4.** to raise the nap on cloth by rubbing it with a teasel. Also spelled teasle, teazel, or teazle.

teat *n.* (Anglo-Saxon *tit*, teat) a nipple or pap; a protrusion from the breast or udder of a female mammal that allows her young to suckle milk. **2.** a similarly shaped protrusion, such as a nozzle or the point of a drill.

technetium (Tc) *n.* (Greek *technētos*, artificial + *ium*, element) element 43, atomic weight 98; an unstable radioactive element with a half-life of 4.2 million years (or less, depending on the isotope) that is produced synthetically and recognized in certain stars. It is a silvery-gray metal that tarnishes slowly in moist air and is an excellent corrosion inhibitor for steel.

technology-inherent risk *n.* a potential problem associated with food safety and/or environmental effects of new technology; e.g., if a genetically modified organism is released into the environment.

technology-transcending risk *n.* political and social risks associated with new technology; e.g., the introduction of genetically modified organisms into a new setting.

tectonic *adj.* (Greek *tektonikos*, pertaining to construction) **1.** dealing with structures and the architectural and engineering principles used to build them. **2.** pertaining to the arrangement of rock layers in the Earth's mantle. **3.** relative to or produced by the forces and rock movements that occur in the crust of the Earth.

tectonic plate *n.* any of the large crustal masses that float on the underlying asthenosphere that makes up most of the Earth's mass. Six or more major plates are recognized (the African, American, Antarctic, Eurasian, Indian, and Pacific plates). Fault zones mark the boundaries of the plates, both on land and under the sea. Earthquakes occur, rocks fold, mountains and troughs form, and volcanoes may erupt when these plates shift locations.

teff *n.* (Amharic *t'ef*, teff) a fragrant grass (*Eragrostis tef*, family Poaceae) native to Ethiopia. It is grown as an ornamental and for its edible seeds.

tektite *n.* (Greek *tēktos*, molten + *ite*, rock) a glassy stone believed to have formed from the impact of a meteorite striking the Earth's surface.

tektosilicate *n.* (Greek *tēktos*, molten + Latin *silex*, flint + *ate*, salt) a framework silicate; a mineral such as quartz or feldspar with a three-dimensional linkage of silica tetrahedra.

telegraph plant *n.* a tropical Asian trefoil (*Desmodium motorium*, family Fabaceae) whose leaves move by noticeable jerks when in the sunlight.

tell *n.* (Arabic *tall*, mound) **1.** an archaeological site composed of layers of debris from structures of various ages, with the oldest layers on the bottom. See glacis. **2.** a story or tale.

tellurium (Te) *n.* (Latin *tellus*, earth + *ium*, element) element 52, atomic weight 127.60; a member of the oxygen-sulfur family, similar to selenium but heavier. It is used in ceramics, blasting caps, and certain alloys.

telmatology *n.* the study of marshes, swamps, and other wetlands and the habitats they provide. See wetland.

telotaxis *n.* (Greek *telos*, far + *taxis*, order) the ability of social insects such as bees and ants to move toward or away from a stimulus or to maintain a constant angle to the source of a stimulus, such as the sun.

temperate zone *n.* **1.** the two bands around the Earth between the tropics and the polar regions: between 23° 27' and 66° 32' of either north or south latitude, that is, between either the Tropic of Cancer and the Arctic Circle in the north or between the Tropic of Capricorn and the Antarctic Circle in the south. **2.** any area on Earth with a warm summer and a cold winter, where the mean annual temperature is less than 68°F (20°C) and the mean temperature of the warmest month is more than 50°F (10°C). Also called variable zone.

temperature inversion *n.* the reverse of the usual trend in the troposphere; increasing air temperature with height. A temperature inversion drastically reduces mixing of the air at different elevations.

tempest *n.* (Latin *tempestas*, season or storm) **1.** a violent wind, usually accompanied by precipitation (rain, sleet, hail, or snow). **2.** agitation, confusion, and uproar.

temporal *adj.* (Latin *temporalis*, of time) **1.** relating to time; temporary or transitory. **2.** pertaining to mortal life; worldly; e.g., temporal goods.

tendril *n.* (French *tendrille*, a little shoot) a slender, often twisting or coiling growth produced by certain climbing plants that allows them to attach to other plants or structures.

Tennessee Valley Authority *n.* a research center established in 1933 at Muscle Shoals, Alabama, to develop electricity, irrigation, flood control, and navigation of the Tennessee River and its tributaries. It later expanded to include the National Fertilizer and Environmental Research Center that was reorganized in 1993 as the National Environmental Research Center and renamed in 1994 as the TVA Environmental Research Center with five areas: Atmospheric Sciences; Agricultural Research and Practices; Waste Management and Remediation; Biotechnology; and Environmental Site Remediation.

tensiometer *n.* (Latin *tensio*, stretching + *metrum*, measure) **1.** a device that measures the tension in a wire, chain, beam, or other member being stretched. **2.** a device that measures the tension in a liquid; e.g., a device that measures soil moisture tension. A typical tensiometer has a porous ceramic cup at the end of a tube with a vacuum gauge on the side. The tube is buried in the soil, and water passes through the ceramic cup to equilibrate with the soil water. The tension read on the vacuum gauge with such a device cannot exceed the external atmospheric pressure, but it can indicate when to turn an irrigation system on or off.

tent caterpillar *n.* the larva of any of several moths (*Malacosoma* sp., family Lasiocampidae) that are serious pests found in North America from the Rocky Mountains eastward. The larvae gather in the forks of the limbs of various orchard and shade trees, spinning a large tentlike web that serves as a nest; they emerge during the day to feed on the foliage.

teosinte *n.* (Nahuatl *teocentli*, divine maize) a wild Central American plant (*Zea mexicana* = *Euchlaena mexicana*, family Poaceae) that is regarded as the probable principal progenitor of modern corn.

tepal *n.* a coined word used to indicate part of a flower that has not differentiated into discrete sepals and petals, e.g., in a tulip or lily.

tepary bean *n.* **1.** a twining or bushy bean (*Phaseolus acutifolius*, var. *latifolius*, family Fabaceae) grown by the Native Americans of southwestern U.S.A.

2. either of two species of small-seeded beans known as gram (*P. aureus* and *P. munga*) cultivated extensively in India and southeastern Asia.

tephra *n.* (Greek *tephra*, ashes) particulate airborne matter ejected from a volcano as part of an eruption that settles as dust, ash, cinders, lapilli, blocks broken from the cone, bombs of congealed magma, or pumice.

tequila *n.* (Nahuatl *Tuiquila*, Tequila, Mexico) **1.** an agave plant (*Agave tequilana*, family Agavaceae). **2.** a strong, distilled alcoholic beverage made from the sap of the plant. See mescal, pulque.

teratology *n.* (Greek *teratos*, monster + *logos*, word) the scientific study of malformed organisms.

terbium (Tb) *n.* (Swedish *Ytterby*, a Swedish village + *ium*, element) element 65, atomic weight 158.92534; a soft, malleable, ductile, silvery gray rare earth metal that occurs in various minerals along with other rare earth metals. It is used in lasers and solid-state devices.

tercel *n.* (French *tiercelet*, third little) a male falcon, especially one that is about a third smaller than a female of the same species; e.g., the common falcon (*Falco peregrinus*, family Falconidae) or other similar species. Also called tiercel.

terebra *n.* (Latin *terebra*, a borer) **1.** the boring body part used by a female hymenopterous insect to drill holes and deposit her eggs. **2.** a univalve marine organism with a turreted shell shaped like an auger (*Terebra* sp.).

teredo *n.* (Greek *terēdōn*, wood-boring worm) a wormlike bivalve marine organism (*Teredo* sp.) that bores into submerged wood such as ship hulls.

terminal moraine *n.* an accumulation of glacial drift, usually in the form of a line of hills that mark the outermost part of the area once covered by a major glacier. Also called an end moraine.

terminal velocity *n.* the velocity of a falling object when friction resulting from its rate of fall equals the gravitational attraction that causes it to fall; e.g., the terminal velocity of a 5-mm raindrop is about 30 ft/sec (9 m/sec).

terminus *n.* (Latin *terminus*, bound or limit) **1.** a boundary or something that marks a boundary, as a stone or other marker designated for that purpose. **2.** either end of a line of travel, as a bus, train, or airline terminal where the carrier reverses direction. **3.** an objective or goal.

termitarium *n.* (Latin *termitis*, wood borer + *arium*, a place for) a mound of cemented soil built by a colony of termites as their home. Also called a termitary. See termite.

termite *n.* (Latin *termitis*, termite) any of about 2000 species of soft-bodied, pale, antlike social insects (order Isoptera) that feed on wood, mostly in warm, humid areas. Some species are very destructive to buildings when they can find wood in contact with soil or can build a tunnel as a connection between the soil and

A soldier beside a termitarium

the wood frame of a structure. The eastern subterranean termite (*Reticulitermes flavipes*, family Rhinotermitidae) is probably the most damaging in the U.S.A. and is most common in the southern states.

tern *n.* (Danish *terne*, tern) any of a large group of slender aquatic birds (family Laridae) with pointed bills, long narrow wings, and forked tails. Terns are similar to gulls but are more slender. Most species are white with black markings, but a few are partly or entirely black. They eat mostly insects and small fish.

ternate *adj.* (Latin *ternatus*, in groups of three arranged in groups of three, especially three leaflets in a plant leaf.

terp *n.* a mound of earth and debris used to raise a building site above flood level, as in The Netherlands.

terpene *n.* (German *terpen*, turpentine) an organic compound containing between two and eight isoprene units linked either linearly or cyclically. Terpenes are major components of some essential oils that give plants characteristic odors or flavors; e.g., lemon oil, mint oil, caraway oil, camphor, and turpentine. Terpenes are also components of molecules that provide color (carotene and xanthophyll), vitamins (A, E, and K), enzymes (e.g., the coenzyme Q family), hormones (e.g., abscisic acid and the giberellins), etc.

terrace *n.* (Latin *terracia*, terrace) **1.** a relatively level raised area with a steep slope or a vertical drop-off on one or more sides and sloping upward to a higher area on another side; e.g., a stream terrace representing a former floodplain that rises above the current floodplain and extends to the valley wall or to the foot of another terrace. **2.** an earthen structure built to conserve soil or serve as a building site with a similar form to that of a stream terrace, often in a set of several terraces built on a hillside. **3.** an unroofed flat wooden or masonry structure (as a deck or porch) or a paved area extending from the side of a house or other building and standing above the surrounding grounds. **4.** the flat roof of a house, especially one that is used as a living or work area. **5.** a row of houses or a street on a line at or near the top of a hill.

Bench terraces on slopes as steep as 90%

terracing *n.* (Latin *terracia*, terrace + *ing*, the act of) **1.** the practice of building terraces for soil and/or water conservation. **2.** a set of terraces.

terrace interval *n.* either the vertical or the horizontal spacing between adjoining terraces, especially where one of these intervals is kept constant throughout a set of terraces.

terra firma *n.* (Latin *terra firma*, solid earth) firm dry land, as opposed to water, air, or loose solids that may shift and offer poor support.

terrain *n.* (French *terrain*, ground) **1.** a tract of land under observation, especially with regard to its physical features (slope, roughness, wet spots, etc.) and its suitability for a particular use. **2.** terrane. See terrane.

terra incognita *n.* (Latin *terra incognita*, unknown land) **1.** land that has not been explored and is therefore unknown. **2.** a field of knowledge that has not been explored and is therefore not understood.

terrane *n.* (French *terrain*, ground) a geologic rock formation or set of related formations, or an area where they can be found and studied. Also spelled terrain.

terrapin *n.* (Algonquian *torope*, tortoise) **1.** any of several edible turtles (family Emydidae or other similar families) with webbed feet, a horny beak, and a carapace covered with horny shield plates; native to fresh or brackish waters of North America, especially the diamondback terrapin (*Malaclemys terrapin*) of the Atlantic and Gulf Coasts. **2.** terrapin flesh used as food.

terrarium *n.* (Latin *terra*, earth + *arium*, place for) **1.** a small enclosure where small land animals are kept in a simulated natural environment. **2.** a small gardenlike display of plants grown in a glass enclosure.

terrestrial *adj.* (Latin *terrestris*, earthy) **1.** pertaining to things on Earth, as distinguished from any other planet or heavenly body. **2.** pertaining to things on land, as distinguished from a water environment. **3.** living on land, as distinguished from living in the water, in the air, or in trees. **4.** rooted in soil rather than rooted in water or living as a parasite.

terrestrial tide *n.* small changes in the absolute elevation of the Earth's surface caused by the gravitational attraction of the moon and the sun; similar to ocean tides but smaller.

terricolous *adj.* (Latin *terricola*, earth dweller) living in or on the soil; e.g., a mole or a cave dweller.

terrigenous *adj.* (Latin *terrigenus*, of earth) **1.** produced by the Earth, especially coming from dry land. **2.** formed of material that came from dry land; e.g., a layer of volcanic ash that settles to the sea floor and forms a deep sea deposit.

territoriality *n.* (Latin *territorialis*, of land + *itas*, condition) **1.** territorial status, quality, or condition; e.g., the territoriality of the area is sometimes established by conquest. **2.** the behavior of certain animals that mark a certain terrain as their own (e.g., a bear leaving claw marks on trees or a dog urinating to leave a scent) and defend it, especially against others of their own kind. **3.** the tendency of humans to designate land as belonging to a certain individual, group, or nation. Also called territorial imperative. **4.** claiming the right to be the decision maker in a certain type of matters; e.g., the territoriality of the legislative, executive, and judicial branches of government.

territorial waters *n.* the water area adjacent to a nation that is considered to be a part of that nation's jurisdiction. Traditionally, territorial waters are within 3 mi (4.8 km) of the coastline, but more recently they have been claimed out to 12 mi (19.3 km) by some nations and even to 200 mi (321.8 km) by a few nations.

territory *n.* (Latin *territorium*, surrounding land) **1.** an area of land; a region or district. **2.** the land area and territorial waters governed by a particular nation or ruler. **3.** a separate area under the control of a nation; e.g., the American colonies were established as territories of European nations. **4.** a part of a nation that does not have full statehood status; e.g., Puerto Rico is a territory of the U.S.A. that is now officially called the Commonwealth of Puerto Rico. **5.** a field, region, or sphere of action and responsibility, as a sales territory assigned to an agent. **6.** the area that an animal marks and defends as its own. **7.** a specialty; a topic or subject that is someone's expertise, as George Washington Carver made peanuts his territory.

tertiary treatment *n.* treatment to remove dissolved solutes (e.g., plant nutrients such as nitrates and phosphates) from liquid wastes by either chemical treatment of the effluent or application of the effluent to land so that it will filter through soil. Tertiary treatment follows primary treatment to remove materials that will float or sink and secondary treatment using microbes to oxidize much of the organic matter. See sewage treatment.

tesla (T) *n.* the international unit of magnetic flux density equal to 1 weber per square meter. Named after Nikola Tesla (1856–1943), a U.S. electrical engineer, physicist, and inventor, who was born in Croatia.

testa *n.* (Latin *testa*, covering) **1.** the outer coat or integument of a seed, usually a hard covering. **2.** the hard outer covering of certain animals, such as the shell of a mollusk or a sea urchin. Also called a test.

test animal *n.* a laboratory animal used to determine the relative toxicity of a pesticide or other toxin or for other health-related or environmental purposes. The data obtained are extrapolated to humans on a relative-weight basis plus a margin for safety.

testicle *n.* See testis.

testis (pl. **testes**) *n.* (Latin *testis*, testis) either of a pair of male reproductive glands (gonads) located in the scrotum that produce testosterone and sperm. Also called testicle.

testosterone *n.* (Latin *testis*, testis + *sterol* from cholesterol + Greek *one*, chemical) the principal male sex hormone ($C_{19}H_{28}O_2$) produced in the testes and responsible for the development and maintenance of secondary male characteristics, such as a deeper voice and more muscle mass and bone tissue than a neutered male or a female. It can also be produced synthetically.

tetanus *n.* (Greek *tetanos*, spasm) an infectious disease caused by bacteria (*Clostridium tetani*, family Bacillaceae) that can infect both humans and animals and produce a toxin. Symptoms appearing after about two weeks of incubation include muscle spasms and rigidity (especially in the lower jaw and neck), seizures, and respiratory paralysis. Also called lockjaw.

tetany *n.* (Latin *tetania*, spasm) a syndrome that occurs with faulty calcium metabolism (sometimes caused by a deficiency of vitamin D). It is characterized by sharp bending of wrists and ankles,

muscle twitching and cramps in the arms and legs and sometimes is accompanied by convulsions. See grass tetany.

tetraethyl lead *n.* a colorless liquid, $(C_2H_5)_4Pb$, used as an antiknock agent in gasoline for internal combustion engines until it was phased out because of its environmental toxicity to humans. See lead.

tetrahedral *adj.* (Greek *tetraedros*, four sided) arranged, shaped like, or pertaining to a tetrahedron.

tetrahedron (pl. **tetrahedra**) *n.* (Greek *tetra*, four + *hedron*, sides) **1.** a solid figure with four sides that are all equilateral triangles. **2.** an arrangement of four ions centered on the four points of a tetrahedron. An oxygen tetrahedron has enough space in the middle to hold a silicon ion (Si^{4+}) or to crowd in an aluminum ion (Al^{3+}). Such tetrahedra form the basic structural unit of silicate minerals, often with an octahedron formed where two tetrahedra occur side by side with their points in opposite directions. See octahedron.

A tetrahedral arrangement of four spheres with a tetrahedron inside formed by connecting the centers of the spheres

tetraploid *adj.* (Greek *tetra*, four + *ploos*, fold + *eidos*, form) having four haploid sets of chromosomes rather than the usual two sets. Tetraploid plants produced by doubling the chromosome number (e.g., by treating cells with colchicine) generally grow more slowly and have larger cells and larger blossoms than their diploid progenitors. See polyploid.

tetrazolium *n.* **1.** a complex ion ($CH_3N_4^+$) that reacts much like an ammonium ion (NH_4^+). **2.** a stain made from any of several tetrazolium compounds. If there is any metabolic activity, it acts as a terminal electron acceptor and is used to distinguish between living and dead tissue.

tetrose *n.* (Greek *tetra*, four + *ose*, sugar) a four-carbon sugar ($C_4H_8O_4$). There are four possible aldehyde forms (D and L forms of erythrose and threose) and two possible ketone forms, but these rarely occur in nature.

Texas longhorn *n.* a breed of rugged cattle that can survive the heat, insects, diseases, coyotes, wolves, and low-quality forage in the semiarid Texas environment. Texas longhorns have slender, twisted horns that span about 40 in. (1 m). They are descendants of cattle brought to Santo Domingo by Christopher Columbus in 1493 on his second trip and by Francisco Vasquez de Coronado (1510–1554).

Raised extensively, especially in Texas, in the 19th century, they became almost extinct in the 20th century before being rescued and increased again.

Texas longhorns

texture *n.* (Latin *textura*, weave) **1.** the process or technique of weaving. **2.** the physical character of a surface, especially with regard to repeated small-scale variations in height that can be detected by feel; e.g., the surface character of a fabric as influenced by fine or coarse threads and weaving pattern. **3.** the effect created by the size, shape, and arrangement of component parts, as the texture of a work of art, of a room, or of a rock. **4.** the distribution of particles sizes in a soil, especially when evaluated as percentages of sand, silt, and clay in a soil sample. See soil texture.

thallium (Tl) *n.* (Greek *thallos*, green stalk + *ium*, element) element 81, atomic weight 204.3833; a heavy metal in the same group as aluminum (Al). Usually present in lead ores, it is used in some alloys and in making rat poison.

thallium sulfate *n.* a poisonous chemical (Tl_2SO_4) used to kill rats and ants. Accidental poisoning of humans and animals is known as thallotoxicosis and can cause death.

thallophyte *n.* (Greek *thallos*, young shoot + *phyton*, plant) a primary division (Thallophyta) in some older plant classification systems consisting of a group of organisms with no differentiation into leaves, stems, and roots; including bacteria, algae, fungi, and lichens.

thallus *n.* (Greek *thallos*, young shoot) the body of a thallophyte; a plant with no leaves, stems, or roots.

thalweg or **talweg** *n.* (German *thal*, valley + *weg*, way) **1.** the line marking the lowest part of a valley; the line that joins the low points of all cross sections of a valley, whether under water or not. Also called valley line. **2.** a subsurface stream that follows the same general course as a valley above it. **3.** the center of the principal navigable channel of a river that is the border between two states.

Thamnidiacea *n.* a group of fungi (family Thamnidiacea, order Mucorales) that live mostly on dung. A few species are parasitic.

thanatology *n.* (Greek *thanatos*, death + *logos*, word) **1.** the study of death and its causes, e.g., in forensic medicine. **2.** the study of the effects of death and dying, especially with regard to alleviating the suffering of the terminally ill and those associated with them.

thanatophidia *n.* (Greek *thanatos*, death + *ophis*, snake) deadly serpents considered collectively.

thatch *n.* (Anglo-Saxon *thaec*, thatch or roof) **1.** a roof made from organic material, such as palm leaves, reeds, or straw, or the plant material used to make it. **2.** a dense layer of dead grass leaves, stems, and roots on the soil surface beneath the growing grass. **3.** a dense growth that resembles thatch, especially a thick growth of hair. *v.* **4.** to apply thatch to a roof. **5.** to remove thatch from a lawn. Also called dethatch.

theodolite *n.* (Latin *theodelitus*, theodolite) a survey instrument designed to measure horizontal and vertical angles as well as elevations. It was invented about 1571, probably by Leonard Digges, an English mathematician. See transit, dumpy level.

Theophrastus *n.* a Greek philosopher and natural scientist (372?–287? B.C.) credited with being the founder of botany for his writings relating plants to their environment.

therm *n.* (Greek *thermē*, heat) **1.** a unit of heat equal to 100,000 British thermal units (BTU). **2.** a unit of heat equal to a large calorie, 1000 large calories, or (rarely) a small calorie.

thermae *n.* (Greek *thermai*, hot baths) **1.** the public baths of the ancient Greeks and Romans. **2.** hot springs or hot baths.

thermal *adj.* (Greek *thermē*, heat + Latin *alis*, like) **1.** pertaining to heat and its effects or to thermae. **2.** serving to reduce heat loss from a body, as a thermal blanket. *n.* **3.** a rising air current caused by a warm area on the surface, especially when no cloud is formed.

thermal pollution *n.* the discharge of heated water from an industrial process or nuclear power plant when it causes adverse effects on plants, fish, animals, or humans by raising the water temperature in a stream or lake.

thermal spring *n.* a spring with a water temperature that is higher than the mean annual soil temperature. Also called a hot spring.

thermic *adj.* (Greek *thermē*, heat + *ikos*, caused by) **1.** thermal; pertaining to or caused by heat. **2.** a modifier used in Soil Taxonomy to name soil subgroups that have mean annual temperatures between 15° and 22°C (59° to 71.6°F); between mesic and hyperthermic.

thermocline *n.* (Greek *thermē*, heat + *klinein*, to recline) a water layer with a relatively steep temperature gradient (at least 0.67°F/ft or 1.2°C/m of depth) between the warmer epilimnion of mixed water above and the colder, stagnant hypolimnion in the bottom of a lake or deeper in the sea.

thermograph *n.* (Greek *thermē*, heat + *graphos*, recorder) **1.** a recording thermometer. **2.** an image based on temperature rather than light. It is used for the early identification of breast cancer, to sense the presence of warm-blooded animals or humans, to make a satellite picture of the heat radiated from the surface of the Earth, etc.

thermography *n.* (Greek *thermē*, heat + *graphia*, writing) the process or technique of making thermograph images.

thermokarst *n.* (Greek *thermē*, heat + German *karst*, karst) karstlike features (sinkholes, caverns, underground streams, etc.) in an area where localized melting has occurred in permafrost; e.g., where vegetation has been killed or a structure built. See karst, permafrost.

thermolysis *n.* (Greek *thermē*, heat + *lysis*, loosening) **1.** the dissipation of heat from a body. **2.** dissociation caused by heating a substance.

thermometer *n.* (Greek *thermē*, heat + *metron*, measure) an instrument for measuring temperature, usually based on the difference in expansion rates between a liquid and its glass enclosure, the bending of a bimetallic strip when the two surfaces expand at different rates, or the electrical conductivity of certain metal alloys in a thermistor. See Celsius scale, Fahrenheit, Kelvin scale, Rankine scale.

thermonuclear *adj.* (Greek *thermē*, heat + French *nucléaire*, nuclear) pertaining to or using a nuclear reaction that releases heat, as a thermonuclear power plant or a thermonuclear bomb.

thermophile *n.* (Greek *thermē*, heat + *philos*, loving) an organism that thrives at high temperatures (e.g., between 50° and 60°C or between 122° and 140°F).

thermophyte *n.* (Greek *thermē*, heat + *phyton*, a plant) a plant that grows best at relatively high temperatures; e.g., cotton.

thermotropism *n.* (Greek *thermē*, heat + *tropos*, turning + *ismos*, action) the action of an organism that responds to heat by moving or growing toward or away from the heat source.

theroid *adj.* (Greek *ther*, wild beast + *oid*, like) like a wild animal; beastly. See feral.

therology *n.* (Greek *thēr*, wild beast + *logos*, word) the scientific study of wild animals.

therophyte *n.* (Greek *théros*, summer + *phyton*, a plant) an annual plant; a plant that germinates from a seed, grows, matures, and produces seed in one growing season and survives any unfavorable season as seeds.

thicket *n.* (Anglo-Saxon *thiccet*, thicket) a thick growth of shrubs, bushes, and/or small trees; dense enough to resist passage.

thigmotropism *n.* (Greek *thigma*, touch + *tropos*, turning + *ismos*, condition) the response of a plant or animal to touching or being touched; e.g., a tendril wrapping around a support or the bursting open and scattering of seeds from the pod of an impatiens plant that is touched when ripe.

thiobacillus *n.* (Greek *theion*, sulfur + Latin *bacillum*, little rod) any of several species of rod-shaped autotrophic bacteria (*Thiobacillus* sp., family Pseudomonadineae) that use carbon dioxide as their sole source of carbon and obtain energy by oxidizing sulfur or sulfides; commonly found in soil, sewage, mine spoils, etc.

thixotrophy *n.* (Greek *thixis*, touching + *tropos*, turning) the characteristic of certain gels (e.g., some wet Vertisols or other soils high in clay) to change suddenly from a solid to a fluid by agitation or pressure, a serious problem when buildings or roads are constructed on such clays.

thorax *n.* (Greek *thōrax*, chest) **1.** the upper body cavity of a human, above the abdomen; the cavity that contains the heart, lungs, liver, etc. **2.** the corresponding part of an animal. **3.** the part of the body that encloses this cavity. **4.** the body segment of an insect where the legs are attached, between the head and the abdomen.

Thoreau, Henry David *n.* an early U.S. naturalist (1817–1862) who, in his book *Walden*, published in 1854, advocated protecting the environment against unrestricted human development

thorium (Th) *n.* (Norse *Thorr*, thunder + *ium*, element) element 90, atomic weight 232.0381; a radioactive heavy metal that is considered the source of much of the Earth's internal heat and can be used for nuclear power. Thorium dioxide (ThO_2) is used to make Welsbach incandescent mantles for gaslights.

Thornthwaite, Charles Warren *n.* U.S. climatologist and geographer (1899–1964) who devised a well-known system for classifying climates.

Thornthwaite's climatic classification *n.* a well-known system for classifying climates, originally published in 1831 by C. W. Thornthwaite, that divides the world into five zones based on vegetation and humidity, especially on ratios of precipitation to evaporation—wet rain forest, humid forest, subhumid grassland, semiarid steppe, and arid desert—and into six zones based on vegetation and temperature—tropical, mesothermal, microthermal, taiga, tundra, and frost.

threadworm *n.* any of several tiny roundworms (class Nematoda) that live in the small intestines of animals and humans and cause diarrhea, mostly in tropical and subtropical countries. The larvae are expelled in the feces, develop in the soil, penetrate the skin on contact, and are carried in the blood stream to the lungs, where they cause bleeding. Threadworms move from the lungs via the esophagus to the small intestines, where they may cause parasitic dysentery and anemia. See filaria, pinworm.

threatened species *n.* any species of plant or animal whose population is so sparse in some parts of its natural range that it is likely to disappear from those parts, but the existence of the species is not endangered because it is still adequately maintained in other places. See endangered species.

Three Mile Island *n.* an island in the Susquehanna River southeast of Harrisburg, Pennsylvania, that was the site of an accident in a nuclear-powered electric generating plant on March 28, 1979. Leakage of radioactive gases into the atmosphere led to a major re-evaluation of the U.S. nuclear energy program.

three-way cross *n.* the progeny from three inbred lines of the same species. Two inbred lines (e.g., of corn) are crossed to make a vigorous hybrid; this hybrid is then pollinated by the third inbred line to produce seed that farmers use to grow field corn.

thremmatology *n.* (Greek *thremmatos*, a nursling + *logos*, word) the branch of science dealing with the breeding and nursing of domestic plants and animals to develop and maintain desired characteristics.

thresher or **threshing machine** *n.* (Anglo-Saxon *threscan*, to thresh + *er*, doer) a machine that threshes ripe grain (especially the small grains, sorghum, dry peas, and other seeds of similar size), usually by passing it through a cylinder that knocks the grains loose from the head and then separates the grain from straw and chaff by passing it over a set of sieves with an air current blowing up through them.

threshing *n.* (Anglo-Saxon *threscan*, to thresh + *ing*, action) **1.** a beating. **2.** the process of separating ripe seed from the plant where it grew.

threshold *n.* (Anglo-Saxon *therscwold*, trample part) **1.** a door sill; the piece immediately beneath a closed door; the part one crosses to enter a building or room. **2.** any entry point, as the threshold of a new career. **3.** a minimum amount to cause something to be noticed, as the threshold of pain or pleasure. **4.** the minimum concentration of a particular pest to make it economically profitable to use a control measure.

thrips *n.* (Greek *thrips*, woodworm) any of about 500 species of tiny insects (order Thysanoptera) that damage almost any species of crop plants by sucking the plant sap. Some species are predators on mites, other small insects, and insect eggs.

thrive *v.* (Norse *thrifa*, to grasp) to grow and develop well; to prosper or flourish; as one hopes that the crop will thrive, the livestock will thrive, the children will thrive, and the business will thrive.

throughfall *n.* (Anglo-Saxon *thurh*, through + *feallan*, to fall) raindrops that fall through a canopy and strike bare soil, thus causing splash erosion, either without being intercepted or by falling from a high tree branch. See raindrop erosion, splash erosion.

throwaway *adj.* (Anglo-Saxon *thrawan*, to twist or turn + *aweg*, away) **1.** a consumer item that is designed to be used only once or for a short time and then discarded, as throwaway paper plates, etc. **2.** a person or group that uses throwaway goods, as we live in a throwaway society. *n.* **3.** a homeless child. **4.** a handbill or circular intended to be given away to as many people as possible. **5.** words, phrases, or sentences that can be eliminated because they don't mean much.

throwback *n.* (Anglo-Saxon *thrawan*, to twist or turn + *baec*, rear) **1.** the action of throwing something back, or something that is thrown back, as the small fish was a throwback. **2.** an organism that exhibits characteristics of an ancestral type, as a throwback to an earlier generation.

thrush *n.* (Anglo-Saxon *thrysce*, thrush) **1.** any of several small to medium-sized songbirds (*Turdus* sp., subfamily Turdinae, family Muscicapidae) that typically spend much time on the ground and migrate at night rather than in the daytime. See robin. **2.** any of several similar species that may be mistaken for a true thrush.

thuja *n.* (Latin *thuia*, thuja) any of several trees (*Thuja* sp., family Cupressaceae) with fragrant scale-like leaves, including giant arborvitae (*T. plicata*), also known as western red cedar, and arborvitae (*T. occidentalis*), a white cedar with leaves that contain a toxin. Thuja wood makes durable fenceposts, and there are many ornamental varieties.

thulium (Tm) *n.* (Latin *Thule*, Scandinavia + *ium*, element) element 69, atomic weight 168.934; a rare heavy metal of the rare earth group lanthanide series. It is a soft, ductile, malleable, silver-gray metal.

thunder *n.* (Anglo-Saxon *thunor*, thunder) **1.** the sound produced by the explosive expansion and contraction of air heated by a stroke of lightning. **2.** any loud sound resembling that caused by lightning. **3.** a loud, authoritative utterance, especially a denunciation. *v.* **4.** to make a noise like thunder. **5.** to utter a loud, forceful, often threatening, declaration.

thunderbolt *n.* (Anglo-Saxon *thunor*, thunder + *bolt*, blow or burst) **1.** a flash of lightning and the sound of thunder produced by it. **2.** a sudden, astounding event, as the news struck like a thunderbolt (or thunderclap).

thunderclap *n.* (Anglo-Saxon *thunor*, thunder + *claeppan*, to clap) **1.** a single booming sound of thunder. **2.** a sudden, startling occurrence.

thundercloud *n.* (Anglo-Saxon *thunor*, thunder + *clud*, mass of rock) a storm cloud that delivers thunder and lightning; usually a cumulonimbus cloud.

thunderhead *n.* (Anglo-Saxon *thunor*, thunder + *heafod*, hood or head) a mass of cumulus clouds that indicates a thunderstorm is coming.

thunderstorm *n.* (Anglo-Saxon *thunor*, thunder + *storm*, storm) a storm that includes thunder, lightning, and usually wind and rain; usually intense but of short duration.

thyme *n.* (Latin *thymum*, thyme) a tiny perennial shrub (*Thymus vulgaris* and related species, family Lamiaceae) with small, aromatic leaves that are used for seasoning foods. Dried leaves and flowers are used to provide scent in sachets, floral arrangements, and baths.

thymus *n.* (Greek *thymos*, thymus) an edible, butterfly-shaped, ductless glandular body at the base of the neck of an animal. It helps produce T cells for the immune system but gradually degenerates after puberty. The thymus of a calf or lamb is known as sweetbread when used as human food.

thyroxine or **thyroxin** *n* a hormone ($C_{15}H_{11}I_4NO_4$) produced in the thyroid gland that regulates general metabolic activity. A deficiency of iodine causes hypothyroidism and leads to cretinism (inadequate mental development) in children or young animals and to goiter in adults. Excess thyroxine causes hyperthyroidism.

thysanuran *n.* (Greek *thysanos*, fringe + *oura*, tail) a bristletail insect (order Thysanura). See bristletail, firebrat, silverfish.

tick *n.* (Anglo-Saxon *ticia*, tick) **1.** any of a large number of wingless, eight-legged (six-legged in the larval stage), bloodsucking parasites that bury their heads in the skin (especially hairy parts) of people, dogs, cats, cows, and most wild animals, often falling onto a person or animal when they brush against a bush or tree. Ticks are classified in two families: soft ticks (family Argasidae) and hard ticks (family Ixodidae). Some ticks transmit diseases, such as lyme disease and Rocky Mountain spotted fever. **2.** any of a group of degenerate two-winged parasitic insects. **3.** a light clicking sound, especially at regular intervals, as the sound of a watch or clock. **4.** a small mark, often used to indicate that an item has been counted or checked. **5.** a cloth covering filled with cotton or other soft substance to make a mattress or pillow. *v.* **6.** to make a clicking sound, especially at regular intervals, as a clock ticks. **7.** to mark with ticks. **8.** to list a group of items in sequence, as when one ticks off a list of priorities or accomplishments.

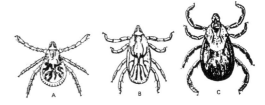

A larva (left), nymph (center), and engorged adult female (right) of the American dog tick (*Dermacentor variabilis*, family Ixodidae)

tick paralysis *n.* a progressive weakness (paralysis) beginning with the legs and progressing upward that has affected children and domestic animals. It is caused by a neurotoxin secreted by a female tick (*Dermacentor, Haemaphysalis, Ixodes,* and *Rhipicephalus* sp.).

tickseed *n.* **1.** any of several plants whose seed resembles a tick; e.g., coreopsis. **2.** tick trefoil.

tick trefoil *n.* any of several legumes (*Desmodium* sp., family Fabaceae) with trifoliate leaves, jointed pods, and stickers that cause the pods to adhere to wool, hair, or clothing.

tidal bore *n.* a wave that moves rapidly upstream in the lower reaches of a river; caused by a rapid rise in sea level at high tide, impacting a narrow river channel.

tidal day *n.* the interval between a high or low tide one day and the corresponding tide the next day at the same location; about 24 hr and 49 min. Also called a lunar day.

tidal flat *n.* a nearly level area that is covered by water at high tide but exposed at low tide; usually muddy or marshy. Also called tideland.

tidal interval *n.* the time between successive high tides (or between successive low tides), about 12 hr and 24 min.

tidal marsh *n.* a wetland that is subject to flooding by high ocean tides; an area at the ecological contact of saltwater, freshwater, and dry land, making it an environment conducive to great biological diversity.

tidal pool *n.* a basin near a sea coast that is covered with water at high tide and holds a pool of water when the tide goes out.

tidal power *n.* electricity generated from the power of incoming and outgoing tides. Water at high tide is caught in reservoirs, funneled through turbines that drive generators, and emptied into reservoirs that are drained at low tide.

tidal range *n.* the difference in elevation of the water surface between low and high tides, ranging from less than a foot (30 cm) in open sea to more than 50 ft (15 m) in a few enclosed bays.

tidal wave *n.* **1.** a tsunami. **2.** a large wave (similar to a tsunami) caused by strong winds blowing over a large area of ocean. **3.** (figurative) a widespread sentiment that becomes a strong public opinion.

tidal wind *n.* a gentle breeze in an inlet that blows onshore with rising tide and offshore with ebbing tide.

tide *n.* (Anglo-Saxon *tid*, time or season) **1.** the twice daily rise and fall of sea level (and of land elevation to a lesser extent) caused mostly by the gravitational attraction of the moon and strengthened by that of the sun when the sun, moon, and Earth are aligned (a spring tide) or weakened when they form a right angle (a neap tide). See tidal day, tidal interval. **2.** anything that rises and falls periodically. **3.** the peak or climax of something, as the tide of an illness or fever. **4.** a strong current or tendency, as the tide of public opinion.

tide gate *n.* a barrier that opens to let the tide flow one way but closes to prevent it from flowing the other way.

tideland *n.* (Anglo-Saxon *tid*, time or season + *land*, land) **1.** a tidal flat. **2.** a shallow area near the coast, within the territorial waters of a nation.

tiger *n.* (Latin *tigris*, tiger) **1.** a large predatory cat (*Panthera tigris*) with black stripes on a tawny coat; native to most of Asia. **2.** any of several other similar large cats, such as the cougar or jaguar. **3.** a bold, courageous person or animal. **4.** any of several other animals with stripes like those of a tiger. **5.** any of several ferocious fish, as certain sharks.

tiger lily *n.* **1.** an ornamental plant (*Lilium tigrinum*, family Liliaceae) with purplish black spots on its orange blossoms; native to China and especially popular in English gardens. **2.** any of several other similar lilies.

tiger moth *n.* any of several large moths with streaks on their wings that resemble the stripes on a tiger's coat.

tile drain *n.* a lined underground channel used to drain wet land; usually made of short lengths of baked clay or concrete tile (short pipes that are often called drain tile) or long lengths of corrugated plastic pipe with holes for water to enter. It is used mostly as part of a tile drainfield to remove water from the subsoil, but some tile drains have surface inlets to remove surface water, e.g., from the channel behind a terrace ridge.

tile drainage system *n.* a set of tile drains used to remove water and increase aeration in the soil, usually in a cropped field. The system pattern may be random (going from one wet spot to the next in a rolling area), regular (in a rectangular or other systematic pattern in a relatively level area), or interceptor (a single line across the slope). See tile drainfield.

tile drainfield *n.* **1.** a tile drainage system used to remove water and increase aeration in the soil of a large area (often an entire field). A design is needed to plan the pattern, depth, size, and gradient of the tile lines. Most installations are now made by trenching machines that use a rotating laser beam as a reference to control depth; older machines used a pointer on the machine and a cord stretched between posts to serve as a guide. See tile outlet. **2.** a similar set of tile lines used to dispose of the effluent from a septic tank by allowing it to seep into the soil.

tile outlet *n.* the place where a tile line empties into a ditch or pond. It generally requires a solid pipe about 20 ft (6 m) long that empties into the water or middle of a ditch so it will not erode the bank; the pipe generally needs to have a gate or grill on the end of it to prevent small animals from entering.

till *v.* (Dutch *telen*, to breed or cultivate) **1.** to prepare soil for planting a crop, usually by plowing or other stirring action that kills weeds and loosens the soil. **2.** to perform other types of cultivation; e.g., harrowing to smooth and compact plowed soil prior to planting, tillage to break a soil crust so seedlings can emerge, or emergency tillage to control wind erosion. **3.** to work, especially any farming operation (to till the land) or other form of nurture, as a missionary tills the field by preaching the gospel. *n.* **4.** a layer of unsorted soil and rock materials (stones, sand, silt, and clay in any proportion) deposited by the melting of a glacier. Also called glacial till or ground moraine. See moraine. **5.** a place where money or valuables are kept, especially a cash register.

tillable *adj.* (Dutch *telen*, to breed or cultivate + French *able*, able) suitable for tillage, as tillable land. Also called arable.

tillage *n.* (Dutch *telen*, to breed or cultivate + French *age*, pertaining to) any practice or system of manipulating the soil physically to make it more suitable for growing a crop, including plowing, harrowing, cultivating, subsoiling, etc.

tillage pan *n.* a compacted soil layer resulting from tillage, usually immediately beneath the plow layer. Also called a plow pan.

tiller *n.* (Anglo-Saxon *telgor*, twig or shoot) **1.** a new stalk that grows from the base of the original stem of a plant of the grass family. **2.** a young tree. **3.** a person who tills the soil; a farmer or gardener. **4.** a machine used to till the soil, especially one with rotating tines. **5.** a handle or lever, especially one used to steer a ship or vehicle. **6.** a small drawer, especially one used to hold money. *v.* **7.** to produce tillers from the base of a plant. A crop of wheat, barley, or other small grain will tiller if there is enough moisture and the stand is thin.

tillering *n.* (Anglo-Saxon *telgor*, twig or shoot + *ing*, making) the process of producing tillers from the base of a plant. Also called stooling.

tilth *n.* (Anglo-Saxon *tilth*, culture) **1.** the workability of a soil. A soil in good tilth is mellow, porous, and easy to till. **2.** the process of tilling land. **3.** the state of being tilled.

timber *n.* (Anglo-Saxon *timber*, timber) **1.** a stand of growing trees, especially one ready to be harvested and made into lumber. **2.** land covered by such trees. **3.** wood suitable for construction use, especially as beams for sturdy bracing or framing purposes. **4.** a wooden rib in a ship. **5.** the quality of a person, as a man of leadership timber. *v.* **6.** to fell trees (often with the warning call Timber!). **7.** to brace or support with sturdy beams.

timber beetle *n.* any of several species of beetles (e.g., *Lymexylon sericeum*) whose larvae bore into the wood of trees. Also called bark beetle.

timber grouse *n.* any of several grouse that inhabit wooded areas; e.g., the ruffed grouse.

timberline or **tree line** *n.* (Anglo-Saxon *timber*, timber + *line*, cord) **1.** the upper limit of tree growth on a mountain. **2.** the polar limit of tree growth in a cold climate.

timber wolf *n.* a large gray wolf (*Canis lupus occidentalis*) native to mountainous areas in northern North America; a predator that can attack cattle and other large animals.

time of concentration *n.* the time it takes for water that has fallen as raindrops or melted from snow to reach a gauging station on a stream.

time-of-travel (TOT) *n.* the length of time it takes for water in an aquifer to travel a specified distance, a concept used to determine the source water protection area around a well that needs to be monitored to avoid pollution.

time zone *n.* any of 24 bands wherein all clocks should indicate the same time, from the South Pole to the North Pole. Each time zone should theoretically be 15° wide, but the boundaries actually used are crooked because they are adjusted for social and political reasons. The Greenwich Meridian running through Greenwich, England, is arbitrarily taken as the zero point for longitude all around the Earth, and the International Date Line passing through the Pacific Ocean (180° from Greenwich) is used as the boundary between the new day to the west and the old day to the east of it as the midnight line for the other boundary sweeps around the Earth.

timothy *n.* a short-lived perennial forage grass (*Phleum pratense*, family Poaceae) named after Timothy Hanson who planted it in the Carolinas about 1720. Timothy was native to Europe but was first cultivated in the U.S.A. and was the most important hay grass in the U.S.A. during most of the 19th century. Timothy grows 2 to 4 ft (0.6 to 1.2 m) tall and has a dense cylindrical inflorescence that is commonly about 1 cm in diameter and 2 to 4 in. (5 to 10 cm) long.

tin *n.* (Anglo-Saxon *tin*, tin) **1.** Sn (from Latin *stannum*, tin), element 50, atomic weight 118.710; a ductile metal with a low melting point. It is used in tin plating, tinfoil, alloys, and soft solder. **2.** a shallow metal pan, especially one used in an oven for baking. **3.** a metal can used as a sealed container for storing sterilized food. *adj.* **4.** made of or covered with tin, as a tin can or a tin roof. **5.** cheap, worthless, or counterfeit, as tin jewelry. *v.* **6.** to coat with tin or with solder.

tinaja *n.* (Spanish *tinaja*, tinaja) **1.** a large ceramic vase used to hold water or other liquid, especially one that is unglazed so enough water seeps out to cool it by evaporation. **2.** a cavity in a rock that holds water after a rain.

tinamou *n.* (French *tinamu*, tinamou) any of several ratite birds (*Tinamus* sp., family Tinamidae) native to Central and South America; e.g., the great tinamou (*T. brasiliensis*), a bird about 18 in. (45 cm) long that inhabits forested areas in the northern part of South America.

tinder *n.* (Anglo-Saxon *tynder*, tinder) **1.** any easily flammable solid substance that can be ignited by a spark; e.g., dry, charred cloth or wood. *adj.* **2.** dry enough to ignite easily, as the woods are tinder dry this summer.

tinea *n.* (Latin *tinea*, a gnawing worm) **1.** any of several fungal skin diseases, especially ringworm. See ringworm. **2.** any of several small moths (*Tinea* sp., family Tineidae), including clothes moths (*T. pillionella*).

tineid *n.* (Latin *Tineidae*, moth family) any of several small lepidopterous moths (*Tinea, Scardia, Incurvaria, and Lampronia* sp.). See tinea, clothes moth.

tipburn *n.* a drying and browning of the tips and edges of leaves of potatoes, lettuce, and other plants as a result of hot, dry weather, insect damage, or disease.

Tiros *n.* any of a series of U.S. infrared observational weather satellites that take pictures of cloud cover. Tiros I was launched on April 1, 1960.

tissue analysis *n.* a laboratory test for determining the concentration of an element (usually a nutrient) in plant tissue. It can be compared to an optimum concentration as a basis for making a fertilizer recommendation.

tissue test *n.* **1.** a qualitative test for the presence of a nutrient dissolved in plant sap. It usually is used in the field to determine if the plant is adequately nourished, but is not recommended as a quantitative basis for making a fertilizer recommendation. **2.** tissue analysis.

titanium *n.* (Latin *titan*, the sun + *ium*, element) element 22, atomic weight 47.88; a metallic element that burns readily in air, forming titanium dioxide (TiO_2); the 9th most abundant of all elements in the crust of the Earth. It is used to make lightweight alloys with aluminum, molybdenum, manganese, iron, etc.

titmouse *n.* (Anglo-Saxon *tit*, little + *mase*, a bird) any of several small, crested songbirds (*Purus* sp., family Paridae) about 5 in. (13 cm) long with gray backs and off-white breasts often fringed with brown, especially the tufted titmouse (*P. bicolor*) in eastern U.S.A. and the plain titmouse (*P. inornatus*) in western U.S.A.

toad *n.* (Anglo-Saxon *tadige*, toad) **1.** any of several small, tailless amphibians (*Bufo*, sp., family Bufonidae) that resemble frogs but have warty skin and smaller, less powerful hind legs than frogs have. Toads eat insects and live on land most of the time but become aquatic during the breeding season. **2.** any of several other similar animals; e.g., frogs and certain lizards are sometimes called toads. **3.** an object of disdain or disgust, as that is just a toad (sometimes applied to a person).

toadstool *n.* (Anglo-Saxon *tadige*, toad + *stol*, throne or seat) **1.** a fruiting body with a short stalk and a rounded cap (resembling an umbrella) produced by a fungus (*Agaricus* sp. and related genera, class Basidiomycetes, order Agaricales). **2.** such a fruiting body that is poisonous to humans, as distinguished from edible forms known as mushrooms. **3.** any of several other fungal fruiting bodies, including puffballs and coral fungi.

tobacco *n.* (Spanish *tabaco*, tobacco) **1.** any of several plants (especially *Nicotiana tabacum* and wild *N. rustica*, family Solanaceae) with large leaves that are used to make tobacco products. Tobacco requires 40 to 45 in. (100 to 115 cm) of annual precipitation or irrigation equivalent and a frost-free period of at least 100 days. **2.** the dried leaves of the plant, either intact (e.g., for making cigars) or ground to make cigarettes or pipe tobacco. **3.** any of the products made from tobacco, or all such products collectively; e.g., tobacco consumption can cause cancer and is estimated to cause the deaths of 2.5 million people per year.

toddy palm *n.* any of several Asian palm trees (e.g., *Caryota urens* or *Borassus flabellifer*) with sap that is used to make a beverage.

toeslope *n.* a smooth, gentle slope between a footslope and a flat area such as a valley floor; the lowest part of the sloping deposit from the side of a hill or other higher area.

tokay *n.* (Malay *toke*, tokay) a relatively large gecko (*Gecko gecko*, family Gekkonidae) that grows as long as 14 in. (35 cm); native to southeastern Asia. It is sometimes kept as a pet.

tolerable soil loss (*T* value) *n.* the maximum rate of permissible soil erosion for a particular soil; the maximum rate that can occur without eventually decreasing the crop-production potential of the soil. It is usually taken as a value between 1 and 5 tons/ac (2 to 11 metric tons/ha) of annual soil loss, depending on soil depth, underlying material that may be converted into soil, etc., but some authorities consider these rates too high because of probable long-term effects.

tolerance *n.* (Latin *tolerantia*, tolerance) **1.** willingness to accept others even if they have characteristics and opinions that differ from one's own; an attitude of fairness, objectivity, and permissiveness toward others. **2.** the ability to endure harmful or disagreeable elements in the environment, as pesticide tolerance or pain tolerance. **3.** permissible limit, as the maximum legal concentration of pesticide residues on food or the allowable variation in a characteristic such as the weight, hardness, or a dimension of a product or part.

tolu *n.* **1.** a South American tree (*Myroxylon balsamum*, family Fabaceae) named after Tolú, Colombia. **2.** a fragrant yellowish brown balsam produced by the tree and used in medicines and perfumes.

tomatillo *n.* (Spanish *tomatillo*, little tomato) a plant (*Physalis ixocarpa*, family Solanaceae) with yellow blossoms and sticky blue berries that develop in a purple-veined calyx. Also called ground cherry or Mexican ground cherry.

tomato *n.* (Nahuatl *tomatl*, tomato) **1.** any of several bushy or viny annual plants (*Lycopersicon esculentum* and related species, family Solanaceae) with deeply serrated leaves; native to South America. It is a popular garden crop with smooth, round fruit that turns from green to red when it ripens. **2.** the red fruit produced by the plants that may be eaten raw, cooked as a vegetable, or processed to make catsup and other condiments. **3.** (slang) a pleasing young woman (can be either flattering or offensive).

ton *n.* (Anglo-Saxon *tunne*, a large vessel) **1.** a short ton; a unit of weight equal to 2000 lb (907.2 kg); the most common ton in U.S. usage. **2.** a long ton; a unit of weight equal to 2240 lb (1016.06 kg); commonly used in Great Britain. Also called a shipping ton. **3.** a metric ton; a unit of weight equal to 1000 kg (2204.6 lb). Also spelled tonne. **4.** a unit of volume used for ships, equal to 100 ft^3 (2.8317 m^3) of internal volume, 40 ft^3 (1.1327 m^3) of freight storage capacity, or 35 ft^3 (0.9911 m^3) of water displacement (approximately 1 long ton of seawater). **5.** (slang) a large amount of anything, as I have a ton of work to do.

tongue *n.* (Anglo-Saxon *tunge*, tongue) **1.** the movable, flexible muscular structure attached to the floor of the mouth in most vertebrates that aids in chewing, swallowing, the perception of taste, and making sounds. **2.** an animal tongue cooked for food. **3.** the ability to speak or the action of speaking. **4.** a language or dialect, as one's mother tongue. **5.** a movable, usually flexible, piece shaped like a tongue, as the tongue of a shoe. **6.** a pole or other long piece at the front of an implement or wagon used to attach one or more animals, a tractor, or some other vehicle to pull the implement or wagon. **7.** a narrow strip that extends outward from a larger area, as a tongue of land in a lake. **8.** an organ or part that resembles a tongue, as the proboscis or ligula of an insect.

tonka bean *n.* **1.** a tall tropical tree (*Dipteryx odorata*, family Fabaceae) with fragrant seeds; native to South America. **2.** the seeds from the tree. It is used as a substitute for vanilla to flavor candies and tobacco, as a perfume, and as a source of coumarin.

tonne *n.* (French *tonne*, ton) a metric ton (1000 kg). See ton.

tonoplast *n.* (Greek *tonos*, stetching + *plastos*, formed) the membrane that surrounds a vacuole in a plant cell and separates it from the cytoplasm.

toothache tree *n.* prickly ash (*Zanthoxylum americanum* and *Z. carolinianum*, family Rutaceae), a small tree or shrub with spines at the base of its aromatic, pinnate leaves. Its bark contains a bitter alkaloid that has been used as a tonic reputed to help a person feel better, forget a toothache, overcome rheumatism, etc.

topaz *n.* (Latin *topazus*, yellow topaz) **1.** an aluminum silicate ($Al_2SiO_4F_2$) that represents number 8 on Mohs' scale of hardness; orthorhombic crystals that may be white, colorless, or a transparent pale blue, yellow, or yellowish brown color. Yellow varieties are used as gemstones. **2.** certain other yellow gems or crystals that resemble topaz, as yellow sapphire or yellow quartz. **3.** either of two brightly colored hummingbirds (*Topaza* sp.) with long bills and tails; native to South America.

topdress *v.* to apply fertilizer on the soil surface, usually on a perennial crop such as pasture or hay or after an annual crop has started to grow.

topi *n.* (Swahili *topi*, topi) **1.** a large East African antelope (*Damaliscus lunatus*, family Bovidae) with a dark brown coat and bluish black and yellow markings. The topi is considered to be the fastest hoofed animal. Also called damalisk. **2.** a pith helmet, lightweight headgear worn for protection from the sun in tropical areas.

topographic map *n.* a map that shows the shape of the land surface, usually by means of contour lines.

topography *n.* (Greek *topographia*, topography) **1.** the shape of the land surface and the hills, valleys, streams, lakes, roads, bridges, etc., that are present. **2.** the science or technique of making topographic maps that show these features. **3.** a detailed description of a tract of land, estate, city, district, or other area. **4.** a detailed study of a place, system, or body part showing the size, shape, placement, and relationships of its component parts.

topology *n.* (Greek *topos*, place + *logos*, word) **1.** a mathematical study of the properties of geometric forms that maintain the same configuration even if bent or stretched. **2.** a topographical study of the shape of the surface of an area, object, entity, or concept, as the topology of the mind.

toposequence *n.* (Greek *topos*, place + Latin *sequens*, sequence) a group of soils that occur in the same landscape with differences related mostly to topographic position and its effects on drainage. Also called a catena if the parent material is the same for all of them.

topsoil *n.* (Anglo-Saxon *top*, top + Latin *solum*, soil) **1.** the upper part of the soil, generally darker and easier to work than the underlying subsoil; sometimes equated with the A horizon, the plow layer, or a mollic epipedon. **2.** any dark-colored soil material that is good for growing plants, especially if it is easy to work and suitable for placement over other material so a lawn, garden, or other plant cover can be established. *v.* **3.** to cover an area with topsoil. **4.** to strip the topsoil from an area.

tor *n.* (Anglo-Saxon *torr*, a tower or rock) a rocky peak or hilltop with little or no soil.

torch thistle *n.* a cactus (*Cereus* sp.) that was sometimes used by Native Americans to make torches.

torchwood *n.* **1.** any of several subtropical trees (*Amyris* sp., family Rutaceae) with dense, resinous wood. **2.** wood that contains enough resin to make a good torch.

tornado *n.* (Spanish *tronado*, thunderstorm) **1.** a violent funnel-shaped storm extending downward from a cumulonimbus cloud with strong winds up to 200 or 300 mph (320 to 480 km/h) whirling around an intense low-pressure center between two air masses moving in opposite directions or at very different speeds. The funnel is large at the top, with its narrow part extending down toward the ground. When it reaches the ground, its low pressure and high wind velocity can lift heavy objects, tear buildings apart, and leave a band of devastation that is typically a few hundred feet wide but sometimes is up to a mile (1.6 km) wide as it moves across the land, commonly at 25 to 40 mph (40 to 65 km/h). Sometimes called a twister; a similar storm over water is called a waterspout. See funnel cloud. **2.** (loosely) any whirling storm, especially if violent. **3.** a burst of activity or strong emotion.

torrent *n.* (Latin *torrens*, seething) **1.** a large, rushing flow of water. **2.** very heavy rainfall. **3.** a large flow of anything, as a torrent of tears, a torrent of mail, or a torrent of words.

torrential *adj.* (Latin *torrent*, seething + *ialis*, action) **1.** fast and violent, like a torrent. **2.** pertaining to a torrent. **3.** the results of a torrent, as torrential damage. **4.** overwhelming; copious. **5.** passionately emotional, possibly leading to violence.

torric *adj.* See soil moisture regime.

torrid *adj.* (Latin *torridus*, dried up) **1.** subject to intense heat from the sun, as a torrid desert. **2.** oppressively hot, as a torrid climate. **3.** very passionate, as a torrid love story or song.

torrid zone *n.* that part of the Earth's surface on both sides of the equator between the Tropic of Cancer and the Tropic of Capricorn. Also called the tropics.

tortoise *n.* (Greek *tartarouchos*, evil demon) **1.** a turtle, especially one that lives on land. **2.** a person or thing that moves very slowly. **3.** a testudo, a movable domed shelter used by Roman soldiers for protection from above.

This tortoise is about 10 in. (25 cm) long.

total digestible nutrients (TDN or **t.d.n.)** *n.* an evaluation of the energy value of livestock feeds obtained by summing the digestible protein, fiber, and carbohydrate plus fat times 2.25. Feed containing TDN equal to 0.75% of the live weight of an animal is recommended as a maintenance daily ration. One pound (454 g) of TDN is considered equal to 1.814 million small calories or 1.814 therms.

tote road *n.* an unpaved, temporary road used to haul logs or supplies.

toucan *n.* (Brazilian *tucano*, the sound of the bird's cry) **1.** any of several brightly colored, perching, fruit-eating birds (*Rhamphastos* sp., family Rhamphastidae) with large downward-curving bills; native to tropical America. **2.** Tucana, a constellation in the southern hemisphere.

tourmaline *n.* (Sinhalese *tōramalli*, carnelian) a trigonal cyclosilicate mineral containing aluminum, boron, silicate groups (Si_6O_{18}), hydroxyl ions, and variable amounts of sodium, magnesium, iron, and other elements. Tourmaline is usually black, but iron-free varieties can be colorless, blue, green, or pink, and some are used as gems and in optical instruments.

township *n.* **1.** an area 6 miles (9.65 km) square divided into 36 sections of 640 acres each (with some adjustments as needed at correction lines). Since the rectangular survey system was adopted in 1875, the U.S. General Land Office has surveyed townships in tiers or townships east or west of a principal meridian and ranges north or south of a baseline in 30 of the 50 states; a similar system is used in much of Canada. **2.** a unit of local government consisting of one or more survey townships or subdivisions thereof (common in eastern and midwestern states of the U.S.A. and in Canada). **3.** a small town and its surrounding district in England or Australia. **4.** an area designated for nonwhites in the Union of South Africa when it was segregated.

toxalbumin *n.* (Latin *toxicum*, poison + *albus*, white + *in*, chemical) any of several toxic high-molecular-weight albumin proteins produced by certain pathogenic bacteria and plants (abrin, ricin, phallin) and found in snake venom.

toxaphene *n.* an amber chemical ($C_{10}H_{10}Cl_8$) that is insoluble in water; used as an agricultural pesticide to kill insects or rodents.

toxic *adj.* (Greek *toxikon*, poison) **1.** poisonous (originally the toxicant used on poisonous arrows). **2.** pertaining to poisons or their effects, as toxic conditions.

toxicant *n.* (Latin *toxicum*, poison + *antem*, agent) a poison; any agent that injures or kills plants, animals, or humans, e.g., by interfering with the action of an enzyme, reacting with a cell to block its proper function, or some secondary action such as causing the body to activate a defense system. Note: Almost any substance becomes poisonous at some concentration; a toxicant has negative effects at a relatively low concentration.

toxic equivalent *n.* the amount of poison per unit of body weight necessary to kill a human or animal.

toxicity *n.* (Latin *toxicum*, poison + *itas*, condition) the ability to damage or kill, or the degree of expression of that ability. Toxicity may be described as follows:

- acute toxicity: injuring or killing after a single exposure;
- chronic toxicity: injury or death following long-term exposure;
- dermal toxicity: able to pass through unbroken skin and cause injury or death;
- oral toxicity: causing injury or death if swallowed.

toxicology *n.* (Greek *toxikon*, poison + *logos*, word) the scientific study of toxic substances, their effects, methods of detection, antidotes, and related topics.

toxicosis *n.* (Greek *toxikon*, poison + *osis*, condition) any diseased condition caused by poisoning. Exogenic toxicosis is caused by the ingestion of toxicants; retention toxicosis is caused by failure to excrete noxious wastes from the body.

toxic release *n.* any accident or intentional action that allows toxic material to escape into the air, water, soil, or other part of the environment where it may damage plants, animals, or people.

toxic residue *n.* any poisonous substance left on the surface of a plant, animal, or object or in soil or water

following the application of a pesticide or other material.

toxic waste *n.* any waste product that contains one or more toxicants such as heavy metals, pesticide residues, or radioactive substances.

toxin *n.* (Greek *toxikon*, poison) any organic toxicant produced by an organism; generally a large molecule that causes the immune system of a person or animal to produce an antibody.

toxoid *n.* (Greek *toxikon*, poison + *eidēs*, resembling) a toxin that has been treated so it can no longer cause a disease but still retains its ability to stimulate the production of antibodies and generate immunity when injected in the body of an animal or human.

trace *n.* (Latin *tractus*, drawn) **1.** any surviving sign or evidence of something that has happened; e.g., something was spilled, an animal or person passed through, or someone built something or developed a concept. **2.** a very small amount of something that is detectable but often too small to measure accurately; e.g., a trace of precipitation is often defined as less than 0.005 in. (0.127 mm). **3.** a faintly discernible expression of a quality, as a trace of irony in his story. **4.** a faint trail or path. **5.** a line drawn by a recording instrument, as a trace of the daily temperature. **6.** the line on a screen indicating the passage of a subatomic particle. **7.** either of the two long straps that attach a harnessed animal to a wagon or implement. **8.** a bar or rod that connects two moving parts with hinges or pins at each end (e.g., a piece that allows a wheel or gear to drive a shaker). Sometimes called a connecting rod or con rod. *v.* **9.** to follow a trace left by an animal or person. **10.** to follow or describe the development of a group, invention, or concept; e.g., to trace it to its origin. **11.** to copy an image by placing a transparent sheet over it and duplicating its lines. **12.** to mark a route, as to trace one's way on a map. **13.** to sketch, especially with faint lines. **14.** to form letters slowly and carefully. **15.** to record data by drawing a line on a graph, as a seismograph traces an earthquake. **16.** to outline, as when one traces out a plan to be developed in detail later.

trace element *n.* See micronutrient.

tract (Latin *tractus*, drawing out or extent) *n.* **1.** an area of land, water, or mineral deposit; e.g., a housing tract or a mining tract. **2.** a part of the body, as the digestive tract. **3.** an interval of time. **4.** a pteryla; a site on a bird's skin where a feather grows. **5.** a small booklet, especially one dealing with a religious or political topic.

trade wind *n.* a wind that blows steadily toward the equator from the northeast in the tropics north of the equator and from the southeast in the tropics south of the equator.

trafficability *n.* (French *trafic*, traffic + Latin *abilitas*, ability) **1.** the capability of the soil or terrain to bear traffic. Trafficability often varies according to moisture conditions; loose, dry dune sand or wet soils high in clay may not support wheeled vehicles. **2.** the ability to maneuver a vehicle, especially over difficult terrain. **3.** the market potential of a product.

tragacanth *n.* (Latin *tragacantha,* a goat thorn) **1.** goat's thorn, a plant (*Astragalus tragacantha*, family Fabaceae) that yields a gummy juice. **2.** any of several related plants (*Astragalus* sp.). **3.** the white or reddish gum produced from the juice of *Astragalus* plants and used to make pharmaceuticals, to print calico, and to stiffen certain kinds of cloth.

trail *n.* (French *trailler*, to drag along) **1.** a pathway for people, animals, or light vehicles; e.g., a bicycle trail, especially one that crosses remote or rough terrain. **2.** the track, scent, or other mark left by the passage of a person, animal, or vehicle. **3.** something that follows behind, as the long train of a wedding dress; a line of people, animals, or vehicles; or a cloud of dust or smoke. *v.* **4.** to follow a trail, either an established pathway or the marks left by the passage of a person or animal. **5.** to follow behind something that leads the way, especially to follow at a considerable distance (to trail behind). **6.** to hang down loosely and/or drag behind. **7.** to show where something has passed, as condensed moisture (a contrail) trails behind an airplane or dust trails behind a moving car. **8.** to gradually change course, as to trail to the left. **9.** to gradually lose strength or volume, as a voice trails down to a whisper. **10.** to grow laterally, as a vine trails across the ground.

trailblazer *n.* **1.** a pathfinder, one who removes or cuts away obstacles and marks a path for others to follow, e.g., to cross a wilderness. **2.** one who begins the study of a new area in science or does something new in some other human endeavor; e.g., a trailblazer in computer design.

trailing arbutus *n.* a sweet-smelling, vinelike plant (*Ericacea repens*, family Ericaceae) with leathery oval leaves and clusters of pink or white flowers; native to northeastern U.S.A. and Canadian woods. It is a suffruticose plant.

trailing fuchsia *n.* a small shrub (*Fuchsia procumbens*, family Onagraceae) with long-stalked leaves and trailing orange and purple blossoms; native to New Zealand. It often is used in hanging baskets.

trait *n.* (Latin *tractus*, a stretch of space or time) **1.** a characteristic that distinguishes or identifies the personality of a person or animal, as a behavioral trait. **2.** a stroke or line drawn with a pen or pencil.

trample *v.* (German *trampeln*, to tramp on) **1.** to flatten by walking, as to trample the grass. **2.** to violate or break down, as to trample the rights of helpless people.

tranquil *adj.* (Latin *tranquillus*, quiet, calm) **1.** serene, placid, free from disturbance; not noisy, agitated, or emotional. **2.** quiet; not moving, as tranquil water.

transcription *n.* (Latin *transcriptio*, copying) **1.** the process of making a full copy; e.g., the transcription of notes into a typed copy. **2.** the conversion of a musical score for use by a different medium or instrument. **3.** the copy produced by any such process. **4.** the process that forms a strand of RNA as the complement of a segment of DNA.

transduction *n.* (Latin *trans*, across + *inductio*, inducing) **1.** the process of transferring energy from one system to another. **2.** the transfer of genetic material from one bacterium to another by a bacteriophage.

transect *n.* (Latin *trans*, across + *sectus*, cut) a line or cross section across an area being studied to describe the area, make a survey, or make an environmental assessment.

trans fat *n.* transformed fat produced when vegetable oils are hydrogenated to prevent spoilage and to change them into solids or semisolids, as in stick margarine and shortening. Trans fat comprises up to 10% of the calories in a typical U.S. diet.

transferase *n.* (Latin *trans*, across + *ferre*, to carry + *ase*, enzyme) any enzyme that catalyzes the transfer of a functional group from one substrate to another. See enzyme.

transfer RNA (tRNA) *n.* a small RNA molecule that functions in protein synthesis by picking up a free amino acid from the cytoplasm and transferring it to a ribosome that is generating a protein molecule.

transformation *n.* (Latin *transformatio*, change of shape) **1.** the process of changing the form, appearance, character, or disposition of a being or object; e.g., a change of costume or scenery in a dramatic presentation or the metamorphosis of an insect. **2.** the state of the being or object after such a change. **3.** a mathematical manipulation of a set of numbers or symbols. **4.** the transfer of genetic material from a donor cell to a recipient cell. **5.** a pathological change in tissue that affects the function or use of the tissue or part. **6.** a change of matter from one state (solid, liquid, or gas) to another. **7.** a change from one form of energy to another, as a transformation from potential energy to kinetic energy or heat energy. **8.** the conversion of one form of electricity to another; e.g., a change in voltage.

transgene *n.* (Latin *trans*, across + *genes*, born) a gene that has been inserted into a chromosome by recombinant DNA technology.

transgenic *adj.* (Latin *trans*, across + *gene*, stock + *icus*, of) having one or more genes that have been inserted by recombinant DNA technology, whether the inserted genes come from another species, the same species, or even the same plant.

transhumance *n.* (French *transhumer*, to move livestock seasonally + Latin *antia*, act or process) the seasonal movement of livestock and herders between seasonally distinct pastures; e.g., using mountain pasture in the summer and lowland pasture in the winter or dryland pasture during favorable periods and irrigated pasture or hay during dry periods. See nomad.

transit *n.* (Latin *transitus*, gone across) **1.** a passage or crossing, as he made a transit across the area. **2.** transportation, especially a public conveyance, as a city transit system, or the movement of goods or people that are in transit. **3.** a change or transition. **4.** the passage of one heavenly body relative to another, as the transit of Venus through one's field of view or between Earth and the sun. **5.** a survey instrument with a telescope mounted on a tripod that can be used to measure horizontal and vertical angles as well as elevations (a form of theodolite). See dumpy level, theodolite.

transitory *adj.* (Latin *transitorius*, fleeting) brief; fleeting; passing quickly; not lasting or enduring.

translocation *n.* (Latin *trans*, across + *locatio*, placing) **1.** a change in the location of a material or thing, as the translocation of a nutrient, a sugar, or a pesticide in a plant or the movement of a factory or business to a new location. **2.** the transfer of a segment of genetic material to a new site, often on another chromosome.

translucent *adj.* (Latin *translucens*, shining through) allowing light to pass through in a diffused form that obscures images on the other side; e.g., frosted glass is translucent but not transparent.

transmembrane *adj.* (Latin *trans*, across + *membrana*, membrane) occurring across a membrane, as the passage of ions or the development of an osmotic or electric potential.

transmigrant *n.* or *adj.* (Latin *transmigrans*, migrating across) one who transmigrates.

transmigrate *v.* (Latin *transmigrare*, to migrate across) **1.** to leave one place, cross an area, and settle in a new location. **2.** to emigrate from one country and immigrate to another.

transmission *n.* (Latin *transmissio*, sending across) **1.** an act of sending or conveying something across a distance or to another being or place. **2.** the thing, entity, or message that is so sent or conveyed. **3.** a mechanism that transmits power and movement in a machine, especially one that can change velocity. **4.** an electromagnetic radio or television signal that carries a message.

transmittance *n.* (Latin *transmittere*, to send across + *antia*, act or process) the ability to convey or to allow something to pass through; e.g., the percentage or fraction of light that passes through a piece of glass or the amount of radiation that passes through the Earth's atmosphere.

transonic or **transsonic** *adj.* (Latin *trans*, across + *sonus*, sound + *icus*, of) near the speed of sound. See sonic boom, sound barrier.

transparent *adj.* (Latin *transparens*, appearing through) **1.** allowing light to pass through without distortion, as opposed to opaque and as distinguished from translucent. **2.** sheer, open enough to permit objects to be seen through the material or object. **3.** easily understood; not deceiving.

transpiration *n.* (Latin *transpiratio*, breathing across) **1.** the evaporation of water from a plant through stomata, mostly from the leaves. **2.** sweating, the loss of water through the pores of the skin of humans and animals. **3.** the passage of a gas under pressure through a capillary tube or a porous material.

transpiration coefficient or **transpiration ratio** *n.* the number of units of water transpired by a plant to produce one unit of dry matter. The water transpired by most plants weighs 200 to 1000 times as much as the dry matter they produce.

transpiration efficiency *n.* the amount of biomass produced by a plant that transpires a given amount of water (commonly expressed as grams of biomass per kilogram of water). Called evapotranspiration efficiency when water loss by evaporation is included in the calculation.

transplant *v.* (Latin *trans*, across + *plantare*, to plant) **1.** to move something (e.g., a tree or a plant) to a new location and plant it there. The best time to transplant depends on the species, but it is often in the late fall because the roots can grow and establish themselves for absorbing water and nutrients while the top part of a plant is dormant and therefore does not need much water. **2.** to surgically implant an organ (usually from the body of an accident victim) into the body of a person or animal. **3.** to move a person, family, or group to a new location, usually to either a different country or a contrasting environment. *n.* **4.** a plant, animal, organ, or person that has been transplanted. **5.** the process or act of transplanting.

transported soil *n.* soil that is not residual; soil that has been moved to its present location by gravity or by a transporting agent such as water, wind, or ice.

transport protein *n.* a plasma membrane component that controls the movement of a specific ion or molecule through the membrane.

transposition *n.* (Latin *transpositio*, a change of position) **1.** an action that changes the position of something; e.g., the transfer of a gene from one chromosomal site to another. **2.** the state produced by such a transfer, or the new form so produced. **3.** a change in an image; e.g., the reversal from a negative to a positive image. **4.** the mathematical interchange of two elements or groups of elements in a set.

transposon *n.* (French *transpose*, change position + *on*, elementary particle) a segment of DNA (e.g., a gene) that can move from one chromosomal site to another, sometimes producing a new genetic effect. See jumping gene.

trash *n.* (Norwegian *trask*, rubbish) **1.** discarded material, refuse, rubbish, unwanted trimmings, etc. See solid waste. **2.** empty, foolish, or offensive words, ideas, speech, or written material. **3.** disreputable persons, individually or collectively. **4.** the refuse left from sugar cane after the juice has been squeezed out. **5.** a hindrance or constraint; e.g., a leash and collar on a dog. *v.* **6.** to discard something, as to trash the branches trimmed from a tree. **7.** to vandalize a place, making it look trashy by breaking and scattering items. **8.** to criticize a person, group, or idea with a derogatory evaluation. **9.** to strip away the outer leaves of sugar cane in preparation for harvest. **10.** to hinder or restrain, as with a leash.

trashy fallow *n.* fallow with plant residues left on the surface and/or protruding from the surface, e.g., by tillage that cuts beneath the surface without inverting the soil.

travertine *n.* (Italian *travertino*, an Italian limestone) light-colored calcium carbonate ($CaCO_3$) deposited by evaporation of water, e.g., at the mouth of a hot spring or as dripstone in limestone caves forming stalactites and stalagmites. Porous forms are called tufa, sinter, or spring deposit; denser metamorphosed varieties are called onyx marble and can be polished for use as an ornament. See karst, stalactite, stalagmite.

treacle *n.* (Latin *theriaca*, antidote) **1.** a medicine formerly used as an antidote for poisonous bites of venomous beasts. **2.** molasses, especially that drained from vats where sugar has been refined. **3.** feigned or unrestrained sentimentality.

tree *n.* (Anglo-Saxon *trēow*, tree or wood) **1.** any woody perennial plant more than 10 ft (3 m) tall and supported by a single stem. A group of trees covering an area may be called a grove, a woodlot, or a forest; timber consists of trees of commercial size and quality. **2.** (loosely) any large bush or plant (e.g., a banana plant). **3.** something with a support and crosspieces that resemble the branches of a tree; e.g., a clothes tree. **4.** a diagram with branches in its design, as a family tree or a data tree. **5.** a support of some kind, as a shoe tree or a beam in a structure. *v.* **6.** to drive an animal up a tree. **7.** to put a person in a difficult situation.

tree farm *n.* a commercial enterprise that grows trees on a sustained basis, replacing harvested trees with new plantings, usually on a rotational basis so that timber, Christmas trees, or other tree products are produced continuously.

tree fern *n.* any of several mostly tropical ferns (family Cyatheaceae) that have foliage at the top of an upright stem as tall as a tree. One attractive species (*Cyathea medullaris*) provides a mucilaginous pulp that is used for food in Polynesia and New Zealand.

tree frog *n.* any of several frogs or toads that live in trees (mostly family Hylidae). Tree frogs have adhesive disks at the ends of their toes.

tree-of-heaven *n.* See Chinese sumac.

tree ring *n.* an annual growth ring as seen in a cross section of a tree. Each year the tree grows a relatively soft layer of wood with large cells in the spring and a denser layer with small cells in the summer.

tree sparrow *n.* **1.** a small Eurasian sparrow (*Passer montanus*) related to the house sparrow but smaller; common in Great Britain. **2.** American tree sparrow (*Spizella arborea*, family Embarizinae). It frequents thickets and weedy fields during the winter in most of the U.S.A. and migrates into Canada and Alaska in the summer. This sparrow is marked with a rusty cap and a dark breast spot.

tree squirrel *n.* a true squirrel (*Sciurus* sp.) that has a long, bushy tail and lives mostly in trees; common in wooded areas and parks.

tree-tip mound *n.* a knoll of soil material deposited from the root system of a fallen tree. See cradle knoll.

tree-tip pit *n.* a depression left where the soil was lifted by the root system of a fallen tree.

trefoil *n.* (French *trifoil*, triple leaf) **1.** any of several leguminous plants (*Trifolium* sp., family Fabaceae) with trifoliate leaves and small purple, yellow, or white blossoms, including the common clovers; widely used as forage crops. **2.** any of several similar plants. **3.** a design with three lobes used for ornamentation or for an emblem; e.g., the official emblem of the Girl Scouts of America.

trehalose *n.* an edible nonreducing disaccharide sugar ($C_{12}H_{22}O_{11}$) present in yeasts, lichens, and fungi such as mushrooms; used to identify certain bacteria.

trematode *n.* (Greek *trematoda*, having holes) any of several parasitic flatworms (flukes of class Trematoda), including internal and external human and animal parasites. See fluke.

trembles *n.* (Latin *tremulus*, trembling) poisoning of sheep and cattle by eating plants, such as white snakeroot in eastern U.S.A. or jimmyweed in southwestern U.S.A., that contain tremetol, a toxic substance ($C_{13}H_{14}O_2$) that can cause muscular tremors. Consuming milk, butter, or flesh from affected animals causes potentially fatal milk sickness in humans.

tremetol See trembles.

tremor *n.* (Latin *tremor*, a trembling) **1.** a trembling, shaking, or quivering motion of a body or limbs, often caused by disease, fear, or debility. **2.** a shaking of the ground; an earthquake or aftershock. **3.** a quavering of the voice caused by nervousness, uncertainty, or excitement.

trench *v.* (French *trencher*, to cut) **1.** to cut into, especially to open a channel or ditch. **2.** to extend into, as to trench upon another's land or rights. *n.* **3.** a long, narrow cut, as a deep furrow or a ditch. **4.** such an excavation made deep enough to protect troops in a war. **5.** a deep valley, especially one in the ocean floor. **6.** something that resembles a furrow or trench, as a deep wrinkle in one's face.

trench plow *n.* a large plow that opens a deeper furrow than most moldboard plows; used to make ditches or for deep plowing.

trench silo *n.* a horizontal silo consisting of a trench excavated into a sloping area where excess water will not be a problem, usually lined with wood or concrete walls and floor. It commonly is 6 to 8 ft (1.8 to 2.4 m) deep, 15 to 25 ft (4.5 to 7.5 m) wide, and as long as needed to provide the desired capacity. Material to be ensiled is placed inside and covered with plastic or canvas (often with a layer of soil on top) to seal the enclosure.

trend *v.* (Anglo-Saxon *trendan*, to turn) **1.** to move, incline, or extend in a certain direction, as to trend toward the south or toward the right. **2.** to have a certain tendency, as their opinion or policy trends toward liberalism. *n.* **3.** the general direction that something moves or extends, as the trend of a coastline. **4.** the general course of a conversation, discussion, or policy or the course of events. **5.** the current style in fashion.

treponema (pl. **treponemae**) *n.* (Latin *treponema*, a turning thread) a group of anaerobic spirochetes (*Treponema* sp., family Treponemataceae), including several that are parasitic pathogens on humans or animals, such as those that cause treponematosis.

treponematosis or **treponemiasis** *n.* (Latin *treponema*, a turning thread + *osis*, disease) any of several infectious diseases caused by treponemae, including bejel, pinta, syphilis, yaws, etc.

tribe *n.* (Latin *tribus*, tribe) **1.** one of the three groups of ancient Romans, of Latin, Etruscan, or Sabine origin. **2.** a close-knit community of ancient or primitive people who share a common ancestry; usually a group of related clans; e.g., any of the 12 tribes of ancient Israel. **3.** any definable group of people who have the same occupation and similar ideas, often used in a derogatory sense, as a tribe of troublemakers or idiots. **4.** a similar group of plants or animals used in some taxonomic systems. **5.** a group of livestock that are all descended from one female ancestor through continuous female lines. **6.** (slang) a large family.

tribulosis *n.* (Latin *tribulare*, to afflict + *osis*, condition) a disease of sheep caused by feeding on puncturevine (*Tribulus terrestris*, family Zygophyllaceae), especially in South Africa. Also known as bighead disease.

tributary *n.* (Latin *tributarius*, of tribute) **1.** an incoming stream; a stream that flows into a larger stream or body of water. **2.** a person, group, or nation that pays tribute to another. *adj.* **3.** flowing into a larger body. **4.** subsidiary to another; contributing support. **5.** paying or paid as tribute. **6.** obliged to pay tribute.

trichina *n.* (Greek *trichinos*, hairy) a very small parasitic nematode (*Trichinella spiralis*). The adults live in the intestines and produce larvae that encyst in the muscles of humans, pigs, or rats, causing trichinosis.

trichinosis *n.* (Greek *trichinos*, hairy + *osis*, disease) a disease caused by trichina nematodes that infest the intestines of mammals and are transmitted to humans in raw meat, especially pork. Symptoms include diarrhea, nausea, colic, fever, muscular stiffness, swelling of the face, sweating, and insomnia. Also known as trichiniasis.

trichlorophenoxyacetic acid (2,4,5-T) *n.* a selective herbicide ($C_8H_5Cl_3O_3$) that is most effective against woody vegetation and is somewhat effective against broadleaved herbaceous plants. Its use is now restricted because it and/or its contaminants can cause birth defects. See agent orange.

trichogramma *n.* a tiny hymenopterous chalcid wasp (*Trichogramma* sp., family Chalcididae) that parasitizes the eggs of many insect pests, especially lepidopteran hosts. Trichogrammas are the most utilized beneficial insect parasites in the world. Many species and strains or races are produced in insectaries and colonized in many countries, especially Russia and China, and used on millions of hectares of crops. Four popular species and the pests they attack are

Wasp (*T. evanescens*) parasitizing an egg

- *T. evanescens:* European corn borer.
- *T. minutum:* codling moth on apples and walnuts.
- *T. platneri:* avocado worms.
- *T. preteosum:* cabbage loopers on cotton, corn, tomatoes, and many other vegetables.

trichomoniasis *n.* (Latin *trichomonad*, a form of protozoa + Greek *iasis*, condition) a sexually transmitted disease caused by flagellate protozoa (*Trichomonas* sp.). Examples: a human form (*T. vaginalis*) that causes a frothy vaginal discharge in women, a bovine form (*T. foetus*) that makes bulls sterile, and a poultry form (*T. gallinae*) that affects the digestive tracts of chickens, turkeys, guineas, and other poultry.

trickle irrigation *n.* See drip irrigation.

trickling filter *n.* small stones in a tank or a pile where sewage effluent can be sprayed over them. Microbes thrive on the stone surfaces and rapidly decompose the organic materials carried in the effluent.

trilobite *n.* (Greek *trilob*, three-lobed + *ite*, fossil) any of an extinct group of marine arthropods (class Trilobita) that were abundant during the Paleozoic Era. The flattened oval bodies of trilobites ranged from about 1 in. (2.5 cm) to 2 ft (60 cm) long.

A model of a trilobite

triose *n.* (Latin *tria*, three + *ose*, sugar) a three-carbon sugar ($C_3H_6O_3$), either the D or L form of glyceraldehyde or dihydroxyacetone. See aldehyde, ketone, sugar.

triple point *n.* the temperature and pressure conditions when the solid, liquid, and vapor (gas) states of a substance coexist in equilibrium. The triple point of water occurs at a temperature of 273.16°K (0.01°C) and a saturation vapor pressure of 6.11 millibars (4.6 mm of mercury or 0.09 lb/in.^{2}).

tristeza *n.* (Latin *tristitia*, tristeza) a viral disease that causes wilting, chlorosis, and root destruction of certain citrus trees.

triticale *n.* (Latin *triticum*, wheat + *secale*, rye) a hybrid small grain (*Triticosecale*, family Poaceae) developed as a cross between wheat (*Triticum aestivum*) and rye (*Secale cereale*), with the chromosome number of the progeny doubled. It combines the high yield potential of wheat with the hardiness and disease resistance of rye, but its flour is not as good as wheat flour for making bread.

tritium (T or **^{3}H)** *n.* (Greek *tritos*, third + *ium*, element) hydrogen with an atomic mass of 3, formed in the upper atmosphere by the action of cosmic rays. Each tritium atom has two neutrons along with one proton in the nucleus. Tritium is radioactive, with a half-life of 12.3 years as it changes to helium (^{3}He). See deuterium.

trochophore *n.* (Greek *trochos*, wheel + *phoros*, bearer) a ciliated larva of certain aquatic invertebrates; e.g., mollusks, rotifers, and annelid worms that swim freely in a marine environment.

trochus *n.* (Greek *trochos*, wheel) **1.** any of several gastropods (*Trochus* sp., family Trochidae) whose shells have a flattened base and a pyramidal shape.

troglobiont *n.* (Greek *trōglē*, gnawed hole + *biont*, living) any creature that lives in a cave.

troglodyte *n.* (Greek *trōglodytēs*, cave dweller) **1.** a prehistoric person who lived in a cave. **2.** a person of primitive character and behavior. **3.** a person who lives in isolation, unacquainted with events in the rest of the world. **4.** an ape, chimpanzee, or gorilla. **5.** an animal that lives underground.

trogon *n.* (Greek *trōgōn*, gnawing) any of several brightly colored birds (*Trogon* sp., family Trogonidae) with short bills and long tails; native to tropical and subtropical areas in America.

trophallaxis *n.* (Greek *trophē*, nourishment + *allaxis*, exchange) **1.** the symbiotic relationship among members of a colony of social insects such as ants or termites that exchange food and/or glandular secretions. **2.** the symbiotic exchange of food between adults and larvae of social insects such as bees and wasps.

trophic *adj.* (Greek *trophikos*, pertaining to nutrition) involved in or pertaining to nutrition and nutritional processes.

trophic level *n.* the position of an organism in a food chain. For example, a primary producer (a plant) at the base, an herbivore (e.g., a cabbage worm feeding on cabbage) at the second trophic level, a carnivore (e.g., a bird that eats the cabbage worm) at the third level, and a top carnivore (e.g., a cat that eats the bird) as the final consumer. Note: The cycle continues with detritivores and decomposers consuming any remnants and the dead bodies of those not otherwise consumed.

trophic structure *n.* the relationships among species in a community in regard to nutrition; the food source for each species in the community from the producers to the final consumers.

trophobiosis *n.* (Greek *trophē*, nourishment + *biosis*, way of life) a symbiotic relationship between some ant species and aphids wherein the ants care for the aphids and the aphids supply secretions that the ants use for food. See ant cow, honeydew.

trophogenic *adj.* (Greek *trophē*, nourishment + *genes*, kind + *ikos*, of) **1.** pertaining to food, eating habits, or related behavior. **2.** pertaining to the upper part of a lake where there is enough light for photosynthesis.

tropholytic *adj.* (Greek *trophē*, nourishment + *lytikos*, causing decomposition) pertaining to the depths of a lake where there is little light, and decomposition of organic materials is the principal source of energy for living organisms.

tropic or **tropics** *n.* (Latin *tropicus*, of a turn) that part of the Earth on either side of the equator between the Tropic of Cancer and the Tropic of Capricorn. Also called tropical zone or torrid zone.

tropical depression *n.* a storm that develops in the tropics with a low-pressure center and cyclonic winds with velocities between 23 and 40 mph (37 to 65 km/h).

tropical disease *n.* any disease that occurs mostly or entirely in the tropics. About 10% of the world's population is afflicted by one or more of the six major tropical diseases (malaria, schistosomiasis, filariasis, trypanosomiasis, leishmaniasis, and leprosy).

tropical disturbance *n.* a minor storm that develops in the tropics with cyclonic winds with velocities of less than 23 mph (37 km/h).

tropical fish *n.* any of several small, usually brightly colored fish native to the tropics that are commonly kept in home aquariums.

tropical rainforest *n.* a general name applied to the high-rainfall lowlands lying between the Tropic of Capricorn and the Tropic of Cancer; generally composed of a wide variety of tree species and understory vegetation and inhabited by many kinds of animals and birds. The largest such area is in the upper Amazon River Basin in Brazil. See rainforest.

tropical soil *n.* any soil characteristic of the tropics, especially a soil with low fertility and reddish colors resulting from iron accumulation under intense weathering. See Oxisol.

tropical storm *n.* a storm that develops in the tropics, especially one that develops over tropical ocean waters and has cyclonic winds with velocities between 40 and 74 mph (65 to 120 km/h).

tropic bird *n.* any of several tropical sea birds (*Phaethon* sp., family Phaethontidae) with webbed feet, mostly white plumage marked with black, and two very long central tail feathers.

Tropic of Cancer *n.* an imaginary line parallel to the equator indicating the northernmost latitude reached by the sun's vertical rays (on June 21 or 22), at about 23.5° north latitude.

Tropic of Capricorn *n.* an imaginary line parallel to the equator indicating the southernmost latitude reached by the sun's vertical rays (on December 21 or 22), at about 23.5° south latitude.

The tropics are between the Tropic of Cancer and the Tropic of Capricorn.

tropism *n.* (Greek *tropos*, turn + *ismos*, action) response of an organism that turns either toward (positive) or away from (negative) a stimulus such as light, especially an oriented growth response of a plant. See tropotaxis.

tropopause *n.* (Latin *tropus*, turn + *pausa*, stopping) the boundary at the top of the troposphere and the bottom of the stratosphere where the air temperature reaches a low of about −57°C, occurring at an altitude of about 6 miles (10 km) over the poles and about 12 miles (20 km) over the equator (or a little higher in the summer and lower in the winter).

tropophyte *n.* (Greek *tropos*, turn + *phyton*, plant) any plant adapted to marked environmental changes, e.g., summer and winter temperatures or seasons of high rainfall followed by low rainfall.

troposphere *n.* (Greek *tropos*, turn + *sphaira*, sphere) the atmosphere from the surface of the Earth to the tropopause at the bottom of the stratosphere; the part of the atmosphere where clouds and storms occur and the air mixes. Temperature decreases as altitude increases in the troposphere. See tropopause.

tropotaxis *n.* (Greek *tropos*, turn + *taxis*, arrangement) movement of an animal directly toward or away from a stimulus; accomplished by comparing the sensations from sensors on each side of the body. See tropism.

trough *n.* (Anglo-Saxon *trōh*, trough) **1.** any container to hold feed or water for animals, usually a long, narrow rectangular box with an open top. **2.** a similar container used for various other purposes. **3.** a channel to catch and carry water, as an eaves trough to catch water than runs off a roof. **4.** a depression between two high points, as the low part between two ocean waves or between two ridges on land. **5.** an elongated low area in the bottom of the ocean that is shallower and wider than a trench. **6.** an elongated area of low atmospheric pressure. **7.** the low point in an economic cycle.

true altitude *n.* the elevation above mean sea level, as distinguished from an elevation above an arbitrary local reference level.

true bearing *n.* **1.** a direction relative to the north–south axis of the Earth, usually expressed as a horizontal angle in degrees, with north as the zero reading. See azimuth. **2.** a magnetic compass direction corrected for the declination. See agonic line.

true north *n.* the direction toward the North Pole.

truffle *n.* (French *truffe*, truffle) any of several edible, potatolike, underground, ascomycetous fungi (mostly *Tuber* sp., family Tuberaceae) that grow in a symbiotic relationship on the roots of some oak (*Quercus* sp.) or birch (*Betula* sp.) tree species. In some European countries, hogs and dogs are trained to sniff out ripe truffles.

trumpet creeper *n.* **1.** a vigorous climbing vine (*Campsis radicans* and related species, family Bignoniaceae) with trumpet-shaped orange and scarlet blossoms and elliptical leaflets in pinnate leaves; native to southeastern U.S.A. **2.** trumpet flower.

trumpet flower *n.* **1.** a hardy, woody, climbing ornamental vine (*Tecoma stans*, family Bignoniaceae) with large, trumpet-shaped red blossoms; native to southern U.S.A. Also called trumpet creeper or trumpet vine. **2.** a frost-sensitive annual plant (*Datura metel*, family Solanaceae) with large, trumpet-shaped, usually white (but some are yellow, blue, or red) blossoms up to 10 in. (25 cm) long and 4 in. (10 cm) across. It grows 3 to 5 ft (0.9 to 1.5 m) tall and has oval leaves up to 8 in. (20 cm) long.

trypanosome *n.* (Greek *trypanon*, borer + *sōma*, body) any of several minute, flagellate, parasitic protozoans (*Trypanosoma* sp.) that live in the blood and cause many diseases in vertebrate animals and humans.

trypanosomiasis *n.* (Greek *trypanon*, borer + *sōma*, body + *iasis*, disease) any infection caused by a trypanosome; e.g., African sleeping sickness and Chagas' disease that affect humans or nagana and similar diseases that affect various mammals, causing fever, anemia, and erythema. Trypanosomiasis is transmitted by at least 24 species of tsetse flies (*Glossina* sp.) that are prevalent in most lowland, humid, tropical areas of Africa and South America. Trypanosomiasis is the most important obstacle to animal and human habitation in humid, tropical Africa. See Chagas' disease, nagana, tsetse.

tsetse *n.* (Tswana *tsetse*, fly) any of several small bloodsucking flies (*Glossina mortisans* and related species, family Muscidae) of tropical climates that are vectors of trypanosomiasis. Also called tsetse fly, tzetze, tzetze fly.

tsunami *n.* (Japanese *tsunami*, harbor wave) a giant ocean wave caused by an earthquake, an undersea volcano, or violent winds. A tsunami can be very destructive when it moves onshore. Often called a tidal wave, even though it is not related to the tides.

tuba *n.* (Latin *tuba*, trumpet) **1.** a brass wind instrument with a low range. Its notes are controlled by valves. **2.** a funnel cloud.

tuber *n.* (Latin *tuber*, swelling or bump) **1.** a fleshy growth produced by an underground stem with eyes where sprouts can grow and reproduce the plant asexually; e.g., a potato. **2.** a tubercle.

tubercle *n.* (Latin *tuberculum*, a small swelling) **1.** a small, firm swelling that protrudes from the skin; a nodule that is larger than a pimple. **2.** a small abnormal growth on a body part or bone; e.g., a bone spur. **3.** a translucent gray skin nodule that is the characteristic lesion of tuberculosis, containing giant cells and connective tissue. **4.** a swelling on a plant root. **5.** a small knob on a lichen frond.

tubercle bacillus *n.* **1.** the bacterium (*Mycobacterium tuberculosis*) that causes tuberculosis in humans. **2.** related species (e.g., *M. bovis*, and *M. avium*) that cause tuberculosis in animals.

tuberculosis (TB) *n.* (Latin *tuberculum*, tubercle + *osis*, disease) an infectious disease caused by a tubercule bacillus that produces tubercles in body tissues, especially in the lungs (pulmonary phthisis), and may spread throughout the body. Tuberculosis causes fever and chills, emaciation, and other symptoms, depending on the site of the infection. There are many forms of tuberculosis, and any part of the body may be affected. It can be cured with proper antibiotic treatment. Formerly called consumption. See tubercule bacillus.

tufa *n.* (Italian *tufa*, porous stone) the porous calcareous deposits formed by evaporation of water from springs, lakes, or groundwater. Also called travertine.

tuff *n.* (French *tuf*, tuff) a porous rock composed of volcanic ash or cinders, usually in stratified layers compressed enough to bond together. Further compression and stronger bonding makes it a welded tuff.

tularemia *n.* (named after Tulare County, California, where it was first identified) a disease of rabbits and other rodents that can also affect people. It is caused by a bacterium (*Francisella tularensis*) transmitted by the bite of deer flies, fleas, lice, and ticks (see *Dermacentor*). Symptoms include chills and fever, headache, muscular pain, and inflammation of lymph glands. Also called rabbit fever or alkali disease.

tule *n.* (Spanish *tule*, a bulrush) either of two large bulrushes (*Scirpus acutus* or *S. validus*, family Cyperaceae) that grow in wetlands of southwestern U.S.A. See wetland.

tulip *n.* (French *tulipe*, tulip) **1.** any of several cool-season plants (*Tulipa gesnerana* and related species, family Liliaceae) that grow from bulbs and produce long, broad, pointed leaves and a vertical stem with one blossom at the top; native to the Middle East, but commonly associated with Holland. **2.** the cup-shaped blossoms produced by such plants in a wide variety of colors.

tulip tree *n.* **1.** a tall deciduous tree (*Liriodendron tulipifera*, family Magnoliaceae) of eastern North America that sheds old branches, leaving a tall trunk with a rounded crown of foliage at the top. It produces large, tulip-shaped, fragrant, greenish yellow blossoms with orange markings inside. Also known as tulip poplar or yellow poplar. **2.** the Chinese tulip tree (*L. chinense*); similar to *L. tulipfera* but not as tall.

tumblebug *n.* (Anglo-Saxon *tumbian*, to dance + *budda*, beetle) any of several beetles (*Canthon* sp. or *Phanaeus* sp., family Scarabaedae) that deposit their eggs in balls of dung and bury them in the soil. This action not only provides a place and nourishment for their young, it also buries the manure, reduces volatilization of nitrogen, and makes the manure more useful as fertilizer. Also known as dung beetle.

tumbleweed *n.* (Anglo-Saxon *tumbian*, to dance + *weod*, weed) any of several weeds with a somewhat rounded top that breaks off when mature and can be blown by the wind in a rolling, tumbling movement that scatters seeds across the landscape. Tumbleweeds often accumulate along fences or in sheltered areas; e.g., amaranth (*Amaranthus* sp., family Amaranthaceae) or Russian thistle (*Salsola kali*, family Chenopodiaceae).

tumulus (pl. **tumuli**) *n.* (Latin *tumulus*, mound) **1.** a mound of earth or stones that indicates human activity, especially one that covers an ancient burial. Also called a barrow or cairn. **2.** a domelike swelling in an old lava flow. **3.** a mound built by wild bees as a nest.

tuna *n.* (Spanish *atún*, tuna) **1.** any of several large fish (e.g., *Thunnus thynnus*, family Scombridae) that grow up to 14 ft (4.3 m) long in temperate and tropical seas; a popular food fish that is also taken for game. Also called tunny, especially in Great Britain. **2.** the pink flesh of the fish, often used to make sandwiches, salads, or casseroles. Also called tuna fish. **3.** a New Zealand eel (*Anguilla aucklandii*). **4.** any of several prickly pear cacti with edible fruit, especially the tall upright Mexican types (*Opuntia tuna* or *O. ficus-indica*). **5.** the edible, spiny, seed-filled fruit of these cacti. Also called Barbary fig.

tundra *n.* (Russian *tundra*, marshy plain) a treeless plain vegetated with lichen-moss, sedges, and small shrubs. Alpine tundra occurs at high altitudes, and Arctic tundra occurs at high latitudes. Arctic tundra has permafrost and stays wet even though annual precipitation may be as low as 10 in. (250 mm). See permafrost, krummholz.

tungsten (W) *n.* (Swedish *tung sten*, heavy stone) element 74, atomic weight 183.84, melting point 3422°C, the highest of any known metal; a silver to white metal that retains tensile strength at high temperatures but oxidizes in air, especially when heated. It is resistant to most acids and is used to make lamp filaments, electrical contact points, alloys for tool steel, etc. Also known as wolfram.

tunnel *n.* (French *tonnel*, a small cask or vessel) **1.** an underground passage; e.g., an animal burrow or a route for a road to pass through a hill. **2.** a horizontal passage in a mine. *v.* **3.** to excavate such a passage; e.g., to tunnel through a mountain or under a river. **4.** to make tunnels in something to hollow out its interior.

tunny *n.* See tuna.

tupelo *n.* (Creek *topilwa*, swamp tree) **1.** any of several tall trees (*Nyssa* sp., family Nyssaceae) with ovate leaves, tiny white blossoms, and purple berries; native to swampy areas in eastern and midwestern U.S.A. **2.** the soft, light wood from such trees. It is used for boxes, crates, furniture, and pulpwood.

turbidimeter *n.* (Latin *turbidus*, disturbed + *metrum*, a measure) an instrument for measuring the turbidity of a liquid or gas; sometimes used to estimate the content of suspended solids.

turbidity *n.* (Latin *turbidus*, disturbed + *itas*, condition) cloudiness or obscurity in a fluid that is clear when pure; e.g., that caused by clay particles or plankton in water or by dust or smoke in the atmosphere.

turbidity current *n.* a flow of fluid caused by the increased density related to carrying suspended particles; e.g., a flow of muddy water across the bottom of a lake or sea. Also called a density current.

turbot *n.* (French *tourbout*, turbot) **1.** a large, edible European flatfish (*Psetta maxima*) with a body shaped like a thin vertical disk with fins. It can weigh up to 30 or 40 lb (14 to 18 kg). **2.** any of several other similar flatfishes.

turbulence *n.* (Latin *turbulentia*, restlessness) **1.** the condition of variable speed and mixing action characteristic of a fluid moving in turbulent flow. See laminar flow, turbulent flow, Reynolds' number. **2.** disorder; violent commotion; unrest and unpredictability, as the turbulence of a mob of frightened people.

turbulent flow *n.* the churning, mixing flow of a fluid moving at a velocity above its Reynolds' number. Turbulent flow has higher energy than laminar flow, because it not only has higher velocity but it also has cross currents, mixing action, and erosive power that laminar flow does not have.

turf *n.* (Anglo-Saxon *turf*, turf) **1.** the matted roots and soil of a dense growth of grass. Also called sod or sward. **2.** artificial turf, a tough green carpet usually underlain by a layer of asphalt as a substitute for natural turf on a football field. **3.** peat, either a peat deposit or a chunk cut therefrom, e.g., to be dried and used as fuel. **4.** a race track for horse racing. **5.** an area controlled or dominated by a person or group, as being on one's home turf. **6.** a person's field of expertise, as the field of ecology is her turf. *v.* **7.** to cover an area with sod.

turfgrass *n.* (Anglo-Saxon *turf*, turf + *graes*, grass) any grass that produces a good sod suitable for a lawn, golf course, football field, or other similar use. The principal natural turfgrasses of the U.S.A. are Kentucky bluegrass, bentgrasses (several species), bermudagrasses, blue gramagrass, buffalograss, carpetgrass, centipedegrass, crested wheatgrass, St. Augustinegrass, tall fescuegrass, and zoysia.

turgid *adj.* (Latin *turgidus*, swollen) **1.** swollen; filled under moderate pressure, as the normal state of healthy cells of plants, animals, and people. (Antonym: flaccid.) **2.** pompous or bombastic, as a turgid speech.

turgor *n.* (Latin *turgor*, swelling) **1.** the hydrostatic pressure inside a cell resulting from the osmotic concentration being higher inside the cell than outside of it. **2.** any swollen or distended condition.

turicata *n.* a Mexican tick (*Argas turicata*, family Argasidae) that is irritating and dangerous to cattle, hogs, and humans; a carrier of the schizomycete bacteria that cause relapsing fever. Also called relapsing fever tick.

turkey *n.* **1.** any of several large, gallinaceous birds (especially *Meleagris gallopavo*, family Meleagrididae) native to North America, including both wild and domesticated poultry with a bare head, various-colored plumage, and a spreading tail. **2.** the flesh of such birds used as food; a popular main dish for special occasions such as Thanksgiving and Christmas dinners. **3.** (slang) a person who is not very smart, clever, or popular.

turkey buzzard *n.* See turkey vulture.

turkey gnat *n.* a small black fly (*Simulium meridionale*, family Simuliidae) that sucks blood from the heads of turkeys and chickens.

turkey louse *n.* a feather-eating louse (*Lipeurus caponis*, family Philopteridae) that infests turkeys and chickens.

turkeymullein *n.* an annual weed (*Eremocarpus setigerus*, family Euphorbiaceae) native to western U.S.A. It produces almost-black seeds that are nutritious for turkeys, but its foliage is hazardous for sheep because they cannot digest the hairs that cover the leaves and stems.

turkey vulture *n.* a large bird (*Cathartes aura*, family Cathartidae) that eats dead animals; native to the Americas. Also called turkey buzzard. See scavenger.

turmeric *n.* (French *terre mérite*, good earth) **1.** an East Indian plant (*Curcuma longa*, or *C. domestica*, family Zingiberaceae) with an aromatic root. **2.** a spice and orange-yellow food coloring made by drying and powdering the root of the plant. It often is used with such foods as rice, salads, and meats.

turnip *n.* (Anglo-Saxon *turnian*, turn + *naep*, mustard) a short-season, cool-day, biennial garden plant (*Brassica rapa*, family Brassicaceae). Its hairy leaves are used as greens and the fleshy roots (both the white turnip and the Swedish turnip also known as rutabaga) are eaten either raw or as cooked vegetables.

turpentine *n.* (Latin *terebinthinus*, of the turpentine tree) **1.** any oleoresin obtained from a coniferous tree (especially the longleaf pine, *Pinus palustris*, family Pinaceae) that yields a volatile oil and a resin when distilled. **2.** the watery, nearly colorless, volatile oil obtained by distillation. It is used as a solvent and a thinner for oil-based paints, varnishes, etc., and in making certain medicines. Properly called oil of turpentine or spirits of turpentine. *v.* **3.** to obtain turpentine from a tree. **4.** to apply turpentine to an object or surface.

turquoise or **turquois** *n.* (French *turqueise*, Turkish) **1.** a hydrated copper-aluminum phosphate mineral [$CuAl_6(PO_4)_4(OH)_8\cdot 4H_2O$] that forms a bluish or greenish semiprecious gemstone. **2.** a greenish-blue characteristic of the mineral.

turreted *adj.* (French *tourete*, little tower) **1.** having one or more turrets (small towers) or turretlike parts. **2.** having a long whorl that tapers to a point, as a conch shell.

turtle *n.* (French *tortue*, tortoise) **1.** any of about 250 species of reptiles (order Chelonia) with soft bodies covered by a shell consisting of a bony carapace above and a plastron beneath. The head, tail, and four legs extend outside the shell for movement, eating, and other functions, but most species can withdraw completely inside the shell when they feel threatened. **2.** any of these animals that spends much of its life in either salt or freshwater, as distinguished from a tortoise that spends its life on land (but the distinction is not consistently maintained). **3.** the flesh of a turtle used as food. **4.** a turtledove.

turtledove *n.* (French *tortue*, tortoise + Anglo-Saxon *dufe*, dove) **1.** any of several Old World doves (*Streptopelia* sp., family Columbidae) with a soft cooing voice and long tails. **2.** a mourning dove.

tusk *n.* (Anglo-Saxon *tucs*, tusk) **1.** either of a pair of long pointed teeth protruding from the mouth of an elephant, walrus, or wild boar; or a single tooth in a narwhal. Tusks are used to dig for food and for fighting. **2.** any long, pointed tooth or other similar projection. *v.* **3.** to dig with a tusk. **4.** to gore an opponent with a tusk.

tussock *n.* (Anglo-Saxon *tuske*, a tuft) **1.** a tuft of grass, sedge, or twigs. **2.** any grass that grows in tussocks. **3.** a tussock moth. **4.** a tuft of hair.

tussock moth *n.* any of several moths (*Hemerocampa* sp., family Lymantriidae) whose larvae have tussocks of hair on their bodies and that feed on the leaves of many species of deciduous trees.

TVA *n.* See Tennessee Valley Authority.

T **value** *n.* See tolerable soil loss.

twig *n.* (Anglo-Saxon *twigge*, twig) **1.** a small branch on a tree or bush. **2.** any small offshoot from a larger stem or branch. **3.** a small dry branch that breaks off easily and can be used to help start a fire. **4.** a small branch of a blood vessel or nerve.

twig borer *n.* any of several beetles or larvae that bore into young twigs on a tree.

twilight *n.* (German *zwielicht*, the light between) **1.** subdued (diffused) light just after sunset or, less often, just before sunrise. See dawn, dusk. **2.** the period of time after sunset until darkness. **3.** a period of gradual decline following a time of glory.

twister *n.* See tornado.

2,4-D *n.* See dichlorophenoxyacetic acid.

2,4,5-T *n.* See trichlorophenoxyacetic acid.

two-way plow *n.* **1.** a moldboard plow with two sets of plowshares, one left-hand and one right-hand, used to plow back and forth across a field. Most are used in irrigated fields to avoid producing back furrows and dead furrows. See moldboard plow. **2.** a disk plow made so its disks can turn to the right or left to plow back and forth across a field.

type genus *n.* the genus that has been designated as a reference to represent its family (or a higher group in the classification system).

type species *n.* the species that has been designated as the most typical example of its genus.

typhoon *n.* (Chinese *tai fung*, great wind) a violent tropical cyclone, especially one that develops over the ocean near Indonesia and Southeast Asia. Similar storms originating in the Atlantic or eastern Pacific areas are called hurricanes; those near India or Australia are called cyclones. See hurricane, cyclone.

U

U *n.* the chemical symbol for uranium.

uarthritis *n.* gout caused by excess uric acid ($C_5H_4N_4O_3$) in the system of a human, bird, or reptile.

ubac *n.* a north-facing mountain slope in the Alps where the shade causes the timberline and snow line to be lower than on the more sunny south-facing slopes.

uberous *adj.* (Latin *uber*, fruitful + *osus*, full of) prolific; abundant; plentiful; copious; highly productive.

uberty *n.* (Latin *ubertas*, fruitfulness) fertility; productivity; abundance.

ubiquitous *adj.* (Latin *ubique*, everywhere + *itas*, condition + *ous*, having) seeming to be present everywhere, especially at the same time.

udder *n.* (Anglo-Saxon *uder*, udder) a mammary gland, especially one with more than one teat, as in cows, mares, ewes, and sows.

udic *adj.* (Latin *udus*, humid) a soil moisture regime used in Soil Taxonomy to indicate soils that are neither dry nor saturated for any extensive period during the year. See ustic, soil moisture regime.

udometer *n.* (Latin *udus*, humid + *metron*, meter) a rain gauge.

ulexite *n.* a hydrous borate mineral ($NaCaB_5O_9 \cdot 8H_2O$) that occurs in arid regions and is an important source of boron; named for George L. Ulex, a German chemist.

uliginous *adj.* (Latin *ūliginosus*, full of moisture) **1.** living in swampy or marshy ground; e.g., a frog; sphagnum moss. **2.** a muddy, slimy condition, as that of some bogs and lake bottoms. See wetland.

ulmaceous *adj.* (Latin *ulmus*, elm + *aceus*, belonging to) belonging to the elm family (Ulmaceae) of trees.

ulmin *n.* an amorphous brown substance produced during the decomposition of organic matter. Also called humin.

Ultisol *n.* (Latin *ultimus*, last + *solum*, soil) a soil order in Soil Taxonomy for soils that have been intensely leached, have argillic horizons, and typically are reddish or yellowish, especially in the subsoil. Ultisols are the dominant soils in southeastern U.S.A. and represent about 13% of U.S. and 6% of world soils.

ultrabasic *adj.* (Latin *ultra*, beyond + *basis*, lowest part) containing no quartz; having less that 45% SiO_2 equivalent in its composition. Ultrabasic rocks are high in nesosilicates (usually olivene).

ultramundane *adj.* (Latin *ultra*, beyond + *mundanus*, worldly) **1.** outside the Earth or the solar system. **2.** outside the realm of physical things.

ultrasonic *adj.* (Latin *ultra*, beyond + *sonus*, sound + *icus*, characteristic of) pertaining to, using, or characteristic of ultrasound.

ultrasonics *n.* the science that deals with ultrasound and its effects on humans.

ultrasound *n.* (Latin *ultra*, beyond + *sonus*, sound) acoustic waves with a frequency higher than the range audible to the human ear; above 20,000 Hz (cycles per second). It is used for medical examinations of internal body organs and structures, for some types of therapy, and for dispersion of clay particles for mechanical analysis.

ultraviolet radiation *n.* electromagnetic wave energy with wavelengths shorter than those of the visible spectrum; shorter than 4000 angstrom units. Sunlight contains ultraviolet radiation that can convert ergosterol to vitamin D_2, tan the skin, and cause sunburn.

ultravirus *n.* (Latin *ultra*, beyond + *virus*, slimy liquid) any virus so small that it can pass through the pores of a fine filter (includes most viruses). Also called filterable virus.

ululate *v.* (Latin *ululatus*, howling) **1.** to howl like a dog or wolf or to hoot like an owl. **2.** to utter a wavering, wailing cry, often a cry of lamentation.

umbel *n.* (Latin *umbella*, sunshade) a type of flower group (family Apiaceae) with blossoms on short pedicels radiating from a common center into an umbrella shape.

Umbelliferae *n.* the older name for plants of family Apiaceae.

umber *n.* (Latin *umbra*, shade) **1.** a natural brown earthy material high in iron and manganese oxides that is used in its natural state as a brown pigment or heated to produce burnt umber, a reddish brown pigment. **2.** either the dark dusky brown or the dark reddish brown of these pigments. **3.** (in northern Britain) shade or shadow. **4.** the European grayling fish (*Thymallus thymallus*).

umbra *n.* (Latin *umbra*, shade) **1.** shade or shadow. **2.** a characteristic or thing that always accompanies a person or thing. **3.** the cone of full shadow cast by a planet or moon, as the umbra of the moon makes a solar eclipse on Earth. **4.** the dark part in the center of a sunspot. **5.** a ghost, phantom, or other apparition of a being that is not physically present.

umbrella bird *n.* any of several birds (*Cephalopterus ornatus* and related species), native to South and Central America, with a notable bluish black crest that curves forward toward the beak and a long tuft of feathers on the neck. Also called dragoon bird.

umbrella pine *n.* an evergreen tree (*Sciadopitys verticillata*), native to Japan, with leaves that grow in whorls shaped like an umbrella. It is used as an ornamental.

umbrella plant *n.* **1.** a type of sedge (*Cyperus alternifolius*, family Cyperaceae), native to Africa, that has umbrella-shaped clusters of leaves at the tops of its vertical stems. **2.** wild buckwheat; any of several plants (*Eriogonum* sp., family Polygonaceae), native to western U.S.A. and Mexico, with white, yellow, or red umbels.

umbrella tree *n.* **1.** a magnolia tree (*Magnolia tripetala*, family Magnoliaceae), native to America, that has large oval pointed leaves in clusters shaped like an umbrella and produces foul-smelling white blossoms and red fruit. **2.** any of several other trees whose leaves form an umbrella shape.

umbric epipedon *n.* a thick, dark epipedon similar to a mollic epipedon in Soil Taxonomy except that its base saturation is less than 50%.

unavailable water *n.* soil water that is so tightly adsorbed by soil colloids that plant roots cannot absorb it fast enough to meet plant needs; i.e., water held at tensions greater than the permanent wilting point.

uncinate *adj.* (Latin *uncinatus*, having hooks) having one or more hooks or barbs, especially at the end.

unconformity *n.* (Latin *un*, not + *conformare*, to make the same + *itas*, condition) **1.** incongruous; inconsistent; lack of agreement with the surroundings or norm. **2.** a missing period in the geologic record at a site; a place with a discontinuity where sedimentation was interrupted by a period of erosion, leaving an age difference between adjoining rock layers. **3.** the interface surface between nonconforming strata (usually seen as a line in an exposure).

unconsolidated *adj.* (Latin *un*, not + *consolidare*, to make solid) somewhat porous and loose rather than compacted, compressed, or cemented; e.g., sediments that have never been covered by a heavy overburden or glacier.

underbrush *n.* (German *unter*, under + French *brosse*, bush) mostly shrubs, vines, and small trees growing beneath a canopy of taller trees. See undergrowth.

undercurrent *n.* (German *unter*, under + French *curant*, running) **1.** a flow of air, water, or other fluid that is beneath the surface, often beneath another current that flows in a different direction. **2.** an underlying feeling or tendency, as an undercurrent of opposition in a crowd.

underdrainage *n.* removal of water that has passed through the soil; e.g., by tile drainage, mole drainage, or deep ditches that intercept the water table.

underfit stream *n.* a stream that is smaller than usual relative to the size of its valley, often an indication that there once was a larger stream there (perhaps during glacial times) or a stream that has changed course.

underfur *n.* (German *unter*, under + French *fourrure*, fur lining) a layer of fine, soft hair covering the skin beneath the longer, coarser outer hair on certain animals such as beavers, otters, and seals.

underground *adj.* (German *unter*, under + *grund*, base) **1.** beneath the land surface, as underground water or an underground burrow, cellar, or subway. **2.** in hiding; secret, as an underground spy organization or an underground newspaper. *n.* **3.** the region beneath the land surface. **4.** a secret internal organization that opposes the government in power.

underground stem *n.* any plant stem that grows beneath the soil surface, often having a reproductive function; e.g., a rhizome, tuber, or corm.

undergrowth *n.* (German *unter*, under + *gruoen*, to grow) underbrush and other plants that grow beneath a canopy of taller trees.

undersea *adj.* (German *unter*, under + *see*, sea) beneath the surface of the sea, especially near the bottom; e.g., undersea life is abundant in shallow areas near the coast.

undershot *adj.* (German *unter*, under + *schuss*, shot) **1.** having the lower jaw thrust forward so the lower front teeth pass in front of the upper teeth, like those of a bulldog. **2.** driven from beneath, as an undershot waterwheel.

understory *n.* or *adj.* (German *unter*, under + Latin *historia*, picture) underbrush; the trees and shrubs that grow beneath the canopy of taller trees, as understory vegetation.

undertow *n.* (German *unter*, under + Anglo-Saxon *togian*, to pull) an undercurrent in a sea or lake, especially one flowing away from the shore. Also called rip current.

undifferentiated group *n.* a soil map unit that indicates the presence of any one or more of its group of designated soils. Its use is limited to soils that have very similar use and management in the survey area.

undulating *adj.* (Latin *undulatus*, waved + *ing*, instance of) having a wavy surface; not quite smooth or level, but less rough than rolling.

undulation *n.* (Latin *undulatus*, waved + *tion*, state or thing) a gentle wave form, usually one of a series; e.g., a surface undulation.

unfledged *adj.* (Anglo-Saxon *un*, not + *flycge*, fledged) **1.** not having enough feathers to fly (said of a young bird). **2.** young and immature; lacking experience.

unguiculate *n.* or *adj.* (Latin *unguiculatus*, fingernail) **1.** having nails or claws on the feet (and hands). **2.** any mammal that has nails or claws rather than hooves. See ungulate.

unguis *n.* (Latin *unguis*, a nail, claw, or hoof) **1.** a hard nail, claw, or hoof on the foot (or hand) of an animal or person. **2.** a clawlike projection at the base of the petals of certain flowers.

ungulate *n.* or *adj.* (Latin *ungulatus*, having hooves) any mammal with hooves (order Ungulata), such as cattle, horses, sheep, goats, and hogs. See unguiculate.

unicorn plant *n.* a flowering plant (*Proboscidea louisianica*, family Martyniaceae) with white to red blossoms marked with purple blotches and a long, curved, woody, beaklike capsule.

unidentified flying object (UFO) *n.* any object or light that is seen moving across the sky and cannot be explained; assumed by some to be of extraterrestrial origin.

unifactorial *adj.* (Latin *unus*, one + *factor*, maker + *alis*, act of) pertaining to or controlled by a single gene.

uniflagellate *adj.* (Latin *unus*, one + *flagellatus*, whipped) having a single flagellum, as certain bacteria.

uniflorous *adj.* (Latin *unus*, one + *florus*, flowered) having a single blossom.

unifoliate *adj.* (Latin *unus*, one + *foliatus*, leafy) having a single leaf.

unifoliolate *adj.* (Latin *unus*, one + *foliolum*, little leaf) having individual leaves rather than groups.

uniformitarianism *n.* (Latin *uniformitas*, uniform state + *arius*, having + *isma*, action or state) the concept that the same geologic processes act today as in the past, so "The present is the key to the past," a concept formulated by James Hutton (1726–1797), a Scottish geologist.

unijugate *adj.* (Latin *unijugus*, having one yoke + *ate*, to produce) having one pair of leaves in a pinnate leaf.

unilobed *adj.* (Latin *unus*, one + *lobus*, hanging down) having a single lobe, as the maxilla of an insect.

uninucleate *adj.* (Latin *unus*, one + *nucleatus*, having a kernel) having only one nucleus in a cell.

uniparental *adj.* (Latin *unus*, one + *parentalis*, of parent) having a single parent, as an organism produced by parthenogenesis.

uniparous *adj.* (Latin *uniparus*, single birth) **1.** giving birth to a single offspring at a time, as is typical of cattle, horses, and people. **2.** producing one egg at a time. **3.** producing a single axis at each branch point of a plant.

unique agricultural land or **unique farmland** *n.* land that does not qualify as prime farmland but has some special combination of soil quality, location, climate, etc., that makes it especially suitable for producing specific high-value food and/or fiber crops; e.g., a sloping sandy soil that is a good site for orchards because it has good air and water drainage. See prime cropland.

unisexual *adj.* (Latin *unus*, one + *sexualis*, of sex) **1.** pertaining to one sex only. **2.** having the organs of only one sex, as most animal and certain plants whose flowers are either staminate (male) or pistillate (female).

United Nations Environment Programme (UNEP) *n.* a program established in 1972 by the United Nations to safeguard the world environment. The UNEP operates Earthwatch to monitor environmental conditions globally and to register potentially toxic chemicals so it can predict potential environmental crises and issue warnings.

United States Agency for International Development (USAID) *n.* the agency responsible for economic and technical aid programs provided by the U.S. government to help other nations and to promote the international interests of the U.S.A. Much of its work is done through contracts with universities and groups established for that purpose. It operates under the direction of the Secretary of State.

United States customary units *n.* the set of units for weights and measures customarily used in the U.S.A.; sometimes called the foot-pound-second system. Note: Most of the U.S. customary units have an origin in English units, but there are differences, for example in the size of the gallon and other liquid measures.

United States Department of Agriculture (USDA) *n.* a department established in 1862 as part of the U.S. government. Now headed by the Secretary of Agriculture, a cabinet-level position, it coordinates several agencies that provide educational, technical, and economic assistance to farmers and other land users, manage government-owned land, combat hunger, inspect food products, and research agricultural topics.

United States Environmental Protection Agency (EPA) *n.* See Environmental Protection Agency.

United States Geological Survey (USGS) *n.* a bureau founded in 1871 as part of the U.S. Department of the Interior. The USGS collects, monitors, analyzes, and provides scientific data about natural resources and environmental conditions, issues, hazards, and problems. Topographic maps of the U.S.A. are one of its most-used products, made available to the public in a form often called quad sheets.

universal gas constant (R) *n.* the constant in the ideal gas law that relates the volume of a gas to the number of moles present and the conditions of temperature and pressure, equal to 0.082058 liter-atmospheres per mole degree K, 8.3145 joules per mole degree K, or 1.987 calories per mole degree K.

universal soil loss equation (USLE) *n.* a technique published in 1965 by Walter H. Wischmeier and Dwight L. Smith of the USDA Agricultural Research Service for estimating the probable soil loss from a specific field based on precipitation, soil erodibility, length and steepness of slope, crop and soil management factors, and conservation practices in use. It is now used mostly in a revised form called RUSLE.

universal time (UT) *n.* Greenwich mean time; mean solar time at the meridian of Greenwich, England. Formerly used as the international reference for time, UT terminology is still used, but the actual time standard is now atomic time. See atomic time.

universe *n.* (Latin *universum*, all in one) **1.** everything that is known or supposed to exist; the Earth, sun, moon, stars, and everything in between. **2.** the world, especially as known to humankind. **3.** the entirety of any kind of system or activity.

universology *n.* (Latin *universum*, all in one + *logia*, word) the science that studies the universe.

univoltine *adj.* having only one generation per year.

unsanitary *adj.* (Latin *in*, not + *sanitas*, health + *aris*, like) dirty; polluted; likely to spread disease.

unsaturated fatty acid *n.* a fatty acid with one (monounsaturated) or more (polyunsaturated) double bonds in its chain. Double bonds are more common in fatty acids from plants than in animal fats. They lower the melting point of the fatty acid as compared with a saturated fatty acid of the same chain length.

unsaturated flow *n.* water movement through soil that contains air in its larger pores. It is mostly capillary flow in response to differences in matric potential. See vadose water.

unsex *v.* (Latin *in*, not + *sexus*, sex) to castrate, spay, or caponize; to surgically remove or chemically deactivate the testes of a male or the ovaries of a female human, animal, or bird.

unstratified *adj.* (Latin *in*, not + *stratificare*, layered) not separated into layers; either uniform throughout or gradational without sudden changes, as unstratified rock or an unstratified atmosphere.

upas *n.* (Javanese *upas*, dart poison) **1.** a large tree (*Antiaris toxicaria*, family Moraceae) native to tropical areas in Asia, Africa, and the Philippine Islands. **2.** the sap of the tree, which is used as a dart poison.

updraft *n.* (Anglo-Saxon *up*, up + *dragan*, draw or pull) an upward flow of a gas, especially an upward air current in the atmosphere.

upheaval *n.* (Anglo-Saxon *up*, up + *hebban*, to lift + *al*, process) **1.** a pushing upward, as the upheaval that produces a mountain chain. **2.** the area that has been pushed up. **3.** a violent social disruption.

uphill *adj.* (Anglo-Saxon *up*, up + *hyll*, hill) **1.** increasing in elevation, as an uphill route. **2.** located at the top, as an uphill site. **3.** difficult or tiring, as an uphill struggle.

upkeep *n.* (Anglo-Saxon *up*, up + *coepan*, watch out for) **1.** the cost of providing food, water, shelter, and other life necessities for an animal or person. **2.** the items so provided. **3.** the cost of maintaining property or things in good condition and repair, as the upkeep of a building, machine, or garden. **4.** the process of providing such maintenance.

upland *n.* (Anglo-Saxon *up*, up + *land*, land) **1.** land at a higher elevation than its surroundings; e.g., on a hilltop or hillside rather than a valley floor. **2.** land that does not flood.

upland cotton *n.* the type of cotton (*Gossypium hirsutum*, family Malvaceae) usually grown in the U.S.A. and representing half or more of the world crop [either the short-staple type with fibers up to 1 in. (2.5 cm) long or the long-staple type with fibers over 1 in. long], as distinguished from Egyptian cotton, sea-island cotton (*G. barbadense*), pima cotton, and Asian cotton.

upland moccasin *n.* a snake (*Ancistrodon atrofuscus*) found in upland areas of southern U.S.A. It resembles a rattlesnake but has no rattles.

upland plover *n.* a large sandpiper (*Bartramia longicauda*, family Scolopacidae) with spotted plumage.

It is native to upland areas of eastern and central U.S.A., central Canada, and part of Alaska. Also called upland sandpiper.

upriver *adj.* (Anglo-Saxon *up*, up + French *rivière*, river) upstream; toward the source of a river; in the inland area rather than near the coast.

upstream *adj.* (Anglo-Saxon *up*, up + *stream*, stream) **1.** nearer the source of a stream. **2.** the early steps in a process, as distinguished from the later parts considered downstream. **3.** in the direction opposite to the normal flow; e.g., opposite to the direction of transcription or synthesis of DNA, RNA, or protein.

uptake *n.* (Anglo-Saxon *up*, up + *tacun*, taking) **1.** absorption; taking in; e.g., uptake of water and plant nutrients by plant roots. **2.** learning or catching on to a new idea or method, as being quick on the uptake. **3.** a pipe or conduit used to exhaust gases from a furnace, engine, or area.

upwarp *n.* (Anglo-Saxon *up*, up + *wearp*, throw) an upheaval that lifts a large area with gently sloping sides. See downwarp.

upwelling *n.* (Anglo-Saxon *up*, up + *wella*, boil + *ing*, instance of) **1.** an upward flow of water from the depths of a sea or lake; e.g., replacing water drawn away by a surface current. **2.** a gradual increase, as an upwelling of public opinion.

upwind *adj.* (Anglo-Saxon *up*, up + *wind*, wind) **1.** facing the wind; the direction opposite that of the air movement, as turning upwind. **2.** located in or moving toward the direction of the wind source.

uranic *adj.* (Greek *ouranos*, heaven + *ikos*, of) having to do with the heavens; astronomical; celestial.

uranite *n.* (Latin *Uranus*, the 7th planet + *ite*, mineral) a dense, black mineral (UO_2, often with inclusions of radium, thorium, lead, etc.) that is the principal ore of uranium.

uranium (U) *n.* (Latin *Uranus*, the 7th planet + *ium*, element) element 92, atomic weight 238.0289; the heaviest naturally occurring element. A white, radioactive metal that has 14 isotopes (^{238}U is most abundant; ^{235}U is used for atomic energy) and occurs in several minerals, including uraninite and carnotite, it is used in nuclear fuels and in nuclear bombs. Depleted uranium (with the most radioactive isotopes removed) is used in guidance devices, shielding, and armor-piercing ammunition.

uranography *n.* (Greek *ouranographia*, heavens mapping) the branch of astronomy that describes and maps the positions of stars in the heavens. Also called uranology.

uranology *n.* (Greek *ouranos*, heaven + *logos*, word) **1.** an older name for uranography. **2.** a book dealing with stars and the heavens.

Uranus *n.* (Greek *Ouranos*, heaven) **1.** the third largest planet in the solar system, the seventh platet from the sun, located between Saturn and Neptune. **2.** the Greek god that personified the heavens.

urban *adj.* (Latin *urbanus*, of the city) **1.** having to do with a city or town (generally a settled area with more than 2500 inhabitants). **2.** characteristic of a city or town, or of life therein. **3.** having the habits of one who dwells in a city.

urban heat island *n.* the area of warmer temperature that encompasses an urban area as a result of the high level of activity, burning of fossil fuels, and replacement of green vegetation with buildings and streets.

urbanization *n.* (Latin *urbanus*, of the city + *izare*, to make or become + *tion*, act or state) **1.** the process of making an area or a person urban. **2.** the process of becoming urban.

urban plume *n.* the zone of increased atmospheric pollution that drifts downwind from an urban area.

urban renewal *n.* improvement of living and working conditions in an urban environment by slum clearance, restoration of rundown buildings, construction of new housing and facilities, improvement of services, etc.

urban runoff *n.* surface water from streets, parking lots, sidewalks, roofs, lawns, etc., that constitute an urban area, usually carried by a system of storm sewers and emptied into a local stream without treatment. Such waters usually contain pollutants such as oil and grease, particulates, domestic animal effluent, lawn pesticides, and miscellaneous factory wastes.

urban sprawl *n.* the spreading of an urban area onto adjoining land, especially when new housing areas and shopping centers expand onto good cropland.

urbiculture *n.* (Latin *urbs*, city + *cultura*, cultivation) the lifestyle of people living in cities.

urd *n.* (Hindi *urd*, a pulse crop) a legume (*Vigna mungo*, family Fabaceae) that is widely grown in tropical Asian areas for its edible seeds and as livestock forage. Also called gram or black gram.

urea *n.* (Greek *ouron*, urine) a white, crystalline, organic chemical [$CO(NH_2)_2$] that is readily soluble in water, occurs naturally in all human and animal urine, and can be synthesized from ammonia (NH_3) and carbon dioxide (CO_2). Urea is the highest analysis (45% N) and most popular solid form of nitrogen fertilizer made in the U.S.A. (second only to anhydrous ammonia) but is subject to volatilization losses if left on the soil surface for an extended time. It is also used as a livestock feed supplement because ruminants can convert some of the nitrogen in urea to protein.

urease *n.* (Greek *ouron*, urine + *ase*, enzyme) an enzyme that catalyzes the hydrolysis of urea to ammonium carbonate.

urechitoxin *n.* a bitter, white, poisonous glycoside ($C_{13}H_{20}O_5$) contained in Savannah flower (*Urechites suberecta*) of tropical America.

ureide *n.* (Greek *ouron*, urine + *ide*, compound) a form of nitrogen produced by nitrogen-fixing bacteria and transported to other parts of the host plant.

urena *n.* (Malayalam *urinna*, urena) any of several tropical plants or shrubs (*Urena* sp., family Malvaceae) that bear clusters of small yellow flowers.

uric acid *n.* a crystalline organic acid ($C_5H_4N_4O_3$) present in animal and human urine in small amounts and in larger amounts in reptile urine. Its salts (e.g., monosodium urate) crystallize into urinary calculi and cause gout when present in the blood. **2.** a purified form of this acid used for the synthesis of organic compounds.

urn *n.* (Latin *urna*, earthen vessel) **1.** a large decorative vase, especially one with a flared base. **2.** a container for the ashes of a cremated body. **3.** a large container with a spigot; used for brewing, holding, and dispensing coffee. **4.** a sporangium; a cavity filled with spores in the capsule of a moss.

Uroglena *n.* any of several flagellate, plantlike organisms (*Uroglena* sp., family Ochromonadidaceae) that form spherical clusters in water and impart a fishy odor to drinking water.

uropygial gland *n.* a gland with an opening near the base of the tail of many birds. Birds use its oily secretion to preen their feathers. Also called oil gland or preen gland.

Ursa Major *n.* (Latin *ursa major*, great bear) a conspicuous northern constellation that includes the Big Dipper. A line through the two outer stars of the Big Dipper points toward Polaris, the North Star. See Ursa Minor.

Ursa Minor *n.* (Latin *ursa minor*, little bear) the northernmost constellation. It includes the Little Dipper with Polaris, the North Star, at the end of its handle. See Ursa Major.

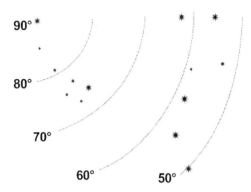

A line through the pointer stars of Ursa Major (top right) leads to Polaris at the handle end of Ursa Minor (top left).
Note: The rotation of the Earth causes our view of the entire group to rotate around Polaris.

urushiol *n.* (Japanese *urushi*, lac tree + *ol*, alcohol or phenol) a mixture of several oleoresins that are the principal irritant oils in poison ivy, poison oak, poison sumac, Japanese lac tree, and related plants.

USDA *n.* See United States Department of Agriculture.

USGS *n.* See United States Geological Survey.

USLE *n.* See universal soil loss equation.

usnea *n.* (Arabic *ushnah*, lichen) any of several lichens (*Usnea* sp.) that produce pale green or gray mosslike growths on rocks and trees.

ustic *adj.* (Latin *ustus*, burnt) a soil moisture regime used in Soil Taxonomy to indicate a soil that at a depth of 50 cm (20 in.) is dry for a total of between 90 and 180 days in most years (drier than udic and moister than aridic) but is not continuously dry in the summer as in a xeric soil moisture regime.

uterus *n.* (Latin *uterus*, womb) womb; the portion of the reproductive system in a female mammal where a fertilized egg implants and develops.

utricle *n.* (Latin *utriculus*, a little bag) **1.** a small baglike structure; e.g., an air-filled cavity in seaweed. **2.** a thin pericarp surrounding one or more seeds. **3.** the larger of the two cavities in the membranous labyrinth of the inner ear.

UV *n.* ultraviolet.

uva *n.* (Latin *uva*, grape) **1.** a small fruit with a thin skin over a pulpy interior that contains several seeds; e.g., a grape. **2.** a raisin (dried grape).

uvala *n.* a depressional valleylike landscape feature overlying limestone bedrock. It is usually formed by the coalescence of several sinkholes.

V *n.* **1.** vanadium. **2.** volt.

vaccine *n.* (Latin *vaccinus*, from cows) **1.** a suspension of killed or altered bacteria, viruses, or rickettsiae administered to a human or animal as a vaccination. **2.** a suspension of the cowpox virus used to produce immunity to smallpox. *adj.* **3.** having to do with vaccination or with vaccinia. **4.** having to do with or coming from cows.

vaccinia *n.* (Latin *vaccinus*, from cows + *ia*, disease) cowpox, a contagious disease of cows that can be transmitted to people by contact. The cowpox virus is used as a smallpox vaccine for people.

vacuole *n.* (Latin *vacuus*, empty + *olum*, small) **1.** a bubblelike cavity in a plant or microbial cell separated from the cytoplasm by a membrane that allows an exchange of water, nutrients, waste materials, etc., with the cytoplasm and serves as a storage site. **2.** a small cavity or vesicle in the tissue of any organism.

vadose water *n.* (Latin *vadosus*, shallow) water held by partially filled voids in the soil and rock material between the soil surface and the groundwater table. Also called suspended water. See unsaturated flow, pellicular water, phreatic.

vagile *adj.* (Latin *vagus*, wandering + *ilis*, suitable for) able to move itself or migrate.

vagina *n.* (Latin *vagina*, sheath) **1.** the sheath that wraps around the stem at the base of the leaves of certain plants; e.g., those of the small grains and many other grasses. **2.** the passage in certain female mammals that connects the uterus to the vulva.

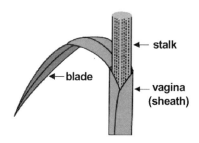

A grass leaf forms a vagina where it wraps around a stalk or stem.

vagrant *n.* (Latin *vagari*, to wander) **1.** a vagabond, one who wanders from one place to another without employment or permanent residence. **2.** a beggar or tramp with no apparent means of support. *adj.* **3.** characteristic of the life of a vagrant. **4.** nomadic, roaming from place to place. **5.** moving or growing aimlessly; e.g., a vagrant leaf driven by the wind or a vagrant vine that meanders across a surface.

vale *n.* (Latin *vallis*, valley) **1.** a valley. **2.** a small trough or channel. **3.** mortal life, especially when wretched, as living in a vale of tears.

valerian *n.* (Latin *valeriana*, valerian) any of several herbaceous plants and shrubs (*Valeriana officinalis* and related species, family Valerianaceae), native to Eurasia, with small white, pink, or lavender blossoms and fruits that dry but do not open at maturity. **2.** a drug made from the dried rhizomes of the plant for use as a sedative and antispasmodic.

valinomycin *n.* an antibiotic ($C_{36}H_{60}N_4O_{12}$) produced by a soil bacillus (*Streptomyces fulvissimus*); a cyclic peptide that increases the transport of potassium ions across cell membranes.

valley *n.* (Latin *vallis*, a vale) **1.** the low area eroded from a landscape by a river or stream, including the alluvial bottomland, any stream terraces that may be present, and the relatively steep valley walls leading to the adjoining hills, mountains, or other nearby upland area. **2.** all of the land associated with a river, as the valley of the Nile. **3.** any low area flanked by higher ground; e.g., a rift valley, a sunken area formed by crustal movement. **4.** an inside junction of two parts of a roof where water flows in from two directions. **5.** a low point in a process or wave motion, especially one plotted on a graph. **6.** a period of depression, fear, or gloom, as a valley of despair.

valley train *n.* an alluvial deposit of sand and gravel produced by meltwater flowing through a narrow valley leading away from the margin of a glacier.

valley wind *n.* a wind that blows upstream in a valley or from a valley up into the adjoining hills or mountains during a warm day. See frost pocket, mountain wind.

value *n.* (French *value*, worth) **1.** the fair price of an item; the amount it is worth, usually expressed in terms of money or other goods. **2.** the purchasing power of money or goods. **3.** the inherent worth of a person, place, thing, or action. **4.** the lightness or darkness of a tone or color. In the Munsell system, value is equal to the square root of the percentage of light reflected. See hue, chroma. **5.** the number assigned to a symbol, as the value of x, y, or z in an equation, or the value of x^2 is 4 when x equals 2. *v.* **6.** to calculate or assign a monetary worth to something, as they value the painting at a thousand dollars. **7.** to rate the worth of something, as we value our health. **8.** to esteem highly, as a person values a good friend.

VAM *n.* vesicular-arbuscular mycorrhizae. See endomycorrhiza.

vampire bat *n.* **1.** any of several small bats (family Desmodontidae) native to tropical America that live on a diet of animal blood and can transmit rabies and other diseases. **2.** any of several similar bats mistakenly accused of sucking blood.

vanadinite *n.* (Icelandic *Vanadis*, a goddess + *ite*, mineral) an ore [$Pb_5(VO_4)_3Cl$] of lead and vanadium; a soft orange to brownish red secondary mineral in the oxidized zone of sulfide deposits.

vanadium (V) *n.* (Icelandic *Vanadis*, a goddess + Latin *ium*, element) element 23, atomic weight 50.9415; a soft, ductile metal that occurs naturally in a number of minerals and is used in certain steel alloys, ceramics, and as a catalyst. Both the metal and its compounds are toxic.

Van Allen belt *n.* either of two doughnut-shaped belts of intense radiation named for James A. Van Allen (1914–), the U.S. physicist who designed the space probes that discovered them in 1958. They are part of the magnetosphere encircling the Earth. The inner belt is composed of high-energy protons and is centered at an altitude of about 2000 mi (3200 km), whereas the outer belt is composed of high-energy electrons and other charged particles at an altitude between 9000 and 12,000 mi (14,500 to 19,000 km). A third belt composed of ions of oxygen, nitrogen, and neon was found inside the inner Van Allen belt in 1993.

vanda *n.* (Sanskrit *vandā*, mistletoe) any of several epiphytic orchid plants (*Vanda* sp.), native to tropical regions in the Eastern Hemisphere, with white, lilac, greenish, or blue blossoms.

vanilla *n.* (Spanish *vainilla*, little pod) **1.** any of several climbing, viny, epiphytic orchid plants (*Vanilla planifolia* and related species, family Orchidaceae) native to tropical regions. See Mexican vanilla. **2.** the cured immature podlike fruit, about 10 in. (25 cm) long, of such orchids, usually hand pollinated. Also called vanilla bean. **3.** an extract (usually in alcohol) from the crushed beans that is used as a flavoring in foods, confections, beverages, perfumes, and pharmaceuticals. Also called vanilla extract. *adj.* **4.** containing vanilla or having a vanilla flavor; e.g., vanilla ice cream.

vapor or **vapour** *n.* (Latin *vapor*, steam) **1.** a visible suspension of tiny droplets or particles in the atmosphere; e.g., a cloud, smoke, dust, or a mist pesticide. **2.** a mist of fine droplets produced by a sprayer. **3.** a medicine that is converted to a vapor so it can be inhaled. **4.** something transitory and probably meaningless, as much idle talk.

vapor barrier *n.* a sheet of plastic or other impervious material installed in a wall, ceiling, or other location to block the passage of water vapor, especially in locations with a marked temperature gradient.

vaquero *n.* (Spanish *vaquero*, cattle raiser) a person who raises or herds cattle; a cowboy or cattle rancher, especially in Latin America and southwestern U.S.A.

vara *n.* (Spanish *vara*, a rod or stick) **1.** a long, slender pole, stick, or branch. **2.** a unit of length in Latin America. It varies from one country to another between 83.5 and 110 cm (32 to 43 in.).

variant *adj.* (Latin *varians*, varying) **1.** deviating from a reference or norm. **2.** changing, exhibiting diversity. **3.** having alternatives, not definitive, as variant wording. **4.** not accepted by some. *n.* **5.** a person or thing that deviates from the usual or the defined norm; e.g. a soil variant that resembles a defined soil series except for one or two characteristics. **6.** a spelling or pronunciation that differs from that commonly used for the word involved.

varied thrush *n.* a songbird (*Ixoreus naevius*, family Muscicapidae) that resembles a robin with a prominent black bar across the chest of the male or a fainter gray bar on the females. It is common in humid coniferous forests of western U.S.A. and Canada.

variety *n.* (Latin *varietas*, various) **1.** a group that is varied or diversified, as a variety of colors or a variety of sizes. **2.** the variation within such a group. **3.** a unique form of some thing, as a variety of pastry or a variety of work. **4.** a subdivision of a species, especially one produced with human guidance and selection. Also called a cultivar or race.

variety meat *n.* something other than the usual cuts of meat, usually organ meat such as heart or liver. See giblets.

varmint *n.* (Latin *vermis*, worm) **1.** vermin, an undesirable predatory animal; e.g., a coyote or bobcat. **2.** an undesirable, obnoxious person; e.g., a sneaky thief or a noisy ruffian.

varnish tree *n.* any of several trees that produce a resinous exudate, either naturally or from an incision, that can be used to make a varnish or lacquer; e.g., the lacquer tree (*Rhus verniciflua*, family Anarcadiaceae).

varve *n.* (Swedish *varv*, a layer) a layer of alluvial sediment deposited in one year at the bottom of a lake, including a fine-textured sublayer deposited by slow-moving water in the winter and a coarser-textured sublayer deposited by faster-moving water in the spring and/or summer. It is a common feature in glacial lake deposits. Varves are sometimes counted to estimate the number of years represented by a deposit; the weather may be interpreted as a dry or a wet year according to the thickness of each varve.

vascular plant *n.* a plant that has xylem tissue that carries water and mineral nutrients from the roots to the leaves and phloem tissue that carries dissolved sugars and other organic products from the leaves to the roots. Most plants living now are vascular plants (bryophytes are the exceptions).

veal *n.* (Latin *vitellus*, little calf) **1.** a small calf, especially one that will be slaughtered for meat. **2.** the flesh of such a calf used as food.

vector *n.* (Latin *vector*, one that conveys) **1.** something (often an insect) that transmits a disease from one plant, animal, or person to another; e.g., the female *Anopheles* mosquito carries the protozoan that causes malaria; the elm bark beetle transmits Dutch elm disease. **2.** a virus, bacterium, or plasmid used in genetic engineering to insert a strand of DNA into a host cell. **3.** a mathematical quantity that has both magnitude and direction, often indicated by a line with an arrow point. **4.** the course followed by an aircraft, missile, or other flying object.

vedalia *n.* a predacious Australian lady beetle (*Rodolia cardinalis*) introduced in the U.S.A. and other countries to control certain scale insects that attack citrus crops.

vegeculture *n.* (Latin *vegetare*, to quicken + *cultura*, cultivation) agriculture based on raising starchy root crops (e.g., manioc, taro, yams, and sweet potato), especially in the early evolution of agriculture in tropical settings, including tropics in South America, Africa, and southeastern Asia. See granoculture.

vegetable *n.* (Latin *vegetabilis*, able to live and grow) **1.** any plant, as distinguished from animal and microbial life and from nonliving things. **2.** any plant that supplies food, especially for use as a salad or main course (excluding dessert). **3.** the part of such a plant that is eaten, whether it be a seed, fruit, flower, leaf, stem, root, tuber, bulb, or some combination thereof eaten either raw or cooked. **4.** a person who has lost awareness and conscious control of activity; e.g., one who is brain dead but still breathing. *adj.* **5.** consisting of vegetables, as a vegetable diet. **6.** dealing with or pertaining to plants, as the vegetable kingdom. **7.** inactive, as a vegetable existence.

vegetable oil *n.* any of a large group of oils obtained from a wide variety of plants, including corn oil, cottonseed oil, olive oil, soybean oil, etc. Most vegetable oils are deemed more healthy than animal oils for use as food because the vegetable oils are generally less saturated (have more double bonds in their chains).

vegetal *adj.* (Latin *vegetare*, to quicken + *alis*, like) **1.** pertaining to or being like plants or vegetable matter. **2.** asexual in reproduction.

vegetate *n.* (Latin *vegetatus*, quickened) **1.** to grow, exist, or behave like a plant; to remain in one place, passive, and unthinking. **2.** to develop an abnormal enlarged growth. **3.** to produce a large amount of leaves and stems but only a few or no flowers and fruit. Note: Excess nitrogen can cause some plants to vegetate.

vegetated waterway *n.* See grassed waterway.

vegetation *n.* (Latin *vegetatio*, process of vegetating) **1.** plant life, either in general or all forms present in a particular area. **2.** the process of establishing plants in an area. **3.** the process of becoming settled and not moving any more, as the vegetation of a formerly active individual. **4.** the dull existence of a being that has ceased to be active. **5.** an abnormal, morbid growth of the body of a person or animal.

vegetative practice *n.* any method of erosion control that relies on vegetation to protect the soil by either intercepting raindrops so they cannot impact the soil or by producing a root mat that binds the soil together; e.g., a grassed waterway that protects a low area from intermittent runoff, or a crop rotation that provides vegetative cover a larger percentage of the time than more intensive cropping would provide. See mechanical practice.

vegetative propagation *n.* reproduction of plants without the use of seeds; e.g., by budding, grafting, rooting cuttings, or growth from stolons or rhizomes. Also called vegetative reproduction or asexual reproduction.

vein *n.* (Latin *vena*, vein) **1.** a blood vessel that carries blood to the heart. See artery. **2.** a vascular bundle in a plant leaf, usually showing as a ridge on the surface of the leaf. **3.** a riblike part in the framework of an insect wing. **4.** a geologic layer containing a valuable mineral, either as a primary deposit (e.g., a coal vein) or as a secondary deposit (e.g., sulfides filling cracks or fissures). Also called a lode if it is metallic. **5.** a contrasting, irregular streak in a piece of wood, stone, or plastic. **6.** a subterranean stream course or the water that flows in it, especially if it can be tapped by a well. **7.** a unique quality in a person's character, mood, or communication, as an optimistic vein or a humorous vein in one's writing.

velamen *n.* (Latin *velamen*, covering) **1.** a velum, a membranous covering. **2.** the corky layer covering the aerial roots of epiphytic orchids that absorb water from the atmosphere.

veld or **veldt** *n.* (Dutch *veld*, field) a grassland with only a few scattered bushes and few or no trees. The term is used in South Africa for grasslands at high elevations and for savanna vegetation at low elevations. See savanna.

velour *n.* (French *velous*, shaggy) **1.** a velvety fur such as that of beaver; often used for hats. **2.** a velvety fabric similar to the fur, made from any of several natural or synthetic fibers; used for upholstery, drapery, and clothing. Also spelled velours.

velum *n.* (Latin *velum*, sail or covering) **1.** a membranous partition in the body of an organism. **2.** the soft palate (the soft part at the back of the roof of the mouth). **3.** a thin layer of clouds either above or pierced by cumulus clouds.

velvet bean or **velvetbean** *n.* an annual legume (*Mucuna deeringiana*, family Fabaceae) with purplish blossoms and hairy pods that is grown in warm climates for livestock feed, soil cover, and soil improvement or as an ornamental.

velvetbean caterpillar *n.* a tropical caterpillar (*Anticarsia gemmatalis*, family Nactuidae) that feeds on leaves of velvet bean, soybeans, cowpeas, peanuts, kudzu, black locust, alfalfa, etc., usually in summer or fall because the insect is killed by winter weather and flies in each year from areas that do not freeze.

velvetleaf *n.* **1.** a warm-season annual herb (*Abutilon theophrasti*, family Malvaceae) about 5 ft (1.5 m) tall, grown for its yellow flowers and beautiful heart-shaped velvety leaves, but also known as a troublesome weed, especially in southern gardens, corn and soybean fields, and waste areas. A native of India, its coarse fiber is used there for making ropes and bags. Also called American jute, buttonweed, chingma abutilon, or Indian mallow. **2.** any of several other similar plants.

Velvetleaf

velvet plant *n.* a composite plant (*Gynura aurantiaca*, family Asteraceae) with a dense cover of purplish hairs on its leaves and stems. Native to tropical areas in the Old World, it is often grown as a house plant.

venation *n.* (Latin *vena*, vein + *ation*, condition) the pattern of veins on a leaf or on the wings of an insect; e.g., parallel, palmate, or pinnate venation.

venereal disease (VD) *n.* **1.** any of several contagious diseases transmitted by sexual contact that include syphilis, gonorrhea, chlamydia, and genital herpes virus infection. Also called sexually transmitted disease (STD). **2.** any of several similar diseases that afflict animals.

venin *n.* (French *venin*, poison) any of several toxic compounds found in snake venom.

venison *n.* ((French *veneison*, venison) deer meat or the flesh of similar wild animals used as food.

venom *n.* (Latin *venenum*, poison) **1.** a poisonous secretion of a snake, spider, bee, or several other insects. Venom is injected through a bite or sting to stun a victim. **2.** a poisonous attitude or effect, as the venom of spite, hate, or jealousy. **3.** (archaic) any poison.

venomous *adj.* (Latin *venenosus*, poisonous) **1.** having venom and a means of injecting it, as a venomous snake or insect. **2.** hateful, spiteful, malignant, or malicious.

ventifact *n.* (Latin *ventus*, wind + *factum*, made) a pebble or stone that has been grooved, shaped, and/or smoothed by the erosive action of wind-driven sand.

Venus *n.* (Latin *Venus*, love) **1.** the second planet from the sun, located between Mercury and Earth, and nearly the same size as Earth. It is the brightest planet in the sky, often seen before the stars show in the evening. **2.** an ancient Italian goddess of love and beauty. **3.** a beautiful woman or a statue or statuette representing a beautiful woman. **4.** any of several conches or conch shells (*Venus* sp., family Veneridae).

Venus's flower basket *n.* a glass sponge (*Euplectella* sp.) with a cylindrical skeleton composed of siliceous spicules. It is found in deep waters near the Philippine Islands and Japan.

Venus's flytrap or **Venus' flytrap** *n.* a perennial herb (*Dionaea muscipula*, family Droseraceae) with white blossoms. Native to swampy areas of North and South Carolina, its leaves have sensitive hairs that cause them to snap shut when touched to trap and digest an insect.

verdant *adj.* (Latin *viridis*, green + *antem*, having) **1.** covered with green vegetation. **2.** green. **3.** inexperienced, immature; e.g., a verdant beginner; a greenhorn.

verdin *n.* (French *verdin*, yellowhammer) **1.** a small bird (*Auriparus flaviceps*, family Remizidae) with a long tail that inhabits desert areas of southwestern U.S.A. and Mexico. It is marked with a gray body, yellow head and throat, and chestnut patches on the wing shoulders, and builds a compact, spherical nest made of thorny twigs. **2.** a green passerine songbird (*Chloropsis* sp.) native to southeast Asia.

verdure *n.* (Latin *viridis*, green + *ura*, state of being) **1.** flourishing green vegetative cover, especially a vigorous growth of grass. **2.** vigorous growth in general; flourishing conditions.

verge *n.* (Latin *virga*, rod or twig) **1.** outer margin or edge, as the verge of a forest. **2.** the decisive point for a major change, as the verge of a discovery or on the verge of disaster. **3.** a narrow strip of grass between a sidewalk or walkway and a street, field, or garden area. **4.** a decorative border around an area, especially around a circular area. **5.** the border of a jurisdiction. **6.** a rod or staff that signifies authority. *v.* **7.** to border, as a property that verges on an estate. **8.** to almost reach, as to verge on greatness.

verjuice *n.* (Latin *viridis*, green + *jus*, juice) **1.** sour, acidic juice squeezed from crab apples, unripe grapes, or other similar unripe fruit for use in cooking. **2.** a sour disposition.

vermicide *n.* (Latin *vermis*, worm + *cida*, killer) a drug or pesticide for killing worms, especially intestinal worms.

vermicomposting *n.* (Latin *vermis*, worm + *compositus*, put together + Anglo-Saxon *ing*, action) the use of worms, especially earthworms, to hasten the breakdown of organic residues to make a suitable organic compost for gardening. The worm casts are rich in nitrogen, phosphorus, and other nutrients and contain polysaccharides that stabilize soil structure.

vermicular *adj.* (Latin *vermiculus*, little worm + *ar*, connected with) **1.** having to do with or done by worms. **2.** having the form of a worm. **3.** resembling the wavy markings commonly made by worms.

vermicule *n.* (Latin *vermiculus*, little worm) **1.** a small worm, grub, larva, or maggot. **2.** a small structure shaped like a worm.

vermiculite *n.* (Latin *vermiculus*, little worm + *itus*, mineral) any of a group of platy, micaceous minerals $[(Mg,Fe,Al)_3(Si,Al)_4O_{10}(OH)_2$ + enough interlayer $Mg^{2+}(H_2O)_6$ to balance the excess charge] with tetrahedral layers on both sides of each octahedral layer. It is derived from weathering or alteration of micas. Soil and rock vermiculite has a layer spacing of 1.4 to 1.5 nm, but when heated at high temperature the interlayer water evaporates and the mineral expands into accordionlike pieces with up to 20 times the original volume. This expanded vermiculite is used in lightweight concrete, as insulation, and in greenhouses as a component of rooting media.

vermiculture *n.* (Latin *vermis*, worm + *cultura*, cultivation) raising earthworms; e.g., to sell to fishermen, for vermicomposting, or for some other earthworm product or by-product.

vermiform *adj.* (Latin *vermiformis*, shaped like a worm) having a long, slender, often sinuous shape, like a worm.

vermin *n.* (Latin *vermis*, worm) **1.** any of a number of small, troublesome insect or animal pests; e.g., bedbugs, flies, lice, mice, and rats. See varmint. **2.** (British) any animal or bird that hunts and kills small animals or birds; e.g., a coyote or weasel. **3.** an obnoxious, contemptuous person, or such persons collectively.

verminosis *n.* (Latin *vermis*, worm + *osis*, disease) infestation with worms.

verminous *adj.* (Latin *verminosus*, wormy) **1.** being like vermin. **2.** caused by or pertaining to vermin, as a verminous disease. **3.** infested with vermin.

vermivorous *adj.* (Latin *vermis*, worm + *vorus*, eating) worm-eating, as robins are vermivorous birds.

vernal *adj.* (Latin *vernalis*, belonging to spring) **1.** pertaining to the spring season and springtime events; e.g., vernal growth of plants and vernal migration of birds. **2.** characteristic or suggestive of spring, as mild vernal temperatures and vernal green colors. **3.** youthful, as the vernal freshness of youth.

vernal equinox *n.* (Latin *vernalis*, belonging to spring + *equinox*, equal night) the time when Earth's equatorial plane points directly toward the sun, marking the beginning of spring (March 21 or 22 in the Northern Hemisphere or September 22 or 23 in the Southern Hemisphere). See autumn, equinox.

vernalization *n.* (Latin *vernalis*, belonging to spring + *izatio*, act or process) subjecting seeds or seedlings to a period of cold temperatures to cause changes that are essential to germination, development, and flowering. Also called jarovization or yarovization.

vernation *n.* (Latin *vernatio*, greening) the arrangement of leaves within a leaf bud before it opens.

versant *n.* (Latin *versans*, turning) **1.** the general slope of a mountainside. **2.** the general slope of a plateau or other large region.

versatile *adj.* (Latin *versatilis*, turning) **1.** capable of turning freely in all directions, such as the antenna of an insect or an anther in a flower. **2.** able to move either forward or backward, as a versatile toe. **3.** having many uses, as a versatile tool. **4.** capable of doing many things well, as a versatile worker.

versicolor *adj.* (Latin *versus*, turning + *color*, color) **1.** changing color, as a versicolor chameleon. **2.** having several colors, as a versicolor bouquet.

vertebra (pl. **vertebrae**) *n.* (Latin *vertebra*, a joint) a segment of the spinal column consisting of a bone with a hole for the spinal cord and several projections where muscles attach.

vertebrate *adj.* (Latin *vertebratus*, jointed) **1.** having vertebrae; having a skeleton that includes a spinal column as a backbone. *n.* **2.** an animal with vertebrae; a member of the subphylum of chordate animals (Vertebrata) that includes mammals, birds, fish, amphibians, and reptiles.

vertical erosion *n.* movement of soil material downward through underlying soil and rock by solution processes, by transport in suspension in water flowing through cracks and channels, or by collapse into underground channels and cavities. See piping, Vertisol, karst, thermokarst.

vertical mulching *n.* placement of crop residues or other mulching material in vertical slits cut into the soil by a subsoiler. It is used to increase water penetration into soils with low permeability.

verticil *n.* (Latin *verticillus*, little vertex or whirl) **1.** a whorl of leaves or blossoms, all attached in a circle around the stem. **2.** a circle of hairs radiating from a center on the skin of an animal.

verticillium wilt *n.* a very prevalent disease (*Verticillium albo-atrum*, family Moniliaceae) that attacks a wide variety of plants (many tree species, shrubs, and garden crops such as tomato, potato, beets, and chrysanthemums) during the summer in temperate climates and is caused by a soil-borne fungus (*Verticillium* sp.). Symptoms include a yellowing of the plants and withering near the base of the stems.

vertiginous *adj.* (Latin *vertiginosus*, dizzy) **1.** spinning, as a whirlwind has a vertiginous current of air. **2.** dizzy; affected by vertigo. **3.** likely to cause vertigo, as a vertiginous climb. **4.** unstable; changeable, as a vertiginous personality or a vertiginous economy.

vertigo *n.* (Latin *vertigo*, dizziness) **1.** a sensation of dizziness or tilting for no apparent reason. It can affect humans or animals. **2.** any of several marsh or land snails (*Vertigo* sp.) with fusiform shells, usually with dentate openings.

Vertisol *n.* (Latin *vertic*, a turn + *solum*, soil) **1.** a self-mixing soil with granular surface soil that falls and is washed into wide shrinkage cracks that open when it dries. Rewetting causes the soil to expand and churn, gradually mixing the soil to a depth of at least 20 in. (50 cm).

A Vertisol with cracks reaching to a depth of 150 cm (5 ft.)

Formerly called grumusol. **2.** an order in Soil Taxonomy for mineral soils that have at least 30% clay, form large cracks when dry, and have gilgai microrelief, slickensides, or tilted wedge-shaped structural units in their subsoils. Vertisols occupy about 1% of the U.S. and 2% of world soils. Their high clay content make tillage difficult, and the large volume changes make them problem areas for building roads or foundations for structures.

vesical *adj.* (Latin *vesicalis*, bladderlike) **1.** pertaining to a bladder, especially the urinary bladder. **2.** elliptical; shaped like a bladder.

vesicle *n.* (Latin *vesicula*, little bladder) **1.** a small bladder or cavity, especially one filled with liquid. **2.** a blister. **3.** a small cyst. **4.** a small cavity in a rock that was filled with gas when the rock solidified.

vesicular *adj.* (Latin *vesicularis*, like a vesicle) **1.** shaped or formed like a vesicle. **2.** pertaining to or characteristic of a vesicle. **3.** having or consisting of vesicles.

vesicular-arbuscular mycorrhizae (VAM) *n.* See endomycorrhiza.

vesicular exanthem or **vesicular exanthema** *n.* **1.** an infectious viral disease that causes blisters on the snout, in the mouth, and on the feet of swine. **2.** a disease that causes blisters or nodules in the mucous membranes of the sex organs of horses.

vesicular stomatitis *n.* a disease that produces blisters on the lips, on the snout, and in the mouth of a cow, horse, or hog and causes symptoms similar to those of foot-and-mouth disease.

vespertine or **vespertinal** *adj.* (Latin *vespertinus*, of evening) **1.** occurring in the evening, as vespertine stillness. **2.** blossoming in the evening, as evening primrose and certain other flowers. **3.** most active in the evening; e.g., mosquitoes and lightning bugs.

vestige *n.* (Latin *vestigium*, footprint) **1.** a trace or remnant of something that once was present, as a vestige of past glory or a vestige of an old custom that is mostly forgotten. **2.** a trace or small amount of something, as a vestige of woodland. **3.** an organ or structure that is degenerate or poorly developed and useless, representing an earlier stage in the development of an organism or species.

vetch *n.* (French *veche*, vetch) **1.** any of several viney leguminous plants (especially *Vicia* sp., family Fabaceae) with pinnate leaves that end in tendrils. It is used as animal fodder and for soil improvement. **2.** the small beanlike seed or fruit of such a plant.

vetchling *n.* (French *veche*, vetch + *ling*, small or young) any of several climbing plants (*Lathyrus* sp., family Fabaceae) that resemble vetch but have an angular or winged stem.

veteran's preference *n.* an advantage assigned to veterans (especially those of World War II) when applying for certain jobs or to purchase a farm. Farmers Home Administration gave preference to veterans of the U.S. armed forces for help in becoming farm owners, and the Bureau of Reclamation gave preference to veterans for land in new irrigation projects.

veterinarian *n.* (Latin *veterinarius*, pertaining to beasts of burden) a doctor for animals; one who practices veterinary medicine.

veterinary medicine *n.* the study and treatment of the medical and surgical needs of animals, especially farm animals and pets.

veterinary science *n.* the field of study that deals with animal well-being (especially domestic animals), including veterinary medicine, animal anatomy, physiology, nutrition, and breeding (but not the products, marketing, and economic aspects included in animal science).

vetiver *n.* (Tamil *vettivēru*, vetti root) **1.** a robust tropical grass (*Vetiveria zizanioides*, family Poaceae) that grows as tall as 8 ft (2.4 m) and is well suited for erosion control on low-fertility soils in the tropics and subtropics. It is tolerant of heavy metals, high or low pH, some salt, and mild frost. Its seeds are sterile, but it is easily reproduced by cuttings. Also called sevendara, khus-khus (cuscus), or khas-khas. **2.** the aromatic roots of the plant that are used to make hangings and screens and containing an oil that is used to make perfume.

This planting of vetiver grass has trapped enough soil to make a small terrace across the slope.

viability *n.* (Latin *vita*, life + *abilitas*, ability) **1.** the quality of being viable; e.g., a seed lot with a viability of 95%. **2.** the ability to survive, as the viability of a species of plants or animals. **3.** the ability to succeed, as this plan has a high degree of viability.

viable *adj.* (Latin *vita*, life + *abilis*, able) **1.** able to survive; capable of living; e.g., a fertile egg, a developing embryo, or a fetus capable of living outside the uterus. **2.** a seed capable of growing into a plant. **3.** workable; able to succeed, as a viable plan.

viaduct *n.* (Latin *via*, way + *ducta*, conduit) a bridge with a number of pillars or towers (usually masonry, but now sometimes made of steel) supporting several short spans that carry a road or railroad across a valley or other low area. (A similar structure that carries water is an aqueduct.)

vibrio *n.* (Latin *vibrare*, to vibrate) any of several curved or S-shaped motile bacteria (*Vibrio* sp., family Pseudomonadaceae). Most live in water, but some species are pathogens that live in humans or animals; e.g., the Asiatic cholera organism (*V. cholerae*).

vibriosis *n.* (Latin *vibrare*, to vibrate + *osis*, disease) a contagious venereal disease, caused by bacteria (*Vibrio fetus*), that induces abortions in cattle, sheep, and goats.

vicinage *n.* (Latin *vicinus*, near + *aticum*, place of) **1.** the nearby region or vicinity. **2.** a particular neighborhood or its inhabitants. **3.** proximity; nearness, as living in the same vicinage.

vicissitude *n.* (Latin *vicissitudo*, turning or changing) **1.** a change or variation in course, setting, or conditions, as the vicissitudes of life (often alternating or fluctuating between two or more states). **2.** a regular, repeated succession (e.g., seasonal changes that recur annually). **3.** a mutation or other comparable change.

vicuna or **vicuña** *n.* (Quecha *wikuña*, vicuna) a South American ruminant (*Vicugna vicugna*, family Camelidae) similar to a guanaco or llama, but smaller. It produces a soft, fine, shaggy wool that is used to make garments and other articles with a soft nap.

vigor *n.* (Latin *vigor*, force or energy) **1.** usable strength or force. **2.** power; vitality; physical and/or mental energy. **3.** strong, energetic activity. **4.** healthy growth in a plant or growth and activity in an animal. **5.** validity; binding legal force.

vile *adj.* (Latin *vilis*, cheap or base) **1.** morally degraded; sinful; wicked. **2.** polluted; degrading; disgusting. **3.** cheap; of inferior quality; having little or no value. **4.** unpleasant, as vile weather.

village *n.* (Latin *villaticus*, belonging to a country home) **1.** a cluster of homes in a rural area, either incorporated or unincorporated. It is smaller than a town but larger than a hamlet. **2.** the people who live in such a place. **3.** a similar grouping of animal life, as a prairie dog village or a beaver village. *adj.* **4.** pertaining to a village or to life in a village.

vine *n.* (Latin *vinea*, vine) **1.** a grape plant. **2.** any plant that has long, slender stems that trail across the ground or other surface, wrap around a support, or cling to a support by tendrils. **3.** the stem of such a plant.

vinegar *n.* (French *vin*, wine + *aigre*, sour) **1.** the impure, dilute (4% or more) acetic acid (CH_3COOH) product of fermentation of wine, cider, beer, etc., by vinegar bacteria. It is used as a condiment, preservative, or cleaner. Often called white vinegar if clear or yellow vinegar if colored. **2.** a medicine dissolved in acetic acid. **3.** a bitter tone in a statement, as a touch of vinegar in his words. **4.** pep; energy; high spirits, as the horse was full of vinegar.

vinegar bacteria *n.* any of several bacteria (*Acetobacter* sp. and *Acetomonas* sp.) that carry out the secondary fermentation that converts ethyl alcohol (C_2H_5OH) to acetic acid (CH_3COOH).

vinegar eel *n.* a tiny nematode (*anguillula aceti*) that commonly grows in vinegar and other acidic environments. Also called vinegar worm.

vinegar fly *n.* any of several small flies (*Drosophila* sp., family Drosophilidae) whose larvae can grow in vinegar and often feed on fruit and other plant material that is fermenting as it decays.

vinegar tree *n.* a sumac tree (*Rhus typhina*, family Anacardiaceae) with sour berries that are sometimes used to make vinegar.

vinegarweed *n.* a weedy mint plant (*Tricostema lanceolatum*, family Lamiaceae) with clusters of blue flowers that grows in dry, sandy soils along the west coast of the U.S.A.

vine hopper *n.* a grape hopper, a sap-sucking insect (*Erythroneura* sp.) that infests grape vines.

vine maple *n.* a low-growing maple (*Acer circinatum*, family Aceraceae) whose vinelike branches form dense thickets near the west coast of North America.

vine mildew *n.* a fungal growth that is parasitic on grape vines.

vinery *n.* (Latin *vinarium*, of wine) **1.** a greenhouse or other place where vines (especially grape vines) are grown. **2.** vines collectively.

vine snake *n.* a slender snake (*Oxybelis* sp.) that lives in trees from Arizona to Bolivia.

vine sorrel *n.* a climbing evergreen plant (*Cissus acida*, family Vitaceae), native to the West Indies, that resembles a grape vine.

vine weevil *n.* a weevil that lives on grape vines and causes great damage.

vineyard *n.* (Anglo-Saxon *wingeard*, vineyard) **1.** a grape plantation, especially one that produces grapes to be made into wine. **2.** a work area, especially for the missionary outreach of a church.

vinyl chloride *n.* chlorinated ethylene ($CH_2 = CHCl$), a highly flammable, explosive, carcinogenic gas used as a refrigerant and as an ingredient for the manufacture of polyvinyl chloride (PVC), other plastics, and various organic compounds. Workplace health standards limit its concentration in the air, e.g., to 1 part per million for a long-term average and 5 parts per million for a short-term maximum.

violet *n.* (Latin *viola*, violet) **1.** a low-growing plant (*Viola* sp., family Violaceae) with fragrant five-petal blossoms that may have any of several bright colors. **2.** any of several other similar plants. **3.** a flower from any of these plants. **4.** a reddish blue with a dominant wavelength between 400 and 450 nm.

viper *n.* (Latin *vipera*, producing live young) **1.** any of several poisonous snakes, including the true vipers (family Viperidae) and the pit vipers (family Crotalidae). **2.** any of several other venomous or supposedly venomous snakes. **3.** a malicious or treacherous person; one who is likely to act or speak spitefully.

viperfish *n.* any of several species of fish (*Chauliodus* sp., family Chauliodontidae) with large mouths and long, fanglike teeth; found in the deep seas.

viremia *n.* (Latin *virus*, slime or poison + Greek *haima*, blood) the presence of viruses in the blood of a human or animal.

vireo *n.* (Latin *vireo*, greenfinch) any of several small songbirds (*Vireo* sp., family Vireonidae), native to North America, with gray or olive-green wings and backs and white or yellow bellies that eat crawling insects and larvae from the foliage of trees.

virga *n.* (Latin *virga*, rod or streak) precipitation that does not reach the ground because the water droplets or ice particles evaporate as they fall.

virgin *n.* (Latin *virgo*, maiden) **1.** a person (especially a young woman) or animal that has never engaged in sexual intercourse. **2.** one who is inexperienced or uninformed in any area. *adj.* **3.** being or having the characteristics of a virgin. **4.** new, fresh, unaffected by previous use or occurrences, as virgin wool or virgin snow. **5.** pure; unadulterated, as virgin gold. **6.** never before tilled, cropped, or harvested, as virgin land or virgin forest. **7.** unfertilized, as a virgin egg. **8.** made from new materials rather than recycled, as virgin steel or virgin paper. **9.** made from the first pressing (before heat and higher pressure have been applied), as virgin olive oil.

Virginia creeper *n.* a climbing vine (*Parthenocissus quinquefolia*, family Vitaceae) with palmate leaves and bluish-black berries that is native to North America. Also called American ivy or ivy vine.

viricide or **virucide** *n.* (Latin *virus*, slime or poison + *cida*, killer) any agent that inhibits or kills viruses.

virion *n.* (Latin *virus*, slime or poison + *on*, functional unit) an individual virus particle; a nucleic acid, either DNA or RNA, covered with a protective coat of protein and, in some, with an outer envelope.

viroid *n.* (Latin *virus*, slime or poison + *oid*, resembling) a protein that causes plant disease. It is similar to a virus but composed of a single strand of RNA without a protein coat. See virion.

virology *n.* (Latin *virus*, slime or poison + *logia*, word) the study of viruses and the diseases they cause.

virulent *adj.* (Latin *virus*, slime or poison + *ulentus*, full of) **1.** exceedingly pathogenic or poisonous; deadly. **2.** malicious; hostile; bitterly spiteful.

virus *n.* (Latin *virus*, slime or poison) any of a large number of very small infectious agents that replicate only within host cells of bacteria, plants, animals, or humans. An individual virus particle is called a virion.

virusoid *n.* (Latin *virus*, slime or poison + *oid*, resembling) a small fragment of RNA associated with the virion of certain plant viruses.

viscera *n.* (Latin *viscera*, inner parts) **1.** the internal bodily organs, especially those located in the abdomen and thorax, including the heart, liver, lungs, kidneys, intestines, etc. **2.** bowels; intestines.

viscosity *n.* (Latin *viscosus*, viscous + *itas*, condition) **1.** the viscous condition. **2.** internal friction; natural resistance to flow. **3.** a value on a scale that evaluates the resistance of a fluid to flow. This property is temperature dependent in most fluids.

vitals *n.* (Latin *vitalia*, life factors) the body organs that are essential to the life of a person or organism, including the brain, heart, liver, stomach, and lungs.

vital signs *n.* **1.** characteristics that indicate the presence of life; the temperature, pulse rate, and respiration of a person or animal. **2.** corresponding features of an organization or other entity, as the vital signs of the environment or of a business.

vitamin *n.* (Latin *vita*, life + *amine* made of ammonia) a general term coined by Casimir Funk (1884–1967), a Polish biochemist, for several unrelated organic substances that occur in foods and are required in small amounts to regulate various metabolic processes in human and animal bodies.

vitamin A *n.* retinol ($C_{20}H_{30}O$), a fat-soluble vitamin obtained from liver, eggs, and milk, or produced from carotene. It is essential to vision, for new cell growth, and in the functioning of the immune system. Also called vitamin A_1 to distinguish it from dehydroretinol ($C_{20}H_{28}O$), a component of fish oils that is called vitamin A_2.

vitamin B_1 *n.* thiamine ($C_{12}H_{17}ON_4SCl$), a water-soluble vitamin required for growth, digestion, fertility, lactation, carbohydrate metabolism, and normal functioning of the nervous system. Deficiencies cause beriberi, edema (body swelling), and heart and nerve problems. Thiamine is abundant in pork, soybeans, beans, and enriched whole-grain breads and cereals.

vitamin B_2 *n.* riboflavin. It is formed from ribose ($C_{17}H_{22}N_4O_6$), helps the body obtain energy from proteins and carbohydrates, and is essential for growth. A deficiency causes scaly skin inflammation, cracked lips, and poor vision. It is abundant in leafy vegetables, liver, cheese, lean meat, eggs, and enriched whole-grain breads.

vitamin B_3 *n.* niacin or nicotinic acid (C_6H_5NO). It is needed for the metabolism of carbohydrates and is present in proteins from liver, lean meat, peas, beans, and whole-grain or enriched breads and cereals. Deficiencies cause pellagra.

vitamin B_6 *n.* pyridoxine, pyridoxal, or pyridoxamine, three forms of vitamin B_6 that are used by the human body in the utilization of protein, for growth, and for other body functions. Deficiencies may result in weight loss, dizziness, mouth sores, and nervousness. It is present in liver, red meats, whole-grain cereals, potatoes, sweet potatoes, corn, and green vegetables.

vitamin B_{12} *n.* cyanocobalamin, an essential vitamin for the development of red blood cells, bone marrow cells, and the nervous system. Deficiencies cause pernicious anemia and degeneration of the spinal cord. Good sources include organ meats, other lean meats, fish, eggs, milk, and shellfish. Note: Vegans need to supplement their diets with vitamin B_{12} because it is deficient in all plant products.

vitamin C *n.* ascorbic acid, the least stable of the vitamins, and a growth promoter that aids in the healing of wounds, tooth formation, and bone development. Deficiencies cause scurvy. Food sources include most garden vegetables, especially turnip greens, green peppers, members of the cabbage family, citrus fruits, and tomatoes. It is also used as a food preservative.

vitamin D *n.* calciferol, the sunshine vitamin, produced by the action of ultraviolet light (as in sunshine) on vitamin D_2 (ergosterol, $C_{28}H_{43}OH$, from plant sources) or vitamin D_3 (cholecalciferol, $C_{27}H_{43}OH$, from animal sources). Deficiencies cause rickets. All forms of vitamin D aid in the absorption of calcium and phosphorus in bone formation. Saltwater fish and egg yolks are good natural food sources, and milk and margarine are often fortified with vitamin D.

vitamin E *n.* the tocopherols, eight closely related viscous oils that are antioxidants that help prevent destruction of essential substances such as vitamin A. A rare form of anemia in infants responds to vitamin E medication, but no clinical effects have been associated with vitamin E deficiency in adults. See white-muscle disease.

vitamin H *n.* biotin (actually a member of the vitamin B complex), which takes part in the metabolism of proteins, carbohydrates, and fats. Deficiencies are rare but can cause mild skin disorders, depression, anemia, muscle pain, and sleeplessness. Milk, eggs, and meats supply biotin, and it is manufactured by bacteria in the intestinal tract.

vitamin K *n.* either vitamin K_1 (phytonadione, $C_{31}H_{46}O_2$, obtained from plants) or vitamin K_2 (menaquinone, $C_{41}H_{56}O_2$, formed by bacteria in the intestinal tract), which promotes clotting of blood and occurs naturally in alfalfa, spinach, cabbage, and egg yolk.

vitamin K_3 *n.* menadione, a yellow crystalline chemical ($C_{11}H_8O_2$) that is used as a supplement to control bleeding. Deficiencies cause bleeding and injury to the liver. Green and leafy vegetables, egg yolks, and liver are rich in this vitamin. Also used as a fungicide.

vitamin L *n.* this vitamin is necessary for lactation in rats and is found in beef-liver extract and yeast.

vitamin M *n.* folic acid or folacin ($C_{19}H_{19}N_7O_6$), actually a member of the vitamin B complex. It helps the body make red blood cells and convert food to energy. Deficiencies cause anemia. Foods rich in folic acid include liver, navy (white) beans, and green and leafy vegetables.

vitular or **vituline** *adj.* (Latin *vitulus*, a calf) having to do with a calf or calves.

vivarium *n.* (Latin *vivarius*, concerning living creatures) **1.** an enclosure that simulates a natural environment for keeping living animals and plants, especially for research or observation; e.g., a zoo, laboratory, park, pond preserve, terrarium, or warren. **2.** a large tank for keeping fish, underwater animals, and plants; an aquarium.

viverrine *n.* or *adj.* (Latin *viverra*, ferret + *inus*, like) any of several long-bodied, short-legged carnivores (family Viverridae) with sharp muzzles, bristly fur, and long tails; e.g., a civet, ferret, or genet.

viviparous *adj.* (Latin *viviparus*, bearing living young) **1.** giving birth to living young rather than laying eggs that hatch; e.g., mammals, snakes, and certain fish. **2.** sprouting in place on the parent plant before dropping to the ground; e.g., mangrove seeds are viviparous. See oviparous.

vixen *n.* (Anglo-Saxon *fyxen*, female fox) **1.** a female fox. **2.** a malicious, bad-tempered, quarrelsome woman.

vlei *n.* (Dutch *vlei*, marsh) a marsh or swamp in a poorly drained valley, usually including a pond or other area of open water, at least during the wet season. Also spelled vley.

void *n.* (French *voide*, empty) **1.** empty space, as the void between stars. **2.** open space, as that between soil particles, also called pore space. Soil voids hold water and air. **3.** an opening, as a void in the wall. **4.** an unknown place, as he disappeared into the void. **5.** a place where something is missing, as the child's absence left a void in her life. *adj.* **6.** empty. **7.** meaningless; not legally effective. **8.** unoccupied, as a void position or office. **9.** not present; e.g., a void (empty) set in mathematics or a void (missing) suit in one's hand in a card game. *v.* **10.** to nullify, as to void a check. **11.** to empty, especially to void excrement. **12.** to evacuate, as to void a room of its occupants or to void the air from a container.

volatile *adj.* (Latin *volatilis*, flying) **1.** able to evaporate, especially a liquid that evaporates rapidly. **2.** likely to change suddenly, as a volatile situation, a volatile disposition, or a volatile stock market. **3.** transient, as the volatile beauty of a sunset. **4.** flying or able to fly. *n.* **5.** a liquid that can evaporate rapidly.

volatile oil *n.* an oil obtained by distillation, especially one obtained from plant tissue; e.g., vanillin or oil of wintergreen.

volatility *n.* (Latin *volatilis*, flying + *itas*, character) the tendency of a substance to evaporate.

volatilization or **volatilisation** (British) *n.* (Latin *volatilis*, flying + *izatio*, process) the process of becoming a gas; evaporation.

volcanic *adj.* (Latin *volcanus*, volcano + *icus*, like) **1.** coming from or having to do with a volcano, as volcanic ash or a volcanic eruption. **2.** having volcanoes, as a volcanic area. **3.** unstable, volatile, as a volcanic temper.

volcanic ash *n.* material composed of angular glass particles up to 5 mm (0.2 in.) in diameter expelled into the atmosphere by a volcano and deposited as a layer over the landscape downwind from the volcano. The silt-sized portion is also called volcanic dust.

volcanic block *n.* an angular stone produced by the cooling of a mass of magma blown from a volcano during an eruption.

volcanic bomb *n.* a rounded stone produced by the cooling of a mass of magma blown from a volcano during an eruption.

volcanic breccia *n.* rock formed of volcanic blocks embedded in volcanic ash.

volcanic dust *n.* dust-sized particles of volcanic ash; angular glass particles less than 0.25 mm in diameter. Volcanic dust from a major eruption may circulate in the atmosphere for several years and reduce the amount of solar radiation reaching the Earth, thus causing global cooling.

The 1980 volcanic eruption of Mount Saint Helens killed 30 people and spewed large quantities of ash, carbon dioxide, sulfur dioxide, etc. into the atmosphere.

volcaniclastic *adj.* (Latin *volcanus*, volcano + Greek *klastos*, broken pieces) composed of rock fragments of volcanic origin, including volcanic ash, cinders, volcanic blocks and bombs, volcanic breccia, welded tuff, etc.

volcano *n.* (Latin *volcanus*, volcano) **1.** an opening in the Earth's crust where steam, volcanic ash, cinders, and/or lava are vented. **2.** a generally cone-shaped mountain or hill, often with a crater at the top. It is designated as active when one or more vents (either in the crater or on the side of the cone) are emitting steam, ash, cinders, or lava.

volcanology *n.* (Latin *volcanus*, volcano + *logia*, word) the geologic study of volcanoes and associated phenomena.

vole *n.* (Norwegian *vollmus*, field mouse) any of several small rodents (*Microtus* sp., family Cricetidae) that resemble a mouse or rat with short legs and tail. Voles are often used in human and animal nutrition research.

volt (**V** or **v**) *n.* the SI unit for electromotive force. A difference in electric potential that will cause a current of 1 amp to flow through a conductor with 1 ohm of resistance. It is named after Alessandro Volta (1745–1827), an Italian physicist.

volume weight *n.* an older term equal to the bulk density of a soil in grams per cubic centimeter but without any units; the weight of a volume of oven-dry soil divided by the same volume of water.

volvox *n.* (Latin *volvere*, to turn) any of several freshwater, flagellated green algae (*Volvox* sp., family Volvocaceae) that form a spherical colony that rolls in the water.

von Liebig, Justus *n.* See Liebig.

vortex *n.* (Latin *vertex*, turning) **1.** a whirling mass of water that draws objects toward a conical depression in its center. **2.** a whirling mass of air, especially one that is visible like a whirlwind, funnel cloud, or tornado. **3.** a whirling mass of anything; e.g., a vortex of flames. **4.** a large amount of violent activity, as the vortex of war.

vug or **vugh** *n.* (Cornish *vooga*, cave) **1.** a small cavity in a rock; e.g., a geode lined with crystals. **2.** a relatively large void in a soil; usually irregular in size and shape and surrounded by soil with normal pore space.

vulpicide *n.* (Latin *vulpes*, fox + *cida*, killer) **1.** killing a fox in any way other than using hounds for a fox hunt. **2.** a person who kills one or more foxes without using hounds.

vulpine *adj.* (Latin *vulpinus*, like a fox) **1.** having the characteristics of a fox. **2.** cunning or crafty, like a fox.

vulture *n.* (Latin *vultur*, vulture) **1.** any of several large Old World birds (family Accipitridae) with dark plumage and bare, usually red, heads. They feed mostly on carrion (dead animals). See scavenger. **2.** any of several similar American birds (family Cathartidae). See turkey vulture. **3.** a greedy, unscrupulous person who takes advantage of others.

W

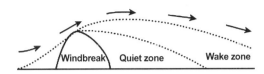

W *n.* the chemical symbol for tungsten (wolfram in German).

wacke *n.* (German *wacko*, gravel, stone) See graywacke.

waddle *v.* (German *watteln*, waddle) **1.** to sway from side to side while walking with short steps, as a duck walks. **2.** to move with any similar swaying motion. *n.* **3.** the action of swaying and walking with short steps.

wadi *n.* (Arabic *wādī*, river) a water course that is dry much of the year except during the rainy season; a term common in northern Africa and the Arabian Peninsula. Also spelled wady, ouady, oued.

wading bird *n.* any of several large birds with long legs, long necks, and long bills that wade in shallow water and eat fish, frogs, etc.; e.g., cranes, flamingos, herons, ibises, storks.

wagon *n.* (Dutch *wagen*, wagon) **1.** a four-wheeled vehicle, usually one pulled by a tractor, animals, or a person, and usually carrying some kind of cargo. **2.** a station wagon, an automobile with an enclosed space in the back suitable for carrying either passengers or cargo. **3.** a police vehicle designed to transport prisoners.

wahoo *n.* (Dakota *wanhu*, arrow wood) **1.** a large shrub or small tree (*Euonymus atropurpureus*, family Celastraceae) with serrated elliptical leaves and pendulous fruit capsules. This native of eastern U.S.A. is grown for its red fall foliage and scarlet fruit. It is toxic to sheep and goats, causing nausea, prostration, and cold sweat. Also called eastern wahoo or burningbush. **2.** any of several other American shrubs or trees, such as winged elm (*Ulnus alata*) or linden (*Tilia heterophylla*). **3.** a large mackerel (*Acanthocybium solanderi*), a fast-swimming oceanic fish that often leaps from the water. Bluish on the back and silvery below, it is valued as a food and game fish.

wakame *n.* (Japanese *wakame*, wakame) a brown seaweed (*Undaria pinnatifida*) with pinnate blades and a short stipe; used extensively in Chinese, Japanese, and Korean foods.

wake zone *n.* the partially sheltered zone on the downwind side of a barrier above and downwind from the quiet zone that offers the most protection.

waldsterben *n.* (German *waldsterben*, forest death) a forest ecosystem that is dying or in severe decline because of acid rain and other forms of air pollution.

walking catfish *n.* an Asian catfish (*Clarias batrachus*) that can survive on land and use its fins to cross from one body of water to another nearby body of water.

walking fish *n.* any of several fish that can survive on land for a short time and use their pectoral fins to move about; e.g., the walking catfish (*Clarias batrachus*), the frogfish (*Antennarius* sp., family Antennaridae).

wallaba *n.* (Arawak *wallaba*, wallaba) **1.** any of several trees (*Eperua* sp., family Fabaceae) native to the northern part of South America. **2.** the hard, dense, durable, deep red wood of such trees; used for building construction, shingles, and posts.

wallaby *n.* (Dharuk *walaba*, wallaby) any of several small to medium-sized herbivorous marsupials (*Macropus* sp., *Petrogale* sp., *Thylogale* sp., etc.) that are closely related to kangaroos but are smaller.

Wallace, Henry A. *n.* a pioneer corn breeder who helped establish hybrid corn and Pioneer Seed Corn, published *Wallace's Farmer,* and became secretary of agriculture in 1933 and then vice president in 1941 under President Franklin D. Roosevelt.

wall cloud *n.* a rotating cloud that extends downward from cumulonimbus clouds and usually includes a funnel cloud that may become a tornado.

wall creeper *n.* any of several birds (family Certhiidae) that inhabit mountain cliffs, building ruins, etc., and feed on insects.

walleye *n.* (Norse *valdeygthr*, wall-eyed) **1.** a large game fish (*Stizostedian vitreum*) found in lakes and rivers of northeastern U.S.A. and eastern Canada. Also called walleyed pike or jack salmon. **2.** any of several other fish with similar eyes that appear to stare. **3.** an eye that turns outward (opposite of a cross-eye). **4.** an eye with an abnormal white iris or cornea.

wall fern *n.* a fern (*Polypodium vulgare*, or *P. virginianum*, family Polypodiaceae) that forms a dense mat of stems creeping across a wall or cliff.

wallflower *n.* **1.** any of several perennial European plants (*Cheiranthus cheiri* and related species, family Brassicaceae) that can cling to a wall or cliff. Its sweet-scented blossoms are usually yellow or orange but can range to brownish red or purple. **2.** a person who stands or sits by the wall and watches, especially a young woman at a dance without a partner.

wallow *v.* (Anglo-Saxon *wealwian*, to roll around) **1.** to lie or roll in mud, dust, snow, or other soft substance, as a pig wallows in the mud. **2.** to self-indulge, as to wallow in luxury or in sentiment. **3.** to move uncertainly and erratically, as a damaged boat wallows toward shore. **4.** to surge or billow, as smoke wallows from an opening. *n.* **5.** a place where animals wallow or have wallowed. **6.** the action of wallowing.

wall rock *n.* **1.** the rock face next to a fault. **2.** rock hanging from a wall in a mine.

wall wasp *n.* any wasp that nests in walls, especially the common European wasp (*Odynerus murarius*).

walnut *n.* (Anglo-Saxon *wealh-hnutu*, foreign nut) **1.** any of several large deciduous trees (*Juglans* sp., family Juglandaceae) with deeply furrowed brown bark and large compound leaves. The trees, including the English walnut (*J. regia*) and black walnut (*J. nigra*), produce nuts and timber. See juglone for black walnut toxicity. **2.** the oval nut from the trees. The double-lobed seeds are inside a hard shell encased in a tough husk and are eaten by people and squirrels. **3.** the hard wood from the trees; used for furniture, woodwork, plywood, and novelties. **4.** a reddish brown color like that of the heartwood of black walnut.

walrus *n.* (Danish *hvalros*, whale horse) a large marine carnivore (*Odobenus rosmarus*) up to 11 ft (3.3 m) long that inhabits arctic seas. It is noted for its tough hide, thick blubber, thick mustache, twin tusks up to 3 ft (0.9 m) long pointing downward from its upper jaw, and strong flippers.

walrus bird *n.* the pectoral sandpiper, an arctic bird.

Walton, Izaak *n.* See Izaak Walton.

wandering dune *n.* a sand dune that moves gradually in whatever direction the wind blows. Such a moving dune needs stabilizing vegetation.

wandering Jew *n.* any of several trailing or creeping plants with smooth stems and leaves and red, white, or blue blossoms (e.g., *Zebrina pendula or Tradescantia fluminensis*).

wandoo *n.* (Nyungar *wando*, wandoo) **1.** a white gum tree (*Eucalyptus redunca*, family Myrtaceae) native to western Australia. **2.** the dense, durable wood of the tree, prized for making wheels.

waning moon *n.* the period when the sunlit portion of the moon as seen from Earth is diminishing from a full moon to the next new moon. See waxing moon, moonarian.

wapatoo *n.* (Chinook *wapatoo*, wapatoo) an arrowhead plant (*Sagittaria latifolia*, family Alismataceae). Also known as common arrowhead.

wapiti *n.* (Shawnee *waapiti*, white rump) elk.

warble *n.* (Swedish *varbulde*, boil) **1.** a lump in the skin on the back of an animal containing the larva of a fly. See cattle grub, heel fly, warble fly. **2.** a lump on a horse's back caused by the rubbing of a saddle. **3.** a bird's song, or an imitation thereof. See warbler. *v.* **4.** to sing like a bird with whistles, trills, and/or quavers.

warble fly *n.* any of several species of flies (*Hypoderma bovis and H. lineatum*, families Oestridae and Hypodermatidae) whose larvae form raised places under the skin of cattle. The larvae later emerge, thus forming a hole in the skin and decreasing its value as leather.

warbler *n.* (French *werbler*, singer) **1.** any of a large number of small, mostly bright-colored, migratory songbirds of both American (family Parulidae) and Eurasian (family Sylviidae) origin. Warblers are insect eaters that have slender, pointed bills and melodious songs. **2.** a person or thing that warbles somewhat like a bird.

warm-blooded *adj.* See homeothermic, endotherm.

warm front *n.* the leading edge of an advancing mass of air that is warmer than the air ahead of it. The warmer air normally rises and moves over the cooler air, often resulting in precipitation.

warm-season plant *n.* **1.** any of several perennial plants (e.g., switchgrass or Bermuda grass) that grow best during the summer (most have optimum air temperatures between 80° and 95°F, or 27° to 35°C) and are dormant during the colder weather. **2.** an annual plant (e.g., corn or cotton) that germinates after the soil has warmed up in mid- to late spring and grows during the summer. See cool-season plant.

warning coloration *n.* an easily recognizable coloration pattern of an animal (e.g., the stripes of a skunk) that alerts enemies to stay away so the animal will not have to use its defense mechanism. Also called aposematic coloration.

warren *n.* (French *warenne*, preserve) **1.** a place where rabbits or game animals live and breed. **2.** any overcrowded living area.

wart *n.* (Anglo-Saxon *wearte*, wart) **1.** a small, hard, usually harmless growth protruding from the skin of a person or animal. **2.** a similar protuberance on the surface of a plant.

wart hog *n.* any of several wild African swine (e.g., *Phacochoerus aethiopicus*) about 30 in. (75 cm) tall and 4.5 ft (1.35 m) long, with large, curved tusks and warty protuberances on its face.

wart snake *n.* a nonvenomous snake (especially *Acrochordus javanicus*) with warty scales on its skin; native to India, southeast Asia, and northern Australia.

wasabi *n.* (Japanese *wasabi*, wasabi) **1.** an herb (*Eutrema wasabi*, family Brassicaceae) with greenish roots. **2.** a condiment similar to horseradish made by grating the roots of the plant; used in Japanese dishes.

Washington lupine *n.* a hardy perennial plant (*Lupinus polyphyllus*, family Fabaceae) of the U.S. West Coast that produces long spikes of white, yellow, or blue flowers. It contains alkaloids that are toxic to sheep, causing nervousness, convulsions, and frothing at the mouth. See lupine.

wasp *n.* (Anglo-Saxon *waesp*, wasp) **1.** any of a large number of hymenopterous insects (families Vespidae, Sphecidae, and others) with long, slender bodies, narrow waists, and two pairs of membranous wings. The females of most species have stingers that can inject venom containing histamine ($C_5H_9N_3$) to paralyze or kill an insect and lay eggs in its body. After the eggs hatch, the young wasps consume the insect. Some wasps are pests that may attack humans, but most are beneficial predators on aphids and other insects. Some have been raised for biological control of pests such as cotton boll weevils and tobacco budworm. **2.** an annoying, insolent person.

A wasp (*Microphitis croceipes*) inserting her eggs into a tobacco budworm (*Heliothis virescens*)

wastage *n.* (French *waster*, to lay waste + *age*, related to) **1.** loss of raw materials as scraps, as remnants, or by wastefulness during a production process. **2.** losses during distribution and marketing, e.g., by spoilage or damage. **3.** losses during use by wear, decay, or breakage. **4.** any process that causes waste, as the wastage of erosion. **5.** the material lost by any of the above. **6.** the gradual removal of ice and snow by melting and vaporization.

waste *v.* (French *waster*, to lay waste) **1.** to consume needlessly or excessively; to use more of a resource or commodity than is needed or beneficial. **2.** to squander; to use something without obtaining an appropriate benefit. **3.** to fail to use, as to waste an opportunity. **4.** to destroy gradually; to wear away, as by erosion. **5.** to destroy by violence, as an enemy wastes a city. **6.** (slang) to kill. **7.** to dwindle, as untended resources waste away. *n.* **8.** discarded materials; garbage; trash. See wastage, clinical waste, demolition waste, hazardous waste, household waste, industrial waste, solid waste, yard waste. **9.** needless consumption or expenditure. **10.** unused resources or time. **11.** material gradually lost by wear and erosion. **12.** an area that has been devastated, as by fire, earthquake, drought, or war.

waste-derived fuel *n.* any waste material used as fuel; e.g., paper, plastics, etc., sorted from garbage and mixed with coal to generate electricity, or similar use of sawdust or agricultural residues. Waste-derived fuel typically has only half the heat energy value of coal, but it also generally has a low sulfur content.

waste disposal *n.* an organized process for processing waste; e.g., garbage collection and a sewer system and treatment plant.

wasteland *n.* **1.** barren, unproductive land; areas where little or no vegetation is present. **2.** an area that has been devastated by a drought, flood, hurricane, war, or other catastrophe.

waste management *n.* the planning and providing for suitable handling of wastes at all stages of production, sorting, transport, and disposal.

waste separation *n.* any process that sorts waste materials into various types according to potential hazards and value, including special handling of clinical and hazardous wastes, recycling of containers, paper, etc., and mechanical sorting that removes metals for recycling and combustible materials for power production.

waste stabilization *n.* any process used to reduce the mobility of hazardous components in waste materials; e.g., converting liquids to solids, mixing them with solids that will absorb and hold them, heat or chemical treatment to reduce the hazard, etc.

wastewater *n.* **1.** unprocessed sewage or sewage effluent. **2.** water that has been used in an industrial process (e.g., for cooling, washing, or flushing) and is not being recycled. **3.** runoff water from an irrigated field.

water *n.* (Anglo-Saxon *waeter*, water) **1.** the oxide of hydrogen (H_2O); a clear, odorless and tasteless, inorganic substance that is normally liquid between 32°F and 212°F (0 to 100°C). Water is essential to and the largest percentage constituent of all plant, animal, and human life. Water is also a major factor in weather, weathering processes, and erosion, and it is used in cleaning, manufacturing, transportation, and other processes. **2.** any body or stream of water, as a lake pond, stream, ocean, etc. **3.** any liquid composed mostly of water but containing solutes, etc.; e.g., perspiration, tears, urine. *v.* **4.** to supply with water, as to water a plant, an animal, a ship, a city, or a region. **5.** to dilute with water, as to water down the soup, or (figuratively) to water down the information in a report.

water application efficiency *n.* the fraction of the water from a source that is stored in soil in the root zone for use by the current crop, usually expressed as a percentage.

water arum *n.* See wild calla.

water balance *n.* **1.** the ratio of water taken into a body as compared to that leaving the body of a person, animal, reservoir, etc. **2.** the condition of a body with a nearly constant water content; when the water balance ratio is 1.0 over a period of time.

water beetle *n.* any of several beetles that can survive underwater (e.g., *Captotomus interrogatus*).

water bird *n.* any of a variety of birds that enjoy a water habitat, including wading birds, ducks and geese, gulls, etc.

waterbuck *n.* any of several African antelopes (*Kobus* sp.) that inhabit marshy areas.

water buffalo *n.* a large mammal (*Bos bubalus*, family Bovidae) that resembles a large beef cow with large, strong horns; native to the jungles of India, Africa, and the Philippine Islands. It is tolerant of wet environments and other adverse conditions and is easily domesticated and used as a draft animal (e.g., in rice fields) as well as a source of rich milk. Also called water ox or carabao.

water bug *n.* any of several aquatic insects that either can survive underwater or are able to walk on the surface of water.

water celery *n.* an acrid-smelling annual herb (*Ranunculus scleratus*, family Ranunculaceae) that grows in wet areas. Also called cursed crowfoot.

water chestnut *n.* **1.** an Old World aquatic plant (*Trapa natans* and related species, family Trapaceae). Also called water caltrop. **2.** the edible nutlike fruit of the plant.

water chinquapin *n.* **1.** an American lotus plant (*Nelumbo lutea*, family Nymphaeaceae) with pale yellow blossoms. **2.** the edible seed of the plant.

water clarification *n.* the process of removing suspended particles that cause turbidity in water, e.g., by treating the water with alum to cause colloidal particles to flocculate and settle out.

water clover *n.* a freshwater fern (*Marsilea quadrifolia*) that grows with its roots in the soil beneath shallow water around the edges of quiet lakes and ponds and produces four-lobed cloverlike leaves that float on the water. Also called pepperwort or water shamrock.

water conservation *n.* **1.** the use of practices that reduce water consumption in homes, offices, and factories; e.g., flow restrictors in showers, small-flush-volume toilets, restrictions on watering plants or washing cars, recycling water in factories, etc. **2.** carefully controlled irrigation that applies water uniformly and only in the amount needed for the crop. **3.** measures that reduce runoff and promote infiltration of water into the soil.

watercourse *n.* **1.** a natural stream or river channel, whether carrying water or as a dry stream bed. **2.** a canal, ditch, or other channel used to convey water. **3.** the stream of water conveyed in any stream, ditch, or other channel.

water cress *n.* **1.** a free-floating plant (*Nasturtium officinale*, family Brassicaceae) that grows best in flowing, nonpolluted but nutrient-rich waters; native to Europe. **2.** the pungent leaves of the plant; used in salads, soups, and garnishes.

water crowfoot *n.* any of several aquatic herbs (*Ranunculus* sp., family Ranunculaceae), especially *R. aquatilis* with its showy white blossoms. See water celery.

water culture *n.* **1.** hydroponics. **2.** aquaculture.

water cycle *n.* See hydrologic cycle.

water deficit *n.* a deficiency of water for a particular need or purpose; e.g., the water a plant needs for transpiration at a given time or during a season, or the water needed to bring a dry soil surface to field capacity.

water deposit *n.* any body of material deposited by water; e.g., an alluvial soil, a sedimentary rock, or an evaporite deposit.

water eagle *n.* See osprey.

waterfowl *n.* **1.** any water bird. **2.** water birds collectively.

water garden *n.* a pond or other body of water used to grow water plants ranging from a few water lilies to a variety of ornamental and/or vegetable plants, often with one or more decorative fountains and/or rock gardens included.

water harvesting *n.* a system that collects runoff from hills and channels it to small areas where crops or trees are planted. Water is stored in the soil rather than in a reservoir. Farmers in the Negev Desert in Israel grow crops and trees this way. The Hopi and Papago tribes of Native Americans in southwestern U.S.A. use a similar system in an area that averages 9 to 13 in. (230 to 330 mm) of annual precipitation.

waterhemlock *n.* any of several aquatic plants (*Cicuta* sp., family Apiaceae) with strong-smelling leaves, small white blossoms, and poisonous roots that cause diarrhea, vomiting, difficulty breathing, and convulsions in livestock that eat them. See cowbane.

water hole *n.* **1.** a pool of water that remains in a depression when other water is not available, especially that in a depression in an otherwise dry stream bed. **2.** a spring, well, oasis, or other source of water in the desert. **3.** an opening in the ice covering a pond, lake, or stream in the winter.

water hyacinth *n.* a floating aquatic plant (*Eichhornia crassipes*, family Pontederiaceae) with feathery roots that grows so thick in lakes and rivers that it often blocks the passage of boats; native to tropical America but common in southern U.S.A.

water level *n.* **1.** the upper surface of any body of water, e.g., that of a pond, stream, or groundwater table. **2.** the water line on an object partially immersed in water, e.g., on a post or a boat.

water lift *n.* any of several kinds of devices used to lift water for household use, irrigation, or other purposes. Some are driven by the water (e.g., a water wheel), others by human or animal power.

water lily *n.* **1.** any of several aquatic plants (*Nymphaea* sp., family Nymphaeaceae) with large floating leaves and showy blossoms. **2.** the blossom from such a plant. **3.** any of several other similar aquatic plants (e.g., *Nuphar* sp.).

A water lily in bloom

water locust *n.* a spiny tree (*Gleditsia aquatica*, family Fabaceae) with pinnate leaves, yellowish green bell-shaped blossoms, and slender seed pods; grows in wet areas. Also called swamp locust.

waterlogged *adj.* **1.** saturated with water, as waterlogged soil. See wetland. **2.** filled or flooded with water, as a waterlogged float or a waterlogged boat.

watermelon *n.* **1.** a prostrate annual plant (*Citrullus lanata*, family Cucurbitaceae) whose vines reach lengths of 10 to 15 ft (3 to 4.5 m). **2.** the large green (often striped) gourd produced by the plants. Its sweet, juicy, red flesh often is eaten as a salad or dessert.

water moccasin *n.* **1.** a thick-bodied, dark olive to black, semiaquatic pit viper (*Agkistrodon piscivorous*, family Viperidae) that grows to lengths of 3 to 6 ft (0.9 to 1.8 m) in swamps and on riverbanks along the U.S. Gulf Coast. It commonly feeds on fish, frogs, and small mammals, and its bite can be fatal to larger animals and humans. Also called cottonmouth for the white interior of its mouth. **2.** any of several other aquatic snakes that appear similar to the water moccasin but are harmless.

water of crystallization *n.* water molecules that occupy specific sites in a crystal structure and affect the physical properties (volume, hardness, color, etc.) of the mineral but can be driven off by heat (usually in the range between 100°C and 200°C), leaving the rest of the chemical intact as an anhydrite; e.g., gypsum ($CaSO_4 \cdot 2H_2O$) and blue vitriol ($CuSO_4 \cdot 5H_2O$) contain water of crystallization in their structures.

water of hydration *n.* water bound to a molecule, ion, or particle by hydrogen bonding. It does not change the internal structural arrangement of the molecule, ion, or particle.

water pollution *n.* contamination of water with dissolved and/or suspended material that degrades its quality and usefulness. The largest mass of such pollutants is sediment, but acids, pesticides, toxins, decomposing organic materials, and excessive levels of plant nutrients are more serious. According to the U.S. Environmental Protection Agency, the major sources of pollution in surface waters of the U.S.A. are

Agriculture	40.0%
Municipal	10.8%
Hydro/Habitat modification	9.6%
Resource extraction	9.0%
Storm sewers and runoff	8.0%
Industrial	6.0%
Silviculture	6.0%
Construction	4.0%
Land disposal	3.0%
Unknown	3.6%

water power *n.* the power of flowing or falling water, especially when used to drive machinery (e.g., via a waterwheel) or to generate electricity.

waterproof *adj.* not penetrable by water, either by nature or by treatment with a waterproofing substance.

water rail *n.* an Old World bird (*Rallus aquaticus*, family Rallidae) with olive-brown plumage, long legs, and a long, red bill.

water rat *n.* **1.** any of several rats that inhabit wet areas. **2.** a muskrat (*Ondatra zibethica*). **3.** a thief or tramp who frequents waterfront areas.

water-repellent *adj.* resistant to the entrance of water but subject to wetting by long-continued exposure, as a water-repellent fabric or garment. Also called water-resistant.

water resource region *n.* the drainage basin of a major river. The U.S.A. is divided into 18 water resource regions in the 48 conterminous states plus one each in Alaska, Hawaii, and the Caribbean.

water right *n.* a legal right to use water from a stream, lake, or other body of water, as governed by state law. Water rights in the U.S.A. are based on the riparian rights doctrine in the eastern states and the prior appropriations principle in most of the western states, with a few states using a combination of the two. See riparian rights, prior appropriations.

watershed *n.* the land area that supplies runoff to a particular stream. Also called catchment area or drainage basin.

water snake *n.* any of several harmless snakes (*Natrix* sp.) that live in or near freshwater and feed on small water animals.

waterspout *n.* **1.** the equivalent of a tornado that touches the surface of a large body of water and sucks up enough water to show its funnel shape. **2.** a vertical pipe that drains water from an eaves trough. Also called a rainspout or downspout.

water table *n.* the upper level of the saturated zone in soil or the underlying rock material. See aquifer.

water thrush *n.* either of two wetland warblers (*Seirus noveboracensis* or *S. motacilla*, family Muscicapidae) native to North America.

water tupelo *n.* a tall tree (*Nyssa aquatica*, family Nyssaceae) that grows in swampy areas of southeastern U.S.A.

water use efficiency *n.* the amount of crop or other product produced per unit of water used; e.g., the number of grams of biomass produced divided by the number of grams of water used in evapotranspiration, or the crop yield in bushels or tons per acre divided

by the total depth of rainfall (plus irrigation water if applied) in inches.

waterwheel or **water wheel** *n.* **1.** a wheel that turns by the weight and/or velocity of flowing water; either an overshot waterwheel that receives water at the top and releases it at the bottom or an undershot waterwheel that is turned by water flowing beneath it; often used to produce power, e.g., for a flour mill. **2.** a wheel with buckets mounted along its rim; used to lift water.

water witch *n.* (water dowser) a person who claims to be able to find underground water by using a divining rod. Also called a dowser or diviner. See divining rod.

watt (W) *n.* the SI unit of power equal to 1 joule per second; named after James Watt (1736–1819), a Scottish engineer and inventor. The amount of power that causes a current of 1 ampere to flow through an electric potential difference of 1 volt; watts = amperes × volts.

watt-hour *n.* the amount of energy represented by a current of 1 watt flowing for 1 hr; equal to 3600 joules. Note: Electricity is usually sold in units of kilowatt-hours (equal to 1000 watt-hours).

wattle *n.* (Anglo-Saxon *watul*, covering) **1.** a cluster of slender branches wrapped and tied together in a bundle perhaps 6 in. (15 cm) in diameter and 5 ft (1.5 m) long, more or less, preferably made from live willow branches or others that will grow in a wet environment; used to stabilize soil on steep banks. **2.** a group of rods or stakes with slender branches interwoven to form a fence, wall, or roof. **3.** a fleshy lobe hanging from the chin or throat of certain birds (chickens, turkeys, etc.)

wave *n.* (Anglo-Saxon *wafian*, to wave one's hand) **1.** a swell or ridge that moves across the surface of a liquid, especially across the surface of a sea or lake, usually as one of a series. **2.** a ripple or undulation that moves across a surface, e.g., that of a field of standing grain. **3.** a side-to-side, up-and-down, or to-and-fro motion, as a wave of the hand. **4.** a fluctuation in the intensity of energy, as a sound wave or light wave. **5.** an increase above the former level, as a heat wave, cold wave, or crime wave. **6.** a widespread opinion or feeling, as a wave of anger, panic, or relief. *v.* **7.** to move back and forth, as to wave one's hand or a flag. **8.** to flutter, as tree branches wave or a flag waves in the breeze. **9.** to follow a series of S-curves, as a path or a road waves through a valley. **10.** to signal a greeting or a farewell, as with a wave of the hand.

wave base *n.* the bottom of the water layer that is stirred sufficiently by wave action to cause agitation of suspended sediments; about half of the spacing between wave crests. Also called wave depth.

wave crest *n.* the ridge or line joining the highest points on a wave.

wave height *n.* the vertical difference in elevation between a wave crest and the adjacent wave trough.

wavelength or **wave length** *n* **1.** the horizontal distance between two successive wave crests or troughs. **2.** the corresponding distance between two electromagnetic waves, e.g., light waves or radio waves; the inverse of the frequency of such waves (the frequency in hertz × wavelength = the speed of the waves, which is 299,792,458 m/sec).

wave trough *n.* the line joining the low points between two successive wave crests.

wax *n.* (Anglo-Saxon *weax*, wax) **1.** any of a large number of esters of long-chain alcohols and long-chain fatty acids. Wax is a combustible substance that is insoluble in water and is easily molded when warm but hardens when cold. It is used to make candles and models and to provide a shiny, water-resistant protective surface on shoes, furniture, etc. **2.** a number of complex waxlike compounds in plant and animal tissue that occur on the surfaces of stems, leaves, skin, hair, and feathers. *v.* **3.** to coat with wax. **4.** to increase (opposite of wane). *adj.* **5.** made of wax.

wax bean *n.* **1.** a type of string bean with long, slender, yellow pods. **2.** the pods eaten as a vegetable. Also called butter bean.

waxing moon *n.* the period when the sunlit part of the moon as seen from Earth is enlarging, between a new moon and the next full moon. See waning moon, moonarian.

wax insect *n.* any of several scale insects that secrete wax; e.g., the small, white, Chinese wax insect (*Ericerus pela*) bred commercially in China for the production of candle wax.

wax myrtle *n.* an aromatic shrub (*Myrica cerifera*, family Myrtaceae) that bears small wax-coated berries; native to southeastern U.S.A. It sometimes is used to make scented candles.

wax palm *n.* **1.** an Andean palm tree (*Ceroxylon alpinum* or *andicola*, family Arecaceae) that produces wax in its stems and fronds; used to make candles. **2.** a Brazilian palm tree (*Copernicia cerifera* or *prunifera*, family Arecaceae) that produces wax. Also called carnauba.

Wb *n.* the symbol for weber.

wean *v.* (Anglo-Saxon *wenian*, to accustom) **1.** to train a child or young animal to eat other food and no longer suckle on its mother's milk. **2.** to train a person or animal to overcome any form of dependence or habit, as to wean someone from constantly watching television.

weaner *n.* or *adj.* (Anglo-Saxon *wenian*, to accustom + *er*, one who) **1.** a young animal that has recently been weaned; e.g., a weaner pig. Also called weanling. **2.** a device used to wean a young animal by covering its mouth so it cannot suckle.

weasel *n.* (Anglo-Saxon *wesle*, weasel) **1.** a common name for at least a dozen small carnivores (*Mustela* sp., family Mustelidae) with long, slender bodies whose reddish brown summer coats turn white in winter. The white fur is called ermine and is prized for making fur coats.

weather *n.* (Anglo-Saxon *weder*, weather) **1.** the state of the atmosphere at any given time and place regarding temperature, wind, precipitation, relative humidity, cloudiness, and barometric pressure. See climate, meteorology. **2.** stormy weather; strong wind and rain or snow, or stormy conditions collectively. **3.** a weather forecast, especially one that is broadcast, as it is time now to do the weather. *v.* **4.** to change or be changed by long exposure to atmospheric conditions, as a rock weathers or an exposed board weathers. **5.** to endure and survive, as a ship weathers a storm or a person weathers great difficulties.

weather-beaten *adj.* marked by features that indicate long exposure to weather; e.g., discoloration from wind, rain, and sun, wrinkling by shrinkage, erosion by windblown sand.

weathercast *n.* a forecast and explanation of the weather, usually for the next few days, especially as part of a television or radio newscast.

weathercaster *n.* a meteorologist who delivers weathercasts. Also called weatherman, weatherwoman, or weatherperson.

weather front *n.* the leading edge of a weather system; a warm front, cold front, or storm front. A front usually is marked by a change in wind direction, a reversal of a warming or cooling trend, the coming or leaving of a storm or group of storms, and often by a bank of clouds or a change in type of clouds.

weathering *n.* (Anglo-Saxon *weder*, weather + *ung*, action) **1.** the chemical and physical processes that cause rocks to disintegrate and minerals to decompose or change, including solution, hydration, hydrolysis, carbonation, and oxidation-reduction reactions. Weathering changes rocks into parent material and parent material into soil. **2.** the evidence that such processes have been at work, as the weathering on the surface of a rock. **3.** similar changes in other materials.

weathering zone *n.* the soil and rock material near the Earth's surface that show the effects of weathering.

weather map *n.* a map that illustrates a weathercast, commonly showing high and low pressure centers, weather fronts, and other present or predicted conditions, based on observations from many sites, satellite images, etc. Also called weather chart.

weatherproof *adj.* **1.** able to withstand all types of weather conditions. *v.* **2.** to make something weatherproof, e.g., by providing a waterproof covering and/or insulation against heat and cold.

weather radar *n.* a radar used to detect precipitation and the movement of storm systems as input for weathercasts. See Doppler effect.

weather satellite *n.* a satellite that takes high-altitude photographs of the Earth to show cloud patterns and weather systems as inputs for weathercasts. Also called meteorological satellite.

weather station *n.* a site equipped to collect meteorological data, usually by automatic measurement and recording of temperature, precipitation, humidity, and wind speed.

weather system *n.* a large air mass with weather properties different from those of adjacent air masses, especially one that involves wind and precipitation. In temperate regions, weather systems tend to move from west to east, while in tropical and polar regions they tend to move from east to west.

weathervane or **weather vane** *n.* a device that is turned by the wind and aligns its long axis with the direction the wind is blowing.

weaverbird *n.* any of several African and Asian birds (family Ploceidae) that live in colonies and weave intricate nests. Also called weaver finch.

web *n.* (Anglo-Saxon *webb*, web) **1.** something woven, as a flat fabric or a carpet without any pile. **2.** a spiderweb made of silken fibers radiating from a center and joined by concentric circular fibers; used to trap flying insects. **3.** similar silken structures made by various insect larvae, such as webworms and tent caterpillars. **4.** an intricate set of events or circumstances, as a food web, a web of evidence, or the web of life. **5.** a set of circumstances that trap someone, as a web of intrigue. **6.** the membranous skin that joins the digits on the feet of certain aquatic birds and animals. **7.** a similar thin part joining sturdier structural parts of certain machine frames or building frameworks. **8.** an interlinked network of radio stations, human services, or other organizations.

Webber's brown fungus *n.* a fungus (*Aegerita webberi*) used in the biological control of certain citrus flies. See white-fringe fungus.

weber (Wb) *n.* the unit for magnetic flux or magnetic pole strength in the SI system; the flux required to produce an electromotive force of 1 volt in a single turn of wire when it is reduced to zero in 1 sec. Named for Wilhelm E. Weber (1804–1891), a German physicist.

weed *n.* (Anglo-Saxon *weod*, weed) **1.** a plant out of place; an unwanted plant, especially one that displaces desired plants or contaminates the product of a crop or one that has toxic effects to people (e.g., poison ivy) or to forage-eating animals (e.g., *Astragalus* sp.). **2.** an inferior tree in a forest or an inferior animal, such as a lame horse. **3.** (informal) a cigarette or cigar. **4.** a tobacco or marijuana plant. **5.** anything that grows rapidly, as that boy is growing like a weed. *v.* **6.** to remove weeds, as to weed a garden. **7.** to eliminate those who do not do well, as to weed out the poor performers.

weedkiller *n.* a chemical that kills weeds. See herbicide.

weeping willow *n.* a medium-sized Asian willow tree (*Salix babylonica*, family Salicaceae) that has long, slender branches that hang straight down. It often is used as an ornamental tree that grows well in moist soil and tolerates city smoke.

weever *n.* (French *wivre*, serpent) **1.** either of two small marine fish (*Trachinus draco* or *T. vipera*, family Trachinidae, the greater weever and lesser weever, respectively) that have poisonous dorsal spines. They inhabit areas near shrimp beds in bottom areas of temperate seas. **2.** any fish of family Trachinidae.

weevil *n.* (Anglo-Saxon *wifel*, beetle) any of many beetles, especially the snout beetle (family Curculionideae), whose larvae are very destructive of many fruits and grains.

weir *n.* (Anglo-Saxon *wer*, weir or dam) **1.** a small dam across a stream. **2.** a net, fence, or woven structure installed in a stream channel to catch fish. **3.** a notched structure extending across an open water channel designed to regulate and/or measure water flow rate using a formula based on the depth of water above the weir notch in a still basin above the weir and a free fall of water below the weir. See Parshall flume.

welded tuff *n.* a porous volcanic rock formed by consolidation of volcanic ash or other similar material, usually showing stratification. Also called ignimbrite.

well *n.* (Anglo-Saxon *wella*, well) **1.** a spring where water flows from the earth, as a well of water. **2.** a deep hole drilled, driven, or dug vertically to reach

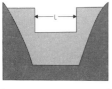

Rectangular weir
$Q = 3.33 L' H^{1.5} ft^{0.5}/sec$
$Q = 1.84 L' H^{1.5} m^{0.5}/sec$
$L' = L - 0.2 H$
H = water depth above notch

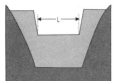

Cipoletti weir
(1:4 side slopes)
$Q = 3.37 L H^{1.5} ft^{0.5}/sec$
$Q = 1.86 L H^{1.5} m^{0.5}/sec$
H = water depth above notch

$90°$ Triangular weir
$Q = 2.48 H^{2.5} ft^{0.5}/sec$
$Q = 1.37 H^{2.5} m^{0.5}/sec$
H = depth above bottom of V

The three most common types of weirs are rectangular, Cipoletti, and 90º triangular.

underground water that can be removed by pumping or bailing. **3.** a deep hole drilled or driven into the earth to obtain petroleum or natural gas. **4.** a deep hole drilled or driven for the purpose of extracting a soluble salt such as sodium chloride (NaCl) or potassium chloride (KCl).

welter *v.* (Anglo-Saxon *weltan*, to roll) **1.** to roll and toss, as ocean waves welter. **2.** to roll about, relax, tumble, or wallow, as pigs welter in the mud. **3.** to be drenched, especially in blood. **4.** to be deeply involved or entangled, as to welter in confusion, despair, or sin. *n.* **5.** a confused muddle or jumble, as a welter of miscellaneous parts. **6.** a condition of turmoil, as the welter that followed an explosion.

westerlies *n.* the prevailing winds that blow from west to east in the two bands around the Earth between about 35° and 65° north or south latitude.

western azalea *n.* a beautiful flowering, deciduous shrub (*Rhododendron occidentale*, family Ericaceae) that is poisonous to sheep and goats; common on rangeland in western U.S.A.

western pine beetle *n.* a very damaging insect (*Dendroctonus brevicomis*, family Scolytidae) that has killed large amounts of ponderosa pine and Coulter pine in western U.S.A., especially the less vigorous and older trees.

western ragweed *n* a perennial weed (*Ambrosia psilostachya*, family Asteraceae) whose wind-borne pollen causes hay fever; native to North America. Also called perennial ragweed. See ragweed.

western red cedar *n.* **1.** a large arborvitae (*Thuja plicata*, family Cupressaceae) with reddish bark,

small scalelike leaves 1/16 to 1/8 in. (1.5 to 3 mm) long, and small cones 1/2 in. (13 mm) long; native to the Washington and Oregon coast and inland along both sides of the U.S.–Canadian border to the continental divide. **2.** the wood of the tree, widely used in cedar shingles and for poles and posts.

wet-bulb thermometer *n.* See psychrometer.

wether *n.* (Anglo-Saxon *wether*, wether) a male sheep or goat castrated before it reaches maturity, generally used to carry the bell to keep the flock together, hence, the name bell wether.

wetland *n.* an area that has water-loving vegetation growing on soils that show evidence of wetness in their profiles, including bogs, marshes, swamps, and other wet areas. Wetlands provide habitat for water-loving plants, birds, and animals; they are also important water reservoirs and serve as sites that purify water by trapping sediments, using plant nutrients, and breaking down certain types of pollutants. Much natural wetland has been drained for agricultural and construction purposes, but further drainage is now restricted by the U.S. Food Security Act of 1985.

A muskrat home in the middle of a wetland

wetland vegetation *n.* any plant life that thrives in saturated soil and/or shallow water. There are an estimated 5000 species of plants (trees, shrubs, grasses, herbs, ferns, and algae) that grow in wetlands of the U.S.A.

wettable powder *n.* a powdered material that will mix with water to form a suspension, not a solution. Wettable powders are used to apply certain pesticides.

wetting agent *n.* an additive (e.g., a detergent) that reduces the surface tension of water and allows water and water solutions to spread out and coat a surface more evenly. Wetting agents are useful for cleaning and in the application of pesticides that otherwise might run off smooth surfaces or waxy leaves. Also called a surface-active agent or surfactant.

whale *n.* (Anglo-Saxon *hwael*, whale) **1.** any of several large, warm-blooded marine mammals (families Balaenidae and Physeteridae, order Cetacea) that typically grow to lengths of 25 to 100 ft (8 to 30 m) and weigh up to 150 tons. Whales are insulated against cold water by a thick layer of blubber; they breathe through one or two blowholes, usually located on top of the head. Whales swim by wriggling their bodies, flippers, and flukes; they can locate objects underwater by echolocation. See dolphin, porpoise. **2.** anything that is unusually big or fine, as that was a whale of a show or a whale of a good time. *v.* **3.** to go after whales; to capture whales for their blubber, flesh, whalebone, etc.

whale shark *n.* a large tropical shark (*Rhincodon typus*), commonly 30 to 60 ft (9 to 18 m) long, that has a sievelike structure over its gills to catch plankton for food.

whale sucker *n.* a large blue sucker (*Remora australis*) that attaches itself to the body of a whale or dolphin.

wharf *n.* (Anglo-Saxon *hwearf*, embankment) a place for mooring and loading or unloading a ship or boat; usually a platform projecting into or along the shoreline of a harbor or stream. Also called a pier.

wharf rat *n.* **1.** a large brown rat (*Mus decumanus*) that commonly lives around wharves. **2.** a vagrant who loiters around wharves, often pilfering from cargo.

wheat *n.* (Anglo-Saxon *hwaete*, wheat) **1.** any of a group of 22 wild and 13 cultivated grass species (*Triticum* sp. and *Aegilops* sp., family Poaceae, especially *T. aestivum*) with a diploid set of 14 chromosomes, though some species are tetraploid (28 chromosomes) or hexaploid (42 chromosomes); native to the Middle East and central Asia. **2.** any of a large number of domestic varieties of wheat that came from einkorn (*T. boeoticum*) and two wheat grasses (*A. speltoides* and *A. squarrosa*). **3.** the grain from such plants. Wheat is the leading grain crop of the world and is used to make bread, cake, cereal, etc. *adj.* **4.** producing or processing wheat, as a wheat field. **5.** made with wheat, as wheat flour or wheat bread. **6.** containing some whole wheat flour instead of all white flour, as wheat bread (also called brown bread).

wheat berry *n.* an individual kernel of wheat.

wheat flour *n.* flour made by grinding wheat. It usually is divided into white flour made mostly from the endosperm and whole wheat flour that includes the wheat germ and bran. Specialty flours are made from selected varieties of wheat.

wheat germ *n.* the embryo from a kernel of wheat (or usually many such embryos collectively). Wheat germ is a nutritious food containing a concentration of vitamins.

wheat germ meal *n.* a commercial byproduct of flour milling that contains the wheat germ and some bran and middlings. It is used as a high-protein supplement in human foods, poultry feeds, and calf meals. Also called wheat germ oil meal.

wheatgrass *n.* (Anglo-Saxon *hwaete*, wheat + *graes*, grass) any of several hardy, drought-resistant, mostly perennial grasses (*Agropyron* sp., family Poaceae) commonly used as forage for livestock. Examples include bluebunch wheatgrass (*A. spicatum*), crested wheatgrass (*A. cristatum*), intermediate wheatgrass (*A. intermedium*), slender wheatgrass (*A. trachycaulum*), and western wheatgrass (*A. smithii*).

wheat rust *n.* any of several fungi (*Puccinia* sp.) that cause disease on wheat plants. Symptoms include rust spots on stems and/or leaves; the disease can cause large yield losses, but only in years when an epidemic occurs.

wheel-track planting *n.* an early form of reduced tillage whereby planters were mounted on a tractor to plant seeds in the soil that had been compacted by the wheels; usually performed on plowed land with no secondary tillage. See plow-planting.

whelk *n.* (Anglo-Saxon *wioluc*, to turn) **1.** any of several large marine gastropods (family Buccinidae) with spiral shells, especially those (*Buccinum undulatum*) eaten as food in Europe. **2.** a pimple or pustule.

whelp *n.* (Anglo-Saxon *hwelp*, whelp) **1.** a young dog. **2.** the young of any of several other animals, including bears, coyotes, lions, seals, tigers, wolves. **3.** a derogatory term for a young person, especially an impudent boy, as you little whelp. **4.** any of a set of vertical slats placed around a capstan on a ship where a hawser is wrapped around them, or similar slats on a winch or windlass. **5.** any one of the teeth on a sprocket wheel. *v.* **6.** to give birth to puppies (or other whelps), as by a female dog.

whey *n.* (Anglo-Saxon *hwaeg*, whey) the liquid part remaining from curdled milk after the curd has been removed to make cheese. Whey is often dried and used as a concentrated ingredient for food products; it has also been used experimentally to reclaim sodic soils.

whiff *n.* **1.** a small puff or gust of air, often carrying an odor, vapor, or smoke, as a whiff of perfume or a whiff of fresh air. **2.** a single breath of such a body of air, as she caught a whiff of the onion smell. **3.** a trace of anything, as a whiff of scandal or a whiff of bad temper. **4.** a flatfish (*Citharichthys* sp.) that has both eyes on the left side of the head; e.g., the horned whiff (*C. cornutus*) of the Atlantic Ocean from New England to Brazil.

whiffle *v.* **1.** to blow in gusts or puffs, as a shifting wind. **2.** to roll and toss, as a leaf or sheet of paper in a shifting breeze.

whippoorwill or **whip-poor-will** *n.* a North American bird (*Caprimulgus vociferus*, family Caprimulgidae) with rounded wings and variegated brown plumage; seen mostly at dusk in wooded areas bordering on fields. Named for its call.

whipsnake *n.* any of several long, slender New World snakes (*Masticophis* sp.) with a slender, whiplike tail.

whiptail *n.* a disease of cauliflower caused by a molybdenum deficiency that causes the leaves to be long, slender, and cupped; most prevalent on acid soils.

whirligig beetle *n.* a predatory black beetle (family Gyrinidae) that runs rapidly across the surface of water. Whirligig beetles are usually seen in large groups running, swimming, or resting near each other.

whirlpool *n.* (Anglo-Saxon *hwyrfepol*, whirlpool) water flowing rapidly in a circular pattern, forming a funnel-shaped depression in the center, where objects may be pulled downward.

whirlwind *n.* (Norse *hvirfilvindr*, whirlwind) a small rotating windstorm of air rotating inward and upward, usually moving across an open field or expanse of water on a hot, dry day. It is usually less than 100 yards (90 m) across.

white ant *n.* a termite.

white bass *n.* an edible freshwater fish (*Morone chrysops*) with a silvery back, yellow front, and sides streaked with black. It is found in the Great Lakes and the Mississippi River system.

whitecap *n.* the frothy white foam on the top of a breaking wave.

white cedar *n.* **1.** any of several coniferous trees, especially northern (or eastern) white cedar (*Thuja occidentalis*, family Cupressaceae), also known as arborvitae, and Atlantic (or southern) white cedar (*Chamaecyparis thyoides*, family Cupressaceae). **2.** the lightweight wood from any of these trees; used as poles, posts, railroad ties, and lumber.

white clover *n.* a low-growing clover (*Trifolium repens*, family Fabaceae) with small white or light pink blossoms. It is an important pasture legume, especially in pastures that are closely grazed, and it also occurs in lawns.

whitedamp or **white damp** *n.* a mixture of gases in a coal mine that includes toxic levels of carbon monoxide (CO). See coal-mine gases.

white dwarf *n.* a late phase of a small star that has collapsed into a compact mass about the size of a planet and will eventually become a cold, dark black dwarf.

white fir *n.* **1.** a large evergreen tree (*Abies concolar*, family Pinaceae) of western U.S.A. with soft, flat, pale bluish green needles 1.5 to 2.5 in. (4 to 6 cm) long and upright cones 3 to 5 in. (8 to 12 cm) long. **2.** the wood from such trees; used for building construction, general millwork, and pulpwood. **3.** any of certain other similar trees or their wood.

whitefish *n.* (Anglo-Saxon *white*, white + *fisc*, fish) **1.** any of several fish (family Coregonidae) that resemble trout but are more scaly and have smaller mouths. Whitefish are found in northern waters of North America and Eurasia. **2.** a marine food fish (*Caulolatilus princeps*) from the area near the California coast. **3.** any of several fish with a silvery color, including certain minnows, young bluefish, and carp. **4.** the white whale (*Delphinapterus leucas*), also called beluga.

whitefly *n.* (Anglo-Saxon *white*, white + *fleoge*, fly) any of several homopterous insects (family Aleyrodidae) whose bodies and wings are dusted with a white, powdery wax. Whiteflies are widely distributed in tropical regions. They obtain nourishment by sucking sap from plants and can be serious crop pests.

white-fringed beetle *n.* any of several weevils (*Graphognathus* sp., family Curculionidae) of South American origin whose larvae feed on the roots and lower stems of a wide variety of plants and cause serious damage.

white-fringe fungus *n* a fungus (*Fusarium aleyrodis*) used in the biological control of several species of citrus flies and purple scale. See Webber's brown fungus.

white grub *n.* a curled, white larva of the June beetle (or June bug or May beetle) (family Scarabaeidae). The larvae live in the soil and feed on roots of many plants (many trees, garden plants, grasses, etc.); they damage lawns by eating grass roots and attracting moles that make tunnels through the lawn as they eat the grubs. Also called grub worms. See June beetle.

white interval *n.* a range of wavelengths of infrared radiation useful for remote sensing because it can pass through the atmosphere without undue absorption. These are customarily identified by the letters A to F; A = 0.36 to 1.2 microns wavelength;

B = 1.5 to 1.8 microns; C = 2.0 to 2.4 microns; D = ? (seldom used); E = ~4.9 microns; F = 8 to 14 microns.

white lead *n.* basic lead carbonate [$2PbCO_3 \cdot Pb(OH_2)$] formerly used in powdered form as a white paint pigment, but prohibited for such use since 1978 because of lead toxicity; especially hazardous to small children who eat paint chips.

white-muscle disease *n.* muscular degeneration in lambs, calves, and other young animals associated with a deficiency of selenium in the mother's diet during gestation.

white noise *n.* **1.** a constant background sound that is generally unobtrusive and may mask unwanted sounds; e.g., a constant hum or the sound of a gentle rain. **2.** background radiation of uniform intensity over a wide range of frequencies, e.g., of radio waves.

white oak *n.* **1.** a large deciduous hardwood tree (*Quercas alba*, family Fagaceae) with deeply serrate five- to nine-lobed leaves 4 to 9 in. (10 to 23 cm) long that are green on top but pale beneath and turn deep red in fall; native to eastern U.S.A. and Canada. It is the state tree of Connecticut and Maryland. **2.** the wood from the tree; used for lumber and especially for making tight oaken barrels. **3.** any of certain other oaks (especially *Q. garryana* or *Q. lobata* of western North America) or their wood.

whiteout *n.* a blinding condition that makes everything look white caused by a blending of heavy cloud cover or blowing snow with a snow-covered landscape such that the light from the sky is equal to the reflected light from the snow.

white phosphorus *n.* the most common of several allotropic forms of elemental phosphorus, with the phosphorus atoms arranged in tetrahedral molecules. White phosphorus is very poisonous and is normally stored under water because it self-ignites in air. Heating white phosphorus in the absence of air causes it to polymerize into red phosphorus, which is stable in air.

white pine *n.* **1.** any of several large five-needle pine trees (*Pinus* sp., family Pinaceae), mostly of northern U.S.A. and Canada. Named for a white stripe that occurs on the side of each needle and for the light color of the wood, especially eastern white pine (*P. strobus*), the state tree of Maine and Minnesota, and western white pine (*P. monticola*), the state tree of Idaho. **2.** the soft, light-colored, usually straight-grained wood from such trees; a highly valued wood used for construction, matchsticks, boxes, and millwork.

white pine blister rust *n.* a destructive fungal disease (*Cronartium ribicola*, family Melampsiraceae) of white pine that depends on alternate hosts such as gooseberries and currants (*Ribes* sp.). It was introduced from Europe in the early part of the 20th century. The fungus girdles and quickly kills seedlings and slowly kills mature trees. The most practical way to protect the trees is the eradication of all ribes in their vicinity.

white pine cone beetle *n.* a beetle (*Conophthorus corniperda*, family Scolytidae) that occurs in northern U.S.A. and Canada. Its larvae eat the seeds and other parts of the cones of eastern white pine.

white snakeroot *n.* a perennial herbaceous plant (*Eupatorium rugosum*, family Asteraceae) that contains tremetol, a poison to most mammals. Milk from a poisoned animal may also be toxic to suckling young or to humans who drink it. White snakeroot is grown in flower gardens for its clusters of white flowers that persist all summer. See trembles, tremetol.

white-tailed deer *n.* an abundant deer (*Odocoileus virginianus*) in North America with a white posterior and belly.

white water *n.* water that contains enough bubbles to appear white, as the white caps on some waves and parts of certain rapids where the current dashes against rocks.

whole-wheat *adj.* made from the entire wheat kernel (e.g., whole-wheat flour or whole-wheat bread), as distinguished from white flour and its products made from the endosperm.

whooping crane *n.* a rare large crane (*Grus americana*, family Gruidae); a tall, stately bird with white plumage (marked with reddish brown when young) except for black wingtips, dark legs, long dark bill, and a bare red cap. It eats small rodents, frogs, and insects. See sandhill crane.

whorl *n.* (Anglo-Saxon *hwyrfel*, whirling spindle) **1.** a flywheel that helps regulate speed, as on a spinning wheel. **2.** a circular arrangement of similar parts, as a whorl of petals in a blossom or a whorl of leaves on a plant. **3.** one complete turn in a spiral shell. **4.** anything with a spiral shape. **5.** the part of a fingerprint where the lines form a complete circle around an axis or center.

wiesenboden *n.* an intrazonal soil in the system of soil classification defined in *Soils and Men*, the 1938 USDA Yearbook of Agriculture; a meadow soil with poor drainage and dark color.

wigwam *n.* a rounded structure made with a framework of poles covered with bark, skins, or mats; used as a dwelling by Native Americans; a more permanent residence than a teepee.

wild *adj.* (Anglo-Saxon *wilde*, wild or bewildered) **1.** living in a natural state; not controlled by or nurtured by humans; as a wild animal or a wild flower. See feral. **2.** uninhabited, as wild country. **3.** uncivilized; uncultured; as a wild man or a wild tribe. **4.** unrestrained, often ferocious, as a wild storm. **5.** likely to become violent, as a wild appearance or wild with rage. **6.** undisciplined, often lawless or immoral, as a wild gang or a wild youth. **7.** unreasonable or unlikely to succeed, as a wild idea or a wild scheme. **8.** disorderly, as wild hair or a wild crowd. **9.** without much basis, as a wild guess. **10.** missing the target, as a wild throw or a wild shot. **11.** eager; enthusiastic; as wild about a project or an idea.

wild black cherry *n.* a medium to large tree (*Prunus serotina*, family Rosaceae) native to the eastern half of the U.S.A. and the adjacent part of Canada. Its wood is highly prized for making furniture. Its ripe berries are readily eaten by birds, but the leaves, flowers, and seeds contain amygdalin, a glucoside that produces hydrocyanic acid that is toxic to livestock. See chokecherry.

wild boar *n.* an Old World swine (*Sus scrofa*) that is believed to be the wild ancestor of domestic hogs.

wild calla *n.* an aquatic herb (*Calla palustris*, family Araceae) with creeping roots, heart-shaped leaves, tiny green blossoms, and red berries. It grows in cool wetlands of the north temperate zone and is toxic to livestock. Also called water arum.

wild carrot *n.* See Queen Anne's lace.

wildcat *n.* (German *wildkatte*, wildcat) **1.** any of several North American lynx (*Lynx* sp.), a wild medium-sized cat. **2.** a wild Old World cat (*Felis sylvestris*), yellowish gray with black stripes, that is closely related to domestic cats. **3.** any of several other wild felines. **4.** a feral cat. **5.** a savage, quick-tempered person. **6.** an exploratory oil well in a new area. **7.** a risky business undertaking. *adj.* **8.** reckless or insecure, as a wildcat company. **9.** unauthorized, as a wildcat product or a wildcat strike. **10.** unregulated; not on a regular schedule, as a wildcat train. *v.* **11.** to search an unproven area for oil or valuable mineral deposits. **2.** to promote or participate in a wildcat strike.

wildebeest *n.* (Dutch *wildebeest*, wild beast) See gnu.

wilderness *n.* (Anglo-Saxon *wilddeornes*, land of wild beasts) **1.** an area that is wild, uncultivated, and uninhabited by humans, whether barren, desert, grassland, or forest. **2.** any desolate area of land or water. **3.** an area of forest, grassland, or desert set aside and officially recognized in accordance with the U.S. Wilderness Act of 1964 for recreation and scientific study; administered by the U.S. Forest Service, Fish and Wildlife Service, National Park Service, and Bureau of Land Management. Motorized vehicles are not allowed, nor are groups of more than ten people.

wildfire *n.* (Anglo-Saxon *wildfyr*, wild fire) **1.** a fire that has gotten out of control and is spreading rapidly. **2.** a material that readily produces such a fire, especially the highly flammable material also known as Greek fire that was formerly used in warfare. **3.** a kind of heat lightning that does not produce thunder. **4.** a bacterial disease of soybeans and tobacco (*Pseudomonas tabaci*, family Pseudomonadaceae) that produces brown leaf spots fringed with yellow.

wild flooding *n.* a method of surface irrigation sometimes used on rough pasture or hay land, or perhaps a field of small grain. A ditch supplies water to the high point of the area and it is released to flood across the land; small ridges and channels made with a shovel are used to guide it to cover areas that would otherwise remain dry.

wildflower or **wild flower** *n.* **1.** any flowering plant that grows without being planted or tended by humans, whether in fields, pastures, roadsides, woodlands, or elsewhere. **2.** the blossoms produced by such plants.

wild fowl or **wildfowl** *n.* any wild bird, especially a game bird such as a duck, goose, partridge, or pheasant.

wild goose *n.* any nondomestic goose, especially a Canada goose (*Branta canadensis*) or the British greylag (*Anser ferus*).

wild leek *n.* a bulbous perennial herb (*Allium tricoccum*, family Liliaceae) that grows wild and resembles an onion. Cows that eat wild leeks will give off-flavor milk. Also called ramp. See wild onion.

wildlife *n.* any nondomesticated animal, including amphibians, birds, fish, reptiles, etc., especially those that humans enjoy watching or hunting.

wildlife management *n.* any set of practices intended to favor wildlife or limit the population of certain species, including plantings that provide food and shelter, regulation of hunting and fishing, restocking depleted populations, etc.

wildlife refuge *n.* an area set aside and maintained by a government agency, private organization, or individual as a haven or sanctuary for wild animals and/or plants, especially those that are threatened or endangered.

wild oats *n.* a weedy grass (*Avena fatua*, family Poaceae) that came from Europe and is difficult to eradicate from fields of wheat and other small grains. The seeds have a twisted awn that unwinds when wet and tends to bury the seed so it will have a better chance of producing a plant.

wild onion *n.* an annual weed (*Allium canadense*, family Liliaceae) that gives milk an unpopular onion flavor when the cows eat it. Also called Canada garlic.

Wild oats

wild rice *n.* **1.** an annual ricelike plant (*Zizania aquatica*) that grows in shallow lakes or ponds, especially in Minnesota, but extending from there into Canada, eastward to Maine, and south to Louisiana and Florida. Also known as Canada rice. **2.** two Asiatic wild rice species (*Oryzafatua* and *O. spontanea*). **3.** the grain of such plants used as food.

wild river *n.* a river or large stream in a wilderness area where there are no dams, bridges, or other artificial structures.

wildwood *n.* (Anglo-Saxon *wilde wudu*, wildwood) a forested area that is seldom entered by humans and remains in its natural state.

williwaw or **willywaw** *n.* **1.** a strong, cold wind that blows from a mountain toward the sea in far northern or southern latitudes, especially in the Straits of Magellan at the southern tip of South America. **2.** a sudden squall. **3.** a violent commotion.

will-o'-the-wisp *n.* **1.** a fleeting light sometimes seen over a marshy area and attributed to phosphorescent gases released by decomposing organic matter. Also called *ignis fatuus* or Jack-o-lantern. **2.** a wishful objective that keeps one seeking after a goal that cannot be attained.

willow *n.* (Anglo-Saxon *welig*, willow) **1.** any of several shrubs and small to medium-sized trees (*Salix* sp., family Salicaceae) with long, narrow pointed leaves and tiny blossoms arranged in catkins. Many species produce long, slender, pliable branches that may be used for weaving wickerwork. **2.** the soft wood from such trees.

willow oak *n.* **1.** a deciduous tree (*Quercus phellos*, family Fagaceae) of southern U.S.A. that grows up to 80 ft (25 m) tall with a narrow crown and short branches with long, slender, glossy leaves similar to willow leaves. **2.** the hard, heavy wood from such trees; used for building construction.

Wilson's warbler *n.* a small yellow songbird (*Wilsonia pusilla*) with black markings. It eats insects and frequents thickets in most of North America, especially in Canada and western U.S.A.

wilt *v.* (German *welk*, withered) **1.** to droop and wither from lack of water associated with dryness, heat, disease, damage, or age of a flower or plant. **2.** to lose strength and vigor, as working in a hot room can cause a person to wilt. **3.** to lose courage, as his resolve may wilt if the opposition is strong. **4.** to cause something to wilt. *n.* **5.** the act or condition of wilting. **6.** any of several diseases that cause plants to wilt. **7.** a viral disease that liquefies the body tissues of certain caterpillars.

wilting point *n.* the water content in the root zone when growing plants wilt and will not recover even if placed in a saturated atmosphere. Also called permanent wilting percentage.

wind *n.* (Anglo-Saxon *wind*, wind) **1.** air movement between areas of unequal pressure, especially rapid horizontal air movement. **2.** any air current, whether natural or forced by a fan or bellows. **3.** a stream of air used to produce music in an organ or any of several wind instruments. **4.** respiration of a human or animal, as the blow knocked the wind out of him. **5.** the strength needed to continue, as he caught his second wind and continued running. **6.** gas produced in the intestinal tract and released as flatulence. **7.** meaningless speech.

windblown *adj.* **1.** showing the effects of wind, as windblown hair. **2.** shaped by strong prevailing winds, as windblown trees. **3.** moved by the wind, as windblown sand.

windbreak *n.* **1.** a line of trees, or a few lines of trees and bushes, placed to protect a living area for people and animals, e.g., upwind from a farmhouse and set of farm buildings. **2.** a shelterbelt. **3.** any structure that reduces the wind velocity across a protected area.

windburn *n.* **1.** skin irritation caused by exposure to a desiccating wind that dries the skin of a human or animal. **2.** damage caused by drying of plant parts, especially on evergreen trees and other greenery exposed to desiccating wind in winter when the soil is frozen or too cold to provide much water.

wind chill factor or **wind chill index** *n.* an estimate of how cold it feels (the apparent temperature) when bare skin is exposed to a combination of wind and low temperature; used as an indication of how much clothing a person needs for protection. See sensible temperature, heat index.

Wind speed	Actual temperature, °F						
(mph)	+30	+20	+10	0	-10	-20	-30
	Apparent temperature, °F						
10	+16	+3	-9	-22	-34	-46	-58
20	+4	-10	-24	-39	-53	-67	-81
30	-2	-18	-33	-49	-64	-79	-93
40	-5	-21	-37	-53	-69	-84	-100

wind direction *n.* the direction the wind comes from, not the direction it is moving.

wind energy *n.* energy derived from wind, usually by means of a windmill used to generate either electricity or mechanical power, e.g., a windmill that pumps water. Wind energy is clean and free, but it is intermittent, it takes space, and the equipment may be expensive. Also called wind power.

wind erosion *n.* removal of soil by wind, as in a dust storm. Wind erosion removes more soil than water erosion in many arid locations.

windfall *n.* **1.** a toppled plant, especially a tree, caused by strong wind. Windfalls can mix soil by lifting it with the roots when a tree topples; over time, it falls from the roots and is redistributed by wind and water. Also called wind throw. **2.** profit or good fortune that comes unexpectedly, usually in considerable amount. *adj.* **3.** to occur unexpectedly as a windfall; e.g., windfall profits.

windmill *n.* **1.** a mechanism that uses power from the wind turning a large wheel to drive a grinder that mills grain, a pump that lifts water, or some other mechanical device. It is usually mounted on a tower so it reaches fast-moving air. **2.** a wind-driven generator that produces electricity. See wind energy.

A windmill that pumps water

window *n.* (Norse *vindauga*, a window) **1.** an opening in a wall that allows light and/or air to enter a room, structure, or vehicle. **2.** a frame holding glass panes, often arranged in sashes that can be opened or closed, that is installed in an opening in a wall. **3.** an open or transparent section in anything, as a window in an envelope. **4.** a portion of a computer screen that is designated for a particular purpose. **5.** an interval of time when something can be done; e.g., a launch window for a space mission. **6.** an interval where something can be done; e.g., a wavelength interval for infrared radiation that can pass through the atmosphere without undue absorption; a white interval. See white interval.

windowpane *n.* **1.** a piece of window glass installed in a frame for use in a window. **2.** a flounder (*Scophthalmusaquosus*) with a thin translucent body, found in the Atlantic Ocean off the coast of North America.

wind-pollinated *adj.* pollinated by airborne pollen carried to the blossoms by the wind.

wind power *n.* See wind energy.

wind rose *n.* a circular diagram that resembles a wheel with various lengths of spokes around a circle; used to illustrate the distribution of wind speed and direction at a specific location for a year.

windrow *n.* **1.** a line made by raking hay into a low, fluffy row so it can dry in the field. **2.** a similar line of cut grain, often tied in sheaves, left to dry in the field. **3.** a line of debris resting against a fence or other barrier where the wind has blown it. *v.* **4.** to make a windrow of hay, grain, or other material.

wind scale *n.* See Beaufort wind scale.

wind shadow *n.* a partially sheltered area on the downwind side of a landscape feature, structure, or object where the wind speed is markedly reduced.

wind shear *n.* **1.** the differential in wind velocity in a unit distance measured across the direction of the wind. **2.** a sudden large change in wind velocity that causes a hazard to aircraft.

windsock *n.* a tapered sleeve made of canvas or other cloth fastened at its larger end to a ring that can rotate freely; used to indicate the direction and approximate strength of the wind.

windstorm *n.* a strong wind that has little or no precipitation to accompany it.

windswept *adj.* exposed to frequent strong winds, as a windswept plain or a windswept beach.

windthrow *n.* a tree uprooted by the wind. Also called windfall.

wind tide *n.* **1.** the rise in water level caused by a strong wind blowing across a large body of water. See seiche. **2.** the difference in water level between the leeward side and the windward side of a body of water when a strong wind is blowing across it. Also called wind setup. **3.** a storm surge.

wind velocity *n.* the speed and direction of the wind.

windward side *n.* the side impacted directly by the wind; also called the upwind side. The opposite side is called the lee side.

wine *n.* (Latin *vinum*, wine) **1.** fermented grape juice; used as an alcoholic beverage (usually 14% alcohol or less), in cooking, and in certain religious ceremonies. The unfermented juice is usually called grape juice. A distilled liquor made from wine is called brandy. **2.** either fermented or unfermented juice of various fruits (grapes, currants, gooseberries, etc.) used as a beverage or in cooking. **3.** a dark reddish color characteristic of red wines. **4.** an enthusing feeling or spirit, as the wine of success. *v.* **5.** to drink wine. **6.** to supply with wine. *adj.* **7.** having a dark red color like that of red wine. **8.** having to do with wine, as a wine cellar.

wine palm *n.* any of several palms (e.g., the coquito) with sap that is made into wine.

wing *n.* (Danish *vinge*, wing) **1.** either of the two feathered forelimbs of a bird commonly, but not necessarily, used for flying. **2.** the forelimb of a bird used as food. **3.** either of the surfaces formed by membranes between the fingerlike digits of the forelimbs of a bat; used for flying. **4.** any of the flight surfaces of an insect. **5.** any similar surface on a flying squirrel, flying fish, or other flying or gliding animal. **6.** any one of the main supporting surfaces of an airplane. **7.** a float shaped like a wing, as a water wing. **8.** an extension from a main part, as a wing of a building or a wing on a table or chair. **9.** a group of people in a position to the right or the left of the main body of an organization. **10.** a group of military aircraft. *v.* **11.** to fly. **12.** to provide with wings. **13.** to wound a flying bird, especially in the wing. **14.** to wound a person, especially in the arm.

wingspread *n.* the total width of a bird when its wings are fully extended. Also called wingspan.

winnow *v.* (Anglo-Saxon *windwian*, to winnow) **1.** to separate grain from chaff by exposing it to wind or an air current. **2.** to drive something away (e.g., dust or leaves) by blowing or fanning. **3.** to sort or separate some from a group, as to winnow the group of political candidates or to winnow through a collection of articles. **4.** to fly as a bird does by flapping its wings. **5.** to stir the air by flapping wings, arms, fans, or other objects.

winter *n.* (Anglo-Saxon *winter*, winter or year) **1.** the coldest part of the year; normally December, January, and February in the Northern Hemisphere or June, July, and August in the Southern Hemisphere. **2.** the period from the winter solstice to the vernal equinox. **3.** any period of cold weather. **4.** the colder half of the year, as in winter and summer. **5.** a year (when used to express age), as a person of 80 winters. **6.** a late, dreary, or adverse stage in the life of a person, animal, or thing. *adj.* **7.** pertaining to the cold season, as winter snowfall or a winter sunset. **8.** useful during the winter season, as winter fruits and vegetables. **9.** growing both before and after the cold season, as winter wheat. *v.* **10.** to pass the winter season, as to winter in a warmer climate. **11.** to care for animals or plants during the winter, as to winter them indoors.

winter barley *n.* any variety of barley that is normally planted in the fall and harvested the next summer.

wintergreen *n.* (Dutch *wintergroen*, evergreen) **1.** a small, creeping, evergreen shrub (*Gaultheria procumbens*, family Ericaceae) with white bell-shaped blossoms, bright red berrylike fruit, and aromatic leaves. **2.** oil of wintergreen, an oil extracted from the leaves of the plant for use as a flavoring in foods, gums, and medicines. **3.** the flavor of this oil or of a product flavored with it. **4.** any of several related plants (*Gaultheria* sp., *Pyrola* sp., and *Chimaphila* sp.).

winterhardy *adj.* able to survive through a winter, as a winterhardy plant.

winter oats *n.* any variety of oats that is normally planted in the fall and harvested the next summer.

winter rye *n.* any variety of rye that is normally planted in the fall and harvested the next summer.

winter solstice *n.* the time of year when the North Pole (for the Northern Hemisphere) or the South Pole (for the Southern Hemisphere) points most directly away from the sun. It occurs on the shortest day of the year, on or about December 21 (Northern Hemisphere) or June 21 (Southern Hemisphere). See solstice.

winter wheat *n.* any variety of wheat that is normally planted in the fall and harvested the next summer.

wireworm *n.* **1.** any of the slender, hard-bodied larvae of the click beetles (family Elateridae). The larvae infest seeds, underground stems, and roots of crops such as Irish potatoes, carrots, beets, onions, corn, tobacco, etc. **2.** a millipede. **3.** a roundworm (*Haemonchus contortus*) that is a stomach parasite of sheep and cattle, especially young animals, causing weakness, anemia, and loss of flesh. Also known as stomach worm or twisted stomach worm.

wisent *n.* (German *wisunt*, bison) the European bison (*Bison bonasus*). Also called zubr.

wisp *n.* **1.** a slender bundle of straw, hay, hair, or other loose material. **2.** a twisted bundle of straw or paper used as a torch or to light a fire. **3.** a thin puff of smoke or steam. **4.** a very slender person, as a wisp of a girl. **5.** something barely discernible, as a wisp of a smile. **6.** a will-o'-the-wisp.

wisteria *n.* any of several climbing woody vines (*Wisteria* sp., family Fabaceae) with pinnate leaves and clusters of white, purple, or rose flowers. All parts are toxic to humans. Named after Caspar Wistar (1761–1818), a U.S. anatomist.

witches'-broom *n.* a growth pattern with clusters of long, slender branches or twigs that grow on a tree or shrub as a result of a fungus, virus, insect, or mistletoe, or as a result of severe pruning. See decapitation, coppice forest.

witch grass *n.* a weedy, hairy grass (*Panicum capillare*, family Poaceae) with a bushlike compound panicle; native to North America.

witch hazel *n.* **1.** a small shrub (*Hamamelis virginiana*, family Hamamelidaceae) with deeply toothed oval leaves and small yellow flowers. **2.** an alcohol solution containing an extract from the leaves or bark of the shrub; used on skin bruises and inflammations or as an astringent.

witch moth *n.* any of several large dull-toned moths (*Erebus* sp., family Noctuidae) that fly at night and are attracted to lights; native to an area from southern U.S.A. into South America, including the West Indies.

witchweed *n.* a parasitic weed (*Striga asiatica*, family Scrophulariaceae) native to Asia and South Africa that has recently invaded southeastern U.S.A. Witchweed is an obligate parasite only on the roots of plants of the grass family, especially corn, sorghum, and sudangrass. It grows 2 to 15 in. (5 to 38 cm) tall and produces red blossoms with a yellow eye and large numbers of tiny sporelike seeds.

withe or **withy** *n.* (Anglo-Saxon *withig*, willow) **1.** a tough, flexible branch from a willow or vine; used for wrapping bundles or tying things together. **2.** a twisted rope or leash made of such branches. **3.** a tool handle that is flexible enough to absorb shock. **4.** a partition dividing a chimney into separate flues.

witness corner *n.* a marker located at a known distance and direction from a point (usually a corner) on a property boundary where the boundary point itself is not accessible for placement of a marker.

witness tree *n.* a tree used as a witness corner; also called a bearing tree. A tree that marks the actual boundary is called a boundary tree.

wold *n.* (Anglo-Saxon *weald*, forest or meadow) **1.** a treeless rolling plain at a high elevation. **2.** a forest. **3.** an open, hilly area, especially in parts of England.

wolf *n.* (Anglo-Saxon *wulf*, wolf) **1.** any of several large, doglike, carnivorous animals (*Canis* sp., family Canidae) that usually hunt in packs and can kill large or small animals. Once common throughout the Northern Hemisphere, wolves now are found mostly in remote or sparsely inhabited areas. **2.** any of several similar wild animals. **3.** the pelt or fur of such an animal. **4.** the constellation Lupus. **5.** a man who makes unwanted advances to women. **6.** a cruel person who takes advantage of others.

wolfram (**W**) *n.* (German *wolf*, wolf + *ram*, dirt) tungsten.

wolfsbane *n.* any of several poisonous plants (*Aconitum* sp., including the variety *A. napellus* known as aconite or monkshood, family Ranunculaceae) with large, showy blue, white, or yellow blossoms having hooded sepals.

wolf spider *n.* any of several spiders (family Lycosidae) that live in natural crevices and tunnels and do not spin webs. Also known as ground spider or jumping spider. The spiders are used in China to control insect pests in rice fields.

wolf tree *n.* a large tree in a woodlot or forest that occupies a lot of space but usually has little commercial value. Often it is a tree that was left when the more valuable trees were harvested from a previous stand.

wolverine *n.* (Anglo-Saxon *wulf*, wolf + *ine*, small) a stocky carnivorous mammal (*Gulo luscus*, family Mustelidae) with thick, shaggy black fur and white markings; native to Canada and northern U.S.A. Also called glutton.

wombat *n.* (Austral *wombat*, wombat) any of several herbivorous, stocky marsupials (family Vombatidae) native to Australia; a burrowing animal about the size of a badger.

wood *n.* (Anglo-Saxon *wudu*, wood) **1.** the xylem of a tree or shrub; the fibrous material beneath the bark of the stems and branches. Wood is composed mostly of about ⅔ cellulose and ⅓ lignin. **2.** the trunks and large branches of trees that are used as poles or timbers, sawed into lumber, used to make paper, or burned as fuel. **3.** such material used for any purpose. *adj.* **4.** made of wood; wooden.

wood ash *n.* the solid residue left from burning wood, commonly about 0.5% of the mass of the wood. Wood ashes have some fertilizer value for phosphorus and potassium (typically about 2% P_2O_5 and 6% K_2O equivalents) and a liming equivalent of about one-third of their weight in limestone. They also have been used as a potassium source for making soap.

woodchuck *n.* (Algonquian *ockqutchaun*, woodchuck) a burrowing rodent (*Marmota monax*) that has coarse reddish brown fur, lives in underground colonies, often in stony ground, and hibernates in the winter. It becomes a nuisance by making dens in pastures and orchards and by eating vegetables from gardens. Also called groundhog. See prairie dog.

wood duck *n.* a common duck (*Aix sponsa*, family Anatidae) that nests in trees in areas where open woodland adjoins a lake or stream. It has a large head with a crest extending toward the back, a short neck, and a square tail. Colors vary from dull browns and blues in females to bright blues, browns, and reds in males at certain seasons.

wood ibis *n.* See wood stork

woodland *n.* (Anglo-Saxon *wuduland*, woodland) **1.** land with trees as the dominant vegetation. *adj.* **2.** pertaining to woodland or forest.

woodpecker *n.* any of several birds (family Picidae) that use their sturdy beaks to drill holes in trees and wooden structures in search of insects to eat or to open space for a nest. Woodpeckers have stiff tail feathers that help brace them while they climb and drill. Many have bright black, white, and red markings.

wood pitch *n.* the final residue from the destructive distillation of wood or coal; a black resinous substance used to make plastics, caulking compounds, and insulating materials.

wood preservative *n.* any material used to impregnate wood and make it resistant to fungal decay and attack by termites. Generally it is one of three types of materials: coal tar creosote (see creosote), pentachlorophenol, or water-soluble salts such as copper or zinc chlorides.

wood pulp *n.* wood that has been broken down into loose fibers by chemical and mechanical treatment; used to make paper.

woods *n.* **1.** woodland. **2.** the trees and understory vegetation growing on an area of woodland.

woodsia *n.* any of several small ferns (*Woodsia* sp.) with short, sturdy stalks and lance-shaped fronds; common in rocky ledges and crevices of high mountain and arctic regions. Named after Joseph Woods (1776–1864), an English botanist.

wood sorrel *n.* any of several plants (*Oxalis* sp., family Oxalidaceae) with trifoliate, heart-shaped leaves and white, pink, red, yellow, or purplish blossoms with five petals that form a cone-shaped tube.

wood stork *n.* the only North American stork (*Mycteria americana*, family Ciconiidae); a large bird, about 3 ft (0.9 m) long with a wingspan of about 5.5 ft (1.65 m), that is mostly white with bare black head, neck, and legs and a long, sturdy yellow bill. It lives around swamps, marshes, and ponds and eats fish, reptiles, and amphibians. Also called wood ibis.

wood tar *n.* a dark, syruplike substance obtained from wood by dry distillation; used as a preservative for wood, ropes, etc., or further distilled to produce creosote, oils, and wood pitch.

wood thrush *n.* a large thrush (*Hylocichla mustelina*, family Muscicapidae) that is about the size and build of a robin and has a rusty head, a brown back, and a white front with large black or brown spots. This songbird eats insects and berries and frequents forest understory.

wood tick *n.* See American dog tick.

woody plant *n.* any perennial tree, shrub, or vine that has woody stems and branches and produces aboveground buds that survive over winter.

wool *n.* (Anglo-Saxon *wull*, wool) **1.** the fleece of sheep, llamas, alpacas, or certain other animals; a dense, curly, hairlike, often oily covering that holds air and insulates well. **2.** yarn, fabric, clothing, or other product made from such material. **3.** a similar fibrous covering on a caterpillar or plant stem. **4.** any of several substances obtained from plants or produced synthetically and treated to simulate wool for making woolen products. **5.** certain other filamentous materials, such as glass wool, rock wool, or steel wool. **6.** a short, thick growth of human hair. *adj.* **7.** made of wool. Also called woolen, as a wool suit or woolen goods.

wool grass *n.* a bulrush (*Scirpus eriophorum*, family Cyperaceae) with woolly spikes; native to eastern U.S.A.

woolly aphid *n.* any of several common plant lice (family Aphidae) that secrete a waxy thread, forming a cottony mass where it grows in wounds on trunks and branches of trees or shrubs; e.g., the woolly apple aphid (*Eriosoma lanigerum*, family Aphidae) that infests apple, pear, hawthorn, elm, and mountain ash or the woolly elm aphid (*E. americanum*). Seriously infested trees may be stunted or killed.

woolly croton *n.* an annual herb (*Croton capitatus*, family Euphorbiaceae) native to North America. It sometimes poisons livestock that eat it in hay.

woolly loco *n.* a locoweed (*Astragalus mollissimus*, family Fabaceae) that grows from South Dakota to Arizona. Both green and dried foliage are poisonous to livestock, causing dullness, loss of appetite, dragging feet, shagginess, and loss of weight. Also called Texas loco, purple locoweed, and woolly crazyweed.

woollypod milkweed *n.* a perennial herb (*Asclepias eriocarpa*, family Asclepiadaceae) that grows in California and is very poisonous to cattle and sheep, causing loss of appetite, diarrhea, rapid heartbeat, and low body temperature. Also called broad-leaved milkweed.

wooton loco *n.* a locoweed range plant (*Astragalus wootoni*, family Fabaceae)that grows in western Texas and New Mexico and is poisonous to all livestock in both its green and dry stages.

worker bee *n.* a sterile female bee that gathers nectar and pollen, brings in water, and protects the hive. See honeybee.

workhorse *n.* **1.** a horse used for pulling loads; generally a large, heavy-built horse, as distinguished from a saddle horse used for riding or a race horse. **2.** a person who is a very diligent, reliable worker.

World Meteorological Organization *n.* an agency of the United Nations with headquarters in Geneva, Switzerland, established in 1951 to standardize meteorological information and improve its exchange among nations.

Worldwatch Institute *n.* an independent, not-for-profit organization dedicated to reporting on global environmental problems and concerns. It has published its report, *State of the World*, annually since 1984. The general tone of the reports is pessimistic, emphasizing the loss of essential resources, political and disease problems, and the environmental problems of the world.

World Weather Watch *n.* a global environmental monitoring system consisting of an international network of ground-, sea-, and space-based weather stations, satellites, computers, etc., that makes weather data available to all nations.

World Wildlife Fund (WWF) *n* an environmental group organized to promote wildlife protection.

worm *n.* (Anglo-Saxon *wyrm*, worm or snake) **1.** any of a large number of crawling invertebrates with elongated, soft, legless bodies that are either round or flattened in cross section (phyla Annelida, Nematoda, Nemertea, or Platyhelminthes). Examples: angleworms, tapeworms, and many insect larvae. **2.** any of several similar elongated soft-bodied creatures with very short limbs, including many insect larvae and some adult insects. **3.** an object with a spiral shape, as a worm gear, a worm auger, or a spiral-shaped condenser for a still. **4.** something that gnaws or forces its way in and gradually consumes an organism or thing. **5.** an illicit computer program that destroys stored information, similar to a virus. **6.** a contemptible person. *v.* **7.** to move forward slowly and or with stealth, as though crawling like a worm. **8.** to move or work insidiously, as to worm one's way into a favored position. **9.** to get rid of worms, as to worm a dog.

wormseed *n.* **1.** the wormseed goosefoot (*Chenopodium ambrosioides*, family Chenopodiaceae). See goosefoot. **2.** the fruit of such plants, used to make Mexican tea or American wormseed for killing intestinal worms. **3.** Levant wormseed (*Artemisia cina*, family Asteraceae). Also known as santonica. **4.** the dried, unopened buds of flowers of this plant.

wormwood *n.* (Anglo-Saxon *wermod*, wormwood) **1.** any of several aromatic bushes (*Artemisia* sp.) with white or yellow blossoms, especially the bush (*A. absinthium*) that yields the bitter dark green oil used to make absinthe. **2.** a bitter and very unpleasant experience.

woundwort *n.* **1.** any of several British plants (*Stachys* sp., family Lamiaceae) with hairy stems and leaves and clusters of small reddish blossoms. **2.** kidney vetch.

wourali plant *n.* a twining woody vine (*Strychnos toxifera*, family Loganiaceae) with long reddish hairs, ovate leaves, and large round fruit; the usual source of curare.

wrangler *n.* (Spanish *caballerango*, stable worker) **1.** a person in charge of wild horses or other livestock and the rangeland where they roam. **2.** a person who argues and disputes at any opportunity.

wren *n.* (Anglo-Saxon *wrenna*, wren) any of a large number of small brown songbirds (family Troglodytidae) with a light-colored front, a slender bill, and a short, barred tail that cocks upward. Most wrens have a white or light-colored stripe passing from the upper bill, over the eye, to the back of the head.

WWF *n.* World Wildlife Fund.

WWW *n.* **1.** World Weather Watch. **2.** world wide web.

X

X *n.* **1.** the Roman numeral for 10. **2.** a sex chromosome. See X chromosome.

xanthan *n.* a natural, water-soluble gum produced by fermentation of glucose by certain bacteria (*Xanthomonas campestris*); used as a stabilizer in certain foods. Also called xanthan gum.

xanthein *n.* (French *xanthéine*, xanthein) water-soluble material that gives flowers a yellow color, as distinguished from xanthin.

xanthene *n.* a yellow substance ($C_{13}H_{10}O$) that is soluble in ether; used as a fungicide and in organic synthesis.

xanthin *n.* (French *xanthine*, xanthin) **1.** yellow material in flowers that is not soluble in water, as distinguished from xanthein. **2.** xanthine.

xanthine *n.* (Greek *xanthos*, yellow + *ine*, chemical) in blood, urine, and certain plants, a nitrogenous substance ($C_5H_4N_4O_2$) that oxidizes to uric acid. **2.** any of a group of drugs (xanthine, caffeine, theobromine, and theophylline) that stimulate the central nervous system; used as heart stimulants. See caffeine.

xanthopyll *n.* (Greek *xanthos*, yellow + *phyllon*, leaf) **1.** a yellow carotenoid pigment ($C_{40}H_{56}O_2$) in green leaves that is masked by chlorophyll until cold autumn weather causes the chlorophyll to disintegrate, revealing the yellow of xanthophyll. Also called lutein. See carotene. **2.** any of a group of red or orange terpenes found in plants that provide much of the red and orange in fall foliage.

xanthopterin *n.* (Greek *xanthos*, yellow + *pteron*, wing) a yellow pigment ($C_{19}H_{18}O_6N_6$) found in butterfly and moth wings, in the bodies of wasps and hornets, and in the urine of mammals.

X chromosome *n.* a sex chromosome. In humans and most animals, two X chromosomes produce a female, whereas an X and a Y chromosome produce a male. In birds, the opposite is true. In many insects, there is no Y chromosome, but one X chromosome produces a male, and two of them produce a female.

xenia *n.* (Greek *xenia*, hospitality) the influence of pollen from another species on some part of a plant other than the embryo, e.g., on the shape or color of the seed or fruit.

xenobiosis *n.* (Greek *xenos*, stranger or guest + *biōsis*, way of life) communal living between two species (e.g., certain species of ants) that live together but raise their young separately.

xenobiotic *n.* or *adj.* (Greek *xenos*, stranger or guest + *biōtikós*, of life) a synthetic (anthropogenic) substance in the environment, especially one that is harmful to life. Examples: chlorofluorocarbons; plastics; pesticides.

xenogamy *n.* (Greek *xenos*, stranger or guest + *gamia*, marriage) cross-pollination of plants.

xenogenesis *n.* (Greek *xenos*, stranger or guest + *genesis*, origin) **1.** the alternation of generations, as in certain insects, so that the offspring are different than the parents but like the parents of the parents. Also called heterogenesis. **2.** the supposed production of a new species unlike their ancestry.

xenology *n.* (Greek *xenos*, stranger or guest + *logos*, word) **1.** the science that studies the relationships of parasites to their hosts. See parasitology. **2.** the dating of early events in our planetary system on the basis of excess xenon 129 in meteorites. Iodine 129 has a half-life of 17 million years as it decays to xenon 129.

xenon (Xe) *n.* (Greek *xenon*, strange) element 54, atomic weight 131.29; a noble gas that is a large enough atom to produce compounds with fluorine, oxygen, cesium, and the alkali metals. All of its compounds are toxic, even though xenon gas is not toxic. Xenon is used in certain lightbulbs, as a filler in electronic tubes, and as an anesthetic.

xenoparasite *n.* (Greek *xenos*, stranger or guest + *parasitos*, parasite) an organism that is normally harmless to a host but becomes a parasite when the host is in a weakened condition.

xeric *adj.* (Greek *xēros*, dry) **1.** deficient in water. **2.** living in a water-deficient environment. **3.** a soil moisture regime defined in Soil Taxonomy as being typical of a Mediterranean climate where soils are

An Arizona landscape with xeric climate and vegetation

normally dry throughout the profile for at least 45 consecutive days in the summer and moist throughout for at least 45 consecutive days in the winter in at least 6 years out of 10. See udic, ustic, soil moisture regime.

xeriscaping *n.* using plants and techniques that minimize the need to irrigate for landscaping in a dry environment.

xeroderma *n.* (Greek *xēros*, dry + *derma*, skin) **1.** dry skin. **2.** a disease that makes the skin dry, scaly, and itchy.

xerophilous *adj.* (Greek *xēros*, dry + *philos*, loving) capable of thriving in a hot, dry climate, as some plants, animals, and people.

xerophyte *n.* (Greek *xēros*, dry + *phyton*, plant) a xerophilous plant; one that can grow under desert conditions. Also called eremophyte.

xerosere *n.* (Greek *xēros*, dry + Latin *series*, order) the succession of ecological communities that occur in a dry setting, beginning with sterile soil.

xerosis *n.* (Greek *xērosis*, dry condition) abnormal dryness of the eyes or skin from environmental conditions, age, or other causes.

xerotes *n.* (Greek *xērotēs*, dryness) a dry condition of a body, place, or thing.

xerothermic *adj.* (Greek *xēros*, dry + *thermē*, heat) **1.** both dry and hot, as a xerothermic climate. **2.** adapted to a hot, dry environment, as a xerothermic plant or animal.

X-linked *adj.* pertaining to a gene located on the X chromosome or to a trait controlled or affected by such a gene.

X ray *n.* **1.** electromagnetic radiation similar to light but with much shorter wavelength (0.1 to 10 nm) and greater penetrating power. The absorption of X rays depends on density of tissues. Also called Roentgen ray. **2.** a radiographic image made with X rays, e.g., of some portion of a body. Since bones are denser than flesh, they absorb more X rays and appear lighter in tone on an X-ray negative. *v.* **3.** to use X rays to examine the internal structure of a body, object, or container (e.g., luggage at an airport). **4.** to use X rays to treat an illness. Also called X-ray therapy.

X-ray astronomy *n.* the study of stars and other celestial bodies based on their emission of X rays.

X-ray crystallography *n.* the process used to determine the internal structure of a crystal from X-ray diffraction.

X-ray diffraction *n.* **1.** a technique used to determine the spacing between layers of atoms in a crystal by measuring the angle of diffraction of X rays passing through the crystal. **2.** an angle measured by this technique.

X-ray star *n.* a star that emits a strong X-ray signal.

xylem *n.* (Greek *xylon*, wood + *ema*, structural element) the supporting and conducting tissue in woody plants that provides stiffness and carries water with dissolved nutrients upward from the plant roots to leaves and other growing parts of plants; generally the interior, woody part of a stem. See phloem.

xylocarp *n.* (Greek *xylon*, wood + *karpos*, fruit) any fruit that is hard and woody.

xylocopa *n.* (Greek *xylon*, wood + *koptein*, to cut) carpenter bees (*Xylocopa* sp. and *Ceratina* sp.) are small bees that tunnel through plant stems and wood and make their nests in the tunnels.

xylophagous *adj.* (Greek *xylophagos*, wood eating) feeding on wood, as termites, certain larvae, or fungi.

Y *n.* yttrium.

yak *n.* (Tibetan *gyag*, yak) a large, shaggy, cold-tolerant, wild ox (*Bos grunniens*, family Bovidae), native to Tibet and central Asian mountains, which has been domesticated for work and milk production. See zho.

yam *n.* (Portuguese *inhame*, yam) **1.** any of several climbing tropical vines (*Dioscorea* sp., family Dioscoreaceae) that bear a large starchy root. **2.** the starchy tuberous root of such plants, used extensively as human food. **3.** (loosely) a sweet potato. See sweet potato.

yamp *n.* (Native American *yamp*, yamp) **1.** an umbelliferous plant (*Carum gairdneri*, family Apiaceae) that grows in western U.S.A. **2.** the edible tubers from such plants.

yampee *n.* **1.** a South American vine (*Dioscorea trifida*, family Dioscoreaceae) with large leaves. Also called cush-cush yam, aja, or mapuey. **2.** the edible tubers produced by such plants.

yard *n.* (Anglo-Saxon *geard*, enclosure) **1.** a tract of ground adjacent to a home or public building, usually grassy and used for pleasure. **2.** a work area for a business, usually having a fence but no roof; e.g., a lumberyard or a brickyard. **3.** an exercise area for a school or prison. **4.** an enclosure to hold livestock, as a stockyard or a farmyard. **5.** a railroad switching area. **6.** a unit of measure equal to 3 ft or 36 in. (0.9144 m) in the system of customary units used in the U.S.A. and some other English-speaking countries. **7.** a long spar that supports the top of a sail on a sailing ship.

yardang *n.* a long, irregular, sharp-crested ridge shaped by wind in a desert region, usually oriented parallel to the prevailing wind.

yard waste *n.* clippings from bushes and trees and other waste materials from the yards and gardens around homes.

yarn *n.* (Anglo-Saxon *gearn*, yarn) **1.** coarse, twisted thread made from either natural (e.g., cotton or wool) or synthetic (e.g., nylon or rayon) fibers that is used for crocheting, knitting, weaving, etc. **2.** a strand that is twisted with other strands to make a rope. **3.** a continuous strand or fiber made of glass, plastic, or metal. **4.** a tall tale, usually an entertaining story with embellishments and exaggerations that make it difficult to believe.

yarovization *n.* See vernalization.

yarrow *n.* (Anglo-Saxon *gearuwe*, yarrow or healer) **1.** a hardy perennial plant (*Achillea millefolium*, family Asteraceae) with a strong odor and taste, fernlike leaves, and clusters of small white or pink blossoms. Native to Eurasia, it has become a weed in North America on roadsides and untilled areas. Also called milfoil or thousand leaf. **2.** any of several related plants (*Achillea* sp.).

yaupon *n.* (Catawba *yopún*, yop bush) a deciduous holly shrub or small tree (*Ilex vomitoria*, family Aquifoliaceae) of southern U.S.A. Its dried leaves can be made into a tea. Small red berries from female plants are eaten by birds in late fall and winter. Also called emetic holly.

yaws *n.* an infectious tropical disease caused by a spirochete (*Treponema pertenue*) closely related to syphilis, but nonvenereal. It occurs mostly in children. The first stage produces lesions on the lower extremities, followed by skin elevations over the entire body that turn into gummy lesions similar to those caused by syphilis accompanied by destructive bone lesions. Also called frambesia. See pian.

Y chromosome *n.* a sex chromosome that pairs with an X chromosome to produce a male in humans and most animals, or a female in birds. See X chromosome.

yean *v.* (Anglo-Saxon *eanian*, to bring forth lambs) to give birth to young (sheep or goats).

yeanling *n.* (Anglo-Saxon *eanian*, to bring forth lambs + *ling*, little) a young sheep or goat; a lamb or a kid.

year *n.* (Anglo-Saxon *gear*, year) **1.** one complete revolution of the Earth around the sun. See solar year, sidereal year, lunar year. **2.** a period of 365 or 366 days. The Gregorian calendar now in use divides the year into 12 months and begins each new year on January 1. **3.** any period of approximately 365 days or 12 months, as a fiscal year, or a school year, or this should be done in a year. **4.** the comparable period for some other planet, as a Martian year.

yearling *n.* (Anglo-Saxon *gear*, year + *ling*, little) **1.** an animal that is between its first and second birthdays. **2.** a horse that has passed its second but not its third January 1. **3.** an event or condition that occurred or began between one and two years ago, as a yearling marriage.

yeast *n.* (Anglo-Saxon *gist*, yeast) any of a large number of unicellular fungi (phylum Ascomycota) that seldom form mycelia; usually spherical, oval, or cylindrical cells that are considerably larger than bacteria, have internal structures that can be seen under a microscope, and reproduce by budding.

2. any of these fungi (*Saccharomyces* sp.) that can decompose carbohydrates aerobically into water and carbon dioxide, or anaerobically into alcohol and carbon dioxide with the aid of the enzyme, zymase; e.g., brewer's yeast (*S. cerevisiae*) used in baking bread and in brewing beer and other alcoholic liquors. **3.** a yellow mass of such fungi prepared for use in baking or brewing or to be eaten as a dietary supplement that is a natural source of protein and B-complex vitamins. See leaven. **4.** certain of these unicellular fungi that can be pathogenic and cause yeast infection.

yellow fever *n.* a deadly infectious disease caused by a virus (*Charon evagatus*) that is transmitted by mosquitoes (*Aedes aegypti* and others) and causes fever, liver damage, and jaundice in monkeys and humans in tropical and subtropical North and South America and Africa.

yellow gum *n.* **1.** any of several eucalyptus trees (e.g., *Eucalyptus melliodora*, family Myrtaceae) with yellowish bark that are native to Australia. **2.** any of several tupelos.

yellowhammer *n.* **1.** a European bunting (*Emberiza citrinella*, family Emberizidae). Male birds of this species have bright yellow markings. **2.** a common flicker (*Colaptes auratus*, family Picidae) with yellow under its wings and tail.

yellow jacket *n.* a social wasp or hornet (*Vespa* sp., family Vespidae) with bright yellow and black bands around its body. It builds gray nests of paperlike material with interior cells like those of a honeycomb and seeks nectar from flowers and sometimes from sweet drinks such as soda. The females can inflict a very painful sting, but the males do not sting.

yellow pine *n.* **1.** ponderosa pine. **2.** the light yellowish wood of ponderosa pine or of certain other North American pine trees.

yellow poplar *n.* **1.** tulip tree. **2.** the light-colored wood of such trees that is used for furniture, boxes, cabinets, construction, and pulpwood.

yerba *n.* (Spanish *yerba*, herb) any small plant or herb, especially yerba maté.

yerba buena *n.* an aromatic trailing plant (*Satureja douglasii*, family Lamiaceae), native to the coastal area of California, with egg-shaped leaves and solitary white blossoms. It spreads by rooting at the tips of its branches when they touch the ground and was formerly used as an antihelmintic and anticatarrhal drug.

yerba maté *n.* See maté.

yersinia *n.* **1.** a bacterial pathogen (*Yersinia pestis*, family Enterobacteriaceae); a gram-negative bacillus that causes bubonic plague in humans, rats, ground squirrels, and other rodents. See bubonic plague. **2.** a related common pathogen (*Y. pseudotuberculosis*) that attacks humans, rodents, birds, guinea pigs, and rabbits.

yersiniosis *n.* an intestinal disease caused by a gram-negative bacillus (*Yersinia enterocolitica*, family Enterobacteriaceae) that produces symptoms resembling appendicitis in children, young adults, and animals.

yew *n.* (Anglo-Saxon *eow*, yew) **1.** any of several evergreen trees or shrubs (*Taxus* sp., family Taxaceae) with dark-green needles and red berries. Yews are popular as ornamentals, but their wood, bark, needles, and seeds are poisonous to humans and animals. See ground hemlock. **2.** the close-grained, elastic wood of such trees or large shrubs. A favorite wood for bows used in archery.

yield *v.* (Anglo-Saxon *gieldan*, to pay) **1.** to produce as a return for invested time, effort, seed or other materials, or money, as a wheat field yields grain or bonds yield interest. **2.** to bear, bring forth, or produce. **3.** to give up, concede, or surrender, as to yield one's turn. **4.** to move in response to pressure, as a wall that yields but does not fall. *n.* **5.** an amount of return, commonly expressed in bushels per acre, quintals per hectare, tons per acre or hectare, percentage on an investment, or percentage of a theoretical potential, etc. **6.** an act of concession or response to pressure.

ylang-ylang *n.* (Tagalog *ilang-ilang*, ylang-ylang) **1.** a small tree (*Cananga odorata*, family Annonaceae) native to Java and the Philippine Islands. Its fragrant, drooping flowers are the source of an oil that is used to make perfume. **2.** the oil obtained from the flowers, or a perfume made from it.

yoke *n.* (Anglo-Saxon *geoc*, yoke) **1.** a sturdy crosspiece (usually wooden) that fits across the necks of two animals (especially oxen) and is held in place by a U-shaped piece around each of their necks. It is tied to a load for the animals to pull. **2.** a pair of animals used to pull a load with a yoke. **3.** a sturdy crosspiece (usually wooden) shaped to fit across the shoulders of a person and fitted to carry a bucket or

A yoke suitable for use with two oxen

other load at each end. **4.** a symbol of oppression, indicating the imposition of a heavy burden, e.g., high taxes. **5.** a piece of a machine that clamps onto two parts and holds them together. **6.** a U-shaped control used to pilot an airplane. **7.** a piece of a shirt or coat that fits across the neck and shoulders or a trousers piece that fits around the hips. **8.** a U- or Y-shaped support piece in various structures or mechanisms. *v.* **9.** to connect two animals or things together with a yoke. **10.** to hook onto a load by means of a yoke.

yolk *n.* (Anglo-Saxon *geolca*, yellow part) **1.** the interior part of an egg, usually yellow or orange and roughly spherical in shape, surrounded by the white (or clear) part. The embryo in a fertilized egg develops at the outer margin of the yolk and is nourished by both the yolk and the white of the egg. **2.** grease and other nonwool materials included in the fleece of a sheep when it is sheared.

young soil *n.* a soil that has relatively little profile development, e.g., loss of highly soluble components and color changes from accumulation of organic matter in the upper part and perhaps some red or brown subsoil colors, but no noticeable changes in texture and only weak soil structure.

young topography *n.* a landscape that has recently been rejuvenated; e.g., a plain that has been uplifted and is beginning to be eroded, so there are broad, nearly level uplands cut by narrow, steep-sided valleys. Swamps, lakes, and other wet areas may exist on the uplands where the new drainage system is not yet fully integrated.

ytterbium (Yb) *n.* (Swedish *Ytterby*, a quarry near Stockholm + *ium*, element) element 70, atomic weight 173.04; a soft, malleable, ductile rare earth metal with a bright silvery luster. It is tarnished by air and water.

yttrium (Y) *n.* (Swedish *Ytterby*, a quarry near Stockholm + *ium*, element) element 39, atomic weight 88.90585; a rare earth metal with a silvery luster that is unstable in air. It is used to make red phosphors for television tubes and as a component of various alloys.

yucca *n.* (Spanish *yuca*, yucca) a succulent, perennial, evergreen plant (*Yucca* sp., family Agavaceae) with pointed, rigid, swordlike leaves that produces a single cluster of spectacular white blossoms that reach upward as high as 5 to 12 ft (1.5 to 3.6 m). It is native to warm, dry climates of America and is the state flower of New Mexico.

Yucca in bloom

yucca moth *n.* a white moth (*Tegeticula alba* and related species, family Prodoxinae) that pollinates the yucca plant. Its larvae develop in the ovaries of the yucca plant. This is an obligatory form of mutualism needed by both the moth and the plant.

Z

zamarra *n.* (Basque *zamar*, sheepskin) a sheepskin coat like those worn by Spanish shepherds.

zamia *n.* (Latin *zamiae*, pine nuts) any of several palmlike shrubs or trees (*Zamia* sp., family Cycadaceae) with a sturdy trunk capped by a crown of pinnate leaves and oblong cones; native to tropical and subtropical America.

zanja *n.* (Spanish *zanja*, excavation) a canal excavated between camellones (raised cropping beds) in parts of Mexico. Zanjas are a source of soil for building camellones, a major reservoir that supplies water to the crops via capillary rise from the water table, and they may be used for irrigation during the dry season. Organic matter from aquatic plants and sediment that accumulates in the zanjas are used to maintain fertility in the camellones. See camellone.

Zea mays *n.* corn in the U.S.A.; maize throughout most of the world.

zebra *n.* (Portuguese *zebra*, wild ass) **1.** any of several African mammals (*Equus* sp.) that resemble small horses with black or dark-brown stripes on a white background. **2.** a zebra butterfly. **3.** (slang) a football official wearing a striped shirt.

A zebra

zebra butterfly *n.* a tropical butterfly (*Heliconius charithonius*) with yellow stripes on black wings.

zebra fish *n.* a small freshwater fish (*Brachydanio rerio*) that is popular in home aquariums for its striking pattern of luminous bluish black horizontal stripes on a silvery gold background.

zebra mussel *n.* a fast-breeding freshwater bivalve mollusk (*Dreissena polymorpha*, family Dreissenidae) with a rough brown outer shell about a half-inch (1.3 cm) long. It attaches itself in clusters to solid objects and has become a major pest in northeastern U.S.A. by clogging waterways and plugging inlet pipes for municipal water systems.

zebra swallowtail *n.* a butterfly (*Papilio marcellus*) with a deeply forked tail and black and greenish white stripes on its wings. See swallowtail.

zebrawood *n.* **1.** a tree (*Connarus guianensis*) native to tropical America. **2.** the striped hard wood from the tree. It is used for making furniture. **3.** the similar striped wood of several other trees.

zebu *n.* (French *zébu*, zebu) a type of domesticated cattle (*Bos indicus*, family Bovidae) developed in India and used as a beef animal, beast of burden, or milk producer. It is heat tolerant, with a conspicuous hump over its shoulders and a large dewlap beneath its neck. See brahman.

zeitgeber *n.* (German *zeitgeber*, time giver) an environmental occurrence that serves as a trigger for a plant to change its growth habit; e.g., a day length or temperature that triggers a response in the plant and makes it begin to flower, mature, or enter some other phase of growth.

zenith *n.* (Latin *cenith*, way or path) **1.** a point in the heavens directly above a person's head (opposite of nadir). **2.** the highest point in a landscape. **3.** the greatest attainment in a person's career.

zenithal *adj.* (Latin *cenith*, way or path + *alis*, like) **1.** pertaining to the zenith. **2.** located near the place or time of a zenith; e.g., zenithal rains in the tropics when the sun is nearly overhead.

zeolite *n.* (Swedish *zeolit*, boil) any of a group of hydrous aluminosilicates of sodium, potassium, calcium, barium, and strontium that easily absorb or release water without changing their structure. They are naturally occurring secondary minerals in basic volcanic rocks and are used as molecular sieves and in water-softening techniques by ion exchange. Some zeolites have negative charge and are used for cation exchange, whereas others have positive charge and are used for anion exchange.

zephyr *n.* (Latin *zephyrus* west wind) **1.** a west wind. **2.** a gentle breeze. **3.** a soft, lightweight yarn, fabric, or garment.

zero *n.* (Arabic *sifr*, cipher) **1.** the numerical symbol for none or nothing, 0. **2.** the beginning point on a scale, or the point where the numbers change from positive to negative. **3.** the temperature when a thermometer reads 0; the freezing point of water on the Celsius scale or 32ºF colder than the freezing point of water. See absolute zero. **4.** the setting for an artillery piece to make a direct hit on a target in the absence of wind. **5.** a cloud ceiling within 50 ft (15 m) of ground level, too low for pilots to land or navigate by visible markers.

zero gravity *n.* apparent weightlessness when a body is in free fall or when the centrifugal force on a body in orbit equals the gravitational attraction.

zero population growth *n.* stable population numbers; the condition when deaths plus emigration equal births plus immigration. See family planning.

zero tillage *n.* see no-till.

zho *n.* a cross between a bull yak (*Bos grunniens*) and a common cow (*Bos taurus*) that is used in northern India, Nepal, Tibet, and Bhutan as a domestic animal. See yak.

zinc (Zn) *n.* (German *zink*, zink) element 30, atomic weight 65.39, usual charge +2; a reactive heavy metal that is explosive as a powder and an essential micronutrient for plants, animals, and humans. It is a constituent of certain enzymes that help transport carbon dioxide through the blood to the lungs and function in protein metabolism, and also is used in several alloys as a protective coating (galvanizing) on sheet iron, and as a component of paints, cosmetics, plastics, batteries, insect and rat poisons, etc.

zinnia *n.* **1.** any of several fast-growing annual flowering plants (*Zinnia elegans* and related species, family Asteraceae) native to Mexico and nearby areas; named after J. G. Zinn (1727–1759), a German botanist. **2.** the blossoms of such plants with many different colors marked with rays.

zircon *n.* (Arabic *zarqūn*, gold color) a common accessory mineral ($ZrSiO_4$) in igneous rocks and pegmatites, commonly with 1% to 4% of the zircon replaced by hafnium, thorium, uranium, and some of the rare earth elements. It is the principal source of zirconium and hafnium, especially from deposits on the coast of Queensland, Australia, and in Florida. Also called jargon, hyacinth, jacinth, or ligure.

zirconium (Zr) *n.* (Arabic *zargūn*, gold color + Latin *ium*, element) element 40, atomic weight 91.224; a lustrous grayish white metal that is highly resistant to corrosion. It is used to clad fuel elements in atomic reactors, and in certain alloys, vacuum tubes, lamp filaments, explosive primers, etc.

Zn *n.* the chemical symbol for zinc.

zodiac *n.* (Latin *zodiacus*, zodiac) **1.** an imaginary band in the heavens extending about 8° on each side of the ecliptic (the sun's yearly path), encompassing the apparent paths of the moon and most of the planets. Twelve major constellations within the zodiac were chosen by Hipparchus (a Greek astronomer) in the 2nd century B.C. as signs of the zodiac, each representing 30° along the ecliptic. **2.** an astrological figure representing the 12 signs of the zodiac. **3.** a circuit or round trip.

zoea *n.* (Latin *zoea*, an animal) any of several free-swimming larvae of certain crustaceans (e.g., crabs) with lateral and medial eyes and spiny projections on the carapace that covers its body.

zoic *adj.* (Greek *zōē*, life + *ikos*, having to do with) pertaining to animal life or characterized by animal life.

zonal soil *n.* a soil with a developed profile that reflects the dominant influence of the climate and vegetation of its zone. It was used in the 1938 U.S. system of soil classification, but is now an obsolete term. See azonal, intrazonal, Soil Taxonomy.

zone *n.* (Latin *zona*, girdle) **1.** an encircling band, belt, girdle, or stripe that differs from the adjoining parts in content, color, texture, structure, or other noticeable characteristics. **2.** a climatic band around the Earth, including the torrid zone near the equator, the temperate zones next to it, and the frigid zones near the North and South Poles. See climatic zone. **3.** any area with distinct characteristics and use, as a school zone, residential zone, time zone, war zone, etc. **4.** a section of a city; e.g., the area served by a post office. **5.** the area that is charged a certain rate of postage, freight, or other service charge, usually a band with a certain range of distances from the point of origin. **6.** a section on the surface of a sphere between two parallel lines. **7.** a designated part of the body, as the abdominal zone. *v.* **8.** to divide an area into zones, e.g., by a zoning ordinance that designates residential, commercial, and industrial zones in a city. **9.** to mark with a band or wrap a band around an object, thus creating a zone.

zone of eluviation *n.* the A horizon, the part of a soil that has been eluviated (where soluble salts have been leached away and part of the fine clay has been moved downward into the zone of illuviation).

zone of illuviation *n.* the B horizon, the part of a soil that has accumulated fine clay or other materials carried down from an overlying zone of eluviation.

zoochlorella *n.* (Greek *zōion*, animal + *chlōros*, green + *ella*, little) **1.** a small green granule found in the body of certain hydras, protozoa, or other tiny invertebrate animals. **2.** a unicellular green alga that lives symbiotically inside the cells of certain freshwater invertebrate animals.

zooecology *n.* (Greek *zōion*, animal + *oikos*, house + *logos*, word) the ecological study of relationships between animals and their environment.

zooid *n.* (Greek *zōion*, animal) an individual animal or protist in a colony of its type.

zoology *n.* (Greek *zōion*, animal + *logos*, word) **1.** the biological science that studies animals. **2.** the animal population of an area.

zoon (pl. **zoa**) *n.* (Greek *zōion*, animal) **1.** any complete individual that is part of a compound organism. **2.** any or all of the organisms developed from a single fertilized egg.

zoonosis *n.* (Greek *zōion*, animal + *osis*, disease condition) any disease that can be transmitted from vertebrate animals to humans. Examples are rabies, anthrax, tuberculosis, tularemia, brucellosis, leptospirosis, and ornithosis.

zoophagous *adj.* (Greek *zōion*, animal + *phagos*, eating) carnivorous; animal-eating.

zoophagous parasite *n.* any parasite that thrives in or on animals.

zoophyte *n.* (Greek *zōion*, animal + *phyton*, plant) any animal that appears to be a plant; e.g., a barnacle, coral, or sponge.

zooplankton *n.* (Greek *zōion*, animal + *planktos*, wandering) the tiny animals or animal-like creatures (e.g., protozoa) that live in water and are included in plankton.

zoosemiotics *n.* (Greek *zōion*, animal + *sēmeiōtikos*, sign) the study of sounds and signals that animals use to communicate.

zoospore *n.* (Greek *zōion*, animal + Latin *spora*, seeding) a motile, flagellated, asexual spore; a type of resting and reproductive cell produced by many green algae, some brown algae, and certain fungi. Also called swarm spore.

zootoxin *n.* (Greek *zōion*, animal + *toxikon*, poison) a toxic substance of animal origin; e.g., snake venom, bee venom, or black widow spider venom.

zooxanthella *n.* (Greek *zōion*, animal + *xanthos*, yellow + *ella*, little) a small particle of yellow pigment such as those found in the bodies of yellow-green algae (zooxanthellae).

zooxanthellae *n.* a type of microscopic yellow-green algae that lives symbiotically with certain marine invertebrates (protozoa, sea anemone, corals, and clams).

zosteraceae *n.* See eelgrass.

zoysia *n.* any of several low-growing grasses (*Zoysia matrella* and related species, family Poaceae) that are used as lawn grasses in warm climates. Native to tropical Asia, it is named after Karl von Zoil (?–1800), a German botanist.

Zr *n.* the chemical symbol for zirconium.

zubr *n.* (Polish *zubr*, bison) the European bison (*Bison bonasus*). Also called wisent.

zugunruhe *n.* (German *zug*, a pull or migration + *unruhe*, unrest) the migratory drive in many birds and in some animals.

Zuider Zee *n.* a shallow arm of the North Sea in the Netherlands that has been mostly reclaimed and used for growing crops. The undrained part is now called the IJsselmeer. See polder.

zwitterion *n.* (German *zwitterion*, hermaphroditic ion) an ion that carries a positive charge at one site and a negative charge at another; e.g., a free amino acid in solution (the amino group becomes $-NH_3^+$ and the carboxyl group becomes $-COO^-$).

zygogenesis *n.* (Greek *zygon*, yoke + *genesis*, formation) formation of zygotes as a form of reproduction.

zygomycete *n.* (Latin *zygomycētēs*, zygomycete) any of a large group of fungi (class Zygomycetes) that reproduces sexually by forming zoospores.

zygospore *n.* (Greek *zygon*, yoke + Latin *spora*, seeding) a cell formed by certain algae and fungi by the merging of two similar gametes.

zygote *n.* (Greek *zygōtos*, yoked) a cell formed by the merger of two gametes; the parent cell that divides repeatedly and becomes a new individual.

zymase *n.* (Greek *zymē*, leaven + *ase*, enzyme) any of the group of enzymes formed by yeast that catalyzes the (anaerobic) fermentation of sucrose into alcohol and carbon dioxide. See yeast.

zymogen *n.* (Greek *zymē*, leaven + *genēs*, birth) any organic molecule that is readily converted into an enzyme; a precursor for an enzyme.

zymology *n.* (Greek *zymē*, leaven + *logos*, word) enzymology; the science that deals with enzymes and fermentation by enzymatic activity.

ISBN 0-8138-0283-0